D0769289

HANDBOOK

OF

BIOLOGICAL

PHYSICS

VOLUME 2

Transport Processes
in Eukaryotic and
Prokaryotic Organisms

HANDBOOK

OF

BIOLOGICAL

PHYSICS

Series Editor: A.J. Hoff

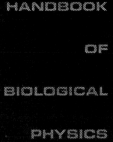

HANDBOOK

OF

BIOLOGICAL

PHYSICS

VOLUME 2

Transport Processes in Eukaryotic and Prokaryotic Organisms

Editors:

W.N. Konings

Department of Microbiology
Groningen Biomolecular Sciences and Biotechnology Institute
University of Groningen, Biology Centre
Kerklaan 30
9751 NN Haren, The Netherlands

H.R. Kaback

Howard Hughes Medical Institute
Departments of Physiology and Microbiology & Molecular Genetics
Molecular Biology Institute
University of California Los Angeles
Los Angeles, CA 90024-1662, USA

J.S. Lolkema

Department of Microbiology
Groningen Biomolecular Sciences and Biotechnology Institute
University of Groningen, Biology Centre
Kerklaan 30
9751 NN Haren, The Netherlands

N·H

1996

ELSEVIER

Amsterdam - Lausanne - New York - Oxford - Shannon - Tokyo

ELSEVIER SCIENCE B.V.
Sara Burgerhartstraat 25
P.O. Box 211, 1000 AE Amsterdam, The Netherlands

ISBN 0-444-82442-1

© 1996 Elsevier Science B.V. All rights reserved.

No part of this publication may be reproduced, stored in a retrieval system or transmitted in any form or by any means, electronic, mechanical, photocopying, recording or otherwise without the prior written permission of the publisher, Elsevier Science B.V., Copyright & Permissions Department, P.O. Box 521, 1000 AM Amsterdam, The Netherlands.

Special regulations for readers in the U.S.A. – This publication has been registered with the Copyright Clearance Center Inc. (CCC), 222 Rosewood Drive, Danvers, MA 01923. Information can be obtained from the CCC about conditions under which photocopies of parts of this publication may be made in the U.S.A. All other copyright questions, including photocopying outside the U.S.A., should be referred to the copyright owner, Elsevier Science B.V., unless otherwise specified.

No responsibility is assumed by the publisher for any injury and/or damage to persons or property as a matter of products liability, negligence or otherwise, or from any use or operation of any methods, products, instructions or ideas contained in the material herein.

This book is printed on acid-free paper.

Printed in The Netherlands.

General Preface

Biological Physics is "The physics of the life processes". Aspects of the life processes are studied within the laws of nature, which are assumed to be equally valid for living and dead matter, with physical concepts and methods. A multidisciplinarian approach brings together elements from biology – knowledge of the problem that is attacked – and from the physical sciences – the techniques and the methodology for solving the problem. In principle, Biological Physics covers the physics of all of biology, including medicine, and therefore its range is extremely broad. There is a need to bring some order to the growing complexity of research in Biological Physics so as to present the experimental results obtained in its manifold (sub)fields, and their interpretation, in a clear and concise manner. With this in mind, the Handbook of Biological Physics has been set up – a series of interconnected monographs each devoted to a certain subfield that is covered in depth and with great attention to clarity of presentation. The Handbook is conceived in such a way that interrelations between fields and subfields are made clear, areas are identified in which a concentrated effort might solve a long-standing problem and, ideally, an evaluation is presented of the extent to which the application of physical concepts and methodologies, often with considerable effort in terms of personal and material input, has advanced our understanding of the biological process under examination.

Individual volumes of the Handbook will be devoted to an entire "system" unless the field is very active or extended (as, e.g., for membrane or vision research), in which case the system will be broken down into two or more subsystems. Depending on the subject, there will be an emphasis on physical chemistry approaches (emphasis on structure at the molecular level) and biophysical approaches (emphasis on mechanisms). The guiding principle of planning the individual volumes is that of going from simple, well-defined concepts and model systems on a molecular level, to the highly complex structures and working mechanisms of living matter. Each volume will contain an introductory chapter defining the place of each of the other chapters. Generally, the volume will end with a closing chapter indicating which areas need further discussion and will provide an outlook into the future of the field.

The first volume of the Handbook – Structure and Dynamics of Membranes – dealt with the morphology of biomembranes and with different aspects of lipid and lipid–protein model membranes (Part A), and with membrane adhesion, membrane fusion and the interaction of biomembranes with polymer networks such as cytoskeleton (Part B). This second volume continues the discussion of biomembranes, now as the barrier that separates the inside of the cell from the outside world, or that divides the cellular inner space in distinct compartments. In the living cell a multitude of transport processes occur, molecules of all sorts moving from the environment into the cell, from the cell outwards, and to and from the different compartments inside the cell. Transport across the biomembrane barriers is mediated by a host of transport proteins, many of which have become known only in recent years. The structure–function relationships of these transport proteins are now being elucidated with a great many biophysical and biochemical techniques, including molecular genetics, X-ray analysis, and a variety of spectroscopic methods. Particular attention is now given to the dynamic aspects of the structure of the proteins, conformational changes, etc.

Part 1 of this volume deals with primary transport phenomena. A number of ion pumps are treated in detail, and multidrug resistance and export systems are discussed. Part 2 covers secondary transport, including substrate transport, symporters and antiporters, exchangers and neurotransmitters. In addition, three chapters present as many views on homeostasis. Part 3 is devoted to phosphotransferase systems, while Part 4 gives an in-depth treatment of channels and porines. Finally, Part 5 discusses transport of macromolecules such as DNA and proteins across the bacterial cytoplasmic membrane, and protein transport across the outer and inner membrane of mitochondria.

Planned volumes

With the first two volumes of the Handbook, the foundation has been laid for covering the wide expanse of Biological Physics by a series of interconnected monographs. The "bottom up" approach adopted for individual volumes is also the guideline for the entire series. Having started with two volumes treating the molecular and supramolecular structure of the cell, the Handbook will continue with several volumes on cellular and supracellular systems. Finally, a number of volumes is planned on the biophysics or suborganismal systems and whole organisms. Volumes planned are:

- Vision – molecular aspects and the retina, perception, pattern recognition, imaging
- Neuro-informatics – neural modelling, information processing
- Photosynthesis and electron transport
- Fluid dynamics and chaos
- Motion and contractile systems
- The vestibular system
- Hearing
- Electro-reception and magnetic field effects

Further volumes will be added as the need arises.

We hope that the present volume of the Handbook will find an equally warm welcome in the Biological Physics community as its first volume, and that those who read these volumes will communicate their criticisms and suggestions for the future development of the project.

Leiden, Spring 1996
Arnold J. Hoff
Editor

Preface to Volume 2
Transport Processes in Eukaryotic and Prokaryotic Organisms

With the rapid development of molecular biology, it has become possible to manipulate almost any protein in a fashion that was virtually unimaginable only fifteen years ago. Individual amino-acid residues can be mutagenized selectively, portions of molecules can be deleted and one molecule, or a portion thereof, can be fused to another to produce chimeras of any variety. Thus, in many ways, the modern molecular biologist is in a position to "play God". However, if the problem in question involves membrane proteins, God wears a blindfold. That is, there is a large barrier which is yet to be overcome, particularly with the class of membrane proteins discussed in this volume, and that barrier is the difficulty inherent in obtaining high-resolution structural information about hydrophobic membrane proteins. Thus far, only a handful of membrane proteins have been crystallized in a form that yields a high-resolution 3-dimensional structure from X-ray diffraction, porin (see Rosenbusch, Chapter 26), photosynthetic reaction centers and terminal oxidases from *Paracoccus denitrificans* and beef heart mitochondria. In addition, high-resolution structures for bacteriorhodopsin (Henderson et al., 1990) and the plant light-harvesting complex (Kuhlbrandt et al., 1994) have been elucidated from electron diffraction studies.

Since the techniques of molecular biology allow manipulation of proteins at the 2 Å level of resolution, it is clear that structural information at the same level of resolution is required to obtain mechanistically relevant information. This is not to say that a structure will yield the mechanism directly without additional information. Since many of the membrane transport proteins discussed in this volume probably undergo widespread conformational changes, dynamic as well as static information will be required. In this respect, it is becoming clear that one particularly useful application of molecular biology is the engineering of membrane proteins for various biophysical approaches that have the capability of yielding dynamic information at high resolution (see Chapter 10).

In summary, therefore, although our knowledge of membrane transport proteins has increased dramatically during the past few years, the molecular mechanisms by which these energy transducing machines operate are likely to remain obscure until static and dynamic aspects of the structures of the proteins involved are resolved at a meaningful level of resolution.

References

Deisenhofer, J. and Michel, H. (1989) Science **245**, 1463–1473.
Henderson, R., Baldwin, J.M., Ceska, T.A., Zemlin, F., Beckmann, E. and Downing, K.H. (1990) J. Mol. Biol. **213**, 899–929.
Iwata, S., Ostermeier, C., Ludwig, B. and Michel, H. (1995) Nature **376**, 660–669.
Kuhlbrandt, W., Wang, D.N. and Fujiyoshi, Y. (1994) Nature **367**, 614–621.
Tsukihara, T., Aoyama, H., Yamashita, E., Tomizaki, T., Yamaguchi, H., Shinzawa-Itoh, K, Nakashima, R., Yaono, R. and Yoshikawa, S. (1995) Science **269**, 1069–1074.

Wil N. Konings
H. Ronald Kaback
and Juke S. Lolkema

Contents of Volume 2

PART 1. PRIMARY TRANSPORT
Ion pumps

Multidrug resistance and export systems

PART 2. SECONDARY TRANSPORT
Substrate transport and exchangers

Neurotransmitters

Homeostasis

Contributors to Volume 2

L.M. Amzel, Department of Biophysics and Biophysical Chemistry, Johns Hopkins University School of Medicine, Baltimore, MD 21205, USA

Y. Anraku, Department of Biological Sciences, Graduate School of Science, University of Tokyo, Hongo, Bunkyo-ku, Tokyo 113, Japan

E.P. Bakker, Abteilung Mikrobiologie, Universität Osnabrück, D-49069 Osnabrück, Germany.

M.F. Bauer, Adolf Butenandt Institute of Physiological Chemistry, University of Munich, Goethestrasse 33, D-80336 Munich, Germany

H. Bénédetti, Laboratoire d'Ingénérie et Dynamique des Systèmes Membranaires, CNRS, 31 Chemein J. Aiguier, B.P. 71, 13402 Marseille Cedex 20, France

M.A. Blight, Institut de Génétique et Microbiologie, URA 1354 CNRS, Bâtiment 409, Université Paris XI, 91405 Orsay cedex 05, France

H. Boer, Department of Biochemistry, Groningen Biomolecular Sciences and Biotechnology Institute, University of Groningen, Nijenborgh 4, 9747 AG Groningen, The Netherlands

H. Bolhuis, Department of Microbiology, Groningen Biomolecular Sciences and Biotechnology Institute, University of Groningen, Kerklaan 30, 9751 NN Haren, The Netherlands

M. Bonhivers, Laboratoire des Biomembranes, URA CNRS 1116, Université Paris Sud, 91405 Orsay Cedex, France

I.R. Booth, Department of Molecular and Cell Biology, University of Aberdeen, Marischal College, Aberdeen, AB9 1AS, UK

J. Broos, Department of Biochemistry, Groningen Biomolecular Sciences and Biotechnology Institute, University of Groningen, Nijenborgh 4, 9747 AG Groningen, The Netherlands

N. Carrasco, Department of Molecular Pharmacology, Albert Einstein College of Medicine, Bronx, NY 10461, USA

A. Carruthers, Department of Biochemistry and Molecular Biology, Program in Molecular Medicine, University of Massachusetts Medical School, 373 Plantation Street, Worcester, MA 10605, USA

S.L. Chan, MRC Group in Membrane Biology, Departments of Medicine and Biochemistry, University of Toronto, Toronto, Canada

D. Chapman, Department of Protein and Molecular Biology, Royal Free Hospital School of Medicine, Rowland Hill Street, London NW3 2PK, UK

G. Dai, Department of Molecular Pharmacology, Albert Einstein College of Medicine, Bronx, NY 10461, USA

N. Demaurex, Division of Cell Biology, Hospital for Sick Children, 555 University Ave., Toronto, Canada

S. Dey, Department of Biochemistry, Wayne State University, School of Medicine, 540 E. Canfield Ave., Detroit, MI 48201, USA

P. Dimroth, Mikrobiologisches Institut, Eidgenössische Technische Hochschule, ETH-Zentrum, Schmelzbergstr. 7, CH-8092 Zürich, Switzerland

A.J.M. Driessen, Department of Microbiology, Groningen Biomolecular Sciences and Biotechnology Institute, University of Groningen, Kerklaan 30, 9751 NN Haren, The Netherlands

P. Fafournoux, I.N.R.A.-U.N.C.M. Theix, 63122 St Genès-Champanelle, France

G.P. Ferguson, Department of Molecular and Cell Biology, University of Aberdeen, Marischal College, Aberdeen, AB9 1AS, UK

C.V. Franklund, Department of Microbiology, School of Medicine, University of Virginia, Charlottesville, VA 22908, USA

M. Futai, Department of Biological Science, Institute of Scientific and Industrial Research, Osaka University, Ibaraki, Osaka 567, Japan

V. Géli, Laboratoire d'Ingénérie et Dynamique des Systèmes Membranaires, CNRS, 31 Chemein J. Aiguier, B.P. 71, 13402 Marseille Cedex 20, France

S. Grinstein, Division of Cell Biology, Hospital for Sick Children, 555 University Ave., Toronto, Canada

P. Gros, Department of Biochemistry, McGill University, 3655 Drummond, Montreal, Que., Canada H3G 1Y6

M. Hanna, Department of Biochemistry, McGill University, 3655 Drummond, Montreal, Que., Canada H3G 1Y6

P.I. Haris, Department of Protein and Molecular Biology, Royal Free Hospital School of Medicine, Rowland Hill Street, London NW3 2PK, UK

K.J. Hellingwerf, Department of Microbiology, E.C. Slater Institute, University of Amsterdam, 1018 WS Amsterdam, The Netherlands

I.B. Holland, Institut de Génétique et Microbiologie, URA 1354 CNRS, Bâtiment 409, Université Paris XI, 91405 Orsay cedex 05, France

K. Jahreis, Universität Osnabrück, Fachbereich Biologie/Chemie, D-49069 Osnabrück, Germany

M.A. Jones, Department of Molecular and Cell Biology, University of Aberdeen, Marischal College, Aberdeen, AB9 1AS, UK

H.R. Kaback, Howard Hughes Medical Institute, Departments of Physiology and Microbiology & Molecular Genetics, Molecular Biology Institute, University of California Los Angeles, Los Angeles, CA 90024-1662, USA

R.J. Kadner, Department of Microbiology, School of Medicine, University of Virginia, Charlottesville, VA 22908, USA

B.I. Kanner, Department of Biochemistry, Hadassah Medical School, The Hebrew University, Jerusalem, Israel

J.H. Kim, Division of Cell Biology, Hospital for Sick Children, 555 University Ave., Toronto, Canada

W.N. Konings, Department of Microbiology, Groningen Biomolecular Sciences and Biotechnology Institute, University of Groningen, Biology Centre, Kerklaan 30, 9751 NN Haren, The Netherlands

J.T. Lathrop, Department of Microbiology, School of Medicine, University of Virginia, Charlottesville, VA 22908, USA

M. Lebendiker, Division of Microbial and Molecular Ecology, The Institute of Life Sciences, The Hebrew University of Jerusalem, 91904 Jerusalem, Israel

M. Lee, Department of Genetics, University of Washington, Seattle, WA 98195, USA

J.W. Lengeler, Universität Osnabrück, Fachbereich Biologie/Chemie, D-49069 Osnabrück, Germany

L. Letellier, Laboratoire des Biomembranes, URA CNRS 1116, Université Paris Sud, 91405 Orsay Cedex, France

O. Levy, Department of Molecular Pharmacology, Albert Einstein College of Medicine, Bronx, NY 10461, USA

J.S. Lolkema, Department of Microbiology, Groningen Biomolecular Sciences and Biotechnology Institute, University of Groningen, Biology Centre, Kerklaan 30, 9751 NN Haren, The Netherlands

P.C. Maloney, Department of Physiology, Johns Hopkins Medical School, Baltimore, MD 21205, USA

C. Manoil, Department of Genetics, University of Washington, Seattle, WA 98195, USA

D. McLaggan, Department of Molecular and Cell Biology, University of Aberdeen, Marischal College, Aberdeen, AB9 1AS, UK

W. Meijberg, Department of Biochemistry, Groningen Biomolecular Sciences and Biotechnology Institute, University of Groningen, Nijenborgh 4, 9747 AG Groningen, The Netherlands

S. Miller, Department of Molecular and Cell Biology, University of Aberdeen, Marischal College, Aberdeen, AB9 1AS, UK

S. Mordoch, Division of Microbial and Molecular Ecology, The Institute of Life Sciences, The Hebrew University of Jerusalem, 91904 Jerusalem, Israel

L.S. Ness, Department of Molecular and Cell Biology, University of Aberdeen, Marischal College, Aberdeen, AB9 1AS, UK

W. Neupert, Adolf Butenandt Institute of Physiological Chemistry, University of Munich, Goethestrasse 33, D-80336 Munich, Germany

Y. Nikolaev, Department of Molecular and Cell Biology, University of Aberdeen, Marischal College, Aberdeen, AB9 1AS, UK

H. Omote, Department of Biological Science, Institute of Scientific and Industrial Research, Osaka University, Ibaraki, Osaka 567, Japan

E. Padan, Department of Microbial and Molecular Ecology, The Institute of Life Sciences, The Hebrew University of Jerusalem, 91904 Jerusalem, Israel

R. Palmen, CHU Rangueil, Bacteriologie, University Paul Sabatier, Toulouse, France

B. Poolman, Department of Microbiology, Groningen Biomolecular Sciences and Biotechnology Institute, University of Groningen, Biology Centre, Kerklaan 30, 9751 NN Haren, The Netherlands

M. Popov, MRC Group in Membrane Biology, Departments of Medicine and Biochemistry, University of Toronto, Toronto, Canada

J. Pouysségur, Centre de Biochimie-CNRS, Université de Nice, Parc Valrose, 06108 Nice, France

M. Putman, Department of Microbiology, Groningen Biomolecular Sciences and Biotechnology Institute, University of Groningen, Kerklaan 30, 9751 NN Haren, The Netherlands

R.A.F. Reithmeier, MRC Group in Membrane Biology, Departments of Medicine and Biochemistry, University of Toronto, Toronto, Canada

G.T. Robillard, Department of Biochemistry, Groningen Biomolecular Sciences and Biotechnology Institute, University of Groningen, Nijenborgh 4, 9747 AG Groningen, The Netherlands

B.P. Rosen, Department of Biochemistry, Wayne State University, School of Medicine, 540 E. Canfield Ave., Detroit, MI 48201, USA

J.P. Rosenbusch, Department of Microbiology, Biozentrum, University of Basel, Klingelbergstr. 70, CH-4056 Basel, Switzerland

G. Rudnick, Department of Pharmacology, Yale University School of Medicine, New Haven, CT 06510, USA

M.H. Saier, Jr., Department of Biology, University of California at San Diego, La Jolla, CA 92093-0116, USA

G.A. Scarborough, Department of Pharmacology, School of Medicine, University of North Carolina, Chapel Hill, NC 27599, USA

M. Schleyer, Abteilung für Angewandte Genetik der Mikroorganismen, Universität Osnabrück, D-49069 Osnabrück, Germany.

A. Schlösser, Abteilung für Angewandte Genetik der Mikroorganismen, Universität Osnabrück, D49069, Osnabrück, Germany

S. *Schuldiner*, Department of Microbial and Molecular Ecology, The Institute of Life Sciences, The Hebrew University of Jerusalem, 91904 Jerusalem, Israel

G.K. *Schuurman-Wolters*, Department of Biochemistry, Groningen Biomolecular Sciences and Biotechnology Institute, University of Groningen, Nijenborgh 4, 9747 AG Groningen, The Netherlands

S. *Stumpe*, Abteilung Mikrobiologie, Universität Osnabrück, D-49069 Osnabrück, Germany.

D. *Swaving-Dijkstra*, Department of Biochemistry, Groningen Biomolecular Sciences and Biotechnology Institute, University of Groningen, Nijenborgh 4, 9747 AG Groningen, The Netherlands

R. *ten Hoeve-Duurkens*, Department of Biochemistry, Groningen Biomolecular Sciences and Biotechnology Institute, University of Groningen, Nijenborgh 4, 9747 AG Groningen, The Netherlands

S. *Tötemeyer*, Department of Molecular and Cell Biology, University of Aberdeen, Marischal College, Aberdeen, AB9 1AS, UK

H.W. *van Veen*, Department of Microbiology, Groningen Biomolecular Sciences and Biotechnology Institute, University of Groningen, Kerklaan 30, 9751 NN Haren, The Netherlands

C.M. *Wood*, Department of Molecular and Cell Biology, University of Aberdeen, Marischal College, Aberdeen, AB9 1AS, UK

J.J. *Ye*, Department of Biology, University of California at San Diego, La Jolla, CA 92093-0116, USA

R. *Yelin*, Division of Microbial and Molecular Ecology, The Institute of Life Sciences, The Hebrew University of Jerusalem, 91904 Jerusalem, Israel

H. *Yerushalmi*, Division of Microbial and Molecular Ecology, The Institute of Life Sciences, The Hebrew University of Jerusalem, 91904 Jerusalem, Israel

R.J. *Zottola*, Department of Biochemistry and Molecular Biology, Program in Molecular Medicine, University of Massachusetts Medical School, 373 Plantation Street, Worcester, MA 10605, USA

CHAPTER 1

ATP-coupled Pumps for Heavy Metals and Metalloids

S. DEY and B.P. ROSEN

Department of Biochemistry, Wayne State University,
School of Medicine, 540 E. Canfield Ave., Detroit, MI 48201, USA

© *1996 Elsevier Science B.V.*
All rights reserved

Handbook of Biological Physics
Volume 2, edited by W.N. Konings, H.R. Kaback and J.S. Lolkema

Contents

1. Introduction

Efflux-mediated resistances to toxic metals are ubiquitous; heavy metal and metal-loid translocating ATPases have been identified in prokaryotes, both gram positive and gram negative, and in all types of eukaryotes, including fungi, plants, protozo-ans and animals. The chemical properties of the metals and metalloids vary consid-erably, as do the true substrates recognized by the transport systems. For example, cadmium is a cation that can be transported by members of the P-type family of ATPases. At the same time, cadmium reacts with thiolates as a soft metal, and members of the ABC-type family of ATPases most likely recognize thiol conjugates of cadmium rather than the metal ion. As(+3) and Sb(+3), the trivalent (reduced) forms of ionized arsenic and antimony, are metalloids or semi-metals that are oxyanions in solution but can react with thiolates as soft metals. Thus arsenicals and antimonials are most likely extruded as the glutathione conjugate in eukaryotic cells, while the ArsA ATPase probably transports the oxyanions arsenite and antimonite (although, as discussed below As(+3) and Sb(+3) allosterically activate the ATPase though reaction with thiol groups of cysteine residues in the catalytic subunit). In this review the members of those three evolutionarily unrelated families of translocat-ing ATPases that transport heavy metals or metalloids will be discussed.

2. Heavy metal transport by P-type ATPases

Within the family of P-type ATPases a subgroup of related proteins is found that transport Cu^+/Ag^+ or Cd^{2+}/Zn^{2+} [1]. Copper and zinc are metals that are required cofactors for some enzymes but are at the same time toxic in high concentrations. Thus opposing pairs of transport systems must exist that catalyze uptake and extrusion of these ions. Solioz has recently shown that there are two P-type ATPases, CopA and CopB, in the gram positive bacterium *Enterococcus hirae* [2,3]. CopA most likely is the uptake system, while CopB is the efflux pump. The CopB protein has been shown to catalyze ATP-driven Cu^+ and Ag^+ transport in membrane vesicles, demonstrating that this protein is a transport ATPase [3]. Genes with high sequence similarity to CopA and CopB are associated with inherited metabolic disorders in copper metabolism Menkes and Wilson diseases [1,4]. While these pro-teins have not been shown to catalyze copper transport, that is a reasonable inference.

The plasmid-borne cadmium resistance operon of *Staphylococcus aureus* en-codes a gene, *cadA*, for a P-type ATPase [5]. The *cadA* gene could be expressed in *Bacillus subtilis*, and everted membrane vesicles prepared from those cells accumu-lated $^{109}Cd^{2+}$ in an ATP-dependent process [6]. The CadA protein could be identified

3

on sodium dodecyl sulfate-polyacrylamide gel electrophoresis [7]. When expressed in *E. coli*, the *cadA* gene product was shown to catalyze uptake of both Cd^{2+} and Zn^{2+} into everted membrane vesicles (K. Tsai and A. Linet, personal communication). Cadmium transport was inhibited by micromolar concentrations of either Cu^+ or Cu^{2+}. Thus the CadA protein is an ATP-dependent pump for cadmium and zinc and perhaps copper. Related genes have been identified in a number of microorganisms, but none of the protein products have been identified [1].

In all P-type ATPases there is a proline residue in a putative membrane spanning region that may be part of the cation translocating domain of the protein. In the copper and cadmium P-type ATPases this conserved proline residue is flanked with either cysteines or a cysteine–histidine pair in the CopB protein. Although there are no data about their role, it has been postulated that these cysteines are in the translocation pathway [5]. However, it would seem that binding of a heavy metal to a cysteine pair would be an essentially irreversible phenomenon, inconsistent with a transport function. Alternately, the conserved proline residue located between the two cysteines could result in the two cysteines thiolates being too far apart to allow for them to be liganded to the metal simultaneously. In that case, a serial association might be involved in the translocation event.

Another region unique to the heavy metal translocating P-type ATPases is the presence of a cysteine containing motif (GXXCXXC) near the N-terminus of each of the proteins except for CopB. In the CadA protein this sequence is $GFTC_{23}ANC_{26}$. These two cysteines were individually altered to glycine and serine residues (K. Tsai and A. Linet, personal communication). Everted membrane vesicles prepared from cells expressing each of the mutants exhibited uptake of $^{109}Cd^{2+}$ at 5–30% wild type levels. These results suggest that the cysteines are not essential for the catalytic activity of these P-type ATPases but are required for maximal activity. They could form a portion of the substrate binding domain of the enzymes; alternatively, the N-termini could be activation domains. Most ATPases do not hydrolyze ATP in the absence of substrate. As described below, the catalytic subunit of the arsenite-translocating ATPase has a low basal rate of ATP hydrolysis in the absence of arsenite or antimonite. The enzyme is allosterically activated by binding of As(+3) to three cysteines that are not in the translocation pathway; thus arsenite is both substrate and activator but at different sites. Since binding of heavy metals to cysteine pairs is not easily reversible, a reasonable hypothesis is that the metal ion bound to the N-terminal cysteines in the copper and cadmium ATPases activates the enzymes but is not itself translocated.

3. Heavy metal transport by ABC-type ATPases

In mammalian cells the P-glycoprotein, the *mdr* gene product, is responsible for resistance to chemotherapeutic drugs [8,9]. When amplified this protein catalyzes extrusion of a large number of structurally unrelated drugs. The superfamily of proteins that includes the P-glycoprotein includes members found in most, probably all, organisms and is related to drug resistance in many cases [8,9]. There are a number of proteins encoded by genes that have diverged considerably from the main branch of P-glycoproteins and constitute a distinct subgroup of the ABC transporters superfamily. One subgroup has members that are associated with heavy metal resistances in different organisms. This subgroup includes the cystic fibrosis transmembrane regulator (CFTR) [10]; the multidrug resistance associated protein (MRP) found in multidrug resistant small cell carcinoma cell lines [11]; the YCF1 protein involved in cadmium resistance in yeast [12], the PgpA protein found in a variety of *Leishmania* species [13,14], and the recently described β cell high affinity sulfonylurea receptor, a regulator of insulin secretion [15]. The sequence similarities of PgpA, MRP and YCF1 are noteworthy, as overexpression of MRP and PgpA are associated with resistance to arsenite and antimonite, and YCF1 may transport the cadmium-glutathione complex $Cd(GS)_2$ (D. Thiele, personal communication). Another yeast ABC transporter, HMT1 [16], distantly related to YCF1 and PgpA, was shown to transport Cd^{2+} into the vacuole as a complex with phytochelatin, a derivative of glutathione [17]. Thus it is likely that the heavy metal transporters that are members of the ABC family transport thiolate complexes and not the free metals.

The treatment of choice for all forms of leishmaniasis depends on Sb(+5)-containing drugs such as sodium stibogluconate (Pentostam). Strains selected for arsenite (As(+3)) resistance exhibit cross resistance to antimonite (Sb(3+)) and to Pentostam [18,19]. Although amplification of the *pgpA* gene of *Leishmania* species is associated with low level arsenite resistance, high level resistance requires an efflux pump that is unrelated to the *ltpgpA* gene of *L. tarentolae* [20,21]. Recently everted plasma membrane vesicles from that organism have been shown to accumulate the glutathione complex of arsenite [22]. The transport system requires ATP and is inhibited by antimonite. The affinity of the pump is approximately 10^{-4} M for $As(GS)_3$. Although glutathione is present in *Leishmania* at about 4 nmol/10^8 cells, the other major intracellular reductant in *Leishmania* is the dithiol trypanothione, which contains two glutathione moieties crosslinked through the glycines by a spermidine and is also present in about equal amounts [23]. Thus the *in vivo* substrate might be a conjugate with either glutathione, trypanothione or both.

We propose a model for Pentostam resistance in *Leishmania* in which there are a minimum of three components (Fig. 1). The first step is reduction of Sb(5+)/As(5+) to Sb(3+)/As(3+). Pentostam contains oxidized Sb(5+), but most likely it must first be reduced to be cytotoxic. We would postulate the existence of a reductase similar to the plasmid-encoded arsenate reductases [24,25]. The second step would be conjugation of the reduced semimetal to the thiol of glutathione and/or trypanothione. Although no glutathione S-transferase activity has been reported in *Leishmania* species, one of the major intracellular reductants is trypanothione, so that perhaps additional efforts will be required to identify the proper conjugating enzyme. The final step in resistance is active extrusion by a glutathione(trypanothione)-linked pump that is perhaps similar to MRP in function if not in evolution. As we have shown, this pump is not PgpA. In addition to the *pgpA* gene, four other *pgp* genes have been identified in *L. tarentolae*, *pgpB* to *pgpE* [26]. We propose that one of these or an additional *pgp* gene is responsible for the high

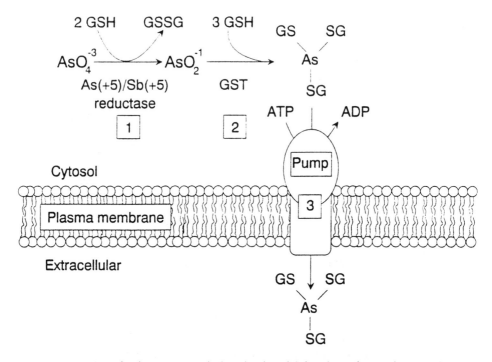

Fig. 1.   Mechanism of resistance to arsenicals and antimonials in eukaryotic organisms. A minimum of three components are required: (1) Reduction of Sb(5+)/As(5+) to Sb(3+)/As(3+). (2) Conjugation of the reduced semimetal to a thiol, most likely glutathione. (3) Active extrusion of the thiol conjugate by an ATP-coupled pump.

level resistance, and that the event(s) leading to resistance do not require gene amplification. Other ABC transporters with structural similarities with PgpA have been isolated from *L. tarentolae* (M. Ouellette, personal communication), and one of these could potentially correspond to the pump.

The fact that there are potentially multiple steps involved in resistance is a unifying factor, explaining why high level resistance requires multiple steps of selection, why arsenite resistance sometimes correlates with increased levels of glutathione S-transferase π in mammalian cells [27], and suggests a relationship between this metal resistance and the P-glycoprotein and MRP family.

4. The Ars arsenite-translocating ATPase

4.1. The ars operon

In contrast to the glutathione-linked pumps found in eukaryotes, resistance to arsenite and antimonite in bacterial cells has been found to be mediated by an oxyanion pump. In *Escherichia coli* arsenite resistance has been shown to be independent of the intracellular levels of reduced glutathione. Mutation in the genes

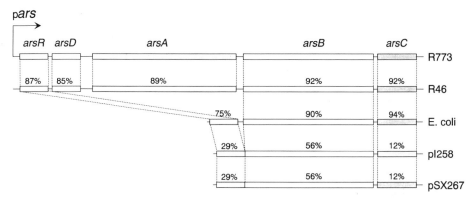

Fig. 2. Bacterial *ars* operon. In the top line the five genes of the *ars* operon of plasmid R773 are shown with the direction of transcription indicated by the arrow, starting with the promoter, p_{ars}. Genes are indicated by boxes, with the intergenic spaces as single lines. The genes of homologous *ars* operons of plasmid R46 (unpublished), the *E. coli* chromosomal operon [34], and staphylococcal plasmids pI258 [32] and pSX267 [33] are aligned below. The similarities of the gene products to the R773 proteins are given as % identity.

for glutathione synthesis (*gshA* and *gshB*) or glutathione reduction (*gor*) had no effect on arsenite resistance [28]. The arsenical pump of the *E. coli* plasmid R773 is encoded by the metalloregulated operon [29–31]. The operon consists of two regulatory genes (*arsR and arsD*) and three structural genes (*arsA, B, and C*) (Fig. 2). Resistance to arsenite and antimonite is conferred by a membrane-bound oxyanion translocating ATPase consisting of two different subunits, the ArsA and the ArsB proteins. The ArsA subunit is a 63 kDa anion-stimulated ATPase that remains anchored to the inner membrane by interaction with the ArsB protein (Fig. 3). The ArsB protein is an intrinsic membrane protein that most likely forms the anion conducting pathway. Expression of a third structural gene, the *arsC* gene, expands the resistance spectrum to allow for arsenate resistance. The 16 kDa ArsC protein functions as an arsenate reductase that reduces arsenate to arsenite and perhaps channels the product directly into the active site of the pump. Another transmissible plasmid, R46, has recently been shown to have a highly homologous *ars* operon with the same five genes (unpublished) (Fig. 2).

Similar arsenite resistance operons has been characterized from the plasmids of Gram-positive organisms (Fig. 2). The two operons, one from plasmid pI258 of *Staphylococcus aureus* [32] and the other from pSX267 of *Staphylococcus xylosus* [33] code for only three genes, *arsRBC*. A third operon recently been identified in the chromosome of *E. coli* [34] has been shown to be responsible for low level arsenical resistance [35]. This chromosomal operon is closely related in sequence to the R773 *ars* operon but, too, has only the *arsRBC* genes. The lack of the gene for the ATPase subunit, the *arsA* gene, in these operons raises a critical question about the energetics and molecular mechanism of the two efflux systems. This difference between the arsenical resistance systems will be considered in more detail below.

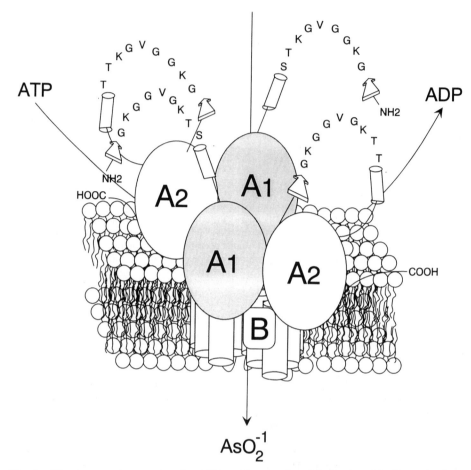

Fig. 3.   The Ars oxyanion-translocating ATPase. The complex of the ArsA and ArsB proteins forms an oxyanion-translocating ATPase that catalyzes extrusion oxyanions of arsenic and antimony in the +3 oxidation state. The ArsA protein has two homologous halves, A1 (N-terminal) and A2 (C-terminal). Two subunits of the ArsA protein are indicated to reflect its structure in solution. The ArsA protein is the catalytic subunit, exhibiting oxyanion-stimulated ATPase activity. The primary sequence of the two phosphate loops of the nucleotide binding domains in the A1 and A2 halves are shown. The ArsB protein is an inner membrane protein in *E. coli* and serves both as the membrane anchor for the ArsA protein and as the anion conducting subunit of the pump.

4.2. The ArsA protein

The sequence of the ArsA protein predicted from the nucleotide sequence suggests that it arose through gene duplication and fusion of a gene half the size of the existing *arsA* gene [29]. Both the N-terminal or A1 half of the ArsA protein and the C-terminal or A2 half contain a consensus sequence for a nucleotide binding site (Fig. 4A). Both sites have been shown by site directed mutagenesis to be required for resistance, ATPase activity and arsenite transport [36,37]. It is interesting that

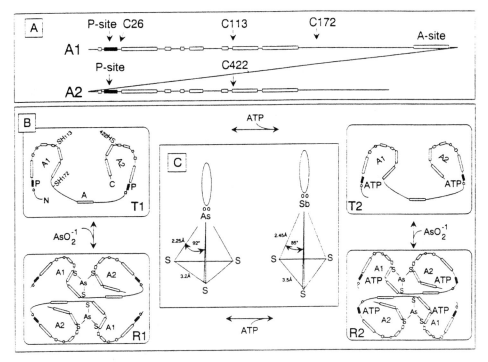

Fig. 4.   Metalloregulation of the ArsA ATPase. A: The ArsA protein consists of two homologous halves, A1 and A2. The aligned boxes in the two halves indicate the regions of greatest sequence similarity. Each half has a consensus sequence for the binding site of phosphoryl groups of ATP (P-site). The A-site is the region that interacts with adenine of ATP bound to the A1 half. The location of the four cysteine residues of the ArsA protein are indicated. B: ArsA ATPase activity is allosterically regulated by binding of the As(+3) or Sb(+3). The 63 kDa ArsA has independent binding sites for metalloid and ATP. In the absence of either nucleotide or As(+3)/Sb(+3) the monomer exists in an inactive T state. Binding of either ATP or semi-metal each produces a unique and different confor-mational change in the protein. In solution the monomer exists in equilibrium with dimer. As(+3) or Sb(+3) bind as soft metals to the thiolates of Cys113 and Cys172 in the A1 half and to Cys422 in the A2 half of each monomer, producing a more compact structure that favor the subunit interactions that stabilize the catalytically active R state. When the ArsA protein is membrane-bound as a subunit of the pump, it is postulated to be at all times a dimer, and the effect of substrate binding is to produce the conformational change from the inactive T state to the active R conformation, promoting catalysis. C: The structure of the soft metal-thiol cage in the ArsA protein is postulated from the bond angles and distances found by crystallographic analysis of small molecules containing As–S or Sb–S bonds. Both structures contain tricoordinately liganded sulfur thiolates, with the unpaired electron pair of the metal forming the apex of a tetrahedron.

all of the known homologues of the ArsA protein are soluble ATPases, none of which have transport functions [38]. The homologues include plasmid replication proteins, division protein and the NifH subunit of nitrogenase. The closest homo-logues are found in eukaryotes: in both *Caenorhabditis elegans* [39] and *Arabadop-sis* [40] reading frames of unknown function but related to the ArsA protein were identified in genome sequencing projects.

Although functionally a component of the oxyanion pump, overexpressed ArsA protein can be purified from cytosol. Purified ArsA protein exhibits arsenite- and antimonite-stimulated ATPase activity [41]. $Mg^{2+}ATP$ was the only nucleotide hydrolyzed, and inhibitors of other types of ATPases did not inhibit ArsA activity, including N,N'-dicyclohexylcarbodiimide, azide, vanadate, and nitrate [42]. The optimal pH range was 7.5 to 7.8. The only two oxyanions that stimulated ATPase activity were arsenite and antimonite, the substrates of the pump.

The ArsA protein forms an adduct with [^{32}P]ATP in a light activated reaction [41]. For reasons that are not understood, only the A1 site forms an adduct [36,37]. The glycine rich sequence is thought to interact the phosphoryl groups of ATP, while photoadduct formation is thought to be through the adenine ring. Thus identification of the site of the ^{32}P label would indicate which residues are near the adenine ring in the A1 ATP binding site. A sequence containing residues 283–304 of the ArsA protein has been identified as containing the label [43]. Interestingly, this sequence is over 260 residues away from the phosphate binding loop. When the two halves of the ArsA protein are aligned with each other, this sequence, termed the 'A-site' (for adenine interacting site) corresponds to a region of nonalignment at the end of the ArsA1 half of the protein (Fig. 4A). This suggests that the A site is a linker connecting the A1 and A2 halves of the protein.

As discussed above, the activity of the ArsA protein is low in the absence of the substrate of the pump, thus preventing ATPase hydrolysis in the absence of transport. The increase in activity appears to be an allosteric activation, where As(+3) or Sb(+3) bind at a site on the ArsA protein to activate and a separate site on the ArsB protein to be transported. There are several lines of evidence that these semimetals act as allosteric activators by homeotropic interaction of two ArsA subunits. First, genetic results suggest interaction of subunits [44]. Mutant *arsA* genes were expressed separately or in combination. Any combination of mutants in which the *arsA1* sequence was unaltered co-expressed with a wild type *arsA2* sequence resulted in resistance, even though the individual constructs did not. Thus two defective ArsA proteins could interact to form a functional ArsA complex. Second, using both chemical crosslinking and light scattering, purified ArsA protein has been shown to form an As(+3)/Sb(+3)-dependent homodimer [45]. These results suggest that soluble ArsA protein is in an equilibrium of inactive monomer and active dimer (Fig. 4B). Activation of the ArsA protein by arsenite or antimonite correlates with formation of an ArsA homodimer. Therefore the monomer is considered to be in the T form with a low basal activity, where as the dimer is the activated R form of the enzyme. The conformation that favors dimerization preferentially binds arsenite or antimonite; thus by mass action binding of the activator increases the concentration of dimer, resulting in activation.

What is the nature of the activator binding site? Are the metalloids bound as oxyanions through ionic bonds? Do they form hydrogen bonds with backbone atoms as do the phosphate and sulfate binding proteins [46]? Alternatively, they could react as soft metals. In the reduced form As(+3) can react with thiol groups. For example, in the reaction of arsenite with dithiolthreitol As(+3) is covalently bonded to two sulfur thiolates and the oxygen of one hydroxyl to form a tricoordinate pyramidal structure in which the free electron pair forms the apex of the pyramid.

Two lines of evidence suggested involvement of cysteine residues in activation of ATPase activity of the ArsA protein. First the ATPase activity of the ArsA protein is extremely sensitive to sulfhydryl reagents such as *N*-ethylmaleimide or methyl methanethiosulfonate [47,48]. The ArsA protein also reacts with fluorescent maleimides such as 2-(4′-maleimidoanilino)naphthalene-6-sulfonic acid that react specifically with cysteine residues, and preincubation with arsenite or antimonite decreased the rate of reaction [47].

The second line of evidence comes from site directed mutagenesis of the cysteinyl residues in the ArsA protein [48]. There are only four cysteines: Cys26, 113, 172, and 422 (Fig. 4A). Cys113 and 422 occupy equivalent positions in the two homologous halves of the ArsA protein. The codon for each cysteine was changed to serine codons. Mutants C113S, C172S, and C422S, when co-expressed with the wild type ArsB protein, give reduced level of resistance to arsenite and antimonite indicating loss of function. Cells expressing a C26S mutant had normal resistance. Each of the four altered proteins was purified and shown to have a structure grossly the same as the wild type, as determined by circular dichroism. The C26S protein had normal ATPase activity, while the other were defective. Each mutant ArsA protein had a K_m for ATP in the same order of magnitude as the wild type protein. The C113S, C172S and C422S enzymes had reduced affinity for antimonite and arsenite. The Vmax of the activated enzymes ranged from very low for the C113S and C422S enzymes to near normal for the C172S enzyme. These results suggest a mechanism of activation by formation of a tricoordinate complex between Sb(+3) or As(+3) and the cysteine thiolates 113, 172 and 422 (Fig. 4B).

Cysteines 113, 172 and 422 are quite distant from each other in the primary sequence. To produce the proposed tricoordinate As-trithiolate species, the sulfur atoms must be less then 4 Å from each other and from the arsenic atom. To test this model a molecular ruler was used. The compound dibromobimane forms a covalent complex with thiolates. It becomes fluorescent if both functional groups react with sulfur thiolates. To do so, the thiols must be less than 6 Å but more than 3 Å from each other. Mutant *arsA* genes were constructed with cysteine codons altered to serines substitutions in each of the four positions and in various combinations of the four mutations (H. Bhattacharjee and B.P. Rosen, unpublished). Purified wild type ArsA protein and each of the proteins with single substitutions reacted with dibromobimane to yield highly fluorescent products. Each ArsA protein from a double mutant containing a C26S alteration still formed a fluorescent product. Only when two of the three cysteines in the putative reactive triad of Cys 113, 172 or 422 were altered together did the resulting ArsA protein no longer form a fluorescent product with dibromobimane. These results indicate that the thiolate sulfur atoms of each of the three cysteine residues are within 3–6 Å of each other in the folded ArsA protein and strongly support a model in which they form the base of the pyramid that contains As(+3) or Sb(+3) as a central soft metal (Fig. 4C), thus accounting for metalloactivation. Note that although the overall geometry is similar between the complexes of the two semimetals, the bond angles and distances may be sufficiently different to account for the dramatically greater activation by Sb(+3) compared to As(+3).

4.3. The ArsB protein

The ArsB protein is 45 kDa integral membrane protein localized in the inner membrane. Topological analysis of the ArsB protein using gene fusions to three different reporter genes demonstrated that the protein has 12 membrane spanning α-helices, with 6 periplasmic and 5 cytoplasmic loop regions and the N-, and C-terminus oriented to the cytoplasmic side of the inner membrane (Fig. 5) [49].

The ArsB protein functions as membrane anchor for the ArsA protein [50]. Interaction between these two proteins has been characterized using *in vitro* reconstitution of purified ArsA protein with everted membrane vesicles containing the ArsB protein [51]. Interestingly, binding exhibits a sigmoidal dependence on the concentration of ArsA protein. In the presence of arsenite or antimonite, binding is hyperbolic, and the concentration of ArsA protein required for half-maximal saturation reduces 10-fold. As discussed above, in solution purified ArsA protein is active as a dimer, with As(+3) or Sb(+3) acting as an allosteric activator by promoting dimerization [45] (Fig. 4B). The metalloid-dependent conversion from sigmoidal to hyperbolic dependence on ArsA concentration is consistent with the membrane-bound form of the ArsA protein being a dimer. In the absence of metalloid the concentration of dimer would increase with increasing protein concentration; if only an ArsA dimer could bind to the ArsB protein, a sigmoidal dependence would follow. In the presence of metalloid the concentration of ArsA dimer would be substantially increased at low protein concentration. Binding would become hyperbolic, with an increase in the affinity of ArsB subunit for the ArsA subunit. However, in its physiological membrane bound state the ArsA subunit is firmly membrane bound, requiring chaotropic treatment to remove it [50]. We proposed that the complex contains a minimum of two ArsA subunits even in the absence of metalloid (Fig. 3). In the absence of the metalloid the two ArsA subunits exist in an inactive T state. Reaction of C113, C176 and C422 with either As(+3) or Sb(+3) induces an allosteric change in the membrane-bound form to the active R state of the ArsA dimer. If the metalloid binding site on the ArsA protein is an allosteric site, then it is likely that the ArsB protein contains a separate substrate site for oxyanion binding and subsequent translocation.

4.4. Arsenite transport in vitro

The ArsB protein has been overexpressed in a chimeric form with a portion of the ArsA protein genetically engineered to the NH_2-terminus of the former [52]. The chimeric protein, like the wild type ArsB protein, interacts with the ArsA protein in the inner membrane and confers some resistance to arsenite and antimonite. Everted membrane vesicles containing both the chimeric ArsB protein and the ArsA protein accumulates arsenite in an ATP-dependent manner [53]. The requirement for ATP could be replaced with other nucleoside triphosphates or by nonhydrolyzable ATP analog, ATPγS. ATP-dependent transport by the ArsA–ArsB complex was insensitive to CCCP. Transport in vesicles prepared from *unc* strains could not be driven by oxidation of lactate or NADH, suggesting that transport by ArsA–ArsB enzyme complex is obligatorily coupled to ATP. Like all other transport ATPases, arsenite

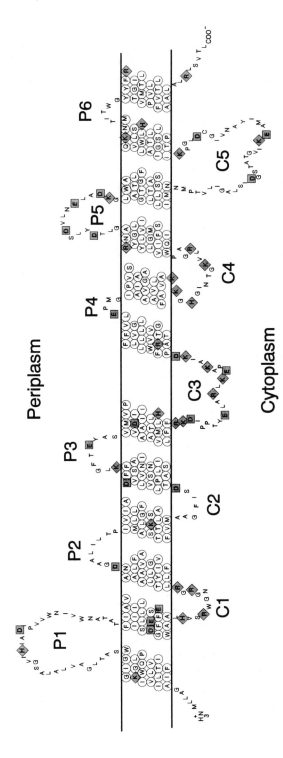

Fig. 5. Topological model of the ArsB protein. The model proposes 12 membrane-spanning α-helices joined by 6 periplasmic loops (P1–P6) and five cytoplasmic loops (C1–C5), with the N- and C-termini in the cytosol [49]. Acid (□) and basic (○) residues are indicated. The distribution of positive charges is in general agreement with the inside positive rule.

transport exhibited an absolute requirement for divalent cation specifically Mg^{2+} or Mn^{2+}. Inhibitors of P-type and F-type ATPases such as vanadate and NaN_3 respectively had no effect on arsenite uptake in everted membrane vesicles. The sulfhydryl reagent, N-ethylmaleimide completely inhibited arsenite transport under similar condition. Antimonite, the other substrate of the resistance pump, inhibited arsenite transport with an apparent K_i 10-fold less than the K_m for arsenite. The fact that ATP-dependent arsenite transport could be measured in everted membrane vesicles free from cytosolic components is further evidence that arsenite transport and resistance are independent of glutathione in *E. coli*.

4.5. Dual mode of energy coupling of the ArsB protein

The closest relatives of the R773 ArsB proteins are the homologues from other *ars* operons, including the ArsB proteins of the gram positive plasmids pI258 [32] and pSX267 [33] (both of which exhibit 56% identity) and the ArsB proteins encoded by the *E. coli* chromosome [34] (90% identity) and plasmid R46 (unpublished) (92% identity). Two other open reading frames of unknown function with slight but significant sequence similarity to the ArsB protein have recently been identified. One potentially encodes a 45 kDa hydrophobic protein in *Mycobacterium leprae* (L. Oskam, C. Hermans, G. Jarings, P.R. Klatser and R.A. Hartskeerl, unpublished), and the second is involved in the human disease type II oculocutaneous albinism [54].

Interestingly, the *ars* operons of the two staphylococcal plasmids and the *E. coli* chromosome do not encode for the catalytic subunit of the pump, the ArsA protein, yet they confer resistance and catalyze arsenite efflux from cells in an energy dependent process [32,35]. When the *arsA* gene was deleted from the R773 ars operon the ArsB protein alone conferred an intermediate level of resistance to arsenite and antimonite [55]. Similarly expression of an *arsA* gene in trans with the staphylococcal plasmid pI258 *arsB* or with a gene fusion consisting of the sequence of the first eight membrane spanning domains of the pI258 ArsB protein and the last four membrane spanning domains of the R773 ArsB protein increased arsenite resistance to the same level as in cells expressing the R773 *arsAarsB* genes [56]. Therefore both the *E. coli* plasmid R773 and the staphylococcal plasmid pI258 encoded ArsB proteins have the ability to function either as a subunit of the Ars oxyanion-translocating ATPase or independently as an oxyanion translocator. In the absence of the catalytic ATPase subunit, the ArsB protein might extrude arsenite, functioning as a secondary carrier coupled to the outside positive membrane potential. However, the possibility of recruiting a chromosomally encoded ATPase subunit by the ArsB protein could not be ruled out. To distinguish between these two possibilities the energetics of arsenite exclusion from cells of *E. coli* catalyzed by the R773 ArsB protein alone was compared with that in the presence of the ArsA subunit [55]. A primary pump would show dependency on chemical energy, while a secondary porter requires electrochemical energy. The studies were performed in *unc* strain that is unable to equilibrate chemical and electrochemical energy. De-energized cells of an *unc* strains supplied with glucose generates both ATP and electrochemical energy. Addition of KCN inhibits respiration, preventing generation of electrochemical energy, while ATP levels remain high through substrate level phosphorylation. Thus intracellular conditions can be established in which chemical energy is present in the absence of electrochemical energy. An ATP-dependent pump should still function under such conditions while a secondary carrier should not. In contrast, addition of respiratory chain substrates such as succinate allows for generation of electrochemical energy. In an *unc* strain this energy cannot be used to drive oxidative phosphorylation, so ATP levels remain low, establishing intracellular conditions where only electrochemical energy is available.

Under conditions where only chemical energy was available (glucose + cyanide), only cells with both the ArsA and ArsB proteins actively extruded arsenite [55]. Cells supplied with succinate (electrochemical energy only) did not extrude arsenite. In contrast the presence of ATP was not sufficient to drive arsenite exclusion from cells expressing the ArsB protein alone. ArsB-mediated exclusion of arsenite required intracellular conditions that generated electrochemical energy, inhibited by KCN and reversed by protonophores that dissipate the electrochemical gradient. These results clearly show that transport by the ArsB protein in the absence of the ArsA protein is coupled to electrochemical energy. Although relative contribution of ΔpH and $\Delta\psi$ can only be confirmed in an *in vitro* system, under the conditions used the majority of the driving force should have been in the form of a $\Delta\psi$. The simplest hypothesis is that anion transport by the ArsB protein is coupled to the positive exterior membrane potential, although coupling with other ions cannot be excluded from the data.

In contrast to the ArsA protein, where activation results from the reaction of the soft metal As(+3) with cysteine thiolates, the ArsB protein contains only a single cysteine residue that can be mutated to serine without effect on resistance or transport [57]. Thus its catalytic mechanism cannot involve thiol chemistry, and the ArsB protein is most likely a true oxyanion carrier protein. Since As(+3) and Sb(+3) are semimetallic elements that straddle the metallic series, their interaction with proteins by nonmetal oxyanion chemistry is just as likely as soft metal chemistry.

When the ArsA protein interacts with the ArsB protein it could induce conformational changes that occlude the ion conducting pathway. This is consistent with the finding that an inactive ArsA protein reverts the arsenite resistance conferred by the ArsB protein [55]. ATP hydrolysis by the ArsA subunit perhaps opens the pathway for ion translocation through conformational change of the complex. Like other solute translocating ATPases the free energy of ATP hydrolysis would also contribute directly to the translocation process of the anion by a mechanism not yet fully understood for any transport ATPase. Therefore the arsenite transport system is either an obligatory ATP-coupled pump or a secondary carrier coupled to electrochemical gradient, depending on the subunit composition of the transport complex (Fig. 6). Similarly the H^+-translocating F_0F_1 ATPase is an ATP-dependent H^+ pump, but when the complex is dissociated, the two portions function independently [58]. Dissociated F_1 is a soluble ATPase. By itself the F_0 is a proton conducting pathway which transports protons into the cell, responding to the $\Delta\psi$. Even though that is not the physiological role of the F_0, it points out the fact that other membrane complexes can work either as carriers or as a pump subunits.

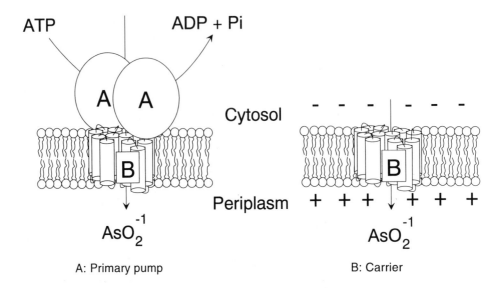

Fig. 6. Dual energy coupling of the ArsB protein. The ArsB protein is capable of functioning in two modes: (A) as an obligatory ATP-coupled pump or (B) as a $\Delta\psi$-driven carrier protein. When the ArsA protein is absent, the ArsB protein translocates arsenical and antimonial oxyanions, with energy derived from the proton pumping respiratory chain or F_0F_1 ATPase. When the ArsA protein is bound to the ArsB protein, the complex is an anion-translocating ATPase that is unable to utilize $\Delta\psi$.

If the ArsB protein is sufficient for arsenite resistance, why did the *arsA* gene evolve? One possibility is that the ArsA-ArsB complex provides a more effective resistance mechanism; ATP levels drop more slowly than the membrane potential under stressful conditions such as presence of toxic compounds in the environment. This is evident from the higher level of arsenite resistance conferred by the ArsB protein when the ArsA subunit is present [55]. Although the origin of the *arsA* gene is unknown, the presence of natural and man-made arsenicals and antimonials in the environment [59,60] could have exerted a selective pressure for the evolution of a more efficient pump in the some organisms.

5. Evolution of ion pumps

The ability of transport proteins to serve either as carriers or as pump subunits suggests a hypothesis for the evolution of transport ATPases. Multisubunit pumps could not have evolved in a single step. More likely the genes for the subunits evolved separately for a variety of different functions. As the extrinsic and intrinsic components of the membrane complexes evolved affinity for each other, the complexes eventually became functional to produce transport ATPases such as the Ars ATPase and the F_0F_1 (Fig. 7). The bacterial solute transport ATPases of the ABC family such as the maltose pump of *E. coli* [61] may similarly have evolved from

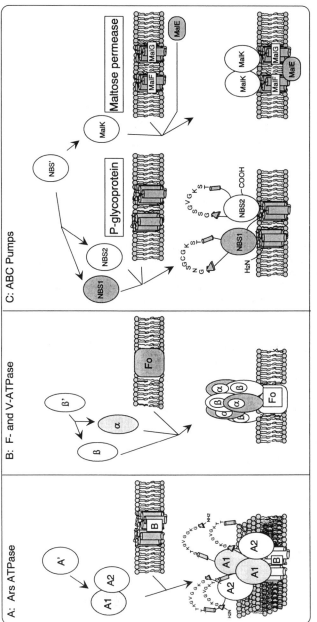

Fig. 7. Model for the evolution of primary pumps. Primary pumps evolved from the separate evolution of the genes for membrane proteins and soluble ATPases. Membrane proteins such as the ArsB protein, the F_0, or the ancestor of the membrane components of the P-glycoprotein or maltose permease may have evolved as secondary carriers for protons, ions or organic solutes or as channels. Soluble proteins with ATPase activity evolved to catalyze reactions involved in intracellular functions. The genes for the smaller nucleotide binding proteins ancestral to the N- and C-terminal halves of the ArsA protein or the P-glycoprotein or to the α and β subunits of the F_1 would have duplicated (some extant pumps simply use multiple copies of the same gene product, for example, the MalK protein). The duplicated genes would then evolve separate functions, sometimes fused to produce larger proteins with multiple nucleotide-binding sites, e.g., the *arsA* gene, or other times remaining separate within a single operon, e.g., the genes for the α and β subunits of F_1 and eventually becoming functional solute-translocating ATPases such as the oxyanion pump (A) and the H^+-translocating F_0F_1 (B). In members of the ABC superfamily, gene fusions created single genes for large multifunctional proteins in eukaryotes such as the P-glycoprotein or remained as a multisubunit complex of separate proteins such as the maltose permease (C). If the genes for the membrane components of the ABC superfamily arose separately, they could have had different original functions, for example secondary sugar transport for the MalF/G complex or chloride channel activity for the P-glycoprotein.

the independent evolution of the *malK* gene for the soluble ATPases and the *malFG* genes for the intrinsic membrane subunits. The *malE* gene for the binding protein is dispensable and most likely a later addition to the bacterial members of the ABC transporter family. The genes for the eukaryotic members of the ABC family such as the P-glycoprotein [8,9] and cystic fibrosis transmembrane regulator (CFTR) protein [10] are single large genes that contain all of the coding information for the pump complex. It is not unusual for large eukaryotic genes to encode multifunctional proteins equivalent to the products of operons of separate genes in prokaryotes.

What is clear it that the ABC family is a superfamily only in terms of the portions of the proteins involved in energy utilization, in particular around the ATP binding sites [62]. The membrane subunits of the bacterial pumps fall into a number of subfamilies that appear to be unrelated to the membrane sectors of the larger eukaryotic proteins [62,63]. For example, the integral membrane proteins of the Nod or Cys systems are related to each other but unrelated to the analogous membrane constituents of the Cys or Nod systems, even though the subunits with the ATP binding sites are closely related. Similarly, analysis of the membrane proteins of the maltose permease demonstrates no relationship to the membrane sectors of the P-glycoprotein. In turn neither of them show a relationship to the *drrB* gene product, the putative membrane sector of the *Streptomyces* daunorubicin efflux system [64]. Different membrane proteins may have formed association with a family of soluble ATPase to give rise to a heterogeneous group of solute ATPases such as the ABC superfamily. The fact that the ArsA protein, the α and β subunits of the F_1, and the catalytic subunits of the ABC proteins all have the same consensus nucleotide binding site [65] suggests that this portion of the catalytic process may be common among transport ATPases. On the other hand, the mechanism of solute translocation through the intrinsic membrane proteins may be as unrelated as their sequences, even among the members of the supposed ABC superfamily. Finally, evolution of pumps through association with different types of membrane proteins could also account for the observation that some such as the P-glycoprotein are pumps, while others such as CFTR are ion channels members, and others such as the ArsB protein or F_0 can have dual modes of energy coupling.

References

1.   Silver, S. and Ji, G. (1994) Environ. Hlth. Perspec. **102**, 107–113.
2.   Odermatt, A., Suter, H., Krapf, R. and Solioz, M. (1993) J. Biol. Chem. **268**, 12775–12779.
3.   Solioz, M. and Odermatt, A. (1995) J. Biol. Chem. **270**, 9217–9221.
4.   Bull, P.C. and Cox, D.W. (1994) Trends Genet. **10**, 246–252.
5.   Nucifora, G., Chu, L., Misra, T.K. and Silver, S. (1989) Proc. Natl. Acad. Sci. (USA) **86**, 3544–3548.
6.   Tsai, K.J., Yoon, K.-P. and Lynn, A.R. (1992) J. Bacteriol **174**, 116–121.
7.   Tsai, K.J. and Lynn, A.R. (1993) Arch. Biochem. Biophys. **305**, 267–270.
8.   Gottesman, M.M. and Pastan, I. (1993) Annu. Rev. Biochem. **62**, 385–427.
9.   Juranka, P.F., Zastawny, R.L., and Ling, V. (1989) FASEB J. **3**, 2583–2592.
10. Grzelczak, Z., Zielenski, J., Lok., S., Plavsic, N., Chou, J.-L., Drumm, M.L., Iannuzzi, M.C., Collins, F.S. and Tsui, L.-C. (1989) Science **245**, 1066–1073.
11. Cole, S.P.C., Bhardwaj, G., Gerlach, J.H., Mackie, J.E., Grant, C.E., Almquist, K.C., Stewart, A.J.,

Kurz, E.U., Duncan, A.M.V., Deely, R.G. (1992) Science **258**, 1650–1654.

12. Szczypka, M.S., Wemmie, J.A., Moye-Rowley, W.S. and Thiele D.J. (1994) J. Biol. Chem. **269**, 22853–22857.

13. Beverley, S.M. (1991) Ann. Rev. Microbiol. 45, 417–444.

14. Ouellette, M., Légaré, D. and Papadopoulou, B. (1994) Trends Microbiol. **2**, 407–411.

15. Aguilar-Bryan, L., Nichols, C.G., Wechsler, S.W., Clement, J.P. IV, Boyd, A.E. III, Gonzalez, G., Herrera-Sosa, H., Nguy, K., Bryan, J. and Nelson, D.A. (1995) Science **268**, 423–426.

16. Ortiz, D.F., Kreppel, L., Speiser, D.M., Scheel, G., McDonald, G. and Ow, D.W. (1992) EMBO J. **11**, 3491–3499.

17. Ortiz, D.F., Ruscitti, T., McCue, K.F. and Ow, D.W. (1994) J. Biol. Chem. **270**, 4721–4728.

18. Callahan, H.L. and Beverley, S.M. (1991) J. Biol. Chem. **266**, 18427–18430.

19. Ouellette, M. and Borst, P. (1991) Res. Microbiol. **142**, 737–746.

20. Papadopoulou, B., Roy, G., Dey, S., Rosen, B.P. and Ouellette, M. (1994) J. Biol. Chem. **296**, 11980–11986.

21. Dey, S., Papadopoulou, B., Roy, G., Grondin, K., Dou, D., Rosen, B.P., and Ouellette M. (1994) Molec. Biol. Parasitol. 67, 49–57.

22. Dey, S., Ouellette M., Lightbody, J., Papadopoulou, B. and Rosen, B.P. (1996) Proc. Natl. Acad. Sci. USA **93**, 2192–2197.

23. Fairlamb, A.H., Blackburn, P., Ulrich, P., Chait, B.T., Cerami A. (1985) Science **227**, 1485–1487.

24. Ji, G. and Silver, S. (1992) Proc. Natl. Acad. Sci. **89**, 9474–9478.

25. Gladysheva, T.B., Oden, K.L. and Rosen, B.P. (1994) Biochemistry **33**, 7287–7293.

26. Légaré, D., Hettema, E and Ouellette, M. (1994) Molec. Biochem. Parasitol. **68**, 81–91.

27. Wang, H.F. and Lee, T.C. (1993) Biochem. Biophys. Res. Commun. **192**, 1093–1099.

28. Oden, K.L., Gladysheva, T.B. and Rosen, B.P. (1994) Molec. Microbiol. **12**, 301–306.

29. Chen, C.M., Misra, T., Silver, S. and Rosen, B.P. (1986) J. Biol. Chem. **261**, 15030–15038.

30. Rosen, B.P., Bhattacharjee, H. and Shi, W.P. (1995) J. Biomembr. Bioenerg. **27**, 85–91.

31. Dey, S. and Rosen, B.P. (1995) in Drug Transport in Antimicrobial and Anticancer Chemotherapy, ed N.H. Georgopapadakou, Dekker, pp. 103–132.

32. Ji, G. and Silver, S. (1992) J. Bacteriol. **174**, 3684–3694.

33. Rosenstein R., Peschel, P., Wieland B. and Götz, F. (1992) J. Bacteriol. **174**, 3676–3683.

34. Sofia, H.J., Burland, V., Daniels, D.L., Plunkett III, G. and Blattner, F.R. (1994) Nucl. Acids Res. **22**, 2576–2586.

35. Carlin, A., Shi, W., Dey, S. and Rosen, B.P. (1995) J. Bacteriol. **177**, 981–986.

36. Karkaria, C.E., Chen, C.M. and Rosen, B.P. (1990) J. Biol. Chem. **265**, 7832–7836.

37. Kaur, P. and Rosen, B.P. (1992) J. Biol. Chem. **267**, 19272–19277.

38. Kaur, P. and Rosen, B.P. (1992) Plasmid **27**, 29–40.

39. Coulson, A., Waterston, R., Kiff, J., Sulston, J. and Kohara, Y. (1988) Nature **335**, 184–186.

40. Newman, T., de Bruijn, F.J., Green, P., Keegstra, K., Kende, H., McIntosh, L., Ohlrogge, J., Raikhel, N., Somerville, S., Thomashow, M., Retzel, E. and Somerville, C. (1994) Plant Physiol. **106**, 1241–1255.

41. Rosen, B.P., Weigel, U., Karkaria, C. and Gangola, P. (1988) J. Biol. Chem. **263**, 3067–3070.

42. Hsu, C.M. and Rosen, B.P. (1989) J. Biol. Chem. **264**, 17349–17354.

43. Kaur, P. and Rosen, B.P. (1994) Biochemistry **33**, 6456–6461.

44. Kaur, P. and Rosen B.P. (1992) J. Bacteriol. **175**, 351–357.

45. Hsu, C.M., Kaur, P., Karkaria, C.E., Steiner, R.F. and Rosen, B.P. (1991) J. Biol. Chem. **266**, 2327–2332.

46. Luecke, H. and Quiocho, F.A. (1990) Nature **347**, 402–406.

47. Ksenzenko, M.Y., Kessel, D.H. and Rosen, B.P. (1993) Biochemistry, **32**, 13362–13368.

48. Bhattacharjee, H., Li, J. Ksenzenko, M.Y. and Rosen, B.P. (1995) J. Biol. Chem. **270**, 1–6.

49. Wu, J.H., Tisa, L.S. and Rosen, B.P. (1992) J. Biol. Chem. **267**, 12570–12576.

50. Tisa, L.S. and Rosen, B.P. (1990) J. Biol. Chem. **265**, 190–194.

51. Dey, S., Dou, D., Tisa, L.S. and Rosen, B.P. (1994) Arch. Biochem. Biophys. **311**, 418–424.

52. Dou, D., Owalabi, J.B., Dey, S. and Rosen, B.P. (1992) J. Biol. Chem. **267**, 25768–25775.

53. Dey, S., Dou, D. and Rosen, B.P. (1994) J. Biol. Chem. **269**, 25442–25446.

54. Rinchik, E.M., Bultman, S.J., Horsthemke, B., Lee, S.T., Strunk, K.M., Spritz, R.A., Avidano, K.M., Jong, M.T. and Nicholls, R.D. (1993) Nature **361**, 72–76.
55. Dey, S. and Rosen, B.P. (1995) J. Bacteriol. **177**, 385–389.
56. Dou, D., Dey, S. and Rosen, B.P. (1994) Antonie van Leeuwenhoek **65**, 359–368.
57. Chen, Y., Dey, S. and Rosen, B.P., in preparation.
58. Futai, M. and Kanazawa, H. (1983) Microbiol. Rev. **47**, 285–313.
59. Abernathy, J.R. (1983) in Arsenic, eds W.H. Lederer and R.J. Fensterheim, pp. 57–62, Van Nostrand Reinhold Co., New York.
60. Alden, J.C. (1983) in Arsenic, eds W.H. Lederer and R.J. Fensterheim, pp. 63–71, Van Nostrand Reinhold Co., New York.
61. Shuman H.A. and Panagiotidis, C.H. (1993) J. Biomembr. Bioenerg. **25**, 613–620.
62. Reizer, J., Reizer, A and Saier, M.H. (1992) Prot. Sci. **1**, 1326–1332.
63. Fath, J.M. and Kotler, R. (1993) Microbiol. Rev. **57**, 995–1017.
64. Guilfoile, P.G. and Hutchinson, C.R. (1991) Proc. Natl. Acad. Sci. (USA) **88**, 8553–8557.
65. Walker, J.E., Saraste, M., Runswick, M.J., and Gay, N.J. (1982) EMBO J. **1**, 945–951.

CHAPTER 2

Sodium Ion Coupled F_1F_0 ATPases

P. DIMROTH

Mikrobiologisches Institut, Eidgenössische Technische Hochschule,
ETH-Zentrum, Schmelzbergstr. 7, CH-8092 Zurich, Switzerland

© 1996 Elsevier Science B.V.
All rights reserved

Handbook of Biological Physics
Volume 2, edited by W.N. Konings, H.R. Kaback and J.S. Lolkema

Contents

1. Introduction

ATP is the central compound of energy metabolism in all living cells. As the membrane of these cells is typically impermeable to ATP, the ATP molecule is permanently synthesized within the cell from ADP and inorganic phosphate with the consumption of an energy source such as light or chemical energy. Numerous enzymes within the cell hydrolyze the ATP molecule again into ADP and inorganic phosphate and couple the free energy of this reaction to do work in either of three forms: chemical work in biosynthetic reactions, mechanical work in the various systems performing motion or osmotic work in the numerous transport processes. The efficiency of this cell energy cycle is enormous. Under resting conditions, an adult may turn over half of its body weight in ATP per day and this amount may increase under working conditions to several times the body weight [1,2].

Most of the ATP synthesis in aerobic cells is performed by F-type or F_1F_0 ATP synthases (also called F_1F_0 ATPases because the direction of the reaction can be reverted). These enzymes are located in the inner membrane of mitochondria, in the thylakoid membrane of chloroplasts or in the cytoplasmic membrane of bacteria (for recent reviews see Refs. [1,3–8]). The membrane-bound location of the F_1F_0 ATPases is essential for the energy converting function of these catalysts. In photophosphorylation or oxidative phosphorylation light or chemical energy, respectively, is converted into an electrochemical proton gradient ($\Delta\mu H^+$) which is exploited by the ATPase to drive the endergonic synthesis of ATP from ADP and inorganic phosphate.

It was formerly believed that energization of ATP synthesis can only be performed *via* a $\Delta\mu H^+$, because proton movement across the ATPase was regarded to be an intrinsic part of the catalytic mechanism of ATP synthesis [9]. This view had to be modified, when it was discovered that in the anaerobic bacterium *Propionigenium modestum* a Na^+ cycle is operative that directly drives ATP synthesis *via* an ATPase of the F_1F_0 type [10–12]. The Na^+ gradient has the same function in ATP synthesis by the *P. modestum* ATPase as the proton gradient by other ATPases. The altered specificity of the *P. modestum* enzyme to use Na^+ as coupling ion has distinct advantages in studying several aspects of this enzyme experimentally. Unlike H^+ concentration, the Na^+ concentration can be varied over a broad range without denaturing the protein. Na^+ movements across the membrane can be easily monitored with the radioactive isotope $^{22}Na^+$ and with this technique, exchange of internal and external Na^+ ions or even counterflow of radioactive Na^+ against a concentration gradient of the unlabeled alkali ion can be measured. Moreover, a vectorial catalyst performing Na^+ pumping across the membrane is expected to contain a Na^+ binding site. Amino acids contributing liganding groups to this site may be detected by mutagenesis, if such mutants exhibit altered properties, or even complete loss of Na^+ binding, or a change of ion binding specificity to another cation. These specific advantages of the *P. modestum* ATPase to use Na^+ as its principle coupling ion have in the past been used to study the mechanism of ion translocation and to determine the Na^+ (cation) binding site of F_1F_0 ATPases [13].

2. General features of F_1F_0 ATPases

Despite the wide distribution of the F_1F_0 ATPases in nature, structure and function of these catalysts have been highly conserved. The usual physiological function of these enzymes is to synthesize ATP under the consumption of $\Delta\mu H^+$ ($\Delta\mu Na^+$ in some cases). Only in some anaerobic bacteria is the direction reversed and $\Delta\mu H^+$ is generated at the expense of ATP hydrolysis.

Under the electron microscope, the appearance of the ATPases from such different sources as bacteria, mitochondria and chloroplasts is essentially the same. A globular headpiece is connected *via* a stalk to a broad basepiece, the membrane integral sector of the molecule [14–16]. In biochemical experiments, the molecule usually dissociates into two parts, the water-soluble F_1 moiety and the integral membrane sector, called F_0. The headpiece is predominantly F_1, the basepiece is F_0 and the stalk is derived from both F_1 and F_0 components.

The F_1 moiety consists of five different subunits, α, β, γ, δ, ϵ, which in the case of the *Escherichia coli* F_1 have molecular masses (in kDa) of α (55.2), β (50.1), γ (31.4), δ (19.6) and ϵ (14.9), respectively. These subunits are arranged in the stoichiometry $\alpha_3\beta_3\gamma\delta\epsilon$, yielding a molecular mass of the *E. coli* F_1 of 382 kDa [3]. Sequence comparison of different ATPases revealed that the β and α subunits are the most highly conserved, with about 70% similarity between the mitochondrial and *E. coli* β subunits, whereas the smaller subunits are less conserved [3–8].

The F_0 moiety of *E. coli* consists of three different subunits, a, b, c, with molecular masses of 30.3, 17.3 and 8.3 kDa, respectively. The stoichiometry of these subunits in the F_0 moiety is a b_2 $c_{10\pm1}$, yielding a molecular mass of about 148 kDa. Subunits in addition to the three basic F_0 subunits have been found in the mitochondrial and chloroplast enzymes [3–8].

The isolated F_1 moiety has been shown to catalyze the hydrolysis of ATP to ADP and inorganic phosphate, and the F_0 moiety catalyzes proton (Na^+) conduction across the membrane. When F_1 and F_0 are coupled, the F_1F_0 complex functions as a reversible, H^+ (Na^+)-translocating ATPase or ATP synthase [3–8].

A breakthrough for understanding the function of the F_1F_0 ATPase came recently from resolving the structure of F_1 from beef heart mitochondria at 2.8 Å resolution [17]. The structure shows a globular hexagonal arrangement of alternating α and β subunits, positioned like the segments of an orange around the central γ-subunit. The N-terminal parts of the α and β subunits consist of β-barrel structure and form something like a crown on top of the particle, opposite to the F_0 moiety that keeps the whole assembly together. The γ-subunit forms an extended antiparallel coiled coil structure in the center of the $\alpha_3\beta_3$ complex, with the C-terminal residue emerging into a dimple of the crown structure. In the middle of the α and β

subunits, a nucleotide binding site is apparent near the interface of each alternating $\alpha\beta$ pair, where the three catalytical binding sites are formed preferentially from amino acids of the β-subunits and the three non catalytical binding sites are preferentially formed from amino acids of the α-subunits. Importantly, the nucleotide binding sites on each $\alpha\beta$ pair are in one of three defined conformational states: at any one time, one catalytic site is occupied with ATP (in the crystal with the analogue AMP-PNP), the next is occupied with ADP, and the third site is empty. These different conformations are apparently caused by different interactions of the asymmetrical γ-subunit with certain parts of the α- and β-subunits. The asymmetric features of the F_1 structure are in perfect agreement with the binding change mechanism, which was formulated based on kinetic experiments [5,9,18,19]. According to this model, each catalytic site can occur in either of three different conformations, an o (open) conformation with no nucleotides bound, a l (loose) conformation with loosely bound ADP and phosphate and a t (tight) conformation with tightly bound ATP. At any one time, all three conformations exist in parallel on the three catalytic sites of the β-subunits, and during ATP synthesis, the three conformations change simultaneously so that each site undergoes a continuous change between all three different conformations. The supposed driving force for these conformational changes is the energy of the electrochemical proton (Na^+) gradient. ATP synthesis is accomplished by converting the site with the l conformation, containing loosely-bound ADP and phosphate, into the t conformation. Here, tightly-bound ATP is synthesized from ADP and phosphate at an equilibrium constant near 1. After a new conformational change, which brings the t conformation into an o conformation, ATP is released from its binding to the protein. These continuous and simultaneous conformational changes of the three catalytic sites could be explained best in accordance to the structure, if the asymmetric γ-subunit would rotate relative to the α and β subunits [17]. By this rotation of the γ-subunit, a similar contact would be formed with each of the three β-subunits, one after the other, so that each one of them would go through the same set of conformational changes during a complete turn of the γ-subunit. The rotational model resembles the flagellar motor, where the flagellar rotation is driven by the $\Delta\mu H^+$ [20]. Likewise, proton movement across the F_0 part of the ATPase should be linked to the rotation of the γ-subunit, if the rotational model is correct. The mechanism of this rotation is neither understood, however, for the flagellar motor, nor for the F_1F_0 ATPase.

3. Discovery of a Na^+-translocating F_1F_0 ATPase

Propionigenium modestum, a strictly anaerobic bacterium, grows from the fermentation of succinate to propionate and CO_2. The pathway of energy metabolism is shown in Fig. 1 [10]. It involves the intermediates succinyl-CoA, (*R*) and (*S*) methylmalonyl-CoA and propionyl-CoA and operates at a total free energy change of only -20 kJ/mol. This amount of energy is not sufficient to synthesize 1 mol of ATP from ADP and inorganic phosphate, because the free energy expense for ATP synthesis under *in vivo* conditions is about 70–80 kJ/mol [21]. Thus, three to four decarboxylation reactions have to be coupled with one ATP synthesis reaction.

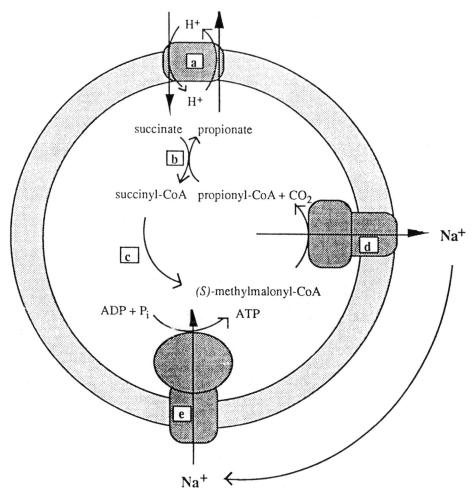

Fig. 1.   Energy metabolism of *Propionigenium modestum* with a Na^+ ion cycle coupling the exergonic decarboxylation of (S)-methylmalonyl-CoA to endergonic ATP synthesis. (a) Succinate uptake system (the mechanism of the transport is not known); (b) succinate propionyl-CoA:CoA transferase; (c) methylmalonyl-CoA mutase and methylmalonyl-CoA epimerase; (d) methylmalonyl-CoA decarboxylase; (e) ATPase. Stoichiometries for the coupling ions are unknown and were therefore not taken into account.

An important enzyme in the energy metabolism is the membrane-bound methyl-malonyl-CoA decarboxylase. This is a biotin-containing Na^+ pump that converts the free energy of the decarboxylation reaction into a $\Delta\mu Na^+$ [10,22,23]. The electro-chemical Na^+ ion gradient thus generated is the only free energy source for ATP synthesis in this bacterium. To cope with the problem of using $\Delta\mu Na^+$ for ATP synthesis, *P. modestum* possesses a unique type of F_1F_0 ATPase that uses Na^+ as the physiological coupling ions [10–12]. The coupling of the ATPase to Na^+ translocation was first demonstrated with inverted bacterial vesicles which catalyzed an active accumulation of Na⁺ within the interial volume of these vesicles upon ATP hydrolysis [10]. The transport was completely abolished by the Na⁺-translocating ionophore monensin, but was not significantly affected by the uncoupler carbonylcyanide-*p*-trifluoromethoxyphenylhydrazone (CCFP), indicating that a proton gradient is not an intermediate for the accumulation of Na⁺. The energetic link between methyl-malonyl-CoA decarboxylation and ATP synthesis *via* a Na⁺ circuit was also demonstrated with these vesicles. The synthesis of ATP, which was dependent on the decarboxylation reaction, was completely abolished in presence of the Na⁺ ionophore monensin [10].

The action of the *P. modestum* ATPase as a primary Na⁺ pump has been clearly demonstrated with proteoliposomes containing the purified enzyme. The rate of

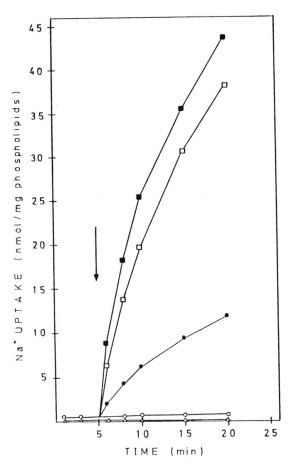

Fig. 2. Kinetics of Na⁺ transport into proteoliposomes containing the purified ATPase from *P. modestum* and effect of ionophores on Na⁺ uptake. The Na⁺ transport was initiated by ATP addition (arrow). Ionophores were contained as follows: (●) control; (■) 30 μM valinomycin; (□) 50 μM CCCP; (Δ) 50 μM monensin; (O) control in the absence of ATP [12].

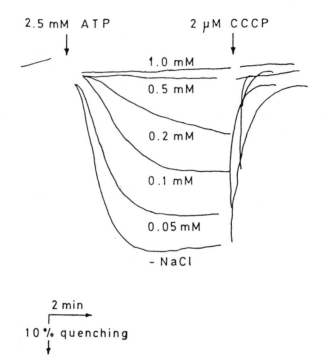

Fig. 3.   Proton pumping by proteoliposomes containing the purified ATPase from *P. modestum* and effect of Na$^+$ on this activity. ACMA fluorescence quenching was initiated by adding ATP (arrow) to reaction mixtures containing the concentrations of NaCl indicated. The quenching was released by adding 2 μM CCCP [24].

ATP-driven Na$^+$ transport into these proteoliposomes increased about four-fold after dissipation of the rate-determining membrane potential with either valinomycin (in the presence of K$^+$) or the uncoupler carbonylcyanide-*m*-chloro-phenylhydrazone (CCCP) [12] (Fig. 2). Stimulation rather than inhibition of the Na$^+$ transport rate by the uncoupler excludes a Na$^+$ transport mechanism *via* a H$^+$ gradient as an intermediate and demonstrates that the *P. modestum* ATPase pumps Na$^+$ ions directly. Another interesting feature of the *P. modestum* ATPase is a switch from Na$^+$ to H$^+$ translocation at unphysiologically low (< 1 mM) Na$^+$ concentrations [24]. The proton pump activity was demonstrated with reconstituted proteoliposomes by the ATP-dependent quenching of the fluorescence of 9-amino-6-chloro-2-methoxyacridine (ACMA) which is indicative for the formation of a ΔpH (internally acidic) across the membrane. No ΔpH was formed in the presence of the uncoupler CCCP or by blocking the ATPase with dicyclohexylcarbodiimide (DCCD). The effect of Na$^+$ ions on the quenching of ACMA fluorescence is shown in Fig. 3 [24]. The rate and extent of ΔpH formation were at maximum in the absence of Na$^+$ and declined gradually to zero by increasing the Na$^+$ concentration from 0 to 1 mM. Half maximal proton-pumping activity was observed at 0.2 mM NaCl. The half maximal rate of Na$^+$ transport was at 0.4 mM NaCl [24]. These results suggest competition of Na$^+$ and H$^+$ for a common binding site on the enzyme. At pH 7.0 and Na$^+$ concentrations of about 10^{-4} M, this site may be occupied with either Na$^+$ or H$^+$, and both cations are translocated, whereas at Na$^+$ concentrations of 10^{-3} M and above, only the alkali ion is transported. Whereas the affinity of the ATPase for protons exceeds its affinity for Na$^+$ by about three orders of magnitude, the V_{max} for Na$^+$ transport is about 10 times the V_{max} for H$^+$ transport. The efficiency of the *P. modestum* ATPase to function as a Na$^+$ pump is therefore based primarily on a high V_{max} and not on a low K_m for this alkali ion. However, as *P. modestum* requires 0.35 M NaCl for optimum growth, the physiological coupling ion is certainly Na$^+$ and not H$^+$.

4. Relationship of the *P. modestum* ATPase to H$^+$-translocating F_1F_0 ATPases

After purification of the *P. modestum* ATPase, it became evident that its subunit composition was very similar to that of H$^+$-translocating F_1F_0 ATPases, e.g. the *E. coli* enzyme [12]. The structural relationship was also indicated from the cross reactivity of the α and β subunits of the *P. modestum* ATPase with antiserum raised against the *E. coli* enzyme [25]. The genes of the *P. modestum* enzyme are arranged on the chromosome as an operon with the same sequence of genes as in the *E. coli atp* operon [26,27]. Sequence analyses revealed a clear relationship between the two enzymes, yielding 71% identity for the β subunits, 54% for the α subunits and between 10 and 38% for the remaining subunits [28,29].

This structural relationship correlates with biochemical properties of the enzyme function, e.g. the effect of various inhibitors such as DCCD, venturicidin, or azide [12]. It is not to be doubted, therefore, that the principal catalytic events leading to ATP synthesis are the same for Na$^+$- or H$^+$-translocating ATPases. This excludes any mechanism that specifically involves protons as coupling ions and may not be extended to Na$^+$ as coupling ions. One such proposal, which predicted the direct specific participation of the coupling protons in the chemical mechanism of ATP synthesis at the catalytic sites of F_1, must therefore be dismissed [9]. The existence of the Na$^+$-translocating ATPase of *P. modestum* also argues against a proton translocation mechanism across the membrane-bound F_0 moiety *via* a network of hydrogen bonds, where a proton adding the chain of hydrogen bonds at one end leads to the release of another proton at the other end (proton wire) [30].

5. Evidence for a Na$^+$ binding site in the *P. modestum* ATPase

ATPase activity measurements with the F_1F_0 complex from *P. modestum* revealed a significant activation by Na$^+$ ions and at 10 times higher concentrations a similar activation by Li$^+$ ions (Fig. 4) [10–12,31]. This activation was only observed with the F_1F_0 complex, but not with the isolated F_1 moiety. Upon reconstitution of F_1 with F_0, the activation of ATPase activity by Na$^+$ ions was recovered [11]. These results indicated a Na$^+$ binding site on the F_0 moiety of the *P. modestum* ATPase that must be occupied to activate the ATPase activity of the F_1F_0 complex. Activation of ATP

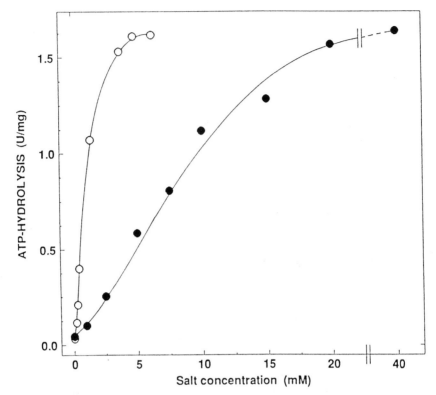

Fig. 4.   Activation of the *P. modestum* ATPase by NaCl and LiCl. ATPase activity measurements were performed at pH 8.0 with the NaCl (O) or LiCl (●) concentrations indicated [31].

hydrolysis at the F_1 moiety by binding the coupling ion to the F_0 moiety is in accord with a coupling between the chemical and the vectorial reaction and indicates long-ranging conformational changes that are transmitted from the Na^+-binding sites on the F_0 moiety into the catalytic sites of the F_1 moiety.

The isolated F_1F_0 ATPase of *P. modestum* catalyzed the hydrolysis of ATP in a Na^+ and pH-dependent fashion. The pH profile in the absence of Na^+ ions and in presence of saturating Na^+ concentrations is shown in Fig. 5 [32]. The maximal ATPase activity observed with saturating Na^+ concentrations followed a bell-shaped curve with a broad pH optimum around pH 7.5 and rather sharp decreases below pH 7.0 and above pH 8.5. In contrast, ATPase activity in the absence of Na^+ ions was much smaller, and significant activities were observed only at pH values below neutrality. The pH-optimum of the Na^+-independent ATPase was at pH 6.0–6.5, and at increasing pH values the activity dropped continuously to reach nondetectable levels at pH 9.7. The degree of activation by Na^+ ions increased continuously from about 1.5-fold at pH 5.5 to about 20-fold at pH 7.5 and above. This remarkable activation of the enzyme by Na^+ ions indicates that a reaction step involving binding of Na^+ ions to the F_0 moiety is rate limiting for the ATPase activity. The pH profile of maximal ATPase activity at saturating Na^+ concentrations

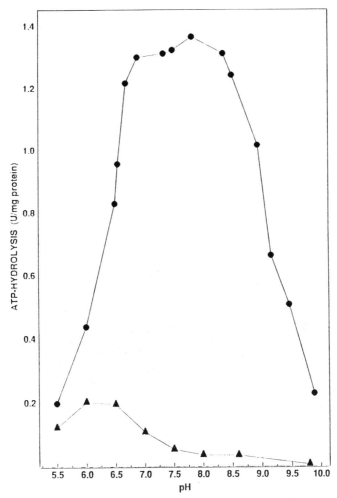

Fig. 5.   pH-dependence of ATPase activity in the presence of saturating Na⁺ concentrations (8 mM
NaCl) (●) or without NaCl addition (endogenous Na⁺ concentration 0.03 mM (▲) [32].

therefore reflects the pH profile of the Na⁺ binding active site on F_0, not the pH
dependence of the ATP hydrolyzing catalytic center on F_1. A plot of log V_{max} against
pH indicated pK values of 6.8 and 8.7 that may therefore be the pK values of
ionizing groups at the Na⁺ binding site [32]. Activation profiles of the ATPase by
Na⁺ ions were very different at acidic or alkaline pH values. At pH 9.0, there was
virtually no ATPase activity in the absence of Na⁺ ions. With increasing Na⁺
concentrations, the ATPase activity increased with strong positive cooperativity (n_H
= 2.6) and reached a maximum at about 1 mM NaCl. These results indicate the
presence of at least three Na⁺ binding sites that must be occupied for maximal
ATPase activity. At pH 6.5, ATPase activity was already observed in the absence
of Na⁺ ions. Upon Na⁺ addition, the activity increased with negative cooperativity

($n_H = 0.6$) reaching its maximum at about 12 mM NaCl. The negative cooperativity at acidic pH values may indicate that more than one site must be occupied with Na^+ to obtain full activation. If most of the sites are occupied with protons, the binding of several Na^+ ions to the same ATPase becomes less favorite, resulting in negative cooperativity. In summary, these results suggest a competition between Na^+ and H^+ for a common binding site. The binding of protons to this site increases at the more acidic pH values and causes partial activation of the ATPase. At the more alkaline pH values, Na^+ is the favorite cation to bind to the coupling ion binding site, and its binding causes the maximal activation of ATPase activity [32].

6. Mechanism of Na^+ translocation through the F_0 part of the *P. modestum* ATPase

To understand the catalytic mechanism of F_1F_0 ATPases and the coupling between the chemical and the vectorial reaction, detailed knowledge about the mechanism of ion translocation through F_0 is indispensable. A widely accepted model for H^+ translocation across F_0 from proton-translocating ATPases is that of a pore, through which the protons flow in response to an electrical or concentration gradient and which becomes blocked upon F_1 binding [4,33]. As Na^+ fluxes are much easier to determine than proton fluxes, studies on the mode of Na^+ transport by the *P. modestum* ATPase were undertaken. Therefore, the F_1 part of reconstituted proteoliposomes containing F_1F_0 was removed, and Na^+ uptake into the F_0-liposomes was determined. Upon application of a K^+ diffusion potential (inside negative), Na^+ was taken up into these F_0-liposomes. The velocity of Na^+ uptake was dependent on the external Na^+ concentration, following typical saturation kinetics with K_T of about 0.6 mM [34]. As this values is about the same as that for Na^+ pumping by the F_1F_0 complex, the Na^+ binding properties of F_0 are not significantly influenced by the binding of F_1. It is also evident from these results that a Na^+ binding step is involved in the Na^+ translocation and thus that F_0 contains (a) Na^+ binding site(s). In the absence of Na^+ ions, protons were translocated by the F_0 liposomes. Upon Na^+ addition, the rate of proton transport gradually decreased, and with 2 mM NaCl proton conduction through F_0 was completely abolished. The same NaCl concentration on the inside of the proteoliposomes, however, was without effect on proton translocation, indicating that the accessibility of the Na^+ binding site in F_0 is strictly oriented with respect to the applied membrane potential [34]. With an inside negative membrane potential, Na^+ binds from the outside and is released to the inside, but no Na^+ transport occurs in the opposite direction.

Of considerable interest was the observation that the rate of Na^+ translocation downhill its concentration gradient was very low if no membrane potential was applied [34,35]. Even at $\Delta\psi = -40$ mV, the Na^+ transport into the proteoliposomes remained slow. However, on further increasing the potential, the rate of Na^+ transport increased rapidly to reach about 10 times higher rates at -115 mV than at -40 mV. This behavior could indicate a gated channel with a threshold of > -40 mV for opening, or alternatively, a transporter with a membrane potential-dependent step in the reaction cycle.

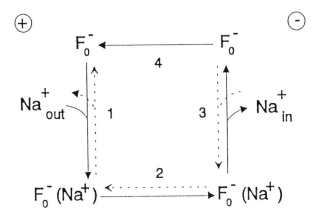

Fig. 6.   Schematic representation of reactions involved in Na⁺ influx and counterflow as catalyzed
by the F_0 moiety of the *P. modestum* ATPase. Na⁺ counterflow proceeds by the reversible reactions
1, 2 and 3 (solid and dashed lines) in the absence of a membrane potential. Na⁺ uptake (solid lines)
in addition requires the membrane potential-dependent reorientation of the unloaded carrier (step 4).
At membrane potentials of about −100 mV, the velocity of step 4 must be so high that counterflow does
not occur. F_0 catalyzes Na⁺ efflux, if the membrane potential is reversed (outside negative) [34].

It was important, therefore, to determine whether in the absence of $\Delta\psi$ F_0 catalyzed an exchange of internal and external Na⁺ ions. The results obtained indicated that the exchange was not only catalyzed at equal Na⁺ concentrations on both sides of the membrane, but also if ²²Na⁺ had to be translocated against a Na⁺ concentration gradient (counterflow) [34]. ²²Na⁺ counterflow could be accomplished by a channel mechanism, if a diffusion potential was generated by the downhill flux of unlabeled Na⁺ ions, which could then drive ²²Na⁺ uptake against the Na⁺ concentration gradient. The observation that ²²Na⁺ counterflow was 10 times faster than Na⁺ uptake in the absence of internal Na⁺ and at $\Delta\psi = -40$ mV, however, is not compatible with the channel mechanism.

We therefore explain these data by a transporter-type model, as shown in Fig. 6 [34]. It is assumed that the binary complex of F_0 and Na⁺ can move freely over the membrane and that Na⁺ can bind or dissociate on both sides. This leads to the observed exchange of Na⁺ in the absence of a membrane potential. In the presence of a potential, the model predicts that the unloaded carrier moves its (empty) Na⁺ site from the negative towards the positive surface of the membrane. If this process is faster than the binding of Na⁺ to the empty site or the reorientation of the binary complex, only unidirectional Na⁺ transport occurs and exchange is prevented, exactly what was found experimentally.

7. Location of the Na⁺ binding site

As described above, there is compelling evidence that the Na⁺ binding site is located on the F_0 part of the *P. modestum* ATPase. This location is in accord with the function of F_0 to translocate Na⁺ ions across the membrane. The cooperative binding of Na^+ with $n_H = 2.6$ (see above) indicates that at least three Na^+ binding sites must exist simultaneously. As only the c subunit is present in a high enough copy number to meet this requirement, this subunit was the best candidate for Na^+ binding.

The c subunit of different F_1F_0 ATPases is known to be the target for the reaction with DCCD. This compound has been shown to specifically modify a conserved acidic amino acid residue within the C-terminal membrane-spanning helix of this protein [36,37]. Extensive mutational analyses were performed with the *E. coli* c subunit. If the DCCD-reactive amino acid (asp61) was mutated to asparagine or glycine, the proton translocating activity of the ATPase was completely abolished [36]. Interestingly, in the double mutation D61G, A24D, the function of ATP-driven proton translocation was partially recovered [38]. An explanation of this remarkable observation could be provided, when the structure of the c subunit was partially solved by NMR spectroscopic techniques [39–42]. It has been observed that after the usual purification of this protein by extraction with chloroform/methanol (2:1) [39], the structure was largely retained in the solvent mixture chloroform/methanol/H_2O (4:4:1) [39]. These features include the unique chemical reactivity of Asp61 with DCCD [40]. Furthermore, the structure of two anti-parallel helices that are in close contact to one another was retained in the solvent. These two membrane-spanning helices are connected by a hydrophilic loop that makes essential contacts to F_1 subunits. In the structure, the A24 methyl and the D61 carboxyl group are in close contact to each other. Loss of function by mutating D61 to G and partial recovery of this function by the additional mutation of A24 to D, therefore, suggests that a charged residue at the location of these two residues in the double helix is essential for function.

The polar loop contains the conserved sequence, Arg41–Gln42–Pro43. Mutations Q42E or R41K were shown to have an uncoupled phenotype with near normal F_1 and F_0 functions [42,43]. This part of the loop region seems therefore very important for the coupling between the chemical events in F_1 with the vectorial events in F_0. Interestingly, suppressor mutations for the uncoupled Q42E subunit c mutants were found in subunit ε of the F_1 part. These results indicate that the polar loop and subunit ε function in the transmission of the conformational change that couples H^+ transport to ATP synthesis, either through a direct physical contact, or indirectly *via* another of the stalk subunits [45].

The c subunit of the *P. modestum* ATPase is assumed to have a similar gross structure as the *E. coli* c subunit. Alignment of the sequences indicates that the *P. modestum* c subunit extends the *E. coli* c subunit by 4 amino acids at the N-terminus and by 6 amino acids at the C-terminus and that the conserved acidic residue in the C-terminal helix is glutamate 65 [28]. As this residue was a key candidate for a Na^+ binding site, we studied the kinetics of ATPase inactivation by DCCD, which is known to specifically modify the conserved acidic amino acid in the c subunit [36]. The second order rate constant for this reaction was 1×10^5 M^{-1} min^{-1} at pH 5.6 and 0°C which is in the range of second order rate constants (k_{cat}/K_m) of many enzymes [31]. The reaction was roughly 10^7 times faster than the reaction of DCCD with acetic acid at the same pH and temperature. This high rate of inactivation is in accord with the remarkable specificity for modification of a single amino acid residue in the protein [32].

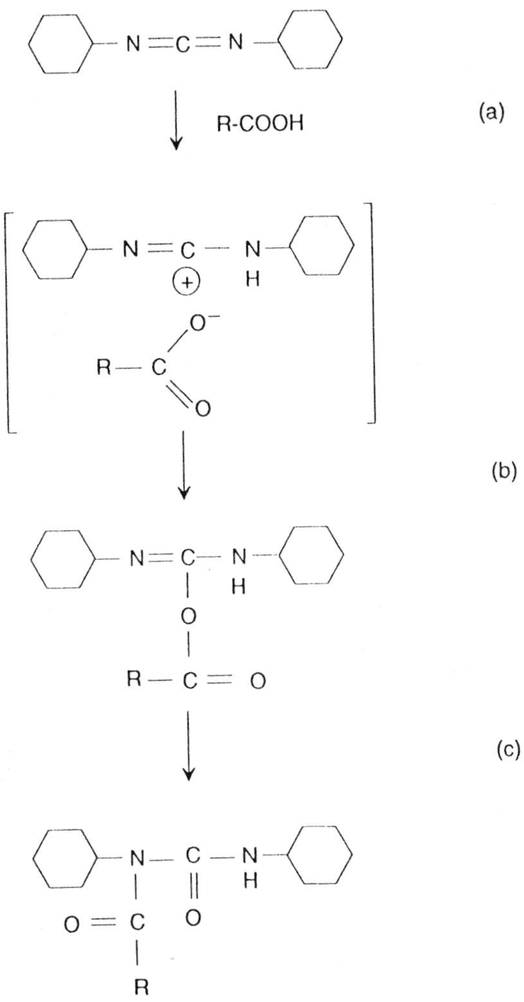

Fig. 7. Hypothetical reaction of DCCD with the protonated carboxyl group of glutamate 65 of subunit c [32].

The rate of ATPase inactivation by DCCD increased with decreasing pH, following a titration curve with pK = 7.0. These results indicate that the DCCD-reactive glutamate residue of subunit c must be protonated in order to get modified and that the carboxyl group of this glutamate residue dissociates with a pK of 7.0. A hypothetical mechanism for the reaction of the carboxyl group of glutamate 65 of the c subunit with DCCD is shown in Fig. 7 [31].

Another important observation was a specific protection of the ATPase by Na^+ ions from inactivation by DCCD [31,32]. This protection was observed at all pH values investigated and led to a shift of the titration curve into the more acidic range. In the presence of 0.5 mM or 1 mM NaCl, the pK of Glu65 thus decreased to 6.2 or 5.8, respectively. These results have been interpreted by a competition of H$^+$ and Na$^+$ for binding to cGlu65, which therefore appears to be the binding site for Na$^+$ ions or protons, if Na$^+$ ions are absent. The protective effect of Na$^+$ ions on the modification by DCCD can be explained by shifting the equilibrium between the protonated and depronotated form of glutamate 65 to the side of the latter, because the concentration of the 'free', deprotonated form of Glu65 is reduced by binding Na$^+$ ions. Na$^+$ ions, therefore, protect from the modification by DCCD by diminishing the amount of the DCCD-reactive protonated form of glutamate 65.

Complementary to the results from ATPase inactivation kinetics were results from [^{14}C]DCCD labeling studies. It was shown that subunit c of the *P. modestum* ATPase was specifically labeled upon incubation with this compound and that this labeling was enhanced at acidic pH values and was prevented in the presence of Na$^+$ ions [32]. These results are in accord with the idea that Na$^+$ binds at the DCCD-reactive site, i.e. glutamate 65 of subunit c. This conclusion was further substantiated by DCCD labeling studies with subunit c, isolated with chloroform/methanol/H$_2$O (4/4/1). After transferring the protein into an aqueous buffer containing dodecylmaltoside, it became labeled after incubation with [^{14}C]DCCD, and the presence of Na$^+$ prevented this reaction [46]. Thus, the isolated c subunit retains the Na$^+$ binding site, which when occupied prevents the modification of glutamate 65 by DCCD.

Intrigued by the very high rate and specificity of the reaction of DCCD with glutamate 65 of subunit c, we considered the possibility that parts of the DCCD structure might resemble the structure of an amino acid side chain that could come into close contact with Glu65 during the catalytic cycle. The protonated DCCD, which might represent the transition state, shares some structural elements with arginine (Fig. 8). The rationale for the DCCD reactivity and specificity could thus be the resemblance at the active site of protonated DCCD and arginine. Supporting evidence for this hypothesis was obtained in studies of the effect of ethylisopropylamiloride (EIPA) on the ATPase. EIPA which has a guanidinium group like arginine protected the enzyme competitively from the modification by DCCD [31].

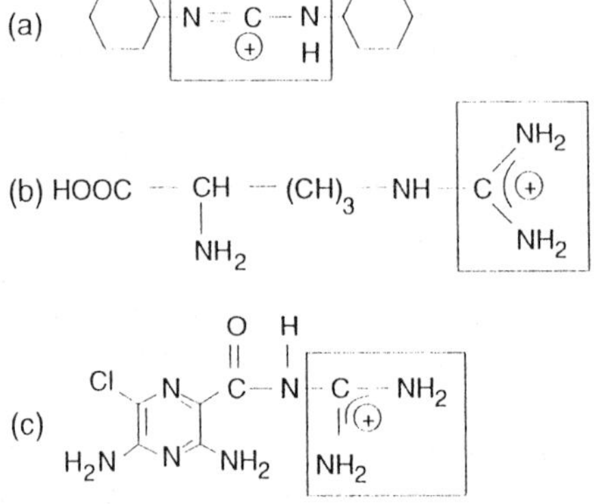

Fig. 8.   Structures of protonated DCCD (a), arginine (b), and EIPA (c). Similar structural elements are boxed.

EIPA like DCCD thus seems to bind near the reactive glutamate-65 residue of subunit c. Such reasoning immediately leads to a transport mechanism for Na^+ (or H^+), where the cation is bound to cGlu65 in one step and is released from this site by ion exchange with an arginine side chain in a later step. In the highly hydrophobic environment of Glu65, the forces attracting a cation to the glutamate anion should be very strong. An ion exchange-type of mechanism would therefore be an ideal means to release the bound Na^+ (or H^+) from the binding to the carboxylate. In addition, this mechanism avoids an energetically unfavorable generation of an unbalanced negative charge in a medium of low dielectric, because this charge would always be compensated by binding Na^+ (or H^+), or arginine. Conserved arginine residues that are essential for function exist in the polar loop region of the c subunit and in a membrane-spanning region of the a subunit.

8. Construction of *P. modestum/E.coli* ATPase hybrids and properties of these chimeras

The construction of *P. modestum/E. coli* ATPase hybrids and determination of their specificity with respect to the coupling ions is a powerful tool to identify those parts of the enzyme, by which these ions are recognized. In one approach to construct such hybrids *in vitro*, a complex consisting of the F_1 moiety of *E. coli* and the F_0 moiety of *P. modestum* (EF_1PF_0) was formed [25]. This hybrid ATPase exhibited the same coupling ion specificity as the homogeneous *P. modestum* ATPase, indicating that the F_0 portion is exclusively responsible for recognition and translocation of the coupling ions. This observation strongly argues against a participation of the coupling protons in the chemical synthesis of ATP at the F_1 catalytic site (see above). The coupling ratio of the EF_1PF_0 ATPase hybrid was only 0.007 Na^+ transported per ATP hydrolyzed [25]. The sophisticated mechanism of coupling was therefore highly impaired in the hybrid enzyme.

In another approach, we tried to complement *E. coli* mutants with defects of certain subunits of the ATPase with the genes on plasmids encoding the corresponding subunits from *P. modestum*. Although subunits a or c were synthesized in the respective *E. coli* mutant, they were unable to substitute for the defective *E. coli* subunits, so that a functional ATPase was not formed [47].

Other expression studies were performed in the *unc* deletion mutant strain *E. coli* DK8. Combinations of ATPase genes from *P. modestum* and *E. coli* were constructed on plasmids and used to transform *E. coli* DK8. Several of these constructs are shown in Fig. 9. Transformants with plasmid pBWF018, expressing an *E. coli* F_1/*P. modestum* F_0 hybrid, exhibited considerable ATPase activities. The enzyme was not functional as an ATP synthase, however, because the coupling between ATP hydrolysis and Na^+ transport was impaired [48]. The reason for this uncoupled phenotype is probably the imperfect interaction between *E. coli* F_1 and *P. modestum* F_0, which is apparent from the weak binding of the F_1 moiety to the membrane. Plasmid pUGP2 contains the whole *atp* operon from *P. modestum* under the control of a *lac* promoter. The defect in the *E.coli* ATPase in strain DK8 could not be complemented by this plasmid. Although all ATPase subunits were expressed,

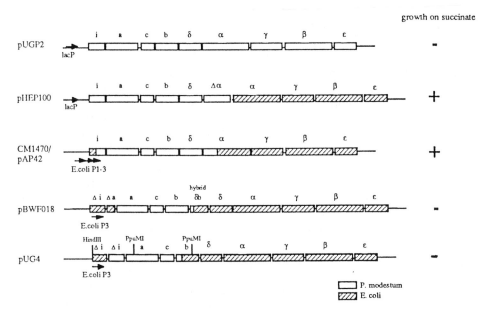

Fig. 9.   Survey of hybrid *P. modestum/E. coli unc* operons encoded on plasmids (pUG2, pHEP100, pBWF018, pUG4) and on the chromosome (CM1470/pAP42, equivalent to strain PEF42).

there was no detectable ATP hydrolyzing activity, indicating that the F_1 part was not functional [48]. The reason for the missing function could be the lack of an additional *P. modestum* gene product (assembly factor), which cannot be replaced by the corresponding *E. coli* protein. An example for the requirement of an assembly factor for the *in vivo* synthesis of a functional F_1F_0 ATPase is known for yeast, where a 30.2 kDa gene product of the ATP10 gene assures the correct assembly of the F_0 complex [49]. Plasmid pUG4 harbors the genes for subunits a and c from *P. modestum* and those for subunits α, β, γ, δ and ε from *E. coli*. The N-terminal hydrophobic part of subunit b, which is supposed to interact with the membrane integral subunits a and c derived from *P. modestum*, and the hydrophilic C-terminal part, which probably interacts with the F_1 subunits, derived from *E. coli*. The hybrid ATPase expressed from pUG4 was unable to complement the ATPase deletion of strain DK8 [48]. The ATP hydrolysis activity of the transformed mutant strain was similar to that of the wild type, but the F_1 moiety was only loosely bound to the membrane, indicating an improper interaction between the F_1 and F_0 moieties. It was also shown that subunit b was accessible to proteases and was partially degraded in the *E. coli* cells. A catalytically active and properly assembled F_0 complex was not formed with the hybrid b subunit. After solubilization of the ATPase from the membrane with Triton X-100 and partial purification, neither a nor c subunits were detectable, whereas the b subunit and a proteolytic fragment of it could be clearly identified. It was not surprising, therefore, that the F_0 moiety expressed from pUG4 was not functional as a proton conductor in response to a membrane potential across the membrane [48].

Interestingly, a functional hybrid ATPase was synthesized from plasmid pHEP100 which carries the *P. modestum* genes for subunits a, b, c and δ and the *E. coli* genes for subunits α, β, γ and ε [50,51]. The hybrid enzyme synthesized in *E. coli* was a fully functional ATP synthase, because the transformed deletion mutant could grow on succinate minimal medium. Growth on this carbon source is dependent on oxidative phosphorylation. A distinct difference of *E. coli* DK8 pHEP100 from the wild type was its Na$^+$-dependence for growth on succinate. This result indicates that H$^+$ and Na$^+$ cycles are operating together for ATP synthesis: the ΔμH$^+$ created by the respiratory chain is converted into ΔμNa$^+$ by the Na$^+$/H$^+$ antiporter and the thus formed ΔμNa$^+$ is used by the hybrid ATPase for ATP synthesis.

The hybrid ATPase has been isolated and its biochemical properties have been studied [50,51]. These proved to be very similar to those of the *P. modestum* wild-type ATPase. The hybrid enzyme was specifically activated by Na$^+$ or Li$^+$ ions with very similar activation profiles at pH 6.0 or pH 8.5, respectively, as the *P. modestum* ATPase. The enzyme was inhibited by DCCD and protected from this inhibition in the presence of Na$^+$ ions. After reconstitution of the hybrid ATPase into proteoliposomes, it catalyzed Na$^+$ transport upon ATP hydrolysis. ATP-dependent proton transport in the absence of Na$^+$ ions and the abolition of this activity by Na$^+$ addition was also demonstrated. Furthermore, the proteoliposomes were able to catalyze ATP synthesis from ADP and inorganic phosphate, if ΔμNa$^+$ of proper direction and magnitude was applied. The hybrid enzyme was thus a tightly coupled Na$^+$ translocating ATPase, that could not be distinguished functionally from the homogeneous *P. modestum* ATPase. It is interesting that a functional hybrid ATPase was only formed, if the δ and the F$_0$ subunits were of a *P. modestum* origin and if the remaining F$_1$ subunits came from *E. coli*, but not with a *P. modestum* F$_0$ and an *E. coli* F$_1$ portion. This indicates important interactions between the δ and the F$_0$ subunits, which could be involved in transmitting the conformational changes linked to ion movement through F$_0$ into the catalytic center of the F$_1$ moiety.

A functional ATPase hybrid consisting of the *P. modestum* subunits a, b, c and δ, the E. *coli* subunits β, γ, ε and a hybrid α subunit deriving in its N-terminal part from *P. modestum* and in its C-terminal part from *E. coli* has also been obtained [51]. Therefore, the *E. coli* mutant strain CM1470 that lacks the genes for the ATPase subunits a, c, b, δ and part of the α-subunit was transformed with plasmid pAP42 that harbors the *P. modestum* genes for the ATPase subunits that were lacking on the *E.coli* chromosome. After selection for growth on succinate, positive clones were obtained that contained a functional hybrid ATPase with properties very similar to those of the homogeneous *P. modestum* enzyme. Analysis of the appropriate gene region indicated that the DNA fragment from *P. modestum* was integrated into the genome of the *E.coli* deletion mutant by site-specific homologous recombination. Two recombination events had occurred, one within the *unc* gene and one in the gene for the α-subunit. The new *E. coli* strain was named PEF42. The ATPase hybrid expressed by these *E. coli* cells was similar to that synthesized from the plasmid pHEP100, with the exception that the former one contained a hybrid α subunit, whereas in the latter the complete α-subunit derived from *E. coli*.

   The Na$^+$ requirement of *E. coli* clones synthesizing a functional *P. modestum/E. coli* ATPase hybrid for growth on succinate offered the possibility to select for mutants that could grow without NaCl addition. After *in vitro* random mutagenesis of the F$_0$ genes of pHEP100, transformation of the mutated plasmid into *E. coli* DK8 and selection for Na$^+$-independent growth on succinate, several mutant clones with the desired phenotype were obtained [52]. Sequencing of the gene for the c subunit revealed a double mutation Phe84Leu Leu87Val in all c subunit mutants. In some of the clones, additional mutations of the c subunit were found, but some contained the double mutation only. The double mutation was then introduced by site directed mutagenesis and homologous recombination into the chromosome of the parent strain PEF42, resulting in *E. coli* strain MPC8487 [52]. As this strain could grow on succinate without Na$^+$ addition, its ATPase must function with H$^+$ as coupling ion, showing that the double mutation is the only reason for the Na$^+$-independent phenotype. The conclusion was confirmed by biochemical analyses of the mutated hybrid ATPase isolated from strain MPC8487. Unlike to the parent enzyme, the mutated ATPase was not activated by Na$^+$ ions, but retained the activation by Li$^+$ ions. In addition, the new enzyme exhibited improved properties for proton coupling, as shown by an about four-fold higher specific ATPase activity in the cell membranes from the mutant strain as compared with the parent strain (in the absence of Na$^+$ ions). Proton pumping was also directly demonstrated after reconstitution of the mutant ATPase into proteoliposomes. Unlike to the *P. modestum* or parent hybrid ATPase, proton pumping was not affected by Na$^+$ ions, but it was still impaired by Li$^+$ addition. These results indicated that in the mutant ATPase Na$^+$ ions were no longer able to bind and compete with H$^+$ for binding and translocation. On the other hand, the cation binding site could still accommodate binding of the smaller Li$^+$ ions. Li$^+$ transport into the proteoliposomes containing the ATPase from stain MPC8487 has in fact been demonstrated [52].

   These results can be interpreted in terms of the hairpin-like structure of subunit c and the location of the binding site for the coupling ions at glutamate 65. Protons can bind from one side of the membrane to this site and can be released to the other side of the membrane, if there are the appropriate access and release channels and if the carboxylate undergoes appropriate pK changes to coordinate binding and release of the proton during the translocation cycle. The binding of Na$^+$ or Li$^+$, however, requires not only the carboxylate group, but also additional liganding groups in the correct geometry to accommodate complexation of the respective metal ion. Such ligands could be the hydroxyl groups of serine 66 and threonine 67 that follow glutamate 65 in the sequence of the *P. modestum* c subunit, or the amide group of glutamine 32, that may be positioned opposite to glutamate 65 on the N-terminal α-helix. The cavity size which is determined by these ligands is essential to discriminate between different metal ions just as the cavity size within a crown ether or an ionophore (e.g. valinomycin) determines which metal ions can be bound. The *P. modestum* c subunit exceeds the length of the *E. coli* c subunit by 4 residues at the N-terminus and by 6 residues at the C-terminus. The Phe84Leu Leu89Val double mutation which leads to the loss of the Na$^+$ binding site is within the extended C-terminal part. The extensions may therefore be important for proper functioning of the *P. modestum* ATPase as a Na^+ coupled enzyme. Conceivably, interaction of the Phe84 and Leu89 within the extended C-terminal tail of the *P. modestum* c subunit with amino acids in their vicinity on the same or the opposite helix is important for forming the binding pocket around Glu65 with the correct positioning of appropriate ligands for selective Na^+ (or Li^+) binding. The altered amino acids (Leu84 and Val89) might interact differently with amino acid residues in their vicinity which could result in a change of the geometry of the metal ion binding pocket at Glu65. If the size of the cavity would become smaller, one could explain why Na^+ ions with an ionic diameter of approximately 1.90 Å could no longer bind, whereas the smaller Li^+ (ionic diameter approximately 1.30 Å) would still fit into the site and could therefore still be transported by the mutant ATPase.

How could one envisage Na, Li^+, or H^+ translocation by the F_0 moiety of the *P. modestum* ATPase across the membrane? A simple model that could accommodate transport of metal ions or protons exists of water-filled access and release channels connected to the binding site at Glu65 of the c subunit. Metal ions moving through the access channel would strip off their water shell and become liganded at cGlu65 if they would fit properly into the binding cavity. The geometry at this site, therefore, determines the coupling ion specificity of the ATPase and is the structural basis of the so called selectivity filter. Subsequently, a conformational change disconnects the bound metal ion from the access channel and connects it to the release channel. Upon lowering the binding affinity, perhaps by ion exchange with an arginine residue moving into the vicinity of cGlu65 [32], the metal ion is released from its binding site, becomes rehydrated, and diffuses through the release channel to the other surface of the membrane. The same principal mechanism can also accommodate H^+ translocation. Protons could either compete with Na^+ or Li^+ transport, if an appropriate metal ion binding site is present, as in the *P. modestum* ATPase, or could be transported exclusively, if a metal ion binding site is missing, as in the *E. coli* ATPase. For proton transport, it is sufficient to deliver these ions from one side of the membrane through an access channel to its carboxylate binding site and release them to the opposite surface through a release channel following a conformational change with a decrease of the pK of the carboxylic acid.

It is obvious from this mechanism that an ATPase with a Na^+ binding site like that from *P. modestum* can use protons as alternative coupling ions. In contrast, a proton translocating ATPase like that from *E. coli* cannot use Na^+ or another metal ion as alternative coupling ion, because no metal ion binding site exists in this enzyme. One might even predict that any cation pump operating by this basic principle can substitute under certain conditions its specific metal ion by a proton. For the Na^+/K^+ ATPase it has indeed been demonstrated that protons can substitute for Na^+ or for K^+, if either of these alkali ions is absent [53]. Another interesting observation is that the Ca^{2+} ATPase of the sarcoplasmic reticulum catalyzes H^+ transport in the opposite direction of Ca^{2+} transport [54]. Still another variation of this theme has been recently discovered for the oxaloacetate decarboxylase Na^+ pump. It was observed that protons compete with Na^+ for binding to the same site, from where they move into the opposite direction of Na^+ pumping, to catalyze the decarboxylation of the acid-labile carboxybiotin residue [54a].

Recently, the c subunit of the *E. coli* ATPase was mutated around the essential acidic residue (Asp61) in such a way that the *E. coli* residues were replaced by the corresponding ones from the *P. modestum* c subunit [55]. By this series of mutations (cVal60Ala, cAsp61Glu, cAla62Ser, cIle63Thr/Ala), a Li^+ binding site (but not a Na^+ binding site) was created. A remarkable difference of this Li^+ binding c subunit mutant and that of the *P. modestum* ATPase is that Li^+ binding to the *E. coli* mutant leads to an inactivation of the enzyme, whereas Li^+ binding to the *P. modestum* c subunit mutant initiates a catalytic cycle that upon ATP hydrolysis leads to Li^+ translocation across the membrane. There are apparently structural restraints in the *E. coli* mutant that prevent the proper conformational changes for release of the Li^+ ions from its binding pocket and delivery to the other surface of the membrane.

In summary, there is now convincing evidence that the conserved acidic amino acid residue in the C-terminal helix of the c subunit undergoes binding and release of the coupling ion during its translocation across the membrane. If metal ions serve as the coupling ions, they must be bound to this site, which requires additional ligands in the vicinity of the carboxylate group to create a binding pocket with the correct geometry to accommodate binding of a specific metal ion. These restraints which apply for Na^+ or Li^+ translocation by the ATPase do not apply for H^+ translocation, for which the carboxylate itself is sufficient to bind the proton and release it after the appropriate conformational change.

9. Are there other Na^+-translocating F_1F_0 ATPases?

The Na^+-translocating F_1F_0 ATPases have apparently not found a wide distribution [56]. The only other example that is known is the ATPase from *Acetobacterium woodii* [57]. During acetate formation from CO_2 and H_2 in this anaerobic bacterium, a primary Na^+ pump operates that couples one of the reactions leading from methylene tetrahydrofolate to acetyl-CoA with the extrusion of Na^+ ions. The Na^+ gradient is then used for ATP synthesis by a Na^+-translocating F_1F_0 ATPase. This enzyme seems to be very similar to that from *P. modestum* [56–58]. Although only 6 instead of the usual 8 subunits were seen on SDS-PAGE of the purified ATPase [56], this does not mean that the *A. woodii* ATPase has an unusual subunit composition. Further experiments are certainly required to clarify this discrepancy.

Additional members of the Na^+-translocating F_1F_0 ATPases family may be expected in other bacteria that dispose of a primary Na^+ pump and create an electrochemical Na^+ ion gradient that could be directly used by a Na^+-translocating ATPase for ATP synthesis. Various anaerobic bacteria are known, that like *P. modestum* contain a Na^+-translocating decarboxylase (for review see Refs. [13,23]). An example is *Klebsiella pneumoniae* which expresses an oxaloacetate decarboxylase Na^+ pump during anaerobic growth on citrate [61–63]. These bacteria do not contain a Na^+-translocating F_1F_0 ATPase. The Na^+ gradient established by the decarboxylase is used for citrate uptake [64], catalyzed by the CitS protein which has recently been purified [65] and characterized [65a], and to drive NADH synthesis by $\Delta\mu Na^+$-driven reversed electron flow [66]. The genes encoding the enzymes of the citrate fermentation pathway citrate lyase [67], oxaloacetate decarboxylase [68–70], and the Na^+-dependent citrate carrier (CitS) [71] are clustered on the chromosome and are coregulated *via* a two component regulatory system [72]. *Veillonella parvula* which grows anaerobically on lactate contains a methylmalonyl-CoA decarboxylase Na^+ pump, but no F_1F_0 ATPase has been found [22,73,74]. The function of the Na^+ gradient in this organism is therefore obscure. *Acidaminococcus fermentans* growing anaerobically on glutamate has a glutaconyl-CoA decarboxylase Na^+ pump without known physiological function [75,76]. *Malonomonas rubra* ferments malonate to acetate and CO_2 and couples this decarboxylation to Na^+ pumping [77,78,78a]. The Na^+ gradient thus established is the only energy form available in *M. rubra* to synthesize ATP like in *P. modestum*. Whether a Na^+ ATPase is operating in *M. rubra*, or whether ATP synthesis occurs by a H^+-coupled ATP synthesis mechanism after converting $\Delta\mu Na^+$ into $\Delta\mu H^+$ has to be established.

Vibrio alginolyticus is an aerobic bacterium that contains a NADH:ubiquinone oxidoreductase functioning as a Na^+ pump [79]. The genes for this Na^+ pump have recently been cloned and sequenced [80–82]. The enzyme was purified in a state competent to catalyze the vectorial reaction and has been characterized biochemically [83,83a]. The F_1F_0 ATPase from this organism has also been cloned, sequenced, and studied biochemically [84]. From these studies it became quite clear that this ATPase is coupled to proton translocation and not to Na^+ translocation, as has been previously suggested [85]. There have been reports in the literature that *E. coli* exposes over a Na^+ translocating ATPase under specific growth conditions [86], but there has also been scepticism about the validity of these findings, and independent confirmation from another laboratory is certainly required.

Some alkaliphilic bacteria grow well at pH around 10, keeping the internal pH well below 9 [87]. These organisms contain H^+-translocating respiratory complexes. The cell interior is acidified relative to the environment by the Na^+/H^+ antiporter with the free energy of $\Delta\psi$. The antiporter action produces an inversed ΔpH ($H_{in}^+ > H_{out}^+$) and a ΔpNa^+ oriented from the outside to the inside. As a consequence, the $\Delta\mu H^+$ is unusually low for an aerobic bacterium, whereas the $\Delta\mu Na^+$ is quite high. From a bioenergetic point of view, a Na^+-translocating ATPase was therefore especially demanding for the alkaliphilic Bacilli. Nevertheless, work in two laboratories has clearly established that these bacteria synthesize ATP by a H^+ translocating F_1F_0 ATPase [88–90]. The mechanism of ATP synthesis by the alkaliphilic Bacilli has been debated: based on measured $\Delta\mu H^+$ values in alkaliphilic Bacilli growing at pH 10 of > -100 mV, our group favors a conventional ATP synthesis mechanism, perhaps with a higher H^+ to ATP stoichiometry [91,92], while the group of Krulwich, measuring $\Delta\mu H^+$ values of around -50 mV, favors an ATP synthesis mechanism with a direct connection between the respiratory H^+ pumping complexes and the ATPase, so that the pumped protons do not equilibrate with the bulk [87,93]. A clear example for a bulk energization of the membrane, arguing against a direct, more localized proton movement from the respiratory chain to the ATPase, exists in *E. coli* PEF42 which links respiratory H^+ pumping *via* a Na^+/H^+ antiporter to Na^+-coupled ATP synthesis (see above, Chapter 8).

In conclusion, the occurrence of Na^+-translocating F_1F_0 ATPases in nature seems to be restricted to a few anaerobic bacterial species that contain a primary Na^+ pump. The presence of a Na^+ pump in a bacterium, however, is not a sufficient criterion for a Na^+ coupled ATP synthesis mechanism, and many of them do in fact synthesize their ATP by a conventional H^+-translocating F_1F_0 ATPase. Not even the alkaliphilic bacteria, where a N^+-translocating ATP synthesis mechanism would be most demanding from bioenergetical considerations, have adopted a Na^+ translocating F_1F_0 ATPase.

10. Concluding remarks

The Na^+-translocating F_1F_0 ATPases are close relatives to the H^+-translocating counterparts. Despite their rare occurrence, they are favorite study objects for the catalytic mechanism of ion translocation across the membrane-bound F_0 moiety. Studies with the *P. modestum* ATPase have considerably contributed to our knowledge of these events. A major contribution for understanding the ATP synthesis mechanism comes from the high resolution structure of F_1 from bovine heart mitochondria. Comparatively little is still known about the molecular details of ion translocation through F_0 and about the coupling events by which this transport promotes ATP synthesis at the catalytic sites of the F_1 moiety. High resolution structural information on the F_0 part is certainly required to understand in full the function of this fascinating molecular machinery.

Abbreviations

    ACMA, 9-amino-6-chloro-2-methoxyacridine
    CCCP, carbonylcyanide-*m*-chloro-phenylhydrazone
    CCFP, carbonylcyanide-*p*-trifluoromethoxyphenylhydrazone
    DCCD, dicyclohexylcarbodiimide
    EIPA, ethylisopropylamiloride
    EF_1, F_1 part of the *E.coli* ATPase
    EF_0, F_0 part of the *E. coli* ATPase
    PF_1, F_1 part of the *P. modestum* ATPase
    PF_0, F_0 part of the *P. modestum* ATPase

References

1.   Pedersen, P.L. and Amzel, M.L. (1993) J. Biol. Chem. **268**, 9937–9940.
2.   Erecinska, M. and Wilson, D. (1978) Trends Biochem. Sci. **3**, 219–222.
3.   Senior, A.E. (1990) Ann. Rev. Biochem. Biophys. Chem. **19**, 7–41.
4.   Fillingame, R.H. (1990) in: T.A. Krulwich (ed.), The Bacteria, Vol. XII. Academic Press, New York, pp. 345–391.
5.   Penefsky, H.S. and Cross, R.L. (1991) Adv. Enzymol. **64**, 173–214.
6.   Hatefi, Y. (1993) Eur. J. Biochem. **218**, 754–767.
7.   Issertel, J.P., Dupuis, A, Garin, J., Lunardi, J., Michel, L. and Vignais, P.V. (1992) Experientia **48**, 351–362.

8. Walker, J.E., Fearnley, I.M., Lutter, R., Todd, R.J. and Runswick, M.J. (1990) Phil. Trans. R. Soc. **326**, 367–378.
9. Mitchell, P. (1974) FEBS Lett. **43**, 189–194.
10. Hilpert, W., Schink, B. and Dimroth, P. (1984) EMBO J. **3**, 1665–1670.
11. Laubinger, W. and Dimroth, P. (1987) Eur. J. Biochem. **168**, 475–480.
12. Laubinger, W. and Dimroth, P. (1988) Biochemistry **27**, 7531–7537.
13. Dimroth, P. (1994) Antonie van Leeuwenhoek **65**, 381–395.
14. Gräber, P., Bottcher, B. and Boekema, E.J. (1990) in: Bioelectrochemistry III, eds G. Milazzo and M. Blank, pp. 247–276. Plenum Press, New York.
15. Boekema, E.J., Berden, J.A. and van Heel, M.G. (1986) Biochim. Biophys. Acta **851**, 353–360.
16. Wilkens, S. and Capaldi, R.A. (1994) Biol. Chem. Hoppe-Seyler **375**, 43–51.
17. Abrahams, J.P., Leslie, A.G.W., Lutter, R. and Walker, J.E. (1994) Nature **370**, 621–628.
18. Boyer, P.D. (1993) Biochim. Biophys. Acta **1140**, 215–250.
19. Cross, R.L. (1981) Ann. Rev. Biochem. **50**, 681–714.
20. Macnab, R.M.A. (1992) Rev. Genet. **26**, 131–158.
21. Thauer, R.K., Jungermann, K. and Decker, K. (1977) Bacteriol. Rev. **41**, 100–180.
22. Hilpert, W. and Dimroth, P. (1983) Eur. J. Biochem. **132**, 579–587.
23. Dimroth, P. (1987) Microbiol. Rev. **51**, 320–340.
24. Laubinger, W. and Dimroth, P. (1989) Biochemistry **28**, 7194–7198.
25. Laubinger, W., Deckers-Hebestreit, G., Altendorf, K. and Dimroth, P. (1990) Biochemistry **29**, 5458–5463.
26. Kaim, G., Ludwig, W., Dimroth, P. and Schleifer, K.H. (1992) Eur. J. Biochem. **207**, 463–470.
27. Krumholz, L.R., Esser, U. and Simoni, R.D. (1992) FEMS Microbiol. Lett. **91**, 37–42.
28. Ludwig, W., Kaim, G., Laubinger, W., Dimroth, P., Hoppe, J. and Schleifer, K.H. (1990) Eur. J. Biochem. **193**, 395–399.
29. Gerike, U. and Dimroth, P. (1993) FEBS Lett. **316**, 89–92.
30. Nagle, J.F. and Morowitz, H.J. (1978) Proc. Natl. Acad. Sci. USA **75**, 298–302.
31. Kluge, C. and Dimroth, P. (1993) J. Biol. Chem. **268**, 14557–14560.
32. Kluge, C. and Dimroth, P. (1993) Biochemistry **32**, 10378–10386.
33. Schneider, E. and Altendorf, K. (1987) Microbiol. Rev. **51**, 477–497.
34. Kluge, C. and Dimroth, P. (1992) Biochemistry **31**, 12665–12672.
35. Kluge, C., Laubinger, W. and Dimroth, P. (1992) Biochem. Soc. Trans. **20**, 572–577.
36. Hoppe, J. and Sebald, W. (1984) Biochim. Biophys. Acta **768**, 164–200.
37. Sebald, W., Machleidt, W. and Wachter, E. (1980) Proc. Natl. Acad. Sci. USA **77**, 785–789.
38. Miller, M.J., Oldenburg, M. and Fillingame, R.H. (1990) Proc. Natl. Acad. Sci. USA **87**, 4900–4904.
39. Girvin, M.E. and Fillingame, R.H. (1993) Biochemistry **32**, 12167–12177.
40. Girvin, M.E. and Fillingame, R.H. (1994) Biochemistry **33**, 665–674.
41. Girvin, M.E. and Fillingame, R.H. (1995) Biochemistry **34**, 1635–1645.
42. Nordwood, T.J., Crawford, D.A., Steventon, M.E., Driscoll, P.C. and Campbell, I.D. (1992) Biochemistry **31**, 6285–6290.
43. Mosher, M.E., White. L.K., Hermolin, J. and Fillingame, R.H. (1985) J. Biol. Chem. **260**, 4807–4814.
44. Fraga, D., Hermolin, J., Oldenburg, M., Miller, M.J. and Fillingame, R.H. (1994) J. Biol. Chem. **269**, 7532–7537.
45. Zhang, Y., Oldenburg, M. and Fillingame, R.H. (1994) J. Biol. Chem. **269**, 10221–10224.
46. Kluge, C. and Dimroth, P. (1994) FEBS Lett. **340**, 245–248.
47. Gerike, U. and Dimroth, P. (1994) Arch. Microbiol. **161**, 495–500.
48. Gerike, U., Kaim, G. and Dimroth, P. (1995) Eur. J. Biochem. **232**, 596–602.
49. Ackerman, S.H. and Tzagoloff, A. (1990) J. Biol. Chem. **265**. 9952–9959.
50. Kaim, G. and Dimroth, P. (1994) Eur. J. Biochem. **222**. 615–623.
51. Kaim, G. and Dimroth, P. (1993) Eur. J. Biochem. **218**, 937–944.
52. Kaim, G. and Dimroth, P. (1995) J. Mol. Biol. **1253**, 726–738.
53. Polvani, C. and Blostein, R. (1988) J. Biol. Chem. **263**, 16757–16763.
54. Levy, D., Seigneuret, M., Bluzat, A. and Rigaud, J.L. (1990) J. Biol. Chem. **265**, 19524–19534.
54a. Di Berardino, M. and Dimroth, P. (1996) EMBO J. **15**, 1842–1849.

55.   Zhang, Y. and Fillingame, R.H. (1995) J. Biol. Chem. **270**, 87–93.
56.   Hoffmann, A., Laubinger, W. and Dimroth, P. (1990) Biochim. Biophys. Acta **1018**, 206–210.
57.   Heise, R., Müller, V. and Gottschalk, G. (1992) Eur. J. Biochem. **206**, 553–557.
58.   Reidlinger, J. and Müller, V. (1994) Eur. J. Biochem. **223**, 275–283.
59.   Spruth, M., Reidlinger, J. and Müller, V. (1995) Biochim. Biophys. Acta **1229**, 96–102.
60.   Reidlinger, J., Mayer, F. and Müller, V. (1994) FEBS Lett. **356**, 17–20.
61.   Dimroth, P. (1982) Eur. J. Biochem. **121**, 443–449.
62.   Dimroth, P. and Thomer, A. (1993) Biochemistry **31**, 1734–1739.
63.   Di Berardino, M. and Dimroth, P. (1995) Eur. J. Biochem. **231**, 790–801.
64.   Dimroth, P. and Thomer, A. (1986) Biol. Chem. Hoppe-Seyler **367**, 813–823.
65.   Pos, M., Bott, M. and Dimroth, P. (1994) FEBS Lett. **347**, 37–41.
65a. Pos, K.M. and Dimroth, P. (1996) Biochemistry **35**, 1018–1026.
66.   Pfenninger-Li, X.D. and Dimroth, P. (1992) Mol. Microbiol. **6**, 1943–1948.
67.   Bott, M. and Dimroth, P. (1994) Mol. Microbiol. **14**, 347–356.
68.   Schwarz, E., Oesterhelt, D., Reinke, H., Beyreuther, K. and Dimroth, P. (1988) J. Biol. Chem. **263**, 9640–9645.
69.   Laußermair, E., Schwarz, E., Oesterhelt, D., Reinke, H., Beyreuhter, K. and Dimroth, P. (1989) J. Biol. Chem. **264**, 14710–14715.
70.   Woehlke, G., Laußermaier, E., Schwarz, E., Oesterhelt, D., Reinke, H., Beyreuther, K. and Dimroth, P. (1992) J. Biol. Chem. **267**, 22804–22805.
71.   Van der Rest, M., Siewe, R.M., Abee, T., Schwarz, E., Oesterhelt, D. and Konings, W.N. (1992) J. Biol. Chem. **267**, 8971–8976.
72.   Bott, M., Meyer, M. and Dimroth, P. (1995) Mol. Microbiol. **18**, 533–546.
73.   Huder, J. and Dimroth, P. (1993) J. Biol. Chem. **268**, 24564–24571.
74.   Huder, J. and Dimroth, P. (1995) J. Bacteriol. **177**, 3623–3630.
75.   Buckel, W. and Semmler, R. (1983) Eur. J. Biochem. **136**, 427–434.
76.   Bendrat, K., and Buckel, W. (1993) Eur. J. Biochem. **211**, 697–702.
77.   Hilbi, H., Dehning, I., Schink, B. and Dimroth, P. (1992) Eur. J. Biochem. **207**, 117–123.
78.   Hilbi, H. and Dimroth, P. (1994) Arch. Microbiol. **162**, 48–56.
78a. Berg, M., Hilbi, H. and Dimroth, P. (1996) Biochemistry **35**, 4689–4696.
79.   Tokuda, H. and Unemoto, T. (1984) J. Biol. Chem. **259**, 7785–7790.
80.   Beattie, P., Tan, K., Bourne, R.M., Leach, D., Rich, P.R. and Ward, F.B. (1994) FEBS Lett. **356**, 333–338.
81.   Hayashi, M., Hirai, K. and Unemoto, T. (1994) FEBS Lett. **356**, 330–332.
82.   Hayashi, M., Hirai, K. and Unemoto, T. (1995) FEBS Lett. **363**, 75–77.
83.   Pfenninger-Li, X.D. and Dimroth, P. (1995) FEBS Lett. **369**, 173–176.
83a. Pfenninger-Li, X.D., Albrecht, S.P.J., van Belzen, R. and Dimroth, P. (1996) Biochemistry, in press.
84.   Krumholz, L.R., Esser, U. and Simoni, R.D. (1990) J. Bacteriol. **172**, 6809–6817.
85.   Dibrov, P.A., Lazarova, R.L., Skulachev, V.P. and Verkhovskaya, M.L. (1986) Biochim. Biophys. Acta **850**, 458–465.
86.   Avetisyan, A.V., Bogachev, A.V., Murtasina, R.A. and Skulachev, V.P. (1993) FEBS Lett. **317**, 267–270.
87.   Krulwich, T.A. and Guffanti, A.A. (1992) J. Bioenerg. Biomembr. **24**, 587–599.
88.   Hicks, D.B. and Krulwich, T.A. (1990) J. Biol. Chem. **265**, 20547–20554.
89.   Hoffmann, A. and Dimroth, P. (1990) Eur. J. Biochem. **194**, 423–430.
90.   Hoffmann, A. and Dimroth, P. (1991) Eur. J. Biochem. **196**, 493–497.
91.   Hoffmann, A. and Dimroth, P. (1991) Eur. J. Biochem. **201**, 467–473.
92.   Dimroth, P. (1992) Biochim. Biophys. Acta **1101**, 236–239.
93.   Guffanti, A.A. and Krulwich, T.A. (1992) J. Biol. Chem. **267**, 9580–9588.

CHAPTER 3

F-Type H$^+$ ATPase (ATP Synthase):
Catalytic Site and Energy Coupling

M. FUTAI and H. OMOTE

Department of Biological Science, Institute of Scientific and Industrial Research,
Osaka University, Ibaraki, Osaka 567, Japan

© *1996 Elsevier Science B.V.*
All rights reserved

Handbook of Biological Physics
Volume 2, edited by W.N. Konings, H.R. Kaback and J.S. Lolkema

Contents

1. Introduction

The F-type H^+ ATPase has no phosphorylated enzyme intermediate, and functions as an ATP synthase in the membranes of bacteria, mitochondria, or chloroplasts [1–3]. The enzyme (F_0F_1) has a conserved basic subunit structure: a minimal complex consisting of catalytic sector F_1, $\alpha_3 \beta_3 \gamma \delta \varepsilon$ (Fig. 1), and membrane sector F_0, $a b_2 c_{10-12}$ (Fig. 2). The F_0F_1 was highly purified recently from bovine heart and rat liver [4,5]. *Escherichia coli* F_0F_1 was purified by column chromatography [6] or density gradient centrifugation [7]. A rapid one-step purification method from an overproducing *E. coli* strain was also established recently [8]. The F_1 sector can be solubilized and purified as a soluble protein by conventional methods. Chloroplast or cyanobacterial F_0 has an additional subunit (*IV* or *b'*) [9], and mitochondrial F_0 has more complicated subunits. Bovine F_0 consists of subunit *a* (subunit 6), *b*, *c*, *d*, *e*, F_6 and A6L, and two more subunits, *f* and *g*, were detected recently [10]. Although organellar F_0 is complicated, *E. coli* F_1 binds functionally to bovine F_0 (mitochondrial membranes depleted of F_1) [11]. *Bacillus* PS3 F_1 could bind functionally to chloroplast F_0 [12], whereas the chloroplast *c* subunit could not replace that of *E. coli* genetically [13]. The mitochondrial enzyme has an inhibitor protein (IF_1) which binds to the β subunit and inhibits ATP hydrolysis in a low electrochemical gradient [14,15]. The ATP synthesis in F_0F_1 is driven by a transmembrane electrochemical proton gradient, which is established primarily by the electron transfer chain. The gradient is required for ATP release from, and ADP+Pi (Phosphate) binding to, the catalytic site in the β subunit. The catalytic cooperativity between the three β subunits in the F_1 sector has been established. The unique feature of this complicated enzyme is the energy coupling between catalysis and the electrochemical proton gradient, although the coupling mechanism is still at an early stage of understanding. Protons transported through F_0 cause conformational changes of different F_1 subunits and drive ATP synthesis in the β subunit. In the reverse reaction, ATP hydrolysis causes conformational changes of different domains of the β subunit, followed by their transmission to other F_1 subunits and finally to proton pathways in F_0. The higher-ordered structure of the bovine F_1 sector ($\alpha_3 \beta_3 \gamma$) has been solved recently by X-ray diffraction [16,17]. Extensive studies on the *E. coli* enzyme have been carried out taking advantage of the easy genetic manipulation of the organism. As the *E. coli* and bovine enzymes exhibit high homology, it became possible to discuss the roles of amino acid residues or domains, with reference to the higher-ordered structure of the bovine enzyme. In this article, we discuss the present status of the understanding of enzyme structure, catalysis and energy coupling, referring mainly to the *E. coli* enzyme. However, it is far beyond our ability to write a comprehensive review in this limited space. We have tried to

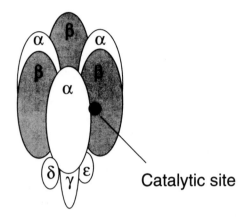

Fig. 1.   A conceptual model of the F_1 sector. A model of F_1 ($\alpha_3\beta_3\gamma\delta\varepsilon$) is shown. The X-ray structure of the $\alpha_3\beta_3\gamma$ complex has been solved [17], but structural information on the δ and ε subunits is still limited.

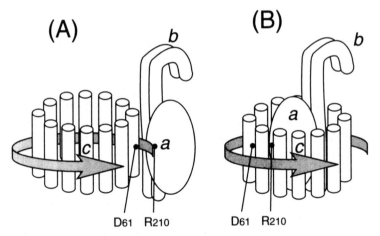

Fig. 2.   A conceptional model of the F_0 sector. Two types of F_0 sector are possible, as judged from the subunit structure and stoichiometry. Altendorf and coworkers supported a similar model to that shown in Fig. 2A based on electron microscopy studies [136]. Further effort is required to obtain the actual structure by X-ray diffraction because no crystal structure has been obtained so far.

produce a critical review and hope to visualize the interesting questions which remain to be answered in the near future. Most of the excellent early studies not cited in this article have been reviewed previously from different aspects [1–3,18–34].

2. Structure of F-type ATPase (F₀F₁)

The primary structure of the enzyme was first derived from the DNA sequence of the *E. coli unc* (*atp*) operon more than a decade ago [18,20]. The cistrons for all subunits are in the operon, in the following order; *a*, *c*, *b*, δ, α, γ, β, ε which is advantageous for further studies on the enzyme from this organism. Eukaryotic enzymes are coded by nuclear and organellar genomes: mammalian mitochondrial A6L and subunit 6 (corresponding to subunit *a*) are encoded by mitochondrial DNA, and other subunits are coded by the nuclear genome [35]. Chloroplast γ and δ and subunit II are coded by the nuclear genome [36]. DNAs or cDNAs for eukaryotic enzymes were cloned and sequenced, and the sequences of all subunits from chloroplasts and mitochondria are known. The catalytic β subunits of 57 different species are known at present (Swiss Prot, release 30). The identity of the α or β subunits among different species is higher than those of other subunits. Furthermore, 103 residues are identical when the *E. coli* α and β sequences are aligned, indicating that the two subunits have a common ancestor. The identities of other subunits are lower but essential domains are conserved among different species. The isolation and sequencing of *E. coli* genes stimulated genetic studies on F_0F_1 from this organism involving analysis of random and then site directed mutations or mutation/suppression [1,18].

Similarities between F_0F_1 and vacuolar ATPases became evident at the primary structure level [37,38]. Vacuolar ATPases found in endomembranes have membrane extrinsic (V_1) and intrinsic (V_0) sectors, and the catalytic *A* and *B* subunits of V_1 are homologous to the α and β subunits of F_0F_1, respectively. Analogy of the vacuolar subunit *D* to the γ subunit has been suggested recently [39]. Vacuolar ATPase has a membrane-embedded proteolipid (16,000 molecular weight) whose amino and carboxyl terminal halves show homology with subunit *c* (8,000 molecular weight) of F_0F_1 [38,40], suggesting that the two proteins have the same ancestral polypeptide and that the vacuolar proteolipid was derived after gene duplication [39]. Nelson and coworkers recently showed the presence of a vacuolar subunit similar to the *b* subunit [41]. The archae ATP synthase or H^+ ATPase (A-ATPase) is similar to the vacuolar ATPase [42–44]. These similarities suggest that studies on F_0F_1 may also contribute to the understanding of other H^+-ATPases.

The F_1 sectors of bovine [17] and rat [45] mitochondria were crystallized and analyzed by X-ray diffraction. The model of bovine F_1 at 6 Å resolution already showed an asymmetric organization of subunits [16]. The higher ordered structure a (flattened sphere of 80 Å high and 100 Å across) of bovine F_1 was solved at 2.8 Å resolution (as schematically shown in Fig. 3) [17]: the α and β subunits are arranged alternatively around the central α helices of the amino and carboxyl terminal domains of the γ subunit [17]. The α and β subunits have similar structures with three distinct domains: the domain farthest from F_0 an amino terminal six-stranded β barrel; the central domain, an α–β domain forming a nucleotide binding site; and the carboxy terminal domain closest to F_0, a bundle of α helices (7 and 6 helices for the α and β subunits, respectively). Consistent with the mechanism of the enzyme described below, the three β subunits differ in conformation and bound-nucleotide (empty, ADP-bound and ATP-bound forms). The crystal structure of rat liver F_1 has been studied and seems to be significantly different in detail from the beef F_1 structure, although high resolution results have not been reported yet [45]. The difference between the two crystal structures, if any, will be interesting from the functional point of view. Attempts to simplify the system were carried out,

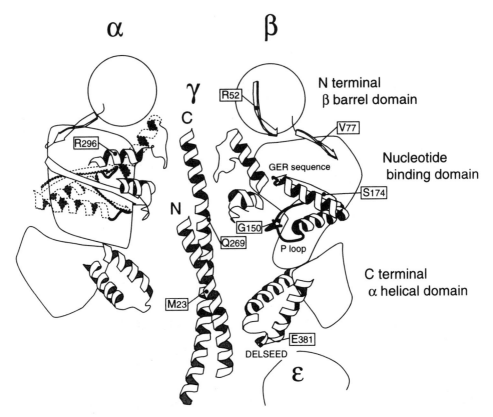

Fig. 3.   A model of the F_1 ($\alpha_3\beta_3\gamma$) complex from X-ray diffraction studies. The structure of the *Escherichia coli* $\alpha_3\beta_3\gamma$ complex cross-sectioned is shown schematically following the X-ray structure reported by Abrahams et al. [17]. Domains discussed in the text are also shown: GER sequence, Gly-Glu-Arg (positions 180–182); P-loop, Gly-Gly-Ala-Gly-Val-Gly-Lys-Thr-Val (positions 149–147); DELSEED, Asp-Glu-Leu-Ser-Glu-Glu-Asp (positions 380–386).

and ATPase complexes with $\alpha_3\beta_3\gamma$ [46,47], $\alpha_3\beta_3$ [48] or $\alpha\beta$ [49,50] were reconstituted. The X-ray structure of the thermophilic $\alpha_3\beta_3$ complex was also solved recently [Y. Shirakibara, personal communication]. Therefore, the results of mutation studies on the *E. coli* enzyme can be interpreted with reference to the higher-ordered structure of the bovine enzyme. However, it is still desirable to solve the higher-ordered structure of the *E. coli* enzyme. Efforts have been directed to crystallize mitochondrial or bacterial F_0F_1, although no success has been reported.

3. Catalysis in the β subunit

3.1. Mechanism of ATP synthesis/hydrolysis

ATP synthesis has been studied with membrane vesicles or F_0F_1 reconstituted into liposomes. The *E. coli* F_0F_1 purified from overproducing strains reconstituted into

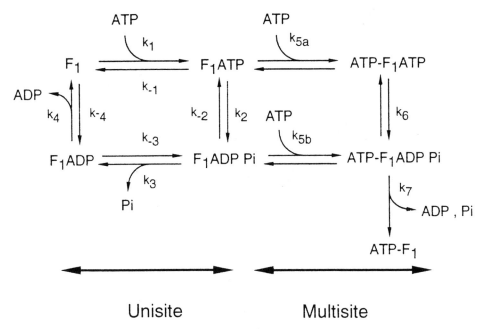

Fig. 4. Kinetics of ATP hydrolysis by purified F_1. The kinetic mechanism of ATP hydrolysis by purified F_1 is shown schematically. In multisite catalysis, three catalytic sites participate in the reactions, although two sites are shown in the figure.

liposomes gave a turnover number of about 30 s^{-1} when energized by an artificial electrochemical gradient of protons [51]. This value is about 2-fold higher than ATP synthesis in membranes energized by respiration.

The kinetics of ATP hydrolysis by the purified bovine or *E. coli* F_1 sector have been verified [52–54] (Fig. 4): unisite (single site) and multisite (steady state) catalysis measured in the presence of a sub-stoichiometric amount of ATP (ATP/F_1 < 1) and excess ATP (ATP/F_1 > 10^6), respectively, clearly showed the presence of cooperativity in multisite catalysis. The higher multisite rate is due to cooperativity between the three catalytic sites in the β subunits. Using the intrinsic fluorescence of the βTrp-331 of βTyr-331→Trp mutant, the requirement of Mg^{2+} for cooperative catalytic site binding was clearly shown [55][1].

In unisite catalysis, ATP binds to a single catalytic site and is hydrolyzed at a rate 10^5–10^6-fold lower than that of multisite catalysis. The association constant, K_1 (k_{+1}/k_{-1}), for ATP binding/release is about 10^9 M^{-1} (*E. coli* value). The equilibrium constant (K_2) in unisite catalysis between $F_1 \cdot$ATP and $F_1 \cdot$ADP·Pi is almost unity (K_2 ≈ 0.5), indicating that enzyme-bound ATP can be formed from enzyme-bound ADP and Pi with very little or no free energy change [52]. These results also suggest that

---

1 Amino acid residues are shown by three letter symbols, with abbreviations for corresponding subunits. Example is: βTry-331, Try residue at position 331 of the β subunit. Mutations are shown with amino acid changes indicated by arrows: βTyr-331→Trp, Tyr-331 of the β subunit changed to Trp.

F_1-bound ATP undergoes hydrolysis/synthesis a number of times, as demonstrated by the incorporation of oxygen of $[^{18}O]$ H_2O into Pi by Boyer and coworkers [56]. The formation of F_1-bound ATP was actually demonstrated by the addition of ADP+Pi to F_1 [57–59].

These observations support the original proposal of Paul Boyer on the "binding change mechanism" of oxidative phosphorylation [60]. The mechanism predicts that the energy of an electrochemical proton gradient is not required for the synthesis of F_1-bound ATP from F_1-bound ADP+Pi but is required for the dissociation of product ATP from and binding of Pi to the catalytic sites. The rate of dissociation of unisite ATP from mitochondrial F_0F_1 was actually accelerated five orders of magnitude in the presence of NADH, and about six orders of magnitude in the presence of NADH and ADP [61]. Catalytic cooperativity in ATP synthesis was also shown [34,62].

Inhibitors showed different sensitivities to uni- and multisite catalysis. Azide [63] or DCCD [64] is a strong inhibitor of multisite but not of unisite catalysis, indicating that they inhibit the catalytic cooperativity required for multisite catalysis but not the chemical reaction itself. N-ethylmaleimide treatment of chloroplast F_1 (alkylation to Cys-89 of the γ subunit) did not affect the unisite activity, but inhibited the multisite activity by more than 90% [65]. Soteropoules et al. suggested that the alkylation of chloroplast Cys-89 alters the interactions between catalytic sites [66]. Anion effects on unisite catalysis are lacking in bacterial or mitochondrial F_1, whereas the thylakoid F_1 unisite catalysis is inhibited by sulfite [65]. Differences among enzymes from different origins were also found: free inorganic phosphate activated mitochondrial unisite catalysis 3-fold, but it affected *E. coli* unisite catalysis only slightly [67].

3.2. βLys-155 and βThr-156 in the phosphate-binding loop are catalytic residues

Amino acid residues essential for catalysis or nucleotide binding have been studied by affinity labeling, chemical modification, and mutagenesis. The results of such studies are discussed below and *E. coli* F_1 amino acid numbers are used unless otherwise mentioned. In Table 1, important residues are shown with both *E. coli* and bovine numbering, together with the experimental findings.

The importance of the glycine-rich sequence or phosphate-binding loop (P-loop), Gly-X-X-X-X-Gly-Lys-Thr/Ser, in the enzyme has been determined by affinity labeling and mutagenesis. One of the early random mutants, βAla-151→Val, in the *E. coli* β P-loop (Gly-Gly-Ala-Gly-Val-Gly-Lys-Thr, *E. coli* β positions 149–156) showed reduced multisite catalysis [68]. Mutation and affinity labeling results indicated that βLys-155 [69–72] and βThr-156 [71] are essential for catalysis. The P-loop of the bovine F_1 is in the central region of the β subunit, and extends in the direction of the carboxyl and amino terminal α helices of the γ subunit [17] (Fig. 3). The P-loop is conserved in many nucleotide binding proteins [73], and forms a similar loop structure between the α helix and β sheet in the p21 *ras* protein (74), elongation factor Tu [75], and adenylate kinase [76]. The P-loops actually form parts of the catalytic sites.

Table 1

Important amino acid residues in the α, β and γ subunits. The mutant results are for the *E. coli* enzyme. Mutations are shown by one letter amino acid symbols: e.g. αR296C, Arg-296 of the α subunit replaced by Cys. The results of affinity labeling and chemical modification studies of bovine (B) and *E. coli* (E) enzymes are shown. Information only from the bovine crystal is also shown. See the text for references. Abbreviations were used for chemical and affinity agents: AP3PL, adenosine triphosphopyridoxal; DCCD, dicyclohexyl-carbodiimide; FSBA, fluorosulfonyl benzoyl adenosine.

Amino acid residue		Experimental finding
E. coli	Bovine	
αArg-296	αArg-304	αR296C suppressed βS174F
αSer-347	αSer-344	catalytic site, crystal
αSer-373	αSer-370	αS373F mutant, low multisite
αArg-376	αArg-373	αR376C mutant, low multisite, catalytic site (B)
βArg-52	βArg-59	βR52C mutation suppressed γ frameshift
βGly-149	βGly-156	βG149C was suppressed by βS174F, βG172E, βE192V, βV198A
βGly-150	βGly-157	βG150D suppressed γ frameshift
βLys-155	βLys-162	catalytic residue, from AP3 PL binding (E), mutant, crystal
βThr-156	βThr-163	catalytic residue, from mutant and crystal
βSer-174	βSer-181	βS174F was suppressed by βG149C
βGlu-181	βGlu-188	catalytic residue, mutant and crystal
βArg-182	βArg-189	essential for multisite and ATP synthesis catalytic residue
βGlu-185	βGlu-192	mutant and crystal
βGlu-192	βGlu-199	DCCD binding site (B,E)
βAsp-242	βAsp-256	Mg^{2+} binding, mutant (E), crystal
βArg-246	βArg-260	βR246H or βR246C mutant, low multisite
βTyr-297	βTyr-311	8-azido ATP binding (B)
βTyr-331	βTyr-345	2-azido ATP binding (B), catalytic site from crystal
βTyr-354	βTyr-368	2-azido ADP binding (B), FSBA binding (B)
γMet-23	γMet-23	γM23K, γM23R low energy coupling
γArg-242	γGln-228	γR242C suppress γM23K
γGln-269	γThr-255	γQ269R suppress γM23K
γThr-273	γArg-259	γT273S suppress γM23K

A synthetic peptide (50 residues) containing the P-loop of rat mitochondrial F_1 could bind ATP tightly (Kd = 17.5 μM) [77]. The binding of one mole of AP3-PL (adenosine triphosphopyridoxal) per mole of F_1 in the presence of $MgCl_2$ completely inhibited the F_1-ATPase activity [78]: the analog covalently bound to βLys-155 [70], indicating that βLys-155 is located near the β/γ phosphate moiety of ATP. βLys-155 is also known as a 7-chloro-4-nitrobenzoxadiazole binding site [79]. Mutants of βLys-155 (βLys-155 → Ala, Ser or Thr, or βThr-155/βLys-156) exhibited no oxidative phosphorylation and membrane ATPase activities [71]. Furthermore, the purified βAla-155 or βSer-155 mutant F_1 exhibited neither unisite nor multisite catalysis activity, although these enzymes bound ATP under both sets of catalytic conditions (Fig. 4). Similar results were obtained with the βGln-155 mutant [72]. The corresponding bovine residue is located near bound ATP in the crystal enzyme [17], supporting that it is a catalytic residue.

The results of replacement of the entire P-loop of the β subunit with those of other proteins suggested the importance of βThr-156 [80]. The enzyme with the p21 *ras* sequence (<u>Gly</u>-Ala-Gly-Gly-Val-<u>Gly</u>-<u>Lys</u>-<u>Ser</u>) exhibited about half the membrane ATPase activity and was capable of oxidative phosphorylation, although three amino acid residues are different between the β subunit and the p21 *ras* P-loop. On the other hand, the enzyme with the P-loop of adenylate kinase (<u>Gly</u>-Gly-Pro-Gly-Ser-<u>Gly</u>-<u>Lys</u>-Gly-<u>Thr</u>) or Gly insertion between βLys-155 and βThr-156 showed no ATP synthesis or ATPase activity. These enzymes were inactive, possibly because the direction of the essential hydroxyl moiety was changed due to the presence of an extra Gly between the essential βLys-155 and βThr-156 residues. Consistent with this notion, βThr-156 → Ala, Cys or Asp, or βThr-155/βLys-156 and βAla-156/βThr-157 mutants were defective in ATP synthesis, and exhibited <0.1% of the membrane ATPase activity of the wild-type [71]. Similar to the p21 *ras* mutant, the βSer-156 mutant was active in ATP synthesis and exhibited ~1.5-fold higher membrane ATPase activity. The purified βAla-156 and βCys-156 enzymes lacked both multisite (<0.02% of the wild-type level) and unisite (<1.5% of the wild-type level) catalysis, and no detectable rates of ATP binding (k_{+1}) [71]. The mutant and wild-type enzymes bound aurovertin (to the β subunit), but its fluorescence in the mutant and wild-type enzymes showed different responses to Mg^{2+}. These results suggest that βThr-156, possibly its hydroxyl moiety, is essential for catalysis and may contribute to Mg^{2+} binding. Consistent with these results, mutagenic analysis of the yeast indicated that the residue corresponding to βLys-155 is invariant and that the residue corresponding to βThr-156 can only be replaced by Ser [81].

The α subunit also has a P-loop, and bovine residues corresponding to αLys-175 and αThr-176 are near the α/β phosphate moiety of ATP and Mg^{2+} bound in the non-catalytic site [17]. However, mutation studies indicated that they are not absolutely essential for catalysis [82,83]. We do not discuss the three non-catalytic (or non-exchangeable) nucleotide binding sites in the α subunits, leaving them for recent reviews [3,25].

3.3. βGlu-181 and βArg-182 in the GER sequence and βGlu-185 are essential

Mutations were introduced systematically in the region near the P-loop by replacing conserved hydrophilic residues [84]. Most of the mutants could grow by oxidative phosphorylation, although their multisite catalysis was decreased to varying degrees depending on the residues introduced. However, mutants at two positions, 181 and 182 (βGlu-181 → Gln, Asp, Asn, Thr, Ser, Ala or Lys; βArg-182 → Lys, Ala, Glu or Gln), could not grow by oxidative phosphorylation and exhibited only negligible membrane ATPase activities [84]. Purified βAla-181 and βGln-181 F_1 showed no multisite and very slow unisite catalysis (~1% of the wild-type level): these defects could be attributed to the decreased rates $(k_{+2}$ and $k_{-2})$ of unisite catalysis. The results with the βGlu-181 → Gln enzyme essentially confirmed the previous results of Senior and Al-Shawi [72]. The βAsp-181 enzyme showed detectable multisite and unisite catalysis rates (27 and 21%, of the wild-type levels, respectively) [84]. These results suggest that βGlu-181, possibly its carboxyl group, is essential for the catalytic steps. It is noteworthy that DCCD selectively binds to the corresponding glutamate residue of thermophilic bacterial F_1 and inactivates the enzyme [85]. Mutation and reconstitution studies with the thermophilic F_1 suggested that the carboxyl moiety is absolutely required for catalysis [86]. On the other hand, βGlu-192 (*E. coli* or mitochondrial DCCD binding residue, Ref. [87]) is not essential and can be changed to other residues. Mutant studies also suggested that βArg-182 is essential for substrate binding and catalysis, although the βLys-182 mutation had less drastic effects: the βLys-182 enzyme showed 1 and 85% of the multisite and unisite wild-type rates, respectively. βGlu-181 and βArg-182 are in the Gly-Glu-Arg sequence (β180-182) conserved in both the β subunit of F_1 and the A subunit of vacuolar type ATPases [42,88]. Recently, the Gly-Glu-Arg sequence was also found in other ATP binding proteins [89,90].

Early observations indicated that the purified β subunit with the βGlu-185 → Lys or Glu mutation could not assemble with the α and γ subunits to form a minimal catalytic $\alpha_3\beta_3\gamma$ complex [91]. This residue is located near the catalytic site in the higher ordered structure [17], and we replaced it with all possible residues. The βAsp-185 mutant showed positive growth by oxidative phosphorylation, 20% of the wild-type membrane ATPase level, and similar unisite catalysis relative to the wild type. However, all other mutants could not grow by oxidative phosphorylation, and exhibited essentially no membrane ATPase activities (≤0.1% of the wild type level) [92]. The purified βGln-185 and βCys-185 F_1 exhibited no multisite catalysis, although they showed substantial unisite catalysis. Interestingly, the βCys-185 enzyme was activated by iodoacetate in the presence of $MgCl_2$, and the enzyme with a βS-carboxymethyl-185 residue (in the three β subunits) exhibited about one-third of the wild-type multisite catalysis. These results taken together suggest that the carboxyl moiety of βGlu-185 is essential for the cooperative catalysis of the enzyme.

The residues forming the adenine-binding site were defined mainly using bovine F_1. βTyr-331 (bovine Tyr-345) (2-azido ATP binding site) (93) and βTyr-297 (bovine Tyr-311) (8-azido ATP binding site) [94] were shown to be in the hydrophobic region interacting with the adenine moiety in the bovine higher ordered structure [17]. On the other hand, βTyr-354 (bovine Tyr-368), a site for 2-azide-ADP [95], and 5′-*p* fluorosulfonyl benzoyl adenosine [96] form a non-catalytic nucleotide binding site together with the α subunit, as indicated by the bovine structure. It is also indicated by the bovine F_1 structure that adenine in the α subunit non-catalytic site is hydrogen-bonded with side chains, whereas adenine in the β subunit is surrounded by a hydrophobic site [17].

Summarizing mutant studies, chemical modification, affinity labeling, and higher ordered structure (Table 1), we propose the roles of the β subunit residues shown in Fig. 5. βLys-155 may bind the β and γ phosphate of ATP, and βThr-156 may bind Mg^{2+}. βGlu-185 is also close to the Mg^{2+} binding site, and a key residue responsible for conformational transmission to other catalytic sites for catalytic cooperativity. βGlu-181 is a candidate for the general base necessary for nucleophilic attack by the water molecule. The βArg-182 and βGlu-181 residues function in nucleotide binding. This model is essentially consistent with the crystal structure

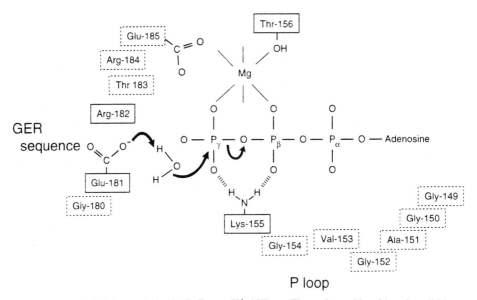

Fig. 5.   A model of the catalytic site in F-type H⁺-ATPase. The amino acid residues in solid boxes are essential for catalysis. Mutations of the residues in dotted boxes lowered multisite catalysis, but are not essential residues. The carboxyl moiety of βGlu-185 was shown to be essential only for multisite catalysis. A possible mechanism of ATP hydrolysis is also shown. See text and Refs. [71,92,99] for details.

of bovine F_1 [17]. A bound water hydrogen-bonded to the residue corresponding to βGlu-181 was observed in the bovine β subunit and this residue is positioned to activate water molecules. The guanidium of αArg-376 (bovine Arg-373) could stabilize the negative charge. The crystal structure also indicates that αSer-347 (bovine Ser-344) is near the ATP γ-phosphate. βArg-246 [54,97] and βAsp-242 [72] residues contribute to catalysis and are located in the catalytic site.

3.4. Domain/domain functional interaction(s) among α/β subunit pairs

Conformational transmission between multiple catalytic sites is essential for cooperativity of the enzyme. H⁺ transported through F_0 changes the conformation of the δ, ε and γ subunits, and finally that of the β subunit to drive ATP synthesis. In the reverse direction, β subunit conformational changes caused by ATP hydrolysis are transmitted to other subunits and finally drive H⁺ transport through the F_0 proton pathway. Mutation/suppression studies can be one approach to understand functional interaction(s) between different domains. The defect due to the mutation in the domain near the catalytic site may be suppressed by second site-mutations in other domains, if the two domains are functionally related.

Among residues in the P-loop, βLys-155 and βThr-156 are essential, as described above, but other residues, such as βGly-149 and βGly-150, can be replaced by Ser or Ala without loss of activity [98]. However, the βCys-149 mutant could not grow by oxidative phosphorylation and was defective in multisite catalysis. The defect of the βCys-149 mutation was suppressed by amino acid replacements of the βGly-172, βSer-174, βGlu-192, or βVal-198 residue [99]. In the reverse direction, the βSer-174 → Phe mutation was suppressed by βGly-149 → Ser or βGly-149 → Cys [98]. The βSer-174 → Phe mutation was also suppressed by other mutations [100]. Bovine residues corresponding to those that suppress the βCys-149 mutant are located near the outer surface of the β subunit and not actually close to βGly-149 which faces the two α helices of the γ subunit at the center of the F_1 molecule [17] (Fig. 3). Thus, the defect of the βCys-149 mutant may be in the conformational transmission from the P-loop to other domain(s), and was suppressed by substitutions at the outer domain of the β subunit. These results may suggest that the conformational transmission between the P-loop and the outer domain is essential during multisite catalysis.

Three copies of the catalytic β subunit synthesize or hydrolyze ATP in a cooperative manner, as described above. As each β subunit is located next to the α subunit in the hexagonal arrangement of the α/β subunit pairs and the different β subunits do not interact directly with each other (Fig. 1), conformational change(s) of one β subunit during catalysis should be transmitted to other β subunit(s) through the nearby α subunit. The requirement of such functional α/β interaction for the catalytic cooperativity is supported by the early α subunit mutants mapped around position 370 such as αSer-373 → Phe [101,102] and α Arg-376 → Cys [103]: their multisite catalysis activities were very low, but they exhibited normal unisite catalysis. The bovine crystal structure [17] indicates that the α subunit region including αSer-373 is close to the α/β subunit interface.

Functional α/β subunit interaction(s) was shown clearly by the results of mutation/suppression studies. The defect of the βSer-174 → Phe mutant [101] was suppressed by the second mutation in the α subunit (αArg-296 → Cys) [104]. The αCys-296/βPhe-174 mutant enzyme was coupled efficiently, although its membrane ATPase activity was essentially the same as that of the βPhe-174 single mutant. These results indicate the importance of functional α/β interaction, possibly between the regions near αArg-296 and βSer-174, for energy coupling. However, the two bovine residues corresponding to αArg-296 and βSer-174 are not located close by in the higher ordered structure [17] (Fig. 3). As described above, mutations in the domain near βSer-174 suppressed βGly-149 → Cys of the P-loop sequence, suggesting functional interaction(s) between the two regions [99]. Thus it may be possible to assume that the conformational change around βGly-149 in the P-loop may be transmitted to the region around αArg-296 through the domain containing βSer-174.

4. Roles of the γ subunit in catalysis and energy coupling

4.1. Roles of the γ subunit in catalysis

This subunit is essential for the assembly of the *E. coli* catalytic complex: the γ subunit is required to form a minimal $\alpha_3\beta_3\gamma$ complex having ATPase activity

[46,47]. Analysis of a series of γ subunit mutants lacking the carboxyl terminal indicated that γGln-269 → end had normal F_0F_1 assembly, while γGln-261 → end affected the assembly of the entire enzyme [105]. The F_1 sector was lost through a deletion in the amino terminus (between γLys-21 and γAla-28) [106]. The *E. coli* [105] or yeast [107] mutant lacking the γ subunit had no stable F_1 attached to F_0. These results suggest the central role of the γ subunit in the assembly of the catalytic sector. A negatively stained specimen labeled with maleimidogold established that the central mass of F_1 contains the amino terminal of the γ subunit [108].

The amino and carboxyl terminal α helices of the γ subunit are both positioned in the center of the higher-ordered structure of bovine F_1 [17] (Fig. 3), suggesting that the γ subunit specifically interacts with the α and β subunits, and such interactions may be critical for enzyme catalysis. The role of the γ subunit in catalysis was first suggested by dithiothreitol activation of the illuminated chloroplast enzyme [109]. Detailed studies of the chloroplast F_1 indicated that the two cysteine residues in the chloroplast γ subunits form a disulfide bond which is inaccessible to dithiothreitol in the dark [109]. Upon illumination, the disulfide becomes accessible to reduction by dithiothreitol or the thioredoxin/feredoxin system, and results in the activation of the enzyme [110,111]. The results of cDNA sequencing of the spinach [112], *Arabidopsis thaliana* [113] and *Clamidomonas reinhardtii* [114] enzymes, and peptide mapping of the spinach enzyme [115] indicated that a disulfide bond (spinach γCys-199 ~ γCys-205) in the unique regulatory domain was reduced during the activation [110]. Such a domain was not found in bacterial or mitochondrial enzymes (Fig. 6) [112].

Compared with the α and β subunits, the γ subunit is less conserved: when the known γ subunit sequences are aligned to obtain maximal identity, 28 of the 286 residues in the *E. coli* subunit are conserved and they are candidates for functional residues [116] (Fig. 6). Eighteen of them are within 50 residues of the carboxyl terminus. Studies with termination mutants indicated that 10 residues (γLeu-278 – γVal-286) within the carboxyl terminus are important, but dispensable for catalysis [117]. On the other hand, the γGln-269 → end could not grow by oxidative phosphorylation and had low membrane ATPase activity. Stimulated by these results, we replaced hydrophilic residues between γGlu-269 and γLeu-278. The γGln-269 → Glu, Arg, or Leu mutant [117] showed reduced growth by oxidative phosphorylation and low membrane ATPase activity, consistent with the finding that bovine residues corresponding to γGln-269 and βAsp-305 form a hydrogen bond in the higher ordered structure [17]. The γThr-273 → Val or Gly and γGlu-275 → Lys or Gln mutations also showed decreased membrane ATPase activity [117]. These results suggest that the carboxyl terminal residues are required for normal ATPase multisite catalysis, especially for the β/γ interaction. Similarly, the yeast γ subunit mutant (Ala → Val) corresponding to *E. coli* γAla-250 had only negligible ATPase activity [107].

Of special interest was that the γGln-269 → Leu, γGlu-275 → Lys and γThr-277 → end mutants had similar ATPase activities (about 14% of the wild type), but γGln-269 → Leu membranes formed a much weaker electrochemical proton gradient than the others [117]. These results suggest that the γ subunit carboxyl terminus also participates in energy coupling and ATPase activity.

```
                                                                                  *
Ec   AGAKEIRSKI ASVQNTQKIT KAMEMVAASK MRKSQDRMAA SRPYAETMRK VIGHLA-HGN L-EYKHPYLE DRDVKRVGYL
mit  ATLKDITRRL KSIRNIQKIT KSMKMVAAAK YARAERELKP ARVYGVGSLA LYEKAD---- ---IKTP--E DK--KK--HL
chl  ANLRELRDRI GSVKNTQKIT EAMKLVAAAK VRRAQEAVVN GRPFSETLVE VLYNMNEQLQ TEDVDVP-LT KIRTVKKVAL
                                         B

Ec   VV--STDRGL CGGLNINLFK KLLAEMKTWT DKGVQCDLAM IGSKGVSFFN SVGGNVVAQV TGMGDN--PS LSELIGPVKV
mit  IIGVSSDRGL CGAIHSSVAK QMKSEAANLA AAGKEVKIIG VGDKIRSILH RTHSDQFLVT FKEVGRRPPT FGDASVIALE
chl  MVV-TGDRGL CGGFNNMLLK KAESRIAELK KLGVDYTIIS IGKKGNTYFI RRPEIPVDRY FD-GTNL-PT AKEAQAIADD

Ec   MLQAYDEGRL DKLYIVSNKF INTMSQVPTI SQLLPLPASD ---------- ---------- ----DDLKH
mit  LLNSGYE--F DEGSIIFNRF RSVISYKTEE KPIFSLDTI- ---------- ---------- ---SSAESMS
chl  VFSLFVSEEV DKVEMLYTKF VSLVKSDPVI HTLLPLSPKG EICDINGKCV DAAEDELFRL TTKEGKLTVE RDMIKTETPA
                                                                A

Ec   KSWDYLYEPD PKALLDTLLR RYVESQVYQG VVENLASEQA ARMVAMKAAT DNGGSLIKEL QLVYNKARQA SITQELTEIV SGAAAV  286
mit  ----IYDDI DADVLRNYQE YSLANIIYYS LKESTTSEQS ARMTAMDNAS KNASEMIDKL TLTFNRTRQA VITKELIEII SGAAAL  272
chl  FSPILEFEQD PAQILDALLP LYLNSQILRA LQESLASELA ARMTAMSNAT DNANELKKTL SINYNRARQA KITGEILEIV AGANACV  323
                                                                                                   C
```

Fig. 6. Alignment of the sequences of γ subunits from chloroplasts, mitochondria and bacteria. The amino acid sequences of the γ subunits from *E. coli* (Ec), bovine mitochondria (mit) and spinach chloroplasts (chl) are shown. The positions of identical residues in all the γ subunits so far sequenced are indicated by asterisks. A unique region for the chloroplast subunit (A) is underlined. Conserved amino (B) and carboxy terminals (C) are shown. The chloroplast cysteine residues forming the disulfide are found in Region A.

4.2. Roles of the γ subunit in energy coupling

Four residues between γLeu-19 and γVal-26 are identical among the γ subunits so far sequenced (Fig. 6), and deletion of γLys-21 to γAla-27 affected F_1 assembly [106], suggesting the importance of the amino terminal domain for subunit/subunit interaction. We introduced mutations systematically into the amino terminal half of the subunit to further examine the role of the amino terminus (Table 1). Most replacements, except γMet-23 → Arg or Lys, had little effect on growth through oxidative phosphorylation [118]. The γMet-23 → Arg or Lys mutant grew only very slowly by oxidative phosphorylation, although other mutants, γMet-23 → Asp and γMet-23 → Asn, were similar to the wild type. Surprisingly, the membranes of γArg-23 and γLys-23 had substantial ATPase activity levels (100 and 65% of wild-type activity, respectively), but formed a much lower electrochemical gradients of protons than the wild type (32 and 17%, respectively). These results suggest that substitutions with positively charged side chains at position 23 reduce the energy coupling efficiency between catalysis and proton transport.

Eight mutations (γArg-242 → Cys, γGln-269 → Arg, γAla-270 → Val, γIle-272 → Thr, γThr-273 → Ser, γGlu-278 → Gly, γIle-279 → Thr, and γVal-280 → Ala) suppressed the defect of γMet-23 → Lys [119]. Similarly, γGln-269 → Arg and γThr-273 → Ser showed reduced growth by oxidative phosphorylation as single mutations, whereas these mutations in combination with γMet-23 → Lys caused substantially increased activity. Furthermore, growth by oxidative phosphorylation or ATP-dependent proton transport of a strain carrying γMet-23 → Lys, γGlu-269 → Arg, or γThr-273 → Ser was temperature sensitive, whereas double mutants γMet-23 → Lys/γGln-269 → Arg, and γMet-23 → Lys/γThr-273 → Ser, were thermally stable. Further studies could define domains critical for both the catalytic function and energy coupling: second-site mutations that suppressed the primary mutations, γGln-269 → Glu and γThr-273 → Val, were mapped at the amino terminal region (residues 18, 34 and 35), and near the carboxyl terminus with changes at residues 236, 238, 242 and 246 [120].

These results led us to conclude initially (before the crystal structure was available) that γMet-23, γArg-242, and the region between γGln-269 and γVal-280 are close to each other, and interact to mediate efficient energy coupling [119,120]. As expected, the amino and carboxyl terminal helices of the γ subunit are in close proximity and the residue corresponding to γMet-23 in the amino terminal α helix is actually near the residue corresponding to γArg-242 in the carboxyl terminal α helix. However, the carboxyl terminal region between γGln-269 and γVal-280 is not in the domain directly interacting with γMet-23. These results suggest that the two regions functionally interact through long range conformational transmission. The three γ subunit segments defined by suppressor mutagenesis, γ21–35, γ236–246, and γ269–280, constitute a critical domain for such conformation transmission in energy coupling [120].

4.3. β/γ Subunit interactions in energy coupling and catalysis

γ Subunit conformational changes have been indicated by crosslinking and tryptic digestion. The γ subunit with the γSer-8 → Cys mutation crosslinked with the β subunit with a bifunctional reagent [121]. The site of crosslinking in the β subunit in the presence of Mg²⁺ ADP was localized in the βVal-145–βLys-155 segment, which contains the P-loop. The γ subunit conformation during ATPase catalysis has been studied using a fluorescent maleimide bound to a Cys residue incorporated by site directed mutagenesis. The fluorescence probe bound to γCys-106, responded to ATP hydrolysis in the catalytic site [122]. Further studies indicated that ATP binding causes fluorescence enhancement and this change is reversed upon bond cleavage to yield an enzyme ADP·Pi complex [123]. Thus the γ subunit changes in conformation upon ATP binding and possibly changes to a different form after ATP hydrolysis.

An interesting frameshift mutant having unrelated 16 carboxyl terminal residues could not grow by oxidative phosphorylation and had low membrane ATPase activities [117]. Two methods led to the prediction that the mutant carboxyl terminus became much longer and forms a long β-strand which extends about 60 Å from γThr-277 (Fig. 7). This β-strand may interact with the upper β-barrel or a part of the catalytic domain of the β subunit, causing defective β/γ subunit conformational

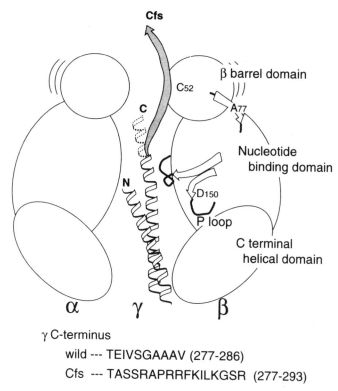

γ C-terminus

wild --- TEIVSGAAAV (277-286)

Cfs --- TASSRAPRRFKILKGSR (277-293)

Fig. 7.   A model of F₁ with the γ frameshift mutation and suppression by the β subunit mutations. The γ frameshift has an unrelated sequence of 16 residues (7 residues longer than the wild type), as shown at the bottom of the figure. It was predicted that the carboxyl terminal region (cfs) of the γ frameshift forms a β-strand of about 60 Å shaded arrow). The βR52C (with or without βV77A) and βG150D mutations restored membrane ATPase activity and ATP synthesis. Mutant residues are located following the bovine F₁ structure.

transmission. The defect of this mutant may suggest that the γ subunit actually rotates during catalysis, because the movement of the mutant γ may be inhibited by the interaction of the long β-strand with the β subunit.

Surprisingly, the βArg-52 → Cys or βGly-150 → Asp mutation could suppress the deleterious effects of the γ frameshift mutation (Fig. 7), suggesting that the altered β/γ interaction in the γ frameshift was restored by the β subunit mutation [124]. Thus, the βArg-52 → Cys suppressed the γ frameshift, possibly by restoring the deleterious interactions between the putative carboxyl terminal β-strand of the γ frameshift and the β-barrel, because the βArg-52 residue corresponds to the bovine residue located in the β-barrel domain. This notion was further supported by the suppressing effect of βVal-77 → Ala. Bovine Ile-84 corresponding to βVal-77 is located in the β sheet connecting the β-barrel with the nucleotide binding domain. Following the rotation model, the frameshift γ subunit could rotate, overcoming its defective interaction with the β-barrel upon the β subunit mutations. The frameshift mutant may provide further approaches to understand the role of the γ subunit. In this regard, we plan to introduce a fluorescent probe to the carboxyl terminus of the γ frameshift or to crosslink the γ carboxyl domains with the β barrel. The defective energy coupling of the γ frameshift was suppressed by the βAsp-150 mutation of the P-loop possibly through a change in the mode of conformational transmission between the β catalytic site and the γ subunit.

Aggeler et al. introduced the βGlu-381 → Cys and βGlu-384 → Cys mutations, and found that they could form a disulfide bond, possibly with γCys-87 [125]. The crosslinking was highest with ATP at the catalytic sites and much lower with ADP. The two β subunit residue are in the "DELSEED" (Asp-Glu-Leu-Ser-Glu-Glu-Asp) region that interacts tightly with a segment of the γ subunit (Fig. 3). This region is also the binding site for inhibitors such as quinacrine mustard [126] and dequalinium [127]. Aurovertin binds to βArg-398 near the DELSEED sequence [128]. The DELSEED region forms a loop interacting with the γ subunit. A conformational change of this region of the chloroplasts subunit was suggested by pyridoxal-5'-phosphate binding [129].

The DELSEED region has also been implicated in the binding of the ε subunit [130,131]. Cys mutation studies [125,132] indicated that εCys-108 (wild-type, Ser) forms a disulfide with βCys-381 or βCys-383 in the β subunits that interact with the γ subunit. The β/ε disulfide formation was also nucleotide-dependent, being highest with ADP and lower with ATP. The disulfide bond formed with either β/γ or β/ε inhibits ATPase activity in proportion to the disulfide bonds formed. The disulfide bond formation between the β and ε subunits is consistent with the earlier finding that the ε subunit and a inhibitory monoclonal antibody interact with the carboxyl terminal region of the β subunit [133]. The crosslinking of the two subunits was induced by 1-ethyl-3-(3-dimethylaminopropyl) carbodiimide [130, 131]. Catalytically, the ε subunit affects the affinity of the β subunit for aurovertin and inhibits product release from the catalytic site in unisite catalysis. These results suggest a model in which the β subunit movement is relative to the γ or ε subunit, and in which such movement is essential for catalytic cooperativity.

5. Proton pathway and its interaction with the F_1 subunits

5.1. Subunit assembly of F_0

As described above, *E. coli* F_0 is composed of the *a*, *b* and *c* subunits with a stoichiometry of $a_1\,b_2\,c_{10-12}$ [2,134] (Fig. 1), and all isolated subunits are necessary to reconstitute functional F_0 [135]. Two different structure models are possible for F_0: 10–12 copies of the *c* subunit form an independent ring which interacts with the complex of subunits *a* and *b* (Fig. 2a); 10–12 copies of subunit *c* form a ring around two *b* and one *a* subunit (Fig. 2b). Conclusive evidence for either model has not been obtained so far. Altendorf and coworkers proposed a similar model to that in Fig. 2a on the basis of analysis of a negatively-stained specimen [136].

Each subunit of F_0 traverses through membranes: subunit *a* crosses the membrane several times (perhaps 6 times), whereas subunit *b* may cross the membrane with a single transmembrane helix, leaving most of the protein exposed to the cytoplasm and interacting with the F_1 subunit. Cryoelectron microscopic analysis of the complex of F_1 and the truncated *b* subunit suggests that the polar domain interacts with a different β subunit than the one interacting with the ε subunit [137]. The *c* subunit forms a hairpin (two α helices) structure with a cytoplasmic loop [138,139]. Close interactions between F_0 subunits were suggested by suppression of *b* subunit mutations with second mutations in subunit *c* or *a* [140,141].

The process of assembly of these subunits has attracted attention since the cloning of the operon. Extensive studies with expression plasmids were carried out [142,143]. Hermolin and Fillingame showed recently that subunits *b* and *c* are inserted into the membrane independently of the other two F_0 subunits [144]. Subunit *a* was not inserted into membranes that lacked either subunit *b* or *c*, indicating that subunit *a* insertion is dependent on the co-insertion of subunit *b* or *c*. Their conclusion is based on analysis of chromosomal mutants defective in each F_0 subunit, and they pointed out that previous results obtained with multicopy plasmids may not reflect the normal assembly.

The eukaryotic F_0 has a more complex structure: chloroplast F_0 has four subunits with two different subunits, similar to the *E. coli b* subunit (*b*, subunit I; *b'*, subunit II) [9], whereas bovine F_0 has a more complex structure, as described above [10]. However, bacterial F_1 could bind to eukaryotic F_0 and forms functional F_0F_1, as discussed above.

5.2. Proton pathway of F_0: residues in subunits a and c

Genetic studies suggested that subunits *a* and *c* form the proton pathway of F_0 [134]. NMR analysis of the purified subunit *c* indicated that this protein is folded as two interacting α helices in a chloroform–methanol–water solution, supporting a hairpin model crossing the membrane [138]. The carboxyl moiety of *c*Asp-61 of the *E. coli c* subunit corresponds to a glutamate residue in mitochondria, chloroplasts or other bacterial subunits [134]. The carboxyl moiety is in the center of the second membrane spanning α helix and thought to undergo a protonation/deprotonation cycle during proton translocation [134,145]. Proton transport activity was still maintained in a mutant (cAla-24 $\rightarrow$ Asp/cAsp-61 $\rightarrow$ Gly) in which the aspartate was moved genetically from position 61 (in the second transmembrane helix) to 24 (in the first helix) [146]. Consistently, NMR studies indicated that cAsp-61 was in van der Waals contact with cAla-24 [145].

Na$^+$ translocating F_0F_1 was first found in _Propionigenium modestum_ [147] and later in other bacteria [148] (See Chapter II by P. Dimroth). The differences between Na$^+$ and H$^+$ transporting F_0F_1 are interesting to study transport through F_0. The transport properties are due to the F_0 sector: the hybrid ATPase (α, γ, β and ε from _E. coli_; a, b, c, δ from _P. modestum_) could support Na$^+$ dependent growth of _E. coli_ through oxidative phosphorylation and catalyzed Na$^+$ transport _in vitro_ [149]. The _in vitro_ reconstitution of _P. modestum_ F_0 and _E. coli_ F_1 resulted in a complex capable of Na$^+$ transport [150]. The _P. modestum_ c subunit exhibits 19% identity with that of _E. coli_ [151]. The binding of Na$^+$ or Li$^+$ to the cation binding site in _P. modestum_ F_0 was shown to activate ATPase activity [147], and both Na$^+$ and Li$^+$ inhibit proton transport through F_0 [152].

The sequence around the presumed _E. coli_ protonation site (Val-Asp-Ala-Ile, residues 60–63) is different from that of _P. modestum_ (Ala-Glu-Ser-Thr, residues 64–67). An _E. coli_ cAsp-61 corresponds to Glu-65 of _P. modestum_. An _E. coli_ mutant with four amino acid replacements (cVal-60 $\rightarrow$ Ala, cAsp-61 $\rightarrow$ Glu, cAla-62 $\rightarrow$ Ser, and cIle-63 $\rightarrow$Thr) could bind Li$^+$, as indicated by Li$^+$ inhibition of ATP-driven H$^+$ transport, F_0F_1 ATPase activity and H$^+$ transport through F_0 [151]. All four replacements (introducing the _P. modestum_ sequence) were necessary for Li$^+$ binding. It is of interest that cAsp-61 should be changed to a glutamate together with other residues, consistent with the presence of a glutamate at the corresponding position of the mitochondrial or chloroplast enzyme. From these studies, Zhang and Fillingame suggested that X-Glu-Ser/Thr-Y is a structural motif for monovalent cation binding. In this regard, it is of interest to note that the 16 kDa proteolipid of Na$^+$ translocating vacuolar ATPase of _Enterococcus_ has an X-Glu-Thr-Y motif in the middle of the fourth transmembrane α helix [153], whereas the proteolipid subunit of other H$^+$ translocating vacuolar ATPases has X-Glu-Val/Gly-Y at the corresponding position [40,154]. The role of the glutamate residue (in the center of the fourth transmembrane helix) in H$^+$ translocation was suggested by mutational studies on yeast vacuolar ATPase [155].

Extensive mutagenesis studies on the a subunit suggested that this subunit plays an essential role in proton transport. The overall similarity of this subunit is lower than that of the catalytic subunit, but it exhibits significant similarity near the carboxyl terminus [134]. The aArg-210 residue could not be replaced by Lys, Glu, Ile, Val or Gln without loss of ATP synthesis and proton translocation dependent on ATP-hydrolysis, indicating that this residue is essential for proton transport [156,157]. Mutational studies also suggested functional roles of aGlu-219 and aHis-245 [157–159]. Single mutants (aGlu-219 $\rightarrow$ His and aHis-245 $\rightarrow$ Glu) were defective, but the double mutant had substantial activity. Similarly, the residues corresponding to aGlu-219 and aHis-245 of the mitochondrial subunit 6 are His and Glu, respectively. Na$^+$ transporting F_0F_1 has Met at position 219 and Asp at the 245 position [160,161]. These results suggest that His and Glu can be moved to positions 219 and 245, respectively, possibly because these two residues are localized nearby in the membrane spanning α helices.

*a*Gln-252 is a strictly conserved residue located in the center of a hydrophobic stretch of 21 residues presumed to be an α helix. This Gln can be replaced by Glu without loss of activity [162]. With this mutation present, *a*Glu-219 can be replaced by Asp, Lys, Gly, Ala or Ser, suggesting that the functional Glu at position 219 can be moved to position 252 [161]. At positions homologous to *E. coli a*Gly-218, *a*Glu-219 and *a*His-245, alkalophilic bacteria have Lys, Glu and Gly, respectively, and chloroplast enzymes have Asp, Glu and Gly, respectively [163]. *a*Gly-218 → Asp, *a*Gly-218 → Lys, and *a*His-245 → Gly did not grow by oxidative phosphorylation and were defective in H^+ transport, although they had substantial membrane bound ATPase [163]. However, the chloroplast and alkalophile-type double mutants as to the two (*a*Gly-218 → Lys/*a*His-245 → Gly and *a*Gly-218 → Asp/*a*His-245 → Gly, respectively) showed restored ATP synthesis and H^+ transport [163]. Thus evidence supports a close spatial interaction between *a*Glu-218 and *a*His-245.

The *c*Ala-24 → Asp /*c*Asp-61 → Gly mutant grew much slower than the wild-type by oxidative phosphorylation, although it could grow to some degree [146]. The additional mutations introduced to *a*Ala-217, *a*Ile-221 or *a*Leu-224 significantly increased the growth by oxidative phosphorylation [164]. These residues are located in the α-helix that includes *a*Arg-210. This helix was proposed to interact with the transmembrane helix of subunit *c* during protonation/deprotonation. A detailed model of H^+ translocation through *a*Arg-210, *a*Glu-219 and *c*Asp-61 was proposed recently [161]. It is possible that *a*Arg-210 lowers the pKa of *c*Asp-61 transiently during sequential protonation of multiple *c* subunits [164]. The participation of residues in ten or twelve *c* subunits and one *a* subunit may be consistent with the rotation of the *c* subunit during H^+ transport coupled catalysis.

The residues in the polar region of the *c* subunit have a conserved sequence, Arg-Glu-Pro (positions 41–43). Mutations in this region gave enzymes with an uncoupled phenotype or with defective assembly [165–167], consistent with the proposal that this region plays a key role in energy coupling between the F_1 and F_0 sectors [165–167]. Recently, Watts and coworkers reported that this region is in direct contact with the γ and ε subunit [168].

5.3. F_0F_1 Subunit interaction

Interactions between subunits of F_0 and F_1 during the transport of protons through F_0 should drive ATP synthesis in the β subunit of F_1 through successive conformational changes. The catalytic sector $\alpha_3\beta_3\gamma$ complex is connected to the F_0 sector by a stalk (~45 Å long and a diameter of 25–30 Å), as visualized by electron microscopy [169]. Both the δ and ε subunits are required for the $\alpha_3\beta_3\gamma$ complex to bind to F_0 [47]. The cytoplasmic domain of subunit *b* [170] and the amino terminus of the *a* subunit are required [171] for binding F_1. Thus it is reasonable to assume that these subunits form the bacterial stalk. The mitochondrial stalk appears to be more complex, and is formed by the γ, δ, ε OSCP, F_6, *b* and *d* subunits [172].

The α subunit mutant, αGly-29 → Asp, showed low growth by oxidative phosphorylation, a proton permeable membrane and DCCD insensitive ATPase activity, suggesting that αGly-29 lies in a region of the α subunit important for binding to F_0 [173]. The δ subunit (176 residues) in F_1 is sensitive to proteases, but is resistant in F_0F_1 [174]. Mutagenesis experiments showed more drastic effects on carboxyl terminal truncation: δCys-140 of F_1 or F_0F_1 was labeled with $[^{14}C]$ NEM and formed a disulfide bond with an α subunit residue when treated with $CuCl_2$ [175]. These results suggest that this subunit interacts with an $\alpha_3\beta_3\gamma$ complex with amino and carboxyl terminals associated with the F_1 sector, and that the carboxyl terminal is essential for proper F_0F_1 interaction. The isolated δ subunit has an elongated shape [176], and mutagenesis experiments suggested that its overall structure is necessary, but that individual residues may not have strict functional roles [177–179].

In contrast to the δ subunit, the ε subunit (138 amino acid residues) has been suggested to regulate enzyme activity [28,180]. Studies on a series of truncated mutants (lacking 70–68 residues) suggested that the carboxyl terminal half of the subunit is not necessary for the binding of F_1 to F_0 [181]. Consistent with these results, the ε subunit of *Chlorobium limicola* has 88 residues [182]. Similarly, F_1 lacking 15 ε amino terminal residues could bind to F_1 in a functionally competent manner, whereas F_1 lacking ε16 amino terminal residues showed defective energy coupling, although mutant F_1 still showed binding to F_0 [183]. It is of interest that the ε subunit lacking 15 residues from the amino terminus and 4 residues from the carboxyl terminus was defective. These results suggest that both terminal regions affect the conformation of the functional domain of the enzyme.

Binding of the ε subunit inhibits the off-rate of Pi from F_1 during multisite catalysis [180]. This subunit also affects Aurovertin D binding to the β subunit [180]. Tryptic digestion experiments clearly indicated that the conformation of this subunit and its binding to F_1 are controlled by the presence of Pi: the cleavage of the ε subunit was fast in the presence of adenine nucleotide with or without Mg^{2+} but became slower when Pi was additionally present. The half maximal concentration of Pi (50 μM) is similar to that for high affinity binding of Pi to F_1.

The ε subunit was shown to bind to the γ subunit with an affinity of ~3 nM [184]. The subunit is also crosslinked with one β subunit forming an ester linkage between βGlu-381 and εSer-108 [131]. The β/ε interaction was shown to be altered by a detergent, lauryldimethylamine oxide [185]. Mutation studies indicated that εHis-38 lies near γ, although it does not directly interact with it. Recent studies also indicated that the cAsp-44 → Cys mutant subunit could form a crosslink with the γ (γ202-286) and ε subunits, indicating that the two subunits span the full length of the stalk between the β and c subunits [168]. It is interesting that the γ subunit spans from the β-barrel of the β subunit to the c subunit.

Mutational studies indicated that the a, b and c subunits are necessary to form a proton pathway and the F_1-binding site. Cleavage of subunit b with trypsin abolishes F_1 binding, although proton conduction through F_0 is not inhibited by proteolytic digestion [170]. The wild type b subunit has been purified and shown to be capable of reconstitution [134]. The b subunit lacking the amino terminal 24 residues was purified and shown to be highly α-helical [186]. This mutant subunit inhibited binding of F_1 to F_0, indicating that the mutant subunit interacts with a subunit of F_1.

6. Summary and perspectives

F-type H^+ ATPase is a proton pump coupling H^+ transport and the chemical reaction, "ATP $\leftrightarrow$ ADP + Pi". This enzyme is conserved among eukaryotic and prokaryotic cells, and synthesizes ATP in most cases. The higher ordered structure of the catalytic sector $\alpha_3\beta_3\gamma$ complex was established by X-ray diffraction. The catalytic site and proton pathway have been studied extensively by means of mutagenesis and chemical approaches. The roles of the γ and other subunits forming a stalk region in energy coupling have been extensively studied. The γ subunit seems to play a key role in the coupling.

To further understand the catalytic mechanism, determination of the detailed structure of the β subunit with all side chains is necessary. The crystallization of F_0F_1 and/or F_0 is the next step for elucidation of the mechanism and structure of the proton pathway. The next obvious question is what is the energy coupling between the transport and chemical reaction; "successive conformation transmission through subunit rotation" is a fascinating model of energy coupling. The conformational transmission between different subunits during catalysis are discussed above and can be summarized as follows: (1) The $\beta \rightarrow \alpha \rightarrow \beta \rightarrow \alpha$ subunit conformational transmission is required for the cooperativity of the multiple catalytic sites. (2) The β subunit interacts with the γ and ε subunits for catalysis and energy coupling. (3) The γ and ε subunits interact with the c subunit for energy coupling. During ATP synthesis, proton transport may cause F_0 subunit conformational changes which are transmitted to the γ and ε subunits, followed by the β subunit. Such successive conformational changes could be analyzed by using appropriate fluorescence probes. The rotation mechanism seems to be inevitable when the subunit stoichiometry of $F_0 F_1$ is considered (Fig. 8). It is reasonable to assume that the γ subunit (one copy) in the F_1 complex interacts successively with the three β subunits. The rotation is also consistent with the structure and function of the F_0 sector because Asp-61 of the c subunits (10–12 copies) and residues in the a subunit (one copy) participate in proton transport. It is not difficult to visualize a single rotational mechanism driving the catalysis. Downhill movement of protons through F_0 rotates the a subunit and stalk region as shown by the shaded arrow in Fig. 8a. It is necessary in this model that $\alpha_3\beta_3$ assembly is fixed to the bilayer by a unknown mechanism. Two types of rotations may be considered in the model shown in Fig. 8b: rotation of $\alpha_3\beta_3$ and the ring assembly of the c subunit. We may need more structural information on F_0 to plan experiments to show rotation of the enzyme experimentally.

Abbreviations

AP$_3$-PL, adenosine triphosphopyridoxal
DCCD, dicyclohexylcarbodiimide
FSBA, fluorosulfonyl benzoyl adenosine
Pi, inorganic phosphate

(a)   (b)

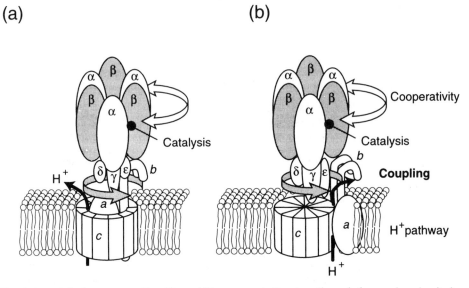

Fig. 8. A model of energy coupling. Down-hill movement of protons through the *a* and *c* subunits in F₀ rotate the stalk region including the γ subunit (shown by a shaded arrow), and the rotation drives ATP synthesis. The catalytic cooperativity is also indicated by an open arrow. See models of F₁ and F₀ in Figs. 1 and 2, respectively.

Acknowledgements

The work carried out in our laboratory was supported in part by grants from the Japanese Ministry of Education, Science and Culture, and the Human Frontier Science Program. We are very grateful to our collaborators, whose names appear in the references.

References

1. Futai, M., Noumi, T. and Maeda, M. (1989) Annu. Rev. Biochem. **58**, 111–136.
2. Fillingame, R.H. (1990) in The Bacteria, ed T.A. Krulwich, pp. 345–391, Academic Press, New York.
3. Senior, A.E. (1990) Annu. Rev. Biophys. Chem. **19**, 7–41.
4. Lutter, R., Saraste, M., van Walraven, H.S., Runswick, M.J., Finel, M., Deatherage, J.F. and Walker, J.E. (1993) Biochem. J. **295**, 799–806.
5. Yoshihara, Y., Nagase, H., Yamane, T., Oka, H., Tani, H. and Higuti, T. (1991) Biochemistry **30**, 6854–6860.
6. Schneider, E. and Altendorf, K. (1982) Eur. J. Biochemistry **126**, 149–153.
7. Foster, D.L. and Fillingame, R.H. (1979) J. Biol. Chem. **254**, 8230–8236.
8. Moriyama, Y., Iwamoto, A., Hanada, H., Maeda, M. and Futai, M. (1991) J. Biol. Chem. **266**, 22141–22146.
9. Cozens, A.L. and Walker, J.E. (1987) J. Mol. Biol. **194**, 359–383.
10. Collinson, I.R., Runswick, M.J., Buchanan, S.K., Fearnley, I.M., Skehel, J.M., van Raaij, M.J., Griffiths, D.E. and Walker, J.E. (1994) Biochemistry **33**, 7971–7978.
11. Zanotti, F., Guerrieri, F., Deckers-Hebestreit, G., Fiermonte, M., Altendolf, K. and Papa, S. (1994) Eur. J. Biochem. **222**, 733–741.

12. Baeuerlein, E., Galmiche, J.M., Stephane, P., Zhao, R. and Girault, G. (1994) FEBS Lett. **338**, 152–156.
13. Burkovski, A., Deckers-Hebestreit, G. and Altendorf, K. (1994) Eur. J. Biochem. **225**, 1221–1228.
14. Mimura, H., Hashimoto, T., Yoshida, Y., Ichikawa, N. and Tagawa, K. (1993) J. Biochem. **113**, 350–354.
15. Walker, J.E. (1994) Curr. Op. Struct. Biol. **4**, 912–918.
16. Abrahams, J.P., Lutter, R., Todd, R.J., van Raaij, M.J., Leslie, A.G.W. and Walker, J.E. (1993) EMBO J. **12**, 1775–1780.
17. Abrahams, J.P., Leslie, A.G.W., Lutter, R. and Walker, J.E. (1994) Nature **370**, 621–628.
18. Futai, M. and Kanazawa, H. (1983) Microbiol. Rev. **47**, 285–312.
19. Kagawa, Y. (1984) in: Bioenergetics, New Comprehensive Biochemistry, pp. 149–186, Elsevier, Amsterdam.
20. Walker, J.E., Saraste, M. and Gay, N.J. (1984) Biochim. Biophys. Acta **768**, 164–200.
21. Vignais, P.V. and Lunardi, J. (1985) Annu. Rev. Biochem. **54**, 977–1014.
22. Futai, M., Noumi, T., Miki, J. and Maeda, M. (1987) Chemica Scripta **27B**, 89–96.
23. Walker, J.E., Cozens, A.L., Dyer, M.R., Fearnley, I.M., Powell, S.J. and Runswick, M.J. (1987) Chemica Scripta **27B**, 97–105.
24. Senior, A.E. (1988) Physiol. Rev. **68**, 177–231.
25. Penefsky, H.S. and Cross, R.L. (1991) Adv. Enzymol. **64**, 173–213.
26. Allison, W.S., Jault, J.-M., Zhuo, S. and Paik, S.R. (1992) J. Bioenerg. Biomembr. **24**, 469–477.
27. Amzel, L.M., Bianchet, M.A. and Pedersen, P.L. (1992) J. Bioenerg. Biomembr. **24**, 429–433.
28. Capaldi, R.A., Aggeler, R., Gogol, E.P. and Wilkens, S. (1992) J. Bioenerg. Biomembr. **24**, 435–439.
29. Duncan, T.M. and Cross, R.L. (1992) J. Bioenerg. Biomembr. **24**, 453–461.
30. Fillingame, R.H. (1992) J. Bioenerg. Biomembr. **24**, 485–491.
31. Gromet-Elhanan, Z. (1992) J. Bioenerg. Biomembr. **24**, 447–452.
32. Kagawa, Y., Ohta, S., Harada, M., Kihara, H., Ito, Y. and Sato, M. (1992) J. Bioenerg. Biomembr. **24**, 441–445.
33. Boyer, P.D. (1993) Biochim. Biophys. Acta **1140**, 215–250Ê.
34. Hatefi, Y. (1993) Eur. J. Biochem. **218**, 759–767.
35. Fearnley, I.M. and Walker, J.E. (1986) EMBO J. **5**, 2003–2008.
36. Hennig, J. and Herrmann, R.G. (1986) Mol. Gen. Genet. **203**, 117–128.
37. Forgac, M. (1989) Physiol. Rev. **69**, 765–796.
38. Nelson, N. (1992) J. Bioenerg. Biomembr. **24**, 407–414.
39. Nelson, H., Mandiyan, S. and Nelson, N. (1995) Proc. Natl. Acad. Sci. USA **92**, 497–501.
40. Mandel, M., Moriyama, Y., Hulmes, J.D., Pan, Y.-C.E., Nelson, H. and Nelson, N. (1988) Proc. Natl. Acad. Sci. USA **85**, 5521–5524.
41. Supeková, L., Supek, F. and Nelson, N. (1995) J. Biol. Chem. **270**, 13726–13732.
42. Inatomi, K., Eya, S., Maeda, M. and Futai, M. (1989) J. Biol. Chem. **264**, 10954–10959.
43. Denda, K., Konishi, J., Oshima, T., Date, T. and Yoshida, M. (1989) J. Biol. Chem. **264**, 7119–7121.
44. Ihara, K. and Mukohata, Y. (1991) Arch. Biochem. Biophys. **286**, 111–116.
45. Bianchet, M., Ysern, X., Hullihen, J., Pedersen, P.L. and Amzel, L.M. (1991) J. Biol. Chem. **266**, 21197–21201.
46. Futai, M. (1977) Biochem. Biophys. Res. Commun. **79**, 1231–1237.
47. Dunn, S.D. and Futai, M. (1980) J. Biol. Chem. **255**, 113–118.
48. Kagawa, Y., Ohta, S. and Otawara-Hamamoto, Y. (1989) FEBS Lett. **249**, 67–69.
49. Ohta, S., Harada, M., Ito, Y., Kobayashi, Y., Sone, N. and Kagawa, Y. (1990) Biochem. Biophys. Res. Commun. **171**, 1258–1263.
50. Andralojc, P.J. and Harris, D.A. (1992) FEBS Lett. **310**, 187–192.
51. Fischer, S., Etzold, C., Turina, P., Deckers-Hebestreit, G., Altendorf, K. and Gräber, P. (1994) Eur. J. Biochem. **225**, 167–172.
52. Grubmeyer, C., Cross, R.L. and Penefsky, H.S. (1982) J. Biol. Chem. **257**, 12092–12100.
53. Duncan, T.M. and Senior, A.E. (1985) J. Biol. Chem. **260**, 4901–4907.
54. Noumi, T., Taniai, M., Kanazawa, H. and Futai, M. (1986) J. Biol. Chem. **261**, 9196–9201.
55. Weber, J., Wilke-Mounts, S. and Senior, A.E. (1994) J. Biol. Chem. **269**, 20462–20467.

56. O'Neal, C.C. and Boyer, P.D. (1984) J. Biol. Chem. **259**, 5761–5767.
57. Feldman, R.I. and Sigman, D.S. (1982) J. Biol. Chem. **257**, 1676–1683.
58. Sakamoto, J. and Tonomura, Y. (1983) J. Biochem. **93**, 1601–1614.
59. Yoshida, M. (1983) Biochem. Biophys. Res. Commun. **114**, 907–912.
60. Boyer, P.D. (1989) FASEB J. **3**, 2164–2178.
61. Souid, A.-K. and Penefsky, H.S. (1995) J. Biol. Chem. **270**, 9074–9082.
62. Matsuno-Yagi, A. and Hatefi, Y. (1985) J. Biol. Chem. **260**, 14424–14427.
63. Noumi, T., Maeda, M. and Futai, M. (1987) FEBS Lett. **213**, 381–384.
64. Tommasino, M. and Capaldi, R.A. (1985) Biochemistry **24**, 3972–3976.
65. Zhang, S. and Jagendorf, A.T. (1995) J. Biol. Chem. **270**, 6607–6614.
66. Soteropoulos, P., Ong, A.M. and McCarty, R.E. (1994) J. Biol. Chem. **269**, 19810–19816.
67. Hanada, H., Noumi, T., Maeda, M. and Futai, M. (1989) FEBS Lett. **257**, 465–467.
68. Hsu, S.Y., Noumi, T., Takeyama, M., Maeda, M., Ishibashi, S. and Futai, M. (1987) FEBS Lett. **218**, 222–226.
69. Tagaya, M., Noumi, T., Nakano, K., Futai, M. and Fukui, T. (1988) FEBS Lett. **233**, 347–351.
70. Ida, K., Noumi, T., Maeda, M., Fukui, T. and Futai, M. (1991) J. Biol. Chem. **266**, 5424–5429.
71. Omote, H., Maeda, M. and Futai, M. (1992) J. Biol. Chem. **267**, 20571–20576.
72. Senior, A.E. and Al-Shawi, M.K. (1992) J. Biol. Chem. **267**, 21471–21478.
73. Saraste, M., Sibbald, P.R. and Wittinghofer, A. (1990) Trends Biochem. Sci. **15**, 430–434.
74. Pai, E.F., Kabsch, W., Krengel, U., Holmes, K.C., John, J. and Wittinghofer, A. (1989) Nature **341**, 209–214.
75. La Cour, T.F.M., Nyborg, J., Thirup, S. and Clark, B.F.C. (1985) EMBO J. **4**, 2385–2388.
76. Dreusicke, D., Karplus, P.A. and Schulz, G.A. (1988) J. Mol. Biol. **199**, 359–371.
77. Chuang, W.J., Abeygunawardana, C., Pedersen, P.L. and Mildvan, A.S. (1992) Biochemistry **31**, 7915–7921.
78. Noumi, T., Tagaya, M., Miki-Takeda, K., Maeda, M., Fukui, T. and Futai, M. (1987) J. Biol. Chem. **262**, 7686–7692.
79. Andrews, W.W., Hill, F.C. and Allison, W.S. (1984) J. Biol. Chem. **259**, 14378–14382.
80. Takeyama, M., Ihara, K., Moriyama, Y., Noumi, T., Ida, K., Tomioka, N., Itai, A., Maeda, M. and Futai, M. (1990) J. Biol. Chem. **265**, 21279–21284.
81. Shen, H., Yao, B.-Y. and Müeller, D.M. (1994) J. Biol. Chem. **269**, 9424–9428.
82. Rao, R., Pagan, J. and Senior, A.E. (1988) J. Biol. Chem. **263**, 15957–15963.
83. Jounouchi, M., Maeda, M. and Futai, M. (1993) J. Biochem. **114**, 171–176.
84. Park, M.-Y., Omote, H., Maeda, M. and Futai, M. (1994) J. Biochem. **116**, 1139–1145.
85. Yoshida, M., Poser, J.W., Allison, W.S. and Esch, F.S. (1981) J. Biol. Chem. **256**, 148–153.
86. Amano, T., Tozawa, K., Yoshida, M. and Murakami, H. (1994) FEBS Letters **348**, 93–98.
87. Yoshida, M., Allison, W.S., Esch, F.S. and Futai, M. (1982) J. Biol. Chem. **257**, 10033–10037.
88. Bowman, E.J., Tenney, K. and Bowman, B.J. (1988) J. Biol. Chem. **263**, 13994–14001.
89. Eichelberg, K., Ginocchio, C.C. and Galán, J.E. (1994) J. Bacteriol. **176**, 4501–4510.
90. Amano, T., Yoshida, M. and Nishikawa, K. (1994) FEBS Lett. **351**, 1–5.
91. Noumi, T., Azuma, M., Shimomura, S., Maeda, M. and Futai, M. (1987) J. Biol. Chem. **262**, 14978–14982.
92. Omote, H., Nga, P.L., Park, M.-Y., Maeda, M. and Futai, M. (1995) J. Biol. Chem. **270**, 25656–25660.
93. Garin, J., Boulay, F., Issartel, J.P., Lunardi, J. and Vignais, P.V. (1986) Biochemistry **25**, 4431–4437.
94. Hollemans, M., Runswick, M.J., Fearnley, I.M. and Walker, J.E. (1983) J. Biol. Chem. **258**, 9307–9313.
95. Cross, R.L., Cunningham, D., Miller, C.G., Xue, Z., Zhou, J.-M. and Boyer, P.D. (1987) Proc. Natl. Acad. Sci. USA **84**, 5715–5719.
96. Bullough, D.A. and Allison, W.S. (1986) J. Biol. Chem. **261**, 5722–5730.
97. Parsonage, D., Duncan, T.M., Wilke-Mounts, S., Kironde, F.A.S., Hatch, L. and Senior, A.E. (1987) J. Biol. Chem. **262**, 6301–6307.
98. Iwamoto, A., Omote, H., Hanada, H., Tomioka, N., Itai, A., Maeda, M. and Futai, M. (1991) J. Biol. Chem. **266**, 16350–16355.
99. Iwamoto, A., Park, M.Y., Maeda, M. and Futai, M. (1993) J. Biol. Chem. **268**, 3156–3160.

100. Miki, J., Fujiwara, K., Tsuda, M., Tsuchiya, T. and Kanazawa, H. (1990) J. Biol. Chem. **265**, 21567–21572.
101. Noumi, T., Futai, M. and Kanazawa, H. (1984) J. Biol. Chem. **259**, 10076–10079.
102. Maggio, M.B., Pagan, J., Parsonage, D., Hatch, L. and Senior, A.E. (1987) J. Biol. Chem. **262**, 8981–8984.
103. Soga, S., Noumi, T., Takeyama, M., Maeda, M. and Futai, M. (1989) Arch. Biochem. Biophys. **268**, 643–648.
104. Omote, H., Park, M.-Y., Maeda, M. and Futai, M. (1994) J. Biol. Chem. **269**, 10265–10269.
105. Miki, J., Takeyama, M., Noumi, T., Kanazawa, H., Maeda, M. and Futai, M. (1986) Arch. Biochem. Biophys. **251**, 458–464.
106. Kanazawa, H., Hama, H., Rosen, B.P. and Futai, M. (1985) Arch. Biochem. Biophys. **241**, 364–370.
107. Paul, M.-F., Ackermann, S., Yue, J., Arselin, G., Velours, J. and Tzagoloff, A. (1994) J. Biol. Chem. **269**, 26158–26164.
108. Wilkens, S. and Capaldi, R.A. (1992) Arch. Biochem. Biophys. **299**, 105–109.
109. Ketcham, S.R., Davenport, J.W., Warncke, K. and McCarty, R.E. (1984) J. Biol. Chem. **259**, 7286–7293.
110. Dann, M.S. and McCarty, R.E. (1991) Plant Physiol. **99**, 153–160.
111. Hartman, H., Syvanen, M. and Buchanan, B.B. (1990) Mol. Biol. Evol. **7**, 247–254.
112. Miki, J., Maeda, M., Mukohata, Y. and Futai, M. (1988) FEBS Lett. **232**, 221–226.
113. Inohara, N., Iwamoto, A., Moriyama, Y., Shimomura, S., Maeda, M. and Futai, M. (1991) J. Biol. Chem. **266**, 7333–7338.
114. Yu, L.M. and Selman, B.R. (1988) J. Biol. Chem. **263**, 19342–19345.
115. Moroney, J.V., Fullmer, C. and McCarty, R.E. (1984) J. Biol. Chem. **259**, 7281–7285.
116. Nakamoto, R.K., Shin, K., Iwamoto, A., Omote, H., Maeda, M. and Futai, M. (1992) Ann. N.Y. Acad. Sci. **671**, 335–344.
117. Iwamoto, A., Miki, J., Maeda, M. and Futai, M. (1990) J. Biol. Chem. **265**, 5043–5048.
118. Shin, K., Nakamoto, R.K., Maeda, M. and Futai, M. (1992) J. Biol. Chem. **267**, 20835–20839.
119. Nakamoto, R.K., Maeda, M. and Futai, M. (1993) J. Biol. Chem. **268**, 867–872.
120. Nakamoto, R.K., Al-Shawi, M.K. and Futai, M. (1995) J. Biol. Chem. **270**, 14042–14046.
121. Aggeler, R., Cai, S.X., Keana, J.F.W., Koike, T. and Capaldi, R.A. (1993) J. Biol. Chem. **268**, 20831–20837.
122. Turina, P. and Capaldi, R.A. (1994) J. Biol. Chem. **269**, 13465–13471.
123. Turina, P. and Capaldi, R. (1994) Biochemistry **33**, 14275–14280.
124. Jeanteur-De Beukelar, C., Omote, H., Iwamoto-Kihara, A., Maeda, M. and Futai, M. (1995) J. Biol. Chem. **270**, 22850–22854.
125. Aggeler, R., Haughton, M.A. and Capaldi, R.A. (1995) J. Biol. Chem. **270**, 9185–9191.
126. Bullough, D.A., Ceccarelli, E.A., Verburg, J.G. and Allison, W.S. (1989) J. Biol. Chem. **264**, 9155–9163.
127. Zhuo, S., Paik, S.R., Register, J.A. and Allison, W.S. (1993) Biochemistry **32**, 2219–2227.
128. Lee, R.S.-F., Pagan, J., Satre, M., Vignais, P.V. and Senior, A.E. (1989) FEBS Lett. **253**, 269–272.
129. Komatsu-Takaki, M. (1995) Eur. J. Biochem. **228**, 265–270.
130. Mendel-Hartvig, J. and Capaldi, R.A. (1991) Biochemistry **30**, 1278–1284.
131. Dallmann, H.G., Flynn, T.G. and Dunn, S.D. (1992) J. Biol. Chem. **267**, 18953–18960.
132. Haughton, M.A. and Capaldi, R.A. (1995) J. Biol. Chem. **270**, 20568–20574.
133. Tozer, R.G. and Dunn, S. (1987) J. Biol. Chem. **262**, 10706–10711.
134. Schneider, E. and Altendorf, K. (1987) Microbiol. Rev. **51**, 477–497.
135. Schneider, E. and Altendorf, K. (1985) EMBO J. **4**, 515–518.
136. Birkenhäger, R., Hoppert, M., Deckers-Hebestreit, G., Mayer, F. and Altendorf, K. (1995) Eur. J. Biochem. **230**, 58–67.
137. Wilkens, S., Dunn, S.D. and Capaldi, R.A. (1994) FEBS Lett. **354**, 37–40.
138. Girvin, M.E. and Fillingame, R.H. (1993) Biochemistry **32**, 12167–12177.
139. Girvin, M.E., Hermolin, J., Pottorf, R. and Fillingame, R.H. (1989) Biochemistry **28**, 4340–4343.
140. Kumamoto, C.A. and Simoni, R.D. (1986) J. Biol. Chem. **261**, 10037–10042.
141. Kumamoto, C.A. and Simoni, R.D. (1987) J. Biol. Chem. **262**, 3060–3064.

142. Klionsky, D.J. and Simoni, R.D. (1985) J. Biol. Chem. **260**, 1200–1206.
143. Aris, J.P., Klionsky, D.J. and Simoni, R.D. (1985) J. Biol. Chem. **260**, 11207–11215.
144. Hermolin, J. and Fillingame, R.H. (1995) J. Biol. Chem. **270**, 2815–2817.
145. Girvin, M.E. and Fillingame, R.H. (1995) Biochemistry **34**, 1635–1645.
146. Miller, M.J., Oldenburg, M. and Fillingame, R.H. (1990) Proc. Natl. Acad. Sci. USA **87**, 4900–4904.
147. Laubinger, W. and Dimroth, P. (1987) Eur. J. Biochem. **168**, 475–480.
148. Reidlinger, J. and Müller, V. (1994) Eur. J. Biochem. **223**, 275–283.
149. Kaim, G. and Dimroth, P. (1994) Eur. J. Biochem. **222**, 615–623.
150. Laubinger, W., Deckers-Hebestreit, G., Altendorf, K. and Dimroth, P. (1990) Biochemistry **29**, 5458–5463.
151. Zhang, Y. and Fillingame, R.H. (1995) J. Biol. Chem. **270**, 87–93.
152. Kluge, C. and Dimroth, P. (1992) Biochemistry **31**, 12665–12672.
153. Kakinuma, Y., Kakinuma, S., Takase, K., Konishi, K., Igarashi, K. and Yamamoto, I. (1993) Biochem. Biophys. Res. Commun. **195**, 1063–1069.
154. Hanada, H., Hasebe, M., Moriyama, Y., Maeda, M. and Futai, M. (1991) Biochem. Biophys. Res. Commun. **176**, 1062–1067.
155. Noumi, T., Beltrán, C., Nelson, H. and Nelson, N. (1991) Proc. Natl. Acad. Sci. USA **88**, 1938–1942.
156. Cain, B.D. and Simoni, R.D. (1989) J. Biol. Chem. **264**, 3292–3300.
157. Eya, S., Maeda, M. and Futai, M. (1991) Arch. Biochem. Biophys. **284**, 71–77.
158. Lightowelers, R.N., Howitt, S.M., Hatch, L., Gibson, F. and Cox, G.B. (1988) Biochim. Biophys. Acta **933**, 241–248.
159. Cain, B.D. and Simoni, R.D. (1988) J. Biol. Chem. **263**, 6606–6612.
160. Esser, W., Krumholz, L.R. and Simoni, R.D. (1990) Nucleic Acids Res. **18**, 5887.
161. Vik, S.B. and Antonio, B.J. (1994) J. Biol. Chem. **269**, 30364–30369.
162. Hartzog, P.E. and Cain, B.D. (1993) J. Bacteriol. **175**, 1337–1343.
163. Hartzog, P.E. and Cain, B.D. (1994) J. Biol. Chem. **269**, 32313–32317.
164. Fraga, D., Hermolin, J. and Fillingame, R.H. (1994) J. Biol. Chem. **269**, 2562–2567.
165. Fraga, D. and Fillingame, R.H. (1989) J. Biol. Chem. **264**, 6797–6803.
166. Fraga, D., Hermolin, J., Oldenburg, M., Miller, M.J. and Fillingame, R.H. (1994) J. Biol. Chem. **269**, 7532–7537.
167. Miller, M.J., Fraga, D., Paule, C.R. and Fillingame, R.H. (1989) J. Biol. Chem. **264**, 305–311.
168. Watts, S.D., Zhang, Y., Fillingame, R.H. and Capaldi, R.A. (1995) FEBS Lett. **368**, 235–238.
169. Gogol, E.P., Lücken, U. and Capaldi, R.A. (1987) FEBS Lett. **219**, 274–278.
170. Perlin, D.S., Cox, D.N. and Senior, A.D. (1983) J. Biol. Chem. **258**, 9793–9800.
171. Eya, S., Noumi, T., Maeda, M. and Futai, M. (1988) J. Biol. Chem. **263**, 10056–10062.
172. Walker, J.E. and Collinson, I.R. (1994) FEBS Lett. **346**, 39–43.
173. Maggio, M.B., Parsonage, D. and Senior, A.E. (1988) J. Biol. Chem. **263**, 4619–4623.
174. Gavilanes-Ruiz, M., Tommasino, M. and Capaldi, R.A. (1988) Biochemistry **27**, 603–609.
175. Tozer, R.G. and Dunn, S.D. (1986) Eur. J. Biochem. **161**, 513–518.
176. Sternweis, P.C. and Smith, J.B. (1977) Biochemistry **16**, 4020–4025.
177. Jounouchi, M., Takeyama, M., Chaiprasert, P., Noumi, T., Moriyama, Y., Maeda, M. and Futai, M. (1992) Arch. Biochem. Biophys. **292**, 376–381.
178. Hazard, A.L. and Senior, A.E. (1994) J. Biol. Chem. **269**, 418–426.
179. Hazard, A.L. and Senior, A.E. (1994) J. Biol. Chem. **269**, 427–432.
180. Dunn, S.D., Zadorozny, V.D., Tozer, R.G. and Orr, L.E. (1987) Biochemistry **26**, 4488–4493.
181. Kuki, M., Noumi, T., Maeda, M., Amemura, A. and Futai, M. (1988) J. Biol. Chem. **263**, 17437–17442.
182. Xie, D.-L., Holger, L., Hauska, G., Maeda, M., Futai, M. and Nelson, N. (1993) Biochim. Biophys. Acta **1172**, 267–273.
183. Jounouchi, M., Takeyama, M., Noumi, T., Moriyama, Y., Maeda, M. and Futai, M. (1992) Arch. Biochem. Biophys. **292**, 87–94.
184. Dunn, S.D. (1982) J. Biol. Chem. **257**, 7354–7359.
185. Dunn, S.D., Tozer, R.G. and Zadorozny, V.D. (1990) Biochemistry **29**, 4335–4340.
186. Dunn, S.D. (1992) J. Biol. Chem. **267**, 7630–7633.

CHAPTER 4

The Neurospora Plasma Membrane Proton Pump

G.A. SCARBOROUGH

*Department of Pharmacology, School of Medicine,
University of North Carolina, Chapel Hill, NC, USA*

© *1996 Elsevier Science B.V.*
All rights reserved

Handbook of Biological Physics
Volume 2, edited by W.N. Konings, H.R. Kaback and J.S. Lolkema

Contents

1. Perspective: the field of membrane transport

An understanding of the molecular mechanism of membrane transport has been a major goal in the field of bioenergetics for many decades. A great many membrane transport molecules have been identified and characterized (e.g. Refs. [1–32]) and many of these have also been solubilized and purified and reconstituted into liposomes in fully functional form. Numerous investigations of these various membrane transporters have yielded a wealth of knowledge about their biochemical properties. More recently, effective techniques for site-directed mutagenesis of membrane transport molecules have emerged, and with this approach the specific functions of a variety of amino acid residues in several membrane transport molecules have been tentatively assigned (e.g. Refs. [33–46]). But in spite of the vast amount of information that has been accumulated regarding the function of membrane transport catalysts, we still know very little about how they really work. Scant progress has been made in developing a generally acceptable model for the molecular mechanism of membrane transport, and thus major questions remain as to the events that transpire as the transported solutes pass through the membrane barrier. As is especially evident in the work of Fersht and his collaborators [47–49], in order to understand how an enzyme works, it is first necessary to know its molecular structure in fine detail at several stages of its catalytic cycle, which is conventionally done by X-ray crystallography. It is then necessary to define the specific functions of a variety of individual amino acid residues in the molecule, which is most rigorously done by site-directed mutagenesis. This requires the ability to grow diffraction quality crystals of the enzyme in at least two of its conformational states and the availability of a high-yield system for expressing recombinant forms of the enzyme with specifically altered residues. Unfortunately, whereas numerous effective systems for expressing mutated transport molecules are available, few useful crystals of these proteins have ever been obtained, and accordingly, the required information regarding the structures of membrane transport molecules is in seriously short supply. It is thus generally agreed that what is desperately needed for progress toward understanding the molecular mechanism of membrane transport is structural information about the membrane transporters at the molecular level of dimensions.

The only available approaches to obtaining this information are X-ray diffraction and electron microscopic analyses of three-dimensional (3D) and two-dimensional (2D) crystals. Currently, the most powerful approach is X-ray diffraction analysis of 3D crystals, as evidenced by the spectacular solution of the structure of the bacterial photosynthetic reaction center by Deisenhofer et al. [50]. Unfortunately, although this approach held great promise for membrane proteins early on, subsequent progress has been slow. This trend may, however, be changing with recent reports of the structures of two different cytochrome c oxidases obtained by X-ray diffraction analyses of 3D crystals [51,52].

The only other available approach for obtaining structural information about transport molecules is electron crystallographic analysis of 2D crystals. The exciting solution of the general structure of bacteriorhodopsin by high-resolution electron microscopy of 2D crystals [53] provided a firm foundation for obtaining structural information about other membrane proteins with this approach. Significant progress in this regard with 2D and tubular crystals of several integral membrane proteins, including the erythrocyte anion transporter [54], the Ca^{++}-ATPase of sarcoplasmic reticulum [55], and the nicotinic acetylcholine receptor [56], has indeed been made. Although the level of structural resolution obtained with these systems has not yet equaled that obtained with bacteriorhodopsin, the recent report by Kühlbrandt et al. [57] describing the structure of the plant light-harvesting complex LHC-II to 3.4 Å resolution by electron crystallography of 2D crystals indicates a promising future for this approach as a means for solving the structures of membrane transport proteins. This is particularly true considering the fact that it has proved somewhat easier to grow 2D crystals of integral membrane proteins than to grow 3D ones.

The primary goal in this laboratory is an understanding of the molecular mechanism of membrane transport catalyzed by the proton-translocating ATPase from the plasma membrane of *Neurospora crassa*, which is an archetype of the so-called P-type ATPase family of ion-translocating ATPases [32]. With the foregoing considerations in mind, we have been working for several years now on obtaining crystals of the H^+-ATPase that are useful for structural studies, and on developing an effective H^+-ATPase expression system with which to define key amino acid residues by site-directed mutagenesis. Encouragingly, in the time that has transpired since our work on this enzyme was last reviewed [58], both of these objectives have been accomplished. In this article, the experimental history of the ATPase is briefly reviewed, and then our more recent progress toward understanding the structure and molecular mechanism of this transporter is described.

2. Salient features of the H^+-ATPase molecule elucidated in earlier studies

The existence of a protonic potential generator fueled by ATP hydrolysis in the plasma membrane of neurospora was first suggested decades ago on the basis of electrophysiological studies carried out in the laboratory of C.L. Slayman [59]. The development in this laboratory of a procedure for confidently isolating neurospora plasma membranes [60] made possible the direct demonstration of the existence of a plasma membrane ATPase [60–62] and demonstration that it acts as an electrogenic [63] proton translocator [64]. Biochemical studies carried out with the isolated plasma membranes defined the hydrolytic moiety of the H^+-ATPase as a polypeptide with a molecular mass of about 100 kDa [65,66] and established that its catalytic mechanism proceeds by way of an aspartyl–phosphoryl–enzyme intermediate [65,

67]. These studies established the kinship of this enzyme with animal cell counterparts such as the Na$^+$/K$^+$-, H$^+$/K$^+$-, and Ca^{++}-ATPases of plasma membranes, and the Ca^{++}-ATPase of muscle sarcoplasmic reticulum, thus placing it in the P-type ionmotive ATPase family. The gene sequence of the H$^+$-ATPase later confirmed this conclusion [68,69]. The H$^+$-ATPase was subsequently detergent-solubilized and purified [66,70], and a large-scale procedure for the facile isolation of 50–100 mg quantities of the ATPase was developed [71,72]. A procedure for reconstituting the purified ATPase in fully functional form into artificial phospholipid vesicles was also worked out [73], and using this reconstituted system it was possible to demonstrate that one copy of the 100 kDa hydrolytic moiety alone is capable of efficient ATP-hydrolysis-driven proton translocation [73,74].

The structure of the H$^+$-ATPase molecule has been probed by a variety of techniques. Biochemical studies have shown that the H$^+$-ATPase is a hexamer as isolated [75] and that the functional properties of this form of the enzyme are the same as those of the ATPase in its monomeric membrane-bound state [76]. An accurate estimation of the secondary structure composition of the ATPase by circular dichroism was made possible by the unique, largely lipid- and detergent-free nature of the purified hexameric ATPase preparation [76], and importantly, an essentially identical secondary structure composition was obtained for the ATPase in its membrane-bound state by infrared attenuated total reflection spectroscopy [77]. Thus, the secondary structure elements that the ATPase comprises are reasonably certain. Protein chemistry procedures for fragmenting the ATPase molecule by trypsinolysis and purifying virtually all of the numerous hydrophilic and hydrophobic fragments produced have been developed [78,79], and with this technology and the purified, reconstituted ATPase proteoliposomes, the transmembrane topography of virtually all of the 919 amino acid residues in the H$^+$-ATPase molecule has been defined [80–83]. The cysteine chemistry of the ATPase has also been defined by protein chemistry techniques [84]. An interesting structural feature of the H$^+$-ATPase molecule revealed by these studies is the presence of an intramembrane disulfide bridge connecting the N- and the C-terminal hydrophobic segments of the molecule. Finally, key features of the conformational dynamics of the ATPase as it proceeds through its catalytic cycle have been elucidated on the basis of the profound effects of ATPase ligands on the sensitivity of the molecule to trypsinolysis [85]. These will be described in more detail below.

3. Recent progress

3.1. Transmembrane topography of the H$^+$-ATPase

In a limit digest of the reconstituted H$^+$-ATPase proteoliposomes, three large membrane-embedded fragments are found comprising approximately residues 99–173, 272–355, and 660–891, with molecular masses of about 7, 7.5, and 21 kDa, respectively [82]. The N- and C-termini of all three of the fragments are on the cytoplasmic side of the membrane. Assuming a predominantly helical structure, which appears to be a valid assumption in light of recent data obtained with the related Ca++-ATPase [86], hydropathy analyses of these fragments indicated that the first fragment probably contains only two transmembrane helices, that the second fragment could contain from two to four transmembrane helices, and that the third fragment could contain as many as six transmembrane helices. In our original model of the transmembrane topography of the ATPase formulated from this data [82], the H+-ATPase was represented with twelve transmembrane helices, a decision based primarily on the hexameric symmetry of the isolated ATPase molecule, and the vast preponderance of the twelve putative helix motif in membrane transport molecules [26,87–89]. This model has now been revised on the basis of recent evidence that the second fragment may contain only two transmembrane helices [90] and to bring it into better agreement with current models for several other P-type ATPases. Figure 1 shows the revised model, which in addition to the change in helix number has also been aligned according to the consensus alignment suggested by Goffeau and Green [22]. The transmembrane helices, M1–M10, are indicated as are the four major cytoplasmic loops, C1–C4. Most models for the transmembrane topography of the P-type ATPases are now quite similar. For the related Ca++-ATPase, some information is emerging regarding the possible three-dimensional arrangement of the transmembrane helices [91], and it is expected that the H+-ATPase will prove to be similar. In this regard, it should be mentioned that for the H+-ATPase, putative helix M2 and either putative helix M9 or M10 must be in juxtaposition because they are covalently connected by a disulfide bridge [84]. A predominantly four-helix bundle arrangement of adjacent transmembrane helixes has been suggested on the basis of thermodynamic considerations [58].

3.2. The nature of the H+-ATPase conformational changes

Early experiments carried out with the H+-ATPase in isolated plasma membranes showed that the enzyme is protected against inhibition by mercurials or trypsin by its substrate, MgATP [61]. Subsequently, the differential sensitivity of the ATPase to tryptic degradation in the presence or absence of MgATP served in the identification of the hydrolytic moiety of the ATPase, and during these studies, it was noted that the potent ATPase inhibitor, orthovanadate, enhances the protective effects of MgATP against tryptic degradation [65]. It was also shown later that the combination of MgATP and vanadate confers stability to the ATPase during detergent solubilization and purification [70]. These various results strongly indicated that the ATPase undergoes significant conformational changes during its catalytic cycle, and because these conformational changes are almost certainly intimately related to the transport mechanism, they were investigated in more detail in a study of the effects of a variety of ATPase ligands and ligand combinations on the sensitivity of the ATPase to degradation by trypsin [85]. The results of these experiments indicated that the ATPase undergoes a clearly definable conformational change upon binding its substrate, MgATP, which is most cleanly dissociated from subsequent conformational changes by the use of the competitive, nonphosphorylating ligand, MgADP. Following the substrate binding conformational change, the ATPase undergoes additional conformational changes as it proceeds through the transition

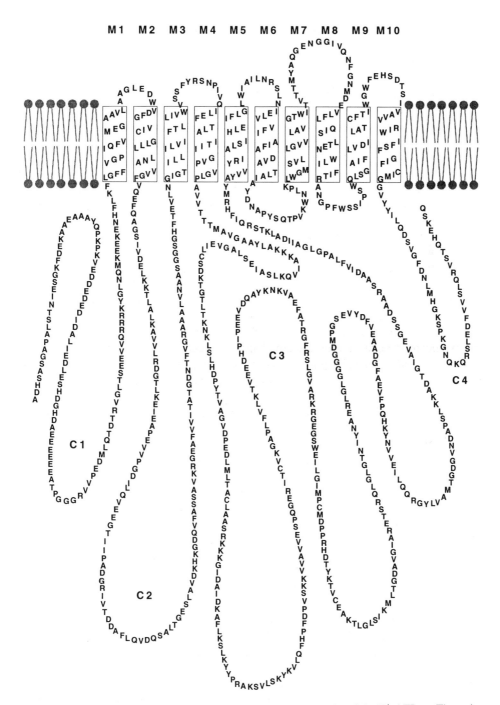

Fig. 1. Current working model for the transmembrane topography of the H⁺-ATPase. The major cytoplasmic loops, C1–C4, are indicated as are the putative transmembrane helices, M1–M10.

states of the enzyme phosphorylation and dephosphorylation reactions. It then undergoes a final conformational change to its unliganded conformation upon product release. The implications of these findings as to the catalytic mechanism of the ATPase, in particular the fundamental importance of transition state binding as the driving force for the overall reaction cycle, have been explained and emphasized on numerous occasions before [58,85,92–94] and will not be reiterated here.

Although it seemed likely that the conformational changes that the ATPase undergoes during its catalytic cycle are substantial, ligand protection of only a few key amino acid residues could conceivably explain the experimental results. The ligand-induced ATPase conformational changes were thus recently further explored using attenuated total reflection Fourier transform infrared spectroscopy [95]. The results clearly showed that the secondary structure components of the H^+-ATPase, i.e., α-helix, β-sheet, turns, and random coil, do not change when the H^+-ATPase undergoes these ligand-induced conformational changes, in agreement with our earlier circular dichroism studies [76]. But importantly, the hydrogen/deuterium (H/D) exchange rates of roughly 175 surface amide linkages in the ATPase polypeptide chain out of a total of about 350 are drastically reduced as the ATPase proceeds from its unliganded conformation to its substrate binding conformation (i.e. bound to MgADP). Similarly, the H/D exchange rates of about 130 residues are reduced in the transition state of the enzyme dephosphorylation reaction (i.e. bound to Mg-vanadate in the presence of ATP). Since virtually all enzymes [96], particularly those that catalyze phosphoryl-transfer reactions [97–103], possess structures with at least two, discrete, relatively rigid structural domains separated by a deep cleft that closes during catalysis, these results constitute strong evidence for a similar behavior of the H^+-ATPase during its catalytic cycle. The proposed cleft closure presumably occludes about 175 surface residues in the ATPase molecule from contact with the aqueous environment in the MgADP bound form and about 130 in the Mg-vanadate bound form. Referring to the model of Fig. 1, the cytoplasmic loops must fold in a way that produces at least two domains and a cleft, and upon ligand binding, numerous interdomain surface residues become occluded when the cleft closes.

The majority of the surface residues of the H^+-ATPase molecule that are occluded as a result of these ligand-induced conformational changes are probably present on the cytoplasmic side of the membrane, because this is the part of the molecule with by far the largest surface area, and because this is the side of the membrane from which the various ligands approach the catalytic center or active site. Thus, the act of occlusion provides a mechanism whereby proton binding residues normally in contact with the aqueous medium on the cytoplasmic side of the membrane can be isolated from it. If the same act of occlusion or any subsequent step in the catalytic cycle presents these residues to the aqueous medium on the other side of the membrane, either directly or through a channel in the molecule, the prerequisites for gated, transmembrane proton movements have been met. This must be the way the ATPase accomplishes the access change required for a mediated transport mechanism and is especially pleasing because it invokes protein conformational dynamics that are already solidly established to exist. The most important question that remains then is the mechanism by which the energy input enhances the release of the protons to the *trans*-ATP side of the membrane. As discussed before [94], this may simply be a kinetic process in which the deprotonated enzyme returns more rapidly to the unliganded state with the proton-binding residues again exposed to the cytoplasmic side of the membrane, or it may involve actual changes in the affinity of the proton-binding sites followed by the access change, as is more often proposed. As also discussed before [58,92–94], the energy input may be direct, with close contact between the active site and the proton-binding sites, or it may be more indirect, with reactions at the active site acting at a distance on the proton-binding sites. Clearly, the location of the active site of the ATPase with respect to the membrane and the proton-binding sites remains a central, as yet unanswered, question regarding the molecular mechanism of active proton translocation catalyzed by the H^+-ATPase, and there is not as yet, a consensus on this important matter [58].

3.3. A high-yield H^+-ATPase expression system

The development of a suitable expression system for defining the function of key residues of the H^+-ATPase by site-directed mutagenesis proved to be a challenging assignment. Expression of the H^+-ATPase in *E. coli* could be demonstrated, but proteolysis was a major, unsolvable problem. Functional H^+-ATPase was expressed in COS cells, but the amounts produced were very low. The amount of the H^+-ATPase produced in the baculovirus/Sf9 insect cell system was enormous, amounting to about a third of the total membrane protein, but the expressed H^+-ATPase molecules were totally inactive. This appeared to be a folding problem because the expressed H^+-ATPase molecules did not assume their normal hexameric form [75] upon solubilization with lysolecithin. After extensive efforts to revive the ATPase expressed in the insect cells, it was concluded that something necessary for proper biosynthesis of the H^+-ATPase in neurospora may be missing in the insect cells. This prompted attempts to express the H^+-ATPase in yeast, owing to its close evolutionary relatedness with neurospora. Two neurospora H^+-ATPase cDNAs differing only in a few bases preceding the coding region were cloned into a high copy number yeast expression vector under the control of the constitutive promoter of the yeast plasma membrane H^+-ATPase, and the resulting plasmids were used to transform *Saccharomyces cerevisiae* strain RS-72, which requires a plasmid-encoded functional plasma membrane H^+-ATPase for growth in glucose medium [104]. Encouragingly, both plasmids supported growth of the cells, indicating that the neurospora ATPase is expressed in functional form in yeast [105]. Western blots of membranes from the transformants confirmed that the neurospora ATPase is expressed in the yeast cells, with production in the range of several percent of the yeast membrane protein. Importantly, when the expressed, recombinant neurospora ATPase molecules are solubilized from the membranes with lysolecithin and subjected to our usual glycerol gradient centrifugation procedure for purifying the native H^+-ATPase, they migrate to a position indistinguishable from that of the native ATPase molecule and display a comparable specific ATPase activity, indicating that the great majority of the recombinant neurospora ATPase molecules produced in yeast fold in a natural manner in fully functional form. The availability of such an assay for proper folding of the H$^+$-ATPase allows us to discriminate site-directed mutants that are inactive because they fold improperly from those that fold properly and are inactive for more interesting reasons. Few heterologous expression systems provide such a convenient indicator of proper folding of the expressed product. This expression system should thus be extremely useful for identifying residues important for the function of the H$^+$-ATPase molecule. Moreover, because the amounts produced are comparable to that produced by neurospora itself, a variety of biochemical experiments with the recombinant ATPase molecules, including crystallization, are entirely feasible.

As part of these studies, an unexpected insight regarding translation initiation in yeast was realized. It was noticed that the two H$^+$-ATPase expression plasmids differing in only two bases immediately upstream from the neurospora ATPase coding sequence produced ATPase molecules that differed significantly in their molecular mass and expression level. The smaller of the ATPase molecules was produced in smaller amounts and had the same molecular mass as the native ATPase. The larger was produced to higher levels and was several thousand daltons larger. This was eventually traced to translation initiation in the more productive construct at an upstream, in frame, non-ATG codon, probably an ATT. Thus, initiation at this codon is considerably more efficient than initiation at the neurospora ATG codon provided in the less productive construct. The upstream initiation site is present in the less productive construct, but it is out of frame with the H$^+$-ATPase coding sequence. These results suggest that special attention should be paid to the bases immediately upstream of the coding sequence when planning cloning strategies for new genes in this and probably other yeast expression systems. Interestingly, the N-terminal extension resulting from initiation at the upstream start site in the more productive H$^+$-ATPase expression plasmid has no deleterious effect on the ATPase whatsoever, suggesting that if needed, the placement of currently popular "flags" or "tags" at the N-termini of the expressed H$^+$-ATPase molecules should be readily tolerated.

Our first mutagenesis studies using this system involved an investigation of the cysteine residues in the H$^+$-ATPase molecule [106]. Each of the eight cysteine residues was changed to a serine or an alanine residue, producing strains C148S and C148A, C376S and C376A, C409S and C409A, C472S and C472A, C532S and C532A, C545S and C545A, C840S and C840A, and C869S and C869A, respectively. With the exception of C376S and C532S, all of the mutant ATPases are able to support the growth of yeast cells to different extents indicating that they are functional. The C376S and C532S enzymes appear to be non-functional. After solubilization of the functional mutant ATPase molecules from isolated membranes with lysolecithin, all behaved similar to the native enzyme when subjected to glycerol density gradient centrifugation, indicating that they quantitatively fold in a natural manner. The kinetic properties of these mutant enzymes were also similar to the native ATPase with the exception of C409A, which has a substantially higher K_m. These results clearly indicate that none of the eight cysteine residues in the

H$^+$-ATPase molecule is essential for ATPase activity, but that cys^{376}, cys^{409}, and cys^{532} may be in or near important sites. They also demonstrate that the disulfide bridge between cys^{148} and cys^{840} or cys^{869} plays no obvious role in the structure or function of the H$^+$-ATPase. Mutagenesis studies of numerous other residues in the ATPase molecule are currently under way. Moreover, a suitable combination of the above-mentioned cys to ala and cys to ser mutations has produced a single cysteine mutant with the cysteine in the active site near asp^{378}. This and other mutants like it should prove to be extremely valuable in future studies of the H$^+$-ATPase structure as will be mentioned again below.

3.4. Large, single 3D crystals of the H$^+$-ATPase

We have been trying for some time now to obtain crystals of the H$^+$-ATPase that are suitable for structural analysis by X-ray diffraction techniques. Early efforts in this regard were largely unsuccessful, but in the last several years, considerable success in crystallizing the H$^+$-ATPase has been realized [107]. The key to obtaining large, 3D crystals of the H$^+$-ATPase was the recognition of the fundamentally chimeric nature of detergent complexes of integral membrane proteins and the development of an experimental approach for finding conditions for crystallizing both parts of the chimera at the same time. The detergent complex of a typical integral membrane protein contains regions of the molecule normally exposed to the aqueous milieu on the cytoplasmic and exocytoplasmic surfaces of the membrane, referred to in the original treatment [107] as the "protein surfaces". The complex contains, in addition, a torus of detergent molecules surrounding and solubilizing the highly hydrophobic region of the molecule normally embedded in the lipid bilayer, termed the "detergent micellar collar". Detergent complexes of integral membrane proteins are thus, in most cases, chimeras comprising surface regions with potentially very different physicochemical properties. It is a simple matter to bring solutions of such detergent complexes of integral membrane proteins to a state of moderate insolubility using conventional protein precipitants as is done for crystallizing soluble proteins, but rarely do membrane protein crystals ever form when this is tried. It seemed possible that it is the chimeric nature of detergent–membrane protein complexes that precludes their crystallization. That is, whereas it might be possible to empirically establish conditions for crystallizing one of the surfaces of the complex, it is unlikely that those conditions would be suitable for crystallizing the other. It was thus reasoned that the key to crystallizing membrane protein–detergent complexes may be the attainment of conditions in which the protein surfaces are moderately supersaturated and the detergent micellar collar is also at or near its solubility limit so that it is not overly or underly willing to coalesce into a solid state.

These considerations suggested that simple attainment of supersaturation using protein precipitants and detergents chosen more or less at random was not likely to be a very successful approach to crystallizing the H$^+$-ATPase, and numerous early attempts to crystallize it using a variety of protein precipitants lent support to this notion. It was therefore decided to approach the problem more systematically in an attempt to find conditions in which the extent of supersaturation of the protein surfaces of the ATPase and the detergent micellar collar were properly matched. Preliminary experiments had indicated that the ATPase is most stable when the detergent used to maintain its solubility is dodecylmaltoside (DDM). The first step was thus to screen a variety of commonly used protein precipitants for those that were able to induce the aggregation of pure DDM micelles. The concentration at which any precipitant induced DDM micellar aggregation was hoped to be close to the concentration at which it might induce insolubility of the detergent micellar collar of the DDM–ATPase complex. Of the nine precipitants tried, seven, all polyethylene glycols (PEGs), were able to induce DDM micelle insolubility. The seven PEGs were then tested for their effect on the solubility of the DDM–ATPase complex, at a concentration slightly below that necessary to induce DDM micellar aggregation. Three of the PEGs caused extensive precipitation of the ATPase at this concentration and were therefore set aside. The other four PEGs did not induce precipitation at the concentration employed and were subsequently used at this concentration for crystallization trials in which the protein concentration was varied. Encouragingly, crystalline plates of the ATPase were obtained for each of the four PEGs tried, indicating that the overall approach may be valid. Unfortunately, the crystals obtained were visibly flawed, suggesting that the proper balance of protein surface and DDM micelle insolubility had not yet been reached. The ionic strength of the crystallization trials was then raised, which was known from other experiments to render the protein surfaces of the ATPase less soluble while having little effect on the DDM micellar aggregation point. For one of the PEGs, PEG 4000, this brought on a new, well-formed hexagonal crystal habit. Subsequent optimization of the initial conditions has yielded large, single, hexagonal 3D crystals of the H^+-ATPase up to about $0.4 \times 0.4 \times 0.15$ mm in size. X-ray diffraction by the ATPase crystals to about 8 Å has been observed thus far, and improvement on this is anticipated. Although the complete unit cell dimensions of the H^+-ATPase crystals cannot be calculated from the available data, the space group is hexagonal or rhombohedral with $a = b \cong 167$ Å. Because the bimodal solubility concept described above and the experimental approach designed to deal with it worked so well after many negative attempts to crystallize the H^+-ATPase, it is hoped that what has been learned from these studies may be generally useful for the crystallization of the detergent complexes of other integral membrane proteins.

3.5. 2D crystals of the H^+-ATPase

Electron microscopy of 3D crystals has often been used to obtain intermediate level resolution images of protein molecules, and this information can sometimes aid in the interpretation of higher resolution X-ray diffraction data [108]. For this reason, an exploration of the H^+-ATPase crystals by electron microscopic techniques was initiated. The first experiments indicated the presence of small 3D crystals in crystallization mixtures similar to those used for obtaining the large, single ATPase crystals, and the above unit cell dimensions of $a = b = 167$ Å were quickly confirmed. Soon after, however, an even more important discovery was made. Whereas 3D H^+-ATPase crystals grow in the drops, large 2D H^+-ATPase crystals

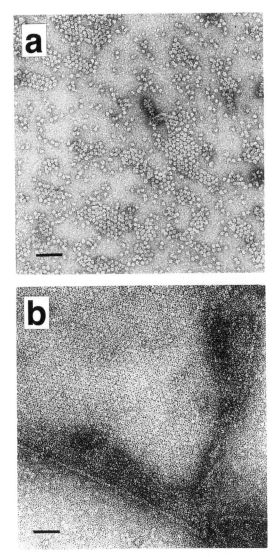

Fig. 2.   Electron micrographs of negatively stained 2D crystals of the H⁺-ATPase. (a) Small 2D crystals
at an early stage of growth; (b) a more fully developed, large single 2D crystal.

readily grow at or near the surface of such drops [109]. The 2D H⁺-ATPase crystals
can be picked up by touching the surface of the drops with a carbon-coated electron
microscope grid and can then be viewed with or without negative stain at low
temperature in a cryoelectron microscope. Figure 2a shows a negative stain image
of the 2D H⁺-ATPase crystals at the very beginning of their formation. The small
circular protomeric units are H⁺-ATPase hexamers just beginning to coalesce into a
hexagonal lattice. Figure 2b shows a negatively-stained large 2D ATPase crystal. In

this micrograph, some of the edges of the crystal have curled and can be seen end-on (straight lines), and their uniform thickness of about 100 Å indicates that it is a purely 2D crystal.

The conventional method for growing 2D crystals of integral membrane proteins involves preparing mixtures containing the protein of interest and a suitable lipid, both solubilized by an appropriate detergent [110]. The detergent is then removed by dialysis and the 2D crystals form as proteolipid complexes. The precipitant-induced 2D crystals of the DDM complex of the H^+-ATPase thus represent an entirely new approach to obtaining 2D crystals of an integral membrane protein useful for structure analysis. Only time will tell if this approach is useful for obtaining crystals of other integral membrane proteins, but it has already proved to be extremely effective for obtaining quality 2D crystals of the H^+-ATPase useful for structural work in a relatively short period of time.

3.6. Emerging structure of the H⁺-ATPase

Images such as those shown in Fig. 2b contain a great deal of information about the structure of the H^+-ATPase molecule, and images of unstained, frozen-hydrated 2D crystals of the H^+-ATPase contain even more. Hundreds of such micrographs have been taken since the discovery of the 2D ATPase crystals, and the best identified by optical diffraction of their negatives in a laser diffractometer. Selected images were then digitized in a flatbed densitometer, Fourier transformed, and processed as described [110]. The inverse Fourier transform of the first image processed yielded the projection map shown in Fig. 3. This projection map extends to a resolution of about 22 Å. The ringlike nature of the H^+-ATPase hexamers is clearly evident in the map. The unit cell is hexagonal with a = b = 167 Å. In projection, the individual monomers viewed from the cytoplasmic aspect are shaped like a boot. Their dimensions in this view are thus quite similar to those of the Ca^{++}-ATPase [55]. Further improvements in specimen preparation and merging of the data from several images have yielded a projection map that extends to 10.3 Å [109].

The ability of the H^+-ATPase monomers to form homogeneous hexamers, and the marked stability of the hexamers during the ATPase purification procedure, raise interesting questions as to the forces that hold the monomers together in the hexamers and the possible physiological role of the hexameric ATPase rings. If there is bonding between the monomers in the hexamers via the cytoplasmic domains, it must be relatively weak, because when the ATPase hexamers are reconstituted into liposomes, single monomer liposomes are formed [74]. It therefore appears that the six monomers are held together primarily via hydrophobic interactions originating from the normally membrane-embedded region of the molecule, which, as mentioned above probably contains ten transmembrane helices. It is clear that the hexamer is not required for functional integrity of the ATPase, because it has been convincingly demonstrated that monomers are fully functional for proton transport [74]. However, hexagonal arrays of ringlike particles very similar in diameter and appearance to the neurospora H^+-ATPase hexamers have been found in the yeast plasma membrane [111], which has a highly homologous

Fig. 3. Projection map of the H⁺-ATPase structure at 22 Å resolution.

H⁺-ATPase. This lends support to the notion that ATPase hexamers may form *in vivo* for a specific but as yet unknown function.

4. Prospectus

With the past progress and recent advances outlined above, the neurospora H⁺-AT-Pase is one of the more likely of the membrane transport systems currently under investigation to eventually yield its molecular mechanism. This transporter can be isolated in large quantities, it is remarkably stable during prolonged periods of storage, and many of its biochemical properties, including certain key features of its conformational dynamics are already known. A high-yield expression system for performing site-directed mutagenesis studies of the H⁺-ATPase is now available, and procedures for preparing both 2D and 3D H⁺-ATPase crystals have been worked out. From this foundation, a variety of future lines of investigation should move us closer to an understanding of how this enzyme works. Residues important for key ATPase functions are beginning to be identified in our site-directed mutagenesis studies, and the delineation of many new key residues is likely to be realized from this approach. Even more importantly, the long-awaited structure of the H⁺-ATPase is beginning to emerge. A 10.3 Å projection map is in hand and extension of the methods used to obtain this map should yield a 3D structure of comparable resolution. Methods for manipulating and improving the quality of both the 2D and 3D crystals are currently being worked out, and it should also be possible to use the phase information derived by image processing of the 2D ATPase crystals for phasing X-ray diffraction data from the 3D crystals, because the symmetry and unit cells of the 2D and 3D crystals appear to be identical. It is thus expected that the current structural resolution will improve. Only a few more Ångstroms resolution may be needed in order to begin to delineate the path of the polypeptide chain. Moreover, since crystals of the ATPase can also be grown in the presence of MgATP and vanadate, the structure of the ATPase in at least one of its closed conformations may also eventually be known. Also, with the availability of single-cysteine mutant forms of the H^+-ATPase, it may be possible to localize the active site and other key sites in the ATPase molecule by cryoelectron crystallography of 2D crystals of these enzymes labeled with gold-maleimide [112]. The neurospora plasma membrane proton pump is thus ripe for further exploration, and the prospects are good that major advances in our understanding of the structure and molecular mechanism of membrane transport catalyzed by this enzyme will be forthcoming in the not-too-distant future.

Acknowledgements

This work was supported primarily by United States Public Health Service National Institutes of Heath Grant GM24784.

References

1. Mueckler, M., Caruso, C., Baldwin, S.A., Panico, M., Blench, I., Morris, H.R., Allard, W.J., Lienhard, G.E. and Lodish, H.F. (1985) Science **229**, 941–945.
2. Aquila, H., Link, T.A. and Klingenberg, M. (1987) FEBS Lett. **212**, 1–9.
3. Kaback, H.R. (1988) Ann. Rev. Physiol. **50**, 243–256.
4. Baldwin, S.A. and Henderson, P.J.F. (1989) Ann. Rev. Physiol. **51**, 459–471.
5. Maiden, M.C.J., Davis, E.O., Baldwin, S.A., Moore, D.C.M. and Henderson, P.J.F. (1987) Nature **325**, 641–643.
6. Hediger, M.A., Turk, E. and Wright, E.M. (1989) Proc. Natl. Acad. Sci. USA **86**, 5748–5752.
7. Yazyu, H., Shiota-Niiya, S., Shimamoto, T., Kanazawa, H., Futai, M. and Tsuchiya, T. (1984) J. Biol. Chem. **25**,. 4320–4326.
8. Botfield, M.C. and Wilson, T.H. (1989) J. Biol. Chem. **264**, 11649–11652.
9. Eckert, B. and Beck, C.F. (1989) J. Biol. Chem. **264**, 11663–11670.
10. Hopfer, U. (1987) in: Physiology of the Gastrointestinal Tract, ed L.R. Johnson, pp. 1499–1526, Raven Press, New York.
11. Kopito, R.R. and Lodish, H.F. (1985) Nature **316**, 234–238.
12. Stroobant, P. and Scarborough, G.A. (1979) Proc. Natl. Acad. Sci. USA **76**, 3102–3106.
13. Rosen, B.P. (1982) in: Membrane Transport of Calcium, ed E. Carafoli, pp. 187–216, Academic Press, New York.
14. Krulwich, T.A. (1983) Biochim. Biophys. Acta **726**, 245–264.
15. Poolman, B. and Konings, W.N. (1993) Biochim. Biophys. Acta **1183**, 5–39.
16. Gabizon, R. and Schuldiner, S. (1985) J. Biol. Chem. **260**, 3001–3005.
17. Senior, A.E. (1990) Ann. Rev. Biophys. Biophys. Chem. **19**, 7–41.
18. Jorgensen, P.L. and Andersen, J.P. (1988) J. Membr. Biol. **103**, 95–120.
19. Anderson, J.P. and Vilsen, B. (1990) Current Opinion in Cell Biology **2**, 722–730.
20. Carafoli, E. (1991) Physiol. Rev. **71**, 129–153.
21. Rabon, E.C. and Reuben, M.A. (1990) Ann. Rev. Physiol. **52**, 321–344.
22. Goffeau, A. and Green, N.M. (1990) in: Monovalent Cations in Biological Systems, ed C.A. Pasternak, pp. 155–169, CRC Press, Inc., Boca Raton, FL.
23. Bowman, B.J. and Bowman, E.J. (1986) J. Membr. Biol. **94**, 83–97.
24. Meadow, N.D., Fox, D.K. and Roseman, S. (1990) Ann. Rev. Biochem. **59**, 497–542.

25. Ames, G.F.-L. (1986) Ann. Rev. Biochem. **55**, 397–425.
26. Silver, S., Nucifora, G., Chu, L. and Misra, T.K. (1989) TIBS **14**,76–80.
27. Endicott, J.A. and Ling, V. (1989) Annu. Rev. Biochem. **58**, 137–171.
28. Forgac, M. (1989) Physiol. Rev. **69**, 765–796.
29. Capaldi, R.A. (1990) Ann. Rev. Biochem. **59**, 569–596.
30. Stoeckenius, W. and Bogomolni, R.A. (1982) Ann. Rev. Biochem. **51**, 587–616.
31. Lanyi, J. (1990) Physiol. Rev. **70**, 319–330.
32. Pedersen, P.L. and Carafoli, E. (1987) TIBS **12**, 146–150.
33. Persson, B., Roepe, P.D., Patel, L., Lee, J. and Kaback, H.R. (1992) Biochemistry **31**, 8892–8897.
34. Lee, J.A., Puttner, I.B. and Kaback, H.R. (1989) Biochemistry **28**, 2540–2544.
35. Lolkema, J.S., Puttner, I.B. and Kaback, H.R. (1988) Biochemistry **27**, 8307–8310.
36. Zhang, Y. and Fillingame, R.H. (1994) J. Biol. Chem. **269**, 5473–5479.
37. Fillingame, R.H., Oldenburg, M. and Fraga, D. (1991) J. Biol. Chem. **266**, 20934–20939.
38. Fraga, D. and Fillingame, R.H. (1989) J. Biol. Chem. **264**, 6797–6803.
39. Hazard, A.L. and Senior, A.E. (1994) J. Biol. Chem. **269**, 418–426.
40. Weber, J., Lee, R.S., Wilke-Mounts, S., Grell, E. and Senior, A.E. (1993) J. Biol. Chem. **268**, 6241–6247.
41. Parsonage, D., Wilke-Mounts, S. and Senior, A.E. (1988) FEBS Lett. **232**, 111–114.
42. Clarke, C.M., Loo, T.W. and MacLennan, D.H. (1990) J. Biol. Chem. **265**, 14088–14092.
43. Maruyama, K., Clarke, D.M., Fujii, J., Inesi, G., Loo, T.W. and MacLennan, D.H. (1989) J. Biol. Chem. **264**, 13038–13042.
44. Clarke, D.M., Maruyama, K., Loo, T.W., Leberer, E., Inesi, G. and MacLennan, D.H. (1989) J. Biol. Chem. **264**, 11246–11251.
45. Maruyama, K. and MacLennan, D.H. (1988) Proc. Natl. Acad. Sci. USA. **85**, 3314–3318.
46. Skerjanc, I.S., Toyofuku, T., Richardson, C. and MacLennan, D. H. (1993) J. Biol. Chem. **268**, 15944–15950.
47. Fersht, A.R., Shi, J.-P., Knill-Jones, J., Lowe, D.M., Wilkinson, A.J., Blow, D.M., Brick, P., Carter, P., Waye, M.M.Y. and Winter, G. (1985) Nature **314**, 235–238.
48. Wells, T.N.C. and Fersht, A.R. (1985) Nature **316**, 656–657.
49. Fersht, A.R., Leatherbarrow, R.J. and Wells, T.N.C. (1986) TIBS **11**, 321–325.
50. Deisenhofer, J., Epp, O., Miki, K., Huber, R. and Michel, H. (1985) Nature **318**, 618–624.
51. Iwata, S., Ostermeier, C., Ludwig, B., and Michel, H. (1995) Nature **376**, 660–669.
52. Tsukihara, T., Aoyama, H., Yamashita, E., Tomizaki, T., Yamaguchi, H., Shinzawa-Itoh, K., Nakashima, R., Yaono, R., and Yoshikawa, S. (1995) Science **269**, 1069–1074.
53. Henderson, R., Baldwin, J.M., Ceska, T.A., Zemlin, F., Beckmann, E. and Downing, K.H. (1990) J. Mol. Biol. **213**, 899–929.
54. Wang, D.N., Kühlbrandt, W., Sarabia, V.E. and Reithmeier, R.A. F. (1993) EMBO J. **12**, 2233–2239.
55. Toyoshima, C., Sasabe, H. and Stokes, D.L. (1993) Nature **362**, 469–471.
56. Unwin, N. (1993) J. Mol. Biol. **229**, 1101–1124.
57. Kühlbrandt, W., Wang, D.N. and Fujiyoshi, Y. (1994) Nature **367**, 614–621.
58. Scarborough, G.A. (1992) in: Molecular Aspects of Transport Proteins, ed J.J.H.H.M. de Pont, pp. 117–134. Elsevier Science Publishers B.V., Amsterdam.
59. Slayman, C.L. (1970) Am. Zool. **10**, 377–392.
60. Scarborough, G.A. (1975) J. Biol. Chem. **250**, 1106–1111.
61. Scarborough, G.A. (1977) Arch. Biochem. Biophys. **180**, 384–393.
62. Bowman, B.J. and Slayman, C.W. (1977) J. Biol. Chem. **252**, 3357–3363.
63. Scarborough, G.A. (1976) Proc. Natl. Acad. Sci. USA **73**, 1485–1488.
64. Scarborough, G.A. (1980) Biochemistry **19**, 2925–2931.
65. Dame, J.B. and Scarborough, G.A. (1980) Biochemistry **19**, 2931–2937.
66. Bowman, B.J., Blasco, F. and Slayman, C.W. (1981) J. Biol. Chem. **256**, 12343–12349.
67. Dame, J.B. and Scarborough, G.A. (1981) J. Biol. Chem. **256**, 10724–10730.
68. Addison, R. (1986) J. Biol. Chem. **261**, 14896–14901.
69. Hager, K.M., Mandala, S.M., Davenport, J.W., Speicher, D.W., Benz, E.J. Jr. and Slayman, C.W. (1986) Proc. Natl. Sci. USA **83**, 7693–7697.

70. Addison, R. and Scarborough, G.A. (1981) J. Biol. Chem. **256**, 13165–13171.
71. Smith, R. and Scarborough, G.A. (1984) Anal. Biochem. **138**, 156–163.
72. Scarborough, G.A. (1988) Meth. Enzymol. **157**, 574–579.
73. Scarborough, G.A. and Addison, R. (1984) J. Biol. Chem. **259**, 9109–9114.
74. Goormaghtigh, E., Chadwick, C. and Scarborough, G.A. (1986) J. Biol. Chem. **261**, 7466–7471.
75. Chadwick, C.C., Goormaghtigh, E. and Scarborough, G.A. (1987) Arch. Biochem. Biophys. **252**, 348–356.
76. Hennessey, J.P., Jr. and Scarborough, G.A. (1988) J. Biol. Chem. **263**, 3123–3130.
77. Goormaghtigh, E., Ruysschaert, J.-M. and Scarborough, G.A. (1988) in: The Ion Pumps: Structure, Function, and Regulation. pp. 51–56, Alan R. Liss, Inc.
78. Rao, U.S., Hennessey, J.P., Jr. and Scarborough, G.A. (1988) Anal. Biochem. **173**, 251–264.
79. Hennessey, J.P., Jr. and Scarborough, G.A. (1989) Anal. Biochem. **176**, 284–289.
80. Hennessey, J.P., Jr. and Scarborough, G.A. (1990) J. Biol. Chem. **265**, 532–537.
81. Scarborough, G.A. and Hennessey, J.P., Jr. (1990) J. Biol. Chem. **265**, 16145–16149.
82. Rao, U.S., Hennessey, J.P., Jr. and Scarborough, G.A. (1991) J. Biol. Chem. **266**, 14740–14746.
83. Rao, U.S., Bauzon, D.D. and Scarborough, G.A. (1992) Biochim. Biophys. Acta **1108**, 153–158.
84. Rao, U.S. and Scarborough, G.A. (1990) J. Biol. Chem. **265**, 7227–7235.
85. Addison, R. and Scarborough, G.A. (1982) J. Biol. Chem. **257**, 10421–10426.
86. Corbalan-Garcia, S., Teruel, J.A., Villalain, J. and Gomez-Fernandez, J.C. (1994) Biochemistry **33**, 8247–8254.
87. Maloney, P.C. (1989) Phil. Trans. R. Soc. Lond. B **326**, 437–454.
88. Maloney, P.C., Ambudkar, S.V., Anantharam, V., Sonna, L.A. and Varadhachary, A. (1990) Microbiol. Rev. **54**, 1–17.
89. Henderson, P.J.F. (1990) J. Bioenerg. Biomembr. **22**, 525–569.
90. Lin, J. and Addison, R. (1994) J. Biol. Chem. **269**, 3887–3890.
91. Stokes, D.L., Taylor, W.R. and Green, N.M. (1994) FEBS Lett. **346**, 32–38.
92. Scarborough, G.A/ (1982) Ann. N.Y. Acad. Sci. **402**, 99–115.
93. Scarborough, G.A. (1985) in: Environmental Regulation of Microbial Metabolism, eds I.S. Kulaev, E.A. Dawes and D.W. Tempest, pp. 39–51. Academic Press.
94. Scarborough, G.A. (1985) Microbiol. Rev. **49**, 214–231.
95. Goormaghtigh, E., Vigneron, L., Scarborough, G.A. and Ruysschaert, J.-M. (1994) J. Biol. Chem. **269**, 27409–27413.
96. Schulz, G.E. and Schirmer, R.H. (1979) in: Principles of Protein Structure. Springer-Verlag. New York.
97. Steitz, T.A., Fletterick, R.J., Anderson, W.F. and Anderson, C.M. (1976) J. Mol. Biol. **104**, 197–222.
98. Stuart, D.I., Levine, M., Muirhead, H. and Stammers, D.K. (1979) J. Mol. Biol. **134**, 109–142.
99. Sachsenheimer, W. and Schulz, G.E. (1977) J. Mol. Biol. **114**, 23–36.
100. Evans, P.R. and Hudson, P.J. (1979) Nature **279**, 500–504.
101. Banks, R.D., Blake, C.C.F., Evans, P.R., Haser, R., Rice, D.W., Hardy, G.W., Merrett, M. and Phillips, A.W. (1979) Nature **279**, 773–777.
102. Pickover, C.A., McKay, D.B., Engelman, D.M. and Steitz, T.A. (1979) J. Biol. Chem. **254**, 11323–11329.
103. Anderson, C.M., Zucker, F.H. and Steitz, T.A. (1979) Science **204**, 375–380.
104. Villalba, J.M., Palmgren, M.G., Berberian, G.E., Ferguson, C. and Serrano, R. (1992) J. Biol. Chem. **267**, 12341–12349.
105. Mahanty, S.K., Rao, U.S., Nicholas, R.A. and Scarborough, G.A. (1994) J. Biol. Chem. **269**, 17705–17712.
106. Mahanty, S.K. and Scarborough, G.A. (1996) J. Biol. Chem. **271**, 367–371.
107. Scarborough, G.A. (1994) Acta. Cryst. D **50**, 643–649.
108. McPherson, A. (1982) Preparation and Analysis of Protein Crystals. John Wiley and Sons, New York.
109. Cyrklaff, M., Auer, M., Kühlbrandt, W. and Scarborough, G.A. (1995) EMBO J. **14**, 1854–1857.
110. Kühlbrandt, W. (1992) Quart. Rev. Biophys. **25**, 1–49.
111. Kübler, O., Gross, H. and Moor, H. (1978) Ultramicroscopy. **3**, 161–168.
112. Milligan, R.A., Whittaker, M. and Safer, D. (1990) Nature **348**, 217–221.

CHAPTER 5

Structure and Function of the Yeast Vacuolar Membrane H^+-ATPase

Y. ANRAKU

Department of Biological Sciences, Graduate School of Science,
University of Tokyo, Hongo, Bunkyo-ku, Tokyo 113, Japan

© *1996 Elsevier Science B.V.*
All rights reserved

Handbook of Biological Physics
Volume 2, edited by W.N. Konings, H.R. Kaback and J.S. Lolkema

Contents

1. Introduction

The yeast vacuolar membrane H⁺-ATPase [1–4] is the first member of a well defined "vacuolar type" ATPase family that has been successively identified in various endocytic and exocytic membrane compartments of eukaryotic cells [5–14]. The vacuolar type H⁺-ATPases characterized thus far are multisubunit complexes composed of an integral membrane V_0 sector and a peripherally associated V_1 sector with similar subunit compositions and conserved functional motifs [8,9, 13]. Current biochemical and genetic studies have revealed that the yeast enzyme consists of at least thirteen subunits with six composing the V_1 sector (69, 60, 54, 42, 32, and 27 kDa), five composing the V_0 sector (100,13, 17, 17', and 23 kDa), and two composing a V_1-V_0 junction core (36 and 14 kDa). The nt sequences of *VMA* genes (for *vacuolar membrane ATPase*) encoding these thirteen subunits have been determined and their roles in expression and assembly of the enzyme complex elucidated.

Early studies on deletion mutants of the *VMA* genes for major subunits of the H⁺-ATPase have shown that the mutants exhibit a common, characteristic set of growth phenotypes [15]. They grow well in YPD medium, but cannot grow on YPD plates containing 100 mM $CaCl_2$ and YP plates containing non-fermentable carbon sources such as 3% glycerol or 3% succinate, clearly showing a Pet⁻ *cls* phenotype [16]. This Pet⁻ phenotype (inability of utilizing non-fermentable carbon sources) has turned out to cosegregate with the *cls* phenotype (calcium sensitive growth), and, like the *vma* deletion mutants, all the Pet⁻ *cls* mutants so far examined have shown Vma⁻ defects, i.e. loss of vacuolar H⁺-ATPase activity and vacuolar acidification *in vivo* [7,8,15]. Later such Pet⁻ *cls* mutants were found to show pH sensitivity for growth in YPD medium at pH 7.0–7.5 [17,18]. During the course of genetic screens for *vma* mutations, two unique *vma* mutants, *vma12* and *vma21*, have been isolated and characterized. The products of *VMA12* [19] and *VMA21* [20], both being a membrane protein located in the vacuolar and ER membranes, respectively, were found to be essential components, but not as a subunit of the H⁺-ATPase, each of them being required for assembly of the enzyme complex.

2. Discovery

Biochemical studies on the chemiosmotic processes in yeast vacuoles [21] have been initiated in 1981 by the work of Ohsumi and Anraku [22]. They established a simple method for separating intact vacuoles of high purity from the budding yeast *Saccharomyces cerevisiae*, which includes spheroplasting of cells followed by differential separation of vacuoles in the crude lysate by flotational centrifugation in discontinuous Ficoll-400 gradients and allows 28-fold enrichment of α-mannosidase, a marker of the vacuolar membrane [1,23]. The purified vacuole fraction contains less than 0.05% each of the marker enzyme activities of succinate dehydrogenase for mitochondria, NADH-cytochrome c reductase for microsomes, and chitin synthetase for plasma membranes [1]. Vacuoles in this preparation are intact as they contain considerable amounts of arginine, polyphosphates and alkaline phosphatase, the markers of the vacuolar sap [22]. Vacuolar membrane vesicles with right-side-out orientation have been obtained after a brief homogenization of the intact vacuoles in a buffer solution with 5 mM $MgCl_2$ followed by centrifugation.

Kakinuma et al. [1] found that the vacuolar membrane vesicles thus prepared have an unmasked Mg^{2+}-ATPase activity with optimal pH of 7.0. The enzyme requires Mg^{2+} but not Ca^{2+} for its activity and hydrolyzes ATP, GTP, UTP, and CTP in this order. The K_m value for ATP was determined to be 0.2 mM. ADP and AMP are not hydrolyzed by the enzyme. The activity is inhibited by DCCD, the potent inhibitor of a family of H^+-translocating ATPases and stimulated 3- and 1.5-fold, respectively, by the protonophore uncoupler SF6847 and the K^+/H^+ antiporter ionophore nigericin.

3. Generation of a proton motive force across the vacuolar membrane

ATP hydrolysis-dependent uptake of protons into the purified vacuolar membrane vesicles was demonstrated directly by the change in quenchings of 9-aminoacridine and quinacrine fluorescence [1]. The electrochemical potential differences of protons across the vacuolar membrane generated upon ATP hydrolysis was determined by a flow-dialysis method with ^{14}C-methylamine for measuring the formation of ΔpH and with ^{14}C-KSCN for measuring the membrane potential. The proton motive force ($Δ_p$) thus calculated is 180 mV, with contributions of 1.7 pH units, interior acid, and of a membrane potential of 75 mV, interior positive [1]. Based on this and other evidence, the Mg^{2+}-ATPase of the yeast vacuolar membrane was proposed to be a new DCCD-sensitive, H^+-translocating ATPase [1–3].

4. Enzymatic properties as vacuolar type H^+-translocating ATPase

4.1. Purification of the H^+-ATPase from yeast vacuoles

The vacuolar membrane H^+-ATPase can be solubilized routinely by the zwitterionic detergent ZW3-14 and purified to near homogeneity by glycerol density gradient centrifugation [2]. Detergents such as cholate, Sarkosyl, Triton X-100, and Tween 80 are not effective for solubilization. Prior treatment of isolated vacuolar membrane vesicles with 1 mM EDTA is recommended before solubilization of the enzyme and does not affect the H^+-ATPase activity. This EDTA-wash removes most of the acid and alkaline phosphatases, which are marker enzymes of the vacuolar sap and associated loosely with the vacuolar membranes. The partially purified enzyme in the presence of phospholipids has the same pH optimum (pH 6.9) and K_m value for ATP hydrolysis (0.21 mM) as the native, membrane-bound enzyme [2]. ADP is not hydrolyzed and inhibits the enzyme activity noncompetitively, with a K_i value of 0.31 mM. The activity of the partially purified enzyme is not inhibited by antiserum against yeast mitochondrial F_1-ATPase and yeast mitochondrial F_1-ATPase inhibitor protein [2].

The original vacuolar membrane H^+-ATPase preparation, which showed a specific activity of 16–18 units/mg of protein in the presence of 0.1 mg/ml of asolectin, appeared to have 69, 60, and 17 kDa polypeptides as major components. Kane et al. [4] found that the same purification protocol results in the preparation of up to eight-subunit H^+-ATPase (11 units/mg of protein): Those are 100, 69, 60, 42, 36, 32, 27, and 17 kDa polypeptides. It was also noted that a similar collection of eight polypeptides is immunoprecipitated from solubilized vacuolar membranes by a monoclonal antibody against the 69 kDa subunit of the H^+-ATPase, suggesting that all eight of these components are candidate subunits of the enzyme [12,13].

4.2. Bafilomycin A₁ as the specific inhibitor of the enzyme

In 1988, Bowman et al. [24] discovered that bafilomycin A_1 is a potent specific inhibitor for "vacuolar type" ATPases. This information immediately brought about profound impact in expanding biochemical and physiological researches on vacuolar type H^+-ATPases in many organelles, cells, and tissues (see Ref. [25]).

Kane et al. [4] reported that the activity of the yeast vacuolar membrane H^+-ATPase in the membranes as well as in a purified form is inhibited by bafilomycin A_1. The I_{50} value for inhibition of the ATPase activity of yeast vacuolar membranes is 0.6×10^{-3} μmol of bafilomycin A_1/mg of protein, which is very similar to the I_{50} value of 0.4×10^{-3} μmol of bafilomycin A_1/mg of protein calculated for *Neurospora* vacuolar membranes. An I_{50} value (1.7×10^{-3} μmol/mg of protein) was determined for the partially purified yeast vacuolar membrane H^+-ATPase and, under this experimental condition, greater than 90% of the ATPase activity was inhibited by 1 μM bafilomycin A_1 [4].

4.3. Comparison of inhibitor sensitivities among three types of H⁺-ATPases

The vacuolar membrane H^+-ATPase is a nitrate-sensitive enzyme and not inhibited at all by azide and vanadate, which are specific inhibitors for the mitochondrial F_1F_0-ATPase and the plasma membrane P-ATPase, respectively [2] (Table 1). The nitrate sensitivity depends on the presence of MgATP [4]: in the presence of 5 mM MgATP, 100 mM KNO_3 inhibited 71% of the H^+-ATPase activity of vacuolar membrane vesicles, with simultaneous removal of the 69 and 60 kDa subunits from the membranes to a similar extent. It has been suggested that nitrate inactivates the enzyme in a manner of a conformation-specific disassembly of the enzyme [4].

DCCD is a potent inhibitor for the vacuolar membrane H^+-ATPase as well as the F_1F_0-type H^+-ATPases from mitochondria (Table 1). The K_i values for DCCD of the partially purified enzyme and H^+-ATPases in vacuolar membranes, submitochondrial particles, and plasma membranes from *S. cerevisiae* have been determined to

Table 1
Effects of inhibitors on the activities of three H^+-ATPases of *Saccharomyces cerevisiae*

Inhibitor	mM	Relative activity (%)[a] of H^+-ATPase			
		Purified vacuolar enzyme	Vacuolar membrane	Mitochondrial membrane	Plasma membrane
None		100	100	100	100
Bafilomycin A_1 [b]	0.001	≤10	ND^c	ND	ND
Sodium azide	2	95	110	4	105
Sodium vanadate	0.1	95	96	100	16
KNO_3	50	57	55	96	100
$CaCl_2$	0.1	101	98	91	82
$CuCl_2$	0.5	10	12	99	1
DCCD	0.001	38	63	12	86
Oligomycin	0.047	96	74	10	74
NBD-Cl	0.1	23	27	6	79
Tributyltin	0.1	14	45	15	33
SITS	0.004	36	44	69	23
Miconazole	0.2	106	109	46	23
DES	0.1	30	48	95	16
Querecetin	0.1	37	67	100	30

[a] Enzyme assays were carried out [2] with 1 mM ATP, 1 mM $MgCl_2$, 25 mM MES/Tris, and the inhibitor indicated at pH 6.9 (vacuolar membrane H^+-ATPase), pH 8.9 (mitochondrial H^+- ATPase) or pH 6.0 (plasma membrane H^+-ATPase).
[b] See Ref. [4] for details.
[c] Not determined [4].

be 0.8, 2, 0.2, and 8 µM, respectively [2]. Table 1 summarizes the inhibitor specificities of the vacuolar membrane H^+-ATPase in comparison of those other two ATPases: the activity is strongly inhibited by NBD-Cl, tributyltin and moderately by SITS, which are inhibitors of F_1F_0 type H^+-ATPases from mitochondria and chloroplasts, but it is not inhibited by oligomycin, which inhibits the mitochondrial F_1F_0-ATPase. It is not affected by miconazole, which is an inhibitor of E_1E_2-type ATPases from plasma membranes, but it is moderately inhibited by DES and quercetin, which are inhibitors of Na^+/K^+-ATPases.

5. Reaction mechanism of ATP hydrolysis

The reaction mechanism of partially purified vacuolar membrane H^+-ATPase has been studied under steady-state (multi-cycle hydrolysis of ATP; 5 mM ATP/34 nM enzyme) and non-steady-state (single-cycle hydrolysis of ATP; 40 nM {γ-^{32}P}ATP/42 nM enzyme) conditions for ATP hydrolysis [3]. The results indicate that the mechanism is similar to those for the mitochondrial and bacterial F_1F_0-ATPases [26–30], suggesting that it has three catalytic sites for ATP hydrolysis in a holoenzyme complex. Interestingly, NBD-Cl inhibits the enzyme activity under the two kinetic conditions equally, whereas DCCD inhibits only the activity under steady state conditions [3]. It is also noted that NBD-Cl inactivates the catalytic sites and results in inhibition of the formation of an enzyme-ATP complex. On the other hand, DCCD does not affect the binding of ATP to a high-affinity catalytic site of the enzyme [3].

Kinetic analyses under similar conditions as above have also been conducted using vacuolar membrane vesicles with right-side-out orientation [31]. Here again, the occurrence of single-cycle hydrolysis of ATP at the NBD-Cl sensitive catalytic site and its cooperative stimulation by excess ATP were demonstrated. DCCD does not inhibit the activity of single-cycle hydrolysis of ATP but strongly inhibits that of multi-cycle hydrolysis of ATP coupled with H^+ translocation across the vacuolar membrane.

To estimate a minimal molecular mass of the H^+-ATPase bound to the vacuolar membrane for the single-cycle hydrolysis of ATP, radiation inactivation analysis has been carried out [31]. When vacuolar membrane vesicles were exposed to γ rays from ^{60}Co, the activities catalyzing single-cycle and multi-cycle hydrolysis of ATP both decreased as a single exponential function of the radiation dosage, and the susceptibility to irradiation of the H^+-ATPase for the former reaction was 4–5 fold lower than that for the latter reaction. If we apply the target theory and assume the molecular mass of a holo complex of the H^+-ATPase to be about 830 kDa[1], the functional molecular mass for single-cycle hydrolysis of ATP is calculated to be 170–210 kDa.

6. Genetic phenotypes and biochemical defects of yeast *vma* mutants

Taiz and his coworkers first cloned and sequenced a cDNA encoding the carrot 69 kDa polypeptide, a catalytic subunit of vacuolar type H^+-ATPase [32]. Independently, Bowman and Bowman and their collaborators reported isolation and sequencing of two genes from *Neurospora crassa*, *vma1* and *vma2*, named for *v*acuolar *m*embrane *A*TPase [33,34]. These initial contributions have provided wide breakthroughs for molecular and genetic studies of vacuolar type H^+-ATPases.

Anraku and coworkers have studied growth phenotypes of the chromosomal *VMA1*, *VMA2*, and *VMA3* disrupted mutants of *S. cerevisiae* [15,17,35,36]. The three *vma* null mutants can grow well in YPD medium, indicating that each *VMA* gene is dispensable for growth. However, they all show a Pet⁻ *cls* phenotype [15]: the *vma* null mutants cannot grow on YPD plates containing 100 mM $CaCl_2$ and on YPG plates. The Pet⁻ phenotype is unique [37] and the calcium-sensitive *cls* phenotype seems to be related with biochemical Vma⁻ defects (Table 2) since the *vma* null mutants have defects of vacuolar membrane H^+-ATPase activity, ATP-dependent Ca^{2+} uptake into isolated vacuoles and vacuolar acidification *in vivo*, and show higher levels of intracellular free calcium concentration [15,38].

---

1   The molecular mass of the yeast vacuolar membrane H^+-ATPase was estimated as about 830 kDa assuming that the enzyme contains three copies each of the *VMA1*, *VMA2*, *VMA3*, *VMA11*, and *VMA16* gene products and one copy each of the remaining eight *VMA* gene products (see Fig. 1).

Table 2
Vma⁻ defects in *vma* null mutants and Pet⁻ *cls* mutants

Mutant	H⁺-ATPase activity[a]	Ca²⁺ uptake activity[b]	Vacuolar acidification[c]	{Ca²⁺}i[d] (nM)
Wild-type	0.67	58	+	150 ± 80
vma1::URA3	0.03	<0.5	−	
vma2::TRP1	<0.01	<0.5	−	
vma3::URA3	0.04	<0.5	−	
cls7/vma3	<0.01	0.9	−	900 ± 100
cls8/vma1	<0.01	0.7	−	900 ± 100
cls9/vma11	<0.01	0.7	−	900 ± 100
cls10/vma12	<0.01	0.8	−	900 ± 100
cls11/vma13	<0.01	0.5	−	900 ± 100

[a] nmol Pi/min/mg of protein: for assay see Ref. [15].
[b] nmol Ca²⁺/min/mg of protein: for assay see Ref. [15].
[c] Vacuolar acidification in intact cells was quantified by measuring accumulation of quinacrine in the vacuoles: vacuoles were (+) or were not (−) stained with quinacrine [15].
[d] Intracellular free calcium concentration: for assay, see Refs. [15,38].

By that time, Ohya et al. [16] had isolated several Pet⁻ *cls* mutants (*cls7–cls11*) of *S. cerevisiae*, which each has a single recessive chromosomal mutation. The complementation analysis between *vma1–vma3* and *cls7–cls11* mutants demonstrated that *vma1* and *vma3* do not complement *cls8* and *cls7*, respectively, and that *vma2* complements all five *cls* mutations. The results indicate that *VMA1* and *VMA3* are identical with *CLS8* and *CLS7*, respectively, and *CLS9*, *CLS10*, and *CLS11* are a new family of *VMA* genes and are named henceforth *VMA11*, *VMA12*, and *VMA13*, respectively (Table 2) [15].

The *vma* mutants also show pH-conditional growth phenotypes: The mutants can grow in YPD medium of pH 5.0–5.5 but cannot grow in YPD medium of pH 7.0–7.5 [17,18], showing the Vma⁻ defects. Preston et al. [39] reported that *vph* mutations (for *v*acuolar *pH* defective) cause a defect of vacuolar acidification in vivo, the phenotypes of which are also useful for isolation of the vacuolar membrane H⁺-ATPase activity-defective *vph* mutants [40–42].

7. Structure of the yeast vacuolar membrane H⁺-ATPase

7.1. V₁ peripheral sector

By the year 1995, thirteen *VMA/VPH* genes that encode subunits of the enzyme and six *VMA* genes that are required for expression of the enzyme activity and for regulation of the enzyme assembly on the vacuolar membrane have been identified (Fig. 1 and Table 3). Based on currently available genetic and biochemical information [19,43,44], I propose a structure model for the yeast vacuolar membrane H⁺-ATPase as shown in Fig. 1.

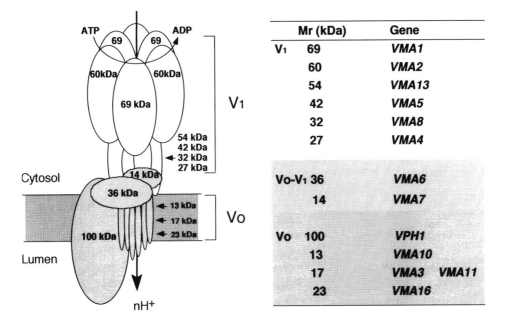

	Mr (kDa)	Gene	
V₁	69	*VMA1*	
	60	*VMA2*	
	54	*VMA13*	
	42	*VMA5*	
	32	*VMA8*	
	27	*VMA4*	
Vo-V₁	36	*VMA6*	
	14	*VMA7*	
Vo	100	*VPH1*	
	13	*VMA10*	
	17	*VMA3*	*VMA11*
	23	*VMA16*	

Fig. 1.   Structural model for the yeast vacuolar membrane H⁺-ATPase and the *VMA* genes encoding the respective subunits.

Table 3
VMA genes that encode non-subunit Vma proteins

Gene	Product	Location	Reference
VMA12/VPH2	25 kDa hydrophobic Vma12p	ER and vacuole	[19] and K.J. Hill and T.H. Stevens (pers. commun.)
VMA14	140 kDa hydrophilic Vma14p	Cytosol	M. Kawasaki, R. Hirose, R. Hirata, Y. Ohya, A. Tzagoloff and Y. Anraku (unpublished)
VMA15/CDC55	60 kDa hydrophilic Vma15p	Cytosol	[75] and R. Hirata, R. Hirose, Y. Ohya, A. Tzagoloff and Y. Anraku (unpublished)
VMA21	8.5 kDa hydrophobic Vma21p	ER	[20]
VMA22	21 kDa hydrophilic Vma22p	ER	K.J. Hill and T.H. Stevens (pers. commun.)
VMA23	not known	not known	[18]

Vma1p, the hydrophilic 69 kDa polypeptide, is the catalytic subunit of the enzyme [3]. Evidence that supports this conclusion is: (1) the ATP analogue 8-azido ATP binds to Vma1p specifically in an ATP-inhibitable manner and, (2) NBD-Cl, which is known to interact selectively with the β subunit (a catalytic polypeptide) of the mitochondrial F_1F_0-H^+-ATPase, binds covalently to the subunit in an ATP-protectable manner and inhibits the formation of an ATP-enzyme complex [3]. Consistent with these results, the deduced sequence of Vma1p shows about 25% sequence identity over 400 amino acid residues with β subunits of F_1F_0-ATPases [36] and contains the consensus sequences for nucleotide-binding domains [45]. This fact suggests that the catalytic subunits of the two classes of H^+-ATPases share a similar structure and mechanism of ATP hydrolysis [3,6,30,46].

Hirata et al. [35] first demonstrated critically that *VMA1* contains a nested genetic element, the *VDE* ORF (open reading frame), in the locus and expresses two functional proteins, Vma1p and *VDE* endonuclease (*VMA1*-derived endonuclease [47]). Thus, the single *VMA1* translational product (a 120 kDa polypeptide/1,071 amino acid residues) catalyzes a self protein splicing in which the internal *VDE* domain (C284-N737) is excised out to produce the 50 kDa *VDE* endonuclease and the N- (M1–G283) and C- (C738–D1,071) domains are ligated instantaneously by a transpeptidation reaction to yield the 69 kDa Vma1p [48]. Based on this novel and protean self catalysis [36,48–50], Anraku and Hirata [51] proposed that the *VMA1* translational product be named *VMA1* protozyme (for *prōtos en zymē*). Ten protozymes that share a common mechanism in protein splicing have been found in six organisms covering three major phylogenic trees (see Refs. [51–53]).

Vma2p, which is encoded by the *VMA2* gene [54–56], is present in an equimolar amount with the major subunit Vma1p in the partially purified enzyme [2,4] and in the vacuolar membrane [57]. This subunit does not bind 8-azido ATP and NBD-Cl under conditions in which they bind to Vma1p [3]. Thus, according to the proposed structure for "vacuolar type" H^+-ATPases [58–63], which is analogous to the structure of the F_1F_0-H^+-ATPases, the yeast enzyme contains three sets of the 69–60 kDa heterodimer in a holoenzyme complex (Fig. 1). If this structure model is correct, Vma2p functions as a regulatory subunit like the α subunit of the F_1F_0-H^+-ATPases.

Vma13p was not originally found in a partially purified enzyme preparation [4]. However, Ho et al. [43] later indicated that this subunit can be co-purified with the active enzyme complex. Interestingly, in a *vma13* null mutant, major peripheral subunits including Vma1p, Vma2p, Vma4p and Vma5p, and integral Vph1p and Vma3p subunits are all present in an inactive complex and assembled onto the vacuolar membrane. This suggests that Vma13p, as a regulatory component, is essential for activity, but not assembly, of the enzyme complex [43].

Vma4p [18,64], Vma5p [18,65], Vma7p [66,67], and Vma8p [68,69] are hydrophilic polypeptides encoded by the respective *VMA* genes. Like Vma1p, Vma2p and Vma13p, they all are removed from the membrane by treatment with chaotropic anions and by cold inactivation. These properties are thus consistent to the notion that they are components of the V_1 peripheral sector of the enzyme complex [44,57]. At present, exact roles of these four subunits are not known, however, a null mutant lacking one of the four *VMA* genes shows a defect of assembling the V_1 peripheral sector on the vacuolar membrane and loses the enzyme activity, suggesting that they have a structural role in assembly.

7.2. V_0 integral sector

VMA3 encodes a hydrophobic polypeptide (160 amino acid residues) with four putative membrane-spanning domains [36,70]. It is a major 17 kDa proteolipid in the vacuolar membrane and is co-purified always associated with the enzyme activity [56]. The subunit is the DCCD-binding component of the complex [2] and functions as a part of a channel for proton translocation in the H^+-ATPase complex. The C-terminal half of yeast Vma3p shows significant identity (35%) to 8 kDa proteolipids of mitochondrial and chloroplast F_1F_0-ATPases and has a conserved glutamic acid residue (E137) in the fourth membrane-spanning domain [36].

Vma11p (164 amino acid residues) was found to be an isoform of Vma3p and contains a conserved glutamic acid residue (E145) [17]. There is a surprising coincidence in the predicted amino acid compositions of the two proteolipids with extensive sequence identity (57% in 150 amino acids). The disruption of either one of the *VMA3* and *VMA11* genes causes loss of the enzyme activity and leads to defective assembly of the V_0 integral sector. In addition, *VMA11* and *VMA3* on multicopy plasmids do not suppress the null mutations of *vma3* and *vma11*, respectively [17]. This result indicates that the two genes do not share functions, but function independently. Interestingly, a mutant Vma11p, in which the conserved E145 is replaced by L145, was fully active in assembly of both V_1 peripheral and V_0 integral sectors onto the vacuolar membrane although the complex does not show any enzyme activity (R. Hirata, A. Takatsuki and Y. Anraku, in collaboration with the T.H. Stevens lab, unpublished). This fact suggests that Vma11p has a structural role in assembly as well as a functional role of participating in proton translocation of the enzyme complex.

Recently, Hirata et al. (R. Hirata, A. Takatsuki and Y. Anraku, unpublished) uncovered that *PPA1/VMA16* may encode a 23 kDa hydrophobic polypeptide that is required for expression of the yeast vacuolar H^+-ATPase activity. *PPA1* was reported to be a gene residing adjacent to the *MAS2* gene *in S. cerevisiae* but its function was unknown [71]. It is now known that a *vma16* null mutant shows typical Vma$^-$ defects: Loss of the H^+-ATPase activity and inability of vacuolar acidification and the mutant vacuolar membrane loses the V_0 integral sector as observed with the null *vma3* and *vma11* mutants.

VMA10 is the gene most recently identified that is also required for expression of the yeast vacuolar membrane H^+-ATPase [72]. It is an intron-containing gene and encodes a novel 13 kDa subunit of the enzyme. This peptide was carefully separated and newly identified as the subunit that is co-purified with the active enzyme complex. Vma10p is a basic protein with isoelectric point of pH 9 and not liberated from the vacuolar membrane by cold inactivation in the presence of MgATP [72]. A *vma10* null mutant shows typical Vma$^-$ defects and a defect in assembly of the V_1 peripheral sector on the vacuolar membrane. Since Vma10p exhibits a significant sequence identity (24%) with subunit *b* of bacterial F_1F_0-ATPases, the authors proposed that the subunit ia a new member of the V_0 integral sector [72].

VPH1 encodes a large hydrophobic integral membrane protein (840 amino acid residues) with six putative membrane-spanning domains and is indispensable for assembly and activity of the H⁺-ATPase [41]. Vph1p seems to be a constituent component of vacuolar type H⁺-ATPases in animal cells, however, its counterpart is not present in F_1F_0-ATPases [41].

7.3. V_1–V_0 junction core

Stevens and coworkers found unique properties of Vma6p [44] and Vma7p [67], which themselves are genuine subunits of the yeast vacuolar membrane H⁺-ATPase but also participate in stabilizing the V_0 integral sector, suggesting that the two subunits may reside in a putative junction site between the V_1 peripheral and V_0 integral sectors (Fig. 1).

Vma6p (345 amino acid residues) encodes a 36 kDa hydrophilic polypeptide in nature and is indeed stripped from the vacuolar membrane by treatment with strong chaotropic agents such as alkaline sodium carbonate (pH 11.5) and 5 M urea. However, unlike the typical peripheral subunits Vma1p and Vma2p, it appears to be tightly associated with the membrane even in the absence of the V_1 peripheral sector and reveals a unique role in stabilizing the components of the V_0 integral sector. It was found that in *vma6* null mutant cells 17 kDa proteolipids are totally absent in the vacuolar membranes and Vph1p exists only in a lower amount than in wild-type cells. In addition, Vma6p is hardly detectable in the membranes from a *vma3* null mutant, suggesting that this subunit is destabilized in the vacuolar membrane lacking the V_0 integral sector. Although no direct evidence has yet been available, Bauerle et al. [44] proposed that Vma6p does likely associate with the vacuolar membrane via direct interaction with one or more of the integral V_0 subunits. There are accumulating evidence that Vma6p and its mammalian counterpart, the 38 kDa bovine clathrin-coated vesicle H⁺-ATPase subunit [73] may be involved in the regulated assembly of the V_1 subunits onto the membrane sector, or alternatively, may function to prevent the passage of protons through V_0 pores in an integral subcomplex [44,74].

Vma7p, one of the new members of the V_1 peripheral sector that was not detected in a partially purified enzyme [2,4], is a 14 kDa hydrophilic subunit of the enzyme [66,67]. Vacuolar membranes isolated from a *vma7* null mutant show no H⁺-ATPase activity and contain greatly reduced levels of the components of the V_0 subcomplex though the mutant cells express normal levels of the V_1 subunits: The null *vma7* cells contain 10-fold lower level of Vph1p and no 17 kDa proteolipids remains in the mutant vacuolar membranes [67]. Furthermore, the level of Vma6p in the mutant cells is comparable to that in wild-type cells, but the amount associated with the vacuolar membrane is significantly reduced. Thus, Vma7p is needed both for assembly and stability of the V_0 integral sector and Vma6p, and may function to form a putative junction bridge between the V_0 integral and V_1 peripheral sectors in the enzyme complex as illustrated in Fig. 1.

8. Regulation of the activity and assembly of the enzyme complex

Genetic screenings for Vma⁻ defects have identified six *vma* mutations that result in a default of assembly of the enzyme (Table 3) [15,18], and among them, Vma12p [19,42] and Vma21p [20] have been well characterized. *VMA12* encodes a 25 kDa hydrophobic polypeptide with two putative membrane-spanning domains and the product appeared to associate with the vacuolar membrane. A *vma12* null mutant shows the typical Vma⁻ defects, although Vma12p is not itself a subunit of the active enzyme complex [19]. Western blotting analyses indicated that none of the V_1 peripheral components including the 69, 60, 42, and 27 kDa subunits resides in the vacuolar membrane from the null *vma12* cells, whereas the cellular levels of these subunits seem to be normal. The 100 and 17 kDa V_0 integral subunits were found to be absent in the mutant vacuolar membranes. These observations indicate that Vma12p is not a component of the active enzyme complex and instead is required during the process of assembly and/or targeting of the enzyme complex to the vacuolar membrane [19].

Vma21p is an 8.5 kDa hydrophobic polypeptide with two membrane-spanning domains and, again, not a subunit of the active H^+-ATPase complex [20]. In a *vma21* null mutant, the 100 kDa V_0 integral subunit was found to be present at greatly reduced level and none is fractionated with the mutant vacuolar membranes. Pulse-chase experiments indicated that this obvious destabilization of Vph1p is due to an increased rate of turnover within the ER in the mutant. Interestingly, Vma21p resides solely in the ER membrane and contains a dilysine motif at the carboxy terminus. Since its location is restricted in the ER membrane and the *vma21* defect only appears to cause a limited destabilization of the Vph1p subunit, but not of other vacuolar marker proteins such as alkaline phosphatase, the authors anticipate that Vma21p may function for stable assembly of the V_0 integral sector in the ER [20].

VMA14 encodes a non-subunit assembly factor with a molecular mass of 140 kDa hydrophilic polypeptide (M. Kawasaki, R. Hirose, R. Hirata, Y. Ohya, A. Tzagoloff and Y. Anraku, unpublished.) and *VMA15* (R. Hirata, R. Hirose, Y. Ohya, A. Tzagoloff and Y. Anraku, unpublished) was found to be identical to *CDC55* that encodes a 60 kDa protein with significant sequence identity (50%) to the B subunit of rabbit skeletal muscle type 2A protein phosphatase (75). Vma22p was recently characterized as a novel 21 kDa hydrophilic protein that is associated with the ER membrane and required for assembly of the vacuolar membrane H^+-ATPase complex (K.J. Hill and T.H. Stevens, personal communication).

9. Distribution of vacuolar type H^+-ATPases

As launched in the dedicated volume "V-ATPases" of *The Journal of Experimental Biology* (see Ref. [25]), bafilomycin A_1-sensitive, vacuolar type H^+-ATPases are now known to be ubiquitously distributed in eukaryotic vacuo-lysosomal and endomembraneous organelles including fungal and plant vacuoles, animal lysosomes, brain clathrin coated vesicles, Golgi bodies, adrenal chromaffin granules and brain synaptic membrane vesicles. Subfamilies of the vacuolar type H^+-ATPase also exist in plasma membranes of archaebacteria and in plasma membranes of neutrophils, macrophages, chicken osteoclasts, bovine kidney brush-border and intercalated cells, and goblet cell apical membranes in the larval midgut of *Manduca sexta*. Recently, the existence and function of other vacuolar type H^+-ATPases have been reported in vanadocyte of the ascidian *Ascidia sydneiensis samea* [76] and in rat retinal pigment epithelium [77]. A new subfamily of Na^+-translocating vacuolar type ATPase was discovered in the eubacterium *Enterococcus hirae* [78] and the *ntp* operon that encodes eleven subunits of the Na^+-ATPase has been characterized in details [79–81].

10. Conclusion and perspectives

I have described in this article the updated information of the structure and function of the vacuolar membrane H^+-ATPase in *Saccharomyces cerevisiae* as a model for many vacuolar type H^+-ATPases in eukaryotic cells. Taking a greater advantage of yeast molecular biological strategies and techniques, a large collection of *vma* mutants has been isolated and characterized, which helps understanding much complicated issues, that had never been thought about in the immediate past, e.g., of expression and assembly of the enzyme complex and its diverse roles in cell physiology. Thus it is needed to know more about the whole structure of the holoenzyme complex and the roles of each subunit in function and regulation. There is a striking similarity and resemblance in the structure and mechanism for ATP hydrolysis between the vacuolar type H^+-ATPases and F_1F_0 H^+-ATPases [11], the former seeming to be a big sister of the latter. Yet, however, the vacuolar type H^+-ATPase is a proton pump and not itself an ATP synthase. This urgent question must be solved in near future. Emerging evidence [19,20] strongly suggests that yeast cells are equipped with a specific vacuolar targeting machinery for delivery and assembly of the enzyme complex. Obviously, elucidation of the mechanism for this particular intracellular protein transport provides a first milestone to understand why vacuolar type H^+-ATPases are destined to be delivered to diverse organelles in different organisms.

It is now commonly recognized that eukaryotic vacuolar type H^+-ATPases perform a diversity of functions in establishing and maintaining pH and cation homeostasis in the lumens of organelles, in the cytosol, and even in the extracellular milieu. The importance of *being acid* in these cellular compartments by the function of vacuolar type H^+-ATPases [5,21,82–84 and see also Ref. [25]) has recently been highlighted where not only normal vegetative growth of cells [15], but also cell transformation [85,86] and differentiation [87], metamorphosis [88], and apoptosis [89] are regulated by the function and components of this unique family of vacuolar type H^+-ATPases. These attractive observations are inviting us to explore new insights of the vacuolar membrane H^+-ATPase-dependent organelle functions in eukaryotic cell systems.

Abbreviations

DCCD, *N,N'*-dicyclohexylcarbodiimide
DES, diethylstilbesterol
ER, endoplasmic reticulum
NBD-Cl, 7-chloro-4-nitrobenzo-2-oxa-1,3-diazole
ORF, open reading frame
SF6847, 3,5-di-*tert*-butyl-4-hydroxybenzylidenemalononitrile
SITS, 4-acetamide-4'-isothiocyanatostilbene-2,2'-disulfonic acid
YPD, 1% yeast extract-2% polypeptone-2% glucose
YPG, 1% yeast extract-2% polypeptone-3% glycerol
ZW3-14, N-tetradecyl-*N,N'*-dimethyl-3-ammonio-1-propane sulfonate

Acknowledgements

The original work from my laboratory described in this article was carried out in collaboration with Drs. Y. Ohsumi, Y. Kakinuma, N, Umemoto, R. Hirata, Y. Ohya and other coworkers, whose names appear in the references. The author thanks Dr. Tom Stevens for information.This study was supported in part by a Grant-in-Aids for Scientific Research on Priority Areas from the Ministry of Education, Science, and Culture of Japan, and a grant from the Human Frontier Science Program Organization in Strasbourg, France.

References

1. Kakinuma, Y., Ohsumi, Y. and Anraku, Y. (1981) J. Biol. Chem. **256**, 10859–10863.
2. Uchida, E., Ohsumi, Y. and Anraku, Y. (1985) J. Biol. Chem. **260**, 1090–1095.
3. Uchida, E., Ohsumi, Y. and Anraku, Y. (1988) J. Biol. Chem. **263**, 45–51.
4. Kane, P.M., Yamashiro, C.Y. and Stevens, T.H. (1989) J. Biol. Chem. **264**, 19236–19244.
5. Anraku, Y. (1987) in: Plant vacuoles, ed B. Marin, pp. 255–265, Plenum Press, New York and London.
6. Anraku, Y., Umemoto, N., Hirata, R. and Wada, Y. (1989) J. Bioenerg. Biomembr. **21**, 589–603.
7. Anraku, Y., Umemoto, N., Hirata, R. and Ohya, Y. (1992) J. Bioenerg. Biomembr. **24**, 395–405.
8. Anraku, Y., Hirata, R., Wada, Y., and Ohya, Y. (1992) J. Exp. Biol. **172**, 67–81.
9. Forgac, M. (1992) Physiol. Rev. **69**, 765–796.
10. Nelson, N. and Taiz, L. (1989) Trends Biochem. Sci. **14**, 113–116.
11. Nelson, N. (1992) Biochim. Biophys. Acta **1100**, 109–124.
12. Kane, P.M. and Stevens, T.H. (1992) J. Bioenerg. Biomembr. **24**, 383–393.
13. Stevens, T.H. (1992) J. Exp. Biol. **172**, 47–55.
14. Stone, D.K., Crider, B.P., Sudhof, T.C. and Xie, X.-S. (1989) J. Bioenerg. Biomembr. **21**, 605–620.
15. Ohya, Y., Umemoto, N., Tanida, I., Ohta, A., Iida, H. and Anraku, Y. (1991) J. Biol. Chem. **266**, 13971–13977.
16. Ohya, Y., Ohsumi, Y. and Anraku, Y. (1986) J. Gen. Microbiol. **132**, 979–988.
17. Umemoto, N., Ohya, Y. and Anraku, Y. (1991) J. Biol. Chem. **266**, 24526–24532.
18. Ho, M.N., Hill, K.J., Lindorfer, M.A. and Stevens, T.H. (1993) J. Biol. Chem. **268**, 221–227.
19. Hirata, R., Umemoto, N., Ho, M.N., Ohya, Y., Stevens, T.H. and Anraku, Y. (1993) J. Biol. Chem. **268**, 961–967.
20. Hill, K.J. and Stevens, T.H. (1994) Mol. Biol. Cell **5**, 1039–1050.
21. Wada, Y. and Anraku, Y. (1994) J. Bioenerg. Biomembr. **26**, 631–637.

22. Ohsumi, Y. and Anraku, Y. (1981) J. Biol. Chem. **256**, 2079–2082.
23. Yoshihisa, T., Ohsumi, Y. and Anraku, Y. (1988) J. Biol. Chem. **263**, 5158–5163.
24. Bowman, E.J., Siebers, A. and Altendorf, K. (1988) Proc. Natl. Acad. Sci. USA **85**, 7972–7976.
25. Harvey, W.R. and Nelson, N. (1992) V-ATPases (eds.) J. Exp. Biol. **172**, 1–485.
26. Grubmeyer, C., Cross, R.L. and Penefsky, H.S. (1982) J. Biol. Chem. **257**, 12092–12100.
27. Cross, R.L., Grubmeyer, C. and Penefsky, H.S. (1982) J. Biol. Chem. **257**, 12100–12105.
28. Duncan, T.M. and Senior, A.E. (1985) J. Biol. Chem. **260**, 4901–4907.
29. Noumi, T., Taniai, M., Kanazawa, H. and Futai, M. (1986) J. Biol. Chem. **261**, 9196–9201.
30. Futai, M., Noumi, T. and Maeda, M. (1988) J. Bioenerg. Bioimembr. **20**, 42–58.
31. Hirata, R., Ohsumi, Y. and Anraku, Y. (1989) FEBS Lett. **244**, 397–401.
32. Zimniak, L., Dittrich, P., Gogarten, J.P., Kibak, H. and Taiz, L. (1988) J. Biol. Chem. **263**, 9102–9112.
33. Bowman, E.J., Tenney, K. and Bowman, B.J. (1988) J. Biol. Chem. **263**, 13994–14001.
34. Bowman, B.J., Allen, R., Wechser, M.A. and Bowman, E.J. (1988) J. Biol. Chem. **263**, 14002–14007.
35. Hirata, R., Ohsumi, Y., Nakano, A., Kawasaki, H., Suzuki, K. and Anraku, Y. (1990) J. Biol. Chem. **265**, 6726–6733.
36. Umemoto, N., Yoshihisa, T., Hirata, R. and Anraku, Y. (1990) J. Biol. Chem. **265**, 18447–18453.
37. Galons, J.P., Tanida, I., Ohya, Y., Anraku, Y. and Arata, Y. (1990) Eur. J. Biochem. **193**, 111–119.
38. Iida, H., Yagawa, Y. and Anraku, Y. (1990) J. Biol. Chem. **265**, 13391–13399.
39. Preston, R.A., Murphy, R.F. and Jones, E.W. (1989) Proc. Natl. Acad. Sci. USA **86**, 7027–7031.
40. Preston, R.A., Reinagel, P.S. and Jones, E.W. (1992) Genetics **131**, 551–558.
41. Manolson, M.F., Proteau, D., Preston, R.A., Stenbit, A., Roberts, B.T., Hoyt, M.A., Preuss, D., Mulholland, J., Botstein, D. and Jones, E.W. (1992) J. Biol. Chem. **267**, 14294–14303.
42. Bachhawat, A.K., Manolson, M.F., Murdock, D.G., Garman, J.D. and Jones, E.W. (1993) Yeast **9**, 175–184.
43. Ho, M.N., Hirata, R., Umemoto, N., Ohya, Y., Takatsuki, A., Stevens, T.H. and Anraku, Y. (1993) J. Biol. Chem. **268**, 18286–18292.
44. Bauerle, C., Ho, M.N., Lindorfer, M.A. and Stevens, T.H. (1993) J. Biol. Chem. **268**, 12749–12757.
45. Walker, J.E., Saraste, M., Runswick, M.J. and Gay, N.J. (1982) EMBO J. **1**, 945–951.
46. Futai, M., Noumi, T., and Maeda, M. (1989) Annu. Rev. Biochem. **58**, 111–136.
47. Gimble, F.S. and Thorner, J. (1992) Nature **357**, 301–306.
48. Hirata, R. and Anraku, Y. (1992) Biochem. Biophys. Res. Commun. **188**, 40–47.
49. Kane, P.M., Yamashiro, C.T., Wolczyk, D.F., Neff, N., Goebl, M. and Stevens, T.H. (1990) Science **250**, 651–657.
50. Cooper, A.A., Chen, Y., Lindorfer, M.A. and Stevens, T.H. (1993) EMBO J. **12**, 2575–2583.
51. Anraku, Y. and Hirata, R. (1994) J. Biochem. **115**, 175–178.
52. Xu, M.-Q., Southworth, M.W., Mersha, F.B., Hornstra, L.J. and Perler, F.B. (1993) Cell **75**, 1371–1377.
53. Xu, M.-Q., Comb, D.G., Paulus, H., Noren, C.J., Shao, Y. and Perler, F.B. (1994) EMBO J. **13**, 5517–5522.
54. Nelson, H., Mandiyan, S. and Nelson, N. (1989) J. Biol. Chem. **264**, 1775–1778.
55. Yamashiro, C.Y., Kane, P.M., Wolczyk, D.F., Preston, R.A. and Stevens, T.H. (1990) Mol. Cell Biol. **10**, 3737–3749.
56. Anraku, Y., Hirata, R., Umemoto, N. and Ohya, Y. (1991) in: New era of bioenergetics, ed Y. Mukohata, pp. 138–168, Academic Press, Tokyo and New York.
57. Kane, P.M., Kuehn, M.C., Howald-Stevenson, I. and Stevens, T.H. (1992) J. Biol. Chem. **267**, 447–454.
58. Arai, H., Terres, G., Pink, S. and Forgac, M. (1988) J. Biol. Chem. **263**, 8796–8802.
59. Puopolo, K. and Forgac, M. (1990) J. Biol. Chem. **265**, 14836–14841.
60. Moriyama, Y. and Nelson, N. (1989) J. Biol. Chem. **264**, 3577–3582.
61. Moriyama, Y. and Nelson, N. (1989) J. Biol. Chem. **264**, 18445–18450.
62. Bowman, B.J., Dschida, W.J., Harris, T. and Bowman, E.J. (1989) J. Biol. Chem. **264**, 15606–15612.
63. Taiz, S.L. and Taiz, L. (1991) Bot. Acta **104**, 117–121.
64. Foury, F. (1990) J. Biol. Chem. **265**, 18554–18560.
65. Beltran, C., Kopecky, J., Pan, Y-C.E., Nelson, H. and Nelson, N. (1992) J. Biol. Chem. **267**, 774–779.

66. Nelson, H., Mandiyan, S. and Nelson, N. (1994) J. Biol. Chem. **269**, 24150–24155.
67. Graham, L.A., Hill, K.J. and Stevens, T.H. (1994) J. Biol. Chem. **269**, 25974–25977.
68. Nelson, H., Mandiyan, S. and Nelson, N. (1995) Proc. Natl. Acad. Sci. USA **92**, 497–501.
69. Graham, L.A., Hill, K.J. and Stevens, T.H. (1995) J. Biol. Chem. **270**, 15037–15044.
70. Nelson, H. and Nelson, N. (1989) FEBS Lett. **247**, 147–153.
71. Apperson, M., Jensen, R.E., Suda, K., Witte, C. and Yaffe, M.P. (1990) Biochem. Biophys. Res. Commun. **168**, 574–579.
72. Supekova, L., Supek, F. and Nelson, N. (1995) J. Biol. Chem. **270**, 13726–13732.
73. Wang, S.-Y., Moriyama, Y., Mandel, M., Hulmes, J.D., Pan, Y.-C.E., Danho, W., Nelson, H. and Nelson, N. (1988) J. Biol. Chem. **263**, 17638–17642.
74. Zhang, J., Myers, M. and Forgac, M. (1992) J. Biol. Chem. **267**, 9773–9778.
75. Healy, A.M., Zolnierowicz, S., Stapleton, A.E., Goebl, M., DePaoli-Roach, A.A. and Pringle, J.R. (1991) Mol. Cell Biol. **11**, 5767–5780.
76. Uyama, T., Moriyama, Y., Futai, M. and Michibata, H. (1994) J. Exp. Zool. **207**, 148–154.
77. Deguchi, J., Yamamoto, A., Yoshimori, T., Sagawa, K., Kato, K., Moriyama, Y., Futai, M., Uyama, T. and Tashiro, Y. (1994) Invest. Ophthalmol. Vis. Sci. **35**, 568–579.
78. Kakinuma, Y. and Igarashi, K. (1990) FEBS Lett. **271**, 97–101.
79. Kakinuma, Y., Igarashi, K., Konishi, K. and Yamato, I. (1991) FEBS Lett. **292**, 64–68.
80. Takase, K., Yamato, I. and Kakinuma, Y. (1993) J. Biol. Chem. **268**, 11610–11616.
81. Takase, K., Kakinuma, S., Yamato, I., Konishi, K., Igarashi, K. and Kakinuma, Y. (1994) J. Biol. Chem. **269**, 11037–11044.
82. Mellman, I., Fuchs, R. and Helenius, A. (1986) Annu. Rev. Biochem. **55**, 663–700.
83. Anraku, Y., Ohya, Y. and Iida, H. (1991) Biochim. Biophys. Acta **1093**, 169–177.
84. Tanida, I., Hasegawa, A., Iida, H., Ohya, Y. and Anralu, Y. (1995) J. Biol. Chem. **270**, 10113–10119.
85. Andresson, T., Sparkowski, J., Goldstein, D.J. and Schlegel, R. (1995) J. Biol. Chem. **270**, 6830–6837.
86. Nakamura, H., Moriyama, Y., Futai, M. and Ozawa, H. (1994) Arch. Histol. Cytol. **57**, 535–539.
87. Lee, B.S., Underhill, D.M., Crane, M.K. and Gluck, S.L. (1995) J. Biol. Chem. **270**, 7320–7329.
88. Sumner, J-P., Dow, J.A.T., Earley, F.G.P., Klein, U., Jager, D. and Wieczorek, H. (1995) J. Biol. Chem. **270**, 5649–5653.
89. Gottlieb, R.A., Giesing, H.A., Zhu, J.Y., Engler, R.L. and Babior, B.M. (1995) Proc. Natl. Acad. Sci. USA **92**, 5965–5968.

CHAPTER 6

Structure and Function of HlyB, the ABC-transporter Essential for Haemolysin Secretion from *Escherichia coli*

I.B. HOLLAND and M.A. BLIGHT

Institut de Génétique et Microbiologie, URA 1354 CNRS,
Bâtiment 409, Université Paris XI, 91405 Orsay cedex 05, France

© 1996 Elsevier Science B.V.
All rights reserved

Handbook of Biological Physics
Volume 2, edited by W.N. Konings, H.R. Kaback and J.S. Lolkema

Contents

1. Introduction

1.1. Background

Haemolysin is a generic term referring to protein toxins capable of lysing erythrocytes. The secretion of a haemolysin into the external medium by *Escherichia coli* was first identified by Lovell and Rees [1]. Subsequently both chromosomally encoded human and plasmid encoded animal uropathogenic *E. coli* isolates have been identified to secrete haemolysin [2–6]. Several determinants have been subcloned from human [7,8] and animal [9] isolates including those from *Proteus* sp. [10] and *Actinobacillus pleuropneumoniae* [11–19]. The haemolysins belong to a large family of RTX toxins (Repeat in Toxin) which possess multiple glycine rich repeats near the C-terminus (reviewed in Ref. [20]).

The genetic (see, e.g., Refs. [8,21–33]) and biochemical analysis (see, e.g., Refs. [24,30,34–68]) has been extensive and has yielded important information concerning the genetic regulation and function of the *E. coli* haemolysin secretion system. Investigations upon the *in vitro* and *in vivo* role of haemolysin in pathogenesis have been equally instructive (see, e.g., Refs. [3,4,54,69–94]), demonstrating the co-ordinate regulation of the *hly* operon with other pathogenic determinants and the *in vitro* and *in vivo* action of the toxin upon different cell lines. A summary of the biochemistry and pathogenesis of haemolytic *E. coli* is not within the scope of this review and the reader is referred to the above bibliography.

1.2. Escherichia coli haemolysin secretion

Secretion of the large 110 kDa haemolytic toxin (HlyA) by *E. coli*, proceeds by the so-called Type I secretion pathway in Gram negative bacteria [95]. This secretion mechanism requires the products of the four contiguous genes of the *hly* operon, *hly C,A,B* and *D* (Fig. 1a) [8,9,22,25,96,97]. HlyA is activated by fatty acylation in the cytoplasm by HlyC (20kDa) [34] and the endogenous acyl carrier protein (ACP) [64,98], followed by secretion and the binding of calcium within the C-terminal glycine rich repeat domain (RTX region) [30,41,42]. Secretion of HlyA does not require activation [34] by fatty acylation whereas erythrocyte lytic activity requires both acylation and calcium binding.

Translocation of HlyA to the medium is achieved through the transenvelope translocator (Fig. 1b) comprising HlyB (66kDa), HlyD (53kDa) [22,24,99] and endogenous TolC (54kDa) [100]. Both HlyB and HlyD are members of two important families of transport and energy-coupling polypeptide families, the ABC (ATP Binding Cassette) [101] (see below) and MFP (Membrane Fusion Protein) [102] families respectively. Membrane localization and topology analyses of HlyB

(a)

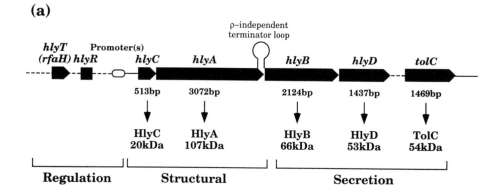

(b)

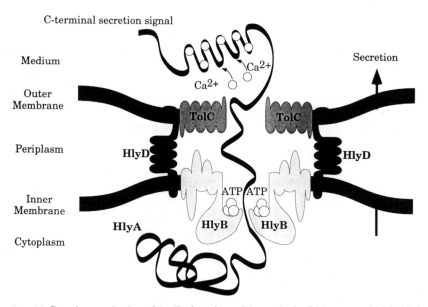

Fig. 1.   (a) Genetic organisation of the *Escherichia coli* haemolysin (*hly*) operon. (b) Model for the
proposed general transenvelope organisation of the Hly translocator.

[50,51] and HlyD [50,55] indicate that both are inner membrane proteins possessing
8 and 1 transmembrane domains respectively.

Secretion of HlyA is apparently independent of the classical General Export
Pathway involving SecA and SecY [39,43] and specificity is achieved through the

uncleaved C-terminal secretion signal of HlyA [35,36,39]. No evidence has yet been presented that HlyA specifically partitions to the membrane bilayer as a prerequisite for translocator interaction. The C-terminal secretion signal has been successfully used for the efficient secretion of many heterologous fusion proteins [37] including mini-antibodies, β-lactamase, prochymosin, β galactosidase, dihydrofolate reductase, the shiga toxin B subunit, *P. haemolytica* glycoprotease, interleukin 1a and a human lipase (see Ref. [103]). The HlyA secretion signal is quite distinct from other targeting signals and the precise functional structure has so far proved to be elusive. The signal is not hydrophobic and appears highly redundant, with many amino acid substitutions having no major effects on function. Genetic studies indicate that the HlyA signal sequence may contribute at least two functions, with residues −15 to −46 involved in translocator recognition [32,53,62] and the extreme C-terminus perhaps playing a role in final folding and release of HlyA to the medium [66] (this laboratory, unpublished data). We have proposed from such studies that secretion signal-translocator interactions may depend upon a small number of critical residues, dispersed throughout the signal region, as part of an extended open structure, somewhat similar to the docking of the peptide antigens with MHC class I molecules [104,105].

Detailed reviews of this type I secretion pathway have been recently published [106,107].

2. Properties of HlyB

2.1. Identification and membrane localization

From sequence analysis, HlyB was first identified 10 years ago [22] as the prototype membrane ABC (ATP Binding Cassette) or Traffic ATPase protein [101]. This family of proteins is characterised by an N-terminal region possessing 6 to 8 transmembrane domains (TMDs) and a cytoplasmically localised C-terminal ATP binding domain. The importance of the ABC proteins within cellular transport processes has become increasingly evident during this decade, with their identification across the phylogenetic scale and their involvement in diverse export and import processes. A provocative example of their importance is the recent completion of the genome sequence from *Haemophilus influenzae* [108] where ABC proteins are encoded by approximately 1% of the genes.

The structure of HlyB deduced from its DNA sequence, indicated an N-terminal hydrophobic domain of approximately 45 kDa and a C-terminal, approximately 35 kDa ATPase (the ABC ATPase), with highly conserved Walker A and B motifs [101]. In this review HlyB will be termed an "ABC-transporter" in distinction to the large family of closely related proteins such as HisP, which are composed solely of the ABC ATPase and require separately encoded integral membrane proteins (permeases) for function. Also in distinction to the HisP family, HlyB-like ABC transporters are normally involved in export rather than import mechanisms. Many homologues of HlyB have now been recognised in both pro- and eukaryotes involved in the secretion of an enormous variety of compounds. As a general term, we prefer "allocrite" rather than substrate, for compounds which are recognised and secreted without modification by a specific transporter [109].

The sequence of HlyB predicts a polypeptide of 79.9 kDa, however, the largest product identified by SDS-PAGE corresponds to a molecular weight of 66 kDa or 69 kDa according to the conditions used [65]. This indicates an aberrant mobility for HlyB in SDS-PAGE but the basis for this is not clear. Several early studies also indicated a 46 kDa protein encoded by HlyB and expressed in mini or maxi cells [22,24,50]. This may be a stable breakdown product or could reflect an alternative translational start or transcriptional termination of *hlyB* mRNA. However, neither the relationship of the 46 kDa protein to either the N- or C-terminal domains of HlyB, nor its physiological significance *in vivo* have been established. HlyB was localized to the inner cytoplasmic membrane in *E. coli* mini cells using both sucrose equilibrium gradient analysis [50] and sarkosyl separation of inner and outer membranes [24]. A close homologue of HlyB, PrtD, responsible for protease secretion in *Erwinia chrysanthemi,* when overproduced in *E. coli* was also localized to the cytoplasmic membrane [110].

A major problem with HlyB studies in *E. coli* has been the failure so far to overproduce the intact protein sufficient either for the production of antibodies or for its specific *in vivo* localization. As an exception to this, Juranka et al., using an HlyB tagged at the C-terminus (residue 704) with the Mdr (P-glycoprotein) C494 monoclonal antibody epitope, localized HlyB to the sarkosyl soluble, inner membrane fraction of the envelope [52]. Attempts to overproduce the HlyB protein have not so far indicated problems with toxicity but rather strict regulation of expression, particularly at the translational level, which limits over expression in *E. coli*. Thus, even when *hlyB* mRNA is overproduced >250-fold, stainable or radiolabelled amounts of the HlyB product cannot be detected in whole cells or in isolated membrane fractions. In contrast, intact HlyB can be easily overproduced when fused downstream of LacZ [65] and the ATPase domain alone or fused to Glutathione S-transferase (GST) can be overproduced in milligram quantities without difficulty [65,111].

2.2. *Topological organization and structure*

The determination of the topological organization of the membrane domain of HlyB and some of its homologues has proved to be extremely contentious and cannot yet be presumed to have been established satisfactorily. The membrane domain of such proteins has relatively low levels of hydrophobic amino acids and relatively small amounts of predicted strongly helical regions. This probably indicates the presence of extensive external or internal loops between transmembrane domains and or a relatively novel organization within the membrane bilayer, with alternatives to α-helices not excluded. An indication of the difficulties of topological predictions is illustrated in Fig. 2 where the number and position of possible TMDs, predicted for HlyB, SecY and bacteriorhodopsin by different algorithms, are compared.

Figure 2 illustrates that the use of several algorithms for the prediction of TMDs is successful for proteins such as bacteriorhodopsin and SecY yet relatively poor

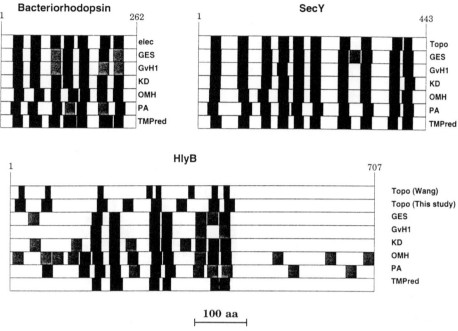

Fig. 2.   Comparison of different hydropathy prediction algorithms for bacteriorhodopsin, SecY and HlyB. Strong and weak hyrdophobic TMD predictions are represented by black and grey boxes respectively. Horizontal boxes represent each prediction method, where "elec" is the electron diffraction structure for bacteriorhodopsin [147], "Topo" the SecY topology [148], "Topo (Wang)" the HlyB β-lactamase probe topology [50], "Topo (this study)" the HlyB topology presented in Fig. 3, "GES" the GES scale [149], "GvH1" the Von Heijne scale [112], "KD" the Kyte and Doolittle scale [150], "OMH" the Eisenberg scale [151], "PA" the Persson-Argos scale [152] and "TMPred" the TMPred scale [153]. The GES, GvH1, KD, OMH, and PA predictions were calculated using the TopPredII program [154]. Numbers represent amino acid positions within the sequences.

for ABC transporters such as HlyB. The number of correctly predicted TMDs with respect to either the known topology of bacteriorhodopsin (electron diffraction analysis) or the experimental topology of SecY is high. However, HlyB clearly illustrates the difficulty in the use of predictive algorithms in that many TMDs are predicted to be "putative" and their relative positions are variable, even being predicted within the C-terminal ABC ATPase.

We have reevaluated and essentially combined the experimental topological data (Fig. 3a) of [50] and [51] to yield a new topological model for HlyB. The model presented in Fig. 3b is based entirely upon the experimental data, initially positioning TMDs at the mid-point between the closest fusions (either β-lactamase, PhoA or LacZ) which define the shift from the cytoplasmic to periplasmic face of the inner membrane. Following the assignment of the TMD mid-points, their precise positions within the bilayer were adjusted to maximize hydrophobic amino acid distribution and to minimize positive charges within an arbitrarly chosen 21 residue TMD. Furthermore, prolines within TMDs were, where possible, positioned within

(a)

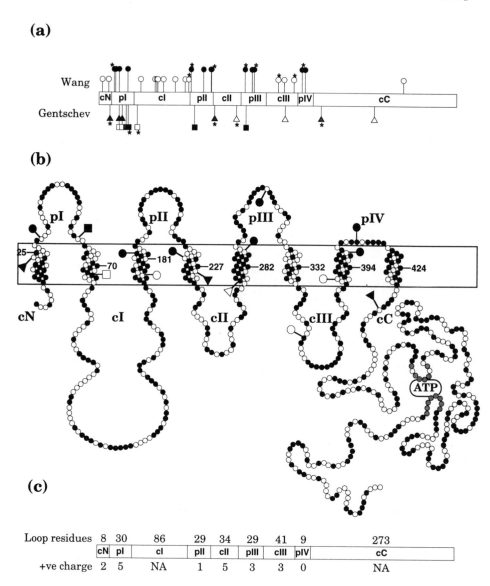

(b)

(c)

Loop residues	8	30	86	29	34	29	41	9	273
	cN	pI	cI	pII	cII	pIII	cIII	pIV	cC
+ve charge	2	5	NA	1	5	3	3	0	NA

Total +ve cytoplasm = 10

Total +ve periplasm = 9

Fig. 3.   (a) Experimental topology model for *E. coli* HlyB from the results of Wang et al. using a β-lactamase probe (circles) [50] and Gentschev et al. using β-galactosidase (triangles) and alkaline phosphatase (squares) probes [51]. Note, for simplicity only fusions closest to TMD mid points are indicated in (b). The horizontal bar represents HlyB from amino acids 1 to 707. Fusions between HlyB and the respective topology probe are as indicated with localisations indicated as filled fusions (black) indicating periplasm, open fusions (white) cytoplasm and grey fusions an uncertain cytoplasmic localisation (β-galactosidase activity between 30–40% wild-type activity). Periplasmic and cytoplasmic domains are as indicated, pI–pIV and cI–cIII respectively, with the cytoplasmically localised

14 residues of either side of the membrane as is observed within 3 out of 7 of the bacteriorhodopsin TMDs (helices B, C and F). The model was further refined in an attempt to fulfill the positive inside rule [112] (Fig. 3c). Thus positive charges were localized maximally and minimally to cytoplasmic and periplasmic loops, respectively, where loops <70 amino-acids only were considered.

The model clearly fits the experimental data well, yet when compared with the predicted hydrophobic TMDs in Fig. 2, still shows relatively poor agreement. Only TMDs 3, 5, 7 and 8 agree with the prediction, albeit with at least a 10 residue shift in position. TMDs 1 and 2 are poorly predicted and TMDs 4 and 6 are shifted completely out of alignment with the prediction. Due to the close fit between the experimental data and the new topological assignment, we are inclined to suggest that the hydropathic prediction is inaccurate for HlyB. On the other hand, given the peculiarities of structure of the membrane domain of the HlyB family, compared with SecY etc., we would be prudent to consider that fusion analysis with such proteins might equally give misleading results.

The difficulties in reconciling the sequence of HlyB with an acceptable topological model for the protein, presumably reflects the specific functional role of the membrane domain, and hence its particular structure. This function might be expected to retain some features common to the whole HlyB family of proteins, despite the highly variable range of allocrites transported. Recalling that the respective subfamilies, of ABC transporters and ABC ATPases represented by HlyB and HisP, are nevertheless involved in contrasting, export and import processes, it is interesting to analyze the membrane domain involved in these processes in more detail. First, the integral membrane proteins which constitute the permease together with HisP, MalK and close relatives, are more typical membrane proteins, highly hydrophobic, relatively easy to model, with at least 5 predicted TMDs. Moreover, this group of membrane proteins all contain the "binding-protein-dependent transport system inner-membrane component" first identified as the "EAA" motif (Prosite 13.0 entry PDOC00364 [113]) [114,115] prior to the last 2 TMDs in a cytoplasmic loop. This motif is not exclusive to the prokaryotic permease systems, having been identified in at least 2 eukaryotes: mouse Nramp, involved in natural resistance to parasites [116] and *Aspergillus nidulans,* CrnA, in the nitrate permease [117]. Moreover, this motif is also associated with proteins such as the *Saccharomyces cerevisiae* uracil permease [118] and the *Drosophila melanogaster* Na$^+$/K$^+$

N- and C-terminii (cN and cC respectively). Vertical lines within the sequence box indicate mid-points between fusions delineating a change from cytoplasm to periplasm (or *vice versa*). (b) Membrane topology model for *E. coli* HlyB. Amino acids (1–707) are represented by small circles where open (white) and closed (black) circles are hydrophilic and hydrophobic respectively (as defined in by Eisenberg [155]. Topology probe fusions delineating membrane boundaries are marked with symbols as described in (a). The mid-points between boundary fusions are numbered. The ATP binding site Walker A and B motifs are indicated by grey circles. (c) Residue and charge distributions within HlyB periplasmic and cytoplasmic loops. The number of residues within each loop are indicated above the bar (representing HlyB) with the number of positive charges within each loop indicated below the bar. NA indicates loops with greater than 65 residues and therefore not applicable to the positive-inside rule [112]. The total number of positive charges in the cytoplasm and periplasm are as shown.

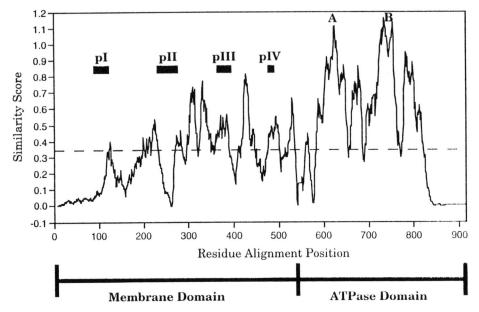

Fig. 4.   Similarity profiles between amino acid sequence alignments of 5 ABC proteins: *E. coli* HlyB, *H. sapiens* Mdr1, *H. sapiens* CFTR, *H.sapiens* TAP1 and TAP2. Mdr1 and CFTR sequences are represented twice as both the N- and C-terminal SF2 moieties (hence 7 sequences in total). The membrane and ATPase domains are as indicated together with the Walker sites (A and B). Positions equivalent to the periplasmic loops (pI–pIV) of HlyB are as shown (black bars).

ATPase which do not interact with ABC ATPases [119]. Both the Nramp and CrnA proteins are predicted to possess at least 10 TMDs with the "EAA" motif occuring prior to the seventh TMD in the cytoplasm [116].

The precise function of the "EAA" motif within the membrane proteins of the permease family remains to be elucidated. However, it seems clear that the overall structural organization of the membrane domain is completely different from that of the HlyB family which also lack the "EAA" motif. Moreover, when the membrane domains of several of the HlyB-like ABC transporters including P-glycoprotein, TAPs and CFTR, are compared, as shown in the similarity plot in Fig. 4, distal sequences are clearly related, although sequences close to the N-terminus are clearly quite different. We presume that these differences between the "EAA" group of membrane proteins and the HlyB-like transporter, reflect the directionality of transport of the different HlyB and HisP families.

We conclude from these results that ABC or Traffic ATPases have associated with at least two different types of membrane protein during evolution. In the HlyB-P-glycoprotein family, conservation of this ancestral protein is most clearly seen in the region corresponding to the two most distal TMDs and surrounding sequences. In contrast, at the N-terminus, most clearly seen within HlyB itself, which contains an approximately 100 residue extension and two additional putative

TMDs, compared with most members of the family, the membrane domains can be quite different. It is tempting to speculate that the N-terminal or proximal portion of the membrane domain therefore plays a major role in allocrite specificity or other specific functions, whilst the distal region expresses a highly conserved function, that of coupling ATPase action to transport.

Studies by G. Ferro-Luzzi Ames and her colleagues have elegantly shown that the functioning of the HisP ATPase requires intimate interaction with the integral membrane proteins HisQ, M. Indeed, the data clearly suggest that a domain of HisP is inserted across the membrane in the presence of the HisQM proteins. [120,121]. Ames and co-workers have proposed that a highly conserved region of HisP, the 15-mer (LSGGQRQRIAIARAL in HlyB) located between the Walker A and B motifs of the HisP and HlyB-P-glycoprotein families, ("Linker" peptide motif Prosite 13.0 entry PDOC 00185 "ABC Transporter family" [113]) [122], and upstream sequences may be involved in insertion of HisP into the membrane and therefore in coupling energy directly to transport. We might anticipate the same function for HlyB.

2.3. HlyB and ABC protein classification

Classification of the super-family into sub-families based upon similar functions and/or substrates, topological structures, interaction with other proteins and host organism is now a large and burgeoning task, since the number of ABC proteins identified to date is at least 216 (SwissProt database version 31.0, 43470 entries, with 0.5% being ABC proteins). The most recent attempt at classification [123] may already require further analysis to include the many ABC proteins identified since publication.

Although it is not within the scope of this review to attempt a classification of the entire ABC super-family to date, we shall present arguments for sub-family classification based upon membrane domain organization and in particular on the conservation of the sequence of the ATPase.

First, classification of the ABC proteins upon the basis of membrane and ATPase domain organisation reveals at least 4 sub-families. The permease sub-family (sub-family, SF1) where the membrane and ATPase domains reside within separate polypeptides; the HlyB sub-family (SF2) with an intimate association of the two domains within the same protein; the TAP sub-family (SF3) where two independent SF2 polypeptides (TAP1 and TAP2) are required for transporter function and finally the P-glycoprotein sub-family (SF4) where two SF2 proteins are joined together in tandem to yield a single ABC transporter. The SF4 group may be sub-divided into SF4a (P-glycoproteins) and SF4b which includes proteins such as CFTR and yeast STE6, possessing insertions within the ABC ATPase. This apparent evolution from SF1 to SF4 also suggests that a two-fold symmetry exists whereby translocator function depends upon at least a dimeric ABC protein. This appears to be the case for the histidine permease with a stoichiometry of 2 HisP (ATPase) to 2 membrane domains (HisQ and HisM) [124] and similarly for the maltose permease with a stoichiometry of MalFGK$_2$ [125]. The combination of the separate membrane and ATPase domains of SF1 proteins to yield an SF2 fusion, normally yields an N-terminal membrane and C-terminal ATPase domain in tandem (e.g. HlyB). However, this topological organisation may not be implicit to function since the *Drosophila melanogaster* White and Brown proteins, required for eye pigment transport, have the reverse topology of the ATPase followed by the membrane domain [126,127].

In view of the above discussion it is therefore anticipated that HlyB also functions minimally as a dimer, however, direct evidence for this has not yet been obtained.

Figure 5a illustrates a simple phylogenetic analysis of 200 amino acids within the ATPase domain for a small number of ABC proteins. This domain is chosen since it is the most highly conserved at the primary sequence level, unlike the membrane domain.The analysis includes representative members from the SF1 to SF4 sub-families and clearly demonstrates that each sub-family clusters upon separate branches (Fig. 5b). Prokaryotic protein transporters (HlyB, LktB, AaltB, CyaB) that require accessory transmembrane proteins for competent translocator function (HlyD and TolC for *E. coli* HlyA transport), clearly group together. Furthermore, the mammalian P-glycoproteins also group together and, moreover, their N and C-terminal tandem repeats (each one HlyB-like) cluster as two separate groups. This may reflect functional differences between the two halves of the tandem ABC transporters.

2.4. ATPase activity of HlyB

Several proteins of the HlyB-P-glycoprotein superfamily have now been purified and shown to have ATPase activity *in vitro*. Previous studies, for example, with *E. coli* MalK [125] or *Salmonella typhimurium* HisP [128], apparently indicated that ATPase activity *in vitro* absolutely required the presence of inner membrane vesicles containing the integral membrane protein components, the periplasmic allocrite binding protein and the allocrite itself. Kinetic and inhibition analysis in these studies yielded tentative ATPase activities, varying five fold between HisP ($V_m = 0.064$ μM ATP min^{-1} mg^{-1} protein; $K_m = 0.20$mM) and MalK ($V_m = 0.341$ μM min^{-1} mg^{-1}). Moreover, several mutants of HisP and mutations affecting the membrane permease required for maltose transport, were isolated, which resulted in a constitutive, substrate independent, ATPase activity, of the ABC, Traffic ATPase. Such findings led to the simple model that allocrite binding to the transport complex activated the ATPase and that hydrolysis of ATP was then in some way coupled to transport activity [120].

This simple model has been complicated by the fact that several ABC transporters have now been isolated intact or as fusion proteins, in the absence of membranes or allocrite, and have been shown to have ATPase activity. The resulting ATPase measurements have however yielded enormous variations, at least 40 fold for V_m and at least 300 fold for K_m. Thus, purified *Salmonella typhimurium*

(a)

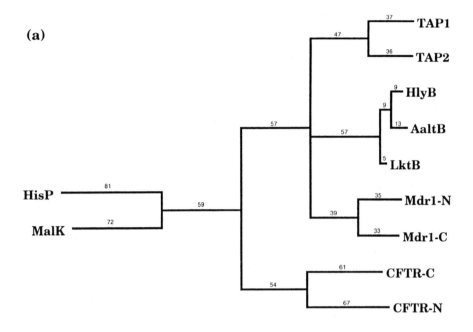

(b)

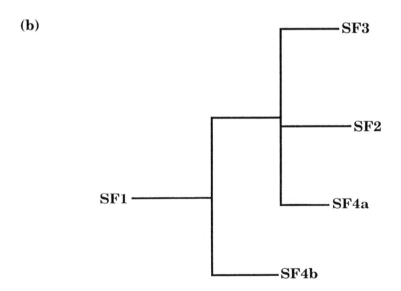

Fig. 5. (a) Phylogenetic analysis of an alignment of a conserved 200 amino acid region within the ATPase domains (ABC ATPase) of several ABC proteins using a bootstrapped (100 replicates) Branch-and-Bound algorithm (program PAUP 3.1.1). Relative ancestral distances are as indicated. (b) Representation of (a) indicating the division of the ABC proteins into 4 major sub-families (SF1-4) and two minor sub-families (SF4a and SF4b). See text for details.

MalK, was reported to have a V_m of 1.3 ± 0.3 μM min^{-1} mg^{-1} and a $K_m = 0.07$ mM [129], with Chinese hamster P-glycoprotein giving a $V_m = 0.321$ μM min^{-1} mg^{-1}, and a $K_m = 0.94$ mM [130]. However, a purified fusion protein between Maltose Binding Protein and a Chinese hamster P-glycoprotein ATPase domain (C-terminal SF2 moiety), gave a $V_m = 0.024$ μM min^{-1} mg^{-1} and $K_m = 20$ mM. Furthermore, partial purification and reconstitution of human P-glycoprotein (Mdr1) into proteoliposomes yielded a vinblastine stimulated $V_m = 8.2$ μM min^{-1} mg^{-1} and vinblastine independent $K_m = 0.285$ mM [131]. An example of a purified intact ABC protein of the SF2 class is the *Erwinia chrysanthemi* PrtD protein [132]. Isolation of PrtD in laurylmaltoside demonstrated a low ATPase activity ($V_m = 0.025$ μM min^{-1} mg^{-1} and $K_m = 0.012$ mM). Attempts to demonstrate ATPase activity for HlyB have been restricted to either a Glutathione-S-transferase fusion protein with the HlyB C-terminal 240 amino acids of the ATPase domain [58] or the same domain over-expressed and refolded from 8M urea (this laboratory, unpublished results). In the former case ATPase kinetics yielded $V_m = 1.0$ μM min^{-1} mg^{-1}, $K_m = 0.2$ mM. The latter purified HlyB ATPase resulted in no detectable ATPase activity, despite binding and purification by dye affinity chromatography on agarose red as has been demonstrated successfully for MalK [133].

Determination of the class of ATPase has in these investigations indicated at least two types based upon inhibitor studies: the P-type ATPases as described above for HisP, HlyB and Mdr and the V-type as suggested for ArsA [134] and CFTR [135]. If inhibitor classification can be relied upon, then there may be several classes of ATPase within the ABC super-family and further divisions into specific sub-families should be considered.

It is clear from the above examples that greatly differing values of V_m and K_m have been calculated for related ABC proteins. Moreover, completely differing effects of substrate upon the various ATPase activities have been described. The GST-HlyB ATPase activity was not affected by the addition of substrate in the form of intact HlyA or the HlyA C-terminal 200 amino acids [58] whereas *E. chrysanthemi* PrtD ATPase activity was actually inhibited by the addition of substrate in the form of either the PrtG or PrtB C-terminal secretion signals [132]. These variable results may not necessarily be surprising, given the range of allocrites involved and the sub-families examined (SF1, SF2 and SF4). However, we might also suspect that in many cases purification protocols and assay conditions are far from optimal and greater efforts may be required to obtain preparations, purified to homogeneity with *in vivo* levels of activity. Such problems are not new to the comparative biochemical study of enzyme activities, but until these are solved the relevance of *in vitro* determined activities to *in vivo* translocation systems, with respect to the regulation of ATPase activity and its coupling to transport, will be suspect.

Nevertheless, *in vitro* analysis of the ATPase activities of "purified" intact and fusion ABC proteins, have yielded important information with respect to the simple diagnostic analysis of the effects of point mutations upon ATP binding and ATP hydrolysis.

3. Genetic analysis of HlyB

3.1. Objectives

In the complete absence of HlyB, haemolysin secretion is completely blocked and active toxin accumulates in the cytoplasm [22,35,136]. It is presumed therefore that HlyB is involved in the initial step in HlyA translocation across the inner membrane, including recognition of the HlyA, C-terminal secretion signal. Further, it is assumed that the specificity for recognition or docking with the translocator can only be provided by the membrane domain of HlyB. However, secretion of HlyA also specifically requires HlyD, which is probably involved in most of the stages in the transport pathway and a role in initiation of translocation for HlyD cannot be ruled out [137]. Direct evidence for interaction between HlyB and HlyD is lacking, although recent evidence (our unpublished data) shows that HlyD is destabilized in the absence of HlyB.

Several recent studies have been concerned with the detailed genetic analysis of HlyB function in the HlyA secretion pathway. The ultimate goal of such studies is three-fold: the identification of the basis of allocrite specificity — initial recognition and subsequent transport — the mechanism of ATPase activation; the mechanism of coupling of ATPase activity to transport of HlyA. A modest start has been made in this respect and the tools are being developed for a much more detailed and extensive analysis in the future. The results to date are summarized below.

3.2. Initial mutagenesis studies

The presumed domains within the HlyB molecule, important in relation to both structure and function, are illustrated in Fig. 6, where the major mutations identified so far by three different groups are also indicated (Table 1). Koronakis et al. first showed that as expected, highly conserved residues within the Walker A motif are essential for secretion of HlyA to the medium [138]. Juranka et al., using insertional mutagenesis at HpaII sites with 15 bp linkers encoding KpnI sites [139], found that all 6 insertions within HlyB completely blocked function. However, insertions at the extreme N- or C-termini retained function. Nevertheless, further attempts to insert an epitope DNA sequence at these new terminal sites, reduced or completely abolished function. In this study with an epitope tagged version of HlyB available for use in some of the constructs, it was possible to demonstrate that at least two of the insertion mutants were unstable and the cause presumably of loss of function. Thus, this study indicated that all the regions scanned (except extreme termini) were essential for function or stability and that this approach could not be exploited for topological studies with inserted epitopes. In addition, Juranka et al. described a site directed mutation (M^{286}-V) (see Fig. 6, Table 1), which resulted in an approximately 4-fold reduction in the level of detectable HlyB protein using an epitope tagged construct [139]. This mutation correspondingly reduced secretion levels of HlyA but the reason for the reduced level of HlyB was not established. This result however provided some evidence that HlyB is probably the limiting factor in HlyA secretion.

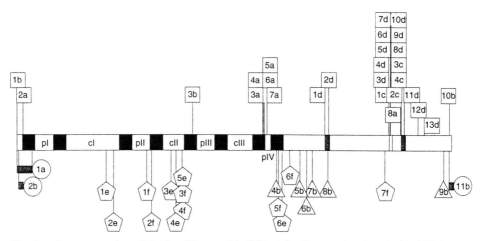

Fig. 6.   A representation of the data illustrated in Table 1 for the mutational analysis of HlyB. The HlyB sequence is represented by a horizontal bar (residues 1–707) with the TMDs predicted from this study (see Fig. 3) shown as black boxes. The periplasmic (pI–pIV) and cytoplasmic loops (cI–cIII) are as indicated together with the Walker A and B ATP binding sites (grey boxes) and "Linker" peptide motif (white box). Mutations are categorised as point (squares), insertion/deletion (circles), KpnI linker insertion (triangles) and suppressors (pentagons). Individual mutations are as numbered (see Table 1).

Table 1

Mutations localised within HlyB. The label and number index the mutations illustrated in Fig. 6. Mutation types are: point, pt; Rep, replacement and Ins, insertion. Defect refers to the phenotype observed as: able or unable to secrete HlyA, sec$^+$ and sec$^-$ respectively; ts, temperature sensitive; Supp^r, suppressor of HlyA signal sequence deletions. *, conditional lethal mutation; †, bind but do not hydrolyse ATP; f, possible role in substrate specificity.

Label	Reference	Number	Mutation	Type	Defect
a	Blight et al. [60]	1	21 aa Cro (Met^1)	Rep	sec$^+$
		2	Gly^{10}-Arg	pt	ts sec$^-$
		3	Ile^{401}-Thr	pt	sec$^-$
		4	Ser^{402}-Pro, Asp^{404}-Lys	pt	sec$^-$
		5	Asp^{404}-Gly	pt	sec$^-$
		6	Ser^{406}-Arg	pt	sec$^+$
		7	*Gly^{408}-Asp	pt	ts sec$^-$
		8	Pro^{624}-Leu	pt	ts sec$^-$
b	Juranka et al. [52]	1	Asp^2-Val, Ser^3-Pro	pt	sec$^+$
		2	Epitope C494 (Val^2)	Ins	sec$^-$
		3	Met^{286}-Val	pt	sec$^+$/sec$^-$
		4	Linker (Pro^{423})	Ins	sec$^-$
		5	Linker (Pro^{462})	Ins	sec$^-$
		6	Linker (Ile^{473})	Ins	sec$^-$
		7	Linker (Pro^{482})	Ins	sec$^-$
		8	Linker (Ser^{607})	Ins	sec$^-$
		9	Linker (Pro^{694})	Ins	sec$^-$

Label	Reference	Number	Mutation	Type	Defect
		10	Leu^{704}-Val, Gln^{705}-Pro	pt	sec^+
		11	Epitope C494 (Val^{704})	Ins	sec^+
c	Koronakis et al. [138]	1	Gly^{605}-Ala	pt	sec^+/sec^-
		2	Gly^{608}-Arg	pt	sec^-
		3	Lys^{625}-Ile	pt	sec^-
		4	Lys^{625}-Arg	pt	sec^+
d	Koronakis et al. [67]	1	Gly^{502}-Glu	pt	sec^-
		2	$^{\dagger}Lys^{508}$-Ile	pt	sec^-
		3	Gly^{605}-Ala	pt	sec^+/sec^-
		4	Gly^{605}-Arg	pt	sec^+/sec^-
		5	$^{\dagger}Ser^{607}$-Asn	pt	sec^-
		6	$^{\dagger}Gly^{608}$-Arg	pt	sec^-
		7	$^{\dagger}Gly^{609}$-Asp	pt	sec^-
		8	Pro^{624}-Leu	pt	sec^-
		9	Lys^{625}-Arg	pt	sec^+
		10	Lys^{625}-Ile	pt	sec^-
		11	$^{\dagger}Asp^{630}$-His	pt	sec^-
		12	Cys^{652}-Arg	pt	sec^+
		13	His^{662}-Leu	pt	sec^-
e	Zhang et al. [140]	1	Ala^{146}-Val	pt	sec^+, $Supp^r$
		2	Leu^{158}-Phe	pt	sec^+, $Supp^r$
		3	Thr^{251}-Ile	pt	sec^+, $Supp^r$
		4	$^{f}Asp^{259}$-Asn	pt	sec^+, $Supp^r$
		5	Ala^{269}-Val	pt	sec^+, $Supp^r$
		6	$^{f}Asp^{433}$-Asn	pt	sec^+, $Supp^r$
f	Sheps et al. [141]	1	Arg212-Lys	pt	sec^+, $Supp^r$
		2	Thr^{220}-Ile	pt	sec^+, $Supp^r$
		3	Ala^{269}-Val	pt	sec^+, $Supp^r$
		4	Ala^{269}-Thr	pt	sec^+, $Supp^r$
		5	Leu^{427}-Phe	pt	sec^+, $Supp^r$
		6	Gly^{445}-Ser	pt	sec^+, $Supp^r$
		7	Val^{599}-Ile	pt	sec^+, $Supp^r$

3.3. Suppressors of HlyA signal sequence mutations

V. Ling's group have attempted to identify regions in HlyB which interact with the HlyA secretion signal, assuming that at some stage this targets HlyA to the translocator, and in particular to the HlyB molecule. The approach involved the isolation of extragenic suppressors within *hlyB* which restored secretion of HlyA lacking the C-terminal 29 residues, i.e. at least two thirds of the signal sequence, or an internal deletion which completely removes almost half of the N-terminal portion of the signal sequence (residues −30 to −58). Both deletions produce very small haloes on blood agar and suppressors were selected by larger haloe size after hydroxylamine or MNNG mutagenesis of *hlyB*, at a frequency of approximately 10^{-4} or less. The position of 11 such suppressors was determined by sequencing, and 10 mapped to the membrane domain in three relatively clustered regions, residues 146–158, 212–269, 427–445, largely irrespective of the HlyA deletion used. However, mutations at A^{269} were isolated with both HlyA deletions and most mutations increased the level of secretion of both of the HlyA signal deficient mutants. Similarly, most suppressor mutations appeared to increase secretion of the defective secretion signals about 4-fold but were not allele specific. Moreover, the HlyB suppressor mutants were not defective in the secretion of wild type HlyA.

These investigations [140,141] constitute an important and interesting attempt to use genetics to identify HlyA signal interactions with HlyB. However, there are several limitations to these studies which preclude the conclusion that specific regions of HlyB, interacting with the HlyA secretion signal, have been identified (allele specific mutations). The use of severely deleted signals, necessary because single point mutations usually have little effect on HlyA secretion [62], eliminates the chances of isolating allele specific mutations and will enhance the possibility of picking up by-pass mutations, possibly involving later stages of recognition and/or translocation. Even the possibility that the efficiency of alternative HlyA secretion signals are actually being enhanced by the presumed suppressors, cannot be excluded. In the absence of any quantitative measurement of HlyB levels in the mutants it is also impossible to exclude, for example, that some mutations result in more efficient translation of *hlyB* mRNA (see above). Suppressor mutations of an HlyA, triple point mutant secretion signal [62], isolated in this laboratory, map outside *hlyB* and probably result in enhanced HlyB expression (unpublished data). In summary, the clustering of several mutations in loops between the TMDs (cI, pII, cII and cC, the ABC ATPase; see Table 1 and Fig. 3b), which suppress secretion signal mutations, is intriguing but does not in itself indicate specific regions of HlyB:HlyA interaction. Some of the mutations in fact map to the periplasmic side of the membrane (loop pII) in the topology model of Wang et al. [50] or the revised model shown in Fig. 3. Clearly, uncertainties with respect to the precise topological structure of HlyB further increase the difficulty of interpretation. In fact, there are precedents for periplasmic located suppressors, whose long distance effects on structure or effects on assembly cannot be discounted. In the case of signal sequence suppressors in the *E. coli* SecY protein, several suppressor mutations have been mapped to the external loop pI [142]. The properties of the HlyB suppressors, although as yet much more limited in number, are in fact remarkably reminiscent of SecY signal sequence suppressors. For that system it has been proposed [142] that a cluster of mutations in TMD7, reduces in particular the proof reading activity of SecY, removing a tolerance barrier to incorrect signals, rather than modifying a docking peptide.

The studies of Ling and collaborators, however, realized two important additional observations. Thus, $D^{259}N$ and $D^{433}N$ HlyB mutations (loop cII and at the border of TMD8 and the ATPase) respectively; see Fig. 6), whilst allowing normal secretion of wild type HlyA, show reduced secretion of an HlyA hybrid with the C-terminus replaced by the leukotoxin (LktA) secretion signal [140]. These mutants therefore display an altered allocrite specificity which could indicate a reduced ability to initiate transport and perhaps recognition of molecules carrying the LktA secretion signal. In P-glycoprotein, a motif ("ERGA") has been identified as being important for drug resistance specificity [143]. The "ERGA" motif is located immediately upstream (1 amino acid) of the "Linker peptide" [122], proposed to couple ATPase activity with allocrite translocation. In HlyB, the "ERGA" motif encompasses amino acids 601–604, intriguingly close to $V^{599}I$ [141], which suppressed the HlyA secretion signal deleted at its N-terminus. It will be important to determine in this case whether for example, HlyB ATPase activity is now constitutive as in certain HisP mutants and that this substitutes in some way for the absence of the N-terminal of the HlyA signal sequence. In summary, mutations in HlyB and P-glycoprotein that appear to be involved in allocrite specificity have been localised to different domains, either HlyB cytoplasmic loop cII and immediately after TMD8 in the ABC ATPase or in the P-glycoprotein ATPase domain.

Koronakis et al. have recently isolated several mutations in the ATP binding region of HlyB by site directed mutagenesis which clearly affect the ATPase of the GST-HlyB fusion *in vitro* [67]. Notably, mutation of highly conserved residues in the ATP binding fold and in the "LSGG-15mer Linker peptide" (see Fig. 6), blocked secretion function *in vivo* and ATPase activity *in vitro*. Mutations in less conserved residues correspondingly had less effect on ATPase activity *in vitro* and protein secretion *in vivo*. In all mutants showing reduced or completely lacking ATPase activity, ATP was nevertheless still bound, as found previously for mutations in the equivalent region of HisP [122]. These results therefore clearly demonstrated that the ATPase activity of HlyB is absolutely essential for secretion of HlyA although the precise point in the secretion pathway, which is consequently blocked in the mutants, remains to be established.

3.4. ATPase and temperature sensitive mutations

Other genetic studies have shown that the N-terminus of HlyB is dispensable for secretory functions. Thus, the first 25 amino acids of HlyB can be replaced by 21 residues of the λ Cro protein without detectable loss of function [60]. This replacement would interrupt the first TMD shown in Fig. 5 and Table 1. As pointed out above, the equivalent N-terminal region of many other, even close homologues of HlyB, is absent and therefore presumably this region may have a unique function in the case of HlyB. Nevertheless, Blight et al., also isolated a temperature sensitive (ts) mutation within the N-terminus of HlyB, $G^{10}R$ [60], which completely blocks secretion at 42°C. Other evidence indicated that this mutation did not destabilize HlyB (see below) and in order to accommodate this finding with the dispensibility of this region, it seems most likely that this radical mutation is *cis* dominant, producing a structural change which interferes with other distal regions of the assembled translocator. Two additional ts mutations isolated in this study are also of some interest. The $P^{624}L$ substitution is located in a relatively well conserved residue in the short region between the Linker peptide and the Walker B motif. Importantly, an identical substitution at this residue in HisP suppresses the requirement for the histidine periplasmic binding protein, apparently because the ATPase activity is now constitutive [122,144]. The substitution, $G^{408}D$, in HlyB, in the short

loop region (pIV), apparently conserved between pro- and eukaryotic ABC transporters between TM7 and TM8, also blocks secretion at high temperature. Moreover, this mutation inhibited the growth of the host cells, particularly in minimal
medium (see also below). Blight and colleagues [60] using site directed mutagenesis also demonstrated that 4 out of 5 substitutions in the pIV, 9 residue periplasmic
loop, greatly reduced secretion. This confirmed the functional importance of this
region in, we may speculate, the coupling of the hydrolysis of ATP to 'activation'
of the membrane domain of HlyB for initial translocation of haemolysin.

4. Does the HlyB, HlyD, TolC translocator form channels in the *E. coli* envelope?

One possible model of HlyA secretion involves translocation of an unfolded HlyA
molecule through a transenvelope channel constituted by the HlyB (inner membrane), HlyD (transenvelope) and TolC (outer membrane) transport complex (see
Fig. 1b). Since HlyB and HlyD have not so far been purified for reconstitution
analysis, direct determination of channel activity of these proteins has not been
possible. We have tried to approach this question indirectly based on the previous
observation [145] that the presence of the Hly system and TolC renders the cells
hypersensitive to large antibiotics, for example vancomycin, that are normally
unable to effectively penetrate the outer membrane. We have shown that hypersensitivity to vancomycin specifically requires the presence of HlyA with a proficient
secretion signal [60]. Moreover, in the wild type pathogenic strain LE2001 which
carries an Hly determinant [8], sensitivity to vancomycin exactly parallels the
narrow window of secretion of HlyA during late exponential phase (A. Pimenta,
this laboratory, unpublished data). These results suggest that active transport of
HlyA, presumably requiring the opening or activation of a transenvelope channel,
in turn facilitates entry of vancomycin. Wandersman and Delepelaire [145] have
previously shown that selection for vancomycin resistance in Hly$^+$ strains, leads to
mutations in either TolC, HlyB or HlyD, demonstrating that all elements of the
translocator are required for the apparent increased drug entry. This phenomena of
allocrite *import* is similar to that reported for the tetracycline efflux transporter,
whose expression results in a large variety of compounds being able to enter into
cells [146]. Sensitivity to vancomycin may therefore be an instructive guide to the
activity of the HlyA translocator. Indeed, amongst mutants (HlyB or HlyD), defective for HlyA secretion, some as expected also acquire resistance to vancomycin,
whilst others, in contrast, retain hypersensitivity to the drug [60,137]. Such drug
hypersensitivity when it occurs is a useful indicator, in the absence of an effective
HlyB antibody, that mutant HlyB molecules have assembled into the envelope. In
the case of the ts mutants described above, P^{624}L and G^{10}R, the cells remained
hypersensitive to the drug at 42°C although secretion was completely blocked. In
control experiments, expression of the mutant HlyB proteins in the absence of
HlyA, did not increase sensitivity to the drug. The results indicated therefore that
whilst translocation of HlyA *per se* was blocked, the transenvelope channel may
nevertheless be activated in these mutants, allowing vancomycin entry.

In the case of $G^{408}D$, at 42°C both secretion activity and vancomycin sensitivity are lost. However, at 30°C the cells display even greater sensitivity to the drug than with wild type HlyB. These results indicate that transenvelope channel activation is blocked at 42°C whilst at 30°C channel activation and concomitant HlyA secretion may be poorly coupled, aborting or preventing HlyA transport, whilst allowing increased drug uptake. The toxicity of this mutation, even at 30°C, with respect to cell growth, at least in minimal medium, also indicates that the altered HlyB molecule may be disrupting general membrane function, and the analysis of the basis of this effect may reveal something of the normal function of HlyB.

5. Perspectives

Efficient purification of both intact HlyB and the ATPase domain are now crucial to further detailed analysis of its function and we expect that this will in principle be possible with current technology being used successfully for other ABC transporters. Over-production of HlyB in *E. coli* for this purpose, however, appears to be extremely difficult and purification may have to depend on simply the use of large scale cultures and conventional protein purification. Rigorous procedures will however be necessary to ensure that the purified protein has an ATPase activity reflecting its *in vivo* activity.

Many interesting mutants of HlyB are now available and further exploitation of such mutants must include the analysis of the effects of these mutations on different steps in the secretion pathway, which now seems feasible from comparable studies with HlyD (Pimenta et al., this laboratory, submitted). In particular, the use of HlyA translocation intermediates would facilitate the dissection of translocator function. Such studies must also include analysis of the coordination of ATP utilisation and optimisation of *in vitro* conditions, which hopefully detect stimulation of HlyB ATPase activity in the presence of the HlyA secretion signal, and hence the possibility to identify ATPase constitutive mutants, as in the case of HisP. The precise topology of HlyB and its homologues remains uncertain and the application of additional techniques, genetical and chemical, particularly to the proximal half of the membrane domain, will be required to ensure greater precision.

Initial genetic suppressor studies have raised the possibility that the HlyB membrane domain may have a proof reading function. More efforts are, however, required to identify presumptive signal docking regions in HlyB and or HlyD.

Purification of HlyB and reconstitution of the secretion pathway *in vitro*, is now the key to many aspects of HlyB function. This combined with the wealth of genetical mutagenesis techniques available and the relative simplicity of *in vivo* functional analysis, will ensure rapid future progress in dissecting its full repertoire of activities.

Acknowledgements

We are extremely grateful to the CNRS and Université Paris XI for continued support. We are also very pleased to acknowledge the important support of the

EU-Biotechnology Program (Contract No. B102-CT93-0145), INSERM and ARC (Association pour la Recherche contre le Cancer) for haemolysin studies in Orsay.

References

1. Lovell, R. and Rees, T.A. (1960) Nature **188**, 755–756.
2. Welch, R.A., Dellinger, E.E., Minshew, B. and Falkow, S. (1981) Nature **294**, 665–667.
3. Cavalieri, S.J. and Snyder, I.S. (1982) Infect. Immun. **37**, 966–974.
4. Hacker, J., Schroter, G., Schrettenbrunner, A., Hughes, C. and Goebel, W. (1983) Zentralbl. Bakteriol. Parasitenkd. Infeksionsfr. Hyg. **254**, 370–378.
5. Waalwijk, C., Bosch, J.F.v.d., McLaren, D.M. and Graff, J.D. (1982) FEMS Microbiol. Lett. **14**, 171–175.
6. Linggood, M.A. and Ingram, P.L. (1982) J. Med. Microbiol. **15**, 23–30.
7. Welch, R., Hull, R. and Falkow, S. (1983) Infect. Immun. **42**, 178–186.
8. Mackman, N. and Holland, I.B. (1984) Mol. Gen. Genet. **196**, 129–134.
9. Goebel, W. and Hedgpeth, J. (1982) J. Bacteriol. **151**, 1290–1298.
10. Koronakis, V., Cross, M., Senior, B., Koronakis, E. and Hughes, C. (1987) J. Bacteriol. **169**, 1509–1515.
11. Gygi, D., Nicolet, J., Frey, J., Cross, M., Koronakis, V. and Hughes, C. (1990) Mol. Microbiol. **4**, 123–128.
12. Chang, Y., Young, R. and Struck, D.K. (1991) J. Bacteriol. **173**, 5151–5158.
13. Kamp, E.M., Popma, J.K., Anakotta, J. and Smits, M.A. (1991) Infect. Immun. **59**, 3079–3085.
14. McWhinney, D.R., Chang, Y.-F., Young, R. and Struck, D.K. (1992) J. Bacteriol. **174**, 291–297.
15. Chang, Y.F., Shi, J., Ma, D.P., Shin, S.J. and Lein, D.H. (1993) DNA Cell Biol. **12**, 351–362.
16. Frey, J., Beck, M., Stucki, U. and Nicolet, J. (1993) Gene **123**, 51–8.
17. Jansen, R., Briaire, J., Kamp, E.M., Gielkens, A.L. and Smits, M.A. (1993) Infect. Immun. **61**, 947–954.
18. Jansen, R., Briaire, J., A.B.M. van Geel, Kamp, E.M., Gielkens, A.L.J. and Smits, M.A. (1994) Infect. Immun. **62**, 4411–4418.
19. Tascon, R.I., Vazquez, B.J., Gutierrez, M.C., Rodriguez, B.I. and Rodriguez, F.E. (1994) Mol. Microbiol. **14**, 207–216.
20. Welch, R.A., Forestier, C., Lobo, A., Pellett, S., Thomas, W.J. and Rowe, G. (1992) FEMS Microbiol. Immunol. **5**, 29–36.
21. Juarez, A. and Goebel, W. (1984) J. Bacteriol. **159**, 1083–1085.
22. Felmlee, T., Pellet, S. and Welch, R.A. (1985) J. Bacteriol. **163**, 94–105.
23. Felmlee, T., Pellett, S., Lee, E.Y. and Welch, R.A. (1985) J. Bacteriol. **163**, 88–93.
24. Mackman, N., Nicaud, J.-M., Gray, L. and Holland, I.B. (1985) Mol. Gen. Genet. **201**, 529–536.
25. Mackman, N., Nicaud, J.-M., Gray, L. and Holland, I.B. (1985) Mol. Gen. Genet. **201**, 282–288.
26. Nicaud, J.-M., Mackman, N., Gray, L. and Holland, I.B. (1985) Mol. Gen. Genet. **199**, 111–116.
27. Hess, J., Wels, W., Vogel, M. and Goebel, W. (1986) FEMS Microbiol. Lett. **34**, 1–11.
28. Ludwig, A., Vogel, M. and Goebel, W. (1987) Mol. Gen. Genet. **206**, 238–245.
29. Koronakis, V., Cross, M. and Hughes, C. (1988) Nucl. Acids Res. **16**, 4789–4800.
30. Ludwig, A.L., Jarchau, T., Benz, R. and Goebel, W. (1988) Mol. Gen. Genet. **214**, 553–561.
31. Vogel, M., Hess, J., Then, I., Jaurez, A. and Goebel, W. (1988) Mol. Gen. Genet. **212**, 76–84.
32. Stanley, P., Koronakis, V. and Hughes, C. (1991) Mol. Microbiol. **5**, 2391–2403.
33. Bailey, M.J.A., Koronakis, V., Schmoll, T. and Hughes, C. (1992) Mol. Microbiol. **6**, 1003–1012.
34. Nicaud, J.-M., Mackman, N., Gray, L. and Holland, I.B. (1985) FEBS Lett. **187**, 339–344.
35. Gray, L., Mackman, N., Nicaud, J.-M. and Holland, I.B. (1986) Mol. Gen. Genet. **205**, 127–133.
36. Nicaud, J.M., Mackman, N., Gray, L. and Holland, I.B. (1986) FEBS Lett. **204**, 331–335.
37. Mackman, N., Baker, K., Gray, L., Haigh, R., Nicaud, J.-M. and Holland, I.B. (1987) EMBO J. **6**, 2835–2841.
38. Felmlee, T. and Welch, R.A. (1988) Proc. Natl. Acad. Sci. USA **85**, 5269–5273.

39. Gray, L., Baker, K., Kenny, B., Mackman, N., Haigh, R. and Holland, I.B. (1989) J. Cell Science Suppl. **11**, 45–47.
40. Oropeza-Wekerle, R.-L., Müller, E., Kern, P., Meyermann, R. and Goebel, W. (1989) J. Bacteriol. **171**, 2783–2788.
41. Boehm, D.F., Welch, R.A. and Snyder, I.S. (1990) Infect. Immun. **58**, 1951–1958.
42. Boehm, D.F., Welch, R.A. and Snyder, I.S. (1990) Infect. Immun. **58**, 1959–1964.
43. Gentschev, I., Hess, J. and Goebel, W. (1990) Mol. Gen. Genet. **222**, 211–216.
44. Hardie, K.R., Issartel, J.-P., Koronakis, E., Hughes, C. and Koronakis, V. (1991) Mol. Microbiol. **5**, 1669–1679.
45. Hess, J., Gentschev, I., Goebel, W. and Jarchau, T. (1990) Mol. Gen. Genet. **224**, 201–208.
46. Oropeza-Wekerle, R.L., Speth, W., Imhof, B., Gentschev, I. and Goebel, W. (1990) J. Bacteriol. **172**, 3711–3717.
47. Pellett, S., Boehm, D.F., Snyder, I.S., Rowe, G. and Welch, R.A. (1990) Infect. Immun. **58**, 822–827.
48. Kenny, B., Haigh, R. and Holland, I.B. (1991) Mol. Microbiol. **5**, 2557–2568.
49. Koronakis, V., Hughes, C. and Koronakis, E. (1991) EMBO J. **10**, 3263–3272.
50. Wang, R., Seror, S.J., Blight, M.A., Pratt, J.M., Broome-Smith, J. and Holland, I.B. (1991) J. Mol. Biol. **217**, 441–454.
51. Gentschev, I. and Goebel, W. (1992) Mol. Gen. Genet. **232**, 40–48.
52. Juranka, P., Zhang, F., Kulpa, J., Endicott, J., Blight, M., Holland, I.B. and Ling, V. (1992) J. Biol. Chem. **267**, 3764–3770.
53. Kenny, B., Taylor, S. and Holland, I.B. (1992) Mol. Microbiol. **6**, 1477–1489.
54. Oropeza-Wekerle, R.-L., Muller, S., Briand, J.-P., Benz, R., Schmid, A. and Goebel, W. (1992) Mol. Microbiol. **6**, 115–121.
55. Schülein, R., Gentschev, I., Mollenkopf, H.-J. and Goebel, W. (1992) Mol. Gen. Genet. **234**, 155–163.
56. Thomas, W.D., Wagner, S.P. and Welch, R.A. (1992) J. Bacteriol. **174**, 6771–6779.
57. Koronakis, V., Hughes, K. and Koronakis, E. (1993) Mol. Microbiol. **8**, 1163–1175.
58. Koronakis, V. and Hughes, C. (1993) Cell Biology **4**, 7–15.
59. Stanley, P.L.D., Diaz, P., Bailey, M.J.A., Gygi, D., Juarez, A. and Hughes, C. (1993) Mol. Microbiol. **10**, 781–787.
60. Blight, M.A., Pimenta, A.L., Lazzaroni, J.-C., Dando, C., Kotelevets, L., Séror, S. and Holland, I.B. (1994) Mol. Gen. Genet. **245**, 431–440.
61. Jarchau, T., Chakraborty, T., Garcia, F. and Goebel, W. (1994) Mol. Gen. Genet. **245**, 53–60.
62. Kenny, B., Chervaux, C. and Holland, I.B. (1994) Mol. Microbiol. **11**, 99–109.
63. Schülein, R., Gentschev, I., Schlör, S., Gross, R. and Goebel, W. (1994) Mol. Gen. Genet. **245**, 203–211.
64. Stanley, P., Packman, L.C., Koronakis, V. and Hughes, C. (1994) Science **266**, 1992–1996.
65. Blight, M.A., Menichi, B. and Holland, I.B. (1995) Mol. Gen. Genet. **247**, 73–85.
66. Chervaux, C. and Holland, I.B. (1995) J. Bacteriol. **178**, 1232–1236.
67. Koronakis, E., Hughes, C., Milisav, I. and Koronakis, V. (1995) Mol. Microbiol. **16**, 87–96.
68. Ostolaza, H., Soloaga, A. and Goñi, F.M. (1995) Eur. J. Biochem. **228**, 39–44.
69. Benz, R., Janko, K. and Lauger, P. (1979) Biochim. Biophys. Acta. **551**, 238–247.
70. Hughes, C., Muller, D., Hacker, J. and Goebel, W. (1982) Toxicon **20**, 247–252.
71. Waalwijk, C., MacLaren, D.M. and de Graaff, J. (1983) Infect. Immun. **42**, 245–249.
72. Cavalieri, S., Bohach, G. and Snyder, I. (1984) Microbiol. Rev. **48**, 326–343.
73. Tschape, H. and Prager, R. (1984) Z. Urol. Nephrol. **77**, 407–413.
74. O'Hanley, P., Low, D., Romero, I., Lark, D., Vosti, K., Falkow, S. and Schoolnik, G. (1985) N. Engl. J. Med. **313**, 414–420.
75. Smith, H.W. and Huggins, M.B. (1985) J. Gen. Microbiol. **131**, 395–403.
76. Bhakdi, S., Mackman, N., Nicaud, J.-M. and Holland, I.B. (1986) Infect. Immun. **52**, 63–69.
77. Bhakdi, S. and Tranum-Jensen, J. (1986) Microb. Pathol. **1**, 5–14.
78. Menestrina, G., Mackman, N., Holland, I.B. and Bhakdi, S. (1987) Biochem. Biophys. Acta **905**, 109–117.
79. Menestrina, G. (1988) FEBS. Lett. **232**, 217–220.
80. Benz, R., Schmid, A., Wagner, W. and Goebel, W. (1989) Infect. Immun. **57**, 887–895.

81. Eberspacher, B., Hugo, F. and Bhakdi, S. (1989) Infect. Immun. **57**, 983–988.
82. Hacker, J., Bender, L., Ott, M., Wingender, J., Lund, B., Marre, R. and Goebel, W. (1990) Microb. Pathog. **8**, 213–25.
83. Bhakdi, S. and Martin, E. (1991) Infect. Immun. **59**, 2955–2962.
84. Blum, G., Ott, M., Cross, A. and Hacker, J. (1991) Microb. Pathog. **10**, 127–136.
85. Grimminger, F., Scholz, C., Bhakdi, S. and Seeger, W. (1991) J. Biol. Chem. **266**, 14262–14269.
86. O'Hanley, P., Lalonde, G. and Ji, G. (1991) Infect. Immun. **59**, 1153–1161.
87. Hacker, J., Ott, M., Blum, G., Marre, R., Heesemann, J., Tschape, H. and Goebel, W. (1992) Int. J. Med. Microbiol. Virol. Parasitol. Infect. Dis. **276**, 165–175.
88. Hughes, C., Stanley, P. and Koronakis, V. (1992) Bioessays **14**, 519–525.
89. Ostolaza, H., Bartolome, B., Ortiz de Zarate, I., de la Cruz, F. and Goñi, F.M. (1993) Biochim. Biophys. Acta **1147**, 81–88.
90. Mobley, H.L., Island, M.D. and Massad, G. (1994) Kidney. Int. Suppl. S129–136.
91. Morschhauser, J., Vetter, V., Emody, L. and Hacker, J. (1994) Mol. Microbiol. **11**, 555–566.
92. Walmrath, D., Ghofrani, H.A., Rosseau, S., Schutte, H., Cramer, A., Kaddus, W., Grimminger, F., Bhakdi, S. and Seeger, W. (1994) J. Exp. Med. **1807**, 1437–1443.
93. Zychlinsky, A., Kenny, B., Menard, R., Prevost, M.C., Holland, I.B. and Sansonetti, P.J. (1994) Mol. Microbiol. **11**, 619–627.
94. Blum, G., Falbo, V., Caprioli, A. and Hacker, J. (1995) FEMS Microbiol. Lett. **126**, 189–195.
95. Salmond, G.P.C. and Reeves, P.J. (1993) TIBS **18**, 7–12.
96. Mackman, N. and Holland, I.B. (1984) Mol. Gen. Genet. **193**, 312–315.
97. Mackman, N., Nicaud, J.-M., Gray, L. and Holland, I.B. (1986) Curr. Top. Microbiol. Immunol. **125**, 159–181.
98. Issartel, J.-P., Koronakis, V. and Hughes, C. (1991) Nature **351**, 759–761.
99. Wagner, W., Vogel, M. and Goebel, M. (1983) J. Bacteriol. **154**, 200–210.
100. Wandersman, C. and Delepelaire, P. (1990) Proc. Natl. Acad. Sci. USA **87**, 4776–4780.
101. Higgins, C.F., Hiles, I.D., Salmond, G.P.C., Gill, D.R., Downie, J.A., Evans, I.J., Holland, I.B., Gray, L., Buckel, S.D., Bell, A.W. and Hermodson, M.A. (1986) Nature **323**, 448–450.
102. Dinh, T., Paulsen, I.T. and Saier-JR, M.H. (1994) J. Bacteriol. **176**, 3825–3831.
103. Blight, M.A. and Holland, I.B. (1994) TIBTECH **12**, 450–455.
104. Fremont, D.H., Matsumura, M., Matsumura, E.A., Peterson, P.A. and Wilson, I.A. (1992) Science **257**, 919–926.
105. Uebel, S., Meyer, T.H., Kraas, W., Kienle, S., Jung, G., Wiesmüller, K.-H. and Tampé, R. (1995) J. Biol. Chem. **270**, 18512–18516.
106. Genin, S. and Boucher, C.A. (1994) Mol. Gen. Genet. **243**, 112–118.
107. Blight, M.A., Chervaux, C. and Holland, I.B. (1994) Curr. Opinion Biotech. **5**, 468–474.
108. Fleischmann, R.D., Adams, M.D., White, O., Clayton, R.A., Kirkness, E.F., Kerlavage, A.R., Bult, C.J., Tomb, J.-F., Dougherty, B.A., Merrick, J.M., McKenney, K., Sutton, G., Fitzhugh, W., Fields, C., Gocayne, J.D., Scott, J., Shirley, R., Liu, L.-I., Glodek, A., Kelley, J.M., Weidman, J.F., Phillips, C.A., Spriggs, T., Hedblom, E., Cotton, M.D., Utterback, T.R., Hanna, M.C., Nguyen, D.T., Saudek, D.M., Brandon, R.C., Fine, L.D., Fritchman, J.L., Fuhrmann, J.L., Geoghagen, N.S.M., Gnehm, C.L., McDonald, L.A., Small, K.V., Fraser, C.M., Smith, H.O. and Venter, J.C. (1995) Science **269**, 496–512.
109. Blight, M.A. and Holland, I.B. (1990) Mol. Microbiol. **4**, 873–880.
110. Delepelaire, P. and Wandersman, C. (1991) Mol. Microbiol. **5**, 2427–2434.
111. Koronakis, V., Hughes, C. and Koronakis, E. (1993) Mol. Microbiol. **8**, 1163–1175.
112. von Heijne, G. (1992) J. Mol. Biol. **225**, 487–494.
113. Bairoch, A. and Bucher, P. (1994) Nucl. Acids Res. **22**, 3583–3589.
114. Dassa, E. and Hofnung, M. (1985) EMBO J. **4**, 2287–2293.
115. Kerppola, R.E. and Ames, G.F.-L. (1992) J. Biol. Chem. **267**, 2329–2336.
116. Vidal, S.M., Malo, D., Vogan, K., Skamene, E. and Gros, P. (1993) Cell **73**, 469–485.
117. Unkles, S.E., Hawker, K.L., Grieve, C., Campbell, E.I., Montagne, P. and Kinghorn, J.R. (1991) Proc. Natl. Acad. Sci. USA **88**, 204–208.
118. Jund, R., Weber, E. and Chevallier, M.-R. (1988) Eur. J. Biochem. **171**, 417–424.

119. Lebovitz, R.M., Takelysu, K. and Fambrough, D.M. (1989) EMBO J. **8**, 193–202.
120. Ames, G.F.-L. and Lecar, H. (1992) FASEB **6**, 2660–2666.
121. Baichwal, V., Liu, D. and Ames, G.F.-L. (1993) Proc Natl Acad Sci USA **90**, 620–624.
122. Shyamala, V., Baichwal, V., Beall, E. and Ames, G.F.-L. (1991) J. Biol. Chem. **266**, 18714–18719.
123. Fath, M.J. and Kolter, R. (1993) Microbiol. Rev. **57**, 995–1017.
124. Kerppola, R.E., Shyamala, V., Klebba, P. and Ames, G.F.-L. (1991) J. Biol. Chem. **266**, 9857–9865.
125. Davidson, A.L., Shuman, H.A. and Nikaido, H. (1992) Proc. Natl. Acad. Sci. USA **89**, 2360–2364.
126. O'Hare, K., Murphy, C., Levis, R. and Rubin, G.-M. (1984) J. Mol. Biol. **180**, 437–455.
127. Dreesen, T.D., Johnson, D.H. and Henikoff, S. (1988) Mol. Cell. Biol. **8**, 5206–5215.
128. Ames, G.F.-L., Nikaido, K., Groarke, J. and Petithory, J. (1989) J. Biol. Chem. **264**, 3998–4002.
129. Morbach, S., Tebbe, S. and Schneider, E. (1993) J. Biol. Chem. **268**, 18617–18621.
130. Shapiro, A.B. and Ling, V. (1994) J. Biol. Chem. **269**, 3745–3754.
131. Ambudkar, S.V., Lelong, I.H., Zhang, J., Cardarelli, C.O., Gottesman, M.M. and Pastan, I. (1992) Proc. Natl. Acad. Sci. USA **89**, 8472–8476.
132. Delepelaire, P. (1994) J. Biol. Chem. **269**, 27952–27957.
133. Schneider, E., Linde, M. and Tebbe, S. (1995) Protein Expr. Purif. **6435**, 10–14.
134. Tisa, L. and Rosen, B. (1990) J. Bioenerg. Biomembr. **22**, 493–507.
135. Anderson, M.P., Gregory, R.J., Thompson, S., Souza, D.W. and Paul, S. (1991) Science **253**, 202–205.
136. Koronakis, V., Koronakis, E. and Hughes, C. (1989) EMBO J. **8**, 595–605.
137. Pimenta, A.L., Jamieson, L. and Holland, I.B. (1996) Mol. Gen. Genet., submitted.
138. Koronakis, V., Koronakis, E. and Hughes, C. (1988) Mol. Gen. Genet. **213**, 551–555.
139. Juranka, P.F., Zastawny, R.L. and Ling, V. (1989) FASEB J. **3**, 2583–2592.
140. Zhang, F., Sheps, J.A. and Ling, V. (1993) J. Biol. Chem. **268**, 19889–19895.
141. Sheps, J.A., Cheung, I. and Ling, V. (1995) J. Biol. Chem. **270**, 14829–14834.
142. Osborne, R.S. and Silhavy, T.J. (1993) EMBO J. **12**, 3391–3398.
143. Beaudet, L. and Gros, P. (1995) J. Biol. Chem. **270**, 17159–17170.
144. Petronilli, V. and Ames, G.F.-L. (1991) J. Biol. Chem. **266**, 16293–16296.
145. Wandersman, C. and Delepelaire, P. (1990) Proc. Natl. Acad. Sci. USA **87**, 4746–4780.
146. Griffith, J.K., Cuellar, D.H., Fordyce, C.A., Hutchings, K.G. and Mondragon, A.A. (1994) Mol. Membr. Biol. **11**, 271–277.
147. Henderson, R. and Unwin, P.N. (1975) Nature **257**, 28–32.
148. Akiyama, Y. and Ito, K. (1987) EMBO J **6**, 3465–3470.
149. Engelman, D.M., Steitz, T.A. and Goldman, A. (1986) Ann. Rev. Biophys. Biophys. Chem. **15**, 321–353.
150. Kyte, J. and Doolittle, R.F. (1982) J. Mol. Biol. **157**, 105–132.
151. Sweet, R.M. and Eisenberg, D. (1983) J. Mol. Biol. **171**, 479–488.
152. Persson, B. and Argos, P. (1994) J. Mol. Biol. **237**, 182–192.
153. Hofmann, K. and Stoffel, W. (1993) Biol. Chem. Hoppe-Seyler **347**, 166–171.
154. Claros, M.G. and von Heijne, G. (1994) Comput. Appl. Biosci. **10**, 685–686.
155. Eisenberg, D., Schwarz, E., Komaromy, M. and Wall, R. (1984) J. Mol. Biol. **179**, 125–142.

CHAPTER 7

The P-Glycoprotein Family and Multidrug Resistance: An Overview

P. GROS and M. HANNA

Department of Biochemistry, McGill University,
Montreal, Quebec, Canada

© *1996 Elsevier Science B.V.*
All rights reserved

Handbook of Biological Physics
Volume 2, edited by W.N. Konings, H.R. Kaback and J.S. Lolkema

Contents

1. General introduction

A major limitation to the successful chemotherapeutic treatment of cancer is the natural or acquired resistance of tumor cells to cytotoxic drugs. While some types of tumors respond poorly to drug treatment, others, such as leukemias, are initially responsive to treatment only to become resistant to drug treatment upon relapse. The latter phenomenon is puzzling since tumor cells can acquire simultaneous resistance to structurally and functionally unrelated compounds to which they have not been previously exposed. This phenomenon of multidrug resistance or MDR affects a large group of natural products such as Vinca alkaloids and anthracyclines, two of the most active classes of anti-cancer drugs. Dramatic progress has recently been made in the understanding of the molecular basis of this phenomenon, including the identification of the proteins involved, and the genes that encode them. In addition, the biochemical and genetic dissection of the protein (P-glycoprotein) and gene (*MDR*) involved in MDR has allowed a much better understanding of the mechanism of action of P-glycoprotein, and has provided *in vitro* and *in vivo* model systems for the rational design and testing of new drugs or modulators capable of by-passing or blocking its action. In this review, we will summarize the current status of knowledge on three major aspects of P-glycoprotein mediated multidrug resistance: (1) the structure/function relationships identified in P-glycoproteins; (2) the mechanistic basis of drug transport by P-glycoproteins; (3) the physiological function and substrates of these proteins in normal tissues.

2. Characteristics of the multidrug resistance phenotype *in vitro*

Since the characterization of cellular changes associated with the onset of multidrug resistance is inherently difficult to carry out *in vivo* in clinical specimens, the elucidation of the genetic, biochemical, and pharmacological bases of this phenomenon have relied on *in vitro* tissue culture models. Multidrug resistant cell lines have been obtained by empirical protocols using step-wise selection in culture medium containing a single cytotoxic agent [1–4] that invariably lead to the low frequency appearance of drug resistant colonies. These cells are then expanded, and the exercise is repeated several times in increasing concentration of drug. A large number of independently derived, highly resistant human and rodent cell lines from different tissue origins have been obtained after exposure to unrelated cytotoxic drugs (reviewed in Refs. [5,6]). A remarkable feature of these cell lines is that their phenotypic characteristics are strikingly similar, despite the diversity of cell lines and drug selection protocols used, suggesting a common underlying mechanistic basis (reviewed in Refs. [7–9]).

These MDR cell lines all display a very similar profile of cross-resistance to structurally and functionally unrelated cytotoxic drugs. This group of drugs (known as MDR drugs) includes Vinca alkaloids (vinblastine, vincristine), anthracyclines (Adriamycin), colchicine, actinomycin D, etoposides (VP-16, VM-21), topoisomerase inhibitors (amsacrine), taxol, and many others, including small peptides such as valinomycin and gramicidin D (for examples see Refs. [10–12]). There are very few structural or functional characteristics that are common amongst MDR drugs: all are very hydrophobic molecules that enter the cell by passive diffusion across the membrane lipid bi-layer or they exert their cytotoxic effect by inserting into the lipid bi-layer itself. In general, they are biplanar, amphiphilic, and usually contain a basic nitrogen atom bearing a positive charge at neutral pH [13,14]. Much effort has been directed towards the identification of structural determinants common to MDR drugs. A comprehensive study of derivatives of the anthracyclines ellipticine and olivacine indicated that lipid solubility, a basic nitrogen atom, and at least two aromatic rings were essential requirements [15]. More detailed studies of analogs of colchicine revealed that not only the chemical composition of the molecule but also its size were important [16]. Colchicine is a plant alkaloid composed of 3 rings, two of which are aromatic, that form the phenyl-tropolone backbone of the molecule with 4 methoxy groups and an acetamido group attached to the periphery. Both the nitrogen atom of the acetamido group, and an overall size expressed as the calculated molar refractivity of less than 9.7 were found to be essential structural parameters defining an MDR analog [16]. A second characteristic of MDR cell lines is that cellular resistance is associated with a decrease in intracellular accumulation of drugs. This decrease is ouabain insensitive and therefore seems independent of an intact membrane potential but is strictly ATP-dependent, as depletion of intracellular ATP pools abrogates resistance [17,18]. This decreased accumulation is linked to a concomitant increase in an ATP and temperature dependent cellular drug efflux, suggesting that an ATP-dependent transport mechanism is at work in MDR cells [19,20]. Thirdly, although low level resistance is usually a stable phenotype, high levels of resistance are often unstable and are lost upon culture of the cells in drug-free medium. These high levels of resistance are frequently associated with karyotypic alterations characteristic of gene amplification such as the appearance of small double minute chromosomes [21], large chromosomal homogeneously staining regions [22,23] or aberrantly banding regions [24,25], suggesting that overexpression of a protein or group of proteins from amplified gene copies is required for high levels of resistance. Although variations in the expression of a number of polypeptides was found in MDR cells, e.g., small calcium binding protein [26 and refs. therein], the most ubiquitous marker of MDR is the expression of a membrane phosphoglycoprotein of molecular weight 160,000 to 210,000 designated P-glycoprotein (P-gp) [2,27]. Detailed catalogs of P-gp positive MDR cells have been published [5,6]. Subsequently, it was demonstrated that P-gp can bind ATP [28,29] and drug analogs [30–36], and has ATPase activity [37]. These findings suggest that this protein may function as a broad specificity drug efflux pump to reduce intracellular drug accumulation in MDR cells. This proposition has proven very difficult to demonstrate experimentally, and the mechanism of action of P-gp has remained a matter of considerable controversy. A final characteristic of the MDR phenotype in P-gp positive cells is that it can be blocked or reversed by a large group of structurally heterogeneous molecules which include calcium channel blockers (verapamil; [38]), calmodulin inhibitors (trifluoperazin; [39]), immunomodulators (cyclosporin A, FK506; [40]), quinidines [41], reserpines [14], piperazine analogs [42], and many others (reviewed in Ref. [8]). These compounds cause a dramatic increase in intracellular drug accumulation and a concomitant increase in cytotoxicity in P-gp positive MDR cells. Although the mechanism of action of these so-called modulators or reversal agents remains poorly understood, they all appear to compete for 'drug binding sites' on P-gp [14,33,36,43–46]. While some of these compounds appear to be themselves transported by P-gp (verapamil; [47]) others are clearly not (progesterone; [48]). Several of these compounds are under clinical evaluation for their activity in drug resistant cancers.

3. P-glycoproteins and *mdr* genes

Several convergent approaches have been used to isolate the genes that encode P-gps. The group of V. Ling produced a monoclonal antibody against Hamster P-gp (C219), which was then used to screen expression cDNA libraries constructed from a highly multidrug resistant Hamster cell line [49,50]. A second approach, used by our group and that of M.M. Gottesman relied on cloning large segments of genomic DNA amplified in common in independently derived MDR cell lines [51–53]. A 120 kb fragment of such a genomic domain was cloned, found to contain a transcription unit which coded for a mRNA whose level of expression was proportional to the degree of drug resistance in P-gp positive cells [54], leading to the ultimate isolation of full length cDNA for P-gp [55,56]. Finally, P-gp was identified as one of several cellular mRNA transcripts overexpressed in drug resistant cells which were then cloned by differential cDNA hybridization [57,58]. An important result from these experiments is that P-gp was not a single protein but rather forms part of a small gene family with three members in rodents (*mdr1, mdr2, mdr3*) and two members in human (*MDR1, MDR2*). Nucleotide and predicted amino acid sequence analysis of the prototype P-gp revealed characteristic features of an integral membrane protein [55,56,58–63]. It is composed of 1276 amino acids (mouse) grouped in two symmetrical halves, each encoding a large highly hydrophobic region (250 residues) and a hydrophilic region (300 residues). Hydropathy profiling indicate that each hydrophobic half can be further divided into 6 putative membrane associated (TM) domains, while each hydrophilic region contains a motif for ATP binding; This motif originally described by Walker [64] is formed by two consensus sequences, one highly conserved including an invariant lysine residue thought to have an interaction with the gamma phosphate of ATP (P loop), and an hydrophobic pocket believed to interact with the adenine moiety of ATP. Another feature of P-gp is the presence of a cluster of deduced N-linked glycosylation signals in the segment separating TM1 and TM2, suggesting that it may be glycosylated. Therefore, the analysis of the deduced amino acid sequence of P-gp predicts structural features

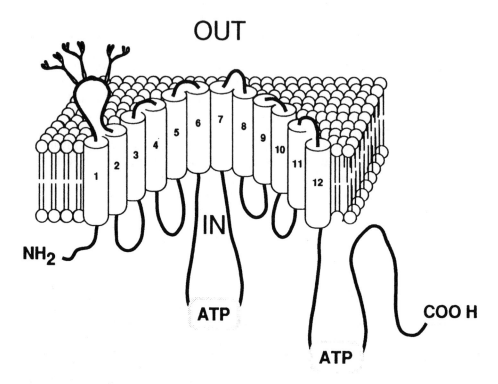

Fig. 1. Proposed membrane topology of P-glycoprotein.

which are in agreement with its previous biochemical analysis. Assuming that P-gp is an integral membrane protein with an even (12) number of TM domains, the combined positioning of: (1) the two predicted ATP binding sites on the intracellular face of the membrane, (2) the predicted N-linked glycosylation sites on the extracellular side of the membrane, together yield a topological model for P-gp shown in Fig. 1.

Alignment of the two symmetrical halves of P-gp reveal that they are 38% identical and 62% homologous, suggesting that they arose at least in part from the duplication of a common ancestor. The region of highest homology is in the predicted ATP binding sites. Homology is lower within the predicted TM domains, and disappears at the extreme amino terminus of each P-gp half [61]. A similar pattern of sequence conservation is also detected amongst members of the P-gp family from the same species. The three mouse *mdr* genes map on the same segment of approximately 500 kb on chromosome 5, and have arisen by two successive gene duplication events from a common ancestor, with the most recent one producing *mdr1* and *mdr3* [65]. The situation is somewhat different in humans, where only two genes (*MDR1* and *MDR2*) exist on the same genomic fragment on chromosome 7, presumably originating from a single gene duplication event [60, 66]. The three mouse proteins share amongst them between 73% and 83% sequence identity for a total homology varying between 85% and 92% [61]. The regions of highest sequence homology between the three proteins are once again the predicted ATP binding domains (over 95% homology), followed by the membrane associated regions (70%) and finally the amino terminus and linker regions (20%).

The formal demonstration that P-gp is indeed responsible for MDR was achieved in transfection studies. Indeed, full length human and mouse *mdr* cDNAs placed under the control of appropriate promotor/enhancer elements could induce the appearance of drug resistant cells in populations of transfected cells plated in drug containing medium [67–69]. Moreover, P-gps encoded by distinct *mdr* genes were found to be functionally distinct: although human *MDR1* and mouse *mdr1* and *mdr3* could confer MDR in this transfection assay [61,67–69], neither human *MDR2* [70] nor mouse *mdr2* could do so [71,72]. In addition, comparison of the drug resistance profiles encoded by mouse *mdr1* and *mdr3* and human *MDR1* revealed both quantitative and qualitative differences, suggesting differences in substrate specificity [73]. Finally, a comparison of the biochemical and pharmacological characteristics of multidrug resistant transfected cell clones stably overexpressing heterologous P-gps indicated that they were identical to those of P-gp positive MDR cells [29,74].

Recent experiments in animals carrying mutations at *mdr* genes have shown that P-gp encoded by *mdr3* can also act on MDR drugs *in vivo* [75]. One of the major sites of P-gp expression in normal tissues is the endothelial cells of the blood brain barrier [76,77], and in the mouse this P-gp is encoded by the *mdr3* gene [78–80]. Recently, gene targeting techniques using homologous recombination in embryonal stem cells, were utilized to introduce a germ line mutation at the *mdr3* locus in mouse [75], and the consequence of this mutation on fitness and response to drug exposure were analyzed. Although these mice were viable and fertile, they displayed a 100-fold increased susceptibility to the centrally neurotoxic peptide ivermectin, and to the MDR drug vinblastine (3-fold). In addition, the mutant mice show profound alterations in the pharmacokinetics and tissue distribution of vinblastine infused intravenously (1 mg/kg, 4 h after i.v. infusion), the most pronounced being a 22-fold increase in brain vinblastine concentration, together with a reduced rate of elimination [75]. These findings indicate that P-gp can act on MDR drugs *in vivo*, in particular within the endothelial cells of the blood brain barrier. These findings may also explain the apparent lack of response of certain brain tumors to chemotherapy, and provide an explanation for some of the side effects observed in patients treated with combinations of cytotoxic drugs and P-gp inhibitors.

4. Structural characteristics of P-glycoprotein

The predicted intracellular location of the two nucleotide binding (NB) sites of P-gp, and the extracellular location of the predicted glycosylated loop have been established by a number of epitope mapping studies using specific antibodies directed against synthetic P-gp peptides of known sequence or against fusion proteins containing P-gp sub-fragments [81,82]. The exact topology of individual TM domains of P-gp has been more difficult to solve and has been tackled by several experimental schemes: (a) *In vitro* methods in which truncated P-gps are fused to an indicator molecule, whose polarity is deduced biochemically after insertion in heterologous membranes, and (b) *In vivo* methods in which discrete tags are inserted in key locations, and whose position with respect to the membrane is deduced using antibodies or other reagents directed against the tag, after expression of the full length and functional protein in a recipient cell. These different approaches have resulted in different topological models for P-gp. Zhang and Ling have used a cell free translation system and pancreatic microsomes to map the topology of P-gp where the first three, four, five or six TM domains were fused to a reporter N-linked glycosylation signal and an epitope tag. They observed two orientations for the N-terminal half of P-gp. One is that predicted by hydropathy, while the other has the predicted TM3 as an intracellular segment and TM5 as extracellular. A similar analysis of the topology of the C-terminal half by this group suggested that predicted TM8 would be extracellular while predicted TM10 would be intracellular [83–85]. Independently, Skach and Lingappa have expressed in *Xenopus* oocytes chimeric P-gps in which the first seven, eight or ten TMs were fused to the C-terminal portion of bovine prolactin as an epitope tag. They found the loop between TM8 and TM9 (predicted to be intracytoplasmic) to be extracellular. Similar results were obtained using 'half molecule' C-terminal constructs [86,87]. Beja and Bibi [88] investigated P-gp topology using alkaline phosphatase (PhoA) fusions expressed in *E. coli* cells. Alkaline phosphatase is active only when translocated to the periplasmic space, and fusion to an intracellular domain gives low PhoA activity while fusion to an extracellular domain gives high activity. In a study of the N-terminal half of P-gp, they found predicted TM4 and part of the predicted intracellular loop which follows TM4 to be extracellular. In a subsequent study of the TM7–TM8 segment [89], they observed that the loop between TM7 and TM8 (predicted extracellular) would be partly in the membrane and partly cytoplasmic, implicating (1) the presence of an additional TM domain between TM7 and TM8, (2) an inside-to-outside orientation for TM8 as opposed to outside-to-inside predicted by the hydropathy plot, and (3) a possible extracytoplasmic location for the C-terminal NB site. Therefore, results from studies of truncated P-gps in different experimental systems have produced inconsistent and sometimes even contradictory topological models. Loo and Clarke studied the topology of human MDR1 P-gp as a full length, functional cystein-less P-gp [90], in which single cystein residues were inserted in key locations according to the original hydropathy plot. The SH group of the cystein serves as a probe for domain accessibility to either membrane permeant or impermeant sulfhydryl-modifying reagents. They showed that the loops between TM5-6, TM7-8 and TM11-12 were extracellular, whereas the loops between TM2-3, TM4-5, TM8-9, TM10-11 and the C-terminus end were cytoplasmic, a topology identical to that proposed by hydropathy. Finally, Kast et al. [91] introduced a hemagglutinin (HA) epitope tag into six different locations in the N-terminal half of P-gp, followed by expression of the functional proteins in transfected cells. Intracellular or extracellular localization of the HA tag was determined in intact or permeabilized cells by immunofluorescence and confocal microscopy, using an anti-HA monoclonal antibody. While tags inserted in the

TM1-2 and TM5-6 intervals were accessible to the anti-HA antibody in intact cells, tags inserted in the TM2-3 and TM6-7 intervals were only accessible in permeabilized cells indicative of a intracytoplasmic localization. More recently, the same group [206] demonstrated that the TM7-8, TM9-10, and TM11-12 are extracellular, while the TM4-5, TM10-11 and the TM12 C-terminal segments were intracellular. The topology model based on these results is also in agreement with the original model deduced by hydropathy. Therefore, the two experimental systems which maintained and monitored the functional integrity of the protein gave results that agree with each other and that fit with the topology predicted from the hydropathy plot.

5. The ABC super family of membrane transport proteins

P-gp belongs to a large super family of structurally related proteins that together form the ABC (ATP Binding Cassette) family of membrane traffic ATPases [92] that spans both the prokaryotic and eukaryotic kingdoms. In Gram-negative bacteria such as *Escherichia coli* and *Salmonella typhimurium*, these transporters carry out high affinity import or export of various types of substrates [93]. They are composed of several subunits including a soluble receptor expressed at high level in the periplasmic space, and a membrane associated complex consisting of two hydrophobic membrane anchors and two peripheral or associated ATP binding units. A large number of these transporters have been identified and transport in an ATP dependent fashion such varied substrates as sugars (*MalK*, [94]), amino acids (*HisP*, [95]), short peptides (*OppD*, [96]), ions (*PstB*, [97]) and several others [92]. P-gps share a more significant homology to another group of bacterial transporters which include the *ChvA* gene of *Agrobacterium tumefaciens* [98] and the *HlyB* gene of *E. coli* [99], which are involved in the secretion/export of β-1,2 glycan and the hemolytic enzyme hemolysin A, respectively. Sequence homology is particularly strong (35% identity) with *HlyB* and overlaps a region that includes the nucleotide binding sites but also the transmembrane region [59]. P-gp homologs have also been identified in lower and higher eukaryotes. These proteins share with P-gp approximately 60% overall homology, but display near identical hydropathy profiles resulting in very similar predicted secondary structure. These include the *pfmdr1* gene of *Plasmodium falciparum* [100] in which mutations have been associated with the onset of chloroquine resistance in the malarial parasite [101]. Members of this family have also been described in other eukaryotic parasites such as *Entamoeba histolytica* (*ehpgp*, emetine resistance; [102]), *Leishmania tarantole* (*ltpgp*, arsenate resistance; [103]), *Leishmania donovani* (*ldmdr1*, multidrug resistance; [104]) and others. Of particular interest is the yeast homolog of P-gp, STE-6, which is responsible for the trans-membrane transport of the farnesylated dodecapepetide mating pheromone **a** factor [105,106]. The structural homology between P-gp and STE-6 also translates into functional homology as the mouse *mdr3* gene can complement a null allele at the *STE-6* locus, restoring mating in a sterile *ste6* mutant [107]. Likewise, it was recently shown that the *pfmdr1* gene can also complement a null mutation at *STE-6* [108]. *mdr*-like genes of yet unknown function have also been identified in the fly (*Drosophila melanogaster*, *Mdr49* and

Mdr65) [109], worm (*Caenorhabditis elegans, pgp-1-4*) [110] and plant (*Arabidopsis thaliana*, atpgpl) [111].

In humans, the ABC family includes several transporters in which alterations cause significant pathologies. Cystic fibrosis, is caused by alterations in Cl⁻ transport across the membrane of epithelial cells, due to mutations in the Cystic Fibrosis Transmembrane Conductance Regulator gene (*CFTR*) [112]. CFTR is a member of the ABC family, and functions as a cyclic AMP responsive outwardly rectifying chloride channel [113]. The deletion of a single phenylalanine (ΔF508) accounts for nearly 70% of all cases of CF; this mutation was shown to be a temperature sensitive mutation resulting in failure of CFTR to mature and be properly targeted to the plasma membrane [114–116]. The biochemical analysis of other CFTR mutations indicates that mutations in the membrane region have a milder phenotype than those in the nucleotide binding sites, and are sometimes associated with altered ion selectivity of the channel [117]. Two additional ABC transporters, designated *TAP-1* and *TAP-2*, have been mapped within the Class II region of the major histocompatibility complex (MHC). These genes encode two proteins half the size of P-gp (1 NB site, 6 TM domains), and deletion or altered expression of these genes in T-lymphocytes abrogate their capacity to present antigens in association with class I MHC molecules. The TAP-1 and TAP-2 proteins function as a heterodimer unit to actively transport antigenic peptides across the membrane of the reticuloendothelium for association with class I molecules and surface antigen presentation [118–120]. Two other ABC transporters have been identified in the peroxisomal membrane, ALDP [121] and PMP70 [122]. Deletion and point mutations in ALDP segregate in families with the X-linked disorders adrenoleukodystrophy and adrenomyeloneuropathy [121,123–125], whereas PMP70 mutations are linked to Zellweger Syndrome [126], an autosomal inborn error of peroxisome biogenesis. ALDP and PMP70 are each a 'half transporter' size with one hydrophobic and one hydrophilic domain, sharing 38.5% sequence identity between them, and are postulated to work as a heterodimer in the peroxisomal membrane. Finally, the gene encoding pancreatic beta cells high affinity sulfonylurea binding protein (sulfonylurea receptor, SUR) was recently cloned and found to be an ABC transporter (127). SUR has been proposed to regulate a K⁺ channel, and mutations in SUR are associated with persistent hyperinsulinemic hypoglycemia of infancy [128]. In addition, another member of the ABC family was recently cloned by differential cDNA cloning from a P-gp-negative, doxorubicin resistant isolate of a small cell lung carcinoma cell line. This gene and protein were designated MRP for Multidrug Resistance associated Protein. Subsequent transfection experiments showed that overexpression of this gene was indeed capable of conferring drug resistance. However, the primary amino acid sequence of MRP, its predicted secondary structure and the encoded MDR phenotype are quite different from those of P-gp [129–131].

6. Interaction of P-glycoprotein with drug molecules

As mentioned above, the extraordinary structural diversity of P-gp substrates is one of the most puzzling aspects of P-gp mediated drug resistance, and much effort has been devoted to the identification of the protein domains associated with drug binding and recognition. This has been achieved by classical biochemical studies of photolabeled P-gp peptides and by the genetic analyses of P-gp mutants showing altered substrate specificity.

One of the few characteristics common to MDR drugs is their high degree of hydrophobicity. This intuitively suggested that MDR drugs may be recognized by P-gp within the lipid bi-layer, and that the predicted TM domains of the protein play an important role in initial interaction with the various substrates. Elegant studies by Raviv et al. [132] suggested that this was indeed the case. 5-[^{125}I]iodonaphthalene-1-azide (INA) is a membrane specific photoactive probe which labels P-gp in intact MDR cells upon excitation with ultraviolet but not with visible light. Conversely, the fluorescent drug daunomycin is excited with visible light but emits UV light. When both daunomycin and INA were incubated with intact P-gp positive cells, and the cells irradiated with visible light, P-gp was labeled by INA by energy transfer from daunomycin suggesting close proximity of INA, daunomycin and P-gp in the plasma membrane. Photoactive radiolabeled analogs of drug molecules and P-gp modulators have been synthesized (reviewed in Ref. [34]), and these have been used in epitope mapping studies to identify P-gp peptides or discrete amino acids potentially forming or lining a drug binding site. The most frequently used photoactive analogs of MDR drugs or P-gp modulators are [^{3}H]-NABV (vindesine/vinblastine), and the P-gp modulators [^{3}H]-azidopine, a calcium channel blocker, and [^{125}I]-iodoarylazidoprazosin, an α1-adrenergic receptor ligand [34]. Several groups have independently identified a major and a minor drug photoaffinity labeling site on P-gp: both sites are located within the membrane associated portion of the protein. One maps within the amino terminal half (minor site, 25% of label) and another at the homologous position in the carboxy terminal half (major site, 75% label) [81,82,133]. Further epitope mapping studies with a series of anti-peptide antibodies demonstrated that prazosin and azidopine bound to the same sites on P-gp, and identified a 5 and a 4-kDa proteolytic fragments mapping immediately downstream of the last membrane domains of each half of P-gp [134]. These results suggest that both P-gp halves contribute to the formation of a single drug binding site, and that the P-gp segments located near the C-terminal portion of the TM region of each half are in close proximity to the substrate molecule. The notion that the two halves of P-gp must cooperate in the formation of a single transport site is supported by independent observations with other ABC transporters: (1) the TAP1 and TAP2 molecules (half size of P-gp) must be expressed in the same cell to produce peptide transport, and (2) independent expression of each half of STE-6 in yeast cells does not yield a functional transporter, while simultaneous expression of both halves does [135].

The importance of the membrane associated portion of P-gp in substrate recognition and binding is also supported by the study of a large number of P-gp mutants with altered substrate specificity. Mouse Mdr1 and Mdr2 are functionally distinct as the former can confer drug resistance while the latter cannot. In a series of Mdr1/Mdr2 chimeras where reciprocal exchange of homologous domains had been carried out [71], it was demonstrated that either of the two NB sites of Mdr2 can be substituted into Mdr1 without loss of drug resistance activity. However, the replacement of as few as two predicted TM domains of Mdr1 by those of Mdr2 resulted in a non-functional protein, suggesting that while the NB sites encode a common mechanistic aspect of transport, the TM regions and their associated loops may encode the functional difference between the two proteins, including possible differences in substrate specificity. Site-directed mutagenesis of individual TM segments has provided direct evidence that these regions are responsible for substrate specificity. A single serine to phenylalanine substitution in predicted TM11 of P-gps encoded by mouse *mdr1* (Ser 941) and *mdr3* (Ser 939) was shown to have a drastic effect on the overall activity, but more importantly on the substrate specificity of the two P-gps [136].This mutation was shown to cause reduced drug binding to the mutant proteins and also to alter sensitivity to the modulatory effect of verapamil and progesterone [137]. Additional substitutions at that site [138] were shown to have opposite effects on substrate specificity, with the same mutation decreasing resistance to certain drugs (ADM, COL) while increasing resistance to another (Actinomycin D). Thus, residue 939/941 in TM11 was shown to be an important determinant for P-gp interaction not only with drug substrates but also with modulators of its activity. Some of these interactions appear to be common to the two drug-resistant isoforms, while others seem to involve additional, non-overlapping determinants in Mdr1 and Mdr3 [42]. Of all the TM domains of P-gp, TM11 is the one that shows the strongest amphipathic character when projected in a α-helical configuration: segregating a hydrophilic and a hydrophobic face of the helix. To analyze possible functional implications of this structural arrangement, TM 11 was systematically mutagenized by alanine scanning [207]. While mutations at 8 residues had no effect on Mdr3 function, mutations at 11 others affected function, with two of the alanine substitutions (Y949A, F953A) causing a dramatic loss of function. The distribution of mutation-sensitive and insensitive residues was not uniform on TM11: while seven of the mutation-insensitive residues mapped to the hydrophobic face of the TM11 helix, a large proportion of the mutation-sensitive ones including Y949 and F953 mapped on the other, less hydrophobic, face of the TM11 helix. These results suggest that the amphipathic character of TM11 is important for substrate interaction or in the formation of a more complex substrate binding site. It is interesting to note that mutations in TM11 of another ABC transporter, the Pfmdr1 protein of *P. falciparum* are associated with the onset of chloroquine resistance in the malarial parasite [100,101].

In support of an important role in substrate binding for the C-terminal P-gp segment is the observation that replacement of the TM11–TM12 interval (extracellular loop) of MDR1 by that of MDR2 increases resistance to actinomycin-D but not to other drugs, while replacement of TM12 of MDR1 by that of MDR2 impaired resistance to actinomycin-D, doxorubicin and vincristine, but not to colchicine [139]. Selection for increasingly high levels of actinomycin D resistance in P-gp positive cells was associated with a Gly^{338}–Ala^{339} to Ala^{338}–Pro^{339} replacement within TM6 of the hamster Pgp1 protein [140]. Also, an alanine scanning mutagenesis of TM6 of human MDR1 identified Val338Ala, Gly341Val, Ala342Leu as mutations affecting drug resistance profiles, while independent mutations at

Ser[344] caused a complete loss of function [141]. These results are in agreement with the previously reviewed biochemical studies on the mapping of P-gp drug photolabeling sites and suggest that TM6 may play an important role in drug–protein interactions [134,142]. Finally, systematic mutagenesis of all proline and all phenylalanine residues found in the TM domains identified mutations in TM4 (Pro[223]), TM6 (Phe[335]), TM10 (Pro[866]), and TM12 (Phe[978]) that alter substrate specificity or overall activity of the protein [143,144]. Interestingly, the pairs of affected residues in TM4/10 and TM6/12 mapped at the exact homologous position in each half of P-gp.

Mutagenesis experiments have also suggested an important role for the predicted intracellular and extracellular loops of P-gp either for direct interaction with substrates or in signal transduction to or from TM domains to the NB sites. Several mutations in these regions that affect substrate specificity have been identified. A Gly185Val substitution in the predicted TM2–TM3 intracellular loop was found to cause increased colchicine resistance but decreased vinblastine resistance [145]. Labeling experiments using photoactivatable analogs of these two drugs showed decreased colchicine binding and increased vinblastine binding to the mutant when compared to the wild type protein [146]. A two step mechanism of drug transport was proposed by these authors based on two distinct binding sites, one involved in initial drug binding ('on' rate), and another involved in drug release ('off' rate). Systematic mutagenesis of twenty glycines in cytoplasmic loops of human MDR1, identified Gly[141] and Gly[187] (TM2-3 intracellular loop), Gly[288] (TM4-5 intracellular loop); and Gly[812], Gly[830] (TM8-9 intracellular loop), as residues where mutations alter the drug resistance profile [147]. Finally, the introduction of the TM11-TM12 extracellular loop of MDR2 into the corresponding region of MDR1 causes an increase in resistance to actinomycin-D, colchicine and doxorubicin, but not to vincristine and is associated with a reduction in photolabeling by [^{125}I]iodoaryl azidoprazosin [139].

7. Interaction of P-glycoprotein with ATP and ATPase activity

The reduced drug uptake and increased drug release characteristic of P-gp expressing cells are strictly ATP dependent, cannot be sustained by non-hydrolyzable ATP analogs and can be blocked by poisons of mitochondrial respiration such as sodium azide and rotenone. These characteristics initially suggested that P-gp may bind ATP and have ATPase activity. Predicted amino acid sequence of P-gp identified two consensus sites for ATP binding, originally described by J.E. Walker in a series of ATP binding proteins and ATPases, and composed of two consensus sequence motifs [64]. The A motif of sequence G-X-X-G-X-G-K-T(S) is believed to form a flexible loop (P loop) with the lysine residue interacting with the terminal phosphate group of ATP and playing a key role in ATP hydrolysis. The B motif of sequence R/K-(X)3-G-(X)3-L-(Hyd)4-(D) is believed to form a hydrophobic pocket possibly forming hydrophobic interactions with the adenine moiety of ATP. Since P-gp can be photolabeled by 8-azido ATP, it has been proposed that the two identified consensus sites play a key role in ATP binding and possibly ATP hydrolysis by this protein. Some of the unresolved questions about the function of these sites are: (1) are both sites catalytically equivalent, and is ATP binding and hydrolysis at both sites required for drug transport; (2) what is the nature and the amino acid residues involved in signaling from and to the TM regions (drug binding) and the predicted NB sites (ATP binding/hydrolysis) to effect transport.

In an early study, the highly conserved lysine and glycine residues of the P-loop (Walker A motif) were mutated in each or both of the NB sites to arginine and alanine, respectively [148]. All mutations abolished Mdr1 drug resistance activity, indicating that both NB sites need to be intact for P-gp function. Since all mutants retained the ability to bind 8-azido ATP, it was proposed that a step subsequent to ATP binding (possibly ATP hydrolysis) was impaired in these mutants. Similar results were obtained in parallel mutagenesis of the same residues of the yeast STE6 protein [135]. Additional studies performed in Mdr1/Mdr2 chimeric molecules showed that both NB sites of Mdr2 could be vertically exchanged (N → N, C → C) in Mdr1 without loss of function, suggesting that the enzymatic activity of these segments is highly conserved amongst functionally distinct P-gps [71]. However, similar experiments showed that the two NB sites could not be horizontally exchanged (C → N), as duplication of the C-terminal NB site (NBD2) into the N-terminal half (NBD1) of P-gp resulted in a non-functional protein [149]. The study of additional chimeric molecules identified two small segments near the Walker B motif which caused a dramatic reduction in resistance to Adriamycin, colchicine and actinomycin-D, but not vinblastine. Site directed mutagenesis identified the ERGA (NBD1) to DKGT (NBD2) substitution upstream of the Walker B, and the threonine 578 (NBD1) to cystein (NBD2) substitution downstream of the Walker B motif as responsible for the effect. The fact that certain mutations in NB sites can alter the drug resistance profile, while mutations at other positions affect overall activity, suggest that the former residues may be involved in transducing a signal (such as drug binding to the TM domains) from the TM domains to the NB sites to possibly effect the catalytic activity of the NB sites.

The ATPase activity of P-gp has been intensively studied (reviewed in Refs. [150,151]). Early studies detected ATPase activity of P-gp purified by immunoprecipitation [37] in the range of 50 nmol of ATP hydrolyzed/min/mg protein. Recent progress in P-gp purification made by several groups together with studies in P-gp enriched plasma membranes, and *in vitro* reconstitutions in liposomes have allowed a detailed analysis of the characteristics of P-gp ATPase activity [150,151]. P-gp has been purified to homogeneity using Zwittergent 3-12 extraction, anion exchange and immunoaffinity chromatography [152], or octylglucoside solubilization followed by chromatography on Reactive Red 120 agarose [153], or CHAPS detergent extraction followed by lentil lectin affinity chromatography [154]. The P-gp ATPase activity has a Vmax of 0.32 to 2.1 μmol/min/mg, and a K_m for MgATP of 0.4 to 0.94 mM with a preference for ATP over other nucleotides. The question of the catalytic activity of either or both predicted NB sites has been addressed by a number of studies, but a clear picture has yet to emerge. The observations that (1) either half of P-gp expressed independently in transfected cells has intrinsic ATPase activity [155], and (2) inhibition of P-gp ATPase activity *in vitro* by either

8-azido ATP or N-ethyl maleimide [156,157] gives a stoichiometry of 2 molecules of inhibitors/molecule of P-gp together suggest that both NB sites are functionally equivalent and that ATP hydrolysis at both sites is required for P-gp function. Alternatively, more complex cooperative interactions between the two sites, with ATP binding and/or hydrolysis at one or both sites, are supported by the observations that (1) mutations within either one of the NB site abrogate P-gp mediated drug resistance [148], (2) inhibition of P-gp ATPase activity by NBD-Cl gives a stoichiometry of 1 molecule inhibitor/P-gp [158], (3) orthovanadate-induced inhibition of P-gp ATPase shows trapping of 1 ADP molecule/P-gp [159], with equal labeling of the two predicted NB sites [208]. A unique feature of the ATPase activity of P-gp is that it can be further modulated by both substrates and inhibitors (reviewed in Refs. [150,151]), and a correlation between stimulation of ATPase activity and transport has been observed in certain substrate series [160]. On the other hand, while certain compounds such as verapamil and vinblastine stimulate ATPase, others like cyclosporin A do not stimulate the ATPase activity but can nevertheless inhibit stimulation by verapamil [161]. Finally, other compounds completely inhibit P-gp ATPase activity (quoted in Ref. [151]). These observations have led to the concept that substrate binding to the TM domains of P-gp stimulates ATP hydrolysis by the NB sites. While certain modulators that are transported by P-gp would mimic substrates and stimulate ATPase activity, others that are not transported would simply bind to TM domains and would uncouple the stimulatory signal to the NB sites. This signal transduction hypothesis is supported by the observations that (1) a mutation near TM3 of human MDR1 (G185V) which affects substrate specificity [145] by modulating drug binding [146], also alters the pattern of drug-stimulatable ATPase activity of the mutant protein when compared to wild type [162]; and (2) certain mutations in the NB sites affect the substrate specificity and drug resistance profile of the protein [149,163].

8. Mechanism of transport

The precise mechanism by which P-gp transports an extraordinary large number of structurally unrelated substrates has been the subject of much debate and controversy over the past five years. Several seemingly incompatible mechanisms of action have been proposed, and convincing experimental data supporting each of these models has been obtained. The models can be grouped into two general types of mechanisms: (1) the intuitively appealing and generally favored model is a direct transport mechanism in which P-gp would recognize drugs within the lipid bilayer and would either transport them outside the cell or would create an intramembranous gradient by transporting drugs from the inner to the outer leaflet of the bilayer, both resulting in net overall decreased accumulation and increased release from the cells; (2) a second and opposing view suggests an indirect mechanism of action in which P-gp would act to modify the intracellular milieu or membrane electrochemical gradient ($\Delta\psi$, ΔpH), thereby indirectly affecting either the partitioning of the drugs on either side of cellular membranes (cell membrane or sub-cellular vesicular compartments) or altering binding of weak bases (such as MDR drugs) to their cellular targets. Several mechanisms of action have been proposed for P-gp to explain such an indirect mechanism of action.

In support of a direct drug transport mechanism are the observations that (1) P-gp can bind photoactivable drug analogs, suggesting direct contact between the substrate and protein [34], (2) single amino acid substitutions can alter the substrate specificity of P-gp [62,136,143–145) by modulating drug transport and drug binding to the protein [137,146], (3) the recently identified normal physiological substrate of Mdr2 and the unique mechanism of trans-membrane transport of this substrate [164,165], (4) P-gp shares homology with a number of prokaryotic and eukaryotic membrane proteins implicated in the ATP-dependent transport of various types of substrates across the membrane (see above); (5) Transport studies in intact cells and plasma membrane vesicles from P-gp expressing cells indeed suggest that P-gp mediates increased ATP-dependent drug binding and/or transport into these vesicles [166–168]. On the other hand, the identification of additional functions associated with P-gp expression in certain MDR cell lines have fueled the argument for an indirect mechanism of action. For example, an increased ATP release from P-gp positive MDR cells or *mdr* transfectants has been noted, prompting the authors to suggest that P-gp may be an ATP channel [169]. This channel-like activity could be blocked by an anti-P-gp antibody and abolished by mutations in the NB domain previously shown to impair drug resistance [169]. It was proposed that the ATP electrochemical gradient, rather than ATP hydrolysis, would provide the driving force for drug efflux. The same group has also proposed that another ABC transporter, the human chloride channel CFTR protein would also function as an ATP channel [170]. Another hypothesis stems from the observation that P-gp positive cells have an elevated intracellular pH (pHi) [171,172]. Roepe et al. later noted that the relative steady state of drug efflux showed a linear relationship to the pHi, but not to the amount of P-gp expressed in a panel of MDR cell lines. A model was proposed in which P-gp may act directly or indirectly to modulate the pHi of the cell resulting in reduced sequestration or retention of drugs in the resistant cells [173–175]. While these models provide an attractive explanation for the very broad substrate specificity of P-gp, they are difficult to reconcile with other biochemical and genetic data reviewed above. One additional controversial proposed function for P-gp is that, in addition to transporting drugs, it may also be a volume activated Cl⁻ channel, or a regulator of such a channel (reviewed in Refs. [180,181]). While some reports find an ATP-dependent, chloride-selective channel which is inhibited by P-gp substrates, modulators and anti-Pgp-antibody, others find no correlation between P-gp expression and the appearance of volume-activated chloride currents [180,181 and refs. therein]. Finally, it must be stressed that these additional properties associated with P-gp in MDR cell lines have been in some instances difficult to reproduce or, at least, seem to be specific to some but not all P-gp positive MDR cells [176–179].

We have chosen to express the three mouse P-gp isoforms in a yeast heterologous expression system in which the activity and mechanism of action of the three proteins could be easily monitored and manipulated by biochemical and genetic means. Several years ago, we showed that expression of mouse Mdr3 could partially complement a null mutant at the *ste6* gene and could restore **a** factor transport and mating in this otherwise sterile mutant, suggesting that the yeast would be a proper host to study P-gp function [107]. Mdr3 expressed in the membrane of these cells was shown to be stable and to bind photoactive drug analogs, and mutations known to affect drug binding in mammalian cells were also shown to have the same effect on Mdr3 expressed in yeast. Mdr3 expression was shown to reduce in an ATP-dependent, and verapamil-sensitive fashion, the accumulation of [³H]-vinblastine and [³H]-FK506 in spheroplasts prepared from these cells. Finally, Mdr3 expression conferred cellular resistance to the anti-fungal macrolide peptides FK506 and FK520 in yeast cells [182]. The effect of P-gp expression on drug transport was initially studied in inside-out-vesicles prepared from yeast plasma membranes. In these studies, drug transport into these vesicles was shown to be verapamil, and vanadate sensitive but was insensitive to the presence of the protonophore CCCP [183]. Further studies on the mechanism of P-gp action were carried out using P-gp inserted in the membrane of secretory vesicles (SVs) purified from the *sec6-4* mutant [184]. This mutant is defective in the last step of the secretory pathway (plasma membrane fusion), and therefore accumulates at the non-permissive temperature large amounts of unfused secretory vesicles containing newly synthesized membrane proteins transiting to the plasma membrane. These SV have major advantages over traditionally prepared membrane vesicles since (1) they can be easily isolated in large amounts; (2) they are tightly sealed, uniform in size and, more importantly, of the same polarity; and (3) they are fully functional organelles in which the endogenous proton translocating PMA1 ATPase maintains a strong electrochemical gradient [185] which can be manipulated to study transport parameters [183,184]. The three mouse P-gps can be expressed in the membrane of these vesicles and expression of Mdr1 and Mdr3 but not Mdr2 in these SVs causes a significant accumulation of VBL in these SVs (inside-out orientation) that can be completely abolished by verapamil. The role of ΔpH and $\Delta\psi$ in drug transport by P-gp was evaluated in these SVs where these parameters can be easily manipulated; First, it was observed that P-gp can concentrate the membrane permeant lipophilic cation TPP$^+$ inside these vesicles, even in the presence of a strong ΔpH (interior positive); Second, it was observed that P-gp can concentrate drugs into such vesicles, even when the total electrochemical gradient ($\Delta,\tilde{u}_{H^+}$) is dissipated by incubation of the vesicles with nigericin in a buffer system containing the membrane permeant anion nitrate; Finally, elimination of the strong electrochemical potential generated across the SV membrane by the yeast endogenous PMA1 H+/ATPase (in a PMA1 temperature sensitive mutant) was found to be without effect on the Mdr3-mediated drug accumulation [183,184]. Hence drug transport by P-gp is neither dependent nor coupled to proton movements, and is independent of an intact electrochemical potential. In addition colchicine accumulation in Mdr3 SVs occurred against a substrate concentration gradient, reaching an intravesicular colchicine concentration of seven times that of the medium concentration (184). Thus, P-gp can transport drugs against the electrical membrane potential and up a substrate concentration gradient, in a manner which is independent of proton movements and therefore excludes a proton-symport or antiport mechanism for

P-gp-mediated drug transport. Finally, P-gp has been purified to homogeneity and reconstituted in liposomes where its mechanism of action can be studied in complete isolation [186,187], and where the possible effect of either membrane potential or pH gradient can be unequivocally established. Two such studies have been recently published. In one case, vectorial colchicine transport by P-gp was clearly demonstrated in reconstituted proteoliposomes [186]. In another study, P-gp was purified and reconstituted in proteoliposomes and P-gp function was monitored using the fluorescent dye Hoechst 33342 as substrate [187]. These authors were able to demonstrate substrate transport into P-gp liposomes which was inhibited by NEM (50 μM), vanadate (50 μM), and was ATP-dependent and could not be sustained by a non-hydrolyzable ATP analog (AMP-PNP). Measurements of intraliposomal pH during transport reaction detected no large change in pH [187]. Therefore, the most recent analyses of P-gp function in either heterologous systems or using purified protein reconstituted in liposomes clearly demonstrate a direct transport mechanism of action for P-gp.

9. Distribution and possible function of P-gp in normal tissues

The large body of data on the functional role of P-gp in multidrug resistance, together with the identification of its cellular and subcellular sites of expression in normal tissues, have led to several hypotheses regarding its function in normal tissues. Although seemingly very distinct, these proposed functions may prove not to be mutually exclusive. The first one is modeled on the activity of P-gp in causing MDR, and favors a detoxifying role for this protein of toxins normally present in the environment. The second, is that MDR drugs are actually fraudulent substrates for P-gp, and that the protein acts on yet to be defined normal cellular products. This is an important aspect of P-gp biology, and again, evidence for and against each of these models has been obtained. In addition, understanding the physiological role of P-gp is more than a strictly academic question, as inhibitors of P-gp function used to modulate drug resistance in P-gp positive tumors *in vivo* will undoubtedly have deleterious effects on the normal physiological events controlled by P-gp.

Expression of the various P-gp isoforms and their cellular mRNAs has been shown to be tightly regulated in an organ and cell specific fashion. The highest levels of human *MDR1* protein and mRNA expression are found in the adrenal gland, kidney, jejunum, colon, and endothelial cells of the blood-brain barrier, whereas human *MDR2* is strongly expressed in liver. In normal mouse tissues, *mdr1* is expressed in the pregnant uterus, placenta, adrenal glands, and kidney, while *mdr2* is expressed primarily in liver but also in muscle and *mdr3* is found in lung, intestine and in brain [66,76–79,172,188,189]. Immunohistochemical analysis with specific anti-P-gp antibodies reveal that P-gps are expressed in a polarized manner on the apical membrane of secretory epithelial cells lining luminal spaces such as the glandular epithelial cells of the endometrium in the pregnant uterus, the biliary canaliculi of hepatocytes, the brush border of renal proximal tubules, pancreatic ductules, columnar epithelium of the intestine, and in endothelial cells of the blood-brain barrier and in the testes [76,77,188,190–192]. Expression of P-gp at the surface of epithelial cells lining the luminal space of the intestine, kidney and liver suggests that P-gp may play a protective role at these sites. Additional experimental evidence in support of this proposal include (1) the altered drug distribution and pharmacokinetics displayed by mutant mice bearing a null allele at *mdr3* [75], (2) endothelial cells from the blood brain barrier can carry out unidirectional drug transport *in vitro* [193], (3) the cardiac glycoside digoxin which is eliminated by glomerular filtration and tubular excretion is a P-gp substrate [194], and (4) pluripotent stem cells of the hemopoietic system can transport the P-gp substrate: fluorescent dye Rhodamine 123 [195]. On the other hand, expression of P-gp at other sites such as the pancreatic ductules, the endometrial glands of the pregnant uterus and the adrenal cortex together with the observation that neither mouse nor human *MDR2* can confer drug resistance in transfection experiments, suggest that P-gp may act on different types of physiological substrates at those sites. It has been proposed that in the uterus and the adrenal gland, P-gp may play a role in hormone transport; recent findings that P-gp can transport steroid hormones such as cortisol, aldosterone, dexamethasone [48] and corticosterone [196] would tend to support this notion. Moreover, since some of the MDR drugs recognized by P-gp have fairly large peptide backbones (actinomycin D, valinomycin, gramicidins), together with the observation that P-gp can transport small peptides [197], including the yeast mating pheromone **a** factor [107] raise the possibility that it may act on small or large physiological peptides. The structural homology detected between P-gp and the CFTR chloride channel suggested the possibility that P-gp may also have ion conducting properties. The group of Higgins (reviewed in Refs. [180,181]) observed that expression of the *MDR1* gene in NIH-3T3 cells resulted in the appearance of small chloride currents that could be inhibited by P-gp modulators verapamil and forskoline. This P-gp associated chloride conductance was found to be regulated by cell volume. Interestingly, the Cl$^-$ channel activity of P-gp was unaffected by mutations in the nucleotide binding sites known to abolish drug transport. These authors proposed that P-gp may be a bi-functional protein capable of acting on both drug molecules and ions in normal tissues and drug resistant cells. A remarkable pattern of complementarity between the expression profiles of CFTR and P-gp was noted in certain normal epithelia [198]. This complementarity was particularly striking in the ileum, where P-gp was expressed in the epithelial cells of the villus, whereas CFTR expression was restricted to the crypt epithelia. A similar complementarity was noticed in the luminal and glandular epithelia of the uterus during pregnancy.

10. Function of P-gp in liver

Interestingly, it is the study of the enigmatic Mdr2 class of P-gp (which does not confer multidrug resistance) that recently provided the most direct clue as to the natural substrate and physiological role of P-gp in normal tissues. Immunological studies with isoform specific antibodies have shown that mouse Mdr2 and human MDR2 expression is largely limited to liver, where it is found only at the canalicular membrane and not the sinusoidal membrane of hepatocytes, and epithelial cells lining the lumen of the bile canaliculi and biliary ductules (see above). It has been proposed that P-gp may be involved at that site in the trans-membrane transport of a bile constituent [191]. Smit et al. [165] have used homologous recombination in embryonal stem cells to create a mouse bearing a null mutation ($mdr2^{-/-}$) at the $mdr2$ locus. Histological examination of this mutant revealed a severe liver pathology that included hepatocyte damage, strong inflammatory response, and more strikingly, destruction of the bile canaliculi. The most obvious biochemical consequence of the mutation was found to be at the level of lipid content of the bile; while heterozygous mice for the mutation ($mdr2^{+/-}$) showed a 50% decrease in the amount of phosphatidylcholine (PC) in the bile, PC was completely absent from the bile of animals homozygous for the mutation ($mdr2^{-/-}$). This observation led the authors to propose that Mdr2 may participate in the translocation of PC across the canalicular membrane into the bile. This possibility was tested directly in yeast SVs stably expressing mouse Mdr2: for this, a fluorescent lipid tag was inserted into the membrane of these vesicles and the asymmetric distribution of the fluorescent lipid tag was monitored over time in the presence or absence of Mdr2 [164]. Mdr2 expression in secretory vesicles caused a time and temperature dependent enhancement of PC translocation from the outer leaflet to the inner leaflet of the membrane of these vesicles. The Mdr2-mediated translocation was specific since expression of Mdr3 in these vesicles was without effect on the membrane distribution of PC. The increased Mdr2-mediated PC translocation was strictly ATP and Mg^{2+} dependent, was abrogated by the ATPase inhibitor vanadate and the P-gp modulator verapamil, but was insensitive to the presence of excess MDR drugs colchicine and vinblastine. These experiments demonstrated that Mdr2 is a lipid transporter and that it functions as a lipid flippase or translocase to move lipids from one leaflet of the membrane to the other [164].

The complexing of bile salts by PC and cholesterol into mixed micelles is thought to greatly reduce the otherwise cytotoxic detergent effect of bile salts traveling within the biliary tree [199], a protection lacking in $mdr2^{-/-}$ mutant mice [165]. However, the observation that the rate of secretion of bile salts and biliary lipids are intimately related has suggested that bile salts may also play a direct positive regulatory role in the trans-hepatic movement of biliary lipids [200]. The mechanism responsible for such an effect of bile salts remains unclear, but has alternatively been proposed to involve direct solubilization of PC from the canalicular membrane [200,201], modulation of either phospholipid biosynthesis [202] or of intracellular trafficking of PC-rich vesicles from the Endoplasmatic Reticulum or the Golgi apparatus [203], or finally through direct activation of a PC transporter (Mdr2) in the canalicular membrane [164,165]. A possible modulatory activity of bile acids on Mdr2 PC translocase was directly monitored in yeast SVs stably expressing Mdr2 [204]. First, it was established that SVs can readily accumulate the bile acid taurocholate (TC), and that this accumulation is mediated by the strong electrochemical gradient H^+ produced by the endogenous yeast PMA1 ATPase [205]. Second, loading of the SVs under polarizing conditions with the primary bile salt taurocholate was found to result in an apparent enhancement of Mdr2-mediated PC translocation activity within the membrane of these vesicles.

Reducing the intravesicular TC concentration by dissipating the electrochemical H^+ gradient across the SV membrane with either nitrate or nigericin eliminated the enhancing effect of TC [204]. Three lines of evidence suggested that the enhanced Mdr2-mediated PC translocation activity was not caused by a regulatory effect of TC on Mdr2 but rather reflected the formation of TC/PC aggregates or micelles in the lumen of SVs. First, significantly higher detergent concentrations were required to reveal the fluorescence of NBD-PC molecules translocated in Mdr2-SV under conditions of TC stimulation than under control conditions; second, the non-micelle-forming bile salt taurodehydrocholate did not cause enhancement of PC translocation in Mdr2-SVs; third, enzyme marker studies indicated that TC behaved as a potent lipid solubilizer directly extracting PC molecules out of the bilayer without causing leakage or otherwise affecting the overall integrity of the membrane of SVs. This resulted in the formation of intravesicular aggregates or even mixed micelles, and provoked the apparent stimulation of Mdr2 activity [204].

An important question that emerges from these findings is whether the flippase mechanism observed for Mdr2 in the transport of PC may also apply to drug transport by the two other members of the family Mdr1 and Mdr3. The high degree of sequence homology between the three proteins (approximately 90%) certainly argues for a common mechanism of action. This notion of sequence/function conservation is also supported by the observation that mouse Mdr3 can complement the biological activity of its yeast homolog STE-6 [107], although the two proteins share only 50% homology. Therefore, it is possible that Mdr1 and Mdr3 may also be flippases for yet to be identified lipids, and that drug molecules inserted in the membrane may be recognized as fraudulent lipid substrates. Of particular interest is the case of the P-gp substrate gramicidin D. Gramicidin D is a linear peptide which forms a homodimer spanning the entire length of the lipid bilayer and kills cells by forming a transmembrane ion channel. It is easy to envision how a flippase-translocase activity for P-gp could disrupt or abolish the formation of such a channel thereby causing cellular resistance to this drug without the necessity of 'transmembrane-transport'. It is interesting to note that cellular levels of drug resistance conveyed by either *mdr1* or *mdr3* are highest for gramicidin amongst all drugs tested in transfectants [137].

References

1. Biedler, J.L. and Riehm, H. (1970) Cancer. Res. **30**, 1174–1184.
2. Bech-Hansen, N.T., Till, J.E. and Ling, V. (1976) J. Cell. Physiol. **88**, 23–32.
3. Fojo, A.T., Whang-Peng, J., Gottesman, M.M. and Pastan, I. (1985) Proc. Natl. Acad. Sci. USA **82**, 7661–7665.
4. Lemontt, J.F., Azzaria, M. and Gros, P. (1988) Cancer Res. **48**, 6348–6353.
5. Beck, W.T. and Danks, M.K. (1991) in: Molecular and Cellular Biology of Multidrug Resistance in Tumor Cells, ed I.B. Roninson, pp. 3–55. Plenum, New York.
6. Sugimoto, Y. and Tsuruo, T. (1991) in: Molecular and Cellular Biology of Multidrug Resistance in Tumor Cells, ed I.B. Roninson, pp. 57–72. Plenum, New York.
7. Endicott, J.A. and Ling, V. (1989) Ann. Rev. Biochem. **58**, 137–171.
8. Roninson, I.B. (1991) Molecular and Cellular Biology of Multidrug Resistance in Tumor Cells. Plenum, New York.

9. Gottesman, M.M. and Pastan, I. (1993) Ann. Rev. Biochem. **62**, 385–427.
10. Gupta, R.S. (1983). Cancer. Res. **43**, 1568–1574.
11. Tsuruo, T., Iida, H., Ohkochi, E., Tsukagoshi, S. and Sakurai, Y. (1983) Gann. **74**, 751–758.
12. Conter, V. and Beck, W.T. (1984) Cancer Treat. Rep, **68**, 831–839.
13. Zamora, J.M., Pearce, H.L. and Beck, W.T. (1988) Mol. Pharmacol. **33**, 454–462.
14. Pearce, H.L., Safa, A.R., Bach, N.J., Winter, M.A., Cirtain, M.C. and Beck, W.T. (1989) Proc. Natl, Acad. Sci. USA **86**, 5128–5132.
15. Chevalier-Multon, M.-C., Jacquemin-Sablon, A., Besselièvre, R., Husson, H-P. and Le Pecq, J-B. (1990) Anti-Cancer Drug Des. **5**, 319–335.
16. Tang Wai, D., Arnold, L., Brossi, A. and Gros, P. (1993) Biochemistry **32**, 6470–6476.
17. Ling, V. and Thompson, L.H. (1974) J. Cell. Physiol. **83**, 103–116.
18. Di Marco, A. (1978) in: Fundamentals in Cancer Chemotherapy Antibiotics Chemotherapy **23**, 216–227. Karger, Basel.
19. Dano, K. (1973) Biochim. Biophys. Acta **323**, 466–483.
20. Skovsgaard, T. (1978) Cancer Res. **39**, 4722–4727.
21. Howell, N., Belli, T.A., Zaczkiewicz, L.T. and Bell, J.A. (1984) Cancer. Res. **44**, 4023–4029.
22. Grund, S.H., Patil, S.R., Shah, H.O., Pauw, P.G. and Stadler, J.K. (1983) Mol. Cell. Biol. **3**, 1634–1647.
23. Meyers, M.B., Spengler, B.A., Chang, T.D., Melera, P.W. and Biedler, J.L. (1985) J. Cell. Biol. **100**, 588–597.
24. Teeter, L.D., Atsumi, S.I., Sen, S. and Kuo, T. (1986) J. Cell. Biol. **103**, 1159–1166.
25. Slovak, M.L., Hoeltege, G.A. and Ganapathi, R. (1986) Cancer. Res. **46**, 4171–4177.
26. Richert, N., Akiyama, S., Shen, D., Gottesman, M.M. and Pastan, I. (1985) Proc. Natl. Acad. Sci. USA **82**, 2330–2333.
27. Juliano, R.L. and Ling, V. (1976) Biochim. Biophys. Acta **455**, 152–162.
28. Cornwell, M.M., Tsuruo, T., Gottesman, M.M. and Pastan, I. (1987) FASEB J. **1**, 51–54.
29. Schurr, E., Raymond, M., Bell, J. and P. Gros. (1989) Cancer Res. **49**, 2729–2734.
30. Cornwell, M.M., Safa, A.R., Felsted, R.L., Gottesman, M.M. and Pastan, I. (1986) Proc. Nat. Acad. Sci. USA **83**, 3847–3850.
31. Safa, A.R., Glover, C.J., Meyers, M.B., Biedler, J.L. and Felsted, R.L. (1986) J. Biol. Chem. **261**, 6137–6140.
32. Safa, A., Metha, N.D. and Agresti, M. (1989) Biochem. Biophys. Res. Commun. **162**, 1402–1408.
33. Safa, A.R. (1988) Proc. Nat. Acad. Sci. USA **85**, 7187–7191.
34. Safa, A.R. (1993) Cancer. Invest. **11**, 46–56.
35. Safa, A.R., Agresti, M., Tamai, I., Mehta, N.D. and Vahabi, S. (1990) Biochem. Biophys. Res. Commun. **166**, 259–266.
36. Safa, A.R., Glover, C.J., Sewell, J.L., Meyers, M.B., Biedler, J.L. and Felsted, R.L. (1987) J. Biol. Chem. **262**, 7884–7888.
37. Hamada, H. and Tsuruo, T. (1988) J. Biol. Chem. **263**, 1454:1458.
38. Tsuruo, T., Iida, H., Tsukagoshi, S. and Sakurai, Y. (1981) Cancer Res. **41**, 1967–1972
39. Tsuruo, T., Iida, H., Tsukagoshi, S. and Sakurai, Y. (1982) Cancer Res. **42**, 4730–4733.
40. Slater, L.M., Sweet, P., Stupecky, M. and S. Gupta. (1986) J. Clin. Invest. **77**, 1405–1408.
41. Tsuruo, T., Iida, H., Kitatani, Y., Yokota, K., Tsukagoshi, S. and Sakurai, Y. (1984) Cancer Res. **44**, 4303–4307.
42. Kajiji, S., Dreslin, J., Grizzuti, K. and Gros, P. (1994) Biochemistry. **33**, 5041–5048.
43. Cornwell, M.M., Pastan, I. and Gottesman, M.M. (1987).J. Biol. Chem. **262**, 2166–2170.
44. Akiyama, S. I., Cornwell, M.M., Kuwano, M., Pastan, I. and Gottesman, M.M. (1988) Mol. Pharmacol. **33**, 144–147.
45. Kamiwatari, M., Nagata, Y., Kikuchi, H., Yoshimura, A., Sumizawa, T., Shudo, N., Sakoda, R., Seto, K. and Akiyama, S.I. (1989) Cancer Res. **49**, 3190–3195.
46. Tamai, I. and Safa, A.R. (1990) J. Biol. Chem. **265**, 16509–16513.
47. Yusa, K and Tsuruo, T. (1989) Cancer. Res. **49**, 5002–5006.
48. Ueda, K., Okamura, N., Hirai, Y., Tanigawara, Y., Saeki, T., Kioka, N., Komano, T. and Hori, R. (1992) J. Biol. Chem. **267**, 24248–24252.

49. Kartner, N., Evernden-Porelle, D., Bradley, G and Ling, V. (1985) Nature **316**, 820–823.

50. Riordan, J.R., Deuchars, K., Kartner, N., Alon, N., Trent, J. and Ling, V. (1985) Nature **316**, 817–819.

51. Roninson, I.B. (1983) Nucleic. Acids. Res. **11**, 5413–5423.

52. Roninson, I.B., Abelson, H., Housman, D.E., Howell, N. and Varshavsky, A. (1984) Nature, **309**, 626–628.

53. Roninson, I.B., Chin, J.E., Choi, K., Gros, P., Housman, D.E., Fojo, A., Shen, D.W., Gottesman, M.M. and Pastan, I. (1986) Proc. Natl. Acad. Sci. USA **83**, 4538–4542.

54. Gros, P., Croop, J., Varshavsky, A. and Housman, D.E. (1986) Proc. Natl. Acad. Sci. USA **83**, 337–341.

55. Gros, P., Croop, J.M and Housman, D.E. (1986) Cell **47**, 371–380.

56. Chen, C.-J., Chin, J.E., Ueda, K., Clark, D.P., Pastan, I., Gottesman, M.M. and Roninson, I.B. (1986) Cell. **47**, 381–389.

57. De Bruijn, M.H.L., Van Der Bliek, A.M., Biedler, J.L. and Borst, P. (1986) Mol. Cell. Biol. **6**, 4717–4722.

58. Van der Bliek, A.M., Van der Velde-Koerts, T., Ling, V. and Borst, P. (1986) Mol. Cell. Biol. **6**, 1671–1678.

59. Gros, P., Raymond, M., Bell, J. and Housman, D.E. (1988) Mol. Cell. Biol. **8**, 2770–2778.

60. Ng, W.F., Sarangi, F., Zastawny, R.L., Veinot-Drebot, L. and Ling, V. (1989) Mol. Cell. Biol. **9**, 1224–1232.

61. Devault, A. and Gros, P. (1990) Mol. Cell. Biol. **10**, 1652–1663.

62. Devine, S.E., Hussain, A., Davide, J.P. and Melera, P.W. (1991) J. Biol. Chem. **266**, 4545–4555.

63. Hsu, S.I.H., Cohen, D., Kirschner, L.S., Lothstein, L., Hartstein, M. and Horwitz, S.B. (1990) Mol. Cell. Biol. **10**, 3596–3606.

64. Walker, J.E., Sraste, M., Runswick, M.J. and Gay, N.J. (1982) EMBO J. **1**, 945–951.

65. Raymond, M., Rose, E., Housman, D.E. and Gros, P. (1989) Mol. Cell. Biol. **10**, 1642–1651.

66. Chin, J.E., Soffir, R., Noonan, K.E., Choi, K. and Roninson, I.B. (1989) Mol. Cell. Biol. **9**, 3808–3820.

67. Gros, P., Ben Neriah, Y., Croop, J.M. and Housman, D.E. (1986) Nature **323**, 728–731.

68. Ueda, K., Cardarelli, C., Gottesman, M.M. and Pastan, I. (1987) Proc. Nat. Acad. Sci. USA **84**, 3004–3008.

69. Guild, B.C., Mulligan, R.C., Gros, P. and Housman, D.E. Proc. Natl. Acad. Sci. USA **85**, 1595–1599, 1988.

70. Schinkel, A.H., Roelofs, M.E.M. and Borst, P. (1991) Cancer Res. **51**, 2628–2635.

71. Buschman, E. and Gros, P. (1991) Mol. Cell. Biol. **11**, 595–603.

72. Buschman, E. and Gros, P. (1994) Cancer. Res. **54**, 4892–4898.

73. Tang-Wai , D., Kajiji, S., DiCapua, F., de Graaf, D., Roninson, I.B and Gros, P. (1995) Biochemistry **34**, 32–39.

74. Hammond, J., Johnstone, R.M. and Gros, P. (1989) Cancer Res. 49. 3867–3871.

75. Schinkel, A.H., Smit, J.J.M., van Tellingen, O., Beijnen, J.H., Wagenaar, E., van Deemter, Mol., C.A.A.M., van der Valk, M.A., Robanus-Maandag, E.C., te Riele, H.P.J., Berns, A.J.M. and Borst, P. (1994) Cell **77**, 491–502.

76. Cordon-Cardo, C., O'Brien, J.P., Casals, D., Rittman-Grauer, L., Biedler, J.L., Melamed, M.R. and Bertino, J.R. (1989) Proc. Nat. Acad. Sci. USA **86**, 695–698.

77. Thiebaut, F., Tsuruo, T., Hamada, H, Gottesman, M.M., Pastan, I. and Willingham, M.C. (1989) J. Histochemi. Cytochem. **37**, 159–164.

78. Arceci, R. J., Croop, J., Horwitz, S.B., Housman, D.E. (1988) Proc. Natl. Acad. Sci. USA **85**, 4350–4354.

79. Croop, J. M., Raymond, M., Haber, D., Devault, A., Arceci, R.J., Gros, P. and Housman, D.E. (1989) Mol. Cell.Biol. 9. 1346–1350.

80. Teeter, L.D., Becker, F.F., Chisari, F.V., Li., D. and Kuo, M.T. (1990) Mol. Cell. Biol. **10**, 5728–5735.

81. Yoshimura, A., Kuwazuru, Y., Sumizawa, T., Ichikawa, M., Ikeda, S.I., Ueda, T. and Akiyama, S.I. (1989) J. Biol. Chem. **264**, 16282–16291.

82. Bruggemann, E.P., Germann, U.A., Gottesman, M.M. and Pastan, I. (1989) J. Biol. Chem. **264**, 15483–15488.

83. Zhang, J.T. and Ling, V. (1991) J. Biol. Chem. **266**, 18224–18232.

84. Zhang, J.T. and Ling, V. (1993) Biochem. Biophys. Acta. **1153**, 191–202.

85. Zhang, J.T., Duthie, M. and Ling, V. (1993) J. Biol. Chem. **268**, 15101–15110.

86. Skach, W.R (1994) Cancer. Res. **54**, 3202–3209.

87. Skach, W.R., Calayag, M.C. and Lingappa, V.R. (1993) J. Biol. Chem. **268**, 6903–6908.

88. Bibi, E. and Béjà, O. (1994) J. Biol. Chem. **269**, 19910–19915.

89. Beja, O. and Bibi, E. (1995) J. Biol. Chem. **270**, 12351–12354.

90. Loo, T.W. and Clarke, D.M. (1995) J. Biol. Chem. **270**, 843–848.

91. Kast, C., Canfield, V., Levenson, R. and Gros, P. (1995) Biochemistry **34**, 4402–4411.

92. Higgins, C.F. (1992) Ann. Rev. Cell. Biol. **8**, 67–113.

93. Higgins, C.F., Hydes, I.D., Salmond, G.P.C., Gill, D.R., Downie, J.A., Evans, I.J., Holland, I.B., Gray, L., Buckel, S.D., Bell, A.W.and Hermondson, M.A. (1986) Nature **323**, 448–450.

94. Gilson, E., Higgins, C.F., Hofnung, M., Ames, G.F.L. and Nikaido, H. (1982) J. Biol. Chem. **257**, 9915–9918.

95. Higgins, C.F., Haag, P.D., Nikaido, K., Ardeshir, F., Garcia, G. and Ames, G.F.L. (1982) Nature **298**, 723–727.

96. Ames, G.F.-L. (1986) Ann. Rev. Biochemi. **55**, 397–425.

97. Surin, B.P., Rosenberg, H. and Cox, G.B. (1985) J. Bact. **161**, 189–198.

98. Cangelosi, G.A., Martinetti, G., Leigh, J.A., Lee, C.C., Theines, C. and Nester, E.W. (1989) J. Bact. **171**, 1609–1615.

99. Felmlee, T., Pellett, S. and Welch, R.A. (1985) J. Bact. **163**, 94–105.

100. Foote, S.J., Thompson, J.K., Cowman, A.F. and Kemp, D.J. (1989) Cell **57**, 921–930.

101. Foote, S.J., Kyle, D.E., Martin, R.K, Oduola, A.M.J., Forsyth, K., Kemp, D.J. and Cowman, A.F. (1990) Nature **345**, 255–258.

102. Samuelson, J., Ayala, P., Oroczo, E. and Wirth, D. (1990) Mol. Biochem. Parasitol. **38**, 281–290.

103. Ouellette, M., Fase-Fowler, F. and Borst, P. (1990) EMBO J. **9**, 1027–1033.

104. Henderson, D.M., Sifri, C.D., Rodgers, M., Wirth, D.F., Hendrickson, N. and Ullman, B. (1992) Mol. Cell. Biol. **12**, 2855–2865.

105. McGrath, J.P. and Varshavsky, A. (1989) Nature **340**, 400–404.

106. Kuchler, K., Sterne,R.E. and Thorner, J. (1989) EMBO J. **8**, 3973–3984.

107. Raymond, M., Gros, P., Whiteway, M. and Thomas, D.Y. (1992) Science **256**, 232–234.

108. Volkman, S.K., Cowman, A.F. and Wirth, D.F. (1995) Proc. Natl. Acad. Sci. USA **92**, 8921–8925.

109. Wu, C.T., Budding, M., Griffin, M.S. and Croop, J. (1991) Mol. Cell. Biol. **11**, 3940–3948.

110. Lincke, C.R., The, I., van Groenigen, M and Borst P. (1992) J. Mol. Biol. **228**, 701–711.

111. Dudler, R. and Hertig, C. (1992) J. Biol. Chem. **267**, 5882–5888.

112. Riordan, J.R., Rommens, J.M., Kerem, B-S., Alow, N., Rozmahel, R., Grzelczak, Z., Zielenski, J., Lok, S., Plavsic, N., Chou, J.L., Drumm, M.L., Ianuzzi, M.C., Collins, F.S. and Tsui, L.C. (1989) Science **245**, 1066–1073.

113. Rich, D.P., Anderson, M.P., Gregory, R.J., Cheng, S.H., Paul, S., Jefferson, D.M., McCann, J.D., Klinger, K.W., Smith, A.E. and Welch, M.J. (1990) Nature **347**, 358–363.

114. Cheng, S.H., Gregory, R.J., Marshall , J., Paul, S., Souza, D.W., White, G.A., O'Riordan, C. R. and Smith, A.E. (1991) Cell **63**, 827–834.

115. Denning, G.M., Anderson, M.P., Amara, J. F., Marshall, J., Smith, A.E. and Welsh, M.J. (1992) Nature **358**, 761–764.

116. Denning, G.M., Ostedgaard, L.S. and Welsh, M.J. (1992) J. Cell Biol. **118**, 551–559.

117. Sheppard, D.N., Rich, D.P., Ostedgaard, L.S., Gregory, R.J., Smith, A.E. and Welsh, M.J. (1993) Nature **362**, 160–164.

118. Kelly, A., Powis, S.H., Kerr, L-A., Mockridge, I., Elliott, T., Bastin, J., Uchanska-Ziegler, B., Ziegler, A., Trowsdale, J. and Townsend, A. (1992) Nature **355**, 641–644.

119. Spies, T., Cerundolo, V., Colonna, M., Cresswell, P., Townsend, A. and DeMars, R. (1992) Nature **355**, 644–646.

120. Powis, S.J., Deverson, E.V., Coadwell, W.J., Ciruela, A., Huskisson, N.S., Smith, H., Butcher, G.W. and Howard, J.C. (1992) Nature **357**, 211–215.

121. Mosser, J., Douar, A-M., Sarde, C-O., Kioschis, P., Feil, R., Moser, H., Poustka, A-M., Mandel, J-L.

and Aubourg, P. (1993) Nature **361**, 726–730.

122. Kamijo, K., Taketani, S., Yokota, S., Osumi, T. and Hashimoto, T. (1990) J. Biol. Chem. **265**, 4534–4540.

123. Fanen, P., Guidoux, S., Sarde, C.O., Mandel, J.L., Goossens, M. and Aubourg, P. (1994) J. Clin. Invest. **94**, 516–520.

124. Braun, A., Ambach, H., Kammerer, S., Rolinski, B., Stockler, S., Rabl, W. et al. (1995) Am. J. Human. Genet. **56**, 854–861.

125. Vorgerd, M., Fuchs, S., Tegenthoff, M. and Malin, J.P. (1995) J. Neurol. Neurosur. Psychiatry **58**, 229–231.

126. Gartner, J. and Vallle, D. (1992) Nature Genet. **1**, 16–23.

127. Aguillar-Bryan, L., Nichols, C.G., Wechsler, S.W., Clement, J.P., Boyd, A.E., Gonzalez G., et al. (1995) Science **268**, 413–426.

128. Thomas, P.M., Cote, G.J., Wohlik, N., Mathew, P.M., Aguillar-Bryan, L., Gagel, R.F., et al. (1995) Science **268**, 426–429

129. Mirski, S.E.L., Gerlach, J. and Cole, S.P. (1987) Cancer Res. **47**, 2594:2598.

130. Cole, S.P., Bhardwaj, G., Gerlach, J.H., Mackie, J.E., Grant, C.E., Almquist, K.C., et al. (1992) Science **258**, 1650–1654.

131. Grant, C.E., Valdimarsson, G., Hipfner, D.R., Almquist, K.C., Cole, S.P.C. and Deeley R.G. (1994) Cancer. Res. **54**, 357–361.

132. Raviv, Y., Pollard, H.B., Bruggemann, E.P., Pastan, I. and Gottesman, M.M. (1990) J. Biol..Chem. **265**, 3975–3980.

133. Greenberger L.M., Lisanti, C.J., Silva, J. T. and Horwitz, S. B. (1991) J. Biol. Chem. **266**, 20744–20751.

134. Greenberger, L.M. (1993) J. Biol. Chem. **268**, 11417–11425.

135. Berkower, C. and Michaelis, S. (1991) EMBO J. **10**, 3777–3785.

136. Gros, P., Dhir, R., Croop, J.M. and Talbot, F. (1991) Proc. Natl. Acad. Sci. USA **88**, 7289–7293.

137. Kajiji, S., Talbot, F., Grissuti, K., Van Dyke-Phillips, V., Safa, A., and Gros, P. (1993) Biochemistry **32**, 4185–4194.

138. Dhir, R., Grizzuti, K., Kajiji, S., and Gros, P. (1993) Biochemistry **32**, 9492–9499.

139. Zhang, X., Collins, K. and Greenberger, L.M. (1995) J. Biol. Chem. **270**, 5441–5448.

140. Devine, S.E., Ling, V. and Melera, P.W. (1992) Proc. Natl. Acad. Sci. USA **89**, 4564–4568.

141. Loo, T.W. and Clarke, D.M. (1994) Biochemistry **33**. 14049–14057.

142. Morris, D.I., Greenberger, L.M., Bruggemann, E.P., Cardarelli, C., Gottesman, M.M., Pastan, I., et al. (1994) Mol. Pharmacol. **46**, 329–337.

143. Loo, T.W., and Clarke, D.M. (1993).J. Biol. Chem. **268**, 3143–3149.

144. Loo, T.W., and Clarke, D.M. (1993) J. Biol. Chem. **268**, 19965–19972.

145. Choi, K., Chen, C.J., Kriegler, M. and Roninson, I.B. (1988) Cell **53**, 519–529.

146. Safa, A.R., Stern, R.K., Choi, K., Agresti, M., Tamai, I., Mehta, N.D. and Roninson, I.B. (1990) Proc. Natl. Acad. Sci. USA **87**, 7225–7229.

147. Loo, T.W. and Clarke, D.M. (1994) J. Biol. Chem. **269**, 7243–7248.

148. Azzaria, M., Schurr, E. and Gros, P. (1989) Mol. Cell. Biol. **9**, 5289–5297.

149. Beaudet, L. and Gros, P. (1995) J. Biol. Chem. **270**, 17159–17170.

150. Scarborough, G.A. (1995) J. Bioenerget. Biomembr. **27**, 37–41.

151. Shapiro, A.B. and Ling V. (1995) J. Bioenerget. Biomembr. **27**, 7–13.

152. Shapiro, A.B. and Ling, V. (1994) J. Biol. Chem. **269**, 3745–3754.

153. Urbatsch, I.L., Al-Shawi, M.K. and Senior, A.E. (1994) Biochemistry. **33**, 7069–7076.

154. Sharom, F.J., Yu, X., Chu, J.W.K. and Doige, C.A. (1995) Biochem. J. **308**, 381–390.

155. Loo, T.W. and Clarke, D.M. (1994) J. Biol. Chem. **269**, 7750–7755.

156. Al Shawi, M.K., Urbatsch, I.L. and Senior, A.E. (1994) J. Biol. Chem. **269**, 8986–8992.

157. Al-Shawi, M.K. and Senior, A.E. (1993) J. Biol. Chem. **268**, 4197–4206.

158. Senior, A.E., Al-Shawi, M.K. and Ursbatsch, I.L. (1995) J. Bioenerget. Biombembr. **27**, 31–6.

159. Urbatsch, I.L., Sankaran, B., Weber, J. and Senior, A.E. (1995) J. Biol. Chem. **270**, 19383–19390.

160. Homolya, L., Hollo, Z., Germann, U.A., Pastan, I., Gottesman, M.M. and Sarkadi, B. (1993) J. Biol. Chem. **268**, 21493–21496.

161. Doige, C.A., Yu, X. and Sharom, F.J. (1993) Biochim. Biophys. Acta **1146**, 65–72.
162. Rao, U.S. (1995) J. Biol. Chem. **270**, 6686–90.
163. Hoof, T., Demmer, A., Hadam, M.R., Riordan, J.R. and Tummler, B. (1994) J. Biol. Chem. **269**, 20575–20583.
164. Ruetz, S. and Gros, P. (1994) Cell **77**, 1071–1081.
165. Smit, J.J., Schinkel, A.H., OudeElferink, R.P., Groen, A.K., Wagenaar, E., Van Deemter, L., Mol, C.A., Ottenhofer, R., Van der Lugt, M.A., Van Roon, M.A., Van der Valk, M.A., Offerhaus, G.J., Berns, A.J. and Borst, P. (1993) Cell **75**, 451–462.
166. Cornwell, M.M., Gottesman, M.M. and Pastan, I. (1986) J. Biol. Chem. **261**, 7921–7928.
167. Horio, M., Gottesman, M.M. and Pastan, I. (1988) Proc. Natl. Acad. Sci. USA **85**, 3580–3584.
168. Kamimoto, Y., Gatmaitan, Z., Hsu, J. and Arias, I.M. (1989) J. Biol. Chem. **264**, 11693–11698.
169. Abraham, E.H., Prat, A.G., Gerweck, L., Seneveratne, T., Arceci, R.J., Kramer, R., Guidotti, G. and Cantiello, H.F. (1993) Proc. Natl. Acad. Sci. USA **90**, 312–316.
170. Reisin, I.L., Prat, A.G., Abraham, E.H., Amara, J.F., Gregory, R.J., Ausiello, D.A. and Cantiello, H.F. (1994) J. Biol. Chem. **269**, 20584–20591.
171. Keizer, H.G. and Joenje, H. (1989) J.Natl. Cancer. Inst. **81**, 706–9.
172. Thiebaut, F., Currier, S.J., Whitaker, J., Haugland, R.P., Gottesman, M.M., Pastan, I. and Willingham, M.C. (1990) J. Histochem. Cytochem. **38**, 685–690.
173. Roepe, P.D. (1992) Biochemistry **31**, 12555–12564.
174. Roepe, P.D., Young, L., Cruz, J. and Carlson, D. (1993) Biochemistry **32**, 11042–11056.
175. Roepe, P.D., Weisburg, J.H., Luz, J.G., Hoffman, M.M. and Wei, L.Y. (1994) Biochemistry **33**, 11008–15.
176. Altenberg, G.A., Young, G., Horton, J.K., Glass, D., Belli, J.A. and Reuss, L. (1993) Proc. Natl. Acad. Sci. USA **90**, 9735–9738.
177. Rasola, A., Galietta, L.J., Gruenert, D.C. and Romeo, G. (1994) J. Biol. Chem. **269**, 1432–1436.
178. Dong, Y., Chen, G., Duran, G.E., Kouyama, K., Chao, A.C., Sikic, B.I., et al. (1994) Cancer. Res. **54**, 5029–5032.
179. Borrel, M.N., Pereira, E., Fiallo, M. and Garnier–Suillerot, A. (1994) Eur. J. Biochem. **223**, 125–33.
180. Higgins, C.F. (1995) J. Bioenerg. Biomembr. **27**, 63–70.
181. Higgins, C.F. (1995) Cell **82**, 693–6.
182. Raymond, M., Ruetz, S., Thomas, D.Y. and Gros, P. (1994) Mol. Cell. Biol. **14**, 277–286.
183. Ruetz, S., Raymond, M. and Gros, P. (1993) Proc. Natl. Acad. Sci. USA **90**, 11588–11592.
184. Ruetz, S. and Gros, P. (1994) J. Biol. Chem. **269**, 12277–12284.
185. Nakamoto, R.K., Rao, R. and Slayman, C.W. (1991) J. Biol. Chem. **266**, 7940–7949.
186. Sharom, F.J., Yu, X. and Doige, C.A. (1993) J. Biol. Chem. **268**, 24197–24202.
187. Shapiro, A.B. and Ling, V. (1995) J. Biol. Chem. **270**, 16167–16175.
188. Thiebaut, F., Tsuruo, T., Hamada, H., Gottesman, M.M., Pastan, I. and Willingham. M.C. (1987) Proc. Natl. Acad.Sci. USA **84**, 7735–7738.
189. Fojo, A., Ueda, K., Slamon, D.J., Poplack, D.G., Gottesman, M.M. and Pastan, I. (1987) Proc. Natl. Acad. Sci. USA **84**, 265–269.
190. Bradley, G., Georges, E. and Ling, V. (1990) J. Cell. Physiol. **145**, 398–408.
191. Buschman, E., Arceci, R.J., Croop, J.M., Che, M., Arias, I.M., Housman, D.E. and Gros, P. (1992) J. Biol. Chem. **267**, 18093–18099.
192. Georges, E., Bradley, G., Gariepy, J. and Ling, V. (1990).Proc. Natil. Acad. Sci. USA **87**, 152–156.
193. Tatsuta, T., Naito, M., Oh-hara, T., Sugawara, I.and Tsuruo, T.(1992) J. Biol. Chem. **267**, 20383–20391.
194. Tanigawara, Y., Okamura, N., Hirai, M., Yasuhara, M., Ueda, K., Kioka, N., Komano, T. and Hori, R. (1992) J. Pharmacol. Exp. Ther. **263**, 840–845.
195. Chaudhary, P.M. and Roninson, I.B. (1991). Cell **66**, 85–94.
196. Wolf, D.C. and Horwitz, S.B. (1992) Int. J. Cancer **52**, 141–146.
197. Sharma, R.C., Inoue, S., Roitelman, J., Schimke, R.T. and Simoni, R.D. (1992) J. Biol. Chem. **267**, 5731–5734.
198. Trezise, A.E.O., Romano, P.R., Gill, D.R., Hyde, S.C., Sepœlveda, F.V., Buchwald, M. and Higgins, C.F. (1992) EMBO J. **11**, 4291–4303.

199. Sagawa, H., Tazuma, S. and Kajiyama, G. (1993) Am. J. Physiol. **264**, G835–G839.
200. Coleman, R. (1987) Biochem. J. **244**, 249–261.
201. Lowe, P.J., Barnwell, S.J. and Coleman, R. (1984) Biochem. J. **222**, 631–637.
202. Balint, J.A., Beeler, D.A., Kyriakides, E.C. and Treble, D.H. (1971) J. Lab. Clin. Med. **77**, 122–133.
203. Hayakaka, T., Sakisaka, S., Meier, P.J. and Boyer, J.L. (1990) Gastroenterology **99**, 216–228.
204. Ruetz, S. and Gros, P. (1995) J. Biol. Chem. **270**, 25388–25395.
205. St. Pierre, M. V., Ruetz, S., Epstein, L. F., Gros, P. and Arias, I. M. (1994) Proc. Natl. Acad. Sci. USA **91**, 9476–9479.
206. Kast, C., Canfield, V., Levenson, R. and Gros, P. (1996) J. Biol. Chem. **271**, 2940–2948.
207. Hanna, M., Brault, M., Kwan, T., Kast, C. and Gros, P. (1996) Biochemistry **35**, 3625–3635.
208. Urbatsch, I.L., Sankaran, B., Bhagat, S. and Senior, A.E. (1995) J. Biol. Chem. **270**, 26956–26961.

CHAPTER 8

Multidrug Resistance in Prokaryotes: Molecular Mechanisms of Drug Efflux

H.W. VAN VEEN, H. BOLHUIS, M. PUTMAN and W.N. KONINGS

Department of Microbiology, Groningen Biomolecular Sciences and Biotechnology Institute,
University of Groningen, Haren, The Netherlands

© *1996 Elsevier Science B.V.*
All rights reserved

Handbook of Biological Physics
Volume 2, edited by W.N. Konings, H.R. Kaback and J.S. Lolkema

Contents

1. General introduction

The introduction of antibiotics into clinical use has brought about a spectacular decline in the prevalence of infectious diseases in humans. However, the emergence of resistant bacteria is changing this situation. Diseases such as tuberculosis and pneumonia, evocative of a former era when pathogenic bacteria were often deadly killers, are now re-emerging as major threats to public health. In addition, the constant pressure of antibiotics in hospital environments has selected for drug resistant bacterial species which may not possess strong virulence but can infect debilitated patients [1–3].

Toxic compounds have always been part of the natural environment in which bacteria dwell. The development of strategies for life in this habitat has been crucial for survival of the organisms. As a result, bacteria have developed versatile mechanisms to resist antibiotics and other cytotoxic drugs [4,5]. Some of these resistance mechanisms are rather specific for a given drug or a single class of drugs. Examples are the enzymatic degradation or inactivation of drugs [5,6], and the alteration of the drug target [7]. In addition, more general mechanisms exist in which the entrance of a variety of cytotoxic compounds into the cell is prevented by the barrier and active transport function of the cell envelope which encloses the cytoplasm [8].

The envelope of Gram-negative bacteria consists of the cytoplasmic (or inner) membrane, a peptidoglycan layer, and the outer membrane that contains lipopolysaccharides [9,10]. Gram-positive bacteria are not enclosed by an outer membrane. Instead, their cytoplasmic membrane is directly surrounded by a thick peptidoglycan layer. Both the peptidoglycan layer of certain Gram-positive bacteria (e.g. mycobacteria [11]) as well as the outer membrane of Gram-negative bacteria are able to serve as a barrier against rapid penetration of drugs. However, these barriers cannot prevent the drugs from exerting their action once they have entered the cell [12–15]. Additional mechanisms are therefore needed to achieve significant levels of drug resistance.

One of the major mechanisms to lower the cytoplasmic drug concentration is based on direct efflux of drugs across the cytoplasmic membrane. For this purpose, prokaryotes have developed drug efflux systems which are embedded in the cytoplasmic membrane [16]. Although some of these transporters are relatively specific for a given substrate, others can handle an extraordinary wide range of structurally unrelated compounds [17]. Such systems resemble the mammalian multidrug resistance P-glycoprotein that mediates multidrug resistance in tumour cells [18,19]. Similarly, multidrug transporters in prokaryotes can frustrate the drug-based treatment of bacterial infections in mammals and plants. The recent increase in the clinical use of antimicrobial agents has been accompanied by a subsequent increase in the frequency of bacterial resistance based on active efflux [4,5,20].

Bacteria have a plethora of interesting facets to their lifestyles, some of which are unique, while others provide us with exciting insights into cellular behaviour in general. This review shows that the multidrug efflux systems encountered in prokaryotic cells are very similar to those observed in eukaryotic cells. A study of the factors which determine the substrate specificity and energy coupling to drug translocation in prokaryotes therefore has significance for the general field of multidrug resistance. Accordingly, three issues will be dealt with in this review. First, an overview of the various classes of prokaryotic multidrug transporters will be presented. Evidently, emphasis will be put on recent developments in the field. A number of excellent reviews have been devoted to this subject in the past [8,16,17]. Second, the current understanding of the regulation of bacterial multidrug resistance will be summarized. Third, the present knowledge of the molecular mechanisms involved in drug transport processes will be discussed. It is hoped that this brief introduction will encourage the reader to delve more deeply into the wealth of scientific material which follows.

2. Classification of prokaryotic drug extrusion systems

Currently, over 30 drug transporters have been discovered in prokaryotes. On the basis of bioenergetic and structural criteria, these transport systems can be divided into: (i) secondary drug transporters, and (ii) ATP-binding cassette (ABC)-type drug transporters.

The secondary drug transporters are single membrane proteins which mediate the extrusion of drugs in a coupled exchange with protons (or sodium) ions [21]. In this exchange process, the ion motive force can drive the efflux of drugs against a concentration gradient. On the basis of similarities in size and secondary structure, the secondary drug transporters can be divided into two subgroups. The members of these subgroups are referred to as TEXANs (Toxin EXtruding ANtiporters) and Mini TEXANs [22]. TEXANs are about 45–50 kDa with a hypothesized topology of 12–14 transmembrane α-helical domains [23]. This subgroup includes all members of the drug resistance branch of the Major Facilitator superfamily of transporters [24], and some members of the Resistance-Nodulation-Division family of membrane proteins [25]. Mini TEXANs are functionally similar to TEXANs, but are much smaller (12–15 kDa) and form only 4 putative membrane spanning domains [26].

The ABC-type drug transporters belong to the ABC superfamily, the members of which contain a highly conserved ATP-binding cassette [27–30]. They utilize the energy of ATP hydrolysis to pump cytotoxic substrates out of the cell [31]. Typically, ABC transporters require the function of four protein domains [27,32]. Two of these domains are highly hydrophobic and each consist of 6 putative transmembrane segments. The hydrophobic domains form the pathway through which substrate crosses the membrane. The other two domains are peripherically located at the cytoplasmic face of the membrane, bind ATP and couple ATP hydrolysis to the translocation of substrate [33–35].

3. Tetracycline carrier, a paradigm for prokaryotic drug transporters

Drug resistance of bacteria caused by active efflux of drugs across the cell envelope began to attract the attention of scientists around 1980, when resistance to tetracycline in *Escherichia coli* was shown to be linked to the energy-dependent extrusion of tetracycline [36]. Currently, eight classes of tetracycline efflux proteins have been identified whose amino acid sequences are similar, but not identical [5,37–39]. The most prevalent of these tetracycline efflux proteins among bacterial pathogens is encoded by the *tetA*(class B) gene from transposon Tn10 among Gram-negative bacteria and the *tetA*(class K) gene from *Staphylococcus aureus* among Gram-positive bacteria. The TetA(B) protein has become a paradigm for secondary drug efflux systems in prokaryotes.

The transport mechanism of TetA(B) has been studied extensively in cells and membrane vesicles of *E. coli* [36,40,41]. Tetracycline is largely anionic at neutral pH, but this speciation is affected by divalent cations which form a metal-tetracycline chelate. Interestingly, tetracycline accumulation in inside-out membrane vesicles is strongly dependent on divalent cations in the following order $Co^{2+} > Mn^{2+} > Mg^{2+} > Cd^{2+} > Ca^{2+}$ [42]. This order is consistent with the increasing order of the dissociation constants for metal chelate complexes of tetracycline. Equimolar cotransport of tetracycline and divalent cations, tetracycline-dependent uptake of Co^{2+}, and inhibition by Mg^{2+} of Co^{2+} uptake in the presence of tetracycline indicate that a metal-tetracycline chelate is the transported solute [42]. This monocationic complex is extruded by means of an electroneutral exchange with a proton [43]. The transport mechanism of TetA(K) has been presumed to be similar to that of TetA(B) [44–47]. Interestingly, some chromosomally-encoded tetracycline carriers (e.g. TetA(L) in *Bacillus subtilis*) mediate cation/H^+ antiport in addition to tetracycline/H^+ antiport [48–50].

TetA(B) is an integral membrane protein whose topography suggest 12 membrane-spanning segments (in α-helical configuration) which are connected by hydrophilic loops protruding into the cytoplasm or periplasmic space. This secondary structure model is proposed on the basis of the results of limited proteolysis [51], site-directed antibody binding [52], and PhoA-fusion analysis [53]. A relatively large cytoplasmic loop in the middle of the protein separates TetA(B) in an N- and C-terminal half. Both halves are evolutionary related, presumably by a gene duplication event, and are well conserved among TetA proteins of various classes [54]. The N- and C-terminal halves of TetA(B) represent separate domains in the drug transporter [55]. The functional interaction between these domains is suggested by (i) the intragenic complementation between mutations in the first and second halves of the *tetA*(B) gene [56-58], (ii) the ability of hybrid tetracycline efflux transporters to confer resistance [59] and (iii) the functional reconstitution of tetracycline resistance upon co-expression of the N- and C-terminal halves of TetA(B) as separate polypeptides [60]. TetA(B) is homologous to a large number of transport proteins, including drug transporters, sugar-H^+ symporters and the glucose facilitators of mammalian cells, and is a member of the Major Facilitator superfamily [24].

4. Overview of multidrug efflux systems

While some of the known drug transporters, like TetA(B) and others [61–66], show high selectivity for a particular toxin, bacteria have developed many so-called multidrug transporters which are much less selective, reminiscent of the multidrug resistance P-glycoproteins of mammalian cells [67,68]. It is increasingly recognized that these multidrug transporters play a major role in the resistance of clinically relevant pathogenic bacteria. For example, enteric pathogens (e.g. *Escherichia* [69,70], *Salmonella* [71], *Klebsiella* [72], *Serratia* [73], and *Proteus* [74]) and other 'hospital' bacteria (e.g. methicillin-resistant *S. aureus* [75–77]) have acquired efflux-mediated resistance to antiseptics and disinfectants. Consequently, these drug resistant bacteria are difficult to control in hospital environments due to the limited number of effective agents that are clinically available [78,79].

4.1. Secondary transporters

A number of secondary multidrug transporters has been characterized in Gram-positive bacteria (Table 1). Among these are the chromosome-encoded TEXANs LmrP of *Lactococcus lactis* [80,81] (Fig. 1), Bmr of *B. subtilis* [82], and NorA of *S. aureus* [83–85] which are structurally similar to TetA(B) (22–25% sequence identity) but are not dedicated to the efflux of a single drug. Instead, NorA mediates the extrusion of quaternary amine compounds such as the disinfectant benzalkonium, whereas LmrP and Bmr show specificity for an exceptionally wide range of amphiphilic, cationic drugs including (i) aromatic dyes, e.g. ethidium bromide, (ii) quaternary amines, and (iii) derivatives of tetraphenylphosphonium. Bmr and LmrP are inhibited by reserpine, a well-known inhibitor of the human multidrug resistance P-glycoprotein [86] and the vesicular neurotransmitter transporters [22]. Overexpression of LmrP and Bmr proteins in drug resistant mutants of *L. lactis* and *B. subtilis*, or expression of these proteins from high-copy plasmid vectors results in increased resistance to these drugs. In contrast, the lack of LmrP and Bmr expression through gene inactivation results in increased drug sensitivity. Thus, LmrP and Bmr-mediated drug efflux across the cytoplasmic membrane is vital for *in vivo* resistance of the organisms to amphiphilic cations [81,82].

In Gram-negative bacteria, transport from the cytoplasm to the external medium requires the translocation of solutes across the cytoplasmic and outer membrane [87]. For this purpose, some transporters are connected to a periplasmic lipoprotein, termed accessory protein, which forms a complex with a porin in the outer membrane (e.g. TolC [88,89]). Examples of such transporters are the chromosome-encoded TEXANs EmrB [90–92], AcrB [93,94] and EnvD [95] in *E. coli*, MexB in *Pseudomonas aeruginosa* [96,97], and MtrD from *Neisseria gonorrhoeae* [98]. Although EmrB is predicted to contain 14 transmembrane α-helices, it shares some sequence homology with TetA(B) [90]. EmrB has specificity for moderately hydrophobic compounds such as CCCP [91] and thiolactomycin [92]. MexB, AcrB, EnvD, and MtrD have specificity for basic dyes, detergents and antibiotics such as tetracycline. The proteins share extensive sequence similarity (up to 78%), are

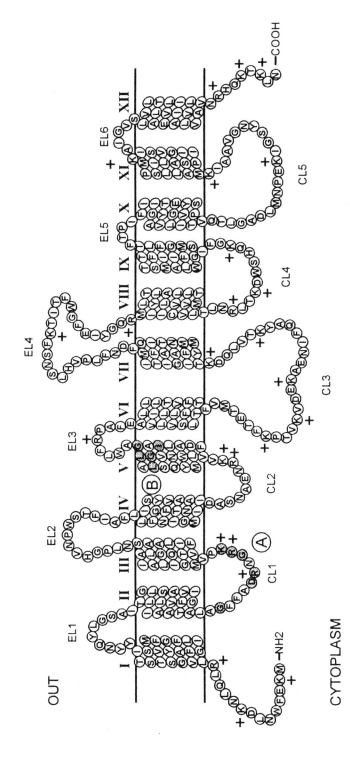

Fig. 1. Secondary structure model of the TEXAN LmrP of *L. lactis*. The model is based on hydropathy profiling of the amino acid sequence and the distribution of positively charged Arg and Lys residues (+) according to the 'positive inside rule' by Von Heijne [194]. This model is very similar to that proposed for TetA(B). Bmr and NorA. LmrP contains a sequence Motif A in the first cytoplasmic loop (CL1) and a Motif B in transmembrane α-helix V (indicated by shaded residues) which are conserved in TetA(B) and its homologs.

Table 1
Overview of multidrug transporters in prokaryotes

Transporter family	Protein	Organism	Reference
ABC-type[a]	LmrA	*L. lactis*	80,135
TEXAN MF[b]	LmrP	*L. lactis*	80,81
	Bmr	*B. subtilis*	82
	NorA	*S. aureus*	83–85
	QacA	*S. aureus*	113,114
	EmrB	*E. coli*	90–92
RND[c]	AcrB	*E. coli*	93,94
	EnvD	*E. coli*	95
	MexB	*P. aeruginosa*	96,97
	MtrD	*N. gonorrhoeae*	98
Mini TEXAN	QacC	*S. aureus*	115,117
	QacE	*K. aerogenes*	26
	EmrE	*E. coli*	119

[a] ATP-binding cassette transporter family [27].
[b] Major Facilitator superfamily [24].
[c] Resistance-Nodulation-Division family [25].

predicted to contain 12 transmembrane α-helices, but are unrelated to TetA(B). Instead, MexB, AcrB, EnvD, and MtrD are members of the Resistance-Nodulation-Division family of membrane proteins [25].

Together with the accessory protein MexA and the outer membrane protein OprK, MexB confers resistance to a wide range of antimicrobial agents [96]. These agents do retain some common features (such as an aromatic ring), and most exhibit the ability to bind cations, including iron [99–103]. In this regard, they resemble the catechol-containing chromophore of pyoverdine [104], an iron chelator which is possibly excreted via MexAB-OprK under iron-limiting growth conditions [96]. Thus, MexAB-OprK-mediated drug efflux may result from the structural similarities between the drugs and the physiological substrate of the transporter.

The genes *acrAB* were first identified because mutations at these loci led to hypersusceptibility to acridines used for plasmid curing [105,106]. Subsequent studies showed that mutations at the same loci also determined susceptibility to other basic dyes, detergents, and certain antibiotics [107]. Originally, mutations in *acrAB* were thought to contribute to the integrity of *E. coli* membranes [108–111]. The proposal that *acrAB* encodes a drug-efflux pump has been supported by the structural homology to *mexAB* and several other drug efflux systems [94].

Most substrates of AcrAB and MtrCD, such as cationic dyes, are not found in the natural environment of *E. coli* and *N. gonorrhoeae*. It seems therefore likely that these transporters have evolved for other purposes. In this regard, it is noteworthy that the natural environment of enteric bacteria is enriched in bile salts and fatty acids [112], and that these compounds are substrate for AcrAB and MtrCD. Thus, AcrAB, MtrCD and related transporters contribute to the intrinsic resistance of the enteric bacteria against these naturally occurring hydrophobic compounds.

Resistance to antiseptics and disinfectants in *S. aureus* has emerged on plasmids. Two different types of resistance determinants have been described which have been cloned, sequenced and expressed in *E. coli*. The QacA and QacB proteins are TEXANs with a putative topology of 14 transmembrane α-helices [113,114], as predicted for EmrB. In contrast, the Mini TEXAN QacC is almost five times smaller, and contains 4 putative transmembrane α-helices [115,116]. This topological model has recently been confirmed by genetic fusion using alkaline phosphatase and β-galactosidase as reporters of subcellular localization [117].

Due to their presence on broad-host plasmids and transposons, Mini TEXANs are widely spread among different bacterial species. The *qacC* gene is identical in sequence to the staphylococcal *smr* gene [118]. The Mini TEXAN family includes *qacE* which was initially characterized in *Klebsiella aerogenes* [26], and *emrE* [119] (formerly known as *ebr* [120,121] or *mvrC* [122]) which is found on the chromosome of *E. coli*. Similar systems are present in *Acinetobacter calcoaceticus* [123] and *Arthrobacter globiformis* [124]. Mini TEXANs are true multidrug transporters which are able to extrude a wide variety of amphiphilic cationic compounds, such as ethidium and phosphonium ions. Interestingly, the transmembrane distribution of tetraphenylphosphonium is widely used in bioenergetic studies to quantitate the membrane potential across biological membranes. It is obvious that the active tetraphenyl phosphonium extrusion via Mini TEXANs and other multidrug transporters can seriously interfere with this approach [118,125,126].

TEXANs and Mini TEXANs are located in the bacterial cytoplasmic membrane, across which a proton motive force (Δp, interior negative and alkaline) exists. This Δp can be generated by primary H^+ extrusion systems such as the respiratory chain or the F_1F_0 H^+-ATPase [127]. It is generally assumed that the Δp is used by (Mini) TEXANs as the driving force for drug efflux through the exchange of a drug molecule and protons. Most of our current knowledge about the energetics of secondary drug transport originates from studies on the TEXAN LmrP, and the Mini TEXANs EmrE and Smr. LmrP-mediated drug efflux in cells of *L. lactis* is strongly inhibited by protonophore CCCP (carbonyl cyanide *m*-chlorophenylhydrazone) suggesting a role of the Δp in drug extrusion [80]. The dependency of drug transport on the Δp was studied in greater detail using ionophores which selectively dissipate the components of the Δp. Dissipation of the transmembrane pH gradient (ΔpH) by nigericin partially inhibited drug extrusion compared to a control in which LmrP activity was totally blocked by reserpine. On the other hand, the simultaneous dissipation of the ΔpH and the transmembrane potential ($\Delta \psi$) by nigericin plus valinomycin inhibited drug efflux up to levels observed in the presence of reserpine [80]. The involvement of both ΔpH and $\Delta \psi$ in the energization of LmrP-mediated drug efflux has been confirmed in transport studies using membrane vesicles of *E. coli* in which LmrP is heterologously expressed [128]. Thus, LmrP catalyzes an electrogenic drug/nH^+ ($n \geq 2$) antiport reaction. The proposed mechanism for the Mini TEXANs Smr and EmrE is strikingly similar to that of LmrP. Both proteins have been overexpressed, purified and reconstituted into proteoliposomes [119, 129]. With these *in vitro* experimental systems, it has been demonstrated that the purified proteins function as independent multidrug transporters. An artificially imposed Δp was shown to be a driving force, and an (electrogenic) drug/proton antiport was suggested as the molecular mechanism of drug transport.

4.2. ABC-type transporters

A number of dedicated ABC-type antibiotic export systems have been found in prokaryotes. In *Streptomyces* strains, these transporters mediate the excretion of various antibiotics to ensure self-resistance to the antibiotics that they produce. Well-known examples are amongst others DrrAB [130], TnrB [131], Ard1 [132], and OleC [133] which are implicated in the efflux of daunorubicin/doxorubicin, tetranasin, antibiotic A201A, and oleandomycin, respectively. ABC transporters dedicated to the efflux of specific drugs have also been detected in other bacteria such as *Bacillus licheniformis* in which the BcrABC proteins confer resistance to bacitracin [134].

Remarkably few prokaryotic ABC-type efflux systems have been characterized that are able to transport multiple drugs (Table 1). To date, most multidrug transporters in bacteria utilize the proton motive force, rather than ATP as the driving force and act via a drug/H$^+$ antiport mechanism. The first example of an ABC-type multidrug transporter has been found in *L. lactis* [80,135]. The gene encoding this transporter, termed *lmrA*, has been cloned and sequenced. It encodes a 589-amino-acid membrane protein, the hydropathy analysis of which suggests the presence of an N-terminal hydrophobic domain with 6 putative transmembrane regions and a C-terminal hydrophilic domain (Fig. 2). The latter domain contains an ATP-binding cassette.

Comparison of LmrA and various ABC transporters clearly identifies the lactococcal transporter as a homolog of the human multidrug resistance P-glycoprotein. This 1280-residue membrane protein is predicted to contain two homologous halves, each with 6 transmembrane regions and an ABC domain. Amino acid alignment of LmrA and each half of P-glycoprotein indicates that they share 32% identity with an additional 16% conservative substitutions for an overall similarity of 48%. Interestingly, the sequence identity includes particular regions which, in P-glycoprotein, have been implicated as determinants of drug recognition and binding. Indeed, the specificity of both transporters for amphiphilic organic cations appears to be identical, whereas the activity of both systems is reversed by a similar set of inhibitors. Thus, LmrA represents a naturally occurring, functional 'half-molecule' of the human multidrug resistance P-glycoprotein. On the basis of the general domain organization of ABC transporters, two ATP binding domains and two hydrophobic domains [27,32], the lactococcal export system is predicted to function as a homodimeric complex.

Although true P-glycoprotein homologs have not yet been found in bacteria other than *L. lactis*, there is little doubt that many more systems remain to be identified. Recently, physiological evidence has been obtained for the presence of

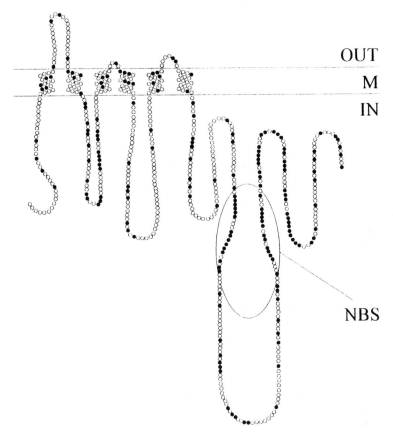

Fig. 2.   Secondary structure model of the ABC-type multidrug transporter LmrA of *L. lactis*. LmrA is predicted to contain 6 transmembrane α-helices and a large cytoplasmic loop encoding an ATP-binding domain. This nucleotide binding site (NBS), identified on the basis of homology to other ABC proteins, is indicated. The structure of LmrA is very similar to that of the mammalian multidrug resistance P-glycoprotein which is predicted to contain 12 transmembrane segments in two homologous halves, each with 6 transmembrane segments and a large cytoplasmic loop encoding an ATP-binding site. In view of the twelve-transmembrane model proposed for P-glycoprotein, LmrA may function as a homodimeric complex. Residues identical between LmrA and the C-terminal half of P-glycoprotein are depicted as closed circles. Out and In refer to the outside and inside of the cytoplasmic membrane (M), respectively.

an ATP-dependent multidrug transporter in *Enterococcus hirae* [136] and *Enterobacter cloacae* [137]. In *E. coli*, the *mdl* gene is predicted to encode a member of the ABC transporter superfamily closely related to the human multidrug resistance P-glycoprotein [138]. Functional studies on this *E. coli* protein have not yet been performed.

Besides the TEXAN LmrP and the P-glycoprotein homolog LmrA, *L. lactis* also possesses an ATP-dependent multidrug extrusion system with specificity for organic anions (Fig. 3). This system was detected in cells of *L. lactis* due to its ability to extrude the fluorescent pH indicator BCECF (2′,7′-bis-(2-carboxyethyl)-5(and 6)-carboxy-

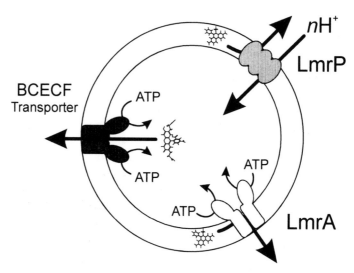

Fig. 3.   Multidrug transporters in *L. lactis*. In addition to the multidrug transporters LmrP and LmrA, *L. lactis* contains a BCECF transporter. This system may be a functional homolog of the mammalian multidrug associated protein (MRP), and may play a role in a glutathione S-transferase-dependent detoxification pathway for electrophilic drugs, and in the excretion of oxidized glutathione during oxygen stress.

fluorescein) [139,140]. BCECF efflux is directed against a concentration gradient and strictly correlates with the cellular ATP concentration. In addition, BCECF efflux is strongly decreased in the presence of ortho-vanadate, a well known inhibitor of P-type ATPases and ABC transporters. Most convincingly, a UV mutant with a strongly reduced efflux rate could be isolated from a BCECF-loaded and lactose-energized cell population by selection of highly fluorescent cells in a flow cytometer-cell sorter [140]. Studies in inside-out membrane vesicles of *L. lactis* have indicated that the BCECF transporter is not only able to transport carboxyfluorescein derivatives, but also glutathione conjugates such as dinitrophenyl-glutathione [141]. Interestingly, the substrate specificity of the BCECF transporter of *L. lactis* is remarkably similar to that of the human multidrug resistance-associated protein (MRP) [142–145]. The overexpression of this ABC transporter in human cancer cells results in a P-glycoprotein-independent multidrug resistance phenotype. In hepatocytes, MRP plays a role in a glutathione S-transferase-dependent detoxification pathway for electrophilic drugs, and in the excretion of oxidized glutathione (GSSG) during oxygen stress [146]. The BCECF transporter of *L. lactis* may have functions similar to those of MRP. Although *L. lactis* contains a manganese-dependent superoxide dismutase to detoxify oxygen radicals, the enzyme is not essential for the aerotolerance of this fermentative bacterium [147]. An alternative protection mechanism against oxygen damage may be based on the detoxification of oxygen radicals by reduced glutathione [148]. This compound is present at high concentrations in *L. lactis* and other facultative anaerobic bacteria [149]. During oxygen stress, optimal levels of reduced versus oxidized glutathione may be maintained through the excretion of GSSG via the BCECF transporter.

5. Regulation of multidrug resistance

The expression of drug-specific efflux transporters is usually induced by the drugs themselves through their interaction with DNA-binding regulatory proteins. The best understood example is the regulation of bacterial tetracycline-efflux transporter TetA(B), where the binding of tetracycline to a specific repressor protein (TetR) causes the repressor to dissociate from the promoter of the transporter gene thereby allowing transcription [38,150,151]. Studies on the regulation of the expression of Bmr, EmrAB, and MtrCD suggest that specific regulator proteins also play a role in the regulation of the expression of multidrug transporters.

Drugs recognized by Bmr enhance the expression of this multidrug transporter. In drug resistant mutants, Bmr can be overexpressed by gene amplification. In wild type cells, however, the regulation of gene expression is dependent upon expression of the regulatory protein BmrR which is encoded downstream of the *bmr* gene [152]. BmrR shows sequence homology and functional similarity with several known bacterial transcription activator proteins, and contains a conserved N-terminal DNA binding domain. Upon binding of drugs, the protein binds to the *bmr* promoter and induces Bmr expression. Interestingly, the *B. subtilis* genome encodes a second multidrug efflux transporter, Blt, which shares 51% amino acid sequence identity with Bmr [153]. Both transporters have a similar drug specificity but are differentially expressed in response to drugs. The expression of Blt is regulated by BltR, a transcriptional activator. Although the DNA-binding domains of BmrR and BltR are related, their putative drug-binding domains are dissimilar. Apparently, the expression of Bmr and Blt is mediated by regulatory proteins which, like their regulated transporters, are capable of recognizing structurally diverse compounds.

Like Bmr and Blt, the expression of EmrAB in *E. coli* is enhanced in the presence of the drugs which are recognized by this multidrug efflux system. However, regulation of the expression of EmrAB is dependent on a negative regulator, EmrR, which is encoded upstream of *emr*AB [154]. Whether the transcriptional repression caused by EmrR is due to direct binding to the *emrAB* promoter is not known. In contrast to BmrR and BltR, EmrR does not contain known DNA-binding motifs.

MtrCD expression in *N. gonorrhoeae* is negatively regulated by a repressor protein, MtrR, which is homologous to TetR [98,155]. Mutations that inactivate MtrR result in an increase in the level of MtrCD expression. Interestingly, the MtrR protein was found to be homologous to previously unidentified open reading frames (*acr*R and *env*R) that are upstream of the *E. coli acr*AB and *env*CD operons [155]. Mutations in these regulatory genes may result in increased levels of AcrAB and EnvCD proteins, and hence, in increased resistance of *E. coli* to antimicrobial agents.

Besides the specific regulatory mechanisms that affect the expression of single multidrug efflux systems, global regulatory mechanisms have evolved in bacteria that affect the expression of (dedicated) drug efflux systems and other proteins involved in the intrinsic resistance of the cell. The *mar* regulon of *E. coli* [156, 157] and various other bacteria [158–161] is involved in chromosomally mediated multiple antibiotic and superoxide resistance. The inducing agents of this regulon are often used in clinical situations: the antibiotics tetracycline and chloramphenicol, and aromatic weak acids including salicylate and acetylsalicylate. The agents stimulate the transcription of the *mar*RAB operon [156]. Specifically, expression of *mar*A, encoding a putative transcriptional activator [162], elevates the expression of about 10 unlinked genes that affect the outer membrane permeation [156, 163], antibiotic efflux [164,165], and the reducing potential of the cell [166]. Although salicylate appears to induce the *mar* operon by binding to the negative regulator MarR, tetracycline and chloramphenicol do not bind to the regulator. The latter compounds must therefore induce the *mar* operon by an indirect mechanism in which MarB may be involved [167]. Besides the *mar*RAB-dependent pathway for multiple antibiotic and superoxide resistance, the *sox*RS genes provide *E. coli* with a *mar*-independent pathway [156,157,168–171]. SoxS is a transcriptional activator homologous to MarA (39% identical residues), which elevates the transcription of the operons of the *mar* regulon [162]. SoxR is a sensor of superoxides that negatively regulates the *sox*RS gene [171]. The dual regulation by *mar*RAB and *sox*RS of a common set of genes may enable different environmental signals to trigger responses [157].

6. Structure/function relationships in (multi)drug transporters

Drug efflux proteins perform two basic functions: the translocation of drugs and the conversion of the (electro)chemical energy of the driving force into that of the drug gradient. In the absence of three-dimensional structural information, the identification of structure/function relationships in drug transporters has relied on the scrutiny of the primary sequence and on biochemical and genetic analysis.

Multiple alignment of the amino acid sequences of the TEXANs LmrP, Bmr, NorA, and TetA(B) reveals the presence of two conserved sequence motifs that are located at similar positions within the putative secondary structure of the proteins. Motif A [GXXXDRXGR(K/R)] is found in many members of the Major Facilitator superfamily [24,172,173] and is present in the cytoplasmic loop between transmembrane segment (TMS) 2 and 3. Motif B (GpilGPvlGG), the drug extrusion consensus sequence, is found at the end of TMS 5, and is typical for LmrP (Fig. 1) and other drug export systems of the Major Facilitator superfamily [23,24,174].

Residues important for catalytic function of secondary drug transporters in prokaryotes have been studied most extensively in TetA(B). Functional studies of site-directed mutants of TetA(B) showed that the negatively charged aspartate (Asp[66]) and the positively charged arginine (Arg[70]) present in Motif A are essential for transport [172–175]. Possibly, Motif A is of structural importance to the transporters, but it may also be involved in the conformational changes required for opening and closing of the translocation pathway. Motif A is also present in a modified form in the cytoplasmic loop between TMS 8 and 9 (Motif A') [176]. None of the residues in this region are essential for transport. It has been proposed that Motif A forms an active leaflet in TetA(B) whereas Motif A' forms a silent counterpart in the second half of the molecule [177].

TetA(B) contains four charged residues in putative transmembrane segments (Asp^{15}, Asp^{84}, His^{257} and Asp^{285}). His^{257} and Asp^{285} mutants have tetracycline transport without proton coupling, pointing to the involvement of these residues in metal–tetracycline coupled proton transport [178,179]. Substitution of Asp^{15} and Asp^{84} in TMS 1 and 3, respectively, by neutral residues affects the apparent affinity of TetA(B) for metal–tetracycline [179,180]. Using random mutagenesis, two residues have been identified in the second half of TetA(B) that play a role in the drug specificity (Trp^{231} and Leu^{308}) [181]. These residues are located within TMS 7 and 9, which are the equivalents of TMS 1 and 3 in the second half of the TetA(B) protein. TMS 1, 3, 7 and 9 may be involved in the recognition of tetracycline through their participation in a pore-like structure. However, it cannot be excluded that the various mutations in these transmembrane domains affect a distantly located drug binding site via intramolecular interactions. Currently, the role of Motif B in drug recognition and binding by TetA(B) is unclear.

Similar to the observations on TetA(B), residues within the transmembrane segments of TEXANs and Mini TEXANs can play a role in the substrate specificity or transport function of these transporters. The ability of Bmr to bind reserpine is strongly affected by substitutions of Val^{286} which is located within TMS 8 [182]. The substrate specificity of the mini TEXAN QacC is altered by mutations in TMS 2 (Pro^{31} and Cys^{42}) [117]. Glu^{13} mutants (TMS 1) of Smr may have impaired drug/proton antiport activity [118]. Mutations in TMS 3 (QacC Tyr^{59} and QacC/Smr Trp^{62}) result in low levels of protein expression, suggesting that these residues are involved in the stability or structure of the Mini TEXANs [117,129].

7. Molecular models of multidrug transport

Several models have been postulated for the pump function of multidrug transporters to explain their broad specificity for chemically unrelated compounds (Fig. 4). Drug translocation may involve substrate transport from the cytoplasm to the exterior which would require an enormous flexibility of an 'enzyme-like' substrate recognition site (conventional transport hypothesis) [183,184]. Alternatively, multidrug transporters could recognize the lipophilic drugs by their physical property to intercalate into the lipid bilayer (vacuum cleaner hypothesis) [185], and transport drugs from the inner leaflet to the outer leaflet of the lipid bilayer or to the exterior (flippase hypothesis) [186,187].

On the other hand, the extraordinary broad substrate specificity of multidrug transporters has led to the proposal of alternative, indirect mechanisms via which such proteins reduce the intracellular drug concentration. In one hypothesis, multidrug transporters may affect the electrochemical proton gradient across the plasma membrane [188,189]. Many of the transported drugs are weak bases having a

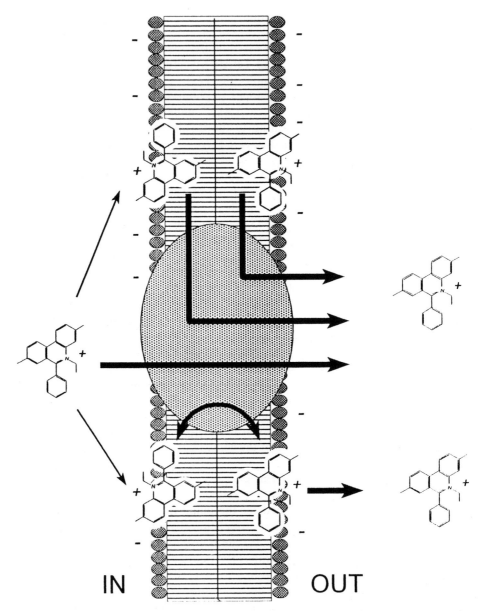

Fig. 4. Molecular models of substrate binding and drug extrusion by multidrug transporters. See text
for further explanation.

positive charge at neutral pH. An increase in the transmembrane pH gradient
(interior alkaline) and/or decrease in the transmembrane membrane potential (interior negative), observed in human multidrug resistant cancer cells [189–191],
would result in a decrease in the cellular drug concentration and, hence, in increased

drug resistance. Currently, no information is available with respect to changes in the magnitude or composition of the proton motive force in multidrug resistant bacteria. However, the observation of LmrA-mediated drug transport in the absence of a proton motive force favours a direct drug extrusion mechanism [135].

Cells expressing LmrA were found to actively extrude the hydrophobic acetoxymethyl ester of BCECF [192]. Similar observations on the human multidrug resistance P-glycoprotein were taken as evidence for the transport of hydrophobic compounds from the plasma membrane [193]. The transport mechanism of LmrA has been studied in greater detail by using the amphiphilic membrane probe TMA-DPH. Intercalation of TMA-DPH in the phospholipid bilayer is based on a fast partitioning of the probe in the outer leaflet, followed by a slower transbilayer movement of the probe from the outer to the inner leaflet of the membrane. The effect of the partitioning of TMA-DPH in the membrane on the drug pumping activity of LmrA suggests that LmrA transports TMA-DPH from the inner leaflet of the membrane [192]. Thus, LmrA recognizes drugs by their hydrophobic properties as suggested by the 'flippase' and 'vacuum cleaner' models.

8. Concluding remarks

Significant progress has been made towards the understanding of the cell biological mechanisms through which prokaryotes develop multidrug resistance. The majority of bacterial multidrug transporters characterized thus far, operates via a secondary transport mechanism. A number of genes encoding TEXANs and Mini TEXANs have been cloned and sequenced. Analysis of these primary sequences has revealed a striking similarity in the overall structure, suggesting that the proteins may function via a similar mechanism. Transport studies in membrane vesicles of *L. lactis* and *E. coli* have demonstrated the drug/proton antiport mechanism of the TEXAN LmrP. Smr and EmrE have been purified and characterized upon reconstitution into proteoliposomes. These studies have established that a single multidrug resistance protein is able to function as a drug pump, and have shown the Δp-dependence of drug transport via the Mini TEXANs.

A new development in the field has been the discovery of the ABC-type multidrug transporter LmrA in *L. lactis*. Interestingly, the substrate specificity of LmrA shows overlap with that of the human multidrug resistance P-glycoprotein, the TEXANs LmrP and Bmr, and Mini TEXANs Smr, EmrE and QacC. On the other hand, the substrate specificity of these multidrug transporters is very different from that of the TetA(B) and other transporters dedicated to the efflux of a specific drug or class of drugs. The molecular basis for these (dis)similarities is not well understood.

To obtain structure/function information of these proteins, mutant proteins can be isolated on the basis of sequence homology, properties of certain amino acids in relation to their location in putative secondary structure, or by selection for a specific phenotype. Although these studies are only beginning to emerge, the first results suggest that the residues important for substrate recognition are located in the hydrophobic transmembrane segments, rather than the hydrophilic loops.

Interestingly, most drugs recognized by multidrug transporters have the ability to intercalate into the phospholipid bilayer due to their amphiphilic nature. In their 'flippase model', Higgins and Gottesman [186] suggest that this partitioning of hydrophobic drugs in the membrane could be a first step in specificity of multidrug transporters. The subsequent interaction between the drugs and a fairly non-specific binding site on the transport protein would be a second determinant of drug specificity. Evidence has been presented that P-glycoprotein and LmrA recognize drugs by their hydrophobic properties. Thus, a major physiological function of multidrug transporters could involve the efflux of hydrophobic compounds which are produced by metabolism or which are encountered in the natural environment.

Acknowledgements

We thank Bert Poolman and Arnold Driessen for valuable discussions. Arnold Driessen is gratefully acknowledged for providing Fig. 3. Work in the authors laboratory is supported by the European Community (BIO2CT-930145) and the Dutch Cancer Society (RUG 96-1218).

References

1. Cohen, M.L. (1992) Science **257**, 1050–1054.
2. Bloom, B.R. and Murray, C.J.L. (1992) Science **257**, 1055–1064.
3. Neu, H.C. (1992) Science **257**, 1064–1073.
4. Lyon, B.R. and Skurray, R.A. (1987) Microbiol. Rev. **51**, 88–134.
5. Speer, B.S., Shoemaker, N.B. and Salyers, A.A. (1992) Clin. Microbiol. Rev. **5**, 387–399.
6. Davies, J. (1994) Science **264**, 375–382.
7. Spratt, B.G. (1994) Science **264**, 388–393.
8. Nikaido, H. (1994) Science **264**, 382–388.
9. Nikaido, H. and Vaara, M. (1985) Microbiol. Rev. **49**, 1–32.
10. Beveridge, T.J. and Graham, L.I. (1991) Microbiol. Rev. **55**, 684–705.
11. Nikaido, H., Kim, S.-H., Rosenberg, E.Y. (1993) Mol. Microbiol. **8**, 1025–1030.
12. Nikaido, H. (1989) Antimicrob. Agents Chemother. **33**, 1831–1836.
13. Pugsley, A.P. and Schnaitman, C.A. (1978) J. Bacteriol. **133**, 1181–1189.
14. Jaffé, A., Chabbert, Y.A. and Derlot, E. (1983) Antimicrob. Agents Chemother. **23**, 622–625.
15. Curtis, N.A.C., Eisenstadt, R.L., Turner, K.A. and White, A.J. (1985) J. Antimicrob. Chemother. **15**, 642–650.
16. Levy, S.B. (1992) Antimicrob. Agents Chemother. **36**, 695–703.
17. Lewis, K. (1994) Trends Biochem. Sci. **19**, 119–123.
18. Gottesman, M.M. (1993) Cancer Res. **53**, 747–754.
19. Shapiro, A.B. and Ling, V. (1995) J. Bioenerg. Biomembr. **27**, 7–13.
20. Bergogne-Bérézin, E. and Joly-Guillou (1991) in: The Biology of Acinetobacter, eds K.J. Towner, E. Bergogne-Bérézin and C.A. Fewson, pp. 83–115, Plenum Press, New York.
21. Poolman, B. and Konings, W.N. (1993) Biochim. Biophys. Acta **1183**, 5–39
22. Schuldiner, S., Shirvan, A. and Linial, M. (1995). Physiol. Rev. **75**, 369–392.
23. Paulsen, I.T., and Skurray, R.A. (1993) Gene **124**, 1–11.
24. Marge, M. and Saier, M.H. (1993) Trends Biochem. Sci. **18**, 13–20.
25. Saier, M.H., Tam, R., Reizer, A. and Reizer, J. (1994) Mol. Microbiol. **11**, 841–847.
26. Paulsen, I.T., Littlejohn, T.G, Rådström, P., Sundström, L., Sköld, O., Swedberg, G. and Skurray, R.A. (1993) Antimicrob. Agents Chemother. **37**, 761–768.
27. Higgins, C.F. (1992) Annu. Rev. Cell Biol. **8**, 67–113.

28. Hyde, S.C., Emsley, P., Hartshorn, M.J., Mimmack, M.M., Gileadi, U., Pearce, S.R., Gallagher, M.P., Gill, D.R., Hubbard, R.E. and Higgins, C.F. (1990) Nature **346**, 362–365.
29. Mimura, C.S., Holbrook, S.R. and Ames, G.F.-L. (1991) Proc. Natl. Acad. Sci. USA **88**, 84–88.
30. Yoshida, M. and Amano, T. (1995) FEBS Lett. **359**, 1–5.
31. Senior, A.E., Al-Shawi, M.K. and Urbatsch, I.L. (1995) J. Bioenerg. Biomembr. **27**, 31–36.
32. Fath, M.J. and Kolter, R. (1993) Microbiol. Rev. **57**, 995–1017.
33. Dean, D.A., Davidson, A.L. and Nikaido, H. (1989) Proc. Natl. Acad. Sci. USA **86**, 4254–4260.
34. Ames, G.F.-L., Nikaido, H., Groake, J. and Petithory, J. (1989) J. Biol. Chem. **246**, 3398–4002.
35. Bishop, L., Agbayani, R., Ambudkar, S.V., Maloney, P.C. and Ames, G.F.-L. (1989) Proc. Natl. Acad. Sci. USA **86**, 6953–6957.
36. McMurry, L., Petrucci, R.E. and Levy, S.B. (1980) Proc. Natl. Acad. Sci. USA **77**, 3974–3977.
37. Chopra, I., Hawkey, P.M. and Hinton, M. (1992) J. Antimicrob. Chemother. **29**, 245–277.
38. Hillen, W. and Berens, C. (1994) Annu. Rev. Microbiol. **46**, 345–369.
39. Jorgensen, R.A. and Reznikoff, W.S. (1979) J. Bacteriol. **138**, 705–714.
40. Kaneko, M., Yamaguchi, A. and Sawai, T. (1985) FEBS Lett. **193**, 194–198.
41. Nelson, M.L., Park, B.H., Andrews, J.S., Georgian, V.A., Thomas, R.C. and Levy, S.B. (1993) J. Med. Chem. **36**, 370–377.
42. Yamaguchi, A., Udagawa, T. and Sawai, T. (1990) J. Biol. Chem. **265**, 4809–4813.
43. Yamaguchi, A., Iwasaki-Ohba, Y., Ono, N., Kaneko-Ohdera, M. and Sawai, T. (1991) FEBS Lett. **282**, 415–418.
44. McMurry, L.M., Park, B.H., Burdett, V. and Levy, S.B. (1987) Antimicrobiol. Agents Chemother. **31**, 1648–1650.
45. Hoshino, T., Ikeda, T., Tomizuka, N. and Furukawa, K. (1985) Gene **37**, 131–138.
46. Ishiwa, H. and Shibahara, H. (1985) Jpn. J. Genet. **60**, 485–498.
47. Mojumdar, M. and Khan, S.A. (1988) J. Bacteriol. **170**, 6119–6122.
48. Guffanti, A.A. and Krulwich, T.A. (1995) J. Bacteriol. **177**, 4557–4561.
49. Cheng, J., Guffanti, A.A. and Krulwich, T.A. (1994) J. Biol. Chem. **269**, 27365–27371.
50. Guay, G.G., Tuckman, M., McNicholas, P. and Rothstein, D.M. (1993) J. Bacteriol. **175**, 4927–4929.
51. Eckert, B. and Beck, C.F. (1989) J. Biol. Chem. **264**, 11663–11670.
52. Yamaguchi, A., Adachi, K. and Sawai, T. (1990) FEBS Lett. **265**, 17–19.
53. Allard, J.D. and Bertrand, K.P. (1992) J. Biol. Chem. **267**, 17809–17819.
54. Rubin, R.A., Levy, S.B., Heinrikson, R.L. and Kezdy, F.S. (1990) Gene **87**, 7–13.
55. Curiale, M. and Levy, S.B. (1982) J. Bacteriol. **151**, 209–215.
56. Curiale, M., McMurry, L.M. and Levy, S.B. (1984) J. Bacteriol. **157**, 211–217.
57. Hickman, R.K. and Levy, S.B. (1988) J. Bacteriol. **170**, 1715–1720.
58. McNicholas, P., McGlynn, M., Guay, G.G. and Rothstein, D.M. (1995) J. Bacteriol. **177**, 5355–5357.
59. Rubin, R.A. and Levy, S.B. (1990) J. Bacteriol. **172**, 2303–2312.
60. Rubin, R.A. and Levy, S.B. (1991) J. Bacteriol. **173**, 4503–4509.
61. George, A.M. and Levy, S.B. (1983) J. Bacteriol. **155**, 531–540.
62. Bissonnette, L., Champetier, S., Buisson, J.-P. and Roy, P.H. (1991) J. Bacteriol. **173**, 4493–4502.
63. Jones, C.S., Osborne, D.J. and Stanley, J. (1992) Lett. Appl. Microbiol. **15**, 106–108.
64. Hansen, L.M., McMurry, L.M., Levy, S.B. and Hirsh, D.C. (1993) Antimicrob. Agents Chemother. **37**, 2699–2705.
65. Sloan, J., McMurry, L.M., Lyras, D., Levy, S.B. and Rood, J.I. (1994) Mol. Microbiol. **11**, 403–415.
66. Bentley, J., Hyatt, L.S., Ainsley, K., Parish, J.H., Herbert, R.B., and White, G.R. (1993) Gene **127**, 117–120.
67. Endicott, J.A. and Ling, V. (1989) Annu. Rev. Biochem. **58**, 137–171.
68. Gottesman, M.M. and Pastan, I. (1993) Annu. Rev. Biochem. **62**, 385–427.
69. Kern, W.V., Andriof, E., Oethinger, M., Kern, P., Hacker, J. and Marre, R. (1994) Antimicrob. Agents Chemother. **38**, 681–687.
70. Heisig, P. and Tschorny, R. (1994) Antimicrob. Agents Chemother. **38**, 1284–1291.
71. Piddock, L.J.V., Griggs, D.J., Hall, M.C. and Jin, Y.F. (1993) Antimicrob. Agents Chemother. **37**, 662–666.
72. López-Brea, M. and Alarcón, T. (1990) Eur. J. Clin. Microbiol. Infect. Dis. **9**, 345–347.

73. Fujimaki, K., Fujii, T., Aoyama, H., Sato, K.-I., Inoue, Y., Inoue, M. and Mitsuhashi, S. (1989) Antimicrob. Agents Chemother. **33**, 785–787.

74. Ishida, H., Fuziwara, H., Kaibori, Y., Horiuchi, T., Sato, K. and Osada, Y. (1995) Antimicrob. Agents Chemother. **39**, 453–457.

75. Kaatz, G.W., Seo, S.M. and Ruble, C.A. (1991) J. Infect. Dis. **163**, 1080–1086.

76. Yoshida, S., Kojima, T., Inoue, M. and Mitsuhashi, S. (1991) Antimicrob. Agents Chemother. **35**, 368–370.

77. Tankovic, J., Desplaces, N., Duval, J. and Courvalin, P. (1994) Antimicrob. Agents Chemother. **38**, 1149–1151.

78. Acar, J.F. and Francoual, S. (1990) J. Antimicrob. Chemother. **26**, 207–213.

79. Neu, H.C. (1992) Annu. Rev. Med. **43**, 465–486.

80. Bolhuis, H., Molenaar, D., Poelarends, G., Van Veen, H.W., Poolman, B., Driessen, A.J.M. and Konings, W.N. (1994) J. Bacteriol. **176**, 6957–6964.

81. Bolhuis, H., Poelarends, G., Van Veen, H.W., Poolman, B., Driessen, A.J.M. and Konings, W.N. (1995) J. Biol. Chem. **270**, 26092–26098.

82. Neyfakh, A.A., Bidnenko, V.E. and Chen, L.B. (1991) Proc. Natl. Acad. Sci. USA **88**, 4781–4785.

83. Yoshida, H., Bogaki, M., Nakamura, S., Ubukata, K. and Konno, M. (1990) J. Bacteriol. **172**, 6942–6949.

84. Kaatz, G.W., Seo, S.M. and Ruble, C.A. (1993) Antimicrob. Agents Chemother. **37**, 1086–1094.

85. Ng, E.Y.W., Trucksis, M. and Hooper, D.C. (1994) Antimicrob. Agents Chemother. **38**, 1345–1355.

86. Pearce, H.L., Safa, A.R., Bach, N.J., Winter, M.A., Cirtain, M.C. and Beck, W.T. (1989) Proc. Natl. Acad. Sci. USA **86**, 319–335.

87. Thanassi, D.G., Suh, G.S.B. and Nikaido, H. (1995) J. Bacteriol. **177**, 998–1007.

88. Wandersman, C. and Delepelaire, P. (1990) Proc. Natl. Acad. Sci. USA **87**, 4776–4780.

89. Wandersman, C. (1993) Trends Genet. **8**, 317–321.

90. Lomovskaya, O. and Lewis, K. (1992) Proc. Natl. Acad. Sci. USA **89**, 8938–8942.

91. Lewis, K., Naroditskaya, V., Ferrante, A. and Fokina, I. (1994) J. Bioenerg. Biomembr. **20**, 639–646.

92. Furukawa, H., Tsay, J.-T., Jackowski, S., Takamura, Y. and Rock, C.O. (1993) J. Bacteriol. **175**, 3723–3729.

93. Ma, D., Cook, D.N., Alberti, M., Pon, N.G., Nikaido, H. and Hearst, J.E. (1993) J. Bacteriol. **175**, 6299–6313.

94. Ma, D., Cook, D.N., Alberti, M., Pon, N.G., Nikaido, H. and Hearst, J.E. (1995) Mol. Microbiol. **16**, 45–55.

95. Klein, J.R., Henrich, B.H. and Plapp, R. (1991) Mol. Gen. Genet. **230**, 230–240.

96. Poole, K., Krebes, K., McNally, C. and Neshat, S. (1993) J. Bacteriol. **175**, 7363–7372.

97. Poole, K. Heinrichs, B. and Neshat, S. (1993) Mol. Microbiol. **10**, 529–544.

98. Hagman, K.E., Pan, W., Spratt, B.G., Balthazar, J.T., Judd, R.C. and Shafer, W.M. (1995) Microbiology **141**, 611–622.

99. Bochner, B., Huang, H.–C., Schieven, G.L. and Ames, B.N. (1980) J. Bacteriol. **143**, 926–933.

100. Ettner, N., Hillen, W. and Ellestad, G.A. (1993) J. Am. Chem. Soc. **115**, 2546–2548.

101. Moss, M.L. and Mellon, M.G. (1942) Ind. Eng. Chem. **14**, 862–865.

102. White, J.R. and Yeowell, H.N. (1982) Biochem. Biophys. Res. Commun. **106**, 407–411.

103. Yeowell, H.N. and White, J.R. (1982) Antimicrob. Agents Chemother. **22**, 961–968.

104. Demange, P., Wendenbaum, S., Bateman, A., Dell, A. and Abdallah, M.A. (1987) in: Iron Transport in Microbes, Plants and Animals, eds G. Winkelmann, D. Van Der Helm and J.B. Neilands, pp. 167–187, VCH Verlagsgesellschaft mbH, Weinheim, Germany.

105. Nakamura, H. (1965) J. Bacteriol. **90**, 8–14.

106. Nakamura, H. (1974) J. Gen. Microbiol. **89**, 85–93.

107. Nakamura, H. (1968) J. Bacteriol. **96**, 987–996.

108. Coleman, W.G., and Leive, L. (1979) J. Bacteriol. **139**, 899–910.

109. Hase, A., Funatsuki, K., Kawakami, M. and Nakamura, H. (1990) Plant Cell Physiol. **31**, 1053–1057.

110. Sedgwick, E.G. and Bragg, P.D. (1992) Biochim. Biophys. Acta **1099**, 45–50.

111. Sedgwick, E.G. and Bragg, P.D. (1992) Biochim. Biophys. Acta **1099**, 51–56.

112. Lentner, C. (1981) in: Geigy Scientific Tables, 8th ed., vol. 1, pp. 139–146, Ciba-Geigy Corp., Basle.

113. Tennent, J.M., Lyon, B.R., Midgley, M., Jones, I.G., Purewal, A.S. and Skurray, R.A. (1989) J. Gen. Microbiol. **135**, 1–10.
114. Rouch, D.A., Cram, D.S., DiBerardino, D., Littlejohn, T.G. and Skurray, R.A. (1990) Mol. Microbiol. **4**, 2051–2062.
115. Littlejohn, T.G., DiBerardino, D., Messerotti, L.J., Spiers, S.J. and Skurray, R.A. (1990) Gene **101**, 59–66.
116. Littlejohn, T.G., Paulsen, I.T., Gillespie, M.T., Tennent, J.M., Midgley, M., Jones, I.G., Purewal, A.S. and Skurray, R.A. (1992) FEMS Microbiol. Lett. **95**, 259–266.
117. Paulsen, I.T., Brown, M.H., Dunstan, S.J. and Skurray, R.A. (1995) J. Bacteriol. **177**, 2827–2833.
118. Grinius, L., Dreguniene, G., Goldberg, E.B., Liao, C.-H. and Projan, S.J. (1992) Plasmid **27**, 119–129.
119. Yerushalmi, H., Lebendiker, M. and Schuldiner, S. (1995) J. Biol. Chem. **270**, 6856–6863.
120. Purewal, A.S., Jones, I.G. and Midgley, M. (1990) FEMS Microbiol. Lett. **68**, 73–76.
121. Purewal, A.S. (1991) FEMS Microbiol. Lett. 82, 229–232.
122. Morimyo, M., Hongo, E., Hama-Inaba, H. and Machida, I. (1992) Nuc. Acids Res. **20**, 3159–3165.
123. Midgley, M., Iscandar, N.S. and Dawes, E.A. (1986) Biochim. Biophys. Acta **856**, 45–49.
124. Midgley, M., Iscandar, N.S., Parish, M. and Dawes, E.A. (1986) FEMS Microbiol. Lett. **34**, 187–190.
125. Midgley, M. (1986) J. Gen. Microbiol. 132, 3187–3193.
126. Midgley, M. (1987) Microbiol. Sci. **4**, 125–127.
127. Harold, F.M. (1986) The Vital Force: a Study of Bioenergetics, Freeman, New York.
128. Bolhuis, H., Van Veen, H.W., Brandts, J.-R., Poolman, B., Driessen, A.J.M. and Konings, W.N. (1996) Submitted for publication.
129. Grinius, L.L. and Goldberg, E.B. (1994) J. Biol. Chem. **269**, 29998–30004.
130. Guilfoile, P.G. and Hutchinson, C.R. (1991) Proc. Natl. Acad. Sci. USA **88**, 8553–8557.
131. Linton, K.J. Cooper, H.N., Hunter, I.S. and Leadlay, P.F. (1994) Mol. Microbiol. **11**, 777–785.
132. Barrasa, M.I., Tercero, J.A., Lacalle, R.A. Jimenez, A. (1995) Eur. J. Biochem. **228**, 562–569.
133. Rodriguez, A.M., Olano, C., Vilches, C., Méndez, C. and Salas, J.A. (1993) Mol. Microbiol. **8**, 571–582.
134. Podlesek, Z., Comino, A., Herzog-Velikonja, B., gur-Bertok, D., Komel, R. and Grabnar, M. (1995) Mol. Microbiol. **16**, 969–976.
135. Van Veen, H.W., Venema, K., Bolhuis, H., Oussenko, I., Kok, J., Poolman, B., Driessen, A.J.M. and Konings, W.N. (1996) PNAS, in press.
136. Midgley, M. (1994) FEMS Microbiol. Lett. **120**, 119–124.
137. Aspedon, A. and Nickerson, K.W. (1993) Can. J. Microbiol. **40**, 184–191.
138. Allikmets, R., Gerrard, B., Court, D. and Dean, M. (1993) Gene **136**, 231–236.
139. Molenaar, D., Abee, T. and Konings, W.N. (1991) Biochim. Biophys. Acta **1115**, 75–83.
140. Molenaar, D., Bolhuis, H., Abee, T., Poolman, B. and Konings, W.N. (1992) J. Bacteriol. **174**, 3118–3124.
141. Van Veen, H.W. and Müller, M. (1995) Unpublished data.
142. Müller, M., Meijer, C., Zaman, G.J.R., Borst, P., Scheper, R.J., Mulder, N.H., De Vries, E.G. and Jansen, P.L.M. (1994) Proc. Natl. Acad. Sci. USA **91**, 13033–13037.
143. Leier, I., Jedlitschky, G., Buchholz, U., Cole, S.P.C., Deeley, R.G. and Keppler, D. (1994) J. Biol. Chem. **269**, 27807–27810.
144. Zaman, G.J.R., Flens, M.J., Van Leusden, M.R., De Haas, M., Mülder, H.S., Lankelma, J., Pinedo, H.M., Scheper, R.J., Baas, F., Broxterman, H.J. and Borst, P. (1994) Proc. Natl. Acad. Sci. USA **91**, 8822–8826.
145. Feller, N., Broxterman, H.J., Währer, D.C.R. and Pinedo, H.M. (1995) FEBS Lett. **368**, 385–388.
146. Ishikawa, T. (1992) Trends Biochem. Sci. **17**, 463–468.
147. Sanders, J.W., Leenhouts, K.J., Haandrikman, A.J., Venema, G. and Kok, J. (1995) J. Bacteriol. **177**, 5254–5260.
148. Meister, A. (1988) J. Biol. Chem. **263**, 17205–17208.
149. Fahey, R.C., Brown, W.C., Adams, W.B. and Worsham, M.B. (1978) J. Bacteriol. **133**, 1126–1129.
150. Hinrichs, W., Kisker, C., Düvel, M., Müller, A., Tovar, K., Hillen, W. and Saenger, W. (1994) Science **264**, 418–420.
151. Kisker, C., Hinrichs, W., Tovar, K., Hillen, W. and Saenger, W. (1995) J. Mol. Biol. **247**, 260–280.

152. Ahmed, M., Borsch, C.M., Taylor, S.S., Vázquez-Laslop, N. and Neyfakh, A.A. (1994) J. Biol. Chem. **269**, 28506–28513.

153. Ahmed, M., Lyass, L., Markham, P.N., Taylor, S.S., Vázquez-Laslop, N. and Neyfakh, A.A. (1995) J. Bacteriol. **177**, 3904–3910.

154. Lomovskaya, O., Lewis, K. and Matin, A. (1995) J. Bacteriol. **177**, 2328–2334.

155. Pan, W. and Spratt, B.G. (1994) Mol. Microbiol. **11**, 769–775.

156. Cohen, S.P., Levy, S.B., Foulds, J. and Rosner, J.L. (1993) J. Bacteriol. **175**, 7856–7862.

157. Rosner, J.L. and Slonczewski, J.L. (1994) J. Bacteriol. **176**, 6262–6269.

158. Sawai, T., Hirano, S. and Yamaguchi, A. (1987) FEMS Microbiol. Lett. **40**, 233–237.

159. George, A.M., Hall, R.M. and Stokes, H.W. (1995) Microbiology **141**, 1909–1920.

160. Zhanel, G.G., Karlowsky, J.A., Saunders, M.H., Davidson, R.J., Hoban, D.J., Hancock, R.E.W., McLean, I. and Nicolle, L.E. (1995) Antimicrob. Agents Chemother. **39**, 489–495.

161. Morshed, S.R., Lei, Y., Yoneyama, H. and Nakae, T. (1995) Biochem. Biophys. Res. Commun. **210**, 356–362.

162. Gambino, L., Gracheck, S.J. and Miller, P.F. (1993) J. Bacteriol. **175**, 2888–2894.

163. Hooper, D.C., Wolfson, J.S., Bozza, M.A. and Ng, E.Y. (1992) Antimicrob. Agents Chemother. **36**, 1151–1154.

164. McMurray, L.M., George, A.M. and Levy, S.B. (1994) Antimicrob. Agents Chemother. **38**, 542–546.

165. Seoane, A.S. and Levy, S.B. (1995) J. Bacteriol. **177**, 530–535.

166. Ariza, R.R., Cohen, S.P., Bachhawat, N., Levy, S.B. and Demple, B. (1994) J. Bacteriol. **176**, 143–148.

167. Martin, R.G. and Rosner, J.L. (1995) Proc. Natl. Acad. Sci. USA **92**, 5456–5460.

168. Greenberg, J.T., Monach, P., Chou, J.H., Josephy, P.D. and Demple, B. (1990) Proc. Natl. Acad. Sci. USA **87**, 6181–6185.

169. Amabile-Cuevas, C.F. and Demple, B. (1991) Nucl. Acid Res. **19**, 4479–4484.

170. Wu, J. and Weiss, B. (1992) J. Bacteriol. **174**, 3915–3920.

171. Hidalgo, E. and Demple, B. (1994) EMBO J. **13**, 138–146.

172. Yamaguchi, A., Ono, N., Akasaki, T. Noumi, T. and Sawai, T. (1990) J. Biol. Chem. **265**, 15525–15530.

173. Yamaguchi, A., Someya, Y. and Sawai, T. (1992) J. Biol. Chem. **267**, 19155–19162.

174. Griffith, J.K., Baker, M.E., Rouch, D.A., Page, M.G.P., Skurray, R.A., Paulsen, I.T., Chater, K.F., Baldwin, S.A. and Henderson, P.J.F. (1992) Curr. Opin. Cell Biol. **4**, 684–695.

175. Yamaguchi, A., Nakatani, M. and Sawai, T. (1992) Biochemistry **31**, 8344–8348.

176. Yamaguchi, A., Kimura, T., Someya, Y. and Sawai, T. (1993) J. Biol. Chem. **268**, 6496–6504.

177. Yamaguchi, A., Kimura, T. and Sawai, T. (1994) J. biochem. **115**, 958–964.

178. Yamaguchi, A., Adachi, K., Akasaki, T., Ono, N. and Sawai, T. (1991) J. Biol. Chem. **266**, 6045–6051.

179. Yamaguchi, A., Akasaka, T., Ono, N., Someya, Y., Naketani, M. and Sawai, T. (1992) J. Biol. Chem. **267**, 7490–7498.

180. McMurray, L.M., Stephan, M. and Levy, S.B. (1992) J. Bacteriol. **174**, 6294–6297.

181. Guay, G.G., Tuckman, M. and Rothstein, D.M. (1994) Antimicrob. Agents Chemother. **38**, 857–860.

182. Ahmed, M., Borsch, C.M., Neyfakh, A.A. and Schuldiner, S. (1993) J. Biol. Chem. **268**, 11086–11098.

183. Altenberg, G.A., Vanoye, C.G., Horton, J.K. and Reuss, L. (1994) Proc. Natl. Acad. Sci USA **91**, 4654–4657.

184. Mülder, M., Van Grondelle, R., Westerhoff, H.V. and Lankelma, J. (1993) Eur. J. Biochem. **218**, 871–882.

185. Raviv, Y., Pollard, H.B., Bruggeman, E.P., Pastan, I. and Gottesman, M.M. (1990) J. Biol. Chem. **265**, 3975–3980.

186. Higgins, C.F. and Gottesman, M.M. (1992) Trends Biochem. Sci. **17**, 18–21.

187. Higgins, C.F. (1994) Cell **79**, 393–395.

188. Simon, S., Roy, D. and Schindler, M. (1994) Proc. Natl. Acad. Sci. USA **91**, 1128–1132.

189. Roepe, P.D., Weisburg, J.H. Luz, J.G., Hoffman, M.M. and Wei, L.-Y. (1994) Biochemistry 33, 11008–11015.

190. Roepe, P.D., Wei, L.-Y., Cruz, J. and Carlson, D. (1993) Biochemistry **32**, 11042–11056.
191. Roepe, P.D. (1992) Biochemistry **31**, 12555–12564.
192. Bolhuis, H., Van Veen, H.W., Molenaar, D., Poolman, B., Driessen, A.J.M. and Konings, W.N. (1995) EMBO J., in press.
193. Homolya, L., Holló, Z., Germann, U.A., Pastan, I., Gottesman, M.M. and Sarkadi, B. (1993) J. Biol. Chem. **268**, 21493–21496.
194. Von Heijne, G. (1986) EMBO J. **5**, 3021–3027.

CHAPTER 9

Molecular Genetic Analysis of Membrane Protein Topology

M. LEE and C. MANOIL

Department of Genetics, Box 357360,
University of Washington, Seattle, WA 98195, USA

© 1996 Elsevier Science B.V.
All rights reserved

Handbook of Biological Physics
Volume 2, edited by W.N. Konings, H.R. Kaback and J.S. Lolkema

Contents

1. Introduction

A fundamental aspect of integral membrane protein structure is the transmembrane topology of the polypeptide chain, i.e., the disposition of the chain relative to the lipid bilayer in which it is embedded. A variety of biochemical techniques have been employed to elucidate topological structure, including proteolysis, side chain covalent modification, and epitope recognition [1]. However, these techniques often are unable to provide a full topological description of a membrane protein because of their reliance on the natural occurrence of residues that are susceptible to proteolytic cleavage, have reactive side chains or are accessible to antibody binding. These limitations have been largely circumvented by the development of molecular genetic methods which can be used in conjunction with the more traditional approaches.

Two molecular genetic strategies for analyzing membrane protein topology have been developed. The first relies on the properties of gene fusion-encoded hybrid proteins in which enzymes whose activities reflect subcellular localization are attached at different sites in a membrane protein. The second strategy depends on the introduction of sequences (e.g., protease cleavage sites) whose subcellular disposition can be easily assayed. In this chapter, we will review both approaches, with special reference to work that has analyzed *Escherichia coli* lactose permease as a model. The reader is also referred to several excellent earlier reviews [1–3].

2. Construction of topology models based on amino acid sequences

The first step in analyzing the topology of a membrane protein is usually the formulation of a model or models based on the amino acid sequence of the protein. The construction of such models is generally based on two assumptions. The first assumption is that membrane-spanning sequences correspond to apolar α-helical sequences that are sufficiently long to span the bilayer. The second assumption is that the spanning sequences are oriented in accordance with the 'positive inside' rule [4], that is, the favored orientation maximizes the exposure of positive residues in hydrophilic loops facing the cytoplasmic side of the membrane.

The major difficulties encountered in formulating topology models using these rules are: (1) some spanning sequences contain polar residues which can make their identification problematic; (2) large extramembranous domains do not follow the positive inside rule [5], and (3) non-helical spanning sequences, such as β-strands and sequences which insert into the membrane without spanning it, may contribute to a protein's membrane association [6–9]. Nevertheless, most of the integral membrane protein structures that have been determined are in reasonable accord with these general rules [7,10,11].

3. Assay of membrane protein topology using gene fusions

The logic of using gene fusions to analyze membrane protein topology is illustrated in Fig. 1 [12]. Fusions of a gene encoding a protein whose enzymatic activity reflects its subcellular location (e.g., *E. coli* alkaline phosphatase (AP) lacking its signal sequence) are generated at multiple sites in the gene for a membrane protein, and the activities of the resulting hybrid proteins are compared to the previously formulated models. In principle, the activity of each hybrid protein reveals whether the junction site normally faces toward or away from the cytoplasm. A number of different proteins have been used as subcellular location 'sensors' in this type of analysis. In bacteria, the most commonly used sensor proteins are alkaline phosphatase and β-lactamase (which require export to the periplasm for activity) and β-galactosidase (which requires cytoplasmic localization for maximal activity) [2]. The combined use of periplasmic-active and cytoplasmic-active fusions provides positive activity signals for domains on both sides of the membrane, allowing a particularly thorough analysis (e.g., see Ref. [13]).

A gene fusion analysis of topology generally consists of three steps: (1) the generation of gene fusions, (2) the assay of hybrid protein enzymatic activity, (3) a comparison of results to topology models.

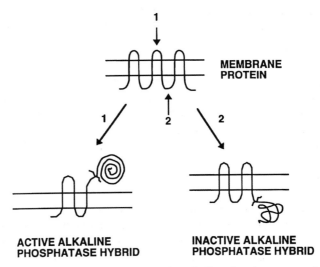

Fig. 1.   Analysis of membrane protein topology using alkaline phosphatase gene fusions. The top panel represents a membrane protein with six membrane-spanning segments. Gene fusions with fusion junctions corresponding to periplasmic positions (e.g., position 1) form hybrids with an exported alkaline phosphatase (AP) attached to N-terminal membrane protein sequences (bottom, left) which exhibit AP activity. Fusions at cytoplasmic positions (e.g., position 2) generate a cytoplasmically disposed AP which is inactive. Cytoplasmic AP appears to be inactive because it cannot form disulfide bonds necessary for folding [47].

3.1. Generation of gene fusions

Two methods have been commonly used for the generation of gene fusions: the insertion of transposon derivatives that 'automatically' generate the fusions, and *in vitro* mutagenesis using oligonucleotide-based methods. Both approaches employ membrane protein genes carried on plasmids. The two methods have different advantages and limitations. The generation of fusions by transposition is technically simple but suffers from the disadvantage that the junction sites are determined by the specificity of transposon insertion [14]. Although this difficulty is minimized for transposon derivatives with a relatively low insertion specificity (e.g., derivatives based on Tn5) [15], it may still be difficult to identify fusions with junctions in small extramembranous domains, which are common in transport proteins. The generation of fusions with predefined junctions using site-directed mutagenesis [16] or Polymerase Chain Reaction-based methods [17] is more laborious than generating the fusions by transposition, but has the advantage that junction positions can be specified by the oligonucleotides employed. This advantage is especially useful in distinguishing between closely related models for a structure (e.g., see Ref. [18]). Additional methods for generating fusions based on the interconversion of different fusion types ('fusion switching') [19] and the generation of nested deletions in appropriate plasmid constructs [20] have also been developed. The fusion switching technique makes it possible to compare the activities of hybrid proteins with different sensor proteins attached at exactly the same site in the membrane protein. The junction sites of fusions generated using any of these methods can be identified simply by DNA sequence analysis using oligonucleotide primers complementary to sequences encoding the 'sensor' proteins [21].

3.2. Assay of hybrid protein activity

The enzyme activities of hybrid proteins are typically determined by assay of permeabilized cells using chromogenic substrates (alkaline phosphatase or β-galactosidase) [21] or by means of a bioassay (β-lactamase) [22]. Since gene fusions with different junction positions may be expressed at different levels, it is essential that rates of hybrid protein synthesis are examined to correct for such differences [23].

3.3. Evaluation of models

The final step in a topology analysis is a comparison of results with models predicted from the sequence. As an example, the results of an AP gene fusion analysis of *E. coli* lactose permease are shown (Fig. 2). The fusions were generated by a combination of transposon Tn*phoA* insertion and *in vitro* mutagenesis [18,24]. The activities of most fusions are consistent with the topology model, with periplasmic fusions expressing high alkaline phosphatase activities and cytoplasmic fusions expressing lower activities.

Many of the lactose permease–alkaline phosphatase fusion junctions fall into the putative membrane-spanning sequences. A study of fusions with junctions within spanning sequences oriented with their N-termini facing the cytoplasm

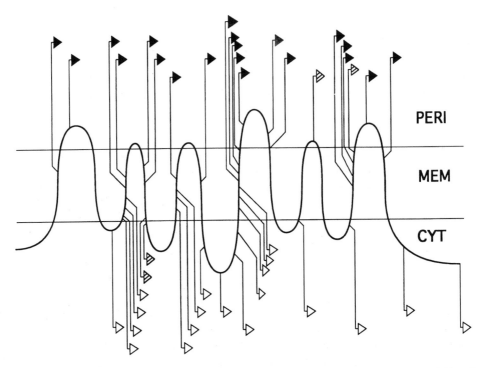

Fig. 2.   Alkaline phosphatase gene fusion analysis of lactose permease. The positions and activities of AP gene fusions are shown [18,24] on the twelve-span model for the topology of lactose permease. Highly active fusions are represented as filled arrowheads, partially active fusions as cross-hatched arrowheads, and inactive fusions as open arrowheads. PERI, periplasm; MEM, membrane; CYT, cytoplasm.

(spanning sequences 3, 5, 7 and 11) indicates that fusions carrying more than 9–11 residues of the spanning sequence express maximal AP activities. It thus appears that only about half of a spanning sequence suffices for the protein export function of a membrane-spanning sequence. This conclusion is consistent with a variety of other genetic studies of protein export signals [25].

Fusions with junctions in spanning sequences in the opposite orientation (N end periplasmic) tend to show elevated alkaline phosphatase activities unless they contain not only the entire apolar sequence but C-terminal charged residues as well. This requirement evidently accounts for the partial AP activities of the fusions in cytoplasmic segment 3 (indicated by cross-hatched arrowheads). Detailed studies of analogous alkaline phosphatase fusions to MalF protein first implied a special role for positively charged residues in the cytoplasmic localization of such domains [26]. An amphipathic sequence plays an apparently analogous role in the *E. coli* serine chemoreceptor [27]. Two other classes of anomalous fusions have been defined and are discussed below (Section 3.4).

The overall success of gene fusion methods in analyzing membrane protein topology implies that the subcellular localization of a site in a membrane protein does not, in general, require C-terminal sequences. The information dictating a particular topology must therefore be present in sequences N-terminal to a given site or redundantly determined by N-terminal and C-terminal sequences. In addition, the folding defects caused by the loss of C-terminal membrane protein sequences in hybrid proteins evidently does not usually prevent the remaining N-terminal sequences from adopting the wild-type topology.

3.4. Anomalous behavior in topology studies

Three types of anomalous behavior have been noted in topology studies involving alkaline phosphatase fusions. As mentioned in the previous section, alkaline phosphatase fusions with junctions near the N-terminal end of a cytoplasmic loop in a membrane protein can exhibit anomalously high activity. The unexpectedly high activity of such fusions may be due to loss of C-terminal hydrophilic sequences such as positive residues required to orient the preceding transmembrane sequence. A recent study of one such fusion site in MalF indicated that β-lactamase fusions may be less prone to this type of anomaly than AP fusions [28].

A second anomaly occasionally occurs when periplasmic domain AP fusions exhibit unusually low activities (e.g., the fusion to the fifth periplasmic domain of lactose permease) (Fig. 2) [18,29]. These cases involve fusions with junction points C-terminal to spanning sequences carrying charged residues. The charged residues reduce the overall hydrophobicity of the spanning sequence and may decrease the sequence's ability to promote export. In such cases, C-terminal spanning sequences in the unfused protein may be required for efficient export of the hydrophilic domain in question. Since such charged residues in spanning sequences are often important in the biological function of membrane proteins, anomalously low alkaline phosphatase activities may commonly be observed for fusions with junctions in regions of functional importance (e.g., regions in a transport protein involved in binding hydrophilic substrates during passage across the membrane).

A third anomaly in gene fusion analysis of topology is encountered for membrane proteins whose (unprocessed) N-termini face the periplasm [13,30]. Hybrid proteins with junctions in these N-terminal domains lack transmembrane sequences and are, in general, not exported.

3.5. Alkaline phosphatase sandwich fusions

The anomalies in gene fusion analysis of membrane protein topology are thought to arise at sites where C-terminal sequences are required for normal subcellular localization. In an attempt to circumvent this problem, a novel type of alkaline phosphatase fusion ('sandwich fusion') was developed in which the alkaline phosphatase sequence is inserted *into* the membrane protein sequence rather than *replacing* C-terminal membrane protein sequences [31]. One such sandwich fusion with an insertion site at the N-terminal end of a MalF cytoplasmic domain did indeed show behavior expected of cytoplasmically situated alkaline phosphatase (i.e., low activity). A disadvantage of using alkaline phosphatase sandwich fusions is that they can be quite toxic to cells producing them (Boyd, personal communication). An alternative approach that also leaves C-terminal membrane protein sequences intact with derivatives less prone to cause toxicity involves the insertion of short sequences whose subcellular locations can be easily determined (see Section 4).

3.6. Analysis of mammalian membrane proteins in E. coli

The topologies of several mammalian membrane proteins expressed in *E. coli* have been analyzed using gene fusions. The first protein to be analyzed was the β-subunit of sheep kidney Na,K-ATPase, a protein with a very simple topology containing a single spanning sequence oriented with its N-terminal end facing the cytoplasm [32]. The properties of a set of β-lactamase gene fusions were consistent with this topology, implying normal membrane insertion in *E. coli.*

Three topologically complex mammalian membrane proteins have also been analyzed using alkaline phosphatase gene fusions. In all three cases, there was auxiliary evidence that at least some of the expressed protein molecules were functional when expressed in *E. coli* implying that they had attained their normal topologies and folded structures. For all three proteins, the fusion studies showed results compatible with the established mammalian cell topologies at some sites of the proteins but not at others. For example, in the analysis of the human β_2-adrenergic receptor, the fusion analysis correctly placed five extramembranous segments situated in the C-terminal part of the membrane-associated domain, but showed results inconsistent with the normal topology for the remaining three extramembranous segments in the N-terminal part of the protein [33]. The inconsistent behavior of the N-terminal fusions is probably due to the fact that the N-terminal region of the protein (which lacks a cleavable signal sequence) normally faces away from the cytoplasm, an arrangement known to give rise to anomalous results (see above). An AP fusion analysis of the first half of the mouse multidrug resistance (MDR) protein (corresponding to six transmembrane sequences) was also compatible with the proposed topology for most of the structure [34]. Interestingly, the analysis worked well for the first translocated segment, which contains the protein's only normally glycosylated site. Since the glycosylation does not occur in *E. coli*, the success of the fusion analysis indicates that the absence of glycosylation does not alter the topology in this region of MDR. Analysis of AP fusions to the C-terminal region of MDR led to a proposal for a major revision of the protein's topology [35]. In a third case, the human erythrocyte glucose transporter (GLUT1), an AP fusion analysis yielded results similar to those found for the MDR protein, with fusions at the N-terminal half of the protein behaving largely as predicted from topology models, but with inconsistencies in the C-terminal region (Traxler and Beckwith, unpublished). The reasons for the difficulties in the analysis of MDR and GLUT1 have not been explored. However, in neither case was the proportion of functional molecules in the expressed forms determined, and it is possible that the proteins are inserted into the membrane of *E. coli* in a mixture of topologies, and that the activities of the hybrid proteins reflect this heterogeneity.

4. Analysis of membrane protein topology using inserted sites

Most of the complications that arise in the analysis of membrane protein topology using gene fusions are due to the loss of C-terminal sequences in the hybrids analyzed. Several alternative approaches that do not suffer from this limitation are based on the insertion of easily identified target sites into the polypeptide. Examples include sequences subject to proteolysis, antibody binding, vectorial labeling, or endogenous modification. In these methods, site-directed mutagenesis is used to insert the target sequence (often after the elimination of interfering endogenous sites) into a specific segment of the protein, followed by a determination of the inserted sequence's subcellular location (Fig. 3). The overall topology of a membrane protein can be determined by assaying inserts at a number of different positions.

4.1. Proteolysis

There is a substantial literature describing the use of proteolysis at naturally-occurring susceptible sites in a membrane protein to define the subcellular locations of the sites. However, the utility of this assay is generally limited by the rarity of naturally occurring protease-sensitive sites. For example, lactose permease was analyzed using chymotrypsin, trypsin, thermolysin, and papain treatments [36,37]. In each case, only a single cytoplasmic segment, about halfway along the length of the polypeptide, was efficiently cleaved.

Molecular genetic methods have made it possible to overcome the limited distribution of naturally-occurring cleavage sites by the introduction of new sites. For example, in the case of leader peptidase (a bitopic membrane protein with one

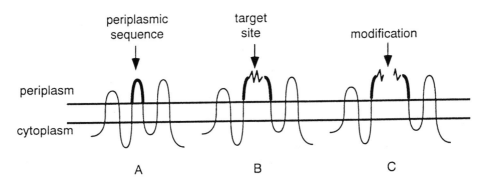

Fig. 3. Analysis of membrane protein topology using introduced target sequences. (A) A model of a protein's membrane topology is formulated and a specific segment of the protein (such as the periplasmic segment depicted as a bold line) is assigned a testable subcellular location. (B) Site-directed mutagenesis is used to alter the segment to include a target site that can be modified in a compartment-specific fashion, (e.g., a protease-sensitive site). (C) The assay is used to test the location of the segment. If the target site is modified (e.g., cleaved by a protease), then the segment bearing the site must be exposed to the compartment that contains the modifying agent.

cytoplasmic segment), a trypsin-sensitive site was introduced into the cytoplasmic segment using a combination of deletion and insertion mutations. Peptide sequencing was used to prove that cleavage occurred at the expected site.

In such constructions, it is crucial to identify the sites where cleavage occurs because a mutation generating a new cleavage site may also uncover sites elsewhere in the protein that are normally resistant to proteolysis. An example of such behavior was observed in studies of derivatives of the outer membrane protein OmpA [38]. OmpA mutants were constructed with trypsin cleavage sites at each of the three hydrophilic segments exposed to the periplasm. These mutant proteins were resistant to extracellular trypsin activity but were cleaved when both sides of the outer membrane were exposed to protease. One mutant, however, cleaved both at the newly introduced site and at a second position, indicating that a cryptic site had been uncovered. The mutation might cause misfolding which makes the cryptic site accessible to trypsin, or trypsin cleavage at the introduced site might make the cryptic site accessible.

In such an approach, it is also important to demonstrate that mutations which introduce cleavage sites do not alter the topology. This difficulty was observed in studies in which lactose permease was modified by the insertion of a sequence capable of both biotinylation by endogenous enzymes and by cleavage by Factor Xa protease [39]. The insertion was introduced into periplasmic segments, but assays of both biotinylation and Factor Xa cleavage demonstrated that the mutant segments were retained in the cytoplasm.

In a different study, lactose permease was mutagenized to introduce a trypsin cleavage site in the sixth periplasmic segment (Lee and Manoil, unpublished observations). The cleavage site was introduced by mutagenizing the lactose permease gene using a degenerate mixture of oligonucleotides encoding a lysine residue (recognized by trypsin) with a variety of adjacent sequences. This family of lactose permease derivatives containing potential cleavage sites was then screened for an insertion that both retained transport activity and was efficiently cleaved by trypsin in spheroplasts. This approach provides a simple systematic method to identify protease-sensitive derivatives of membrane proteins whose topology and folding are not significantly altered from that of wild-type.

4.2. Antibody binding

The subcellular location of a segment within a membrane protein can also be identified after introducing an epitope by analyzing antibody binding. This method was used to study the large extramembranous domains in the α1 subunit of the human acetylcholine receptor (AChR) [40].

A recurring problem in the design of target site assays is finding regions within a membrane protein that will tolerate an insert. The authors of the AChR study showed that certain regions of the protein are poorly conserved between species and also have naturally occurring epitopes that are recognized in both native or denatured AChR. They reasoned that these regions probably do not have a tightly constrained conformation in native AChR, and are thus likely to tolerate insertions without causing major alterations in the protein's structure.

The study, which employed two different monoclonal antibodies, found that small insertions (five or eight residue epitopes) were poorly recognized by antibody compared to longer insertions or insertions in which the epitope was inserted in a tandem repeat. The use of extender sequences appears to have improved accessibility to the target sites, recognition of the target sites, or both.

4.3. Vectorial labeling

A number of studies of membrane protein topology have utilized sulfhydryl reagents to modify selectively cysteine residues introduced into a protein by site-directed mutagenesis. This approach was introduced by Falke et al. in an analysis of the Tar chemoreceptor [41], and has also recently been used to analyze the MotA protein [42]. Cysteine residues situated in periplasmic segments of the proteins were rapidly modified by membrane-impermeant maleimide derivatives, whereas cysteine residues in cytoplasmic segments were slowly modified. For both proteins, the effects of the mutations on membrane insertion of the altered proteins could be assessed by analyzing swarming behavior.

An extension of this approach was utilized in the analysis of the *E. coli* glucose 6-phosphate transporter [43]. In this study, transporter mutants containing cysteine residues introduced at different sites were modified with *p*-chloromercuribenzosulfonate, a sulfhydryl agent that is similar to substrate glucose 6-phosphate in both size and charge. The analysis allowed the authors not only to demonstrate the location of particular hydrophilic loops, but also to probe the structure of the pore through which glucose 6-phosphate enters cells.

Cysteine residues have also been introduced into bacteriorhodopsin and modified with spin label reagents whose subcellular locations could be assayed using electron paramagnetic resonance. It was possible to distinguish residues buried inside the folded protein from residues exposed on the protein surface (to either the membrane or the aqueous environment) [44]. An extension of this method can distinguish between the membrane and aqueous environments and also provides a metric for how deeply a given surface residue is buried within the membrane [45].

4.4. Endogenous modification

In eukaryotic cells, the glycosylation activity found in the lumen of the endoplasmic reticulum and Golgi can be used to modify target sites in a compartment-specific manner. In principle, potential glycosylation sites introduced into translocated sites but not cytoplasmic sites of a membrane protein are modified. The twelve transmembrane sequence model of the cystic fibrosis transmembrane conductance regulator (CFTR) was tested using this method [46]. The two known glycosylation sites in the fourth translocated segment were eliminated, and novel glycosylation sites were individually inserted into each of the other five putative translocated domains. When expressed in cells, all of the mutant proteins with sites in putative translocated domains were efficiently modified, a result which supports the CFTR topology model. All of the mutant proteins showed iodide efflux activities comparable to the wild type, implying that the modifications did not dramatically alter the topology or folding of the proteins.

4.5. General considerations

The assays described in this section are all complicated by the need for the inserted target site to be accessible to the modifying or binding agent. A target site within a hydrophilic segment may be poorly detectable due to steric hindrance from the segment itself, from neighboring hydrophilic segments, or from the lipid bilayer. The addition of 'extender' sequences has been useful in both the proteolytic and epitope binding assays to increase accessibility of target sites. In addition, the placement of a variety of different sequences on both sides of a target site makes it possible to screen for conformations that are efficiently recognized. A second concern in the use of such assays is that the introduction of target sites may alter the topology of the membrane protein being analyzed. However, if the mutant protein containing the introduced site exhibits a specific activity (e.g., for transport) which is the same as that of wild-type, then the topology of the mutant derivative is likely to be essentially normal.

5. Conclusion

A variety of molecular genetic approaches are available for the analysis of membrane protein topology in prokaryotes and eukaryotes. Methods based on the generation and analysis of hybrid proteins are simple and generally reliable, but are subject to several difficulties due to the loss of C-terminal membrane protein sequences in the hybrid proteins generated. These difficulties can be overcome through the use of methods that involve the insertion of easily identifiable target sequences into the membrane protein sequence without loss of C-terminal sequences. However, such methods are somewhat laborious and are subject to technical difficulties of their own due to the requirements that the inserted sequences be accessible to external agents and that they not significantly alter the membrane topologies of the proteins modified.

Abbreviations

AchR, al subunit of the human acetylcholine receptor
AP, alkaline phosphatase
CFTR, cystic fibrosis transmembrane conductance regulator
GLUT1, human erythrocyte glucose transporter
MDR, mouse multidrug resistance protein

References

1. Jennings, M. (1989) Annu. Rev. Biochem. **58**, 999–1027.
2. Traxler, B., Boyd, D. and Beckwith, J. (1993) J. Membr. Biol. **132**, 1–11.
3. Boyd, D. (1994) in: Membrane Protein Structure, ed S. White. pp. 144–163. Oxford University Press, New York.
4. von Heijne, G. (1994) Annu. Rev. Biophys. Biomol. Struct. **23**, 167–192.
5. von Heijne, G. (1986) EMBO J. **5**, 3021–3027.

6. Cowan, S. and Rosenbusch, J. (1994) Science **264**, 914–916.
7. Kuhlbrandt, W., Wang, D. and Fujiyoshi, Y. (1994) Nature **367**, 614–621.
8. Picot, D., Loll, P. and Garavito, R. (1994) Nature **367**, 243–249.
9. Hucho, F., Goeme-Tschelnokow, U. and Strecker, A. (1994) TIBS **19**, 383–387.
10. Rees, D., Komiya, H. and Yeates, T. (1989) Annu. Rev. Biochem. **58**, 607–633.
11. Henderson, R., Baldwin, J., Ceska, T., Zemlin, F., Beckman, E. and Downing, K. (1990) J. Mol. Biol. **213**, 899–929.
12. Manoil, C. and Beckwith, J. (1986) Science **233**, 1403–1408.
13. Whitley, P., Amaber, T., Ehrmann, M., Haardt, M., Bremer, E. and von Heijne G. (1994) EMBO J. **13**, 4653–4661.
14. Manoil, C. and Beckwith, J. (1985) Proc. Natl. Acad. Sci. USA **82**, 8129–8133.
15. Berg, D., Schmandt, M. and Lowe, J. (1983) Genetics **105**. 813–828.
16. Boyd, D., Manoil, C. and Beckwith, J. (1987) Proc. Natl. Acad. Sci. USA **84**, 852–8529.
17. Boyd, D., Traxler, B. and Beckwith, J. (1993) J. Bacteriol. **175**, 553–556.
18. Calamia, J. and Manoil, C. (1990) Proc. Natl. Acad. Sci. USA **87**, 4937–4941.
19. Manoil, C. (1990) J. Bacteriol. **172**, 1035–1042.
20. Sugiyama, J., Mahmoodian, S. and Jacobson, G. (1991) Proc. Natl. Acad. Sci. USA **88**, 9603–9607.
21. Manoil, C. (1991) Meth. Cell Biol. **34**, 61–75.
22. Broome-Smith, J., Tadayyon, M. and Zhang, Y. (1990) Mol. Microbiol. **4** 1637–1644.
23. San Millan, J., Boyd, D., Dalbey, R., Wickner, W. and Beckwith, J. (1989) J. Bacteriol. **171**, 5536–5541.
24. Ujwal, M., Jung, H., Manoil, C., Altenbach, C., Hubbell, W. and H.R. Kaback (1995) Biochemistry **34**, 14909–14917.
25. Lee, E. and Manoil, C. (1994) J. Biol. Chem. **269**, 28822–28828.
26. Boyd, D. and Beckwith, J. (1990) Cell **62**, 1031–1033.
27. Seligman, L. and Manoil, C. (1994) J. Biol. Chem. **269**, 19888–19896.
28. Prinz, W. and Beckwith, J. (1994) J. Bacteriol. **176**, 6410–6413.
29. Allard, J. and Bertrand, K. (1992) J. Biol. Chem. **267**, 17809–17819.
30. Lewis, M., Chang, J. and Simoni, R. (1990) J. Biol. Chem. **265**, 10541–10550.
31. Ehrmann, M., Boyd, D. and Beckwith, J. (1990) Proc. Natl. Acad. Sci. USA **87**, 7574–7578.
32. Zhang, Y. and Broome-Smith, J. (1990) Gene **96**, 51–57.
33. LaCatena, R., Cellini, A., Scavizi, F. and Tocchini-Valentini, F. (1994) Proc. Natl. Acad. Sci. USA **91**, 10521–10525.
34. Bibi, E. and Beja, O. (1994) J. Biol. Chem. **269**, 19910–19915.
35. Beja, O. and Bibi, E. (1995) J. Biol. Chem. **270**, 12351–12354.
36. Stochaj, U., Biesler, B. and Ehring, R. (1986) Eur. J. Biochem. **158**, 423–428.
37. Goldkorn, T., Rimon, G. and H.R. Kaback (1983) Proc. Natl. Acad. Sci. USA **80**, 3322–3326.
38. Reid, G., Koebnik, R., Hindennach, I., Mutschler, B. and Henning, U. (1994) Mol. Gen. Genet. **243**, 127–135.
39. Zen, K., Consler, T. and Kaback, H.R. (1995) Biochemistry **34**, 3430–3437.
40. Anand, R., Bason, L., Saedi, M., Gerzanich, V., Peng, X. and Lindstrom, J. (1993) Biochemistry **32**, 9975–9984.
41. Falke, J., Dernburg, A., Sternberg, D., Zalkin, N., Milligan, D. and Koshland, D.E. (1988) J. Biol. Chem. **263**, 14850–14858.
42. Zhou, J., Fazzio, R. and Blair, D. (1995) J. Mol. Biol. **251**, 237–242.
43. Yan, R.-T. and Malony, P. (1993) Cell **75**, 37–44.
44. Altenbach, C., Marti, T., Khorana, H. and Hubbell, W. (1991) Science **248**, 1088.
45. Altenbach, C., Greenhalgh, D., Khorana, G. and Hubbell W. (1993) Proc. Natl. Acad. Sci. USA **91**, 1667–1671.
46. Chang, X.-B., Hou, Y.-X., Jensen, T. and Riordan, J. (1994) J. Biol. Chem. **269**, 18572–18575.
47. Derman, A. and Beckwith, J. (1991) J. Bacteriol. **173**, 7719–7722.

CHAPTER 10

The Lactose Permease of *Escherichia Coli*: Past, Present and Future

H.R. KABACK

Howard Hughes Medical Institute,
Departments of Physiology and Microbiology and Molecular Genetics,
Molecular Biology Institute, University of California Los Angeles,
Los Angeles, CA 90024-1662, USA

© *1996 Elsevier Science B.V.*
All rights reserved

Handbook of Biological Physics
Volume 2, edited by W.N. Konings, H.R. Kaback and J.S. Lolkema

Contents

1. General introduction

The molecular mechanism of energy transduction in biological membranes is an important unsolved problem in the life sciences. Although the driving force for a variety of seemingly unrelated phenomena (e.g., secondary active transport, oxidative phosphorylation, rotation of the bacterial flagellar motor) is a bulk-phase, transmembrane electrochemical ion gradient, the molecular mechanism(s) by which free-energy stored in such gradients is transduced into work or into chemical energy remains enigmatic. Nonetheless, gene sequencing and analyses of deduced amino-acid sequences indicate that many biological machines involved in energy transduction, secondary transport proteins in particular [1,2], fall into families encompassing proteins from archaea to the mammalian central nervous system, thereby suggesting that the members may have common basic structural features and mechanisms of action. In addition, many of these proteins have been implicated in human disease (e.g. diabetes mellitus, glucose/galactose malabsorption, some forms of drug abuse, stroke, antibiotic resistance), as well as the mechanism of action of certain psychotropic drugs.

This chapter discusses selected observations with a specific transport protein, the lactose (lac) permease as a paradigm for a huge number of proteins that catalyze similar reactions in virtually all biological membranes. It should be stated at the outset that structural information at high resolution is particularly difficult to obtain with hydrophobic membrane proteins [3]. The great majority of membrane proteins, lac permease in particular, have yet to be crystallized and it is becoming increasingly apparent that structural information is a prerequisite for solving their mechanism of action.

2. The past

The β-galactoside or lac transport system in *Escherichia coli* is the most extensively studied secondary transport system. It was first described in 1955 by Cohen and Rickenberg [4] and is part of the lac operon, which enables the organism to utilize the disaccharide lactose. In addition to regulatory loci, the lac operon contains three structural genes: (i) the Z gene encoding β-galactosidase, a cytosolic enzyme that cleaves lactose once it enters the cell; (ii) the Y gene encoding lac permease which catalyzes transport of β-galactosides through the plasma membrane; and (iii) the A gene encoding thio-β-galactoside transacetylase, an enzyme that catalyzes the acetylation of thio-β-galactosides with acetyl-CoA as the acetyl donor and has an unknown physiological function.

As postulated by Mitchell (reviewed in Refs. [5,6]), supported by the experiments of West [7] and West and Mitchell [8,9] and demonstrated conclusively in bacterial membrane vesicles (reviewed in Refs. [10–12]), accumulation of a wide variety of solutes against a concentration gradient is driven by a proton electrochemical gradient ($\Delta\bar{\mu}_H^+$; interior negative and/or alkaline). More specifically, β-galactoside accumulation in *E. coli* is catalyzed by lac permease, which carries out the coupled stoichiometric translocation of a β-galactoside with H^+ (i.e., β-galactoside/H^+ symport or cotransport). Physiologically, the permease utilizes free energy released from downhill translocation of H^+ to drive accumulation of β-galactosides against a concentration gradient. In the absence of $\Delta\bar{\mu}_H^+$, however, lac permease catalyzes the converse reaction, utilizing free energy released from downhill translocation of β-galactosides to drive uphill translocation of H^+ with generation of a $\Delta\bar{\mu}_H^+$ the polarity of which depends upon the direction of the substrate concentration gradient.

The *lacY* gene was the first gene encoding a membrane protein to be cloned into a recombinant plasmid [13] and sequenced [14]. By combining overexpression of *lacY* with the use of a highly specific photoaffinity probe [15] and reconstitution of transport activity in artificial phospholipid vesicles (i.e., proteoliposomes) [16], the permease was solubilized from the membrane, purified to homogeneity [17–19] and shown to catalyze all the translocation reactions typical of the transport system *in vivo* with comparable turnover numbers [20,21]. Thus, the product of the *lacY* gene is solely responsible for all the translocation reactions catalyzed by the β-galactoside transport system.

2.1. Lac permease contains 12 transmembrane domains in α-helical conformation

Circular dichroic measurements with purified lac permease demonstrate that the protein is ca. 80% helical, an estimate consistent with the hydropathy profile of the permease which suggests that approximately 70% of its 417 amino acid residues are found in hydrophobic domains with a mean length of 24 ± 4 residues. Based on these findings, it was proposed (22) that the permease has a hydrophilic N terminus followed by 12 hydrophobic segments in α-helical conformation that traverse the membrane in zigzag fashion connected by hydrophilic domains (loops) with a 17-residue C-terminal hydrophilic tail (Fig. 1).

Support for general features of the model and evidence that the N and C termini, as well as the loops between helices IV and V and helices VI and VII, are exposed to the cytoplasmic face of the membrane were obtained from laser Raman [23] and Fourier transform infrared spectroscopy (J. LeCoutre, H. Jung, K. Gerwert and H.R. Kaback, unpublished information), immunological studies [24–32], limited proteolysis [33,34] and chemical modification [35]. However, none of these approaches is able to differentiate between the 12-helix motif and other models containing 10 [23] or 13 [36] transmembrane domains.

Unequivocal support for the topological predictions of the 12-helix model was obtained by analyzing an extensive series of lac permease-alkaline phosphatase (*lacY-phoA*) fusion proteins [37]. In addition, it was shown that the alkaline phosphatase activity of fusions engineered at every third amino-acid residue in

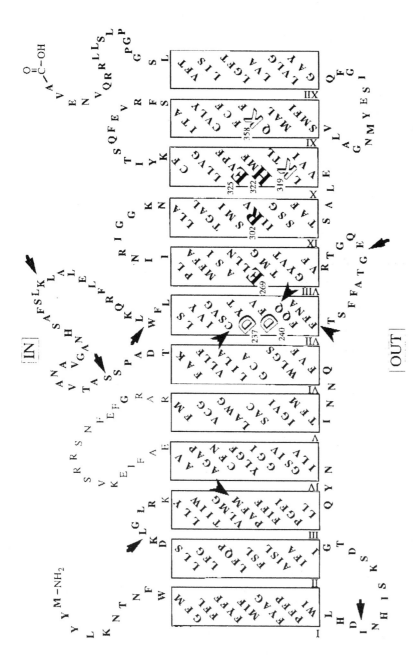

Fig. 1. Secondary-structure model of lac permease. The single amino-acid code is used, and hydrophobic segments are shown in boxes as transmembrane α-helical domains connected by hydrophilic loops. Small black letters indicate residues that yield active permease when replaced with Cys or in some instances other amino acids; large black letters indicate residues that yield inactive permease when replaced with a number of different residues; large open raised letters indicate residues that are charge-paired [Asp237 (helix VII)-Lys358 (helix VII) and Asp240 (helix VII)-Lys319 (helix X)]; small gray letters indicate residues that have not been mutagenized. Gray arrowheads indicate active split permease constructs; black arrowheads signify inactive constructs.

putative helices III and V increases abruptly as the fusion junction proceeds from the 8th to the 11th residue. Thus, approximately half a transmembrane domain is needed to translocate alkaline phosphatase through the membrane to the external surface. When fusions are constructed at alternate amino-acid residues in putative helices VII and XI [38] or at each residue in helices IX and X (M.J. Ujwal, E. Bibi, C. Manoil and H.R. Kaback, unpublished information), the data are in excellent agreement with the model shown in Fig. 1. In addition, expression of contiguous, non-overlapping permease fragments with discontinuities in cytoplasmic or periplasmic loops leads to functional complementation, while fragments with discontinuities in transmembrane domains do not, and this approach has been used to approximate the boundaries of helix VII [39,40].

2.2. Lac permease is functional as a monomer

One particularly difficult problem to resolve with hydrophobic membrane proteins is their functional oligomeric state. Notwithstanding strong evidence that lac permease is functional as a monomer [41,42], certain paired in-frame deletion mutants are able to complement functionally [43]. Although cells expressing the deletions individually do not catalyze active transport, cells simultaneously expressing specific pairs of deletions catalyze transport up to 60% as well as cells expressing wild-type permease, and it is clear that the phenomenon occurs at the protein and not the DNA level. Remarkably, complementation is observed *only* with pairs of permease molecules containing large deletions and *not* with missense mutations or point deletions. Although the mechanism of complementation is unclear, it is probably related to the phenomenon whereby independently expressed N- and C-terminal fragments of the permease interact to form a functional complex [43]. In any case, the observation that certain pairs of deletion mutants can interact rekindled concern regarding the oligomerization state of wild-type permease.

Recently [44], a fusion protein was engineered that contains two lac permease molecules covalently linked in tandem (permease dimer). Permease dimer is inserted into the membrane in a functional state, and each half of the dimer exhibits equal activity. Thus, inactivating point mutations in either half of the *lacY* tandem repeat lead to 50% loss of transport. Furthermore, the activity of a permease dimer composed of wild-type permease and a mutant devoid of Cys is inactivated ca. 60% by N-ethylmaleimide (NEM). In order to test the caveat that oligomerization between dimers might account for the findings, a permease dimer was constructed that contains two different deletion mutants which complement when expressed as untethered molecules. Since this construct does not catalyze lactose accumulation, it is unlikely that permease dimers oligomerize. The experiments provide strong support for the conclusion that lac permease is functional as a monomer.

2.3. Site-directed mutagenesis reveals that few amino-acid residues are important for activity

By using site-directed mutagenesis with wild-type permease or a functional mutant devoid of Cys residues (C-less permease) [45] individual residues in the permease

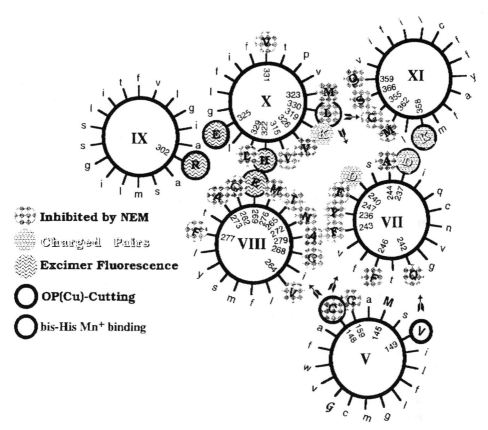

Fig. 2.   Helix packing model of helices V and VII–XI in lac permease viewed from the periplasmic surface. In addition to the symbol key given at the left of the figure, positions 330 (helix X) and 148 and 149 (helix V) are shown with arrows indicating cleavage sites in neighboring helices.

that are important mechanistically have been identified (reviewed in Refs. [46–48]. Almost all of the 417 residues in C-less permease have been replaced with Cys or other residues, and *remarkably*, only 4 residues have been shown to be obligatory for active lactose transport thus far [49–53] (Fig. 1). Among the residues mutated, none of the 8 Cys [45,54–61], 6 Trp [62], 12 Pro [63,64] 14 Tyr [65][1] or 36 Gly residues [66] and only 1 out of 4 His residues [67–70] is obligatory for activity. On the other hand, Glu269 (helix VIII) [71,72], Arg302 (helix IX) [73,74], His322

---

1   Although Tyr → Phe replacements indicated originally that 3 Tyr residues are important for activity, Cys-scanning mutagenesis has revealed that Tyr26, Tyr236 and Tyr336 can be replaced with Cys in either the Cys-less or wild-type background with retention of significant activity. Similarly, replacement of Thr348 or Ala177 with Cys in wild-type permease results in highly significant activity.

(helix X) [67–70] and Glu325 (helix X) [75,76] are essential for substrate accumulation and/or high-affinity binding (reviewed in Ref. [47]). Moreover, differences
in the properties of the mutants suggest that Arg302, His322 and Glu325 may
function in a H^+ translocation pathway, although it is possible that the residues also
form part of a coordination site for H_3O^+ (discussed in Ref. [12]). It is also
noteworthy that certain active Cys-replacement mutants are altered by alkylation,
and these mutants cluster on helical faces so as to line a cavity in the C-terminal
half of the permease (Fig. 2). In any event, very few residues are mandatory for
substrate binding or transport, and it is unlikely that individual amino-acid replacements, particularly with Cys, cause global conformational changes.

2.4. Site-directed fluorescent labeling and mutagenesis yield helix packing in the C-terminal half of the permease

The observations described above highlight a need for static and dynamic
information at high-resolution in order to define the transport mechanism, and
recent studies with permease molecules engineered specifically for spectroscopic
approaches [45,62] are providing important new insight. Since none of the individual native Cys residues in the permease are required for activity, all 8 Cys residues
were replaced simultaneously to yield a C-less permease that retains at least 50%
of wild-type activity [45]. Molecules containing 1 or 2 Cys replacements at any
desired position(s) can then be tagged with thiol-specific biophysical probes after
solubilization and purification. By using this approach, proximity relationships
between transmembrane helices in the C-terminal half of the permease have been
established [77]. Pairs of charged amino-acid residues in transmembrane domains
were replaced with Cys in a C-less construct containing a biotin-acceptor domain
in the middle cytoplasmic loop to facilitate purification by monovalent avidin
affinity chromatography [78]. As a fluorophore, N-(1-pyrenyl)maleimide (PM) was
selected, since PM can form an excited state dimer (excimer) that exhibits a unique
emission maximum if two conjugated ring systems are within about 3.5 Å and
correctly oriented. The findings obtained indicate that: (i) transmembrane domain
X is in α-helical conformation; (ii) helix IX (Arg302) and helix X (Glu325) are in
close proximity; and (iii) helix VIII (Glu269) is in close proximity to helix X
(His322) (Fig. 2).

Second-site suppressor analysis [79] and site-directed mutagenesis combined
with chemical modification [80–82] demonstrate that Asp237 (helix VII) interacts
with Lys358 (helix XI), probably forming a salt bridge. Individual replacement of
Asp237 or Lys358 in C-less permease with Cys or Ala abolishes active lactose
transport, while simultaneous replacement of both charged residues with Cys
and/or Ala or *reversal* of the residues leads to fully active permease. Remarkably,
mutant D237C is restored to full activity by carboxylmethylation which recreates a
negative charge at position 237, and mutant K358C is restored to full activity by
treatment with ethylammonium methanethiosulfonate which recreates a positive
charge at position 358. It has also been shown by site-directed mutagenesis [80,82]

and second-site suppressor analysis [83] that Asp240 (helix VII) interacts with Lys319 (helix X). Individual replacement of either Asp240 or Lys319 in C-less permease with neutral amino acid residues inactivates, but double neutral mutants retain significant activity. In contrast to Asp237-Lys358, however, the polarity of the interaction between Asp240 and Lys319 is important, as reversal of the residues inactivates. In any case, the findings indicate that helix VII (Asp237 and Asp240, respectively) is in close proximity to helices XI (Lys358) and X (Lys319). It is also apparent that none of these residues plays a direct role in the mechanism. However, the interaction between Asp237 and Lys358 appears to be important for folding of the permease and insertion into the membrane (81). In any event, taken together with the conclusions derived from site-directed pyrene labeling, the observations describe the packing of helices VII–XI (Fig. 2), the first tertiary structure information regarding this class of membrane proteins [75] (reviewed in Refs. [46,47]).

2.5. Ligand binding induces widespread conformational changes

Excimer fluorescence can also be used to study dynamic aspects of permease folding [84]. Excimers formed *between* transmembrane domains are markedly diminished by denaturants, while the excimer observed *within* helix X is unaffected, indicating that tertiary interactions are disrupted with little effect on secondary structure. Consistently, interacting helices do not exhibit excimers in detergent, but only after reconstitution into membranes. One of the excimers described also exhibits ligand-induced alterations. Excimer fluorescence due to the interaction between helices VIII (E269C) and X (H322C) is quenched by Tl^+, and the effect is markedly and specifically attenuated by permease substrates. The reactivity of single Cys residues placed in many transmembrane domains is also dramatically altered in the presence of ligand, implying that transport involves widespread changes in tertiary structure [49,85–88]. In one Cys replacement mutant (V315C), the effect of ligand is mimicked by imposition of $\Delta\bar{\mu}_H^+$, providing a preliminarily suggestion that either ligand binding or $\Delta\bar{\mu}_H^+$ may cause the permease to assume the same conformation [49,85]. As discussed above, many of the Cys-replacement mutants that are inactivated by alkylation cluster on helical faces (Fig. 2), indicating that the surface contours of the helices may be important. This surmise coupled with the indication that very few residues are essential in a mechanism that involves widespread conformational changes is encouraging, as it suggests the possibility that a relatively low resolution structure (i.e., helix packing) and localization of the translocation pathway will provide important mechanistic insights.

2.6. The Saga of Cys148

Lac permease is irreversibly inactivated by N-ethylmaleimide (NEM) and protection is afforded by certain substrates [89]. On the basis of these findings, it was postulated that a Cys residue is at or near the substrate-binding site in the permease, and it was shown sometime later [90] that the substrate-protectable Cys is at position 148. Since Cys148 can be replaced with Gly [54,56] or Ser [57,61] with little or no effect on activity, it is apparent that this residue is not essential, and the possibility arises that substrate protection may be due to a long-range conformational effect.

Recently [91], Cys148 was replaced with hydrophobic (Ala, Val, Ile, Phe), hydrophilic (Ser, Thr) or charged (Asp, Lys) residues, and the properties of the replacement mutants were analyzed. Although Cys148 is not essential for transport, the size and polarity of the side chain at this position modifies transport activity and substrate specificity. Thus, small hydrophobic side-chains (Ala, Val) generally increase the apparent affinity of the permease for substrate, while hydrophilic side-chains (Ser, Thr, Asp) decrease the apparent affinity and bulky or positively charged side-chains (Phe, Lys) virtually abolish activity. In addition, hydrophilic substitutions (Ser, Thr, Asp) severely decrease the activity of the permease towards galactose with little effect on the transport of more complex substrates (e.g. lactose). Based on these and other observations, it is concluded that Cys148 is located in a sugar binding site of lac permease and probably interacts hydrophobically with the galactosyl moiety of the substrate. The non-galactosyl moiety increases affinity via nonspecific interactions.

Using purified permease molecules with single-Cys residues at positions 145, 146, 147 or 148 and 2-(4'-maleimidylanilino)naphthalene-6-sulfonic acid (MIANS), it was shown [86] that Cys148 is highly accessible and reactivity is blocked by ligand. Importantly, reactivity of Cys148 is blocked by galactose, but not by glucose. Thus, the specificity of the permease is directed exclusively towards the asymmetry at the fourth position of the galactose moiety. Permease with Cys145 (positions 148 and 145 are presumably on the same face of helix V) behaves in a similar, but much less dramatic fashion. In contrast, permease with Cys146 or Cys147 exhibits similar accessibility, but the reactivity of Cys147 is enhanced by ligand, while the reactivity of Cys146 is unaffected. Studies with I^- indicate that MIANS at positions 145 or 148 is accessible to the collisional quencher, while MIANS at positions 146 or 147 is not accessible. Strikingly, I^- quenching of MIANS at position 145 is diminished in the presence of ligand. The results indicate that Cys148 is in a substrate binding site in the permease with Met145 on the periphery.

3. The present

Clearly, the most pressing problem with regard to polytopic membrane proteins is a need for high-resolution structural information. Therefore, a major thrust of this laboratory over the past 3 years and in the immediate future has been towards crystallization of wild-type lac permease, as well as engineered chimeric forms of the protein that contain polar domains on the surfaces [92,93]. This aspect of the research is not covered here. However, since it is already apparent that lac permease is difficult to crystallize, development of other techniques to obtain structural information are essential. Towards this end, new approaches have been developed that combine the use of protein engineering with biophysical and biochemical techniques in order to obtain proximity relationships within the permease. It should be emphasized that these approaches are also being utilized to obtain dynamic information regarding conformational changes in the permease which will be essential to determining its mechanism of action.

3.1. Metal–spin label interactions yield distances within proteins

The magnetic dipolar interaction between a nitroxide spin label and a paramagnetic metal ion has been exploited to measure distances in natural metalloproteins [94], and Leigh [95] has developed a quantitative treatment for determining the distance-dependent line broadening of a nitroxide center by paramagnetic ions. Thus far, however, use of this technique has been limited to natural metalloproteins. In order to evaluate this potentially powerful method for general use, molecular genetic techniques have been used to introduce both a paramagnetic metal ion binding site and a nitroxide side-chain at selected positions in T4 lysozyme, a protein with a well-known crystallographic structure [96,97]. Thus, a His-X_3-His metal ion binding motif [98] was introduced at the N terminus of the long inter-domain helix of T4 lysozyme (K65H/Q69H) of three mutants, each containing a single Cys residue at position 71, 76 or 80. The mutants were purified from the soluble fraction of *E. coli*, concentrated and labeled with the thiol-specific (1-oxyl-2,2,5,5-tetramethyl-pyrroline-3-methyl) methanethiosulfonate (methanethiosulfonate spin label). After removal of the unreacted spin label, EPR spectra were recorded in the presence or absence of Cu(II) using a loop gap resonator [99] under four conditions: room temperature, −20°C, 35% sucrose or 35% ficoll. The correlation of the nitroxide distances from the Cu(II) site and the calculated distances using Leigh's analysis agree under each of the four conditions, yielding distances of 11, 13 and 18 Å for mutants V71C, R76C and R80C, respectively, which correspond strikingly well with the distances observed in the X-ray structure of lysozyme.

The metal–nitroxide approach has several advantages over other spectroscopic techniques for distance determinations. Room temperature capability allows detection of time-dependent conformational changes. In addition, metal–nitroxide interactions have the potential to probe longer distances than nitroxide–nitroxide interactions [100]. Finally, the method does not suffer from the orientation ambiguity and background problems inherent in fluorescent energy transfer.

Two or 6 contiguous His residues can be inserted into all but 3 of the hydrophilic domains of lac permease without seriously compromising function [101]. By using either radioactive nickel or EPR spectroscopy with Cu(II) or Mn(II), it was demonstrated that permease with 6 His residues in either the first or second cytoplasmic loop bind metal with a stoichiometry of unity and a K_D in the µM range [102]. In order to test the feasibility of the metal–nitroxide approach with respect to distance measurements in lac permease, 3 individual Cys residues were introduced into helix IV in place of Ile103, Gly111 or Gly121 in C-less permease containing 6 contiguous His residues in the periplasmic loop between helices III and IV and a biotin acceptor domain in the middle cytoplasmic loop to facilitate purification [96,102]. The proteins were expressed, solubilized and purified by monomeric avidin affinity chromatography after spin labeling on the column and concentrated. EPR spectroscopy was then carried out as described above in the presence and absence of Cu(II). The results demonstrate that positions 103, 111 and 121 are 8, 14 and >25 Å, respectively, from the metal binding site in the loop. As demonstrated by molecular dynamics simulation [102], the distances measured are entirely consistent with the contention that transmembrane domain IV in the permease is indeed in α-helical conformation.

Distances of >25 Å are an approximation, as measurement of distance by Cu(II) broadening of a strongly immobilized nitroxide spectrum is relatively insensitive at distances greater than 20 Å [97]. However, greater distances (at least 30 Å) are practical to measure using the Leigh analysis at S band microwave frequencies with deuterated spin labels [103] and/or a paramagnetic ion such as Gd^{3+} (which has a large magnetic moment) is used as the probe. Site-specific labeling of Cys148 with a sulfhydryl-specific lanthanide chelator [104] might be useful for this purpose. In addition, fast relaxing metal ions [Ni(II)] together with T_{1e} measurements of the nitroxide may permit the determination of distances up to 50 Å [100].

3.2. Engineering metal binding sites within the transmembrane helices of lac permease

As discussed above (Fig. 2), site directed excimer fluorescence indicates that Glu269 (helix VIII) and His322 (helix X) lie in close proximity in the tertiary structure of lac permease. Since the simplest design of a metal-binding site is based on the ability of bis-His residues to chelate divalent metal ions [105], a lac permease mutant was constructed in which Glu269 was replaced with His [106]. If position 269 is in close approximation to His322, it was reasoned that His replacement for Glu269 should result in bis-His residues at positions 269 and 322 which might form a divalent metal-binding site. Thus, wild-type and E269H permease containing a biotin acceptor domain in the middle cytoplasmic loop were purified by affinity chromatography, and binding of Mn(II) was measured by EPR spectroscopy. The amplitude of the Mn(II) EPR spectrum is reduced by the E269H mutant, while no change is observed in the presence of wild-type permease. The E269H mutant contains a single binding site for Mn^{2+} with a K_D of about 43 µM, and Mn(II) binding is pH dependent with no binding at pH 5.0, stoichiometric binding at pH 7.5 and a mid-point at about 6.3. It has also been shown that permease with His residues in place of Asp237 (helix VII) and Lys358 (helix XI), two residues thought to be salt-bridged (see Refs. [46,47]), binds Mn(II) stoichiometrically in a manner similar to that described for E269H permease [107]. On the other hand, permease with His residues in place of Asp240 (helix VII) and Lys319 (helix X), two residues that interact in a manner different from that of Asp237 and Lys358 [80,82], does not bind Mn(II). Furthermore, replacement of Arg302 (helix IX) with His generates a Mn(II) binding site with a K_D approximating that of E269H permease, indicating that positions 269 and 322 are in close proximity, and introduction of a third His residue in place of Glu325 results in a binding site with higher affinity (ca. 12 µM) [108]. Finally, permease with His residues in place of Met145 and Val149 on the same face of helix V also binds Mn(II) with a K_D in the micromolar range (J. Wu and H.R. Kaback, unpublished information). Clearly, the results provide strong confirmation of the helix packing model shown in Fig. 2 and provide a novel approach for studying helix proximity, as well as solvent accessibility, in polytopic membrane proteins. Furthermore, the ability to design metal-binding sites within the transmembrane domains of the permease will be particularly useful with respect to the use of metal–spin label interactions to study helix packing.

3.3. Site-directed peptide bond cleavage

A class of chemical probes devised largely for footprinting of DNA generate reactive oxygen species that cleave surrounding structural elements by oxidative degradation [109,110]. Several metal chelates bound to protein [111] are effective reagents for localized cleavage of polypeptide chains by reactive oxygen species [112]. When EDTA-Fe, for example, is attached covalently to a protein, cleavage occurs close to the attachment site in the 3-D structure, suggesting that the backbone segments that are in close proximity to the site at which EDTA-Fe is attached can be identified. Diffusible hydroxyl radicals [113] and reagent-bound metal–peroxide species [112] have been proposed to explain the reactions, as these highly reactive species attack the polypeptide backbone and side chains leading to chemical modification and protein fragmentation. Another metal chelate, *o*-phenanthroline-Cu [OP(Cu)] which reacts with DNA via a fundamentally different mechanism from EDTA-Fe, has been used by Sigman and co-workers [114,115], and it has been shown recently [116] that this reagent provides a valuable chemical approach for determining proximity relationships in lac permease.

Biotinylated lac permease with given single-Cys replacements was purified by avidin affinity chromatography, and the single-Cys residues were modified with 5-(bromoacetamino)-1,10-phenanthroline while the protein was bound to the affinity resin. After elution, the extent of modification was determined by measuring the residual free thiol content [86], and it is clear that G147C, single-Cys148, V149C or L330C permease treated as described are devoid of free thiols.

Experiments with purified L330C permease in dodecylmaltoside demonstrate that after derivatization with OP(Cu) and incubation in the presence of ascorbate with or without peroxide (115), major cleavage products with M_rs of about 19 and 6–8 kDa are observed on immunoblots with anti-C-terminal antibody (see Fig. 2). Remarkably, the same cleavage products are observed with the protein in the native membrane. It is difficult to assign the site(s) of cleavage *a priori* based on the M_rs observed on immunoblots because of the anomalous electrophoretic behavior of hydrophobic proteins like lac permease. Therefore, as a standard, the C-terminal half of the permease which contains the last 6 transmembrane helices (C6) was expressed and shown to exhibit an M_r of about 20 kDa. Comparing the M_r of C6 with the cleavage products from L330C-OP(Cu) permease, the large fragment migrates slightly faster than the C6 standard, indicating that the cleavage site is near the N terminus of helix VII. The relative M_r of the smaller fragment is consistent with a cleavage site near the N terminus of helix XI.

In an effort to determine the cleavage sites more precisely, the fragments were separated electrophoretically and subjected to amino acid analysis and N-terminal microsequencing. Although the N termini are blocked, due presumably to the oxidative nature of the cleavage reaction, the amino acid composition of the 19 and 6–8 kDa fragments is consistent with the general assignment of the cleavage sites as given. In this regard, it is important to note that position 330 is close to the cytoplasmic end of helix X (Fig. 1). Thus, the fragments should include a major portion of the transmembrane helix in which cleavage occurs (e.g. the 19 kDa fragment should include most of the last 6 transmembrane domains and therefore exhibit an M_r close to that of the C6 standard).

After incubation of OP(Cu)-labeled single-Cys148 (helix V) permease with ascorbate, major cleavage products that react with anti-C-terminal antibody are observed at about 19 kDa (with a slightly smaller M_r than the C6 standard) and at 15–16 kDa (see Fig. 2). The larger fragment probably reflects a cleavage site near the N terminus of helix VII, while the relative M_r of the 15–16 kDa fragment suggests a cleavage site near the C terminus of helix VIII. When the cleavage reagent is moved 100° to position 149 [V149C-OP(Cu) permease], a single major cleavage product is observed at 19 kDa, consistent with cleavage at the N terminus of helix VII. However, when the reagent is moved 100° in the other direction to position 147 [G147C-OP(Cu) permease], no cleavage product is observed. The results are consistent with the interpretation that helix V is in close proximity to helices VII and VIII with position 148 in the interface between the helices, position 149 facing helix VII and position 147 facing the lipid bilayer (Fig. 2).

3.4. Spin–spin interaction and fluorescence energy transfer confirm that helix V is in close proximity to helices VII and VIII

Experiments utilizing spin–spin interaction and fluorescence energy transfer indicate that these techniques represent useful additional approaches for confirming proximity relationships. Thus, permease containing a biotin acceptor domain and paired Cys residues at positions 148 (helix V) and 228 (helix VII), 148 and 226 (helix VII) or 148 and 275 (helix VIII) was affinity purified and labeled with a sulfhydryl-specific nitroxide spin label. Spin–spin interactions are observed with the 148/228 and 148/275 pairs, indicating close proximity between appropriate faces of helix V and helices VII and VIII. Little or no interaction is evident with the 148/226 pair, in all likelihood because position 226 is on the opposite face of helix VII from position 228. Furthermore, the interaction is stronger in the 148/228 pair than in the 148/275 pair. Broadening of the electron paramagnetic resonance spectra in the frozen state was used to estimate distance between the 148/228 and the 148/275 pairs. The nitroxides at positions 148 and 228 or 148 and 275 are within approximately 9 Å or 14 Å, respectively. Finally, Cys residues at positions 148 and 228 are crosslinked by dibromobimane, a bifunctional crosslinker that is about 5 Å long, while no crosslinking is detected between Cys residues at positions 148 and 275 or 148 and 226. The results provide strong support for a structure in which helix V is in close proximity to both helices VII and VIII and oriented in such a fashion that Cys 148 is closer to helix VII.

4. The future

The experimental approaches and methods described above are feasible for studying both static any dynamic aspects of the permease, and the techniques are widely applicable. It should be emphasized in addition, that almost 400 unique permease mutants containing single-Cys residues have been constructed, many of which have a biotin acceptor domain or a 6-His tail [93] for rapid purification. The few remaining residues in C-less lac permease that have not been mutated are being replaced with Cys, and the mutants will be characterized with respect to activity and expression. Thus, a library of unique permease molecules are already on hand, each of which contains a unique Cys residue and can be purified and labeled as desired. The expediency of these manipulations, as well as the ability to insert metal binding sites and the biotin acceptor domain into any of the mutants, is greatly facilitated by the availability of a cassette *lacY* gene (EMBL X-56095) that contains unique restriction sites approximately every 100 bp. Furthermore, the great majority of the single-Cys mutants are active, even after alkylation. It is also noteworthy that there is ample evidence indicating that the permease maintains close to native conformation in dodecylmaltoside [86–88].

4.1. Use of paramagnetic metal–spin label interactions to determine helix packing

As discussed above [96,97], collaborative efforts with Wayne L. Hubbell demonstrate with the long interdomain helix of T4 lysozyme that precise distance information is obtained by metal–spin label interactions up to a distance of about 25 Å. It has also been shown that a similar approach can be used to measure distances between an engineered metal binding site in the second periplasmic loop and nitroxide-labeled Cys residues at different positions in transmembrane helix IV of lac permease [102]. Although the studies utilize nitroxide line broadening to assess distance, it is highly relevant that the distance capability of metal–spin label interactions can be extended up to about 50 Å by other types of EPR measurements.

Permease mutants containing a single metal binding site and a single Cys residue are being constructed in order to obtain helix packing in lac permease. Three sets of constructs with a metal binding site and a single-Cys residue in each transmembrane domain at approximately the same depth in the permease will be used: (i) One set will contain a metal binding site in a periplasmic loop (six contiguous His residues) in 12 different mutants containing a single-Cys replacement near the periplasmic end of each transmembrane helix. (ii) A second set will contain a metal binding within the transmembrane domains (bis-His residues between helices or between turns of one helix) of 12 different mutants containing a single-Cys replacement near the middle of each transmembrane helix. (iii) A third set will contain a metal binding site in a cytoplasmic loop (six contiguous His residues) in 12 different mutants containing a single-Cys replacement near the cytoplasmic end of each transmembrane helix.

Each mutant will also contain a biotin acceptor domain at the C terminus in order to facilitate rapid purification by avidin affinity chromatography (the biotin acceptor domain has a factor Xa protease site at the N terminus, and the domain can be removed if necessary [78]). The proteins will be labeled with nitroxide spin label, EPR measurements will be carried out with stoichiometric concentrations of Cu(II), and helix packing will be determined by triangulation. In a parallel series of experiments, the purified proteins with bound Cu(II) will be reconstituted into phospholipid vesicles prior to making the EPR measurements. In addition, the distance measurements can be carried in the absence and presence of a high affinity ligand. Since the dimensions of the permease approximate 44 × 50 Å [42,117)] it may be necessary to make the distance measurements from two metal binding sites inserted independently on the surfaces and within the permease. However, the range of metal ion–nitroxide interaction can likely be increased to at least 50 Å, making this unnecessary.

4.2. Confirmation and extension of conclusions derived from metal–spin label measurements of helix packing

(1) Once helix packing is determined from metal–spin label interactions, paired Cys replacements will be introduced into neighboring helices. The construction of such double mutants is accomplished readily, since it is a simple operation to carry out 'cutting and pasting' with the cassette *lacY* gene encoding appropriate single-Cys mutants with the biotin acceptor domain (see Ref. [75]). The double mutants will be expressed, solubilized and purified by affinity chromatography and studied by:

(a) Site-directed excimer fluorescence. As outlined above, site-directed excimer fluorescence with pyrene has proven to be very useful for obtaining proximity relationships between the helices in the C-terminal half of lac permease. However, in order to observe excimer fluorescence, two pyrene moieties must be within 3.5 Å and in the correct orientation. Therefore, it is important to be able to make an 'educated guess' as to where to place paired Cys residues. In the C-terminal half of the permease, this is possible because there are charged residues in the transmembrane domains that have been postulated to interact. However, there are no charged residues in transmembrane domains in the N-terminal half of the protein (Fig. 1). Therefore, even with helix packing information from metal–spin label measurements, it may not be straightforward to predict which residues are on complementing helical faces. For this reason, although it has been clearly demonstrated that site-directed excimer fluorescence can yield information of the type desired, the requirements for excimer formation are sufficiently stringent that either site-directed spin-spin interaction or peptide bond cleavage will be used prior to site-directed excimer fluorescence, as these techniques detect interactions up to about 10 Å.

(b) Site-directed spin–spin interactions. Detection of spin–spin interaction between nitroxide free radicals at positions 148 (helix V) and 275 (helix VIII) is described above, and the observation supports other data indicating that helices V and VIII are in close proximity. In order to pursue this approach, single and paired Cys residues in adjoining helices, as determined by metal–spin label interaction, will be labeled with nitroxide spin label after purification, and EPR measurements will be carried out at room temperature and –90°C. Comparison of the spectra from

the summed single-labeled proteins to the spectrum of the double-labeled protein will reveal whether the two nitroxides are in close proximity [118]. In addition, the measurements can be carried out in the absence and presence of a high-affinity ligand.

(c) Site-directed peptide bond cleavage. Recent experiments carried out in collaboration with David S. Sigman [116] are encouraging with respect to the use of site-directed peptide bond cleavage using OP(Cu) as a means of studying helix proximity in lac permease. Using this approach, evidence has been obtained that confirms conclusions indicating that helix X is in close proximity to helices VII and XI. Moreover, the approach has shown that helix V is close to helices VII and VIII (Fig. 2). At the present time, these conclusions are based on the relative size of the fragments produced from single-Cys permease molecules labeled with OP(Cu) after incubation with ascorbate because the N termini of the fragments are blocked, making N-terminal microsequencing impossible. However, the amino-acid composition of the fragments is consistent with the assignment of the cleavage sites, and more importantly, attempts are being made to regenerate the N termini by chemical means. In any case, the use of this approach *in conjunction* with the others described will provide further information about helix packing.

(2) *Engineering metal binding sites.* Another approach that will be used to "fine tune" helix packing data obtained from metal–spin label interactions is engineering Mn(II) binding sites between neighboring helices by introduction of bis-His residues [106–108]. This novel approach confirms most of the conclusions derived from site-directed excimer fluorescence and provides the first direct evidence that Asp237 (helix VII) and Lys358 (helix XI), residues thought to be salt bridged, are indeed in close physical proximity. The introduction of paired His residues by site-directed mutagenesis, followed by cutting and pasting appropriate DNA restriction fragments is readily achieved using the cassette *lacY* gene, as is expression of the constructs, solubilization and purification. Mn(II) binding will be determined by EPR. As indicated for site-directed excimer fluorescence, the geometric requirements needed to form a metal binding site from two His residues are relatively stringent. Therefore, this approach, like site-directed excimer fluorescence, will be used only after metal–spin label measurements and either site-directed spin–spin interaction or peptide bond cleavage.

(3) *Fluorescence energy transfer.* A technique that has been used to measure distances within proteins is resonance energy transfer between an appropriate donor (e.g. tryptophan) and an appropriate acceptor (e.g. anilinonaphthalenesulfonate) [119]. This approach has been used to provide evidence that helix V (Cys148) is near helix VII (Trp223). The experiments utilize a permease mutant containing a single native Trp residue at position 223 as the donor to MIANS-labeled Cys148. In order to apply this methodology to lac permease in a more general manner, two approaches are being taken:

(a) A completely functional lac permease in which all 6 Trp residues are replaced with Phe has been described [62]. A biotin acceptor domain (with its single Trp residue replaced by Phe) has been inserted into Trp-less permease, and the fluorescent properties of the native Trp residues has been characterized individually

by constructing a series of 6 mutants, each with a single native Trp residue, and purifying the proteins [120]. The excitation and emission spectra of each native Trp residue was studied, as well as accessibility to hydrophilic or hydrophobic collisional quenchers. The results support the topological location of the Trp residues in the permease as depicted in Fig. 1. Since Cys148 can be labeled specifically with MIANS in each of the mutants, resonance energy transfer will be measured between each native Trp residue and MIANS at position 148. Alternatively, Cys148 will be labeled with methanethiosulfonate spin label, and quenching of Trp fluorescence will be studied in each mutant. It is also possible to introduce Trp residues at any position in the Trp-less permease and measure energy transfer or fluorescence quenching with appropriate probes at position 148. However, in these constructs which contain all of the native Cys residues, the measurements are limited to proteins with the acceptor covalently bound to Cys148.

(b) Another more generalized approach to resonance energy transfer involves the use of a permease devoid of both Cys and Trp residues that has been constructed recently (C. Weitzman and H.R. Kaback, unpublished information). By using this construct with the Trp-less biotin acceptor domain, a single Trp and a single Cys residue will be introduced, respectively, into neighboring helices (e.g. IX and X). The protein will then be purified, labeled with MIANS, and resonance energy transfer will be studied both in dodecylmaltoside and after reconstitution into phospholipid vesicles. The efficiency of energy transfer by donor quenching will be calculated from the decrease in fluorescence intensity of the single Trp at the emission maximum brought about by the presence of acceptor MIANS at the single Cys. Alternatively, the Cys residue will be reacted with nitroxide spin label, and quenching of Trp fluorescence will be examined. Both types of measurements are less restrictive than site-directed excimer fluorescence, and their feasibility will be tested by using residues known to be in close proximity (Fig. 2).

4.3. Localization of the substrate translocation pathway

Lac permease recognizes a large variety of α- and β-galactosides containing both ether and thioether linkages [121,122]. In general, α-linked sugars have a higher affinity for the permease than β-linked sugars. Competitive inhibition studies of lactose transport by a variety of sugars shows that the relative importance of the stereochemistry of the galactosyl hydroxyl groups follows the order C3–OH > C4–OH > C6–OH > C2–OH > C1–OH, and that the preferred size for the non-galactosyl moiety is hexose, phenyl ring > methyl group > no aglycone > disaccharide > trisaccharide [122]. Thus, binding of the galactosyl moiety is specific whereas that of the non-galactosyl part is non-specific [86,91].

Since very little is known about the substrate binding site(s) in lac permease, it is not possible to design rationally a ligand that will bind to the permease in a predictable manner. Therefore a series of spin-labeled ligands (spin probes) are being synthesized, and binding to the permease will be studied in order to identify ligands that can report on the environment of the protein in the vicinity of the nitroxide. Ligands that bind with high affinity ($K_D \leq 10^{-6}$ M) can then be used to determine their distance from either metal binding sites incorporated into the permease or from spin-labeled Cys residues in the permease. A few spin-labeled galactosides have been synthesized [123] but their use in EPR studies has not yet been reported. A tritiated, spin-labeled galactoside was found to be actively transported by lac permease [124], but the nitroxide was probably reduced in intact cells.

Although purified lac permease can be reconstituted into proteoliposomes, the permease retains the ability to bind ligands specifically in dodecylmaltoside [86–88]. Therefore, both reconstituted permease and permease in dodecylmaltoside will be employed in EPR experiments. Binding of spin-probes to the permease will be studied to obtain the following information.

Initially, an approximation of affinity of the permease for the spin-labeled galactosides can be obtained by means of transport assays (i.e., determination of K_i) with intact cells or right-side-out membrane vesicles or by measuring the ability of a given spin-labeled galactoside to block MIANS labeling of single-Cys148 permease [77]. Titration of purified lac permease with increasing concentrations of spin probes that exhibit appropriately low K_is will be performed, and K_Ds and number of binding sites will be measured [19]. The EPR spectrum of a free spin probe in aqueous solution has the typical three line nitroxide spectrum. Spectral identification of the bound species is dependent upon its K_D, and frequently, a mixed spectrum of bound and free species is observed. When the K_D is fairly large ($K_D \geq 10^{-3}$ M), the extent of binding is measured by an almost stoichiometric decrease in the spectral amplitude of the free species upon binding to the protein [125]. This allows a fairly accurate measurement of K_D since the much broader bound-species spectrum contributes very little to the observed spectral peak height. Spin probes with a K_D ($\leq 10^{-6}$ M) can be observed as the protein-bound form. The K_D in such cases is measured by adding a ligand (such as NPG or TDG) that competes for the binding site and measuring the spectral intensity of the liberated free species [125,126].

The polarity of the environment of the nitroxide can be directly determined from the value of the splitting of the outer hyperfine extrema in the absence of motion ($2A_{zz}^o$) [127]. This can be determined in frozen solutions or by recording the spectrum at two microwave frequencies [127].

In order to probe the steric environment of a nitroxide ligand, the mobility of the nitroxide bound to the permease will be studied. In the anticipated slow motional regime ($10^{-7} > \tau_c > 10^{-8}$), this is simply measured by the parameter $S \equiv A_{zz}/A_{zz}^o$, where A_{zz} and A_{zz}^o are the splitting of hyperfine extrema at the experimental temperature and in the absence of motion, respectively [128]. If the spectra reflect motion in the range $10^{-9} < \tau_c < 10^{-8}$, the motion will be analyzed by spectral simulation techniques [128].

It is not possible to predict the behavior of the various spin probes used in these EPR studies due in part to the lack of detailed information about the substrate binding site(s) in the permease. EPR spectral parameters can be interpreted to determine the number, polarity and steric environment of the binding site(s) and the dissociation constant of a given spin-labeled ligand for the binding sites. Furthermore, these studies can provide a basis to select and design ligands that can be used for additional EPR studies to derive more structural and dynamic information about the permease.

One general and two more specific approaches can be used to localize the substrate translocation pathway in the permease. First, the position of a spin labeled galactoside with an appropriately high affinity can be localized within the tertiary structure of the permease by using metal–spin label interactions as described above. Thus, functional permease molecules containing metal binding sites in periplasmic, cytoplasmic loops or within the permease will be purified and distance between the bound ligand and the metal binding site can be determined as described by [96,97, 102]. Once the general location of the binding site is determined by triangulation, focus will shift to specific single-Cys replacement mutants. For example, site-directed fluorescence studies (86) demonstrate that Cys148 is in a substrate binding site and that Met145, located on the same helical face as Cys 148, is in close proximity. Thus, the reactivity of M145C permease with MIANS is mildly inhibited by ligand, and the accessibility of MIANS-labeled M145C permease to I⁻ is blocked by ligand. In contrast, modification of Cys148, which is in the binding site, abolishes ligand binding. Therefore, permease containing a single Cys at position 145 will be derivatized with a thiol-specific nitroxide spin label and reconstituted into proteoliposomes or studied in dodecylmaltoside. Binding of a spin-labeled galactoside with a $K_D \leq 10^{-6}$ M to permease spin-labeled at position 145 will be studied by EPR. Dipolar interaction between two nitroxides depends on $1/r^3$ and is reflected in the ratio of the difference in intensities of the hyperfine extrema to the intensity of the central line in the EPR spectrum [129]. The distance between the nitroxide on the galactoside and that on the permease can be approximately determined from this ratio [100], but more accurately from spectral simulation [118]. Another focused approach will involve labeling M145C permease with a thiol-specific fluorophore [86] and monitoring the quenching of fluorescence by the bound spin-labeled galactoside. Fluorescence quenching will be observed if the fluorophore and the nitroxide are separated by a distance of up to about 10 Å [130]. Clearly, with the availability of a library of permease molecules containing a single-Cys residue at each position in the protein which can be readily purified by affinity chromatography, this approach can be extended as desired to residues in the area of interest in the molecule.

5. Summary and concluding remarks

The lac permease of E. coli is providing a paradigm for secondary active transporter proteins that transduce the free energy stored in electrochemical ion gradients into work in the form of a concentration gradient. This hydrophobic, polytopic, cytoplasmic membrane protein catalyzes the coupled, stoichiometric translocation of β-galactosides and H⁺, and it has been solubilized, purified, reconstituted into artificial phospholipid vesicles and shown to be solely responsible for β-galactoside transport as a monomer. The lacY gene which encodes the permease has been cloned and sequenced, and based on spectroscopic analyses of the purified protein and hydropathy profiling of its amino-acid sequence, a secondary structure has been proposed in which the protein has 12 transmembrane domains in α-helical configuration that traverse the membrane in zigzag fashion connected by hydrophilic loops with the N and C termini on the cytoplasmic face of the membrane. Unequivocal support for the topological predictions of the 12-helix model has been obtained by analyzing a large number of lac permease-alkaline phosphatase (*lacY-phoA*) fusions. Extensive use of site-directed and Cys-scanning mutagenesis indicates that very few residues in the permease are directly involved in the transport mechanism. Interestingly, however, Cys148 which is not essential for activity is located in a binding site and probably interacts weakly and hydrophobically with the galactosyl moiety of the substrate.

Second-site suppressor analysis and site-directed mutagenesis with chemical modification have provided evidence that helix VII is close to helices X and XI in the tertiary structure of the permease. Experiments in which paired Cys replacements in C-less permease were labeled with pyrene, a fluorophore that exhibits excimer fluorescence when two of the unconjugated ring systems are in close approximation, indicate that His322 and Glu325 are located in an α-helical region of the permease and that helix IX is close to helix X. Based on certain second-site suppressor mutants, it was suggested that helix VIII (Glu269) is close to helix X (His322), and site-directed pyrene-maleimide labeling experiments provide strong support for this postulate. Taken together, the findings lead to a model describing helix packing in the C-terminal half of the permease, and recent experiments using engineered metal binding sites (i.e., bis-His residues in transmembrane helices) and site-directed peptide bond cleavage have confirmed and extended the model. Other findings indicate that widespread conformational changes in the permease result from either ligand binding or imposition of $\Delta\bar{\mu}_H^+$.

Since many membrane transporters appear to have similar secondary structures based on hydropathy profiling, it seems likely that the basic tertiary structure and mechanism of action of these proteins have been conserved throughout evolution. Therefore, studies on bacterial transport systems which are considerably easier to manipulate than their eukaryotic counterparts have important relevance to transporters in higher order systems, particularly with respect to the development of new approaches to structure–function relationships. Although it is now possible to manipulate membrane proteins to an extent that was unimaginable only a few years ago, it is unlikely that transport mechanisms can be defined on a molecular level without information about tertiary structure. As indicated by many of the experiments outlined here, site-directed spectroscopy will be particularly useful for obtaining both static and dynamic information.

Remarkably few amino-acid residues appear to be critically involved in the transport mechanism, suggesting that relatively simple chemistry drives the mechanism. On the other hand, the use of site-directed fluorescence and chemical labeling indicate that widespread conformational changes are involved. Furthermore, certain active Cys-replacement mutants are altered by alkylation, and these mutants cluster on helical faces. As a whole, the results suggest that the permease is comprised of 12 loosely packed helices in which surface contours are important for sliding motions that occur during turnover. This surmise coupled with the indication that few residues are essential to the mechanism is encouraging in that it suggests the possibility that a relatively low resolution structure (i.e., helix packing) plus localization of the critical residues and the translocation pathway will provide important insights.

References

1. Henderson, P.J. (1990) J. Bioenerg. Biomembr. **22**, 525–569.
2. Marger, M.D. and Saier, M.H., Jr. (1993) TIBS **18**, 13–20.
3. Deisenhofer, J. and Michel, H. (1989) Science **245**, 1463–1473.
4. Cohen, G.N. and Rickenberg, H.V. (1955) Compt. Rendu. **240**, 466–468.
5. Mitchell, P. (1963) Biochem. Soc. Symp. **22**, 142–168.
6. Mitchell, P. (1968) Chemiosmotic coupling and energy transduction. Glynn Research Ltd, Bodmin, UK.
7. West, I.C. (1970) Biochem. Biophys. Res. Commun. **41**, 655–661.
8. West, I.C. and Mitchell, P. (1972) J. Bioenerg. **3**, 445–462.
9. West, I.C. and Mitchell, P. (1973) Biochem. J. **132**.
10. Kaback, H.R. (1976) J. Cell Physiol. **89**, 575–593.
11. Kaback, H.R. (1983) J. Membr. Biol. **76**, 95–112.
12. Kaback, H.R. (1989) Harvey Lect. **83**, 77–103.
13. Teather, R.M., Müller-Hill, B., Abrutsch, U., Aichele, G. and Overath, P. (1978) Molec. Gen. Genet. **159**, 239–248.
14. Buchel, D.E., Gronenborn, B. and Muller-Hill, B. (1980) Nature **283**, 541–545.
15. Kaczorowski, G.J., Leblanc, G. and Kaback, H.R. (1980) Proc. Natl. Acad. Sci. USA **77**, 6319–6323.
16. Newman, M.J. and Wilson, T.H. (1980) J. Biol. Chem. **255**, 10583–10586.
17. Newman, M.J., Foster, D.L., Wilson, T.H. and Kaback, H.R. (1981) J. Biol. Chem. **256**, 11804–11808.
18. Foster, D.L., Garcia, M.L., Newman, M.J., Patel, L. and Kaback, H.R. (1982) Biochemistry **21**, 5634–5638.
19. Viitanen, P., Newman, M.J., Foster, D.L., Wilson, T.H. and Kaback, H.R. (1986) Meth. Enzymol. **125**, 429–452.
20. Matsushita, K., Patel, L., Gennis, R.B. and Kaback, H.R. (1983) Proc. Natl. Acad. Sci. USA **80**, 4889–4893.
21. Viitanen, P., Garcia, M.L. and Kaback, H.R. (1984) Proc. Natl. Acad. Sci. USA **81**, 1629–1633.
22. Foster, D.L., Boublik, M. and Kaback, H.R. (1983) J. Biol. Chem. **258**, 31–34.
23. Vogel, H., Wright, J.K. and Jahnig, F. (1985) EMBO J **4**, 3625–3631.
24. Carrasco, N., Tahara, S.M., Patel, L., Goldkorn, T. and Kaback, H.R. (1982) Proc. Natl. Acad. Sci. USA **79**, 6894–6898.
25. Seckler, R., Wright, J.K. and Overath, P. (1983) J. Biol. Chem. **258**, 10817–10820.
26. Carrasco, N., Viitanen, P., Herzlinger, D. and Kaback, H.R. (1984) Biochemistry **23**, 3681–3687.
27. Herzlinger, D., Viitanen, P., Carrasco, N. and Kaback, H.R. (1984) Biochemistry **23**, 3688–3693.
28. Carrasco, N., Herzlinger, D., Mitchell, R., DeChiara, S., Danho, W., Gabriel, T.F. and Kaback, H.R. (1984) Proc. Natl. Acad. Sci. USA **81**, 4672–4676.
29. Seckler, R. and Wright, J.K. (1984) Eur. J. Biochem. **142**, 269–279.
30. Herzlinger, D., Carrasco, N. and Kaback, H.R. (1985) Biochemistry **24**, 221–229.
31. Danho, W., Makofske, R., Humiec, F., Gabriel, T. F., Carrasco, N. and Kaback, H.R. (1985) in: Use of Site-directed Polyclonal Antibodies as Immunotopological Probes for the lac Permease of *Escherichia coli*, eds C.M. Deber, V.J. Hruby and K.D. Kopple. Pierce Chem Co, pp. 59–62.
32. Seckler, R., Moroy, T., Wright, J.K. and Overath, P. (1986) Biochemistry **25**, 2403–2409.
33. Goldkorn, T., Rimon, G. and Kaback, H.R. (1983) Proc. Natl. Acad. Sci. USA **80**, 3322–3326.
34. Stochaj, U., Bieseler, B. and Ehring, R. (1986) Eur. J. Biochem. **158**, 423–428.

35. Page, M.G. and Rosenbusch, J.P. (1988) J. Biol. Chem. **263**, 15906–15914.
36. Bieseler, B., Heinrich, P. and Beyreuther, C. (1985) Ann. NY Acad. Sci. **456**, 309–325.
37. Calamia, J. and Manoil, C. (1990) Proc. Natl. Acad. Sci. USA **87**, 4937–4941.
38. Ujwal, M.L., Jung, H., Bibi, E., Manoil, C., Altenbach, C., Hubbell, W.L. and Kaback, H.R. (1995) Biochemistry, **34**, 14909–14917.
39. Bibi, E. and Kaback, H.R. (1990) Proc. Natl. Acad. Sci. USA **87**, 4325–4329.
40. Zen, K.H., McKenna, E., Bibi, E., Hardy, D. and Kaback, H.R. (1994) Biochemistry **33**, 8198–8206.
41. Dornmair, K., Corin, A.F., Wright, J.K. and Jahnig, F. (1985) EMBO J. **4**, 3633–3638.
42. Costello, M.J., Escaig, J., Matsushita, K., Viitanen, P.V., Menick, D.R. and Kaback, H.R. (1987) J. Biol. Chem. **262**, 17072–17082.
43. Bibi, E. and Kaback, H.R. (1992) Proc. Natl. Acad. Sci. USA **89**, 1524–1528.
44. Sahin-Tóth, M., Lawrence, M. C. and Kaback, H.R. (1994) Proc. Natl. Acad. Sci. USA **91**, 5421–5425.
45. van Iwaarden, P.R., Pastore, J.C., Konings, W.N. and Kaback, H.R. (1991) Biochemistry **30**, 9595–9600.
46. Kaback, H.R., Jung, K., Jung, H., Wu, J., Privé, G.G. and Zen, K. (1993) J. Bioenerg. Biomembr. **25**, 627–636.
47. Kaback, H.R., Frillingos, S., Jung, H., Jung, K., Privé, G.G., Ujwal, M.L., Weitzman, C., Wu, J. and Zen, K. (1994) J. Exp. Biol. **196**, 183–195.
48. Kaback, H.R. (1995) in: The Lactose Permease of Escherichia coli: An update, ed T.E. Andreoli, J. Hoffman, D.D. Fanestil and S.G. Schultz, Physiology of Membrane Disorders.
49. Sahin-Tóth, M. and Kaback, H.R. (1993) Protein Sci. **2**, 1024–1033.
50. Dunten, R.L., Sahin-Toth, M. and Kaback, H.R. (1993) Biochemistry **32**, 12644–12650.
51. Sahin-Tóth, M., Persson, B., Schwieger, J., Cohan, M. and Kaback, H.R. (1994) Protein Sci. **3**, 240–247.
52. Frillingos, S., Sahin-Toth, M., Persson, B. and Kaback, H.R. (1994) Biochemistry **33**, 8074–8081.
53. Weitzman, C. and Kaback, H.R. (1995) Biochemistry, **34**, 9374–9379.
54. Trumble, W.R., Viitanen, P.V., Sarkar, H.K., Poonian, M.S. and Kaback, H.R. (1984) Biochem. Biophys. Res. Commun. **119**, 860–867.
55. Menick, D.R., Sarkar, H.K., Poonian, M.S. and Kaback, H.R. (1985) Biochem. Biophys. Res. Commun. **132**, 162–170.
56. Viitanen, P.V., Menick, D.R., Sarkar, H.K., Trumble, W.R. and Kaback, H.R. (1985) Biochemistry **24**, 7628–7635.
57. Sarkar, H.K., Menick, D.R., Viitanen, P.V., Poonian, M.S. and Kaback, H.R. (1986) J. Biol. Chem. **261**, 8914–8918.
58. Brooker, R.J. and Wilson, T.H. (1986) J. Biol. Chem. **261**, 11765–11771.
59. Menick, D.R., Lee, J.A., Brooker, R.J., Wilson, T.H. and Kaback, H.R. (1987) Biochemistry **26**, 1132–1136.
60. van Iwaarden, P.R., Driessen, A.J., Menick, D.R., Kaback, H.R. and Konings, W.N. (1991) J. Biol. Chem. **266**, 15688–15692.
61. Neuhaus, J.-M., Soppa, J., Wright, J. K., Riede, I., Bocklage, H., Frank, R. and Overath, P. (1985) FEBS Lett. **185**, 83–88.
62. Menezes, M.E., Roepe, P.D. and Kaback, H.R. (1990) Proc. Natl. Acad. Sci. USA **87**, 1638–1642.
63. Lolkema, J.S., Puttner, I.B. and Kaback, H.R. (1988) Biochemistry **27**, 8307–8310.
64. Consler, T.G., Tsolas, O. and Kaback, H.R. (1991) Biochemistry **30**, 1291–1298.
65. Roepe, P.D. and Kaback, H.R. (1989) Biochemistry **28**, 6127–6132.
66. Jung, K., Jung, H., Colacurcio, P. and Kaback, H.R. (1995) Biochemistry **34**, 1030–1039.
67. Padan, E., Sarkar, H.K., Poonian, M.S. and Kaback, H.R. (1985) Proc. Natl. Acad. Sci. USA **82**, 6765–6768.
68. Puttner, I.B., Sarkar, H.K., Poonian, M.S. and Kaback, H.R. (1986) Biochemistry **25**, 4483–4485.
69. Puttner, I.B. and Kaback, H.R. (1988) Proc. Natl. Acad. Sci. USA **85**, 1467–1471.
70. Puttner, I.B., Sarkar, H.K., Lolkema, J.S. and Kaback, H.R. (1989) Biochemistry **28**, 2525–2533.
71. Ujwal, M.L., Sahin-Toth, M., Persson, B. and Kaback, H.R. (1994) Mol. Membr. Biol. **1**, 9–16.
72. Franco, P.J. and Brooker, R. J. (1994) J. Biol. Chem. **269**, 7379–7386.

73. Menick, D.R., Carrasco, N., Antes, L.M., Patel, L. and Kaback, H.R. (1987) Biochemistry **26**, 6638–6644.
74. Matzke, E.A., Stephenson, L.J. and Brooker, R.J. (1992) J. Biol. Chem. **267**, 19095–19100.
75. Carrasco, N., Antes, L.M., Poonian, M.S. and Kaback, H.R. (1986) Biochemistry **25**, 4486–4488.
76. Carrasco, N., Puttner, I.B., Antes, L.M., Lee, J.A., Larigan, J.D., Lolkema, J.S., Roepe, P.D. and Kaback, H.R. (1989) Biochemistry **28**, 2533–2539.
77. Jung, K., Jung, H., Wu, J., Privé, G.G. and Kaback, H.R. (1993) Biochemistry **32**, 12273–12228.
78. Consler, T.G., Persson, B.L., Jung, H., Zen, K.H., Jung, K., Prive, G.G., Verner, G.E. and Kaback, H.R. (1993) Proc. Natl. Acad. Sci. USA **90**, 6964.
79. King, S.C., Hansen, C.L. and Wilson, T.H. (1991) Biochim. Biophys. Acta **1062**, 177–186.
80. Sahin-Tóth, M., Dunten, R.L., Gonzalez, A. and Kaback, H.R. (1992) Proc. Natl. Acad. Sci. USA **89**, 10547–10551.
81. Dunten, R.L., Sahin-Toth, M. and Kaback, H.R. (1993) Biochemistry **32**, 3139–3145.
82. Sahin-Tóth, M. and Kaback, H.R. (1993) Biochemistry **32**, 10027–10035.
83. Lee, J.L., Hwang, P.P., Hansen, C. and Wilson, T.H. (1992) J. Biol. Chem. **267**, 20758–20764.
84. Jung, K., Jung, H. and Kaback, H.R. (1994) Biochemistry **33**, 3980–3985.
85. Jung, H., Jung, K. and Kaback, H.R. (1994) Protein Sci **3**, 1052–1057.
86. Wu, J. and Kaback, H.R. (1994) Biochemistry **33**, 12166–12171.
87. Wu, J., Frillingos, S., Voss, J. and Kaback, H.R. (1994) Protein Sci. **3**, 2294–2301.
88. Wu, J., Frillingos, S. and Kaback, H.R. (1995) Biochemistry **34**, 8257–8263.
89. Fox, C.F. and Kennedy, E.P. (1965) Proc. Natl. Acad. Sci. USA **51**, 81–89.
90. Beyreuther, K., Bieseler, B., Ehring, R. and Müller-Hill, B. (1981) in: Identification of Internal Residues of Lactose Permease of *Escherichia coli* by Radiolabel Sequences of Peptide Mixtures, ed M. Elzinga. Humana, Clifton, NY, pp. 139–148.
91. Jung, K., Jung, H. and Kaback, H.R. (1994) Biochemistry **33**, 12160–12165.
92. Privé, G.G., Verner, G.E., Weitzman, C., Zen, K.H., Eisenberg, D. and Kaback, H.R. (1994) Acta Cryst. **D50**, 375–379.
93. Privé, G.G. and Kaback, H.R. (1996) J. Bioenerg. Biomembr. **28**, 29–34.
94. Eaton, S.S. and Eaton, G.R. (1988) Coord. Chem. Rev. **83**, 29–72.
95. Leigh, J.S. (1970) J. Chem. Phys. **52**, 2608–2612.
96. Voss, J., Hubbell, W.L. and Kaback, H.R. (1995) Biophys. J. **68**, A423.
97. Voss, J., Salwinski, L., Kaback, H.R. and Hubbell, W.L. (1995) Proc. Natl. Acad. Sci. USA **92**, 12295–12299.
98. Arnold, F.H. and Haymore, B.L. (1991) Science **252**, 1796–1797.
99. Altenbach, C., Marti, T., Khorana, H.G. and Hubbell, W.L. (1990) Science **248**, 1088–1092.
100. Likhtenshtein, G.I. (1993) Biophysical labeling methods in molecular biology. Cambridge University Press, New York.
101. McKenna, E., Hardy, D. and Kaback, H.R. (1992) Proc. Natl. Acad. Sci. USA **89**, 11954–11958.
102. Voss, J., Hubbell, W.L. and Kaback, H.R. (1995) Proc. Natl. Acad. Sci. USA **92**, 12300–12303.
103. Park, J.H. and Trommer, W.E. (1989) in: Advantages of N-15 and Deuterium Spin Probes for Biomedical Electron Paramagnetic Resonance Investigations, eds L.J. Berliner J. and Reuben. Plenum Press, New York, Vol. 8, pp. 547–595.
104. Selvin, P.R. and Hearst, J.E. (1994) Proc. Natl. Acad. Sci. USA **91**, 10024–10028.
105. Regan, L. (1993) Ann. Rev. Biophys. Biomol. Struct. **22**, 257–287.
106. Jung, K., Voss, J., He, M., Hubbell, W.L. and Kaback, H.R. (1995) Biochemistry **34**, 6272–6277.
107. He, M.M., Voss, J., Hubbell, W.L. and Kaback, H.R. (1995) Biochemistry **34**, 15661–15666.
108. He, M.M., Voss, J., Hubbell, W.L. and Kaback, H.R. (1995) Biochemistry **34**, 15667–15670.
109. Dervan, P.B. (1986) Science **232**, 464–471.
110. Tullius, T.D. (1989) Ann. Rev. Biophys. Biophys. Chem. **18**, 213–237.
111. Ermacora, M.R., Delfino, J.M., Cuenoud, B., Schepartz, A. and Fox, R.O. (1992) Proc. Natl. Acad. Sci. USA **89**, 6383–6387.
112. Rana, T. M. and Meares, C. F. (1991) Proc. Natl. Acad. Sci. USA **88**, 10578–10582.
113. Bateman, R.C., Jr., Youngblood, W.W., Busby, W.H., Jr. and Kizer, J.S. (1985) J. Biol. Chem. **260**, 9088–9091.

114. Sigman, D.S. (1990) Biochemistry **29,** 9097–9105.
115. Sigman, D.S., Kuwabara, M.D., Chen, C.H. and Bruice, T W. (1991) Meth. Enzymol. **208,** 414–433.
116. Wu, J., Perrin, D., Sigman, D. and Kaback, H. (1995) Proc. Natl. Acad. Sci. USA **92,** 9186–9190.
117. Li, J. and Tooth, P. (1987) Biochemistry **26,** 4816–4823.
118. Farahbakhsh, Z.T., Huang, Q.L., Ding, L.L., Altenbach, C., Steinhoff, H.J., Horwitz, J. and Hubbell, W.L. (1995) Biochemistry **34,** 509–516.
119. Stryer, L. and Haugland, R.P. (1967) Proc. Natl. Acad. Sci. USA **58,** 719–726.
120. Weitzman, C., Consler, T.G. and Kaback, H.R. (1995) Protein Sci. **4,** 2310–2318.
121. Sandermann, H., Jr. (1977) Eur. J. Biochem. **80,** 507–515.
122. Olsen, S.G. and Brooker, R.J. (1989) J. Biol. Chem. **264,** 15982–15987.
123. Gnewuch, T. and Sosnovsky, G. (1986) Chem. Rev. **86,** 203–238.
124. Struve, W.G. and McConnell, H.M. (1972) Biochem. Biophys. Res. Commun. **49,** 1631–1637.
125. Weiner, H. (1969) Biochemistry **8,** 526–33.
126. Berliner, L.J. (1978) Meth. Enzymol. **49G,** 418–480.
127. Griffith, O.H. and Jost, P.C. (1976) in: Lipid Spin Labels in Biological Membranes, ed L.J. Berliner. Academic Press, New York, Vol. 1, pp. 453–523.
128. Schneider, D.J. and Freed, J.H. (1989) in: Calculating Slow Motional Magnetic Resonance Spectra: A Users' Guide, eds L.J. Berliner and J. Reuben. Plenum Press, New York, Vol. 8, pp. 1–76.
129. Lassman, G., Ebert, B., Kuznetsov, A.N. and Damerau, W. (1973) Biochim. Biophys. Acta **310,** 298–304.
130. Weiner, J.H., Shaw, G., Turner, R.J. and Trieber, C.A. (1993) J. Biol. Chem. **268,** 3238–3244.

CHAPTER 11

Secondary Transporters and Metabolic Energy Generation in Bacteria

J.S. LOLKEMA, B. POOLMAN and W.N. KONINGS

Department of Microbiology, Groningen Biomolecular Sciences and Biotechnology Institute,
University of Groningen, Biology Centre, Kerklaan 30, 9751 NN Haren, The Netherlands

© *1996 Elsevier Science B.V.*
All rights reserved

Handbook of Biological Physics
Volume 2, edited by W.N. Konings, H.R. Kaback and J.S. Lolkema

Contents

1. Introduction

Nutrients provide bacterial cells with the two essential components for sustaining live: matter and free energy. The transduction of matter and free energy between catabolism and anabolism is indirect which ensures high metabolic flexibility. Nutrients are degraded to more general applicable molecules that serve as multipurpose 'building blocks' in biosynthesis. Two major forms of free energy, usually referred to as metabolic energy, are intermediate between catabolism and anabolism: ATP and ion gradients across the cell membrane. ATP represents a pool of chemical free energy ($\Delta G'_{ATP}$) that is released upon hydrolysis. The ion gradients represent free energy stored in an electrochemical gradient of either H^+ or Na^+ ($\Delta\tilde{\mu}_{H^+}$ and $\Delta\tilde{\mu}_{Na^+}$, respectively). The gradients exert a force on the ions (the ion motive force) that consists of two components, the chemical gradient ΔpX and the electrical potential gradient $\Delta\psi$, according to the following equation:

$$\frac{\Delta\tilde{\mu}_{X^+}}{F} = -\frac{2.3\ RT}{F}\ \Delta pX + \Delta\psi$$

in which X^+ represents either H^+ or Na^+ and the forces are expressed in electrical field units (V). Normal physiological conditions are such that the force on the ions is directed inwards, which usually means that the cell is alkaline, low Na^+ and negative inside relative to outside.

The free energy in the ATP pool and the H^+ and Na^+ ion gradients is termed *metabolic energy* and is available to the cell for energy consuming processes. Generation of metabolic energy refers to metabolic processes by which the free energy in these pools increases. The different pools of free energy are convertible through the action of reversible ATPases/synthases that couple the transport of either H^+ or Na^+ to the hydrolysis/synthesis of ATP. The electrochemical proton and sodium ion gradients can be interconverted by Na^+/H^+ antiporters (Fig. 1). The interconversion between the different pools is an important mechanism to fill the pools, i.e. in one organism catabolism results in a high proton motive force (pmf) that is used to drive ATP synthesis, whereas in other organisms ATP hydrolysis is coupled to the formation of an electrochemical proton gradient. The interconversion does not result in an overall increase of free energy in the pools. In fact, since the interconversion takes place at a certain rate, the free energy as a whole decreases. The total free energy in the pools increases as a result of ATP synthesis by substrate level phosphorylation and the translocation of proton and sodium ions across the cell membrane catalysed by chemiosmotic pumps at the expense of some source of chemical energy or light (Fig. 1) [1]. The ion pumps are termed primary transporters and, by analogy, we propose to call this mechanism of ion motive force

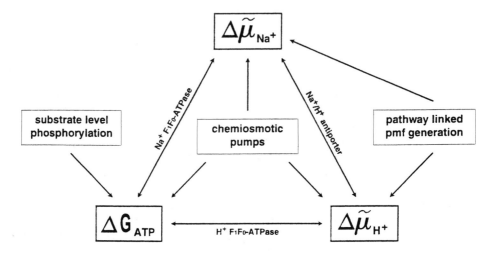

Fig. 1.   Generation and interconversion of metabolic energy. By definition, metabolic energy is the free energy stored in the ATP pool, and the Na^+ and H^+ ion gradients across the cytoplasmic membrane.

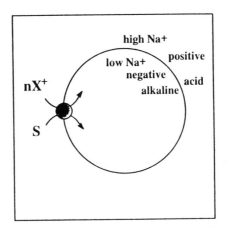

Fig. 2.   A symporter and the energetic state of the cytoplasmic membrane. The secondary transporter catalyses the coupled uptake of solute S with H^+, Na^+ or both. X^+ specifies H^+ or Na^+. Under physiological conditions, the H^+ and Na^+ gradients across the membrane are directed inwards and the membrane potential is positive outside relative to inside.

generation *primary metabolic energy generation*. Less well known is a different mechanism of pmf generation that involves the action of secondary transporters, rather than primary ion pumps, which is termed *secondary metabolic energy generation*. The pmf is generated as a result of both metabolic and transport steps and involves the activity of a complete pathway. Secondary metabolic energy generation, which is particularly important for (facultative) anaerobes, is the subject of this chapter.

Secondary transporters convert the electrochemical energy from one gradient into another one (for a recent review, see Ref. [2]). Mechanistically, this is achieved by coupling fluxes of different species, usually one or more cations and a solute (Fig. 2). The transporters can be classified in three groups: (i) *uniporters* catalyse the translocation of a single solute across the cytoplasmic membrane, (ii) *symporters* couple transport of two or more solutes in the same direction, and (iii) *antiporters* couple the movement of solutes in opposite direction. Usually, secondary transport is looked at as a metabolic energy consuming process. The energetic state of the membrane is such that a force is exerted on the cations to move in (Fig. 2), which allows the cells to use the ion motive force to concentrate a nutrient that is present at low concentration in the medium inside the cell (cation/substrate symport) or to excrete unwanted products into the medium (cation/product antiport). The cations move down their electrochemical gradient and the solute moves against its concentration gradient. In addition to the magnitude of the cation gradient(s) and the solute gradient, the magnitude and direction of the force on the transport process is determined by the transport stoichiometry and the charge of the substrate. This also determines the relative contribution of the chemical gradient (ΔpX) and the membrane potential ($\Delta\psi$) to the driving force. A transporter that is driven by the membrane potential is electrogenic and translocates net positive charge across the membrane.

2. Secondary metabolic energy generation

2.1. Excretion systems

The most straightforward implementation of secondary transport in metabolic energy generation is a proton or Na^+-ion symporter that functions as an extrusion system (Fig. 3A). A solute that accumulates in the cell as a product of metabolism is transported out of the cell in symport with a proton or Na^+-ion (*end-product efflux*). The solute drives the cation against the ion motive force and free energy is converted from the solute gradient into the ion gradient. The outwardly directed product gradient is sustained by continuous uptake of the substrate from the external medium and internal conversion into the product. Since the cell may have to spent metabolic energy in the uptake of the substrate, end-product efflux has been referred to as 'energy recycling' [3]. This is particularly true for those systems that use an excretion system as a means to recover stored energy as has been observed for the excretion of inorganic phosphate in *Acinetobacter johnsonii* [13,14]. Under conditions of excess carbon and energy source, the cell spends metabolic energy to take up phosphate which subsequently is stored as polyphosphate. When conditions get harsh polyphosphate is hydrolysed which results in the recovery of metabolic energy upon cation linked efflux of the released phosphate. Table 1 summarizes a number of well-established and suggested metabolic energy generating excretion systems. These systems will not be further considered here (for a review, see Refs. [4,5]).

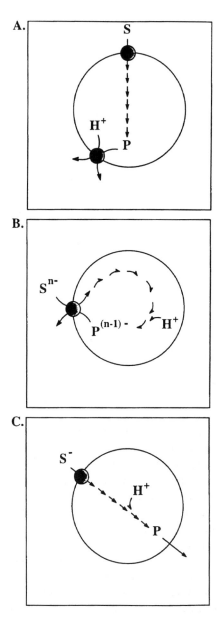

Fig. 3.   Mechanisms of secondary metabolic energy generation. (A) Excretion. Cytosolic reactions maintain an outwardly directed gradient of product P over the membrane. The symporter couples the efflux of the product to the uphill movement of a proton (or Na^+ ion) out of the cell, thereby generating a pmf (or smf). (B) Precursor/product exchange. The transporter couples the uptake of substrate S to the excretion of the metabolic end-product P. Net charge translocation by the transporter results in the generation of a membrane potential, inside negative relative to outside, and proton consumption in the pathway in a pH gradient, inside alkaline. (C) Electrogenic uniport. Carrier mediated uptake of weak acid S followed by conversion into product P that leaves the cell by passive diffusion.

Table 1
Metabolic energy generating excretion systems

Substrate	Cation	Organism	Ref.
Lactate	H$^+$	*Escherichia coli*	6
		Lactococcus lactis	7,8
		Enterococcus faecalis	9,10
Succinate	Na$^+$	*Selenomonas ruminantium*	11
		Escherichia coli	12
MeHPO$_4^-$	H$^+$	*Acetobacter johnsonii*	13
		Escherichia coli	14
Acetate	–	Sulfate reducers	15–17
		Methanogens	18
NH$_4^+$	–	*Ureaplasma* species	19
SO$_4^{2-}$	–	*Thiobacillus* species	19

2.2. Electrogenic precursor/product exchange systems

In these systems the pmf is not generated by direct proton translocation but, instead, is generated indirectly during the conversion of a substrate into the product(s) of a fermentation pathway in a process in which secondary transport plays a central role (Fig 3B). Importantly, the pmf is not generated by F$_0$F$_1$-ATPase subsequent to ATP formation by substrate level phosphorylation, electron transfer, or some other means of chemiosmotic energy conservation. The two components of the pmf, membrane potential and pH gradient, are generated in separate steps in the pathway. The membrane potential is generated by the transport system involved in the coupled uptake and excretion of substrate and product (*precursor/product exchange*). As indicated above, the uptake of nutrients usually requires the input of metabolic energy, i.e., the carriers are driven by the ion gradients. Precursor/product exchangers generate a membrane potential of physiological polarity by net translocation of negative charge into the cell or net positive charge out of the cell (Fig. 4). The force on the coupled precursor uptake/product exit is composed of the precursor and product concentration gradients and counteracted by the membrane potential. Both precursor and product concentration gradients are generated by the metabolic pathway that converts precursor into product. The pH gradient is generated in the conversion of internalized substrate to the product(s) by consumption of a scalar proton. The chemical nature of the pH gradient formation makes that metabolic energy generation is restricted to the pmf as far as the chemical gradient is concerned. The charge translocation contributes to both the pmf and smf. It should be stressed that, whereas the chemiosmotic pumps and the secondary excretion systems generate the pmf in a single step catalysed by a single enzyme or enzyme complex, pmf generation by precursor/product exchange systems involves a minimum of one transport and one metabolic step.

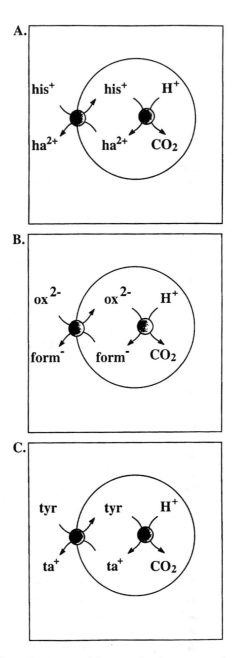

Fig. 4.   Mechanisms of membrane potential generation by precursor/product exchangers. (A) Histidine/histamine exchange in *L. buchneri* [21]. Net positive charge is moved out of the cell by exchange of monovalent histidine (his$^+$) for divalent histamine (ha^{2+}). (B) Oxalate/formate exchange. Net negative charge is moved into the cell by exchange of divalent oxalate (ox^{2-}) for monovalent formate (form$^-$). (C) Tyrosine/tyramine exchange. Net positive charge is moved out of the cell by exchange of neutral tyrosine (tyr) for positively charged tyramine (ta$^+$).

Table 2
Membrane potential generating secondary transport

Transport	Protein	Organism	Ref.
Exchangers/antiporters			
Aspartate/alanine[a]	–	*Lactobacillus* sp.	20
Histidine/histamine	–	*Lactobacillus buchneri*	21
Lysine/alanine[b]	LysI	*Corynebacterium glutamicum*	22
Lysine/cadaverine	CadB	*Escherichia coli*	23
Malate/lactate	MleP	*Lactococcus lactis*	24
Oxalate/formate	OxlT	*Oxalobacter formigenes*	25
Phenylalanine/phenylethylamine[a]	–	*Lactobacillus buchneri*	unpublished
Putrescine/ornithine	PotE	*Escherichia coli*	26
Tyrosine/tyramine[a]	–	*Lactobacillus buchneri*	unpublished
Citrate/lactate	CitP	*Leuconostoc mesenteroides*	27
Uniporters			
Malate	–	*Leuconostoc oenos*	28
Citrate	–	*Leuconostoc oenos*	29

[a] Exchange data in intact cells.
[b] Alanine can be replaced by isoleucine or valine as substrate.

Decarboxylation driven fermentation of dicarboxylic acids and amino acids are representatives of the simplest systems consisting of an electrogenic transporter (Table 2) and a cytoplasmic decarboxylase. The corresponding decarboxylation products, i.e. monocarboxylic acids and amines, leave the cell in exchange for the substrate (Fig. 4). Citrate metabolism in lactic acid bacteria is a more complex system that upon entry of citrate into the cell, involves many metabolic steps that may differ between species and conditions (see below).

2.3. Electrogenic uniporters

Electrogenic uptake of anionic substrates followed by metabolism represents another mechanism of indirect pmf generation that is very similar to the precursor/product exchange systems (Figs. 3C and 3B, respectively). The difference is that the uptake of the substrate is not coupled to the excretion of a product. The substrate is taken up by a uniporter that translocates net negative charge into the cell thereby generating a membrane potential. The end-products leave the cell in their uncharged state, most likely by passive diffusion. The uptake of the substrate is driven only by the gradient of the substrate across the membrane whereas the membrane potential is counteracting transport. A high rate of metabolism of the internalized substrate is necessary to keep the substrate concentration gradient high. Up to now, electrogenic uniporters have been described for malate and citrate uptake in the Lactic Acid Bacterium *Leuconostoc oenos* (Table 2).

2.4. The citrate and malate fermentation pathways in Lactic Acid Bacteria

Citrate and malate fermentation pathways in Lactic Acid Bacteria are related secondary metabolic energy generating pathways that nicely demonstrate the diversity of the systems. They also show how the systems adapt to environmental conditions and integrate with other metabolic pathways in the cell. The malate fermentation pathway is a simple pathway consisting of a transporter and a cytoplasmic decarboxylase that converts malate into lactate plus carbon dioxide (Fig. 5A,C). The process is termed malolactic fermentation. Breakdown of citrate proceeds via citrate lyase that splits the molecule into acetate plus oxaloacetate. The latter is decarboxylated by a cytoplasmic decarboxylase yielding pyruvate plus carbon dioxide. In heterofermentative *Leuconostoc* species and during cometabolism of citrate with a carbohydrate, pyruvate functions as a redox sink for the reducing equivalents produced in the breakdown of the sugar (Fig. 5B,C) [30,31]. Since pyruvate is reduced to lactate we refer to the process as citrolactic fermentation. Both the malolactic and citrolactic fermentation pathways have been implemented in nature in functioning with an electrogenic uniporter and a precursor/product exchanger. In *Leuconostoc oenos*, a bacterium that grows optimally in acidic milieus, the uniporters catalyse transport of the monovalent anions (Fig. 5A,B). These are the protonation states of citrate and malate that are most abundantly present at the low pH values of the medium. The products of the pathways, lactate and acetate readily cross the cytoplasmic membrane in their protonated state by passive diffusion. *Leuconostoc oenos* maintains a cytosolic pH between 5 and 6 [29,32], resulting in a high enough fraction of the protonated end-products to ensure a high rate of excretion. On the other hand, the transporter for malate uptake in *Lactococcus lactis* and the transporter for citrate uptake in *Leuconostoc mesenteroides* are precursor/product exchangers that couple the uptake of the corresponding substrate to the excretion of the product lactate (Fig. 5C,D). Both these organisms grow at more neutral pH values where the divalent anionic states of the substrates are most abundant. A uniporter translocating divalent anions would have the disadvantage that the counteracting force on transport would be twice the membrane potential. Furthermore, the cytosolic pH in these organisms is above 7 [33,34] rendering most of the lactate in the dissociated state which may require a transporter for efficient excretion. The precursor/product exchange mechanism gets around both problems by providing a transport pathway for lactate excretion and by reducing the charge translocation to one unit. In conclusion, these examples suggest that the presence of an electrogenic uniporter or precursor/product exchanger may form an adaptation to the pH in the natural habitat of the organism, which determines the dissociation states of substrates and products.

In *Leuconostoc mesenteroides* citrolactic fermentation is intimately coupled to carbohydrate metabolism. Efficient uptake of citrate occurs only when lactate is produced from pyruvate, which requires the redox equivalents from carbohydrate metabolism [34]. The uniporter mechanism in *Leuconostoc oenos* uncouples both pathways and resting cells metabolize citrate even in the absence of carbohydrate.

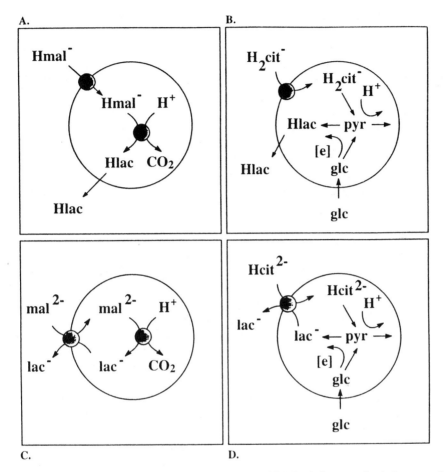

Fig. 5.   The citrate/malate system in Lactic Acid Bacteria. Malolactic fermentation in *Lc. oenos* (A) and *L. lactis* (C) and citrate metabolism in *Lc. oenos* (B) and *Lc. mesenteroides* (D).

Under those conditions pyruvate is converted to acetoin and carbon dioxide, which is a redox neutral process [54]. Energetically, the use of the uniporter in combination with passive efflux of the protonated products and the precursor/product exchanger are equivalent.

A relation between the citrate and malate fermentation pathways is evident from the properties of the transporters in the pathways. Studies with isolated membrane vesicles have shown that the malate uptake system of *Leuconostoc oenos* and the citrate carrier of *Leuconostoc mesenteroides* (CitP) transport malate as well as citrate [27,28]. Moreover, the CitP protein is homologous to the malate transporter (MleP) of *L. lactis* [55].

3. The transporters

3.1. Kinetic mechanism

Two kinetically different mechanisms can be discriminated for a transporter that catalyses coupled translocation of two solutes in opposite directions. In one mechanism, the actual translocation step is an isomerization of a ternary complex of the carrier and the two substrates (Fig. 6A). The carrier molecule contains two physically distinct binding sites, one for each substrate, that face the two opposite sides of the membrane. Translocation follows from the concomitant reorientation of the two binding sites. The distinct binding sites allow antiport activity for structurally non-related substrates. The mitochondrial tricarboxylate transporter is believed to operate via this mechanism, which is often referred to as 'sequential' [35]. In the other mechanism, the transporter contains a single site that is alternately exposed to either side of the membrane (Fig. 6B). The binding site has affinity for substrates that are structurally related and can only reorient when a substrate is bound ('ping-pong' mechanism). Exchange occurs when one substrate is at a high concentration at one side of the membrane and the other at the opposite side. The bacterial arginine/ornithine exchanger is an example of such an antiporter [36]. In both mechanisms there is an obligatory coupling between the two solute fluxes in the two opposite directions. The 'ping-pong' type of exchange mechanism (Fig. 6B) is also observed as a partial reaction of a proton symporter (Fig. 6C). A proton symporter has two binding sites, one for the solute and one for the proton, that are exposed to the same side of the membrane. The catalytic steps involved are (i) binding of substrate and proton to the externally exposed binding sites, (ii) translocation of the ternary complex, (iii) dissociation of the substrate and proton into the

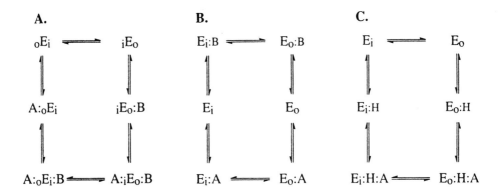

Fig. 6.   Kinetic mechanism for exchange and symport reactions. (A) Exchange proceeding via a ternary complex between the enzyme and the substrates ('sequential' mechanism). (B) Exchange proceeding via a 'ping-pong' mechanism. (C) Proton symport proceeding via a ternary complex between the proton, the solute and the transporter. E, A, B are the enzyme and the two substrates, respectively. Subscript i and o denote the orientation of the binding site to the internal and external medium, respectively.

internal medium, and (iv) isomerization of the 'empty' carrier to yield the binding sites facing the external medium again. In case the internal medium contains a significant concentration of a solute for which the carrier has affinity, this solute may bind to the internally exposed binding site after dissociation of the translocated solute in step (iii). The carrier may return loaded with the newly bound solute, thereby preventing complete turnover. This catalytic mode of transport is called exchange and represents a partial reaction of the catalytic cycle of the carrier. The solute bound to the carrier in the 'back stroke' may be identical to the solute that was transported into the cell (homologous exchange) or a different one for which the carrier has affinity (heterologous exchange). The exchange process omits the catalytic step in which the empty binding sites on the carrier change their orientation from 'in' to 'out'. Therefore, proton symporters catalyse both unidirectional symport (complete turnover) and bidirectional exchange(partial turnover).

The precursor/product exchangers that have been studied so far are not antiporters in the sense that translocation of substrate and product in opposite directions is obligatory coupled. Rather, the systems are proton symporters (or uniporters) that have been optimized to catalyse the partial exchange reaction. Exchange is catalysed much more efficiently than complete turnover, indicating that the reorientation of the binding sites of the carrier is faster with the substrates bound than without. In the case of the simple decarboxylation driven systems, the precursor and product are related chemical structures (see Table 2). Indeed, the citrate carrier of *Leuconostoc mesenteroides* has affinity for citrate, malate and lactate, which are all hydroxy-carboxylic acids [27]. The unidirectional mode of transport has been demonstrated for the malate/lactate exchanger of *Lactoccus lactis* [24], the histidine/histamine exchanger of *Lactobacillus buchneri* [21], and the citrate/lactate exchanger of *Leuconostoc mesenteroides* [27] (Fig. 7). Unfortunately, the question

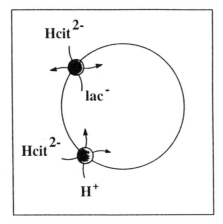

Fig. 7.   Kinetic modes of precursor/product exchangers. The physiological mode of transport is exchange between substrate and metabolic product as demonstrated here for citrate/lactate exchange (top). In addition, the transporters catalyse *in vitro* symport of the substrate with a H^+ (bottom). During both modes of transport a single net negative charge is moved into the cell.

about unidirectional transport was not addressed when the oxalate/formate exchanger was studied after purification and reconstitution into proteoliposomes [37]. The difference between a carrier that has been optimized as an exchanger and one that has been optimized for complete turnover is nicely demonstrated by the catalytic properties of the malate transporters of *Lactococcus lactis* (malate/lactate exchanger) and *Leuconostoc oenos* (malate uniporter). Membrane vesicles from *L. lactis* loaded with malate show very rapid release of malate when diluted into medium containing lactate (exchange mode) compared to dilution into medium without lactate (uniport mode)[24]. The enhancement of release of malate in the presence of lactate is not observed with membrane vesicles from *L. oenos* [28]. The same difference is observed between the citrate carriers of *Lc. mesenteroides* [27] and *Lc. oenos* [29] (Fig. 7).

3.2. Function and structure

The global secondary structure of bacterial secondary solute transporters is believed to be similar (e.g. see Ref. [2]). The majority of the proteins consist of 12 membrane spanning segments (TMS) that traverse the membrane in an α-helical fashion. The TMS are connected by hydrophilic loops that are usually shorter at the periplasmic side than at the cytoplasmic side of the membrane. The information follows from the hydropathy profile of the primary sequences of many transporters and in many cases the predicted topologies have been confirmed by engineering compartment specific reporter enzymes like PhoA, BlaM and LacZ to N-terminal fragments of the transporters [38]. A number of genes coding for transporters involved in secondary metabolic energy generation have been cloned and sequenced. These include the genes for the citrate carrier (CitP) of *Lactococcus lactis* [39], *Leuconostoc lactis* [40] and *Leuconostoc mesenteroides* [41] and the malate carrier (MleP) of *Lactococcus lactis*. The deduced amino acid sequences of the three CitP proteins are 99% identical. CitP and MleP are homologous proteins that share approximately 50% sequence identity [55]. The third known member of this family of homologous proteins is the Na^+-dependent citrate carrier of *K. pneumoniae* (CitS) that shares about 30% identical residues with CitP and MleP [42]. CitS transports citrate obligatory coupled to the translocation of 2 Na^+-ions and 1 H^+ [43]. Since the substrates of CitP and MleP include citrate, malate and lactate, the family is termed the *bacterial hydroxy-carboxylic acid transporters*. Figure 8 shows the hydrophobicity plot of CitP and the average hydrophobicity of the aligned sequence set of the three genes. It is evident that the hydropathy profile is highly conserved in this family. Although computer algorithms predict 12 TMS for the individual members, the predicted secondary structures differ significantly from those predicted for other secondary transporters [44]. The large hydrophilic loop around residue 250 (Fig. 8) is not positioned in the middle of the sequence, and the C-terminal halve is characterized by large periplasmic and short cytoplasmic loops which is opposite to that of most secondary transporters analysed to date. Also, Von Heynes 'positive inside' rule is not very apparent in the C-terminal half of the molecules [45]. The correctness of the predicted model has to be substantiated by experimental evidence.

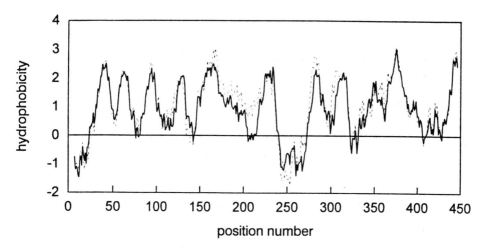

Fig. 8.   Hydropathy profile of the hydroxy-carboxylic acid family of secondary carriers. The family consists of CitP, the citrate carrier of Lactic Acid Bacteria, MleP, the malate carrier of *L. lactis* and CitS, the Na⁺ dependent citrate carrier of *K. pneumoniae*. The solid line is the profile of the complete aligned set. The dotted line is the profile of CitP alone. The window size was 13.

The presence of metabolic energy generating (CitP and MleP) and consuming (CitS) transporters in one family indicates that the carriers involved in secondary metabolic energy generation are normal secondary transporters as was argued above. Apparently, functional differences like substrate and cation specificity as well as mode of energy coupling can easily be accommodated by the same overall three dimensional structure. The different functions, most likely, reflect small details in the structure. The properties of the carriers in the hydroxy-carboxylic acid transporters are summarized in Table 3.

Table 3
Properties of the bacterial hydroxy-carboxylic acid transporters

Carrier	Organism	Substrates	Cation	Net charge translocation	Physiology	
					Transport[a]	Energetics[b]
CitS	*K. pneumoniae*	citrate	Na^+ H^+	positive	symport	dissipating
CitP	*Lc. mesenteroides*	citrate malate lactate	H^+	negative	exchange	generating
MleP	*L. lactis*	malate lactate	H^+	negative	exchange	generating

[a] The mode of transport under physiological conditions.
[b] Effect on proton and/or sodium ion gradient.

4. The pathway

4.1. Coupling between ΔpH and Δψ generation

The secondary metabolic energy generating systems that are driven by decarboxylation generate a proton motive force that is composed of both a membrane potential and a pH gradient. In contrast to a proton pump, the $\Delta\psi$ and ΔpH are generated in separate steps, transport and metabolic conversion, respectively, and the mechanism constitutes a way of indirect proton pumping [37]. Despite the separation of these two events in physically distinct steps, analysis of scalar proton consumption in the metabolic conversion and the requirements for particular protonation states of substrate and product in the transport step, reveal that the formation of $\Delta\psi$ and ΔpH are, in fact, coupled [46]. The decarboxylation driven pathways convert a weak acid substrate, originally outside the cell, into products that leave the cell. The cell functions as a catalyst by taking up the substrate, converting it into products which are subsequently excreted. The overall effect on the pH is independent of the metabolic pathway via which the cell converts substrate into products. Instead, proton consumption is solely determined by the chemical properties of substrate and products. In the kinetic steady state, the conversion of substrate into products will be accompanied by a continuous proton consumption from the medium. For instance, the overall reaction describing malate fermentation (malolactic fermentation) in lactic acid bacteria is

$$H_2mal \rightarrow Hlac + CO_2 \tag{1}$$

in which H_2mal and $Hlac$ are the fully protonated species, i.e. malic acid and lactic acid, respectively. At very low pH the conversion is pH neutral (reaction 1). At alkaline pH values, the effect on the pH depends on the fate of the CO_2 released. Under condition of constant partial carbon dioxide pressure, pCO_2, the production

Opposite: Fig. 9. Proton consumption during malolactic fermentation. Proton consumption during malate fermentation as a result of the overall process (●), the redistribution of malate, lactate and bicarbonate over the different protonated species in the external medium (○), and the conversion of malate into lactate and carbon dioxide inside the cell (▲). Proton consumption is shown for constant pCO_2 (A,B), and under condition of CO_2 accumulation (C,D). A and C represent the electrogenic uniport uptake mechanism (Fig. 10A) and B and D represent the passive diffusion mechanism (Fig. 10B). It is assumed that CO_2 permeates the membrane passively in its uncharged state. Net proton consumption in the overall process (●) was calculated at each pH from the fractions of weak acids in the different protonated species and the overall reaction stoichiometry. Net proton consumption per molecule of malate equals: $2 \times mal^{2-} + Hmal^- - lac^-$ (A,B) and $2 \times mal^{2-} + Hmal^- - lac^- - HCO_3^-$ (C,D) in which mal^{2-}, etc indicate the fraction of malate in the different protonation states. Proton consumption due to redistribution in the external medium is given by $mal^{2-} - H_2mal - lac^-$ (A) and $mal^{2-} - H_2mal - lac^- - HCO_3^-$ (C). Proton consumption in the external medium follows from the same expression as for net proton consumption in the case of passive uptake of protonated malate (B,D). The two pKs for malate are 3.4 and 5.2; the pK for lactate is 3.9; the pK for bicarbonate is 6.4. The second dissociation of bicarbonate (pK = 10.2) was not taken into account.

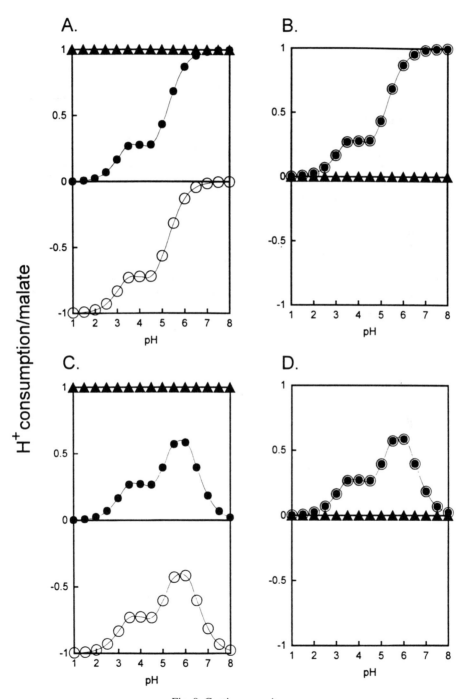

Fig. 9. Caption opposite.

of CO_2 will have no effect on the pH of the medium and results in the consumption of 1 proton per molecule of substrate (reaction 2)

$$mal^{2-} + H^+ \rightarrow lac^- + CO_2 \tag{2}$$

On the other hand, when at alkaline pH values CO_2 accumulates in the system, the reaction becomes pH neutral again (reaction 3)

$$mal^{2-} + H_2O \rightarrow lac^- + HCO_3^- \tag{3}$$

At intermediate, more physiological pH values the alkalinization is determined by the differences in the pKs of malic and lactic acid. The proton consumption per molecule of substrate as a function of medium pH and under constant pCO_2 is shown in Fig. 9A,B; the proton consumption under CO_2 accumulating conditions is shown in Fig. 9C,D. The analysis shows that the fate of CO_2 is only important above pH 6. Experimental evidence for alkalinization of the medium during malate and citrate fermentation by lactic acid bacteria has been described [21,24,29].

The formation of a pH gradient across the membrane during the metabolism of a weak acid is a property of the mechanism by which the cell (the catalysts) catalyzes the overall reaction. The mechanism(s) by which substrate and product(s) are transported into and out of the cell, respectively, determine(s) where the actual proton consumption takes place and, consequently, determines the energetic consequences of the overall alkalinization. In the steady state, alkalinization of the cytoplasm relative to the medium will result in a flux of protons from 'out' to 'in' either by passive diffusion ('leak') or catalyzed by transporters. Similarly, in the steady state, alkalinization of the medium relative to the cytoplasm results in a proton flux from in to out. In either case the medium pH rises, but the pH gradient across the membrane is of opposite sign. Factors that determine the effects on the internal and external pH are (i) the redistribution of substrate and products over their protonation states in the external medium, (ii) the number of protons transported with the substrate into and with the product out of the cell, and (iii) the chemical conversion inside the cell.

Figure 10 shows two conceivable mechanisms by which the cell can catalyze malolactic fermentation that differ in the way malate is transported into the cell. In Fig. 10A malate enters the cell in its monovalent anionic state via a carrier protein as has been proposed for *Leuconostoc oenos* [28]; in Fig. 10B malate enters the cell in its fully protonated state by passive diffusion as has been proposed for *Lactobacillus plantarum* [47]. In both examples lactate leaves the cell in its protonated form. For the following discussion on proton consumption only the net proton stoichiometry of the transport mechanisms is important. The uniporter transporting monoprotonated malate (Hmal$^-$) is energetically equivalent to a symporter transporting deprotonated malate (mal^{2-}) with a proton (H$^+$), and the combination of the uptake of monoprotonated malate (Hmal$^-$) by an uniporter and passive diffusion of lactic acid (Hlac) out of the cell is equivalent to the precursor/product exchanger that couples the uptake of mal^{2-} to the excretion of lac$^-$ (Fig 5). The effect on the external pH is determined by the redistribution of malate and lactate over the different protonation states. When monoanionic malate is transported into the cell

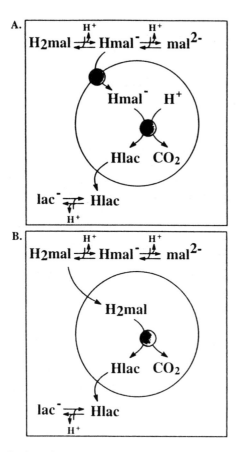

Fig. 10.   Different mechanisms for the malate decarboxylation pathway. (A) Malate is taken up in its monovalent anionic state by an electrogenic uniporter. (B) Malate is entering the cell by passive diffusion of the fully protonated state.

by the carrier the fraction of malate in the H_2mal state will release a proton and the fraction in the mal^{2-} state will take up a proton before being transported. The product lactate leaves the cell in the protonated state and will in part dissociate in the external medium. Redistribution of substrate and product alone results in an acidification of the external medium up to a medium pH of about 7 (open circle in Fig. 9A). Since one proton is needed inside the cell to convert $Hmal^-$ into protonated lactate (reaction 1, triangle in Fig. 9A) the internal pH will rise. The overall result is that in addition to the membrane potential generated by the uniporter, a pH gradient of physiological polarity is generated ($-Z\Delta pH < 0$).

The energetic consequences of the passive uptake of malate (Fig. 10B) are analyzed in Fig. 9B. Obviously, the electroneutral diffusion of the substrate into the cell does not result in a membrane potential. Inside the cell, fully protonated malate is converted into the protonated products without the need for a proton and, therefore, the internal pH is not affected (triangle in Fig 9B). Redistribution of substrate and products results in a net proton consumption on the outside (open circle in Fig. 9B) that follows the net proton consumption in the overall reaction (black circle, Fig. 9B). Consequently, a pH gradient is generated, inside acid $(-Z\Delta pH > 0)$, which is opposite to the physiological pH gradient and, consequently, reduces the pmf across the membrane.

The analysis shows that the generation of membrane potential and pH gradient in secondary metabolic energy generating systems are coupled events. Only in those cases where the uptake of the weak acid results in a membrane potential of physiological polarity, negative inside, a pH gradient is formed that is alkaline inside. In the case the uptake of the substrate is not electrogenic, metabolism results in an inverted pH gradient that lowers the pmf. Such an uptake is metabolic energy consuming and is only feasible when other pmf generating processes take place. Finally, the coupling between the formation of the membrane potential and the pH gradient can also be understood by inspection of the overall reactions (e.g. reaction 1), which show a pH neutral conversion of the fully protonated weak acid into the products. An internal proton is only needed for the metabolism to proceed when a deprotonated state of the substrate enters the cell which, in turn, results in the membrane potential.

4.2. Apparent proton pumping

At low pH and constant pCO_2, and both at low and high pH under accumulating CO_2 conditions, the combined scalar reactions that take place outside and inside the cell in a decarboxylation driven pathway mimic vectorial proton transport by a proton pump, i.e. the proton that disappears inside (triangle in Fig. 9A,C), reappears outside (open circle, Fig. 9A,C) with no net proton consumption (black circle, Fig. 9A,C). At each pH, the internal alkalinization caused by the pathway is clearly shown when the overall reaction is split into partial reactions. For a decarboxylation driven precursor/product exchanger (e.g. Fig. 5C) the partial reactions are as depicted in Scheme 1.

$$RCOO^-_{out} + RH_{in} \rightleftharpoons RCOO^-_{in} + RH_{out}$$

$$RCOO^-_{in} + H^+_{in} \rightleftharpoons RH_{in} + CO_2^{in}$$

$$CO_2^{in} \rightleftharpoons CO_2^{out}$$

$$\underline{\hspace{6cm}} (+)$$

$$RCOO^{in}_{out} + H^+_{in} \rightleftharpoons RH_{out} + CO_2^{out}$$

Scheme 1

The proton that is needed in the decarboxylation of the external carboxylate originates from the inside. Again, the final reaction describing the overall chemistry of the process is the same for a uniporter taking up $RCOO^-$ in combination with passive efflux of the products (RH and CO_2).

For the energetic analysis of the system the overall reaction can be described by any set of partial reactions as long as the sum equals the overall reaction (free energy is an additive property). Scheme 2 shows an alternative description for the above decarboxylation process.

$$H^+_{in} \rightleftharpoons H^+_{out}$$

$$RCOO^-_{out} + H^+_{out} \rightleftharpoons RH_{out} + CO_2^{out}$$

$$\underline{\hspace{10cm}}(+)$$

$$RCOO^-_{out} + H^+_{in} \rightleftharpoons RH_{out} + CO_2^{out}$$

Scheme 2

In this description, the substrate and products are not transported into and out of the cell, respectively, but, instead the proton is transported out of the cell. The formation of the electrochemical proton gradient is immediately evident from the first partial reaction. Scheme 2 shows the analogy between primary and secondary pmf generation. Pathway linked pmf generation behaves energetically like an apparent primary proton pump that couples the translocation of the proton to the decarboxylation reaction that takes place outside the cell (Fig. 11).

5. Efficiency of energy transduction

5.1. Thermodynamic efficiency

The maximal attainable proton motive force in a decarboxylation driven fermentation pathway follows from the expression for the free energy difference over the overall reaction in Schemes 1 and 2. At thermodynamic equilibrium, i.e. when ΔG is zero, it follows (see Appendix):

$$\frac{\Delta\tilde{\mu}_{H^+}}{F} = \frac{\Delta G^{0\prime}}{F} - 7Z + Z\,pH_{out} + Z\log\frac{RH\,CO_2}{RCOO^-}$$

The actual expression for the equilibrium value of the pmf depends on the nature of the substrate RCOOH. RCOOH may be a mono-, di- or tricarboxylic acid (e.g., histidine, malate or citrate, respectively) which determines the pH dependency of

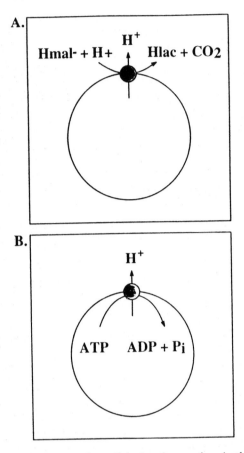

Fig. 11.   Apparent proton pumping by pathway linked pmf generation. Analogy between secondary (malolactic fermentation, A) and primary (F_0F_1-ATPase, B) pmf generation. Pathway linked pmf generation is energetically equivalent to a proton pump that is driven by an external chemical reaction.

the concentration of $RCOO^-$. Related to this is the nature of product RH which may be a weak acid itself and (partially) deprotonate upon leaving the cell. In an open system, the CO_2 concentration in the medium will be constant assuming rapid equilibrium between the air and liquid phase. Figure 12 shows the pH dependency of the equilibrium pmf values for a number of cases. The force on the apparent proton pump is highest at low pH values and increases with the number of carboxylic groups on the substrate. Figure 12 shows the contribution to the equilibrium pmf in addition to the standard free energy $\Delta G^{0'}/F$, which is in the order of $-250\,mV$ for a decarboxylation reaction. Especially at low external pH, the drop of free energy over the driving reaction is large for the pumping of just one proton and the equilibrium pmf is very high. The 'real life' pmf, however, is not determined by the thermodynamics of the system but by the kinetics of pmf formation and consumption, and is known to be much lower than the calculated equilibrium values

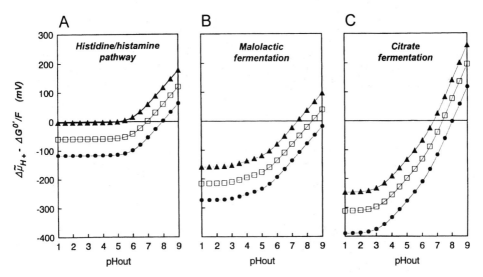

Fig. 12. Equilibrium pmf for pathway linked pmf generating systems with mono- (A), di- (B) and tricarboxylic (C) acids as precursor. The examples used are the pathways for histidine decarboxylation (Fig. 4A), malolactic fermentation (Fig. 5A,C) and citrate metabolism in resting cells of *Lc. oenos*. The pathway products in the latter case are acetic acid, acetoin and carbon dioxide. Plotted are the differences between the equilibrium pmf and the standard free energy difference over the pathway. The initial concentration of the substrate was 50 mM and the forces were calculated at different external pH values for conversions of 0.1 (●), 0.5 (□) and 0.9 (▲). The pCO_2 is assumed to be constant. The pK for histidine is 6; the pKs for malate are 3.4 and 5.2; the pK for lactate was 3.9; the pKs for citrate were 3.1, 4.8 and 6.4 and for acetate 4.8.

[24,27,32]. Thus it seems that the efficiency of energy transduction in a secondary metabolic energy generation is low. Chemiosmotic pumps can increase the efficiency of energy transduction by increasing the proton stoichiometry, either by increasing the number of protons pumped per substrate (e.g., F_0F_1-ATPase pumps 3–4 H^+ per ATP hydrolysed) or by increasing the number of proton translocation sites in a pathway, thereby splitting the free energy drop in smaller parts (e.g., the aerobic respiratory chain). Pathway linked pmf generating systems that metabolize monocarboxylic acids do not allow modulation of the proton stoichiometry. When di- or tricarboxylic acids are metabolized the stoichiometry can, in principle, be increased by modifying the transporter to translocate di- or trivalent anions into the cell. This would require the uptake of two or three scalar protons in the cytosolic reaction(s). Such a mechanism would be equivalent to the pumping of 2 and 3 protons, respectively, in Scheme 2. The consequence of an increased stoichiometry would be that the counteractive force of the membrane potential on the uptake reaction increases considerably and the metabolism may not be capable of overcoming this force efficiently. The advantage of 'pumping' more protons may be diminished by an overall lowering of the flux through the pathway, and a cell is likely to be better off with a fast rate of pmf generation than with an economic system that tries to approach thermodynamic equilibrium.

The systems studied to date all translocate one unit charge over the membrane, which is equivalent to a proton stoichiometry of 1 H^+/substrate. The uptake of citrate in *Lc. mesenteroides* and *Lc. oenos* is indicative in this respect. The citrate permease of *Lc. oenos* transports monovalent citrate (H_2cit^-), the form that is predominantly present in its acidic habitat [29]. *Lc. mesenteroides* grows at higher pH values and the citrate carrier recognizes divalent citrate ($Hcit^{2-}$). However, the divalent form is translocated in symport with a proton or in exchange with monovalent lactate, which again results in the net translocation of one negative charge [27].

5.2. Comparison of primary and secondary metabolic energy generation

Decarboxylases of several bacterial species are intimately associated with the cytoplasmic membrane [48]. Well-studied examples include oxaloacetate decarboxylase of *Klebsiella pneumoniae* and *Salmonella typhimurium*, and the (S)-methyl-malonylCoA decarboxylase of *Propiogenum modestum*. These enzymes have a avidin-sensitive carboxyl transferase activity which results in the formation of an N-carboxybiotin enzyme derivative. The free energy liberated by the decarboxylation of the carboxybiotin enzyme derivative is conserved by pumping Na^+ ions across the membrane. The decarboxylases involved in pathway linked pmf generating systems, i.e. malolactic fermentation, citrate fermentation and oxalate, histidine and lysine decarboxylation, are cytoplasmic rather than integral membrane enzymes [49,50,51,52]. Moreover, unlike the membrane-bound enzymes, the cytoplasmic decarboxylases are not inhibited by avidin which is taken as evidence that these enzymes do not form the N-carboxybiotin derivative. Studies of malate decarboxylase, termed malolactic enzyme, of *Lactobacillus plantarum* and other lactic acid bacteria indicate an requirement for NAD^+ plus Mn^{++}. Malolactic enzyme decarboxylates L-malate to pyruvate with reduction of NAD^+ to NADH, followed by the reduction of pyruvate to L-lactate and reoxidation of NADH to NAD^+ [51]. The intermediates NADH and pyruvate remain bound to the enzyme rather than being released into the medium during enzymatic conversion. Cloning and sequencing of the gene coding for malolactic enzyme of *L. lactis* has revealed homology with malic enzymes that catalyse the conversion of malate into pyruvate with NAD^+ as the electron acceptor [49].

In the membrane-associated decarboxylation processes, a scalar proton is consumed and 2 Na^+ are translocated per molecule decarboxylated as has been shown for the oxaloacetate decarboxylase of *K. pneumoniae* and (S)-methyl-malonylCoA decarboxylase of *P. modestum* [48,53]. These enzymes function in primary metabolic energy generation. At first sight, such a mechanism seems energetically superior over the secondary metabolic energy generating systems described here. However, for a fair comparison between the systems one should consider the pathways as a whole. The first steps in the metabolic pathways for citrate degradation in anaerobically grown *K. pneumoniae* and *Lc. mesenteroides* are identical. Internalized citrate is split into acetate plus oxaloacetate by citrate lyase, followed by decarboxylation of oxaloacetate, yielding pyruvate plus CO_2 (Fig. 13A,C). Both pathways yield metabolic energy in the form of ion gradients, but *Lc. mesenteroides*

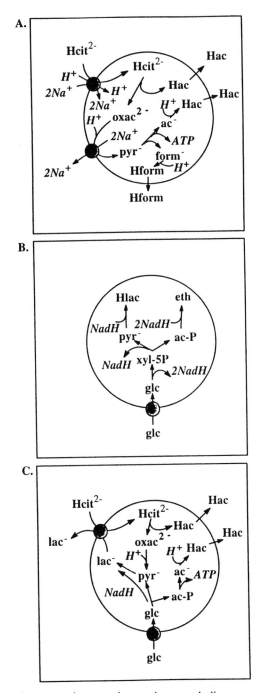

Fig. 13.   Comparison between primary and secondary metabolic energy generation. (A) Citrate metabolism in anaerobically growing *K. pneumoniae*. (B) Heterofermentative glucose metabolism in *Lc. mesenteroides*. (C) Citrate/glucose cometabolism in *Lc. mesenteroides*.

Table 4
Metabolic energy generation by citrate metabolism in *K. pneumoniae* and *L. mesenteroides*[a]

Metabolic step	*K. pneumoniae*	*L. mesenteroides*
Citrate uptake	-2[Na][b], -1[H], -1 charge[c]	$+1$ charge
Decarboxylation	$+2$[Na], $+1$[H], $+2$ charges	$+1$ [H]
Acetate kinase	$+1$ ATP	$+0.5$ ATP
Weak acid efflux	$+2$[H]	$+0.5$[H]
Metabolic energy yield	$+2$[H], $+1$ charge, $+1$ ATP	$+1.5$[H], $+1$ charge, $+0.5$ ATP

[a] The table analyses the involvement of the cations in metabolic energy generation or dissipation in the metabolic steps. Indicated are the number of ions and charges that are translocated or consumed. (–) and (+) indicate consumption and generation of metabolic energy, respectively.
[b] [Na] and [H] represent contribution to the chemical component in the ion motive forces.
[c] Charges are defined as electrogenic translocations contributing to the electrochemical potential of either Na^+ or H^+.

employs a secondary system whereas *K. pneumoniae* uses a combination of a primary Na^+-ion pump and an energy consuming secondary uptake mechanism for citrate. The latter, termed CitS, is driven by the pmf and smf and couples the uptake of divalent citrate with 1 H^+ and 2 Na^+ [43]. Thus, the two Na^+ ions that are pumped out by the oxaloacetate decarboxylase are needed for the uptake of citrate. The proton transported into the cell by the citrate carrier is consumed in the decarboxylation step. Therefore, the pathway up to pyruvate is only slightly energy generating, i.e. one positive charge is translocated out of the cell per citrate metabolized. In *K. pneumoniae* pyruvate is metabolized further via acetylCoA and acetyl-phosphate to acetate which yields 1 ATP by substrate level phosphorylation. The weak acids formed, i.e. acetate and formate, leave the cell, most likely, in their protonated form, thereby transporting two protons out of the cell. Thus, overall, the pathway gains per molecule of citrate consumed, the translocation of one positive charge and two protons out of the cell and one ATP (see Table 4).

In *Lc. mesenteroides* citrate is only metabolized during cometabolism with a carbohydrate and the energetics have to be discussed in the context of the heterofermentative carbohydrate metabolism of *Leuconostoc* ssp. (Fig. 13B). Glucose is converted into xylose-5P which, subsequently, is split into acetyl-P plus glyceraldehyde. The latter is converted to pyruvate. Acetyl-P can be converted to acetate by acetate kinase, yielding ATP, but this would result in severe problems with the redox balance since the conversion of glucose to pyruvate results in 3 NAD(P)H. One of the NAD^+ is regenerated in the reduction of pyruvate to lactate. The other two are regenerated in the reduction of acetyl-P to ethanol, a route that does not yield ATP. Therefore, only 1 ATP is formed per molecule of glucose metabolised. During cometabolism with citrate, pyruvate formed from citrate functions as a redox sink and allows acetate kinase catalysed production of ATP from acetyl-P. Two redox equivalents have to be funnelled to pyruvate for the production of 1 ATP, resulting in an extra metabolic energy gain of 0.5 ATP per molecule of citrate.

Additionally, 0.5 H^+ per molecule of citrate is translocated out of the cell with the acetate produced in the acetate kinase reaction. Together with the secondary metabolic energy conserved in citrate uptake and the proton consumption in the decarboxylation of oxaloacetate, the breakdown of citrate results in a yield, per molecule of citrate, of 1 charge and 1.5 proton translocated out of the cell plus 0.5 ATP (Table 4).

The energetic analysis given above is somewhat simplified since other conversion routes for pyruvate were not considered. Nevertheless, it is clear that energetically the two mechanisms of energy conservation in metabolic breakdown of citrate are roughly equivalent. The yield of metabolic energy in the form of ion gradients is more or less the same. The system in *K. pneumoniae* yields more ATP via substrate level phosphorylation. Depending on the physiological conditions, the criterium for an organism to select for a particular energy conserving system may be related to optimizing metabolic rate rather than energetic efficiency. The *Klebsiella* system seems particularly suitable when the citrate concentration in the medium is low. The combination of the Na^+-ion pump and the secondary transporter for citrate allows the accumulation of citrate in the cell which will speed up internal metabolism. The *Leuconostoc* system will only work efficiently at high external citrate concentrations in the medium, which ensures a high downhill concentration gradient over the membrane.

6. Appendix

6.1. General expression for the equilibrium pmf

The free energy difference over a pmf generating decarboxylation pathway follows most easily from the sum of the free energy differences over the partial reactions depicted in Scheme 2. With all potentials in mV, it follows

$$\frac{\Delta G}{F} = \psi_{out} + Z \log H^+_{out} - \psi_{in} - Z \log H^+_{in} + \frac{\Delta G^0}{F} + Z \log \frac{RH_{out}CO_2^{out}}{RCOO^-_{out}H^+_{out}} \tag{1}$$

Z is equal to 2.3 RT/F. The first four terms equal the electrochemical gradient for protons across the membrane (the pmf)

$$\frac{\Delta\tilde{\mu}_{H^+}}{F} = \Delta\psi - Z\Delta pH = \psi_{in} - \psi_{out} - Z(pH_{in} - pH_{out}) \tag{2}$$

The thermodynamic standard state can be converted to the biochemical standard state as follows

$$\frac{\Delta G^0}{F} = \frac{\Delta G^{0'}}{F} + Z \log 10^{-7} = \frac{\Delta G^{0'}}{F} - 7Z \tag{3}$$

At equilibrium $\Delta G/F$ equals zero which results in the general equation for the equilibrium proton motive force

$$\frac{\Delta\tilde{\mu}_{H^+}}{F} = \frac{\Delta G^{0\prime}}{F} - 7Z + Z\,pH_{out} + Z\log\frac{RH\,CO_2}{RCOO^-} \tag{4}$$

All concentrations in the last term refer to the external medium. The actual expression for the equilibrium pmf depends on the nature of the substrate which may be a mono-, di- or tricarboxylic acid (for instance, histidine, malate and citrate, respectively; Fig. 12).

6.2. Monocarboxylic acids

The concentration of the substrate in the deprotonated state equals

$$RCOO^- = \frac{K}{K + H_{out}^+}\,RCOOH^{total} \tag{5}$$

in which K is the dissociation constant and $RCOOH^{total}$ represents the sum of the concentrations of the different protonation states. Substitution of Eq. (5) in Eq. (4) results in

$$\frac{\Delta\tilde{\mu}_{H^+}}{F} = \frac{\Delta G^{0\prime}}{F} - 7Z + Z\,pH_{out} + Z\,pK + Z\log(K + H_{out}) + Z\log\frac{RH\,CO_2}{RCOOH^{total}} \tag{6}$$

At low pH when $H_{out} \gg K$ the equilibrium pmf is independent of the pH

$$\frac{\Delta\tilde{\mu}_{H^+}}{F} = \frac{\Delta G^{0\prime}}{F} - 7Z + Z\,pK + Z\log\frac{RH\,CO_2}{RCOOH_{total}} \tag{7}$$

and at high pH when $K \gg H_{out}$ the pmf decreases 60 mV per pH unit (see Fig. 12A)

$$\frac{\Delta\tilde{\mu}_{H^+}}{F} = \frac{\Delta G^{0\prime}}{F} - 7Z + Z\,pH_{out} + Z\log\frac{RH\,CO_2}{RCOOH^{total}} \tag{8}$$

6.3. Dicarboxylic acids

Decarboxylation of a dicarboxylic acid results in a monocarboxylic acid and, consequently, both RH and RCOOH in Eq. (4) depend on pH. The fraction of the substrate in the monoanionic state is

$$RCOO^- = \frac{K_1^{RCOOH}\,H_{out}^+}{H_{out}^{+2} + H_{out}^+\,K_1^{RCOOH} + K_1^{RCOOH}\,K_2^{RCOOH}}\,RCOOH^{total} \tag{9}$$

in which the Ks are the two dissociation constants. The fraction of the product in the protonated state is

$$RH = \frac{H^+_{out}}{H^+_{out} + K^{RH}} RH^{total} \tag{10}$$

Substitution in Eq. (4) gives

$$\frac{\Delta\tilde{\mu}_{H^+}}{F} = \frac{\Delta G^{0\prime}}{F} - 7Z + Z \, pK_1^{RCOOH} + Z \, pH_{out} +$$

$$Z \log \frac{(H^{+2}_{out} + H^+_{out} K_1^{RCOOH} + K_1^{RCOOH} K_2^{RCOOH})}{H^+_{out} + K^{RH}} + Z \log \frac{RH^{total} \, CO_2}{RCOOH_{total}} \tag{11}$$

At low pH when $H^+_{out} \gg K_1^{RCOOH}$, K_2^{RCOOH} and K^{RH} this results again in an pH independent equilibrium pmf

$$\frac{\Delta\tilde{\mu}_{H^+}}{F} = \frac{\Delta G^{0\prime}}{F} - 7Z + Z \, pK_1^{RCOOH} + Z \log \frac{RH^{total} \, CO_2}{RCOOH_{total}} \tag{12}$$

and at high pH when $H^+_{out} \ll K_1^{RCOOH}$, K_2^{RCOOH} and K^{RH} in an equilibrium pmf that decreases 60 mV per pH unit (Fig. 12B)

$$\frac{\Delta\tilde{\mu}_{H^+}}{F} = \frac{\Delta G^{0\prime}}{F} - 7Z + Z \, pH_{out} - Z \, pK_2^{RCOOH} + Z \, pK^{RH} + Z \log \frac{RH^{total} \, CO_2}{RCOOH_{total}} \tag{13}$$

6.4. Tricarboxylic acids

Pmf generation by citrate metabolism does not proceed via a single decarboxylation step. In *L. oenos* and below pH 5 citrate is completely converted to acetate, acetoin and carbon dioxide [54]. The substrate is taken up as H_2cit^- and the products are excreted uncharged. The overall reaction equivalent to the ones in Schemes 1 and 2, is

$$H_2cit^-_{out} + H^+_{in} \rightleftharpoons Hac_{out} + 1/2 \text{ acetoin}_{out} + 2 \, CO_2^{out} \tag{14}$$

This modifies general equation (4) to

$$\frac{\Delta\tilde{\mu}_{H^+}}{F} = \frac{\Delta G^{0\prime}}{F} - 7Z + Z \, pH_{out} + Z \log \frac{Hac \, \text{acetoin}^{1/2} \, CO_2^2}{H_2cit^-} \tag{15}$$

with the additional difference that $\Delta G^{0\prime}$ represents the standard free energy difference over the complete pathway. It follows for the pH dependence of the fractions of H_2cit^- and Hac.

$$H_2cit^- = \frac{K_1^{cit} H^{+2}_{out}}{H^{+3}_{out} + H^{+2}_{out} K_1^{cit} + H^+_{out} K_1^{cit} K_2^{cit} + K_1^{cit} K_2^{cit} K_3^{cit}} H_3cit^{total} \tag{16}$$

$$\text{Hac} = \frac{H^+_{out}}{H^+_{out} + K^{Hac}}\, \text{Hac}^{total} \tag{17}$$

Substitution of Eqs. (16) and (17) in the expression for the equilibrium pmf (Eq. (15)) yields

$$\frac{\Delta\tilde{\mu}_{H^+}}{F} = \frac{\Delta G^{0\prime}}{F} - 7Z + 2Z\, pH_{out} + Z\, pK^{cit}_1 +$$

$$Z \log \frac{H^{+3}_{out} + H^{+2}_{out}\, K^{cit}_1 + H^+_{out}\, K^{cit}_1\, K^{cit}_2 + K^{cit}_1\, K_2\, K^{cit}_3}{H^+_{out}\, K^{Hac}} + Z \log \frac{\text{Hac}^{total}\, \text{acetoin}^{1/2}\, CO^2_2}{H_3 cit^{total}} \tag{18}$$

At low pH this results in a pH independent equilibrium pmf (Fig. 12C)

$$\frac{\Delta\tilde{\mu}_{H^+}}{F} = \frac{\Delta G^{0\prime}}{F} - 7Z + Z\, pK^{cit}_1 + Z \log \frac{\text{Hac}^{total}\, \text{acetoin}^{1/2}\, CO^2_2}{H_3 cit^{total}} \tag{19}$$

and at high pH the equilibrium pmf reduces by 120 mV for every pH unit

$$\frac{\Delta\tilde{\mu}_{H^+}}{F} = \frac{\Delta G^{0\prime}}{F} - 7Z + 2Z\, pH_{out} - Z\, pK^{cit}_2 - Z\, pK^{cit}_3 + Z\, pK^{Hac}$$

$$+ Z \log \frac{\text{Hac}^{total}\, \text{acetoin}^{1/2}\, CO^2_2}{H_3 cit^{total}} \tag{20}$$

Abbreviations

LAB, Lactic Acid Bacteria
pmf, proton motive force
smf, sodium ion motive force
TMS, transmembrane segment

References

1. Mitchell, P. (1968) Chemiosmotic Coupling and Energy Transduction. Glynn Research Ltd., Bodmin, UK.
2. Poolman, B. and Konings, W.N. (1993) Biochim. Biophys. Acta **1183**, 5–39.
3. Michels P.A.M., Michels J.P.J., Boonstra J., Konings W.N. (1979) FEMS Microbiol. Lett. **5**, 357–364.
4. Konings, W.N., Poolman, B. and Van Veen, H.W. (1994) Antonie van Leeuwenhoek **65**, 369–380.
5. Konings, W.N., Lolkema, J.S. and Poolman, B. (1995) Arch. Microbiol. **164**, 235–242.
6. Ten Brink B., Konings W.N. (1980) Eur. J. Biochem. **111**, 59–66.

7. Otto R., Lageveen R.G., Veldkamp H., Konings W.N. (1982) J. Bacteriol. **146**, 733–738.
8. Otto R., Sonnenberg A.S.M., Veldkamp H., Konings W.N. (1980) Proc. Natl. Acad. Sci. USA **77**, 5502–5506.
9. Simpson S.J., Bendall M.R., Egan A.F., Rogers P.J. (1983) Eur. J. Biochem. **136**, 63–69.
10. Simpson S.J., Vink R., Egan A.F., Rogers P.J. (1983) FEMS Microbiol. Lett. **5**, 85–88.
11. Michel T.A., Macy J.M. (1990) J. Bacteriol. **172**, 1430–1435.
12. Engel P., Krämer R., Unden G. (1994) Eur. J. Biochem. **112**, 605–614.
13. Van Veen H.W., Abee T., Kortstee G.J.J., Konings W.N., Zehnder, A.J.B. (1994) J. Biol. Chem. **269**, 9509–9514.
14. Van Veen H.W., Abee, T., Kortstee G.J.J., Konings W.N., Zehnder A.J.B. (1994) Biochemistry **33**, 1766–1770.
15. McInerney M.J. and Beaty P.S. (1988) Can. J. Microbiol. **34**, 487–493.
16. Ingvorsen K., Zehnder A.J.B. and Jørgenson B.B. (1984) Appl. Env. Microbiol. **47**, 404–408.
17. Schönheit P., Kristjansson J.K. and Thauer R.K. (1982) Arch. Microbiol. **132**, 285–288.
18. Smith M.R. and Lequerica J.L. (1985) J. Bacteriol. **164**, 618–625.
19. Kelly D.P. (1988) Oxidation of sulphur compounds. In: The Nitrogen and Sulphur Cycle, eds J.A. Cole and S.J. Ferguson, pp. 64–98. Society Gen. Microbiol., Cambridge University Press, UK.
20. Higuchi, T., Hayashi, H. and Abe, K. (1993) FEMS Microbiol. Lett. **12**, C39.
21. Molenaar, D., Bosscher, J.S., Ten Brink, B., Driessen, A.J.M. and Konings, W.N. (1993) J. Bacteriol. **175**, 2864–2870.
22. Broër, S. and Krämer, R. (1990) J. Bacteriol. **172**, 7241–7248.
23. Meng, S-Y. and Bennett, G.N. (1992) J. Bacteriol. **174**, 2659–2669.
24. Poolman, B., Molenaar, D., Smid, E.J., Ubbink, T., Abee, T., Renault, P.P. and Konings, W.N. (1991) J. Bacteriol. **173**, 6030–6037.
25. Anantharam, V., Allison, M.J. and Maloney, P.C. (1989) J. Biol. Chem. **264**, 7244–7250.
26. Kashiwagi, K., Miyamoto, S., Suzuki, F., Kobayashi, H. and Igarashi, K. (1992) Proc. Natl. Acad. Sci. USA **89**, 4529–4533.
27. Marty-Teysset, C., Lolkema, J.S., Schmitt, P., Divies, C. and Konings, W.N. (1995) J. Biol. Chem. **270**, 25370–25376.
28. Salema, M., Poolman, B., Lolkema, J.S., Loureiro-Dias, M.C. and Konings, W.N. (1994) Eur. J. Biochem. **225**, 289–295.
29. Ramos, A., Poolman, B., Santos, H., Lolkema, J.S. and Konings, W.N. (1994) J. Bacteriol. **176**, 4899–4905.
30. Starrenburg, M.J.C. and Hugenholtz, J. (1991) Appl. Environ. Microbiol. **57**, 3535–3540.
31. Schmitt, P., Divies, C. and Cardona, R. (1992) Appl. Microbiol. Biotechnol. **36**, 679–683.
32. Salema, M., Lolkema, J.S., San Romão, M.V. and Lourero Dias, M.C. (1996), J. Bacteriol. in press.
33. Molenaar, D., Abee, T. and Konings, W.N. (1991) Biochim. Biophys. Acta **1115**, 75–83.
34. Marty-Teysset, C., Posthuma, C., Lolkema, J.S., Schmitt, P., Divies, C. and Konings, W.N. (1996) J. Bacteriol. **178,** 2178–2185.
35. Bisaccia, F., DePalma, A., Dierks, T., Krämer, R. and Palmieri, F. (1993) Biochim. Biophys. Acta **1142**, 139–145.
36. Driessen, A.J.M., Molenaar, D. and Konings, W.N. (1989) J. Biol. Chem. **264**, 10361–10370.
37. Ruan, Z.-S., Anantharam, V., Crawford, I.T., Ambudkar, S.V., Rhee, S.Y., Allisson, M.J. and Maloney, P.C. (1992) J. Biol. Chem. **267**, 10537–10543.
38. Hennessey, E.S. and Broome-Smith, J.K. (1993) Curr. Opin. Struct. Biol. **3**, 524–531.
39. David, S., van der Rest, M.E., Driessen, A.J.M., Simons, G. and de Vos, W.M. (1990) J. Bacteriol. **172**, 5789–5784.
40. David, S. (1992) Genetics of mesophilic citrate fermenting lactic acid bacteria. Ph.D. thesis, Univ. of Wageningen, The Netherlands.
41. Lhotte, M.E., Bandell, M., Dartois, V., Prévost, H., Lolkema, J.S., Konings, W.N., Divies, C. (1996) submitted.
42. van der Rest, M.E., Siewe, R.M., Abee, T., Schwarz, E., Oesterhelt, D. and Konings, W.N. (1992) J. Biol. Chem. **267**, 8971–8976.
43. Lolkema, J.S., Enequist, H. and van der Rest, M.E. (1994) Eur. J. Biochem. **220**, 469–475.

44. Lolkema, J.S., Speelmans, G. and Konings, W.N. (1994) Biochim. Biophys. Acta 211–215.
45. Von Heijne, G. (1986) EMBO J. **5**, 3021–3027.
46. Lolkema, J.S., Poolman, B. and Konings, W.N. (1995) J. Biomembr. Bioenerg. **27**, 467–473.
47. Olsen, E.B., Russel, J.B. and Henick-kling, T. (1991) J. Bacteriol. **173**, 6199–6206.
48. Dimroth, P. (1987) Microbiol. Rev. **51**, 320–340.
49. Anasany, V., Dequin, S., Blondin, B. and Barre, P. (1993) FEBS Lett. **332**, 74–80.
50. Baetz, A. and Allison, M.J. (1989) J. Bacteriol. **171**, 2605–2608.
51. Caspritz, G. and Radler, F. (1983) J. Biol. Chem. **258**, 4907–4910.
52. Hugenholtz, J., Perdon, L. and Abee, T. (1993) Appl. Environ. Microbiol. **59**, 4216–4222.
53. Dimroth, P. (1991) Bioessays **13**, 463–468.
54. Ramos, A., Lolkema, J.S., Konings, W.N. and Santos, H. (1994) Appl. Environ. Microbiol. **61**, 1303–1310.
55. Bandell, M., Ansanay, V., Rachidi, N., Dequin, S. and Lolkema J.S. (1996) submitted.

CHAPTER 12

Pi-Linked Anion Exchange in Bacteria: Biochemical and Molecular Studies of UhpT and Related Porters

P.C. MALONEY

Department of Physiology, Johns Hopkins Medical School,
Baltimore, MD 21205, USA

© 1996 Elsevier Science B.V.
All rights reserved

Handbook of Biological Physics
Volume 2, edited by W.N. Konings, H.R. Kaback and J.S. Lolkema

Contents

1. Principles and overview

This chapter concerns transport systems that carry out an exchange (antiport) reaction, one of several kinds of 'secondary' transport — that is, a reaction not explicitly associated with chemical or photochemical energy transductions (see Ref. [1] for a formal definition). Membrane carrier proteins involved in such secondary transport are associated with many events in biology, ranging from the accumulation of nutrients and extrusion of wastes by bacteria to the cycling of neurotransmitters at synaptic or synaptosomal membranes. But whether one considers microbiology or neuroscience, we now believe these carrier proteins all share certain basic elements of structure and mechanism. This in turn suggests that progress in understanding any single example might have an impact on others. And because bacterial systems are so tractable to genetic manipulation, a discussion of microbial transporters should have special relevance to the larger community. Certainly, the study of bacterial transport — indeed, a bacterial anion exchange — was important to the early development of this field.

1.1. Origins: GLUT, UhpT and LacY

This general area of study was established in 1952 when Widdas [2] presented an analysis of glucose transport by the mammalian placenta, a reaction now attributed to the GLUT1 isoform of the glucose transport protein. Before Widdas' work, those interested in the transport of sugars, amino acids or other organic substrates evaluated their findings in terms of coefficients of diffusion and permeability, using a framework designed to model the distribution of ions across membranes. While this approach was central to an understanding of ion transport, these factors proved far less powerful in driving an understanding of sugar or amino acid transport, for such coefficients varied in unexpected ways, at times even showing a dependence on the internal concentration of transported solute (see Refs. [2,3]). Widdas offered a new perspective. His 'carrier' hypothesis [2] suggested that transport must be initiated by the binding of substrate to a catalytic membrane element, the carrier. Consequently, the kinetics of glucose transport would have to include terms related to substrate binding. He also proposed that the carrier–substrate complex traveled across the membrane, implying that transport kinetics should also have terms describing the mobility or permeability of the complex, much as in the former paradigm. At the opposite surface, the complex would dissociate, the unoccupied carrier would return to the original membrane surface, and the process would begin again. This new mechanistic view, and simple extensions of it (Fig. 1) [4], are as successful now as they were in the 1950s in describing the kinetics of carrier-mediated

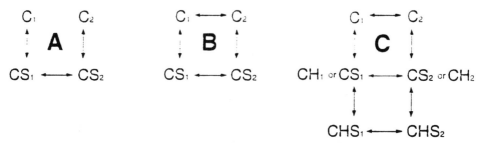

Fig. 1.   Kinetic schemes describing carrier-mediated transport. (A) Antiport. (B) Uniport. (C) Symport. In model C, if the binary species reorient (central arrow), transport of the individual cosubstrates is not coupled in a 1-for-1 stoichiometry.

transport. Of course, it is not likely that carriers diffuse through the membrane. Instead, we see the diffusion and permeability constants of the carrier and carrier–substrate complex as rate constants describing the alternate exposure of substrate binding centers to one or the other membrane surface. Accordingly, substrates are now seen as passing *through* rather than being carried *by* transport proteins.

The second formative event in this field occurred only a year later, in 1953, when Mitchell's laboratory presented a study of phosphate (Pi) exchange in the bacterium, *Micrococcus pyogenes* (now, *Staphylococcus aureus*) [5,6]. This reaction is now assigned to a transporter resembling UhpT, one of several anion exchange proteins found in *Escherichia coli* and *Salmonella typhimurium* [7] (and elsewhere; see below). The most significant aspect of this early work was its experimental proof that a membrane carrier acted as a catalyst. Specifically, Mitchell showed that Pi exchange could be inhibited by mercury, and that inhibition was achieved with far fewer mercuric ions than there were Pi molecules available to the exchange process. Clearly, then, the mercury-sensitive element (a protein or protein–lipid complex) was a catalyst, as Widdas required.

Work with a bacterial system also represented the third important step in this field, when in 1956 Monod and his collaborators described the lactose carrier (LacY) in *E. coli* [8], bringing genetic tools in the study of membrane transport. LacY spent its formative years in the shadow of β-galactosidase but has now emerged as the best-studied example of a membrane carrier [9]. The final event of the 1950s was the insight by Crane, in 1959, that a carrier might combine with more than one substrate at a time (Fig. 1) (see Ref. [10]) thereby explaining the Na-dependence of glucose transport by (the SGLT1 porter of) intestinal tissue. Following his own thoughts concerning the role of protons in membrane biology, in 1963 Mitchell suggested that the mechanism of LacY function was a H^+/lactose cotransport [11]. Both Crane's and Mitchell's suggestions proved to be correct.

The strongest effect of these studies in the 1950s was to set up a theoretical framework whose credentials could be tested by biochemistry and biophysics. Those tests have been applied, and we now view these ideas (Fig. 1) as the most accurate shorthand to use in describing 'carrier-mediated' events. Whether considering the capture of nutrients or the recycling of neurotransmitters, an adequate model can almost always be reached by some combination or elaboration of the these prototypical systems, GLUT1, UhpT, LacY and SGLT1.

1.2. Common biochemical and molecular themes

This early work (above) also introduced to the field model systems exemplifying the three main biochemical mechanisms of carrier-mediated transport (Fig. 1): uniport (e.g., GLUT1), antiport (UhpT) and symport (LacY, SGLT1). Although these reaction schemes have quite different kinetic consequences, they clearly spring from a common argument. Moreover, we now believe a single protein class encompasses all reaction types and that there may actually be a single, generic structure for all membrane carriers. The evidence driving this speculation comes mainly from analysis of the large number of carrier sequences known in prokaryote and eukaryote cells. Analysis of this data-base confirms early suspicions [12] that most examples have a hydrophobic core containing 10–12 likely transmembrane α-helices (see Fig. 4, below), a structure first suggested for the LacY protein [13], following Kyte and Doolittle's suggestion [14] for identification of transmembrane α-helices.

Several hundred carrier sequences have been examined, and together they define two major groups [15–17]. Most cases (≈ 80% of the total), including examples of each reaction type (Fig. 1) and representatives from bacteria, fungi and animal cells, display the ca. 10–12 transmembrane segments noted above. While this structure most often arises simply by analyzing an amino acid sequence, in a growing number of cases there is independent support from various reporter groups inserted into or fused with the target protein (see later). Those examples in the minority group (the remaining ≈ 20%) are typically from eukaryote organelles (usually mitochondria, a favored experimental system) and show a hydrophobic core of 5–7 transmembrane α-helices [17]. Since at least some members of the majority class can function as monomers (see below), while several among the minority group operate as dimers [18], it is feasible that a single organizational principle characterizes all of these systems, and that the smallest unit of biochemical function has a hydrophobic core of ca. 10–12 transmembrane α-helices. If so, the presence of this common denominator makes it possible that contributions to the general field can continue to arise from diverse sources.

2. UhpT and other phosphate-linked anion exchange proteins

The present state of knowledge about UhpT and other so-called 'Pi-linked' exchange proteins should be considered within this larger framework (above). These antiporters are grouped together in part by phenomenology, since they each use Pi as a substrate, and in part by sequence homology (below). Such mixed criteria reflects that quite different cell types served as models for study at different times. Three systems have been most important, and these are the ones emphasized most strongly in the sections that follow. Thus, Mitchell's work with *S. aureus* established the initial model. In the 1980s, when these carriers were redefined as a group, crucial work was performed with *Streptococcus* (now *Lactococcus*) *lactis*; that gave rise to the experimental definition of Pi-linked antiport and to models of biochemical and physiological function. Now, in the 1990s, the examples in *E. coli* and *S. typhimurium* offer the most instructive targets, and this is where current biochemical and molecular studies are helping to define the translocation pathway.

2.1. Early work

Bacterial transport of phosphate was first described in the early 1950s by Mitchell [5,6] who studied the exchange of internal and external phosphate by *S. aureus*, and who identified the *mono*valent anion of phosphate ($H_2PO_4^{1-}$) as the substrate for exchange. A decade later, however, Harold et al. [19] came to different conclusions about Pi transport by *Enterococcus faecalis*, another gram-positive cell. In this latter case, no exchange reaction could be identified. Instead, only a net influx could be seen. Moreover, the reaction required active metabolism (ATP) and selected in favor of *di*valent phosphate (HPO_4^{2-}). These differing observations reflect the diversity of mechanisms exploited by microorganisms to transport Pi. We now know Escherichia has at least three distinct biochemical mechanisms that take ^{32}Pi into the cell. Two of these, now designated as Pst, an ABC transporter [20], and Pit, a H^+/MgPi symporter [21], are highly specific for Pi itself, while the third type appears to accept Pi almost by chance and with low affinity relative to some organic phosphate. Pi-linked antiporters constitute this last group, and in *E. coli* there are two main examples — the UhpT and GlpT porters — which accept sugar 6-phosphates [22] and glycerol 3-phosphate [23], respectively, as their high affinity substrates (reviewed in Ref. [7]).

   Pst and Pit represent the most common solutions to the problem of net import of Pi by bacteria. Pst-like energetics accommodates the early findings in Enterococcus (above) while Pit-like behavior accounts for the proton-motive force dependence of Pi transport by a number of other cells [7]. Early schemes had also considered Pi and sugar phosphate transport by GlpT and UhpT as further examples of nH$^+$/anion symport, but we now know that both UhpT and GlpT belong to the category of antiport [7]. Thus, a summary of the historical record concludes that Pi transport systems come in two varieties — those designed for net Pi movement (Pit, Pst, ...), and those involved in an *exchange* involving Pi and/or a phosphorylated substrate (GlpT, UhpT, ...).

2.2. Model experiments: the homologous and heterologous exchanges

Pi-linked antiporters each carry out a Pi self-exchange reaction, and because this represents the common functional link, it seems appropriate to give a specific example of the reaction. ^{32}Pi transport into washed cells of *L. lactis* is described by Fig. 2 (from Ref. [24]). In the absence of metabolizable substrates, such cells have reduced levels of ATP, no proton-motive force, and an elevated pool of internal Pi (here, 50 mM). Moreover, this internal Pi pool is maintained for *long* periods, even if outside Pi becomes low (20 μM here); for this reason one expects the cell membrane to be impermeable to Pi. However, the experiment of Fig. 2 shows that

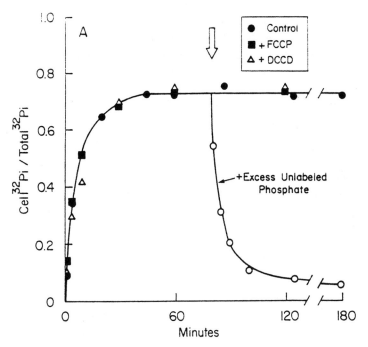

Fig. 2. Phosphate exchange in *L. lactis.* ^{32}Pi transport was estimated by centrifugation through silicone oil after addition of 20 μM ^{32}Pi to washed cells suspended in 300 mM KCl, 20 mM MOPS/K (pH 7). Key: (solid circles), control; (solid squares), 1 μM FCCP present; (open triangles), stock cells had been exposed to 1 mM DCCD for 30 min prior to assay. At the arrow, part of the control suspension received 4 mM KPi. From Ref. [24].

added ^{32}Pi is readily taken up into the internal Pi pool ($\geq 85\%$ of internal ^{32}P was ^{32}Pi in this case) — and just as readily discharged following addition of unlabeled Pi. Because access to the internal pool was unaffected by metabolic inhibitors (FCCP, DCCD), one is presented with an apparent paradox: a membrane that retains Pi against a large concentration gradient while at the same time showing a high permeability to ^{32}Pi, all in the *absence* of energy. Clearly, only if external ^{32}Pi enters in exact exchange for internal Pi is there any simple resolution to this quandary. In fact, this can be demonstrated directly, simply by showing that internal and external Pi pool sizes remain constant despite the transfer of ^{32}Pi [5,24].

Two important aspects of Pi-linked antiport were revealed by examining the Pi self-exchange reaction in *L. lactis* and *S. aureus* [6,24]. Perhaps most unexpected, such work showed that the reaction strongly favors monovalent over divalent Pi. Experimentally, this is inferred from the fact that the Michaelis constant for Pi transport (K_t) is constant with pH in the acid range, but rises as pH moves above the pK_2 for phosphate (p$K_2 = 6.8$) and monovalent Pi becomes depleted [6,24]; other effects of pH were attributed to changes in the maximal velocity [6]. If divalent Pi were a substrate, complex explanations for the pH sensitivity of K_t must be invoked, whereas in the current model the authentic K_t (for monovalent Pi) is pH independent, and pH effects are confined to changes of V_{max}, as is usual in enzymology. Accordingly, in constructing models of these two specific systems — and when extending these models to systems of similar substrate specificity (see Table 1, below) — it has been convenient to assume that when Pi takes part in exchange, it is accepted only as the monovalent anion. This has had an important effect on the structure of biochemical and physiological models outlined later.

That ^{32}Pi is moved across Lactococcal membranes by true exchange (Fig. 2) and that Pi-linked systems also accept organic phosphates suggests antiport as the mechanism of transport for these substrates during heterologous exchange. And, in fact, such mutual exchange can be demonstrated experimentally, but this does not prove that these systems carry out antiport as their exclusive reaction mechanism. For example, symporters often show a substrate exchange that is faster than the reaction of net flux, so one could imagine the exchange of Pi and the organic phosphate as a partial reaction of an ion-coupled symporter. An incidental exchange of this sort would be very different from a mechanistic antiport. It was therefore important to address this possibility directly. The relevant experiments make use of

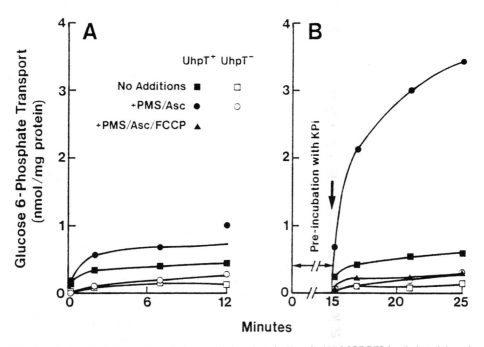

Fig. 3.   Antiport of phosphate and glucose 6-phosphate in *E. coli*. (A) MOPS/K-loaded vesicles of UhpT-positive (UhpT$^+$) or UhpT-negative (UhpT$^-$) strains were placed in MES-based buffer at pH 6, in the absence or presence of 120 μM phenazine methosulfate with 33 mM ascorbate (Asc), as indicated. This was followed after several minutes by 100 μM [^{14}C]glucose 6-phosphate. (B) In parallel tubes, vesicles were preincubated for 15 min with 100 μM KPi, and other supplements as noted in part A, before adding labeled sugar phosphate. The protonophore, FCCP, was used where indicated. From Ref. [25].

membrane vesicles prepared in the *absence* of Pi (using MOPS as the internal buffer), and the work with the *E. coli* UhpT protein gives a useful case study (Fig. 3). When allowed to respire, such MOPS-loaded vesicles accumulate traditional substrates by H^+-coupled systems, and judged by the sustained gradients of substrate (e.g., Pi or proline), oxidation reactions generate a cation-motive force near -140 mV [25]. Nevertheless, these vesicles fail to take up sugar phosphate (Fig. 3). Instead, UhpT function requires a *pre*incubation to establish a pool of internal Pi; only then is there a restoration of the activity found in cells (Fig. 3). This observation — that UhpT requires substrates in *trans* — argues strongly in favor of the idea of antiport.

2.3. Diagnostic features of Pi-linked antiport

Currently known examples of Pi-linked antiport are listed in Table 1. Most information concerning these exchanges comes from a few well-studied systems, each of which uses G6P as its high affinity substrate, and since the example in *E. coli* is known as UhpT (for uptake of hexose phosphate), this abbreviation is often used to encompass all three examples: the one in *S. aureus* (where G3P is also taken), in *L. lactis*, and in *E. coli*. The distribution of Pi-linked exchange is likely to be broader than noted here, as the surveys by Winkler [26] and Dietz [27] document, but the examples shown are those in which some experimental investigation supports the argument. Based on the features of well-studied cases, each family member will exhibit the following properties:

Table 1
Pi-linked anion exchange proteins[a]

Organism	Exchange protein	Primary substrate(s)[b]
Escherichia	UhpT	G6P, M6P, F6P
	UhpC[c]	G6P
	GlpT	G3P
Salmonella	PgtP	PEP > PGA
Bacillus	GlpT	G3P
Staphylococcus	UhpT	G6P > G3P
Lactococcus	UhpT	G6P, M6P, > F6P

[a] The antiporters of Staphylococcus and Lactococcus are not defined genetically, but in Escherichia, Salmonella and Bacillus the UhpT, UhpC, GlpT and PgtP proteins show 30–35% sequence identities [33,36,49–51]; the Salmonella UhpT and UhpC proteins are nearly identical to their counterparts in Escherichia [33]. Modified and simplified from Ref. [7].
[b] The substrates named are glucose 6-phosphate (G6P), mannose 6-phosphate (M6P), fructose 6-phosphate (F6P), glycerol 3-phosphate (G3P), phosphoglyceric acid(s) (PGA), and phosphoenolpyruvate (PEP).
[c] UhpC is not a transporter but a G6P receptor and regulatory protein of high specificity. Occupancy of UhpC by *external* G6P generates the signals required for UhpT expression [52]; this prevents internal G6P from serving as an inducer.

(1) *Catalysis of the Pi self-exchange reaction* (Fig. 2). This is perhaps the simplest biochemical diagnostic for family membership, since this can be reproduced at all levels of analysis, from intact cells to proteoliposomes [7]. Judging from data collected in *S. aureus* [6] and *L. lactis* [24] this exchange in most cases will favor monovalent Pi over divalent Pi, although this is not a strict requirement (see Ref. [7] for discussion of the PgtP protein). However, in all cases so far tested these proteins fail to distinguish between the arsenate and phosphate anions, so the two substitute freely for each other, in all exchange modes.

(2) Each system also catalyzes *the heterologous exchange of Pi and some organic phosphate* (G6P, G3P, etc.), and in both *E. coli* and *S. aureus* one can exclude the operation of symport as a mechanism which might account for sugar phosphate accumulation [25,28]. In no one of these systems has there been a complete kinetic analysis, but in *L. lactis* UhpT [29] and *E. coli* GlpT [23] there is a simple competitive relationship between Pi and the organic phosphate substrate, as might be anticipated for antiport. It is therefore convenient to assume that the two kinds of substrates occupy overlapping binding sites and that the heterologous exchange can be classified as Ping Pong in a biochemical sense; this simplifying assumption is incorporated into several models.

(3) The electrical character of the heterologous reaction has been analyzed in several cases (*L. lactis*, *S. aureus*, *E. coli* UhpT), where *the evidence favors an electroneutral exchange* [7]. This observation, too, has been useful in devising appropriate biochemical models.

(4) Finally, sequence homology among the gram-negative UhpT, GlpT, and PgtP proteins and the gram-positive GlpT example (Table 1), suggest a *common origin and thus a common reaction mechanism*.

To be sure, there are differences among the various examples of Pi-linked exchange, but these differences seem to be mainly in the values of kinetic constants. Thus, the Michaelis constant for Pi transport is spread over about a 10-fold range (250 µM to ≥2.3 mM), and that for the preferred organic phosphates over a similar range but a lower concentration (generally, 20–200 µM). This order of affinity should not be taken to suggest that Pi is an inappropriate or 'unnatural' substrate, however. Instead, its role is essential in several respects — either to serve as the countersubstrate driving accumulation of the organic substrate; as a pathway for net Pi export when internal stores exceed their capacity; or even as the pathway for net Pi accumulation under some circumstances. These varied scenarios depend on three main factors: the transmembrane pH gradient; the relative pool sizes of internal and external Pi; and the net stoichiometry of the exchange of Pi with the organic phosphate substrate (see Ref. [7] for detailed discussion).

2.4. Exchange stoichiometry: implications for biochemistry and cell physiology

Pi-linked antiporters were identified initially by work with gram-positive cells, and those prototypes continue to be the best understood in two respects. In particular, documentation of ionic selectivity is most complete for examples in *S. aureus* and *L. lactis*. And for technical reasons, measurement of exchange stoichiometry is most convincing for the Lactococcal carrier. (The K_t for Pi is relatively low in this example, making it easier to measure heterologous exchange, and contaminating phosphatases are highly sensitive to inhibition by vanadate.) By contrast, analysis using molecular biology is based in the gram-negative systems, especially in UhpT of *E. coli*. Of necessity then, unifying biochemical and molecular models have taken information from different systems, and one should keep in mind the hybrid nature of these general models in evaluating any single example, as in the case of the biochemical model that emerges from considerations of exchange stoichiometry in the gram-positive *L. lactis*.

Work with the UhpT protein of *L. lactis* shows that two Pi anions exchange with a single G6P anion in assays at pH 7 [30]. Since the reaction is electrically neutral [30], and since the Pi self-exchange uses monovalent Pi (see above), this 2-for-1 heterologous antiport is presumed to reflect 2 *monovalent* Pi anions exchanging for the *divalent* G6P anion, as might be expected at pH 7. This exchange is affected in two significant ways by varying assay pH. First, acidification reduces overall function due to a decreased maximal velocity; there is, however, very little change in substrate affinity [29]. Accordingly, the pH-insensitivity in affinity for G6P, over a pH range that spans its own pK_2 (= 6.1), suggests that both monovalent *and* divalent sugar phosphate anions are effective substrates. This stands in contrast to the same sorts of experiments with the inorganic substrate, Pi, since in that case, there was selectivity favoring the monoanion. The second effect of pH is to change the observed heterologous exchange stoichiometry itself: thus, the antiport ratio (Pi/G6P) moves from its upper limit of 2:1 (measured at pH 7–8) to a lower limit near 1:1 (measured at pH 5.2) [29].

These facts are organized by a simple model with important consequences for cell physiology (Fig. 4). In brief, the model proposes that UhpT has a bifunctional active site that accepts either two monovalent anions or a single divalent species, but not both. This is an unusual requirement, but not unprecedented — for example, the ionophore, A23187, uses two carboxylate anions to bind either a pair of H^+ or a single Ca^{2+} (Mg^{2+}) during a 2H:Ca neutral antiport. This constraint ensures an electroneutral exchange with simple Ping Pong kinetics as the antiporter (UhpT or A23187) samples the *cis* and *trans* compartments. Given the preference for monovalent over divalent Pi, one can now understand the finding of a 2:1 (Pi:G6P) exchange stoichiometry at pH 7, where sugar phosphate is mostly divalent. The model also explains a pH-dependent change in stoichiometry, provided that both mono- and divalent sugar phosphates are acceptable, as is predicted by the finding that the affinity for G6P is not much affected by pH. Thus, one imagines two kinds of substrates — monovalent anions (monovalent Pi, monovalent sugar phosphate, and so on), which would exchange in the ratio of 2:2; and divalent anions (typically, sugar phosphates) that would move at 1:1. The overall macroscopic stoichiometry would then be a mixture of these elementary events, weighted according to the relative concentration of each substrate type and the affinity of their interactions with the carrier.

The details of this model depend very much on experiments with *L. lactis*; moreover, this model has not yet been tested by exploring its kinetic predictions.

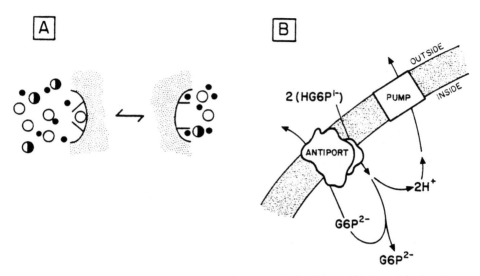

Fig. 4.   A biochemical model and its implications for cell physiology. (A) A model of exchange involving monovalent Pi (solid circles), monovalent sugar phosphate (half-filled circles) or divalent sugar phosphate (open circles). As shown, binding is to the divalent substrate (left figure) or to a pair of monovalent Pi anion (right figure), which are enclosed by overlapping structures (sticks) within the protein. (B) A consequence of the model in part A is that net sugar phosphate accumulation can occur in the presence of a pH gradient. From Ref. [53].

Nevertheless, the general idea does allow one to understand findings elsewhere. For example, Essenberg and Kornberg [31] proposed an $nH^+/G6P$ symport mechanism for the UhpT protein in *E. coli*, and much evidence points to this as a useful *functional* view [7]. In fact, this is entirely consistent with the *L. lactis* biochemical model, given that the *E. coli* cytoplasm is typically more alkaline that the periplasmic space. Thus, this biochemical model predicts a cycle in which net sugar phosphate accumulation arises from the exchange of two monovalent anions and a single divalent species, yielding the net import of $2H^+$ and 1 $G6P^{2-}$ (Fig. 4). Phenotypically, this would appear as a symport with protons, resolving the conflict between thermodynamics (symport) and mechanism (antiport). This physiological perspective also allows one to understand other observations in *E. coli*. It is known that mutants which lack the Pi-specific transporters, Pit and Pst, can nevertheless survive if G6P or G3P is used in the growth media. But this finding is puzzling if UhpT or GlpT operate simply as 2-for-2 or 1-for-1 exchangers, since those reactions could not provide for net Pi entry. On the other hand, an asymmetric self-exchange using sugar phosphate (or Pi) accommodates these findings in a straightforward way by allowing either UhpT or GlpT to serve as the port of net entry of Pi (see Ref. [7] for detailed discussion of this and other scenarios).

3. Biochemical and molecular studies in *E. coli*

3.1. Protein structure and topology

The amino acid sequences of *E. coli* (and *S. typhimurium*) UhpT and GlpT proteins are punctuated by as many as 12 segments whose hydrophobic character could yield a transmembrane α-helix. Guided by these theoretical findings, Kadner [32,33] and Boos [34] and their colleagues have carried out an extensive analysis of topology by studying UhpT and GlpT fusion proteins using a C-terminal reporter group such as alkaline phosphatase (PhoA). This experimental approach is based on the argument [35] that the cytoplasmic or periplasmic localization of PhoA (and other reporters) is determined largely by the orientation of the immediately preceding transmembrane segment. And because recovery of functional PhoA requires its export into the periplasm, the location of PhoA is readily determined by simple enzymatic assay. A large number of UhpT-PhoA and GlpT-PhoA chimeras have been examined [32–34]; a few chimeras between GlpT and β-galactosidase have also been tested [34]. The result of this analysis is the conclusion that UhpT and GlpT have similar topologies, consistent with the similarity in sequence. An example of this topology is shown for the UhpT protein (slightly modified from Kadner's [33] work) (Fig. 5). That model also illustrates the finding [36] that there is an excess of positively charged residues on the cytoplasmic surface, consistent with the 'positive-inside' rule of von Heijne [37]. In addition, the diagram (Fig. 5) illustrates the location of the six cysteine residues within UhpT. The properties of two of these cysteines will be discussed later.

3.2. Monomer or dimer?

It was noted earlier that in analyzing the sequences of a large number of membrane carriers (perhaps 80% of the total), one arrives at a structure resembling that suggested for UhpT (Fig. 5). A number of studies indicate that only a single protein is required for function, but in most cases it is unclear whether the minimal unit of function is the monomer, dimer or some higher aggregate. For UhpT, biochemical work suggests the monomer is functional [38]. A more complete analysis for LacY leads to a similar conclusion [39,40], but in both cases it has proven difficult to arrive at the incontrovertible experiment. In the case of UhpT, it is clear that the *solubilized* protein binds its natural substrates, since addition of various sugar phosphates protects the protein from thermal denaturation or chemical inactivation, with the degree of protection correlating with the rank order of substrate affinity [38]. It is also clear that the solubilized protein is distributed as a monomer in detergent extracts, with or without added substrate. Accordingly, one must conclude that the monomer binds substrate. It has also been found that reconstitution of activity (by a detergent dilution technique) shows a linear dependence on protein concentration, even when the sample is diluted to a point where there should be less than one monomer per proteoliposome. By extension, then, this points to the monomer as the minimal functional unit [38].

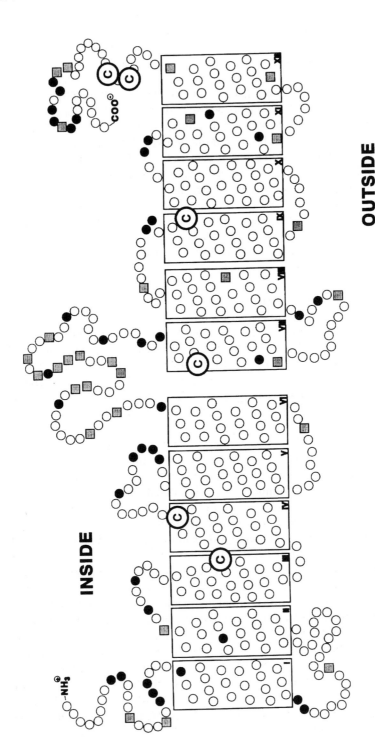

Fig. 5. The topology of UhpT. UhpT topology, slightly modified from Island et al. [33] is shown, with positive charged residues (K,R) indicated by solid circles and negative residues (D,E) given by gray squares. The enlarged circles show the six UhpT cysteine residues at positions 108, 143, 265, 331, 436 and 438. From Ref. [42].

The biochemical finding that UhpT is likely functional as a monomer is at odds with certain genetic observations with its relative, GlpT, where one can obtain mutants that display a dominant-negative phenotype, suggesting an oligomeric structure [41]. Thus, a simple interpretation of biochemical work disagrees with an equally simple interpretation of genetic studies. Since the ability to use genetic manipulations is becoming more and more accessible, it is important to think carefully about such dominant-negative effects. For example, one could imagine such behavior arising during or after assembly by interaction between transmembrane segments of one unit and those of its neighbor, so as to affect the final shape of both. This effect should be most pronounced for over-expressed mutant proteins or when the experimental system has one or several carriers at rather high concentration. Such factors, which are unrelated to transport, may help explain the dominant-negative phenotypes, but further work is clearly needed before one can exclude the occurrence of the dimer (oligomer) as a relevant *in vivo* structure. Of course, the presence of a multimer *in vivo* does *not* conflict with the biochemical finding of the monomer as the minimal unit of function.

3.3. Probing the translocation pathway

Biochemical and biophysical techniques are poorly developed in membrane biology, and one anticipates future work on Pi-linked exchange to emphasize molecular biology as a way to explore protein structure and function. One recent outcome of this approach is development of a line of experiments (below) that locates a portion of the translocation pathway through UhpT. In principle, extensions of this approach could identify a number of the amino acid residues that interact with substrate. With this information in hand, it may be possible to think intelligently about questions of selectivity and stoichiometry that are raised by the biochemical model derived in *L. lactis* (Fig. 4).

Several years ago, experiments with UhpT showed that at least one residue on TM7 must be on the transport pathway [42]. That conclusion arose from a study designed originally to identify mercurial-reactive cysteine(s) (recall Mitchell's original findings [5]) by using site-directed mutagenesis to generate single-cysteine UhpT variants (see Fig. 5). Two of the six cysteines in UhpT (C143 and C265) were shown to be sensitive to the *permeant* SH-reactive probe, *p*-mercuribenzoic acid. Subsequent work then focused on using the related but *impermeant* probe, *p*-mercuribenzosulfonate (PCMBS). This revealed an important difference in the reactivities of these two cysteines. Thus, when intact cells were examined, C143 was insensitive to attack by PCMBS, but if everted membrane vesicles were studied, inhibition was clearly observed [42]. Therefore, it was concluded that C143 was accessible only from the cytoplasmic phase, as might be expected from the likely topology of UhpT (33) (Fig. 5). By contrast, C265 was accessible to the impermeant PCMBS from *both* membrane surfaces — that is, inhibition of UhpT function was found in both intact cells and everted membrane vesicles of the single-cysteine variant containing C265. C265 therefore behaved as if exposed to both sides of the membrane. Further studies of substrate protection showed that the attack by

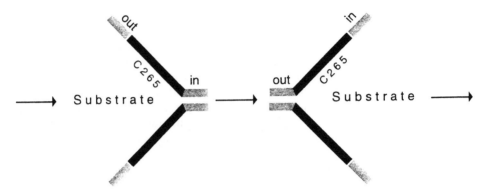

Fig. 6.   The translocation pathway in UhpT. Three regions are shown in this cross-section of the translocation pathway in a membrane carrier. The model is framed as a Ping Pong biochemical mechanism, so that as substrate moves from left to right (outside to inside), one half-turnover is completed. Within the pathway, the peripheral domains, exposed to a single surface ('in' or 'out'), are gray. The core domain, in black, contains at least one residue exposed to both internal and external compartments during the course of substrate transport; C265 of UhpT is an example of such a residue. From Ref. [42].

PCMBS at C265 could be blocked by the UhpT substrate, G6P, whether in cells or vesicles. Taken together, then, these findings suggest C265 lines the substrate translocation pathway extending through UhpT.

With this conclusion in mind, recall that membrane carriers operate to transport a single substrate (or a set of substrates) per turnover (Fig. 1); this results from the stoichiometric combination of carrier and substrate, as Widdas proposed [2], and it is this that distinguishes carriers from channels, where many substrates move as the protein cycles once between its open and closed states. To ensure this relationship between carrier and substrate, the translocation pathway must have three kinds of domains (Fig. 6). Two of these could be said to be 'peripheral', inasmuch as they are exposed to either the inside or the outside aqueous phases, but not to both. By contrast, the third or 'core' domain, lying between these peripheral regions, would become alternately exposed to both surfaces as the carrier orients and reorients to accept and discharge substrate(s). The physical dimensions of these domains are as yet unknown for any carrier, but in principle their sizes might vary considerably, according to requirements set by substrate volume and the kinds and numbers of ligands required for its binding.

An exploration of this model could be highly productive for UhpT, because a cysteine-less version of this antiporter is functional [42], and because the work with C265 establishes a simple experimental test with which to identify residues in the core domain of the pathway — that is the accessibility to impermeant, hydrophilic probes added from either cytosolic or periplasmic surfaces. In principle, then, one should be able to implant new cysteines at new locations and use PCMBS (or other probes) to identify residues and α-helices that surround and delimit the pathway. This cysteine scanning mutagenesis shows promising early results. It appears that the core domain (Fig. 6) extends on TM7 from C265 at least through I276 [43]. And since the pattern of PCMBS accessibility in this region suggests an α-helical structure, this central domain would occupy a 17 Å span in the middle of TM7 [43]. If this dimension is replicated elsewhere, the size of the pathway through UhpT could, in fact, enclose the numbers and kinds of substrates suggested by biochemical modeling (Fig. 4), but it is far too early to raise this as anything but a speculation. More important in the long run is that a simple combination of biochemistry and mutagenesis suggests a general way to probe the translocation pathway.

4. Prospects

Kinetic patterns embodying the carrier hypothesis (Fig. 1) can be mapped onto a single physical structure (Fig. 5), and the current emphasis is on finding systems will look closely at this relationship. In this effort, bacterial models may be especially fruitful since they are among the most susceptible to a molecular analysis. Thus, the next few years should prove especially exciting as we learn how to describe how substrates interact with and move through transporting proteins. Recall there seem to be two kinds of carriers: those of helix number 5–7 that operate as dimers, and those of helix number 10–12 that function as monomers (but which could be viewed as covalent heterodimers). If this generalization holds, one could reasonably argue that substrate(s) move through a transporter at the true dimer interface, or at the equivalent interface found in the larger covalent heterodimer — say, the interface between the N-terminal and C-terminal halves of the molecule [44]. No direct information has tested the idea, but visualization of a pathway could soon come from crystallography, which is most advanced for a red cell antiport protein, band 3 [45]. Alternatively, it may be feasible to use molecular biology to follow clues provided by bacterial examples. Several residues in the pathway of UhpT are now known [42,43], and in LacY and other bacterial carriers (e.g., MelB) residues affecting substrate specificity suggest an equivalent structure [46,47]. At the moment, we arrive at a two dimensional perspective with reasonable confidence. Understanding where substrates bind and move, even in only a few cases, could bring a new rational basis to inferences made for three dimensional structure — perhaps not only for carriers, but also for other transport proteins with similarly structured hydrophobic cores.

Acknowledgements

Work in this laboratory has been supported by grants from the National Institutes of Health (GM24195) and the National Science Foundation (MCB-9220823). We would like to express our gratitude to Dr. R.J. Kadner (University of Virginia, Charlottesville, VA, USA) for his generosity in providing appropriate cell lines with which to study the UhpT protein of *E. coli*.

References

1. Stein, W.D. (1986) Transport and Diffusion Across Cell Membranes. Academic Press, Orlando, FL.
2. Widdas, W.F. (1952) J. Physiol. **118**, 23–39.
3. Widdas, W.F. (1988) Biochim. Biophys. Acta **947**, 385–404.
4. Mitchell, P. (1967) Adv. Enzymol. **29**, 33–87.
5. Mitchell, P. and Moyle, J. (1953) J. Gen. Microbiol. **9**, 257–272.
6. Mitchell, P. (1954) J. Gen. Microbiol. **11**, 73–82.
7. Maloney, P.C., Ambudkar, S.V., Anantharam, V., Sonna, L.A. and Varadhachary, A. (1990) Microbiol. Rev. **54**, 1–17.
8. Rickenberg, H.V., Cohen, G.N., Buttin, G. and Monod, J. (1956) Ann. Inst. Pasteur **91**, 829–857.
9. Kaback, H.R. in: Handbook of Biological Physics, Vol. II. Transport Processes in Membranes, eds W.N. Konings, H.R. Kaback and J.S. Lolkema. Elsevier, Amsterdam, pp. 203–227.
10. Crane, R. (1962) Fed. Proc. **21**, 891–895.
11. Mitchell, P. (1963) Biochem. Soc. Symp. **22**, 142–168.
12. Maloney, PC (1990) Res. Microbiol. **141**, 374–383.
13. Foster, D.L., Boublik, M. and Kaback, H.R. (1983) J. Biol. Chem. **258**, 31–34.
14. Kyte, J. and Doolittle, R.F. (1982) J. Molec. Biol. **157**, 105–132.
15. Marger, MD and Saier, M.H., Jr (1993) Trends Biochem. Sci. **18**, 13–20.
16. Reizer, J., Reizer, A. and Saier, M.H., Jr. (1994) Biochim. Biophys. Acta **1197**, 133–166.
17. Nelson, D.R., Lawson, J.E., Klingenberg, M. and Douglas, M.G. (1993) J. Mol. Biol. **230**, 1159–1170.
18. Aquila, H., Link, T. and Klingenberg, M. (1987) FEBS Lett. **212**, 1–9.
19. Harold, F.M., Harold, R.L. and Abrams, A. (1965) J. Biol. Chem. **240**, 3145–3153.
20. Rosenberg, H. (1987) In: Ion Transport in Prokaryotes eds B.P. Rosen and S. Silver, Academic Press, Inc., New York, pp. 205–248.
21. van Veen, H.W., Abee, T., Kortstee, G.J.J., Konings, W.N. and Zehnder, A.J.B. (1994) Biochemistry **33**, 1766–1770.
22. Winkler, H.H. (1966) Biochem. Biophys. Acta **117**, 231–240.
23. Hayashi, S.-I., Koch, J.P. and Lin, E.C.C. (1964) J. Biol. Chem. **239**, 3098–3105.
24. Maloney, P.C., Thomas, J. and Schiller, L. (1984) J. Bacteriol. **158**, 238–245
25. Sonna, L.A., Ambudkar, S.V. and Maloney, P.C. (1988) J. Biol. Chem. **263**, 6625–6630.
26. Winkler, H.H. (1973) J. Bacteriol. **116**, 1079–1081.
27. Dietz, G.W. (1976) Adv. Enzymol. **44**, 237–259.
28. Sonna, L.A. and Maloney, P.C. (1988) J. Membr. Biol. **101**, 267–274.
29. Ambudkar, S.V., Sonna, L.A. and Maloney, P.C. (1986) Proc. Natl. Acad. Sci. USA **83**, 280–284.
30. Ambudkar, S.V. and Maloney, P.C. (1984) J. Biol. Chem. **259**, 12576–12585
31. Essenberg, R.C. and Kornberg, H.L. (1975) J. Biol. Chem. **250**, 939–945.
32. Lloyd A.D. and Kadner, R.J. (1990) J. Bacteriol. **172**, 1688–1693.
33. Island, M.D., Wei, B.Y. and Kadner, R.J. (1992) J. Bacteriol. **174**, 2754–2762.
34. Gott, P. and Boos, W. (1988) Molec. Microbiol. **2**, 655–663.
35. Calamia, J. and Manoil, C. (1992) J. Molec. Biol. **224**, 539–543.
36. Friedrich, M.J. and Kadner, R.J. (1987) J. Bacteriol. **169**, 3556–3563.
37. von Heijne, G. (1992) J. Molec. Biol. **225**, 487–494.
38. Ambudkar, S.V., Anantharam, V. and Maloney, P.C. (1990) J. Biol. Chem. **265**, 12287–12292.
39. Costello, M.J., Escaig, J., Matsushita, K., Viitanen, P.V., Menick, D.R. and Kaback, H.R. (1987) J. Biol. Chem. **262**, 17072–17082.
40. Sahin-Toth, M., Lawrence, M.C. and Kaback, H.R. (1994) Proc. Natl. Acad. Sci. USA **91**, 5421–5425
41. Larson, T.J., Schumacher, G. and Boos, W. (1982) J. Bacteriol. **152**, 1008–1021.
42. Yan, R.-T. and Maloney, P.C. (1993) Cell **75**, 37–44.
43. Yan, R.-T. and Maloney, P.C. (1995) Proc. Natl. Acad. Sci. USA **92**, 5973–5976
44. Maloney, PC (1994) Curr. Opin. Cell Biol. **6**, 571–582.
45. Wang, D.N., Sarabia, V.E., Reithmeier, R.A. and Kuhlbrandt W. (1994) EMBO J. **13**, 3230–3235.

46. Collins, J.C., Permuth, S.F. and Brooker, R.J. (1989) J. Biol. Chem. **264**, 14698–14703.
47. Poucher, T., Zani, M.L. and Leblanc, G. (1993) J. Biol. Chem. **268**, 3209–3215.
48. Hama, H. and Wilson, T.H. (19940 J. Biol. Chem. **268**, 10060–10065.
49. Eiglmeier, K., Boos, W. and Cole, S.T. (1987) Molec. Microbiol. **1**, 251–258.
50. Goldrick, D., Yu, G.Q., Hong, J.-S.Q. (1988) J. Bacteriol. **170**, 3421–3426.
51. Nilsson, R.P., Beijer, L., Rutberg, G. (1994) Microbiology **140**, 723–730.
52. Kadner, R.J., Island, M.D., Dahl, J.L. and Webber, C.A. (1994) Res. Microbiol. **145**, 381–387.
53. Maloney, P.C., Ambudkar, S.V. and Sonna, L.A. (1987) in: Phosphate Metabolism and Cellular Regulation in Microorganisms, eds A. Torriani-Gorini, F.G. Rothman, S. Silver, A. Wright and E. Yagil. American Society for Microbiology, Washington, D.C.

CHAPTER 13

Structure of the Erythrocyte Band 3
Anion Exchanger

R.A.F. REITHMEIER, S.L. CHAN and M. POPOV

*MRC Group in Membrane Biology, Departments of Medicine and Biochemistry,
University of Toronto, Toronto, Canada*

© *1996 Elsevier Science B.V.*
All rights reserved

*Handbook of Biological Physics
Volume 2, edited by W.N. Konings, H.R. Kaback and J.S. Lolkema*

Contents

1. Introduction

Band 3 is the protein responsible for the rapid exchange of bicarbonate and chloride across the red cell membrane. This process allows the transport of bicarbonate, formed within the red cell by carbonic anhydrase, into the plasma, thereby increasing the CO_2 carrying capacity of the blood. Anion transport is inhibited by a variety of organic anions, most notably by stilbene disulfonates like DIDS. Human Band 3 is present in 1.2×10^6 copies per red cell and consists of a 911 amino acid polypeptide with a single site of N-glycosylation at Asn642. The protein can be divided into two structurally and functionally distinct domains. The amino-terminal cytosolic domain provides the site of attachment of the cytoskeleton and other cytosolic proteins. The carboxyl-terminal membrane domain spans the membrane up to 14 times (Fig. 1) and carries out the anion transport function. Band 3 is a member of a multi-gene family of anion exchangers (AE) and serves as a valuable model for the structure and function of these related proteins.

In this review, the focus is placed on recent structural studies of Band 3 (AE1). Additional information on Band 3 and the AE gene family can be found in other reviews [1–11].

2. Cl⁻/HCO₃⁻ exchangers

2.1. Anion exchanger (AE) gene family

Electroneutral Cl⁻/HCO₃⁻ exchange pathways are present in a wide range of vertebrate tissues. This function is being assigned to a growing number of homologous AE proteins (Table 1), which are encoded by at least three different genes: AE1, AE2, and AE3 [7,9,10,12,13]. The human AE1 gene has been mapped to chromosome 17q21-qter [14], the AE2 gene to chromosome 7q35-7q36 [15] and the AE3 to chromosome 2q36 [16]. The existence of a novel member (AE0) of the AE gene family, localized to human chromosome 22, has recently been reported [17].

AE1 sequences are available for mouse [18], human [19,20], chicken [21–23], and trout [24] erythroid forms, a rat kidney [25] form and three chicken kidney forms [26]. AE2 sequences for human [27,28], mouse [29], rat [30,31] and rabbit [32] are known. Neuronal AE3 for mouse [33] and rat [30] and a shorter rat cardiac AE3 isoform have been identified. The erythroid AE1 proteins range in size up to 929 amino acids, while AE2 and AE3 proteins are larger due to an amino-terminal extension and an insertion (the Z-loop) between TM segments 5 and 6 (Fig. 1).

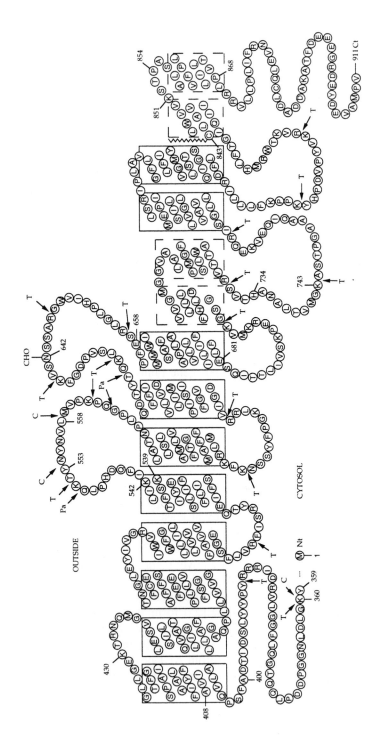

Fig. 1.   Folding model for the membrane domain of human Band 3. Nt, amino-terminus; Ct, carboxy-terminus. T, C, Pa refer to trypsin, chymotrypsin and papain cleavage sites. The topology of only a few of these sites has been established (Table 5). N642 contains the single site of N-glycosylation on Band 3 and C843 is fatty-acylated. A400 to A408 is deleted in SAO Band 3, K430 is the site of reaction of eosin maleimide, K539 and K542 react with H2DIDS and DIDS respectively. E658 is part of the Wr[a] antigen, E681 reacts with Woodward's reagent K, K851 reacts with pyridoxal phosphate and H2DIDS, P854 is part of the Diego antigen, P868 is mutated to L in Band 3 HT. See text for details.

Table 1
Properties of anion exchanger (AE) gene products

Gene	Species	Length	N-glycosylation sites
AE1	Human	911	N642
	Mouse	929	N660
	Chicken	922	N653
	Trout	918	N546, 568
AE2	Human	1240	N858, 867, 881
	Mouse	1237	N855, 866, 878
	Rat	1234	N856, 866, 878
	Rabbit	1237	N855, 864, 878
AE3	Mouse (brain)	1227	N868
	Rat (brain)	1227	N868
	Rat (heart)	1030	N671

2.2. Domain structure and homology

Anion exchangers can be divided into three domains [34] based on sequence similarity: an amino-terminal extension, a cytosolic core domain and a membrane domain (Fig. 2). The amino-terminal extensions of the cytosolic domain in the various gene products vary greatly in length and sequence. The human Band 3 sequence Val59-Glu-Leu-X-Glu-Leu is highly conserved in the AE1, 2 and 3 and defines the beginning of a common cytosolic core. The amino-terminal extensions of AE2 and AE3 are homologous and are much longer (300 residues) than the erythroid extension. In contrast kidney Band 3 (AE1b) is missing the entire amino-terminal extension (see below). The membrane domain is proposed to begin at the perfectly conserved Phe 379 and finish at the carboxyl-terminal Val911 of the human sequence.

Another major difference in the primary sequence of the AE gene family is the presence of an insertion (Z-loop) in the extracellular loop connecting TM segments 5 and 6 (Fig. 1). This insert is present in trout AE1, AE2 and AE3 and contains the sites of N-glycosylation in these proteins. The inserts vary in size (15–24 residues) and they are not related in sequence between the different genes. The published sequence of AE0 [17] covers only the transmembrane domain and it is identical in sequence to AE1 except for an insertion in exon 14 [35], encoding 25 amino acids. The insertion is identical to the third extracellular loop (Z-loop) of the AE2 protein.

The different mouse gene products, AE1, AE2 and AE3, share an overall 35% sequence identity, and AE2 and AE3 are the most similar isoforms (Table 2). The membrane domain is more highly conserved with a 51% identity between the three mouse gene products. This indicates that the anion exchange function has remained their primary function throughout evolution and that extensive protein–protein

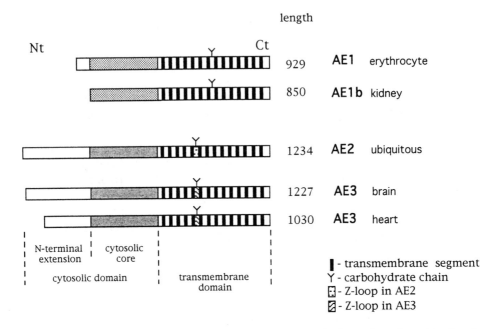

Fig. 2.   Domain organization of rat anion exchanger (AE) gene family products. Length is given in
number of amino acids.

Table 2
Sequence identity for mouse AE proteins

Entire protein	AE1	AE2	AE3
AE1	100	44	42
AE2		100	56
AE3			100
Membrane domain	AE1md	AE2md	AE3md
AE1md	100	67	58
AE2md		100	68
AE3md			100
Cytosolic domain	AE1cd	AE2cd	AE3cd
AE1cd	100	36	39
AE2cd		100	58
AE3cd			100

An alignment of the AE protein sequences was performed using Gene Works™ multiple sequence
alignment software. Identities is the number of perfect matches that occur in the alignment. The identity
value is the percentage of the consensus sequence that the number represents.

interactions occur within this domain. In contrast, the cytosolic domain is less well conserved. The entire AE1 protein is 38% identical between human, mouse, chicken and trout.

2.3. Alternative promoters and transcripts

The diversity of the AE proteins is increased by the use of multiple promoters. The resulting transcripts give rise to proteins which vary in their N-terminal sequence. AE1 is a single copy gene in mouse [36], rat [25], human [14] and chicken [22]. The mouse and human AE1 genes (Fig. 3) are ~20 kb long and consist of 20 exons [35,37,38]. The exon boundaries fall within regions encoding hydrophilic loops connecting TM segments. The AE1 gene contains alternative promoters allowing tissue-specific expression of AE1 isoforms [36]. The erythroid specific promoter region (Fig. 3) is located immediately upstream from exon 1 and is conserved in human, mouse and rat AE1 genes. Exon 2 contains some 5' untranslated region and the first five codons. The first exon (K1) of kidney AE1 transcript is located within intron 3 (Fig. 3). The resulting kidney AE1b mRNA lacks exons 1–3 and uses an initiator codon within exon 4. In mouse and rat kidney AE1b begins at Met80 and human kidney AE1b begins at Met66 [39]. A minor mouse kidney transcript initiates upstream of the erythroid transcription initiation site and encodes a translation product that is identical to erythroid Band 3 [36,40]. The rat AE1 gene also yields two isoforms due to alternative splicing [41]. The two chicken erythroid Band 3 polypeptides differ in length by 33 amino acids at the amino terminus, otherwise they are identical [22,23,26]. In chicken kidney three N-terminal truncated versions are detected [26]. Two chicken kidney transcripts (AE1-3, 5) encode the same truncated protein missing the first 78 amino acids, while AE1-4 also encodes a truncated protein with a novel 21 residue amino-terminal sequence [26].

   Human AE2 cDNA, cloned from kidney mRNA [28] encodes a 1240 amino acid protein. AE2 cDNA has also been cloned from human K562 cells [27], a lymphoid cell line [29], mouse kidney [29], rat choroid plexus [31], rat stomach and brain [30] and from rabbit ileum [32]. In rat tissues a 4.4 kb AE2 transcript was detected in all tissues examined, with stomach containing the highest level [30]. An additional 3.9 kb mRNA was present only in stomach [30]. An unusual, C-terminal truncated isoform of AE2–AE2a, was isolated from guinea pig organ of Corti cDNA library [42]. Due to alternative splicing, 83 bp are deleted which shifted the open reading frame, resulting in a shorter protein with only two TM segments.

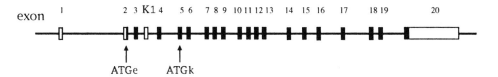

Fig. 3.   Map of the human AE1 gene [35]. The positions of exons 1–20 are indicated, with the protein coding regions shown in solid boxes and the 5' and 3' non-coding regions shown as open boxes. The erythroid (ATGe) and kidney (ATGk) translation start sites and the kidney exon K1 are indicated.

The rat AE3 gene contains 24 exons, and gives rise to at least two mRNA transcripts (4.1 and 3.4 kb in mouse, 4.4 and 3.8 kb in rat): a longer-brain variant and a shorter isoform found in the heart [30,33,43]. Transcription is initiated within intron 6 for the heart isoform. The first 844 nucleotides of the brain transcript are replaced by a different 367 nucleotides in the brain isoform. The first six exons encode the 5' untranslated region and the first 270 codons of the brain isoform. Parts of the human AE3 gene have been sequenced [16,44] and [43], an alternative exon is found in the position corresponding to rat intron 6.

The high degree of sequence variation that occurs at the amino-termini of the AE gene products suggests that this region is involved in specialized functions. These functions may include interaction with cytosolic proteins, subcellular localization or regulation of transport activity. In contrast, the function of the Z-loop is to bear the oligosaccharide chain, thus its particular sequence may not be that important. The highly conserved nature of the membrane domain indicates that all members of this gene family are anion exchangers that operate by a similar if not identical mechanism.

2.4. Mutant Band 3 proteins and diseases

Naturally occurring mutations in human Band 3 are listed in Table 3. Band 3 Memphis I is characterized as an ancient polymorphism. This variant of AE1 has decreased mobility on SDS-PAGE equivalent to 3 kDa and it has been shown to contain a single point mutation Lys56Glu [45,46]. Band 3 Memphis I is relatively common in most populations tested and is particularly prevalent in Japanese (29%), American Indians (25%) and African Americans (15%).

Band 3 Memphis II was distinguished from the Memphis I variant as a sub-population more susceptible to covalent labeling with H_2DIDS. Memphis II variant carries the polymorphism Pro854Leu, as well as Lys56Glu [47]. The Band 3 Memphis II is also associated with the rare Diego (Di^a) blood group antigen. In current models (Fig. 1) Pro854 is located at the extracellular surface in the loop connecting TM segments 13 and 14.

Mutations Glu40Lys (Band 3^{MONTEFIORE}) [48] and Pro327Arg (Band 3^{TUSCALOOSA}) [49] in the cytosolic domain were found in association with atypical spherocytic hemolytic anemia. Both mutants were connected with a deficiency in red blood cell protein 4.2 (88%) and, in Band 3^{MONTEFIORE}, with a spectrin deficiency [50].

The molecular basis for Southeast Asian ovalocytosis (SAO) is deletion of nine amino acids (ΔAla400–Ala408) at the boundary of cytoplasmic and membrane domains of Band 3 [51–53]. SAO erythrocytes have an altered shape and are relatively resistant to invasion by the malarial parasite Plasmodium. The SAO mutation alters the membrane domain of Band 3, resulting in loss of anion transport activity [54,55]. A homozygous SAO mutation is likely lethal due to a lack of anion transport activity in erythrocytes and in the kidney [56]. SAO Band 3 makes up about half of the Band 3 in the mature red cell in heterozygotes. This suggests that during erythroid differentiation the biosynthesis and targeting of SAO Band 3 to the plasma membrane is normal. Band 3 from SAO erythrocytes appeared to have

Table 3
Natural Band 3 mutations

Residue(s)	Name	Description
Lys56Glu	Band 3 Memphis I*	ancient asymptomatic polymorphism
Pro854Leu	Band 3 Memphis II*	more readily labeled with H2DIDS
Glu40Lys	Band 3 Montefiore	atypical spherocytic hemolytic anemia
Pro327Arg	Band 3 Tuscaloosa*	atypical spherocytic hemolytic anemia
ΔAla400→Ala408	SAO Band 3*	malaria resistance, inactive
Glu658Lys	Wright antigen Wr^a	no Band 3 abnormalities
Pro868Leu	Band 3 HT	hereditary acantocytosis, increased anion transport

Band 3 deficiency in hereditary spherocytosis

821aa + 10nt = STOP codon	Band 3 Prague I	absent from erythrocyte plasma membrane
Arg760Gln	Band 3 Prague II*	absent from erythrocyte plasma membrane
Arg760Trp	Band 3 Hradec Kralove	
Arg808Cys	Band 3 Jablonec	
Arg870Trp	Band 3 Prague III	

* Also carries Memphis I mutation.

normal secondary structure as determined by circular dichroism, but with impaired inhibitor binding, a high proportion of tetramers/dimers and no polylactosaminyl type oligosaccharide [54,55,57]. The rotational and lateral mobilities of Band 3 are reduced in SAO red cells, likely due to the higher oligomeric state of Band 3 and an enhanced association of Band 3 with the cytoskeleton [58,59]. SAO Band 3 can form a heterodimer with normal Band 3 [60], which may affect the functioning of the normal subunit. While the transmembrane domain of SAO Band 3 is dramatically changed, the cytosolic domain appears to be normal [55]. Interestingly, although the membrane domain has a normal α-helical content, it does not undergo a usual calorimetric transition, suggesting a denatured protein (55). SAO Band 3 may represent the first example of a 'molten globule' partially folded state in a membrane protein.

The Wright (Wr) blood group collection consists of two antigens, Wr^a and Wr^b. The most common Wr^b has Glu at position 658 in Band 3, while the low-incidence Wr^a carries the mutation Glu658Lys, located in the fourth extracellular loop (Fig. 1), just preceding TM8 [61]. Glu658 may interact with Arg61 of glycophorin A to form the Wr^a antigen [61].

The Band 3^{HT} variant carries the mutation Pro868Leu, located near the carboxyl-terminal end of TM 14 [62]. The mutation is associated with abnormal red cell shape (acanthocytosis), increased anion transport, increased electrophoretic mobility and decreased covalent labeling by H_2DIDS.

Autosomal dominant hereditary spherocytosis can be associated with a deficiency in Band 3. In Band 3 Prague I [63] there is a duplication of 10 bp resulting in a truncated protein in which the carboxyl-terminal 90 amino acids are replaced by a novel sequence of 70 amino acids. This mutant protein is missing the last two transmembrane segments normally found in Band 3. Four different missense hereditary spherocytosis mutations were detected involving substitutions for arginine residues (Arg760Trp, Arg760Gln, Arg808Cys and Arg870Trp) located at the cytoplasmic boundaries of the putative TM helices [50]. These arginines are highly conserved in AE1 from other species, and in AE2 and AE3 proteins. In Prague I and II, the Band 3 proteins corresponding to these mutants are not found in mature red cells. The protein is likely degraded in the endoplasmic reticulum or lost from the plasma membrane during red cell development. A similar situation would account for the decreased level of Band 3 protein in other hereditary spherocytosis mutants.

3. Topology of Band 3

3.1. Hydropathy analysis

Figure 4 shows the analyses of the membrane domain of human AE1 using three different scales designed to detect potential transmembrane segments. The Kyte–Doolittle scale [64] and the MIR scale [65] used a window of 21 amino acids, since this number of residues can span a lipid bilayer as an α-helix. The Persson and Argos scale [66] used a core of 15 residues and two 4-residue flanking regions. All three curves are normalized to the same scale so they may be compared directly. There are 10 peaks in all three curves suggesting that the protein spans the membrane 10 times. However, some of these peaks are broad enough to span a lipid bilayer twice. The anion exchanger proteins were analyzed for TM segments using a neural network method [67]. This procedure predicted 12 TM segments, breaking each of the last two hydrophobic sequences into two TM segments. This splitting of the last two peaks is also visible in the Persson–Argos curve. Moreover, if one uses a smaller window size (e.g., 11 residues) with the Kyte–Doolittle or MIR scales these peaks will also be divided into two, albeit weakly. Since the amino-terminus of Band 3 is cytosolic and the loop between hydrophobic peaks 4 and 5 is known to contain extracellular protease-sensitive sites (Fig. 1), there needs to be an odd number of TM segments corresponding to the first four hydrophobic peaks. Since the second peak is the widest, it likely represents two TM segments. Hydrophobicity analysis of the membrane domains of AE2 and AE3 give a virtually identical profile to AE1 indicating a very similar folding pattern for these homologous proteins.

3.2. Reverse turn and variability analyses

The amino acid sequence of human Band 3 was analyzed for sequences with a high propensity to form reverse turns [68]. In soluble proteins reverse turns are found on the surface of the protein exposed to water and a similar situation is predicted for

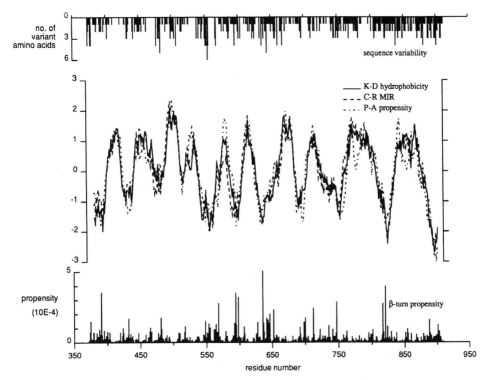

Fig. 4. Sequence analysis of the membrane domain of human Band 3. The top graph gives the sequence variability. Comparison was made between human Band 3 and 9 other AE protein sequences. The length of the bar corresponds to the number of variant amino acids at each position. The middle graph gives the Kyte–Doolittle [64] hydrophobicity plot (solid line), the membrane inclusion ratio [65] plot (dashed line) and the Persson–Argos membrane propensity [66] plot (dash-dotted line). For each scale normalization was performed so that the average value in the region is zero and the vertical axis is in standard deviations from the norm. The lower graph gives the reverse turn propensity of four-residues segments using the Chou–Fasman scale [68].

Band 3. Amino acid sequences that commonly form 4-residue reverse turns are listed in Table 4 and are indicated in Fig. 4. These segments are overwhelmingly located in the hydrophilic loops that connect TM segments. The N-glycosylated loop connecting TM 7 and 8 is particularly rich in reverse turns. Searches of reverse turn propensities in other AE protein sequences show that the turns found in human Band 3 are conserved. No high propensity score was apparent to divide TM 2 and 3 although a cluster of moderate scores was located around Phe464. Similarly, neither of the last two broad hydrophobic peaks can be readily divided into two by a high propensity of reverse turn sequences.

Figure 4 also gives a variability plot for the membrane domain of the AE gene family. There is a general trend that the variability increases as one proceeds towards the carboxyl-terminus. This suggests that the amino-terminal portion of the membrane domain forms a core structure containing extensive and specific protein–

Table 4
Reverse turn propensities in the membrane domain of human Band 3

Sequence	Propensity[1]	Location
G367PDD	4.45	Cytosolic domain
D370PLQ	1.56	
Y390PYY	3.52	
T431RNQ	1.71	TM1/2
E480TNG	1.76	TM3/4
H547PLQ	1.49	TM5/6
L567PNT	2.77	
N593SSY	3.50	TM6/7
F597PGK	3.21	
V634PDG	5.08	TM7/8
V640SNS	1.74	
N642SSA	1.62	
A645RGW	2.02	
H651PLG	2.31	
G699SGF	1.75	TM8/9
G711MGG	2.42	TM9/10
T746PGA	2.85	TM10/11
P815PKY	2.76	TM12/13
H819PDV	4.00	
D887ADD	1.62	Cytosolic C-terminal

[1] The propensity ($\times 10^{-4}$) for forming reverse turns were calculated for each segment of four consecutive amino acids along the Band 3 sequence. The relative propensity of forming a reverse turn for each four residue sequence is the product of the probabilities of finding residue X in the ith position of the reverse turn.

protein interactions. It can also be seen that the variability among the sequences is greatest in the hydrophilic loops connecting the TM segments. Positions where charged residues occur in the hydrophobic regions are conserved. In contrast, hydrophilic loops in AE proteins often contain charged residues that are poorly conserved (Table 5). It is proposed that these residues face the aqueous media. An example of a very poorly conserved hydrophilic region is the Z-loop insertion.

Table 5
Location in the membrane domain of human Band 3 of non-conserved charged residues in AE family

Location	Residues	Location	Residues
Cytosolic domain	S402D,N,H	TM10/11	G742S,R
TM1/2	R432E,K,Q		K743T
	N433H,G,D		A749P,E,D
TM3/4	S477A,R,T,K		A750K,R
	E480K,S,R		Q752E,H,K
	T481S,D,A		Q754E,V
	G483E,N,Q,H,D		E755K
TM4/5	R514K,Q		K757L,R
	R518P	TM11/12	E777G
	Q521R		P778D,A
TM5/6	K542T		S781K,R
	Q545K,E,T		R782M,Y,Q
	D546A,E	TM12/13	D807E,Q
Z-LOOP			K814V,M
	L567Y,V,K,Q		D821A,K,E
TM6/7	K590Q		V822D,E,Q
	S595H,T,V,R		K826Q,T
	K600P,R		K829T,R
TM7/8	Q625E,K,T		H834T,D
	D626G	TM13/14	K851M
	S633V,K,T		A858R
	D636K,R,S,T	Cytosolic C-terminal	R870S
	K639M,E,S		R871M,G
	S644N,T,D,H,E		L876T,R
	A645K		R879T,S,Q
	R656K,Y,A		N880D,E
	S657K,R,T		V881K,L,I,R
	E658P,L		Q884K
TM8/9	K691Q		
	P692K		
	E693A		
	K695M,R		

3.3. N-glycosylation

Human Band 3 contains a single site of N-glycosylation [69,70] at Asn642 [19,20] located on an extracellular loop between TM segments 7 and 8 (Fig. 1). Enzymatic removal of the oligosaccharide chain does not affect the protein structure and does not impair anion transport [71]. Mutagenesis of the consensus glycosylation site resulted in some reduction in cell-surface expression in *Xenopus* oocytes but the non-glycosylated protein was still functional [72]. Trout AE1 does not contain a consensus N-glycosylation site in the loop between TM 7 and 8 but rather contains a site in an insertion in the preceding Z-loop. Interestingly, this is similar to the N-glycosylation pattern in AE2 and AE3. A survey of utilized N-glycosylation sites in multi-span membrane proteins has revealed that N-glycosylation sites are localized to single loops greater than 30 residues in size [73].

Band 3 contains equal amount of two types of complex N-linked oligosaccharide, a short chain and a long chain containing a polylactosaminyl (Gal(b1→4) GlcNAc) structure [74,75]. HEMPAS is an inherited disease that results in the lack of the polylactosaminyl oligosaccharide on Band 3 [76]. Instead, a short high mannose type of oligosaccharide is present on Band 3. This change in sugar structure does not affect anion transport (J.H.M. Charuk, unpublished results).

3.4. Experimental constraints

The high abundance of Band 3 in the erythrocyte membrane and the ability to produce resealed ghosts, and inside-out vesicles have led to extensive studies of Band 3 topology. Table 6 summarizes the sites within the human Band 3 sequence that have been localized to one side of the membrane or the other. The results obtained with small non-penetrating reagents must be interpreted with caution. If Band 3 contains a channel, a reagent applied to the outside of the cell may label a site deep within a channel that is close to the cytosolic side of the membrane. Similarly, vigorous protease treatments alter the structure of the protein and can expose cryptic sites that are normally not accessible in the native protein.

The single site of N-glycosylation at Asn-642 defines a domain of the protein that is exposed to the extracytosolic side of the membrane. Cysteine 843 is palmitoylated [77], a modification that localizes this residue to the cytosolic side of the membrane with the fatty acid intercalated into the inner leaflet of the bilayer (Fig. 1).

Table 6
Topological markers in Band 3

Site	Location	Evidence
Lys 360	Cytosolic	Trypsin cleavage site
Lys430	Extracytosolic	Chemical methylation
Lys539/Lys542	Extracytosolic	(H2)DIDS labelled sites
Tyr553, Leu558	Extracytosolic	Chymotrypsin cleavage sites
Glu630	Extracytosolic	Papain cleavage site
Asn642	Extracytosolic	N-glycosylation site
Gln658	Extracytosolic	Wright antigen site
Lys743	Cytosolic	Trypsin cleavage site
Phe813-Tyr824	Cytosolic	Antibody binding
Cys843	Cytosolic	Palmitoylation site
Pro854	Extracytosolic	Diego antigen site
Cys885	Cytosolic	Maleimide labeling
Carboxyl-terminus	Cytosolic	Antibody binding

Proteolytic cleavage sites have also been localized within the Band 3 sequence (Table 6). There is a region located on the cytosolic side of the membrane that is very sensitive to cleavage by trypsin (Lys360) and chymotrypsin (Tyr359) [78]. An additional cytosolic trypsin-sensitive site has been localized to Lys743 [11,79]. External chymotrypsin cleaves at two sites in the loop between TM 5 and 6 [11]. Extensive treatment of ghost membranes with proteases combined with sequencing has identified accessible cleavage sites (Fig. 1) in Band 3 [80,81].

Both monoclonal [82] and polyclonal antibody [83] binding studies have localized the carboxyl-terminus of Band 3 to the cytosolic side of the erythrocyte membrane. Since the amino-terminus of Band 3 is cytosolic, the polypeptide chain spans the membrane an even number of times. A monoclonal antibody was used to localize an epitope encompassing residues Phe813-Tyr824 to the cytosolic side of the membrane [84]. Monoclonal antibodies also react with the external loop connecting TM 5 and 6 [85] in intact cells. The Diego antigen is due to a point mutation (Pro854Leu) located in the external loop connecting TM 13 and 14.

H_2DIDS when applied from the outside of cells reacts with Lys539 and Lys851 [81]. These residues are therefore accessible from the outside of the cell. Most models (Fig. 1) position Lys539 near the outer end of TM5 while the position of Lys851 with respect to the membrane is more difficult to define. External eosin maleimide reacts with Lys430 which is positioned in the extracellular loop connecting TM 1 and 2 (Fig. 1). Lys430 is also the site of reductive methylation from the outside of the cell [86], confirming the extracellular location of this residue (Fig. 1).

3.5. Models of Band 3 TM segments

Circular dichroism studies have shown that the portions of Band 3 that are embedded in the membrane and thereby protected from proteolysis are α-helical in conformation (87,88). NMR studies of a synthetic peptide corresponding to the first TM segment of Band 3 in organic solvents have confirmed the helical propensity of this sequence [89]. Each TM helix may contain three types of surfaces: a surface that faces the lipid bilayer, a portion that faces a channel or pore and surfaces involved in helix–helix interactions. The high degree of sequence conservation within the membrane domain suggests that helix–helix interactions dominate the features of the TM segments. In fact, proteolytic cleavage of the connecting loops did not result in dissociation of the TM helices [88].

Computer models of each of the TM segments in Band 3 were constructed as α-helices and energy minimized using CHARMm 22. Figure 5 shows each of the TM segments in Band 3 viewed from the side and from the cytosolic end. Of the 14 putative TM segments, TM segments 1, 3 and 5 are highly conserved with variation limited to, at most, 3 residue positions. These segments may not have an extensive interaction with the lipid and may form the core of the protein structure. TM segments 2, 6 and 8 are quite amphipathic and the hydrophobic face is where the variation is the highest. Hence, it is likely that the hydrophobic faces of these TM segments face the lipid.

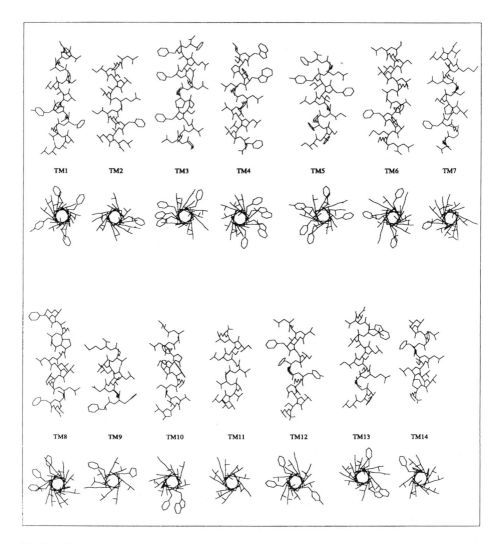

Fig. 5.   Computer-generated models of the 14 putative TM segments of human Band 3. The upper set of helices are displayed in the same orientation as the folding model in Fig. 2 with the cytosolic side of the membrane on the bottom. The lower set of helices are end on, viewed from the cytosolic side of the membrane. The left side of each helix is the most hydrophobic face.

Charged residues found in the middle of TM segments are proposed to face the channel. Interestingly these are mostly anionic in nature [34]. For example, TM2 contains a conserved glutamate at position 439 and TM3 contains a pair of conserved glutamates (Glu472, 473). TM 5 also contains a conserved glutamate (Glu535) located on the same side of the helix as the DIDS-reactive lysines (Lys539, 542). Two conserved aspartates (Asp607, 621) are within TM7. Glu681 in TM8 is also conserved and places a direct role in transport [11]. A negative lining to the channel may prevent substrate anions from interacting with the walls of the channel during translocation. Positive charges tend to cluster near the surface of the membrane, perhaps acting to concentrate anions near the mouth of the channel.

Aromatic residues like tryptophan and tyrosine found near the ends of TM segments may interact with the polar head-groups of the phospholipids. Some TM segments (e.g., TM3, TM8) display phenylalanine residues to one face of the helix (Fig. 5). The high degree of hydrophobicity in most TM segments suggests that they may be able to exist as individual TM segments during biosynthesis [90]. Subsequent helix–helix interactions bring the TM segments together to form the final folded protein. Pairs of helices can be modeled, however further experimental constraints are required before helix–helix interactions can be established with any certainty.

3.6. Biosynthesis and membrane insertion

The Band 3 polypeptide is synthesized on ribosomes bound to the cytosolic side of the ER membrane. The first TM segment likely acts as a signal sequence which binds to the signal recognition particle (SRP) and directs the nascent chain to the translocon in the endoplasmic reticulum [91,92]. The large cytosolic domain retains the amino-terminus of Band 3 to the cytosolic side of the endoplasmic reticulum membrane. At some point TM segments move from the translocon into the lipid bilayer. How this occurs is not known, however the interface region of the lipid bilayer may play a role in helix formation and insertion. The small size of many of the lumenal loops on Band 3 (Fig. 1) suggest that most transmembrane segments (e.g., TM 1 and 2) may insert into the membrane as an anti-parallel helical pair.

Band 3 contains multiple signals for insertion into the endoplasmic reticulum membrane [93]. Truncated Band 3 molecules can insert into the ER membrane in an SRP-dependent fashion. For example, molecules encompassing TM 1–7 or TM 7–14 can insert into the ER membrane. A polypeptide containing TM7 as the only TM segment is able to target to the endoplasmic reticulum membrane and to insert in the proper orientation. Whether multiple TM segments other than TM 1 interact with the SRP machinery in the context of the intact protein is still not known.

Human Band 3 mutants with truncated membrane domains were prepared to provide complementary fragments [94]. One set of mutants contained the first 8 and the last 6 transmembrane segments on separate cRNAs, and another the first 12 and the last 2. When these complementary fragments were expressed together in oocytes, DNDS inhibitable chloride transport reached about 50% of the value for intact Band 3. The fragments can reassemble into a functional transporter and the loops connecting TM8/9 or TM 12/13 need not be intact for functional expression.

4. Structure and function of Band 3

4.1. Cytosolic domain

The amino-terminal cytosolic domain of Band 3 provides the membrane attachment site for a variety of cytosolic proteins including hemoglobin, ankyrin, Band 4.1,

Band 4.2 and glycolytic enzymes [95]. The extreme amino terminal end of Band 3 is responsible for binding to hemoglobin and the structure of a complex between hemoglobin and a peptide corresponding to the amino-terminus of Band 3 has been determined by X-ray diffraction [96]. Binding of denatured hemoglobin to Band 3 causes clustering of Band 3 in the membrane and the creation of binding sites for circulating senescence antibodies [97]. Glycolytic enzymes are inactivated when bound to the acidic amino terminal region of Band 3 and phosphorylation of Tyr8 prevents binding of the enzymes [98].

Kidney Band 3 does not bind ankyrin [99], band 4.1 or aldolase [100]. Ankyrin [101] and glyceraldehyde-3-phosphate dehydrogenase [102] co-localize with kidney Band 3 to the basolateral membrane of intercalated cells. This interaction can not be mediated by the amino-terminal extension as in the erythrocyte. The cytosolic domain of Band 3 has a hinge region located around residues 175–200 that is involved in ankyrin binding. Oxidation or modification of Cys201 and Cys317 inhibit ankyrin binding [103]. However, mutagenesis has clearly shown that the cysteine residues themselves are not required for ankyrin binding [104]. While AE1 and AE3 bind ankyrin, AE2 does not [105]. Interestingly ankyrin co-localizes with the apical Na^+/K^+ ATPase rather than basolateral AE2 in the choroid plexus [106].

The cytosolic domain has been expressed in *E. coli* and this protein displays the features of the native protein [107]. A major breakthrough occurred with the crystallization of the expressed protein [108]. The crystals diffract to under 3 Å and the structure determination is nearing completion (P. Low, personal communication).

4.2. Membrane domain

The cytosolic and membrane domains can be physically separated by chymotrypsin cleavage at Tyr372 in mouse Band 3 and at Tyr359 for chymotrypsin and Lys 360 for trypsin in human Band 3 (Fig. 1). The protease-sensitive sites are located in a very poorly conserved region of the protein that precedes the beginning of the membrane domain. This region is probably a flexible region that links the cytosolic and membrane domains [95,109]. Secondary structure predictions of this region (Table 4) point to a high reverse turn probability consistent with the abundance of proline, glycine, aspartate and serine residues [110]. The membrane domain produced by proteolytic digestion can carry out the anion transport function of Band 3 [111].

The ability to reconstitute purified Band 3 into lipid vesicles with the retention of transport function suggested the use of 2-D arrays for structural studies. Band 3 [112] and its membrane domain [113,114] have been crystallized. The structure of the deglycosylated membrane domain has been determined by electron crystallography. The membrane domain crystallized as a dimer and two different crystal forms were produced: sheets and tubes. The negatively-stained material was studied using optical diffraction to a resolution of 20 Å.

The 2- and 3-D maps are shown in Fig. 6. The 2-D projection map of the sheets and tubes revealed that they differed in the position of one domain. The 3-D map was produced from a tilt series of electron micrographs of the crystalline sheets [114]. Two domains can be distinguished: a lower membrane embedded portion and

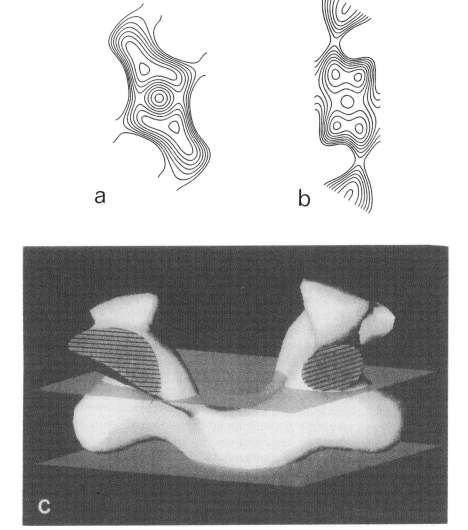

Fig. 6.   Structure of the membrane domain of the Band 3 protein determined from electron microscopy and image reconstruction of two-dimensional crystals. Upper part of the figure shows the projection maps derived from the hexagonal sheets (a) and tubes (b). The lower figure is a three-dimensional map determined from the sheets. The two planes represent the two sides of a lipid bilayer 40 Å thick.

an upper mobile domain. Based on the topology of Band 3 we proposed that the upper domain represent loops of the protein that face the cytosol, but this has yet to be proven. The dimer is 110 Å long, 60 Å wide and 80 Å thick. The upper domain is flexible and is located in different positions in the two crystal forms. The upper domain also provides the contacts that hold the molecules together and there do not appear to be any contacts between dimers in the membrane embedded portion. The upper part of the protein forms two sides of a deep canyon that leads to a depression at the center of the dimer on the cytosolic side of the membrane. At 20 Å resolution using negatively stained material the features of a channel are difficult to visualize. The lower part of the structure contains a central depression and a depression within each monomer. It is proposed that the depressions represent the stilbene disulfonate inhibitor-binding sites or entrances to the sites. The cross-sectional area through the middle of the bilayer is consistent with 14 TM segments per monomer. The presence of a large aqueous channel would reduce this number.

The mobile upper domain may provide a flexible link to the cytosolic domain. The movement of this domain may be involved in anion transport. Movement of a domain embedded in the lipid bilayer would be slower than movement in aqueous media due to the higher viscosity of lipid. If this domain is connected by a single helix to the gate within the anion pore, a conformational change allowing anion passage is possible. Further insight will require a higher resolution structure obtained by cryo-electron microscopy.

4.3. Role of oligomers

An important feature of anion exchangers is their oligomeric structure. Intact Band 3 is mainly dimeric but can form tetramers [1]. The isolated cytosolic domain is dimeric [95] as is the membrane domain [115]. Perhaps Band 3 is dimeric because the anion passage may be located, at least in part, at the interface between the two subunits. Functional reconstitution of Band 3 monomers has however been reported suggesting that a single Band 3 molecule contains a TM passage [116]. Tetramers of Band 3 provide the binding site for ankyrin [117,118]. Allosteric regulation is a feature of oligomeric proteins, including Band 3 [8]. AE1 can operate by a ping-pong mechanism in which the two subunits are uncoupled, allowing one subunit to translocate anions without the other translocating in the opposite direction. Anion exchangers may also operate by a simultaneous mechanism, whereby two anions must be bound to Band 3 on opposite sides of the membrane in order for the translocation of anions to take place. It would be of considerable interest to study the nature of the interactions between subunits in AE2 or AE3 since these proteins are likely under fine regulatory control.

4.4. Inhibitors and chemical modification

The interaction of inhibitors with Band 3 is turning out to be more complex than earlier imagined. The anionic nature of most inhibitors suggested a competitive interaction. Eosin maleimide which reacts with Lys430 [119] inhibits anion transport but does not block access to the external anion binding site [120–123]. Eosin maleimide therefore prevents the anion translocation step, locking the protein into the outward facing conformation. Kinetic evidence for a ternary complex of inhibitors and substrates on Band 3 and allosteric interactions between the inhibitor and substrate binding sites have been reported [124,125].

DIDS or H_2DIDS bind to a cleft in Band 3 that faces the cell exterior but their mode of action is not identical. Mutagenesis experiments suggest that Lys 541 is likely the DIDS reactive site [126], although this has yet to be proven. H_2DIDS crosslinks Lys539 and Lys851 [81]. H_2DIDS crosslinking occurs readily at pH 9.5 [127] while DIDS requires a more alkaline pH (pH = 13). [125,128]. Once bound to Band 3 the inhibitor is not accessible to an anti-H_2DIDS antibody [88]. The binding of inhibitors stabilize Band 3 against denaturation [87,129]. DIDS-labeled Band 3 is more resistant to proteolysis than unlabeled protein and many of the protected peptides are located in cytosolic loops [128]. Proteolysis of H_2DIDS-labeled Band 3 leaves the inhibitor associated with the fragments embedded in the membrane except under harsh denaturing conditions [81,88]. Inhibitor binding can even cause changes in the nature of the oligomeric population of Band 3 [130–132].

Pyridoxal phosphate is transported by Band 3 and can inhibit anion transport by reacting covalently with Band 3. Site-directed mutagenesis studies have indicated that pyridoxal phosphate can react with either Lys558 or Lys869 in mouse Band 3 [133]. This suggests that these two residues are located on the transport pathway. The site of the pyridoxal phosphate reaction in human Band 3 has been localized to Lys851, the equivalent of Lys 869 in mouse [134]. Interestingly, this is the same residue that is the site of H_2DIDS crosslinking.

4.5. Heterologous expression and site-directed mutagenesis

Expression systems used for the AE family include mammalian cultured cells (COS, human embryonic kidney (HEK) 293), chicken embryo fibroblasts, QT6 quail cell line, *Xenopus laevis* oocytes, Sf9 insect cells and yeast. Immunofluorescence staining, pulse-chase labeling and endoglycosidase H digestion have shown that mouse AE1, when expressed in HEK cells, remains in the endoplasmic reticulum (ER) or the cis-Golgi compartment [135]. When mouse AE2 was transfected under the same conditions, 60% of the synthesized protein was processed to the cell surface. Human AE1 was shown to exit the ER of HEK cells when co-transfected with the AE1-binding fragment of ankyrin1 (ANK-90) [136]. This suggests that they may interact in early biosynthetic compartments. Mouse AE1 expressed in HEK cells and reconstituted into liposomes was functional using an SO_4^{2-} transport assay [135,137].

Chicken Band 3 was expressed in chicken embryo fibroblasts and QT6 cells using an avian retrovirus vector SFCV-BIIIR [138]. Band 3 was functionally expressed on the surface, as shown by immunofluorescence staining. Infected cells became highly vacuolated and vacuoles disappeared when cells were grown in pH 8.0 medium or in the presence of 0.4 mM DIDS. Vacuoles were also seen in transformed avian erythroblasts over-expressing Band 3 [139] suggesting that they are created in response to alterations in intracellular pH or ion concentrations.

Cytoplasmic mRNA injections were used to express mouse [140,141] and human [126,142] Band 3 in *Xenopus* oocytes. It was estimated that about 10% of the expressed protein is present in the oocyte plasma membrane. Co-expression of human AE1 and human red cell glycophorin A increased the number of AE1 molecules present at the oocyte surface [142]. A region corresponding to putative TM segments 9–12 of Band 3 was initially proposed for this interaction [94]. New results point to the beginning of TM8 as being important [61]. The role of glycophorin A in erythroid Band 3 synthesis is not established, since glycophorin A-deficient individuals have normal expression levels of Band 3 in their erythrocytes [143].

A cDNA encoding the membrane domain of mouse AE1 [144] is functional when expressed in *Xenopus* oocytes. This confirms earlier protein chemical studies using proteolyzed Band 3 [111]. AE2 and AE3 lacking the cytosolic domain are also functional when expressed in COS cells [31,33].

The IC_{50} for inhibition of anion exchange in human erythrocytes are about 1 μM for H_2DIDS, 0.1 μM for DIDS and 10 μM for DNDS [145]. The IC_{50} values for inhibition of anion transport by H_2DIDS in transfected cells is 142 μM for AE2 and 0.43 μM for AE3 [146]. Recent results [137] using sulfate transport into microsomes isolated from AE1 and AE2 transfected HEK cells have shown that the sensitivity of the two gene products to DNDS (K_i = 2.5, 4 μM for AE1 and AE2 respectively) is about the same [137]. AE2 expressed in oocytes [147,148] was inhibited by DIDS with a K_i of 0.5 μM in low salt and 19 μM under physiological salt concentrations. The IC_{50} for DIDS inhibition of AE2 in Sf9 insect cells is 4 μM [149]. The functional expression of AE1 in yeast [150] opens up the possibility to combine structural and mutagenic studies.

The first target for mutagenesis was the DIDS-reactive lysine residues [126, 151]. Mutation of Lys 539 in human Band 3 showed that this residue is not essential for anion transport and that the mutation did not result in loss of covalent DIDS reaction [126]. The effect of mutation of Lys542 was not tested in this study. In the mouse system, mutation of both Lys 558 (Lys539 in human) and 561 (Lys542 in human) did not prevent reversible inhibition of anion transport by H_2DIDS nor did the mutations eliminate anion transport [133,141]. Lys869 in mouse is the likely site of H_2DIDS crosslinking. This residue is Met in AE2 and therefore a lysine at this position is not essential for anion transport function. Irreversible inhibition by pyridoxal phosphate is prevented by mutation of both Lys 558 and Lys 869 showing that inhibition by pyridoxal phosphate can occur by reaction at either lysine. The affinity of AE1 for DNDS is decreased almost 100-fold (6 μM to 500 μM) by mutation of *both* Lys 558 and Lys 869 in mouse [133]. The K_m value for Cl^- binding, however, decreased to 1/4 of the original value in double mutant. This led to the conclusion that the two lysines are not involved in substrate binding directly, but have an allosteric effect on the substrate binding site. Mutation of Lys449 in mouse AE1 prevents irreversible inhibition of transport by eosin maleimide, confirming the site of reaction of this inhibitor [152]. This mutation however did not impair anion transport.

Mutations of histidine residues in mouse AE1 [153] result in partial (His752Ser, His837Gln/Arg, or complete (His721Gln, His852Gln) inhibition of chloride exchange when expressed in oocytes. Amazingly, the effect of most of these mutations can be reversed by a second mutation (Lys558Asn) at the H_2DIDS-reactive site. This suggests that long-range effects can take place in the Band 3 protein. Mutation of multiple lysines or prolines around His837 led to inhibition of transport that is not reversed by the Lys558Asn mutation. These mutagenesis studies also showed that His752 was the predominant site of diethyl pyrocarbonate reaction, a reaction that modifies inward-facing histidine residues and inhibits anion transport [11].

Anion transport in red cells is inhibited at low pH (pKa = 5.5) and modification of carboxyl groups in Band 3 results in inhibition of monovalent but an acceleration of divalent SO_4^{2-} transport at neutral pH [11,154]. Glu681 in human Band 3 (Glu699 in mouse) is the site of action of Woodward's reagent K and borohydride which reacts from the outside of the cell and reduces the carboxyl group to an alcohol [155]. This perfectly conserved residue is located near the cytosolic end of TM8 (Fig. 1) and it is the site of H^+/SO_4^{2-} co-transport that takes place during Cl^-/SO_4^{2-} exchange. The acceleration of sulfate transport is due to neutralization of the carboxyl group which normally must be protonated for H^+/SO_4^{2-} co-transport to occur. Modification of Glu681 also changes the intracellular pH dependence of anion transport suggesting that this residue is positioned at the permeability barrier in Band 3. Mutation of Glu699 in mouse Band 3 shifts the pH dependence of anion transport from a pKa of 5.8 to 6.8 [156]. Interestingly, mutation of His 752 (His 734 in human) has exactly the same effect. It was suggested [156] that the side-chains of these two residues are hydrogen bonded to one another and participate in a common rate-limiting step in anion transport.

Cys843 of human Band 3 is covalently bound to a fatty acid chain in the erythroid plasma membrane. The thioester bond is usually formed with stearate or palmitate on the cytoplasmic side of the membrane and the fatty acid anchors a portion of protein to the membrane. When Cys861, the mouse equivalent of human Cys843, was mutated to Ser or Met the anion transport and K_i value for DNDS inhibition of Cl^- efflux from *Xenopus* oocytes remained unaffected [80]. Expression of a Band 3 mutant with all five cysteines (Cys201, 317, 479, 843 and 885), replaced by serines did not alter ankyrin binding, the oligomeric state nor the transport function of the protein [104].

5. Beyond the red cell

5.1. Transport in epithelial cells

Epithelial and other polarized cells exploit Cl^-/HCO_3^- exchange in order to catalyze vectorial ion and H_2O movements. To do so, expression of the Cl^-/HCO_3^- exchanger is restricted to the apical or basolateral membrane. The AE1b protein is localized to the basolateral membranes of acid secreting (α-)intercalated cells of the cortical collecting duct of human [157], rat [101,102,158,159] and rabbit [160] kidney and in kidney oncocytoma, an epithelial tumor resembling acid-secreting intercalated cells [39]. α-Intercalated cells in the kidney are involved in HCO_3^- re-absorption, where basolateral AE1b is coupled with an apical H^+-ATPase [161,162]. The β-intercalated cells of the cortical collecting duct in mammalian kidney [162] secrete HCO_3^- and absorb Cl^- via an apical Cl^-/HCO_3^- exchanger. An immortalized bicarbonate-secreting intercalated cell line expressed an apical anion exchanger believed to be AE1 [163].

AE2 protein is located in the brush border membrane of villus enterocytes in the ileum [32] where it functions as a HCO_3^- extruder. Coupled with apical Na^+/H^+-exchanger, it contributes to net NaCl re-absorption. Acid-secreting parietal cells in the stomach secrete hydrochloric acid in the glandular lumen [164,165] against the H^+ gradient. These cells express an AE2 Cl^-/HCO_3^- exchanger on their basolateral membrane [164]. Basolateral localization was also reported for AE2 cloned from choroid plexus epithelium [31,106] where it plays a role in HCO_3^- secretion into cerebrospinal fluid. Osteoclasts are polarized cells that generate a H^+ flux to mobilize bone calcium during the process of bone resorption. Extracellular acidification is generated by an apical vacuolar-type H^+-ATPase, while a basolateral Cl^-/HCO_3^- exchanger protects the cell from alkalization [166]. Parotid acinar cells use basolateral Cl^-/HCO_3^- exchanger to generate Cl^- gradient that drives H_2O across apical membrane [167].

5.2. Intracellular pH regulation

The anion concentration gradients across the plasma membrane in non-erythroid cells drive the exchange of intracellular HCO_3^- for extracellular Cl^-, producing an acidification of the cell. The Cl^-/HCO_3^- exchanger, the acid loader, is activated at elevated pH_i and inhibited by lower pH_i, while acid extruders (alkaline loaders) such as the Na^+/H^+ exchanger are activated when pH_i drops below normal physiological value, maintaining the pH homeostasis [12].

Erythroid and non-erythroid cells show inhibition of monovalent anion exchange when intracellular pH is decreased [168,169]. AE2 and AE3 are inhibited at pH values below 7.0 [146,168], while Band 3 is inhibited below pH 6.0 [170]. The sensitivity of AE2 and AE3 to intracellular pH suggests the presence of a modifier site that may allosterically affect anion transport [146,169]. This is supported by evidence that increased pH_i enhances not only Cl^-/HCO_3^- exchange, but also Cl^-/Cl^- exchange [169]. The domain that regulates pH may be within the unique N-terminal extensions of AE2 and AE3. Removal of the cytosolic domain of AE3 altered the sensitivity of the protein to internal pH [33].

The pH_i dependence of AE2-mediated Cl^-/HCO_3^- exchange was variable in different expression systems, varying from relatively insensitive in Sf9 cells (rat AE2) [149] to activated above resting pH_i for mouse AE2 in COS [31], HEK [146] and CHOP cells [171]. AE2 expressed in oocytes was activated by intracellular alkalinization caused by hypertonic activation of the endogenous Na^+/H^+ exchanger [148].

5.3. Cell growth and differentiation

Chicken erythroblasts transfected with the oncogene v-erb-B express erythroid-specific genes like AE1 and differentiate into erythrocytes [172]. Co-expression of v-erb-A and v-erb-B blocks the differentiation of chicken erythroblasts into erythrocytes and enables them to proliferate under a wide range of media pH and ion (e.g., bicarbonate) concentrations. v-erb-A is an oncogene that is a mutated thyroid hormone receptor $\alpha 1$. In chicken erythroblasts the v-erb-A oncoprotein arrests expression of the avian anion transporter by binding to the regulatory region of the

AE1 gene [173]. Reintroduction of the gene for Band 3 under the control of another promoter into transformed cells results in cells that only grow under a narrow media pH range [172]. Blocking anion transport in v-erb-A/v-erb-B transformed cells with DIDS allows proliferation. It appears that a functional AE1 blocks proliferation and is important for differentiation along the erythroid pathway. Stilbene disulfonates are also mitogenic in B-lymphocytes and this effect may be mediated through inhibition of anion exchangers [174].

6. Conclusions

Human erythrocyte Band 3 is one of the best studied membrane proteins due to its abundance in red blood cells. This has led to many studies of the structure and function of this specialized transporter. The cloning of AE1 led to the discovery of the AE gene family. The wide distribution of AE proteins points to their importance in processes beyond the red cell, such as intracellular pH regulation, transcellular anion transport and mitogenesis. The high degree of sequence similarity in the membrane domain of AE proteins supports the view that the structure and mechanism of erythrocyte Band 3 can serve as a valid model for other members of the AE gene family.

Abbreviations

AE, anion exchanger
DIDS, 4,4'-diisothiocyanostilbene-2,2'-disulfonate
H_2DIDS, 4,4'-diisothiocyanodihyrostilbene-2,2'-disulfonate
DNDS, 4,4'-dinitrostilbene-2,2'-disulfonate
ER, endoplasmic reticulum
HEK, human embryonic kidney
MIR, membrane inclusion ratio
SAO, Southeast Asian ovalocyte
TM, transmembrane

Acknowledgements

The authors thank J.H.M. Charuk, J.W. Vince and H. Kameh for their suggestions.

References

1. Jennings, M.L. (1984) J. Membr. Biol. **80,** 105–117.
2. Passow, H. (1986) Rev. Physiol. Biochem. Pharmacol. **103,** 61–203.
3. Jay, D. and Cantley, L. (1986) Ann. Rev. Biochem **55,** 511–538.
4. Fröhlich, O. and Gunn, R.B. (1986) Biochim. Biophys. Acta **864,** 169–194.
5. Alper, S.L., Kopito, R.R. and Lodish, H.F. (1987) Kidney Int. **32,** S117–S128.
6. Jennings, M.L. (1989) Annu. Rev. Biophys. Biophys. Chem. **18,** 397–430.
7. Kopito, R. R. (1990) Intl. Review of Cytol. **123,** 177–199.
8. Salhany, J.M. (1990) Erythrocyte Band 3 Protein. CRC Press, Boca Raton, FL.
9. Tanner, M.J.A. (1993) Semin. Hematol. **30,** 34–57.

10. Reithmeier, R. A. F. (1993) Curr. Opin. Struct. Biol. **3**, 515–523.
11. Jennings, M.L. (1992) The Kidney: Physiology and Pathophysiology, 2nd Edn. Raven Press, Ltd., New York, pp. 503–535.
12. Alper, S.L. (1991) Annu. Rev. Physiol. **53**, 549–564.
13. Godnich, M.J. and Jennings, M.L. (1995) in: Molecular Nephrology, eds D. Schlöndorf and J.V. Bonventre, pp. 289–307, Marcel Dekker, Inc., New York.
14. Showe, L.C., Ballantine, M. and Huebner, K. (1987) Genomics **1**, 71–76.
15. Palumbo, A.P., Isobe, M., Huebner, K., Shane, S. and Rovera, G. (1986) Am. J. Hum. Genet. **39**, 307–316.
16. Su, Y.R., Klanke, C.A., Houseal, T.W., Linn, S.C., Burk, S.E., Varvil, T.S., Otterud, B.E., Shull, G.E., Leppert, M.F. and Menon, A.G. (1994) Genomics **22**, 605–609.
17. Havenga, M.J.E., Bosman, G.J.C.G.M., Appelhans, H. and Grip, W.J.D. (1994) Mol. Brian Res. **25**, 97–104.
18. Kopito, R.R. and Lodish, H.L. (1985) J. Cell. Biochem. **29**, 1–17.
19. Tanner, M.J., Martin, P.G. and High, S. (1988) Biochem. J. **256**, 703–712.
20. Lux, S.E., John, K.M., Kopito, R.R. and Lodish, H.F. (1989) Proc. Natl. Acad. Sci. **86**, 9089–9093.
21. Cox, J.V. and Lazarides, E. (1988) Mol. Cell. Biol. **8**, 1327–1335.
22. Kim, H.R.C., Yew, N.S., Ansorge, W., Voss, H., Schwager, C., Vennstrom, B., Zenke, M. and Engel, J.D. (1988) Mol. Cell Biol. **8**, 4416–4424.
23. Kim, H.R.C., Kennedy, B.S. and Engel, J.D. (1989) Mol. Cell Biol. **9**, 5198–5206.
24. Hübner, S., Michel, F., Rudloff, V. and Appelhans, H. (1992) Biochem. J. **285**, 17–23.
25. Kudrycki, K.E. and Schull, G.E. (1989) J. Biol. Chem. **264**, 8185–8192.
26. Cox, K.H. and Cox, J.V. (1995) Am. J. Physiol. **95**, F503–F513.
27. Demuth, D.R., Showe, L.C., Ballantine, M., Palumbo, A., Fraser, P.J., Cioe, L., Rovera, G. and Curtis, P.J. (1986) EMBO J. **5**, 1205–1214.
28. Gehrig, H., Muller, W. and Appelhans, H. (1992) Biochim. Biophys. Acta **1130**, 326–328.
29. Alper, S.L., Kopito, R.R., Libresco, S.M. and Lodish, H.F. (1988) J. Biol. Chem. **263**, 17092–17099.
30. Kudrycki, K.E., Newman, P.R. and Schull, G.E. (1990) J. Biol. Chem. **265**, 462–471.
31. Lindsey, A.E., Schneider, K., Simmons, D.M., Baron, R., Lee, B.S. and Kopito, R.R. (1990) Proc. Natl. Acad. Sci. USA **87**, 5278–5282.
32. Chow, A., Dobbins, J.W., Aronson, P.S. and Igarashi, P.(1992) Am. J. Physiol. **263**, G345–G352.
33. Kopito, R.R., Lee, B.S., Simmons, D.M., Lindsey, A.E., Morgans, C.W. and Schneider, K. (1989) Cell **59**, 927–937.
34. Espanol, M.J. and Saier, M.H. (1995) Mol. Membr. Biol. **12**, 193–200.
35. Sahr, K.E., Taylor, W.M., Daniels, B.P., Rubin, H.L. and Jarolim, P. (1994) Genomics **24**, 491–501.
36. Kopito, R.R., Andersson, M.A. and Lodish, H.F. (1987) Proc. Natl. Acad. Sci. USA **84**, 7149–7153.
37. Kopito, R.R., Andersson, M. and Lodish, H.F. (1987) J. Biol. Chem. **262**, 8035–8040.
38. Schofield, A.E., Martin, P.G., Spillett, D. and Tanner, M.J.A. (1994) Blood **84**, 2000–2012.
39. Kollert-Jons, A., Wagner, S., Hubner, S., Appelhans, H. and Drenckhahn, D. (1993) Am. J. Physiol. **93**, F813–F821.
40. Brosius III, F.C., Alper, S.L., Garcia, A.M. and Lodish, H.F. (1989) J. Biol. Chem. **264**, 7784–7787.
41. Kudrycki, K.E. and Shull, G.E. (1993) Am. J. Physiol. **264**, F540–F547.
42. Negrini, C., Rivolta, M.N., Kalinee, F. and Kachar, B. (1995) Biochim. Biophys. Acta **1236**, 207–211.
43. Linn, S.C., Kudrycki, K.E. and Shull, G.E. (1992) J. Biol. Chem. **267**, 7927–7935.
44. Linn, S.C., Askew, R. and Shull, G.E. (1995) Circ. Res. **76**, 584–591.
45. Yannoukakus, D., Vasseur, C., Driancourt, C., Blouquit, Y., Delaunay, J., Wajcman, H. and Bursaux, E. (1991) Blood **78**, 1117–1120.
46. Jarolim, P., Rubin, H.L., Zhai, S., Sahr, K.E., Liu, S.C., Mueller, T.J. and Palek, J. (1992) Blood **80**, 1592–1598.
47. Bruce, L.J., Anstee, D.J., Spring, F.A. and Tanner, M.J.A. (1994) J. Biol. Chem. **269**, 16155–16158.
48. Rybicki, A.C., Qiu, J.J.H., Musto, S., Rosen, N.L., Nagel, R.L. and Schwartz, R.S. (1993) Blood **81**, 2155–2165.
49. Jarolim, P., Palek, J., Rubin, H.L., Prchal, J.T., Korsgren, C. and Cohen, C.M. (1992) Blood **80**, 523–529.

50. Jarolim, P., Rubin, H.L., Brabac, V., Chrobak, L., Zolotarev, A.S., Alper, S.L., Brugnara, C., Wichterle, H. and Palek, J. (1995) Blood **85**, 634–640.

51. Jarolim, P., Palek, J., Amato, D., Hassan, K., Sapak, P., Nurse, G.T., Rubin, H.L., Zhai, S., Sahr, K.E. and Liu, S.C. (1991) Proc. Nat. Acad. Sci. USA **78**, 5829–5832.

52. Schofield, A.E., Tanner, M.J.A., Pinder, J.C., Clough, B., Bayley, P.M., Nash, G.B., Dluzewski, A.R., Reardon, D.M., Cox, T.M., Wilson, R.J.M. and Gratzer, W.B. (1992) J. Mol. Biol. **223**, 949–958.

53. Mohandas, N., Windari, R., Knowles, D., Leung, A., Parra, M., George, E., Conboy, J. and Chasis, J. (1992) J. Clin. Invest. **89**, 686–692.

54. Schofield, A.E., Reardon, D.M. and Tanner, M.J.A. (1992) Nature **355**, 836–838.

55. Moriyama, R., Ideguchi, H., Lombardo, C.R., Dort, H.M.V. and Low, P.S. (1992) J. Biol. Chem. **267**, 25792–25797.

56. Liu, S.-C., Jarolim, P., Rubin, H.L., Palek, J., Amato, D., Hassan, K., Zaik, M. and Sapak, P. (1994) Blood **84**, 3590–3598.

57. Sarabia, V.E., Casey, J.R. and Reithmeier, R.A.F. (1993) J. Biol. Chem. **268**, 10676–10680.

58. Tilley, I., Nash, G.B., Jones, G. and Sawyer, W. (1991) J. Membr. Biol. **121**, 59–66.

59. Liu, S.-C., Palek, J., Yi, S.J., Nichols, P.E., Derick, L.H., Chiou, S.-S., Amato, D., Corbett, J.D., Cho, M.R. and Golan, D.E. (1995) Blood **86**, 349–358.

60. Jennings, M.L. and Gosselink, P.G. (1994) Biochemistry **34**, 3588–3595.

61. Bruce, L.J., Ring, S.M., Anstee, D.J., Reid, M.E., Wilkinson, S. and Tanner, M.J.A. (1995) Blood **85**, 541–547.

62. Bruce, L.J., Kay, J.M.B., Lawrence, C. and Tanner, M.J.A. (1993) Biochem. J. **293**, 317–320.

63. Jarolim, P., Rubin, H.L., Liu, S.-C., Cho, M.R., Brabec, V., Derick, L.H., Yi, S.J., Saad, S.T.O., Alper, S., Brugnara, C., Golan, D.E. and Palek, J. (1994) J. Clin. Invest. **93**, 121–130.

64. Kyte, J. and Doolittle, R.F. (1982) J. Mol. Biol. **157**, 105–132.

65. Landolt-Marticorena, C., Williams, K.A., Deber, C.M. and Reithmeier, R.A.F. (1993) J. Mol. Biol. **229**, 602–608.

66. Persson, B. and Argos, P. (1994) J. Mol. Biol. **237**, 182–192.

67. Rost, B., Casadio, R., Fariselli, P. and Sander, C. (1995) Protein Sci. **4**, 521–533.

68. Chou, P.Y. and Fasman, G.D. (1978) Adv. Enzymol. **47**, 45–148.

69. Drickamer, L.K. (1978) J. Biol. Chem. **253**, 7242–7248.

70. Jay, D.G. (1986) Biochemistry **25**, 554–556.

71. Casey, J.R., Pirraglia, C.A. and Reithmeier, R.A.F. (1992) J. Biol. Chem. **267**, 11940–11948.

72. Groves, J.D. and Tanner, M.J.A. (1993) Mol. Membr. Biol. **11**, 31–38.

73. Landolt-Marticorena, C. and Reithmeier, R.A.F. (1994) Biochem. J. **302**, 253–260.

74. Tsuji, T., Irimura, T. and Osawa, T. (1980) Biochem. J. **187**, 677–686.

75. Fukuda, M., Dell, A., Oates, J.E. and Fukuda, M.N. (1984) J. Biol. Chem. **259**, 8260–8273.

76. Fukuda, M.N. (1990) Glycobiology **1**, 9–15.

77. Okubo, K., Hamasaki, N., Hara, K. and Kageura, M. (1991) J. Biol. Chem. **266**, 16420–16424.

78. Steck, T. L., Ramos, B. and Strapazon, E. (1976) Biochemistry **15**, 1154–1161.

79. Jennings, M.L., Anderson, M.P. and Monaghan, R. (1986) J. Biol. Chem. **261**, 9002–9010.

80. Kang, D., Karbach, D. and Passow, H. (1994) Biochim. Biophys. Acta **1194**, 341–344.

81. Okubo, K., Kang, D., Hamasaki, N. and Jennings, M.L. (1994) J. Biol. Chem. **269**, 1918–1926.

82. Wainwright, S.D., Tanner, M.J.A., Martin, G.E.M., Yendle, J.E. and Holmes, C. (1989) Biochem. J. **258**, 211–220.

83. Lieberman, D.M. and Reithmeier, R.A.F. (1988) J. Biol. Chem. **263**, 10022–10028.

84. Wainwright, S.D., Mawby, W.J. and Tanner, M.J.A. (1990) Biochem. J. **272**, 265–272.

85. Smythe, J.S., Spring, F.A., Gardner, B., Parsons, S.F., Judson, P.A. and Anstee, D.J. (1995) Blood **85**, 2929–2936.

86. Jennings, M.L. and Nicknish, J.S. (1984) Biochemistry **23**, 6432–6436.

87. Oikawa, K., Lieberman, D.M. and Reithmeier, R.A.F. (1985) Biochemistry **24**, 2843–2848.

88. Landolt-Marticorena, C., Casey, J.R. and Reithmeier, R.A. F. (1995) Mol. Membr. Biol. **12**, 173–182.

89. Gargaro, A.R., Bloomberg, G.B., Dempsey, C.E., Murray, M. and Tanner, M.J.A. (1994) Eur. J. Biochem. **221**, 445–454.

90. Popot, J.L. and Engelman, D.M. (1990) Biochemistry **29**, 4031–4037.

91. Braell, W.A. and Lodish, H.F. (1982) Cell **28**, 23–31.
92. Sabban, E., Marchesi, V., Adensik, M. and Sabatini, D.D. (1982) J. Cell Biol. **91**, 637–646.
93. Tam, L.Y., Loo, T.W., Clarke, D.M. and Reithmeier, R.A.F. (1994) J. Biol. Chem. **269**, 32542–32550.
94. Groves, J.D. and Tanner, M.J.A. (1995) J. Biol. Chem. **270**, 9097–9105.
95. Low, P.S. (1986) Biochim. Biophys. Acta **864**, 145–167.
96. Walder, J.A., Chatterjee, R., Steck, T.L., Low, P.S., Musso, G.F., Kaiser, E.T., Rogers, P.H. and Arnone, A. (1984) J. Biol. Chem. **259**, 10238–10246.
97. Waugh, S.M. and Low, P.S. (1985) Biochemistry **24**, 34–39.
98. Low, P.S., Allen, D.P., Zioncheck, T.F., Chari, P., Willardson, B.M., Geahlen, R.L. and Harrison, M.L. (1987) J. Biol. Chem. **262**, 4592–4586.
99. Ding, Y., Casey, J.R. and Kopito, R.R. (1994) J. Biol. Chem. **269**, 32201–32208.
100. Wang, C.C., Moriyama, R., Lombardo, C.R. and Low, P.S. (1995) J. Biol. Chem. **270**, 17892–17897.
101. Drenckhahn, D., Schluter, K., Allen, D.P. and Bennett, V. (1985) Science **230**, 1287–1289.
102. Erocolani, L., Brown, D., Stuart-Tilley, A. and Alper, S. L. (1992) Am. J. Physiol. **262**, F892–F896.
103. Thevenin, B.J.-M., Willardson, B.M. and Low, P.S. (1989) J. Biol. Chem. **264**, 15886–15892.
104. Casey, J.R., Ding, Y. and Kopito, R.R. (1995) J. Biol. Chem. **270**, 8521–8527.
105. Morgans, C.W. and Kopito, R.R. (1993) J. Cell Sci. **105**, 1137–1142.
106. Alper, S. L., Stuart-Tilley, A., Simmons, C.F., Brown, D. and Drenckhahn, D. (1994) J. Clin. Invest. **93**, 1430–1438.
107. Wang, C.C., Badylak, J.A., Lux, S.E., Moriyama, R., Dixon, J.E. and Low, P.S. (1992) Protein Sci. **1**, 1206–1214.
108. Kiyatkin, A.B., Natarajan, P., Munshi, S., Minor, W., Johnson, J.E. and Low, P.S. (1995) Proteins: Struct. Func. Gen. **22**, 293–297.
109. Wang, D.N. (1994) FEBS Lett. **346**, 26–31.
110. Reithmeier, R.A.F. and Deber, C. M. (1992) in: The Structure of Biological Membranes, ed P. Yeagle, pp. 337–393, CRC Press, Boca Raton, FL.
111. Grinstein, S., Ship, S. and Rothstein, A. (1978) Biochim. Biophys. Acta **507**, 294–304.
112. Dolder, M., Walz, T., Hefti, A. and Engel, A. (1993) J. Mol. Biol. **231**, 119–132.
113. Wang, D.N., Kühlbrandt, W., Sarabia, V.E. and Reithmeier, R.A.F. (1993) EMBO J. **12**, 2233–2239.
114. Wang, D.N., Sarabia, V.E., Reithmeier, R.A.F. and Kühlbrandt, W. (1994) EMBO J. **13**, 3230–3235.
115. Casey, J.R. and Reithmeier, R.A.F. (1991) J. Biol. Chem. **266**, 15726–15737.
116. Lindenthal, S. and Schubert, D. (1991) Proc. Natl. Acad. Sci. USA **88**, 6540–6544.
117. Mulzer, K., Kampmann, L., Petrasch, P. and Schubert, D. (1990) Colloid Polym. Sci. **268**, 60–64.
118. Michaely, P. and Bennett, V. (1995) J. Biol. Chem. **270**, 22050–22057.
119. Cobb, C. E. and Beth, A.H. (1990) Biochemistry **29**, 8283–8290.
120. Liu, S. J. and Knauf, P.A. (1993) Am. J. Physiol. **264**, C1155–C1164.
121. Knauf, P.A., Strong, N.M., Penikas, J., R.B. Wheeler, J. and Liu, S.J. (1993) Am. J. Physiol. **264**, C1144–C1154.
122. Liu, D., Kennedy, S.D. and Knauf, P.A. (1995) Biophys. J. **69**, 399–408.
123. Pan, R. and Cherry, R.J. (1995) Biochemistry **34**, 4880–4888.
124. Salhany, J.M., Sloan, R.L., Cordes, K.A. and Schopfer, L.M. (1994) Biochemistry **33**, 11909–11916.
125. Schopfer, L.M. and Salhany, J.M. (1995) Biochemistry **34**, 8320–8329.
126. Garcia, A.M. and Lodish, H.F. (1989) J. Biol. Chem. **264**, 19607–19613.
127. Jennings, M.L. and Passow, H. (1979) Biochim. Biophys. Acta **554**, 498–519.
128. Kang, D., Okubo, K., Hamasaki, N., Kuroda, N. and Shiraki, H. (1992) J. Biol. Chem. **267**, 19211–19217.
129. Appell, K.C. and Low, P.S. (1982) Biochemistry **21**, 2151–2157.
130. Tomida, M., Kondo, Y., Moriyama, R., Machida, H. and Makino, S. (1988) Biochim. Biophys. Acta **943**, 493–500.
131. Schuck, P., Legrum, B., Passow, H. and Schubert, D. (1995) Eur. J. Biochem. **230**, 806–812.
132. Salhany, J.M., Sloan, R.L. and Cordes, K.A. (1990) J. Biol. Chem. **265**, 17688–17693.
133. Wood, P.G., Müller, H., Sovak, M. and Passow, H. (1992) J. Membr. Biol. **127**, 139–148.
134. Kawano, Y., Okubo, K., Tokunaga, F., Miyata, T., Iwanaga, S. and Hamasaki, N. (1988) J. Biol. Chem. **263**, 8232–8238.

135. Ruetz, S., Lindsey, A.E., Ward, C.L. and Kopito, R.R. (1993) J. Cell Biol. **121**, 37–48.
136. Gomez, S. and Morgans, C. (1993) J. Biol. Chem. **268**, 19593–19597.
137. Sekler, I., Lo, R.S., Masrocola, T. and Kopito, R.R. (1995) J. Biol. Chem. **270**, 11251–11256.
138. Fuerstenberg, S., Beug, H., Introna, M., Khazaie, K., Munoz, A., Ness, S., Nordström, K., Sap, J., Stanley, I., Zenke, M. and Vennström, B. (1990) J. Virol. **64**, 5891–5902.
139. Zenke, M., Kahn, P., Disela, C., Vennström, B., Leutz, A., Keegan, K., Hayman, M. J., Choi, H.-R., Yew, N., Engel, J. D. and Beug, H. (1988) Cell **52**, 107–119.
140. Hanke-Baier, P., Raida, M. and Passow, H. (1988) Biochim. Biophys. Acta **940**, 136–140.
141. Bartel, D., Lepke, S., Layh-Schmitt, G., Legrum, B. and Passow, H. (1989) EMBO J. **8**, 3601–3609.
142. Groves, J.D. and Tanner, M.J.A. (1992) J. Biol. Chem. **267**, 11940–11948.
143. Bruce, L.J., Groves, J.D., Okubo, Y., Thilaganathan, B. and Tanner, M.J.A. (1994) Blood **84**, 916–922.
144. Lepke, S., Becker, A. and Passow, H. (1992) Biochim. Biophys. Acta **1106**, 13–16.
145. Cabantchik, Z.I. and Greger, R. (1992) Am. J. Physiol. **31**, C803–C827.
146. Lee, B.S., Gunn, R.B. and Kopito, R.R. (1991) J. Biol. Chem. **266**, 11448–11454.
147. Humphreys, B.D., Jiang, L., Chernova, M.N. and Alper, S.L. (1994) Am. J. Physiol. **94**, C1295–C1307.
148. Humphreys, B.D., Jiang, L., Chernova, M.N. and Alper, S.L. (1995) Am. J. Physiol. **268**, C210–C209.
149. He, X., Wu, X., Knauf, P.A., Tabak, L.A. and Melvin, J.E. (1993) Am. J. Physiol. **264**, C1075–C1079.
150. Sekler, I., Kopito, R. and Casey, J.R. (1995) J. Biol. Chem. **270**, 21028–21034.
151. Passow, H., Wood, P.G., Lepke, S., Müller, H. and Sovak, M. (1992) Biophys. J. **62**. 98–100.
152. Passow, H., Lepke, S. and Wood, P. G. (1992) in Progress in Cell Research, eds E. Bamberg and H. Passow, Vol. 2, pp. 85–98, Elsevier, Amsterdam.
153. Müller-Berger, S., Karbach, D., König, J., Lepke, S., Wood, P.G., Appelhans, H. and Passow, H. (1995) Biochemistry **34**, 9315–9324.
154. Jennings, M. L. (1992) in: The Kidney: Physiology and Pathophysiology, eds D.W. Seldin and G. Giebisch, 2nd Edn, pp. 113–145, Raven Press, Ltd., New York.
155. Jennings, M.L. and Anderson, M.P. (1987) J. Biol. Chem. **262**, 1691–1697.
156. Müller-Berger, S., Karbach, D., Kang, D., Aranibar, N., Wood, P.G., Rüterjans, H. and Passow, H. (1995) Biochemistry **34**, 9325–9332.
157. Drenckhahn, D. and Merte, C. (1987) Eur. J. Cell Biol. **45**, 107–115.
158. Verlander, J.W., Madsen, K.M., Low, P.S., Allen, D.P. and Tisher, C.C. (1988) Am. J. Physiol. **255**, F115–F125.
159. Alper, S.L., Natale, J., Gluck, S., Lodish, H.F. and Brown, D. (1989) Proc. Natl. Acad. Sci. USA **86**, 5429–5433.
160. Schuster, V.L., Fejes-Toth, G., Naray-Feyes-Toth, A. and Gluck, S. (1991) Am. J. Physiol. **260**, F506–F517.
161. Fejes-Tóth, G., Chen, W.-R., Rusvai, E., Moser, T. and Náray-Fejes-Tóth, A. (1994) J. Biol. Chem. **269**, 26717–26721.
162. Schuster, V.L. (1993) Annu. Rev. Physiol. **55**, 267–288.
163. Adelsberg, J.S. v. and Al-Awqati, Q. (1993) J. Biol. Chem. **268**, 11283–11289.
164. Jons, T., Warrrings, B., Jons, A. and Drenckhahn, D. (1994) Histochemistry **102**, 255–263.
165. Muallem, S., Burnham, C., Blissard, D., Berglindh, T. and Sachs, G. (1985) J. Biol. Chem. **260**, 6641–6653.
166. Carano, A., Schlesinger, P.H., Athanasou, N.A., Teitelbaum, S.L. and Blair, H.C. (1993) Am. J. Physiol. **264**, C694–C701.
167. Turner, R.J. and George, J.N. (1988) Am. J. Physiol. **254**, C391–C396.
168. Olsnes, S., Tonnessen, T.I., Ludt, J. and Sandvig, K. (1987) Biochemistry **26**, 2778–2785.
169. Green, J., Yamaguchi, D.T., Kleeman, C.R. and Muallen, S. (1990) J. Gen. Physiol. **95**, 121–145.
170. Cabantchik, Z.I., Knauf, P.A. and Rothstein, A. (1978) Biochim. Biophys. Acta **515**, 239–302.
171. Jiang, L., Stuart-Tilley, A., Parkash, J. and Alper, S.L. (1994) Am. J. Physiol. **267**, C845–C856.
172. Fuerstenberg, S., Leitner, I., Schroeder, C., Schwarz, H., Vennström, B. and Beug, H. (1992) EMBO J. **11**, 3355–3365.
173. Inggvarsson, S. and Vennström, B. (1993) Annals N. Y. Acad. Sci. 426–429.
174. Deane, K.H. and Mannie, M.D. (1992) Eur. J. Immunol. **22**, 1165–1171.

CHAPTER 14

Erythrocyte Sugar Transport

A. CARRUTHERS and R.J. ZOTTOLA

Department of Biochemistry and Molecular Biology, Program in Molecular Medicine, University of Massachusetts Medical School, Worcester, MA 01605, USA

© 1996 Elsevier Science B.V.
All rights reserved

Handbook of Biological Physics
Volume 2, edited by W.N. Konings, H.R. Kaback and J.S. Lolkema

Contents

1. General introduction

In 1952, Wilfred Widdas published a landmark paper in the field of membrane transport. Expanding on earlier studies demonstrating that human erythrocyte sugar transport displays saturation kinetics [1–3], Widdas proposed a 'carrier transfer' mechanism for sugar transport [4] in which the glucose transporter cycles between sugar import- and sugar export-competent states (Fig. 1). This was a fundamental departure from hypotheses proposed at that time for trans-membrane ion conductance via membrane-spanning channels. Unlike the proposed carrier-transfer mechanism, these translocation pathways were thought to be simultaneously accessible to intra- and extracellular substrate [5]. Drawing a parallel between enzyme-kinetics and carrier-mediated transport, Widdas developed a mathematical description of carrier-mediated transport [4,6] that accounts for most properties of human erythrocyte sugar transport. These properties are shared by many carrier-mediated facilitated diffusion (uniport) mechanisms. They include *saturation kinetics, trans-acceleration* (stimulation of unidirectional fluxes by the presence of substrate at the opposite, -trans-side of the membrane), *equilibrative* or *passive* transport (substrates are distributed equally between cell water and interstitium) and *counterflow* (a paradoxical accumulation of labeled substrate when uptake is measured into cells containing high concentrations of unlabeled substrate).

The carrier transfer hypothesis can, with some modification, also account for *antiport, symport* and *primary active* transport systems [7]. The success of this hypothesis has led to its wide acceptance in the field of membrane transport. The emphasis in the glucose transport field has therefore now moved towards the development of strategies that help us to understand how the structure of the glucose transporter and its substrate-liganded intermediates give rise to carrier-transfer kinetics. However, several other hypotheses have been proposed for erythrocyte sugar transport (for reviews see Refs. [8,9]). Some have not withstood experimental challenge. Others are at least as successful as the carrier transfer model for sugar transport but are founded on fundamentally different principles. If more than one hypothesis can account for the properties of the erythrocyte sugar transport system, which model is most useful for understanding the structural basis of transport?

This review examines kinetic and structural properties of the human erythrocyte glucose transport system and, as a framework for discussion, interprets these data within the context of the carrier transfer (simple carrier) hypothesis. We show that unlike other sugar transport systems, the erythrocyte glucose transporter has not been adequately modeled. We agree with Stein's suggestion [7] that the reason for this lies not in the paucity of data but rather in the breadth of studies that have been made. We concur with Naftalin's hypothesis [8] that transport measurements in erythrocytes report the sum of at least two serial processes — transmembrane sugar translocation and intracellular sugar complexation. We show that like most other eukaryotic transport systems, the glucose carrier is a multi-subunit complex. We hypothesize that individual subunits of the carrier complex function as simple carriers but that cooperative interactions between subunits give rise to fundamentally different transport kinetics.

2. Steady-state transport kinetics

Various transport conditions have been developed in order that measurements of Michaelis ($K_{m(app)}$) and velocity (V_{max}) constants may be made unambiguously. These are summarized in Table 1.

2.1. Transport is asymmetric

The first challenge to the simple carrier hypothesis arose when a number of studies indicated that the erythrocyte glucose transporter displays unequal affinities for D-glucose at opposite sides of the membrane. Miller demonstrated that $K_{m(app)}$ for net D-glucose exit (8 mM) is significantly greater than both $K_{m(app)}$ for net D-glucose uptake (1.6 mM) and $K_{m(app)}$ for inhibition of net D-glucose exit by extracellular D-glucose (2 mM) [10,11]. The original carrier hypothesis proposes that $K_{m(app)}$ for D-glucose uptake and exit were identical [4]. Indeed, studies by Harris [33] and later studies by Karlish et al. [24] are consistent with this view demonstrating that net uptake of D-glucose from medium containing saturating D-glucose levels is inhibited half-maximally by an intracellular D-glucose level (2 mM) very similar to $K_{m(app)}$ for net D-glucose uptake. Subsequent, so-called 'zero-trans' measurements, however, confirm that $K_{m(app)}$ and V_{max} for sugar exit into sugar-free medium are significantly (5 to 10-fold) greater than the corresponding values for sugar uptake into sugar-free cells [12–14].

'Asymmetry' in kinetic parameters for exit and entry was resolved when it was recognized that the simple carrier mechanism allows for asymmetric transport parameters provided that the ratio $V_{max}/K_{m(app)}$ is identical for uptake and exit [15,16]. This equality is necessary if the carrier is to behave as a passive transporter [17] and is not specific to the simple carrier hypothesis for transport. Other hypotheses for sugar transport also make this requirement [18]. Inequality in these ratios would indicate an active transport process [7] and red cell sugar transport is passive.

2.2. Transport displays trans-acceleration

An important property of erythrocyte sugar transport is the stimulation of unidirectional sugar exit by extracellular sugar [10,19,20]. This process is called 'trans-acceleration' and is bidirectional. Intracellular sugar also stimulates unidirectional sugar uptake [10]. A related phenomenon is counterflow [10] in which cells con-

Table 1
Sugar transport measurements in human erythrocytes. Adapted from Ref. [8].

Experiment	Procedure	Measures	Parameters
Zero-trans entry	$[Sugar]_o$ varied $[Sugar]_i = 0$	K_m entry V_{max} entry	K_{21}^{zt} V_{21}^{zt}
Zero-trans exit	$[Sugar]_o = 0$ $[Sugar]_i$ varied	K_m exit V_{max} exit	K_{12}^{zt} V_{12}^{zt}
Equilibrium exchange	$[Sugar]_o =$ $[Sugar]_i$ varied	K_m exchange V_{max} exchange	K^{ee} V^{ee}
Infinite-trans entry	$[Sugar]_o$ varied $[Sugar]_i$ saturating	K_m entry V_{max} exchange	K_{21}^{it} V^{ee}
Infinite-trans exit	$[Sugar]_o$ saturating $[Sugar]_i$ varied	K_m exit V_{max} exchange	K_{12}^{it} V^{ee}
Infinite-cis entry	$[Sugar]_o$ saturating $[Sugar]_i$ varied	K_m exit V_{max} entry	K_{21}^{ic} V_{21}^{zt}
Infinite-cis exit	$[Sugar]_o$ varied $[Sugar]_i$ saturating	K_m entry V_{max} exit	K_{12}^{ic} V_{12}^{zt}

taining high concentrations of unlabeled sugar transiently accumulate radiolabeled sugars from medium containing low concentrations of radiolabeled sugar to a level some 10-fold greater than that present outside the cell. In the absence of intracellular sugar, extracellular sugar simply equilibrates between cellular and interstitial water. This accumulation appears paradoxical given the passive nature of the transport system. However, it became apparent that the accumulation of radiolabeled sugar results from dilution of the specific activity of the imported, labelled sugar by the unlabelled intracellular sugar [21].

Trans-acceleration is consistent with the simple carrier hypothesis if the rate-limiting steps in net uptake and exit are not the sugar translocation steps but rather the relaxation steps (see Fig. 1). For example, in net uptake the two limiting conformational changes are sugar import (k_{-1}) in which the externally bound sugar is translocated to the cytosolic surface of the membrane and relaxation (k_0) in which the cytosolic sugar binding site (e1) is lost and the exofacial sugar binding site (e2) is regenerated. In net exit, the two limiting conformational changes are sugar export (k_1) in which the endofacially bound sugar is translocated to the exofacial surface of the membrane and relaxation (k_{-0}) in which the exofacial sugar binding site (e2) is lost and the cytosolic sugar binding site (e1) is regenerated. When intracellular sugar levels are saturating, the exofacial sugar binding site (e2) is regenerated via the export step k_1 and not via relaxation (k_0). Thus if $k_1 > k_0$, the presence of saturating sugar levels within the cell will stimulate unidirectional uptake by a factor approaching k_1/k_0. A similar argument ($k_{-1} > k_{-0}$) accounts for exit stimulation by extracellular sugars.

2.3. Detailed analysis of transport behavior

Kinetic analysis of transport data can indicate the minimum number of carrier-substrate intermediates that exist during a transport cycle and provides information on the rates of their interconversions. While this information does not reveal the structural basis of transport, it can alert us to the existence of a carrier-substrate intermediate that might not otherwise be obvious even if a detailed structure of the transporter were available for analysis. Furthermore, in the absence of any high resolution carrier structure, the kinetic analysis of transport data, the kinetic analysis of transport inhibitions and the kinetic analysis of ligand binding to the transporter remain the most information-rich strategies for understanding transport function.

In the following section we quantitatively compare the properties of human erythrocyte sugar transport with those predicted for a simple carrier mechanism. The basic quantitative properties of the simple-carrier hypothesis for sugar transport are summarized in Tables 2 and 3 with reference to the model shown in Fig. 1. The steady-state solution (transport rate equation) for the simple-carrier mechanism is characterized by 1 affinity parameter and by 4 'resistance' parameters (see Table 3). These parameters can be computed directly from experimental transport data (see Table 2 for details) and allow for assessment of consistency between theory and practice.

A   *Simple Carrier*

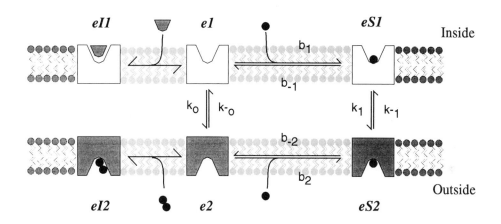

B   *Two Site Carrier*

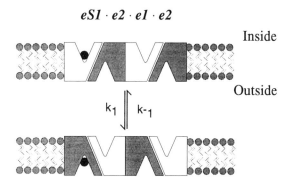

Fig. 1.   (A) The carrier-transfer or simple carrier hypothesis of Widdas [4]. The carrier is shown schematically in the membrane bilayer exposing either an import site (e2) or an export site (e1). The import site can react with D-glucose to form eS2 or with extracellular nontransportable inhibitors (e.g. maltose) to form eI2. The export site can react with intracellular D-glucose to form eS1 or with intracellular nontransportable inhibitors (e.g. cytochalasin B) to form eI1. (B) A variation of this model proposed by Hebert and Carruthers [73] in which glucose transport proteins (GLUT1) of tetrameric GLUT1 expose substrate binding sites in an antiparallel manner. When one subunit undergoes an e1 to e2 conformational change, the adjacent subunit must undergo the antiparallel change and *vice versa*. This transport complex can bind S1 and S2 simultaneously.

Table 2

Michaelis and velocity constants for sugar transport in human red cells measured under conditions summarized in Table 1.

Experiment	Parameter	Simple-carrier solution[a]	D-Glucose[b]	D-Galactose[b]
Zero-trans entry	K_{21}^{zt} [c]	$\dfrac{K\,R_{oo}}{R_{21}}$	1.6±0 [14,28,29]	31.8±4.4 [39]
	V_{21}^{zt} [d]	$\dfrac{1}{R_{21}}$	39.6±6.5	28.6±1.1
Zero-trans exit	K_{12}^{zt}	$\dfrac{K\,R_{oo}}{R_{12}}$	6.5±0.9[e] [11,14,23,23]	74.4±23.9[e] [38]
			26.9±2.8[f] [24–26]	240±57[f] [38]
	V_{12}^{zt}	$\dfrac{1}{R_{12}}$	149.5±12.7[e]	241±66[e]
			186.3±24.1[f]	255±96[f]
Equilibrium exchange	K^{ee}	$\dfrac{K\,R_{oo}}{R_{ee}}$	27.0±3.6 [10,14,27,30–32]	138± 57 [30,31,38]
	V^{ee}	$\dfrac{1}{R_{ee}}$	311.3±20.3	432±44
Infinite-trans entry	K_{21}^{it}	$\dfrac{K\,R_{12}}{R_{ee}}$	1.7 [28]	21±2 [38]
Infinite-cis exit	K_{12}^{ic}	$\dfrac{K\,R_{12}}{R_{ee}}$	1.8±0.1 [10,28,33,34]	19.2±1.5 [40,41]
Infinite-cis entry	K_{21}^{ic}	$\dfrac{K\,R_{21}}{R_{ee}}$	14.3[e] [44]	21.4±16.6[e] [39]
			4.3±1.5[f] [12,25,33,35–37]	24.9±17.6[f] [39]

[a] Each Michaelis ($K_{m(app)}$) and velocity (V_{max}) parameter can be predicted from the steady-state solution to the simple-carrier hypothesis (see Table 3). The solution is provided.

[b] Measured $K_{m(app)}$ and V_{max} parameters are shown for D-glucose and D-galactose transport at 20°C.

[c] $K_{m(app)}$ are in mM.

[d] V_{max} are in mmol sugar per l cell water per minute.

[e] Measured by an initial rate method (see text for details).

[f] Measured by using an integrated Michaelis–Menten equation to obtain $K_{m(app)}$ and V_{max} from the time-course of sugar uptake or exit (see text for details).

The studies from which these data are collated are shown in parentheses.

Table 3
Simple-carrier analysis of human erythrocyte sugar transport

Constant[a]	Simple-carrier solution[b]	D-Glucose[c]	D-Galactose[c]
K [d]	$\dfrac{b_2 k_{-0}}{b_{-2} k_{-1}} = \dfrac{b_1 k_0}{b_{-1} k_1}$	1.36±0.43	11.91±4.83
nR_{00}	$\dfrac{1}{k_{-0}} + \dfrac{1}{k_0}$	6.54±1.32	8.39±0.73
nR_{12}	$\dfrac{1}{k_{-0}} + \dfrac{1}{k_1}$	1.53±0.14	0.96±0.37
nR_{21}	$\dfrac{1}{k_0} + \dfrac{1}{k_{-1}}$	5.75±1.12	7.96±0.32
NR_{ee}	$\dfrac{1}{k_1} + \dfrac{1}{k_{-1}}$	0.73±0.05	0.52±0.07

$$v_{12} = \frac{S_1(K + S_2)}{K^2 R_{00} + K R_{12} S_1 + K R_{21} S_2 + R_{ee} S_1 S_2} \qquad \text{e}$$

[a] Simple-carrier-mediated sugar transport is characterized by 5 constants: one affinity constant K and four resistance parameters R_{00}, R_{12}, R_{21} and R_{ee}; see Refs. [7,16]). K has units of mM. The products (glucose transporter) × resistance parameter $(n\,R)$ were computed assuming $n = 3.8$ µM per 1 cell water [9] and have units of ms.
[b] The steady-state solutions for these constants (see Fig. 1 for details of rate-constants).
[c] The computed values for these affinity and resistance constants based upon the measured rates of D-glucose and D-galactose transport (see Table 2).
[d] K was computed from zero-trans and equilibrium-exchange transport data only.
[e] The steady-state velocity equation for sugar efflux. S_1 is the intracellular sugar concentration while S_2 is the extracellular sugar concentration. Sugar uptake is obtained by interchanging S_1 and S_2 in the numerator.

2.3.1. Asymmetry

Table 2 summarizes the results of a large number of studies by many workers in which the parameters for D-glucose and D-galactose transport at 20°C were measured. V_{max} and $K_{m(app)}$ for zero-trans exit of either sugar are significantly greater than the corresponding parameters for zero-trans uptake of either sugar. This confirms our earlier discussions indicating that transport is asymmetric. This also means that if $k_{-0} = k_0$, k_{-1} cannot be identical to k_1 or *vice versa*. The reason for this is that the simple carrier hypothesis requires that the ratio $K_{m(app)}$exit : $K_{m(app)}$entry is given by

$$\frac{\dfrac{1}{k_0} + \dfrac{1}{k_{-1}}}{\dfrac{1}{k_{-0}} + \dfrac{1}{k_1}}$$

([7] and see Tables 2 and 3).

Zero-trans exit data fall into two categories defined both by the result and by the method of analysis. Exits obtained by methods that measure the initial rate of sugar loss into sugar-free medium [11,14,22,23) indicate a low $K_{m(app)}$ for exit (5–9 mM). This method measures sugar exit from cells loaded with varying sugar levels over a relatively short interval (1–30 s) and assumes that the measured rate is the instantaneous rate of exit uncomplicated by backflux of sugar into the cell. Exits obtained by an integrated rate equation analysis of the time-course of exit [24–26] indicate a significantly higher $K_{m(app)}$ (26 mM) for exit. This method measures the full time course of sugar exit from cells loaded initially with relatively high sugar concentrations. As exit proceeds, intracellular sugar levels fall, the exit site desaturates and the rate of exit falls. This exit time-course therefore contains a full [sugar]/exit velocity data set which can be analyzed using an appropriate form of the integrated Michaelis–Menten equation to obtain $K_{m(app)}$ and V_{max} [24]. This method also requires that exiting sugar does not saturate external sugar binding sites to cause significant sugar backflux. Measurements of V_{max} appear to be independent of the method of analysis.

The initial rate-method is presently viewed to give more accurate determinations of $K_{m(app)}$ for exit [14,27]. The theoretical basis for this conclusion has not been elaborated formally although the use of data obtained by the initial rate method provides a more satisfactory fit to the predictions of the simple carrier hypothesis. If we use the results of initial rate analyses, the ratios $V_{max}/K_{m(app)}$ for D-glucose uptake and exit are identical satisfying the known passive requirement of the transport system. If the results of integrated rate analyses are used, the ratio $V_{max}/K_{m(app)}$ for D-glucose uptake is greater than that for exit and this contravenes the known passive requirement of the transport system. We will return to this point in Section 2.3.4.

Assuming transport is mediated by a simple carrier mechanism, it is possible to use the data and relationships summarized in Table 2 to compute the affinity and resistance parameters for erythrocyte sugar transport (Table 3). From these data we can calculate boundary limits for the various rate constants describing each step of the transport model shown in Fig. 1 [7,42,43]. Table 4 shows the results of these analyses. The simple carrier analysis suggests that transport asymmetry derives from unequal rate constants for relaxation ($k_{-0} > k_0$) and substrate binding ($b_1 < b_2$). Translocation (k_{-1} and k_1) and dissociation constants (b_{-1} and b_{-2}) appear insensitive to direction. As is expected, boundary limits for relaxation constants k_0 and k_{-0} are substrate independent. More interestingly, it seems that the second-order association rate constants (b_1 and b_2) are most sensitive to substrate stereochemistry being an order of magnitude lower for D-galactose than for D-glucose.

The simple carrier hypothesis accurately predicts $K_{m(app)}$ and V_{max} for D-glucose uptake into sugar-free cells, D-glucose exit into sugar-free medium and V_{max} for exchange D-glucose transport (see Table 5). The model provides a less satisfactory fit for $K_{m(app)}$ for equilibrium-exchange D-glucose transport, $K_{m(app)}$ for inhibition of net D-glucose exit by extracellular D-glucose and $K_{m(app)}$ for infinite-cis uptake of D-glucose. Similar problems are also found in analyses of galactose exit from erythrocytes [38,40,41].

Table 4
Computed bounds for the rate-constants of the simple carrier.

Constant[a]	Computed as[b]	D-Glucose[c]	D-Galactose[d]
$k_1, k_{-1}, b_{-1}, b_{-2}$	$\dfrac{1}{nR_{ee}}$	1366	1895
k_0	$\dfrac{1}{n\,R_{21}} < k_0 < \dfrac{1}{n(R_{21} - R_{ee})}$	$173 < k_0 < 199$	$125 < k_0 < 134$
k_{-0}	$\dfrac{1}{n\,R_{12}} < k_{-0} < \dfrac{1}{n(R_{12} - R_{ee})}$	$656 < k_{-0} < 1261$	$1058 < k_{-0} < 2391$
b_1	$\dfrac{1}{n\,K\,R_{21}}$	127,667	10,533
b_2	$\dfrac{1}{n\,K\,R_{12}}$	481,667	88,833

[a] See Fig. 1 for detailed references to rate constants. k_1, k_{-1}, k_0, k_{-0}, b_{-1} and b_{-2} are first order rate constants and are in units of s^{-1}. b_1 and b_2 are second order rate constants and are in units of $molar^{-1}\,s^{-1}$.
[b] The method for computing these constants is indicated and is adapted from Refs. [7,42].
[c] The results for D-glucose and D-galactose transport at 20°C (Table 2 and Table 3) are shown.

2.3.2. High affinity endofacial site

$K_{m(app)}$ for infinite-cis net sugar uptake (saturated uptake of D-glucose into cells loaded with varying levels of D-glucose) is controversial because the result of a majority of studies supports rejection of the simple carrier hypothesis for sugar transport.

The studies of Harris [33], Hankin et al. [12], Lieb and Stein [37] and Melchior and Carruthers [25,35] measure the time course of net D-glucose uptake at 20°C from saturating concentrations of D-glucose. Each analysis indicates that the initial rate of sugar entry falls by one-half at an intracellular D-glucose concentration very similar to that extracellular concentration that inhibits saturated exit by one-half ($\approx$ 2 mM). Based upon the data listed for zero-trans and equilibrium exchange transport data in Table 2, $K_{m(app)}$ for infinite-cis net D-glucose uptake via a simple carrier mechanism is predicted to be 11 mM (Table 5).

The findings of a single study support the simple carrier mechanism suggesting that $K_{m(app)}$ for infinite-cis uptake of D-glucose at 20°C is considerably higher than 2 mM (15 mM) [44]. It has been suggested [44–46] that just as with integrated-rate analyses of sugar exit, the use of an integrated Michaelis–Menten rate equation for infinite-cis uptake to determine $K_{m(app)}$ at the interior of the cell introduces errors which in this instance, lead to underestimates of $K_{m(app)}$. Lieb and Stein [37] have reevaluated their conclusion in the light of this criticism and have developed more robust methods of data analysis. They conclude that there is not more than one chance in twenty ($p <$ 0.05) that $K_{m(app)}$ for infinite-cis entry falls outside the range 0 to 2.8 mM.

Table 5
Predicted sugar transport properties of human erythrocytes assuming transport is mediated by a simple-carrier mechanism

Experiment	Parameter[a]	D-Glucose[b]	D-Galactose[b]
Zero-trans entry	K_{21}^{zt}	1.5	12.5 ± 5.4^{c}
	V_{21}^{zt}	39.6	28.6
Zero-trans exit	K_{12}^{zt}	5.8^{d}	106
	V_{12}^{zt}	149.5	241
Equilibrium exchange	K^{ee}	12.2 ± 1.6^{c}	189
	V^{ee}	311.3	432
Infinite-trans entry	K_{21}^{it}	2.8 ± 0.1^{e}	21.3
Infinite-cis exit	K_{12}^{ic}	2.8 ± 0.1^{e}	21.3
Infinite-cis entry	K_{21}^{ic}	10.7 ± 3.8^{e}	180 ± 14^{e}

[a] See Table 1 for details of the experimental procedure and Tables 2 and 3 for the calculation and use of constants to predict $K_{m(app)}$ and V_{max} parameters for red cell sugar transport. $K_{m(app)}$ parameters are in mM and V_{max} parameters are in mmol sugar per l cell water per min.
[b] Results are calculated for D-glucose and D-galactose transport and are based upon the data listed in Table 2.
[c] This result is significantly less than the experimental observation ($p < 0.05$).
[d] This result is significantly less than the experimental finding obtained by use of the integrated rate-equation approach to determine $K_{m(app)}$ for exit.
[e] This result is significantly greater than the experimental finding ($p < 0.05$).

Ginsburg adopted an empirical approach to this problem [39]. $K_{m(app)}$ for infinite-cis uptake of galactose was measured by two approaches. The initial rate approach measures the rate of saturated net galactose uptake at various preset intracellular galactose concentrations. The results show that net uptake is reduced by half at 25 mM intracellular galactose. Ginsburg also measured $K_{m(app)}$ for infinite-cis uptake of galactose by an integrated Michaelis–Menten rate equation approach. Cells were depleted of sugar and resuspended in medium containing saturating galactose levels. Galactose uptake is initially rapid but falls as the intracellular galactose levels increase causing significant galactose efflux. This net uptake time-course therefore contains a full [intracellular galactose]/net uptake velocity data set which can be analyzed using an appropriate form of the integrated Michaelis-Menten equation to obtain $K_{m(app)}$ for exit and V_{max} for uptake [39]. Ginsburg found that $K_{m(app)}$ for infinite-cis uptake of galactose when measured using the integrated rate equation approach is 22 mM. Since galactose levels transport in red cells is

Table 6
3OMG transport properties of human red cells at 4°C (Cloherty and Carruthers, unpublished)

Experiment	Parameter[a]		b_n
Zero-trans entry	K_{21}^{zt}	0.38±0.13	3
	V_{21}^{zt}	0.18±0.02	
Zero-trans exit	K_{12}^{zt}	4.35±0.62	3
	V_{12}^{zt}	1.62±0.10	
Equilibrium exchange	K^{ee}	22.62±6.17	4
	V^{ee}	9.17±3.44	
Infinite-trans entry	K_{21}^{it}	1.52±0.36	3
Infinite-cis entry	K_{21}^{ic}	0.66±0.15[c]	3
		0.47±0.11[d]	3
		3.35±0.62[e]	2

[a] See Table 1 for details of the experimental procedure. $K_{m(app)}$ parameters have units of mM. V_{max} parameters have units of mM sugar per l cell water per min.
[b] The number of experiments carried out is shown as n.
K_m for infinite-cis uptake was measured by three methods: [c]saturated (20 mM), uni- directional uptake was measured into cells loaded with 0 to 20 mM sugar; [d]cells were loaded with sugar (0 to 20 mM) at an identical specific activity to that present outside (20 mM). Net uptake was then measured over 0–5 min. K_m is computed as that concentration of intracellular sugar that reduces net uptake by 50%; [e]the integrated rate equation approach of Hankin et al. [12] was used.

characterized by 10-fold lower affinity than is D-glucose transport (Table 1), this confirms the presence of a high affinity endofacial site and indicates that this result is independent of methodology. We have recently confirmed this finding for 3-O-methylglucose (3OMG) transport by red cells at ice-temperature (Cloherty and Carruthers, unpublished; see Table 6).

Two infinite-cis D-glucose uptake studies are reported at ice-temperature. One supports the existence of a relatively high affinity binding site for infinite-trans D-glucose exit into galactose [47] while a second study suggests that the affinity of this site exit for glucose is considerably lower [44]. Our own studies at 4°C with the nonmetabolizable but transported sugar 3OMG support the existence of this high affinity sugar exit site (Cloherty and Carruthers, unpublished, see Table 6).

2.3.3. High affinity exofacial site

Infinite-cis D-glucose exit experiments (Sen–Widdas exits) uniformly support the existence of a high affinity exofacial site for D-glucose. Saturated net D-glucose exit into saline containing D-glucose is half-maximally inhibited when the extracellular D-glucose level is ≈ 2 mM (Table 2). Baker and Widdas [48] have reexamined Sen-Widdas exits of 3OMG. To their surprise they discovered that $K_{m(app)}$ for infinite-cis exit of 3OMG is almost 2-fold lower than $K_{m(app)}$ for infinite-trans uptake of 3OMG. According to the simple-carrier hypothesis, these two Michaelis constants are identical. Although the simple-carrier hypothesis predicts a high affinity exofacial sugar binding site, the projected affinity of this site is consistently underestimated by the simple-carrier model (see Table 5).

Baker and Widdas [48] suggest that inhibitions of Sen–Widdas exits by exofacial sugars reflects two processes: (1) sugar import leading to reduced net exit and (2) sugar binding to a high affinity exofacial site that is distinct from the import site and which leads to reduced net exit and increased futile exchange.

2.3.4. Low affinity equilibrium-exchange

$K_{m(app)}$ for exchange D-glucose transport at 20°C (≈ 25–30 mM) is some 2 to 4-fold greater than that predicted by the simple carrier hypothesis (Table 5). This is interesting because this is the one transport condition (intracellular D-glucose and extracellular D-glucose levels are identical) that is least susceptible to experimental error. The measurement of exchange transport requires that tracer exchange is monitored in sugar equilibrated cells. Since this exchange is a simple exponential process [16], its measurement can be achieved using conventional sampling procedures that do not require the use of rapid sampling devices. $K_{m(app)}$ for equilibrium exchange D-glucose transport at ice-temperature [14] is consistent with the predictions of the simple carrier mechanism.

We have discussed previously the requirement of equality in the ratio $V_{max}/K_{m(app)}$ for net sugar uptake and exit. $V_{max}/K_{m(app)}$ for equilibrium exchange D-glucose transport at 20°C (0.2 s^{-1}) is considerably lower than that for zero-trans fluxes (0.4 s^{-1}). According to the simple carrier hypothesis, sugar transport displays simple saturation kinetics under all flux conditions and the relationship $V_{max}/K_{m(app)}$ net exit $= V_{max}/K_{m(app)}$ net entry $= V_{max}/K_{m(app)}$ equilibrium exchange must hold true [16,17]. The result we have just described means either that the transport determinations are compromised or that the hypothesis is incorrect. At least one complete data set [14] obtained at 20°C using a rapid stopped flow apparatus contravenes this fundamental law. This inequality cannot thus be discounted on the basis of lack of temporal resolution in transport determinations.

It should be noted that a fundamentally different hypothesis for erythrocyte sugar transport (the fixed-, or two-site carrier [8,13]) allows equilibrium exchange transport to appear to break the $V_{max}/K_{m(app)}$ equality requirement without contravening the second law of thermodynamics [18]. We discuss this model in Section 2.3.5 below.

The preceding argument assumes $K_{m(app)}$ and V_{max} for net entry are accurate but we discussed previously that $K_{m(app)}$ for exit is either high or low depending on the method of analysis. If we assume that the high $K_{m(app)}$ determined from time-course measurements represents the most accurate determination of exit kinetics, the ratio $V_{max}/K_{m(app)}$ net exit $\approx V_{max}/K_{m(app)}$ equilibrium exchange which are approximately two to three-fold lower than the ratio $V_{max}/K_{m(app)}$ net entry. In this way we might conclude that it is the net entry determinations that are flawed. These considerations suggest very strongly that arguments supporting and refuting the acceptability of experimental transport data are *circulus in probando* and that unrecognized fundamental factors are influencing measurements of erythrocyte sugar transport.

2.3.5. Paradoxical hetero-exchange

Miller [10,21] concluded that erythrocyte heteroexchange sugar transport data are inconsistent with the simple carrier hypothesis for sugar transport. The basis for this conclusion is the demonstration that the extents of trans-acceleration of D-glucose exit produced by extracellular mannose or galactose are inconsistent with the observed rates of self exchange of mannose, D-glucose or galactose and with net mannose or galactose exits. Miller's argument allows for the possibility that trans-location is more rapid than relaxation and assumes: (1) that the rates of translocation of any given sugar are identical in both directions ($k_{-1} = k_1$); (2) that the rate of relaxation is independent of direction ($k_0 = k_{-0}$); (3) that transport is symmetric; and (4) that exchange transport is saturated at the sugar concentrations used (130 mM).

The latter assumption is incorrect. $K_{m(app)}$ for exchange D-galactose transport in human red cells is 140 mM [38]. We also know that transport is asymmetric (see Sections 2.1 and 2.3.1 above) which indicates that $k_0 \neq k_{-0}$.

Miller observed that glucose exit into mannose or galactose is significantly faster than into glucose. He reasoned that this cannot be due to mannose and galactose translocations occurring more rapidly since if this were the case, then mannose–mannose and galactose–galactose exchanges would be faster than glucose–glucose exchange and this is not observed. However, this conclusion is negated when $k_{-1} \neq k_1$.

Assuming only saturated exchange exit occurs, V_{max} for exchange exit is given by

$$\frac{n\, k_1\, k_{-1}}{k_1 + k_{-1}}$$

where n is the concentration of cellular glucose transporters (3.8 $\mu mol \cdot l^{-1}$ cell water; [9]). When $k_{1(glucose)} = k_{-1(glucose)} > k_{1(mannose)} > k_{1(galactose)}$ but $k_{-1(glucose)} < k_{-1(galactose)} < k_{-1(mannose)}$, exofacial mannose and galactose will stimulate glucose exit more than will exofacial glucose. Since Miller measured self-exchange of galactose at a subsaturating concentration (130 mM $\approx K_{m(app)}$) galactose–galactose self-exchange will be slower than saturated glucose–glucose exchange. If relaxation (k_{-0}) occurs some three-fold more slowly than does k_1(glucose), V_{max} for net exit ($n\, k_1\, k_{-0}/(k_1 + k_{-0})$) will increase in the order galactose < mannose < glucose $\leq n\, k_{-0}$. These predictions broadly match the results obtained by Miller [10] and thus cannot be used as they stand to refute the simple carrier mechanism for sugar transport. However, this analysis makes an untested prediction. Galactose exit into D-glucose should occur more slowly than D-glucose exit into galactose or D-glucose. This test has not be made but would be useful for it could confirm or refute the hypothesis that $k_1 < k_{-1}$ for galactose transport.

Naftalin and Rist [49] have examined hetero-exchange sugar transport in rat erythrocytes where 3-O-methylglucose transport is symmetric and where V_{max} for 3-O-methylglucose transport is greater than that for uptake of mannose or 2-deoxy-D-glucose or for exit of mannose. This observation confirms earlier measurements of rat red cell sugar transport [50] demonstrating that unlike sugar transport in human erythrocytes, V_{max} for net sugar uptake in rat cells is strongly dependent on the transported sugar. This result also means that if sugar transport in this tissue is mediated by a simple-carrier, it is not rate-limited solely by substrate-independent relaxation (k_0 or k_{-0}) but also by translocation. Knowing that reduced mannose uptake and exit reflect reduced translocation (k_{-1} and k_1 for mannose are less than k_1 and k_{-1} for 3-O-methylglucose), Naftalin and Rist were at a loss to explain how V_{max} for mannose self-exchange is almost as great as V_{max} for 3OMG self-exchange. They conclude correctly [49] that simple-carrier kinetics cannot account for this behavior and show that a hypothetical carrier mechanism that exposes import and export sites simultaneously is more successful in predicting this result.

2.3.6. Conclusions

The properties of human erythrocyte sugar transport are superficially consistent with the predictions of the simple carrier hypothesis. Close examination, however, reveals systematic, quantitative deviations of experimental from predicted transport behavior that are not always accounted for by use of varying experimental methodologies.

A fundamentally different hypothesis — the fixed- or two-site carrier hypothesis [8,13] — can mimic the predictions of the simple carrier hypothesis [18] and, because this hypothetical carrier mechanism allows equilibrium exchange transport to deviate from simple saturation kinetics at low sugar concentrations, it can additionally give rise to circumstances where a $K_{m(app)}$ for exchange transport can be (inappropriately) measured and $V_{max}/K_{m(app)}$ for exchange transport is considerably lower than the same ratios for net uptake and exit of sugar [18]. This would provide a satisfactory explanation for red cell sugar transport were anomalous exchange transport the only deviation of experimental behavior from predicted kinetics.

However, two additional problems complicate our understanding of erythrocyte sugar transport. The red cell exposes high affinity sugar binding sites both outside and inside the cell. It is not at all certain that these sites are catalytic. Baker and Widdas [48] suggest that these sites are allosteric regulatory sites that increase carrier antiport function at the expense of uniport function. Others suggest that these sites are the result of errors in experimental analysis. While possible, it is also true that the high affinity endofacial site which was first recognized more than 30 years ago in the first infinite-cis net uptake experiments [33], has been demonstrated repetitively in multiple laboratories using divergent techniques.

In recognizing these complexities Naftalin and Holman (1977) and Baker and Naftalin (1979) concluded that the transport mechanism is either considerably more elaborate than believed previously or that what we measure as transport is not just carrier-mediated sugar translocation but rather, is the sum of translocation plus additional intracellular processing of the imported sugar (e.g. binding to an intra-cellular complex). If the latter is true, then the various carrier models hypothesized to account for erythrocyte sugar transport may be fundamentally correct but fail quantitatively because they do not consider the effects of post translocational processing on net sugar movements.

3. Inhibitions of transport

3.1. Transport inhibition is asymmetric

Cytochalasin B, phloretin, maltose and ethylidene-glucose (see Fig. 2) are inhibitors of erythrocyte sugar transport but are not themselves transported by the transport system [51–56].

Widdas and co-workers [13,53,55–57] concluded that cytochalasin B binds at or close to the sugar export site of the glucose carrier and in doing so serves as a competitive inhibitor of sugar exit but as a noncompetitive inhibitor of sugar entry. Phloretin, maltose and ethylidene glucose bind at or very close to the sugar import site and serve as competitive inhibitors of sugar uptake but as noncompetitive inhibitors of sugar exit. Because it is more hydrophobic than glucose, ethylidene glucose can also penetrate cells by transbilayer diffusion and, in doing so, can also act to inhibit transport via low affinity association with the export-site or via a site that exists on the export-competent form of the transporter. These studies indicate that the stereospecificity of nontransportable inhibitor interactions with import- and export-competent forms of the glucose carrier are different and support the view that the transport system is asymmetric. In this instance, the transporter displays asymmetric affinities for inhibitors.

The results of Barnett et al. [54,58,59] support this conclusion. Their studies support the hypothesis that during sugar import, D-glucose enters the translocation pathway with C1 leading. During export, the reverse is true where C4 leads into the translocation pathway. This reinforces the hypothesis that import and export sites do not share identical stereochemistries and suggests the intriguing possibility that the chemistry of exit is, as a first approximation, the reverse chemistry of import.

3.2. Is the transporter a ping-pong carrier?

Because some inhibitors interact only with the export competent form (e1) of the sugar transporter (e.g. cytochalasin B) while others (e.g. maltose, ethylidene glucose) react with the import competent form (e2) of the carrier, it should be possible to test whether e1 and e2 can co-exist. According to the simple carrier hypothesis, transport is a uni uni ping pong process in which e1 and e2 are mutually exclusive and only a single substrate can be bound at any instant. Since e1 and e2 interconvert via the process of relaxation in the absence of transported substrate, addition of cytochalasin B will reduce the fraction of carrier than can exist as e2, exofacial maltose will reduce the fraction of carrier that exists as e1 and these inhibitors will compete for binding.

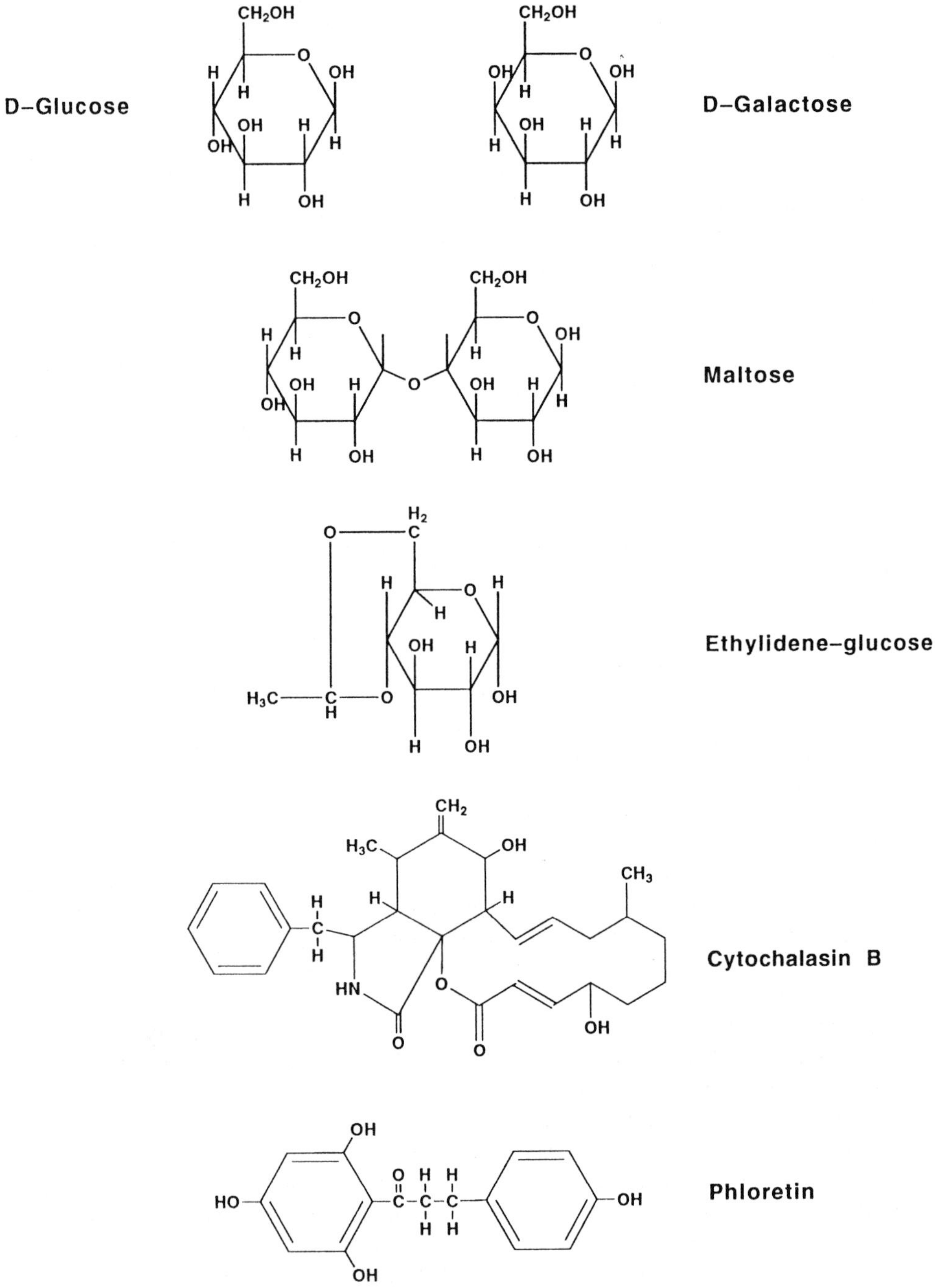

Fig. 2.   Some substrates (D-glucose and D-galactose) and inhibitors (maltose, ethylidene glucose, phloretin and cytochalasin B) of the human red cell sugar transport system.

A carrier that can expose import and export competent states simultaneously, however, should also be capable of binding cytochalasin B and exofacial maltose simultaneously.

3.2.1. Studies of transport inhibition

If e1 and e2 are mutually exclusive, maltose binding to e2 will reduce the availability of e1 and thereby competitively inhibit cytochalasin B binding. Similarly, cytochalasin B will inhibit maltose binding by reducing the availability of e2. These effects are competitive. Thus $K_{i(app)}$ for cytochalasin B inhibition of sugar uptake will be increased competitively by exofacial maltose and $K_{i(app)}$ for maltose inhibition of sugar uptake will be increased competitively by cytochalasin B. If both ligands bind to the transporter simultaneously, exofacial maltose will leave $K_{i(app)}$ for cytochalasin B inhibition of transport unchanged and *vice versa*. This latter conclusion assumes no cooperativity between maltose and cytochalasin B binding. If these ligand binding sites interact with negative cooperativity, exofacial maltose will increase $K_{i(app)}$ for cytochalasin B inhibition of sugar transport. However, this will not be simple linear competitive inhibition. Rather, $K_{i(app)}$ for cytochalasin B inhibition of transport will increase in a saturable fashion with maltose concentration. Similarly, positive cooperativity between maltose and cytochalasin B binding sites would be revealed as increased affinity for ligand binding in the presence of trans-ligand.

This distinction between simple carrier and so-called fixed- or two-site carrier-mediated transport inhibitions resulting from the simultaneous presence of e1 and e2-reactive transport inhibitors is the basis for testing whether the carrier exposes e1 and e2 sites alternately or simultaneously [60]. This insightful strategy examines directly the features that distinguish various hypothetical carrier mechanisms. Unfortunately, the results obtained by Krupka [60] are consistent either with a simple carrier mechanism or with a carrier in which binding of exofacial phloretin and endofacial cytochalasin B show very strong negative cooperativity. Moreover, phloretin inhibitions of transport display a complexity [53] that is difficult to reconcile with the hypothesis that phloretin binds to e2 alone. We therefore re-examined these analyses using a more conventional trans-inhibitor pair to probe the sugar transporter of intact cells. Our findings [61] demonstrate that exofacial maltose and endofacial cytochalasin B inhibitions of D-glucose uptake are consistent with the hypothesis that maltose and cytochalasin B binding sites coexist on the transporter but that these sites interact with negative cooperativity. Thus maltose-saturation of the exofacial site increases $K_{d(app)}$ for cytochalasin B binding to the endofacial site by 3-fold and *vice versa*.

The major assumptions of simple- and two-site carrier inhibition analyses are that the exofacial maltose and endofacial cytochalasin B binding sites correspond to D-glucose import and export sites respectively. The studies of Krupka [62] and Barnett et al. [54] suggest that this may not be true for the cytochalasin B binding site.

Baker and Widdas [48] have shown that $K_{i(app)}$ for inhibition of infinite-cis exit by EG (a nontransported sugar that binds with high affinity to the sugar import site) is almost 3-fold lower than $K_{i(app)}$ for inhibition of infinite-trans 3OMG uptake. The simple carrier hypothesis requires that these inhibitory constants are identical. This result is similar to their observation [48] that extracellular 3OMG inhibits saturated 3OMG net exit half-maximally at a concentration significantly lower that $K_{m(app)}$ for 3OMG uptake into cells containing saturating levels of 3OMG. Baker and Widdas interpret their data as being consistent with two exofacial sugar binding sites — a high affinity site (revealed when the endofacial catalytic sites are saturated) which serves as an allosteric (non catalytic) regulatory domain and a lower affinity catalytic site.

3.2.2. Substrate and inhibitor binding

The simple carrier hypothesis predicts that the glucose carrier cannot bind exofacial maltose and endofacial cytochalasin B simultaneously because this type of transporter can bind only a single substrate at any instant. More complex, multisubstrate binding models relax these restrictions and allow for multiple populations of sugar and ligand binding sites within a single population of carriers. The results of investigations of cytochalasin B binding to human erythrocytes fall into two categories — those that support the simple carrier hypothesis and those that do not.

Theoretical considerations of cytochalasin B binding to a simple carrier (see Fig. 1) reveal the subtleties of this hypothetical transport mechanism [63]. Cytochalasin B is presumed to interact with the e1 conformer at the export site. Addition of a nontransported but binding competent intracellular sugar competes directly with cytochalasin B for binding to e1 and inhibition of binding is simple competitive-inhibition. Addition of an extracellular binding competent but nontransportable sugar prevents e2 relaxation to e1, traps the carrier in the e2 state and thereby reduces available e1 for interaction with cytochalasin B. Again, this inhibition is simple linear competition. Now consider the effect of addition of a transported sugar to one side of the membrane only. Intracellular glucose not only competes directly with cytochalasin B for binding to e1, it also reduces available e1 by promoting the eS1 → eS2 conformational change. Extracellular D-glucose increases e1 by promoting the eS2 → eS1 conformational change. The effects that extra- and intracellular ligands exert on cytochalasin B binding to e1 thus depends upon relative affinities of e1 and e2 for ligands and the relative magnitude of the first order rate constants k_0, k_{-0}, k_1 and k_{-1}. Now consider 4 conditions: (1) D-glucose is absent; (2) saturating extracellular D-glucose is present; (3) saturating intracellular D-glucose is present; and (4) saturating D-glucose is present both inside and outside the cell.

In the absence of sugar, the relative amount of transporter presenting export (e1) sites is given by the ratio $k_{-0}/(k_0 + k_{-0})$. When external sugar is saturating e1 is proportional to $k_{-1}/(k_0 + k_{-1})$. When internal sugar is saturating e1 is proportional to $k_{-0}/(k_1 + k_{-0})$ and when sugar is saturating both inside and outside the cell, e1 is proportional to $k_{-1}/(k_1 + k_{-1})$. Using the boundary estimates for rate constants for red cell D-glucose transport data at 4°C (Table 5) we predict: (1) extracellular D-glucose will have little effect on cytochalasin B binding because most of the carrier (93%) is already in the e1 state; (2) intracellular D-glucose will massively inhibit cytochalasin B binding because saturating intracellular D-glucose reduces available e1 to only 7% of total carrier; (3) when saturating sugar is present both inside and outside the cell, available e1 (e1 + eS1) increases to 52% of total carrier and inhibition of cytochalasin B binding is significantly less (8-fold) than that produced by intracellular D-glucose alone.

This exact experiment has been carried out at ice-temperature using 100 mM D-glucose as the inhibitory sugar [63]. The results show that the combination of intra- and extracellular D-glucose is no less potent than intracellular D-glucose alone in inhibiting cytochalasin B binding to the glucose transporter. This result contradicts the predictions of the simple carrier hypothesis.

In the same studies [63], it was also shown that exofacial maltose, ethylidene glucose and phloretin produce inhibitions of cytochalasin B binding that cannot be accounted for by a simple linear competition. We concluded that the exo- and endofacial ligand binding sites display negative cooperativity when occupied by these transport inhibitors. May [64] has also examined the effects of exofacial maltose on cytochalasin B binding in red cells but finds that inhibition is of the linear competitive type expected for a simple carrier mechanism.

Studies of cytochalasin B binding to glucose transport protein (GLUT1) purified from human erythrocytes or present in unsealed erythrocyte membranes show that cytochalasin B binding sites and sugar binding sites are mutually exclusive [65–69]. The stoichiometry of cytochalasin B binding to GLUT1 is close to unity [70] but falls (reversibly) to 0.5 or less upon removal of reductant [71–73].

Sugar binding to glucose transport protein purified from human erythrocytes or present in unsealed erythrocyte membranes reveals a single [74–76] or multiple [26,73,77,78] sugar binding sites. Sugar binding studies are possible because the intrinsic tryptophan fluorescence of GLUT1 is modified upon binding sugar [74].

The results of Chin et al. [78] are interesting for they show that although cytochalasin B reduces GLUT1 intrinsic fluorescence at 355 nm, D-glucose (but not cytochalasin B) reduces fluorescence at 355 nm but increases fluorescence at 315 nm. This is accounted for by a shift in the GLUT1 emission spectrum to shorter wavelengths that is promoted upon D-glucose binding. What is intriguing however is that increased fluorescence at 315 nm is half-maximal at a glucose concentration of 36 mM while reduced fluorescence at 355 nm is half-maximal at 6 mM. This type of bimodal sugar binding kinetics cannot be explained by the simple carrier mechanism and one must conclude that either one of the sugar binding sites is unassociated with glucose translocation or that the simple carrier hypothesis for sugar transport is incorrect. Hebert and Carruthers [73] have shown that multiple sugar binding sites with widely different affinities for sugar are either lost or are converted to multiple sites with similar affinities for sugar when the purified transporter is exposed to reductant.

3.2.3. Pre-steady-state measurements

Two studies have exploited rapid quenching procedures and stopped-flow fluorimetry to investigate the presteady-state transport and ligand binding properties of the erythrocyte sugar transport system [76,79].

The study of Lowe and Walmsley [79] was designed to measure $e1 \leftrightarrow e2$ turnover (relaxation) in a simple carrier. Red cells were incubated in 150 mM maltose (a non transported but reactive sugar) to trap the hypothetical simple carrier in the external (e2.maltose) conformer. These cells were then transferred to 7.5 mM maltose medium containing 0.1 mM glucose. According to the simple carrier hypothesis, this will lead to a surge of glucose uptake (e2.maltose is reduced due to dissociation of the inhibitor resulting in increased e2 for combination with glucose) which decays as the carrier relaxes to a normal steady state distribution of conformers (e1 > e2 because $k_{-0} > k_0$). Glucose uptake is enhanced by the initial preincubation with maltose because more carrier is initially trapped in the e2 state than would occur in the absence of maltose.

The predicted transient surge in glucose uptake produced upon maltose-preincubation was observed to be greater than that observed in the absence of preincubation with maltose and decayed with a longer half-life than in the absence of maltose. These data are qualitatively consistent with the simple carrier hypothesis for sugar transport [79]. However, subsequent quantitative analysis of these data by Naftalin [80,81] leads to rejection of the simple-carrier hypothesis. The observed increase in uptake is some 32-fold greater than that predicted by the simple carrier hypothesis and the time course of decay or relaxation of the uptake surge is much faster than predicted by the model.

The studies of Appleman and Lienhard [75,76] examined substrate induced quenching of purified glucose carrier intrinsic fluorescence. Using similar rapid mixing methods, they monitored the time course of transporter fluorescence changes promoted upon substrate interaction with the site that quenches GLUT1 fluorescence. The temperature dependence of substrate-induced changes in transporter fluorescence are consistent with the steady-state studies of Lowe and Walmsley [14] and with the hypothesis that transport is mediated by an almost symmetric (asymmetry ≈ 1.6, efflux:influx) simple carrier. This supports previous reconstitution studies that suggest that the purified, reconstituted glucose transporter is intrinsically symmetric [82].

4. Other factors influencing erythrocyte sugar transport

4.1. Cell storage

Weiser et al. [32] have examined the effects of prolonged cold-storage of erythrocytes on D-glucose equilibrium exchange transport by red cells. They observe that $K_{m(app)}$ for exchange transport is doubled by cold-storage. While the reason for this effect is unknown, this observation goes some way to explaining the anomalously high $K_{m(app)}$ for exchange transport. Helgerson et al. [83] have reported that $K_{m(app)}$ for equilibrium-exchange transport in ATP-free erythrocyte ghosts is doubled relative to $K_{m(app)}$ for exchange transport in red cells or in red cell ghosts containing ATP. V_{max} for exchange transport is unchanged by ATP-removal.

4.2. Cellular nucleotide content

Erythrocyte sugar transport asymmetry is significantly reduced or perhaps is even lost in red cell ghosts [23,25,26,36,77,84–87]. Asymmetry can be restored by reconstitution of ghosts with red cell cytosol [25,84]. The cytosolic factor respon-sible for transport modulation appears to be ATP which reduces $K_{m(app)}$ and V_{max} for sugar uptake, reduces $K_{m(app)}$ for equilibrium exchange transport, reduces V_{max} for exit and increases $K_{m(app)}$ for sugar exit. ATP binds to the glucose carrier to modulate the affinity of the carrier for sugars [26,77,88,89] and promotes a conformational change in the carboxyl-terminus of the GLUT1 protein [89]. ATP-binding to the transport protein is inhibited by AMP and by ADP which appear to function as inhibitors of ATP-inhibition of transport [89]. Calcium ions inhibit sugar uptake by red cells and red cell ghosts [83,90,91]. This inhibitory action of Ca^{2+} is reversed by ATP. Unlike ATP-modulation of transport, however, ATP-reversal of Ca^{2+}-inhi-bition of transport appears to require ATP-hydrolysis [83].

These findings suggest that erythrocyte sugar transport may be intrinsically symmetric in the absence of cytosolic ATP but that it is susceptible to direct modulation by ATP and to indirect modulation by Ca^{2+}. The variability in reported parameters for transport might result from variable cellular nucleotide and Ca^{2+} status arising from prolonged cold-storage or prolonged incubation in glucose-free media. The findings of Weiser et al. [32] are consistent with this hypothesis.

4.3. Transporter oligomeric structure

Five independent lines of evidence directly support the view that the erythrocyte glucose transporter is a multisubunit structure. (1) Target inactivation analysis indicates that the D-glucose-inhibitable cytochalasin B binding component of human erythrocytes is either a 124 kDa or a 220 kDa species [92–94]. (2) Reversible chemical cross-linking of human erythrocytes membrane proteins indicates that the sugar transport protein GLUT1 (a 55 kDa species [95–97]) is present in a 220 kDa complex (Hebert and Carruthers, unpublished). Irreversible chemical cross-linking of purified GLUT1 shows that GLUT1 is cross-linked as tetramer in the absence of reductant and as a dimer when reductant is included [72,73]. (3) Shadow analysis of freeze-fracture electron micrographs of reconstituted, reduced GLUT1 indicates a particle size consistent with a GLUT1 dimer [67]. (4) Wild-type, endogenous GLUT1 is co-immunoprecipitated from Chinese Hamster Ovary (CHO) cells when a GLUT1-GLUT4 chimera is immunoprecipitated using a GLUT4-specific antis-erum [98]. (5) Hydrodynamic size analysis of detergent solubilized GLUT1 indi-cates that the detergent/lipid/protein micelles contain more than one copy of GLUT1 [66,72,73,96,99]. When GLUT1 is solubilized from membranes in the presence of reductant, micellular GLUT1 content is two copies per micelle. When GLUT1 is solubilized in the absence of reductant, micellular GLUT1 content is 4 copies per micelle [73].

While the glucose transporter is a GLUT1 oligomer, the importance of trans-porter oligomeric structure to transport function is only now becoming apparent. It is formally possible that although the transporter is a complex of four copies of the glucose transport protein, each protein (or subunit) is independently catalytically active and the transporter's oligomeric structure simply reflects that GLUT1 struc-ture that is most thermodynamically stable within the membrane bilayer. It is also possible that normal transporter function requires obligate coupling between subunits and that individual subunits are incapable of catalytic function in isolation. Studies from this laboratory are beginning to shed some light on the relationship between transporter secondary structure, transporter oligomeric structure and transporter function.

Zoccoli et al. [96] first demonstrated that glucose transport protein cytochalasin B binding capacity is halved if reductant is omitted during solubilization of GLUT1 from erythrocyte membranes. Hebert and Carruthers [72] later showed that purification of GLUT1 in the presence of reductant results in a molar stoichiometry of cytochalasin B binding to GLUT1 protein of almost 1:1 whereas omission of reductant halves the cytochalasin B binding of the purified protein. Conventional wisdom suggested that omission of reductant results in the denaturation of one half of the functional transport protein. However, it was subsequently demonstrated that incubation of erythrocyte membranes either at high pH (24°C) in the presence of reductant or at neutral pH (37°C) in the presence of reductant doubles membrane cytochalasin B binding capacity, results in the loss of negative cooperative interactions between extracellular maltose and intracellular cytochalasin B binding sites and in the loss of multiple affinity D-glucose binding kinetics [72,73]. Assuming GLUT1 is not also denatured in its native environment, this suggests a more fundamental effect of reductant on transporter ligand binding.

Hydrodynamic size analyses indicate that the transporter is a GLUT1 dimer in the presence of reductant and a GLUT1 tetramer in the absence of reductant [72,73]. Analysis of GLUT1 free sulfhydryl content demonstrates that each sodium dodecylsulfate-unfolded subunit of tetrameric GLUT1 exposes only two thiol groups whereas each subunit of dimeric GLUT1 exposes six thiols [71,73]. Since each GLUT1 protein contains 6 cysteine residues, this suggests the potential for as many as 2 intramolecular disulfide bonds or 4 mixed disulfides. It is clear that mixed disulfides, if present, are not formed between subunits because tetrameric GLUT1 is resolved as a monomeric species upon nonreducing sodium dodecylsulfate polyacrylamide gel electrophoresis [71,73]. Assays of tetrameric GLUT1 subunit disulfide bridge content suggest the presence of only a single intramolecular disulfide bridge [71]. Differential alkylation/peptide mapping/N-terminal sequence analyses of tetrameric GLUT1 indicate that the N-terminal and C-terminal halves of GLUT1 are not disulfide-linked and that 3 of the 4 inaccessible thiols of each subunit of tetrameric GLUT1 (cysteine residues 347, 421 and 429) are located within the C-terminal half of the protein [71]. This result is consistent with the effects of serine-substitutions for cysteine at GLUT1 residues 347 or 421 which prevent exposure of tetrameric GLUT1-specific epitopes whereas serine substitutions at cysteine residues 133, 210, 207 or 429 are without effect on the accessibility of tetrameric GLUT1-specific epitopes [71]. These studies are consistent with the hypothesis that each subunit (GLUT1 protein) of the sugar transporter contains a single intramolecular disulfide bridge between cysteines 347 and 421 which in someway stabilizes a subunit structure that promotes GLUT1 tetramerization.

Two contra-indicating studies show that expression of a cysteine-less GLUT1 mutant in *Xenopus* oocytes results in a sugar transporter whose kinetic properties are similar to those of the wild-type protein when expressed in *Xenopus* oocytes [100,101]. Our own studies indicate that transporter expressed in *Xenopus* oocytes, unlike GLUT1 expressed in erythrocytes and CHO cells, neither exposes tetrameric-GLUT1 specific epitopes nor is it sensitive to inhibition by extracellular reductant [71]. This suggests to us that GLUT1 may not be fully post-translationally processed in *Xenopus* and may exist as a GLUT1 dimer in this tissue.

Why do both dimeric and tetrameric GLUT1 expose two cytochalasin B binding sites? We have proposed [71–73] that each GLUT1 protein can function as if a simple carrier but that the conformation (e1 or e2) that each subunit adopts depends upon whether the subunit is present within dimeric or tetrameric forms of the transporter. Specifically we have suggested that dimeric GLUT1 consists of two structurally associated but functionally independent subunits. Cytochalasin B is believed to interact specifically with the e1 form of the simple carrier. Because each subunit of the dimer can undergo e1 ↔ e2 interconversions independently of its neighbor, dimeric GLUT1 can bind 2 mol cytochalasin B per mol dimer.

In tetrameric GLUT1, subunit conformations are arranged in an antiparallel manner (Fig. 1). If one subunit presents an e2 conformation, the adjacent subunit must present an e1 conformation and *vice versa*. When one subunit undergoes an e2 → e1 conformational change, the adjacent subunit must undergo the antiparallel (e1 → e2) translocation. In this way each tetramer can present only two e1 conformers at any instant and can thus bind only 2 molecules of cytochalasin B per 4 molecules of GLUT1. This form of the transporter is comprised of subunits which individually serve as simple carriers but which together provide the transport complex with overall properties that are fundamentally different from those of a simple carrier. The tetramer exposes import and export sites simultaneously whereas each subunit in isolation or each dimer can present only e1 or e2 sites at any instant.

If different laboratories have unknowingly studied different oligomeric (reduced/nonreduced) forms of the transporter, this model could explain why some studies support the view that the erythrocyte sugar transporter is a simple carrier whereas others indicate that the carrier binds substrates at import and export sites simultaneously.

4.4. Compartmentalization of intracellular D-glucose

A number of theoretical and experimental analyses of erythrocyte sugar transport indicate that compartmentalization of intracellular sugar can and does impact seriously the interpretation of experimental transport data [8,9,15,16,18,21,50, 102,103]. *In vitro* measurements suggest that at physiological [Hb], as much as 80% of intracellular glucose is bound nonspecifically to hemoglobin [8]. The results of counterflow and zero-trans sugar exit experiments with human and rat erythrocytes support the hypothesis that intracellular sugars are not uniformly distributed within the cell [50,102].

What are the experimental consequences of compartmentalization of intracellular sugar? We have previously simulated a theoretical situation in which two sugar compartments exist inside the cell [9,18,50]. The first and smaller compartment

(C_m) lies beneath the cell membrane in direct contact with the glucose transport protein. The second, larger compartment (C_i) might represent bulk cell water or hemoglobin-associated D-glucose. D-Glucose exchanges slowly between C_m and C_i. If we arbitrarily simulate transport via a symmetric transport system and estimate net transport rates from the first second of simulated transport, a number of interesting results are found. During sugar uptake, C_m rapidly fills with D-glucose and exchanges only slowly with C_i. As a result [D-glucose] at the cytoplasmic surface of the plasma membrane is seriously underestimated and significant efflux of sugar occurs. The consequence is that V_{max} for sugar uptake is underestimated. During efflux, C_m empties rapidly and further D-glucose exit requires flux of sugar from C_i to C_m. The net effect is that $K_{m(app)}$ for exit is seriously overestimated. The result is an apparent asymmetry in transport (5.6–7.4-fold) regardless of the fact that the translocation process is intrinsically symmetric. These arbitrary simulations illustrate the potential unreliability of transport measurements in human red cells made at intervals as short as even 1 s should an unstirred layer (permeability barrier) exist below the plasma membrane.

Can we test such a model? We suspected that if transport were rate-limited by slow complexation with an intracellular sugar binding species or by slow transport across an intracellular diffusion barrier, then uptake or exit of sugar at very low sugar concentrations would be bi-exponential processes. We therefore examined uptake and exit of 50 μM 3-O-methylglucose (3OMG) in human erythrocytes and human K562 leukemic cells at 4°C [103]. Sugar uptake is a bi-exponential process. In red cells, the first process rapidly fills a small cellular compartment (k = 7.4 ± 1.7 min^{-1}; vol = 29 ± 6%) while the second process slowly fills the bulk cytosol (k = 0.56 ± 0.11 min^{-1}; vol = 71 ± 6%). Sugar exit is a monoexponential process. Several phenomena could give rise to bi-exponential sugar uptake.

Sugar metabolism cannot account for this result because the transported sugar (3OMG) is nonmetabolizable. Were the transporter to contain a high affinity, negative-feedback sugar binding site within a cytosolic GLUT1 domain, sugar uptake would be rapid during initial phases but would slow as intracellular sugar rises and the regulatory site becomes saturated. However, this model also predicts that exit is accelerated as the regulatory site desaturates ($K_{m(app)}$ for exit falls with falling intracellular sugar) and this would result in an hysteresis in the exit progress curve that is not observed experimentally. We therefore reject the kinetic explanation for biphasic sugar uptake.

A more simple explanation describes uptake of sugar into two parallel compartments (i.e. two populations of cell sizes are present in the cell suspension). This seems unlikely, however, since the first-order rate constant for sugar uptake (protein-mediated or leakage) into a sphere of volume v, is given by:

$$k = \text{constant} \cdot vol^{-1/3}$$

For particles of volume v and v/2.5, k varies by only 1.3-fold. The observed difference is 13-fold. Moreover, it can be shown that for any body, the ratio of GLUT1-mediated sugar uptake (transport) to transbilayer diffusion (leakage) is independent of volume and is given by

$$\frac{[GLUT1]\text{k}_{cat}}{P\,N\,K_{m}}$$

where [GLUT1] is surface glucose transporter density, k_{cat} is the turnover number of the transporter, P is the permeability coefficient of the membrane bilayer to sugar, K_m is the Michaelis constant for GLUT1-mediated sugar uptake and N is Avagadro's constant. If two populations of cell sizes give rise to fast and slow components of protein-mediated sugar uptake, sugar leakage should also show fast and slow components. This was not observed and thus permits rejection of this hypothesis. If a second population of cells exists with lower volume and higher sugar transporter cell-surface density, this population would transport sugars more rapidly than the larger cells. Our calculations show that such a population of spherical cells would be characterized by a diameter of 75% of that of the larger cells and would present cell-surface GLUT1 at a density 10-fold higher than that of the larger cells. More importantly, this population of cells would also display elevated rates of sugar exit relative to the larger cells at limiting sugar concentrations and this was not observed. We therefore reject the hypothesis of multiple cell populations of differing GLUT1 content.

If small and large cellular compartments do not exist in parallel, they must be present in series. Since the smaller compartment fills more rapidly than the larger, we conclude that this compartment lies between the glucose transporter and the bulk cell cytosol. The questions we must now address are: (1) Does the compartment extend uniformly across the cytoplasmic surface of the erythrocyte membrane? (2) Why is sugar exit a simple exponential process?

We approached the first question by permeabilizing the red cell membrane using α-toxin of *Staphylococcus aureus*. Our rationale was to bypass the glucose transporter as the major portal for sugar import but in a way as to enable uptake rates that would still be limited by a slower intracellular process. Our studies show that permeabilized erythrocytes fill within 5 seconds of exposure to extracellular sugar. This suggests that the series barrier does not extend uniformly across the endofacial surface of the membrane. Rather, it must be limited (structurally or functionally) to locations coincident with glucose transporter-mediated sugar entry. If the series barrier were an unstirred sugar layer acting to reduce the rate of cytosolic sugar diffusion, bypassing the glucose transporter by cell membrane permeabilization is not expected to impact this phenomenon.

Because of the very high glucose transporter content of human red cells (6% of total erythrocyte surface area is occupied by GLUT1 [73]), we expected to observe uniform cell surface staining when erythrocytes were stained for GLUT1. Contrary to expectations, the GLUT1 staining patterns obtained suggest that GLUT1 distribution is restricted to surface domains of very high local transporter density [103]. This staining pattern contrasts with the uniform cell-surface distribution of the anion transporter noted in the same study and previously in the study of Lieberman and Reithmeier [104]. It is probable, therefore, that the series barrier is coincident with locations of high GLUT1 density.

The series barrier is not formed from macromolecules that are lost upon cellular lysis because multicomponent sugar uptake is still observed in erythrocyte ghosts which lack more than 95% cellular hemoglobin [103]. Human K562 leukemic cells (a pre-erythroid cell line) also shows multi-component sugar uptake which is unaffected by subsequent cell-induction to synthesize hemoglobin [103]. This suggests that sugar binding to hemoglobin (whether free in bulk cytosol or anchored to the membrane) is not responsible for this phenomenon. While hypotonically lysed and washed erythrocytes retain a disproportionate amount of hexokinase I suggesting a membrane association (28% based upon immunoblot analysis), the distribution of this enzyme does not coincide with that of GLUT1 as judged by dual-staining immunofluorescence microscopy [103]. Glycolytic enzymes suggested to associate strongly with the erythrocyte membrane include glyceraldehyde phosphate dehydrogenase and phosphoglycerate kinase [105,106]. GAPDH and bacterial glucokinase have been shown to associate in a nucleotide-dependent fashion with erythrocyte GLUT1 *in vitro* [107,108].

The molecules that associate with erythrocyte GLUT1 are not known. Equilibrium 3OMG space analysis [103] indicates that erythrocytes contain some 1.2×10^6 3OMG binding sites per cell (30 µM; GLUT1:binding site molar ratio = 1:4). The apparent binding constant ($K_{d(app)} = 200$–400 µM) is close to $K_{m(app)}$ for 3OMG net uptake at this temperature [83]. ATP-depletion (e.g. ghost-formation) reversibly slows the rapid (presumed translocation) step slightly but reduces the slower (presumed sugar shunting) step considerably. It is unclear whether the action of ATP on the slow phase of net uptake is related to ATP-hydrolysis dependent reversal of Ca^{2+}-inhibition of sugar transport [83] or to the allosteric (hydrolysis-independent) modulation of transport resulting from nucleotide binding to GLUT1 [89].

We have modeled erythrocyte sugar transport as a four compartment process (Fig. 3). The compartments are the extracellular water (C_o), the endofacial, peri-carrier space (C_m), the sugar binding compartment (C_b) and the bulk, cytosolic water (C_i). Sugar transporters are laterally segregated into domains of very high local transporter density. Below these domains lie the sugar binding complexes (C_b) which may form a complex with cytoplasmic domains of GLUT1 in order that sugar released at the exit site is rapidly bound to the complex. It is also possible that the hypothesized extrinsic 'sugar binding complex' is, in fact, an intrinsic function of the transporter (GLUT1) complex *per se*. If so this could explain the reported multiple affinities shown by GLUT1 for sugars [73].

Sugar bound to C_b (S_b) can dissociate into either C_m or into C_i — the direction of net dissociation being determined by the sugar concentration gradient across the complex (S_m–S_i). C_b thus serves to bind sugars in a saturable fashion and 'transports' bound sugars between C_m and C_i. This transport function is essential for successful model fitting. Simulations of net transport demonstrate that both uptake and exit show marked biphasic kinetics when only a simple binding function is ascribed to C_b (i.e. S_b can dissociate only to S_m).

When extracellular sugar levels are low, the pseudo first-order rate constant for saturable sugar transport ($k_t = V_{max}/K_m$) and the ratio bound to free intracellular sugar [$S_b/(S_m + S_i)$] are greatest. Here, net sugar movement between C_m and C_i is rate-limited by the shunt pathway and by desorption from C_b. At high sugar levels, the pseudo-first-order rate constant for transport falls significantly (GLUT1-mediated transport saturates and $k_t \approx V_{max}/S_o$), the ratio bound to free intracellular sugar falls markedly and the rate of diffusion via the shunt pathway approaches the rate of transport across the cell membrane ($k_4 \approx k_t$). Sugar exit is rate-limited by the availability of S_m. At low sugar concentrations where the bound to free intracellular sugar ratio is greatest, S_m availability is governed by desorption from the complex and by the shunt pathway (k_4; $S_i \rightarrow S_m$). Provided C_m is small (<5% cell water), this compartment will be difficult to measure accurately and sugar exit will appear mono-exponential with apparent first-order rate constant k_4. Biphasic sugar uptake and monophasic sugar exit are modeled accurately by this hypothesis [103]. The sugar binding properties of red cells are rather less well modeled by this scheme. Although the 3OMG space/sugar binding capacity of the red cell is well-approximated by the model, the predicted dissociation constant for 3OMG binding (k_{-3}/k_3 = 1.25 µM) is significantly lower than measured $K_{d(app)}$ ((200–400) ± 140 µM). This theoretical result may lie within experimental error [103] but it is also possible that the measured sugar binding properties of erythrocytes also reflect binding at other sites that do not limit transport and which are not yet considered in the model.

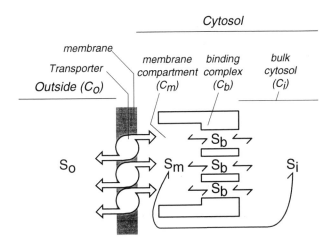

Fig. 3. A model for net sugar transport by human erythrocytes. The upper diagram shows clusters of transporters in the membrane. Transporter/sugar collision frequency is postulated to be limited by increasing medium viscosity. Transporters deliver extracellular sugar (S_o) to a sub-membranous compartment (C_m) of limited free water content at a rate described by the pseudo-first-order rate constant k_t. Sugar within this compartment (S_m) can exit rapidly via transport (k_t), can leak slowly (k_4, a first-order process) into bulk cytosol (C_i) to form Si or can bind rapidly (k_3, a second-order process) and reversibly to a sugar binding complex (C_b) to form bound sugar (S_b). S_b dissociates (k_{-3}, a first-order process) either to S_m or to S_i. S_i can access the sugar exit site either via leakage into C_m k_4) or via binding (k_3) to C_b and subsequent dissociation (k_{-3}) into C_m.

5. Concluding remarks

We conclude that net sugar transport by human erythrocytes is the sum of two sequential processes — sugar translocation and reversible intracellular sugar binding. Sugar binding may occur at either endofacial GLUT1 sites or at a complex in very close association with the glucose transporter. Because of this, steady-state sugar transport measurements report both translocation and binding steps and cannot be used directly to model the translocation process in isolation. This process also illustrates the potential for control of net sugar transport by reversible recruitment of sugar binding complexes to the transporter — a process previously shown to occur in thymocytes, CHO cells and in macrophages [109–112]. These considerations suggest that interpretation of steady state sugar transport data must be carried out cautiously.

It remains formally possible that the transport process can be modeled successfully by one or more of the hypothetical transport mechanisms suggested to date. Exchange transport data, transport inhibition data and ligand binding results suggest to us that the transport system does not function as a simple carrier but rather, can bind export and import ligands simultaneously. This behavior appears to result from cooperative interactions between subunits of the multisubunit transport complex (the GLUT1 tetramer). It is unclear whether individual subunits are functional in isolation from the native (tetrameric GLUT1) transport complex but subunits of the reduced form of the transporter (a GLUT1 dimer) may function independently as simple carriers.

Acknowledgements

This work was supported by NIH grants: DK 36081 and DK 44888.

References

1. Bang, O. and Orskov, S.L. (1937) J. Clin. Invest. **16**, 279–288.
2. Wilbrandt, E.M., Guensberg, E. and Lauener, J. (1947) Helv. Physiol. Acta **5**, C20.
3. LeFevre, P.G. (1948) J. Gen. Physiol. **31**, 405.
4. Widdas, W.F. (1952) J. Physiol. (Lond.) **118**, 23–39
5. Hodgkin, A.L. and Huxley, A.F. (1952) J. Physiol. (Lond.) **117**, 500–544.
6. Widdas, W.F. (1954) J. Physiol. (Lond.) **125**, 163–180.
7. Stein, W.D. (1986) in: Transport and Diffusion across Cell Membranes, pp. 231–305, Academic Press, New York.
8. Naftalin, R.J. and Holman, G.D. (1977) in: Membrane Transport in Red Cells, eds J.C. Ellory and V.L. Lew. pp. 257–300. Academic Press, New York.
9. Carruthers, A. (1990) Physiol. Rev. **70**, 1135–1176.
10. Miller, D.M. (1968) Biophys. J. **8**, 1329–1338.
11. Miller, D.M. (1971) Biophys. J. **11**, 915–923.
12. Hankin, B.L., Lieb, W.R. and Stein, W.D. (1972) Biochim. Biophys. Acta **288**, 114–126.
13. Baker, G.F. and Widdas, W.F. (1973) J. Physiol. (Lond.) **231**, 143–165.
14. Lowe, A.G. and Walmsley, A.R. (1986) Biochim. Biophys. Acta **857**, 146–154.
15. Regen, D.M. and Tarpley, H.L. (1974) Biochim. Biophys. Acta **339**, 218–233.
16. Lieb, W.R. and Stein, W.D. (1974) Biochim. Biophys. Acta **373**, 178–196.

17. Krupka, R.M. (1989) Biochem. J. **260**, 885–891.
18. Carruthers, A. (1991) Biochemistry **30**, 3898–3906.
19. Levine, M., Oxender, D.L. and Stein, W.D. (1965) Biochim. Biophys. Acta **109**, 151–163.
20. Mawe, R.C. and Hempling, H.G. (1965) J. Cell. Comp. Physiol. **66**, 95–102.
21. Miller, D.M. (1968) Biophys. J. **8**, 1339–1352.
22. Brahm, J. (1983) J. Physiol. **339**, 339–354.
23. Jensen, M.R. and Brahm, J. (1987) Biochim. Biophys. Acta **900**, 282–290.
24. Karlish, S.J.D., Lieb, W.R., Ram, D. and Stein, W.D. (1972) Biochim. Biophys. Acta **255**, 126–132.
25. Carruthers, A. and Melchior, D.L. (1983) Biochim. Biophys. Acta **728**, 254–266.
26. Carruthers, A. (1986) Biochemistry **25**, 3592–3602.
27. Wheeler, T.J. (1986) Biochim. Biophys. Acta **862**, 387–398.
28. Lacko, L., Wittke, B. and Geck, P. (1973) J. Cell Physiol. **82**, 213–318.
29. Challiss, J.R., Taylor, L.P. and Holman, G.D. (1980) Biochim. Biophys. Acta **602**, 155–166.
30. Eilam, Y. and Stein, W.D. (1972) Biochim. Biophys. Acta **266**, 161–173.
31. Eilam, Y. (1975) Biochim. Biophys. Acta **401**, 364–369.
32. Weiser, M.B., Razin, M. and Stein, W.D. (1983) Biochim. Biophys. Acta **727**, 379–388.
33. Harris, E.J. (1964) J. Physiol. **173**, 344–353.
34. Sen, A.K. and Widdas, W.F. (1962) J. Physiol. **160**, 392–403.
35. Carruthers, A. and Melchior, D.L. (1985) Biochemistry **24**, 4244–4250.
36. Carruthers, A. (1984) Prog. Biophys. Molec. Biol. **43**, 33–69.
37. Lieb, W.R. and Stein, W.D. (1977) J. Theor. Biol. **69**, 311–319.
38. Ginsburg, H. and Ram, D. (1975) Biochim. Biophys. Acta **382**, 376–396.
39. Ginsburg, H. and Stein, D. (1975) Biochim. Biophys. Acta **382**, 353–368.
40. Ginsburg, H. (1978) Biochim. Biophys. Acta **506**, 119–135.
41. Ginsburg, H. and Yeroushalmy, S. (1978) J. Physiol. (Lond.) **282**, 399–417.
42. Lieb, W.R. (1982) in: Red Cell Membranes. A Methodological Approach, eds J.C.E. and J.D. Young. pp. 135–164. Academic Press.
43. Lieb, W.R. and Stein, W.D. (1974) Biochim. Biophys. Acta **373**, 165–177.
44. Wheeler, T.J. and Whelan, J.D. (1988) Biochemistry **27**, 1441–1446.
45. Foster, D.M. and Jacquez, J.A. (1976) Biochim. Biophys. Acta **436**, 210–221.
46. Wheeler, T.J. and Hinkle, P.C. (1985) Ann. Rev. Physiol. **47**, 503–518.
47. Baker, G.F. and Naftalin, R.J. (1979) Biochim. Biophys. Acta **550**, 474–484.
48. Baker, G.F. and Widdas, W.F. (1988) J. Physiol. **395**, 57–76.
49. Naftalin, R.J. and Rist, R.J. (1994) Biochim. Biophys. Acta **1191**, 65–78.
50. Helgerson, A.L. and Carruthers, A. (1989) Biochemistry **28**, 4580–4594.
51. Bloch, R. (1973) Biochemistry **12**, 4799–4801.
52. LeFevre, P.G. and Marshall, J.K. (1959) J. Biol. Chem. **234**, 3022–3027.
53. Basketter, D.A. and Widdas, W.F. (1978) J. Physiol. (Lond.) **278**, 389–401.
54. Barnett, J.E., Holman, G.D., Chalkley, R.A. and Munday, K.A. (1975) Biochem. J. **145**, 417–429.
55. Baker, G.F. and Widdas, W.F. (1973) J. Physiol. (Lond.) **231**, 129–142.
56. Baker, G.F., Basketter, D.A. and Widdas, W.F. (1978) J. Physiol. (Lond.) **278**, 377–388.
57. Aubby, D.S. and Widdas, W.F. (1980) J. Physiol. (Lond.) **309**, 317–327.
58. Barnett, J.E.G., Holman, G.D. and Munday, K.A. (1973) Biochem. J. **135**, 539–541.
59. Barnett, J.E., Holman, G.D. and Munday, K.A. (1973) Biochem. J. **131**, 211–221.
60. Krupka, R.M. and Devés, R. (1981) J. Biol. Chem. **256**, 5410–5416.
61. Carruthers, A. and Helgerson, A.L. (1991) Biochemistry **30**, 3907–3915.
62. Krupka, R M. (1971) Biochemistry **10**, 1143–1153.
63. Helgerson, A.L. and Carruthers, A. (1987) J. Biol. Chem. **262**, 5464–5475.
64. May, J.M. and Beechem, J.M. (1993) Biochemistry **32**, 2907–2915.
65. Sogin, D.C. and Hinkle, P.C. (1980) Biochemistry **19**, 5417–5420.
66. Sogin, D.C. and Hinkle, P.C. (1978) J. Supramolec. Struct. **8**, 447–453.
67. Hinkle, P.C., Sogin, D.C., Wheeler, T.J. and Telford, J.N. (1979) in: Function and Molecular aspects of Biomembrane Transport, eds E. Quagliariello. pp. 487–494. Elsevier/North-Holland Biomedical Press, New York.

68. Gorga, F.R. and Lienhard, G.E. (1981) Biochemistry **20**, 5108–5113.
69. Baldwin, S.A., Baldwin, J.M., Gorga, F.R. and Lienhard, G.E. (1979) Biochim. Biophys. Acta **552**, 183–188.
70. Baldwin, S.A., Baldwin, J.M. and Lienhard, G.E. (1982) Biochemistry **21**, 3836–3842.
71. Zottola, R.J., Cloherty, E.K., Coderre, P.E., Hansen, A., Hebert, D. N. and Carruthers, A. (1995) Biochemistry **34**, 9734–9747.
72. Hebert, D.N. and Carruthers, A. (1991) Biochemistry **30**, 4654–4658.
73. Hebert, D.N. and Carruthers, A. (1992) J. Biol. Chem. **267**, 23829–23838.
74. Gorga, F.R. and Lienhard, G.E. (1982) Biochemistry **21**, 1905–1908.
75. Appleman, J.R. and Lienhard, G.E. (1985) J. Biol. Chem. **260**, 4575–4578.
76. Appleman, J.R. and Lienhard, G.E. (1989) Biochemistry **28**, 8221–9227.
77. Carruthers, A. (1986) J. Biol. Chem. **261**, 11028–11037.
78. Chin, J.J., Jhun, B.H. and Jung, C.Y. (1992) Biochemistry **31**, 1945–1951.
79. Lowe, A.G. and Walmsley, A.R. (1987) Biochim. Biophys. Acta **903**, 547–550.
80. Naftalin, R.J. (1988) Biochim. Biophys. Acta **946**, 431–438.
81. Naftalin, R.J. (1988) Trends Biochem. Sci. **13**, 425–426.
82. Carruthers, A. and Melchior, D.L. (1984) Biochemistry **23**, 6901–11.
83. Helgerson, A.L., Hebert, D.N., Naderi, S. and Carruthers, A. (1989) Biochemistry **28**, 6410–6417.
84. Hebert, D.N. and Carruthers, A. (1986) J. Biol. Chem. **261**, 10093–10099.
85. Jung, C.Y., Carlson, L.M. and Whaley, D.A. (1971) Biochim. Biophys. Acta **241**, 613–627.
86. Taverna, R.D. and Langdon, R.G. (1973) Biochim. Biophys. Acta **298**, 412–421.
87. Taverna, R.D. and Langdon, R.G. (1973) Biochim. Biophys. Acta **298**, 422–428.
88. Carruthers, A., Hebert, D.N., Helgerson, A.L., Tefft, R.E., Naderi, S. and Melchior, D.L. (1989) Ann. N.Y. Acad. Sci. **568**, 52–67.
89. Carruthers, A. and Helgerson, A.L. (1989) Biochemistry **28**, 8337–8346.
90. Jacquez, J.A. (1983) Biochim. Biophys. Acta **727**, 367–378.
91. May, J.M. (1988) FEBS Lett. **241**, 188–190.
92. Cuppoletti, J., Jung, C.Y. and Green, F.A. (1981) J. Biol. Chem. **256**, 1305–1306.
93. Jarvis, S.M., Ellory, J.C. and Young, J.D. (1986) Biochim. Biophys. Acta **855**, 312–315.
94. Jung, C.Y., Hsu, T.L., Cha, J.S. and Haas, M.N. (1980) J. Biol. Chem. **225**, 361–364.
95. Kasahara, M. and Hinkle, P.C. (1977) J. Biol. Chem. **253**, 7384–7390.
96. Zoccoli, M.A., Baldwin, S.A. and Lienhard, G.E. (1978) J. Biol. Chem. **253**, 6923–6930.
97. Mueckler, M., Caruso, C., Baldwin, S.A., Panico, M., Blench, I., Morris, H.R., Allard, W.J., Lienhard, G.E. and H.F.L. (1985) Science **229**, 941–945.
98. Pessino, A., Hebert, D.N., Woon, C.W., Harrison, S.A., Clancy, B.M., Buxton, J.M., Carruthers, A. and Czech, M.P. (1991) J. Biol. Chem. **266**, 20213–20217.
99. Coderre, P.E., Cloherty, E.K., Zottola, R.J. and Carruthers, A. (1995) Biochemistry **34**, 9762–9773.
100. Wellner, M., Monden, I. and Keller, K. (1994) Biochem. J. **299**, 813–817.
101. Due, A.D., Cook, J.A., Fletcher, S.J., Qu, Z.C., Powers, A.C. and May, J.M. (1995) Biochem. Biophys. Res. Commun. **208**, 590–596.
102. Naftalin, R.J., Smith, P.M. and Roselaar, S.E. (1985) Biochim. Biophys. Acta **820**, 235–249.
103. Cloherty, E.K., Sultzman, L.A., Zottola, R.J. and Carruthers, A. (1995) Biochemistry **34**, 15395–15406.
104. Lieberman, D.M. and Reithmeier, R.A. (1988) J. Biol. Chem. **263**, 10022–10028.
105. Kliman, H.J. and Steck, T.L. (1980) J. Biol. Chem. **255**, 6314–6321.
106. Mercer, R.W. and Dunham, P.B. (1981) J. Gen. Physiol. **78**, 547–568.
107. Lachaal, M., Berenski, C.J., Kim, J. and Jung, C.Y. (1990) J. Biol. Chem. **265**, 15449–15454.
108. Lachaal, M. and Jung, C.Y. (1993) J. Cell. Physiol. **156**, 326–332.
109. Pedley, K.C., Jones, G.E., Magnani, M., Rist, R.J. and Naftalin, R.J. (1993) Biochem. J. **291**, 515–522.
110. Naftalin, R.J. and Rist, R.J. (1990) Biochem. J. **265**, 251–259.
111. Faik, P., Morgan, M., Naftalin, R.J. and Rist, R. (1989) Biochem. J. **260**, 153–156.
112. Naftalin, R.J. and Rist, R.J. (1989) Biochem. J. **260**, 143–152.

CHAPTER 15

The Mediator of Thyroidal Iodide Accumulation: The Sodium/Iodide Symporter

G. DAI and O. LEVY

Department of Molecular Pharmacology,
Albert Einstein College of Medicine,
Bronx, NY 10461, USA

L.M. AMZEL

Department of Biophysics and Biophysical Chemistry,
Johns Hopkins University School of Medicine,
Baltimore, MD 21205, USA

N. CARRASCO

Department of Molecular Pharmacology,
Albert Einstein College of Medicine,
Bronx, NY 10461, USA

© *1996 Elsevier Science B.V.*
All rights reserved

Handbook of Biological Physics
Volume 2, edited by W.N. Konings, H.R. Kaback and J.S. Lolkema

Contents

1. Introduction

The Na^+/I^- symporter (NIS) is a key plasma membrane protein that catalyzes the active accumulation of iodide (I^-) in the thyroid gland, i.e. the first and critical rate-limiting step in the biosynthesis of the thyroid hormones [1,2]. NIS is located in the basolateral membrane of the hormone-producing thyroid follicular cells or thyrocytes [3,4]. In 1995 a cDNA clone that encodes NIS was isolated in our laboratory by functional screening of a cDNA library from a rat thyroid-derived cell line (FRTL-5 cells) in *Xenopus laevis* oocytes [5]. The sequence of NIS comprises 618 amino acids (relative molecular mass 65,196). The hydropathic profile and secondary structure predictions of the protein suggest that the cloned cDNA encodes for an intrinsic membrane protein with 12 putative transmembrane domains. NIS is the first I^- transporting molecule whose cDNA has been cloned. NIS plays a crucial role in the evaluation, diagnosis, and treatment of various thyroid pathological conditions [1,2].

The principal function of the thyroid is to produce the hormones T_3 and T_4 [tri-iodothyronine and thyroxine (or tetra-iodothyronine) respectively] both of which play essential roles in regulating intermediary metabolism in virtually all tissues and in maturation of the nervous system, skeletal muscle and lungs in the developing fetus and the newborn [1,2,6]. Unlike any other hormones, T_3 and T_4 contain iodine as an essential constituent. The ability of thyroid follicular cells to transport I^-, first reported in 1915 [7], is an apparent cellular adaptation to sequester environmentally scarce iodine, thus ensuring adequate thyroid hormone production in most cases. Nevertheless, insufficient dietary supply of iodine is still prevalent among more than 800 million people in many regions of the world, leading to endemic iodine deficiency disorders (IDD) often associated with hypothyroidism [8].

The overall translocation of I^- into the gland for thyroid hormogenesis involves two separate processes, namely I^- accumulation and I^- efflux. I^- accumulation is mediated by NIS, and it consists of the active transport of I^- from the interstitium into the follicular cells across the basolateral plasma membrane. NIS-catalyzed I^- accumulation is a Na^+-dependent active transport process that couples the energy released by the inward "downhill" translocation of Na^+ down its electrochemical gradient to driving the simultaneous inward "uphill" translocation of I^- against its electrochemical gradient [9] (Fig. 1A). The Na^+ gradient acting as the driving force for I^- accumulation is generated by the ouabain-sensitive, K^+_{out}-activated Na^+/K^+ ATPase. Na^+/I^- symport activity in the thyroid is characteristically blocked by the competitive inhibitors perchlorate and thiocyanate.

I^- efflux, on the other hand, is the transfer of I^- from the cytosol of thyrocytes towards the colloid across the apical plasma membrane. I^- efflux is a facilitated diffusion mechanism that has been proposed to be mediated by an I^- channel located in

(A)

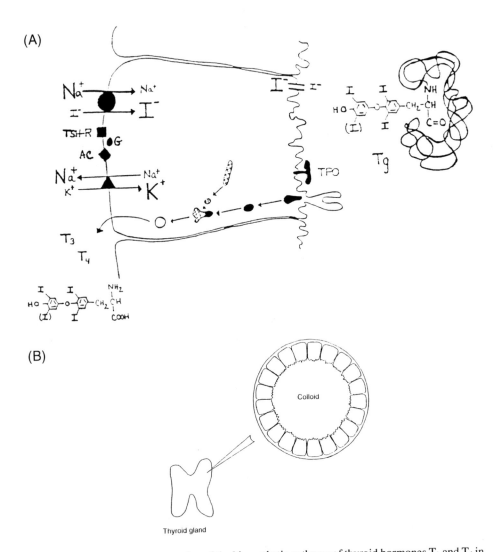

(B)

Fig. 1. (A) Schematic representation of the biosynthetic pathway of thyroid hormones T_3 and T_4 in the follicular cell. The basolateral end of the cell is shown on the left side of the figure, and the apical end on the right. (●) Active accumulation of I^-, mediated by the Na^+/I^- symporter; (▲) Na^+/K^+ ATPase; (■) TSH receptor; (♦) adenylate cyclase; (G) G protein; (=) I^- efflux towards the colloid; (TPO) TPO-catalyzed organification of I^-; (←) endocytosis of iodinated Tg, followed by phago-lysosomal hydrolysis of endocytosed iodinated Tg. (B) Schematic representation of the thyroid gland and a thyroid follicle.

the apical membrane of thyrocytes [10,11]. The colloid, where the large hormone precursor thyroglobulin (Tg) is stored, is located in the follicular lumen, an extracellular compartment. I^- is ultimately required at the cell/colloid interface because this is the site where, to a large extent, hormone biosynthesis takes place (Fig. 1A).

NIS confers to the thyroid gland its most readily distinctive functional attribute, i.e. its ability to actively accumulate I⁻; NIS therefore provides the molecular basis for the thyroidal radioiodide uptake test and for thyroid scintigraphy, two thyroid function tests of considerable value as diagnostic aids in a variety of thyroid pathological conditions [1,2]. For example, the possible existence of thyroid cancer must be ruled out whenever a thyroid nodule is detected. Thyroid nodules that are determined by scintigraphy to accumulate I⁻ equally or more efficiently than the normal surrounding tissue are generally benign, while most thyroid cancers display markedly reduced I⁻ accumulation activity relative to healthy tissue [1,2]. Therefore, it is likely that thyroid cancer can have a deleterious effect on the function of NIS. Still, NIS is sufficiently active in some thyroid cancers and metastases to render them amenable to detection and even destruction with radioiodine [1,2]. Conversely, large doses of radiation reaching the gland via NIS in the form of iodine isotopes can cause thyroid cancer. The most dramatic example of this is the alarming rise in the incidence of thyroid cancer cases in children in Ukraine and Belarus in the wake of the 1986 Chernobyl power plant accident [12–17]. In this instance, ^{131}I in the nuclear fallout was ingested largely through milk and concentrated in the thyroid via NIS.

Surprisingly, the molecular characterization of NIS remained elusive for a long time even as other major thyroid-specific proteins involved in hormogenesis, i.e. Tg, thyroid peroxidase (TPO), and the TSH receptor, were characterized in considerable molecular detail [18–20]. In this chapter, we briefly review the essential microscopic architecture of the thyroid gland, summarize the T_3 and T_4 biosynthetic pathway, survey selected research on the I⁻ transport system of the thyroid that led to the cloning and characterization of NIS from rat thyroid, analyze the NIS molecule, and explore the impact that the characterization of NIS is likely to have on other research areas.

2. Microscopic architecture of the thyroid

Embryologically, the thyroid in mammals develops from the endodermal pharynx as the earliest endocrine structure to appear during gestation [1]. As the rudimentary thyroid continues to develop, it retains a link to the pharyngeal floor through the thyroglossal duct, which in most cases eventually disappears, albeit only partially. The human adult thyroid displays a characteristic bi-lobed shape and weighs approximately 20 g. It is the largest and most richly vascularized solely endocrine gland in humans [1,2]. The microscopic morphology of the human thyroid is also highly characteristic and very similar to other vertebrate thyroids. It consists of an array of follicles of various sizes, each of which is a spheroidal structure made up of a single layer of epithelial follicular cells or thyrocytes surrounding the colloid (Fig. 1B). The thyroid follicular cells produce the colloid, accumulate I⁻, and synthesize the thyroid hormones. The colloid, where Tg is stored, plays an important and rather unique role as an extracellular site of hormogenesis.

A major morphological characteristic of the thyroid follicular cells is their polarity. As illustrated in Fig. 1, these epithelial cells exhibit a basolateral end facing the basement membrane, the interstitial space and the capillary bed, and an apical end facing the colloid. The apical end of the follicular cells displays numerous microvilli that result in a greatly expanded area of exposure to the colloid. This region is known as the cell/colloid interface. A second, less abundant and larger cellular type is also present in the thyroid, mainly in spaces between follicles. These are known as C cells, which produce the calcemia-lowering peptide hormone calcitonin. C cells appear to be of ectodermic origin, derived from the neural crest [1].

Although the follicular structure of the thyroid plays an important role in hormone biosynthesis *in vivo*, it is not required for I^- transport [9,21]. Thus various thyroid-derived systems devoid of follicular structure and polarity have nevertheless been extremely useful in the study of thyroid I^- transport, including cells in culture and membrane vesicles. In contrast, the investigation of the location of the I^- transport systems in the follicular cells does require that the polarity of the cells be preserved. Using cultured porcine thyroid cell monolayers Chambard et al. [3] experimentally showed that both the active I^- accumulation mechanism (i.e. NIS) and the TSH receptor are located only in the basolateral membrane. These findings were the first to clearly prove a functional asymmetry of the membrane domains of thyroid cells, in addition to their morphological distinction. Employing a similar system, Nilsson et al. [10] observed a rapid and temporary TSH-stimulated I^- efflux only in the apical direction, and proposed that a TSH-regulated I^- channel, located in the apical membrane of the follicular cells, mediates TSH-stimulated I^- efflux. The end result of the combined activity of NIS and the I^- channel is to ensure the supply of I^- to the cell/colloid interface for hormogenesis.

3. The T_3 and T_4 biosynthetic pathway

Tg is the most abundant of the thyroid-specific proteins synthesized by the follicular cells, a large (>600 kDa) dimeric glycoprotein that serves as the molecular template for the synthesis of T_3 and T_4 at the cell/colloid interface, and as a storage substrate for the hormones and for accumulated I^-. Tg is by far the principal component of the colloid, where it is found at a concentration of >50 mg/ml [1]. The thyroid hormones are iodothyronines, i.e. the result of two coupled iodotyrosines (the structures of T_3 and T_4 are shown in Fig. 1A). The basic events leading to the biosynthesis of these hormones may be briefly summarized as follows (Fig. 1A): I^- is actively accumulated against an electrochemical gradient across the basolateral plasma membrane of the thyroid follicular cells in a process catalyzed by NIS, and passively translocated across the apical membrane into the colloid via the putative I^- channel. Accumulated I^- that has reached the cell/colloid interface is oxidized and incorporated into some tyrosyl residues within the Tg molecule in a reaction catalyzed by TPO, leading to the subsequent coupling of iodotyrosine residues. This incorporation of I^- into organic molecules is called 'I^- organification', a reaction pharmacologically blocked by such anti-thyroid agents as 6-n-propyl-2-thiouracil (PTU) and 1-methyl-2-mercaptoimidazole (MMI). Iodinated Tg is stored extracellularly in the colloid. In response to demand for thyroid hormones, phagolysosomal hydrolysis of endocytosed iodinated Tg ensues. T_3 and T_4 are secreted into the bloodstream, and non-secreted iodotyrosines are metabolized to tyrosine and I^-, a reaction catalyzed by the microsomal enzyme iodotyrosine dehalogenase. This process facilitates reutilization of the remaining I^- [22]. All steps in the thyroid hormone biosynthetic pathway are stimulated by thyroid stimulating hormone (TSH) from the pituitary. The effect of TSH results from binding of the hormone to the TSH receptor. Unlike hormone biosynthesis in other endocrine glands, hormone production in the thyroid occurs to a large extent in the colloid, an extracellular compartment.

Detailed information on the physical properties, chemical composition, primary sequence, as well as on the biosynthesis and post-translational modifications of Tg has been obtained in the past decade [18,23–25]. Similarly, TPO, the enzyme that catalyzes the organification of I^-, is now known to be an intrinsic, glycosylated heme-containing protein, present on the microvilli of the apical cell membrane [26,27]. Human TPO has been purified and its primary amino acid sequence is known; both its catalytic mechanism and its regulation by TSH have been investigated [26,27]. Moreover, TPO has been identified as the 'microsomal antigen' in Hashimoto's thyroiditis, an autoimmune condition [27]. The TSH receptors from human, dog, and rat thyroids have been cloned and functionally expressed [28–30]. The TSH receptor is a member of a family of plasma membrane receptors with seven putative transmembrane-spanning regions, which are joined together by hydrophilic domains at both sides of the membrane. The intracellular loops interact functionally with membrane-associated guanine nucleotide (G) regulatory proteins during signal transduction [30]. The TSH receptor and the receptors for other glycoprotein hormones display characteristically large extracellular amino termini domains, in correspondence with their respective polypeptide ligands. This contrasts with other members of this receptor family, such as the prototypical β-adrenergic receptor, whose amino termini domains are considerably smaller. Anti-TSH receptor antibodies have been implicated in the pathophysiology of Graves' disease [31].

Even though the I^- transporting ability of the thyroid was first observed in 1915 by Marine and Feiss [7], no molecular information on NIS was available until its cloning in 1995 [5]. The search for the elusive identity of NIS continued in the 1940s, when studies performed in thyroid glands and thyroid slices showed that I^- transport and thyroid hormone biosynthesis were two phenomena that could be pharmacologically distinguished from each other, thus revealing the existence of a special I^- concentrating mechanism in the gland [32,33]. These studies revealed pronounced inhibition of hormone synthesis by antithyroid agents such as sulfonamides and thiouracil, without any effect on I^- accumulation. The prevention of hormone synthesis by these agents is now known to result from inhibition of I^- organification. But a fundamental question had to be asked: what were the properties and the nature of this thyroid I^- transporter?

4. In search of the NIS molecule

4.1. Properties of thyroidal I⁻ accumulation

The thyroid is capable of concentrating I⁻ by a factor of 20–40 with respect to the concentration of the anion in the plasma under physiological conditions [34,35]. I⁻ accumulation in the thyroid was shown early on by many groups (9,21,36,37) to be an active transport process that occurs against an I⁻ electrochemical gradient, stimulated by TSH, and blocked by the well-known 'classic' competitive inhibitors, the anions thiocyanate and perchlorate. The I⁻ transport mechanism of the thyroid displays a considerably higher preference for I⁻ over Cl⁻ or other halides. This has been shown by the early observation that the halides F⁻, Cl⁻ and Br⁻ administered to rats were not concentrated in the thyroid and had no effect on radiolabeled I⁻ previously accumulated in the gland, whereas the administration of unlabeled I⁻ led to rapid discharge of radioiodide [38,39]. Most studies of the I⁻ transport system of the thyroid carried out over the years in thyroid slices, cells in culture and membrane vesicles have been performed in the presence of over 1000-fold higher concentrations of Cl⁻ than I⁻ without indication of Cl⁻ competition for transport. This fact further underscores the high selectivity of the transporter for I⁻.

Iff and Wilbrandt [40] demonstrated in 1963 that I⁻ accumulation in the intact thyroid gland was totally dependent on extracellular Na^+, and they identified a close relationship between the I⁻ trapping system and the Na^+/K^+ ATPase. In 1973, Bagchi and Fawcett [41] suggested that I⁻ is cotransported with Na^+ by a membrane carrier into the cell and that the driving force for the process is the inwardly directed Na^+ gradient generated by the ouabain-sensitive, K^+_{out}-activated Na^+/K^+ ATPase. The chemical nature of the thyroid I⁻ carrier was the subject of some debate in the past [9]. Even endocrinology textbooks published in the 1990s still discuss early data that had been interpreted to suggest that the thyroid I⁻ carrier might be a phospholipid [42,43], although the identity of NIS as a protein has not been in doubt since NIS was expressed in *X. laevis* oocytes in 1989 [44].

4.2. Studies using FRTL-5 cells and membrane vesicles

Further confirmation of the proposed Na^+/I^- symport model has been provided by studies carried out in FRTL-5 cells by Weiss et al. in 1984 [21] and in hog thyroid plasma membrane vesicles by O'Neill et al. in 1986 [36]. FRTL-5 cells (Fisher rat thyroid line) were established by Ambesi-Impiombato et al. [45,46] in 1979. These cells are derived from normal rat thyroids, and are devoid of malignant *in vitro* characteristics: they are untransformed (near-diploid) follicular thyroid cells, do not grow in semi-solid media or in syngenic animals (i.e., they are non-tumorigenic), and show contact inhibition and a normal chromosome morphology. FRTL-5 cells exhibit most of the functional differentiation characteristics of the original thyroid follicular cells, but they do not form the follicular architecture of intact thyroid tissue. FRTL-5 cells were selected in low serum, remain differentiated growing in 5% serum, and depend on the presence of TSH for growth. However, they can be maintained in medium without TSH for as long as 10 days without loss of viability, as evidenced by their retained ability to exclude trypan blue and to be recultured. Withdrawal of TSH from the medium results in a rapid decline in intracellular cAMP content within 3 h. After 24 h and for as long as 10 days, the cells become extremely sensitive to TSH, as indicated by their ability to increase intracellular levels of cAMP upon addition of the hormone. FRTL-5 cells concentrate I^- about 30-fold ($K_m \sim 30 \, \mu M$) [21] and display all the properties of I^- accumulation observed in tissue slices and primary cell culture systems, most significantly Na^+-dependence and inhibition by perchlorate, in addition to their mentioned sensitivity to TSH.

As a result of their characteristics, FRTL-5 cells have been employed extensively in thyroid research. Thyroid membrane vesicles have also proven very useful for the examination of I^- transport. O'Neill et al. [36] prepared plasma membrane vesicles from hog thyroid glands by differential centrifugation of tissue homogenate, thus making it possible to carry out accumulation experiments under well-defined initial ion gradients and bypassing the Na^+/K^+ ATPase. I^- transport ($K_m \sim 5 \, \mu M$) driven by an imposed inward Na^+ gradient was observed in these vesicles in the absence of ATP. As in whole thyroid tissue and FRTL-5 cells, I^- transport was totally Na^+-dependent and inhibitable by thiocyanate and perchlorate. The K_m values for I^- accumulation in FRTL-5 cells and hog membrane vesicles are to be taken as operational values related to the particular experimental conditions used. The different values (30 μM versus 5 μM respectively) may reflect the difference in the species involved (rat versus hog) and/or in the particular experimental conditions. It is significant, however, that either K_m value is far larger than the concentration of free I^- normally present in the blood (6–47 nM) [1,2].

4.3. Electrogenicity of Na^+/I^- symport activity

The fact that the magnitude of the Na^+ gradient generated by the Na^+/K^+ ATPase (~12-fold) is not sufficient to account for the I^- gradient established by the symporter (~30–40-fold) suggests that Na^+/I^- symport activity is electrogenic and, as a corollary, that the stoichiometry is not 1:1. Because Na^+/I^- symport activity involves the translocation in the same direction of two ions of opposite charge across the membrane, the electrogenicity of the process depends on the Na^+/I^- stoichiometry. If the stoichiometry were 1:1, the transport process would not result in a net transfer of charge across the membrane, i.e., it would not be electrogenic. O'Neill et al [36] reported that the Na^+/I^- symporter is electrogenic by observing a slight stimulation of I^- transport upon establishment of a K^+-diffusion membrane potential ($\Delta\psi$) negative inside with respect to the outside. The K^+-diffusion membrane potential was generated by loading the thyroid membrane vesicles with K^+, transferring the vesicles to a medium with a low concentration of K^+, and adding the K^+ ionophore valinomycin. The addition of valinomycin results in an outward K^+ flux which in turn generates a membrane potential, negative inside the vesicles with respect to the outside. That the presence of a negative (inside) membrane potential resulted in a slight stimulation of I^- transport in the vesicles was interpreted as evidence that I^- transport is electrogenic, i.e., that the Na^+/I^- flux stoichiometry is larger than one (more than one Na^+ is translocated per each I^-). The authors also reported that the Na^+ dependence of I⁻ accumulation was sigmoidal, with a Hill coefficient of 1.61. Nakamura et al. [47] have reproduced these findings, obtaining a Hill coefficient of 1.8. While Hill coefficients by themselves do not constitute a definitive index of stoichiometry, the reported Hill coefficient values combined with the observed stimulation of I⁻ transport caused by a K^+-diffusion potential suggest a transport stoichiometry of at least 2 Na^+ per I⁻. It has been proposed that the electrogenic nature of the Na^+/I⁻ symporter may contribute to enhance the I⁻ accumulation efficiency of the thyroid.

We have devised a protocol to prepare functional sealed membrane vesicles (MV) from FRTL-5 cells in which pronounced Na^+/I⁻ symport activity, inhibitable by perchlorate, is measured. Virtually no I⁻ translocation was detected when Na^+ was replaced with choline in the medium [48]. While O'Neill et al. [36] reported that imposition of a $\Delta\psi$ (negative inside) stimulated Na^+/I⁻ symport activity in hog thyroid MV, and suggested on this basis that Na^+/I⁻ symport activity is electrogenic, the observed stimulation was modest and limited to certain valinomycin concentrations, and therefore not conclusive. We used FRTL-5 MV to evaluate the role of $\Delta\psi$ on I⁻ transport because they are highly active and tightly sealed. First, initial rates of Na^+/I⁻ symport activity were measured as a function of varying concentrations of extravesicular Na^+. We observed a clear sigmoidal dependence of I⁻ transport on the Na^+ concentration, suggesting that the process is not electroneutral. The calculated Hill coefficient was 2.0 ± 0.07. The K_m of the Na^+/I⁻ symporter for Na^+ was calculated to be ~75 mM.

To assess the direct effect of $\Delta\psi$ on I⁻ transport, FRTL-5 MV were loaded with K^+ and diluted into a medium without K^+ in the presence of valinomycin to generate a $\Delta\psi$ negative inside with respect to the outside of –70 mV. There was a ~40% increase in Na^+/I⁻ symport activity when a K^+-diffusion gradient was imposed in the presence of valinomycin. When the MV were diluted into a solution without valinomycin, or a solution containing equimolar K^+, a diffusion gradient was not generated, and consequently there was no increase in Na^+/I⁻ symport activity. To explore whether $\Delta\psi$ could act as a sole force in driving I⁻ transport, as $\Delta\psi$ has been shown to act in driving Ca^{2+} translocation catalyzed by the Na^+/Ca^{2+} exchanger [49], FRTL-5 MV were loaded with K^+, Na^+, and ¹²⁵I and diluted into a solution containing equimolar Na^+ and ¹²⁵I in the presence of valinomycin. As a result, a negative $\Delta\psi$ was generated (inside with respect to outside). It was clear that I⁻ was translocated even though there was no Na^+ gradient driving the process. When the polarity of the K^+ gradient was reversed, i.e. when MV were loaded just as above but without K^+, and diluted into a solution that contained K^+, a $\Delta\psi$ that was positive on the inside with respect to the outside was generated. In this case, I⁻ efflux was observed. These results indicate that Na^+/I⁻ symport activity is electrogenic, the stoichiometry of Na^+/I⁻ symport activity is at least 2 Na^+ per I⁻, and $\Delta\psi$ is able to act as the sole driving force for I⁻ translocation (Levy, O. and Carrasco N., unpublished observations).

4.4. Inhibitors

The availability of inhibitors can be of valuable help in protein identification. It has long been known that certain anions such as thiocyanate and perchlorate are competitive inhibitors of I⁻ accumulation in the thyroid [9,34,35]. The antithyroid properties of these anions were first discovered as a side effect of thiocyanate used for the treatment of hypertension, when it was observed that some patients thus treated exhibited goiter and/or hypothyroidism [9]. The mechanism of inhibition of these compounds involves a similarity in size and charge of the anions to I⁻, so that the closer the ionic radius of the inhibitor is to that of I⁻, the lower the K_i value. Moreover, univalency is a requirement for inhibition, since divalent anions fail to inhibit transport [9]. A significant difference between these two inhibitors is that perchlorate is, but thiocyanate is not, concentrated within the thyroid of most species studied. This is due to the fact that while thiocyanate is transported into the cell, it is rapidly metabolized thereafter [9]. However, thiocyanate is concentrated by salivary tissue and gastric mucosa [9]. As perchlorate is 10–100 times more potent than thiocyanate as an inhibitor of I⁻ accumulation in a variety of *in vivo* and *in vitro* systems, it was at one time also used in the treatment of hyperthyroidism, but was eventually withdrawn because of severe secondary effects [9]. Both thiocyanate and perchlorate have been shown to cause the rapid discharge of accumulated I⁻ from PTU-blocked thyroid tissue across the basolateral membrane towards the interstitium. This phenomenon is the basis for the perchlorate discharge test, the purpose of which is to detect defects in intrathyroidal I⁻ organification. In normal subjects the administration of perchlorate blocks the continued accumulation of radioiodide by the thyroid, but causes virtually no release of previously accumulated radioiodide from the gland. In contrast, in patients with an I⁻ organification defect, administration of the inhibitor results in the release of I⁻ from the thyroid. The efficacy of I⁻ organification needs to be evaluated in certain pathological conditions, such as in the presence of organification genetic defects, or when patients are under treatment with antithyroid agents [9].

According to Rocmans et al. [50], perchlorate causes the discharge of I⁻ from the gland towards the interstitium by means of counterflow. Counterflow would occur in this case if either perchlorate (a competitive inhibitor) or unlabeled I⁻ were inwardly translocated by the symporter, causing the simultaneous outward transport of previously accumulated radiolabeled I⁻, also mediated by the symporter. Using MMI-blocked dog thyroid slices, the authors observed that the addition of perchlorate inhibited the influx of radiolabeled I⁻ into the follicles and rapidly discharged previously accumulated radiolabeled I⁻ into the medium. The fact that the addition of unlabeled I⁻ had the same effect as perchlorate strongly suggests that counterflow is at work. This process results from the ability of NIS to translocate its substrates in both directions.

The previously mentioned 'classic' inhibitors of I⁻ transport in the thyroid have been of no value for the identification of NIS, because any chemical modifications to the inhibitor molecules designed to generate affinity chromatography ligands or photoaffinity reagents would interfere with their inhibitory activity. Hence, new compounds have been tested in recent years for their ability to inhibit I⁻ transport. Van Sande et al. [51] have reported that the marine toxin dysidenin, a hexachlorinated tripeptide-like molecule extracted from the sponge *Dysidea herbacea*, inhibited I⁻ accumulation in dog thyroid slices in a 'pseudocompetitive' fashion [51].

Kaminsky et al. [37] have reported that the hallucinogenic drug harmaline and the chemically related convulsive agent TRP-P-2 [3-amino-1-methyl-5H-pyrido(4,3-b)indole acetate] inhibited Na$^+$-dependent I$^-$ accumulation in FRTL-5 cells and calf thyroid membrane vesicles. Harmaline had previously been shown to inhibit other Na$^+$-dependent processes by competitive interaction at the Na$^+$ site [9]. TRP-P-2 (K_i ~ 0.25 mM) was ten-fold more effective as an inhibitor of I$^-$ accumulation than harmaline (K_i ~ 4.0 mM). Inhibition by TRP-P-2 was competitive with respect to Na$^+$ and fully reversible. Although TRP-P-2 should be considered a relatively low affinity inhibitor, its affinity for the Na$^+$ site of NIS was over 100 times higher than that of Na$^+$ (K_m ~ 50 mM). An important structural feature of TRP-P-2 is that it contains a primary amino group that can be derivatized. Hence both TRP-P-2 and dysidenin-related compounds were proposed to be potentially useful for the identification and characterization of NIS. However, none of these inhibitors led to the identification of NIS prior to the isolation of the cDNA encoding the protein.

5. Expression cloning and characterization of NIS

5.1. Expression of Na$^+$/I$^-$ symport activity in X. laevis oocytes

A promising development in the search for the NIS molecule was the expression of perchlorate-sensitive Na$^+$/I$^-$ symport activity in *X. laevis* oocytes by microinjection of poly A$^+$ RNA isolated from FRTL-5 cells [44]. The *X. laevis* oocyte expression system was first used for the functional cloning of the rabbit intestine Na$^+$/glucose cotransporter [52], and thereafter for the functional cloning of numerous other transporters, channels and receptors [53]. A 7-fold increase of I$^-$ accumulation was observed over background 6–7 days after injection. Poly A$^+$ RNA was subsequently fractionated by sucrose gradient centrifugation, and fractions were assayed for their ability to induce I$^-$ accumulation in oocytes. The poly A$^+$ RNA encoding NIS was found in a fraction containing mRNAs that were 2.8–4.0 kb in length [44]. Thus the oocyte system was shown to be of potential value for the possible expression cloning of the cDNA that encodes the symporter, in the absence of oligonucleotides based on protein sequence data or anti-symporter antibodies. Although this cloning strategy proved lengthy, it was eventually successful.

5.2. Isolation of the cDNA clone that encodes rat NIS

We generated several cDNA libraries from poly A$^+$ RNA of FRTL-5 cells, and subjected them to expression screening in *X. laevis* oocytes. The most recently screened cDNA library was constructed in a pSPORT vector, employing a procedure that favors the generation of large inserts [54]. The library was size fractionated and the fraction containing inserts from 2.5–4.5 kb was screened because the poly A$^+$ RNA encoding NIS was found in a fraction containing messages of 2.8–4.0 kb in length [44]. The 2.5–4.5 kb cDNA fraction was plated out in 60 nylon filters (15 cm diameter, Micron Separations Inc.), each containing ~25,000 clones. Replica filters were made, divided into four sections (~6,000 clones each) and cDNA-containing plasmids were isolated from each section. Plasmids were linearized with

Not I. In vitro transcripts were made with bacteriophage T7 RNA polymerase from linearized templates in the presence of an excess of P'-5-(7 methyl)-guanosine-P_3-5'guanosine triphosphate (m^7GpppG) over guanosine 5' triphosphate to allow formation of a 5' cap structure [55]. Templates were degraded with RNAse-free deoxyribonuclease I; transcripts were extracted, precipitated and redissolved in diethylpyrocarbonate (DEPC)-treated water for microinjection into *X. laevis* oocytes. Oocytes were then assayed for Na^+/I^- symport activity 3–9 days after injection. Perchlorate-sensitive Na^+/I^- symport activity in oocytes injected with most of these cRNAs was indistinguishable from the background signal observed in control water-injected oocytes [<1.5 pmol of I^-/oocyte (n = 30 independent experiments)]. However, oocytes injected with cRNAS from group no. 31 (6,000 clones) displayed a modestly higher transport activity (2.84 pmol of I^-/oocyte ± 1.14) than background when assayed on day 5 after injection. Closer scrutiny revealed that 1 out of every 4 oocytes in this group exhibited an only slightly higher but still potentially significant Na^+/I^- symport activity than the basal transport rate, whereas activity in the remaining 3 (out of 4) oocytes was indistinguishable from negative controls. Similar results were obtained when transport assays were conducted on days 8 and 9.

Notwithstanding the modesty of the signal, we decided to examine group no. 31 further. cDNAs from group no. 31 were divided into 12 groups, and the cRNA transcripts from group no. 12 (~500 clones) elicited a considerably higher transport activity (14.42 pmol of I^-/oocyte ± 4.1) which became evident as early as day 4 after injection. cDNAs from group no. 12 were further subdivided into 126 subgroups (containing ~5 clones each), which were then re-distributed in 10 pools. Out of these, pool no. 10 (containing a total of ~50 clones from groups 109–126) elicited an even higher signal (77.89 ± 14 pmol of I^-/oocyte, signal detected 3 days after injection). Further analysis of these subgroups traced the activity to subgroups 109–112 (containing a total of ~10 clones). Transport activity elicited from the pool of subgroups 109–112 was 110 ± 7.4 pmol of I^-/oocyte. After individual clones were separately taken from subgroup no. 111, one clone elicited the strongest signal one day after injection (~192.4 ± 24.4 pmol of I^-/oocyte). This clone was streaked on a plate from which 5 single colonies were taken and analyzed by electrophoresis. Activity elicited in oocytes by microinjected transcripts from each of these colonies was indistinguishable.

In summary, the expression cloning of NIS was carried out by measuring perchlorate-sensitive Na^+/I^- symport activity in oocytes microinjected with cRNAs made *in vitro* from pools containing decreasing numbers of cDNA clones. In addition to control assays in oocytes microinjected with water, Na^+-dependence was ascertained by using choline in place of Na^+, and perchlorate sensitivity was tested by conducting assays in the presence of both Na^+ and perchlorate. Activity in choline or perchlorate assays was virtually indistinguishable from background. Figure 2 shows I^- accumulation in oocytes microinjected with water (control), mRNA from FRTL-5 cells, or cRNA from the NIS clone, which elicited a signal that corresponds to a >700-fold increase in perchlorate-sensitive Na^+/I^- symport activity over background. During the early stages of screening, maintenance and viability of oocytes had to be optimized because activity was detected 6–7 days

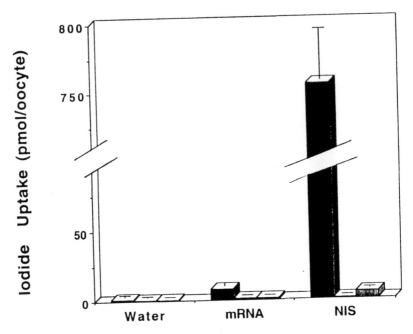

Fig. 2.   Expression of NIS in *Xenopus laevis* oocytes. Oocytes were microinjected with water (lane 1), 50 ng mRNA from FRTL-5 (lane 2), or ~20 ng cRNA transcripts made *in vitro* from the NIS cDNA clone (lane 3). I^- accumulation was assayed, 48 h after microinjection, either in the presence of Na^+ (solid bars), absence of Na^+ (i.e. in the presence of choline, striped bars), or in the presence of both Na^+ and perchlorate (dotted bars) at 45 min, as described [44]. Each point represents the average of data from six oocytes ± SE.

after injection of poly A^+ mRNA from FRTL-5 cells. This is a long latency period compared to the expression of other proteins, which typically takes approximately 2–3 days. The reasons for the long latency period are unknown. A clear correspondence was apparent between the decreasing number of clones in the pools, the rising magnitude of the activity elicited, and the shortening of the latency period for appearance of the signal after injection.

5.3. Time course and kinetic analysis of Na^+/I^- symport activity in oocytes microinjected with the transcript from the NIS cDNA clone

The time course of I^- accumulation was analyzed in oocytes microinjected with the transcript from the NIS cDNA clone, 60 h after injection. I^- accumulation reached nearly 800 pmol of I^-/oocyte at the ~90 min saturation point. Given that the I^- concentration in the transport solution was 50 µM and the functional volume is 0.5 µl/oocyte, the generated I^- concentration gradient was >30-fold [5], i.e. a virtually identical value to that observed in the thyroid gland *in vivo*. A kinetic analysis of Na^+/I^- symport activity was performed in oocytes microinjected with the transcript from the NIS cDNA clone. Initial velocity rates of Na^+/I^- symport activity were determined at the 2 min time points at various I⁻ concentrations (1.25–250 μM). Na⁺/I⁻ symport rates displayed saturation kinetics. The apparent K_m for I⁻ was 36 μM [5], a value consistent with the range of values reported for FRTL-5 cells [21].

5.4. Expression of Na⁺/I⁻ symport activity in COS cells transfected with the NIS cDNA clone

In order to assess the expression of the NIS cDNA clone in mammalian cells, COS cells were transiently transfected with the NIS cDNA in the pSV.SPORT vector. The plasmid contains the SV40 origin of replication and early transcriptional promoter, t-intron region and polyadenylation signals. The NIS cDNA insert was restricted from the original pSPORT vector with *Not* I and *Sal* I, purified and subcloned into the pSV.SPORT. COS cells transfected with the NIS cDNA clone exhibited per-chlorate-sensitive Na⁺/I⁻ symport activity, in contrast to control non-transfected cells or cells transfected with the same plasmid containing an irrelevant insert, neither of which displayed transport activity [5].

The above results provide unequivocal proof that the product of our isolated cDNA clone is sufficient to elicit perchlorate-sensitive Na⁺/I⁻ symport activity in both oocytes and mammalian cells.

5.5. Primary sequence and predicted secondary structure of the NIS molecule

The complete nucleotide sequence of the cloned NIS cDNA and the deduced amino acid sequence are presented in Fig. 2 of Ref. [5]. The nucleotide sequence of the NIS clone indicates that the insert is 2,839 base pairs in length, with a predicted open reading frame of 1854 nucleotides, a 5′ untranslated region of 109 nucleotides and a 3′ untranslated region of 876 nucleotides. Within the 3′ untranslated region, a potential poly A signal was identified at position 2795. The putative initiation codon ATG contains a purine at position 113 and thus represents a reasonable Kozak consensus sequence [56]. Beginning with Met at position 1, a long open reading frame codes for a protein of 618 amino acids (relative molecular mass 65,196). The hydropathic profile and secondary structure predictions [57–59] of the protein suggest that the cloned cDNA encodes for an intrinsic membrane protein with 12 putative transmembrane domains (Fig. 3). The NH₂ terminus has been placed on the cytoplasmic side, given the absence of a signal sequence. The COOH terminus, which has also been predicted to be on the cytoplasmic side, contains a large hydrophilic region of ~70 amino acids within which the only potential cAMP-dependent PKA phosphorylation domain of the molecule is found (positions 549–552). Three potential Asn-glycosylation sites were identified in the deduced amino acid sequence at positions 225, 485, and 497. The last two are located in the 12th hydrophilic sequence, a domain predicted to be on the extracellular face of the membrane. The length of the 12 transmembrane domains in the model ranges from 20 to 28 amino acid residues, except helix V, which contains 18 residues. Only three charged residues are predicted to lie within transmembrane domains, namely Asp 16 in helix I, Glu 79 in helix II, and Arg 208 in helix VI. All three charged residues are located close to the cytoplasmic side of the corresponding domain rather than

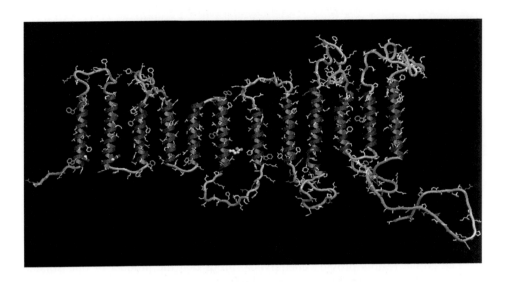

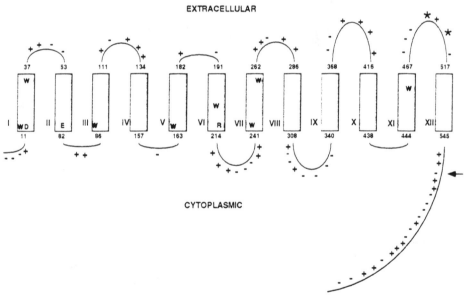

Fig. 3.   (A) Proposed secondary structure model. Initial coordinates were obtained with the program QUANTA (Molecular Simulations Inc., Burlington, Mass.). Regularization of the model was carried out with the program 'O' [69]. Graphics were carried out with the program SETOR [70]. Membrane-spanning helices are depicted in red. The color code for amino acids is: Trp, green; Leu, Ile, Phe, and Tyr, yellow; Asp and Glu, red; Arg, Lys and His, blue; remaining amino acids, grey. In Asp 16 and Glu 79 carbon atoms are depicted white and oxygen atoms purple. In Arg 208 carbon atoms are white and nitrogen atoms blue.   (B) Schematic representation of the putative topology of rat NIS in the membrane. Roman numerals indicate putative membrane-spanning domains. Potential N-linked glycosylation sites are indicated by asterisks. A putative intracellular consensus sequence for cAMP-dependent protein kinase A phosphorylation is indicated with an arrow.

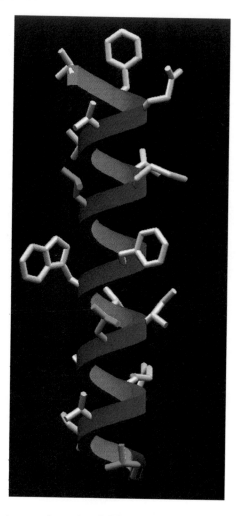

Fig. 4.   Detail of putative transmembrane domain VI. α-Helix backbone is depicted in red, Leu residues in green, and all other amino acid residues in gray.

towards its center. Out of a total of 8 Trp residues found in the membrane, six are located near the extremes of transmembrane domains. The lengths of transmembrane domains and location of Trp residues proposed in the NIS secondary structure model are similar to those found in the *R. viridis* photoreaction center crystal structure [60]. Four Leu residues (positions 199, 206, 213 and 220) appear to comprise a Leu zipper motif (Fig. 4). This motif, which has been proposed to play a role in the oligomerization of subunits in the membrane, has been conserved in all cloned neurotransmitter transporters [61].

Remarkably, NIS falls alongside other anion transporters in a dendrogram representing a cluster analysis of Na⁺-dependent cotransporters on the basis of their amino acid sequences (Fig. 5). A comparison of the predicted amino acid sequence

DENDROGRAM OF SODIUM DRIVEN TRANSPORTERS

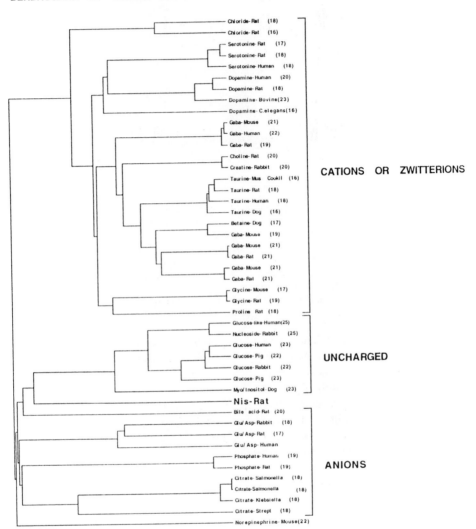

Fig. 5. Dendrogram representing a cluster analysis of members of the Na⁺-dependent cotransporter protein family. A multiple analysis was created using the PILEUP program of the Genetic Computer Group. The dendrogram is a tree representation of clustering relationships among the deduced amino acid sequences of the cDNAs used for the analysis. The numbers in parentheses represent the percent of homology to rat NIS.

of NIS with those of other cloned Na⁺-dependent cotransporters in available databases revealed the highest degree of homology (24.6% amino acid identity) with the human Na⁺/glucose cotransporter. The sequence comparison suggests that the membrane topology of NIS is similar to that of other Na⁺-dependent cotransporters, a notion further reinforced when hydropathy profiles are also compared. For

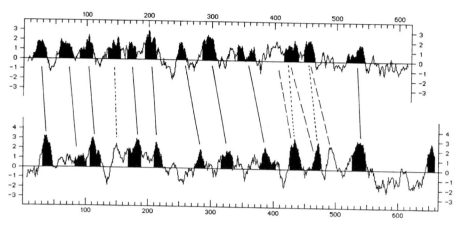

Fig. 6. Hydropathy plots of the deduced amino acid sequences of NIS (top) and the Na$^+$/glucose cotransporter (bottom). The hydropathic analysis was based on the Kyte–Doolittle algorithm [57] with a window of 9 residues. Hydropathy values (positive correspond to hydrophobic and negative to hydrophilic regions) are plotted against amino acid position. Putative membrane-spanning domains shown as filled-in areas under plot lines are according to original models [5,65]. Continuous lines joining domains in the two proteins indicate the nine common domains in the original models; the two alternative models with twelve helices are generated by adding either the three helices joined by the dashed lines or those joined by the dotted lines.

example, a high correspondence is readily apparent between the hydropathy profiles of rat NIS and the human Na$^+$/glucose cotransporter (Fig. 6). Therefore, secondary structure models for these two proteins and for other members of this protein family should probably be the same, given that the models have to be compatible with the corresponding sequences and hydropathic profiles. This implies that the directionality of individual helices be the same in the proteins being compared. Since meeting these requirements considerably narrows the range of possible topologies, any notable differences in proposed model topologies would probably have to be reconciled on the basis of new experimental evidence.

One such striking difference is the existence of a predicted last hydrophobic segment at the COOH terminus of the Na$^+$/glucose cotransporter, occurring immediately after a long hydrophilic segment. A counterpart of this hydrophobic domain is absent in NIS. New experimental evidence is necessary to determine whether this domain in the Na$^+$/glucose cotransporter is actually a transmembrane helix or a long hydrophobic stretch in the COOH terminus. The rest of the transmembrane domains should be the same in both molecules, as alignment of their hydropathy profiles is straightforward (Fig. 6). However, the models proposed for the two molecules are different. One major difference between the models involves transmembrane helix IV of NIS (amino acids 135–156). Although the equivalent residues in the Na$^+$/glucose cotransporter are highly hydrophobic, these residues were not proposed to form a transmembrane helix probably because if such a transmembrane domain were predicted in the Na$^+$/glucose cotransporter it would include two charged residues: Lys 157 and Asp 162. Thus, two possibilities exist for this region that are compatible with the sequences of both proteins: either no transmembrane helix is formed in either protein, or a transmembrane helix exists in both proteins, with the one in the Na$^+$/glucose cotransporter including two charged residues (Fig. 6).

The hydropathy profile of the region spanning from amino acids 380–460 of NIS is very similar to the region 420–520 of the Na$^+$/glucose cotransporter: both contain three hydrophobic domains that can form transmembrane helixes. In the originally proposed models only two of these regions were considered transmembrane helixes in each model: the last two in NIS and the first two in the Na$^+$/glucose cotransporter. However, in principle, all three of these regions can be transmembrane helixes, so that additional possibilities exist in the models. In summary, once the last transmembrane helix proposed in the original models of the Na$^+$/glucose cotransporter is considered as not being present in NIS, for the rest of the structure there are three topological models compatible with the available sequences of both proteins: two with twelve and one with thirteen transmembrane helixes. The first model with twelve helixes is the original model proposed for NIS (Fig. 3), which in the Na$^+$/glucose cotransporter would include the counterpart of transmembrane domain IV of NIS. The second twelve transmembrane helix model lacks transmembrane domain IV of NIS (or its equivalent in the Na$^+$/glucose cotransporter), but includes an additional transmembrane domain, spanning residues 388–413 of NIS (435–460 of the Na$^+$/glucose cotransporter) (Fig. 6). The thirteen transmembrane domain model predicts both amino acid sequences 135–156 and 388–413 in NIS (and the corresponding sequences in the Na$^+$/glucose cotransporter) to be transmembrane domains (Fig. 6). Nevertheless, the latter model is less likely considering that it predicts that the NH$_2$ and COOH termini are on opposite sides of the membrane. As noted above, in the case of NIS both termini appear to be on the cytoplasmic side. Discrimination between these models will require extensive experimentation with both NIS and the Na$^+$/glucose cotransporter family.

A 2.9 kb mRNA transcript that hybridizes with isolated NIS cDNA clone was identified by Northern analysis in both FRTL-5 cells and rat thyroid tissue, but not in such non-thyroid tissues as rat liver, kidney, intestine, brain, or heart, thus suggesting that NIS is primarily expressed in thyroid cells. As expected, the NIS transcript was more readily detectable in the epithelial FRTL-5 cells than in native thyroid tissue.

6. Concluding remarks

In the context of thyroid research, the molecular identification of NIS was surprisingly elusive and long in coming. NIS is one of at least four major proteins that appear to be distinctively thyroidal, the other three being Tg, TPO and the TSH receptor [18–20]. Each of these four proteins mediates a crucial step in thyroid hormogenesis, plays an important role in thyroid function, and is considered a thyroid-specific marker. Yet, even as Tg, TPO and the TSH receptor molecules were characterized in recent decades, NIS remained unidentified for over 50 years. In summary, we have isolated the rat NIS cDNA clone and designed a secondary structure model for the NIS molecule. We have shown that the product of the NIS cDNA clone is sufficient to elicit Na$^+$/I$^-$ symport activity in both oocytes and mammalian cells, suggesting that NIS probably functions as a single subunit or as an oligomer of identical subunits. The data presented here establish that rat NIS is a 618 amino acid (relative molecular mass 65,196) Na$^+$-dependent intrinsic membrane protein with 12 putative transmembrane α-helix domains. Underlining the thyroid specificity of NIS, in Northern blot analyses using the NIS cDNA clone as a probe we observed that a 2.9 kb transcript is present in FRTL-5 cells and in the thyroid gland but not in liver, kidney, intestine, brain, or heart.

The role of NIS as the mediator of I$^-$ accumulation is consistent with its placing alongside other Na$^+$-dependent anion transporters on the basis of clustering relationships between its deduced amino acid sequence and the deduced sequences of other transporters. A comparison of the predicted amino acid sequence of NIS with those of other cloned Na$^+$-dependent cotransporters revealed the highest degree of homology (24.6% amino acid identity) with the human Na$^+$/glucose cotransporter. Given that some of the residues important for function in several membrane transporters are charged amino acids located in putative transmembrane domains [62–64], our model suggests that one or more of the three such residues in NIS, namely Asp 16, Glu 79 and Arg 208, may play a role in Na$^+$/I$^-$ symport activity. Interestingly, Asp 16 in NIS appears to correspond to Asp 28 in the Na$^+$/glucose cotransporter, a residue predicted to be located on the border between the cytosol and the membrane [53]. Although Asp 28 was initially believed to be essential for Na$^+$/glucose cotransport activity, it was subsequently shown that Asp 28 in the Na$^+$/glucose cotransporter is instead required for effective trafficking of the glycosylated protein from the endoplasmic reticulum to the plasma membrane. Such trafficking seems to be prevented in a mutant identified in one family, the glucose/galactose malabsorption mutant, in which Asp 28 is replaced by Asn [65]. Thus, it will be valuable to examine whether Asp 16 in NIS plays a similar role in trafficking.

The cloning and characterization of NIS makes it possible for the first time to raise anti-NIS antibodies to investigate the topology and identify functionally important regions of the NIS molecule, and to elucidate the biogenesis and regulatory mechanisms of NIS by TSH and other hormonal and environmental factors. Moreover, in view of the high degree of homology that exists between a eukaryotic transport protein from a given species and its counterparts in other species, it is now highly likely that the use of the rat NIS cDNA as a probe will result in the isolation of the human NIS cDNA clone. The availability of the human NIS cDNA clone will in turn make it feasible to investigate the expression of NIS in patients in various pathophysiological thyroid states, including hyperthyroidism, hypothyroidism, iodine deficiency, thyroid cancer and congenital lack of iodide transport.

6.1. Use of the NIS clone to investigate the molecular mechanism that underlies congenital lack of I$^-$ transport activity

Congenital lack of thyroidal I$^-$ accumulation is a rare condition that has been reported in several patients over the last few decades [9]. Patients with congenital lack of thyroidal I⁻ accumulation generally lack also the ability to transport I⁻ in such extra-thyroid tissues as salivary glands, gastric mucosa, and choroid plexus [9]. Thus, a genetic link among these I⁻ transport systems has been suggested. The mechanism by which a possible genetic defect leads to the absence of thyroidal I⁻ transport is unknown. Hence, three hypotheses may be proposed: (1) as a result of the defect, no NIS is synthesized at all; (2) NIS is synthesized but is not targeted properly to the plasma membrane; (3) synthesis and targeting of NIS are both unaltered, but the symporter itself is non-functional, either due to a mutation that renders the molecule inactive, or to the absence (or lack of activity) of a separate activating molecule. To examine these hypotheses, thyroid tissue from patients diagnosed with congenital lack of thyroid I⁻ transport may be obtained and anti-NIS antibodies may be used to determine, by immunofluorescence, whether the symporter is present in the plasma membrane. To ascertain whether the symporter molecule bears a mutation(s), RNA from the thyroid tissue samples may be isolated and subjected to RT-PCR to amplify the NIS cDNA. The patients' NIS cDNA sequence would be determined and compared with the control sequence in search of mutations.

The first two hypotheses can readily be addressed with immunofluorescence studies as soon as anti-human NIS antibodies are available. These experiments would reveal whether the symporter is present in the gland and, if it is, whether it is properly targeted to the plasma membrane. Mutation-induced impaired targeting to the plasma membrane of otherwise functional transporters has been reported for two other molecules in addition to the Asp28Asn mutation in the Na⁺/glucose cotransporter described above, namely the defective chloride channel found in patients with cystic fibrosis (CF) and the multidrug resistance protein P-glycoprotein. Impaired targeting of the CF chloride channel to the plasma membrane is observed as a result of the most frequent CF-causing mutation, i.e., deletion of phenylalanine at position 508, which causes retention of the molecule in the endoplasmic reticulum [66]. When the protein bearing the mutation is experimentally targeted to the plasma membrane, as it is when expressed in oocytes, it is at least partially functional. Similarly, some mutations introduced into P-glycoprotein which were initially thought to inactivate this drug efflux pump have been demonstrated to cause, instead, retention of the molecule in the endoplasmic reticulum [67].

If such immunofluorescence studies were to demonstrate the presence of the Na⁺/I⁻ symporter in the plasma membrane, the last hypothesis would be considered the most likely, i.e., such an observation would strongly suggest that NIS in these patients probably bears a mutation that renders it inactive. Therefore, if elucidation of the sequence of the mutated symporter reveals any missing or substituted amino acid residue(s), it would be concluded that this residue(s) is (are) necessary for folding or activity of the symporter.

Given that the characteristics of this congenital condition strongly suggest a genetic link among I⁻ transporters from various tissues, it seems likely that the isolation of the clone that encodes NIS in the thyroid may lead to the characterization of non-thyroid I⁻ transporters and consequently to a far better understanding of I⁻ transport processes in general.

6.2. NIS in autoimmune thyroid disease

An important aspect of the wide pathophysiological impact of Tg, TPO and the TSH receptor is the fact that autoantibodies against all three proteins have been demonstrated in patients suffering from autoimmune thyroid disease. Most notably, anti-TSH receptor antibodies (also called thyroid stimulating antibodies, which mimic TSH action) are a significant causative factor in Graves' disease and its attending hyperthyroidism [1,2], while anti-TPO antibodies are associated with Hashimoto's thyroiditis (goitrous autoimmune thyroiditis), in which antibodies may inhibit the activity of the enzyme and cause hypothyroidism [1,2]. The pathogenesis of autoimmune thyroid disease across the spectrum from Graves' disease to Hashimoto's thyroiditis has been suggested to involve various defects in the development of an organ-specific (in this case thyroid-specific) population of suppressor T-lymphocytes [1,2]. Thus, as NIS has now been identified, it is of considerable interest to explore whether a comparable immunoregulatory defect can lead to the production of anti-NIS autoantibodies in patients with some variety of autoimmune thyroid disease and to investigate what effect, if any, such putative autoantibodies may have on thyroid function.

6.3. NIS and thyroid cancer

Thyroid tumorigenesis by radiation. While cancer of the thyroid is a relatively infrequent condition in the U.S. (0.6% of all cancers in men and 1.6% in women, [1,2]), it has a considerably higher impact on endocrinological practice than would be expected solely on the basis of its incidence. This is so because the possible existence of thyroid cancer must be ruled out whenever a thyroid nodule is detected. There is a high estimated incidence of solitary palpable nodules in U.S. adults (~2 to 4%) [1,2]. The degree of accumulation of I^-, as revealed by scans of the gland, is used as an aid in the differential diagnosis of thyroid nodules. Thyroid nodules that accumulate I^- equally or more efficiently than the normal surrounding tissue are generally benign, while most thyroid cancers display markedly reduced I^- accumulation relative to healthy tissue. The main morphological types of thyroid cancer are the papillary carcinoma (75–85% of all thyroid cancers) and the follicular carcinoma (10–20%). In most instances both types are well-differentiated tumors that originate in the thyroid follicular cells [68]. The reduced I^- accumulation detected in the majority of thyroid cancers suggests that malignant transformation of these cells may have a deleterious effect on the function of NIS. Therefore, it would be highly instructive to ascertain whether any molecular changes can be detected in NIS in cancerous cells, and to correlate identified molecular differences, if any, with NIS function.

In spite of the I^- accumulation decrease of most thyroid cancers, NIS is sufficiently active in some thyroid cancers and metastases to render them amenable to detection and even destruction with I^- radioisotopes [1,2]. In fact, radioactive I^- therapy is used for the ablation of metastases and/or residual thyroid tissue after surgery [1,2], as well as for treatment of hyperthyroidism and other conditions. On the other hand, large doses of radiation reaching the gland via the I⁻ transport system in the form of iodine isotopes can cause thyroid cancer. The most dramatic example of this is the alarming rise in the incidence of thyroid cancer cases in Ukraine and Belarus in the wake of the 1986 Chernobyl power plant accident [12–17]. ^{131}I from the nuclear fallout is believed to be the most likely causative agent. This isotope was ingested largely through milk, mostly by young children, and concentrated in the thyroid via NIS. Virtually all of the identified tumors have been papillary carcinomas. The overall incidence of thyroid cancer in six regions studied around Chernobyl increased from an average of 4 cases per year from 1986 to 1989, to 55 cases in 1991 alone. In the Gomel region of Belarus, the region most exposed to nuclear fallout from the accident, the incidence of thyroid cancer in 1991 was ~80-fold higher than the 'normal' reported incidence elsewhere. Since such a large and rapid increase in cancer incidence is most probably without precedent, a highly unusual and potentially revealing opportunity has thus been created to study cancer-causing mechanisms.

In this context, the NIS cDNA clone may prove useful for the analysis of the mechanism by which radiation is tumorigenic in the thyroid. As the human NIS cDNA clone and anti-human NIS Abs become available, it will be possible to explore the expression of NIS at both the transcriptional and post-transcriptional levels in the glands of thyroid cancer patients exposed to radiation in the areas around Chernobyl. The results of these experiments would be correlated with data obtained from monitoring I⁻ transport, thyroid function, autoantibodies, and other molecular thyroid parameters in these patients. In addition, comparisons could be made between the Chernobyl data and the patterns of NIS expression in survivors from the atomic bombs dropped on Hiroshima and Nagasaki, where a link between radiation dose and incidence of thyroid carcinoma has also been ascertained. Findings from these studies may be of relevance for the understanding and prevention of cancer-causing processes.

Acknowledgements

We thank Dr. G. Castillo for his valuable contribution. This work was supported by the National Institutes of Health grant DK-41544, the Pew Scholars Program in the Biomedical Sciences and the Beckman Young Investigator Award (N.C.). O.L. was supported by the National Institutes of Health Training Program in Pharmacological Sciences GM-07260 and by the National Institutes of Health Hepatology Research Training Grant DK-07218.

References

1.   Werner, S.C. and Ingbar, S. (1991) in: The Thyroid: A Fundamental and Clinical Text, eds L.R. Braverman and R.D. Utiger. pp. 1–1362, J.B. Lippincott, Philadelphia, PA.
2.   DeGroot, L.J. (1995) in: Endocrinology, ed L.J. DeGroot. Grune & Stratton Inc. Orlando, FL.
3.   Chambard, M., Verrier, B., Gabrion, J. and Mauchamp, J. (1983) J. Cell Biol. **96**, 1172–1177.
4.   Nakamura, Y., Kotani, T. and Ohtaki, S. (1990) J. Endocrinol. **126**, 275–281.
5.   Dai, G., Levy, O. and Carrasco, N. (1995) Nature **379**, 458–460.

6.   Pasquinin J.M. and Adamo, A.M. (1994) Dev. Neurosci. **16**, 1–8.
7.   Marine, D. and Feiss, H.O. (1915) J. Pharmacol. Exper. Therap. **7**, 557–576.
8.   Hetzel, B.S., Potter B.J. and Dulberg E.M. (1990) World Rev. Nutr. Diet **62**, 59–119.
9.   Carrasco, N. (1993) Biochim. Biophys. Acta. **1154**, 65–82.
10. Nilsson, M., Bjorkman, U., Ekholm, R. and Ericson, L. (1992) Acta Endocrinol. **126**, 67–74.
11. Golstein, P.E., Sener, A., Beauwens, R. (1995) Am. J. Physiol. **268**, C111–C118.
12. (a) Kaznov, V.S., Demidchik, E.P and Astakhova, L.N. (1992) Nature (London) **359**, 21.
     (b) Baverstock, K., Egloff, B., Pinchera, A., Ruchti, C. and Williams, D. (1992) Nature (Lond.) **359**, 21–22.
     (c) Shigematsu, I. and Thiessen, J.W. (1992) Nature (London) **359**, 680–681.
     (d) Ron, E., Lubin, J. and Schneider, A.B. (1992) Nature (Lond.) **360**, 113.
13. Mettler, F.A., Williamson, M.R., Royal, H.D., Hurley, J.R., Khafagi, F., Sheppard, M.C., Saenger, E.L., Yokoyama, N., Parshin, V., Griaznova, E.A., Tarenenko, M., Chesin, V., Cheban, A. (1992) JAMA **268**, 616–619.
14. Williams, E.D. (1993) Histopathology **23**, 387–389.
15. Nikirov, Y. and Gnepp, D.R. (1994) Cancer **75**, 748–766.
16. Williams, E.D. (1994) Nature (Lond.) **371**, 556.
17. Likhtarev, I.A., Sobolev, B.G., Kairo, I.A., Tronko, N.D., Bogdanova, T.I., Oleinik, V.A., Epshtein, E.V. and Beral, V. (1995) Nature **375**, 365.
18. Mercken, L., Simons, M., Swillens S., Massaer, M.,Vassart, G. (1985) Nature **316**, 647–651.
19. Magnusson, R. P., Gestautas, J., Taurog, A. and Rapoport, B. (1987) J. Biol. Chem. **262**, 13885–13888.
20. Parmentier, M., Libert, F., Maenhaut, C., Lefort, A., Gérard, C., Perret, J., Van Sande, J., Dumont, J. E. and Vassart, G. (1989) Science **246**, 1620–1622.
21. Weiss, S.J., Philp, N.J. and Grollman, E.F. (1984) Endocrinology **114**, 1090–1098.
22. Kohrle, J. (1994) Exp. Clin. Endocrinol. **102**, 63–89.
23. Herzog, V., Neumuller, W., and Holzmann, B. (1987) EMBO. J. **6**, 555–560.
24. Herzog, V. (1985) Eur. J. Cell Biol. **39**, 399–409.
25. Kim, P.S. and Arvan P. (1995) J. Cell Biol. **128**, 29–38.
26. Magnusson, R. P. (1991) Thyroid peroxidase, in: Peroxidases in Chemistry and Biology, eds J. Everse, K.E. Everse and M.B. Grisham. pp. 199–219. CRC Press, Boca Raton, FL.
27. McLachlan, S.M. and Rapoport, B. (1992) Endocr. Rev. **13**, 192–206.
28. Maenhaut, C., Brabant, G., Vassart, G., and Dumont, J.E. (1992) J. Biol. Chem. **267**, 3000–3007.
29. Akamizu, T. Ikuyama, S., and Saji, M., Kosugi, S., and Kozak, C., McBride, O.W. and Kohn, L.D. (1990) Proc. Natl. Acad. Sci. USA **87**, 5677–5681.
30. Ludgate M.E. and Vassart G. (1995) Baillieres Clin. Endocrinol. Metab. **9**, 95–113.
31. Paschke, R., Vassart, G. and Ludgate, M. (1995) Clin. Endocrinol. **42**, 565–569.
32. Shachner, H., Franklin, A.L. and Chaikoff, I.L. (1944) Endocrinology **34**, 159–167.
33. Franklin, A.L. Chaikoff, I.L. and Lerner, S.R. (1944) J. Biol. Chem. **153**, 151–162.
34. Halmi, N.S. (1961) Vitam. Horm. **19**, 133–163.
35. Wolff, J. (1964) Physiol. Rev. **44**, 45–90.
36. O'Neill, B., Magnolato, D., and Semenza, G. (1987) Biochim. Biophys. Acta. **896**, 263–274.
37. Kaminsky, S.M., Levy, O., Garry, M.T. and Carrasco, N. (1992) Eur. J. Biochem. **200**, 203–207.
38. Wyngaarden, J.B., Wright, B.M. and Ways, P. (1952) Endocrinology **39**, 157–160.
39. Williams, R.H., Jaffe, H. and Solomon, B. (1950) Am. J. Med. Sci. **219**, 1–6.
40. Iff, H. W. and Willbrandt, W. (1963) Biochim. Biophys. Acta **70**, 711–752.
41. Bagchi, N. and Fawcett, D.M. (1973) Biochim. Biophys. Acta **318**, 235–251.
42. Vilkki, P. (1962) Arch. Biochem. Biophys. **97**, 425–427.
43. Schneider, P.B. and Wolff, J. (1965) Biochem. Biophys. Acta **94**, 114–123.
44. Vilijn, F. and Carrasco, N. (1989) J. Biol. Chem. **264**, 11901–11903.
45. Ambesi-Impiombato, F.S. and Coon, H.G. (1979) Int. Rev. Cytol. Suppl. **10**, 163–173.
46. Ambesi-Impiombato, F.S., Parks, L.A.M. and Coon, H.G. (1980) Proc. Natl. Acad. Sci. USA **77**, 3455–3459.
47. Nakamura, Y., Ohtaki, S. and Yamazaki, I. (1988) J. Biochem. **104**, 544–549.

48. Kaminsky, S.M., Levy, O., Salvador, C., Dai, G., and Carrasco, N. (1994) Proc. Natl. Acad. Sci. USA **91**, 3789–3793.
49. Reeves, J.P. and Hale, C.C. (1984) J. Biol. Chem. **259**, 7733–7739.
50. Rocmans, P.A., Penel, J.C., Cantraine, F.R. and Dumont, J.E. (1977) Am. J. Physiol. **232**, E343–E352.
51. Van Sande, J., Denebourg, F., Beauwens, R., Brekman, J.C. Daloze, D. and Dumont, J.E. (1990) Am. Soc. Pharmacol. Exp. Ther. **37**, 583–589.
52. Hediger, M.A., Ikeda, T., Coady, M., and Wright, E.M. (1987) Nature **330**, 379–381.
53. Wright, E.M., Loo, D.D.F., Panayatova-Heiermann, M., Lostao, M.P., Hirayama, B.H., Mackenzie, B., Boorer, K. and Zampighi, G. (1994) J. Exp. Biol. **196**, 197–212.
54. Kotexicz, M.L., Sampson, C.M., D'Alessio, J.M. and Gerard, G.F. (1988) Nucl. Acids Res. **16**, 265–277.
55. Krieg, P.A. and Melton, D.A. (1984) Nucleic Acid Res. **12**, 5707–5717.
56. Kozak, M. (1991) J. Biol. Chem. **266**, 19867–19870.
57. Kyte, J. and Doolittle, R.F. (1982) J. Mol. Biol. **157**, 105–132.
58. Garnier, J., Osguthorpe, D.J. and Robson, B. (1978) J. Mol. Biol. **120**, 97–120.
59. Chou, P.Y. and Fasman, G.D. (1974) Biochemistry **13**, 222–245.
60. Deisenhofer, J. and Michel, H. (1989) EMBO J. **8**, 2149–2170
61. Surratt, C.K., Weing, J.-B., Yuhasz, S., Amzel, L.M., Kwon, H.M., Handler, J.S. and Uhl, G.R. (1993) Curr. Opin. Nephrol. Hypertens. **2**, 744–760.
62. Carrasco, N., Antes, L., Poonian, M. and Kaback, H. R. (1986) Biochemistry **25**, 4486–4488.
63. Zhang, Y., Pines, G. and Kanner, B.I. (1994) J. Biol. Chem. **269**, 19573–19577.
64. Pantanowitz, S., Bendahan, A. and Kanner, B.I. (1993) J. Biol. Chem. **268**, 3222–3225.
65. Turk E., Zabel B., Mundlos S., Dyer J., and Wright, E.M. (1991) Nature **350**, 354–356.
66. Cheng, S.H., Gregory, R.J., Marshall, J., Sucharita, P., Souza, D.W., White, G.A., O'Riordan, C.R. and Smith, A. (1990) Cell **63**, 827–834.
67. Loo, T.W. and Clarke, D.M. (1994) J. Biol. Chem. **269**, 28683–28689.
68. Elisei R., Pinchera, A., Romei, C., Gryczynska, Pohl, V., Maenhaut, Fugazzola, L. and Pacini, F. (1994) J. Clin. Endocrinol. Metab. **78**, 867–871.
69. Jones, T.A. (1978) Appl. Crystallog. **11**, 268–272.
70. Evans, S.V. (1993) J. Mol. Graphics. **11**, 134–138.

CHAPTER 16

The Vertebrate Na$^+$/H$^+$ Exchangers
Structure, Expression and Hormonal Regulation

P. FAFOURNOUX

I.N.R.A.-U.N.C.M. Theix,
63122 St Genès-Champanelle,
France

J. POUYSSÉGUR

Centre de Biochimie-CNRS,
Université de Nice, Parc Valrose,
06108 Nice, France

© *1996 Elsevier Science B.V.*
All rights reserved

Handbook of Biological Physics
Volume 2, edited by W.N. Konings, H.R. Kaback and J.S. Lolkema

Contents

1. Introduction

In 1976 Mürer et al. demonstrated the existence of a directly coupled Na^+/H^+ exchange process in brush border membrane vesicles of renal proximal tubule and small intestine [1]. Since that time, the presence of such a transporter has been identified in the plasma membrane of virtually all eucaryotic cells. This process has a central role in cell homeostasis: regulation of intracellular pH, cell Na^+ content, cell volume, transcellular Na^+ and HCO_3^- absorption in epithelial cells (reviewed in Refs. [2,3]). The Na^+/H^+ exchanger is electroneutral and has a tightly coupled 1:1 stoichiometry for exchange of Na^+ and H^+. This exchange reaction is reversible and driven by the transmembrane chemical gradients of Na^+ and H^+, without input of metabolic energy such as ATP hydrolysis. An essential feature of these exchangers is their allosteric activation by intracellular protons through a proton regulatory site whereby H^+ binding exerts an allosteric activation on H^+ extrusion [4]. Moreover, the Na^+/H^+ exchanger is rapidly activated in response to a variety of mitogenic and nonmitogenic signals such as growth factors, oncogenic transformation, sperm, neurotransmitters, hormones, lectins, osmotic change and cell spreading. The Na^+/ H^+ exchanger can be reversibly inhibited by the diuretic drug amiloride and its analogues.

Recently, molecular cloning has provided increasing amount of information about structure, functional features and regulation of the Na^+/H^+ exchangers referred to as NHE (for recent reviews see Ref. [5,6]. The most widely studied NHE isoform, NHE1 is ubiquitously expressed and is involved in a variety of cellular functions by virtue of its ability to govern intracellular pH. Besides NHE1, several other isoforms have been identified mainly in epithelia where they perform more specialized ion transport. Five subtypes of Na^+/H^+ exchangers have been cloned and characterized to date; they define a new gene family of vertebrate transporters. These isoforms share the same structure but exhibit differences with respect to amiloride-sensitivity, cellular localisation, kinetic parameters, regulation by various stimuli and plasma membrane targeting in polarized epithelial cells.

In this chapter, we will present the most recent progress concerning the structure–function relationship of the Na^+/H^+ exchangers and their molecular mechanism of activation in response to extracellular signals.

2. The Na/H exchanger isoforms

2.1. Identification

In 1989, Sardet et al. [7] first isolated a Na^+/H^+ exchanger cDNA clone through genetic complementation of exchanger-deficient cells. The cloning strategy included three steps: (i) selection of a stable exchanger-deficient mouse fibroblast cell line, (ii) transfection of this mutant cell line with human genomic DNA and selection of the cells overexpressing a functional human exchanger and (iii) molecular cloning of the human transfected gene coding for the human Na^+/H^+ antiporter. The strategy for selection of the exchanger-deficient or exchanger-overexpressing cells was based on the Na^+/H^+ exchange reversibility and the toxicity of a high concentration of cytosolic H^+ [8]. By using this strategy for cell selection, a mouse cell line overexpressing the human Na^+/H^+ antiporter was obtained [9]. From this cell line, a 0.8 Kb genomic fragment was isolated and used for cDNA cloning of Na^+/H^+ exchanger. This human exchanger cDNA, now referred to as NHE1 restored a Na^+/H^+ exchange activity when expressed in exchanger-deficient fibroblast [7] or in Sf9 cells [10]. The nucleotide sequence predicted a protein of 815 amino acids. The hydropathy plot revealed that the exchanger contains 10–12 putative transmembrane spanning segments. Recently, NHE1 was shown to have O- and N-linked oligosaccharides that exist only in the first extracellular loop [11]. These oligosaccharides however appear to be dispensable for correct membrane insertion and Na^+/H^+ exchange activity [10,11].

Marked differences in the pharmacological, kinetic and regulatory properties of Na^+/H^+ exchange in various cell types, tissues and species led to surmise the existence of multiple isoforms of exchanger [12]. Recent molecular cloning studies — using the cDNA domain encoding the transmembrane domain of NHE1 as a probe — have confirmed that Na^+/H^+ exchangers constitute indeed a gene family. So far, six vertebrate isoforms: NHE1 [7], βNHE [13], NHE2 [14,15], NHE3 [16,17], NHE4 [16] and NHE5 [18] have been cloned and sequenced.

2.2. Tissue distribution

The first cloned Na^+/H^+ exchanger isoform NHE1 is expressed in virtually all cell types, tissues and species. The other isoforms appear to have a more specialized distribution. NHE2 isoform is predominantly expressed in epithelia: in rabbit, NHE2 message is present in kidney, large intestine, adrenal gland, small intestine and, to a lesser extent, in trachea and skeletal muscle [19]. In rat, NHE2 mRNA was predominantly found in small intestine, colon and stomach as compared to skeletal muscle, brain, kidney, testis, uterus, heart and lung [14]. NHE3 appears to be specifically expressed in a subset of epithelia: namely in intestine, kidney and stomach [16,17]. Tissue expression of NHE4 is decreasing as follows: stomach > small and large intestine >> uterus, brain, kidney, skeletal muscle [16], whereas NHE5, although not yet fully characterized is expressed in brain, testis and skeletal muscle [18].

In polarized epithelial cells, it was found that NHE1 localizes mainly to the basolateral membrane. Using antibodies against NHE1, basolateral localization has been shown in illeal villus, crypt epithelial cell [20] and in multiple nephron segments [21,22]. However, in cultured epithelial cells, the NHE1 isoform either endogenous (MDCK, HT29) or transfected (OK and MDCK cells) is expressed both apically and basolaterally [23]. By contrast, in the same cells in culture, the transfected NHE3 isoform is specifically targeted to the apical membrane [23]. It is presently impossible to determine whether the lack of specificity of NHE1 expression observed in OK and MDCK cells is physiological or corresponds to a loss of functions of the protein trafficking machinery of these cell lines in culture. Nevertheless, these results clearly demonstrate that NHE1 and NHE3 isoforms possess distinct molecular determinants specifying their membrane destination in polarized epithelial cells.

3. Functional studies

3.1. Kinetics and pharmacology

NHE1, NHE2, and NHE3 kinetic properties have been studied in stably transfected antiporter deficient cells [24–27]. For all the expressed exchangers, the dependence of the rate of transport on the extracellular Na^+ concentration follows Michaelis–Menten kinetics, consistent with one single binding site with an apparent K_m in the millimolar range.

In contrast, the concentration dependence of intracellular H^+ on Na^+/H^+ transport is considerably steeper. Such a behaviour is attributed to the existence of a internal H^+ modifier site at which H^+ binding turns on the operation of Na^+/H^+ exchange. The existence of a modifier site is essential in order to understand the mechanism whereby the exchanger exerts its physiological function. When protonated, the modifier site activates Na^+/H^+ exchange, thereby counteracting the acidification of the cytosol. Once the physiological 'set point' pH is reached, the allosteric site terminates exchange, preventing excessive cytoplasmic alkalinization [4]. H^+ kinetic data show similar cooperativity by intracellular H^+ for the three isoforms [25,26], confirming that the presence of a modifier site is a common feature of the Na^+/H^+ exchangers.

While all Na^+/H^+ exchangers are inhibited by amiloride and its analogues, they exhibit quantitative variations in sensitivities to these drugs. Kinetic analyses of amiloride and its more potent 5-amino substituted derivatives revealed that the sensitivity of the Na^+/H^+ exchanger to these compounds decreased in the order NHE1 > NHE2 >> NHE3 [26–28]. Amiloride acts on the exchanger as a competitive inhibitor of Na^+ by acting at or close to the transport site. Recently, an amino acid residue critical for amiloride binding has been identified using the following approach: selection of mutant cells that express an amiloride resistant exchanger [29] and, subsequently, cloning and sequencing their exchanger cDNA [30]. Comparison between sequences of NHE1 cDNA clones isolated from parental and amiloride-resistant cells allowed the identification of one point mutation:leucine to phenylalanine at position 167 (hamster amino acid sequence) which is responsible for the decreased affinity for amiloride observed in the selected cells. It is noteworthy that this amino acid residue is located within a highly conserved part of the fourth transmembrane domain of the antiporter. Subsequently other amino acids located around position Leu167 and within this fourth transmembrane segment were found to be critical for amiloride recognition [31].

3.2. Functional domains of the antiporter

All the NHE molecules can be separated into two large domains: the amino terminal hydrophobic domain and the carboxyl-terminal hydrophilic domain. Deletion of the entire hydrophilic domain (expressed in the antiporter deficient cell line) (i) preserves amiloride-sensitive Na^+/H^+ exchange activity indicating that the cytoplasmic domain is not essential for this function, (ii) markedly shifts the pK value for intracellular H^+ to the acidic range but preserves the allosteric activation by the internal H^+ indicating that the H^+ modifier site must be located within the transmembrane domain, (iii) abolishes growth factor induced cytoplasmic alkalinization [32]. From these studies, a model has been proposed where the antiporter can be separated into two distinct functional domains: the NH2 transmembrane domain that has all the features required for catalyzing amiloride sensitive Na^+/H^+ exchange with a built in H^+ modifier site and the cytoplasmic regulatory domain which determines the pHi set point value and mediates the activation of the exchange by growth factors. Similar results have also been obtained for other NHE isoforms [33,34].

Evidence for the regulatory role of the cytoplasmic domain was also provided by the study of a chimeric molecule between human NHE1 and trout βNHE [35]. Trout exchanger βNHE is activated by cAMP, presumably through direct phosphorylation of its cytoplasmic domain by a cAMP dependent protein kinase [13], whereas mammalian NHE1 does not respond to cAMP. Borgese et al. [35] have shown that cAMP activates a chimera of the N-terminal transmembrane domain of human NHE1 fused to the C-terminal cytoplasmic domain of βNHE. This study demonstrates that the structure of the cytoplasmic domain of the Na^+/H^+ exchanger dictates the nature of the hormonal response, an action that is also dependent upon the cellular context.

3.3. Oligomeric structure

There is substantial evidence that most membrane transport proteins are oligomeric [36]. Concerning the Na^+/H^+ exchanger, the first studies suggested that the protein is present in the membrane as a dimer [37]. Kinetic studies have also suggested that the functional transport unit may be an oligomer [38,39]. Recently, evidence have been presented that the Na^+/H^+ exchangers (NHE1 and NHE3) form stable dimers in the membrane [40,41]. Mobility shift of NHE1 molecule on SDS-PAGE from a monomeric form (110 Kd) to a putative dimeric form (220 Kd) has been observed when cells were exposed to a chemical cross-linker, dissuccinimidyl suberate (DSS) [40]. Utilization of a NHE1 mutant deleted of the cytoplasmic domain demonstrates that the transmembrane domain of the antiporter is sufficient for dimerization. In addition, dimerization is isoform specific because coexpression of NHE1 and NHE3 in the same cell does not lead to the formation of heterodimers. Exploiting the fact that dimerization triggers kinase activity of the receptor tyrosine kinase, a chimera between the NHE1 transmembrane domain and the insulin receptor tyrosine kinase domain was constructed. This chimera was expected to probe the possibility that NHE1 forms spontaneous dimers *in vivo*. When transfected into hamster fibroblasts this chimera generated a functional transporter undergoing autophosphorylation on a tyrosine residue and presenting properties of a constitutively active insulin receptor. These results strongly suggest that NHE1 exists in an oligomeric state in intact cells [40]. Despite the strong evidence for dimeric structure of NHE1 and NHE3, whether individual subunits of NHE1 are the minimum functional unit for Na$^+$/H$^+$ exchange remains obscure. Coexpression of a nonfunctional point mutant molecule (E262I) with an active truncated NHE1 did not lead to a dominant negative effect. This observation would favour the hypothesis that individual subunits of NHE1 function independently within the oligomeric state. However, as discussed in the corresponding article, this conclusion must be taken with caution because the experimental conditions did not permit to reach a sufficient higher level of the inactive NHE1 molecule, a condition required to drive maximally the formation of heterodimers.

4. Regulation

Transporters of the Na$^+$/H$^+$ exchanger family have the capacity to modulate their activity in response to hormones, growth factors and other various extracellular stimuli. The molecular mechanisms of regulation are not fully elucidated yet. The isoform that has been best studied is the ubiquitous form NHE1 which is activated via an increased affinity for internal protons leading to an intracellular alkalinization detectable in the absence of bicarbonate.

4.1. NHE1 regulation

4.1.1. Growth factor activation of NHE1
Phosphorylation of NHE1: Phosphorylation of NHE1 has been postulated as a very likely mechanism for its activation [42,43]. This assumption was based on the fact that phorbol esters and growth factors which are known to activate protein kinases also activate NHE1 [44,45]. After molecular identification of NHE1 and generation of antibodies, it was possible to immunoprecipitate NHE1 from ^{32}P labeled fibroblasts. It was thus shown that NHE1 is phosphorylated in resting cells and that mitogenic stimulation is accompanied by an increase in phosphorylation [46]. Moreover, okadaic acid, a specific serine/threonine protein phosphatase inhibitor could, by itself, trigger activation of NHE1, an effect that correlates well with stimulation of its phosphorylation. Whatever the nature of the stimulus, it was demonstrated that phosphorylation of NHE1 occurs exclusively on serine residues [46]. Particularly, this was observed for stimulation by EGF and thrombin which are known to activate distinct signalling pathways (receptor tyrosine kinase or G-protein coupled receptor, respectively). Moreover, tryptic phosphopeptide maps of NHE1 that was immuno-precipitated from cells treated with EGF or thrombin showed a common pattern of phosphorylation [47]. Taken together, these results suggested that the final step in NHE1 activation is mediated by an unidentified NHE1 kinase that is activated by a pathway integrating all extracellular stimuli.

However, using a set of deletion mutants of the cytoplasmic domain, Wakabayashi et al. [48] demonstrated that all major phosphorylation sites are located in

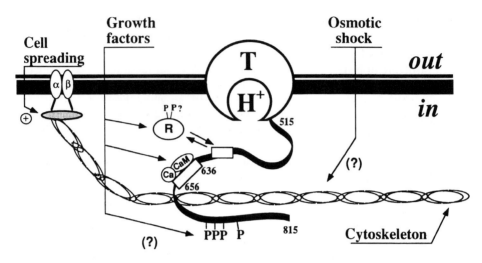

Fig. 1.   Proposed model for NHE1 activation by extracellular stimuli based on available knowledge. Growth factor-induced intracellular alkalinization of NHE1-expressing cells seems to be triggered either through (1) direct phosphorylation of the cytoplasmic domain of NHE1, (2) regulatory protein (R), or/and (3) elevation of intracellular Ca^{++} leading to interaction of the Ca^{++}/Calmodulin complex with a specific region of the cytoplasmic domain. Osmotic shock and cell spreading both activate NHE1 through a possible interaction of cytoskeleton elements with the cytoplasmic tail of NHE1. T represents the transporter domain that consists of 10 to 12 transmembrane segments followed by a cytoplasmic stretch of 300 residues.

the cytoplasmic tail between amino acids 567 and 815 (Fig. 1). Partial deletion of this region (635–815) eliminated all major phosphorylation sites, but the truncated NHE1, although less, was still activatable by all growth factors tested. This finding suggests the existence of an activating mechanism that does not require direct phosphorylation of NHE1. The simplest hypothesis is to postulate the existence of one or multiple regulatory protein(s) that interact with the cytoplasmic domain of NHE1. These results are in good agreement with those of Grinstein et al. [49] who have shown that hyperosmotic shock activates the antiporter without increasing phosphorylation.

Interaction with Ca^{++}/Calmodulin: Bertrand et al. [50] demonstrated that NHE1 is able to bind Ca^{++}/Calmodulin with high affinity. They identified two binding sites located in the cytoplasmic regulatory domain of NHE1. Deletion of the high affinity calmodulin binding region (residues 636–656) and point mutations of positively charged residues within this region reduced Ca^{++}/Calmodulin binding and, concomitantly, the thrombin-induced intracellular alkalinization [51]. In addition, the authors showed that mutation and deletion described above rendered NHE1 constitutively active as demonstrated by the rightward shift in pHi dependence. These results suggest a model in which calmodulin-binding region functions as an autoinhibitory domain for NHE1. Ca^{++}/Calmodulin activates NHE1 by binding to this region and releasing the autoinhibitory domain. This activating mechanism could be critical in the early growth factor response when internal Ca^{++} rise is maximal, whereas the sustained activation of NHE1 could be promoted by a NHE1 accessory protein interacting with the 567–635 region and possibly through the phosphorylation of the cytoplasmic domain (Fig. 1).

4.1.2. Activation of NHE1 by mechanical stimuli and cell spreading
NHE1 is activated in response to mechanical signals such as hyperosmotic shock [45] or cell spreading [52–54]. Hyperosmotic shock activates NHE1 but, in contrast to that observed with mitogens, cell shrinkage-induced alkalinization is not associated with a change in NHE1 phosphorylation state [49] or with a rapid change in cytoplasmic Ca^{++} level [55]. The increase of intracellular Ca^{++} level that is observed in lymphocytes is secondary to cytoplasmic alkalinization [45]. Therefore, an interaction of NHE1 with accessory protein(s) can be postulated to explain its activation by cell shrinkage.

NHE1 can also be activated by cell spreading on dishes coated with fibronectin or by adding fibronectin-coated beads to cells in suspension [54,56]. Such an activation is thought to act via integrin–fibronectin interaction [57]. Integrins are integral proteins known to anchor actin microfilaments to the plasma membrane through a direct interaction with cytoskeletal proteins. In this context it is interesting to note that NHE1 was found to be enriched together with F-actin, vinculin, and talin [58], suggesting direct or indirect interactions with the cytoskeletal elements.

4.1.3. Identification of proteins interacting with NHE1
The putative regulatory protein(s) R (Fig. 1) that interacts with the cytoplasmic domain of NHE1 has not yet been identified. By using the double hybrid technique, we have recently isolated several cDNAs that are good candidates for specific interaction with the cytoplasmic tail of NHE1 (unpublished results). Among these clones, paxillin, myosin light chain, and the human isoform of the chicken myosin-binding subunit of myosin phosphatase -M133- [59] were shown to bind the cytoplasmic domain of NHE1. The physiological relevance of these data are under investigation. However these results are in good agreement with those [58] showing that NHE1 could colocalize with cytoskeletal proteins and particularly with actin in the leading edge and ruffles. The involvement of F-actin has been suggested in regulation of the Na$^+$/H$^+$ exchange in CaCO$_2$ cells [60]. It has also been reported that cytochalasin B (an actin filament disrupter) inhibits hyperosmolarity-induced activation of the exchanger [61] in Barnacle muscle fibers. In addition, activation of NHE1 by growth factors appears to depend on the degree of cell adhesion to the substratum [56], suggesting an interaction of the exchanger with cell cytoskeleton (particularly stress fibers which are not polymerized in non adherent cells).

One of the cDNAs cloned could encode a protein that links NHE1 to the cytoskeleton. That protein could be involved in Na$^+$/H$^+$ exchange regulation. This hypothesis is still under investigation.

4.2. Hormonal regulation of epithelial NHE

The analysis of the hormonal regulation of the different exchangers is particularly complex because, in a variety of tissues, different isoforms can be co-expressed and exhibit different pathways of regulation. To overcome these difficulties, a strategy consisted in studying the regulation of the cloned exchangers in the same recipient cell line defective in endogenous Na^+/H^+ antiporter [26,28]. Such studies have demonstrated that stimulation of NHE2 and NHE3 by growth factors does not involve a change in exchanger affinity for H^+ but induce a change in its V_{max} [25]. NHE3 is stimulated by serum, growth factors and thrombin but is inhibited by phorbol esters whereas NHE2 is activated in response to PKC stimulation [25,33]. These results do not take into account the effects inherent to the cell type or the polarity of epithelial cells on regulation of the exchangers. Indeed, the cellular environment would provide specific regulatory or accessory proteins participating in the regulation of NHE in response to different stimuli. For example, NHE1 is either insensitive [62], stimulated [25], or inhibited [63–65] via the activation of the same kinase (PKC) pathway when expressed in different cell lines.

4.2. Long-term regulation

In addition to the above-described short-term regulation, Na^+/H^+ exchangers can also be regulated on a long-term basis [66]. It has been reported that several pathological conditions (acidosis) are associated with an adaptive increase in the activity of apical Na^+/H^+ exchanger. Moreover, incubation of cultured tubular cells in an acidic medium leads to an increase in Na^+/H^+ exchange activity that is dependent on protein synthesis, and is associated with an increase in NHE mRNA [67]. This acid-induced activation of NHE requires protein kinase C and is associated with an increase in the activity of the transcription factor AP1.

5. Concluding remarks

Biochemical identification of the Na^+/H^+ exchanger and subsequent cloning have provided major information and opened the field of this new transporter family. The generation of specific molecular tools has already permitted to identify several isoforms that are likely to be involved in specialized functions given their restricted expression and unique regulation. Combination of site-directed and random mutagenesis in conjunction with a genetic screen has illuminated key residues involved in ions catalysis and amiloride binding. Such an approach should be further developed to identify the domains involved in H^+ sensing, hormonal regulation and membrane targeting in epithelial cells. Finally, alteration of NHE gene expression in transgenic animal (transgenesis or knock out) could bring valuable information about the physiological role of these exchangers and their possible implication in human diseases.

References

1.   Mürer, H., Hopfer, U. and Kinne, R. (1976) Biochem. J. **154**, 597–604.
2.   Seifter, J.L. and Aronson, P.S. (1986) J. Clin. Invest. **78**, 859–864.
3.   Grinstein, S. ed (1988) Na^+/H^+ Exchange, CRC Press, Boca Raton, FL.

4. Aronson, P.S. (1985) Ann. Rev. Physiol. **47**, 545–560.
5. Bianchini, L. and Pouysségur, J. (1995) in: Molecular of the Kidney in Health and Disease, eds D. Schlondorff and J. Bonventre. Dekker, New York.
6. Noël, J. and Pouysségur, J. (1995) Am. J. Physiol. **268**, C283–C296.
7. Sardet, C., Franchi, A. and Pouysségur, J. (1989) Cell **56**, 271–280.
8. Pouysségur, J. (1985) TIBS **11**, 453–455.
9. Franchi, A., Perucca-Lostanlen, D. and Pouysségur, J. (1986) Proc. Natl. Acad. Sci. USA **83**, 9388–9392.
10. Fafournoux, P., Ghysdael, J., Sardet, C. and Pouysségur, J. (1991) Biochemistry **30**, 9510–9515.
11. Counillon, L., Pouysségur, J. and Reithmeier, R. (1994) Biochemistry **33**, 10463–10469.
12. Clark, J.D. and Limbird, L.L. (1991) Am. J. Physiol. **261**, G945–G953.
13. Borgese, F., Sardet, C., Cappadoro, M., Pouysségur, J. and Motais, R. (1992) Proc. Natl. Acad. Sci. USA **89**, 6765–6769.
14. Wang, Z., Orlowski, J. and Shull, G.E. (1993) J. Biol. Chem. **268**, 11925–11928.
15. Collins, J.F., Honda, T., Knobek, S., Bulus, N.M., Conary, J., Dubois, R. and Ghishan, F.K. (1993) Proc. Natl. Acad. Sci. USA **90**, 3938–3942.
16. Orlowski, J., Kandasamy, R.A. and Shull, G.E. (1992) J. Biol. Chem. **267**, 9331–9339.
17. Tse, C.M., Brant, S.R., Walker, S., Pouysségur, J. and Donowitz, M. (1992) J. Biol. Chem. **267**, 9340–9346.
18. Klanke, C.A., Su, Y.R., Callen, D.F., Wano, Z., Meneton, P., Baird, N., Kandasamy, R.A., Orlowski, J., Otterud, B.E., Leppert, M., Shull, G.E. and Menon, A.G. (1995) Genomics **25**, 615–622.
19. Tse, C.M., Levine, S.A., Yun, C.H.C., Montrose, M.H., Little, P.J., Pouysségur, J., Donowitz, M. (1993) J. Biol. Chem. **268**, 11917–11924.
20. Bookstein, C. De Paoli, A.M., Xie, Y., Niu, P., Musch, M.W., Rao, M.C. and Change, E.B. (1994) J. Clin. Invest. **93**, 106–113.
21. Biemesderfer, D., Reilly, R.F., Exner, M., Igarashi, P. and Aronson, P.S. (1992) Am. J. Physiol. Renal Fluid Electrolyte Physiol. **263**, F833–F840.
22. Krapf, G. and Solioz, M. (1991) J. Clin. Invest. 88, 783-788.
23. Noël, J., Roux, D. and Pouysségur, J. (1996) J. Cell Sci. (in press)
24. Tse, C.M., Ma, A.I., Yang, V.W., Watson, A.J.M., Levine, S., Montrose, M.H., Potter, J., Sardet, C., Pouysségur, J. and Donowitz, M. (1991) EMBO J. **10**, 1957–1967.
25. Levine, S.A., Montrose, M.M., Tse, C.M. and Donowitz, M. (1993) J. Biol. Chem. **286**, 25527–25535.
26. Orlowski, J. (1993) J. Biol. Chem. **268**, 16369–16377.
27. Yu, F.H., Shull, G.E. and Orlowski, J. (1993) J. Biol. Chem. **268**, 25536–25541.
28. Counillon, L., Scholz, W., Lang H.J. and Pouysségur, J. (1993) Mol. Pharmacol. **44**, 1041–1045.
29. Franchi, A., Cragoe, J. and Pouysségur, J. (1986a) J. Biol. Chem. **261**, 14614–14620.
30. Counillon, L., Franchi, A. and Pouysségur, J. (1993) Proc. Natl. Acad. Sci. USA **90**, 4508–4512.
31. Counillon L., Noël J. and Pouysségur J. (1996) submitted.
32. Wakabayashi, S., Fafournoux, P., Sardet, C. and Pouysségur, J. (1992) Proc. Natl. Acad. Sci. USA **89**, 2424–2428.
33. Levine, S.A., Samir, K.N., Yun, C.H.C., Yip, J.W., Montrose, M., Donowitz, M. and Tse, C.M. (1995) J. Biol. Chem. **270**, 13713–13725.
34. Tse, M., Levine, S., Yun, C., Brant, S., Counillon, L., Pouysségur, J. and Donowitz, M. (1993) J. Membr. Biol. **135**, 93–108.
35. Borgese, F., Malapert, M., Fievet, B., Pouysségur, J. and Motais, R. (1994) Proc. Natl. Acad. Sci. USA **91**, 5431–5435.
36. Klingenberg, M. (1981) Nature **290**, 449–454.
37. Béliveau, R., Demeule, M. and Potier, M. (1988) Biochem. Biophys. Res. Commun. **152**, 484–489.
38. Otsu, K., Kinsella, J., Scktor, B. and Froehlich, J.P. (1989) Proc. Natl. Acad. Sci, USA **86**, 4818–4822.
39. Otsu, K., Kinsella, J.L., Koh, E. and Froehlich, J.P. (1992) J. Biol. Chem. **267**, 8089–8096.
40. Fafournoux, P., Noël, J. and Pouysségur, J. (1994) J. Biol. Chem. **269**, 2589–2596.
41. Fliegel, L., Haxorth, R.S. and Dyck, J.R.B. (1993) Biochem. J. **289**, 101–107.
42. L'Allemain, G., Paris, S. and Pouysségur, J. (1984) J. Biol. Chem. **259**, 5809–5815.

43. Pouysségur, J., Chambard, J.C., Franchi, A., Paris, S. and Van Obberghen-Schilling, E.E. (1982) Proc. Natl. Acad. Sci. USA **79**, 3935–3939.
44. Paris, D. and Pouysségur, J. (1984) J. Biol. Chem. **259**, 10989–10994.
45. Grinstein, S., Cohen, S., Goetz, J.D. and Rothstein, A. (1985) J. Cell Biol. **101**, 269–276.
46. Sardet, C., Counillon, L., Franchi, A. and Pouysségur, J. (1990) Science **247**, 723–726.
47. Sardet, C., Fafournoux, P. and Pouysségur, J. (1991) J. Biol. Chem. **266**, 19166–19171.
48. Wakabayashi, S., Bertrand, B., Shigekawa, M., Fafournoux, P. and Pouysségur, J. (1994) J. Biol. Chem. **269**, 5583–5588.
49. Grinstein, S., Woodside, M., Sardet, C., Pouysségur, J. and Rotin, D. (1992) J. Biol. Chem. **267**, 23823–23828.
50. Bertrand, B., Wakabayashi, S., Ikeda, T., Pouysségur, J. and Shigekawa, M. (1994) J. Biol. Chem. **269**, 13703–13709.
51. Wakabayashi, S., Bertrand, B., Ikeda, T., Pouysségur, J. and Shigekawa, M. (1994) J. Biol. Chem. **269**, 13710–13715.
52. Ingber, D.E., Prusty, D., Frangioni, V., Cragoe, E.J., Lechêne, C. and Schwartz, M.A. (1990) J. Cell Biol. **110**, 1803–1811.
53. Novikova, I.Y., Murravyeva, O.V., Cragoe, E.J. and Margolis, L.B. (1993) Biochim. Biophys. Acta **1178**, 267–272.
54. Schwartz, M.A., Both, G. and Lechêne, C. (1989) Proc. Natl. Acad. Sci. USA **86**, 4525–4529.
55. Grinstein, S., Cohen, S., Goetz, J.D., Rothstein, A. and Felfand, E.W. (1985b) Proc. Natl. Acad. Sci. USA **82**, 1429–1433.
56. Schwartz, M.A. and Lechêne, C. (1992) Proc. Natl. Acad. Sci. USA **89**, 6138–6141.
57. Schwartz, M.A. (1992) Trends Cell Biol. **2**, 304–308.
58. Grinstein, S., Woodside, M., Waddell, T.K., Gowney, G.P., Orlowski, J., Pouysségur, J., Wong, D.C.P. and Foskett, J.K. (1993) EMBO J. **12**, 5209–5218.
59. Shimizu, H., Ito, M., Miyahara, M., Ichikawa, K., Okubo, S., Konishi, T. Naka, M., Tanaka, T., Hirano, K., Hartshorne, D.J. and Nakano, T. (1994) J. Biol. Chem. **269**, 30407–30411.
60. Watson, A.J.M., Levine, S., Donowitz, M. and Montrose, M.H. (1992) J. Biol. Chem. **267**, 956–962.
61. Davis, B.A., Hogan, E.M. and Boron, E.F. (1994) Am. J. Physiol. Cell Physiol. **266**, C1744–C1753.
62. Watson, A.J., Levine, S., Donowitz, M. and Montrose, M.M. (1991) Am. J. Physiol. 261, G229-G238.
63. Cano, A., Preisig, P.A., Miller, R.T. and Alpern, R.J. (1993) J. Am. Soc. Nephrol. **4**, 834A.
64. Helmle-Kolb, C., Counillon, L., Roux, D., Pouysségur, J., Mrkic, B. and Mürer, H. (1993) Pflugers Arch. **425**, 34–40.
65. Pollock, A.S., Warnock, D.G. and Strewler, G.J. (1986) Am. J. Physiol. **260**, F217–F225.
66. Horie, S., Moe, O., Tejedor, A. and Alpern, R.J. (1990) Proc. Natl. Acad. Sci. USA **87**, 4742–4745.
67. Horie, S., Moe, O., Yamahi, Y., Cano, A., Miller, R.T. and Alpern, R.J. (1992) Proc. Natl. Acad. Sci. USA **89**, 5236–5240.

CHAPTER 17

Biogenic Amine Transporters of the Plasma Membrane

G. RUDNICK

Department of Pharmacology, Yale University School of Medicine, New Haven, CT 06510, USA

© 1996 Elsevier Science B.V.
All rights reserved

Handbook of Biological Physics
Volume 2, edited by W.N. Konings, H.R. Kaback and J.S. Lolkema

Contents

1. Introduction

The biogenic amine neurotransmitters serotonin (5-hydroxytryptamine, 5-HT), dopamine (DA) and norepinephrine (NE) are inactivated, after their release from nerve terminals, by sequestration within the cells from which they were secreted (Fig. 1). This process, which also occurs with other small neurotransmitters such as GABA and glutamate (see Chapter 19) is catalyzed by neurotransmitter transporters located in the plasma membrane of nerve terminals. With the exception of ACh, all small neurotransmitters are inactivated by this mechanism, and even in the case of ACh, where hydrolysis by cholinesterase is the main inactivation mechanism, the product of hydrolysis (choline) is taken up by a specific transporter [1]. In the nervous system, the transporters function to regulate neurotransmitter activity by removing extracellular transmitter. Inhibitors that interfere with this regulation include antidepressant drugs and stimulants such as the amphetamines and cocaine. For each neurotransmitter, a specific transporter mediates its transport into the nerve terminal, and each transporter is apparently found only on cells which release its cognate transmitter.

Many of these neurotransmitter transporters are related, both structurally and functionally, and constitute a multigene family. These carrier proteins couple the transmembrane movement of Na^+, Cl^-, and in some systems K^+, to the reuptake of neurotransmitters released into the synaptic cleft [2] (Fig. 2). The Na^+ and Cl^--coupled neurotransmitter transporters are a group of integral membrane proteins encoded by a closely related family of recently cloned cDNAs [3–24]. Within this family, transporters for the biogenic amines 5-HT, NE and DA constitute a distinct subclass with structural and functional properties that set them apart from other members of the neurotransmitter transporter gene family.

Each neuron actually contains two distinct transport systems for its specific neurotransmitter (Fig. 1). The first of these systems moves the transmitter (NT) from the extracellular space into the cytoplasm and the second transports cytoplasmic transmitter into the storage organelle, or synaptic vesicle. Both systems move the transmitter uphill, against a concentration gradient, and are therefore coupled to the input of metabolic energy. In each case this energy comes from hydrolysis of cytoplasmic ATP, which is used to generate transmembrane ion gradients. At the plasma membrane, the Na^+/K^+-ATPase generates ion gradients. These are shown in Fig. 1 as arrows in the direction of the downhill movement of the ions. As described below, these gradients provide the driving force for neurotransmitter uptake across the plasma membrane. At the synaptic vesicle membrane, ATP is hydrolyzed by an H^+ pump that acidifies the vesicle interior, generating an H^+ gradient which is used to drive neurotransmitter accumulation by the vesicle.

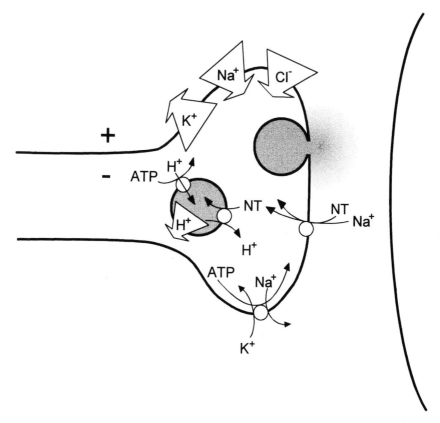

Fig. 1.   Neurotransmitter recycling at the nerve terminal. Neurotransmitter (NT) is released from
the nerve terminal by fusion of synaptic vesicles with the plasma membrane. After release, the
transmitter is transported across the plasma membrane by a Na^+-dependent transporter in the plasma
membrane. Transmitter delivered into the cytoplasm is further sequestered in synaptic vesicles by a
vesicular transporter using the transmembrane H^+ gradient as a driving force. This driving force,
shown by the arrow pointing in the direction of downhill H^+ movement, is generated by an
ATP-dependent H^+ pump in the vesicle membrane. The Na^+ and K^+ gradients across the plasma
membrane are generated by the Na^+/K^+-ATPase. This enzyme also creates a transmembrane electrical
potential (negative inside) that causes Cl^- to redistribute. Neurotransmitter transport across the plasma
membrane is coupled to the Na^+, Cl^- and K^+ gradients generated by the ATPase.

The components responsible for biogenic amine neurotransmitter transport into
and storage within neurons are also found in other tissues, including platelets,
chromaffin cells, mast cells, basophils, and placenta, where they are also utilized
for biogenic amine transport and storage. Recent isolation of cDNA clones for these
proteins [5–12,22] has allowed the examination of various tissues where the amines
are accumulated and stored in secretory granules. These studies have established
the existence of mRNA encoding transporters for serotonin (SERT), norepinephrine (NET) and dopamine (DAT) in cells storing those transmitters. These
biogenic amine transporters are all inhibited by cocaine, and share other structural

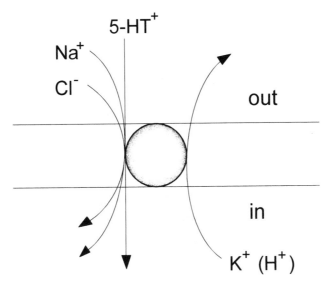

Fig. 2.   Driving forces for 5-HT transport. Na$^+$ and Cl$^-$ on the outside of the cell are transported inward with the cationic form of 5-HT. In the same catalytic cycle, K$^+$ is transported out from the cytoplasm. In the absence of internal K$^+$, H$^+$ ions take the place of K$^+$. The energy released by downhill movement of Na$^+$, Cl$^-$ and K$^+$ provides the driving force for 5-HT accumulation against its concentration gradient.

and mechanistic properties. The vesicular monoamine transporter (VMAT) is also found in all cells secreting biogenic amines (see Chapter 18). Knowledge of the mechanisms for biogenic amine transport and storage in one cell type, therefore, sheds light on amine transport in many other cell types as well.

At the plasma membrane the Na$^+$, K$^+$-ATPase directly creates Na$^+$, (out > in) and K$^+$ (in > out) gradients and, indirectly, through generation of a transmembrane electrical potential ($\Delta\psi$, inside negative), creates a Cl$^-$ gradient (out > in). The transmembrane gradients of Na$^+$, K$^+$, and Cl$^-$ serve, in turn, as driving forces for biogenic amine transport across the plasma membrane. Another component of the plasma membrane, the specific biogenic amine transporter, couples the inward flux of Na$^+$ and Cl$^-$, and the outward flux of K$^+$, to the entry of the neurotransmitter substrate.

The preferred experimental system for studying the mechanism of neurotransmitter transport has been that of isolated plasma membrane vesicles. The 5-HT transporter (SERT) has been studied in membrane vesicles from lysed platelets, which provide a system where internal and external ionic composition can be controlled independently. Preparations of membrane vesicles derived from rat brain synaptic plasma membrane have been used to study the GABA transporter (GAT-1) (see Chapter 19). Membrane vesicles from PC12 cells and placental syncitiotrophoblast cells have been used to study the NE transporter (NET) [25,26]. The advantage of membrane vesicles is that complications of intracellular metabolism or sequestration are minimized, and the accumulation of substrate neurotransmitter within the vesicle lumen is determined by the ion gradients imposed by experimental manipulation. Since both internal and external medium is experimentally variable, the responses of kinetic and thermodynamic transport parameters to ion gradients and membrane potential are readily accessible.

2. Ionic requirements

2.1. The 5-HT Transporter (SERT)

The best studied of the three plasma membrane biogenic amine transporters is the 5-HT transporter. Much of the available data on 5-HT transport comes from studies using platelets. The platelet plasma membrane 5-HT transporter is identical in primary amino acid sequence to the brain 5-HT transporter [9]. In humans and rats, a single gene apparently encodes the 5-HT transporter expressed in all tissues. Although transport studies in intact platelets can yield valuable information, their usefulness is limited by our inability to control the cytoplasmic and intragranular ion composition. To circumvent these difficulties, we have used preparations of vesicles derived from plasma membrane and dense granule membrane as model systems [27]. The availability of membrane vesicles from platelets [28], rat basophilic leukemia cells [29] and placenta [30] has allowed a detailed understanding of the 5-HT transporter's mechanism.

The plasma membrane 5-HT transporter, like other neurotransmitter transporters in this gene family, requires both Na^+ and Cl^- in the external medium for neurotransmitter influx. Early studies by Sneddon and Lingjaerde [31,32] on the 5-HT transporter of intact platelets led to the suggestion that both Na^+ and Cl^- are co-transported with 5-HT. Studies in both intact platelets and in plasma membrane vesicles demonstrated that while Cl^- could be replaced by Br^-, and to a lesser extent by SCN^- or NO_2^-, Na^+ could not be replaced by other cations [31–34].

2.1.1. Coupling to Na+

Evidence that Na^+ and Cl^- were actually co-transported with 5-HT came from studies using plasma membrane vesicles [27]. When a Na^+ concentration gradient (out > in) was imposed across the vesicle membrane in the absence of other driving forces, this gradient was sufficient to drive 5-HT accumulation [33]. Coupling between Na^+ and 5-HT transport follows from the fact that Na^+ could drive transport only if its own gradient is dissipated. Thus, Na^+ influx must accompany 5-HT influx. Na^+-coupled 5-HT transport into membrane vesicles is insensitive to inhibitors of other Na^+ transport processes such as ouabain and furosemide, supporting the hypothesis that Na^+ and 5-HT fluxes are coupled directly by the transporter [33,35]. Many of these results have been reproduced in membrane vesicle systems from cultured rat basophilic leukemia cells [29], mouse brain synaptosomes [36], and human placenta [30].

2.1.2. Coupling to Cl-

The argument that Cl^- is cotransported with 5-HT is somewhat less direct, as it has been difficult to demonstrate 5-HT accumulation with only the Cl^- gradient as a driving force. However, the transmembrane Cl⁻ gradient influences 5-HT accumulation when a Na⁺ gradient provides the driving force. Thus, raising internal Cl⁻ decreases the Cl⁻ gradient, and inhibits 5-HT uptake. External Cl⁻ is required for 5-HT uptake, and Cl⁻ can be replaced only by Br⁻ and, to a lesser extent, by SCN⁻, NO$_3^-$, and NO$_2^-$ [34]. In contrast, 5-HT efflux requires internal but not external Cl⁻ [34]. One alternative explanation for stimulation of transport by Cl⁻ (on the same membrane face as 5-HT) is that Cl⁻ might be required to electrically compensate for rheogenic (charge moving) 5-HT transport. This possibility was ruled out by the observation that a valinomycin-mediated K⁺ diffusion potential (interior negative) was unable to eliminate the external Cl⁻ requirement for 5-HT influx [34].

2.1.3. Coupling to K⁺

Perhaps the greatest surprise during the early studies of 5-HT transport was the discovery that K⁺ efflux was directly coupled to 5-HT influx. Previously, K⁺ countertransport had been invoked for other transport systems to explain stimulation by internal K⁺ [37–39]. In those other cases, however, it became clear that a membrane potential generated by K⁺ diffusion was responsible for driving rheogenic transport processes. Stimulation of 5-HT transport into plasma membrane vesicles by intravesicular K⁺ was also initially attributed to a K⁺ diffusion potential since internal K⁺ stimulates transport but is not absolutely required [33]. Subsequent measurements, however, showed that K⁺ stimulated transport even if the membrane potential was close to zero [40]. When valinomycin was added to increase the K⁺ conductance of the membrane, a K⁺ diffusion potential (inside negative) was generated. Transport was essentially the same whether or not a diffusion potential was imposed in addition to the K⁺ concentration gradient [41]. Thus, the K⁺ gradient did not seem to act indirectly through the membrane potential, but rather, directly by exchanging with 5-HT.

2.1.4. Coupling to H⁺

The results described above argue that K⁺ is countertransported during 5-HT influx. However, transport still occurred in the absence of K⁺. The reason for this became apparent in a study of the pH dependence of 5-HT transport. In the absence of K⁺, internal H⁺ ions apparently fulfill the requirement for a countertransported cation. For example, when the interior of plasma membrane vesicles was acidified to pH 5.5, and the vesicles were diluted into a medium of pH 7.5, 5-HT influx was stimulated, and the stimulation by internal K⁺ was significantly blunted [42]. Conversely, when internal K⁺ was high, internal H⁺ ions could no longer stimulate. Even when no other driving forces were present (NaCl in = out, no K⁺ present), a transmembrane pH difference (ΔpH, interior acid) could serve as the sole driving force for transport. ΔpH-driven 5-HT accumulation required Na⁺ and was blocked by imipramine or by high K⁺ (in = out), indicating that it was mediated by the 5-HT transporter, and not due to non-ionic diffusion [42]. From all of these data, it was concluded that inwardly directed Na⁺ and Cl⁻ gradients, and outwardly directed K⁺ or H⁺ gradients served as driving forces for 5-HT transport (Fig. 2).

2.1.5. Transport stoichiometry

A number of methods have been devised that could be used to evaluate how many Na^+, Cl^-, and K^+ ions are transported with 5-HT. For the 5-HT transporter, the most useful have been thermodynamic measurements. In this method, a known Na^+, Cl^-, or K^+ concentration gradient is imposed across the plasma membrane as a driving force, and the substrate concentration gradient accumulated in response to that driving force is measured at equilibrium. By varying the concentration gradient of the driving ion, and measuring the effect on substrate accumulation, the stoichiometry can be calculated. If more than one solute is coupled to substrate transport, then the above analysis still holds as long as the gradients of all other co- or countertransported solutes (as well as $\Delta\psi$ for a rheogenic process) are held constant. Using this method, a stoichiometry of 1:1 was determined for 5-HT transport with respect to both Na^+ [43] and K^+ [33]. The Cl^- stoichiometry was deduced from the fact that 5-HT transport was not affected by imposition of a $\Delta\psi$ (interior negative), and was, therefore, likely to be electroneutral. Given that 5-HT is transported in its cationic form [42,44], only a $5\text{-}HT^+$:Na^+:Cl^-:K^+ stoichiometry of 1:1:1:1 is consistent with all the known facts. Obviously, the above analysis requires an experimental system, like membrane vesicles, where the composition of both internal and external media can be controlled.

2.1.6. Coupling to $\Delta\psi$

Evidence relating the membrane potential to 5-HT transport has been mixed in other systems. Bendahan and Kanner [29] found that 5-HT transport into plasma membrane vesicles from rat basophilic leukemia cells was stimulated by a K^+ diffusion potential. However, other workers studying plasma membrane vesicles from mouse brain and human placenta concluded that 5-HT transport in these tissues was not driven by a transmembrane electrical potential ($\Delta\psi$, interior negative) [45,46]. Further evidence that K^+ acts directly on the transporter came from measurements of K^+ effects on 5-HT transport which occur even in the absence of a K^+ gradient. The addition of 30 mM internal K^+ increased the transport rate 2.5-fold even when 30 mM K^+ was added simultaneously to the external medium [41].

One might expect that electrogenicity could be easily tested if cells expressing the 5-HT transporter could be directly impaled with microelectrodes. This has been done by Mager et al. [47], using *Xenopus* oocytes injected with 5-HT transporter mRNA, with somewhat surprising results. A simple prediction is that current should flow across the membrane during 5-HT transport if the transporter is electrogenic. In fact, a 5-HT-dependent current has been measured, but closer inspection of its properties suggests that it does not represent electrogenic 5-HT transport but, rather, a conductance that is triggered by transport. The key finding is that the transport-associated current is voltage-dependent. Thus, the inward current increases as the inside of the cell is made more negative. In the same cells, however, [^{3}H]5-HT transport is independent of membrane potential. It is, therefore, very unlikely that the voltage-dependent current represents the voltage-independent transport process.

2.2. The NE transporter (NET)

2.2.1. Coupling to Na⁺, Cl⁻ and K⁺

Although no membrane vesicle systems have been described that transport DA, two plasma membrane vesicle systems have emerged for studying NE transport: the placental brush border membrane [26] and cultured PC-12 cells [25]. Harder and Bonisch [25] concluded that NE transport into PC12 vesicles was coupled to Na^+ and Cl^-, and was electrogenic, but they failed to arrive at a definitive coupling stoichiometry because of uncertainties about the role of K^+. According to their analysis, stimulation of NE influx by internal K^+ resulted either from direct K^+ countertransport as occurs with SERT [41], or from a K^+ diffusion potential which drives electrogenic NE influx, as with GAT-1 [48]. Ganapathy and coworkers [26] studied NET mediated transport of both NE and DA into placental membrane vesicles (DA is actually preferred by NET as a substrate [49]). They reached similar conclusions regarding ion coupling, but also were left with some ambiguity regarding K^+. In fact, the effects of ions on NET-mediated DA accumulation were similar to those observed with SERT-mediated 5-HT transport in the same membranes and the two activities were distinguished only by their inhibitor sensitivities [50]. Part of the difficulty in interpreting and comparing these data stems from the fact that they were obtained in various cell types, with unknown, and potentially very different conductances to K^+.

Two further problems make it difficult to interpret existing data on NET ion coupling. Both previous studies assumed that the cationic form of the catecholamine substrate was transported [25,26]. However, both cationic and neutral forms of catecholamine substrates are present at physiological pH, and there is no *a priori* reason to prefer one over the other. In the case of the vesicular monoamine transporter (VMAT) the ionic form of the substrate is a matter of some controversy. Different investigators have reached opposite conclusions [51–53]. A further problem in previous studies estimating NET stoichiometry is in the use of kinetic, rather than thermodynamic measurements. The number of Na^+ ions cotransported with each catecholamine substrate was estimated from the dependence of transport rate on Na^+ concentration [25,26]. This method is capable of detecting the involvement of multiple Na^+ ions only if they have similar binding affinity and kinetics. If two Na^+ ions (for example) with widely different affinities or binding kinetics are cotransported, the dependence of transport rate on Na^+ may reflect only binding of the lowest affinity or most slowly associating Na^+ ion.

We recently established LLC-PK₁ cell lines stably expressing the biogenic amine transporters SERT, NET and DAT as well as the GABA transporter GAT-1. Using these cell lines, we have characterized and compared the transporters under the same conditions and in the same cellular environment [49]. One attractive advantage of LLC-PK₁ cells is that it has been possible to prepare plasma membrane vesicles that are suitable for transport studies [54]. We took advantage of this property to prepare membrane vesicles containing transporters for GABA, 5-HT and NE, all in the same LLC-PK₁ background. These vesicles should have identical composition except for the heterologously expressed transporter. Moreover, these vesicles are suitable for estimating equilibrium substrate accumulation in response to imposed ion gradients. This property allowed us to define the ion coupling stoichiometry for NET using the known stoichiometries for GAT-1 and SERT mediated transport as internal controls.

The results for SERT and GAT-1 are consistent with previously reported determinations of ion coupling stoichiometry. For NET, accumulation of [^{3}H]DA was stimulated by imposition of Na^+ and Cl^- gradients (out > in) and by a K^+ gradient (in > out). To determine the role that each of these ions and gradients play in NET mediated transport, we measured the influence of each ion on transport when that ion was absent, present at the same concentration internally and externally, or present asymmetrically across the membrane. The presence of Na^+ or Cl^-, even in the absence of a gradient, stimulated DA accumulation by NET, but K^+ had little or no effect in the absence of a K^+ gradient. Stimulation by a K^+ gradient was markedly enhanced by increasing the K^+ permeability with valinomycin, suggesting that net positive charge is transported together with DA. The cationic form of DA is likely to be the substrate for NET, since varying pH did not affect the K_M of DA for transport. We estimated the Na^+:DA stoichiometry by measuring the effect of internal Na^+ on peak accumulation of DA. An increase in internal Na^+ decreased the Na^+ gradient and decreased the peak accumulation of 5-HT, GABA and DA by SERT, GAT-1 and NET, respectively. GAT-1 was more sensitive to internal Na^+ than SERT or NET, which were of approximately equal sensitivity. The results are consistent with known Na^+:substrate stoichiometries of 2 for GAT-1 and 1 for SERT, and suggest a 1:1 cotransport of Na^+ with substrate also for NET. Taken together, the results suggest that NET catalyzes cotransport of one cationic substrate molecule with one Na^+ ion, and one Cl^- ion, and that K^+ does not participate directly in the transport process.

2.3. The DA transporter (DAT)

2.3.1. Coupling to Na^+ and Cl^-

The DA transporter has a different ion dependence than do the 5-HT and NE transporters. While initial rates of DA transport were found to be dependent on a single Cl^-, two Na^+ ions were apparently involved in the transport process [55]. Thus, the initial rate of DA transport into suspensions of rat striatum was a simple hyperbolic function of [Cl^-] but depended on [Na^+] in a sigmoidal fashion. These data are consistent with a Na^+:Cl^-:DA stoichiometry of 2:1:1. These differences in Na^+ stoichiometry have been reproduced with the cloned transporter cDNAs stably expressed in LLC-PK$_1$ cell lines, indicating that they are intrinsic properties of the transporters and not artifacts due to the different cell types used [49].

It is important at this point, to remember that steady-state kinetics do not necessarily provide accurate information on co-transport stoichiometry. It is possible that more than one Na^+ ion is required for DA binding or even translocation (as reflected in the rate measurements), but that only one of those Na^+ ions is actually co-transported. For this reason, it is essential to confirm the stoichiometry by an independent method. For transport systems, thermodynamic measurements of coupling between solute fluxes has proven useful. In the case of the GABA transporter, where both thermodynamic and kinetic data exist, both methods indicate a $Na^+:Cl^-$:GABA stoichiometry of 2:1:1 [56–58]. Thus, the dependence of transport rate on a given ion may suggest a transport stoichiometry, but cannot provide proof for it.

3. Mechanism

It is interesting to consider how the biogenic amine transporters, with molecular weights of 60–80 Kd, are able to couple the fluxes of substrate, Na^+, Cl^-, and K^+ in a stoichiometric manner. The problem faced by a coupled transporter is more complicated than that faced by an ion channel, since a channel can function merely by allowing its substrate ions to flow across the lipid bilayer. Such uncoupled flux will dissipate the ion gradients and will not utilize them to concentrate another substrate. However, the structural similarities between transporters and ion channels may give a clue to the mechanism of transport (Fig. 3). Just as an ion channel is thought to have a central aqueous cavity surrounded by amphipathic membrane-spanning helices, a transporter may have a central binding site which accommodates Na^+, Cl^- and substrate. The difference in mechanism between a transporter and a channel may be that, while a channel assumes open (conducting) and closed (nonconducting) states, a transporter can also assume two states which differ only in the accessibility of the central binding site. In each of these states, the site is exposed to only one face of the membrane, and the act of substrate translocation represents a conformational change to the state where the binding site is exposed on the opposite face (Fig. 3). Thus, the transporter may behave like a channel with a gate at each face of the membrane, but only one gate is usually open at any point in time.

For this mechanism to lead to cotransport of ions with substrate molecules, the transporter must obey a set of rules governing the conformational transition between its two states (Fig. 4). For cotransport of Na^+, Cl^- and 5-HT, the rule would allow a conformational change only when the binding site was occupied with Na^+, Cl^- and substrate or when the site was completely empty. To account for K^+ countertransport with 5-HT, the conformational change could occur when the binding site contained either K^+ or it contained Na^+, Cl^- and 5-HT. This simple model of a binding site exposed alternately to one side of the membrane or the other can explain most carrier-mediated transport. However, it makes specific predictions about the behavior of the transport system.

In particular, this model requires that the substrate is transported in the same step as Na^+ and Cl^- but in a different step than a countertransported ion such as K^+. There is evidence that 5-HT and K^+ are transported in different steps. This comes from efflux and exchange experiments with plasma membrane vesicles. In vesicles containing NaCl, [^{3}H]5-HT efflux is supported by external K^+ in a reversal of the normal influx reaction. In the absence of external K^+, radiolabel efflux is inhibited, but it can be accelerated by external unlabeled 5-HT [41]. This reaction represents an exchange of internal and external 5-HT which does not require the K^+-dependent

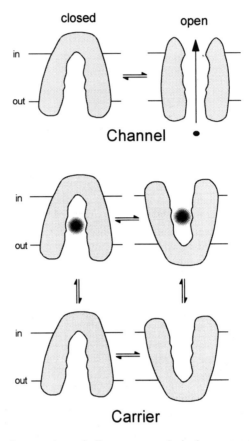

Fig. 3.   Channels and carriers may have similar structures. A single structural model can account for transport by carriers and channels. In a channel (above), one or more "gates", or permeability barriers, open to allow free passage for ions from one side of the membrane to the other. A carrier can be thought of as a channel with two gates. Normally only one is open at a time. By closing one gate and opening another, the carrier allows a solute molecule, bound between the two gates, to cross the membrane.

rate-limiting step for net 5-HT efflux. Thus the steps required for 5-HT binding, translocation and dissociation do not include the steps (where K^+ is translocated) that become rate-limiting in the absence of K^+.

Another prediction of the model for biogenic amine transport is that Na^+, Cl^- and substrate should all be bound to the transporter prior to translocation. Since it has been difficult to measure directly substrate binding to the biogenic amine transporters, the most direct data relating to this point comes from studies of the Na^+ and Cl^- dependence of inhibitor binding. The tricyclic antidepressant imipramine is a high affinity ligand for the 5-HT transporter. Na^+ and Cl^- both increase the affinity of the 5-HT transporter for imipramine [59]. To the extent that imipramine binding reflects the normal process of substrate binding, these results indicate that Na^+ and Cl^- are bound together with 5-HT prior to translocation.

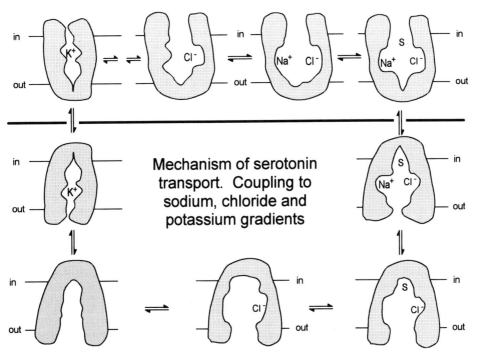

Fig. 4.   Mechanism of 5-HT transport. Starting at the lower left and continuing clockwise, the transporter binds Cl⁻, 5-HT (S), and Na⁺. These binding events permit the carrier to undergo a conformational change to the form in the upper right hand corner. This internal-facing form dissociates Na⁺, Cl⁻ and 5-HT to the cytoplasm. Upon binding internal K⁺, another conformational change allows the carrier to dissociate K⁺ on the cell exterior, generating the original form of the transporter which can initiate another round of transport by binding external Na+, Cl⁻ and 5-HT.

Somewhat different results were obtained using 2β-2-carbomethoxy-3-(4-[^{125}I]iodophenyl)tropane (β-[^{125}I]CIT) as a ligand for the 5-HT transporter [60]. Binding of this cocaine analog is stimulated by Na⁺, but Cl⁻ has no effect. β-CIT binds to the same number of sites as imipramine, and binding of the two ligands is mutually exclusive, indicating that they both bind at or near the substrate binding site on the 5-HT transporter. Using the ability of β-CIT to bind in the absence of Na⁺ and Cl⁻, Humphreys et al. [60] determined that Na⁺ was not absolutely required for binding of any ligand, and that substrates and inhibitors have distinctly different Na⁺ and Cl⁻ requirements for binding affinity. Together with results using a variety of antidepressants [61], these findings suggest that there are subsites within the substrate binding pocket that define distinct but overlapping binding sites for a variety of substrates and inhibitors.

There are interesting similarities and differences in the binding properties of β-CIT to the 5-HT and dopamine transporters in rat striatal membranes and platelet plasma membranes [62]. For example, β-[^{125}I]CIT binding to both these transporters was pH-dependent and Cl⁻-independent. These findings suggest that there is a protonatable residue involved in the binding of cocaine to both the 5-HT and dopamine transporters. One difference between the two transporters is that, while β-[^{125}I]CIT binding to both transporters is stimulated by Na^+, the Na^+ dependence is sigmoidal for DAT and hyperbolic for SERT. This behavior is similar to the Na^+ dependence of transport by the two systems [43,63] and may reflect different Na^+ transport stoichiometries.

4. Channel-like properties

4.1. The 5-HT transporter

While the simple model described above is able to explain much of the transport phenomena observed with the 5-HT transporter, it did not predict the channel properties of the transporter observed by Mager et al. [47]. In these studies, cation conductances were observed under three conditions. There is a resting, uncoupled current in the absence of 5-HT. In the presence of substrate, this current increases and is referred to as a transport current. Finally, a transient current is produced upon hyperpolarization of the membrane in the absence of substrate. All of these currents are dependent upon expression of SERT and are blocked by transport inhibitors. The relationship between these three kinds of conductances is still not clear, but it is becoming more obvious that they are not an obligate part of the transport cycle.

More recent experiments [64] demonstrate discrete single channel behavior of the 5-HT transporter expressed in patches of oocyte membrane. From an estimate of the frequency of single-channel openings, the number of transporters in the patch and the turnover number, it has been estimated that only in 100–1000 turnovers does the transporter conduct as a channel. Although these openings are much less frequent than 5-HT transport events, they are more likely to occur when the transporter is working. The most likely explanation for this behavior is that during transport, there is some probability that the transporter will transiently open up into a conducting state like a channel and, during that time, will allow many ions to cross the membrane. The probability that the 'channel' opens is a function of transport, which is voltage-independent, but the rate at which ions pass through the open channel depends on the applied voltage.

It is difficult to understand if the conductive states of the 5-HT transporter are physiologically significant or if they represent unproductive side reactions of the transporter separate from its catalytic cycle. If one imagines the permeability pathway of the protein to consist of a central binding site for 5-HT, Na^+, Cl^-, and K^+ separated by gates from the internal and external media, then the rules for efficient coupling would predict that both gates should not be open simultaneously. However, the observed ion channel activity could result if these rules were not always obeyed, and there was a significant probability that one or more of the intermediates in the transport cycle was in equilibrium with a form in which both gates were open.

If the currents mediated by SERT do not represent 5-HT transport events, it is not obvious that they will be informative with respect to transport mechanism.

However, the kinetics of the channel openings by SERT have only begun to be analyzed. One possibility is that different states of the transporter will have different kinetics for channel opening and closing. This is almost certainly true, since the presence of 5-HT increases the frequency of channel openings. It may be that each intermediate in the transport cycle has characteristic channel properties and that interconversion between these intermediates (by ligand binding events and conformational changes) could be detected by measuring channel parameters such as unitary conductance, and channel opening and closing rates.

4.2. Other transporters

There is ample evidence that other transporters also exhibit conductances in addition to those expected from transport properties of the carrier. In the case of the NE transporter, HEK-293 cells stably expressing hNET display NET-mediated currents that are not seen in untransfected cells and that are blocked by NET antagonists [65]. Moreover, the currents observed when these cells are transporting NE, are much larger than predicted from the turnover number of NET and its level of expression. Galli et al. [65] concluded that NET also functioned as a ligand-gated ion channel in addition to it's transport function.

The GABA transporter, GAT-1 is a member of the same transporter family as SERT and NET and has been shown to conduct current concomitant with GABA transport. Transient currents believed to represent the interaction of Na^+ with the transporter have also been observed [66]. In addition to these currents, which are consistent with charge movement coupled to GABA flux, Cammack and Schwartz [67] have recently observed single channel cation conductances in cell-attached and excised membrane patches from HEK-293 cells expressing GAT-1. These channels were not observed in the presence of GAT-1 inhibitors or in untransfected cells. In addition, a high frequency current noise was observed in the presence of Na^+ and GABA, suggesting that cation channel activity also occurs during GABA transport.

These channel-like properties of neurotransmitter transporters are not restricted to the family of NaCl dependent transporters represented by SERT, NET, and GAT-1. A glutamate transporter, EAAT-4 has been shown to conduct Cl^- ions in the presence of glutamate [68]. Finally, the Na^+ dependent glucose transporter SGLT, has been demonstrated to catalyze uncoupled flux of Na^+ in the absence of glucose [69]. This phenomenon has been interpreted as a transport leak but it may represent the same channel phenomenon observed with SERT and GAT-1.

5. Reversal of transport

5.1. Efflux and exchange

The 5-HT transporter is quite capable of catalyzing efflux as well as influx. Efflux is stimulated by internal Na^+ and Cl^- and by external K^+, and is inhibited by the tricyclic antidepressant imipramine, which also inhibits 5-HT influx into intact platelets, membrane vesicles, and synaptosomes. If NaCl is present in the external medium, the transporter also catalyzes an exchange reaction between internal labeled 5-HT and external unlabeled 5-HT. The exchange process is independent of external K^+ and provides some of the support that K^+ and 5-HT are transported in separate steps (see above). This exchange phenomenon also may play a role in the pharmacological action of amphetamines, an interesting class of drugs that interact with both plasma membrane and vesicular biogenic amine transport systems.

5.2. Amphetamine action

Amphetamines represent a class of stimulants that increase extracellular levels of biogenic amines. Their mechanism differs from simple inhibitors like cocaine, although it also involves biogenic amine transporters. Amphetamine derivatives are apparently substrates for biogenic amine transporters, and lead to transmitter release by a process of transporter-mediated release from intracellular stores [70-72]. Both catecholamine and 5-HT transporters are affected by amphetamines. In particular, compounds such as *p*-chloroamphetamine and 3,4-methylenedioxymethamphetamine (MDMA, also known as 'ecstasy') preferentially release 5-HT, and also cause degeneration of serotonergic nerve endings [73]. Other amphetamine derivatives, such as methamphetamine, preferentially release catecholamines.

5.2.1. Actions at the plasma membrane transporter
The process of exchange stimulated by amphetamines results from two properties of amphetamine and its derivatives. These compounds are substrates for biogenic amine transporters and they also are highly permeant across lipid membranes. As substrates, they are taken up into cells expressing the transporters, and as permeant solutes, they rapidly diffuse out of the cell without requiring participation of the transporter. The result is that an amphetamine derivative will cycle between the cytoplasm and the cell exterior in a process that allows Na^+ and Cl^- to enter the cell and K^+ to leave each time the protonated amphetamine enters. Additionally, a H^+ ion will remain inside the cell if the amphetamine leaves as the more permeant neutral form (Fig. 5). This dissipation of ion gradients and internal acidification may possibly be related to the toxicity of amphetamines *in vivo*. The one-way utilization of the transporter (only for influx) leads to an increase in the availability of inward-facing transporter binding sites for efflux of cytoplasmic 5-HT, and, together with reduced ion gradients, results in net 5-HT efflux.

5.2.2. Actions at the vesicular membrane
In addition, the ability of amphetamine derivatives to act as weak base ionophores at the dense granule membrane leads to leakage of vesicular biogenic amines into the cytoplasm [72,74]. Weakly basic amines are able to raise the internal pH of acidic organelles by dissociating into the neutral, permeant form, entering the organelle, and binding an internal H^+ ion. Weakly basic amines such as ammonia and methylamine have been used to raise the internal pH of chromaffin granules, for example. While amphetamine derivatives are certainly capable of dissipating transmembrane ΔpH by this mechanism, they are much more potent than simple

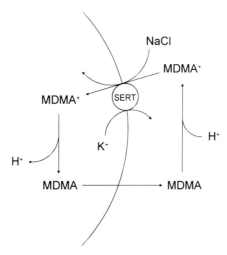

Fig. 5.   Interaction of 3,4-methylenedioxymethamphetamine (MDMA) with the 5-HT transporter. MDMA is a substrate for SERT and, like 5-HT, is transported into cells together with Na^+ and Cl^- and in exchange for K^+. Since it is membrane permeant in its neutral form, MDMA deprotonates intracellularly and leaves the cell, at which time it can reprotonate and serve again as a substrate for SERT. This futile transport cycle may lead to dissipation of cellular Na^+, Cl^- and K^+ gradients and acidification of the cell interior.

amines when tested in model systems. For example, *p*-chloroamphetamine (PCA) and 3,4-methylenedioxymethamphetamine (MDMA, ecstasy) are 5–10 times more potent than NH_4Cl in dissipating ΔpH in chromaffin granule membrane vesicles [71,74]. This result suggests that not only are these compounds crossing the membrane in their neutral form, but also as protonated species. By cycling into the vesicle as an uncharged molecule and back out in the protonated form, an amphetamine derivative could act as a classical uncoupler to increase the membrane permeability to H^+ ions.

In addition to this uncoupling activity, amphetamine derivatives have affinity to the vesicular monoamine transporter, VMAT. Binding of various amphetamine derivatives to VMAT has been observed with both native and heterologously expressed VMAT [75]. Despite the affinity of many amphetamines to VMAT, at least one compound, PCA, has no demonstrable binding to VMAT [76] despite its robust ability to release stored biogenic amines [74]. Thus, the ability of amphetamines to dissipate vesicular pH differences is sufficient to explain their effects on vesicular release.

Sulzer et al. [77] have extended this hypothesis by measuring the effects of intracellularly injected amphetamine and DA. Using the *Planorbis corneus* giant dopamine neuron, they demonstrated that amphetamine could act intracellularly to release DA from the cell. Moreover, injections of DA directly into the cytoplasm led to DA efflux that was sensitive to nomifensine, suggesting that it was mediated by the plasma membrane transporter. According to the hypothesis put forward by Sulzer et al. [77], amphetamine action at the vesicular membrane is sufficient to account for amphetamine-induced amine release. These results would also appear to explain the observation that blockade of plasma membrane transporters prevents the action of amphetamines [70,78].

5.2.3. The amphetamine permeability paradox

If amphetamine derivatives act only by uncoupling at the level of biogenic amine storage vesicles, then classical uncouplers such as 2,4-dinitrophenol should act as psychostimulants like amphetamine. However, no such action has been reported for uncouplers or other weakly basic amines. Moreover, all of the amphetamine derivatives that we have tested bind to plasma membrane amine transporters [71,74, 76,79–81]. What role could the plasma membrane transporters play in amphetamine action? One possibility is that they serve merely to let amphetamine derivatives into the cell. This would account for the specificity of various amphetamine derivatives. For example, the preferential ability of methamphetamine to release dopamine while MDMA and fenfluramine release 5-HT could result from their relative ability to be transported by SERT or DAT. However, it seems unlikely that amphetamines, which are so permeant that they act as uncouplers at the vesicle membrane, are unable to cross the plasma membrane without the aid of a transporter. If this is truly the role of the plasma membrane transporters in amphetamine action, it would seem that the vesicular and plasma membranes have vastly different permeabilities to amphetamines.

An alternative possibility, however, is that the ability of amphetamines to serve as substrates for plasma membrane transporters is important in their action even if the membrane does not constitute a barrier to amphetamine diffusion. In this view, futile cycling of the plasma membrane transporter is induced by transporter mediated influx followed by diffusion back out of the cell. As described above, this process will lead to dissipation of Na^+, Cl^- and K^+ gradients and possibly also acidify the cell interior. As a result of the lower gradients and the appearance of cytoplasmic binding sites following amphetamine dissociation on the cell interior, biogenic amine efflux via the transporter will be stimulated. Clearly, the issues of amphetamine permeation across the plasma membrane and the specificity of amphetamine action need to be explored further.

6. Purification

As a strategy to identify biogenic amine transporters, a number of groups have attempted to purify the 5-HT transporter to homogeneity. The protein has been solubilized in an active form using digitonin [82]. Using this solubilized preparation, two groups achieved significant purification of the 5-HT transporter from platelet and brain using affinity resins based on citalopram, a high affinity ligand for the transporter [83,84]. In neither case was the protein purified to homogeneity or reconstituted in liposomes to recover transport activity. Launay et al. [85] have apparently purified the transporter to homogeneity using affinity columns based on 5-HT itself or 6-fluorotryptamine, but also did not reconstitute transport activity with the purified protein. Reconstitution of the transporter has, however, been demonstrated with a urea-cholate extract of placental membranes [86]. As these studies were in progress, molecular cloning of other Na+-dependent transporter cDNAs was providing an independent approach for studying their structure.

7. Cloning

The identification of cDNAs for Na+-dependent biogenic amine transporters has contributed greatly to our understanding of the molecular structure and function of these important proteins. The first cDNA encoding a biogenic amine carrier was identified by Pacholczyk et al. [5] using an expression cloning strategy in COS cells. A norepinephrine transporter cDNA (NET) was isolated from an SK-N-SH human neuroblastoma library on the basis of its ability to direct the transport of ^{125}I-labeled m-iodobenzylguanidine, a norepinephrine analog. Comparison of the predicted amino acid sequence for NET with that for the Na+-dependent γ-aminobutyric acid (GABA) transporter, GAT-1, previously identified by Guastella et al. [3] revealed that these transporters are members of a multi-gene family. Given the significant degree of homology between the two transporters, several groups designed degenerate oligonucleotide primers and have used them in homology-based polymerase chain reaction strategies for identifying additional members of the family. Several members of this transporter gene family have been identified, including carriers for dopamine [10–13,22] and 5-HT [6,7]. Hydropathy analyses of these transporter sequences are virtually superimposable, and predict proteins with 12 transmembrane domains having both their amino and carboxyl termini on the cytoplasmic side of the plasma membrane (Fig. 6). There is a significant degree of identity across this family of transporters (25%) despite their very different substrate specificities. Homology is highest in the proposed transmembrane domains (TM) and particularly sparse in the extracellular loop connecting TM3 and TM4. This large extracellular loop is the location of from 1–4 potential N-linked glycosylation sites in each transporter sequence.

5-HT Transporter cDNAs have been isolated independently from rat basophilic leukemia (RBL) cells, rat brainstem, human placenta, placenta, platelet and brain, and mouse brain [6,7,87,88]. Transport by the 5-HT transporter expressed in transfected CV-1 and HeLa cells was saturable with a K_m from 300 to 500 nM [6,7]. 5-HT transport was shown to depend on both Na+ and Cl−, which is characteristic of all members of this family examined thus far. As a result, this group of carriers has come to be known as the Na+- and Cl−-dependent neurotransmitter transporter family. Several tricyclic and heterocyclic antidepressants inhibited the heterologously expressed 5-HT transporter with a rank order of potency similar to that observed for the native brain 5-HT transporters [89] indicating that the expressed protein possesses the ability to both transport 5-HT and to bind antidepressants. The anorectic drug fenfluramine and the neurotoxin MDMA were extremely effective at blocking 5-HT uptake [6]. Northern blot analyses reveal a single mRNA species of 3.1 to 3.7 kb in size expressed in rat midbrain and brainstem, as well as several peripheral tissues including rat gut, lung, adrenal, spleen, stomach, uterus and kidney [6,7]. *In situ* hybridization histochemistry with 5-HT transporter probes

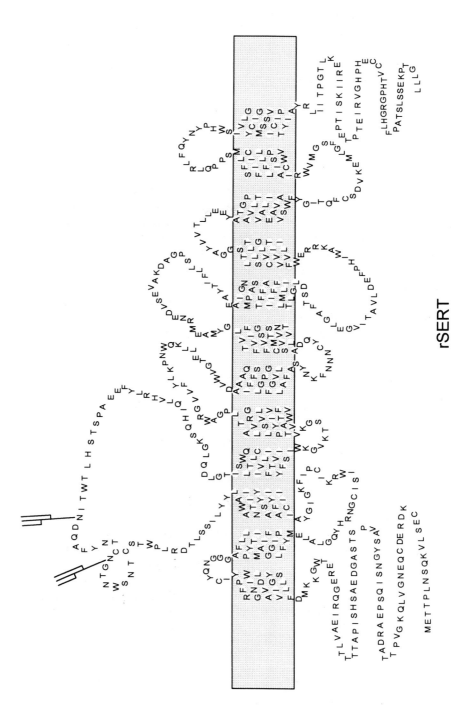

rSERT

Fig. 6.   Diagram of the 5-HT amino acid sequence. Using hydropathy analysis, transmembrane domains were identified. The predicted structure has both N- and C-termini in the cytoplasm, and the large, glycosylated loop on the cell surface.

revealed mRNA in serotonergic neurons of the rat brain, in addition to the lamina propria of the stomach and duodenum, and the chromaffin cells of the adrenal [7]. In each mammalian species examined, only one 5-HT transporter gene has been found, with minor differences between species [7,9]. This contrasts markedly with 5-HT receptors, of which there are many types and subtypes [90].

8. Mutagenesis

8.1. Point mutations

Relatively few mutagenesis studies with biogenic amine transporters have been reported. In DAT, replacement of aspartate at position 79, which resides in the first transmembrane domain, with alanine, glycine, or glutamate reduced transport of DA and MPP^+ and reduced affinity for binding of the cocaine analog CFT, although the number of CFT binding sites was unchanged [91]. Mutants in DAT 7th and 11th hydrophobic putative transmembrane domains increase MPP^+ uptake velocity and affinity ($1/K_D$), respectively. These mutations exert much more modest effects on dopamine uptake and have little impact on cocaine analog binding [92]. Replacement of serine residues in the seventh putative transmembrane domain led to a loss of transport activity but affected cocaine analog binding to a lesser degree [91]. In contrast, serine residues in the eighth transmembrane domain could be replaced with no loss of transport or binding activity [91].

8.2. Glycosylation mutants

All members of the NaCl coupled transporter family contain potential glycosylation sites in the putative extracellular loop region between transmembrane domains 3 and 4. These sites are utilized, as judged by the affinity of the transporters to immobilized lectins [83,93]. Tate et al. [94] compared the expression of SERT in baculovirus-infected insect cells with mutants in which the glycosylation sites were ablated. The glycosylation-deficient mutants were able to bind ligands and transport 5-HT with unchanged affinities, but the apparent level of functional transporters in the plasma membrane was dramatically reduced. These results suggest that glycosylation is not critical for function, but is necessary for efficient and stable delivery of the transporter to the plasma membrane. The same region containing the glycosylation sites contains a pair of highly conserved cysteine residues which are also apparently involved in proper expression of DAT. Replacing either Cys180 or Cys189 in DAT with Ala leads to poor expression and the transporter accumulated in perinuclear regions of the cell [95]. In contrast, replacement of cysteine residues at positions 90, 242, 305, and 345 did not lead to a significant change in transport.

8.3. Chimeras

Early attempts to generate chimeric transporters composed of NET and GAT or NET and SERT sequences demonstrated that functional transporters resulted only when all of the putative transmembrane domains were donated from the same protein [96,97]. The close similarity between NET and DAT sequences (68% amino acid identity) allowed chimeras derived from those two proteins to function even when the junction was between transmembrane domains [98,99]. By examining the substrate K_M and inhibitor K_I values on the chimeric mutants, various regions of the structure were assigned roles in the binding of substrates and inhibitors for the transporters. A similar approach was used with the 5-HT transporters from rat and human, which have different affinities for tricyclic antidepressants and amphetamine [100]. Chimeras between rat and human SERT sequences suggest that this difference resides, at least in part, in the distal regions of the sequence, possibly involving the twelfth putative transmembrane domain. One difficulty with this approach is that the two parent sequences in the chimera must be quite similar to allow the formation of a functional chimera. Thus, while regions of difference between the parent sequences might be associated with various functions, nothing may be learned about the importance of regions that are identical in both donor proteins.

Another use of chimeras has been to examine the role of the hydrophilic domains of the protein, postulated to be cytoplasmic or extracellular loop regions. Replacement of part of the putative loop between transmembrane domains 3 and 4 in SERT with sequence from NET generates a chimeric transporter with very low transport rates (5–10% of wild type) but normal binding properties for Na$^+$, 5-HT, and the inhibitors β-CIT, paroxetine, and desipramine [101]. Moreover, cell surface biotinylation studies indicate that the chimeric mutant is expressed on the cell surface to the same extent as the wild-type transporter. These results suggest that the chimeric transporter is folded and inserted properly into the membrane, but fails to undergo conformational changes necessary for transport. The implication that putative extracellular domains are involved in conformational changes is unexpected, and suggests that the loops connecting transmembrane domains are not merely passive linkers.

References

1. Knipper, M., Kahle, C. and Breer, H. (1991) Biochim. Biophys. Acta **1065**, 107–113.
2. Kanner, B.I. and Schuldiner, S. (1987) CRC Crit. Rev. Biochem. **22**, 1–38.
3. Guastella, J., Nelson, N., Nelson, H., Czyzyk, L., Keynan, S., Miedel, M., Davidson, N., Lester, H. and Kanner, B.I. (1990) Science **249**, 1303–1306.
4. Nelson, H., Mandiyan, S. and Nelson, N. (1990) FEBS Lett. **269**, 181–184.
5. Pacholczyk, T., Blakely, R. and Amara, S. (1991) Nature **350**, 350–354.
6. Hoffman, B.J., Mezey, E. and Brownstein, M.J. (1991) Science **254**, 579–580.
7. Blakely, R., Berson, H., Fremeau, R., Caron, M., Peek, M., Prince, H. and Bradely, C. (1991) Nature **354**, 66–70.
8. Lesch, K., Wolozin, B., Estler, H., Murphy, D. and Riederer, P. (1993) J. Neural. Transm. Gen. Sect. **91**, 67–72.
9. Lesch, K., Wolozin, B., Murphy, D. and Riederer, P. (1993) J. Neurochem. **60**, 2319–2322.
10. Shimada, S., Kitayama, S., Lin, C., Patel, A., Nanthakumar, E., Gregor, P., Kuhar, M. and Uhl, G. (1991) Science **254**, 576–578.
11. Kilty, J., Lorang, D. and Amara, S. (1991) Science **254**, 578–579.
12. Usdin, T., Mezey, E., Chen, C., Brownstein, M. and Hoffman, B. (1991) Proc. Nat. Acad. Sci. USA **88**, 11168–11171.
13. Giros, B., Elmstikawy, S., Bertrand, L. and Caron, M.G. (1991) FEBS Lett. **295**, 149–154.

14. Mayser, W., Betz, H. and Schloss, P. (1991) FEBS Letters **295**, 203–206.

15. Guimbal, C. and Kilimann, M. (1993) J. Biol. Chem. **268**, 8418–8421.

16. Fremeau, R., Caron, M. and Blakely, R. (1992) Neuron **8**, 915–926.

17. Liu, Q.-R., Lopez-Corcuera, B., Nelson, H., Mandiyan, S. and Nelson, N. (1992) Proc. Natl. Acad. Sci. USA **89**, 12145–12149.

18. Clark, J., Deutch, A., Gallipoli, P. and Amara, S. (1992) Neuron **9**, 337–348.

19. Borden, L., Smith, K., Hartig, P., Branchek, T. and Weinshank, R. (1992) J. Biol. Chem. **267**, 21098–21104.

20. Smith, K., Borden, L., Hartig, P., Branchek, T. and Weinshank, R. (1992) Neuron **8**, 927–935.

21. Smith, K., Borden, L., Wang, C., Hartig, P., Branchek, T. and Weinshank, R. (1992) Mol. Pharmacol. **42**, 563–569.

22. Giros, B., Elmestikawy, S., Godinot, N., Zheng, K., Han, H., Yang-Feng, T. and Caron, M. (1992) Mol. Pharmacol. **42**, 383–390.

23. Yamauchi, A., Uchida, S., Kwon, H., Preston, A., Robey, R., Garciaperez, A., Burg, M. and Handler, J. (1992) J. Biol. Chem. **267**, 649–652.

24. Liu, Q.-R., Nelson, H., Mandiyan, S., Lopez-Corcuera, B. and Nelson, N. (1992) FEBS Lett. **305**, 110–114.

25. Harder, R. and Bonisch, H. (1985) J. Neurochem. **45**, 1154–62.

26. Ramamoorthy, S., Leibach, F.H., Mahesh, V.B. and Ganapathy, V. (1992) Am. J. Physiol. **262**, C1189–C1196.

27. Rudnick, G. (1986) in: Platelet Responses and Metabolism, ed H. Holmsen, Vol. II, pp. 119–133, CRC Press, Boca Raton, FL.

28. Barber, A.J. and Jamieson, G.A. (1970) J. Biol. Chem. **245**, 6357–6365.

29. Kanner, B.I. and Bendahan, A. (1985) Biochim. Biophys. Acta **816**, 403–410.

30. Balkovetz, D., Tirruppathi, C., Leibach, F., Mahesh, V. and Ganapathy, V. (1989) J. Biol. Chem. **264**, 2195–2198.

31. Lingjaerde, O. (1969) FEBS Letters **3**, 103–106.

32. Sneddon, J.M. (1969) Br. J. Pharmacol. **37**, 680–688.

33. Rudnick, G. (1977) J. Biol. Chem. **252**, 2170–274.

34. Nelson, P. and Rudnick, G. (1982) J. Biol. Chem. **257**, 6151–6155.

35. Nelson, P. and Rudnick, G. (1981) Biochemistry **20**, 4246–49.

36. O'Reilly, C.A. and Reith, M.E.A. (1988) J. Biol. Chem. **263**, 6115–6121.

37. Eddy, A.A. (1968) Biochem. J. **108**, 195–206.

38. Eddy, A.A., Indge, K.J., Backen, K. and Nowacki, J.A. (1970) Biochem. J. **120**, 845–52.

39. Crane, R.K., Forstner, G. and Eichholz, A. (1965) Biochim. Biophys. Acta **109**, 467–477.

40. Rudnick, G. and Nelson, P. (1978) Biochemistry **17**, 4739–4742.

41. Nelson, P. and Rudnick, G. (1979) J. Biol. Chem. **254**, 10084–10089.

42. Keyes, S. and Rudnick, G. (1982) J. Biol. Chem. **257**, 1172–1176.

43. Talvenheimo, J., Fishkes, H., Nelson, P.J. and Rudnick, G. (1983) J. Biol. Chem. **258**, 6115–6119.

44. Rudnick, G., Kirk, K.L., Fishkes, H. and Schuldiner, S. (1989) J. Biol. Chem. **264**, 14865–14868.

45. Reith, M.E.A., Zimanyi, I. and O'Reilly, C.A. (1989) Biochem. Pharmacol. **38**, 2091–2097.

46. Cool, D.R., Leibach, F.H. and Ganapathy, V. (1990) Biochemistry **29**, 1818–1822.

47. Mager, S., Min, C., Henry, D.J., Chavkin, C., Hoffman, B.J., Davidson, N. and Lester, H. (1994) Neuron **12**, 845–859.

48. Kanner, B., I. (1978) J. Biochem. **17**, 1207–1211.

49. Gu, H., Wall, S. and Rudnick, G. (1994) J. Biol. Chem. **269**, 7124–7130.

50. Ramamoorthy, S., Prasad, P., Kulanthaivel, P., Leibach, F., Blakeley, R. and Ganapathy, V. (1993) Biochemistry **32**, 1346–1353.

51. Knoth, J., Isaacs, J. and Njus, D. (1981) J. Biol. Chem. **256**, 6541–6543.

52. Scherman, D. and Henry, J.-P. (1981) Eur. J. Biochem. **116**, 535–539.

53. Kobold, G., Langer, R. and Burger, A. (1985) Nauyn-Schmiedegerg's Arch. Pharmacol. **331**, 209–19.

54. Brown, C., Bodmer, M., Biber, J. and Murer, H. (1984) Biochim. Biophys. Acta **769**, 471–478.

55. McElvain, J.S. and Schenk, (1992) J. Biochem. Pharmacol. **43**, 2189–2199.

56. Keynan, S. and Kanner, B.I. (1988) Biochemistry **27**, 12–17.

57. Radian, R. and Kanner, B., I. (1983) Biochemistry **22**, 1236.
58. Keynan, S., Suh, Y., Kanner, B. and Rudnick, G. (1992) Biochemistry **31**, 1974–1979.
59. Talvenheimo, J., Nelson, P.J. and Rudnick, G. (1979) J. Biol. Chem. **254**, 4631–4635.
60. Humphreys, C.J., Wall, S.C. and Rudnick, G. (1994) Biochemistry **33**, 9118–9125.
61. Humphreys, C., Beidler, D. and Rudnick, G. (1991) Biochem. Soc. Trans. **19**, 95–98.
62. Wall, S., Innis, R. and Rudnick, G. (1993) Mol. Pharmacol. **43**, 264–270.
63. Friedrich, U. and Bonisch, H. (1986) Naunyn-Schmiedegerg's Arch. Pharmacol. **333**, 246–252.
64. Lin, F., Lester, H. and Mager, S. (1995) Soc. Neurosci. Abstr. **21**, 781.
65. Galli, A., Defelice, L.J., Duke, B.J., Moore, K.R. and Blakely, R.D. (1995) J. Exper. Biol. **198**, 2197–2212.
66. Mager, S., Naeve, J., Quick, M., Labarca, C., Davidson, N. and Lester, H. (1993) Neuron **10**, 177–188.
67. Cammack, J. and Schwartz, E. (1995) Proc. Natl. Acad. Sci. USA.
68. Fairman, W., Vandenberg, R., Arriza, J., Kavanaugh, M. and Amara, S. (1995) Nature **375**, 599–603.
69. Parent, L. and Wright, E. (1993) Soc. Gen. Physiol. Ser. **48**, 263–281.
70. Fischer, J.F. and Cho, A.K. (1979) J. Pharm. Exp. Therap. **208**, 203–209.
71. Rudnick, G. and Wall, S.C. (1992) Proc. Natl. Acad. Sci. USA **89**, 1817–1821.
72. Sulzer, D. and Rayport, S. (1990) Neuron **5**, 797–808.
73. Mamounas, L., Mullen, C., Ohearn, E. and Molliver, M. (1991) J. Comp. Neurol. **314**, 558–586.
74. Rudnick, G. and Wall, S.C. (1992) Biochemistry **31**, 6710–6718.
75. Peter, D., Jimenez, J., Liu, Y., Kim, J. and Edwards, R. (1994) J. Biol. Chem. **269**, 7231–7237.
76. Schuldiner, S., Steiner-Mordoch, S., Yelin, R., Wall, S. and Rudnick, G. (1993) Mol. Pharmacol. **44**, 1227–1231.
77. Sulzer, D., Chen, T., Lau, Y., Kristensen, H., Rayport, S. and Ewing, A. (1995) J. Neurosci. **15**, 4102–4108.
78. Azzaro, A.J., Ziance, R.J. and Rutledge, C.O. (1974) J. Pharmacol. Exp. Ther. **189**, 110–118.
79. Wall, S.C., Gu, H.H. and Rudnick, G. (1995) Mol. Pharmacol. **47**, 544–550.
80. Rudnick, G. and Wall, S. (1993) Mol. Pharmacol. **43**, 271–276.
81. Rudnick, G. and Wall, S.C. (1992) Ann. N.Y. Acad. Sci. **648**, 345–347.
82. Talvenheimo, J. and Rudnick, G. (1980) J. Biol. Chem. **255**, 8606–11.
83. Biessen, E., Horn, A. and Robillard, G. (1990) Biochemistry **29**, 3349–3354.
84. Graham, D., Esnaud, H. and Langer, S.Z. (1992) Biochem. J. **286**, 801–805.
85. Launay, J., Geoffroy, C., Mutel, V., Buckle, M., Cesura, A., Alouf, J. and DaPrada, M. (1992) J. Biol. Chem. **267**, 11344–11351.
86. Ramamoorthy, S., Cool, D., Leibach, F., Mahesh, V. and Ganapathy, V. (1992) Biochem. J. **286**, 89–95.
87. Ramamoorthy, S., Bauman, A., Moore, K., Han, H., Yang-Feng, T., Chang, A., Ganapathy, V. and Blakely, R. (1993) Proc. Natl. Acad. Sci. USA **90**, 2542–2546.
88. Gregor, P., Patel, A., Shimada, S., Lin, C., Rochelle, J., Kitayama, S., Seldin, M. and Uhl, G. (1993) Mammalian Genome **4**, 283–284.
89. Koe, B.K. (1976) J. Pharm. Exp. Therap. **199**, 649–661.
90. Fuller, R. (1992) J. Clin. Psychol. **53**, 36–45.
91. Kitayama, S., Shimada, S., Xu, H., Markham, L., Donovan, D. and Uhl, G. (1992) Proc. Natl. Acad. Sci. USA **89**, 7782–7785.
92. Kitayama, S., Wang, J. and Uhl, G. (1993) Synapse **15**, 58–62.
93. Radian, R., Bendahan, A. and Kanner, B., I. (1986) J. Biol. Chem. **261**, 15437–15441.
94. Tate, C. and Blakely, R. (1994) J. Biol. Chem. **269**, 26303–26310.
95. Wang, J., Moriwaki, A. and Uhl, G. (1995) J. Neurochem. **64**, 1416–1419.
96. Blakely, R., Moore, K. and Qian, Y. (1993) Soc. Gen. Physiol. Ser. **48**, 283–300.
97. Rudnick, G. and Kanner, B. (1992) unpublished observations.
98. Buck, K. and Amara, S. (1993) Abstracts, Soc. Neurosci. **19**, 40.12.
99. Giros, B., Wang, Y., Suter, S., Mcleskey, S., Pifl, C. and Caron, M. (1994) J. Biol. Chem. **269**, 15985–15988.
100. Barker, E., Kimmel, H. and Blakely, R. (1994) Mol. Pharmacol. **46**, 799–807.
101. Stephan, M., Chen, M., Panado, K. and Rudnick, G. (1995) submitted for publication.

CHAPTER 18

From Multidrug Resistance to Vesicular Neurotransmitter Transport

S. SCHULDINER, M. LEBENDIKER, S. MORDOCH,
R. YELIN and H. YERUSHALMI

Division of Microbial and Molecular Ecology, The Institute of Life Sciences,
The Hebrew University of Jerusalem, 91904 Jerusalem, Israel

© *1996 Elsevier Science B.V.*
All rights reserved

Handbook of Biological Physics
Volume 2, edited by W.N. Konings, H.R. Kaback and J.S. Lolkema

Contents

1. General introduction

Multidrug resistance is a major concern in medical and agricultural diseases. In medicine, the emergence of resistance to multiple drugs is a significant obstacle in the treatment of several tumors as well as many infectious diseases. In agriculture, the control of resistance of plant pathogens is of major economic importance. However, these traits existed for aeons long before the drugs and antibiotics were discovered by humans. Resistance to a wide range of cytotoxic compounds is a common phenomenon observed in many organisms throughout the evolutionary scale and probably developed in order to cope with the variety of toxic compounds which are part of the natural environment in which living cells dwell. Only those organisms which have developed through evolution the ability to cope with a wide variety of compounds have been able to survive.

One of the strategies which evolved is removal of toxic substances by transport. Multidrug transporters are the proteins usually responsible for performing this task and have been found in many organisms, from bacteria to man. They can actively remove a wide variety of substrates in an energy dependent process and decrease the concentration of the offending compounds near their target. A great diversity of multidrug transporters are known to us today and we can group them in five different families (Fig. 1) based on structure similarities. In all cases, two basic energy coupling mechanisms have been characterized: those proteins which utilize the hydrolysis of ATP to actively remove toxic compounds and those which utilize the proton electrochemical gradient ($\Delta\tilde{\mu}_{H^+}$) generated by primary pumps. One family (ABC type) is best known for the P-glyco-protein or MDR in its various forms, which confers multidrug resistance to cancerous cells [1–5]. This family includes also many bacterial transport proteins (see Chapter 8). The ABC transporters utilize ATP to actively transport a wide variety of compounds.

All the other families which have been studied at a mechanistic level are antiporters. They exchange various drugs with one or more protons and utilize in this way the proton electrochemical gradient to actively transport a variety of substrates. The TEXANs (**T**oxin **EX**truding **AN**tiporters) include proteins which render fungi and bacteria resistant to antibiotic treatment [6–9] and mammalian proteins located in intracellular organelles which transport neurotransmitters such as dopamine, adrenaline, serotonin and Acetylcholine [10,11]. A third family (Mini-TEXANs) is represented by very small proteins, about 100 amino acids long which render bacteria resistant to a variety of toxic cations [7,12,13]. In another family (RND) it is not clear that all the proteins are transporters and their energy coupling mode has not been studied although they do not have nucleotide binding domains [14]. The last 'family' has a single protein: the organic cation antiporter from

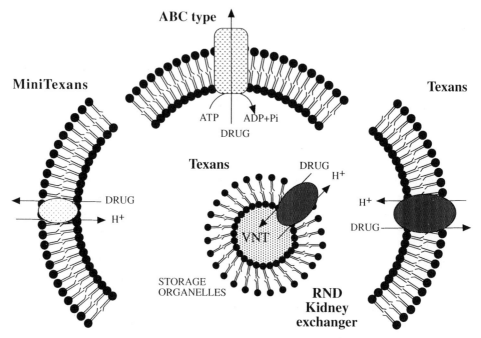

Fig. 1. Families of multidrug transporters. For explanation, see text.

kidney, unrelated to the others in its primary sequence but capable, as all the others, to remove a wide variety of toxic compounds against concentration gradients [15].

A question commonly asked about multidrug transporters is what is their 'real' function. Are these proteins functioning solely for protection of the organism against toxic compounds or do they have a very specific function and just by chance they happen to be also polyspecific. As in many other cases, the answer seems to be a complex one: thus, clearly in proteins functioning in the blood brain barrier [16] or in the kidney [15] there is little doubt that they are protecting the organism against toxic compounds by removing them from the organism or by preventing their passage to the brain. In the case of bacterial proteins such as the *Bacillus* BmrI and *E. coli* EmrA or Mar proteins, whose expression is regulated by multiple xenobiotics [17,18], it also seems reasonable that they have still a major role in protection of the cell as judged from their regulation. Yet, in other cases, it seems that given proteins, have evolved to perform specific roles other than multidrug resistance such as is the case of vesicular neurotransmitter transporters (VNTs), lipid translocators [19] and bacterial amino acid and sugar transporters from the ABC family [5].

This review will focus mainly in novel findings in the study of the vesicular neurotransmitter transporters, proteins which play a central role in neurotransmission. The topic seems to us very timely because of the realization of their functional and structural similarity with bacterial multidrug transporters. Also, their study hints about the possible new modes of protection from toxic substrates in mammalian cells by compartmentalization of the drugs. Their study should link together many different fields, such as pharmacology, microbiology and neurobiology.

The study of VMAT has also led us to search for other, simpler model systems of multidrug transporters, the Mini TEXANs, which we will very briefly describe at the end of this review. The Mini TEXANs turn out to display unique properties and may serve as models for understanding the molecular basis of high-affinity recognition of multiple substrates and the mechanism of ion-coupled transport.

2. TEXANs

The VNTs are part of a family which includes at least 40 proteins from prokaryotes and eukaryotes and has been surveyed in several monographs [7,10,20]. Based on an analysis of the evolutionary relationships four subgroups have been identified. All the proteins present in microorganisms are presumed to be exporters located in the cytoplasmic membrane of these cells which confer resistance to a large list of compounds due to their ability to actively remove them from the cell (Fig. 1). The VNTs, on the other hand, are located in intracellular vesicles. While the bacterial transporters extrude the toxic compounds to the medium, those presently known in mammals, remove neurotransmitters from the cytoplasm into intracellular storage compartments. In both cases, as a result of their functioning, the concentration of the substrates in the cytoplasm is reduced. When the substrate of the VNTs is cytotoxic, such as is the case with MPP^+, a substrate of the vesicular monoamine transporter (VMAT), the removal of the toxic compound from the cytoplasm, away from its presumed target will ameliorate the toxicity of the compound. Indeed, CHO cells expressing VMAT are more resistant to MPP^+ ([21] and see below). As mentioned above, these findings suggest a novel type of drug resistance in eukaryotic cells: protection of the life essential devices by compartmentalization (Fig. 1). Similar strategies have previously been suggested to explain tolerance to high salt and toxic compounds in plants and yeasts [22–24], however no molecular description of these phenomena has yet been proposed. One orphan mammalian protein, more similar in sequence to the bacterial Tet A-Bmr cluster than to the VNTs, has been described thus far [25].

Most transporters of the family have a very broad specificity for substrates. All of the substrates are aromatic compounds, usually bearing an ionizable or permanently charged nitrogen moiety. In some substrates, however, carboxylic groups are also present (i.e. norfloxacin); in others, a phosphonium moiety is present (i.e. TPP^+), and yet in others, no positive charge is present at all (i.e. actinorhodin, uncouplers). Many of the substrates are common to many multidrug transporters, as can be seen for example in Fig. 2A and B in which structural formulas of some of VMAT and EmrE (a MiniTEXAN) inhibitors and substrates are presented. Although the proteins are unrelated in their sequence, there is a large overlap in substrates, albeit they recognize them with different affinities (Fig. 2). All of the compounds described in Fig. 2 also interact with Mdr1. This large overlap in substrate recognition may hint about common solutions to the problem.

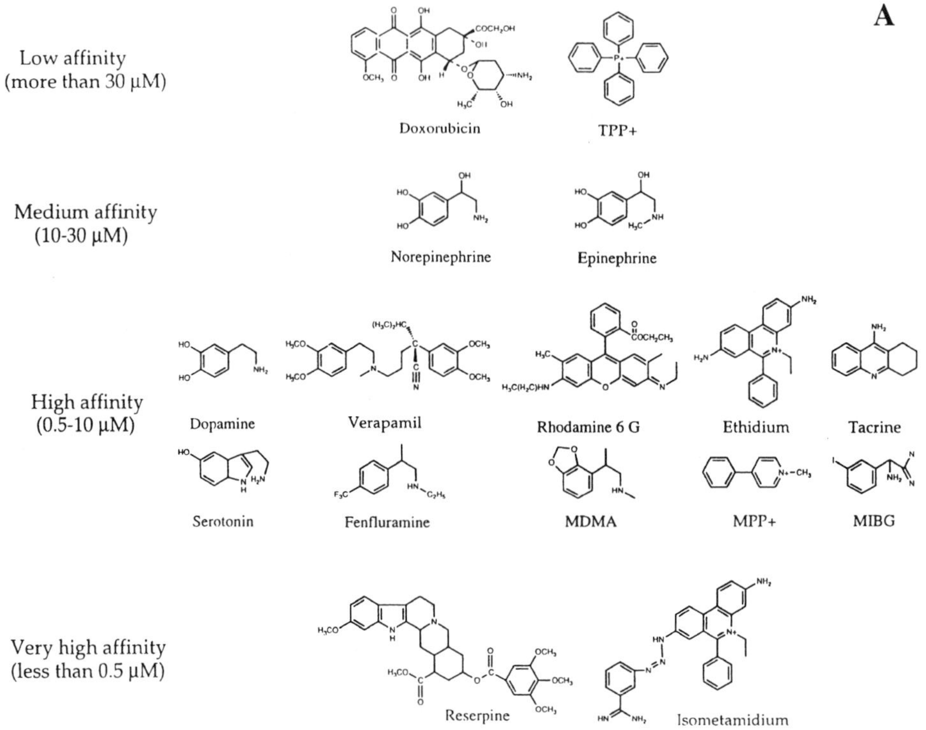

Fig. 2. Substrates and inhibitors of two multidrug transporters. (A) VMAT; *opposite*: (B) EmrE .

All the transporters of the TEXAN family are located in membranes across which H^+ electrochemical gradients exist. The gradients are generated by primary pumps, such as the bacterial respiratory chain or the H^+-translocating ATPases of both bacteria and intracellular storage organelles. The gradient is utilized by the protein through the exchange of a substrate molecule with one or more hydrogen ions. All the neurotransmitter storage vesicles studied thus far, in brain, platelets, mast cells and adrenal medulla, contain a vacuolar type H^+ pumping ATPase, similar in composition to the ATPase of lysosomes, endosomes, Golgi membranes and clathrin coated vesicles [26–28]. In all these organelles, the activity of this proton pump generates an H^+ electrochemical gradient ($\Delta\tilde{\mu}_{H^+}$ acid and positive inside). In synaptic vesicles and neurotransmitter storage organelles the proton electrochemical gradient is utilized by the VNTs, which couple efflux of H^+ ions to the uptake of a neurotransmitter molecule (for review see Ref. [10]). In the case of Tet proteins, it has been suggested that the exchange is between a metal–tetracycline complex and one proton in an electroneutral process. Also BMR mediated drug efflux is apparently driven by a transmembrane pH gradient. While very little is known about the other transporters the fact that they all display sequence similarities, and none of them show any ATP binding domains, in addition to the fact that they are all found in membranes with H^+ gradients suggests that they all are antiporters which exchange one or more H^+ ions with a substrate molecule.

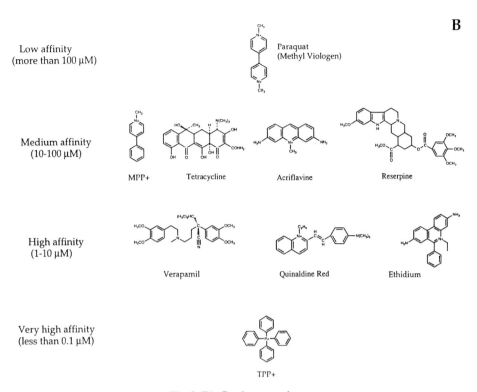

Fig. 2 (B). Caption opposite.

3. Transporters and neurotransmission

Synaptic transmission involves the regulated release of transmitter molecules to the synaptic cleft whereby they interact with postsynaptic receptors which subsequently transduce the information. Removal of the transmitter from the cleft enables termination of the signal, it usually occurs by its reuptake back to the presynaptic terminal or into glial elements in a sodium dependent process. This process assures constant and high levels of neurotransmitters in the neuron and low concentrations in the cleft.

Storage of neurotransmitters in subcellular organelles ensures their regulated release, and is crucial for protecting the accumulated molecules from leakage or intraneuronal metabolism and the neuron from possible toxic effects of the transmitters. In addition, the removal of intraneuronal molecules into the storage system effectively lowers the concentration gradient across the neuronal membrane and thus acts as an amplification stage for the overall process of uptake. Drugs which interact with either transport systems have profound pharmacological effects as they modify the levels of neurotransmitter in the cleft. Drugs that interfere with these activities include the tricyclic antidepressants, stimulants such as amphetamines and cocaine, antihypertensives such as reserpine and neurotoxins such as MPP$^+$.

Plasma membrane transporters have been intensively studied at a mechanistic, biochemical and molecular level. The molecular characterization began with the purification, amino acid sequencing and cloning of the γ-aminobutyric acid (GABA) transporter [29]. Since, it has become clear that these Na^+ and Cl^--coupled transporters represent a group of integral membrane proteins encoded by a closely related family of genes which include the transporters for monoamines, GABA, glycine, proline, choline and taurine [30,31]. A different class of plasma membrane transporters is represented by the Na^+ and K^+-coupled glutamate transporter [32,33, 34,35] (see also Chapter 19).

Vesicular transport has been observed for several classical transmitters, including acetylcholine (ACh), the monoamines, glutamate, GABA and glycine (reviewed in Ref. [10]). The vesicular monoamine transporter (VMAT) and ACh (VAChT) have been the most intensively studied and are the ones for which most molecular information have been obtained and we will review the most salient features of both. In both cases, the key for this knowledge resides in the availability of excellent experimental paradigms for their study and potent and specific inhibitors.

4. Pharmacology of VNTs

The best characterized inhibitors of VNTs are reserpine and tetrabenazine, the two principal agents that inhibit vesicular monoamine transport [36,37]. In recent years, vesamicol, a novel inhibitor of VAChT has been introduced and studied in detail [38–40].

4.1. Molecular mechanism of reserpine action

Reserpine presumably binds at the site of amine recognition as judged by the fact that it inhibits transport in an apparent competitive way, with K_is in the subnanomolar range [41,42], that it binds to the transporter with a K_D similar to its K_i [41,42] and that transport substrates prevent its association in a concentration range similar to the range of their apparent K_ms [41]. Its effect *in vitro* is practically irreversible [43], in line with the *in vivo* effect of the drug which is extremely long lasting and is relieved only when new vesicles replace the ones that were hit [44]. As a result of this action, it depletes monoamine stores, providing considerable information on the physiological role of biogenic amines in the nervous system [45]. Reserpine has been used because it potently reduces blood pressure. However, it frequently produces a disabling effect of lethargy that resembles depression and has limited its clinical utility [46]. This observation has given rise to the amine hypothesis of affective disorders which in modified form still produces a useful framework for considering this group of major psychiatric disorders.

The time course of reserpine binding is relatively slow. This low rate of association is consistent with a similar time course for inhibition of monoamine transport [43]. Reserpine binding is accelerated by $\Delta\tilde{\mu}_{H^+}$ whether generated by the H^+-ATPase [41,47] or artificially imposed [43]. This acceleration is observed also in proteoliposomes reconstituted with the purified protein [48].

In all cases, in the presence or absence of $\Delta\tilde{\mu}_{H^+}$ and in the native as well in the purified protein, two distinct populations of sites have been detected [41,48]. A high affinity site, $K_D = 0.5$ nM, $B_{max} = 7$–10 pmol/mg protein in the native chromaffin granule membrane vesicle preparation (0.3 and 310, respectively for the purified protein) and a low affinity site, $K_D = 20$ nM, $B_{max} = 60$ pmol/mg protein in the native system (30 and 4200, respectively for the purified preparation). Surprisingly the apparent K_D does not change with an imposition of a $\Delta\tilde{\mu}_{H^+}$ even though the on rate increases several fold [41]. It has to be assumed that the off rate changes also accordingly, although the off rate is so slow that it is impossible to measure. In this context, it is important to note that the K_D values obtained under conditions where the concentration of ligand binding sites does not exceed the value of the dissociation constant, is 30 pM, i.e. about ten times higher affinity than previously estimated [42].

The reserpine binding rate is less sensitive than transport to changes in the ΔpH, and is stimulated equally efficiently by ΔpH and $\Delta\psi$ [43]. These findings suggest that fewer protons are translocated in the step which generates the high affinity binding site than in the overall transport cycle. Changes in binding rate probably reflect changes in the availability of reserpine binding sites and translocation of a single H^+ generates the binding form of the transporter. Figure 3 schematically demonstrates a model to explain transport, reserpine binding, and reserpine occlusion by the vesicular monoamine transporter. In this model, the high affinity form of the transporter (TH$^+$) is apparently achieved by either protonation of VMAT or by H^+ translocation. The energy invested in the transporter may be released by ligand binding (THS) and converted into vectorial movement of a substrate molecule across the membrane or directly into binding energy in the case of reserpine (THR). In the case of a substrate a second conformational change results in the ligand binding site being exposed to the vesicle interior, where the substrate can dissociate (TH$_2$S). The second H^+ in the cycle, may be required to facilitate the conformational change or to allow for release of the positively charged substrate from the protein. In the model this second H^+ binding and release is arbitrarily located, since there is no information about the order of the reactions. Interestingly, in this respect, are the findings that the apparent affinity of the transporter for substrates drops when the pH decreases [49,50]. This could reflect a mechanism for releasing the substrate in the acidic lumenal milieu. Substrate and H^+ release regenerates the transporter (T) which can now start a new cycle. In the case of reserpine, however, its structure (bulk of its side chain?) restricts the conformational change so that instead of releasing the ligand on the interior, the complex becomes trapped in a state from which reserpine cannot readily dissociate and which cannot translocate another H^+ to regenerate the high affinity form. It is not known whether the slow binding of reserpine requires also protonation of VMAT. If this is the case, protonation in the absence of $\Delta\tilde{\mu}_{H^+}$ would be the rate limiting step.

Tetrabenazine (TBZ) is another potent inhibitor of the transporter. Radiolabeled dihydrotetrabenazine (TBZOH) has been used in binding studies to characterize the protein [41,51], to study the regulation of its synthesis [52–54] and its distribution in various tissues [55]. Binding of TBZOH to the transporter is not modified by the imposition of a $\Delta\tilde{\mu}_{H^+}$ as shown for reserpine. In addition, binding is not inhibited by

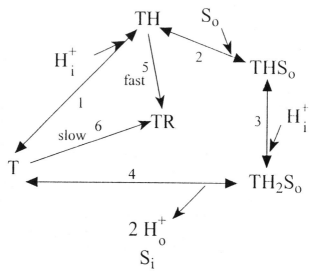

Fig. 3.   Model for substrate transport and reserpine binding. See explanations in text. T, transporter;
S, substrate; R, reserpine; H^+, hydrogen ion. Subscripts denote whether the substrate is inside the
vesicle (i) or outside (o).

reserpine at concentrations which fully inhibit transport. Moreover, transport substrates block binding only at concentrations 100 fold higher than their apparent K_m values. These findings have led Henry and collaborators to suggest that TBZOH binds to a site on the protein which is different from the reserpine and substrate binding site [42,51,56]. It has been suggested that both sites are mutually exclusive, i.e., VMAT exists in two different conformations, each conformation binding only one type of ligand, TBZ or reserpine. According to this interpretation, addition of TBZ would pull the conformational equilibrium toward the TBZ binding conformation, which is unable to bind reserpine. Indeed, under proper conditions (low protein concentration and short incubation times) 50 nM TBZOH inhibits reserpine binding by 70%. ATP, through the generation of $\Delta\tilde{\mu}_{H^+}$, would pull the conformational equilibrium towards the RES binding site [42]. Although elegant and attractive, this model has yet to explain the lack of effect of $\Delta\tilde{\mu}_{H^+}$ on TBZ binding, which should be inhibited, if the two forms were mutually exclusive. Also, the concentrations of reserpine required to inhibit TBZ binding are higher than those required for site occupancy (binding and inhibition of transport).

4.2. Acetylcholine transporter

An important development in the study of VAChT was the discovery of a specific inhibitor, *trans*-2-(4-phenylpiperidino)cyclohexanol, code named AH5183 and now called vesamicol, which blocks neuromuscular transmission and shows unusual characteristics of action. Marshall [38] hypothesized that AH5183 blocks storage by synaptic vesicles and indeed it inhibits ACh storage by purified *Torpedo* synaptic vesicles with an IC_{50} of 40 nM. The drug was the most potent inhibitor found among at least 80 compounds initially screened ([57,58] and see Ref. [40] for review). [^{3}H]vesamicol binding showed an apparent K_D of 34 nM [59–63]. ACh inhibits vesamicol binding only at very high concentrations (20–50 mM) and other high affinity analogues were shown to competitively inhibit binding. However, in all cases where the analogues are transported the inhibition constant is about 20 fold higher than the apparent constants for transport. Non transported analogues show the same efficiency for inhibition of both ACh transport and [^{3}H]vesamicol binding [64]. A kinetic model has been suggested in which it is assumed that vesamicol binds to an allosteric site to form a dead end complex when ACh is not bound. As described above, the existence of two sites with similar properties has been observed also for VMAT. In the latter, TBZ is a potent inhibitor of monoamine accumulation but its binding is inhibited by monoamines only at very high concentrations (see above for detailed discussion and references). In the case of VMAT, both binding sites are present in one protein, since the purified and the recombinant proteins show high sensitivity to TBZ. In the case of VAChT, the existence of a 'receptor' has been postulated which could lie on the same protein or in a separate one. The vesamicol 'receptor' has been extensively studied by Parsons and collaborators, it has been purified [65] and labelled with photoaffinity labels [40,66,67]. The 'receptor' solubilized in cholate and stabilized with glycerol and a phospholipid mixture was purified to yield a specific binding of 4400 pmol/mg protein, a purification factor of about 15. Unfortunately, the purified receptor exhibits very heterogeneous electrophoretic mobility in SDS PAGE with very diffuse stain at about 240 kDa. This is the typical behavior of membrane glycoproteins which are not fully monodispersed because of the detergent used and in addition boiled, treatments known to induce aggregation of membrane proteins.

Recently, glycoproteins from various species which bind vesamicol with high affinity have been expressed in CV1-fibroblasts ([68–70] and see Section 6). In addition the rat VAChT expressed in CV1-fibroblasts catalyzes vesamicol-sensitive ACh accumulation [68]. As will be seen below, the evidence available now clearly demonstrates that vesamicol binds to the vesicular ACh transporter itself.

5. Identification of functional transporters

The only VNT purified in a functional form is the bovine VMAT. The high stability of the complex [^{3}H]reserpine–transporter has been used to label the transporter and follow its separation through a variety of procedures. In these experiments a small amount of Triton X100-extracts from prelabeled membranes were mixed with a four to five-fold higher amount of extract from unlabelled membranes. Purification of the material labeled in this way has revealed the presence of two proteins that differ in pI, a very acidic one (3.5) and a moderately acidic one (5.0) [48]. Reconstitution in proteoliposomes has shown that both catalyze monoamine transport with the expected properties. The more acidic isoform is a glycoprotein of 80 kDa, which has been purified and reconstituted in proteoliposomes. It catalyzes transport of serotonin with an apparent K_m of 2 μM and a V_{max} of 140 nmol/mg/min, about 200-fold higher than the one determined in the native system. Transport is inhibited by reserpine and tetrabenazine, ligands which bind to two distinct sites on the transporter. In addition, the reconstituted purified transporter binds reserpine with a biphasic kinetic behavior, typical of the native system. The results demonstrated that a single polypeptide is required for all the activities displayed by the transporter, i.e. reserpine- and tetrabenazine-sensitive, ΔpH-driven serotonin accumulation, and binding of reserpine in an energy-dependent and independent way. Based on these and additional findings it was estimated that the transporter represents about 0.2–0.5% of the chromaffin granule membrane vesicle and that it has a turnover of about 30 min^{-1}.

The assignment of the activity to the 80 kDa polypeptide and the localization of the tetrabenazine and reserpine binding sites in the same polypeptide is confirmed by several independent approaches, including direct sequencing of the purified protein [71,72] and cloning and analysis of the recombinant protein [21,73,74]. Vincent and Near purified a TBZOH-binding glycoprotein from bovine adrenal medulla using a protocol identical to the one used to purify the functional transporter. The TBZOH-binding protein displays an apparent Mr of 85 kDa [75]. In addition, Henry and collaborators, labelled bovine VMAT with 7-azido-8-[^{125}I] iodoketanserin, a photoactive derivative of ketanserin, which is thought to interact with the TBZ binding site. The labelled polypeptide displayed an apparent Mr of 70 kDa and a pI ranging from 3.8 to 4.6 [76]. In all the cases broad diffuse bands, characteristic of membrane proteins, were detected so that the differences in the Mr's reported (70, 80 and 85 kDa) are probably due to a different analysis of the results and not to innate variations. Sequencing of the labelled protein confirmed that it is identical with the functional transporter [72].

The basis of the difference between the two isoforms has not yet been studied and could be due to either covalent modification (i.e phosphorylation or different glycosylation levels) of the same polypeptide backbone, to limited proteolysis during preparation or to a different polypeptide backbone. Since we know now (see Section 6) that there are at least two types of VMATs: VMAT2, which is sensitive to TBZ, and VMAT1, which is less sensitive, it should be determined whether the activity of the high pI form is less sensitive to TBZ. The sequence of 26 N-terminal amino acids of the purified protein is practically identical to the predicted sequence of the bovine adrenal VMAT2 ([71,72,77] and see below). Antibodies raised against a synthetic peptide based on the described sequences specifically recognize the pure protein on Western blots and immunoprecipitate reserpine binding activity under conditions where the 80-kDa protein alone is precipitated [71].

6. Cloning and functional expression

6.1. VMATs

Several sequences of VMAT and VAChT from various species are now available. Rat VMAT was cloned by Edwards et al. [21] and Erickson et al. [74] practically at the same time using different strategies (reviewed in Refs. [10,78]). Erickson and coworkers [74] used expression cloning in CV-1 cells transfected with c-DNA prepared from rat basophilic leukemia (RBL) cells mRNA. Edwards et al. took advantage of the ability of VMAT1 to render CHO cells resistant to MPP^+ by means of its ability to transport the neurotoxin into intracellular acidic compartments thereby lowering its effective concentration in the cytoplasm [21].

Sequence analysis of the cDNA conferring MPP^+ resistance (VMAT1) shows a single large open reading frame which predicts a 521 amino acid protein. Analysis by the method of hydrophobic moments predicts 12 putative transmembrane segments (TMS). A large hydrophilic loop occurs between membrane domains 1 and 2 and contains three potential sites for N-linked glycosylation. According to the 12 TMS model [21] this loop faces the lumen of the vesicle, and both termini face the cytoplasm.

Biochemical and quantitative evidence for the identity of the cloned cDNA was provided by developing a cell free system in which membranes were assayed for dopamine transport and reserpine binding [21,73].

VMAT1 expressed in MPP^+-resistant CHO cells accounts for about 0.1% of the total cell membrane protein [73] while the bovine vesicular transporter for 0.2–0.5% of the chromaffin granule membrane [48].

A transporter distinct from VMAT1 has been identified in rat brain (VMAT2) and in RBL cells [74]. The predicted protein shows an overall identity of 62% and a similarity of 78% to VMAT1. The major sequence divergences occur at the large lumenal loop located between the first and the second transmembrane domains and at the N- and C-termini [72,77,79,80].

6.2. The two VMAT subtypes differ also in some functional properties

A comprehensive comparison between the functions of rat VMAT1 and VMAT2 has been performed by Edwards and collaborators in membrane vesicles prepared from CHO stable transformed cells lines in which the respective proteins are expressed [81]. According to these studies VMAT2 has a consistently higher affinity for all the monoamine substrates tested. In the case of serotonin, dopamine and epinephrine the apparent K_m of VMAT1 for ATP dependent transport is 0.85, 1.56 and 1.86 μM respectively, while the corresponding K_ms measured for VMAT2 are 4- to 5-fold lower, 0.19, 0.32 and 0.47 μM correspondingly. Although the affinities are slightly different, the rank order for the various monoamines is similar. Also other substrates such as MPP^+ (1.6 vs 2.8 μM, respectively) and methamphetamine (2.7 vs 5.5 μM respectively as estimated from measurements of the ability to inhibit reserpine binding) display a similar pattern. A most striking difference is detected for histamine: 3 μM for VMAT2 and 436 μM for VMAT1. Also, VMAT1 is significantly less sensitive to tetrabenazine, IC_{50} = 3–4 μM [21,81], than either VMAT2 (IC_{50} = 0.3–0.6 μM) [74,81] or the native [37,42] and the purified transporter (IC_{50} = 25 nM) [48] from bovine adrenal medulla. Interestingly, the apparent affinities determined in heterologous expression systems or in proteoliposomes reconstituted with the purified transporter are higher than those determined in chromaffin granules membrane vesicles (see Ref. [82] for references). The turnover number (TO) of the recombinant protein has been calculated based on the V_{max} for serotonin transport and the number of reserpine binding sites. The TO for VMAT1 is 10 min^{-1}, while that for VMAT2 is 40 [81]. A similar analysis of the purified bovine transporter (VMAT2 type) showed a TO of 30 min^{-1} (Section 5 and Ref. [48]). These values are lower than the 135 min^{-1} estimated for intact bovine chromaffin granules [83] but coincide well with the values obtained from brain regions (10–35 min^{-1}) [55] and other estimates (15 and 35) in chromaffin granules [56].

6.3. Tissue distribution

Tissue distribution of rat VMAT subtypes has been studied very intensively using a variety of techniques: Northern analysis [21] [74], *in situ* hybridization [21] and immunohistochemistry [84,85]. From these studies it is concluded that expression of VMAT1 and VMAT2 is mutually exclusive: VMAT1 is restricted to non neuronal cells and VMAT2 to neuronal cells. VMAT1 is expressed in endocrine/paracrine cells: in the adrenal medulla chromaffin cells, in the intestine and gastric mucose in serotonin and histamine-containing endocrine and paracrine cells and in SIF cells of sympathetic ganglia. VMAT2 is expressed in neuronal cells throughout, including in the intestine and stomach. There are two exceptions for this restriction: a subpopulation of VMAT2-expressing chromaffin cells in the adrenal medulla and a population of VMAT2, chromogranin A-positive endocrine cells of the oxyntic mucose of the rat stomach [84].

While the studies in rat are very definitive, the situation is very different in other species. In human pheochromocytoma mRNA for both subtypes is found [84]. The bovine adrenal medulla expresses a VMAT2 type transporter whose message has also been detected in brain [72,77]. VMAT2 is the main adrenal medulla transporter as judged from direct protein purification studies [48,71,72,77]. The purified transporter from bovine adrenal is VMAT2 and accounts for at least 60% of the activity in the gland.

6.4. Subcellular targeting of VMAT

Storage of monoamines differs from that of other classical neurotransmitters. Thus, while the latter are stored in small synaptic vesicles, monoamines in the adrenal medulla are stored with neural peptides in large dense core chromaffin granules. In the CNS, neurons store monoamines in small vesicles that may contain a dense core. The difference in the storage of the monoamines as compared to that of classical transmitters may reflect differential sorting of the VNTs. Sorting of VMAT was studied with immunohistochemical and biochemical tools [86,87].

In heterologous expression of VMAT1 in CHO cells, the transporter is targeted to a population of recycling vesicles and co-localizes with the transferrin receptor. Thus, localization in CHO cells is similar to that of other neuronal vesicle proteins such as synaptophysin and SV2 [86].

In PC12 cells, endogenous VMAT1, occurs principally in large dense core granule (LDCV). In synaptic like microvesicles (SLMVs) and in endosomes only small amounts are found [86].

In the rat adrenal medulla, immunoreactivity for VMAT1 occurs at several sites in the secretory pathway but most prominently in the chromaffin granules, support-ing the results in PC12 cells [86].

In central neurons r-VMAT2s localization was studied in the nuclei of solitary tract, a region known to contain a dense and heterogenous population of monoamin-ergic neurons [87]. VMAT2 localizes primarily to LDCVs in axon terminals. It is also detected in less-prominent amounts in small synaptic vesicles, the trans-Golgi network and other sites of vesicles transport and recycling. Thus, both VMAT1 and VMAT2 are primarily sorted to LDCVs in all the cell types studied.

6.5. Regulation of expression of VMAT

Evidence for regulation of expression of VMAT was first obtained in insulin shocked rats in which an increase in the number of [^{3}H]TBZOH binding sites in the adrenal medulla was detected. The increase was maximal after 4–6 days [52]. A similar increase was observed *in vitro* in bovine chromaffin cells in culture in the presence of carbamylcholine or depolarizing concentrations of potassium ions [54]. The response was mimicked by forskolin and by phorbol esters and was blocked by actinomycin and cycloheximide, suggesting involvement of transcriptional activation.

This suggestion was supported by the detection of an increase in message for VMAT2 after 6 h depolarization [72]. After 5 days the cells contained fewer secretory granules and those left had a higher density, suggesting that they are newly synthesized and immature. While the catecholamine, chromogranin A and cytochrome b561 content decreased, [^{3}H]TBZOH binding sites increased about 1.5-fold. The physiological significance of these findings is not obvious. It has been speculated that this phenomenon may reflect the fact that vesicular uptake might be rate limiting and that in order to accelerate refilling an increase in VMAT is needed. Mahata et al., have suggested, however, that there is no increase in other membrane proteins in the rat granule [88,89]. They have reported no changes in the level of VMAT1 message (the main subtype in rat adrenal) under conditions at which mRNA for the matrix peptide NPY increased [88,89]. Since in the same system the [^{3}H] TBZOH binding sites increased, the latter findings may suggest a novel mode of regulation of activity of preexisting protein.

7. Vesicular acetylcholine Transporters

The powerful genetics of the nematode *Caenorhabditis elegans* has provided important information regarding VAChT. The elegant approach used by Rand and collaborators [90] was based on the analysis of one of the mutants described by Brenner twenty years ago [91]. Mutations in the *unc-17* gene of the nematode result in impaired neuromuscular function, which suggest that cholinergic processes might be defective in the mutant. In addition, *unc-17* mutants were resistant to cholinesterase inhibitors [91], a resistance which may result from decreased syn-thesis or release of the transmitter. Moreover, *unc-17* was found closely linked to *cha-1* gene, which encodes choline acetyl-transferase [92]. The genomic region of *unc-17* was cloned by walking from the *cha-1* gene and thereafter cDNA was isolated from a library [90]. Injection of a cosmid containing the complete coding sequence of the isolated cDNA rescues the mutant phenotypes of *unc-17* animals. A protein with 532 amino acids is predicted from the isolated DNA sequences. This protein (UNC-17=VAChT) is 37% identical to VMAT1 and 39% identical to VMAT2 . The findings strongly suggested that UNC-17 is a vesicular acetylcholine transporter. This was supported by the fact that antibodies against specific peptides stain most regions of the nervous system. Within individual cells, staining was punctate and concentrated near synaptic regions. Double labeling with anti-synaptotagmin showed colocalization of the two antigens. In addition, in *unc-104* mutants, a mutation in a kinesin related protein required for the axonal transport of synaptic vesicles, synaptic vesicles accumulate in cell bodies. In these animals, anti-UNC-17 staining was restricted to neuronal cell bodies. More than 20 alleles, viable as homozygotes have now been identified. Their phenotypes vary from mild to severe. In two of these mutants, staining was dramatically decreased throughout the nervous system. Two other alleles were isolated which are lethal as homozygotes and they seem to represent the null *unc-17* phenotype. This is the first demonstration that the function encoded by a VNT is essential for survival.

Homology screening with a probe from *unc17* allowed for the isolation of DNA clones from *Torpedo marmorata* and *Torpedo ocellata* [70]. The *Torpedo* proteins display approximately 50% identity to UNC17 and 43% identity to VMAT1 and VMAT2. Message is specifically expressed in the brain and the electric lobe. The *Torpedo* protein, expressed in CV-1 fibroblast cells, binds vesamicol with high affinity (K_D= 6 nM). Interestingly, the UNC17 protein expressed in the same cells binds also vesamicol, albeit with a lower affinity (124 nM) [70]. Mammalian VAChTs (human and rat) have been identified [68,69]. The predicted sequences of both proteins are highly similar to those of the *Torpedo* and *C. elegans* counterparts. The rat VAChT has been shown to bind vesamicol with high affinity (K_D = 6 nM). It also catalyzes proton-dependent, vesamicol sensitive, ACh accumulation in transfected CV1 cells [68]. The distribution of rat VAChT mRNA coincides with that reported for choline acetyltransferase (ChAT), the enzyme required for ACh biosynthesis, in the peripheral and central cholinergic nervous systems. The human VAChT gene localizes to chromosome 10q11.2 which is also the location of the ChAT gene. The entire sequence of the human VAChT coding area is contained uninterrupted within the first intron of the ChAT gene locus [68]. Transcription of both genes from the same or contiguous promoters provides a novel mechanism for coordinate regulation of two proteins whose expression is required to establish a phenotype.

8. Structure–function studies: identification of residues/domains with putative roles in structure and function

The only topological information about the VNTs thus far is that obtained from an hydropathic analysis of the protein sequence. The results predict 12 putative transmembrane segments and a large hydrophilic loop between transmembrane domains

1 and 2. The loop contains three potential sites for N-linked glycosylation. Previous studies have demonstrated that all the glycan moieties in glycoproteins of chromaffin granules and AcCh storage face the lumen. Therefore, according to the latter finding and the model, the loop faces the lumen and both termini face the cytoplasm.

Identification of functional residues in VNTs is based on studies using site directed mutagenesis and relatively specific chemical modifiers. These studies are facilitated by the availability of sequences from different species and subtypes. It is usually assumed that residues which play central roles in catalysis are conserved throughout species. The degree of conservation in the group of VNTs is rather high. The members which are farther away in the group (human VMAT2 and the *C. elegans* VAChT) are still 38% identical and 63% similar. In the schematic representation in Fig. 4, an alignment of the nine known sequences of the VNTs is presented. The highest divergence is detected in the N- and C-termini and in the glycosylation loop between putative TMS I and II. The highest identity is observed

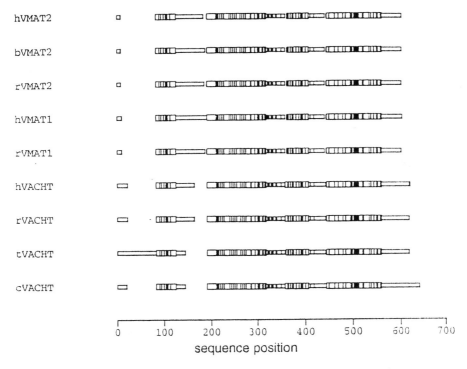

Fig. 4. Schematic representation of a multiple alignment of vesicular neurotransmitter transporters. Multiple alignment was first performed using Pileup (Wisconsin Package, 1994) and then blocks of similarity were identified using MACAW [118]. The best blocks are shown, a vertical line indicates fully conserved residues. The width of the line is proportional to the number of conserved residues. The sequences aligned are as described in Table 1 of Schuldiner, 1995. The sequence of hVMAT1 was kindly made available before publication by J. Erickson (NIH). The least conserved domains are the N- and C-termini and the large glycosylation loop. TM VIII is conserved within the VMATs and the VAChTs but there are only four residues identical in both.

Table 1
Summary of mutants in VMAT

Mutation	Transport	Reserpine binding	Coupling to $\Delta\tilde{\mu}_{H^+}$	Remarks	Subtype and Reference
D33E	Normal	Normal	Normal	Impaired	VMAT2 [96]
D33N	Undetectable	Normal	Normal	Substrate recognition	
G151L	Normal	Normal	Normal		
T154A	Normal	Normal	Normal		
N155Q	Normal	Normal	Normal		VMAT2
N155D	Normal	Normal	Normal		
G158L	Normal	Normal	Normal		
(Stmd3A)	Undetectable	Normal	Normal	Impaired	VMAT2
S180-182A				substrate recognition	
S197, 198, 200, 201A (Stmd4A)	Normal	Normal	Normal		VMAT2
H384C	Normal	Normal	Normal		VMAT1 [103]
H384R					
D404E	Shifted to acidic pH	Normal	Normal		VMAT1 (Mordoch et al.,
D404C	Undetectable	Undetectable	–		unpublished)
D404S	Undetectable	Undetectable	–		
H419C	Undetectable	Normal	None		VMAT1 [103]
H419R	Undetectable	Normal	None		
D431E	Undetectable	Normal	Normal		VMAT1
D431S	Undetectable	Normal	Normal		(Mordoch et al.,
D431C	Undetectable	Undetectable	No expression		unpublished)

in TM I, II and XI where at least 11 amino acids are fully conserved. In TMS XI the conservation is particularly striking since practically all the amino acids conserved are in a contiguous stretch 'SVYGSVYAIAD'.

8.1. Carboxyl residues

Particularly striking are four conserved Asp residues D34, D267, D404 and D431 (numbers of rVMAT1) in the middle of TMS I, VI, X and XI. In addition to their being conserved charged residues in the membrane, biochemical evidence was available that N,N'-dicyclohexylcarbodiimide (DCC) inhibits VMAT mediated transport [93–95]. DCC reacts with a carboxyl residue whose availability is influenced by the occupancy of the tetrabenazine binding site [94]. Reaction with the above carboxyl residue inhibits not only overall transport activity but also TBZ and reserpine binding, suggesting that the residue plays a role in one of the first steps of the transport cycle. As with all chemical modifiers, indirect effects such as steric hindrance by the DCC moiety or indirect effect on the structure of the protein cannot be ruled out at present. Therefore, mutagenesis studies of the roles of these four Asp residues on VMAT should be instructive.

Conservative replacement of Asp33 with Glu in rVMAT2 reduced activity but did not abolish it [96]. Replacement with Asn abolished transport indicating the crucial role of a negative charge at this position. However, VMAT could still bind [^{3}H] reserpine and binding was accelerated by $\Delta \tilde{\mu}_{H^+}$ [96]. Inhibition of reserpine binding in D33N by serotonin differs dramatically from the wild type suggesting interference with substrate recognition. Similarly, replacements in positions 404, 431 located in TMS X and XI had dramatic effects on activity [96a]. In the case of 431, even a conservative replacement with Glu led to transport inhibition. Replacement of Asp 404 with Glu generated a very interesting protein; the pH optimum of transport was shifted by about 1 pH unit to the acidic side. Replacement with either Cys or Ser on both positions yielded proteins which displayed no transport at all. The Asp431 replacements, D431E and D431S, but not D431C which was not expressed at detectable levels, bound [^{3}H] reserpine normally and binding was accelerated by $\Delta \tilde{\mu}_{H^+}$. The Asp404 replacements D404S and D404C showed no binding at all [96a].

An important conclusion from these studies is that replacements of negative charges in the middle of putative transmembrane segments, while having dramatic effects on transport, do not necessarily perturb the protein structure since some of the mutants still bind a high affinity ligand and respond to $\Delta \tilde{\mu}_{H^+}$ (see Table 1). The ability to measure partial reactions in VMAT allows for a more sophisticated analysis of mutagenized proteins than previously possible. Thus, although it does not transport, it was shown that the D33N mutant protein has a lower affinity to serotonin than the wild type. Also, the mutant D404E, shows an acid shift on the overall cycle but not on partial reactions suggesting an effect on a pKa important for the final steps of transport (Fig. 3). It is tempting to speculate a direct role of D404 in the translocation of the second H$^+$ needed for the overall cycle. Direct proof of this contention will need further experimentation.

In this context, it is interesting to point out that only two of the 19 mutants described in Table 1 are completely inactive. One of them, is not expressed to detectable levels.

8.2. His and Arg residues

Also His residues have been suggested to play a role in H$^+$ translocation and sensing in other H$^+$-coupled transporters [97,98]. In VMATs, there is only one His conserved (His419) in loop 10, between TMS IX and X. This His is immediately behind an Arg residue conserved throughout the whole VNT family and very close to the longest conserved sequence stretch (see above). Although present also in the rat, human and *C. elegans* VAChT, it is replaced by a Phe in the Torpedo VAChT. Biochemical evidence suggested a role for His also in VMAT. Phenylglyoxal (PG) and diethyl pyrocarbonate (DEPC) are reagents relatively specific for Arg and His residues, respectively. They both inhibit serotonin accumulation in chromaffin granule membrane vesicles in a dose dependent manner (IC_{50} of 8 and 1 mM, respectively) [99,100]. The inhibition by DEPC was specific for His groups since transport could be restored by hydroxylamine [99]. Neither PG nor DEPC inhibited binding of either reserpine or tetrabenazine, indicating that the inhibition of transport is not due to a direct interaction with either of the known binding sites. Interestingly, however, the acceleration of reserpine binding by a transmembrane H^+ gradient was inhibited by both reagents [99]. The results suggest that either proton transport or a conformational change induced by proton transport is inhibited by both types of reagents. Several other transport systems are sensitive to DEPC and phenylglyoxal (review in [99,101,102].

A more direct analysis of the role of histidines in VMAT has been carried out by site directed mutagenesis of rVMAT1 [103]. Replacement of His419, the only His conserved in VMAT1, with either Cys (H419C) or Arg (H419R) completely abolishes transport as measured in permeabilized CV1 cells transiently transformed with the mutants DNA. In the absence of $\Delta\tilde{\mu}_{H^+}$, reserpine binding to the mutant proteins is at levels comparable to those detected in the wild type. However, acceleration of binding in the presence of $\Delta\tilde{\mu}_{H^+}$ is not observed in either H419C or H419R. These results suggest that His419 plays a role in a step other than binding and may be associated directly with H^+ translocation or in conformational changes occurring after substrate binding.

8.3. Serine residues

In the β-adrenergic receptor and in the dopamine plasma membrane transporter [104] it has been suggested that serines play a role in ligand recognition. Two groups of serine residues occur in VMAT2: in TM III and TM IV. Simultaneous replacement of 4 serines in TM IV (S197, 198, 200 and 201) with alanine does not affect transport activity. On the other hand, mutant VMAT2 in which serines 180, 181 and 182 (in TM III) were replaced with alanine, showed no transport activity. Moreover, binding of [³H] reserpine was at normal levels and it was accelerated by $\Delta\tilde{\mu}_{H^+}$. However, in contrast to wild type, and similar to D33N, binding was not inhibited by serotonin even at concentrations of 500 μM. The results suggest a possible role of Ser 180-182 in substrate recognition [96].

8.4. Mutations in TetA and BMR proteins

Interesting information can also be inferred from studies in homologous proteins performing similar or identical functions. TEXANs have evolved in many living organisms as transport proteins which play central roles in survival. Because of the overall similarities in mechanism (H^+/substrate antiporters), secondary structure and sharing of some of the substrates, the information obtained in studies with other TEXANs deserve some attention. Two proteins, TetA and Bmr, have been studied in detail and a large number of mutants were characterized [105–108].

Comparison to VMAT has suggested a number of residues that may be involved in catalysis. Replacement of N155 in VMAT2 with the Asp found in related bacterial protein does not affect transport activity. Also, replacement of the adjacent potential phosphorylation site (T154) with alanine does not impair function, indicating that phosphorylation at this site is not required for activity. Replacement of Gly 151 and 158 in the conserved motif GXXXXRXG in loop 3 has no effect on transport.

Random mutagenesis of BMR gave rise to four independent mutants exhibiting altered spectra of cross-resistance to various drugs (A. Neyfakh, personal communication). All these mutants were located within the region of transmembrane domains IX–XI. In addition, a reserpine-insensitive mutant located in the same region was identified [108]. Four more site-directed mutations in this region were engineered and all of them changed quantitatively the cross-resistance profile of bacteria expressing Bmr (A. Neyfakh, personal communication). These site-directed mutations were designed based on the following principle: the *S. aureus* homolog NorA, is very similar to Bmr in the cross-resistance profile it confers to *B. subtilis* [109], but different from their homologous Tet transporters. Amino acids conserved between Bmr and NorA were converted into the Tet-specific amino acids. For example, if a certain residue in Bmr and NorA is M, but in TetA, B and C the corresponding residue is G, M was converted in Bmr into G. Using the latter principle control mutations in some other regions of the transporter were made, one in the VIIth TM domain, another in the IVth TM domain. Both these control mutations gave changes in the resistance profile as large as the ones observed with the mutations in the IX–XI region indicating that other residues in different areas of the protein are needed for substrate specificity. It is not clear therefore why all the random mutations showing altered cross-resistance profile, were clustered in the IX–XI region (the probability of such a clustering by pure chance is just 1.2%).

An extensive mutagenesis study has been performed also in the *E. coli* TetA protein [105–107]. These studies highlight residues with putative roles in catalysis in TM domains VIII–IX, as well as in I–III. Other transport proteins, like the *lac* permease from *E. coli*, a 12 TM protein which catalyzes symport of H^+ and β-galactosides but does not belong to the TEXAN superfamily imply a similar conclusion. In addition to important residues in TM IX and X, it was found by Kaback and coworkers that very few of the residues in the other TM areas are essential [101]. In fact, TM II–VII can be completely deleted and the protein still catalyzes facilitated diffusion of lactose [110]. Studies by Wilson and collaborators on the melibiose symporter suggest that the above suggestions may be an oversimplification since more than one domain is involved in substrate recognition [111].

9. The MiniTEXANs, a family with unique properties

It was only during recent years that it became evident that there are more multidrug transporters than previously thought [11,21]. As described in the introduction, we

Table 2
Members of the MiniTEXAN family

Name	Organism	Other names	Length	Known function	Acc. no.
EmrE	*Escherichia coli*	Ebr, Mvrc	110	Multidrug resistance	M62732
Smr	*Staphylococcus aureus*	QacC, Ebr	107	Multidrug resistance	M33479
QacE	*Klebsiella aerogenes*	–	110	Multidrug resistance	X68232
SugE	*Escherichia coli*	–	105	Suppression of GroE mutation	X69949

believe that this diversity is the result of a continuous selection that favors those organisms which have developed one or more strategies to cope with as many toxic compounds as possible.

One of the most interesting families of multidrug transporters, the MiniTEX-ANs, has received quite a lot of attention during the past year. This family includes only a few proteins (Table 2). The genes coding for most of them were identified on the basis of the resistance conferred. One protein in the family (SugE) has no known transport activity.

A unique feature of the proteins in this family is their size: 100–110 amino acids proteins with four putative transmembrane segments catalyze the same activities which have previously been observed in proteins three to nine times larger. The fact that the small proteins of this family are transporters has been previously hypothesized before [12,112,113]. Recently, EmrE and Smr were purified and reconstituted in a functional form and it was clearly demonstrated that a single polypeptide catalyzes the exchange of a toxic cation with protons, in an electrogenic exchange which suggests a possible stoichiometry of $2H^+/$ cation [13,114]. The question has been raised then why are the other transporters so large when the minimal subunit necessary is at the most 110 amino acids. One possibility is that the Mini TEXANs function as a homo-oligomer to form at least the twelve transmembrane segments usually observed in other transporters. On the other hand, it is also possible that only some domains of the larger proteins are needed for transport and the rest functions for regulation of the activity and/or interaction with other proteins.

Another very unique feature of EmrE, is its solubility in chloroform:methanol which provides us with a powerful and unique tool for purification and characterization [13]. Since only a handful of minor proteins is soluble in these organic solvents, after overproduction of EmrE it was possible to extract it highly purified. Most importantly, EmrE was active after this treatment suggesting that the extraction did not induce major irreversible changes in its structure. The reconstituted protein has an apparent K_m for methyl viologen of 247 μM and a V_{max} of 1572 nmol/min/mg protein. The latter represents a turnover number of 14 min^{-1}, which is in the same order of magnitude of many ion-coupled transporters [115]. No quantitative data is available on the native protein but the specificity range of the purified protein is practically identical to the range of resistance reported. The solubility of EmrE in organic solvents does not reside in a specially high proportion

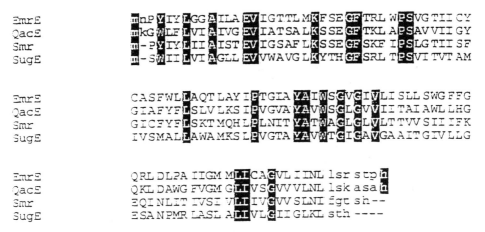

Fig. 5.   Alignment of MiniTEXANs. Multiple alignment was first performed using Pileup (Wisconsin Package, 1994) and then blocks of similarity were identified using MACAW [118].

of hydrophobic amino acids: the percentage in proteins is 49%, which is not different from other classical ion-coupled transporters, which is between 41 to 49.6% in seven randomly chosen H^+-coupled bacterial transporters analyzed based in the sequences available from the databank [13]. On the other hand the percent of charged amino acids is 7.2 while in the twelve transmembrane helix transporters scanned as above it is between 11 and 14.8%. Another important difference is the number of net charges which is only +1 in EmrE. The range in the other transporters is between –7 and +7 with the lowest absolute number of net charges in the group analyzed being 4. It is therefore suggested that the existence of unpaired charges on the protein represent the main energy barrier for solubilization in the highly hydrophobic milieu of the organic solvents. This has been extensively discussed in the case of charge insertion in membranes [116] and has also received some experimental support in experiments in which neutralization of charges of membrane proteins with H^+ or Ca^{++} rendered them soluble in organic solvents [117].

The range of substrates recognized by EmrE has been summarized in Fig. 2. As already discussed there is a large overlap with the specificities displayed by other multidrug transporters.

An alignment of the four proteins is shown in Fig. 5. TMS I and TMS II show the highest conservation. Several of the conserved residues have been mutagenized and activity is lost or modified [113,114].

Although there is no sequence similarity with other multidrug transporters, the MiniTEXANs share many of the properties of their larger analogs: they confer resistance to a variety of toxicants thanks to their ability to actively remove them away from their target. Because of their size and unique properties, the MiniTEX-ANs may provide a very useful model to understand the molecular basis of recognition of multiple substrates with high affinity as displayed by the other multidrug transporters. It also should be a good model to understand structure–function aspects of transport reactions in ion-coupled processes in general.

Abbreviations

DCC, , N,N'-dicyclohexylcarbodiimide
DEPC, diethylpyrocarbonate
LDCV, Large dense core granules
MPP$^+$, N-methyl-4-phenylpyridinium
SLMV, Synaptic like microvesicles
TBZ, , Tetrabenazine
TBZOH, Dihydrotetrabenazine
VAChT, Vesicular acetylcholine transporter
VMAT, Vesicular monoamine transporter
VNT, , Vesicular neurotransmitter transporter
$\Delta\tilde{\mu}_{H^+}$, , Proton electrochemical gradient

Acknowledgements

Work in the authors laboratory was supported by grants from the National Institute of Health (NH16708), the United States-Israel Binational Science Foundation (93-00051), the National Institute of Psychobiology in Israel and the Israeli Academy of Sciences. We thank R. Edwards and J. Erickson for sharing with us unpublished observations.

References

1. Gros, P., Croop, J. and Housman, D. (1986) Cell **47**, 371–380.
2. Higgins, C. (1992) Ann. Rev. Cell Biol. **8**, 67–113.
3. Endicott, J. and Ling, V. (1989) Annu. Rev. Biochem. **58**, 137–171.
4. Gottesman, M.M. and Pastan, I. (1993) Annu. Rev. Biochem. **62**, 385–427.
5. Doige, C.A. and Ames, G.F.L. (1993) Annu. Rev. Microbiol. **47**, 291–319.
6. Neyfakh, A., Bidnenko, V. and Chen, L. (1991) Proc. Natl. Acad. Sci. USA **88**, 4781–4785.
7. Paulsen, I. and Skurray, R. (1993) Gene **124**, 1–11.
8. Lewis, K. (1994) Trends Biochem. Sci. **19**, 119–123.
9. Nikaido, H. (1994) Science **264**, 382–388.
10. Schuldiner, S., Shirvan, A. and Linial, M. (1995) Physiol. Rev. **75**, 369–392.
11. Schuldiner, S. (1994) J. Neurochem. **62**, 2067–2078.
12. Grinius, L., Dreguniene, G., Goldberg, E.B., Liao, C.H. and Projan, S.J. (1992) Plasmid **27**, 119–129.
13. Yerushalmi, H., Lebendiker, M. and Schuldiner, S. (1995) J. Biol. Chem. **270**, 6856–6863.
14. Saier, M., Tam, R., Reizer, A. and Reizer, J. (1994) Mol. Microbiol. **11**, 841–847.
15. Grundemann, D., Gorboulev, V., Gambaryan, S., Veyhl, M. and Koepsell, H. (1994) Nature **372**, 549–552.
16. Schinkel, A.H., Smit, J.J.M., van Tellingen, O., Beijnen, J.H., Wagenaar, E., Deemter, V.L., Mol, C., Valk, V.D.M., Robanus-Maandag, E., Riele, T.H., Berns, A. and Borst, P. (1995) Cell **77**, 491–502.
17. Ahmed, M., Borsch, C.M., Taylor, S.S., Vazquezlaslop, N. and Neyfakh, A.A. (1994) J. Biol. Chem. **269**, 28506–28513.
18. Lomovskaya, O., Lewis, K. and Matin, A. (1995) J. Bacteriol. **177**, 2328–2334.
19. Ruetz, S. and Gros, P. (1995) Cell **77**, 1071–1081.
20. Griffith, J., Baker, M., Rouch, D., Page, M., Skurray, R., Paulsen, I., Chater, K., Baldwin, S. and Henderson, P. (1992) Curr. Opin. Cell Biol. **4**, 684–695.
21. Liu, Y., Peter, D., Roghani, A., Schuldiner, S., Prive, G., Eisenberg, D., Brecha, N. and Edwards, R. (1992) Cell **70**, 539–551.

22. Wink, M. (1993) J. Exp. Bot. **44**, 231–246.
23. Flowers, T.J., Troke, P.F. and Yeo, A.R. (1977) Ann. Rev. Plant. Physiol. **28**, 89–121.
24. Ohya, Y., Umemoto, N., Tanida, I., Ohta, A., Iida, H. and Anraku, Y. (1991) J. Biol. Chem. **266**, 13971–13977.
25. Duyao, M., Taylor, S., Buckler, A., Ambrose, C., Lin, C., Groot, N., Church, D., Barnes, G., Wasmuth, J., Housman, D., MacDonald, M. and Gusella, J. (1993) Human Mol. Genet. **2**, 673–676.
26. Nelson, N. (1992) J. Bioenerg. Biomembr. **24**, 407–414.
27. Gogarten, J.P., Kibak, H., Dittrich, P., Taiz, L., Bowman, E.J., Bowman, B.J., Manolson, M.F., Poole, R.J., Date, T., Oshima, T., Konishi, J., Denda, K. and Yoshida, M. (1989) Proc. Natl. Acad. Sci. USA **86**, 6661–6665.
28. Gluck, S. (1992) J. Bioenerg. Biomembr. **24**, 351–359.
29. Guastella, J., Nelson, N., Nelson, H., Czyzyk, L., Keynan, S., Miedel, M., C, Davidson, N., Lester, A., H and Kanner, B.I. (1990) Science **249**, 1303–1306.
30. Amara, S.G. and Kuhar, M.J. (1993) Annu. Rev. Neurosci. **16**, 73–93.
31. Rudnick, G. and Clark, J. (1993) BBA Rev. Bioenerget. **1144**, 249–263.
32. Storck, T., Schulte, S., Hofmann, K. and Stoffel, W. (1992) Proc. Natl. Acad. Sci. USA **89**, 10955–10959.
33. Pines, G., Danbolt, N., Bjoras, M., Zhang, Y., Bendahan, A., Eide, L., Koepsell, H., Stormmathisen, J., Seeberg, E. and Kanner, B. (1992) Nature **360**, 464–467.
34. Kanner, B. (1993) FEBS Lett. **325**, 95–99.
35. Kanai, Y. and Hediger, M. (1992) Nature **360**, 467–471.
36. Kirshner, N. (1962) J. Biol. Chem. **237**, 2311–2317.
37. Pletscher, A. (1977) Br. J. Pharmacol. **59**, 419–424.
38. Marshall, I. (1970) Br. J. Pharmacol. **38**, 503–516.
39. Marshall, I. and Parsons, S. (1987) Trends Neurosci. **10**, 174–177.
40. Parsons, S.M., Bahr, B.A., Rogers, G.A., Clarkson, E.D., Noremberg, K. and Hicks, B.W. (1993) Acetylcholine Transporter Vesamicol Receptor Pharmacology and Structure, Elsevier, Amsterdam.
41. Scherman, D. and Henry, J.-P. (1984) Molec. Pharm. **25**, 113–122.
42. Darchen, F., Scherman, D. and Henry, J.-P. (1989) Biochemistry **28**, 1692–1697.
43. Rudnick, G., Steiner-Mordoch, S.S., Fishkes, H., Stern-Bach, Y. and Schuldiner, S. (1990) Biochemistry **29**, 603–608.
44. Stitzel, R.E. (1977) Pharm. Rev. **28**, 179–205.
45. Carlsson, A. (1965) Hand. Exp. Pharmacol. **19**, 529–592.
46. Frize, E. (1954) N. Engl. J. Med. **251**, 1006–1008.
47. Weaver, J.A. and Deupree, J.D. (1982) Eur. J. Pharm. **80**, 437–38.
48. Stern-Bach, Y., Greenberg-Ofrath, N., Flechner, I. and Schuldiner, S. (1990) J. Biol. Chem. **265**, 3961–3966.
49. Scherman, D. and Henry, J. (1981) Eur. J. Biochem. **116**, 535–539.
50. Darchen, F., Scherman, D., Desnos, C. and Henry, J.-P. (1988) Biochem. Pharmacol. **37**, 4381–4387.
51. Henry, J.-P. and Scherman, D. (1989) Biochem. Pharmacol. **38**, 2395–2404.
52. Stietzen, M., Schober, M., Fischer-Colbrie, R., Scherman, D., Sperk, G. and Winkler, H. (1987) Neuroscience **22**, 131–139.
53. Desnos, C., Raynaud, B., Vidal, S., Weber, M. and Scherman, D. (1990) Develop. Brain Res. **52**, 161–166.
54. Desnos, C., Laran, M. and Scherman, D. (1992) J. Neurochem. **59**, 2105–2112.
55. Scherman, D. (1986) J. Neurochem. **47**, 331–339.
56. Scherman, D., Jaudon, P. and Henry, J. (1983) Proc. Natl. Acad. Sci. USA **80**, 584–588.
57. Rogers, G. and Parsons, S. (1989) Mol. Pharmacol. **36**, 333–341.
58. Rogers, G., Parsons, S., Anderson, D., Nilsson, L., Bahr, B., Kornreich, W., Kaufman, R., Jacobs, R. and Kirtman, B. (1989) J. Med. Chem. **32**, 1217–1230.
59. Kaufman, R., Rogers, G.A., Fehlmann, C. and Parsons, S.M. (1989) Mol. Pharmacol. **36**, 452–456.
60. Bahr, B. and Parsons, S. (1986a) J. Neurochem. **46**, 1214–1218.
61. Bahr, B. and Parsons, S. (1986b) Proc. Natl. Acad. Sci. USA **83**, 2267–2270.
62. Bahr, B., Noremberg, K., Rogers, G., Hicks, B. and Parsons, S. (1992) Biochemistry **31**, 5778–5784.

63. Bahr, B., Clarkson, E., Rogers, G., Noremberg, K. and Parsons, S. (1992) Biochemistry **31**, 5752–5762.
64. Clarkson, E., Rogers, G. and Parsons, S. (1992) J. Neurochem. **59**, 695–700.
65. Bahr, B. and Parsons, S. (1992) Biochemistry **31**, 5763–5769.
66. Rogers, G. and Parsons, S. (1992) Biochemistry **31**, 5770–5777.
67. Rogers, G.A. and Parsons, S.M. (1993) Biochemistry **32**, 8596–8601.
68. Erickson, J.D., Varoqui, H., Schafer, M.K.H., Modi, W., Diebler, M.F., Weihe, E., Rand, J., Eiden, L., Bonner, T.I. and Usdin, T.B. (1994) J. Biol. Chem. **269**, 21929–21932.
69. Roghani, A., Feldman, J., Kohan, S.A., Shirzadi, A., Gundersen, C.B., Brecha, N. and Edwards, R.H. (1994) Proc. Natl. Acad. Sci. USA **91**, 10620–10624.
70. Varoqui, H., Diebler, M., Meunier, F., Rand, J., Usdin, T., Bonner, T., Eiden, L. and Erickson, J. (1994) FEBS Lett. **342**, 97–102.
71. Sternbach, Y., Keen, J., Bejerano, M., Steiner-Mordoch, S., Wallach, M., Findlay, J. and Schuldiner, S. (1992) Proc. Natl. Acad. Sci. USA **89**, 9730–9733.
72. Krejci, E., Gasnier, B., Botton, D., Isambert, M.F., Sagne, C., Gagnon, J., Massoulie, J. and Henry, J.P. (1993) FEBS Lett. **335**, 27–32.
73. Schuldiner, S., Liu, Y. and Edwards, R. (1993) J. Biol. Chem. **268**, 29–34.
74. Erickson, J., Eiden, L. and Hoffman, B. (1992) Proc. Natl. Acad. Sci. USA **89**, 10993–10997.
75. Vincent, M. and Near, J. (1991) Molec. Pharmacol. **40**, 889–894.
76. Isambert, M., Gasnier, B., Botton, D. and Henry, J. (1992) Biochemistry **31**, 1980–1986.
77. Howell, M., Shirvan, A., Stern-Bach, Y., Steiner-Mordoch, S., Strasser, J.E., Dean, G.E. and Schuldiner, S. (1994) FEBS Lett. **338**, 16–22.
78. Usdin, T., Eiden, L., Bonner, T. and Erickson, J. (1995) TINS **18**, 218–224.
79. Erickson, J. and Eiden, L. (1993) J. Neurochem. **61**, 2314–2317.
80. Vandenbergh, D., Persico, A. and Uhl, G. (1992) Molec. Brain Res. **15**, 161–166.
81. Peter, D., Jimenez, J., Liu, Y.J., Kim, J. and Edwards, R.H. (1994) J. Biol. Chem. **269**, 7231–7237.
82. Johnson, R. (1988) Physiol. Rev. **68**, 232–307.
83. Scherman, D. and Boschi, G. (1988) Neuroscience **27**, 1029–1035.
84. Weihe, E., Schafer, M.-H., Erickson, J. and Eiden, L. (1995) J. Molec. Neurosci. **5**, 149–164.
85. Peter, D., Liu, Y., Sternini, C., de Giorgio, R., Brecha, N. and Edwards, R. (1995) J. Neurosci. **15**, 6179–6188.
86. Liu, Y., Schweitzer, E., Nirenberg, M., Pickel, V., Evans, C. and Edwards, R. (1994) J. Cell Biol. **127**, 1419–1433.
87. Nirenberg, M., Liu, Y., Peter, D., Edwards, R. and Pickel, V. (1995) Proc. Natl. Acad. Sci. USA **92**, 8773–8777.
88. Mahata, S.K., Mahata, M., Fischercolbrie, R. and Winkler, H. (1993) Molec. Brain Res. **19**, 83–92.
89. Mahata, S.K., Mahata, M., Fischercolbrie, R. and Winkler, H. (1993) Neurosci. Lett. **156**, 70–72.
90. Alfonso, A., Grundahl, K., Duerr, J.S., Han, H.P. and Rand, J.B. (1993) Science **261**, 617–619.
91. Brenner, S. (1974) Genetics **77**, 71–94.
92. Rand, J. (1989) Genetics **122**, 73–80.
93. Schuldiner, S., Fishkes, H. and Kanner, B.I. (1978) Proc. Natl. Acad. Sci. USA **75**, 3713–3716.
94. Suchi, R., Sternbach, Y., Gabay, T. and Schuldiner, S. (1991) Biochemistry **30**, 6490–6494.
95. Gasnier, B., Scherman, D. and Henry, J. (1985) Biochemistry **24**, 1239–1244.
96. Merickel, A., Rosandich, P., Peter, D. and Edwards, R. (1995) J. Biol. Chem. **270**, 25798–25804.
96a. Steiner-Mordoch, S., Shirvan, A. and Schuldiner, S. (1996). J. Biol. Chem., in press.
97. Puettner, I.B., Sarkar, H.K., Padan, E., Lolkema, J.S. and Kaback, H.R. (1989) Biochemistry **28**, 2525.
98. Gerchman, Y., Olami, Y., Rimon, A., Taglicht, D., Schuldiner, S. and Padan, E. (1993) Proc. Natl. Acad. Sci. USA **90**, 1212–1216.
99. Suchi, R., Stern-Bach, Y. and Schuldiner, S. (1992) Biochemistry **31**, 12500–12503.
100. Isambert, M. and Henry, J. (1981) FEBS Lett. **136**, 13–18.
101. Kaback, H. (1992) Biochim. Biophys. Acta **1101**, 210–213.
102. Padan, E., Sarkar, H., K, Vitanen, P.V., Poonian, M.S. and Kaback, H.R. (1985) Proc. Natl. Acad. Sci. USA **82**, 6765–8.
103. Shirvan, A., Laskar, O., Steiner-Mordoch, S. and Schuldiner, S. (1994) FEBS Lett. **356**, 145–150.

104. Kitayama, S., Shimada, S., Xu, H., Markham, L., Donovan, D. and Uhl, G. (1992) Proc. Natl. Acad. Sci. USA **89**, 7782–7785.
105. Yamaguchi, A., Akasaka, T., Kimura, T., Sakai, T., Adachi, Y. and Sawai, T. (1993) Biochemistry **32**, 5698–5704.
106. Yamaguchi, A., Oyauchi, R., Someya, Y., Akasaka, T. and Sawai, T. (1993) J. Biol. Chem. **268**, 26990–26995.
107. Yamaguchi, A., Kimura, T., Someya, Y. and Sawai, T. (1993) J. Biol. Chem. **268**, 6496–6504.
108. Ahmed, M., Borsch, C., Neyfakh, A. and Schuldiner, S. (1993) J. Biol. Chem. **268**, 11086–11089.
109. Neyfakh, A. (1992) Antimicrob. Agents Chemother. **36**, 484–485.
110. Bibi, E., Verner, G., Chang, C. and Kaback, H. (1991) Proc. Natl. Acad. Sci. USA **88**, 7271–7275.
111. Hama, H. and Wilson, T. (1993) J. Biol. Chem. **268**, 10060–10065.
112. Purewal, A.S., Jones, I.G. and Midgley, M. (1990) FEMS Microbiol. Lett. **68**, 73–76.
113. Paulsen, I., Brown, M., Dunstan, S. and Skurray, R. (1995) J. Bacteriol. **177**, 2827–2833.
114. Grinius, L. and Goldberg, E. (1994) J. Biol. Chem. **269**, 29998–30004.
115. Taglicht, D., Padan, E. and Schuldiner, S. (1991) J. Biol. Chem. **266**, 11289–11294.
116. Engelman, D., Steits, T. and Goldman, A. (1986) Annu. Rev. Biophys. Chem. **15**, 321–353.
117. Gitler, C. (1977) Biochem. Biophys. Res. Comm. **74**, 178–182.
118. Schuler, G., Altschul, S. and Lipman, D. (1991) Proteins **9**, 180–190.

CHAPTER 19

Structure and Function of Sodium-Coupled Amino Acid Neurotransmitter Transporters

B.I. KANNER

Department of Biochemistry, Hadassah Medical School,
The Hebrew University, Jerusalem, Israel

© 1996 Elsevier Science B.V.
All rights reserved

Handbook of Biological Physics
Volume 2, edited by W.N. Konings, H.R. Kaback and J.S. Lolkema

Contents

1. Introduction

Sodium-coupled neurotransmitter transporters, located in the plasma membrane of nerve terminals and glial processes, serve to keep the extracellular transmitter levels below those which are neurotoxic. Moreover, they help, in conjunction with diffusion, to terminate its action in synaptic transmission. Such a termination mechanism operates with most transmitters, including γ-aminobutyric acid (GABA), L-glutamate, glycine, dopamine, serotonin and norepinephrine. Another termination mechanism is observed with cholinergic transmission. After dissociation from its receptor, acetylcholine is hydrolysed into choline and acetate. The choline moiety is then recovered by sodium-dependent transport as described above. As the concentration of the transmitters in the nerve terminals are much higher than in the cleft — typically by four orders of magnitude — energy input is required. The transporters that are located in the plasma membranes of nerve endings and glial cells obtain this energy by coupling the flow of neurotransmitters to that of sodium. The $(Na^+ + K^+)$-ATPase generates an inwardly-directed electro-chemical sodium gradient which is utilized by the transporters to drive 'uphill' transport of the neurotransmitters (reviewed in Ref. [1–3]).

Neurotransmitter uptake systems have been investigated in detail by using plasma membranes obtained upon osmotic shock of synaptosomes. It appears that these transporters are coupled not only to sodium, but also to additional ions like potassium or chloride (Table 1).

These transporters are of considerable medical interest. Since they function to regulate neurotransmitter activity by removing it from the synaptic cleft, specific transporter inhibitors can be potentially used as novel drugs for neurological disease. For instance, attenuation of GABA removal will prolong the effect of this inhibitory transporter, thereby potentiating its action. Thus, inhibitors of GABA transport could represent a novel class of anti-epileptic drugs. Well-known inhibitors which interfere with the functioning of biogenic amine transporters include anti-depressant drugs and stimulants such as amphetamines and cocaine. The neurotransmitter glutamate — at excessive local concentrations — causes cell death, by activating N-methyl-D-aspartic acid (NMDA) receptors and subsequent calcium entry. The transmitter has been implicated in neuronal destruction during ischaemia, epilepsy, stroke, amyotropic lateral sclerosis and Huntington's disease. Neuronal and glial glutamate transporters may have a critical role in preventing glutamate from acting as an exitotoxin [4,5].

In the last few years major advances in the cloning of these neurotransmitter transporters have been made. After the GABA transporter was purified [6], the ensuing protein sequence information was used to clone it [7]. Subsequently the

Table 1
Comparison of GABA and glutamate transporters

	GABA Tp	Glutamate Tp
Cosubstrates	Na^+, Cl^-	Na^+, K^+ (OH^- or H^+)
Electrogenicity	+	+
Localization	neuronal, glial	neuronal, glial
'Sociology'	belong to large family of transporters for all neurotransmitters excluding glutamate	belong to separate small family of glutamate transporters
Relationship to bacterial transporters	–	glt-P, glutamate transporter dct-A, dicarboxylic acid transporter
Predicted topology	12 TMs amino and carboxyl termini are cytoplasmic	6–10 TMs amino and carboxyl termini are cytoplasmatic
Glycosylation	+	+
Possible regulation	protein kinase C	protein kinase C arachidonic acid
Pore domain	not yet identified	confined to a short conserved stretch in the carboxyl terminal half of the transporter

expression cloning of a norepinephrine transporter [8] provided evidence that these two proteins are the first members of a novel superfamily of neurotransmitter transporters. This result led — using polymerase chain reaction (PCR) and other technologies relying on sequence conservation — to the isolation of a growing list of neurotransmitter transporters (reviewed in Refs. [9–11]). This list includes various subtypes of GABA transporters as well as those for all the above-mentioned neurotransmitters, except glutamate. All of the members of this superfamily are dependent on sodium and chloride and by analogy with the GABA transporter [12] are likely to cotransport their transmitter with both sodium and chloride. Interestingly sodium-dependent glutamate transport is not chloride dependent, but rather sodium and glutamate are countertransported with potassium (Table 1, [13,14]). A few years ago three distinct but highly related glutamate transporters have been cloned [15–17]. Very recently a fourth subtype was cloned, exhibiting a large chloride conductance [18]. These transporters represent a distinct family of transporters.

Here we describe the current status on two prototypes of these distinct families; the GABA and glutamate transporters.

2. Mechanism

The GABA transporter cotransports the neurotransmitter with sodium and chloride in an electrogenic fashion ([1,12], Table 1, Fig. 1, top). The available measurements include tracer fluxes [12] and electrophysiological approaches [19,20].

The latter approach reveals that at very negative (inside) potentials, the chloride dependency is not absolute [20]. At the present time it is not clear what the mechanistic interpretation of this result is. One possibility is that under these conditions, another anion — such as hydroxyl — may take over the role of chloride. In addition, a transient current is observed in the absence of GABA. This transient can be blocked by bulky GABA analogues, which can bind to the transporter, but are not translocated by it [20]. It probably reflects a conformational change of the transporter occurring after the sodium has bound. These measurements also permit determination of the turnover number of the transporter. These estimates of a few cycles per second [20] are in agreement with biochemical ones [6]. It is of interest to note that although both GABA and 5-HT transporters belong to the same transporter superfamily (Table 1), the latter one appears to exhibit quite distinct properties [21]. It appears that the 5-HT transporter is electroneutral, but that a transporter-associated current can be detected. This is probably related to the observation that under some conditions this transporter may act as a channel. This transporter-channel appears to be less sodium selective than the transporter mode which is carrying 5-HT in a coupled fashion [21].

The mechanism of sodium-dependent L-glutamate transport has been studied initially using tracer flux studies employing radioactive glutamate. These studies indicated that the process is electrogenic, with positive charge moving in the direction of the glutamate [13]. This observation suggested that it is possible to monitor L-glutamate transport electrically using the whole-cell patch-clamp tech-

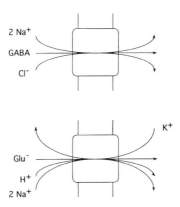

Fig. 1. Schematic representation of GABA and L-glutamate transport. The diagrams illustrate the coupled fluxes of GABA (top) and L-glutamate (bottom) with their cosubstrates. As indicated in the text it is equally possible that an hydroxyl ion is countertransported, rather than the cotransport of a proton (or a hydronium ion).

nique [22]. In addition to L-glutamate, D- and L-aspartate are transportable substrates with affinities in the lower micromolar range. The system is stereospecific with regard to glutamate, the D-isomer being a poor substrate. Glutamate uptake is driven by an inwardly directed sodium ion gradient and at the same time potassium is moving outwards. The potassium movement is not a passive movement in response to the charge carried by the transporter. Rather, it is an integral part of the translocation cycle catalyzed by the transporter. Its role is further described below. Recently, evidence has been presented that another ionic species is countertransported (in addition to potassium), namely hydroxyl ions [23].

The first-order dependence of the carrier current on internal potassium [24], together with the well-known first-order dependence on external L-glutamate and sigmoid dependence on external sodium, suggest a stoichiometry of $3Na^+:1K^+:1$glutamate [13,24]. This stoichiometry implies that one positive charge moves inwards per glutamate anion entering the cell. If a hydroxyl anion is countertransported as well [23], the stoichiometry could be $2Na^+:1K^+:1$glutamate$:1OH^-$, and transport would still be electrogenic. However, the alternative that a proton moves in together with the sodium and the glutamate is possible with an equal probability (Fig. 1, bottom). A stoichiometry of $2Na^+:1$glutamate is also favored by direct experimental evidence obtained by kinetic [25] and thermodynamic methods [26].

The study of the ion dependence of partial reactions of the glutamate transporter has revealed that glutamate transport is an ordered process. First sodium and glutamate are translocated. After their release inside the cell, potassium binds and is translocated outward so that a new cycle can be initiated [14,27]. Using electrophysiological methods it has been possible to monitor binding of sodium to the glutamate transporter [28]. The turnover number of the transporter is in the same order of that of the GABA transporter, namely, a few per second [28,29].

3. Reconstitution, purification and localization

Using methodology which enables one to reconstitute many samples simultaneously and rapidly, one of the subtypes of the GABA [6] as well as of the L-glutamate transporter [29] have been purified to an apparent homogeneity. Both are glycoproteins and both have an apparent molecular mass of 70–80 kDa. The two transporters retain all the properties observed in membrane vesicles. They are distinct not only because of their different functional properties. Also antibodies generated against the GABA transporter [6] react (as detected by immunoblotting) only with fractions containing GABA transport activity and not with those containing L-glutamate transport activity [29]. The opposite is true for antibodies generated against the glutamate transporter [30]. Recently, the glycine transporter has been purified and reconstituted as well. Interestingly, it appears to be a larger protein than the GABA and glutamate transporters — about 100 kDa in size [31]. The serotonin transporter has also been purified, but these preparations, containing a band around 70 kDa, have been shown to be active only in the binding of $[^3H]$-imipramine but not in serotonin transport [32,33]. Immunocytochemical localization studies of the GABA transporter reveal that in most brain areas it is located in the membranes of nerve terminals [34], although in some areas, such as substantia nigra, glial processes were labelled (Table 1).

Using the antibodies raised against the glutamate transporter, the immunocyto-chemical localization of the transporter was studied at the light and electron microscopic level in rat central nervous system. In all regions examined (including cerebral cortex, caudato-putamen, corpus callosum, hippocampus, cerebellum, spinal cord) it was found to be located in glial cells rather than in neurons. In particular, fine astrocytic processes were strongly stained. Putative glutamatergic axon terminals appeared nonimmunoreactive [30]. The uptake of glutamate by such terminals (for which there is strong previous evidence) therefore may be due to a subtype of glutamate transporter different from the glial transporter. Using a monoclonal antibody raised against this transporter, a similar glial localization of the transporter was found [35]. Another glial transporter, as well as a neuronal one, have been identified (see below).

4. The large superfamily of Na-dependent neurotransmitter transporters

Partial sequencing of the purified $GABA_A$ transporter allowed the cloning of the first member of the new family of Na-dependent neurotransmitter transporters [7]. After expression and cloning of the noradrenaline transporter [8], it became clear that it had significant homology with the $GABA_A$ transporter. The use of functional c-DNA expression assays and amplification of related sequences by polymerase chain reaction (PCR) resulted in the cloning of additional transporters which belong to this family, such as the dopamine [36–38] and serotonin [39,40] transporters, additional GABA transporters [41–44]), transporters of glycine [45–47], proline [48], taurine [49,50], betaine [51] and 'orphan' transporters, whose substrates are still unknown. In addition, another family member which was originally thought to be a choline transporter [52], probably is in fact a creatine transporter [53]. A novel glycine transporter cDNA encoding for a 799 amino acid protein has recently been isolated [54]. This is significantly longer than most members of the superfamily. If we take into account that part of the mass of these transporters is constituted by sugar, it could encode for the 100 kDa glycine transporter which was purified and reconstituted [43].

The deduced amino acid sequences of these proteins reveal 30–65% identity between different members of the family. Based on these differences in homology the family can be divided into four subgroups: (a) transporters of biogenic amines (noradrenaline, dopamine and serotonin); (b) various GABA transporters as well as transporters of taurine and creatine; (c) transporters of proline and glycine; and (d) 'orphan' transporters. These proteins share some features of a common secondary structure. Each transporter is composed of 12 hydrophobic putative transmembrane α-helices. The lack of a signal peptide suggests that both amino- and carboxy-terminii face the cytoplasm. These regions contain putative phosphorylation sites, that may be involved in regulation of the transport process (see below). The second extracellular loop between helices 3 and 4 is the largest, and it contains putative glycosylation sites.

Alignment of the deduced amino acid sequences of 13 different members of this superfamily, whose substrates are known (subgroups a–c) revealed that some segments within these proteins share a higher degree of homology than others. The most highly conserved regions (>50% homology) are helix 1 together with the extracellular loop connecting it with helix 2, and helix 5 together with a short intracellular loop connecting it with helix 4 and a larger extracellular loop connecting it with helix 6. These domains may be involved in stabilizing a tertiary structure, that is essential for the function of all these transporters. Alternatively, they may be related to a common function of these transporters, such as the translocation of sodium ions. The region stretching from helix 9 on is far less conserved than the segment containing the first 8 helices. Possibly, this domain contains some residues that are involved in translocating the different substrates. The least conserved segments are the amino- and carboxy-terminii. As was mentioned above, these areas may be involved in regulation of the transport process. The 'orphan' transporters differ from all other members of the family in three regions. They contain much larger extracellular loops between helices 7–8 and helices 11–12.

5. Molecular cloning and predicted structure of glutamate transporters

Transporters for many neurotransmitters were cloned on the assumption that they are related to the GABA [7] and norepinephrine [8] transporters [9–11]. This approach was unsuccessful for the glutamate transporter. Three different glutamate transporters were cloned using different approaches: GLAST [15], GLT-1 [16] and EAAC 1 [17]. The former two appear to be of glial [15,16,30,55], the latter of neuronal origin [17,56]. It is not yet known whether the newly cloned EAAT-4 [18] is neuronal or glial. Indeed, the three transporters are not related to the above superfamily [15–17]. On the other hand, they are very similar to each other, displaying ~50% identity and ~60% similarity. They also appear to be related to the proton-coupled glutamate transporter from *E. coli* and other bacteria (**glt-P** [57]) and the dicarboxylate transporter (**dct-A** [58]) of *Rhizobium meliloti*. In these cases the identities are around 25–30%. Thus, they form a distinct family. They contain between 500–600 amino acids. It has been shown that this family also encodes sodium-dependent transporters which do not use dicarboxylic acids as substrates, but rather neutral amino acids [59,60]. Recently the three human homologues of the rat brain glutamate transporters have been cloned [61], as well as a novel subtype which is characterized by a large substrate-induced chloride current [18]. A similar but smaller current, which is not thermodynamically coupled to glutamate transport, has been observed in several of the other subtypes as well [62].

GLT-1, which encodes the glutamate transporter which was purified [16,29,30], has 573 amino acids and a relative molecular mass of 64 kDa, in good agreement with the value of 65 kDa of the purified and deglycosylated transporter [30]. Hydrophaty plots are relatively straightforward at the amino terminal side of the protein and the three different groups have predicted 6 transmembrane α-helices at very similar positions [15–17]. On the other hand, there is much more ambiguity at the carboxyl side where zero [15], two [16] or four [17] α-helices have been predicted. However, all three groups note uncertainty in assigning transmembrane α-helices in this part of the protein, taking into account alternative possibilities including membrane spanning β-sheets [15]. It is clear that experimental approaches to delineate their topology are badly needed.

6. Regulation of neurotransmitter transport

It is conceivable that the reuptake process is subject to physiological regulation. However, our knowledge of this aspect of neurotransmitter function is rudimentary. It has been shown that arachidonic acid, which may be released via phospholipase A_2, can inhibit several sodium coupled uptake systems [63] including the uptake systems for glycine [64] and glutamate [65]. Glutamate transport in rat brain membrane preparations is inhibited by arachidonate and this compound also inhibits the purified and reconstituted glutamate transporter GLT-1 [66]. However, the situation is more complex as transport mediated by GLT-1 expressed in oocytes is stimulated by arachidonate, while the converse is true for another glutamate transporter [67]. Nieoullon et al. [68] found that *in vivo* electrical stimulation (for 10 min) of frontal cortex increased high affinity glutamate uptake in rat striatum. The increase was due to an increase in affinity. The uptake measurements were done in tissue samples dissected out 20 min after the cessation of stimulation. This increase from basal level could be inhibited by dopaminergic activity [69]. The existence of putative phosphorylation sites [16] indicates that this glial glutamate transporter may be regulated by protein kinase and phosphatases. The finding that glutamate transport activity (V_{max}, but not K_m) is increased in cultured glial cells after incubation of the cells with phorbol esters [70], suggests that the putative phosphorylation sites are physiologically relevant. We have provided evidence that this stimulation of glutamate transport by phorbol esters is due to a direct phosphorylation of the transporter by protein kinase C. A single serine residue (serine 113) located in the loop connecting putative transmembrane helices 2 and 3, appears to be the major site of this phosphorylation [71].

In the case of the GABA transporter, modulation of GABA transport activity by phorbol esters has been reported. However, some unclarity as to this phenomenon is that while one group reported stimulation by the phorbol ester [72], another group found preincubation of oocytes with this compound to inhibit [73].

7. Structure–function relationships in the superfamily of neurotransmitter transporters

It has been shown previously that parts of amino- and carboxyl-terminii of the $GABA_A$ transporter are not required for function [74]. In order to define these domains, a series of deletion mutants was studied in the GABA transporter [75]. Transporters truncated at either end until just a few amino-acids distance from the beginning of helix 1 and the end of helix 12, retained their ability to catalyze sodium and chloride-dependent GABA transport. These deleted segments did not contain any residues conserved among the different members of the superfamily. Once the truncated segment included part of these conserved residues, the transporter's activity was severely reduced. However, the functional damage was not due to impaired turnover or impaired targeting of the truncated proteins [75].

Fragments of the $(Na^+ + Cl^-)$-coupled $GABA_A$ transporter were produced by proteolysis of membrane vesicles and reconstituted preparations from rat brain [76]. The former were digested with pronase, the latter with trypsin. Fragments with different apparent molecular masses were recognized by sequence directed antibodies raised against this transporter. When GABA was present in the digestion medium the generation of these fragments was almost entirely blocked [76]. At the same time, the neurotransmitter largely prevented the loss of activity caused by the protease. The effect was specific for GABA; protection was not afforded by other neurotransmitters. It was only observed when the two cosubstrates, sodium and chloride, were present on the same side of the membrane as GABA [76]. The results indicate that the transporter may exist in two conformations. In the absence of one or more of the substrates, multiple sites located throughout the transporter are accessible to the proteases. In the presence of all three substrates — conditions favoring the formation of the translocation complex — the conformation is changed such that these sites become inaccessible to protease action.

We have investigated the role of the hydrophilic loops connecting the putative transmembrane α-helices connecting GAT-1. Deletions of randomly picked nonconserved single amino acids in the loops connecting helices 7 and 8 or 8 and 9 result in inactive transport upon expression in Hela cells. However, transporters where these amino acids are replaced with glycine retain significant activity. The expression levels of the inactive mutant transporters was similar to that of the wild-type, but one of these, ΔVal-348, appears to be defectively targeted to the plasma membrane. Our data are compatible with the idea that a minimal length of the loops is required, presumably to enable the transmembrane domains to interact optimally with each other [77].

The substrate translocation performed by the various members of the superfamily is sodium- and usually chloride-dependent. In addition, some of the substrates contain charged groups as well. Therefore, charged amino acids in the membrane domain of the transporters may be essential for their normal function. This was tested using the GABA transporter [78]. Out of 5 charged amino acids within its membrane domain only one, arginine-69 in helix 1, is absolutely essential for activity. It is not merely the positive charge which is important, as even its substitution to other positively charged amino acids does not restore activity. The functional damage is not due to impaired turnover or impaired targeting of the mutated protein. The three other positively charged amino acids and the only negatively charged one are not critical [78]. It is possible that the arginine-69 residue may be involved in chloride binding.

The transporters of biogenic amines contain an additional negatively charged residue in helix 1: aspartate-79 (dopamine transporter numbering). Replacement of aspartate-79 in the dopamine transporter with alanine, glycine or glutamate significantly reduced the transport of dopamine, MPP^+ (parkinsonism inducing neurotoxin), and the binding of CFT (cocaine analogue) without affecting B_{max}. Apparently, aspartate-79 in helix 1 interacts with dopamine's amine during the transport process. Serine-356 and serine-359 in helix 7 are also involved in dopamine binding and translocation, perhaps by interacting with the hydroxyls on the catechol [79].

Studies of other proteins indicate that in addition to charged amino acids, aromatic amino acids containing π-electrons are also involved in maintaining the structure and function of these proteins [80]. Therefore, tryptophan residues in the membrane domain of the GABA transporter were mutated into serine as well as leucine [81]. Mutations at the 68 and 222 position (in helix 1 and helix 4, respectively) led to a decrease of over 90% of the GABA uptake. On the basis of the alignments of the transporters of the superfamily, it was postulated that tryptophan-222 is involved in the binding of the amino group of GABA. The replacement of tryptophan-68 to leucine results in an increased affinity of the transporter for sodium (S. Mager, N. Kleinberger-Doron, G. Keshet, N. Davidson, B.I. Kanner and H.A. Lester, unpublished experiments). This strongly suggests the involvement of this residue in sodium binding.

8. Structure–function relationships for the glutamate transporters

In the case of the glutamate transporter GLT-1, in addition to the putative protein kinase C consensus sites, the conserved charged amino acids have been mutated and analyzed. These include a conserved lysine located in helix 5 and a histidine in helix 6. While the former is not critical, the latter — histidine-326 — is [82]. If in fact a proton and not an hydroxyl ion participates in the translocation cycle, histidine-326 is an excellent candidate to serve as the proton binding site. Furthermore, five conserved negatively-charged amino acids are clustered in the carboxyl terminal hydrophobic part of the transporter for which the hydrophaty plot is ambiguous. Three of these residues appear to be critical for the transporter's function and one of them, glutamate-404, appears to be involved in the substrate discrimination of the transporter [83]. A mutation in this residue severely impairs transport of L-glutamate but transport of D- and L-aspartate is almost unaffected. This suggests that the conserved stretch surrounding glutamate-404 may form the translocation pore of glutamate and aspartate. Evidence to support this comes from experiments in which the determinant of the binding of dihydrokainic acid — a glutamate analogue — was found to be located within a stretch of 75 amino acids, in the middle of which is located residue glutamate 404 [84].

Substrate-induced conformational changes in the GLT-1 transporter have been detected, as revealed by the altered accessibility of trypsin sensitive sites to the protease [85]. These experiments indicate that lithium can occupy one of the sodium binding sites and also that there are at least two transporter-glutamate bound states [85].

Acknowledgements

I wish to thank Mrs. Beryl Levene for expert secretarial assistance and Dr. Michael P. Kavanaugh for information on important experimental studies prior to publication. The work from the author's laboratory was supported by the Bernard Katz Minerva Center for Cell Biophysics, and by grants from the US–Israel Binational Science Foundation, the Basic Research Foundation administered by the Israel Academy of Sciences and Humanities, the National Institutes of Health and the Bundesministerium fur Forschung und Technologie.

References

1. Kanner, B.I. (1983) Biochim. Biophys. Acta **726**, 293–316.
2. Kanner, B.I. and Schuldiner, S. (1987) CRC Crit. Rev. Biochem. **22**, 1–39.
3. Kanner, B.I. (1989) Curr. Opin. Cell Biol. **1**, 735–738.
4. Johnston, G.A.R. (1981) in Glutamate: Transmitter in the Central Nervous System, eds P.J. Roberts, J. Storm-Mathisen and G.A.R. Johnston, pp. 77–87. John Wiley and Sons, Chichester/New York/Brisbane/Toronto.
5. McBean, G.J. and Roberts, P.J. (1985) J. Neurochem. **44**, 247–254.
6. Radian, R., Bendahan, A. and Kanner, B.I. (1986) J. Biol. Chem. **261**, 15437–15441.
7. Guastella, J., Nelson, N., Nelson, H., Czyzyk, L., Keynan, S., Miedel, M.C., Davidson, N., Lester, H. and Kanner, B.I. (1990) Science **249**, 1303–1306.
8. Pacholczyk, T., Blakely, R.D. and Amara, S.G. (1991) Nature **350**, 350–353.
9. Uhl, G.R. (1992) Trends Neurosci. **15**, 265–268.
10. Schloss, P., Mayser, W. and Betz, H. (1992) FEBS Lett. **307**, 76–78.
11. Amara, S.G. and Kuhar, M.J. (1993) Ann. Rev. Neurosci. **16**, 73–93.
12. Keynan, S. and Kanner, B.I. (1988) Biochemistry **27**, 12–17.
13. Kanner, B.I. and Sharon, I. (1978) Biochemistry **17**, 3949–3953.
14. Kanner, B.I. and Bendahan, A. (1982) Biochemistry **21**, 6327–6330.
15. Storck, T., Schulte, S., Hofmann, K. and Stoffel, W. (1992) Proc. Natl. Acad. Sci. USA **89**, 10955–10959.
16. Pines, G., Danbolt, N.C., Bjoras, M., Zhang, Y., Bendahan, A., Eide, L., Koepsell, H., Storm-Mathisen, J., Seeberg, E. and Kanner, B.I. (1992) Nature **360**, 464–467.
17. Kanai, Y. and Hediger, M.A. (1992) Nature **360**, 467–471.
18. Fairman, W.A., Vandenberg, R.J., Arriza, J.L., Kavanaugh, M.P. and Amara, S.G. (1995) Nature **375**, 599–603.
19. Kavanaugh, M.P., Arriza, J.L., North, R.A. and Amara, S.G. (1992) J. Biol. Chem. **267**, 22007–22009.
20. Mager, S.J., Naeve, J., Quick, M., Guastella, J., Davidson, N. and Lester, H.A. (1993) Neuron **10**, 177–188.
21. Mager, S., Min, C., Henry, D.J., Chavkin, L., Hoffman, B.J., Davidson, N. and Lester, H.A. (1994) Neuron **12**, 845–859.
22. Brew, H. and Atwell, D. (1987) Nature **327**, 707–709.
23. Bouvier, M., Szatkowski, M., Amato, A. and Atwell, D. (1992) Nature **335**, 433–435.
24. Barbour, B., Brew, H. and Atwell, D. (1988) Nature **335**, 433–435.
25. Stallcup, W.B., Bullock, K. and Baetge, E.E. (1979) J. Neurochem. **32**, 57–65.
26. Erecinska, M., Wantorsky, D. and Wilson, D.F. (1983) J. Biol. Chem. **258**, 9069–9077.
27. Pines, G. and Kanner, B. I. (1990) Biochemistry **29**, 11209–11214.
28. Wadiche, J.I., Arriza, J.L., Amara, S.G. and Kavanaugh, M.P. (1995) Neuron **14**, 1019–1027.
29. Danbolt, N.C., Pines, G. and Kanner, B.I. (1990) Biochemistry **29**, 6734–6740.
30. Danbolt, N.C., Storm-Mathisen, J. and Kanner, B.I. (1992) Neuroscience **51**, 295–310.
31. Lopez-Corcuera, B., Vazquez, J. and Aragon, C. (1991) J. Biol. Chem. **266**, 24809–24814.
32. Launay, J.M., Geoffroy, C., Mutel, V., Buckle, M., Cesura, A., Alouf, J.E. and Da-Prada, M. (1992) J. Biol. Chem. **267**, 11344–11351.
33. Graham, D., Esnaud, H. and Langer, S.Z. (1992) Biochem. J. **286**, 801–805.

34. Radian, R., Ottersen, O.L., Storm-Mathisen, J., Castel, M. and Kanner, B.I. (1990) J. Neurosci. **10**, 1319–1330.
35. Hees, B., Danbolt, N.C., Kanner, B.I., Haase, W., Heitmann, K. and Koepsell, H. (1992) J. Biol. Chem. **267**, 23275–23281.
36. Shimada, S., Kitayama, S., Lin, C.L., Patel, A., Nanthakumar, E., Gregor, P., Kuhar, M. and Uhl, G. (1991) Science **254**, 576–578.
37. Kilty, J.E., Lorang, D. and Amara, S.G. (1991) Science **254**, 578–579.
38. Usdin, T.B., Mezey, E., Chen, C., Brownstein, M.J. and Hoffman, B.J. (1991) Proc. Natl. Acad. Sci. USA **88**, 11168–11171.
39. Hoffman, B.J., Mezey, E. and Brownstein, M.J. (1991) Science **254**, 579–580.
40. Blakely, R.D., Benson, H.E., Fremeau, R.T. Jr., Caron, M.G., Peek, M.M., Prince, H.K. and Bradley, C.C. (1991) Nature **353**, 66–70.
41. Clark, J.A., Deutch, A.Y., Gallipoli, P.Z. and Amara, S.G. (1992) Neuron **9**, 337–348.
42. Borden, L.A., Smith, K.E., Hartig, P.R., Branchek, T.A. and Weinshank, R.L. (1992) J. Biol. Chem. **267**, 21098–21104.
43. Lopez-Corcuera, B., Liu, Q.R., Mandiyan, S., Nelson, H. and Nelson, N. (1992) J. Biol. Chem. **267**, 17491–17493.
44. Liu, Q.R., Lopez-Corcuera, B., Mandiyan, S., Nelson, H. and Nelson, N. (1993) J. Biol. Chem. **268**, 2104–2112.
45. Smith, K.E., Borden, L.A., Hartig, P.A., Branchek, T. and Weinshank, R.L. (1992) Neuron **8**, 927–935.
46. Liu, Q.R., Nelson, H., Mandiyan, S., Lopez-Corcuera, B. and Nelson, N. (1992) FEBS Lett. **305**, 110–114.
47. Guastella, J., Brecha, N., Weigmann, C. and Lester, H.A. (1992) Proc. Natl. Acad. Sci. USA **89**, 7189–7193.
48. Fremeau, R.T. Jr., Caron, M.G. and Blakely, R.D. (1992) Neuron **8**, 915–926.
49. Uchida, S., Kwon, H.M., Yamauchi, A., Preston, A.S., Marumo, F., and Handler, J.S. (1992) Proc. Natl. Acad. Sci. USA **89**, 8230–8234.
50. Liu, Q.R., Lopez-Corcuera, B., Nelson, H., Mandiyan, S. and Nelson, N. (1992a) Proc. Natl. Acad. Sci. USA **89**, 12145–12149.
51. Yamauchi, A., Uchida, S., Kwon, H.M., Preston, A.S., Robey, R.B., Garcia-Perez, A., Burg, M.B. and Handler, J.S. (1992) J. Biol. Chem. **267**, 649–652.
52. Mayser, W., Schloss, P. and Betz, H. (1992) FEBS Lett. **305**, 31–36.
53. Guimbal, C. and Kilimann, M.W. (1993) J. Biol. Chem. **268**, 8418–8421.
54. Liu, Q.R., Lopez-Corcuera, B., Mandiyan, S., Nelson, H. and Nelson, N. (1993) J. Biol. Chem. **268**, 22802–22808.
55. Lehre, K.P., Levy, L.M., Ottersen, O.P., Storm–Mathisen, J. and Danbolt, N.C. (1995) J. Neurosci. **15**, 1835–153.
56. Rothstein, J.D., Martin, L., Levey, A.I., Dykes–Hoberg, M., Jun, L., Wu, D., Nash, N. and Kuncl, R.W. (1994) Neuron **13**, 713–725.
57. Tolner, B., Poolman, B., Wallace, B. and Konings, W.N. (1992) J. Bacteriol. **174**, 2391–2393.
58. Jiang, J., Gu, B., Albright, L.M. and Nixon, B.T. (1989) J. Bacteriol. **171**, 5244–5253.
59. Shafqat, S., Tamarappoo, B.K., Kilberg, M.S., Puranam, R.S., McNamara, J.O., Guadaño-Ferraz, A. and Fremeau, R.T. (1993) J. Biol. Chem. **268**, 15351–15355.
60. Arriza, J.L., Kavanaugh, M.P., Fairman, W.A., Wu, Y.–N., Murdoch, G.H., North, R.A. and Amara, S.G. (1993) J. Biol. Chem. **268**, 15329–15332.
61. Arriza, J.L., Fairman, W.A., Wadiche, J.I., Murdoch, G.H., Kavanaugh, M.P. and Amara, S.G. (1994) J. Neurosci. **14**, 5559–5569.
62. Wadiche, J.I., Amara, S.G. and Kavanaugh, M.P. (1995) Neuron **15**, 721–728.
63. Rhoads, D.E., Ockner, B.K., Peterson, N.A. and Raghupathy, E. (1983) Biochemistry **22**, 1965–1970.
64. Zafra, F., Alcantara, R., Gomeza, J., Aragon, C. and Gimenez, E. (1990) Biochem. J. **271**, 237–242.
65. Barbour, B., Szatkowski, M., Ingledew, N. and Attwell, D. (1989) Nature **342**, 918–920.
66. Trotti, D., Volterra, A., Lehre, K.P., Rossi, D., Gjesdal, ., Racagni, G. and Danbolt, N.C. (1995) J. Biol. Chem. **270**, 9890–9895.

67. Zerungue, N., Arriza, J.L., Amara, S.G. and Kavanaugh, M.P. (1995) J. Biol. Chem. **270**, 6433–6435.
68. Nieoullon, A., Kerkerian, L. and Dusticier, N. (1983) Neurosci. Lett. **43**, 191–196.
69. Kerkerian, L., Dusticier, N. and Nieoullon, A. (1987) J. Neurochem. **48**, 1301–1306.
70. Casado, M., Zafra, F., Aragon, C., and Gimenez, C. (1991) J. Neurochem. **57**, 1185–1190.
71. Casado, M., Bendahan, A., Zafra, F., Danbolt, N.C., Aragon, C., Gimenez, C. and Kanner, B.I. (1993) J. Biol. Chem. **268**, 27313–27317.
72. Corey, J.L., Davidson, N., Lester, H.A., Brecha, N. and Quick, M.W. (1994) J. Biol. Chem. **269**, 14759–14767.
73. Osawa, I., Saito, N., Koga, T. and Tanaka, C. (1994) Neurosci. Res. **19**, 287–293.
74. Mabjeesh, N.J. and Kanner, B.I. (1992) J. Biol. Chem. **267**, 2563–2568.
75. Bendahan, A. and Kanner, B.I. (1993) FEBS Lett. **318**, 41–44.
76. Mabjeesh, N.J. and Kanner, B.I. (1993) Biochemistry **32**, 8540–8546.
77. Kanner, B.I., Bendahan, A., Pantanowitz, S. and Su, H. (1994) FEBS Lett. **356**, 192–194.
78. Pantanowitz, S., Bendahan, A. and Kanner, B.I. (1993) J. Biol. Chem. **268**, 3222–3225.
79. Kitayama, S., Shimada, S., Xu, H., Markham, L., Donovan, D.M. and Uhl, G.R. (1992) Proc. Natl. Acad. Sci. USA **89**, 7782–7785.
80. Sussman, J.L. and Silman, I. (1992) Curr. Opin. Struc. Biol. **2**, 721–729.
81. Kleinberger–Doron, N. and Kanner, B.I. (1994) J. Biol. Chem. **269**, 3063–3067.
82. Zhang, Y., Pines, G. and Kanner, B.I. (1994) J. Biol. Chem. **269**, 19573–19577.
83. Pines, G., Zhang, Y. and Kanner, B.I. (1995) J. Biol. Chem. **270**, 17093–17097.
84. Vandenberg, R.J., Arriza, J.L., Amara, S.G. and Kavanaugh, M.P. (1995) J. Biol. Chem. **270**, 17668–17671.
85. Grunewald, M. and Kanner, B.I. (1995) J. Biol. Chem. **270**, 17017–17024.

CHAPTER 20

Intracellular pH: Measurement, Manipulation and Physiological Regulation

J.H. KIM, N. DEMAUREX and S. GRINSTEIN

Division of Cell Biology, Hospital for Sick Children
555 University Ave., Toronto, Canada M5G 1X8

© 1996 Elsevier Science B.V.
All rights reserved

Handbook of Biological Physics
Volume 2, edited by W.N. Konings, H.R. Kaback and J.S. Lolkema

Contents

1. Introduction

Because most enzymes have well defined pH optima, continuous and accurate regulation of the local [H$^+$] is crucial to the maintenance of many cellular functions. This notion has elicited great interest in the measurement of the pH of the individual subcellular compartments and of the mechanisms underlying its regulation. This chapter will attempt to summarize the methods used to measure and manipulate cellular pH and the current status of the knowledge regarding pH homeostasis in the cytosol and in the lumen of endomembrane compartments.

Researchers interested in pH regulation must not only be able to measure the relevant pH accurately, reliably and with reasonable time resolution but they must also have the ability to alter the pH in a predictable, controlled manner to study the responsiveness of the system. The first section describes the most popular techniques used to measure pH in biological systems and outlines some of the procedures used to alter the intracellular pH. The two subsequent sections detail measurements of cytosolic and organellar pH and the mechanisms underlying pH regulation in mammalian cells.

2. Measurement of intracellular pH

During the past decades, a variety of techniques have been developed that now allow researchers to measure the pH of systems as diverse as whole organs, tissue fragments, cell populations, single cell preparations and even intracellular organelles. Three very different techniques that have found widespread applications in the study of pH homeostasis will be described here: (1) pH-sensitive microelectrodes; (2) fluorescence spectroscopy and (3) nuclear magnetic resonance. Measurements based on the partitioning of radiolabeled weak acids or bases have become less popular and are dealt with briefly in the section describing the pH of organelles; for this particular subject, readers are referred to the comprehensive review of Roos and Boron [1], as well as to more recent reviews on pH measurements [2,3]. Each of the techniques described in this section has unique advantages that makes it best suited for a particular application, explaining its popularity among researchers in a given field. The goal of this section is to provide the reader with a comprehensive introduction, outlining the theoretical basis, technical requirements, and the advantages and limitations of the different methodologies.

2.1. pH Electrodes

Intracellular pH can be measured with microelectrodes whose tip is selectively permeable to H$^+$ ions (reviewed in Ref. [4,5]). This approach provides an immediate estimate of the local H^+ activity with high precision (± 0.01 pH units), as pH electrodes have an almost Nernstian voltage response (≈ 60 mV/pH unit). The electrochemical potential across the H^+-selective membrane is measured by connecting the electrode inside the pipette (which is filled with a solution of known pH) to an external (bath) electrode via a high-impedance voltameter. When the electrode is used to measure extracellular pH, as is done daily by most biologists who care about the pH of their solutions, the voltage output can be directly converted into pH units, provided that the electrode is calibrated by immersion into media of known pH. When a microelectrode is used to measure intracellular pH, however, the electrochemical potential across the electrode tip is in series with the membrane potential of the cell, V_m, which has to be subtracted from the voltage output. Hence, V_m has to be measured independently, either with a separate microelectrode or by using a special double-barreled electrode, with one barrel dedicated to pH measurement and the other to voltage determination.

Regardless of the electrode type, the cell has to be impaled at least once, which restricts the approach to preparations that can withstand invasive procedures. Recently, pH-sensitive suction electrodes have been used to obtain low-resistance access to the cell cytosol, in a configuration similar to the whole-cell patch clamp technique [6]. Although this approach can be applied to all cells amenable to patch-clamp measurements, a second patch or intracellular voltage electrode is still required to measure V_m, which renders the procedure at least as invasive as a single impalement with a double-barreled microelectrode.

Types of pH electrodes. Two types of pH microelectrodes are commonly used: (1) glass-sensitive microelectrodes, usually of the recessed-type developed by Thomas [7], and (2) liquid ion-exchange (LIX) microelectrodes, in which the H^+-selective interface is made of an ionophoretic H^+-binding resin [8,9]. The two types of electrodes can be made of various sizes, but behave differently in terms of selectivity, response time and resistance.

Recessed-tip microelectrodes are composed of a relatively large closed-tip, pH-sensitive glass electrode, encased into a conventional microelectrode which acts as a shield. The cells are impaled with the small tip of the shielding electrode and the cytoplasm equilibrates with the recessed volume, allowing the larger surface of the pH-sensitive electrode to measure the cytosolic pH. These electrodes have near-Nernstian slopes (≥ 57 mV/pH unit) and provide accurate pH measurements over a wide range of pH values (2–9) [7]. Furthermore, they show minimal drifting over time and can be stored and reused for several days. However, due to the recessed volume, their response time is relatively slow, the time constant ranging from 15 to 120 seconds, depending on the electrode construction [4]. They also tend to have a fairly high resistance (100–1000 GΩ), resulting in a noisy signal, and to cause greater damage to small cells than the sharper LIX microelectrodes. Recessed-tip microelectrodes are difficult to manufacture, but a skilled researcher can produce electrodes with reasonably sharp tips, fast response time and low resistance, making them amenable to pH measurements from a wide range of cell types.

LIX microelectrodes use a neutral carrier resin that is selectively permeable to H^+, the most popular being based on the carrier 4-nonadecylpyridine (ETH 1907)

[9]. As is the case with recessed-tip microelectrodes, obtaining a good LIX microelectrode is the critical step, as the surface of the exposed resin as well as the length of the resin column determine the response time and resistance of the pH electrode [5]. Typically, a conventional intracellular electrode is pulled, its inner surface is rendered hydrophobic by high-temperature evaporation of an aminosilane compound (silanation), and the electrode tip is further beveled on a grinding plate. Then, a small column (20–200 μm) of the liquid sensor is aspirated into the electrode tip under microscopic observation and the electrode is backfilled with the reference pH solution. The resulting LIX microelectrodes are stable for several days, show minimal drifting and can be reused several times. They are faster than glass microelectrodes, with response times usually in the range of seconds [5]. Low resistance (and noise) microelectrodes can be obtained by beveling the electrode and minimizing the height of the resin column. LIX microelectrodes, however, sometimes exhibit sub- or super-Nernstian slopes and can even become non-linear at extreme pH values, their pH range being limited by the pK of the neutral carrier. In addition, several organic compounds have been shown to alter the H^+ selectivity of LIX electrodes [9]. A major advantage of LIX microelectrodes is that they can readily be constructed in the double-barrel configuration, either by using theta-shaped glass capillaries or by joining together two separate microelectrodes. Double-barreled LIX microelectrodes with sharp tips (<1 μM) are suitable for intracellular recording from small cells [10] and can have response times as fast as 60 ms when a linearization network is used to improve the frequency response [11]. Due to their relative ease of manufacture and the need for only one impalement, double-barreled LIX microelectrodes have become the electrodes of choice for intracellular pH recording from mammalian cells.

In summary, pH microelectrodes are the most sensitive assay to measure intracellular pH. They provide a direct and simple reading of the local pH, which is not affected by the pH of neighbouring organelles. Another feature (or caveat) is that, by necessity, the membrane potential has to be measured, providing additional information about the physiology and integrity of the cell. A major advantage is that the measurement is quite simple, the electrodes easily calibrated and the required equipment is available in any laboratory equipped for intracellular electrical recordings. Hence, this technique has been readily adopted by electrophysiologists, the only new skills required being in manufacturing pH microelectrodes. The limitations of pH electrodes are their slow response time, lack of spatial resolution and the fact that impalement is an invasive procedure. Additional problems specific to LIX microelectrodes include non-linearity and interference with organic compounds.

2.2. Fluorescence spectroscopy

Fluorescence spectroscopy is widely used to measure intracellular pH, as the technique is non-invasive, very sensitive, and provides better temporal and spatial resolution than either pH electrodes or nuclear magnetic resonance (the advantages of fluorescent indicators are reviewed in Ref. [12]). The new fluorescent pH indicators have a resonance structure that changes with pH, the protonated indicator absorbing or emitting light at a different wavelength than the non-protonated form. As this spectral shift is pH-specific, it allows to correct for fluorescence changes caused by leakage, quenching, or bleaching of the dye, as well as for a variety of sources of optical interference which could be mistaken for pH changes. This is easily accomplished by selecting two different wavelengths where the fluorescence intensity varies differently with pH and calculating the ratio of the corresponding signals. In contrast, fluorophores that redistribute across biological membranes according to the prevailing pH gradients generate a signal that is proportional to the dye concentration and artifacts cannot be corrected by making the ratio. These probes are now used mostly to measure organellar pH and are described in a subsequent section of this chapter.

The new generation of pH-sensitive indicators derive their $[H^+]$ dependence from a titratable group attached to a parent fluorophore, such as fluorescein, pyranine, umbelliferone, or calcein [13]. The useful range of the pH indicators is mostly determined by the pK value of the titratable group, as fluorescence changes linearly with pH only over a ≈ 0.5 pH range around the pK. Thus, most probes designed to measure intracellular pH have pKs in the physiological pH range. The most widely used pH indicator, 2,7-bis-(2-carboxyethyl)-5(6)-carboxyfluorescein or BCECF, introduced by Tsien and colleagues [14], has a pK of 6.97. This dye has a bright fluorescence that increases with pH when it is excited at 490 nm and the emission measured at 535 nm, allowing qualitative pH measurements with standard fluorescein optics. Furthermore, BCECF fluorescence with excitation at 440 nm is pH-independent, so this wavelength can be used to monitor fluorescence drifts due to bleaching, leakage, or quenching of the dye and to correct these by performing ratio measurements.

Another widely used family of dyes are the carboxy-seminaphthorhodafluor, or SNARF derivatives [15], which can be used either as excitation or as emission ratio dyes. Ratio emission dyes allow measurement with microsecond resolution, as the emitted fluorescence light can be split with a dichroic mirror and measured simultaneously by two light detectors. Excitation ratio dyes, in contrast, require sequential illumination of the sample at two different wavelengths, so the time resolution is thus limited by the speed at which the alternate illumination can be achieved.

The dyes can be introduced into cells either by microinjection (which renders the technique invasive) or more commonly by incubation with a membrane-permeant, non-fluorescent form which is cleaved by intracellular non-specific esterases, yielding a membrane-impermeant, fluorescent derivative. This technique is by far the most commonly used, making it possible to 'load' a large number of cells in a relatively short time (typically 10–20 min). However, in this case the dye is not specifically 'targeted' to the cytosol and hydrolysis can occur in intracellular compartments, causing the signal to be potentially contaminated by the pH of organelles. This possibility, which is often overlooked, can pose serious problems in some cell types, but it also offers an alternate approach to measure organellar pH.

Optical configurations. The most attractive feature of fluorescent pH indicators is their versatility. A fluorescence-activated cell sorter can rapidly acquire statistical information from a cell population and can separate cells based on their pH. A conventional spectrofluorimeter allows any researcher to perform rapid, on-line measurements of intracellular pH from populations of adherent or suspended cells, simply by measuring the fluorescence intensity at a fixed excitation and emission wavelength. Although this configuration does not allow for correction of pH-independent fluorescence changes, it provides physiological information at minimal cost. A spectrofluorimeter equipped with alternate illumination or two emission monochromators and detectors can be used for ratio measurements and thus quantitative pH determinations.

A fluorescence microscope equipped with a dynamic light detector (e.g. a photometer or intensified video camera) can be used to perform on-line measurements from single cells. It is thus possible to record from rare, scarce cell types. Moreover, imaging also addresses the problems caused by cellular heterogeneity (the cell(s) of interest can be selected among a population). More importantly, fluorescence microscopy can be coupled to electrophysiological recordings or microinjection. Slow-scan digital cameras, though expensive, offer the best spatial resolution, making it possible to image subcellular structures. They also have a greater dynamic range, facilitating accurate quantification of fluorescence from sources of varying intensity. These cameras are typically used for excitation ratio measurements, as in the configuration described below (Fig. 1), but ratio emission measurements can also be performed using a beam splitter/mirror assembly, which redirect the fluorescence emitted at two different wavelengths onto separate regions of the same imaging chip.

Fluorescence ratio is usually converted to pH by performing an *in situ* titration, according to the method of Thomas [16]. Cation-selective ionophores such as nigericin or monensin are used to equilibrate the transmembrane H^+, K^+ and/or Na^+ gradients, and the cells are sequentially exposed to media of similar ionic composition but different pH. Provided that the medium K^+ and/or Na^+ concentrations approximate the intracellular values, the intra- and extracellular pH can be assumed to be identical and the fluorescence signal can be calibrated to the medium pH. The main inconvenience of ionophores is that they can irreversibly damage the cells, requiring that the calibration be performed at the end of the experiment or in parallel samples. Although *in vitro* calibration can in theory be used, most researchers prefer *in situ* calibration, since the fluorescence of the pH-sensitive probes can be affected by the viscosity, refraction and autofluorescence of the cells.

In summary, fluorescence is the preferred method to measure pH, as it is very sensitive, offers excellent temporal and spatial resolution and is amenable to a wide range of applications. Artifacts caused by bleaching, quenching, or leakage of the dyes are largely eliminated by ratio measurements. Calibration is straightforward, but necessitates incubation of the cells with ionophores. Another limitation is that some pH-sensitive dyes are toxic to cells, especially when a strong illumination is used (phototoxicity). pH measurements can be performed in any laboratory equipped with a fluorimeter, but special optical configurations are required to take advantage of the spectral properties of the dyes and high resolution fluorescence microscopy still necessitates expensive equipment.

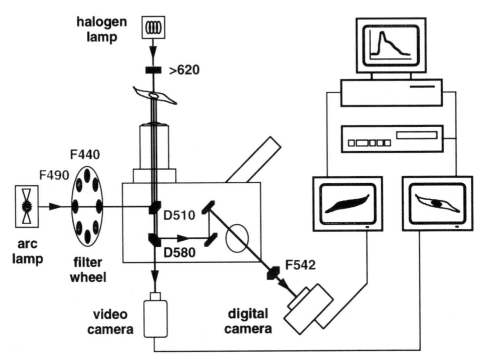

Fig. 1. Optical configuration for simultaneous determination of cellular pH and morphology using light microscopy. Alternate excitation at 490 and 440 nm is provided by a Xenon arc lamp and a computer-controlled shutter/filter wheel assembly (Sutter Corp.), while continuous >620 nm illumination is achieved by filtering the transmitted incandescent source provided with the microscope through a long-pass filter (F620). The transillumination path is equipped with a polarizer and a Wollaston prism, providing differential interference contrast (DIC or Nomarski optics). The 440/490 light is reflected to the cells by an excitation dichroic mirror (D510) and the emitted fluorescence (>510 nm) is separated from the transmitted red light (>620) by an emission dichroic mirror (D565). The red light is directed to a video camera, allowing continuous visualization of the cells, while the fluorescent light is directed onto a barrier filter (F542) and imaged with a slow-scan, cooled (−30°C) charge-coupled device (CCD) camera (Princeton Instruments). The large separation between the fluorescence and DIC signals (542 and 620 nm) prevents cross-talk between the two light paths and allows continuous, simultaneous assessment of cell morphology and intracellular pH. Image acquisition and filter selection is controlled by the Metamorph/Metafluor software (Universal Imaging). Pairs of DIC images (512×512 pixels) and ratio images (max. 1035×1350 pixels) can be acquired at frequencies ranging from 0.1 to 1 Hz, depending on exposure time and image resolution.

2.3. Nuclear magnetic resonance

Nuclear magnetic resonance (NMR) is increasingly used for determinations of intracellular pH since the technology is now widely available and allows *in situ* measurements in intact animals via magnetic resonance imaging (MRI). NMR measures the resonance frequency of nuclei which are magnetically 'visible', i.e. which have a spin value $I \neq 0$, by exposing the sample to an alternating electromagnetic field (see Ref. [17] for a general review on NMR). Each nucleus has a characteristic frequency, or 'signature', which can be resolved on the NMR spectrum. Because the magnetic field imposed on the nucleus is locally affected by the electronic structure of the atom, some nuclei can be effectively used to 'sense' their chemical environment. An ion that alters its electronic structure in a pH-dependent manner will therefore have a resonance frequency that shifts on the NMR spectrum. These shifts are typically quite small, a few Hz on the MHz scale of the NMR spectrum, but become apparent when the spectra are normalized to a reference compound that does not alter its electronic structure as a function of pH. ^{31}P-NMR is generally used to measure intracellular pH [18], as inorganic phosphate, a comparatively abundant species, is titrated from $H_2PO_4^-$ to HPO_4^{2-} in the physiological pH range; the global NMR signal obtained is a weighted average of the two phosphate forms. Shifts in the NMR spectrum of inorganic phosphate (usually expressed in parts per million, or ppm) are often compared to the creatine phosphate signal, that is invariant with pH. Some organic phosphate-containing compounds can also be used for pH measurements. These include 2-deoxy-D-glucose, which is phosphorylated into 2-deoxy-D-glucose-6-phosphate, a species that is pH-sensitive in the physiological range. The signals from exogenous, 'reporter' phosphates are easier to interpret but necessitate loading of cells with the indicator, which renders the technique somewhat invasive.

In summary, ^{31}P-NMR is the method of choice to measure intracellular pH *in vivo* or in perfused organs or tissues. The technique is non-invasive and measures not only intracellular pH but also other physiologically relevant metabolites. Small pH changes can be detected, but the absolute precision of NMR is limited by the lack of accurate calibration and the poor time resolution precludes measurement of rapid pH transients. NMR requires expensive equipment and dedicated operators, but the technology is becoming increasingly available for clinical and biological research.

3. Manipulation of intracellular pH

3.1. Weak electrolytes

Exposure of cells to weak electrolytes can be used to produce a continued acid or base load, as desired. More frequently, however, imposition of transient and well defined pH transients is preferable, to evaluate the cellular responses unencumbered by a continuous stress. The method of choice to acidify cells is the ammonium prepulse technique introduced by Boron [19], which takes advantage of the differential permeabilities of NH_3 and NH_4^+ across biological membranes. Whereas the membrane-permeant NH_3 diffuses within seconds across lipid bilayers, NH_4^+ entry is usually much slower and involves facilitated diffusion through membrane transporters. In most cell types, a 10–30 min incubation is generally sufficient for substantial NH_4^+ entry into the cells. During this incubation or 'pulse' period, a rapid alkalinization is observed as H^+ ions are consumed by the intracellular protonation of the membrane-permeant NH_3. Thereafter, NH_4^+ (a conjugated acid) slowly enters the cells, gradually restoring the pH. The subsequent removal of external NH_4^+/NH_3

produces a massive acidification, as all the NH_4^+ imported by the cell during the prepulse suddenly exits the cell in the form of deprotonated NH_3, leaving behind an excess of intracellular H^+ ions. Clearly, an equivalent acid prepulse method can be designed to obtain a transient alkaline load.

3.2. Ionophores

Proton-transporting ionophores can be used to provide an effective pH 'clamp'. The need to preserve electroneutrality during H^+ translocation implies that counterions must be displaced simultaneously. When using conductive ionophores such as CCCP or FCCP, it must be ensured that counterion conductance suffices to support H^+ transport, which can often be accomplished by introduction of another conductive ionophore such as valinomycin. Alternatively, electroneutral, cation-exchanging ionophores such as monensin or nigericin can be used. Inevitably these measures disrupt other cationic gradients and may affect the biological response of interest. Moreover, the ionophores can irreversibly damage the cells, although toxicity can be minimized by using low concentrations of ionophores. In addition, the intracellular pH can be set to a value different from the external pH by taking advantage of the prevailing K^+ or Na^+ gradients. For instance, when nigericin is used in low $[K^+]$ solutions, extracellular H^+ is exchanged for intracellular K^+, producing a cytosolic acidification [20].

3.3. Micropipettes

The intracellular pH of isolated cells can be manipulated by microinjecting small amounts of concentrated acids or bases [21]. However, the pH 'surge' is not uniformly distributed across the cytoplasm and is difficult to quantify, due to possible leakage at the site of microinjection. An alternative is to use the whole-cell configuration of the patch-clamp technique to equilibrate the intracellular pH with the pH of the pipette solution. Using this configuration, a constant intracellular pH can be established and maintained, but heavily buffered pipette solutions and long equilibration times are required [22]. pH changes can be imposed either by perfusing the pipette with different solutions during the experiment or, more conveniently, by including a permeable weak electrolyte such as NH_4^+ in the pipette solution. This provides a virtually unlimited intracellular H^+ donor/acceptor system, allowing the experimenter to rapidly and reversibly change the intracellular pH by simply changing the external NH_4^+ concentration, by a principle similar to that described above for the NH_4^+ prepulse [23].

4. Regulation of cytosolic pH

4.1. General principles

The cytosolic pH (pH_c) has been measured in a variety of animal cell types using the methods detailed above. In the vast majority of cases the cytosolic pH has been found to range between 7.0 and 7.3 when the cells are bathed at physiological extracellular pH (i.e. 7.3–7.4) [1]. While intra- and extracellular pH are of the same order, this is not a coincidental, passive occurrence. Thermodynamically, the system is far from equilibrium, in that the membrane potential of most mammalian cells is internally negative. The predicted pH_c at equilibrium, calculated using an average potential of non-excitable cells of –60 mV, would be a full pH unit more acidic than the external pH. This passive tendency to acidify is compounded by the ongoing net generation of metabolic acid by the cells themselves. A number of predominant metabolic pathways yield acid equivalents that would normally accumulate in the cytosol. These include the hydrolysis of ATP and GTP, glycolysis, the hexose monophosphate shunt pathway, triglyceride hydrolysis and many others. In view of this continuous acid stress, the maintenance of near-neutral pH_c must result from a dynamic and precisely regulated process.

Short-term homeostasis of pH_c can be favored by cellular buffering, which will tend to mitigate rapid departures from the physiological pH level. The intracellular buffering power of animal cells near physiological pH is substantial, estimated to range from about 25 to 100 mM/pH unit [1]. However, while acute changes of pH_c can be partially counteracted by the cellular buffering capacity, long term regulation requires the continuous operation of transport processes that are 'active'. Transport of H^+ equivalents for the purpose of regulating pH_c need not be 'primary active', i.e. directly requiring ATP hydrolysis for the catalytic (transport) event. In fact, many pH regulating systems in animal cell membranes are 'secondary active'. In such cases, the energy for the translocation of H^+ (equivalents) is provided by coupling to the movement of another chemical species down its electrochemical gradient. Often, the gradient of the 'driver' species is itself established by a primary active process.

In principle, regulation of pH_c can be accomplished either by translocation of acidic or basic equivalents out of the cell across the plasmalemma or by accumulation into cytosolic organelles. However, the importance of organellar accumulation for the purposes of cytosolic pH regulation is dubious, given their finite capacity and the necessity to independently maintain intraorganellar pH. Therefore, it is generally felt that long-term regulation of pHc is primarily the responsibility of plasma membrane sensors and transporters. Transmembrane transfer of acid equivalents is accomplished by a variety of membrane-spanning proteins which have unique functions, structures and pharmacological characteristics. The best characterized systems involved in pH_c regulation in animal cells are discussed next.

4.2. Na⁺/H⁺ exchangers

Na⁺/H⁺ exchangers (NHE), also termed Na⁺/H⁺ antiporters, are integral plasmalemmal proteins with 10–12 membrane spanning domains and a cytoplasmic carboxy-terminus tail. Recent evidence suggest the existence of at least four isoforms of the exchanger, each with specific pharmacological and biochemical properties, particular cellular/tissue localizations, and unique functional characteristics [24,25]. NHE-1 is the best characterized of all the isoforms. It is a near ubiquitous glycoprotein, present in the plasmalemma of virtually all the mammalian cells studied to date, including the basolateral membrane of epithelial cells. NHE-2 and NHE-3 have a much more restricted tissue distribution, appearing predominantly in epithelia, where they are believed to localize to the apical membrane (though discrepant findings have been reported, cf. Ref. [26,27]). NHE-4 is abundant in the gastrointestinal tract, but is found also in kidney and other tissues [24]. Unlike NHE-1, NHE-2 and NHE-3 are seemingly not glycosylated. Less is known about NHE-4 from both the structural and functional standpoints partly because, unlike the other isoforms, NHE-4 has been refractory to heterologous expression in transfected cells.

Typically, mammalian Na^+/H^+ antiporters catalyze the exchange of extracellular Na^+ for intracellular H^+, thereby counteracting the tendency of the cytosol to become acidic. The stoichiometry of the exchange process is one Na^+ for one H^+. As a result the transport cycle is predicted to be electroneutral and insensitive to changes in the membrane potential. This prediction was recently borne out experimentally for NHE-1, NHE-2 and NHE-3 that were transfected into cells lacking endogenous antiport activity [28]. Electroneutrality and voltage insensitivity enable excitable cells to undergo electrical potential changes without jeopardizing pH_c regulation and *vice versa*.

Because of their voltage independence, transport through the NHEs is driven solely by the combined concentration gradients of the substrate ions, i.e. Na^+ and H^+. In theory, therefore, net flux through the exchanger should cease when the Na^+ concentration gradient is balanced by an identical H^+ gradient. Under physiological conditions, however, such equilibrium is never attained and the direction of H^+ transport is constantly outward, driven by the prevailing inward Na^+ gradient (extracellular $[Na^+]$ being over ten-fold higher than the cytosolic $[Na^+]$). This gradient, of course, is generated and maintained by the Na^+/K^+ pump. As a consequence, the continued extrusion of intracellular H^+ is directly fueled by the Na^+ gradient and indirectly by the hydrolysis of ATP. In this sense, the Na^+/H^+ antiport can be considered a 'secondary' active transport mechanism.

NHE activity is greatly stimulated when the cytosol is acidified. This peculiar behavior is partly attributable to the increased availability of internal substrate (i.e. H^+), but is mostly due to the protonation of one or more allosteric H^+ binding site(s) on the internal surface of the exchanger [29]. When protonated, these putative sites are thought to activate Na^+/H^+ exchange, thereby counteracting the acidification of the cytosol. All isoforms studied display this characteristic allosteric behavior, though the threshold pH of activation and the apparent affinity towards H^+ vary somewhat between NHEs.

Cation exchange by the NHEs is characterized by sensitivity to inhibition by amiloride and its analogs. However, the susceptibility of the individual isoforms to the inhibitors varies widely, with NHE-1 being most sensitive and NHE-3 most resistant. For inhibitor HOE694 the sensitivity of the isoforms differs by upwards of three orders of magnitude [30]. As before, characterization of NHE-4 remains to be accomplished.

An interesting feature of the antiporters is their regulation by osmolarity. NHE-1 and NHE-2 have been reported to be stimulated when the medium is made hypertonic [31]. This behavior could account for the compensatory volume increase, which in many cells has been reported to be inhibited by amiloride. Thus, the antiporters may play a dual role in pH and cell volume regulation. Interestingly, NHE-3 differs from the other isoforms in that it is inhibited by increased osmolarity [31]. In the kidney, where NHE-3 predominates, such inhibition would modulate Na^+ and bicarbonate reabsorption in the context of osmoregulation.

4.3. H+-pumping ATPases

Proton pumps hydrolyze ATP to power the translocation of H^+ across the membrane. As such, they can be considered a 'primary active' transport system. In some mammalian cells, vacuolar or V-type proton pumps contribute to the regulation of pH_c. The V-type ATPases from eukaryotic cells have a multiple subunit structure (8–9 subunits ranging in size from 17 kDa to 100 kDa) and can be easily distinguished from other classes of ATPases, such as the mitochondrial or F-type ATPases and the phosphorylated intermediate or P-type ATPases, on the basis of their pharmacological characteristics. V-type pumps are exquisitely sensitive to the macrolide antibiotics bafilomycin and concanamycin, which preclude transport at nanomolar doses. Comparable concentrations have negligible effects on F-type and P-type pumps. The subunit assembly of V-type pumps has been grossly subdivided into a hydrophobic, tightly membrane-associated V_0 domain, and an extrinsic V_1 domain that encompasses the ATP-binding region [32].

The V-type pump itself does not translocate an associated counterion. Neutralization arises from the parallel movement of ions (likely chloride) through independent pathways. As such, this system is electrogenic and sensitive to changes in the transmembrane potential. It has been postulated that, when counterion permeability is limiting, such parallel pathways may be the main determinant of the rate of pumping and therefore a target of regulation.

As discussed below, V-type H^+ ATPases are prominent in a variety of intracellular membranes [32], where they play a crucial role in intraorganellar acidification. In addition, H^+ ATPases have also been detected in the plasma membranes of some tissues, such as some epithelial cells of the urinary tract [33]. Further studies indicated that the pumps reached these membranes through exocytic insertion of endomembrane vesicles. Such insertion is a reversible, tightly regulated process activated by intracellular acidification. Plasmalemmal V-type pumps have also been detected functionally and/or immunochemically in non-epithelial cells such as macrophages, activated neutrophils and osteoclasts [34]. Regulated vesicular traffic is probably also responsible for pump delivery in these cells. In osteoclasts, V-type ATPases play a role in the acidification of the lacunae, where bone resorption occurs, and may also be important for intracellular pH regulation. In neutrophils and macrophages the pumps most likely play a role in the maintenance of pH_c. These phagocytic cells undergo violent bursts of metabolic acid generation that must be neutralized. Furthermore, pH_c homeostasis is complicated when the cells are in acidic environments such as exist within tumors or at sites of infection. Under these conditions, addition of bafilomycin depresses pH_c and impairs normal phagocyte function, highlighting the important role of V-type pumps in pH homeostasis.

4.4. H+ conductive pathways

Experiments using model systems such as planar bilayers have unequivocally demonstrated that H^+ (or OH^-) can permeate through lipid bilayer membranes, albeit at extremely low rates. In addition, H^+ equivalents can traverse biological membranes passively through other components, likely proteins or glycoproteins. Evidence for high conductance, highly H^+-selective, voltage dependent 'channels' has recently been obtained in neurons, epithelial cells, oocytes and phagocytes (reviewed in Refs. [35,36]). The physiological significance of this conductive pathway remains unclear, since its operation would appear to be deleterious under most circumstances. As discussed earlier, at the resting potential and ΔpH prevailing across the membranes of most mammalian cells, a passive conductance would favor net uptake of acid equivalents. This would only exacerbate the tendency for the cytosol to become acidic due to metabolic H^+ production. However, in all the cells studied, the conductance displays certain properties that ensure that flux of H^+ will occur only outward. Specifically, the system is activated only when cells are depolarized. Moreover, extracellular acidification reduces the conductance while acidification of the cytosol activates it. When combined, these properties support activation of the pathway only when the protonmotive force points outward.

The conductive pathway is not ubiquitous. It is interesting that it has been detected in cell types in which membrane depolarization occurs upon activation and is accompanied by a metabolic burst (e.g. nerve cells, phagocytes). Because the channels are activated by depolarization and internal acidification, it is possible that net acid efflux occurs during depolarization, facilitating the maintenance of pH_c. On the other hand, the potentially deleterious entry of acid into repolarized (inside negative) cells seems to be prevented by the rectification properties of the conductance [35].

4.5. HCO3 transporters

An electroneutral anion exchange process, analogous to the cation (Na^+/H^+) antiport, is also important in pH_c regulation in most cells. This anion antiport is independent of the cationic composition of the medium and can transport a variety of halides as well as nitrate, sulfate and phosphate. Under physiological conditions, however, the anion exchangers are believed to transport mainly Cl^- and HCO_3^- [37].

The Cl^-/HCO_3^- exchanger translocates anions with a one-to-one stoichiometry and, like the Na^+/H^+ antiporter, it is therefore electroneutral and insensitive to alterations in transmembrane potential. This type of anion exchange is passive, driven by the concentration gradients of the substrate ions. In most cell types, the intracellular activities of Cl^- and HCO_3^- are lower than those in the external medium, due in part to the electronegativity of the cell interior in combination with anion conductive channels. However, because the concentration of HCO_3^- is a function of pH, a parameter controlled independently of potential, the inward concentration gradient for Cl^- generally exceeds that for HCO_3^-. This imbalance is expected to drive the net influx of Cl^- simultaneously with HCO_3^- efflux through the anion exchanger, promoting cytosolic acidification. It has therefore been postulated that cation-independent anion exchange serves to protect cells from alkalosis.

On the other hand, the spontaneous tendency of the exchanger to acidify the cytosol contravenes the needs of cells near normal pH_c. This deleterious action is curtailed by the peculiar pH_c dependence of the exchanger, which is virtually inactive below physiological pH and is greatly activated upon alkalinization [38]. Thus, its activation profile is opposite to that of the Na^+/H^+ exchanger. The behavior of the Cl^-/HCO_3^- exchanger is consistent with a role in the recovery from alkalosis, since the activated exchange of external Cl^- for cytosolic HCO_3^- would tend to restore the physiological pH_c.

Cl^-/HCO_3^- exchange is inhibited by disulfonic stilbene derivatives. For this reason, disulfonic stilbenes that bind covalently to amino groups of proteins were initially used to label and identify the molecules responsible for anion exchange. Anion exchangers were found to be glycoproteins of approx. 100 kDa that span the membrane multiple times and extend a sizable N-terminal hydrophilic domain into the cytosolic interior. Identification of the red cell exchanger, since renamed AE1, was followed by the discovery of additional isoforms and splice variants. A truncated form of AE1 is expressed in the kidney and a separate isoform, termed AE2 is thought to be ubiquitous. Additional isoforms with more restricted localization were described subsequently [39,40]. The structural and functional characterization of the isoforms is at a preliminary stage.

A second type of Cl^-/HCO_3^- exchange mechanism exists in the plasma membranes of some mammalian cells [41,42]. It differs from the cation-insensitive anion exchanger described above in that Na^+ is an essential requirement for anion exchange to occur. In fact, the cation is cotransported into the cell. Under physiological conditions, this system catalyzes the exchange of extracellular Na^+ plus HCO_3^- for intracellular Cl^-. Since the overall exchange cycle has been found to be electroneutral, it has been inferred that some other cation is extruded from the cell or that an additional anion is transported inward. Measurements of pH indicate that two acid equivalents exit the cell per transport cycle. Three possible ionic combinations have been suggested to account for both the electroneutrality and acid transport stoichiometry: extracellular Na^+ could enter the cell accompanied by one CO_3^{-2} or two HCO_3^- ions, rather than a single HCO_3^-, or alternatively one H^+ could exit the cell along with Cl^- [42]. While mechanistically different, these combinations would be functionally equivalent.

The large inward Na^+ gradient across the plasmalemma is the chief determinant of the directionality of Na^+-dependent Cl^-/HCO_3^- exchange. By driving net HCO_3^- uptake into the cells, this transporter tends to counteract the acidification of the cytosol. For this reason, Na^+-dependent Cl^-/HCO_3^- exchange is considered to be a functional pH regulatory system in the resting state. In fact, its activation occurs at more alkaline pH than the set point of the Na^+/H^+ exchanger, suggesting that the anion exchanger is primarily responsible for pH_c homeostasis in resting cells. Consistent with this view, mutants devoid of Na^+/H^+ exchangers survive and proliferate normally, provided bicarbonate is present in the medium. Despite their functional similarities, Na^+-dependent Cl^-/HCO_3^- exchange and Na^+/H^+ exchange can be readily distinguished pharmacologically. Like the cation-independent Cl^-/HCO_3^- exchanger, the Na^+-dependent anion exchanger is sensitive to stilbene derivatives, but not to amiloride and its analogs. Conversely, the Na^+/H^+ exchanger is sensitive to amiloride and its analogs, but not to the disulfonic stilbenes.

It has been demonstrated that HCO_3^- traverses the membrane not only by electroneutral exchange, but also through conductive paths. These are in all likelihood channels which transport predominantly other anions such as Cl^- [43]. Considering the direction and magnitude of the membrane potential, activation of anion channels is expected to induce net HCO_3^- efflux and consequently a cytosolic acidification. While its physiological significance remains obscure, the predicted acidification has indeed been measured in both secretory cells and in neurons.

4.6. Monocarboxylate-H^+ cotransport

Tissues with few or no mitochondria, such as white muscle and red cells, respectively, depend largely on glycolysis as a source of ATP. Lactate, the end product of this metabolic pathway, must be eliminated from the cells to enable continued glycolytic flux. To this end, cells have developed transport mechanisms for the disposal of lactate and other monocarboxylates [44]. In principle, because protons are also generated during glycolysis, translocation of both products is necessary and could be coordinated. In fact, protonated organic acids can traverse the lipid bilayer of the plasma membrane. However, when the pK_a of the acids and the prevailing pH are taken into consideration, the rates of monocarboxylic acid transport observed experimentally in cells greatly exceed those predicted on the basis of diffusion across the hydrophobic layer. These findings prompted the suggestion that specific monocarboxylate-H^+ cotransporters existed in the membranes of animal cells. This prediction was borne out by the observation that transport was stereospecific and also by pharmacological means, when hydroxycinnamate derivatives were found to block lactate and pyruvate transport [45].

The monocarboxylate-H^+ symporter can transport not only L-lactate, but also a wide range of unbranched aliphatic monocarboxylates, including pyruvate, as well as ketone bodies like β-hydroxybutyrate and acetoacetate. Dicarboxylates, sulfonates, formate and bicarbonate, in contrast, are not accepted. The transport stoichiometry is proposed to be one-to-one so that, as in the case of the antiporters, the symport process is electroneutral and voltage insensitive. The pharmacology of the monocarboxylate-H^+ symporter differs from that of the transporters listed above. It is characteristically inhibited by micromolar doses of α-cyanocinnamate derivatives. In contrast, the monocarboxylate symporter is only moderately sensitive to disulfonic stilbenes, distinguishing it from the inorganic anion transporters. Other potent inhibitors include quercetin, phloretin and niflumic acid [44].

While efforts to identify the symporter protein of red cells by chemical means were under way, the molecule was serendipitously cloned, as a mutant of the mevalonate transporter [46]. In fact, two different isoforms of the monocarboxylate-H^+ symporter have been identified thus far. The cloned symporters, named MCT-1 and MCT-2 [47] encode membrane proteins of close to 500 amino acids, with 10–12 transmembrane domains. The two isoforms are $\approx 60\%$ identical and differ in their tissue distribution, kinetic and pharmacological properties [47]. They are therefore likely to play somewhat different functional roles.

In anaerobic cells, such as erythrocytes or white muscle fibers, release of lactate is substantial. To the extent that protons accompany the monocarboxylate, the acidic burden of the cytosol is relieved and the MCT can be considered a pH regulatory system.

5. Organellar pH homeostasis

Not only is it important for cells to regulate their cytosolic pH, but also the pH inside organelles. As detailed below, the luminal pH of individual organelles can vary greatly and each must be tightly controlled for the resident enzymes to function optimally. In fact, aberrations of the normal pH homeostasis, either through disease, infection, or by pharmacological means, can lead to significant functional changes [48–50]. In the last ten years our understanding of the pH regulation of the exocytic and endocytic pathways has grown rapidly. The existing literature is very extensive so that, rather than being a comprehensive review, this section will mainly highlight the tools currently used for organellar pH measurements and the central features of pH regulation in each compartment.

5.1. Measurement of organellar pH

Three main types of probes have been used for measurement of the luminal pH of subcellular organelles: (i) fluorescent permeant indicators (e.g. acridine orange); (ii) fluorescent impermeant indicators (e.g. fluorescein isothiocyanate (FITC)-conjugated endocytosis markers, liposome fusion) and (iii) non fluorescent permeant indicators (e.g. DAMP, primaquine, [^{14}C]-methylamine).

(i) *Fluorescent permeant indicators.* Acridine orange, 3,6-dimethylaminoacridine (AO), belongs to a unique family of fluorescent amines that permeate through membranes and accumulate in acidic compartments. AO and other members, 9-aminoacridine and atebrin, exhibit the property of fluorescence quenching when accumulation of dye in acidic compartments results in molecular crowding and self-interactions. AO is further characterized by a concentration-dependent shift in its fluorescence spectrum. Therefore, not only is there a concentration dependent quenching of its monomeric fluorescence at 530 nm, but formation of multimeric aggregates of acridine orange, at concentrations greater than 100 μM, produces a new fluorescence emission peak at 640 nm. This fluorescence shift provides a visual tool to identify compartments with greater AO sequestration, correlating with the presence of steeper pH gradients (acidic inside). The use of acridine derivatives, however, is complicated by their binding to DNA and RNA, carcinogenicity and comparatively low sensitivity [51].

(ii) *Fluorescent impermeant indicators.* The measurement of the pH of endocytic vesicles in intact cells is possible through the covalent linkage of pH-sensitive fluorescein derivatives, such as FITC, to macromolecules such as dextran, asialooorosomucoid, or ovalbumin which can be endocytosed by living cells. Through pulse-chase experiments these fluid phase pH indicators are temporally localized to specific compartments of the endocytic pathway. FITC-dextran has been the most popular not only because of its ideal pK$_a$ around 6.4, but also for its resistance to organellar enzymes (particularly lysosomal enzymes), its hydrophilicity (which makes it membrane impermeant) and its stability and nontoxicity in biological systems [52]. Confinement of FITC-dextran to the endosomal compartment requires the inhibition of endosome–lysosome fusion, which can be achieved by lowering the experimental temperature to 18°C or below [53].

A more specific approach to compartmental labeling is based on the measurement of the fluorescence of FITC-conjugated ligands that bind specifically to receptors restricted to endosomes or lysosomes. For instance, FITC-transferrin is known to be confined to the endosome and thus provides more accurate and less time-dependent measures of endosomal pH [54].

More recently, measurements of *trans*-Golgi pH in living cells were reported [55]. The approach used was based on an earlier observation that liposomes of ≈70 nm diameter preferentially fuse with the *trans*-Golgi membrane. Size-fractionated liposomes containing fluid phase, impermeant fluorophores (derivatives of both fluorescein and rhodamine) were microinjected into cultured human skin fibroblasts. The liposomes were allowed to fuse with the trans-Golgi and the pH-sensitive emission of fluorescein was monitored. The pH-insensitive fluorescence of rhodamine was used to normalize the determinations.

Our group has harnessed the unique intracellular pathway traversed by a bacterial toxin, verotoxin, to determine the pH of the Golgi of living cells (unpublished observations). The toxin is internalized via clathrin-coated pits and follows a retrograde pathway to reside temporarily in the Golgi apparatus, before eventually reaching the endoplasmic reticulum. Verotoxin can be labeled with fluorescein isothiocyanate and allowed to bind to its plasma membrane receptors. The fluorophore can then be chased to the Golgi apparatus where it will remain for several hours. Measurements of Golgi luminal pH can then be obtained by fluorescence imaging, essentially as described for the liposomal fusion method.

(iii) *Non-fluorescent permeant indicators.* DAMP, 3-(2,4-dinitro anilino)-3′-amino-*N*-methyldipropylamine, is a weak base that permeates through biological membranes and accumulates in acidic compartments [56]. Standard aldehyde fixation and immunogold detection using an antibody against the dinitrophenol group can provide not only subcellular localization but semi-quantitative measures of organellar pH by electron microscopy. Using this method with control cells and cells defective in the cystic fibrosis gene, it has been postulated that the *trans*-Golgi and a prelysosomal compartment of cells with mutations in the cystic fibrosis gene were less acidic (i.e. contained fewer DAMP particles) than the control counterparts [48]. An alternative probe used for immunological detection is primaquine, for which good antibodies are available [57].

[^{14}C]-Methylamine, also a weak base, has similarly been used to determine organellar pH, following its accumulation in acidic compartments [58,59]. Quantitative measurements of pH can be made by applying this method to isolated organelles. However, in intact cells the identity of the compartments studied is ill defined. Moreover, the use of radioisotopes to measure pH has fallen out of favor because of the terminal nature of the determinations and due to cost and safety concerns.

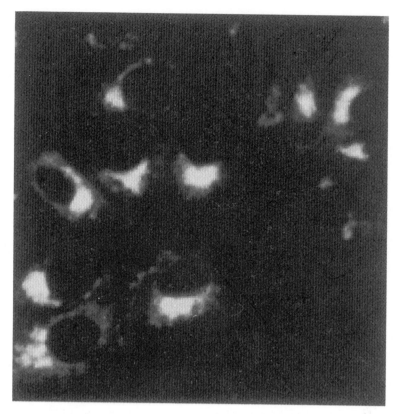

Fig. 2.   Digital image of luminal Golgi staining with FITC-labeled verotoxin. Image of viable Vero cells generated by fluorescence microscopy, detected with a cooled CCD camera.

5.2. Manipulation of organellar pH

The pH of organelles can be altered by interfering with the systems that transport proton (equivalents) into or out of their lumen. This can be accomplished in cell-free systems as well as in intact cells by pharmacological means. Potent inhibitors of pumps, channels and exchangers are available. Of particular interest are the macrolide antibiotics, bafilomycin A_1 and concanamycin, isolated from *Streptomyces* sp., which were recently described to be strong and highly specific inhibitors of the vacuolar type (V) H^+-ATPase [60,61]. They are ineffective towards other members of the H^+-ATPase family, such as F-type or E-type H^+-ATPases. When added to organelles of the endocytic or exocytic pathways, these antibiotics can rapidly dissipate the luminal acidification.

As in the case of the cytosol, the pH of organelles can also be manipulated readily with weak electrolytes and with ionophores that catalyze net translocation of H^+ equivalents.

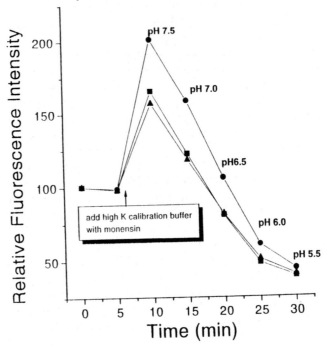

Fig. 3. Measurement of Golgi pH *in situ.* Vero cells were stained with FITC-labeled verotoxin, as in fig. 2. Fluorescence was quantified microscopically using a CCD. Where indicated, fluorescence was calibrated as a function of pH by perfusing the cells in K-rich media of the indicated pH containing 5 μM monensin.

5.3. Calibration

Various methods of calibration are available to correlate physical measurements with absolute pH. With fluorescence measurements in living cells, the most reliable calibrations are those performed *in situ.* The procedure aims to equilibrate the intraorganellar and cytoplasmic pH with a well defined extracellular pH. This equilibration is frequently accomplished using ionophores which increase the permeability to monovalent cations, including H^+. One example is monensin, a well characterized metabolite of *Streptomyces cinammonensis*, that binds H^+, Na^+ and K^+ ions and exchanges them with a 1:1 stoichiometry across membranes [62]. Because the overall monovalent cation concentration is assumed to be equal inside organelles as in the cytoplasm and external medium, addition of monensin is expected to equilibrate the pH of all compartments. Other ionophores often used to increase H^+ permeability for calibration of pH measurements are nigericin, and the conductive protonophores CCCP, and FCCP.

5.4. Regulation of organellar pH

5.4.1. Endosomes

Endosomal acidification was first appreciated during measurements of the fluorescence of FITC-conjugated α-2 macroglobulin through receptor-mediated endocytosis [63]. This notion was subsequently confirmed by labeling the transferrin receptor, known to remain within the prelysosomal compartment, with pH-sensitive dyes. Because the prelysosomal compartment is so heterogeneous, containing early endosomes, intermediary endosomal carrier vesicles and late endosomes, current estimates of the pH value for this compartment are only approximate. Nevertheless, most groups agree that endosomal pH ranges between pH 5 and 6 [52,64].

Protons accumulate in endosomes mostly, if not exclusively, through the vacuolar type H^+-ATPase. The ATP dependence of endosomal acidification was first appreciated while studying purified endosomal fractions prelabeled *in situ* with FITC-conjugated dextran. Endosomal acidification was found to be electrogenic and sensitive to *N*-ethylmaleimide (NEM), as well as bafilomycin and concanamycin. These are hallmarks of V-type ATPases [60,61,65,66]. It is well accepted that a parallel chloride conductance exists in the endosomal membrane and that this pathway compensates for the electrogenicity of the pump, permitting effective net transport of H^+ and consequently luminal acidification. Hence, chloride has been postulated to control the rate and/or extent of acidification. The role of other regulators is debatable. Early studies of the kinetics of endosomal acidification supported the hypothesis that the Na,K-ATPase may antagonize the activity of the proton pump. Due to the directionality of its electrogenic effect, the Na,K-ATPase would generate a luminal positive membrane potential that would oppose H^+ pumping by the V-type pump. The cumulative evidence is contradictory, however. When rat liver endosomes were studied, Na,K-ATPase was detected immunologically. Despite this, *in vitro* studies by some groups failed to detect the presence of a functionally active pump [67]. Other groups have provided evidence for the presence of an amiloride-insensitive Na^+/H^+ exchanger in endosomes [68,69]. The evidence is based on Na^+ dependent proton efflux from endosomes that is independent of K^+ or ATP. Clearly, a more detailed look at the modes of regulation of endosomal acidification is warranted. This should be coupled to a more refined classification of vesicular types based on their function.

5.4.2. Lysosomes

The lysosome is the most acidic compartment in mammalian cells. Over a hundred years ago Metchnikoff demonstrated the acidity of the lysosomal compartment of protozoa following ingestion of litmus paper, which displayed a distinct color change indicative of acidity. There is general consensus that the pH of lysosomes is between 4.5 and 5.5 [63,70]. Like endosomes, the pH gradient of lysosomes is generated by a vacuolar type H^+-ATPase, which transports protons in the presence of Mg^{2+}-ATP. Selective inhibition is possible with bafilomycin and concanamycin, and with the alkylating agent NEM. In contrast the lysosomal pump is unresponsive to oligomycin, azide and vanadate. No other determinants of lysosomal pH regulation have been identified. It is clear, however, that the ionic permeabilities of lysosomes are different from those of endosomes. The lysosomal membrane is tighter to protons, Na^+, Li^+ and Cl^- and more permeable to K^+ and phosphate [71]. Therefore, while the general mechanism of lysosomal acidification is not dissimilar to that of endosomes, there are subtle quantitative differences. These may reflect the difference in the density of proton pumps, differential degrees of pump activation, different isoforms of some of the pump subunits, or differences in counterion permeability.

5.4.3. Specialized compartments

(a) Phagosomes. Mammalian organisms have well differentiated cells that function as a defence against external assault from microorganisms. Phagocytosis is an important mechanism of protection whereby microorganisms can be internalized and digested. Most of the breakdown of ingested products occurs in the phagolysosome, which results from fusion of the phagosome with lysosomes. The pH of the phagosome decreases rapidly once it fuses with lysosomes and/or other endomembrane vesicles, attaining pH levels similar to those reported in lysosomes [51]. The underlying process is similarly thought to involve V-ATPases. Interestingly, the addition of ammonium chloride has been reported to inhibit normal phagosome–lysosome fusion and may in certain cell types promote phagosome–endosome fusion [72].

(b) Secretory granules. Secretory organelles, such as the chromaffin granules of the adrenal medulla, also have an acidic lumen. Chromaffin granules are responsible for the storage of catecholamines and their subsequent release upon stimulation. The granules are rich in vacuolar type H^+-ATPases, which function to provide an acidic environment to the lumen. The acidity generates an appropriate protonmotive gradient that is utilized for catecholamine accumulation in the granule. In addition, the acidic pH protects the catecholamines, which are unstable at alkaline levels. Determinations of pH in secretory granules have yielded values as low as 5.5 [73].

5.4.4. Mitochondria

Mitochondrial pH regulation is complex due to the nature of the energetically active inner membrane, which contains the respiratory chain. This complex series of enzymes acquire redox energy which is converted to a transmembrane protonmotive force. The apparent global stoichiometry is the extrusion of two protons for every electron pair translocated. A proton gradient is therefore generated to produce an inside-alkaline pH in the range of 8.0. The mitochondrion contains in addition a large multi-subunit complex, called the F_1F_0-ATPase, which is related structurally to the V-type ATPase. Under normal circumstances, the F_1F_0-ATPase does not consume but rather produces ATP, utilizing the protonmotive energy generated by the respiratory chain. Other ions, such as calcium, potassium and sodium, also harness the proton electrochemical gradient to translocate across the inner mitochondrial membrane [74–77]. These mechanisms are beyond the scope of this review.

5.4.5. Endoplasmic reticulum

Compared to the endocytic pathway, analyzing the components of the exocytic pathway poses a greater challenge to the cell biologist, as there is no simple procedure for studying them *in situ*. Most of the current understanding of pH regulation in the endoplasmic reticulum, therefore, is based on analysis of cell-free systems. The only available estimates of the native pH of the endoplasmic reticulum, obtained by immunoelectron microscopy using DAMP, suggest that the proton concentration is similar to that of the cytosol [56]. Thus far, the presence of vacuolar type H^+-ATPases in the endoplasmic reticulum has not been documented by immunocytochemistry. Some evidence does exists, however, for the presence of a functioning nucleotide-regulated Cl^-/OH^- exchanger in this compartment. This transporter appears to be active in cell-free experiments employing microsomes derived from the reticulum. It is activated by either ATP or GTP although hydrolysis of the nucleotide is seemingly not required [78]. The contribution of this system to the maintenance of the pH of the endoplasmic reticulum remains unclear.

5.4.6. Golgi apparatus

The Golgi apparatus is a crucial component of the exocytic pathway. It consists of perinuclear stacks of cisternae that can be divided into four main compartments based on ultrastructural and functional analysis [79]. Most proximal to the nucleus is the *cis*-Golgi, which is followed by the medial Golgi, *trans*-Golgi and the *trans*-Golgi network or TGN. Each set of stacks contains specific enzymes that participate in series to perform the necessary post translational modifications, much like a factory assembly line. The TGN has an additional function of sorting products of the exocytic pathway and is an active site for cycling of resident proteins as well as plasma membrane proteins. Golgi resident enzymes have specific pH optima at which they function best. Many of them work better in a slightly acidic milieu. In fact when attempts were made to dissipate pH gradients in living cells using ionophores or permeant amines, normal secretion and membrane traffic were altered [80].

Initial studies investigated the pH regulation of the Golgi apparatus using cell-free systems. The presence of an electrogenic V-type ATPase which is selective for proton translocation was detected [81,82]. As expected, this pump was found to be ATP dependent. Regulation was suggested to depend on a parallel chloride conductance pathway, as in endosomes. These studies were helpful, but failed to reproduce physiologic conditions and were fraught with problems such as contamination with other membranes.

Most recently, studies were reported using a microinjection technique whereby fluorophore-filled liposomes were injected into living cells [55]. Liposomes of a specific size fuse with the *trans*-Golgi and their contents, presumably dispersed uniformly into this compartment, fluoresce with an intensity dictated by the proton concentration in the cistern. Measurements were made with a specialized imaging system coupled to a fluorescence microscope. A steady state pH value for the *trans*-Golgi of 6.17 was calculated in these studies. Some of the limitations of this technique include the rapid dissipation of fluorescence and the need for microinjection.

Our group has employed the technique described above using recombinant B subunit of verotoxin (unpublished observations). FITC-labeled verotoxin was allowed to internalize, reaching the Golgi apparatus within 1–2 hours. Fluorescence measurements were made of selected regions corresponding to the Golgi with a sensitive cooled CCD camera. *In situ* calibrations were performed using the ionophore monensin. The average pH value of the Golgi apparatus calculated with this method was 6.5. The discrepancy between the two techniques may depend on several factors, the most important of which might be the nature of the compartment being measured. The FITC-labeled verotoxin may actually reside in earlier Golgi compartments, which are presumed to be less acidic than the trans-Golgi. This is deduced from the fact that DAMP accumulation and V-ATPase detection were minimal in the earlier (e.g. *cis*, medial) compartments [83].

Studies treating cells with bafilomycin demonstrate a rapid alkalinization of the Golgi, suggesting the presence of an active V-ATPase proton pump. Preliminary experiments have failed to detect Na^+-dependent pH changes, arguing against functional Na^+/H^+ exchange in the Golgi membrane. As further *in situ* experiments are under way, a clearer picture of the mechanisms regulating pH in different compartments of the Golgi complex should emerge shortly.

Acknowledgements

J.H.K. is supported by a Postdoctoral Fellowship from the Canadian Cystic Fibrosis Foundation (CCFF) and the Janssen Ortho Incorporated. N.D. is supported by the Swiss Foundation for Medical and Biological Research. S.G. is cross-appointed to the Department of Biochemistry of the University of Toronto and is an International Scholar of the Howard Hughes Medical Institute (HHMI). Original research in the authors' laboratory is funded by the CCFF, by the HHMI and by the Medical Research Council of Canada.

References

1. Roos, A. and Boron, W.F. (1981) Physiol. Rev. **61**, 296–434.
2. Grinstein, S. and Putnam, R. (1994) Measurement of intracellular pH. in, Methods in Membrane and Transporter Research. R.G Landes Company, Austin, TX, pp. 113–141.
3. Frohlich, O. and Wallert, M.A. (1995) Cardiovasc. Res. **29**, 194–202.
4. Thomas, R.C. (1978) Ion-sensitive Intracellular Microelectrodes. How to Make and Use Them. Academic Press, New York, 110 pp.
5. Ammann, D. (1986) Ion-sensitive Microelectrodes. Principles, Design and Application. Springer Verlag, New York, 346 pp.
6. Rodrigo, G.C. and Chapman, R.A. (1990) Pflugers Arch. **416**, 196–200.
7. Thomas, R.C. (1974) J. Physiol. **238**, 159–180.
8. Ammann, D. Lanter, F. Steiner, R.A., Schulthess, P. Shijo, Y and Simon, W. (1981) Anal. Chem. **53**, 2267–2269.
9. Chao, P., Ammann, D., Oesch, A., Simon, W. and Lang. F. (1988) Pflugers Arch. **411**, 216–219.
10. Aickin, C.C. (1984) J. Physiol. **349**, 571–585.
11. Muckenhoff, K., Schreiber, S., De Santis, A., Okada, Y. and Scheid, P. (1994) J. Neurosci. Meth. **51**, 147–153.
12. Tsien, R.Y. (1989) Fluorescent indicators of ion concentrations, in: Fluorescence Microscopy of Living Cells in Culture, eds Y.L. Wang and D.L. Taylor. Academic Press, pp. 127–156.

13. Haugland, R.P. (1994) Handbook of Fluorescent Probes and Research Chemicals. Molecular Probes Inc., Eugene, OR, 421 pp.
14. Rink, T.J., Tsien, R.Y. and Pozzan, T. (1982) Cell. Biol. **95**, 189–196.
15. Whitaker, J.E., Haugland, R.P. and Prendergast, F.G. (1991) Anal. Biochem. **194**, 330–344.
16. Thomas, J.A., Buchsbaum, R.N., Zimniac, A. and Racker, E. (1982) Biochemistry **18**, 2210–2218.
17. Gillies R.J. (1994) NMR spectroscopy in Biomedicine. Academic Press, New York.
18. Szwergold, B.S. (1992) Annu. Rev. Physiol. **54**, 775–798.
19. Boron, W.F. and De Weer, P.J. (1976) Gen. Physiol. **67**, 91–112.
20. Grinstein, S., Cohen, S., Goetz-Smith, J.D and Dixon, S.J. (1989) Meth. Enzymol. **173**, 777–790.
21. Connor, J.A. and Ahmed, Z. (1983) Cell. Mol. Neurobiol. **4**, 53–66.
22. Demaurex, N., Grinstein, S., Jaconi, M., Schlegel, W., Lew, D.P. and Krause, K.H. (1993) J. Physiol. **466**, 329–344.
23. Grinstein, S., Romanek, R. and Rotstein, O.D.(1994) Am. J. Physiol. **267**, C1152–1159.
24. Orlowski, J., Kandasamy, R. and Shull, G.E. (1992) J. Biol. Chem. **267**, 9331–9339.
25. Fliegel, L. and Frohlich, O. (1993) Biochem. J. **296**, 273–285.
26. Soleimani, M., Singh, G., Gullans, S. and McAteer, J.A. (1994) J. Biol. Chem. **269**, 27973–27978.
27. Mrkic, B., Tse, C., Forgo, J., Helmle-Kolb, C., Donowitz, M. and Murer, H. (1993) Pflugers Arch. **424**, 377–384.
28. Demaurex, N., Orlowski, J., Woodside, M. and Grinstein, S. (1995) J. Gen. Physiol. **106**, 85–111.
29. Aronson, P.S. (1985) Ann. Rev. Physiol. **47**, 545–560.
30. Counillon, L., Scholz, H., Lang, J. and Pouyssegur, J. (1993) Mol. Pharmacol. **44**, 1041–1045.
31. Kapus, A., Grinstein, S., Kandasamy, R. and Orlowski, J. (1994) J. Biol. Chem. **269**, 23544–23552.
32. Forgac, M. (1992) J. Bionereg. Biomembr. **24**, 341–350.
33. Al-Awqati, Q. (1986) Ann. Rev. Cell Biol. **2**, 179–200.
34. Chow, C-W., Demaurex, N. and Grinstein, S. (1995) Curr. Opin. Hematol. **2**, 89–95.
35. Lukacs, G., Kapus, A., Nanda, A., Romanek, R. and Grinstein, S. (1993) Am. J. Physiol. **265**, C3–C14.
36. DeCoursey, T.E. and Cherny, V.V. (1994) J. Membr. Biol. **141**, 203–223.
37. Jennings, M.L. (1989) Annu. Rev. Biophys. Chem. **18**, 397–430.
38. Olsnes, S., Tonnessen, T.I. and Sandvig, K. (1986) J. Cell Biol. **102**, 967–971.
39. Alper, S.L., Kopito, R.R. and Lodish, H.F. (1987) Kidney Int. **32**, S117–S128.
40. Kopito, R.R. (1990) Int. Rev. Cytol. **123**, 177–199.
41. Boron, W.F. (1983) J. Membr. Biol. **72**, 1–14.
42. Chen, L.K. and Boron, W. F. (1991) Kidney Int. **40**, S11–S17.
43. Bretag, A.H. (1987) Physiol. Rev. **67**, 618–645.
44. Poole, R.C. and Halestrap, A.P. (1993) Am. J. Physiol. **264**, C761–C782.
45. Halestrap, A.P. and Denton, R.M. (1974) Biochem. J. **138**, 313–316.
46. Garcia, C.K., Goldstein, J.L., Pathak, P.K., Anderson, R.G. and Brown, M.S. (1994) Cell **76**, 865–873.
47. Garcia, C.K., Brown, M.S., Pthak, P.K. and Goldstein, J.L. (1995) J. Biol. Chem. **270**, 1843–1849.
48. Barasch, J., Kiss, B., Prince, A., Saiman, L., Gruenert, D. and Al-Awqati, Q. (1991) Nature **352**, 70–73.
49. Wang, C., Lamb, R.A. and Pinto, L.H. (1994) Virology **205**, 133–140.
50. Poole, B. and Ohkuma, S. (1981) J. Cell Biol. **90**, 665–669.
51. Steinberg, T.H., Swanson, J.A. (1994) Meth. Enzymol. **236**, 147–60.
52. Geisow, M.J. (1984) Exper. Cell Res. **150**, 29–35.
53. Ohkuma, S. (1989) Meth. Enzymol. **174**, 131–154.
54. Yamashiro, D.J. and Maxfield, F.R. (1984) J. Cell. Biochem. **26**, 231–246.
55. Seksek, O., Biwersi, J. and Verkman, A.S. (1995) J. Biol. Chem. **270**, 4967–4970.
56. Anderson, R.G., Falck, J.R., Goldstein, J.L., Brown, M.S. (1984) Proc. Natl. Acad. Sci. USA **81**, 4838–4842.
57. Hiebsch, R.R., Raub, T.J. and Wattenberg, B.W. (1991) J. Biol. Chem. **266**, 20323–20328.
58. de Duve, C., Ohkuma, S., Poole, B. and Tulkens, P. (1978) in: Microenvironments and Metabolic Compartmentation, eds P.A. Srere and R.W. Estabrook. Academic Press, New York, pp. 371–380.

59. Kopitz, K., Gerhard, C., Hofler, P. and Cantz, M. (1994) Clin. Chim. Acta **227**, 121–133.

60. Bowman, E.J., Siebers, A. and Altendorf, K. (1988) Proc. Natl. Acad. Sci. USA **85**, 7972–7976.

61. Woo, J.T., Shinohara, C., Sakai, K., Hasumi, K. and Endo, A. (1992) J. Antibiot. **45**, 1108–1116.

62. Tartakoff, A.M. (1983) Cell **32**, 1026–1028.

63. Tycko, B. and Maxfield, F.R. (1982) Cell **28**, 643–645.

64. Tycko, B., Keith, C.H. and Maxfield, F.R. (1983) J. Cell Biol. **97**, 1762–1776.

65. Young, G.P., Qiao, J.Z. and Al-Awqati, Q. (1988) Proc. Natl. Acad. Sci. USA **85**, 9590–9594.

66. Forgac, M. (1989) Physiol. Rev. **69**, 765–796.

66. Sipe, D.M., Jesurum, A. and Murphy, R.F. (1991) J. Biol. Chem. **266**, 3469–3474.

68. Hilden, S.A., Ghoshroy, K.B. and Madias, N.E. (1990) Am. J. Physiol. **258**, F1311–1319.

69. Anbari, M., Root, K.V. and Van Dyke, R.W. (1994) Hepatology **19**, 1034–1043.

70. Ohkuma, S. and Poole, B. (1978) Proc. Natl. Acad. Sci. USA **75**, 3327–3331.

71. Piqueras, A.I., Somers, M., Hammond, T.G., Strange, K., Harris, H.W., Jr., Gawryl, M. and Zeidel, M.L. (1994) Am. J. Physiol. **266**, C121–133.

72. Hart, P.D. and Young, M.R. (1991) J. Exp. Med. **174**, 881–889.

73. Mellman, I., Fuchs, R. and Helenius, A. (1986) Ann. Rev. Biochem. **55**, 663–700.

74. Ernster, L. and Schatz, G. (1981) J. Cell Biol. **91**, 227s–255s.

75. Jezek, P., Orosz, D.E., Modriansky, M. and Garlid, K.D. (1994) J. Biol. Chem. **269**, 26184–92610.

76. Papa, S. (1982) J. Bioenerg. Biomembr. **14**, 69–86.

77. Sholtz, K.F., Gorskaya, I.A. and Kotelnikova, A.V. (1983) Eur. J. Biochem. **136**, 129–134.

78. Begault, B. and Edelman, A. (1993) Biochim. Biophys. Acta **1146**, 183–190.

79. Mellman, I. and Simons, K. (1992) Cell **68**, 829–840.

80. Thorens, B. and Vassalli, P. (1986) Nature **321**, 618–620.

81. Moriyama, Y. and Nelson, N. (1989) J. Biol. Chem. **264**, 18445–18450.

82. Marquez-Sterling, N., Herman, I.M., Pesacreta, T., Arai, H., Terres, G. and Forgac, M. (1991) Eur. J. Cell Biol. **56**, 19–33.

83. Anderson, R.G. and Pathak, R.K. (1985) Cell **40**, 635–643.

CHAPTER 21

K⁺ Circulation Across the Prokaryotic Cell Membrane: K⁺-uptake Systems

S. STUMPE, A. SCHLÖSSER, M. SCHLEYER and E.P. BAKKER

Abteilung Mikrobiologie, Universität Osnabrück, D-49069 Osnabrück, Germany

© *1996 Elsevier Science B.V.*
All rights reserved

Handbook of Biological Physics
Volume 2, edited by W.N. Konings, H.R. Kaback and J.S. Lolkema

Contents

1. General introduction

All living cells contain relatively high concentrations of K^+ in their cytoplasm. In prokaryotes the main function of this cell K^+ lies in its role in turgor-pressure homeostasis (with related processes such as osmoadaptation, osmoregulation and volume regulation, for reviews see Refs. [1–4]) and in pH homeostasis [5]. For this purpose the cells transport K^+ across the cell membrane via a variety of both K^+-uptake and K^+-efflux systems. Until recently all of these systems were thought to function as either primary pumps or secondary transport systems. However, within the last year it has been reported that three different types of K^+-accepting channels may exist in the cell membrane of prokaryotes [6–8]. The activity of these transport systems and channels is coordinated in such a manner, that the basal rate of K^+ cycling across the cell membrane is low. Thereby, the cells avoid unnecessary loss of energy due to futile K^+ circulation. Under extreme conditions (e.g. after osmotic shock or a sudden change in pH of the medium) the cells rapidly activate one or more of their K^+-transport systems and depending on the conditions this can lead to either net K^+ uptake or net K^+ exit. This acute stress situation is followed by a phase of adaptation, during which the cells adjust to the new condition and the transmembrane K^+ fluxes gradually return to their original level.

In this chapter, we will describe which K^+-transport genes and K^+-transport systems have been identified in prokaryotes, which roles K^+ and K^+ transport play in the physiology of these organisms and what is known about the various K^+-uptake systems. In separate chapters I.R. Booth et al. (Chapter 30) and L. Letellier and M. Bonhivers (Chapter 27) will describe the properties of K^+-efflux systems and of channels from prokaryotes, respectively [9,10]. More data about the properties of the K^+-transport systems from prokaryotes can be found in a number of reviews [11–17].

2. K^+-transport genes and K^+-transport systems

Figure 1A gives a survey of the genes, that code for products involved in K^+ circulation across the *Escherichia coli* cell membrane [6,7,11,18]. In *Salmonella typhimurium* several *trk*-genes have been labelled sap ([19–21], see also Section 4.2). These genes are given within parentheses in Fig. 1A. The genome of *Haemophilus influenzae* contains the *trkA*, *trkH* (frame HI0725), *trkE* (*sapABCDF*), *kefC* and *mscL* genes (frame HI0625; [22] and our own alignment studies). However, this organism lacks *kdp* genes and *kup* [22]. Almost nothing is known about the pattern of K^+-transport genes in other prokaryotes. However, with the recent acceleration

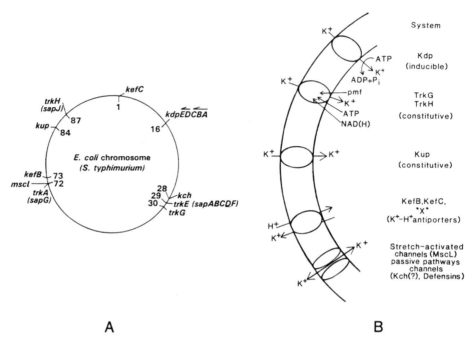

Fig. 1.   K$^+$-transport genes (A) and K$^+$-transport systems (B) in *Escherichia coli*. See the text for details. The *S. typhimurium sap* genes equivalent to the *E. coli trk* genes are given within parentheses in panel A.

in prokaryotic genome-sequencing projects [22], many relevant data are expected to appear within the next two years.

The genes depicted in Fig. 1 (A) encode 4 different K$^+$-uptake systems (Kdp, TrkH, TrkG and Kup), two K$^+$-efflux systems (KefB and KefC [9,15]), one mechano-sensitive channel (MscL [6]), and one putative, but biochemically not yet identified K$^+$ channel (Kch [7]). This list is incomplete with respect to efflux systems and channels: first, the genetically identified KefB and KefC systems are primarily involved in detoxification of glutathione-adducts, rather than in K$^+$ efflux [9,15,23]. Activity measurements show that the cells must contain at least one additional K$^+$/H$^+$-antiporter [24]. In addition, *E. coli* may contain an antiporter that exchanges K$^+$ and Na$^+$ against H$^+$ [25]. Secondly, apart from MscL, *E. coli* cell membranes contain two additional mechano-sensitive channels with smaller pore diameters [26]. Finally, the recently described K$^+$ channel (Kcs) of *Streptomyces lividans* [8] may also be present in other bacteria including *E. coli*.

Although little is known about the pattern of K$^+$-transport systems in prokaryotes other than *E. coli*, one expects it to be similar to that depicted in Fig. 1B. However, some well-documented differences exist, and as far as this concerns K$^+$-uptake systems these differences are discussed in Section 4.4.

3. Function of cell K$^+$ and of transmembrane K$^+$ circulation

3.1. Status of cell K$^+$; turgor pressure

The activities of the K$^+$-transport systems of prokaryotes are coordinated such that under normal conditions the cells contain a K$^+$ concentration of at least several hundreds of mM in their cytoplasm. There has been some debate about which fraction of this cell K$^+$ is bound to macromolecules (e.g. Ref. [27]). Recent experiments with *E. coli* have shown that even in cells grown at low external osmolarity 44% of cell K$^+$ (amounting to 320 mmol/g dry wt. of cells, or approximately 190 mM K$^+$) is free inside the cytoplasm [28]. With increasing medium osmolality free K$^+$ in the cytoplasm increases to a value of 890 mmol/g dry wt of cells (approximately 600 mM) at a medium osmolality of 1.26 Osm [28]. Under these conditions the amount of bound K$^+$ also increases. This is explained by K$^+$ replacing bound putrescine, which is released by the cells at high medium osmolality [29,30]. In prokaryotes other than *E. coli*, too, a considerable fraction of their cell K$^+$ is expected to occur free in the cytoplasm.

At these concentrations K$^+$ is the main osmolyte in the cytoplasm. Thereby it makes a major contribution to the turgor pressure of the cells, which is defined as the difference in osmotic pressure between the cytoplasm and the medium. Turgor pressure is essential for growth. It causes the cytoplasmic membrane to be pressed against the cell wall. In the few bacteria in which turgor pressure has been measured it amounts to values between 1 and 4 Atm [31,32].

Very high turgor pressures (at least 25 Atm) must exist in extremely thermophilic archaea like *Methanothermus fervidus*, *Methanococcus kandleri* or *Pyrococcus woesii*. Even when grown at low medium osmolarity these organisms contain K$^+$ concentrations of 1–2.3 M in their cytoplasm. This K$^+$ forms the counterion to unusual osmolytes like the trianionic cyclic 2,3-diphosphoglycerate [33] or di-myo-inositol-1,1-phosphate [34]. These solutes play a role in the thermostabilization of cytoplasmic enzymes rather than in osmoregulation of the cells [34]. Even so, these archaea must possess extremely strong cell walls in order to withstand the turgor pressure exerted by the high internal solute concentrations.

3.2. Osmoadaptation

3.2.1. Hyperosmotic shock; osmoregulation at high medium osmolality
An increase in the osmolality of the medium ('upshock') will reduce turgor pressure. When it becomes negative water leaves the cells, leading to shrinkage of the cytoplasm and retraction of the cell membrane from the cell wall (plasmolysis). Within certain limits growing prokaryotes can adapt to this new condition by accumulating solutes from the medium and/or by synthesizing and storing solutes in their cytoplasm. Especially the solutes accumulated at high medium osmolality are of interest. Zwitterions like betaine or ectoine (1,4,5,6-tetrahydro-2-methyl-4-pyrimidincarboxylic acid), polyols and/or in some cases carbohydrates have been found in up to molar concentrations inside the cytoplasm of a number of halophilic and halotolerant bacteria [35,36]. At such high concentrations these compounds are

believed to preserve the hydration shell of proteins and are therefore called compatible solutes [37].

With *E. coli* and *S. typhimurium* the earliest detected adaptative response to upshock is the accumulation of K^+ inside the cytoplasm [28,38]. It is caused by the activation of the K^+-uptake systems Trk and Kdp ([39], see Section 4). Low turgor pressure also stimulates the synthesis of the Kdp proteins [40]. This control occurs at the level of transcription (see Section 4.1.1).

Electroneutrality for upshock-stimulated K^+ uptake is initially maintained by H^+ extrusion, leading to a transient alkalinization of the cytoplasm [28,38]. At a later stage the cells extrude up to 45 mM of the divalent cation putrescine [29,30]. About 1 min after upshock the cells start to synthesize glutamate. Its amount is smaller than that of the accumulated K^+ [28,38,41]. Recent results indicate that the difference reflects K^+ binding to sites abandoned by the putrescine molecules leaving the cells [28,30]. The result of these processes is that early after osmotic upshock the cells restore their turgor pressure by accumulating potassium glutamate. However, accumulation of this salt only reflects adaptation of the cells to the acute stress situation. Potassium and glutamate ions are not compatible solutes. During a second phase of adaptation the cells will therefore replace parts of or even their complete osmo-induced potassium glutamate load by compatible solutes. The most favourable situation for the cells is to accumulate betaine (or related compounds like proline) from the medium. For this purpose the cells both synthesize and activate the betaine-uptake systems ProP and ProU [42,43]. ProU is the more important of the two systems and it starts to mediate the uptake of considerable amounts of betaine within minutes after the hyperosmotic shock. This event reverses the direction of K^+ movement across the cell membrane [38,44] and inhibits osmo-induced glutamate accumulation [1]. In the absence of betaine, but with choline present in the medium, the cells can still accumulate betaine. They do this by accumulating choline from the medium and oxidizing it to betaine [1,45]. In the absence of betaine, choline or related compounds in the medium the cells have to resort to a third strategy. Under these conditions they synthesize trehalose and store this compound in their cytoplasm [38,41,46]. Trehalose replaces a major part of the free potassium glutamate accumulated after osmotic upshock [38,47]. A scheme of the different pathways of osmoadaptation in *E. coli* is given in Fig. 2.

Several authors have proposed that the high K^+ concentration occurring in the cytoplasm of upshocked bacteria serves as a second messenger in triggering the expression of genes and in activating enzymes involved in osmoadaptation [3,4]. However, with the possible exception of a function of high K^+ in trehalose synthesis [46], there is no evidence for this notion [1].

Even in the presence of betaine enteric bacteria are unable to grow at NaCl concentrations above 2 M. By contrast, many halophilic bacteria thrive both at moderately low and at very high NaCl concentrations (up to 2.5–4 M). These organisms synthesize compatible solutes such as betaine and ectoine. At high osmolarity they do not appear to accumulate high concentrations of K^+ and it is a matter of debate whether this cation plays a role in osmoregulation of these halophiles at all [12].

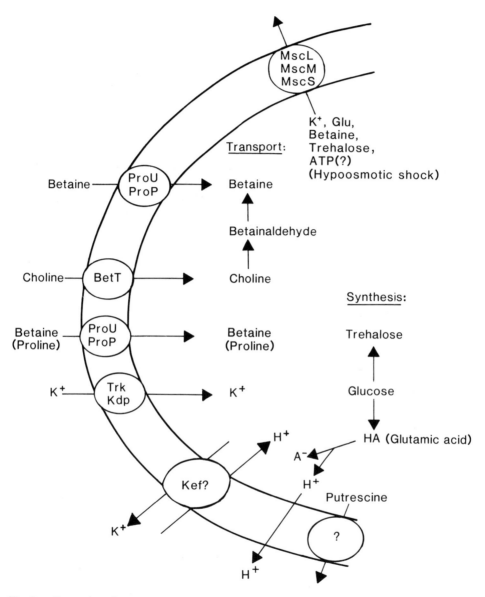

Fig. 2. Osmoadaptation pathways in enteric bacteria. See the text for details. The upper right part of the figure depicts solute efflux through mechano-sensitive channels occurring after hypoosmotic shock.

Extremely halophilic archaea have resorted to a completely different strategy. Species like *Halobacterium salinarium* (formerly *H. halobium* [48]) maintain about 4 M of K$^+$ in their cytoplasm [49]. These organisms can live with this 'non-compatible' ion because they have adapted their proteins by drastically increasing the fraction of negative charges. Hydrated K$^+$ ions are believed to bind to these negatively charged residues, thereby allowing native folding of these proteins [50].

3.2.2. Hypoosmotic shock

Transmembrane K$^+$ movements also play a major role in the adaptation of osmotically down-shocked *E. coli* cells. Under these conditions turgor pressure increases to such an extent that the cells are in danger of bursting. Cell rupture is avoided by the instantaneous release of solutes from the cytoplasm. This extremely rapid efflux is supposed to occur through mechano-sensitive channels [6,26], which open under these conditions [51]. Moderate osmotic downshock of growing *E. coli* cells adapting to high osmolality leads to an almost complete release of their potassium glutamate and of parts of their trehalose (Fig. 2). Under these conditions the cells remain viable, start to reaccumulate potassium glutamate from the medium and resume growth at about 10 min after hypoosmotic shock [47]. Hence, after osmotic downshock the cells resort to the following strategy: By opening their mechano-sensitive channels followed by solute exit the cells reduce their dangerously high turgor pressure to a very low value. Subsequently they adapt permanently to the new situation by reaccumulating some of their solutes and thereby restoring their turgor pressure [47].

3.3. Other functions

The role of K$^+$ transport in pH regulation of the cytoplasm of prokaryotes has been reviewed [5,52,53]. Since hardly any new data are available on this topic, it will not be discussed here. The same applies to the potential role of K$^+$ uptake in the development of the reversed (internally positive) membrane potential of acidophiles at pH values below 3 [52]. A further role for K$^+$ uptake has recently been described with respect to virulence of *S. typhimurium* [19–21]. This function will be described in Section 4.2.6. Finally, in their chapter Booth et al. [9] describe that in the presence of glutathione adducts with toxic compounds, the cells have found a way to extend survival, by exchanging their cell K$^+$ against H$^+$ via the KefB/C systems [23].

4. K$^+$-uptake systems

In this section an overview on the K$^+$-uptake systems Kdp, Trk and Kup is given. Subsequently, K$^+$-uptake systems with properties different from these three systems will be discussed. K$^+$ channels are not included here, since there exists no evidence that they play a role in K$^+$ uptake. For instance, if *E. coli* were to contain a channel with such properties, cells lacking functional Kdp, Trk and Kup systems would grow at K$^+$ concentrations below 10 mM, which is not observed [54,55].

4.1. Kdp

When grown at low K$^+$ concentrations *E. coli* induces the high-affinity K$^+$-uptake system Kdp. With its low K_m-value for K$^+$ of about 2 µM Kdp-containing cells are able to reduce the K$^+$ concentration of the medium to about 50 nM, showing that

Kdp functions as a K$^+$-scavenging system [11]. Uptake of K$^+$ is coupled to the hydrolysis of ATP via a P-Type of ATPase mechanism [56,57]. The genes encoding the proteins of the ATPase complex are organized in the *kdpFABC* operon [16,58–60] (Fig. 3). The expression of this operon is controlled by the *kdpDE*-gene products. KdpD functions as a sensor-kinase. By phosphorylation it communicates with the KdpE protein, which in turn controls expression of the *kdpFABC*-operon (reviewed in Refs. [16,60–62]) (Fig. 3). Below we will discuss signal-sensing by KdpD, control of transcription by KdpE, the properties of the ATPase complex, and of its single subunits, the mechanism by which Kdp transports K$^+$, and the spreading of the Kdp–ATPase among prokaryotes.

4.1.1. Regulation of transcription

The nucleotide sequence of the *kdpDE*-genes [63] suggested that their products belong to the class of two component regulator systems, later renamed as sensor kinase-response regulator systems [64,65]. The sensor kinase is KdpD, which is located in the cell membrane [63]. The protein is predicted to consist of 894 residues, thereby being twice as large as other membrane-bound sensor kinases [61,63–65]. KdpD is composed of three domains: a cytoplasmic, mostly hydrophilic, only partially essential N-terminus of about 400 residues, a hydrophobic region of 100 residues which is predicted to span the membrane four times, and a hydrophilic cytoplasmic C-terminal domain of about 400 residues, containing the active center of the enzyme [61,63,67,68] (Fig. 3A).

Signal recognition by KdpD occurs within or close to the membrane domain. The evidence for this notion is based on the observations that (i) a truncated KdpD protein lacking the membrane region ('KdpD*') is inactive in sensing, but shows *in vitro* still some activity in transducing the signal to KdpE [66], and (ii) mutants impaired in sensing show amino-acid changes in transmembrane helices 3 and 4 of this domain, as well as in the first 25 residues of the subsequent hydrophilic domain [68].

Confusion exists about the nature of the signal sensed by KdpD. Early observations showed that the Kdp system is made under conditions at which the cells lack K$^+$ [54,55]. Subsequent studies indicated that low turgor pressure stimulates the expression of the *kdpFABC*-operon, since hyperosmotic shock exerted by different compounds leads to a transient induction of the operon [40]. The large level of induction observed at low external K$^+$ concentrations was interpreted to mean that the low turgor pressure prevailing under these conditions leads to *kdpFABC* expression [40]. This proposal is attractive, because it links sensing by the KdpD-transmembrane domain with the transmembrane phenomenon of turgor pressure. However, subsequent experiments have indicated that the situation is more complex. First, influencing turgor pressure with the same osmolality of sucrose or NaCl gives rise to quite different levels of induction [44,69,70]. Secondly, *kdp* induction turned out to be strongly pH dependent [70]. The KdpD variants with amino-acid changes in or close to the KdpD-transmembrane domain [68] have given additional information about the induction process. In the mutant strains induction by hyperosmotic shock is normal, but repression by external K$^+$ occurs at much higher concentrations than in the wild type. Moreover, ethanol, chlorpromazine, the local

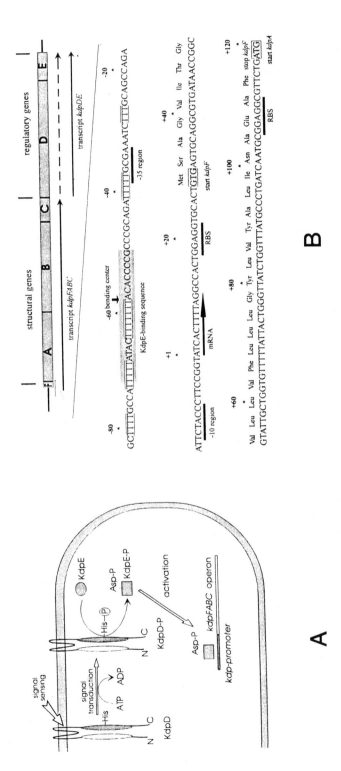

Fig. 3. Control of *kdpFABC*-transcription by the KdpDE proteins. See the text for details.

anaesthetic procaine, or a high $(NH_4)_2SO_4$ concentration also causes *kdp* expression [68]. These observations are interpreted to mean that KdpD senses two parameters: (i) changes in the physicochemical state of the membrane (which may or may not be equivalent to changes in the lateral stretch of the membrane due to changes in turgor pressure), and (ii) the level of external K+ [68]. Clearly, more work is needed to answer the question about the nature of the signal(s) sensed by KdpD.

Signal sensing by KdpD leads to its autophosphorylation. No information is available about how the protein transduces the sensed signal to its catalytic domain. Presumably, the first 150 residues of its C-terminal domain are important in this process. Sequence comparisons with related sensor kinases suggest that the C-terminal 250 residues of KdpD form the catalytic domain and that residue His673 is most likely its autophosphorylation site [61,63,65,71]. Phosphorylated KdpD is able to transfer its phosphate group to presumably residue Asp9 or Asp52 of KdpE [63,66,71].

Mizuno and his coworkers have characterized the binding of (phospho-)KdpE to the promoter region of the *kdpFABC*-operon. DNase-I footprinting showed that KdpE interacts with a region of DNA 72-50 nucleotides 5′ to the transcriptional start site of the operon [59]. The DNA of this part of the promoter is highly curved [72,73] (Fig. 3B). Subsequent results indicated that phospho-KdpE has a much higher affinity for the *kdpFABC*-promoter than has non-phosphorylated KdpE and that *in vitro* transcription only occurs in the presence of phospho-KdpE [73].

In summary, the following scenario for the regulation the *kdpFABC*-operon has emerged: As yet unknown stimuli affect the transmembrane region of KdpD in such a manner that a signal is transduced to the kinase parts of the molecule, leading to its autophosphylation. Phosphorylated KdpD transfers then its phosphate group to KdpE which binds to the promoter region of the *kdpFABC*-promoter in a process that may involve a change in DNA bending of this region [74], thereby promoting transcription of the operon (Fig. 3).

4.1.2. The ATPase complex; mechanism of action

The nucleotide sequence of the *E. coli kdpFABC* operon predicts that the gene products KdpF, KdpA, KdpB and KdpC exhibit M_r-values of 3,090, 59,189, 72,112 and 20,267, respectively [16,58]. The detergent Aminoxid WS35 has been used to solubilize the Kdp–ATPase complex from cytoplasmic membranes of *E. coli* [76], *Alicyclobacillus* (formerly *Bacillus*) *acidocaldarius* [77], and *Rhodobacter sphaeroides* [78]. Initially, the complex has been purified by a number of chromatographic steps in the presence of Aminoxid WS35 [76,77]. Subsequently, the purification protocol has been simplified by employing affinity chromatography on triazine dye containing TSK AF-Red material [78,79].

The Kdp–ATPase complex has long been considered to consist of the KdpA, KdpB and KdpC proteins only [76–80]. However, recent evidence indicates that KdpF may also form part of the complex [81] (see below). The ATPase activity of the complex is with 1–2 U/mg protein [76–79] (turn-over number of about 2 $ATP \cdot complex^{-1} \cdot s^{-1}$) about two orders of magnitude smaller than that of Kdp in intact cells. The reason for this low activity of the complex is not known. The ATPase activity of the complexes isolated from all three species is stimulated by low K^+ concentrations (K_m values of 3.5–25 µM [76–78], suggesting that the reaction cycle of the solubilized enzyme still includes a K^+-dependent step. ATPase activity is inhibited by low concentrations of vanadate [76–79], and the KdpB-subunit in the complex from *E. coli* can be phosphorylated by ATP [82], presumably at residue Asp307 [83]. Together with the extensive sequence homology of the KdpB protein with that of other P-type ATPases [58,84,85], these data suggest that the Kdp system belongs to this family of ATPases.

P-type ATPases are characterized by their property that the enzyme occurs in two conformations, E_1 and E_2 during the reaction cycle. In the first step MgATP transduces energy to the E_1 form by phosphorylating it (E_1-P). This form relaxes to conformation 2 (E_2-P; step 2). The E_2-P form interacts with the transported ion(s), and translocate(s) it (them) across the membrane. This process is driven by the hydrolysis of the mixed anhydride–phosphate bond of the enzyme, giving rise to E_2 (step 3). In the final step E_2 relaxes to the original E_1 form (step 4) (see Refs. [57,85] for more details). According to this scheme one would expect that addition of K^+ leads to rapid dephosphorylation of the Kdp complex. This has indeed been observed for a variant Kdp system [82] with a greatly reduced affinity for K^+ [86]. However, later results showed, that release of phosphate by added K^+ occurs only in the presence of ATP [87], and therefore an extra step, involving K^+ release upon ATP binding to the enzyme has been included in the reaction cycle [62]. Nothing is known about how the Kdp-complex senses changes in turgor pressure [39].

A further question concerns the ions transported via Kdp. Tests for Kdp activity in intact cells are done under conditions at which the K^+ taken up exchanges mainly against extruded Na^+ [55]. The role of Na^+ in this process is to maintain electroneutrality, rather then that it serves as a second substrate for the Kdp system [88]. Unlike K^+, Na^+ does not activate the ATPase of the Kdp complex [82,87]. Moreover, net K^+ uptake via the Kdp-system leads to depolarization of the cell membrane, suggesting that K^+ uptake is electrogenic [89]. Recent experiments with the reconstituted ATPase complex adsorbed to bilayer membranes show unequivocally, that Kdp activity is associated with the formation of an electric field across the membrane (K. Fendler, S. Dröse, K. Altendorf and E. Bamberg, personal communication). These data suggest that K^+ is the only ion transported by the complex. K^+ cannot be replaced by any other ion in tests for Kdp activity [76,90,91].

4.1.3. Topology and function of the single subunits

The structure of Kdp is more complex than that of any other P-type ATPase. This complexity may be necessary for the transport of K^+ with high affinity [16]. For this purpose the function of K^+ transport and energy conversion by ATP hydrolysis has been divided between the KdpA and the KdpB subunits. KdpC and possibly also KdpF may have structural functions. Many conclusions about the structure and topology of the single Kdp subunits have been derived from the nucleotide sequence of the *E. coli kdpFABC*-operon [58], homologous proteins from other P-type ATPases [84,85] and the recently determined nucleotide sequence of the *Clostridium acetobutylicum kdpFABC* operon [92].

Topological analysis of KdpA suggests that it spans the membrane ten times, with both its N- and C-termini being located at the periplasmic side of the membrane [93]. The main evidence that KdpA forms the K^+-translocating subunit of the complex is based on the observation that almost all of the mutations giving rise to lower affinity for K^+ in uptake assays (K_m mutants) map inside the kdpA gene [93]. The residues changed in these KdpA variants cluster at three hydrophilic regions of the protein at the periplasmic side of the membrane and in one hydrophilic loop inside the cytoplasm [93]. Moreover, with one exception, these residues of *E. coli* KdpA are all conserved in the clostridial protein [92,93]. From these data it is concluded that KdpA contains K^+-binding sites at each site of the membrane [93]. Independently, Jan and Jan have proposed that KdpA contains two regions resembling the H5 domain of potassium channels of eukaryotes [94]. H5-domains are relatively hydrophilic loops located at the external side of the cell membrane. They are thought to fold back into the membrane and to be directly involved in channel formation [94]. Two of the periplasmic K^+-binding regions identified in [93] form the center of the putative H5 domains of KdpA [94]. Taken together, these data suggest that the periplasmic K^+-binding site of KdpA is buried inside the membrane and may form part of the K^+-translocation pathway.

The catalytic subunit KdpB contains six putative transmembrane segments, with its C- and N-termini being located in the cytoplasm [16,58]. In analogy with the function of other P-type ATPase domains, the large cytoplasmic domains between the putative transmembrane helices 2 and 3 (140 amino-acid residues) and 4 and 5 (345 amino-acid residues) are thought to be involved in the E_1 to E_2 conformational change and in ATP hydrolysis plus phospho-enzyme formation at residue Asp307 [83], respectively [58,85]. Compared to KdpB the catalytic subunits of the K^+-translocating Na^+/K^+-ATPase and H^+/K^+-ATPase contain at least two additional transmembrane helices at their C-terminus [85]. This might suggest that these helices are involved in ion transport, and that for Kdp this function has been transferred to the KdpA subunit. However, at least two other bacterial P-type ATPases also contain only 6 transmembrane helices, without having additional subunits [95–97]. Apparently, a domain composed of six transmembrane helices is sufficient for the recognition and translocation of a particular ion via a P-type ATPase.

The structure of KdpC resembles that of the β-subunits of the Na^+/K^+-ATPase and H^+/K^+-ATPases, in that its N-terminal domain is predicted to traverse the cell membrane once and that it contains a large hydrophilic C-terminal domain [16]. However, the orientation of KdpC in the membrane is opposite to that of these β-subunits [16]. Hence it is unlikely that the hydrophilic domains of these proteins have similar functions. Despite of its mainly hydrophilic nature KdpC must be buried deep inside the Kdp complex, since it is poorly accessible to hydrophilic reagents added from either side of the membrane. Without KdpC KdpA/B-complexes do not form (A. Siebers and K. Altendorf, personal communication), suggesting that KdpC either has a role in the assembly of the complex or that it connects the KdpA and KdpB subunits [16,60].

The *E. coli* KdpF peptide is encoded by the open reading frame of 29 codons just upstream of the *kdpA* gene [16,58] (Fig. 3B). A slightly shorter frame is also present in the *C. acetobutylicum* operon [92]. The peptides encoded by these *kdpF* genes are highly hydrophobic and thereby resemble the proteolipids of a number of other P-type ATPases (e.g. Ref. [98]). The activity of a protein fusion between KdpF and β-galactosidase indicates that the fusion protein is synthesized under the same conditions as Kdp [60]. In addition, the extent of galactosidase activity of the fusion protein indicates that the N-terminus of KdpF is located outside and that its C-terminus is in or near the cytoplasm [62]. A small peptide of the size expected for KdpF is made in minicells and has recently been detected in the KdpABC complex [81]. Since it is not known how much KdpF is present in the complex, it is too premature to decide whether the protein forms a functional important part of it.

4.1.4. Spreading among prokaryotes

Kdp occurs in a wide variety of bacteria, including many *Enterobacteriaceae* [38,99], various purple phototrophs [78,100], several cyanobacteria [99], the Gram-positives *A. acidocaldarius* [77] and *C. acetobutylicum* [92], and possibly in *Rhizobium* [101] and *Paracoccus denitrificans* [16]. However, for prokaryotes the possession of a Kdp-system is not essential. *kdp* Genes are not present in the genome of *H. influenzae* [22]. Moreover, *Vibrio alginolyticus* [102], *B. subtilis* (W. Epstein, personal communication) and *Enterococcus hirae* [17] do not contain Kdp. With respect to the latter organism it should be mentioned here that a P-type ATPase from this organism, which has long been considered to transport K$^+$ [95] has turned out to transport Cu [96]. Finally, no Kdp-system has been described for any archaeum.

4.2. Trk

At K$^+$ concentrations above 200 μM *E. coli* accumulates K$^+$ mainly via its Trk system(s). Most K-12 strains contain two of these systems (called TrkG and TrkH, see Section 4.2.3). However, this is an exception. Many other *E. coli* strains and all other bacteria are likely to possess only a single Trk system. For *H. influenzae* this has been shown to be the case, since on its chromosome it carries only a single set of *trk* genes [22]. Below we call the system Trk, if its general properties are described and TrkH or TrkG, if referring to one of the two *E. coli* K-12 Trk systems.

Trk has properties very different from Kdp. Its genes lie scattered on the chromosome [18,54] (see Fig. 1A) and are expressed constitutively [54,55]. The system transports K$^+$ with a low affinity (K_m value of about 1 mM), but at a high rate [55]. Since the cells contain only small amounts of Trk proteins [103,104], the turn-over rate of the system amounts to 10^4–10^5 molecules of transported K$^+$ · complex^{-1} · s^{-1}. This value is extremely high for a transport system and at least 100-fold larger than that of Kdp. Like Kdp, Trk is composed of several subunits [18,103] and requires ATP for activity [39,88]. However, ATP is thought to activate the transport process rather than that its hydrolysis drives transport [105]. K$^+$-uptake is linked to transmembrane proton circulation [88], and probably occurs as K$^+$/H$^+$ symport [13]. The TrkA protein binds NAD$^+$ and NADH, suggesting that these dinucleotides also effect the activity of the Trk system [106].

Below we will first describe the properties of the different Trk proteins, then address the question of how Trk transports K^+ across the membrane and finally discuss the role of Trk in virulence of *S. typhimurium*.

4.2.1. Composition

The Trk system consists of at least three proteins: an integral membrane protein (in *E. coli* K-12 TrkH or TrkG, in other bacteria TrkH [18,22,107,108]), a peripheral membrane protein that binds NAD(H) (TrkA [106]) and one or two ATP-binding proteins ('TrkE' which is equivalent to the SapD and SapF proteins from *S. typhimurium* [18,20,109]). A Trk complex has neither been detected nor isolated. If we refer to such a complex, it is in the sense of the composition of the system as given in Fig. 4.

4.2.2. TrkH and TrkG

The products of the *E. coli* trkH and trkG genes, TrkH (483 amino-acid residues) and TrkG, (485 residues), share 41% identical residues throughout their complete sequence [108]. The trkG gene is located within the prophage *rac* region of the chromosome and exhibits for *E. coli* an unusually low G+C content of 36%

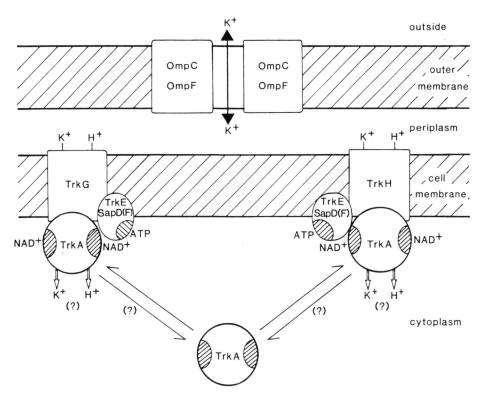

Fig. 4.   The *E. coli* TrkH and TrkG systems. See the text for details. K^+ ions are assumed to permeate freely through the outer membrane porins, indicated with "Omp" in the figure.

[104,108]. Moreover, *E. coli* strains that do not contain *rac* also lack *trkG*. It is therefore assumed that *trkH* is the intrinsic *E. coli* gene and that *trkG* has entered the cells with the prophage [13,108]. The source of *trkG* is not clear. Sequence comparison with TrkH of *H. influenzae* (frame HI0723, 52% identity with *E. coli* TrkH, taking a frame-shift in *H. influenzae trkH* into account [22]), show that *trkG* is likely to have originated from a bacterium more remote from *E. coli* than is *H. influenzae*.

The TrkH and TrkG proteins are predicted to contain 10 membrane spanning regions, and their N- and C-termini to be located at the cytoplasmic side of the cell membrane [13,108] (Fig. 5). Many of the hydrophilic regions connecting the 10 membrane spans are relatively large in TrkH and TrkG. Two large and well conserved periplasmic loops resemble H5 regions of K^+ channels [94] (Fig. 5), suggesting that TrkH (or TrkG) is the K^+-translocating subunit of the complex. This notion is supported by the observations that (i) the K^+/Rb^+-specificity of the system is determined by the TrkG or TrkH subunit [108], and (ii) the exchange of conserved residue Glu468 to Ala in transmembrane helix X of TrkG (TrkH numbering, see Fig. 5) increases the K_m value of this system for K^+ uptake tenfold [110].

The TrkH or TrkG proteins do not form a simple K^+ pore, since (i) their overproduction alone does not lead to an increase of V_{max} for K^+ uptake [104,106], and (ii) deletion of either *trkA* or *trkE* leads to complete inactivation of the Trk system [21,109,111]. We assume that the hydrophilic loops of TrkH (or TrkG) located at the cytoplasmic side of the cell membrane bind TrkA [103] and possibly also TrkE (Figs. 4 and 5), and that the proper interaction between these proteins opens the K^+ pathway through the membrane.

4.2.3. TrkA

TrkA forms a peripheral membrane protein bound to the inner side of the cell membrane. This notion is based on conclusions drawn from the *trkA*-nucleotide sequence [106] and from the observation that always some TrkA is present in the cell fraction of cytoplasmic proteins [103]. Alignment with its own structure showed that TrkA consists of two similar halves [106]. Each half contains a complete dinucleotide-binding domain, similar to that found in the NAD^+-dependent dehydrogenases, LDH, MDH and GAPDH. In addition, each TrkA half contains at its C-terminus a domain of 100 residues resembling the first (and less important) part of the catalytic domain of GAPDH from *E. coli*. Isolated and renatured TrkA binds both $[^{32}P]$-NAD^+ and $[^{32}P]$-NADH [106].

The nucleotide sequence of 4 bacterial *trkA* genes [21,22,106,112] and one archaeal putative *trkA* gene [113] have been reported. Sequence alignments show that the TrkA structure has been conserved well among prokaryotes. All of these proteins are composed of two similar halves and all of them contain two complete NAD^+-binding domains. However, the archaeal *Methanosarcina mazei* gene is with 406 instead of 458 codons shorter than the bacterial *trkA* genes [113].

The dinucleotide-binding fold of TrkA is conserved in degenerated form in two prokaryotic K^+-efflux systems (KefB [I.R. Booth, personal communication] and KefC [21,22,114,115]), in the putative K^+-channel Kch from *E. coli* [7,21], and in

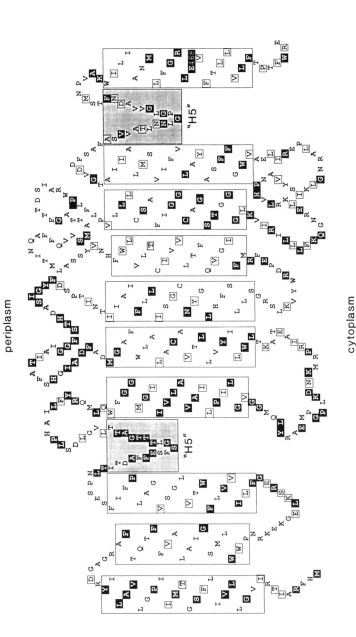

Fig. 5.   Model for the folding of the *E. coli* TrkH-protein in the membrane. Residues identical in *E. coli* and *H. influenzae* TrkH and in *E. coli* TrkG are given in white with a black background. Conservatively exchanged residues have been put inside squares. Conservative exchanges are: D = E, E = Q, D = N, N = Q, K = R, S = T, A = G, G = P, I = L = M = V, F = W = Y. The putative H5 regions [94] have been drawn as folding back into the membrane (grey areas).

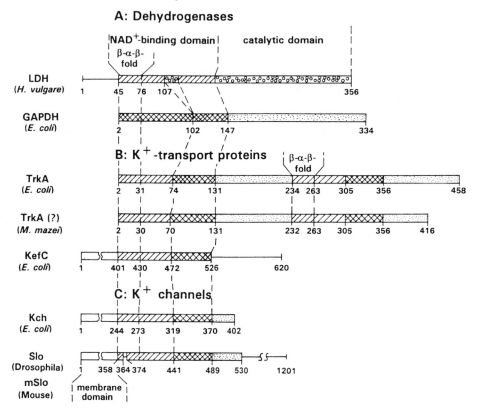

Fig. 6. Degenerated dinucleotide binding folds in K$^+$-transport proteins and K$^+$ channels.

the K$^+$-channels Slo from *Drosophila* and the homologous Mslo from mouse [7,116]. In all of these proteins the domain is attached C-terminal to the membrane domain (Fig. 6). The region aligned for the different proteins comprises up to 160 residues. Homology is not large, with value of % identity and % similarity varying between 11–14% and 24–32%, respectively. The conserved domain is too degenerated to allow for dinucleotide binding to it [106,117], and at present we do not have a clue which function(s) these domains exert in the process of K$^+$ translocation via these transporters or channels.

4.2.4. TrkE

The *E. coli trkE* gene maps inside a region equivalent to the *sapABCDF*-operon from *S. typhimurium* ([20,54] and W. Epstein, personal communication). The products of this operon form a putative ABC-transporter, composed of a periplasmic-binding protein (SapA), two integral membrane proteins (SapB, SapC) and two ATP-binding proteins (SapD, SapF) [20]. Sequence comparisons have suggested that the SapABCDF system transports peptides, but this has not been verified experimentally [20]. Our experiments with the cloned *E. coli trkE* region indicates that this bacterium contains and expresses the complete *sapABCDF* operon [109]. Complementation experiments indicate that neither *sapA* nor *sapF* are required for K^+ uptake via the Trk systems of an *E. coli* strain carrying the *trkE80* allele. At the other hand *sapD* (*trkE*) is indispensable [109]. These data suggest that the earlier observed ATP dependence of K^+ transport via Trk [88,105] can be attributed to the putative ATP-binding protein SapD (TrkE), which must interact with or form part of the Trk complex. Further evidence for this notion was sought with an *E. coli* strain in which the *sapABCDF* operon had been deleted from the chromosome. Complementation studies gave the same result as with the *trkE80* strain, except that SapF supports SapD in its K^+ transport function [109]. From these results we conclude that Trk 'borrows' at least one and possibly two ATP-binding protein(s) from the Sap system.

4.2.5. Mechanism of action

Three models for the mechanism of K^+ transport via Trk are given in Fig. 7. K^+-uptake is assumed to be mediated by the TrkH protein. According to model I this occurs in symport with protons and are TrkA, SapD and possibly SapF involved in the regulation of transport activity. In model II it is assumed that ATP-hydrolysis via SapD (and SapF?) supplies the energy to drive K^+ uptake. According to model III Trk functions as a K^+-translocating redox pump, in which energy coupling occurs via a redox reaction with NAD(H) at the TrkA protein.

We favour model I. High rate, low affinity transport systems function almost always as secondary transport systems. These transport proteins occur in relatively low copy numbers and mediate transport with higher turn-over numbers than do

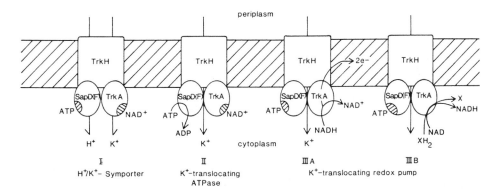

Fig. 7. Alternative models for the mode of action of K^+ transport via the Trk system. See the text for details.

primary pumps. All of these aspects apply to Trk (Section 4.2). The evidence that K$^+$ uptake occurs in symport with protons is indirect and has been reviewed before [13]. In this respect it should be mentioned that K$^+$-uptake systems from two lower eukaryotes [118,119] and from plant cells have been shown to function as H$^+$/K$^+$-symporters [120,121].

Since SapD and SapF ('TrkE') are putative ATP-binding proteins [20], we assume that the action of ATP on the complex occurs via these proteins. According to model I ATP binding to these Sap proteins opens the K$^+$-transport pathway. This type of interaction with ATP would be similar to that known for the CFTR protein from animal cells, where ATP opens the Cl$^-$ channel by binding to the protein [122]. Such a function of ATP would be new for an ABC transporter from a prokaryote. Hence, the possibility that ATP drives K$^+$ transport via its hydrolysis at the SapD/F proteins should also be considered (model II). However, the high turn-over number of the system (which also applies when the calculation is based on the copy numbers of the SapD/F proteins/cell [R. Tewes and E.P. Bakker, unpublished observations]) and the low rate of ATP consumption during K$^+$ uptake via Trk [105] argue against model II.

Little is known about the function of TrkA and about the significance of its *in-vitro* observed NAD(H) binding in K$^+$ transport. In cell metabolism NAD(H) serves much more often as a coenzyme than as an effector of enzymes. Hence, with respect to this dinucleotide the first possibility to be considered is that Trk functions as a K$^+$-translocating redox pump (model III). According to such a mechanism NADH could either donate its electrons via TrkA to the electron chain (model IIIA) or NAD$^+$ could accept electrons from a reduced, not yet identified substrate (model IIIB). The major problem with these models is that besides its complete NAD$^+$-binding domain TrkA does not contain a complete catalytic domain and that the part of this domain present in TrkA is in GAPDH mainly involved in subunit–subunit interaction [106]. Moreover, the high turn-over number of Trk (based on the copy number of TrkA/cell [103]) argues against a role of TrkA as a catalytic subunit in an NAD(H)-dependent reaction (model III). If NAD(H) has a function in K$^+$ transport, we assume it therefore to occur via binding to TrkA (models I and II).

Nothing is known about how the Trk system senses turgor pressure. One possibility would be that changes in membrane stretch alter the packing of the transmembrane helices of TrkH, and thereby influence the activity of the Trk system.

4.2.6. The Trk system and virulence of S. typhimurium

In their search for mutants sensitive to defensin-like antimicrobial peptides, Groisman et al. have identified the *trk* genes from *S. typhimurium* [19–21]. Defensins are small basic peptides, which protect the body against bacteria [123]. Using protamine as a model compound for defensin action two loci within the *sapABCDF* operon ('*trkE*', see above) [20], the *sapG* locus (*trkA*) [21] and the *sapJ* locus (*trkH*, E. Groisman, personal communication) have been identified (Fig. 1A). Mutations in these genes rendered the cells sensitive to several defensins and made them avirulent when administered orally to mice. Moreover, except for the *sapJ* strain, which did not show a large effect, all of these mutant strains exhibited a thousand-fold increase of LD_{50}-value of mice, when administered intraperitoneally [19]. The insight that SapABCDF may function as a peptide transporter [20] led to the hypothesis that this ABC-transporter is involved in the detoxification of defensins, by transporting these compounds into the cytoplasm, where they are degraded by proteases. TrkA has been proposed to support the action of the Sap system [20,21]. Support for the notion that resistance is not related to K^+ transport came from the observations that (i) mutations in some of the relevant genes show differential effects with respect to resistance and to K^+ transport ([21] and E. Groisman, personal communication) and (ii) Kdp cannot not replace Trk in conferring resistance to the toxic peptides [21]. On these grounds an alternative hypothesis, according to which (i) defensins form channels in cell membranes through which K^+ leaves the cells [124,125], and (ii) a high rate K^+-uptake system (like Trk) rescues the cells by reaccumulating K^+ until the defensins are detoxified by proteases, has been rejected [21]. We tested both hypothesis with E. coli K^+ transport mutants. Our results show (i) always a good correlation between the loss of resistance towards protamine and the inhibition of K^+ uptake by the cells, and (ii) that the presence of the *kup* gene on a multiple-copy plasmid partially restores resistance of strains with chromosomal deletions of *trkA*, of *trkH* plus *trkG*, or of *sapA-F* (S. Stumpe and E.P. Bakker, unpublished observations). Under those conditions the Kup system transports K^+ with a rate half as large as that of the TrkH plus TrkG systems [126]. These results suggest that the second hypothesis is correct and that peptide transport via the SapABCDF system does not contribute to the resistance phenomenon.

4.2.7. Ubiquity among prokaryotes

With the possible exception of *Halobacterium salinarium*, all prokaryotes investigated thus far have been shown to posses a high rate, low affinity K^+-uptake system [11,13]. In those cases in which the transport mechanism has been studied in detail, it showed considerable similarity with that of the E. coli Trk systems. This is true for bacteria with small genomes like *Enterococcus hirae* (formarly *Streptococcus faecalis* ATCC 9790) and *Mycoplasma mycoides* var. *Capri*, the extremophile *A. acidocaldarius* and two methanogens (reviewed in Refs. [11,13,17]). Together with the observations that a third methanogen possesses a *trkA*-like gene [113] (see Section 4.2.2) and that in its small genome *H. influenzae* carries a complete set of *trk* genes [22], these data suggest that Trk systems are wide-spread among prokaryotes.

4.3. Kup

Kup forms the third type of K^+-uptake system in E. coli. It is a separate system, since in contrast to Kdp and Trk it accepts Cs^+ and does not disciminate between K^+ and Rb^+ [126]. The *kup* gene (formerly *trkD*) is expressed constitutively [54,127] and the concentration of its product in the cell is equally low as that of the Trk proteins [127]. Kup is not involved in osmoadaption of the cells to elevated NaCl concentrations [38,47]. Since the V_{max} for net K^+ uptake via chromosomically encoded Kup is relatively low, and its K_m value is only slightly lower than that of the Trk systems

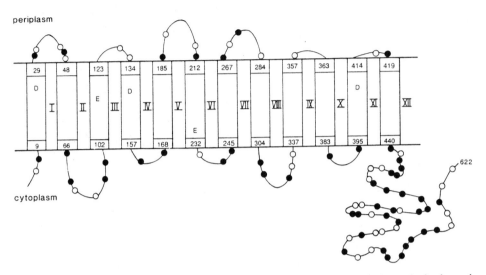

Fig. 8.   Model for the folding of the Kup protein in the membrane. Positively and negatively charged residues are indicated by closed and open circles respectively. Taken with permission from Ref. [126].

[55,126], the question arose why *E. coli* possesses Kup. Experiments with strains deleted for different K⁺-uptake genes showed that the presence of Kup gives *E. coli* a growth advantage at K⁺ concentrations between 0.5 and 5 mM. This effect increases with external pH [127]. Moreover, in cells lacking Kdp this growth advantage is larger and extends to K⁺ concentrations as low as about 50 µM. Since Kdp is not universally present among bacteria (see Section 4.1.4), the function of a Kup-like system is therefore to help cells grow at K⁺-concentrations below 5 mM [127]. It is not known how widely Kup is spread among prokaryotes. Several Enterobacteriaceae contain the gene [13]. However, it is not present in *H. influenzae* [22] and *Vibrio alginolyticus* does not contain a Kup system [102].

The nucleotide sequence of the *kup* gene predicts that the protein consists of two domains. Its 440 residue hydrophobic N-terminal part can span the membrane up to 12 times [128] (Fig. 8). In this respect Kup resembles a large group of secondary transport systems called 'major facilitator superfamily' (MFS) [129]. However, sequence comparisons show that Kup does not belong to this superfamily [128]. The primary sequence of this part of Kup also does not show any similarity to that of any other protein involved in K⁺-translocation across membranes.

The membrane domain of Kup is followed by a C-terminal hydrophilic domain of 182 residues [128] (Fig. 8). With this two-domain structure Kup resembles the proteins KefB/C ([114] and I.R. Booth, personal communication), Kch [7] and BetT [46] from *E. coli* and LacS from *Lactobacilli* [130]. The hydrophilic domains of these proteins are supposed to be involved in regulation of transport activity. However, except for the domain being partially essential for K⁺ uptake [128], no specific function for it has yet been found in Kup [127].

Experiments with intact cells and right-side out membrane vesicles showed that Kup functions as a secondary transport system [127,131]. Na⁺-ions are not required. Most likely, Kup is a K⁺/H⁺-symporter [127].

4.4. Other systems

Several prokaryotes have been reported to synthesize K⁺-uptake systems different from Kdp, Trk or Kup, when grown under (K⁺)-stress conditions. *Vibrio algi-nolyticus* synthesizes a new system with a tenfold lower K_m-value for K⁺ uptake than that of the constitutive Trk system, when grown at low K⁺ concentrations. The new system is different from Kup, since it is inducible, and accepts Rb⁺ poorly. A mutant in which the new system is non-functional has been described [102]. The gene encoding the new system has not yet been cloned.

When grown at low proton-motive force *E. hirae* overproduces a second K⁺-uptake system, KtrII [132]. Its mode of action resembles that of the Na⁺/K⁺-ATPase from higher animal cells, in that the hydrolysis of ATP is coupled to the extrusion of Na⁺ and the uptake of K⁺. However, both the mechanism of energy coupling to transport in KtrII [17] and its structure (Y. Kakinuma, personal communication) are completely different from that of the Na⁺/K⁺-ATPase.

Abbreviations

ABC-Transporter, ATP-binding cassette transport system
GAPDH, NAD⁺-dependent glyceraldehyde-3-phosphate dehydrogenase
LDH, NAD⁺-dependent lactate dehydrogenase
MDH, NAD⁺-dependent malate dehydrogenase
MFS, major facilitator superfamily
M_r, molecular weight

Acknowledgements

We thank E. Limpinsel for expert technical assistance and for preparing the figures, W. Puppe for Fig. 3 and, with other members of the Osnabrück Kdp group, for many helpful discussions, and W. Epstein and K. Altendorf for supplying manuscripts before their publication. The work from our group has been supported by the Deutsche Forschungsgemeinschaft (SFB171, Teilproject C1), by a grant from the Volkswagen Foundation for cooperation between Israel and the State of Lower Saxony, and by the Fonds der Chemischen Industrie.

Note added in proof

KefB/C: In Fig. 1B KefB and KefC are drawn as K⁺/K⁺-antiporters. In their chapter on K⁺-efflux systems Booth et al. describe these proteins as K⁺ channels [9]. The former notion is based on the observation that upon activation of KefB/C in intact cells K⁺ appeared to move away from electrochemical equilibrium, suggesting that K⁺ efflux is coupled to the movement of another ion (i.e. H⁺ influx) [133]. Subsequent patch-clamp experiments with cytoplasmic membranes (cited in Ref. [9])

have shown that after its activation KefC exhibited K$^+$-channel activity. More work is needed to resolve this controversy.

Trk: The *Mycoplasma genitalium* genome does not contain *trk* genes ([134] and our own computer searches). Hence, the contention that the K$^+$-uptake system from a different *Mycoplasma* species is similar to Trk from *E. coli* [13] may not extend to sequence similarities between K$^+$-uptake proteins of the two species. In addition, a pendant for model I of Fig. 7 was found in animal cells: the sulfonylurea receptor from pancreatic β cells turned out to be an ABC-transporter. Substrate binding (and transport?) is thought to close an ATP-dependent K$^+$ channel. It is speculated that the ATP dependence of the channel resides in the ABC transporter [135].

Kup: Bañueles et al. [136] have shown that the fungus *Schwanniomyces occidentalis* possesses a Kup-like high affinity K$^+$-uptake system (HAK1, 32% identity with the Kup primary sequence). The two proteins are predicted to contain four conserved negatively charged residues inside the hydrophobic part of the membrane (residues Asp24, Glu116, Glu229, and Asp408 in Kup, corresponding to residues Asp13, Asp153, Glu264, and Glu447 in HAK1). These residues might play a role in the K$^+$-transport process.

References

1. Csonka, L.N. and Epstein, W. (1995) in: *Escherichia coli* and *Salmonella typhimurium*, Edn 2 , ed F.C. Neidhardt. ASM Press, Washington, in press.
2. Csonka, L.N. and Hanson, A.D. (1991) Annu. Rev. Microbiol. **45**, 569–606.
3. Booth, I.R. and Higgins, C.F. (1990) FEMS Microbiol. Rev. **75**, 239–246.
4. Epstein, W. (1986) FEMS Microbiol. Rev. **39**, 73–78.
5. Booth, I.R. (1985) Microbiol. Rev. **49**, 359–378.
6. Sukharev, S.I., Blount, P., Martinac, B., Blattner, F.R. and Kung, C. (1994) Nature (Lond.) **368**, 265–268.
7. Milkman, R. (1994) Proc. Natl. Acad. Sci. USA **91**, 3510–3514.
8. Schrempf, H., Schmidt, O., Kümmerlen, R., Hinnah, S., Müller, D., Betzler, M., Steinkamp, T. and Wagner, R. (1995) EMBO J. **14**, 5170–5178.
9. Booth, I.R., Jones, M., McLeggan, D., Nikolaev, Y., Ness, L., Wood, C., Miller, S., Tötemeyer, S. and Ferguson, G. (1996) in: Handbook of Biological Physics, Vol. 2: Transport Processes in Membranes, eds W. Konings, H.R. Kaback and J.S. Lolkema, pp. 693–729. Elsevier, Amsterdam.
10. Letellier, L. and Bonhivers, M. (1996) in: Handbook of Biological Physics, Vol. 2: Transport Processes in Membranes, eds W.N. Konings, H.R. Kaback and J.S. Lolkema, pp. 615–636. Elsevier, Amsterdam.
11. Walderhaug, M.O., Dosch, D.C. and Epstein, W. (1987) in: Ion Transport in Prokaryotes, eds B.P. Rosen and S. Silver, pp. 85–130, Academic Press, New York.
12. Bakker, E.P. (1993) in: Alkali Cation Transport Systems in Prokaryotes, ed E.P. Bakker, pp. 205–224. CRC Press, Boca Raton, FL.
13. Bakker, E.P. (1993) in: Alkali Cation Transport Systems in Prokaryotes, ed E.P. Bakker, pp. 253–276, CRC Press, Boca Raton, FL.
14. Epstein, W., Buurman, E., McLaggan, D. and Naprstek, J. (1993) Biochem. Soc. Trans. **21**, 1006–1010.
15. Booth, I.R., Douglas, R.M., Ferguson, G.P., Lamb, A.J., Munro, A.W. and Ritchie, G.Y. (1993) in: Alkali Cation Transport Systems in Prokaryotes, ed E.P. Bakker, pp. 291–308, CRC Press, Boca Raton, FL.
16. Siebers, A. and Altendorf, K. (1993) in: Alkali Cation Transport Systems in Prokaryotes, ed E.P. Bakker, pp. 225–252. CRC Press, Boca Raton, FL.

17. Kakinuma, Y. (1993) in: Alkali Cation Transport Systems in Prokaryotes, ed E.P. Bakker, pp. 277–290, CRC Press, Boca Raton, FL.

18. Dosch, D.C., Helmer, G.L., Sutton, S.H., Salvacion, F.F. and Epstein, W. (1991) J. Bacteriol. **173**, 687–696.

19. Groisman, E.A., Parra-Lopez, C., Salcedo, M., Lipps, C.J. and Heffron, F. (1992) Proc. Natl. Acad. Sci. USA **89**, 11939–11943.

20. Parra-Lopez, C., Baer, M.T. and Groisman, E.A. (1993) EMBO J. **12**, 4053–4062.

21. Parra-Lopez, C., Lin, R., Aspedon, A. and Groisman, E.A. (1994) EMBO J. **13**, 4053–4062.

22. Fleischmann, R.D., Adams, M.D., White, O., Clayton, R.A., Kirkness, E.F., Kerlavage, A.R., Bult, C.J., Tomb, J.-F., Dougherty, B.A., Merrick, J.M., McKenney, K., Sutton, G., FitzHugh, W., Fields, C., Gocayne, J.D., Scott, J., Shirley, R., Liu, L.-I., Glodek, A., Kelley, J.M., Weidman, J.F., Phillips, C.A., Spriggs, T., Hedblom, E., Cotton, M.D., Utterback, T.R., Hanna, M.C., Nguyen, D.T., Saudek, D.M., Brandon, R.C., Fine, L.D., Fritchman, J.L., Fuhrmann, J.L., Geoghagen, N.S.M., Gnehm, C.L., Mcdonald, L.A., Small, K.V., Fraser, C.M., Smith, H.O. and Venter, J.C. (1995) Science **269**, 496–512.

23. Ferguson, G.P., Munro, A.W., Douglas, R.M., McLaggan, D. and Booth, I.R. (1993) Mol. Microbiol. **9**, 1297–1303.

24. Bakker, E.P., Booth, I.R., Dinnbier, U., Epstein, W. and Gajewska, A. (1987) J. Bacteriol. **169**, 3743–3749.

25. Verkovskaya, M.L., Verkovsky, M.L. and Wikström, M. (1995) FEBS Lett. **363**, 46–48.

26. Berrier, C., Besnard, M., Ajouz, B., Coulombe, A. and Ghazi, A., unpublished observations cited in Ref. [1].

27. Wiggins, P.M. (1990) Microbiol. Rev. **54**, 432–449.

28. McLaggan, D., Naprstek, J., Buurman, E.T. and Epstein, W. (1994) J. Biol. Chem. **269**, 1911–1917.

29. Munro, G.F., Hercules, K., Morgan, J. and Sauerbier, W. (1972) J. Biol. Chem. **247**, 1272–1280.

30. Cayley, S.M., Capp, W., Guttman, H.J., and Record, M.T., Jr., unpublished observations cited in Ref. [1].

31. Reed, R.H. and Walsby, A.E. (1985) Arch. Microbiol. **143**, 290–296.

32. Koch, A.L. and Pinette, M.F.S. (1987) J. Bacteriol. **169**, 3654.

33. Huber, R., Kurr, M., Jannasch, H.W. and Stetter, K.O. (1989) Nature (Lond.) **342**, 833–834.

34. Scholz, S., Sonnenbichler, J., Schäfer, W. and Hensel, R. (1992) FEBS Lett. **306**, 239–242.

35. Yancey, P.H., Clark, M.E., Hand, S.C., Bowlus, R.D. and Somero, G.N. (1982) Science **217**, 1214–1222.

36. Galinski, E.A., Pfeiffer, H.-P. and Trüper, H.G. (1985) Eur. J. Biochem. **149**, 135–139.

37. Brown, A.D. (1990) Microbial Water Stress Physiology: Principles and Perspectives. Wiley, Chicester.

38. Dinnbier, U., Limpinsel, E., Schmid, R. and Bakker, E.P. (1988) Arch. Microbiol. **150**, 348–357.

39. Rhoads, D.B. and Epstein, W. (1978) J. Gen. Physiol. **72**, 283–295.

40. Laimins, L.A., Rhoads, D.B. and Epstein, W. (1981) Proc. Natl. Acad. Sci. USA **78**, 464–468.

41. Larsen, P.I., Sydnes, L.K., Landfald, B. and Strom, A.R. (1987) Arch. Microbiol. **147**, 1–7.

42. Cairney, J.C., Booth, I.R. and Higgins, C.F. (1985) J. Bacteriol. **164**, 1218–1223.

43. Cairney, J., Booth, I.R. and Higgins, C.F. (1985) J. Bacteriol. **164**, 1224–1232.

44. Sutherland, L., Cairney, J., Elmore, M.J., Booth, I.R. and Higgins, C.F. (1986) J. Bacteriol. **168**, 805–814.

45. Lamark, T., Kaasen, I., Eshoo, M.W., Falkenberg, P., McDougall, J. and Strom, A.R. (1991) Mol. Microbiol. **5**, 1049–1064.

46. Strom, A.R. and Kaasen, I. (1993) Mol. Microbiol. **8**, 205–210.

47. Schleyer, M., Schmid, R. and Bakker, E.P. (1993) Arch. Microbiol. **160**, 424–431.

48. Larsen, H. (1984) in: Bergey's Manual of Systematic Bacteriology, Vol. 1, eds N.R. Krieg and J.G. Holt, pp. 261–267. Williams and Wilkins, Baltimore/London.

49. Lanyi, J.K. and Silverman, M.P. (1972) Can. J. Microbiol. 18, 993–995.

50. Zaccai, G. and Eisenberg, H. (1991) in: Life under Extreme Conditions. Biochemical Adaptation, ed G. di Prisco, pp. 125–137, Springer, Berlin.

51. Berrier, C., Coulombe, A., Szabo, I., Zoratti, M. and Ghazi, A. (1992) Eur. J. Biochem. **206**, 559–565.

52. Bakker, E.P. (1990) FEMS Microbiol. Rev. **75**, 319–334.
53. Abee, T. and Konings, W.N. (1993) in: Alkali Cation Transport in Prokaryotes, ed E.P. Bakker, pp. 333–348, CRC Press, Boca Raton, FL.
54. Epstein, W. and Kim, B.S. (1971) J. Bacteriol. **108**, 639–644.
55. Rhoads, D.B., Waters, F.B. and Epstein, W. (1976) J. Gen. Physiol. **67**, 325–341.
56. Pedersen, P.L. and Carafoli, E. (1987) Trends Biochem. Sci. **12**, 146–150.
57. Epstein, W. (1990) in: Bacterial Energetics. The Bacteria, Vol. 13, ed T.A. Krulwich, pp. 87–100, Academic Press, Orlando, FL.
58. Hesse, J., Wieczorek, L., Altendorf, K., Reicin, A.S., Dorus, E. and Epstein, W. (1984) Proc. Natl. Acad. Sci. USA **81**, 4746–4750.
59. Sugiura, A., Nakashima, K., Tanaka, K. and Mizuno, T. (1992) Mol. Microbiol. **7**, 1769–1776.
60. Altendorf, K. and Epstein, W. (1993) Cell. Physiol. Biochem. **4**, 160–168.
61. Altendorf, K., Voelkner, P. and Puppe, W. (1994) Res. Microbiol. **145**, 374–381.
62. Altendorf, K. and Epstein, W. (1995) in: Biomembranes, Vol. 6, in press.
63. Walderhaug, M.O., Polarek, J.W., Voelkner, P., Daniel, J.M., Hesse, J.E., Altendorf, K. and Epstein, W. (1992) J. Bacteriol. **174**, 2152–2159.
64. Stock, J.B., Ninfa, A.J. and Stock, A.M. (1989) Microbiol. Rev. **53**, 450–490.
65. Parkinson, J.S. and Kofoid, E.C. (1992) Annu. Rev. Genet. **26**, 71–112.
66. Nakashima, K., Sugiura, A., Momoi, H. and Mizuno, T. (1992) Mol. Microbiol. **7**, 1777–1784.
67. Zimman, P., Puppe, W. and Altendorf, K. (1995) J. Biol. Chem., in press.
68. Sugiura, A., Hirokawa, K., Nakashima, K. and Mizuno, T. (1994) Mol. Microbiol. **14**, 929–938.
69. Gowrishankar, J. (1985) J. Bacteriol. **164**, 434–445.
70. Asha, H. and Gowrishankar, J. (1993) J. Bacteriol. **175**, 4528–4537.
71. Voelkner, P., Puppe, W. and Altendorf, K. (1993) Eur. J. Biochem. **217**, 1019–1026.
72. Tanaka, K.-I., Muramatsu, S., Yamada, H. and Mizuno, T. (1991) Mol. Gen. Genet. **226**, 367–376.
73. Sugiura, A., Nakashima, K. and Mizuno, T. (1993) Biosci. Biotech. Biochem. **57**, 356–357.
74. Perez-Martin, J., Rojo, F. and de Lorenzo, V. (1994) Microbiol. Rev. **58**, 268–290.
75. Nakashima, K., Sugiura, A., Kanamuru, K. and Mizuno, T. (1993) Mol. Microbiol. **7**, 109–116.
76. Siebers, A. and Altendorf, K. (1988) Eur. J. Biochem. **178**, 131–140.
77. Hafer, J., Siebers, A. and Bakker, E.P. (1989) Mol. Microbiol. **3**, 487–495.
78. Abee, T., Siebers, A., Altendorf, K. and Konings, W.N. (1992) J. Bacteriol. **174**, 6911–6917.
79. Siebers, A., Kollmann, R., Dirkes, G. and Altendorf, K. (1992) J. Biol. Chem. **267**, 12717–12721.
80. Laimins, L.A., Rhoads, D.B., Altendorf, K. and Epstein, W. (1978) Proc. Natl. Acad. Sci. USA **75**, 3216–3219.
81. Möllenkamp, T. and Altendorf, K., personal communication
82. Siebers, A. and Altendorf, K. (1989) J. Biol. Chem. **264**, 5831–5838.
83. Puppe, W., Siebers, A. and Altendorf, K. (1992) Mol. Microbiol. **6**, 3511–3520.
84. Serrano, R., Kielland-Brandt, M.C. and Fink, G.R. (1986) Nature (Lond.) **319**, 689–693.
85. Serrano, R. (1988) Biochim. Biophys. Acta **947**, 1–28.
86. Epstein, W., Whitelaw, V. and Hesse, J. (1978) J. Biol. Chem. **253**, 6666–6668.
87. Naprstek, J., Walderhaug, M.O. and Epstein, W. (1991) Ann. N.Y. Acad. Sci. **671**, 481–483.
88. Rhoads, D.B. and Epstein, W. (1977) J. Biol. Chem. **252**, 1394–1401.
89. Bakker, E.P. and Mangerich, W.E. (1981) J. Bacteriol. **147**, 820–826.
90. Damper, P.D., Epstein, W., Rosen, B.P. and Sorensen, E.N. (1979) Biochemistry **18**, 4165–4169.
91. Rhoads, D.B., Woo, A. and Epstein, W. (1978) Biochim. Biophys. Acta **469**, 45–51.
92. Treuner, A. and Dürre, P. (1994) Bioengineering, Vol. 3 10, 29. Abstract Spring Meeting of the German Microbiological Societies VAAM and DGHM.
93. Buurman, E.T., Kim, K.-T. and Epstein, W. (1995) J. Biol. Chem. **270**, 6678–6685.
94. Jan, L.Y. and Jan, Y.N. (1994) Nature (Lond.) **371**, 119–122.
95. Solioz, M., Mathews, S. and Fürst, P. (1987) J. Biol. Chem. **262**, 7358–7362.
96. Odermatt, A., Suter, H., Krapf, R. and Solioz, M. Ann. N.Y. Acad. Sci. 671, 484–486.
97. Nucifora, G., Chu, L., Misra, T.K. and Silver, S. (1989) Proc. Natl. Acad. Sci. USA **86**, 3544–3548.
98. Navarre, C., Catty, P., Leterme, S., Dietrich, S. and Goffeau, A. (1994) J. Biol. Chem. **269**, 21262–21268.

99. Walderhaug, M.O., Litwack, E.D. and Epstein, W. (1989) J. Bacteriol. **171**, 1192–1195.
100. Abee, T., Knol, J., Hellingwerf, K., Bakker, E.P., Siebers, A. and Konings, W.N. (1992) Arch. Microbiol. **158**, 374–380.
101. Kim, S.T. (1985) Arch. Microbiol. **142**, 393–396.
102. Nakamura, T., Suzuki, F., Abe, M., Matsuba, Y. and Unemoto, T. (1994) Microbiol. **140**, 1781–1785.
103. Bossemeyer, D., Borchard, A., Dosch, D.C., Helmer, G.C., Epstein, W., Booth, I.R. and Bakker, E.P. (1989) J. Biol. Chem. **264**, 16403–16410.
104. Schlösser, A. (1993) Ph.D. Thesis, University of Osnabrück.
105. Stewart, L.M.D., Bakker, E.P. and Booth, I.R. (1985) J. Gen. Microbiol. **131**, 77–85.
106. Schlösser, A., Hamann, A., Bossemeyer, D. and Bakker, E.P. (1993) Mol. Microbiol. **9**, 533–543.
107. Schlösser, A., Kluttig, S., Hamann, A. and Bakker, E.P. (1991) J. Bacteriol. **173**, 3170–3176.
108. Schlösser, A., Meldorf, M., Stumpe, S., Bakker, E.P. and Epstein, W. (1995) J. Bacteriol. **177**, 1908–1910.
109. Epstein, W., Nölker, E., Stumpe, S. and Bakker, E.P., unpublished observations.
110. Stumpe, S. (1993) M.Sc. Thesis, University of Osnabrück.
111. Hamann, A. (1991) Ph.D. Thesis, University of Osnabrück.
112. Nakamura, T., Matsuba, Y., Yamamuro, N., Booth, I.R. and Unemoto, T. (1994) Biochim. Biophys. Acta **1219**, 701–705.
113. Macario, A.J.L., Dugan, C.B. and Conway de Macario, E. (1993) Biochim. Biophys. Acta **1216**, 495–498.
114. Munro, A.W., Ritchie, G.Y., Lamb, A.J., Douglas, R.M. and Booth, I.R. (1991) Mol. Microbiol. **5**, 607–615.
115. McKie, J.H. and Douglas, K.T. (1993) Biochem. Soc. Trans. **21**, 540–544.
116. Butler, A., Tsunoda, S., McCobb, D.P., Wei, A. and Salkoff, L. (1993) Science **261**, 221–224.
117. Wierenga, R.K., Terpstra, P. and Hol, W.G.J. (1986) J. Mol. Biol. **187**, 101–107.
118. Rodriguez-Navaro, A., Blatt, M.R., Slayman, C.L. (1986) J. Gen. Physiol. **87**, 649–674.
119. Gläser, H.-U. and Pick, U. (1990) Biochim. Biophys. Acta **1019**, 293–299.
120. Schachtman, D.P. and Schroeder, J.I. (1994) Nature (Lond.) **370**, 655–658.
121. Maathuis, F.J.M. and Sanders, D. (1994) Proc. Natl. Acad. Sci. USA **91**, 9272–9276.
122. Andersen, M.P., Berger, H.A., Rich, D.P., Gregory, R.J., Smith, A.E. and Welsh, M.J. (1991) Cell **67**, 775–784.
123. Lehrer, R.I., Lichtenstein, A.K. and Ganz, T. (1993) Annu. Rev. Immunol. **11**, 105–128.
124. Westerhoff, H.V., Deretic, D., Hendler, R.W. and Zasloff, M. (1989) Proc. Natl. Acad. Sci. USA **86**, 6597–6601.
125. Cociancich, S., Ghazi, A., Hetru, C., Hoffmann, J.A. and Letellier, L. (1993) J. Biol. Chem. **268**, 19239–19245.
126. Bossemeyer, D., Schlösser, A. and Bakker, E.P. (1989) J. Bacteriol. **171**, 2219–2221.
127. Schleyer, M. (1994) Ph.D. Thesis, University of Osnabrück.
128. Schleyer, M. and Bakker, E.P. (1993) J. Bacteriol. **175**, 6925–6931.
129. Marger, M. and Saier, M. (1993) Trends Biochem. Sci. **18**, 13–20.
130. Poolman, B., Modderman, R. and Reizer, J. (1992) J. Biol. Chem. **267**, 9150–9157.
131. Montag, M. (1991) Ms.C. Thesis, University of Osnabrück.
132. Kobayashi, H. (1982) J. Bacteriol. **150**, 506–511.
133. Bakker, E.P. and Mangerich, W.E. (1982) FEBS Lett. **140**, 177–180.
134. Fraser, C.M., Gocayne, J.D., White, O., Adams, M.D., Clayton, R.A., Fleischmann, R.D., Bult, C.J., Kerlavage, A.R., Sutton, G., Kelley, J.M., Fritchman, J.L., Weidman, J.F., Small, K.V., Sandusky, M., Fuhrmann, J.F., Ngyen, D., Utterback, T.R., Saudek, D.M., Phillips, C.A., Merrick, J.M., Tomb, J.F., Dougherty, B.A., Bott, K.F., Hu, P.-C., Lucier, T.S., Petereson, S.N., Smith, H.O., Hutchison III, C.A. and Venter, J.C. (1995) Science **270**, 397–403.
135. Aguilar-Bryan, L., Nichols, C.G., Wechsler, S.W., Clement IV, J.P., Boyd III, A.E., González, G., Herrera-Sosa, H., Nguy, K., Bryan, J. and Nelson, D.A. (1995) Science **268**, 423–426.
136. Bañueles, M.A., Klein, R.D., Alexander-Bowman, S.J. and Rodriguez-Navarro, A. (1995) EMBO J. **14**, 3021–3027.

CHAPTER 22

Bacterial Na⁺/H⁺ Antiporters – Molecular Biology, Biochemistry and Physiology

Bacterial Na$^+$/H$^+$ Antiporters – Molecular Biology, Biochemistry and Physiology

E. PADAN and S. SCHULDINER

Department of Microbial and Molecular Ecology, The Institute of Life Sciences, The Hebrew University of Jerusalem, 91904 Jerusalem, Israel

© 1996 Elsevier Science B.V.
All rights reserved

Handbook of Biological Physics
Volume 2, edited by W.N. Konings, H.R. Kaback and J.S. Lolkema

Contents

1. Introduction

Whatever their origin, (whether eukaryotic or prokaryotic of plant or animal origin), very rigid physiological constraints are imposed on all cells regarding the most ubiquitous ions, Na^+ and H^+. All cells must maintain intracellular pH (pH_{in}) constant and intracellular concentration of $Na^+([Na^+_{in}])$ lower than the extracellular concentration (reviews [1,2]). It is not surprising therefore that in the cytoplasmic membrane of all cells (reviews [1–3]), as yet with only one known exception [4], there are proteins that couple the fluxes of these ions and are involved in their homeostasis. These are the Na^+/H^+ antiporters discovered by P. Mitchell [5,6]. Na^+/H^+ antiporters have since also been found in membranes of many organelles [2].

The Na^+/H^+ antiporters couple between $\Delta\tilde{\mu}_{H^+}$ and $\Delta\tilde{\mu}_{Na^+}$ so that at steady state (Eq. 1) $\Delta\tilde{\mu}_{Na^+} = n\Delta\tilde{\mu}_{H^+}$ ($\Delta\tilde{\mu}_{Na^+}$ = Na^+ electrochemical potential difference, $\Delta\tilde{\mu}_{H^+}$ = H^+ electrochemical potential difference, n= H^+/Na^+ stoichiometry of the antiporter). Thus, the study of the molecular physiology of the Na^+/H^+ antiporter in relation to other systems involved in the cycling of Na^+ and H^+ is the basis for the understanding of H^+ and Na^+ homeostasis in any cell.

Comprehensive reviews regarding the various Na^+/H^+ antiporters have recently been published [1–3,7–10]. This review will focus on novel data obtained in bacteria, mainly through molecular biology and biochemistry of purified antiporter proteins.

2. Molecular nature and properties of the Na^+/H^+ antiporters

2.1. Na^+/H^+ antiporters of E. coli

A clue to the molecular nature of the antiporters was obtained in *E. coli* by combining molecular biology and biochemistry (Table 1 and reviews [1–3]). Two genes were identified, mapped and cloned, *nha*A [11,12] and *nha*B [13]. They encode two membrane proteins, which share no significant homology but have a similar predicted secondary structure of 12 putative transmembrane segments (TMS) (Rothman, Padan and Schuldiner, preliminary results and [13]), characteristic of transporters. Multicopy of either gene specifically affect the $Na^+(Li^+)/H^+$ exchange activity of isolated membrane vesicles. The proof for their being structural genes of Na^+/H^+ antiporters was obtained by a biochemical approach. Each protein was identified, overexpressed, purified and reconstituted, separately, in a functional form in proteoliposomes [14,15]. The turnover number of both purified NhaA and NhaB proteins (Table 1) is very similar to that in the native membrane. Hence a single polypeptide is enough to catalyze the exchange reaction. The

Table 1
Comparison between NhaA and NhaB

Property	Units	NhaA [Ref.]	NhaB [Ref.]
Molecular mass	kDa	41 [11,12,14]	55.5 [13,15]
Number of transmembrane segments		12 [Rothman, Padan and Schuldiner, unpublished results]	12 [13]
Stoichiometry	H^+/Na^+ ratio	2 [16]	1.5 [15]
Cation specificity and affinity		$Li^+ > Na^+$ [11]	$Na^+ > Mn^{++} > Li^+$ [17]
Inhibition by amiloride derivatives		− [18]	+ [17]
pH dependence	change in rate with pH (pH 6.5–8.5)	1800- fold [14,19]	1.8-fold [15]
Turnover number	min^{-1}	48 at pH 6.5 89000 at pH 8.5 [14,16]	5500 [15]
Role		indispensable at high salinity and alkaline pH in the presence of Na^+ [20]	required only when NhaA is inactive or absent [21]

purified functional proteins provided a system to study properties of the antiporters that could not be studied easily or unequivocally in the intact cells or isolated membrane vesicles due to leaks and even ion fluxes mediated by other proteins. In fact, since in *E. coli* two Na^+/H^+ antiporters are present, a reconstituted purified protein is the most suitable system to test properties such as stoichiometry, sensitivity to inhibitors and pH sensitivity of the antiporters (Table 1).

2.1.1. Stoichiometry of the Na^+/H^+ antiporters
Although it was first suggested that the Na^+/H^+ exchange is electroneutral [5], further evidence has indicated an electrogenic activity, $H^+/Na^+ > 1$ [22]. It has also been proposed that below a certain pH_{out} the antiporter is electroneutral, while above it, it is electrogenic [23]. This question was treated extensively in intact cells [24,25]. In these studies, the magnitudes of $\Delta\tilde{\mu}_{H^+}$ and $\Delta\tilde{\mu}_{Na^+}$ were measured at various external pH values using $^{23}Na^+$ and ^{31}P-NMR spectroscopy. From these measurements, the apparent stoichiometry was found to change from 1.1 at $pH_{out} = 6.5$ to 1.4 at $pH_{out} = 8.5$. To explain this phenomenon the authors suggested a change in the relative contribution of two antiporters, one electroneutral and the other electrogenic with a stoichiometry of $2H^+/Na^+$. While the rate of the first one is constant, the rate of the latter device is accelerated upon alkalinization of the medium. Indeed, two different specific Na^+/H^+ antiporters were found in *E.coli* namely NhaA and NhaB [20]. Measuring their stoichiometries therefore, was important both for the understanding of their mechanism as well as for their participation in the cell physiology.

Proteoliposomes reconstituted with purified functional antiporters have been proven to be the preferable system to measure the stoichiometry of the ions. In this system passive and mediated leaks are minimal and ion gradients can be imposed across the proteoliposome membrane and manipulated at will.

The first evidence that both NhaA and NhaB are electrogenic is that imposed $\Delta\psi$, negative inside, in proteoliposomes, enhances the rate of NhaA or NhaB-mediated Na$^+$ efflux [14–16]. A more direct demonstration of the rheogenic nature of the antiporters is that when a Na$^+$ gradient directed outward is imposed, both NhaA [16] and NhaB [15] generate a membrane potential positive inside. The membrane potential is enhanced when the ΔpH, formed by the antiporter activity is collapsed by nigericin.

Estimation of H$^+$ to Na$^+$ stoichiometry was achieved by a thermodynamic approach: the size of the membrane potential generated by the antiporters depends on the magnitude of the Na$^+$ gradient and its stoichiometry [16] according to ΔpNa $= (n - 1)\Delta\psi$ (Eq. 2, derived from 1). From measurements of the size of the membrane potential as determined at steady state at different Na$^+$ gradients, H$^+$/Na$^+$ ratio of 2 for NhaA [16] and 1.5 for NhaB [15] were calculated (Fig. 1A).

Another technique to measure stoichiometry was based on a kinetic approach; direct measurements of initial rates of H$^+$ and Na$^+$ yielded in the case of NhaA a value close to 2 [16] but gave equivocal results in the case of NhaB due to difficulties in estimation of initial rates [15]. A modification of the kinetic approach was therefore undertaken with NhaB [15]. The activity of a rheogenic antiporter involves the electrophoretic movement of permeant ions to compensate for charge translocation. The ratio between the movement of the counter ion and the activity of the antiporter is therefore a measure of the number of net charges transferred in one catalytic cycle. K$^+$ and its analog Rb$^+$, in the presence of valinomycin was used as a counter ion in proteoliposomes reconstituted with NhaB. When conditions that drive Na$^+$ uptake were created, (ΔpH, acid inside), the initial rate of Na$^+$ uptake was stimulated 3–4 fold upon addition of 10 mM KCl and 1 μM valinomycin. This phenomenon was explained by the movement of K$^+$ to compensate for charge translocation which allowed for further uptake of ^{22}Na$^+$. Under these conditions, as expected, an amiloride (a specific inhibitor of NhaB, see Section 2.1.2. and Ref. [17]) sensitive ^{86}Rb$^+$ uptake was observed. This uptake was dependent on the presence of valinomycin and on the Na$^+$ concentration in the assay medium. In these type of experiments initial rates of ^{22}Na$^+$ and ^{86}Rb$^+$ uptake were measured with and without 1 mM amiloride in the presence of 1 μM valinomycin and with different NaCl concentrations. The ratio of the fluxes of Na$^+$/Rb$^+$ was found to be very close to two. This is the ratio that is expected if one charge is translocated per 2 Na$^+$ ions. Such a case is consistent with a H$^+$/Na$^+$ stoichiometry of 3:2 for NhaB [15].

2.1.2. Sensitivity of the Na$^+$/H$^+$ antiporters to inhibitors

The diuretic drug amiloride (Fig. 2) is a specific inhibitor of sodium transporting proteins in several cell types. It is a competitive inhibitor of plasma membrane Na$^+$/H$^+$ antiporter in animal cells as well as epithelial Na$^+$ channels [26]. In algae and plants it has been shown to inhibit tonoplast [27] and plasma membrane [28]

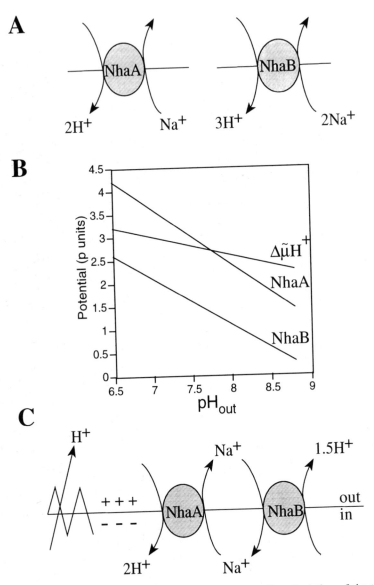

Fig. 1.   H$^+$/Na$^+$ stoichiometry of NhaA and NhaB (A) allows the calculation of the theoretical maximal Na$^+$ gradient (ΔpNa) maintained by NhaA and NhaB at various pH and $\Delta\tilde{\mu}_{H^+}$ values (B), suggesting a possible Na$^+$ cycle at alkaline pH (C). (A) Schematic model of the H$^+$/Na$^+$ stoichiometry of NhaA and NhaB. (B) Calculations are based on ΔpH and $\Delta\psi$ values from Ref. [25] and the steady-state equations ΔpNa = 2ΔpH + $\Delta\psi$ for NhaA and ΔpNa = 1.5ΔpH + 0.5 $\Delta\psi$ for NhaB. The ion electrochemical potentials are given in p units, where a concentration ratio of 10 represents a chemical potential of 1 p unit. For monovalent cations, a $\Delta\psi$ of 59 mV is equivalent to 1 p unit. (C) A possible Na$^+$ cycle at alkaline pH.

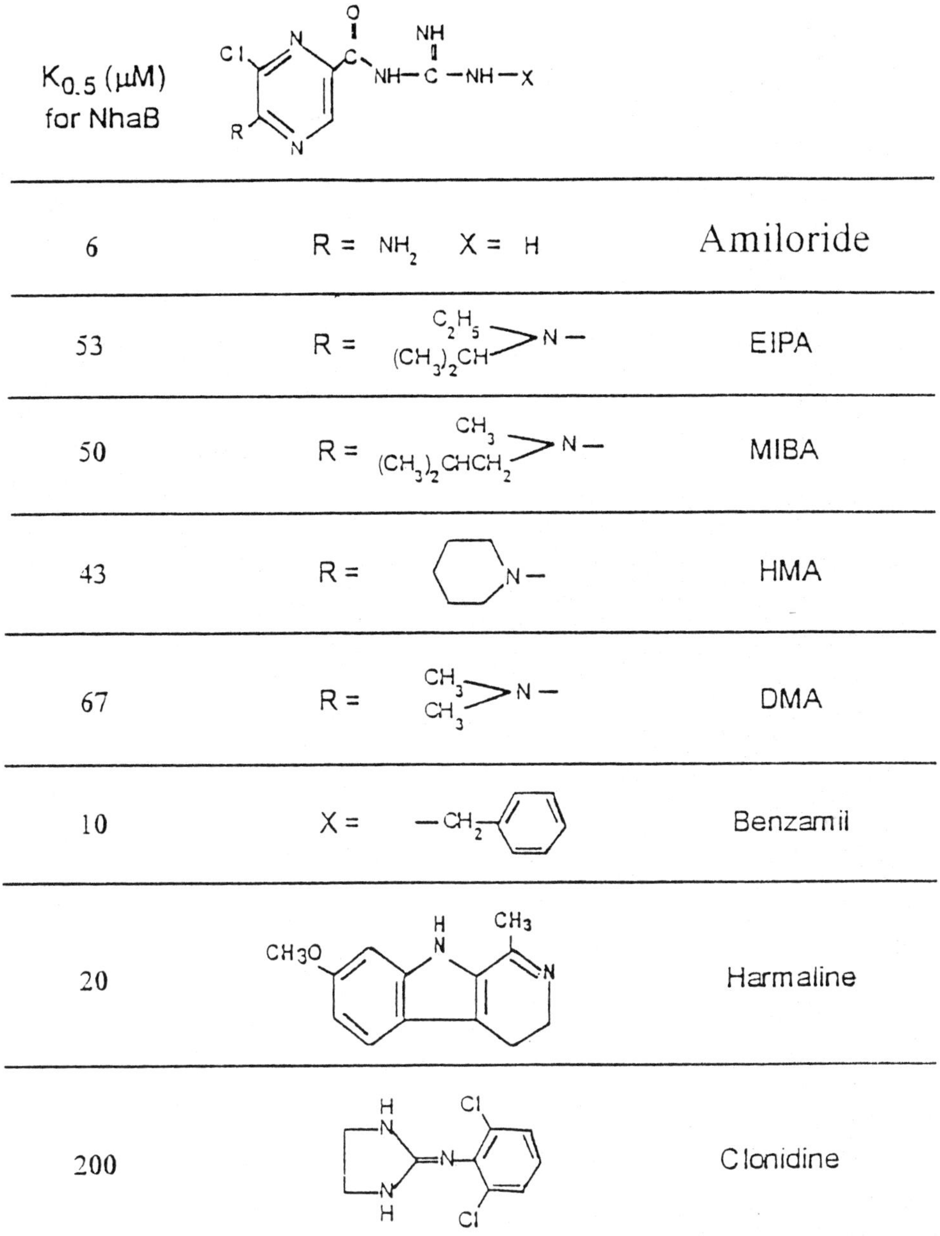

Fig. 2. Amiloride derivatives and their effect on NhaB. The data are from Ref. [17].

Na$^+$/H$^+$ antiport activity. The Na$^+$/H$^+$ antiporter of methanogenic bacteria was also found to be inhibited by both amiloride and harmaline (Fig. 2) [29].

There has been some controversy whether amiloride inhibits also the Na$^+$/H$^+$ antiport activity of *E. coli* [8,30,31]. Since they are weak amines amiloride and some of its derivatives may inhibit due to their uncoupling activity and therefore it has been hard to differentiate between specific and non specific effects of amiloride on ΔpH driven Na$^+$ transport [31,32].

To avoid the possible uncoupling activity of amiloride it is preferable to test the effect of this drug on ΔpH independent reactions, namely passive Na$^+$ efflux and ^{22}Na$^+$:Na$^+$ exchange. No significant inhibition of downhill sodium efflux was detected when catalyzed by purified NhaA in proteoliposomes [14,18]. Amiloride inhibited, the passive Na$^+$ efflux catalyzed by NhaB significantly (t$^{1/2}$ decreased from 25 s to > 120 s at 1 mM amiloride) as well as ^{22}Na$^+$:Na$^+$ exchange and ΔpH driven Na$^+$ uptake (competitive inhibition Ki = 20 μM [17]).

NhaB differs in its sensitivity to amiloride derivatives from that of the mammalian exchanger isoforms. Rather, it resembles the amiloride sensitive sodium channel from epithelial cells or the antiporter of alga *Dunaliella salina* [28], in being more sensitive to amiloride than to its derivatives [17] (Fig. 2).

It has been reported that amiloride inhibits growth of *E. coli* cells at alkaline pH [33,34]. It was suggested that this phenomenon results from the inhibition of Na$^+$/H$^+$ exchange activity, which is responsible for pH regulation. This is probably not the case because of several reasons: (a) NhaA which is supposed to be very active at alkaline pH is amiloride resistant; (b) an *E. coli* strain in which both *nha*A and *nha*B are deleted (Δ*nha*AΔ*nha*B) can grow at alkaline pH [21]; (c) growth of Δ*nha*AΔ*nha*B is also inhibited by amiloride at alkaline pH, indicating that this growth inhibition is not due to inhibition of specific Na$^+$/H$^+$ antiport activity.

In a study aimed to characterize the amiloride binding site of the mammalian antiporter Nhe-1, Pouyssegur and co-workers developed a procedure to isolate Nhe-1 mutants resistant to the amiloride analog 5-N-(methyl propyl) amiloride (MPA) [35]. Two Nhe-1 mutants with decreased sensitivity to MPA were isolated and both had a replacement of Leu-167 to Phe. This Leu is found in a sequence which is conserved among Nhe isoforms: ^{164}VFFLFLLPPI^{173} which is located in the middle of the fourth putative trans-membrane segment. Analysis of the sequence of NhaB reveals the pentamer ^{445}FLFLL^{450} which is identical to 5 amino acid of the putative amiloride binding site [17]. However it should be noted that in contrast to Nhe-1, this sequence resides in the middle of the eleventh putative trans-membrane segment of NhaB and the homology between NhaB and Nhe-1 is not extended beyond that in this region.

Harmaline and clonidine but not cimetidine, compounds known to inhibit the mammalian Na$^+$/H$^+$ antiporters, inhibit NhaB (17 and Fig. 2).

Li$^+$ is a known substrate of both Na$^+$/H$^+$ antiporters of *E. coli* [20] and Mn^{++} is a substrate of mitochondrial Na$^+$/H$^+$ exchanger [36]. Both LiCl and MnCl$_2$ inhibit NhaB-dependent Na$^+$ uptake [15,17]. The K$_{0.5}$ values at a sodium concentration of 100 μM were found to be 1.2 and 5 mM for Mn^{++} and Li$^+$, respectively. These data indicate that, surprisingly, NhaB has a higher affinity for Mn^{++} than to Li$^+$, a situation similar to that observed for the mitochondrial Na$^+$/H$^+$ exchanger [36].

2.1.3. NhaA functions simultaneously as a pH sensor and a titrator
NhaA is highly sensitive to pH, changing its rate more than 1000 fold between pH

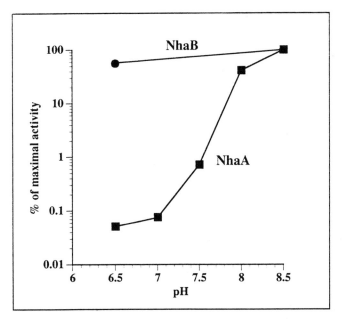

Fig. 3.   pH profile of NhaA and NhaB activity. The data are from Refs. [14,15], given for each antiporter as percent of its maximal activity.

7-8, thus functioning simultaneously as a pH 'sensor' and a 'titrator' [14,22] (Fig. 3). This distinct pH dependence of the Na$^+$/H$^+$ antiporter of *E. coli*, was first demonstrated in right side out membrane vesicles measuring ^{22}Na efflux driven by imposed artificial ΔpH or $\Delta\psi$ [22] and then in purified protein functionally reconstituted in proteoliposome, measuring passive ^{22}Na efflux (Fig. 3) [14].

Since flux of cations via the cation/H$^+$ antiporters affects the ΔpH across the membrane, cation induced changes in the fluorescence of acridine orange, and similar probes measuring ΔpH, has proven a fast and reproducible way to monitor the activity of H$^+$/cation antiporters in isolated membrane vesicles [23,37]. Calculation of the kinetic parameters of an antiporter with this technique is complicated due to the indirect nature of the measurement. However, when measurements are conducted at cation concentrations above the K_m of the antiporter, it most probably reflects the V_{max} of the system and is thus most suitable for comparison of antiporters activity [37].

Indeed with this technique a pronounced pH dependence of NhaA activity was found, similar to that found by direct flux measurements of ^{22}Na [19]. The experimental system used was everted membrane vesicles isolated from ΔnhaAΔnhaB transformed with wild-type *nhaA* bearing plasmid. ΔnhaAΔnhaB is devoid of Na$^+$/H$^+$ antiporters, and is highly sensitive to Na$^+$ (Fig. 4) but the plasmidic *nhaA* restores in it Na$^+$/H$^+$ antiporter activity and Na$^+$ resistance [20,21]. This system therefore has proven most suitable to identify (by site directed or random mutagenesis) residues in NhaA involved in the pH sensitive domain of the protein (Fig. 5A).

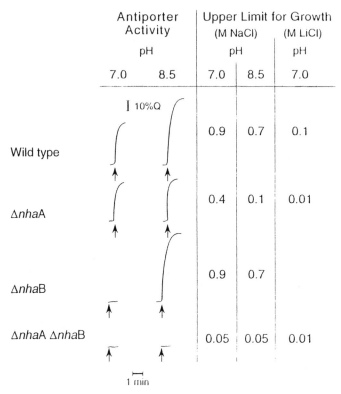

Antiporter Activity		Upper Limit for Growth		
		(M NaCl)		(M LiCl)
pH		pH		pH
7.0	8.5	7.0	8.5	7.0

Wild type — 0.9 | 0.7 | 0.1

ΔnhaA — 0.4 | 0.1 | 0.01

ΔnhaB — 0.9 | 0.7

ΔnhaA ΔnhaB — 0.05 | 0.05 | 0.01

Fig. 4.   Phenotype of the antiporter mutants. The upper limit of the concentrations of NaCl (in LB) [20] and LiCl (in minimal medium) [20] permitting growth of wild-type *E. coli* and its antiporter mutants ΔnhaA, NM81 [20], ΔnhaB, EP431 [21], and ΔnhaAΔnhaB, EP432 [21] is presented. Also shown are the respective antiporter activities in everted membrane vesicles isolated from the cells grown in LB in the presence of 100 mM NaCl except for EP432 which was grown in LBK [21]. The activity is measured according to [11,37] as a change (Q) in ΔpH monitored by acridine orange fluorescence upon addition of 10 mM Na^+ (arrow).

A mutation can be easily introduced into the plasmidic NhaA and its effect tested both *in vitro*, in everted isolated membrane vesicles, and *in vivo*, in both cases with no background of chromosomal encoded NhaA or NhaB Na^+/H^+ antiporters.

Using this approach it was found that none of the eight histidines of NhaA are essential for Na^+/H^+ exchange activity of the protein [19]. However, the replacement of His-226 by Arg (H226R) markedly changed the pH dependence of the antiporter [19] (Fig. 6). A strain deleted of both antiporters genes, *nhaA* and *nhaB*, transformed with multicopy plasmid bearing wild-type *nhaA*, exhibited both Na^+ and Li^+ resistance, throughout the pH range of 6–8.5. In marked contrast, transformants of plasmid bearing H226R-*nhaA* are resistant to Li^+ and to Na^+ at neutral pH, but became sensitive to Na^+ above pH 7.5. Analysis of the Na^+/H^+ antiporter activity of membrane vesicles derived from H226R cells showed that the mutated protein was activated by pH to the same extent as the wild type. However, whereas the

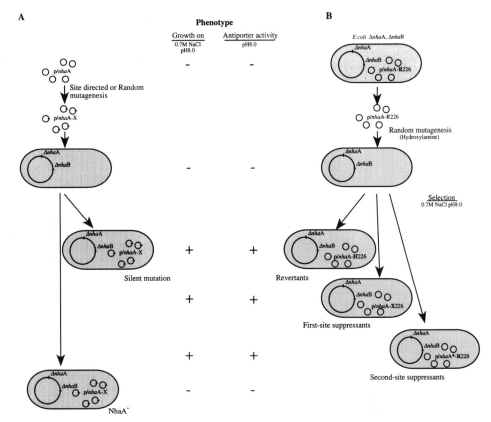

Fig. 5. Scheme for site directed and random mutagenesis of *nha*A (A) and a method for selection of revertants and suppressants of an *nha*A⁻ mutation (B).

activation of the wild-type NhaA occurred between pH 7 and pH 8, that of H226R antiporter occurred between pH 6.5 and pH 7.5. In addition, while the wild-type antiporter remains almost fully active, at least up to pH 8.5, H226R is reversibly inactivated above pH 7.5, retaining only 10–20% of the maximal activity at pH 8.5 (Fig. 6) [19].

Histidine and arginine share certain properties: both most probably bear a positive charge and both are polar residues capable of hydrogen bonding. The question as to which of these properties is important for the pH sensitivity of NhaA was then tested.

The fact that H226R is inactive at alkaline pH has provided a very powerful system to isolate revertants or suppressants to H226R (Fig. 5B). First site suppressants were obtained in which cysteine and serine replace H226 (Fig. 6) [127].

In solution both cysteine and histidine have a pK in the physiological range (Table 5.1 in Ref. [38]), whereas serine does not ionize under these conditions. Both histidine and cysteine have been shown to be ionized in proteins with a pK not very different from that observed in solution [38,39]. Nevertheless it is impossible to

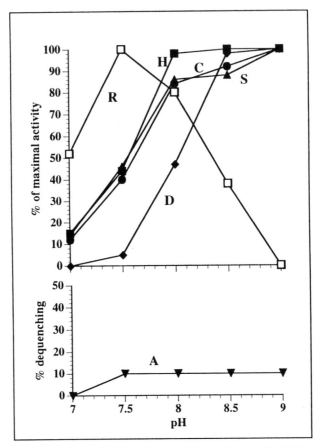

Fig. 6.   pH dependence of the Na$^+$/H$^+$ antiporter activity of *nha*A His-226 mutants. Everted membrane vesicles were prepared and assayed as described [19,127] at the indicated pH values. The data were obtained from EP432 (Δ*nha*AΔ*nha*B) transformed with the wild type (H) or the various mutated plasmids in residue H226. The amino acid replacements are indicated.

predict the pK of an amino acid in a protein neither from its ionization constant in solution, nor from its pK in another protein. It has been suggested that serine is also capable of ionization in a protein of the bacterial photosynthetic reaction center, but only when its electrostatic environment is drastically changed by electron transport [40], an unlikely situation in the case of NhaA. Therefore, we suggest that ionization of residue 226 is not important for pH regulation of the antiporter, rather it is the polarity and/or capacity to form hydrogen bonds, properties shared by histidine, serine and cysteine, which is essential.

    To test the effect of negative charge, aspartate was introduced at position 226 (H226D). As opposed to Arg which shifted the pH profile towards acidic pH, Asp shifted the pH profile towards basic pH (Fig. 6) [127]. When Ala (H226A) was introduced, which is neither polar nor charged, the antiporter showed no activity

(Fig. 6) [127]. In conclusion, a polar group is essential; charge shifts the pH profile, positive, towards acidic pH, negative towards basic pH.

As yet it is not known how these residues exert their effect. Amino acid residues are known to cause micro changes within proteins reflected in the electrostatic microenvironment, stability of acidic or basic residues, density of local protons and even binding of water molecules [38–40]. It is assumed that at a particular pH, a certain proton density, at a given site, is crucial for the pH sensitivity of NhaA. According to this reasoning at a pH below this critical value, a positive charge, at position 226, would be beneficial to reduce the excess proton concentration. At the alkaline range negative charge would be useful to attract the scarce protons. It should be emphasized however, that as long as the structure of the protein is unsolved, short-range steric effects or long range conformational effects, not directly related to the 'pH sensor', cannot be excluded.

It is of interest that the difference between the mutants is not expressed in growth phenotype at the acidic pH range. H226D with the lowest activity at this pH range grows like the wild type. On the other hand, at alkaline pH growth is detected only in the mutants which are substantially activated by pH, thus H226A and H226R stop growing beyond pH 8.4. These results emphasize the physiological importance of NhaA at alkaline pH in the presence of Na⁺.

2.1.4. pH Dependence of the NhaB protein

In contrast to NhaA, whose activity is extremely pH dependent (Section 2.1.3), NhaB activity was considered to be pH independent [20,21]. Indeed, when ammonium gradient driven ^{22}Na⁺ uptake was measured in NhaB-proteoliposomes at external pHs of 7.2, 7.6 and 8.5, the V_{max} found was 107, 67.5 and 87.8 μmol/min/mg, respectively (15 and Fig. 3). However, a ten-fold increase in the K_m of NhaB to Na⁺ was detected (from 1.5 to 16.6 mM) upon decrease of the pH in these experiments.

A similar effect of pH on the K_m of Na⁺/H⁺ antiporter activity was reported by Leblanc and collaborators [22]. These studies were performed in membrane vesicles isolated from cells containing both *nha*A and *nha*B genes. However, the cells were grown under conditions (low Na⁺, neutral pH) in which the level of *nha*A expression is known to be low (Section 3). These results reflect therefore the changes in affinity of NhaB. It has been suggested that the effect of pH on the K_m of NhaB is due to competition of both ions H⁺ and Na⁺, on a common site [22].

2.2. Na⁺/H⁺ antiporters of other bacteria

A comprehensive review of the properties of bacterial Na⁺/H⁺ antiporters, based on their activities in isolated membrane vesicles and intact cells, has recently been published [2]. Here we will focus only on the few antiporters whose genes have recently been cloned. As yet only the *E. coli* NhaA and NhaB proteins have been purified, reconstituted in proteoliposomes in a functional form and properties of the pure proteins studied in detail (Section 2.1).

Two strategies for cloning of Na⁺/H⁺ antiporter genes have been advanced. Each based on one of the two substrates, common to all Na⁺/H⁺ antiporters, Li⁺ and Na⁺.

On a concentration basis, Li^+ is 10 times more toxic than Na^+, both to wild-type *E. coli* (Fig. 4) [11] and fission yeast [41]. Therefore Li^+ provides a screen for cells capable of maintaining low internal Na^+ or Li^+ levels without selecting for osmotolerance. Another advantage of Li^+ selection over that of Na^+ is that it can be applied directly to wild-type cells. Realizing this advantage of Li^+ selection, *nha*A was cloned using a DNA library containing sequences overlapping the *nha*A locus [11]. The *sod*2 gene has been cloned from *Schizosaccharomyces pombe* using a similar approach [41].

Although wild-type *E. coli* cells transformed with multicopy plasmids bearing *nha*A become Li^+ resistant as compared to the wild type, their tolerance to Na^+ is unchanged. This implies that other factors, such as adaptation to increased osmolarity determines the upper level of resistance to Na^+. In contrast to wild-type cells, mutants, ΔnhaA or ΔnhaAΔnhaB (Fig. 4), which are Na^+ sensitive due to the lack of the antiporters are most suitable to apply Na^+ selection and clone by complementation DNA inserts encoding Na^+/H^+ antiporter genes. With this approach and the *E. coli* mutants various antiporter genes have been cloned from very different bacteria including *nha*B [13] and *cha*A [42] from *E coli* and *nha*C from an alkaliphile *Bacillus firmus* OF4 [43]. A similar approach applied to *Enterococcus hirae* yielded the *nap*A antiporter [44]. Utilizing this approach several new Na^+/H^+ antiporters have recently been cloned from bacteria (Section 2.2.1).

2.2.1. Na^+/H^+ antiporters of Vibrio alginolyticus and Vibrio parahaemolyticus

In contrast to *E. coli*, whose growth is independent of Na^+, the marine bacterium *Vibrio alginolyticus* requires 0.5 M NaCl for optimal growth [45]. In addition it has a primary Na^+ extrusion system, which is an electrogenic Na^+-translocating NADH-quinone reductase in the respiratory chain [46,47]. Like *E. coli* it has a Na^+/H^+ antiporter which is driven by a protonmotive force [48] and suggested to play a primary role in Na^+ circulation at neutral pH, when the respiratory linked Na^+ pump is not operating. Many transport systems for carbohydrates and amino acids of *V. alginolyticus* are Na^+-substrate symport systems that utilize $\Delta \tilde{\mu}_{Na^+}$ as a driving force [46]. Also, flagellar rotation, is driven by an influx of Na^+ [49]. Thus, Na^+ circulation across the cell membrane is very important for energy transduction in this organism.

A gene has been cloned from a DNA library from the marine bacterium *Vibrio alginolyticus* that functionally complements the ΔnhaA mutant strain of *E. coli* (Fig. 4) conferring resistance to Na^+ (0.5 M NaCl at pH 7.5) and, concomitantly, increasing Na^+/H^+ antiport activity measured in isolated everted membrane vesicles [45]. The *V. alginolyticus* protein also allowed Na^+ preloaded cells of *E. coli* NM81 (ΔnhaA) and RS1 (ΔnhaA, *cha*A$^-$) [50], to extrude Na^+ at alkaline pH [51]. The extrusion of Na^+ occurred against its chemical gradient in the presence of a membrane-permeable amine and $\Delta \psi$. Thus, the protein is functional as an electrogenic Na^+/H^+ antiporter in *E. coli* cells and in this respect it is functionally similar to *E. coli* NhaA.

The nucleotide sequence of the cloned fragment revealed an open reading frame, which encodes a protein with a predicted 383 amino acid sequence and a molecular mass of 40400 Da. The hydropathy profile is characteristic of a membrane protein with 11 membrane spanning regions. The deduced protein shows no significant sequence similarity with any cation/H⁺ antiporter except for the *E. coli* NhaA. Since it is 58% identical to the *E. coli* NhaA it was designated *nha*Av (Fig. 7) [44]. We suggest, however, designating it Va-*nha*A to follow a more general convention used also with eukaryotic genes.

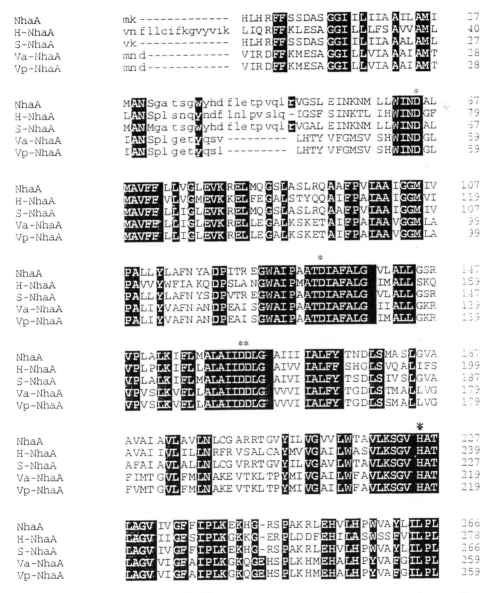

Fig. 7. Multiple alignment of NhaA Na⁺/H⁺ antiporters. The alignment was first performed using Pileup (Wisconsin Package, 1994) and then MACAW [128]. H-NhaA, S-NhaA, Va-NhaA and Vp-NhaA are *nha*A homologues of *H. influenzae* [57], *S. enteriditis* [58], *V. alginolyticus* [45] and *V. parahaemolyticus* [52] respectively. *, conserved Asp residues; ✳, conserved His residues.
(Continued overleaf).

```
NhaA     FAFANAG VSLQGVTLDGLT SILPLGIIAGLLI GKPLGISL   306
H-NhaA   FAFANAG VSFAGIDVNMIS SPLLLAIASGLII GKPVGIFG   318
S-NhaA   FAFANAG VSLQGVTIDGLT SMLPLGIIAGLLI GKPLGISL   306
Va-NhaA  FAFANAG ISLEGVSMSGLT SMLPLGIALGLLV GKPLGIFT   299
Vp-NhaA  FAFANAG ISLEGVSMSGLT SMLPLGIALGLLI GKPLGIFS   299


NhaA     FCWLALRLKL AHLPEGTTYQQIMVVGI LCGIGFTMS IFIA   346
H-NhaA   FSYISVKLGL AKLPDGINFKQIFAVAV LCGIGFTMS MFLA   358
S-NhaA   FCWLALRFKL AHLPQGTTYQQIMAVGI LCGIGFTMS IFIA   346
Va-NhaA  FSWAAVKMGV AKLPEGVNFKHIFAVSV LCGIGFTMS IFIS   339
Vp-NhaA  FSWAAVKLGV AKLPEGINFKHIFAVSV LCGIGFTMS IFIS   339


NhaA     SLAF-GSVDPEL INWAK LGILVGSISSAVIGYSWLRvrlr   385
H-NhaA   SLAFdANAGESV NTLSR LGILLGSTVSAILGYLFLKqttk   398
S-NhaA   SLAF-GNVDPEL INWAK LGILIGSLLSAVVGYSWLRarln   385
Va-NhaA  SLAF-GNVSPEF DTYAR LGILMGSTTAALLGYALLHfslp   378
Vp-NhaA  SLAF-GNVSPEF DTYAR LGILMGSTTAVLGYALLHfslp   378


NhaA     psv--                                      388
H-NhaA   ln---                                      400
S-NhaA   apa--                                      388
Va-NhaA  kkaga                                      383
Vp-NhaA  kkagd                                      383
```

Fig. 7 Continued. Caption on p. 515.

Vibrio parahaemolyticus is also a slightly halophilic marine bacterium, which requires at least 30 mM Na^+ for growth [52] and is very similar in its Na^+ metabolism to *V. alginolyticus* [52]. Also in this bacterium, a Na^+/H^+ antiporter is very important for Na^+ circulation, especially at neutral pH when the primary respiratory linked pump is inactive [53].

Recently, the Na^+/H^+ antiporter activity of *V. parahaemolyticus* has been characterized [54]. It seems that there are two Na^+/H^+ antiporters in the cell membrane of *V. parahaemolyticus*, resembling NhaA and NhaB of *E. coli* (Section 2). One system is pH-dependent (activity increases as pH increases between 7.0 and 9.0) and the other is pH-independent and less active. The K_m values and V_{max} values of the *V. parahaemolyticus* Na^+/H^+ antiporters are considerably larger than those of *E. coli* [52,54]. Since *V. parahaemolyticus* lives in seawater and large quantities of Na^+ enter *V. parahaemolyticus* cells, a low-affinity and high-capacity Na^+ extruding system would be necessary for *V. parahaemolyticus* cells.

A gene encoding a Na^+/H^+ antiporter was cloned from the chromosomal DNA of *Vibrio parahaemolyticus* using the Na^+ and Li^+ sensitive $\Delta nhaA\Delta nhaB$ *E. coli* mutant (Fig. 4) [52]. As expected, membrane vesicles prepared from the original *E. coli* mutant did not show any detectable Na^+/H^+ (and Li^+/H^+) antiport activity, while a high Na^+/H^+ (and Li^+/H^+) antiport activity was observed in membrane vesicles prepared from the transformed cells. The predicted protein encoded by the cloned gene consists of 383 amino acid residues putatively forming 12 TMS and showing high homology (59% identity and 87% similarity) with the NhaA Na^+/H^+ antiporter of *E. coli* and 97.4% identify to that of *V. alginolyticus* Va-NhaA (Fig. 7) [52].

These results imply the presence of NhaA type antiporter in marine bacteria. We suggest designating the *V. parahaemolyticus* gene Vp-*nha*A. The activity of Vp-NhaA increased dramatically when the pH of the assay medium was raised from 7.0 to 8.5, similar to its *E. coli* homolog. Interestingly, however, in contrast to the amiloride-insensitive *E. coli* NhaA (Section 2.1.2), the activity of the pH-dependent Na⁺/H⁺ antiporter of *V. parahaemolyticus* is inhibited by the drug [54].

It should be noted that the activity conferred by the *V. parahaemolyticus* gene expressed in *E. coli*, an heterologous host, differs from the native activity in untransformed *V. parahaemolyticus*. A drastic decrease in K_m values for Li⁺ and Na⁺ was observed in the heterologous expression system (Section 2.2.5).

2.2.2. Search for common denominators among the Na⁺/H⁺ antiporters, the NhaA family

Can we find common molecular denominators among the Na⁺/H⁺ antiporters? There is quite a high degree of sequence conservation among the mammalian Na⁺/H⁺ antiporters [9,55], but there is no significant homology between them and NhaA, and not even between NhaA and NhaB [2,7]. These findings would suggest that there are no universal Na⁺ and H⁺ recognition and exchange sites and that, instead, different sequences accomplish similar functions. However, it is possible that very few conserved residues are adequate to carry out these functions, even when they are dispersed throughout the protein. Identification of the domains necessary for activity regulation, and H⁺ and Na⁺ sensing in the various proteins is certainly one of the future challenges.

NhaB displays a weak but distinct homology to a membrane protein with unknown function in *Mycobacterium leprae* (Accession #L10660, 24% identity, 55% similarity) [17]. In addition there is a limited homology to other transporters found in bacteria and animal cells: the rat sodium dependent sulfate transporter (#L19102: 24% identity, 50% similarity), the product of the mouse pink-eyed dilution gene (#M97900: 21% identity, 53% similarity), and the ArsB protein [56].

Interestingly, NhaB but not NhaA of *E. coli* is inhibited by amiloride (Section 2.1.2). The Vp-NhaA Na⁺/H⁺ antiporter is also inhibited by amiloride. A search for homologous domains in NhaB revealed one sequence (⁴⁴⁵FLFLL⁴⁵⁰) that was very similar to the amiloride binding domain in human Nhe¹⁶⁴(VFFLFLLPPI¹⁷³) (Section 2.1.2.). A VFFLLI sequence was found in TMS2 of Vp-NhaA [52]. Since amiloride appears to compete with Na⁺, this region may be involved in Na⁺ binding. A similar motif was found in the Na⁺/H⁺ antiporter of *Enterococcus hirae* (NapA) [44] and other Na⁺ driven transporters [52]. It is not known, however, whether the activity of these transporters is inhibited by amiloride or not.

The sequence of the entire genome of *Haemophilus influenzae* Rd has recently been published [57]. It includes genes encoding proteins homologous to NhaB, NhaC, NhaA (87%, 62% and 75% similarity, respectively) (Fig. 7). This newcomer joins the growing family of NhaA which includes the Na⁺/H⁺ antiporter from *Salmonella enteriditis* [58], *V. alginolyticus* [45] and *V. parahaemolyticus* [52] and shows substantial homology with the *E. coli* protein.

Pinner et al. [13] reported that in one domain of the *E. coli* NhaB (starting at amino acid position 295), the homology with NhaA (starting at 300) at practically the equivalent position in the protein was quite significant (43% identity and 65% similarity). A diffuse and restricted homology with other Na$^+$-transporting proteins was found in this area (GXXLXXXXA). Between the Gly and Ala, no charged residues were found in any of the sequences. Also, this stretch of the homology overlaps with part of the 'sodium consensus box' previously identified in other Na$^+$-translocating systems [59]. However, replacement in NhaA of the conserved Gly with Cys did not affect activity (P. Dibrov, E. Padan and S. Schuldiner, unpublished results). Similar sequences were found in Vp-NhaA [52]. At present, the functional significance of these domains is not clear.

In the *E. coli* NhaA, His-226 has been assigned a role in the pH sensitivity of the protein (Section 2.1.3). Most interestingly, this histidine is conserved in all the members of the NhaA family (Fig. 7), suggesting that the 'pH sensor' domain is similar in all of them.

There are 9 aspartic acids and 13 glutamic acids in the Va-NhaA protein. Of 22 negatively charged residues, only 4 amino acids are predicted to be in TMS and all of them are aspartic acid (45,51). These 4 amino acids are conserved in NhaA of *E. coli*, *Salmonella enteritides* and Vp-NhaA (Fig. 7). Furthermore, D-111, which is predicted in a loop region between the TMS III and IV is also conserved in NhaA. Each of these conserved aspartic acid residues was replaced by asparagine. *E. coli* NM81 cells containing a plasmid harboring the Va-*nha*A gene mutated at D-125, -155 or -156 could neither grow in a high NaCl medium nor extrude Na$^+$ at alkaline pH against its chemical gradient (Fig. 7) and [51]. It is reasonable to assume that negatively charged amino acid residues localized in putative TMS are important for binding or translocation of cations like H$^+$ and Na$^+$. Nevertheless it should be stressed that intramembrane charges can also be involved in the assembly and folding of the protein [60,61]. Hence the suggestion that these aspartates form the Na$^+$ binding site [51], although appealing, may be premature.

Eight consecutive amino acid residues starting at position 45 in *E. coli* NhaA are missing in Va-*nha*A [45]. Since these two NhaA antiporters possess similar properties, these eight residues probably have no significant role in the antiporter protein.

The expression of *nha*A in *E. coli* has been shown to be regulated by Na$^+$ and by a regulatory protein encoded by *nha*R located just downstream from *nha*A (Section 3). The structural gene corresponding to *nha*R, however, was not found in the upstream and downstream regions of either Va-*nha*A or Vp-*nha*A. Since these marine bacteria require Na$^+$ for optimal growth, it has been suggested that the Na$^+$/H$^+$ antiporter is constitutively expressed for the extrusion of cellular Na$^+$ in marine bacteria [44].

2.2.3. *The chromosomal tetracycline — resistance locus tetB(L) of Bacillus subtilis encodes a protein which confers Na$^+$/H$^+$ antiporter activity*

Tetracycline-resistance determinants that function *via* proteins catalyzing efflux of the antibiotic have long been recognized in numerous plasmids and transposons from both gram-negative [62] and gram-positive bacteria [63]. In addition, an efflux type of *tet* gene exists in single copy near the origin of replication of some

B. subtilis strains that are nonetheless tetracycline-sensitive. This *tet*B gene, which is referred to as *tet*B(L), has remained an unexplained, presumably cryptic oddity [64].

Transpositional disruption of the chromosomal *tet*B(L) locus of *B. subtilis* led to impaired growth at alkaline pH in the presence of Na^+ and reduced rates of electrogenic Na^+ efflux at alkaline pH [65]. The mutant phenotype was reversed by transformation with a plasmid expressing the *tet*B(L) gene, suggesting a physiological role for this locus in Na^+-resistance and Na^+-dependent pH homeostasis. TetB(L) was also inferred to have a modest capacity for K^+ efflux [65]. Energy-dependent tetracycline efflux rates in the wild type were larger than in the transposition mutant, but were not sufficient to confer resistance to the antibiotic. The plasmidic *tet*B(L) conferred only a limited Tet resistance.

To further explore its function, the *tet*B(L) gene was expressed in *E. coli* ΔnhaA (NM81) which is sensitive to Na^+. ΔnhaA is also completely inhibited by tetracycline concentrations as low as 2 μg/ml. Both Na^+ resistance up to 0.6 M and tetracycline resistance (4 μg/ml) was conferred upon transformation of ΔnhaA with the *tet*B(L) bearing plasmid. Furthermore, everted membrane vesicles isolated from the transformants catalyze Na^+/H^+ antiport even more actively than tetracycline/H^+ antiport [65].

These results for the first time raise the possibility of the activity of Tet proteins in Na^+ efflux, and of physiological roles for *tet* gene products that are unrelated to antibiotic resistance. TetB(L) is suggested to function as a Na^+/H^+ antiporter at neutral pH as well as at pH 8.5. Nevertheless, this interesting suggestion will be experimentally proven only when the TetB(L) protein is purified in a functional form.

2.2.4. Na⁺/H⁺ antiporter of extreme alkaliphiles

The pioneering studies of Krulwich and colleagues on the physiology of extreme alkaliphiles, demonstrated that Na^+/H^+ antiporters play a crucial role in pH homeostasis in these organisms [66,67]. Intracellular pH regulation in these bacteria, which at alkaline pH involves acidification of the cytoplasm, requires Na^+ and a mutant lacking the antiporter activity is no longer alkaliphilic. *nha*C was cloned from *B. firmus* OF4 [43] using *E. coli* ΔnhaA as a host and its Na^+ sensitivity for functional complementation. *nha*C confers Na^+ resistance to the *E. coli* mutant and increases Na^+/H^+ activity in isolated everted membrane vesicles, suggesting that it encodes a Na^+/H^+ antiporter. Further evaluation of the role of *nha*C in alkaliphily awaits developing a genetic system in the extreme alkaliphile.

Recently, a series of alkali-sensitive mutants of the alkaliphilic *Bacillus* sp. C-125 were isolated [68]. One of the mutants, 38154 could not grow above pH 9.5 and was unable to sustain a low internal pH in an alkaline environment. A DNA fragment from the wild type parent was cloned by functional complementation of the mutant restoring alkaliphily [69]. Direct sequencing of the mutant's DNA corresponding region, revealed that the mutation resulted in an amino acid substitution from Gly-393 to Arg of the putative ORF1 product. ORF1 is predicted to encode an 804-amino-acid polypeptide with a molecular weight of 89070 and hydrophobicity characteristics of a membrane protein. The N-terminal part of the putative ORF1 product showed amino acid similarity to those of the chain-5 products of eukaryotic NADH quinone oxidoreductases. Membrane vesicles prepared from mutant 38154 did not show membrane potential ($\Delta\psi$)-driven Na^+/H^+ antiporter activity. Antiporter activity was regained by introducing the cloned gene. These results indicate that the mutation in 38154 affects, whether or not directly, the electrogenic Na^+/H^+ antiporter activity. This is the first report which directly shows that a gene encoding a putative Na^+/H^+ antiporter is important in the alkaliphily of alkaliphilic microorganisms.

2.2.5. *Expression of homologous and heterologous antiporter genes in E. coli devoid of its own Na^+/H^+ antiporter genes*

As previously mentioned, heterologous expression of antiporter genes in *E. coli* ΔnhaAΔ*nha*B provided a very powerful tool for cloning of antiporter genes. This expression system was also used for the study of the activity of antiporter proteins [42,45,51,52,65,70]. However, it should be noted that the affinity of the *V. parahaemolyticus* Vp-NhaA antiporter for Li^+ and Na^+ is significantly increased when the protein is expressed heterologously in *E. coli* cells [52]. Since the pH activity relationship was similar in both *E. coli* and *V. parahaemolyticus*, it seems that no significant change in the affinity for H^+ had occurred. Although we do not know the reason(s) for the changes in affinities, several possibilities can be considered. (1) A difference in membrane lipids in *V. parahaemolyticus* and *E. coli* cells may affect the activity of the Na^+/H^+ antiporter. The composition of lipids in the membrane have been shown to change the properties of transporters [71]. (2) The Vp-*nha*A gene or its protein may be specifically modified in some way in *E. coli* or in *V. parahaemolyticus* cells to account for the differences. (3) Na^+/H^+ antiporter activities in the two organisms are regulated by different mechanisms.

3. Regulation of expression of *nha*A

3.1. *A novel signal transduction pathway to Na^+ regulates expression of nhaA*

Since the fluxes of Na^+ and H^+ are coupled through Na^+/H^+ antiporters in every cell, their regulation deserves attention. Furthermore, in cases like *E. coli* where two antiporters function, the regulation of activity and expression of each of the individual elements is crucial for the understanding of both Na^+ and H^+ circulation in the cells, and the integrative operation of the antiporter in these circulations. The only regulation of expression thus far studied has been that of *nha*A in *E. coli*. Two promoters have been mapped by primer extension in the 5′ upstream region and the upstream DNA sequence characterized [72].

A chromosomal translational fusion between *nha*A and *lac*Z (*nha*A′–′*lac*Z) was constructed. The levels of expression were very low unless Na^+ or Li^+ were added. Na^+ and Li^+ ions increased expression in a time-and concentration-dependent manner [72]: the maximal increase was detected when the cells were exposed to 50 to 100 mM of either ion for a period of 2 h. This effect is specific to the nature of the cation and is not due to a change in osmolarity. The effect of the ions is potentiated at extracellular alkaline pH. Northern analysis demonstrated that the regulation is at the level of transcription (N. Dover, S. Schuldiner and E. Padan, unpublished results). The pattern of regulation of *nha*A thus reflects its role in the adaptation to high salinity and alkaline pH in *E. coli* (Section 4). It also implicates the involvement of a novel regulatory gene, in addition to *nha*A and *nha*B.

The NhaR protein, the product of the gene *nha*R, located downstream of *nha*A, regulates expression of *nha*A. Inactivation of chromosomal *nha*R yields cells sensitive to Li$^+$ and Na$^+$ in spite of having intact *nha*A. Furthermore, multicopy *nha*R increases expression of the *nha*A′−′*lacZ* fusion, and this increase is completely Na$^+$-dependent [73]. These results indicate that NhaR is a positive regulator of *nha*A, which functions in *trans* and its effect is Na$^+$-dependent. Accordingly, partially purified NhaR protein binds specifically to the promoter region of *nha*A and retards its mobility in a gel retardation system [73].

On the basis of protein homology NhaR belongs to a large family of positive regulatory proteins called the LysR family which contains in their N-terminal a helix-turn-helix motive that is believed to bind DNA [74,75]. Most importantly, several of these proteins, are part of a signal transduction pathway involved in response to environmental stress [76,77]. We thus conclude that NhaR is part of a signal transduction which is essential for challenging Na$^+$ and pH stress (Fig. 8). Hence *nha*A is regulated both at the activity and at the expression level; the former by pH (Figs. 3 and 8) [14] (Section 2.1.3); the latter by Na$^+$ via NhaR (Fig. 8) [73,79].

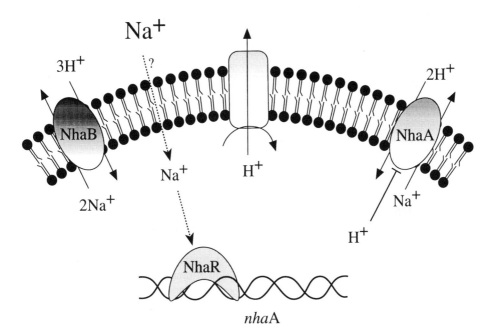

Fig. 8.   Schematic presentation of the Na$^+$ responsive regulation of *nha*A mediated by NhaR. Activity of NhaA is regulated by H$^+$, high pH activates [14]. Expression of *nha*A is regulated by intracellular Na$^+$ via the positive regulator NhaR [73,79].

3.2. The 'Na sensor' in NhaR

The change in extracellular Na⁺-concentration, must reach NhaR, since the effect of NhaR on *nha*A expression is Na⁺-dependent. It is conceivable that a change in intracellular Na⁺ which follows the change of the extracellular concentration of the ion, serves as the immediate signal. If this is the case then, manipulation of intracellular Na⁺ should change the expression levels. Transformation of the *nha*A'–'*lac*Z bearing strain with either *nha*A or *nha*B inhibits the Na⁺-dependent induction [7]. Since these genes share very little homology but a common activity, Na⁺ extrusion, we suggest that an increase in the intracellular Na⁺ serves as the immediate on-signal for the NhaR dependent *nha*A expression (Fig. 8). In agreement with this suggestion, *nha*A–*lac*Z is fully expressed at a lower Na⁺ concentration (10 mM, pH 7.5) in a Δ*nha*AΔ*nha*B mutant than in a Δ*nha*A mutant (100 mM, pH 7.5) (E. Pinner, E. Padan and S. Schuldiner, unpublished results).

Since the induction of *nha*A by intracellular Na⁺ is dependent on NhaR, it is conceivable that a 'Na⁺ sensor'-site exists on NhaR which can be identified by mutations affecting the Na⁺ sensitivity of the expression system. Should such a mutation increase the affinity to Na⁺ of the expression system, then at a given Na⁺ concentration, it may even increase the Na⁺/H⁺ antiporter activity in the membrane above the wild type level. A previously isolated mutation, designated *ant*^{up} was found to increase the Na⁺/H⁺ antiporter activity, thereby conferring Li⁺ resistance to cells which are otherwise Li⁺ sensitive (Nha^{up} phenotype) [11,78]. The Nha^{up} mutation resides in the C-terminus of NhaR and is glu134 to ala134 substitution in the protein [79]. This mutation increases the affinity to Na⁺ of the NhaR mediated *nha*A transcription. Although Na⁺ may indirectly affect NhaR it is tempting to speculate that glu134 is part of the 'Na⁺ sensor' of NhaR.

4. Molecular physiology of the antiporters

4.1. The role of the Na⁺/H⁺ antiporters in homeostasis of Na⁺

A straightforward approach to assess the role of a given gene in a given trait is to construct a mutation lacking the gene and compare its activity to that of the wild type and the mutant retransformed with the gene. Any difference found can then be ascribed to the tested gene, provided that, in its absence, no other systems replace its activity. The physiological importance of the antiporters of *E. coli* has become apparent by construction of deletions mutations, Δ*nha*A, Δ*nha*B, Δ*nha*AΔ*nha*B (Fig. 4). Analysis of Δ*nha*AΔ*nha*B showed that there are no specific Na⁺/H⁺ antiporters other than NhaA and NhaB in the membrane. A K⁺/H⁺ non-specific antiporter [80] was the only remaining activity in Δ*nha*AΔ*nha*B [21] (Fig. 4). Accordingly, the Δ*nha*AΔ*nha*B mutant is most sensitive to Na⁺ having an upper limit for growth of ≤0.05 M. It is also very sensitive to Li⁺, yet in the absence of added Na⁺ (contaminating level of 10 mM) or Li⁺ it grows at the entire pH range, pH 6.5–8.4 [21] (Fig. 4). Hence the sensitivity of the mutant is specific to Na⁺ and Li⁺.

Deletion of each gene separately allowed identification of the relative role of the respective proteins in Na$^+$ metabolism. Similar to $\Delta nhaA\Delta nhaB$ also $\Delta nhaA$ is Na$^+$ sensitive. It cannot withstand sodium concentrations as high as the wild type does (0.9 M NaCl at pH 7, Fig. 4) [20] and it cannot challenge the toxic effects of Li$^+$ ions (0.1 M). In two respects the $\Delta nhaA$ ion sensitivity differs from that of $\Delta nhaA\Delta nhaB$. It is less sensitive than the latter and its Na$^+$ sensitivity is pH-dependent, increasing at alkaline pH (from 0.4 M NaCl at pH 7 to 0.1 M NaCl at pH 8.5). It is thus concluded that *nhaA* is indispensable for adaptation to high salinity, for challenging Li$^+$ toxicity, and for growth at alkaline pH (in the presence of Na$^+$). In the absence of added Na$^+$ $\Delta nhaA$ like $\Delta nhaA\Delta nhaB$ grows at the same rate as the wild-type strain. Hence the phenotype of both mutants exposes a specific sensitivity of the cytoplasm to Na$^+$.

The $\Delta nhaB$ strain shows no impairment in its ability to adapt to high salt or alkaline pH nor in its resistance to Li$^+$ [21]. These findings suggest that NhaA alone can cope with the salt and pH stress, having adequate capacity for these functions. Nevertheless, in the absence of *nhaA*, *nhaB* confers a limited resistance to Na$^+$ since $\Delta nhaA\Delta nhaB$ is more sensitive to Na$^+$ than $\Delta nhaA$. The role of nhaB also becomes apparent in the wild type at low pH and low Na$^+$ when *nhaA* is not induced and the activity of NhaA is shut down [21]. Growth on Na$^+$/symport substrates such as glutamate and proline does not occur in $\Delta nhaB$ under these conditions.

The phenotypes of the deletion mutations provide the most compelling evidence for the major role of Na$^+$/H$^+$ antiporter activity in the Na$^+$ cycle of bacteria. Thus the level of the Na$^+$/H$^+$ antiporter activity in isolated membrane vesicle in each mutant explains the respective phenotype; $\Delta nhaA\Delta nhaB$ has no activity, $\Delta nhaA$ has partial activity (40–60%) and $\Delta nhaB$ has full activity (Fig. 4). Furthermore the Na$^+$ sensitive phenotype of both $\Delta nhaA\Delta nhaB$ and $\Delta nhaA$ is cured upon transformation with *nhaA*. The Na$^+$ and Li$^+$ sensitive phenotype is cured only partially by multicopy *nhaB* (see below).

Based on these results and on the fact that the expression of *nhaA* is highly regulated and increases significantly under the conditions in which it is essential: high salt, alkaline pH (in the presence of Na$^+$ ions), and the presence of toxic Li$^+$ ions (Section 3) we suggest that *nhaB* is the housekeeping antiporter while *nhaA* is the adaptive gene for Na$^+$ tolerance.

Although it confers resistance to Li$^+$, *nhaA* does not increase the limits of the pH or salt that wild-type *E. coli* can cope with, suggesting that factors such as osmotolerance set the upper limits of tolerance.

As described above, the addition of Na$^+$ dramatically affects the growth of $\Delta nhaA$, while its sensitivity to Na$^+$ intensifies with increasing pH [20]. The increase in the detrimental effect of Na$^+$ on $\Delta nhaA$ with pH, has been explained by either the observed increase in intracellular Na$^+$ with pH [25] and/or an increase in Na$^+$ toxicity in the more alkaline cytoplasm. The possibility that Na$^+$ toxicity is not directly related to the antiporters and may occur via competition between excess Na$^+$ and H$^+$ in the active sites of many essential systems has also been raised [7]. In this respect it would be interesting to test whether the Na$^+$ sensitivity of $\Delta nhaA$ $\Delta nhaB$ is indeed pH independent.

The cause of the increased sensitivity of the ΔnhaA strain at alkaline pH has most likely a thermodynamic component [15]. With a stoichiometry of 1.5, NhaB cannot maintain a Na^+ gradient when the driving force reaches 0 (the sum of 1.5 ΔpH + 0.5 $\Delta \psi$) even though $\Delta \tilde{\mu}_{H^+} > 0$ (Eq. 1, Fig. 1 and Section 4.2). In E. coli at acid and neutral pH, the ΔpH is alkaline inside and decreases with increasing pH_{out} [81]. $\Delta \psi$ increases so that $\Delta \tilde{\mu}_{H^+}$ is more or less constant [82,83]. The sum of 1.5 ΔpH + 0.5 $\Delta \psi$ decreases with the alkalinization of the pH_{out} and will reach a value equivalent to 60 mV at a pH_{out} of about 8.1 and will be 0 at pH_{out} of 9 [25]. Therefore, the driving force for NhaB activity is expected to be very low at a pH of about 8.5, and NhaB will not be able to maintain a significant sodium gradient at these pH values. Indeed, we could not complement the growth properties of the ΔnhaA strain (NM81) on 100 mM NaCl at pH 8.5 with a nhaB, which shows that the inability of NhaB to extrude sodium at alkaline pH_{out} is a consequence of a thermodynamic rather than a kinetic barrier [15]. Also multi nhaB does not complement the Li^+ sensitive phenotype.

4.2. The role of the Na^+/H^+ antiporters in pH homeostasis

One of the major roles assigned to the antiporter is in the regulation of intracellular pH (pH_{in}) mainly at alkaline extracellular pH [80–82]. In E. coli, pH_{in} has been shown to be clamped at around 7.8, despite large changes in the extracellular pH [81,82–86]. When the pH of the external medium is rapidly lowered or raised by over one unit, the E. coli internal pH shifts slightly, then recovers [86,87]. During anaerobic growth, cells maintain a constant internal pH of 7.4 at external pH 6.6–7.0 [88]. Many bacteria as well as eukaryotic cells have since been shown to strictly maintain a constant cytoplasmic pH at around neutrality [82–86,89–92]. Relatively small increases in pH_{in} halt cell division and activate the expression of specific genes [93,94] and of regulons [95,96]. It is therefore not surprising that both eukaryotic and prokaryotic cells have evolved several pH_{in} regulative mechanisms to eliminate metabolically induced changes in pH_{in} or to counter extreme environmental conditions [82,91,96–100].

The elucidation of the mechanisms of pH homeostasis in neutrophilic bacteria, including E. coli, have proven to be remarkably elusive [7,81,82,84,86,96,100]. A well documented system is that of Enterococcus hirae (Streptococcus faecalis), in which the proton-translocating ATPase regulates internal pH by excreting H^+ [101,102]. However, this cell is limited in its pH range of growth in the absence of carbonate (pH 6.5–7.9) [102,103]. In E. coli, unc mutants lacking an active H^+/AT-Pase regulate internal pH normally [104].

We proposed that Na^+/H^+ antiporters in conjunction with the primary H^+ pumps are responsible for homeostasis of intracellular pH in E. coli [81,82]. Involvement of K^+/H^+ antiporter has been also suggested both in E. coli [80] and E. hirae [105]. The suggestion of the involvement of Na^+/H^+ antiporters had its most compelling experimental validation in akaliphiles, in which it was shown that Na^+ ions are required for acidification of the cytoplasm and for growth and a mutant lacking antiporter activity lost both its pH homeostasis and growth capacity at alkaline pH

[106–108] (Section 3.2.4). In neutrophiles, such as *E. coli*, there is no direct evidence that supports this contention, since it is not clearly established that Na$^+$ is required for growth at alkaline pH. Nor is the requirement for Na$^+$ easy to demonstrate in some alkaliphiles, presumably due to a very high affinity for Na$^+$ (as low as 0.5 mM), such that the contamination present in most media suffices to support growth [109]. McMorrow et al. [110] have taken special precautions to reduce Na$^+$ to very low levels (5–15 μM) and reported a strict requirement for Na$^+$ (saturable at 100 μM) for growth of *E. coli* at pH 8.5. This range of concentrations of Na$^+$ required for growth is well within the range of the K$_m$ of the NhaA system (100 μM).

As discussed in Section 4.1, a straightforward approach to study the involvement of the Na$^+$/H$^+$ antiporters in a physiological function such as pH homeostasis is the analysis of the phenotypes of the various antiporter mutants and the deduction of the role of the antiporters by a comparison to the wild type phenotype under different conditions pertaining to Na$^+$ and H$^+$. With this approach a novel gene affecting Na$^+$/H$^+$ antiport activity has recently been shown to participate in pH homeostasis in alkaliphilic bacteria [68,69] (Section 2.2.4). In *E. coli* we found that as long as Na$^+$ is withheld from the medium, Δ*nhaA*Δ*nhaB*, and the two respective single mutants grow at the entire pH range of growth [20,21]. Similarly, *E. coli* has been shown to grow with negligible proton motive force, in the presence of carbonyl cyanide m-chlorophenylhydrazone (CCCP), both at alkaline pH [111,112] and at neutral pH [113], a condition under which the Na$^+$/H$^+$ antiporter cannot operate. Hence, the simplistic assumption that cells lacking the antiporter activity are pH-sensitive has been disproved in *E. coli*.

Nevertheless, a mutation, Hit1, which is closely linked to *nha*B has been proposed to affect pH homeostasis, based on its growth inhibitory effect [114] and direct measurements of pH$_{in}$ [115] at alkaline pH. Since a deletion in *nha*B does not yield a similar phenotype [21,111], we conclude that Hit1 contains an additional unidentified mutation, or that *hit*1 modifies NhaB so that the aberrant protein inhibits growth at alkaline pH. The role of NhaB as that of NhaA in pH-regulation is therefore still unclear.

One possibility to explain the growth of Δ*nhaA*Δ*nhaB* at alkaline pH in the absence of Na$^+$ is that although the antiporters are involved in pH regulation, pH-homeostasis is not required in the absence of Na$^+$. On the other hand, assuming that pH-homeostasis is an absolute requirement for growth [81,82], these results suggest that neither NhaA nor NhaB participate in pH homeostasis, or that only in their absence, *kha*A–K$^+$/H$^+$ antiporter [116,117] or an alternative system regulates intracellular pH (Section 4.3).

An alternative mechanism for pH homeostasis in the absence of the antiporters does not exclude the possibility that when present, the antiporters are involved both in pH homeostasis and Na$^+$ extrusion. This question can now be addressed by the new tools developed for the measurement of Na$^+$/H$^+$ antiporter activity, each separately and in combination.

The primary means of H$^+$ extrusion under aerobic conditions is the electron transport chain, which in the absence of permeable ions develop a negligible ΔpH and a large Δψ. Apparently Na$^+$ and K$^+$ are involved in the modulation of Δμ̃$_{H^+}$

components: an electrogenic uptake of K^+, with or without H^+, compensates for charge extrusion and thus permits the development of ΔpH [81,82,117]; an exchange of Na^+ and H^+ has been implied in the generation of an inverted ΔpH at alkaline pH [81]. We know that NhaA is electrogenic, with a stoichiometry of two (Section 2.1.1) and that it is a highly active system: the maximal value of NhaA mediated downhill Na^+ efflux at pH 8.5 is about 400 nmol/min/mg cell protein, and it can increase even more when fully induced [7]. Since the rate of H^+ extrusion through the respiratory chain is about 1000 nmol/min/mg cell protein, and the rate of K^+ transport through the Trk systems is about 500 [118], NhaA can, indeed, quantitatively account for a rapid response to changes in ion content.

The apparent stoichiometry (1.13–1.4) of H^+/Na^+ exchange in respiring E. coli cells was found to be dependent on pH and explained by the presence of two Na^+/H^+ antiporters with different stoichiometries [25]. Both NhaA and NhaB were found to be electrogenic, with a stoichiometry of $2H^+/Na^+$ [16] and 1.5 H^+/Na^+ [15] respectively (Section 2.1.1).

The different stoichiometries imply a different dependence on $\Delta\tilde{\mu}_{H^+}$, the driving force. The total value of $\Delta\tilde{\mu}_{H^+}$ changes slightly throughout the pH range between 6.5 and 8.5. However, its components change drastically; the ΔpH component is reduced with an increase in extracellular pH so as to maintain constant intracellular pH. Thus, it is zero at pH 7.8 and becomes negative above it [81]. Concomitantly the $\Delta\psi$ is increased in a compensatory fashion to maintain a constant $\Delta\tilde{\mu}_{H^+}$ [83]. At steady state the maximal concentration gradient of Na^+ maintained by NhaA is $\Delta pNa = 2 \Delta pH + \Delta\psi$ and that by NhaB is $\Delta pNa = 1.5 \Delta pH + 0.5\Delta\psi$.

Figure 1 shows the change of the ΔpNa maintained by each antiporter at steady state, at various pH values. Throughout the pH range, NhaA maintains a Na^+ gradient higher than that maintained by NhaB. The difference between the two gradients is increased with increasing pH. At pH 8.5 the gradient generated by NhaA is 20-fold higher than that of NhaB.

Hence the velocity, K_m, regulation of the activity, regulation of expression and in particular its stoichiometry make NhaA most suitable to function at alkaline pH and by extruding Na^+ in exchange for H^+ acidify the cytoplasm to regulate pH_{in}. Such an activity entails a recycling of Na^+ so that intracellular concentration of the ion will not be reduced below the K_m of NhaA to limit the activity of the antiporter.

Figure 1 suggests that at alkaline pH the driving force for NhaB, will be the Na^+ gradient maintained by NhaA. Then, NhaB will, rather than excrete Na^+, function in the uptake mode and complete the cycle.

The presence of two antiporters with different stoichiometries raises the problem of futile cycles; the protein with the lower stoichiometry (NhaB) will be able to recycle Na^+ back, but not to extrude it, while the antiporter with the high stoichiometry (NhaA) is active. This necessitates very careful regulation of expression and activity of both antiporters. Indeed, NhaA is regulated at both transcription [72,73] and activity levels [14]. Further work is required to establish the patterns of regulation of NhaB.

These data together with the parameters regarding the regulation of activity and expression of the antiporters will eventually permit construction of a model, simulating the integrative activity of the antiporters in the H$^+$ and Na$^+$ cycles of the cells. Understanding these cycles will provide important clues for our comprehension of the process of adaptation to extreme pH and salt environments.

4.3. Na$^+$ excretion machinery other than the Na$^+$/H$^+$ antiporters in E. coli

The high Na$^+$/H$^+$ antiporter activity in the cytoplasmic membrane due to the Na$^+$/H$^+$ antiporters implies that as long as they are in the membrane it is very hard, if not impossible, to reach a conclusion about the existence of additional Na$^+$ fluxes. Hence, a strain deleted of both antiporters, Δ*nha*AΔ*nha*B due to its Na$^+$ sensitivity, is a most suitable system to apply, selection for Na$^+$ resistance, a most powerful tool for the cloning and expression of genes conferring Na$^+$ resistance. These already include homologous or heterologous antiporters genes (Section 2.2), and even *cha*A a Ca^{++}/H$^+$ antiporter with apparently low affinity for Na$^+$ [42]. Most interestingly, since Δ*nha*AΔ*nha*B bears deletion mutations that cannot revert, it also allows the search for suppression mutations that, restore resistance, and thus unravel novel systems that are not necessarily antiporters [119].

Suppression mutations in Δ*nha*AΔ*nha*B were found to be very frequent. Several were isolated and it was found that they confer Na$^+$ but not Li$^+$ resistance. All map in the same locus which we call MH1 [119].

MH1 does not encode for a Na$^+$/H$^+$ antiporter since such activity was not detected in its membrane. The mutation also affects neither the K$^+$/H$^+$ nor the Ca^{++}/H$^+$ antiport activity. Remarkably, however, MH1 maintains a Na$^+$ gradient of 5–8 (directed inward) in the presence of 50 mM [Na$^+$]$_{out}$ as does the wild-type. Furthermore, up to 350 mM [Na$^+$]$_{out}$ the gradient of MH1 is only slightly lower than that of the wild-type and only beyond it, decreases. Most interestingly, at 400 mM [Na$^+$]$_{out}$ when the [Na$^+$]$_{in}$ concentration of MH1 reaches 90 mM, as compared with 15 mM in the wild-type, a difference in growth rate between MH1 and wild-type becomes apparent. These results imply that the mutation MH1 exposes a Na$^+$ export machinery which at least up to 350 mM [Na$^+$]$_{out}$, is similar in its capacity to that of the wild-type [119].

Δ*unc* strains cannot interconvert phosphate bond energy with the electrochemical proton gradient, allowing conditions to be established in which the source of energy available for transport are both phosphate bond energy and an electrochemical gradient (during metabolism of glucose), only phosphate bond energy (glucose metabolism in the presence of uncouplers or respiratory inhibitors) or only an electrochemical gradient of protons or redox (during respiration of substrates of the electron transport chain). As previously shown in Ref. [120] the driving force for Na$^+$ extrusion in wild-type *E. coli* (in our case TA15) is $\Delta\tilde{\mu}_{H^+}$. Thus TA15Δ*unc* maintains a Na$^+$ gradient like TA15 and the uncoupler CCCP collapses the gradient in TA15Δ*unc* [119]. In contrast to TA15Δ*unc*, MH1Δ*unc* maintains the Na$^+$ gradient even in the presence of uncoupler [119]. It could be argued that MH1Δ*unc* is insensitive to uncouplers. This is highly unlikely since, as described above, the isogenic strain TA15Δ*unc* is uncoupler sensitive. Furthermore, the collapse of the Na$^+$ gradient in MH1 by anaerobiosis is accelerated in the presence of uncouplers.

Hence Na^+ extrusion in MH1Δunc is not driven by $\Delta\tilde{\mu}_{H^+}$. Since only anaerobiosis in the presence or absence of uncouplers collapsed the gradient, we have suggested that the Na^+ extrusion in MH1 can be directly coupled to electron transport, demonstrating the presence of a respiration dependent, $\Delta\tilde{\mu}_{H^+}$ independent, Na^+ extrusion mechanism in *E. coli*. This mechanism became evident in MH1, a mutant devoid of both antiporters and carrying a mutation. Hence the triple mutant, MH1, allowed us to describe a mechanism which in the wild-type, if present, is masked by the activity of two antiporters.

MH1 can be a mutation in a Na^+ pump itself increasing its activity and/or expression. On the other hand, it is also possible that MH1 affects another system which is needed for expression or which limits the activity of the pump. In this respect, $\Delta nhaA\Delta nhaB$ (EP432) in the presence of high K^+, shows a limited Na^+ extrusion capacity [119]. It is possible that this low activity can be increased by induction, under conditions unfavorable to the H^+ cycle (in the presence of protono-phores or at alkaline pH) as suggested before [121–126] or by MH1-like mutations [119].

Abbreviations

pH_{in}, cytoplasmic pH
pH_{out}, extracellular pH
$\Delta\psi$, electrical potential
ΔpH, $pH_i – pH_o$
ΔpNa, $–\log [Na^+]_i/[Na^+]_o$
$\Delta\tilde{\mu}_{Na^+}$, sodium electrochemical gradient
$\Delta\tilde{\mu}_H$, proton electrochemical gradient
MPA, 5-N-(methylpropyl) amiloride
TMS, transmembrane segments

Acknowledgements

The research in the authors' laboratory was supported by grants from the US-Israel Binational Science Foundation to E.P.

References

1. Padan, E. and Schuldiner, S. (1993) J. Bioenerg. Biomembr. **25**, 647–669.
2. Padan, E. and Schuldiner, S. (1994) Biochim. Biophys. Acta **1185**, 129–151.
3. Schuldiner, S. and Padan, E. (1993) Int. Rev. Cytol. **137C**, 229–266.
4. Speelmans, G., Poolman, B., Abee, T. and Konings, W.N. (1993) Proc. Natl. Acad. Sci. USA **90**, 7975–7979.
5. Mitchell, P. (1961) Nature (London) **191**, 144–146.
6. Mitchell, P. and Moyle, J. (1967) Biochem. J. **105**, 1147–1162.
7. Padan, E. and Schuldiner, S. (1992) in: Alkali Cation Transport Systems in Prokaryotes, ed E.P. Bakker, pp. 3–24, CRC Press, Boca Raton, FL.
8. Schuldiner, S. and Padan, E. (1992) in: Alkali Cation Transport Systems in Prokaryotes, ed E.P. Bakker, pp. 25–51, CRC Press, Boca Raton, FL.

9. Tse, M., Levine, S., Yun, C., Brant, S. Counillon, T., Pouyssegur, J. and Donowitz, M. (1993) J. Membr. Biol. **135**, 93–108.
10. Yun, C.H., Tse, C.M., Nath, S., Levine, S.L. and Donowitz, M. (1995) J. Physiol. (Lond.) **482**, Suppl. P1S–6S.
11. Goldberg, B.G., Arbel, T., Chen, J., Karpel, R., Mackie, G.A., Schuldiner, S. and Padan, E. (1987) Proc. Natl. Acad. Sci. USA **84**, 2615–2619.
12. Karpel, R., Olami, Y., Taglicht, D., Schuldiner, S. and Padan, E. (1988) J. Biol. Chem. **263**, 10408–10414.
13. Pinner, E., Padan, E. and Schuldiner, S. (1992) J. Biol. Chem. **267**, 11064– 11068.
14. Taglicht, D., Padan, E. and Schuldiner, S. (1991) J.Biol. Chem. **266**, 11289–11294.
15. Pinner, E., Padan, E. and Schuldiner, S. (1994) J. Biol. Chem. **269**, 26274–26479.
16. Taglicht, D., Padan, E. and Schuldiner, S. (1993) J. Biol. Chem. **268**, 5382–5387.
17. Pinner, E., Padan,. E. and Schuldiner, S. (1995) FEBS Lett. **365**, 18–22.
18. Taglicht, D. (1992) Ph.D. Thesis (Hebrew University).
19. Gerchman, Y., Olami, Y., Rimon, A., Taglicht, D., Schuldiner, S. and Padan, E. (1993) Proc. Natl. Acad. Sci. USA **90**, 1212–1216.
20. Padan, E., Maisler, N., Taglicht, D., Karpel, R. and Schuldiner, S. (1989) J. Biol. Chem. **264**, 20297–20302.
21. Pinner, E., Kotler, Y., Padan, E., and Schuldiner, S. (1993) J. Biol. Chem. **268**, 1729–1734.
22. Bassilana, M., Damiano, E. and Leblanc, G. (1984) Biochemistry **23**, 5288–5294.
23. Schuldiner, S. and Fishkes, H. (1978). Biochemistry **17**, 706–710.
24. Castle, A.M., Macnab, R.M. and Schulman, R.G. (1986) J. Biol. Chem. **261**, 3288–3294.
25. Pan, J.W. and Macnab, R.M. (1990) J. Biol. Chem. **265**, 9247–9250.
26. Kleyman, T.R. and Cragoe, Jr. E.J. (1988) J. Membr. Biol. **105**, 1–21.
27. Barkla, B.J. and Blumwald, E. (1991) Proc. Natl. Acad. Sci. USA **88**, 11177–11181.
28. Katz, A. Kleyman, T.R. and Pick, U. (1994) Biochemistry **33**, 2389–2393.
29. Schonheit, P. and Beimborn, D.B. (1985) Arch. Mircobiol. **142**, 354–360.
30. Mochizuki–Oda, N. and Oosawa, F. (1985) J. Bacteriol. **163**, 395–397.
31. Leblanc, G., Bassilana, M. and Damiano, E. (1988) in: Na⁺/H⁺ Exchange, ed S. Grinstein, pp. 103–117. CRC Press. Boca Raton, FL.
32. Davies, K. and Solioz, M. (1992) Biochemistry **31**, 8055–8058.
33. McMorrow, I. Shuman, H.A., Sze, D., Wilson, D.M. and Wilson, T.H. (1989) Biochim. Biophys. Acta **981**, 21–26.
34. Onoda, T. Oshima, A. Fukunaga, N. and Nakatani, A. (1992) J. Gen. Microbiol. **63**, 1265–1270.
35. Counillon, L. Franchi, A. and Pouyssegur, J. (1993) Proc. Natl. Acad. Sci. USA **90**, 4508–45123.
36. Garlid, K.D., Shariat-Madar, Z., Nath, S. and Jezek, P. (1991) J. Biol. Chem. **266**, 6518–6523.
37. Rosen, B.P. (1986) Methods Enzymol., **125**, 328–336.
38. Fersht, A. (1985) in: Enzyme Structure and Mechanism, ed A. Fersht, pp. 155–175, W.H. Freeman and Co., New York.
39. Antosiewitcz, J., McCammon, J.A. and Gilson, M.K. (1994) J. Mol. Biol. **238**, 415–436.
40. Okamura, M.Y. and Feher, G. (1992) Annu. Rev. Biochem. **61**, 861–896.
41. Jia, Z.P., McCullough, N., Martel, R., Hemmingsen, S. and Young, P.G. (1992) EMBO J. **11**, 1631–1640.
42. Ivey, D.M., Guffanti, A.A., Zemsky, J., Pinner, E., Karpel, R., Padan, E., Schuldiner, S. and Krulwich, T.A. (1993) J. Biol. Chem. **268**, 11296–11303.
43. Ivey, D.M., Guffanti, A.A., Bossewitch, J.S., Padan, E. and Krulwich, T.A. (1991) J. Biol. Chem. **266**, 23483–23489.
44. Waser, M., Hess-Bienz, D., Davies, K. and Solioz, M. (1992) J. Biol. Chem. **267**, 5396–5400.
45. Nakamura, T., Komano, Y., Itaya, E., Tsukamoto, K., Tsuchiya, T. and Unemoto, T. (1994) Biochim. Biophys. Acta **1190**, 465–468.
46. Tokuda, H. and Unemoto, T. (1982) J. Biol. Chem. **257**, 10007–10014.
47. Unemoto, T. and Hayashi, M. (1993) J. Bioenerg. Biomembr. **25**, 385–391.
48. Nakamura, T., Kawasaki, S. and Unemoto, T. (1992) J. Gen. Microbiol. **138**, 1271–1276.
49. Atsumi, T., McCarter, L. and Imae, Y. (1992) Nature **335**, 182–184.

50. Ohyama, T., Imaizumi, R., Igarashi, K. and Kobayashi, H. (1992) J. Bacteriol. **174**, 7743–7749.
51. Nakamura, T., Komano, Y. and Unemoto, T. (1995) Biochim. Biophys. Acta, in press.
52. Kuroda, T., Shimamoto, T., Inaba, K., Tsuda, M. and Tsuchiya, T. (1994) J. Biochem. **116**, 1030–1038.
53. Tsuchiya, T. and Shinoda, S. (1985) J. Bacteriol. **162**, 794–798.
54. Kuroda, T., Shimamoto, T., Inaba, K., Kayahara, T., Tsuda, M. and Tsuchiya, T. (1994) J. Biochem. **115**, 1162–1165.
55. Tse, M., Levine, S., Yun, C., Brant, S., Pouyssegur, J. and Donowitz, M. (1993) Am. Soc. Nephrol. **4**, 969–975.
56. Seung-Taek, L., Nicholls, R.D., Jong, M.T.C., Fukai, K. and Spritz, R.A. (1995) Genomics **28**, 354–363.
57. Fleischmann, R.D., Adams, M.D., White, O., Clayton, R.A., Kirkness, E.F., Kerlavage, A.R., Bult, C.J., Tomb, J., Dougherty, B.A., Merrick, J.M., McKenney, K., Sutton, G., FitzHugh, W., Fields, C., Gocayne, J., Scott, J., Shirley, R., Liu, L., Glodek, A., Kelley, J.M., Weidman, J.F., Philips, C.A., Spriggs, T., Hedblom, E., Cotton, M.D., Utterback, T.R., Hanna, M.C., Nguyen, D.T., Saudek, D.M., Brandon, R.c., Fine, L.D., Fritchman, J.L., Fuhrmann, J.L., Geoghagen, N.S.M., Gnehm, C.L., McDonald; L.A., Small, K.V., Frase, C.M., Smith, H.O. and Venter, J.C. (1995) Science **269**, 496–512.
58. Pinner, E., Carmel, O., Bercovier, H., Sela, S., Padan, E. and Schuldiner, S. (1992) Arch. Microbiol. **157**, 323–328.
59. Deguchi, Y., Yamato, T. and Anraku, Y. (1990) J. Biol. Chem. **265**, 21704–21708.
60. King,S.C., Hansen, C.L. and Wilson, T.H. (1991) Biochim. Biophys. Acta **1062**, 177–186.
61. Kaback, H.R., Jung, K., Jung, H., Wu, J., Prive, G.G. and Zen, K. (1993) J. Bioenerg, Biomembr. **25**, 627–636.
62. Nguyen. T.A., Postle, K. and Bertrand, K.P. (1983) Gene **25**, 83–92.
63. Schwartz, S., Cardoso, M. and Wegener, H.C. (1992) Antimicrob. Agents Chemother. **36**, 580–588.
64. Salyers, A.A., Speer, B.S. and Shoemaker, N.B. (1990) Molec. Microbiol. **4**, 151–156.
65. Cheng, J., Guffanti, A.A. and Krulwich, T.A. (1994) J. Biol. Chem. **269**, 27365–27371.
66. Krulwich, T.A. (1983) Biochim. Biophys. Acta **726**, 245–264.
67. Krulwich, T.A. and Guffanti, A.A. (1989) J. Bioenerg. Biomembr. **21**, 663–677.
68. Kudo, T., Hino, M., Kitada, M. and Horikoshi, K. (1990) J. Bacteriol. **172**, 7282–7283.
69. Hamamoto, T., Hashimoto, M., Hino, M., Kitada, M., Seto, Y., Kudo, T. and Horikoshi, K. (1994) Molec. Microbiol. **14**, 939–946.
70. Strausak, D., Waser, M. and Solioz, M. (1993) J. Biol. Chem. **268**, 26334–26337.
71. Van de Vossenberg, J.L.C.M., Ublink-Kok, T., Elferink, M.G.L., Driessen, A.J.M. and Konings, W.N. (1995) Molec. Microbiol. in press.
72. Karpel, R., Alon, T., Glaser, G., Schuldiner, S. and Padan, E. (1991) J. Biol. Chem. **266**, 21753–21759.
73. Rahav-Manor, O., Carmel, O., Karpel, R., Taglicht, D., Glaser, G., Schuldiner, S. and Padan, E. (1992) J. Biol. Chem. **267**, 10433–10438.
74. Henikoff, S., Haughn, G.W., Calvo, J.M. and Wallace, J.C. (1988) Proc. Natl. Acad. Sci. USA **85**, 6602–6606.
75. Christman, M.F., Storz, G. and Ames, B.N. (1989) Proc. Natl. Acad. Sci. USA **86**, 3484–3488.
76. Schell, M.A. (1993) Annu. Rev. Microbiol. 47, 597–626.
77. Storz, G., Tartaglia, L.A. and Ames, B.N. (1990) Science **248**, 189–194.
78. Niiya, S., Yamasaki, K., Wilson, T.H. and Tsuchiya, T. (1982) J. Biol. Chem. **257**, 8902–8906.
79. Carmel, O., Dover, N., Rahav-Manor, O., Dibrov, P., Kirsch, D., Karpel, R., Schuldiner, S. and Padan, E. (1994) EMBO J. **13**, 1981–1989.
80. Rosen, B.P. (1986) Annu. Rev. Microbiol. **40**, 263–286.
81. Padan, E., Zilberstein, D. and Rottenberg, H. (1976) Eur. J. Biochem. **63**, 533–541.
82. Padan, E., Zilberstein, D. and Schuldiner, S. (1981) Biochim. Biophys. Acta **650**, 151–166.
83. Zilberstein, D., Schuldiner, S. and Padan, E. (1979) Biochemistry **18**, 669–673.
84. Booth, I.R. (1985) Microbiol. Rev. **49**, 395–378.
85. Padan, E. and Schuldiner, S. (1986) Meth. Enzymol. **125**, 337–352.
86. Slonczewski, J.L., Macnab, R.M., Alger, J.R. and Castel, A. (1992) J. Bacteriol. **174**, 7743–7749.
87. Zilberstein, D., Agmon, V., Schuldiner, S. and Padan, E. (1984) J. Bacteriol. **158**, 246–252.

88. Kashket, E.R.. (1983) FEBS Lett. **154**, 343–346.

89. Pouyssegur, J., Sardet, C. Franchi, A., L'Allemain, G. and Paris, S. (1984) Proc. Natl. Acad. Sci. USA **81**, 4833–4837.

90. Grinstein, S., Rotin, D. and Mason, M.J. (1989) Biochim. Biophys. Acta **988**, 73–97.

91. Krulwich, T.A. (1986) J. Membr. Biol. **89**, 113–125.

92. Slonczewski, J.L., Rosen, B.P., Alger, S.R. and Macnab, R.M. (1981) Proc. Natl. Acad. Sci. USA **78**, 6271–6275.

93. Bingham, R.J., Hall, K.S. and Slonczewski, J.L. (1990) J. Bacteriol. **172**, 2184–2186.

94. Olson, E.R. (1993) Mol. Microbiol. **8**, 5–14.

95. Schuldiner, S., Agmon, V., Brandsma, J. Cohen, A., Friedman, E. and Padan, E. (1986) J. Bacteriol. **168**, 936–939.

96. Padan, E. and Schuldiner, S. (1987) J. Membr. Biol. **95**, 189–198.

97. Grinstein, S. (ed.) (1988) Na⁺/H⁺ Exchange. CRC Press, Boca Raton, FL.

98. Grinstein, S. and Furuya, W. (1988) Am. J. Physiol. **23**, C272–C285.

99. Sardet, C., Franchi, A. and Pouyssegur, J. (1989) Cell **56**, 271–280.

100. Slonczewski, J.L. (1992) ASM News **58**, 140–144.

101. Kobayashi, H., Murakami, H. and Unemoto, T. (1982) J. Biol. Chem. **257**, 13246–13252.

102. Kobayashi, H. (1985) J. Biol. Chem. **260**, 72–76.

103. Kakinuma, Y. (1987) J. Bacteriol. **169**, 4403–4405.

104. Kashket, E.R. (1981) J. Bacteriol. **146**, 377–384.

105. Kakinuma, Y. and Igarashi, K. (1995) J. Bacteriol. **177**, 2227–2229.

106. Krulwich, T.A., Guffamti, A.A., Bornstein, R.F. and Hoffstein, T. (1982) J. Biol. Chem. **257**, 1885–1889.

107. Krulwich, T.A., Federbush, J.G. and Guffanti, A.A. (1985) J. Biol. Chem. **260**, 4055–4058.

108. McLaggan, D., Selwyun, M.Y. and Dawson, A.P. (1984) FEBS Lett. **165**, 254–258.

109. Sugiyama, S.H., Matsukura, H. and Imae, Y. (1985) FEBS Lett. **182**, 265–268.

110. McMorrow, I., Shuman, H.A., Sze, D., Wilson, D.M. and Wilson, T.H. (1989) Biochim. Biophys. Acta **981**, 21–26.

111. Ohyama, T., Imaizumi, R., Igarashi, K. and Kobayashi, H. (1992) J. Bacteriol. **174**, 7743–7749.

112. Mugikura, S., Nishikawa, M., Igarashi, K. and Kobayashi, H. (1990) J. Biochem. **108**, 86–91.

113. Kinoshita, N., Unemoto, T. and Kobayashi, H. (1984) J. Bacteriol. **160**, 1074–1077.

114. Thelen, P., Tsuchiya, T. and Goldberg, E.B. (1991) J. Bacteriol. **173**, 6553–6557.

115. Shimamoti, T., Inaba, K., Thelen, P., Ishikawa, T., Goldberg, E., Tsuda, Masaaki. and Tsuchiya, T. (1994) J. Biochem. **116**, 285–290.

116. Brey, R.N., Beck, J.C. and Rosen, B.P. (1978) Biochem. Biophys. Res. Commun. **83**, 1588–1594.

117. Bakker, E.P. and Mangerich, W.E. (1981) J. Bacteriol. **147**, 820–826.

118. Walderhaug, M.O., Dosch, D.C. and Epstein, W. (1987) in: Ion Transport in Prokaryotes, ed B.P. Rosen and S. Silver, p. 85, Academic Press, San Diego.

119. Harel-Bronstein, M., Dibrov, P. Olami, Y., Pinner, E., Schuldiner, S. and Padan, E. (1995) J. Biol. Chem. **270**, 3816–3822.

120. Borbolla, M.G. and Rosen, B.P. (1984) Arch. Biochem. Biophys. **229**, 98–103.

121. Avetisyan, A.V., Dibrov, P.A., Skulachev, V.P. amd Sokolov, M.V. (1989) FEBS Lett. **254**, 17–21.

122. Avetisyan, A.V., Bogachev, A.V., Murtasina, R.A. and Skulachev, V.P. (1992) FEBS Lett. **306**, 199–202.

123. Avetisyan, A.V., Bogachev, A.V., Murtasina, R.A. and Skulachev, V.P. (1993) FEBS Lett. **317**, 267–270.

124. Skulachev, V.P. (1984) Trends Biochem. Sci. 9, 483–485.

125. Skulachev, V.P. (1988) in: Membrane Bioenergetics, ed V.P. Skulachev, pp. 293–326, Springer-Verlag, Berlin.

126. Dibrov, P.A. (1991) Biochim. Biophys. Acta **1056**, 209–224.

127. Rimon, A, Gerchman, Y., Olami, Y., Schuldiner, S. and Padan, E. (1995) J. Biol. Chem. **270**, 1–5.

128. Schuler, G.D., Altschul, S.F., Lipman, D.J. (1991) Proteins: Structure, Function and Genetics **9**, 180–190.

CHAPTER 23

Biophysical Aspects of Carbohydrate Transport Regulation in Bacteria

M.H. SAIER, JR. and J.J. YE

Department of Biology, University of California at San Diego,
La Jolla, CA 92093-0116, USA

© *1996 Elsevier Science B.V.*
All rights reserved

Handbook of Biological Physics
Volume 2, edited by W.N. Konings, H.R. Kaback and J.S. Lolkema

Contents

1. Introduction

Multiple mechanisms of transmembrane sugar transport in bacteria are regulated in response to various sensors of energy availability. These sensors of energy availability include intracellular metabolites, cytoplasmic ATP or phosphoenolpyruvate, the proton motive force, and exogenous sources of energy. Specific regulatory proteins are designed to sense the availability of these potential energy sources via complex sensory transmission mechanisms that involve reception, protein phosphorylation and allosteric protein–protein interactions. In this chapter, two of the best characterized such transmission mechanisms will be examined from a biophysical standpoint.

2. Mechanisms of carbohydrate transport in bacteria

Bacteria take up exogenous sources of carbon and energy via hundreds of solute-specific transport systems that function by seven different mechanisms and belong to several independently evolving protein families or superfamilies (Table 1; see Refs. [1,2] for current reviews). Among these systems are the lactose and melibiose permeases of *E. coli* that function by proton and sodium symport and belong to the major facilitator and sodium:solute symporter superfamilies, respectively (entries 1 and 2 in Table 1). These transporters are driven by chemiosmotic energy in the form of ion gradients and transmembrane electrical potentials. Chemically driven transport systems such as the ABC-type maltose and the PTS-type mannitol permeases of *E. coli* function by ATP-driven active transport and PEP-dependent group translocation, respectively (entries 3 and 4 in Table 1). Facilitative diffusion transporters equilibrate their solutes across the membrane without energy expenditure. The glycerol facilitator of *E. coli* (of the MIP family) is believed to function by a channel-type mechanism while the glucose facilitator of *Zymomonas mobilis* (of the MFS) is thought to function by a carrier-type mechanism (entries 5 and 6 in Table 1) [1–3]. Accumulation of these carbohydrates in the bacterial cell cytoplasm can occur only if they are phosphorylated by ATP-dependent cytoplasmic kinases (glycerol kinase and glucokinase, respectively). Finally, a few permeases specific for phosphorylated compounds (e.g., the hexose-P, phosphoglycerate and glycerol-P permeases of *E. coli*) function by antiport, bringing in the phosphorylated carbohydrate in exchange for inorganic phosphate (entry 7 in Table 1).

Table 1
Representative bacterial transport systems specific for carbohydrates: classification and energy coupling

Representative transport system	Organism	Family	Energy source	Mechanism of transport energy coupling	Ref.
1. Lactose permease	*E. coli*	Major facilitator superfamily (MFS)	Proton motive force	Sugar:H^+ symport	69
2. Melibiose permease	*E. coli*	Sodium:solute symporter superfamily (SSSS)	Sodium motive force	Sugar:Na^+ symport	67
3. Maltose permease	*E. coli*	ATP binding cassette (ABC) superfamily	ATP	Conformational coupling to ATP hydrolysis	70,71
4. Mannitol permease	*E. coli*	Phosphotransferase system (PTS) superfamily	PEP	Group translocation	12
5. Glycerol facilitator	*E. coli*	Major intrinsic protein (MIP) family	None	Passive diffusion through a channel protein	72
6. Glucose facilitator	*Z. mobilis*	Major facilitator superfamily (MFS)	None	Carrier-type facilitated diffusion	73
7. Hexose-P permease	*E. coli*	Major facilitator superfamily (MFS)	P_i concentration gradient	Sugar-P:P_i antiport	69,74

3. Mechanisms of transport regulation in bacteria

Just as there are multiple mechanisms of transport, multiple mechanisms exist for the regulation of the activities of transport systems (Table 2). Integral membrane proteins can exist in alternative conformations which differ in activity. Each such protein conformation exhibits a dipole moment which interacts with an electric field (the membrane potential, $\Delta\psi$). The magnitude of the membrane potential determines the conformation of the protein which exhibits lowest free energy. Consequently, a transport protein can respond to an imposed membrane potential ($\Delta\psi$) (one form of cellular energy) with either a decrease or an increase in activity. Alternatively, it can respond to pH differences (ΔpH) that exist across the membrane as a result of selective protonation/deprotonation phenomena (entry 1 in Table 2).

Transport can also be regulated by competition for exogenous sugar binding (thus sensing external energy availability) (entry 2 in Table 2) or by competition between two functionally similar permeases for common energy sources. In the case of the phosphoenolpyruvate:sugar phospho-transferase system (PTS), competition for

Table 2
Mechanisms of carbohydrate transport regulation in bacteria

Representative transport system	Organism	Family[a]	Energy source sensed	Mechanism of regulation	Ref.
1. Phosphoglycerate permease	*S. typhimurium*	MFS	$\Delta\psi$; ΔpH	Dipole moment alignment; residue protonation	75–78
2. Mannose permease	*S. typhimurium*	PTS	External sugar	Competition for permease binding	79
3. Fructose permease	*S. typhimurium*	PTS	HPr(his-P)	Competition for energy	80,81
4. Mannitol permease	*E. coli*	PTS	Cytoplasmic sugar-Ps (metabolites)	Direct binding	82–84
5. Lactose permease (LacS)	*Streptococcus thermophilus*	SSSS	PTS sugar	Permease regulatory domain phosphorylation	6
6. Lactose permease (LacY)	*E. coli*	MFS	PTS sugar	Allosteric inhibition by IIA^{glc}	10,11
7. Lactose permease (LacP)	*Lactobacillus brevis*	?	Cytoplasmic metabolites	Allosteric inhibition by HPr(ser-P)	62,63

[a] Abbreviations of the transport protein families are provided in Table 1.

the phosphoryl donor for permease energization, phospho HPr, has been shown to provide a mechanism whereby the functioning of one PTS permease can inhibit the activity of another (entry 3 in Table 2). Direct binding of a cytoplasmic metabolite to a permease protein, either at the substrate binding site, or at an allosteric site localized to the cytoplasmic surface of the membrane, provides another means by which the activity of a permease can be regulated in response to a source of energy (in this case, cytoplasmic metabolite levels) (entry 4 in Table 2).

Recently, evidence has been presented to suggest that direct phosphorylation of bacterial permease proteins functions in the control of their activities by mechanisms that are unrelated to energization (see entry 5 in Table 2). The same has been extensively documented in eukaryotes [4]. The lactose permease of *Streptococcus thermophilus* (LacS) is directly phosphorylated by phospho HPr of the PTS on a histidyl residue of a IIA^{glc}-like domain localized to the C-terminus of the permease [5]. This phosphorylation event is believed to regulate the activity of the permease in response to the activity of the PTS, the dominant sugar transport system [6].

The last two entries in Table 2 deal with allosteric regulation of permease function by direct protein–protein interactions. In each case cited, phosphorylation of an allosteric regulatory protein controls its interaction with the permease. In *E. coli* and other enteric bacteria, the free form of the IIA^{glc} protein of the PTS binds to target permeases to inhibit their activities (entry 6 in Table 2). Phosphorylation of histidine in IIA^{glc}, catalyzed by the phosphotransfer proteins of the PTS, Enzyme I and HPr, at the expense of PEP, prevents binding. The structure of the IIA^{glc} protein is understood in 3-dimensions, and this IIA^{glc}-mediated regulatory mechanism is well established. It will be the focus of the next section (Section 4). In *Lactobacillus brevis* and certain other low GC Gram-positive bacteria, the serine phosphorylated form of HPr of the PTS binds to target permeases to inhibit their activities (entry 7 in Table 2). Phosphorylation of serine in HPr is catalyzed by a metabolite-activated, ATP-dependent protein kinase and reversed by a protein-P phosphatase. The 3-dimensional structure of HPr and its serine phosphorylated derivative are available, and the cited regulatory mechanism is well established. This mechanism will be discussed in Section 5.

4. Transport regulation by the IIA^{glc} protein in enteric bacteria

In enteric bacteria, the PTS regulates the uptake, or the cytoplasmic generation, of inducers of non-PTS catabolic operons (inducer exclusion) in processes that are coordinate with inhibition of adenylate cyclase [7]. The mechanism involves a central regulatory protein of the PTS, the glucose-specific Enzyme IIA (IIA^{glc}; 8–10), previously termed Enzyme III^{glc}, the product of the *crr* gene (Fig. 1; for reviews, see Refs. [10–12]).

PTS permeases function as transmembrane signaling devices in one of the first complex sensory transduction mechanisms to be elucidated in detail. When one of the PTS sugars is present in the bacterial growth medium, the IIA^{glc} protein becomes dephosphorylated as the phosphoryl groups are transferred to incoming sugar molecules via the sugar-specific permease proteins. This process is believed to result in both allosteric deactivation of adenylate cyclase (which seems to be positively regulated by phosphorylated IIA^{glc}) and allosteric inhibition of the non-PTS permeases and catabolic enzymes that generate cytoplasmic inducers (which are negatively controlled by free IIA^{glc}) (Fig. 1) [10,11].

Early experiments had established the validity of the main mechanistic features of the process whereby IIA^{glc} regulates the activities of target permeases and catabolic enzymes (see Fig. 1). Consequently, recent efforts have focused on elucidation of the three dimensional structure of the central IIA^{glc} regulatory protein and identification of the sites of the protein–protein regulatory interactions [13–23]. The concerted mechanism of catabolite repression, involving PTS-mediated protein phosphorylation and the control of cAMP synthesis, is less well understood at the molecular level than is transport regulation [11,24–26).

Exactly how the IIA^{glc} protein interacts with HPr and certain targets of allosteric regulation has recently been elucidated. Mutations in the lactose, maltose, and melibiose permeases as well as glycerol kinase that specifically abolish regulation by IIA^{glc}, presumably by destroying the complementarity of the allosteric binding sites on these permeases, have been isolated, and, in some cases, they have been characterized in molecular detail [27] (see Ref. [10] for more recent primary references). The IIA^{glc} binding site in LacY is almost certainly within the central

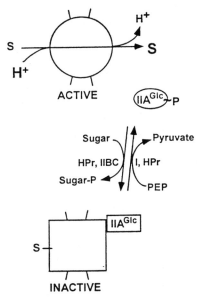

Fig. 1. Model illustrating the phosphotransferase system (PTS)-mediated regulation of the lactose:H$^+$ symport permease of *E. coli*. The negative allosteric effector is the unphosphorylated IIA^{glc} protein of the PTS. Dephosphorylation of the histidyl phosphorylated IIA^{glc} protein (IIA^{glc}~P) is catalyzed by IIBC^{glc} in response to extracellular glucose, or by an Enzyme II complex and HPr in response to other extracellular PTS sugars. Rephosphorylation of IIA^{glc} is accomplished by phosphoryl transfer from phosphoenolpyruvate (PEP) through Enzyme I (I) and HPr. Binding of IIA^{glc} to the permease inhibits transport. S, lactose; sugar, any PTS sugar.

cytoplasmic loop of the permease between transmembrane spanners 6 and 7. The same is probably true of the raffinose permease of *E. coli* [28]. The IIA^{glc} binding site within the maltose permease is localized to the cytoplasmic ATP hydrolyzing subunit of this permease, MalK. That on the melibiose permease may be localized to the C-terminal tail of this protein, but this conclusion is more tentative (see Ref. [10] for primary references).

Examination of the sequences of the MalK, LacY and RafB permeases led to the identification of a consensus sequence for IIA^{glc} binding [28]. This consensus sequence was VGANXSLXSX. Arabinose isomerase of *E. coli* was shown to possess elements of this consensus sequence and was later shown to be regulated by IIA^{glc} *in vitro*. The IIA^{glc} binding consensus sequence is, however, lacking in glycerol kinase, the melibiose permease and adenylate cyclase, proteins that are also allosterically regulated by IIA^{glc} or its histidyl phosphorylated derivative. In these latter cases, IIA^{glc} presumably binds to a different set of amino acyl residues, or to the same residues occurring in different sequence. In the case of glycerol kinase, the binding site involves the C-terminal region of the 50 kDa polypeptide chain of this tetrameric enzyme. The sequence to which IIA^{glc} binds is: PGIET-TERNY (residues 472 to 481; underlined residues provide points of direct contact with IIA^{glc}). This sequence exhibits similarity with the consensus binding sequence noted above [29].

Three-dimensional analyses of IIA^{glc} revealed that this protein exhibits a structure consisting of antiparallel β-strands forming a collapsed β-barrel or a β-sandwich with six strands on either face. The barrel-like structure exhibits Greek key and jelly roll topological features [19]. Two active site histidyl residues (his-75 and his-90 in the *E. coli* protein) are found adjacent and perpendicular to each other (the N-3 atoms are 3.3 Å apart) within a hydrophobic pocket. The N-3 atom of his-90 (the phosphorylation site) is exposed to the aqueous milieu while the N-1 atom of his-90 is probably protonated and hydrogen bonded to the carbonyl oxygen of gly-97. His-75, which is required for phosphoryl transfer between IIA^{glc} and IIB^{glc} but not between HPr and IIA^{glc}, has its N-1 atom hydrogen bonded to the γ-oxygen of thr-73. In the his-90 phosphorylated state of IIA^{glc}, the phosphate may interact with the N-3 atom of his-75 [19,23]. Mutagenic analyses of IIA^{glc} [30] have led to the conclusion that the lactose permease binds to the active site face of IIA^{glc}. The same is established for glycerol kinase [29].

HPr, which interacts with IIA^{glc} during phosphoryl transfer, has the structure of a deformed open-faced sandwich with three α-helices overlying a 4-stranded, skewed, anti-parallel β-sheet [31]. NMR analyses of the complex of these two proteins revealed that HPr binds to a site on the surface of IIA^{glc} that overlaps the active site crevice located on one side of the β-barrel [22,23]. Chemical shift changes induced in HPr upon binding to IIA^{glc} were observed for residues 12 to 22 which include the active site histidyl residue, his-15. The sequence corresponding to residues 12–22 for the *E. coli* HPr is NGLHTRPAAQ. Thus, in all proteins known to interact with IIA^{glc} for which the binding sites have been localized (MalK, LacY, RafB, GlpK and HPr), the G at the second position in the 10 residue binding segment is always conserved. As glycyl residues are of extreme importance for maintenance of 3-dimensional protein structures and are consequently among the most conserved residues found in families of homologous proteins, it is possible that all proteins that bind IIA^{glc}, either for catalytic purposes or for regulatory purposes, exhibit a similar 3-dimensional fold in their binding regions.

Elegant X-ray diffraction studies have led to the conclusion that the tetrameric glycerol kinase binds four molecules of IIA^{glc} in an allosteric regulatory complex that is presumably relevant to the *in vivo* complex [29]. The crystal structure of this complex at 2.6 Å resolution revealed that IIA^{glc} binds to glycerol kinase at a C-terminal site that is distant from the catalytic site of the latter enzyme. This fact suggests that long-range conformational changes mediate the inhibition of glycerol kinase by IIA^{glc}. The two proteins interact largely employing hydrophobic and electrostatic attractive forces. Only one hydrogen bond involving an uncharged group appears to play a role. The phosphorylation site of IIA^{glc} (his-90 in the *E. coli* IIA^{glc}) proved to be buried in a hydrophobic environment formed by the active site region of IIA^{glc} and a 3_{10} helix of glycerol kinase. This fact suggests that phosphorylation prevents IIA^{glc} binding by directly disrupting protein–protein interactions.

The elucidation of the structure of this protein complex established beyond a doubt that IIA^{glc} is both necessary and sufficient to effect the allosteric regulation of its various target systems, the permeases and catabolic enzymes that are inhibited by the nonphosphorylated form of this protein. It seems likely that regulation of the permeases will involve a comparable allosteric complex in which the IIA^{glc} protein binds to a cytoplasmically exposed region of the permease to control its activity. Because the central loop of the lactose permease which binds IIA^{glc} is not required for permease activity [32], it can be anticipated that as for glycerol kinase, IIA^{glc} binding will induce a conformational change in the permease that negatively affects its activity. Since in all cases that have been reported to date, IIA^{glc} diminishes the V_{max} of transport, possibly by blocking transport altogether, IIA^{glc} could either close the channel through which the sugar passes or inhibit the cyclic conformational changes that accompany sugar transport. Binding of IIA^{glc} presumably increases the free energy of translocation. Because IIA^{glc} binding is cooperative with sugar binding, giving rise to transmembrane signaling [11], an inhibitory effect of the protein by decreasing the affinity of the carrier for its substrate is not likely.

5. Transport regulation by HPr(ser-P) in low GC Gram-positive bacteria

Sugar uptake in bacteria is regulated by the various mechanisms outlined in Table 2. Another regulatory device, a vectorial process which controls sugar or sugar-phosphate accumulation by *efflux* of the intracellular sugar from the cell, occurs in Gram-positive bacteria [33–36]. Expulsion of intracellular sugars was first observed during studies on transport of β-galactosides in resting cells of *Streptococcus pyogenes* and *Lactococcus lactis* [37,38]. Like many other lactic acid bacteria, starved cultures of *S. pyogenes* and *L. lactis* use the PTS and cytoplasmic stores of PEP for the uptake and phosphorylation of PTS sugars such as lactose and its non-metabolizable analogue, methyl-β-D-thiogalactopyranoside (TMG). Subsequent addition of a metabolizable PTS sugar (but not a nonmetabolizable sugar analogue) to a bacterial culture preloaded with TMG elicits rapid dephosphorylation of intracellular TMG-phosphate and efflux of the free sugar. The half-time for expulsion of TMG by glucose (15–20 seconds) is more than an order of magnitude shorter than that for accumulation of the sugar phosphate. The intracellular pool of TMG-P is essentially stable in the absence of glucose, and expulsion of TMG is therefore not due to rapid turnover of TMG-P and inhibition of TMG influx by glucose. A cytoplasmic sugar-P phosphatase is apparently activated, and rapid, energy-independent efflux of the sugar follows [33,39,40].

Attempts to define the mechanism of sugar expulsion led to identification of a unique phosphorylated derivative of HPr. A small protein was initially shown to be phosphorylated in cultures of *S. pyogenes* in response to conditions which promote expulsion [39]. In further studies, the low molecular weight protein was phosphorylated *in vitro* employing a crude extract of *S. pyogenes* in the presence of [γ-^{32}P]ATP, and after purification, the isolated protein was identified as a phosphorylated derivative of HPr [41]. This phosphorylated HPr differed from HPr(his$_{15}$~P) and proved to be HPr(ser$_{46}$-P) [41–44].

The streptococcal HPr(ser) kinase has a molecular weight of approximately 60,000, and its activity is dependent on divalent cations as well as on one of several intermediary metabolites, the most active of which are fructose-1,6-bisphosphate and gluconate-6-P [42,45]. Phosphorylation of the seryl residue in HPr is strongly inhibited by inorganic phosphate or by phosphorylation of the active site histidyl residue [42]. The converse is also observed, i.e., ser_{46} phosphorylation strongly inhibits his_{15} phosphorylation.

Since the discovery of the kinase in *S. pyogenes*, protein kinases which phosphorylate HPr have been described in work from several laboratories for a number of low GC (but not high GC) Gram-positive bacteria [46–49]. The phosphorylation of serine-46 in HPr was implicated in the regulation of PTS activity on the basis of *in vitro* studies (50), but the physiological significance of this observation in certain bacteria was questioned in view of *in vivo* results with *B. subtilis* and *S. aureus* [44,51,52]. Recently, *in vivo* work with *B. subtilis* strains that bear either the wild-type or S46A mutant HPr chromosomal genes have reaffirmed the notion that HPr(ser) phosphorylation regulates sugar uptake via the PTS. Moreover, regulation of PTS sugar uptake into membrane vesicles of *L. lactis* and *L. brevis*, due to HPr(ser) phosphorylation, has also been documented (J.J. Ye and M.H. Saier, Jr., unpublished results). These experiments suggest that one consequence of the metabolite-activated kinase-dependent phosphorylation of seryl residue 46 in HPr is the regulation of sugar uptake via the PTS.

The kinetic parameters influenced by HPr(ser) phosphorylation have been defined in *in vitro* PTS-catalyzed sugar phosphorylation reactions [44,53], and the development of methods for the *in vivo* quantification of the four forms of HPr (free HPr, HPr(ser-P), HPr(his-P), and HPr(ser-P, his-P)) have resulted in the unexpected finding that substantial amounts of the doubly phosphorylated form of HPr exists in certain streptococci under certain physiological conditions [54]. The strong inhibition of HPr(his) phosphorylation via the Enzyme I-catalyzed reaction by phosphorylation of ser_{46} has been reported to be relieved by complexation of HPr with an Enzyme IIA such as IIA^{glc} [50], but the significance of this observation has been questioned [44,53]. The mechanism responsible for the appearance of substantial amounts of doubly phosphorylated HPr and the physiological significance of the presence of this derivative have yet to be determined.

Expulsion of PTS-accumulated sugar phosphates from *S. pyogenes* or *L. lactis* cells occurs in a two step process with cytoplasmic sugar-P hydrolysis being rate limiting and preceding efflux (see above). Both inhibition of PTS sugar uptake and stimulation of sugar-P release were recently shown to be dependent on HPr(ser-P) or the HPr(ser-P) mutant analogue, S46D HPr, in membrane vesicles of *L. lactis* [55,56]. It was also shown that in toluenized vesicles, TMG-P hydrolysis is stimulated by intravesicular S46D HPr or by wild-type HPr under conditions that activate the HPr(ser) kinase [55]. This work was extended by demonstrating activation of a peripherally membrane-associated sugar-P phosphatase by HPr(ser-P) in crude extracts of *L. brevis*.

Thompson and Chassy [57] purified a hexose-6-P phosphohydrolase from *L. lactis*, but they did not provide evidence for or against its potential activation by HPr(ser-P). Recent work established that this purified enzyme is not activated by S46D HPr [58,59]. Instead, a small, heat stable sugar-P phosphatase proved to be responsible for initiating efflux. This enzyme was purified to homogeneity. It has a molecular weight of about 9,000 and forms a 1:1 stoichiometric complex with S46D HPr. It is activated more than 10-fold by the binding of either HPr(ser-P) or S46D HPr. The availability of this enzyme in large quantities should allow detailed X-ray crystallographic and nuclear magnetic resonance analyses of the 3-dimensional structure of the complex formed between the enzyme and its phosphorylated allosteric regulatory protein.

In contrast to the homofermentative lactic acid bacteria described above, heterofermentative lactobacilli such as *Lactobacillus brevis* transport glucose, lactose, and their non-metabolizable analogues, 2-deoxyglucose (2DG) and thiomethyl-β-galactoside (TMG), respectively, by active transport energized by the proton motive force [60,61]. They exhibit metabolite-activated, vectorial sugar expulsion and possess an ATP-dependent HPr kinase [34,46]. Thus, [^{14}C]TMG accumulates in the cytoplasm of lactobacilli when provided with an exogenous source of energy such as arginine, but addition of glucose to the preloaded cells promotes rapid efflux that establishes a low cellular concentration of the galactoside. Counterflow experiments have shown that metabolism of glucose by *L. brevis* results in the conversion of the active β-galactoside transport system into a facilitated diffusion system (i.e., from a sugar:H$^+$ symporter to a sugar uniporter; 61, confirmed in Ref. [62]).

Employing electroporation to transfer purified proteins and membrane impermeant metabolites into right-side-out vesicles of *L. brevis*, TMG uptake was recently shown to be inhibited, and pre-accumulated TMG was shown to efflux from the vesicles upon addition of glucose, if and only if intravesicular wild-type HPr of *Bacillus subtilis* was present [62]. Glucose could be replaced by intravesicular (but not extravesicular) fructose-1,6-bisphosphate, gluconate-6-phosphate, or 2-phosphoglycerate, but not by other phosphorylated metabolites, in agreement with the allosteric activating effects of these compounds on HPr(ser) kinase measured *in vitro*. Intravesicular S46D HPr promoted regulation, even in the absence of glucose or a metabolite. HPr(ser-P) appeared to convert the lactose/H$^+$ symporter into a sugar uniporter [62].

2-Deoxyglucose (2DG) uptake and efflux via the glucose:H$^+$ symporter of *L. brevis* were shown to be regulated as for the lactose:H$^+$ symporter [58]. Uptake of the natural, metabolizable substrates of the lactose, glucose, mannose, and ribose permeases was shown to be inhibited by wild-type HPr in the presence of fructose-1,6-diphosphate or by S46D HPr. These last observations clearly suggest metabolic significance in addition to the significance of this regulatory mechanism to the control of inducer levels. Thus, intracellular metabolites activate the kinase to provide a feedback mechanism to limit the overall rate of sugar uptake. The regulatory mechanism is apparently not merely designed to create a hierarchy of preferred sugars.

The direct binding of ^{125}I-[S46D]HPr to *L. brevis* membranes containing high levels of the lactose permease [63] or the glucose permease [64] has recently been demonstrated. The radioactive protein was found to bind to membranes prepared from galactose-grown *L. brevis* cells in the presence (but not in the absence) of one of the substrates of the lactose permease. Membranes from glucose-grown cells did not exhibit lactose analogue-promoted ^{125}I-[S46D]HPr binding, but binding to these

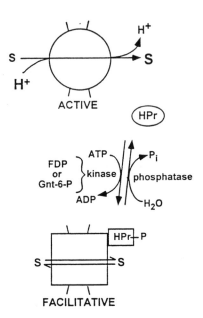

Fig. 2.   Model illustrating the PTS-mediated regulation of the lactose:H$^+$ symport permease of *Lactobacillus brevis*. The negative allosteric effector is the seryl-46 phosphorylated derivative of HPr (HPr-P). Phosphorylation of serine in HPr is catalyzed by an ATP-dependent metabolite-activated protein kinase (kinase). Activating metabolites include fructose-1,6-diphosphate (FDP) and gluconate-6-phosphate (Gnt-6-P). Dephosphorylation of HPr(ser-P) is catalyzed by an HPr(ser-P) phosphatase. Binding of HPr(ser-P) to the permease converts the lactose:H$^+$ symporter to a lactose facilitator, unable to accumulate the sugar against a concentration gradient. S, lactose.

membranes was observed in the presence of glucose or 2-deoxyglucose. The substrate and inducer specificities for binding correlated with those of the two permeases for transport. The apparent sugar binding affinities calculated from the sugar-promoted [125]I-[S46D]HPr binding curves were in agreement with sugar transport K_m values. Moreover, [S46D]HPr but not wild-type or [S46A]HPr effectively competed with [125]I-[S46D]HPr for binding [63,64]. These results establish the involvement of HPr(ser-P) in the proposed regulatory mechanism for PTS-mediated control of non PTS permeases and suggest a direct, allosteric binding mechanism. The proposed regulatory mechanism is depicted in Fig. 2.

X-ray diffraction and NMR studies have established the 3-dimensional structures of several PTS phosphocarrier proteins. High resolution structural data are available for *B. subtilis* HPr, its seryl phosphorylated and mutant derivatives as well as the interactive glucose-specific Enzyme IIA-like protein domain (IIA^{glc}) of the *B. subtilis* Enzyme II^{glc} complex [13,16–18,21,31,65,66]. HPr resembles a skewed open-faced β-sandwich with four antiparallel β-strands underlying three α-helices. The active site H15 residue and the regulatory S46 residue are not in direct contact with each other but are also not distant from each other. Phosphorylation of S46 induces a local conformational change which was initially reported to be transmitted partly through secondary structural elements to the active site H15 residue, thereby providing an explanation for the reciprocal inhibitory effect of S46 phosphorylation on the PTS phosphoryl transfer reaction [13,44,65].

More recently Pullen et al. [68] re-examined the conformational changes induced by seryl phosphorylation of HPr using NMR spectroscopy and solvent denaturation. Phosphorylation of S46 was found to stabilize a short α-helix (helix-B) that exhibits behavior indicative of conformational averaging in unphosphorylated HPr in solution. Backbone amide proton exchange rates of helix-B residues were shown to decrease in the phosphorylated form. Phosphorylation apparently stabilizes the protein to solvent and thermal denaturation, with a $\Delta\Delta G$ of 0.7–0.8 kcal mol^{-1}. Similar stabilization was measured for the HPr mutant protein in which serine-46 was replaced by aspartic acid, indicating that an electrostatic interaction between the negatively charged groups and the N-terminal end of the helix contributes significantly to the stabilization. The results indicate that phosphorylation of serine-46 at the N-terminal end of α-helix B does not cause a conformational change per se but rather stabilizes the helical structure. These observations may prove mechanistically relevant to many targets of protein phosphorylation both in bacteria and in eukaryotes.

6. Conclusions and perspectives

Results currently available clearly indicate that the metabolite-activated protein kinase-mediated phosphorylation of serine-46 in HPr plays a key role in the control of inducer levels and catabolite repression in low GC Gram-positive bacteria. This enzyme is not found in enteric bacteria such as *E. coli* and *S. typhimurium* where an entirely different PTS-mediated regulatory mechanism is responsible for catabolite repression and inducer concentration control. In Figs. 1 and 2, these two mechanistically dissimilar but functionally related processes are compared. In Gram-negative enteric bacteria, an external sugar is sensed by the sugar-recognition constituent of an Enzyme II complex of the PTS (IIC), and a dephosphorylating signal is transmitted via the Enzyme IIB/HPr proteins to the central regulatory protein, IIA^{glc} [10,11]. Targets regulated include permeases specific for lactose, maltose, melibiose and raffinose, and catabolic enzymes such as glycerol kinase and arabinose isomerase that generate cytoplasmic inducers as well as the cyclic AMP biosynthetic enzyme, adenylate cyclase that mediates catabolite repression [10,11]. In low GC Gram-positive bacteria, cytoplasmic phosphorylated sugar metabolites are sensed by the HPr kinase which is allosterically activated. HPr becomes phosphorylated on serine-46, and this phosphorylated derivative binds to its target proteins. These targets include both PTS and non-PTS permeases which are inhibited, and a peripherally, cytoplasmic membrane-bound sugar-P phosphatase which is activated. The net result is to reduce cytoplasmic inducer levels.

The parallels between the Gram-negative and Gram-positive bacterial regulatory systems are superficial at the mechanistic level but fundamental at the functional level. Thus, the PTS participates in regulation in both cases, and phosphorylation of its protein constituents plays key roles. However, the stimuli sensed, the transmission mechanisms, the central PTS regulatory proteins that effect allosteric regulation, and some of the target proteins are completely different. It seems clear that these two transmission mechanisms evolved independently. They provide a prime example of functional convergence.

Acknowledgements

Thanks are given to Mary Beth Hiller for providing expert assistance in the preparation of this manuscript. Work in the author's laboratory was supported by Public Health Service grants 5RO1 AI21702 and 2RO1 AI14176 from the National Institute of Allergy and Infectious Diseases.

References

1. Saier, M.H., Jr. (1994) Bioessays **16**, 23–29.
2. Saier, M.H., Jr. (1994) Microbiol. Rev. **58**, 71–93.
3. Saier, M.H., Jr. (1985) Mechanisms and Regulation of Carbohydrate Transport in Bacteria. Academic Press, New York.
4. Saier, M.H., Jr., Daniels, G.A., Boerner, P. and Lin, J. (1988) J. Membr. Biol. **104**, 1–20.
5. Poolman, B., Modderman, R. and Reizer, J. (1992) J. Biol. Chem. **267**, 9150–9157.
6. Poolman, B., Knol, J., Mollet, B., Nieuwenhuis, B. and Sulter, G. (1995) Proc. Natl. Acad. Sci. USA **92**, 778–782.
7. Saier, M.H., Jr. and Feucht, B.U. (1975) J. Biol. Chem. **250**, 7078–7080.
8. Saier, M.H., Jr. and Reizer, J. (1992) J. Bacteriol. **174**, 1433–1438.
9. Saier, M.H., Jr. (1993) J. Cell. Biochem. **51**, 1–6.
10. Saier, M.H., Jr. (1993) J. Cell. Biochem. **51**, 62–68.
11. Saier, M.H., Jr. (1989) Microbiol. Rev. **53**, 109–120.
12. Postma, P.W., Lengeler, J.W. and Jacobson, G.R. (1993) Microbiol. Rev. **57**, 543–594.
13. Wittekind, M., Reizer, J., Deutscher, J., Saier, M.H., Jr. and Klevit, R.E. (1989) Biochemistry **28**, 9908–9912.
14. Dean, D.A., Reizer, J., Nikaido, H. and Saier, M.H., Jr. (1990) J. Biol. Chem. **265**, 21005–21010.
15. Wilson, T.H., Yunker, P.L. and Hansen, C.L. (1990) Biochim. Biophys. Acta **1029**, 113–116.
16. Fairbrother, W.J., Cavanagh, J., Dyson, H.J., Palmer, A.G., III, Sutrina, S., Reizer, J., Saier, M.H., Jr. and Wright, P.E. (1991) Biochemistry **30**, 6896–6907.
17. Fairbrother, W.J., Gippert, G.P., Reizer, J., Saier, M.H., Jr. and Wright, P.E. (1992) FEBS Lett. **296**, 148–152.
18. Fairbrother, W.J., Palmer, A.G., III, Rance, M., Reizer, J., Saier, M.H., Jr. and Wright, P.E. (1992) Biochemistry **31**, 4413–4425.
19. Liao, D.-I., Kapadia, G., Reddy, P., Saier, M.H., Jr., Reizer, J. and Herzberg, O. (1991) Biochemistry **30**, 9583–9594.
20. Pelton, J.G., Torchia, D.A., Meadow, N.D., Wong, C.-Y. and Roseman, S. (1991) Proc. Natl. Acad. Sci. USA **88**, 3479–3483.
21. Stone, M.J., Fairbrother, W.J., Palmer, A.G., III, Reizer, J., Saier, M.H., Jr. and Wright, P.E. (1992) Biochemistry **31**, 4394–4406.
22. Chen, Y., Reizer, J., Saier, M.H., Jr., Fairbrother, W.J. and Wright, P.E. (1993) Biochemistry **32**, 32–37.
23. Chen, Y., Fairbrother, W.J. and Wright, P.E. (1993) J. Cell. Biochem. **51**, 75–82.
24. Pastan, I. and Adhya, S. (1976) Bacteriol. Rev. **40**, 527–551.
25. Ullmann, A. and Danchin, A. (1983) Advan. Cyclic Nucl. Res. **15**, 1–53.
26. Peterkofsky, A., Seok, Y.-J., Amin, N., Thapar, R., Lee, S.Y., Klevit, R.E., Waygood, E.B. Anderson, J.W., Gruschus, J., Huq, H. and Gollop, N. (1995) Biochemistry **34**, 8950–8959.

27. Saier, M.H., Jr., Straud, H., Massman, L.S., Judice, J.J., Newman, M.J. and Feucht, B.U. (1978) J. Bacteriol. **133**, 1358–1367.

28. Titgemeyer, F., Mason, R.E. and Saier, M.H., Jr. (1994) J. Bacteriol. **176**, 543–546.

29. Hurley, J. H., Faber, H.R., Worthylake, D., Meadow, N.D., Roseman, S., Pettigrew, D.W. and Remington, S.J. (1993) Science **259**, 673–677.

30. Zeng, G.Q., De Reuse, H. and Danchin, A. (1992) Res. Microbiol. **143**, 251–261.

31. Herzberg, O., Reddy, P., Sutrina, S., Saier, M.H., Jr., Reizer, J. and Kapadia, G. (1992) Proc. Natl. Acad. Sci. USA **89**, 2499–2503.

32. Bibi, E. and Kaback, H.R. (1990) Proc. Natl. Acad. Sci. USA **87**, 4325–4329.

33. Reizer, J., Deutscher, J., Sutrina, S., Thompson, J. and Saier, M.H., Jr. (1985) Trends Biochem. Sci. **1**, 32–35.

34. Reizer, J., Peterkofsky, A. and Romano, A.H. (1988) Proc. Natl. Acad. Sci. USA **85**, 2041–2045.

35. Reizer, J., Hoischen, C., Reizer, A., Pham, T.N. and Saier, M.H., Jr. (1993) Prot. Sci. **2**, 506–521.

36. Reizer, J., Romano, A.H. and Deutscher, J. (1993) J. Cell. Biochem. **51**, 19–24.

37. Reizer, J. and Panos, C. (1980) Proc. Natl. Acad. Sci. USA **77**, 5497–5501.

38. Thompson, J. and Saier, M.H., Jr. (1981) J. Bacteriol. **146**, 885–894.

39. Reizer, J., Novotny, M.J., Panos, C. and Saier, M.H., Jr. (1983) J. Bacteriol. **156**, 354–361.

40. Sutrina, S., Reizer, J. and Saier, M.H., Jr. (1988) J. Bacteriol. **170**, 1874–1877.

41. Deutscher, J. and Saier, M.H., Jr. (1983) Proc. Natl. Acad. Sci. USA **80**, 6790–6794.

42. Reizer, J., Novotny, M.J., Hengstenberg, W. and Saier, M.H., Jr. (1984) J. Bacteriol. **160**, 333–340.

43. Deutscher, J., Pevec, B., Beyreuther, K., Kiltz, H.-H. and Hengstenberg, W. (1986) Biochemistry **25**, 6543–6551.

44. Reizer, J., Sutrina, S.L., Saier, M.H., Jr., Stewart, G.C., Peterkofsky, A. and Reddy, P. (1989) EMBO J. **8**, 2111–2120.

45. Deutscher, J. and Engelmann, R. (1984) FEMS Microbiol. Lett. **23**, 157–162.

46. Reizer, J., Saier, M.H., Jr., Deutscher, J., Grenier, F., Thompson, J. and Hengstenberg, W. (1988) CRC Crit. Rev. Microbiol. **15**, 297–338.

47. Hoischen, C., Dijkstra, A., Rottem, S., Reizer, J. and Saier, M.H., Jr. (1993) J. Bacteriol. **175**, 6599–6604.

48. Titgemeyer, F., Walkenhorst, J., Cui, X., Reizer, J. and Saier, M.H., Jr. (1994) Res. Microbiol. **145**, 89–92.

49. Titgemeyer, F., Walkenhorst, J., Reizer, J., Stuiver, M.H., Cui, X. and Saier, M.H., Jr. (1995) Microbiology **141**, 51–58.

50. Deutscher, J., Kessler, U., Alpert, C.A. and Hengstenberg, W. (1984) Biochemistry **23**, 4455–4460.

51. Sutrina, S., Reddy, P., Saier, M.H., Jr. and Reizer, J. (1990) J. Biol. Chem. **265**, 18581–18589.

52. Deutscher, J., Reizer, J., Fischer, C., Galinier, A., Saier, M.H., Jr. and Steinmetz, M. (1994) J. Bacteriol. **176**, 3336–3344.

53. Reizer, J., Sutrina, S.L., Wu, L.-F., Deutscher, J., Reddy, P. and Saier, M.H., Jr. (1992) J. Biol. Chem. **267**, 9158–9169.

54. Vadeboncoeur, C., Brochu, D. and Reizer, J. (1991) Anal. Biochem. **196**, 24–30.

55. Ye, J.-J., Reizer, J., Cui, X. and Saier, M.H., Jr. (1994) J. Biol. Chem. **269**, 11837–11844.

56. Ye, J.-J., Reizer, J. and Saier, M.H., Jr. (1994) Microbiology **140**, 3421–3429.

57. Thompson, J. and Chassy, B.M. (1983) J. Bacteriol. **156**, 70–80.

58. Ye, J.-J., Neal, J.W., Cui, X., Reizer, J. and Saier, M.H., Jr. (1994) J. Bacteriol. **176**, 3484–3492.

59. Ye, J.-J. and Saier, M.H., Jr. (1995) J. Biol. Chem. **270**, 16740–16744.

60. Romano, A.H., Trifone, J.D. and Brustolon, M. (1979) J. Bacteriol. **139**, 93–97.

61. Romano, A.H., Brino, G., Peterkofsky, A. and Reizer, J. (1987) J. Bacteriol. **169**, 5589–5596.

62. Ye, J.-J., Reizer, J., Cui, X. and Saier, M.H., Jr. (1994) Proc. Natl. Acad. Sci. USA **91**, 3102–3106.

63. Ye, J.-J. and Saier, M.H., Jr. (1995) Proc. Natl. Acad. Sci. USA **92**, 417–421.

64. Ye, J.-J. and Saier, M.H., Jr. (1995) J. Bacteriol. **177**, 1900–1902.

65. Wittekind, M., Reizer, J. and Klevit, R.E. (1990) Biochemistry **29**, 7191–7200.

66. Kapadia, G., Reizer, J., Sutrina, S.L., Saier, M.H., Jr., Reddy, P. and Herzberg, O. (1990) J. Mol. Biol. **212**, 1–2.

67. Reizer, J., Reizer, A. and Saier, M.H., Jr. (1994) Biochim. Biophys. Acta **1197**, 133–166.

68. Pullen, K., Rajagopal, P., Branchini, B.R., Huffine, M.E., Reizer, J., Saier, M.H., Jr., Scholtz, M. and Klevit, R.E. (1995) Prot. Sci. **4**, 2478–2486.
69. Marger, M.D. and Saier, M.H., Jr. (1993) Trends Biochem. Sci. **18**, 13–20.
70. Tam, R. and Saier, M.H., Jr. (1993) Microbiol. Rev. **57**, 320–346.
71. Kuan, G., Dassa, E., Saurin, W., Hofnung, M. and Saier, M.H., Jr. (1995) Res. Microbiol. **146**, 271–278.
72. Reizer, J., Reizer, A. and Saier, M.H., Jr. (1993) Crit. Rev. Biochem. Mol. Biol. **28**, 235–257.
73. Weisser, P., Krämer, R., Sahm, H. and Sprenger, G.A. (1995) J. Bacteriol. **177**, 3351–3354.
74. Maloney, P.C., Ambudkar, S.V., Anantharam, V., Sonna, L.A. and Varadhachary, A. (1990) Microbiol. Rev. **54**, 1–17.
75. Reider, E., Wagner, E.F. and Schweiger, M. (1979) Proc. Natl. Acad. Sci. USA **76**, 5529–5533.
76. Saier, M.H., Jr., Grenier, F.C., Lee, C.A. and Waygood, E.B. (1985) J. Cell. Biochem. **27**, 43–56.
77. Poolman, B., Driessen, A.J.M. and Konings, W.N. (1987) Microbiol. Rev. **51**, 498–508.
78. Tsong, T.Y. (1990) Annu. Rev. Biophys. Biophys. Chem. **19**, 83–106.
79. Rephaeli, A.W. and Saier, M.H., Jr. (1980) J. Biol. Chem. 8585–8591. **VOL. NO??**
80. Dills, S.S., Apperson, A., Schmidt, M.R. and Saier, M.H., Jr. (1980) Microbiol. Rev. **44**, 385–418.
81. Kornberg, H. and Lambourne, L.T.M. (1994) Proc. Natl. Acad. Sci. USA **91**, 11080–11083.
82. Saier, M.H., Jr., Newman, M.J. and Rephaeli, A.W. (1977) J. Biol. Chem. **252**, 8890–8898.
83. Saier, M.H., Jr., Feucht, B.U. and Mora, W.K. (1977) J. Biol. Chem. **252**, 8899–8907.
84. Saier, M.H., Jr., Cox, D.F. and Moczydlowski, E.G. (1977) J. Biol. Chem. **252**, 8908–8916.

CHAPTER 24

Domain and Subunit Interactions and Their Role in the Function of the *E. Coli* mannitol transporter, EII^{MTL}

G.T. ROBILLARD, H. BOER, P.I. HARIS*, W. MEIJBERG,
D. SWAVING-DIJKSTRA, G.K. SCHUURMAN-WOLTERS,
R. TEN HOEVE-DUURKENS, D. CHAPMAN* and J. BROOS

*Department of Biochemistry and Groningen Biomolecular Sciences and Biotechnology Institute,
University of Groningen, Nyenborgh 4, 9747 AG Groningen, The Netherlands*

**Department of Protein and Molecular Biology, Royal Free Hospital School of Medicine,
Rowland Hill Street, London NW3 2PF, UK*

© *1996 Elsevier Science B.V.*
All rights reserved

Handbook of Biological Physics
Volume 2, edited by W.N. Konings, H.R. Kaback and J.S. Lolkema

Contents

1. General introduction

The phosphoenolpyruvate-dependent phosphotransferase system (PTS) is responsible for transport and phosphorylation of a variety of carbohydrates in prokaryotes,
for the regulation of the activity of various other transport systems and for the
regulation of the synthesis of the proteins of these systems [1]. The system works
via a series of phosphoryl group intermediates whose number depends on the
carbohydrate being transported and the microorganism in which it is situated. For
this reason it is better to focus on the number of phosphoryl group transfer steps
which occur. There are five steps as given in Scheme I.

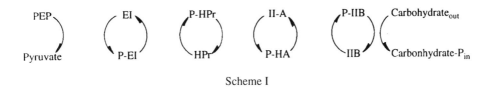

Scheme I

For the discussion in this chapter, the central component is EII, the membrane-
bound protein which catalyzes both phosphorylation and transport. Each PTS
carbohydrate has a specific carrier with a more or less complicated domain architecture. There are two globular cytoplasmic domains, A and B, which catalyze the
phosphoryl group transfer to the carbohydrate after it has been brought across the
membrane by an integral membrane C domain. In some cases there is an additional
membrane-bound D domain. These domains can exist as separate functional proteins, or as fused covalently-linked multi-domain proteins. The domain architecture
of a representative number of phosphotransferase systems is shown in Fig. 1.
   Virtually all domain combinations have been found except a completely covalently linked PTS starting from EI through to IIC. In a very limited number of cases
the rational for domains existing as separate proteins is evident; a protein fulfils
more than one function and to do so must interact with several different partners.
Such is the case for HPr and IIA^{glc}. P-HPr phosphorylates a variety of carbohydrate-
specific IIA proteins or A domains while IIA^{glc} binds to adenylate cyclase, glycerol
kinase and various secondary transport proteins and thereby regulates the synthesis
and transport activity of these transporters. An obvious advantage of the fused
proteins would be to·economize on the amount of protein needed for high rates of
phosphoryl group transfer and accompanying carbohydrate transport. Significant
rate acceleration can be achieved by proximity alone obviating the need for high

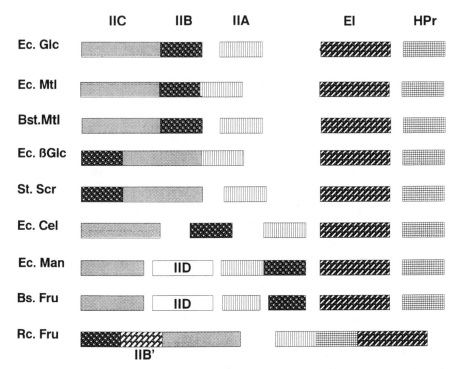

Fig. 1.   Domain architecture of various PTS proteins. Ec = *E. coli*, Bs = *B. Subtilis*, Rc = *R. Capsulata*, St = *S. typhimurium*, Bst = *B. Stearothermophilus*. The shading in the top row is assigned to the domain listed above it. For the other rows the changing position of the shaded domains reflects an altered covalent linkage. For instance, in Ec. βGlc, the B domain is situated in the N-terminal region followed by the C domain and then the A domain.

concentrations to any component or high affinity constants for the interaction of the individual components.

One feature of EII which distinguishes it from all other transport proteins is its ability to chemically modify its substrate by phosphorylation. This gives the experimentalist a significant advantage in studying the activity of the transport protein since one can monitor phosphorylation rather than transport thereby removing the restraint of working with compartmentalized systems. Activity can be measured in vesicles, in detergent or in detergent/lipid mixtures. The common activities measured are phosphorylation and exchange as given in Scheme II.

The purpose of our investigations is to develop a detailed understanding of the molecular events associated with transport via EII. For this we need information on:
  – the role of the association state of the protein in transport and phosphorylation
  – the role of domain interactions in phosphoryl group transfer and energy coupling
  – the process of mannitol binding, the location of the binding sites and the isomerization of these sites in the transport process.

$$PEP + Carbohydrate \rightarrow \rightarrow \rightarrow \rightarrow \rightarrow \rightarrow Carbohydrate\text{-}P + Pyruvate$$

EI, HPr, EII

PHOSPHORYLATION

$$\textit{Carbohydrate} + Carbohydrate\text{-}P \rightleftarrows \textit{Carbohydrate-P} + Carbohydrate$$

EII

EXCHANGE

Scheme II

To address these issues we have subcloned the various domains and domain combinations of EII^{mtl} to give us the following proteins: IIC^{mtl}, IIB^{mtl}, IIA^{mtl}, $IICB^{mtl}$ and $IIBA^{mtl}$. These proteins and mutants thereof are being used for 3D structure determination and to address the above issues.

2. Domain and subunit interactions in phosphorylation and transport

EII^{mtl} and EII^{glc} have been shown to be dimeric enzymes by a variety of methods including crosslinking, gel filtration, SDS gel electrophoresis, radiation inactivation analysis and analytical ultracentrifugation [2–7]. In the case of EII^{mtl}, mannitol binding data and kinetic data on mannitol phosphorylation and exchange as a function of EII^{mtl} concentration demonstrate that the associated state is functionally important [8,9]. Low ionic strength and the absence of Mg^{++} caused the enzyme to become completely inactive in mannitol exchange and significantly less active in mannitol phosphorylation. These effects could be reversed by increasing the EII^{mtl} concentration. Even though dimerization appears to be essential for the enzyme's function, no explanation has yet been provided for the role of the dimer in either phosphorylation or transport. One feasible explanation could be that the individual A, B and C domains interact across the subunit interface rather than within one subunit, i.e. the A domain of subunit 1 phosphorylates the B domain of subunit 2 etc. as shown in the diagram in Fig. 2.

Support for such a phosphorylation route comes from complementation studies in which a functional domain could replace a non-functional domain. This was first observed with the *S. typhimurium* glucose transport system where a mutant lacking functional IIA^{glc} was still able to phosphorylate and transport glucose via EII^{glc}. The IIA^{glc}-like activity was associated with the membrane fraction and was shown to be due to the A domain of EII^{nag}, a protein with a covalently fused A domain which is highly homologous to IIA^{glc} [10,11]. In the case of EII^{mtl}, membrane vesicles containing mutant EII's which were inactive by virtue of mutations in the phosphorylation site on the A domain of one protein and the B domain of another could restore phosphorylation and exchange activity by being combined in the presence of detergent. This indicated that an A domain on one subunit could phosphorylate a B domain on a second subunit even when a non-functional A domain was present

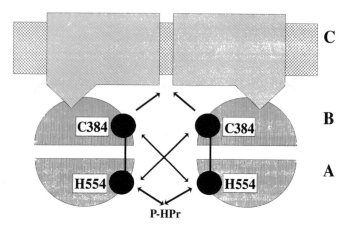

Fig. 2.   Possible phosphorylation routes within the EII^{mtl} dimer. H554 and C384 are the residues which become phosphorylated.

on the second subunit [12]. Further support came with the subcloning and purification of the individual A, B and C domains [13–16]. Each of these proteins restored the activity of a mutant or chimeric protein in which the corresponding domain had been inactivated by virtue of a mutation. Thus functional interactions across the subunit interface are possible, but are they the preferred path or even the mandatory path? Figure 3 presents kinetic complementation studies with purified wild-type and mutant forms of EII^{mtl} which address this issue. The blackened domain in each scheme is the inactive domain. EII^{mtl}-G196D in panel I cannot bind mannitol and is inactive, but in a heterodimer with IIC^{mtl}, phosphorylation and exchange activity are observed. A similar result is presented in panel II, but now with a heterodimer consisting of intact EII^{mtl} mutants, G196D and C384S; even the rates of phosphorylation and exchange are comparable in the two experiments. Clearly transfer across the heterodimer subunit interface between the B domain of the mutant and an active IIC^{mtl} or C domain occurs. If phosphoryl group transfer across the subunit interface is mandatory, then a heterodimer consisting of wild-type EII^{mtl} and an inactive B/C domain double mutant should be inactive yet the heterodimer of wild-type EII^{mtl} and G196D/C384S shows high rates of phosphorylation and exchange (panel III). By comparison of the rates in the three experiments we must conclude that:

(i) phosphoryl group transfer across the subunit interface occurs but that

(ii) phosphoryl group transfer within a single monomeric unit of the dimer is the preferred route;

(iii) only one active A, B and C domain are necessary in a dimer for phosphorylation or exchange activity and they do not have to be situated on the same subunit. Previous experiments have already shown that only one active A domain was necessary.

Since the dimer does not appear to play a role in the phosphoryl group transfer, is it necessary for transport? One could envision, for instance, that two functioning

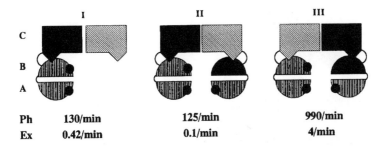

Fig. 3.   Mutant complementation schemes. Panel I: G196D-EII^{mtl} plus IIC^{mtl}. Panel II: G196D-EII^{mtl} plus C384-EII^{mtl}. Panel III: Wild-type EII^{mtl} plus G196D/C384S-EII^{mtl}. Ph = phosphorylation activity, Ex = Exchange activity. The filled circles represent intact phosphorylation sites on the A and B domains.

transporter subunits couple their isomerization states as shown in Fig. 4 such that while one empty site is oriented towards the periplasm for binding mannitol, the other loaded site is oriented towards the cytoplasm for transfer of the phosphoryl group to mannitol. This issue was addressed by creating heterodimers *in vivo* [17]. Cells expressing wild-type EII^{mtl} constitutively were transformed with a plasmid encoding the G196D/C384S double mutant under control of the cI857 temperature-sensitive repressor and phage λ P$_r$ promoter. Growth on glucose led to low levels of wild-type EII^{mtl} and measurable transport. Induction of the double mutant had only a minor effect on the transport rates even though there was a massive over-production of the mutant protein. Since all wild-type EII^{mtl} must be in the heterodimer form under these conditions we have to conclude that the heterodimer with one inactive transport and phosphorylation subunit is still capable of active transport. In conclusion, while EII^{mtl} is active primarily as a dimer, the *raison d'être* for the dimer is not to provide an intersubunit route for phosphoryl group transfer. If it is important for transport, there is no requirement for two transporter subunits both capable of binding mannitol. It is possible that the dimer provides a physical path for movement and facilitates the isomerization of the transporting subunits.

3. Physical Chemical Characterization of A/B domain interactions

The PTS protein domains interact with one another, the most striking example being the influence of the phosphorylation state of the B domain on the conformation of the C domain. Quantitative characterization of these interactions could provide insight into the degree to which they interact and influence each others structure and function.

The thermodynamic parameters ΔG, ΔH and ΔS of unfolding define a protein's stability in solution. When two or more domains interact, it will be reflected in these thermodynamic parameters and the interaction can be quantitated by comparing the values derived from the individual proteins with those from the interacting domains. In recent years a large number of unfolding studies have been performed on

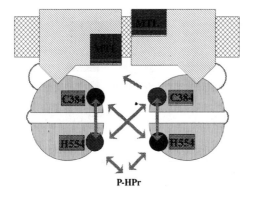

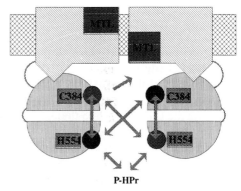

Fig. 4.   Alternating site model for the coupled transport and phosphorylation of mannitol.

proteins containing more than one domain. In principle, two possible observations can be made when unfolding a multidomain protein: (i) a single transition during which all domains unfold cooperatively, or (ii) multiple transitions, indicating that one or more relatively stable intermediates exist on the unfolding pathway. Examples of the former are lysozyme [18,19], phosphoglycerate kinase [20] and ribonuclease A [21]. Multiple transitions were observed in diphtheria toxin [22], pepsinogen [23] and human apotransferrin [24]. The heat denaturation of yeast hexokinase exhibits both types of behavior in the presence or absence of the substrate glucose [25]. The interpretation given to these observations is that multiple transitions indicate the absence of strong interdomain interactions whereas single transitions are an indication that the domains are strongly interacting. Using this model, one can obtain information on the extent of interaction between the domains in a multidomain protein. Information on the energetics of IIBA^{mtl} unfolding have been obtained by studies using either GuHCl or heat as the denaturant [26]. The interaction between the two domains was quantitated by comparing these data to those obtained on separate IIA^{mtl} and IIB^{mtl}. The method was applied to proteins in the unphosporylated and phosphorylated state enabling us to assess the interdomain interactions in IIBA^{mtl} as a function of the phosphorylation state of the enzyme.

The thermal denaturation of IIBA^{mtl} and its individual domains was studied by differential scanning calorimetry. Figure 5 shows typical examples of the excess heat capacity curves as a function of temperature for the three proteins. The IIA^{mtl} and IIB^{mtl} transitions fit a two-state model; the deconvolution analysis of the transition of IIBA^{mtl} is shown in panel C. For each domain there is a good agreement between the observed data and the theoretical fits. The midpoint temperatures at which maximum heat absorption occurs are 64.7°C and 62.7°C for IIA^{mtl} and IIB^{mtl}, respectively. The heat absorption is much more intense for IIA^{mtl} than for IIB^{mtl},

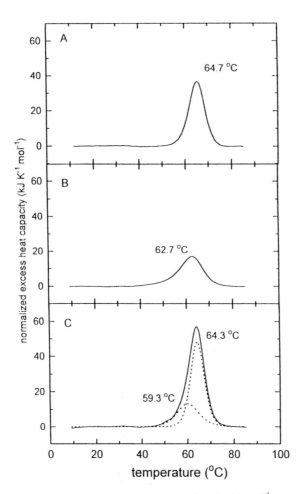

Fig. 5. Excess heat capacity function versus temperature for 94 μM IIA^{mtl} (A), 198 μM IIB^{mtl} (B) and 86 μM IIBA^{mtl} (C). Experimental conditions were 50 mM Hepes, pH 7.6, 10 mM β-mercaptoethanol, the scanning rate was 60°C/h. The experimental curves shown in the figure were obtained by (i) subtracting a buffer-buffer baseline, (ii) normalization to concentration and (iii) subtracting baseline based on how far the transition was from completion. Solid lines are the experimental curves, dotted lines are the theoretical fits to the data.

Table 1
Thermodynamic parameters for the cytoplasmic domains of EII^{mtl} determined by differential scanning calorimetry. Values between brackets are the standard deviations obtained from the fit to the experimental data

Protein	Scan rate ($^\circ$C h^{-1})	T_m ($^\circ$C)	ΔH^{cal} (kJ mol^{-1})	ΔH^{vH} (kJ mol^{-1})	$\Delta H^{cal}/\Delta H^{vH}$
IIAmtl	60	64.7 (0.3)	340 (8)	377 (3)	0.90 (0.02)
IIBmtl	60	62.7 (0.04)	268 (4)	260 (2)	1.03 (0.02)
IIBAmtl	60	59.3	156 (12)	298 (11)	0.53 (0.1)
		64.3	378 (15)	490 (4)	0.77 (0.04)

reflected in the unfolding enthalpies of 377 and 260 kJ/mol, respectively. As will be seen later in the GuHCl unfolding, the transition of IIA^{mtl} is the more cooperative of the two as evidenced by the smaller 8°C width at half height for IIA^{mtl} *versus* 11°C for IIB^{mtl}. The thermal unfolding profile of $IIBA^{mtl}$ exhibits one large peak with maximal heat absorption at 64.3°C. The total enthalpy of unfolding is 531 kJ/mol which, surprisingly, is lower than the sum of the enthalpies of IIA^{mtl} and IIB^{mtl} (637 kJ/mol). Deconvolution analysis showed that the unfolding of $IIBA^{mtl}$ could be best described by two overlapping transitions centered around 59.3 and 64.3°C. The former transition is rather broad and corresponds to a calorimetric enthalpy of unfolding of 156 kJ/mol whereas the latter is sharper and more intense corresponding to a calorimetric enthalpy of 378 kJ/mol. The $\Delta H^{cal}/\Delta H^{vH}$ ratios were 0.53 and 0.77, respectively, indicating that neither transition was of the two-state type. The enthalpies of unfolding and transition midpoint temperatures summarized in Table 1 show that both the T_m and ΔH of IIB^{mtl} are increased relative to those values for the B domain in $IIBA^{mtl}$. Thus, in contrast to what would be expected for two proteins which are produced naturally in the fused state, the interactions of the B domain with the A domain are destabilizing rather than stabilizing.

The influence of phosphorylation on the stability of the individual domains or their interaction could not be probed by DSC because the *in situ* phosphorylation machinery required to maintain the proteins in their phosphorylated form experience thermal inactivation in the same temperature region where the A and B domains unfold. GuHCl-induced unfolding monitored by circular dichroism provided a solution.

The results presented in Fig. 6 for the Gu-HCl-induced unfolding show one transition centered around 1.20 M GuHCl for IIA^{mtl} (Fig. 6A) and 0.95 M GuHCl for IIB^{mtl} (Fig. 6B). It is immediately clear from the slope of the curves in the transitional region that the unfolding of IIA^{mtl} is more cooperative than the unfolding of IIB^{mtl}. The unfolding of $IIBA^{mtl}$ as monitored by the change in elipticity at 222 nm clearly exhibits two stages: the fraction unfolded increases gradually from 0 to 0.2 between 0 and 0.8 M GuHCl and then rises rapidly from 0.2 to 1 between 0.8 and 1.5 M GuHCl. From comparison with the unfolding profiles of the individual domains it is clear that the gradual change at lower GuHCl concentrations corresponds to the unfolding of the B domain and the cooperative transition at

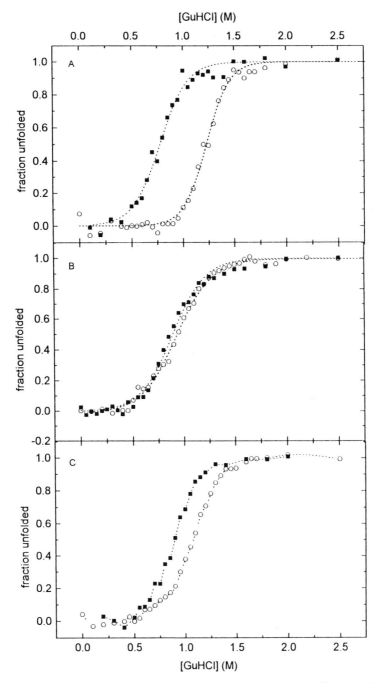

Fig. 6. Circular dichroism at 222 nm of GuHCl-induced unfolding of IIA^{mtl} (A), IIB^{mtl} (B) and IIBA^{mtl} (C) in 10 mM Hepes, pH 7.6, 2 mM β-mercaptoethanol at 25°C. Protein concentrations were 4 μM in each case. The dotted lines through the data are least squares fits to the LEM-equation in (A) and (B) and linear connections of the points in (C). (open circles) unphosphorylated protein; (filled squares) phosphorylated protein.

higher concentration corresponds to the unfolding of the A domain. The GuHCl-induced unfolding of P-IIA^{mtl} in Fig. 6 deviates clearly from the unfolding transition of the unphosphorylated IIA^{mtl} with the midpoint of the transition shifting to 0.4 M lower GuHCl than the midpoint of the IIA^{mtl} transition. The transition also is somewhat less cooperative. The situation is different for phosphorylated IIB^{mtl}; the midpoint of the transition is shifted only marginally, indicating that phosphorylation has a much larger impact on the stability of IIA^{mtl} than IIB^{mtl}. As was observed with IIA^{mtl}, the midpoint of the unfolding transition for IIBA^{mtl} is shifted towards lower GuHCl concentrations upon phosphorylation.

These experiments provide two unexpected results: (i) phosphorylation destabilizes IIA^{mtl} and to a lesser extent IIB^{mtl}; (ii) the interaction of the A and B domains in IIBA^{mtl} destabilize the B domain relative to the separate IIB^{mtl}. The first observation might reflect the loss of interactions between the protein and ordered water as well as the disruption of other polar or ionic interactions in the active site of IIA^{mtl} as a result of introducing the phosphoryl group. The second observation, lower stability of the B domain in IIBA^{mtl} *versus* IIB^{mtl} has also been observed for *E. coli* IIBA^{man} [27]; in both cases it means that the A domain destabilizes the B domain. That nature accepts such a destabilization means that other factors weigh more heavily in keeping the covalent duo together. One such factor could be the proximity advantage which enables high rates of interdomain phosphoryl group transfer with low concentrations of protein. This proximity advantage also means that the binding interactions between the A and B domain do not have to be optimized. Both of these features could be achieved, however, if the A and B domains in IIBA^{mtl} or IIBA^{man} were indifferent to each others presence; destabilization offers no additional advantage. It is possible that, to carry out their natural functions, IIB or the B domains must be in a more flexible, loosely folded, state. This flexibility could result from a decrease number of van der Waal's interactions in the protein interior with a corresponding decrease in ΔH. On the other hand, the destabilization may reflect the artificial experimental situation. In EII^{mtl}, the C domain is covalently attached to the B domain and the complex is situated in the lipid membrane matrix. In EII^{man}, IIBA^{man} is tightly associated with the IIC and IID membrane-bound proteins of this transport system. These interactions could well provide additional stabilizing forces which could more than compensate for the destabilization of the B domain arising from interactions with the A domain. Thermodynamic investigations into the B/C domain interactions are currently in progress to clarify this issue.

4. Kinetic studies of B/C domain interactions

Lolkema et al. [28] demonstrated that when the B domain was phosphorylated, the rate of mannitol transport via the C domain increased by 1000-fold. Roossien and Robillard [29] showed that this same phosphorylation event resulted in changes in the reactivity of C domain residues to chemical modification. Since the phosphoryl group is not transferred to the C domain, the influence of B domain phosphorylation must be communicated to the C domain by a form of conformational coupling.

Table 2
K_ds and $t_{1/2}$ exchange values for mannitol binding

Enzyme	$t_{1/2}$ (s)	K_d (nM)
EII^{mtl}	43	45
IIC^{mtl}	45	139
C384D	11	96
C384S	16	107
C384L	7	151
C384K	5	168
C384H	5	232
C384E	6	263
C384G	8	306

Assuming that the B domain phosphorylation site region is most likely involved in this process, a series of B domain phosphorylation site mutants were prepared to examine their influence on the mannitol binding kinetics of the C domain [30]. Table 2 presents K_d values and exchange half-life values obtained by flow dialysis for mannitol binding to EII^{mtl}, IIC^{mtl} and various C384X mutants.

Membrane vesicles and detergent solutions of IIC^{mtl} and all the phosphorylation site mutants still bind mannitol with a high affinity albeit with some quantitative differences. The parent enzyme binds with the highest affinity suggesting that removal of the A and B domain or changes at the B domain phosphorylation site are felt by the C domain. The kinetics of binding and exchange are also affected in these proteins. Figure 7A shows that when ^{3}H-mannitol is added to EII^{mtl} or IIC^{mtl} containing inverted membrane vesicles present in the dialysis chamber, there is an immediate overshoot followed by a slow decrease in the amount of free mannitol indicating that the rate of mannitol binding is slow [31]. The rate is substantially increased, however, in C384L, C384K and C384E, mutants chosen to examine the effects of charge and polarity; the overshoot which is characteristic of slow binding is completely lacking in these and all other mutants examined (Fig. 7B).

Similar observations were made of the rate of displacement (exchange) of bound ^{3}H-mannitol by unlabelled mannitol. As shown in Table 2, all of the C384 mutations lead to an increased rate of exchange of bound mannitol.

Another striking demonstration of B/C domain interactions was the pH dependence of the mannitol exchange rate. IIC^{mtl} showed no pH dependence between pH 6 and 9 whereas a pH dependence with a pK_a of approximately 8 was observed with EII^{mtl}, suggesting that amino acids in the B domain confer this pH dependence on the C domain. Since the pH dependence was still observed in the mutant enzymes, the effect must be attributed to residues other than the active site cysteine. Both the glutamate and lysine mutant show rate increases over the parent enzyme and the serine mutant. The fact that oppositely charged side chains both result in substantial rate enhancements and that the glutamate and serine mutants both show the same

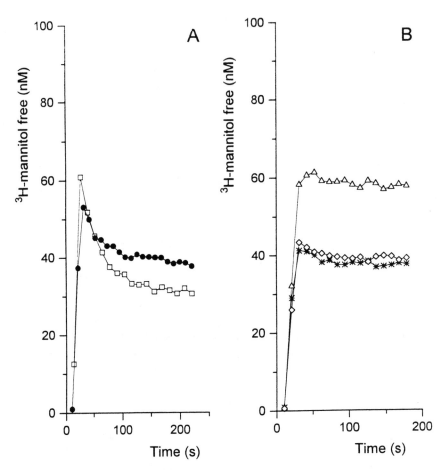

Fig. 7.   Kinetics of binding of mannitol to EII^{mtl} (●) and IIC^{mtl} (□) in panel A and to C384L-EII^{mtl} (◊) C384K-EII^{mtl} (Δ) and C384E-EII^{mtl} (★) in panel B. The rate of binding of mannitol to the enzyme was monitored by adding H^3-mannitol to the upper compartment of a flow dialysis cell containing inside-out membrane vesicles. The amount of free mannitol in the upper compartment was measured by collecting samples of the flow through the lower compartment.

pK$_a$ indicates that there is no simple relationship between the rate changes, pK$_a$ changes and charged state. One might suspect such a relationship if the changes were due to a localized electrostatic interaction between the residue at position 384 and some other residue in the B domain. Most probably, the kinetic changes reflect broader changes in the B domain structure and its interaction with the C domain.

5. FTIR studies on B/C domain interactions

These were undertaken in an effort to specify the nature of the changes occurring in the B/C domain interactions as a result of the conformational coupling event [32].

Since this event is brought about by phosphorylation of the B domain, one must be able to distinguish between phosphorylation-induced changes occurring in the B domain alone *versus* those occurring as a result of alterations in the B/C domain interactions. This can be done by a judicious choice of protein and domain combinations. For this purpose we produced, purified and characterized IICB^{mtl}. Removal of the A domain led to no loss of exchange activity. We had previously produced and kinetically characterized the B and C domains and showed that the phosphorylated B domain had a relatively low affinity for the C domain. Due to this low affinity, hardly any P-IIB^{mtl}:IIC^{mtl} complex would be present in an equimolar solution of P-IIB^{mtl} and IIC^{mtl}. Any changes which did occur upon phosphorylation of such a mixture would be due to the phosphorylation of the B domain only. Additional changes observed upon phosphorylation of the covalently joined IICB^{mtl} could then be assigned to changes due to conformational coupling between the two domains.

The left panel of Fig. 8 presents second derivative FTIR spectra of H$_2$O solutions of IIB^{mtl} (A) and IIC^{mtl} (B) which differ primarily in the position of the main band at 1653 cm^{-1} *versus* 1657 cm^{-1}. The band at 1653 cm^{-1} is the characteristic amide I NH stretch frequency associated with α-helical structure in water-soluble globular proteins. This same band in membrane proteins is often found in the range of 1657–1658 cm^{-1} and in bacteriorhodopsin at an even higher frequency of 1662 cm^{-1} [33,34]. The shift to higher frequencies might be related to the nature of the transmembrane helices or to the lipid or detergent environment surrounding these helices. There is also a strong band in the region of 1630–1634 cm^{-1} indicative of β-sheet structure in both IIC^{mtl} and IIB^{mtl}. Quantitative analysis using a calibration set of 18 soluble proteins to calculate each type of secondary structure[35] revealed 41% α-helix, 28% β-sheet, 17% turns and 14% random coil in IIB^{mtl}. This same analysis could not be applied to IIC^{mtl} because the amide I band is shifted outside of the frequency range of the calibration set. Nevertheless it is clear that IIC^{mtl} consists of significant amounts of α-helix and β-sheet. The right panel of Fig. 8 presents spectra of equimolar amounts of IIC^{mtl} with IIB^{mtl} (C) and P-IIB^{mtl} (D). The spectra are virtually identical with those of IIC^{mtl} alone indicating that, under the conditions of the measurements phosphorylation of IIB^{mtl} brings about no discernable changes in the structure of IIC^{mtl}.

These results are contrasted with those of IICB^{mtl} presented in Fig. 9. Phosphorylation of IICB^{mtl} results in changes in the region 1640–1630 cm^{-1} and addition of perseitol, a non-phosphorylatable mannitol analogue, to P-IICB^{mtl} causes further changes in this same region. Although the changes are not large, they are a clear indication of structural alterations as a result of phosphorylation and substrate binding. Since similar changes did not occur upon phosphorylation of IIB^{mtl} in the equimolar solution of IIB^{mtl} and IIC^{mtl}, the changes reported here for IICB^{mtl} are probably due to the change in B/C domain interactions resulting from conformational coupling upon phosphorylation of the B domain.

Phosphorylation of the B domain does not generate discernable changes in the FTIR spectra and indeed changes were not expected since the phosphorylation site is situated in a loop region sitting atop a twisted four-stranded β-sheet [36]. Figure

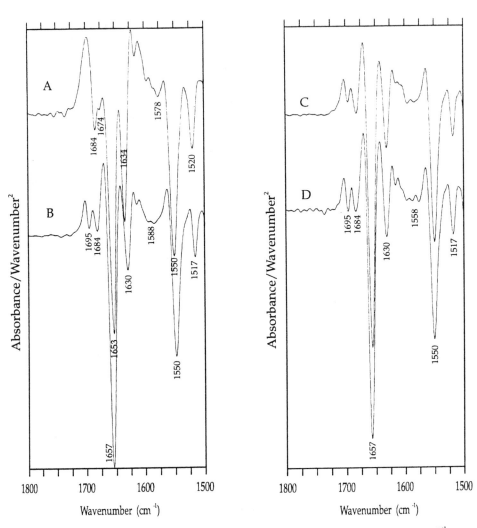

Fig. 8 .   FTIR second derivative spectra of H_2O-containing buffer solutions with (A) IIB^{mtl}, (B) IIC^{mtl}, (C) equimolar solution of IIC^{mtl} and IIB^{mtl}, (D) equimolar solution of $P-IIB^{mtl}$ and IIC^{mtl}. The IIC samples contained, in addition, approximately 100 mM decyl-maltoside.

10 presents the proposed distribution of IIC^{mtl} transmembrane helices [37]; two proposed loop regions account for somewhat less than half of the residues in this domain. It is quite likely that these residues contribute to the β-sheet structure seen in the FTIR spectra. Several reports on EII^{mtl} and EII^{glc} implicate residues in the larger of the two loops at the cytoplasmic face in substrate binding and transport. A variety of mutations in this loop region result in changes in K_m for carbohydrate and/or V_{max} and uncoupling of transport and phosphorylation [38–40]. It is quite likely that this region of the protein is reporting β-sheet structural changes in the FTIR spectra

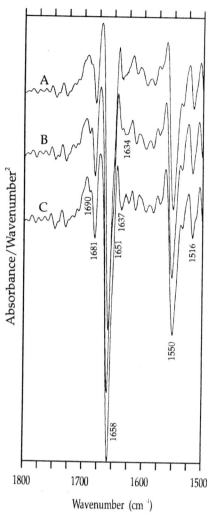

Fig. 9.   FTIR second derivative spectra of (A) IICB^{mtl}, (B) P-IICB^{mtl} and (C) P-IICB^{mtl} plus 10 mM perseitol. All solutions contained approximately 100 mM decyl-maltoside.

upon phosphorylation and substrate binding in response to the conformational coupling event.

6. Fluorescence studies of the C domain and its mannitol interactions

Mannitol binding and the isomerization of the binding site in the transport process are two events which are central to the function of the C domain; any information on these steps can help delineate the kinetics and mechanism of these events. Fluorescence spectroscopy is capable of providing pre-steady state kinetic informa-

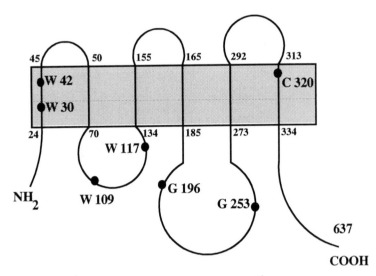

Fig. 10.   Proposed membrane disposition of the C domain of EII^{mtl}. G196 and G253 are two residues
which, when mutated, affect the binding and/or rate of phosphorylation of mannitol.

tion via stopped-flow measurements, structure dynamics information via aniso-
tropy depolarization, and structure information via steady-state and non-radiative
energy transfer measurements. EII^{mtl} possesses four tryptophan residues all of
which are situated in the C domain (see Fig. 10). These can serve as intrinsic
fluorescence reporters to monitor the influence of B domain phosphorylation on the
C domain conformation and to monitor mannitol binding and possibly the isomerization of the binding sites during transport. The information is most valuable when
there is no ambiguity as to which tryptophan is reporting the events. For this reason
a tryptophan minus protein has been constructed, replacing all tryptophans with
phenylalanine and, from there, four single tryptophan mutants were constructed,
each of which involved one of the four natural tryptophan positions in the wild-type
enzyme. All of the mutants including the tryptophan-minus protein retained phosphorylation and exchange activity and mannitol binding affinities comparable to the
parent enzyme.

The spectrum of the trp-minus protein should have shown a maximum between
300 and 305 nm due to tyrosine fluorescence from the 11 tyrosines in EII^{mtl}.
However as Fig. 11 shows the fluorescence spectra of the purified wild-type and
trp-minus EII^{mtl} were virtually identical except for a lower intensity in the trp-minus
protein. DNA sequencing of the trp-minus plasmid confirmed that the protein
definitely lacked tryptophan. One source of the unexpected fluorescence was the
detergent. Figure 12 presents fluorescence excitation and emission spectra of a 1%
solution of various commercially available detergents including crystallization
grade detergents; they are all substantially contaminated with impurities which
fluoresce in the same region as tryptophan. The variation in the excitation spectra
suggest that the impurities are chemically different. They probably arise from

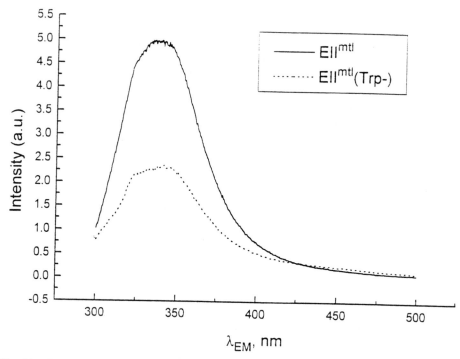

Fig. 11. Fluorescence emission spectra of wild-type EII^{mtl} and Trp-minus-EII^{mtl} each purified with standard preparations of decyl-polyethylene glycol.

stabilizers such as anisole and toluene derivatives, which are added at various stages in the manufacturing process of the precursors to prevent degradation [41]. Other impurities were traced to the plastic tubing used with rotary pumps for column chromatography. Apparently, plasticizers leach out of these polymers when they are brought into contact with detergent-containing solutions. Prior cleaning with organic solvents does not alleviate the problem. Many of these compounds appear to bind strongly to hydrophobic regions of EII^{mtl} and cannot be removed by dialysis. Ionic detergents and some carbohydrate-based detergents can be purified by repeated crystallization but most of these detergents were inhibitory. Since EII^{mtl} is stable in polyethyleneglycol-based detergents over a broad concentration range, a crystallization procedure was developed for pentaethylene glycol monodecyl ether ($C_{10}E_5$) which yielded a fluorescence-free detergent. These procedures enabled us to generate trp-minus EII^{mtl} free of fluorescence intensity in the 325–350 nm region [42].

As a result of these experiences the following precautions are now employed in our laboratory to generate fluorescent contaminant-free purified EII^{mtl}.

(i) All detergents are purified until the fluorescence intensity is clearly lower than the raman peak of water which occurs at 326 nm when an excitation wavelength of 290 nm is used.

(ii) The entire protein purification procedure is done with purified detergents.

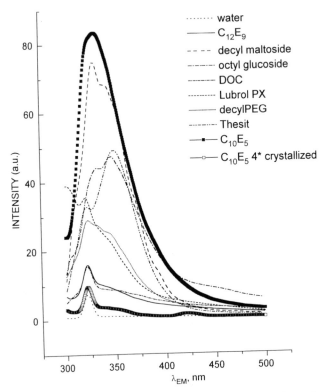

Fig. 12.   Fluorescence emission (A) and excitation (B) spectra of 1% solutions of various commercially available detergents in water. Emission spectra were obtained by excitation at 290 nm. The excitation spectrum for each detergent was recorded at its emission maximum determined from Fig. 12A.

(iii) Enzyme and detergent solutions are never brought into contact with plastic of any kind.

(iv) All glassware is rinsed with spectroscopically pure methanol to remove fluorescence impurities left behind by soap residues.

(v) Stainless steel and teflon tubing and HPLC pumps are used for column chromatography.

Figure 13A compares the fluorescence spectrum of trp-minus and wild-type EII^{mtl} both purified using the above precautions [43]. The trp-minus preparation shows a classical tyrosine fluorescence spectrum with a maximum at 305 nm. Wild-type EII^{mtl} is characterized by a typical tryptophan fluorescence spectrum; tyrosine fluorescence is absent due to non-radiative energy transfer to the tryptophan. When mannitol is added to wild-type EII^{mtl}, an 8% increase is observed, however, it is not possible to assign the increase to one of the four specific tryptophan residues due to energy transfer among the tryptophans. Figures 13B and C show the spectra of the four single tryptophan-EII^{mtl} proteins all normalized to the same concentration and the effect of mannitol on each protein. Each spectrum

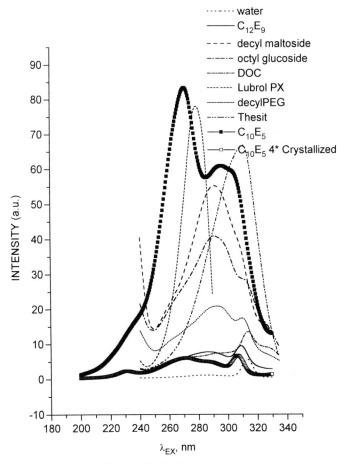

water
$C_{12}E_9$
decyl maltoside
octyl glucoside
DOC
Lubrol PX
decylPEG
Thesit
$C_{10}E_5$
$C_{10}E_5$ 4* Crystallized

INTENSITY (a.u.)

λ_{EX}, nm

Fig. 12 (B) Caption opposite.

now reflects the local environment of the individual tryptophans. The W30 protein is the most blue-shifted indicating a strongly apolar environment; the other three are more red-shifted but do not differ substantially among themselves. Each tryptophan possesses a different quantum yield reflecting the specifics of its local surroundings. Addition of mannitol has a substantial effect only on the intensity of the W30 mutant; it accounts for virtually the entire intensity increase observed upon mannitol addition to the wild-type protein. Further experiments are in progress to determine whether this is due to a direct interaction between W30 and bound mannitol or a conformational change resulting from mannitol binding at a site removed from the tryptophan. The effect observed with W30-EII^{mtl} is interesting of itself, but even more so in as much as mutagenesis studies have shown that residues in the second loop region play a significant role in binding and transport. It is possible that residues in this region are in close contact with residues in the first transmembrane helix.

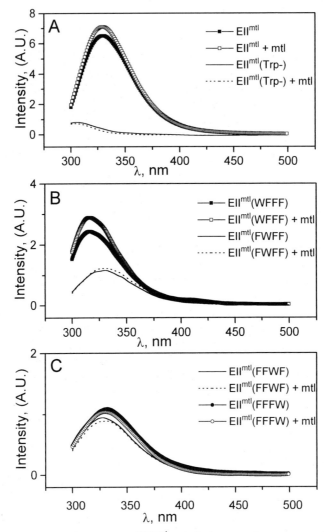

Fig. 13. Fluorescence emission spectra of EII^{mtl} and single tryptophan mutants with and without mannitol. Panel A, wild-type EII^{mtl} and EII^{mtl}(trp–). Panels B and C, the indicated single tryptophan mutant enzymes. For clarity the vertical axis is expanded in panels B and C relative to panel A. All spectra were corrected for buffer background and instrument response.

7. Summary

The issue of subunit and domain interactions and their role in the function of EII^{mtl} in transport and phosphorylation has been addressed with a variety of EII^{mtl} constructs using kinetic and physical chemical methods. We have shown that:

1. Subunit interactions are essential for transport and phosphorylation of mannitol, but that within the associated complex, only one active A, B and C domain

are necessary and they do not have to be situated on the same subunit. The rate of phosphorylation of mannitol is the fastest, however, if the active domains are situated on the same subunit. These observations, together with the high affinities observed during complexation of IIC^{mtl} with wild-type and mutant forms of EII^{mtl}, suggest that the EII dimer possesses an open structure in which the C domains provide the forces for dimerization and the freely mobile A and B domains are capable of transferring their phosphoryl group to their own subunit or neighboring subunit. If these domains are inactive they can be replaced by the free proteins, IIA^{mtl} and IIB^{mtl} but then much higher concentrations are required to achieve significant rates because the affinities between the domains are low and the proximity factor is missing.

2. Phosphorylation of the active site histidine destabilizes the A domain. Interaction between the A and B domains in the covalently-linked complex destabilizes the B domain. The mechanistic basis for this destabilization is currently under investigation.

3. The B and C domain experience a specific interaction upon phosphorylation of the B domain which is reflected in small alterations of the β-sheet secondary structure FTIR bands of the C domain. Perseitol binding has a similar effect.

4. Mannitol binding is reported by the increase in fluorescence of a single tryptophan far removed, in the linear sequence, from other residues in the transporter domain which are involved in binding and transport. This may reflect the proximity of these portions of the protein in the 3D structure.

All aspects of the work presented above continue to be investigated in the confidence that they will provide a more complete picture of the structural and mechanistic relationships associated with enzyme-catalyzed transport.

Abbreviations

Nomenclature: When domains are covalently linked we refer to them as the A, B, C, BA or CB domain. When they are produced separately we refer to them as IIA, IIB, IIC, IIBA or IICB.

EII^{mtl}, EII^{glc}, EII^{man}, the mannitol, glucose and mannose specific Enzyme II, respectively

EII^{mtl}(trp–), tryptophan-minus mutant of EII^{mtl}

EII^{mtl}(WFFF), EII^{mtl}(FWFF), EII^{mtl}(FFWF) and EII^{mtl}(FFFW), EII^{mtl} mutants with tryptophan 30, 42, 109 and 117 each present as a single tryptophan, respectively, while the other three tryptophans had been replaced by phenylalanine

PEP, P-enolpyruvate

PTS, P-enolpyruvate-dependent carbohydrate phosphotransferase system

References

1. Postma, P.W., Lengeler, J.W. and Jacobson, G.R. (1993) Microbiol. Rev. **57**, 543–594.
2. Roossien, F.F., Van Es-Spiekman, W and Robillard G.T. (1986) FEBS Lett. **196**, 284–290.
3. Erni, B. (1986) Biochemistry **25**, 305–312.
4. Erni, B., Trachsel, H., Postma, P.W. and Rosenbusch J.P. (1982), J. Biol. Chem. **257**, 13726–13730.

5. Roossien, F.F. and Robillard, G.T. (1984) Biochemistry **23**, 5682–5685.
6. Lolkema, J.S., Kuiper, H., ten Hoeve-Duurkens, R.H. and Robillard, G.T. (1993) Biochemistry **32**, 1396–1400.
7. Pas, H.H., Ellory, J.C. and Robillard, G.T. (1987) Biochemistry **26**, 6689–6696.
8. Pas, H.H., ten Hoeve-Duurkens, R.H. and Robillard, G.T. (1988) Biochemistry **27**, 5520–5525.
9. Lolkema, J.S. and Robillard, G.T. (1990) Biochemistry **29**, 10120–10125.
10. Vogler A.P., Broekhuizen, C.P., Schuitema, A., Lengeler, J.W. and Postma, P.W. (1988) Mol. Microbiol. **2**, 719–726.
11. Vogler, A.P. and Lengeler, J.W. (1988) Mol. Gen. Genet. **213**, 175–178.
12. van Weeghel, R.P., van der Hoek, Y.Y., Pas, H.H., Elferink, M., Keck, W. and Robillard, G.T. (1991) Biochemistry **30**, 1768–1773.
13. van Weeghel, R.P., Meyer, G.H., Keck, W. and Robillard, G.T. (1991) Biochemistry **30**, 1774–1779.
14. van Weeghel, R.P., Meyer, G., Pas, H.H., Keck, W. and Robillard, G.T. (1991) Biochemistry **30**, 9478–9485.
15. Boer, H., ten Hoeve-Duurkens, R.H., Schuurman-Wolters, G. K., Dijkstra A. and Robillard, G.T. (1994) J. Biol. Chem. **269**, 17863–17871.
16. Robillard, G.T., Boer, H., van Weeghel, R.P., Wolters, G. and Dijkstra, A. (1993) Biochemistry **32**, 9553–9562.
17. Boer, H. and Robillard, G.T. (1996) Biochemistry, in press.
18. Khechinashvili, N.N., Privalov, P.L. and Tiktopulo, E.I. (1973) FEBS Lett. **30**, 57–60.
19. Privalov, P.L. and Khechinaskvili, N.N. (1974) J. Mol. Biol. **80**, 665–684.
20. Hu, C.Q. and Sturtevant, J.M. (1987) Biochemistry **26**, 178–182.
21. Brandts, J.F., Hu, C.Q., Lin, A.-N. and Mas, M.T. (1988) Biochemistry **28**, 8588–8596.
22. Ramsay. G. and Freire, E. (1990) Biochemistry **29**, 8677–8683.
23. Privalov, P.L., Malao, P.L., Khechinaskvili, N.N. Stepanov, V.M., Revina, L.P. (1980) J. Mol. Biol. **152**, 445–464.
24. Lin, L.N., Mason, A.B., Woodworth, R.C. and Brandts, J.F. (1994) Biochemistry **33**, 1881–1888.
25. Takahashi, K., Casey, J.L. and Sturtevant, J.M. (1981) Biochemistry **20**, 4693–4697.
26. Meijberg, W., Schuurman-Wolters, G.K. and Robillard G.T. (1996) Biochemistry **35**, 2759–2766.
27. Markovic-Housley, Z., Cooper, A., Lustig, A., Flukiger, K., Stolz, B. and Erni, B. (1994) Biochemistry **33**, 10977–10984.
28. Lolkema, J.S., ten Hoeve-Duurkens, R.H., Swaving-Dijkstra, D. and Robillard, G.T. (1991) Biochemistry **30**, 6716–6721.
29. Roossien, F.F. and Robillard, G.T. (1984) Biochemistry **23**, 211–215.
30. Boer, H., ten Hoeve-Duurkens, R.H., Lolkema, J.S. and Robillard, G.T. (1995) Biochemistry **34**, 3239–3247.
31. Lolkema, J.S., Swaving-Dijkstra, D. and Robillard, G.T. (1992) Biochemistry **31**, 5514–5521.
32. Boer, H., Haris, P.I., Swaving-Dijkstra, D., Schuurman-Wolters, G.K., ten Hoeve-Duurkens, R.H. Chapman, D. and Robillard, G.T., unpublished data.
33. Haris, P.I. and Chapman, D. (1989) Biochem. Soc. Trans. **17**, 161–162.
34. Haris, P.I. and Chapman, D. (1992) Trends Biochem. Sci. **17**, 328–333.
35. Lee, D.C., Haris, P.I., Chapman, D. and Mitchell, R.C. (1990) Biochemistry **29**, 9185–9193.
36. AB, E., unpublished data.
37. Sugiyama, J.E., Mahmoodian, S. and Jacobson, G.R. (1991) Proc. Nat. Acad. Sci. USA **88**, 9603–9607.
38. Manayan, R., Tenn, G., Yee, H.B., Desai, J.D., Yamada, M., and Saier, M.H., Jr. (1988) J. Bacteriol. **170**, 1290–1296.
39. Buhr, A., Daniels, G.A. and Erni, B. (1992) J. Biol. Chem. **267**, 3847–3851.
40. Ruiter, G.J.G., van Meurs, G., Verwey, M.A., Postma, P.W. and van Dam, K. (1992) J. Bacteriol. **174**, 2843–2850.
41. Ray, W.J. and Puvanthingal, J.M. (1985) Anal. Biochem. **146**, 307–312.
42. Swaving-Dijkstra, D., Broos, J. and Robillard, G.T. (1996) Anal. Biochem., in press.
43. Swaving-Dijkstra, D., Lolkema, J.S., Enequist, H., Broos, J. and Robillard, G.T. (1966) Biochemistry, in press.

CHAPTER 25

Phosphotransferase Systems or PTSs as Carbohydrate Transport and as Signal Transduction Systems

J.W. LENGELER and K. JAHREIS

Universität Osnabrück, Fachbereich Biologie/Chemie,
Osnabrück, Germany

© 1996 Elsevier Science B.V.
All rights reserved

Handbook of Biological Physics
Volume 2, edited by W.N. Konings, H.R. Kaback and J.S. Lolkema

Contents

1. Introduction. The central role of PTSs in bacterial metabolism

Bacterial carbohydrate transport systems are classified primarily according to the energy source which they utilize. Three types of active transport systems can be distinguished as has been described in the preceding chapters of this book. These are: (i) primary transport systems; (ii) secondary transport systems; and (iii) phospho-*enol*pyruvate (PEP)-dependent carbohydrate: phosphotransferase systems, abbreviated hereafter as PTS(s). Primary and secondary systems catalyze the translocation of their substrates in an energy dependent process through the membrane. This causes an accumulation of chemically unchanged substrate molecules at one side, usually the cytoplasmic side, of the inner cell membrane. As indicated by the name, the PTS was considered originally by its discoverer S. Roseman as an unusually complex system to phosphorylate carbohydrates [1]. Genetic studies, mostly from the groups of E.C.C. Lin [2], M.L. Morse [3] and E.D. Rosenblum [4], supported by biochemical studies [5] then hinted at a simultaneous role in transport for a large number of carbohydrates. In contrast to the other active transport systems, PTSs couple the translocation of their substrates through the membrane with the phosphorylation of these substrates at the expense of PEP hydrolysis. Their activity thus causes intracellular accumulation of substrate-phosphate molecules as the first metabolic intermediates.

1.1. The PTSs are a series of carbohydrate transport systems

We have learned many details about the properties of the PTS proteins and their multiple functions, as well as on the genes which encode these proteins, as documented in several recent reviews [6–10]. At present, we know of more than 40 PTS proteins and of a rapidly increasing number of non-PTS proteins which interact with the PTS. This complexity reflects first the central role which the PTS has in transport and phosphorylation of a large number of carbohydrates. Each bacterial cell contains indeed a series of PTS transport systems, e.g. 16 described thus far in *Escherichia coli* and other enteric bacteria. According to their major and inducing substrate, they are called the glucose (Glc)-PTS, the fructose (Fru)-PTS, the mannose (Man)-PTS, the mannitol (Mtl)-PTS, etc. [8]. As indicated schematically in Fig. 1, each PTS includes a substrate-specific Enzyme II (abbreviated hereafter as II^{Glc}, II^{Fru}, II^{Man} or II^{Mtl}) which transports and phosphorylates the various substrates. Each Enzyme II comprises three functional domains named IIA, IIB and IIC, the latter corresponding to the membrane-bound transporter. There are indications (see below) that IIC contains two parts (domains) which are fused normally. If split as

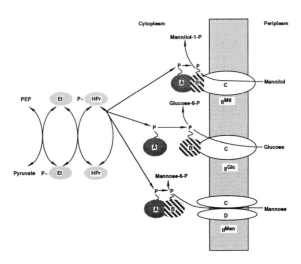

Fig. 1.   Various types of phosphotransferase systems or PTSs. Enzyme I (EI) and histidine protein (HPr) are the general proteins for all PTSs. The Enzymes II (EII) specific for mannitol (II^{Mtl}), for glucose (II^{Glc}), and for mannose (II^{Man}) represent the variety of EIIs found in bacteria. All contain three autonomous domains called IIA (formerly Enzyme III), IIB and IIC. The latter, the membrane-bound transporter, is split in II^{Man} in two parts (IIC and IID). P~ indicates the phosphorylated form of the soluble proteins/domains, while arrows indicate the direction of the biochemical steps (after 8).

in II^{Man}, they will be called IIC and IID. All domains may be fused into a large protein of about 680 amino acids. Such a complex will be described as IICBA^{Mtl} if the domains are fused (starting at the amino terminal end) in the sequence IIC, IIB and IIA. If split into two or more peptides as for II^{Glc} and II^{Man}, they will be described accordingly as IICB^{Glc} IIA^{Glc} or as IIAB^{Man} IIC^{Man} IID^{Man} and in an analogous way for other Enzymes II.

All Enzymes II of a cell are phosphorylated by a PEP-dependent protein kinase, the Enzyme I (EI). In its dimeric form, this kinase autophosphorylates at one histidine residue per subunit and transfers this phosphoryl group to an ancillary phosphocarrier protein, the Histidine-protein (HPr). In each Enzyme II, domain IIA accepts the phosphoryl group from P~HPr at the N3-position of a conserved histidyl residue and donates it to a cysteine (rarely a histidine) residue of its domain IIB. Complexed to the corresponding IIC transporter, the phosphorylated IIB drives the transport of a bound substrate and subsequently liberates the substrate-phosphate molecule into the cytoplasm. The structural genes for Enzymes II are normally clustered together with the genes for the corresponding catabolic enzymes in distinct operons (regulons). Genes *ptsI* and *ptsH* for EI and HPr, however, form the *pts* operon. This arrangement corroborates the hypothesis that each PTS should be considered as an independent transport system which is phosphorylated by the general protein kinase EI through HPr.

1.2. The PTS is more than a series of transport systems

All functions of the various PTSs as transport and phosphotransferase systems for many carbohydrates, however, are not sufficient to explain its unusual complexity. The first hint at an additional role of the PTS in the regulation of carbohydrate metabolism came from the observation that Pts⁻ mutants of *E. coli* and *Salmonella typhimurium* lacking EI and HPr were not only unable to take up PTS-carbohydrates, but also most non-PTS carbohydrates such as lactose, maltose, melibiose, glycerol, pentoses and Krebs cycle intermediates. Subsequent work, primarily by the groups mentioned before [1–5], by P.W. Postma [11], M.H. Saier Jr. [12] and others (for a recent review see Refs. [8,9]), showed that in enteric bacteria the PTS protein IIA^Glc (also IIA^Crr, formerly Enzyme III^Glc) is central to this control. This important protein controls the activity of many carbohydrate uptake systems and the synthesis of all proteins which are a member of the *crp* modulon and depend on 3′-5′, cyclic-adenosine-monophosphate (cAMP) for their synthesis (see Section 3.3).

The cascade of protein kinases and phosphocarrier proteins which constitute the PTS, together with the membrane-bound Enzymes II, strongly resembles other pro- and especially eukaryotic signal transduction systems, in particular those linked to global regulatory networks. All PTS proteins together form a sophisticated signal transduction system through which bacteria monitor their environment and respond to extracellular changes [13]. Prokaryotes sense such changes either directly through an impressive battery of membrane-bound sensors, or indirectly as changes in pools of intracellular molecules which are detected by intracellular sensors. These pools often correlate with the transport capacity of a cell and hence reflect the environment. Most sensors are linked through signal transduction systems to global regulatory networks which control entire metabolic blocs or differentiation processes. Meantime the PTS may be considered as a paradigm for systems in which the transport capacity of a cell is closely linked to cellular control processes [14]. According to this model, it constitutes a signal transduction system which monitors the environment through the transport capacity of its Enzymes II and signals to a global regulatory system which controls many functions involved in the quest of food.

The PTSs and several of their specific functions have been reviewed recently [6–10,13–16] or are described in more detail in this book (Chapters 23 and 24). We will therefore restrict the present description to the proteins and properties mentioned before which (i) enable the transport catalyzed by PTSs to trigger a signal and (ii) enable the system to transduce these signals through the cascade of PTS protein kinases to various target proteins. These are involved in the quest of food which they control through carbon catabolite repression, inducer exclusion and chemotaxis towards PTS-carbohydrates. We will restrict, furthermore, the references to recent reviews and overview articles, and to papers central to the topic of this article.

2. PTSs as carbohydrate transport systems

Soon after its discovery it became clear that the PTS is not only a complex phosphotransferase system, but that it constitutes at the same time a series of carbohydrate transport systems which couple transport and phosphorylation of their substrates ('group translocation'). The economy of the system at the energetic level is nearly optimal because for each PEP molecule hydrolyzed during uptake, two ATP molecules are generated subsequently, one being derived from PEP in the pyruvate kinase reaction. The PTS thus represents a self-priming transport system in which an intermediate in the catabolism of its substrates is used to drive transport and to allow the phosphorylation of these substrates. Furthermore, the high phosphate transfer potential of PEP is retained in the phospho-His and phospho-Cys residues of phosphorylated PTS proteins. This property facilitates the reversible and cost-neutral phosphate transfer among PTS proteins and their communication with other proteins (see Section 3). The versatility of the PTS is enhanced furthermore by a modular construction at the gene and protein level, and the use of units which are often interchangeable and hence facilitate cross-regulation.

2.1. The modular construction of PTSs

All Enzymes II and general PTS proteins have evolved apparently from common ancestors by gene duplications and mutational diversification of the copies. During evolution, the genes have been split furthermore into fragments which encode functional enzyme domains and these fragments are constantly shuffled and reshuffled while being transferred between various organisms. The search for such conserved, but rearranged gene and protein domains has become very helpful to define functional domains and *vice versa*.

DNA and protein sequences for more than 20 different Enzymes II are available. They represent PTSs for 14 different substrates from many Gram-positive and Gram-negative eubacteria and PTSs encoded on chromosomes, plasmids and transposable elements which can be grouped into five genetic families [10] as summarized in Fig. 2. The degree of similarity varies from near identity to a few locally restricted motifs. The least conserved structures, e.g., linker peptides which fuse domains, will deviate already in closely related PTSs, while strongly conserved structures may still be recognizable in barely related systems, e.g., those from different genetic families. Members of the five families, which have been isolated primarily from *E. coli* (Ec), *S. typhimurium* (St), *Klebsiella pneumoniae* (Kp), *Vibrio alginolyticus* (Va), *Bacillus subtilis* (Bs), the genera *Staphylococcus* and *Streptococcus*, *Lactococcus lactis* (Ll) and *Lactobacillus casei* (Lc), are: (i) the glucose/sucrose family with PTSs specific for glucose (Glc), maltose (Mal) and N-acetyl-glucosamine (Nag) on the one hand, for sucrose (Scr, Sac), trehalose (Tre) and β-glucosides (Bgl, Arb, Asc) on the other hand; (ii) the mannitol/fructose family with PTSs specific for mannitol (Mtl, Cmt) or fructose (Fru) and galactitol (Gat); (iii) the lactose/cellobiose family with PTSs specific for lactose (Lac) or cellobiose (Cel); (iv) the mannose/sorbose family with PTSs specific for mannose

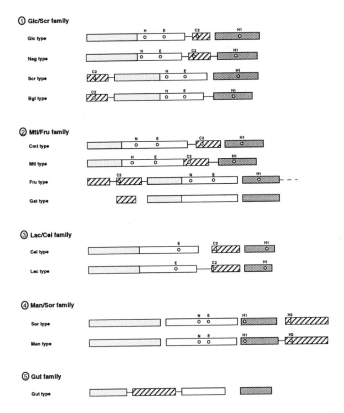

Fig. 2.   Enzymes II and their functional domains ('modules') IIA,B,C and D. All Enzymes II described thus far are a member of one of five genetic families. Their modular construction from four functional domains which are in free form or fused (by linkers, thin line) to larger proteins is evident. Symbols: dark shaded boxes, IIA with the first histidine (H1) phosphorylation site; hatched boxes, IIB with the second histidine (H2) or cysteine (C2) phosphorylation site; lightly shaded boxes, first hydrophobic part of the transporter; white boxes, second part of the transporter with a large hydrophilic loop containing conserved residues as indicated (E, Glu; H, His; N, Asn). Each protein is drawn roughly to scale with its amino-terminus to the left. For FPr (376 residues), only the IIA^{Fru} domain is indicated. The various members of each family are listed in the text (after 10).

(Man), sorbose (Sor) or fructose (Lev); (v) the glucitol family with PTSs specific for glucitol (Gut) (references in Refs. [10,17]).

In Enzymes II, domains IIA, IIB and IIC are either free (see the Gat-, Cel- and Sor-type in Fig. 2) or fused, but in any case they are functionally autonomous [8]. Thus if IIA^{Nag} or IIA^{Mtl}, which are normally fused to their IIBC-domains, are deleted, they can be replaced by another IIA domain (e.g., IIA^{Glc} and IIA^{Bgl} or IIA^{Cmt}, respectively) regardless of whether this IIA is fused to its IIBC part or is a free peptide. IIC domains, on the other hand, form a functional transporter even if expressed in the absence of any matching IIA and IIB. Functional replacements, however, are restricted to members of the same genetic family. A comparison of the arrangements of the various domains within these five genetic families indicates that their organization at the DNA level usually mirrors the organization at the peptide level and that larger functional domains move as entire blocs. In the fused form of Enzymes II, domains are often delineated by characteristic linker peptides of about 20 residues length. Each domain, beginning with the general PTS proteins EI (575 residues) and HPr (85 residues), the hydrophilic domains IIA and IIB (about 100 residues each) that become phosphorylated successively, and ending with the transporter IIC (about 350 residues), catalyzes one major biochemical reaction in the overall process of group translocation (see Section 2.3).

Neither the number of peptides which constitute a PTS nor their arrangement within polypeptides seem essential for function. Thus several families of Enzymes II contain subfamilies in which basically only the arrangement of the domains differs, e.g. sequence IICBA for the Glc/Nag-, and IIBCA for the Scr/Bgl subfamilies of family 1 in Fig. 2. IIBC^{Scr} from enteric bacteria requires the so-called IIA^{Glc} for activity. This example of a natural 'cross-talk' corroborates the conclusion that both subfamilies are members of the same family. As a rule, the genes for domains IIABC of a substrate-specific Enzyme II map in an operon (or regulon), together with the genes for the corresponding metabolic enzymes. This facilitates their coordinated expression and supports the notion that each PTS constitutes a transport unit. An important exception is IIA^{Glc} whose structural gene *crr* (mnemonic for <u>c</u>atabolite <u>r</u>epression <u>r</u>esistance) is part of the *ptsHI crr* operon and not of the *ptsG* operon for its IICB^{Glc} transporter. For historical reasons, this IIA has originally been called Enzyme III (III^{Glc}), then IIA^{Glc}. Because of its involvement also in the transport through II^{Scr} and II^{Tre}, the mapping of its gene in the *pts* operon and its central role in catabolite repression and inducer exclusion which require the coordinated synthesis of this IIA with EI and HPr, it should be called correctly IIA^{Crr} in enteric bacteria and perhaps IIA^{Glc}, IIA^{Scr} or IIA^{Tre} when its role in transport is indicated.

Members of the Man/Sor and of the Gut family seem to be exceptions because their transporter domains appear as split. Assuming that here too only complete functional domains are separated which in the other Enzymes II are fused, we have postulated [10] that each IIC consists of an amino-terminal and a carboxy-terminal functional part (see Section 2.2). These appear to have been split in the Man/Sor family into two peptides (called IIC and IID), perhaps as the consequence of the insertion of a peptide with functions in the injection of phage λ DNA. In the Gut family, however, both parts appear to have been separated by the insertion of IIB^{Gut} between the IIC and IID part, thus forming the somewhat unusual transporter of this family (Fig. 2).

2.2. General structure of a IIC transporter

Uptake through an Enzyme II includes binding of the substrate to one or several binding sites, translocation through the membrane and subsequent phosphorylation. Binding and facilitated diffusion of the substrate down its concentration gradient requires only the transporter IIC. As will be described below (Section 2.3), binding, translocation and phosphorylation are sequential steps which may be uncoupled, e.g., by mutations of the IIC domains.

A set of rules has been described which allow the deduction of a putative 2D-model for the membrane-bound parts of IIC-domains [10]. The rules rely primarily on the well known Kyte and Doolittle hydropathy plots [18], the positive inside rule by v. Heijne [19] and the occurrence of conserved structures and sequence motifs found in all Enzymes II, as outlined in Fig. 2. Limited data from experiments with purified and reconstituted Enzyme II preparations or inside-out and right-side-out vesicles are available (references in Ref. [8]). They all indicate a structure which is nearly insensitive to protease activities, i.e. most parts of the IIC transporter should be buried in the membrane. This contrasts with the hydrophilic domains IIA and IIB which, even if fused to IIC, are accessible to protease treatment, and with peptide linkers which constitute the preferred sites of protease attack.

Gene fusion techniques have been developed which can give a rough estimate on the topology of membrane-bound proteins. Over 40 unique PhoA fusions have been inserted in vitro into II^{Mtl} from *E. coli* using a nested deletion method [20]. They support a 2D model derived from sequence analysis [10] and shown schematically in Fig. 3A. Its most conspicuous features are: (i) the presence of 6 transmembrane structures (helices, TM) which lack almost completely charged and even hydrophilic amino acids, i.e. residues capable of hydrogen bonding. The helices either contain hydrophobic residues which are not conserved (type 1), or they contain a mixture of strongly conserved structural (neutral) and hydrophobic residues (type 2). TM I, II, IV and V of most Enzymes II are of this type 2, while TM III and VI are of type 1; (ii) the presence of three short periplasmic loops of variable length which often lack conserved residues; (iii) for Enzymes II of the IICB type, the carboxy-terminal loop of IIC is fused by a linker to the intracellular domain IIB and the highly conserved TM I together with an amphiphilic structure preceding this TM seems essential for a proper insertion into the membrane. In IIBC types, however, intracellular IIB together with TM I cause the proper insertion at the amino-terminus, while TM VI (and IIA if fused by a linker to IIC) ensure(s) proper insertion at the carboxy-terminus; (iv) a large ($\geq$80 residues) hydrophilic loop 5 connects TM IV and V. It contains an above average percentage of conserved residues (among them many hydrophilic), most of which are essential in substrate binding and phosphorylation (see Section 2.3). This loop is of the size of other soluble domains, e.g., IIA and IIB, and may constitute the central part of the catalytic center and of part IID; (v) a large hydrophilic loop 3 connects TM II and III. Its location (cytoplasmic?) is controversial, at least in IICs from other families.

The topology of IIC^{Glc} from *E. coli* has also been tested through 41 PhoA- and 54 LacZ- fusions which were selected *in vivo* by screening for alkaline phosphatase and β-galactosidase activity, respectively [21]. In the first (amino-terminal) part, the model agrees remarkably well with the IIC^{Mtl} model discussed before and the II^{Scr} model as derived from sequence comparisons [10]. The fusion data, however, place loop 3 into the membrane as shown schematically in Fig. 3B. The two additional membrane-bound structures (TM A and B) have a very low hydropathy value ($\leq$ +2.0) and less than 20 residues in length. They differ consequently from TM I–VI (hydropathy value > +2.0; 20 residues) which may correspond to classical transmembrane helices bridging the membrane once in full length.

(A) Scr/Mtl

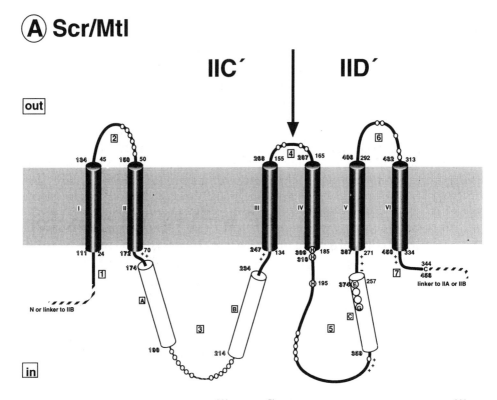

Fig. 3.   2D models as determined for IIC^Mtl and IIC^Glc from *E. coli*. In 3A, the model of IIC^Mtl as supported by a gene fusion study [20] is given. It postulates 6 transmembrane helices (I to VI), 3 periplasmic loops (# 2, 4 and 6) and 4 cytoplasmic loops (# 1, 3, 5 and 7). Also indicated are three weakly hydrophobic structures (A, B, C) in loops 3 and 5, furthermore charged residues (+ or –), conserved histidine (H), glutamate (E) and glycine (G) residues considered as essential in catalysis, and gaps (o – o) in sequence alignments between members of one family or subfamily. The model for IIC^Scr as deduced from sequence alignments [10] is also given with the number of residues for IIC^Scr from pUR400 at the left (light numbers) and those for IIC^Mtl from *E. coli* at the right side (dark numbers). The arrow on top indicates the possible border between parts IIC′ and IID′ as discussed in the text. In 3B, the model of IIC^Glc as supported by a gene fusion study [21] is given. The numbering and symbols are as in 3A. The residues at the left side (light numbers) indicate as in 3A those for IIC^Scr, while the residues at the right side (dark numbers) are those for IIC^Glc from *E. coli*.

Curiously, the IIC^Glc model deviates strongly in its second (carboxy-terminal) part from the IIC^Mtl and the calculated IIC^Scr model. The results place a large part of loop 5 into the membrane (TM C in figure 3B) including residues (e.g. the GITE-motif) which seem essential in substrate binding and phosphorylation and place TM V and TM VI as indicated. Similar fusion studies with IIC^Scr which used artificially created fusion sites in all putative peri-membrane loops (according to the model in Fig. 3B) gave ambiguous results. While the fusions clearly positioned loop 2 into the periplasm, the fusions in the putative loops 3b, 4 and 5a showed intermediate PhoA and LacZ activity, while no stable insertions could be obtained

Ⓑ Scr/Glc

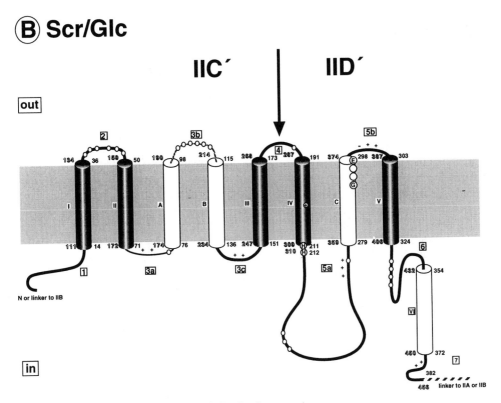

Fig. 3 (B) Caption opposite.

in loops 5b and 6 (K. Rutkowski and J.W. Lengeler, unpublished results).

Considerable evidence (references in Ref. [8]) suggests that a dimer constitutes the active form of an Enzyme II and that all IIC, regardless from which genetic family, make up a translocation unit with a similar 2D structure. We have proposed a transporter model [10] according to which in a dimer (at least) 2×6 transmembrane helices form a hollow structure. Its interior would contain hydrophilic loops which in essence constitute the catalytic center of the transporter for substrate binding and translocation. These loops could be anchored to the interior side of the hydrophobic outside structure through 'transmembrane' structures of low hydropathy and short length such as, e.g., TM A, B and C in Fig. 3B. If this model proves to be correct, the two variants of a IIC transporter shown in Fig. 3 could correspond to a completely (3A) or partially (3B) unfolded form.

2.3. Carbohydrate transport through Enzymes II and group translocation

The isolated IIC domain still catalyzes the binding of its substrates and their translocation through the membrane [22–24]. Hence, IIC must be the real transporter. Furthermore, substrate translocation and phosphorylation considered as typical for group translocation through an Enzyme II correspond to two distinct processes which can be uncoupled [10,25–27]. They can formally be described as facilitated diffusion of a substrate through the membrane-bound facilitator IIC, coupled to subsequent phosphorylation of the intracellular substrate by the soluble phosphotransferases IIA and IIB, as suggested originally by Lin [2].

Results from a series of kinetic studies using whole cells, vesicle preparations and purified Enzyme II preparations, mostly with II^{Mtl} and II^{Glc} (references in Refs. [8,10]), indicate that an Enzyme II-dimer exposes one high-affinity binding site to the periplasmic side [24,28–30]. If the IIC transporter is not complexed to a phosphorylated IIB domain, the substrate is translocated ('facilitated diffusion') slowly to the cytoplasm [23,31]. The entire process probably involves reorientation of the loaded binding site through an 'occluded' site to a low-affinity 'inward-facing' binding site. The mobile state thus corresponds to the shift of a binding site from exposure to one side of the membrane to exposure to the opposite side. Translocation of high-affinity substrates through a non-phosphorylated Enzyme II or an isolated IIC is low, at least 10-fold below the rate required for efficient growth. They are probably 'trapped', e.g., because high-affinity substrates induce a conformational change which locks the binding site in an immobile (or 'occluded') state [29]. This slow transport contrasts with an efficient transport of low-affinity substrates which has been observed for every Enzyme II tested thus far [10]. Their translocation is sufficient to allow fast growth (up to 60 min generation time).

In intact II^{Glc} and in truncated $IICB^{Mtl}$ transporters, so-called uncoupled mutants have been isolated which allow growth on glucose (in a Pts$^-$ strain) [25–27] and on mannitol (through an artificial metabolic pathway for free mannitol) [10]. The 32 mutants characterized thus far in both transporters all carry mutations in loop 5 (or 5a and 5b) in Fig. 3. All mutants show a drastically increased K_M^{app} value (10^2 to 10^3-fold) and their normally high-affinity substrates correspond in these strains to low-affinity substrates. Apparently, low-affinity substrates or uncoupled IIC mutants with decreased affinities cannot lock the transporter in the immobile state. In the absence of a tight coupling between substrate binding and mobilization of the binding site, however, 'active transport' is converted automatically into 'facilitated diffusion' as observed here. It is tempting to speculate that the phosphorylation of a trapped substrate through a phospho-IIB subunit, itself complexed to a IIC domain, converts the high-affinity substrate into a low-affinity substrate-phosphate and thus causes its release into the cytoplasm. 10^3 to 10^4-fold increases in K_M^{app} values between substrates and their phosphates for the corresponding Enzymes II seem typical. An alternative model [23,30,32] assumes that complexing of IIC with its phospho-IIB facilitates reorientation of a loaded binding site from 'outward' to 'inward'. Phosphorylation of the substrate follows and causes dissociation of the substrate into the cytoplasm. In agreement with this model, an increase in the translocation rate by orders of magnitude is observed for high-affinity substrates in the presence of a phospho-IIB complexed to its IIC transporter as determined by in vitro tests. Further, especially *in vivo* tests with intact cells, are needed to decide which model is more accurate.

During transport, ATP-driven (so-called 'ABC-transporters') transport systems apparently transfer a covalently bound high-energy phosphate group from ATP to

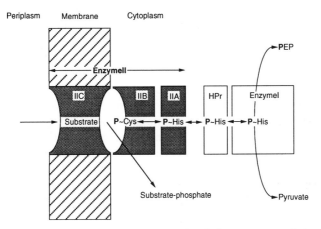

Fig. 4.   Biochemical steps involved in group translocation during transport and phosphorylation of a substrate through a PTS. Indicated are the protein kinase and various phosphocarrier proteins, the residues involved in group transfer, the directionality of the biochemical reactions and the cellular location of the various components.

water. They use the free energy to drive substrate uptake, if necessary against the substrate gradient. PTSs, by contrast, transfer a phospho-group derived from PEP to the substrate (Fig. 4). This remarkably economic procedure not only generates intracellular substrate-phosphate in a PEP-generating ('self-priming') process, but also couples transport directly to a cascade of protein kinases which constitute the central part of a signal transduction pathway as will be described next.

3. The PTS as a central signal transduction system

Complex bacterial activities, e.g., those involved in the quest of food, require coordination of entire metabolic networks [33]. The genes and enzymes involved in carbohydrate uptake and metabolism, in nutrient and energy supply and in cell motility are members of a modulon. The term modulon has been proposed for networks of operons and regulons in which each member is controlled first by individual repressors and activators, and second by a global regulator [34]. This general, pleiotropically active regulator is invariably epistatic, i.e. dominant over individual regulators. These interact with their specific substrates (e.g., β-galactosides induce the *lac* operon, sn-glycerolphosphate the *glp* regulon in *E. coli*). Global systems, in contrast, are coupled to sensors and signal transduction systems which sense global changes in cellular metabolism. Such changes, like feast versus famine and other stress conditions, cause a general alarm. The alarm is often triggered through changes in intracellular pools of metabolites which reflect the activity of metabolic networks. These metabolites or alarmones are frequently derivatives of nucleotides (e.g., cAMP) which are sensed through intracellular sensors [35].

Each molecule transported and phosphorylated via an Enzyme II triggers through group translocation a phospho-relay within the various PTS proteins. Therefore, the PTS with its many Enzymes II (e.g., 16 in *E. coli*), is not only a perfect sensor system to monitor the environment and the metabolic state (feast or famine?) of a cell, but also an optimal signal system to communicate this information to the various global control systems which coordinate the quest of food. Similar to other signal systems, the PTS proteins communicate among themselves and with other proteins through reversible phosphoryl group transfer and through direct protein–protein interactions. Communication is facilitated (i) by the modular construction of the network in which repeating units (domains IIA, B and C, several HPr-like proteins) with similar functions allow a controlled cross-talk (Figs. 1 and 2), and (ii) by the hierarchical construction of the network. The information gathered by the system is integrated and communicated to a global regulatory network which controls most peripheral catabolic pathways and many other functions related to the quest of food, e.g., motility and chemotaxis in enteric bacteria.

3.1. The PTS, a cascade of protein kinases and phosphocarrier proteins

As indicated in Fig. 4 and summarized in Fig. 5 for enteric bacteria, the input side of the PTS as a signal transduction system comprises a series of reversible protein phosphorylation steps which involve phospho-His and phospho-Cys residues. Transport and phosphorylation of any PTS-substrate, a virtually irreversible step, triggers a phospho-relay which causes eventually the dephosphorylation of phospho-EI and requires its PEP-dependent rephosphorylation to reset the system. PEP pools are an important regulatory element in this process. The phosphorylation state of the PTS and in particular of EI, HPr and IIA^{Crr} (IIA^{Glc}) therefore reflects the presence of carbohydrates in the medium (feast), and the glycolytic and gluconeogenic activities as mirrored in PEP pools.

The cascade of protein kinases and phosphocarriers involved in PTS activities are arranged in a hierarchical way (Fig. 5). At the top of this hierarchy is Enzyme I. This PEP-dependent protein kinase corresponds to the catalytic subunit of other, mostly eukaryotic signal transduction systems, also involved in the pleiotropic and global control of enzyme activities and gene expression. Probably to increase its mobility and versatility the relatively large kinase ($M_r \sim 65{,}000–85{,}000$ per monomer) transfers its phosphate group to a His-group of the small (M_r ca. 9,000) phosphocarrier HPr which, as indicated, represents a mobile 'universal joint'. For HPr, IIA^{Crr} and IIA^{Mtl} structures as determined by 2D and 3D nuclear magnetic resonance (NMR) and by X-ray crystallography studies have become available (references in Refs. [8,10]). According to these, amino acids His15 (the residue phosphorylated by EI), Arg17 and Glu85 form the catalytic center of HPr and are located at the surface of the molecule [36]. For both IIAs, the residues involved in binding HPr, are also located at the surface [37–39]. These mostly hydrophobic residues are clustered in IIA^{Crr} within a shallow depression where they surround two essential His residues, one of which (His90 in *E. coli*) accepts the phosphoryl group from phospho-HPr. Although both IIAs interact with phospho-His of HPr, no

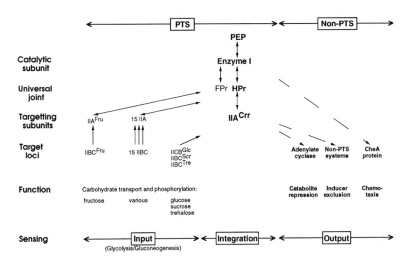

Fig. 5.   The hierarchical structure of the PTS as a signal transduction system with a central role in global control of functions involved in the quest of food. For explanations see text.

consensus motif for the binding site can be found. Apparently, the hydrophobic structure of IIAs is matched by a corresponding structure around His15 of HPr. Both structures together form the binding site and allow phosphoryl group transfer when put together. It is tempting to speculate that similar non-specific (hydrophobic–hydrophobic) contact sites exist for other IIAs and may be typical for proteins at the top of the cascade ('targeting subunits') which have to communicate with many target loci. As expected according to this rule, the consensus sequences for IIB and IIC binding sites ('target loci') are specific (up to 40 structural and hydrophilic residues) and both domains are normally covalently bound to each other [40].

3.2. Signal integration within the PTS

Similar to other sophisticated signal systems involved in the transient response of a sensory system to a stimulus, the PTS must be able to integrate various signals and to adapt to longer lasting stimuli. Central to these processes are the general PTS proteins EI, HPr and IIA^{Crr}. Based on an impressive number of biochemical and genetical experiments, on in vivo as well as in vitro studies with purified proteins, and even on first 3D structures, a picture on the events involved in signal integration by EI, HPr and IIA^{Crr} begins to emerge (references in Refs. [8,9,13,15,16]). Central is the phosphorylation cycle of EI as indicated in Fig. 6. As for other protein kinases, autophosphorylation of EI occurs in the dimeric form. Once phosphorylated, the subunits dissociate into monomers which, at least *in vitro*, phosphorylate HPr in a rapid process [41,42]. In the absence of PTS-dependent transport and a high PEP pool (ca. mM), all IIAs and IIBs (i.e. the targeting subunits and target loci on the 'input' side of Fig. 5) are eventually phosphorylated. This process is facilitated by the reversibility of all phosphorylation reactions involved and by an

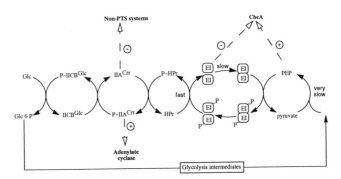

Fig. 6.   Phosphorylation steps involved in signal transduction through the PTS. Upon uptake and phosphorylation of glucose molecules, IIB, IIA, HPr and finally EI monomers are dephosphorylated rapidly (in msec). The process uses up phospho-IIA^{Crr} and prevents the activation (+) of adenylate cyclase ('catabolite repression'). It generates free IIA^{Crr} which inhibits (−) several non-PTS transport systems and catabolic enzymes ('inducer exclusion'). EI monomers dimerize slowly and the dimer autophosphorylates at the expense of PEP. Transiently accumulated free EI then inhibits CheA (−), thus causing positive chemotaxis. Glucose 6-P is converted in a very slow process through glycolysis to PEP which in high amounts activates CheA (+).

excess of HPr and IIA^{Crr} (1–2×10^4 molecules per cell) compared to Enzymes II (2,000 molecules per fully induced cell). The system is in a pre-stimulus equilibrium, the cell is adapted. The actual equilibrium is largely dependent on the intracellular concentrations of essential proteins and molecules and on their kinetic values. The former are not easily determined because they vary up to 20-fold under different conditions. During growth on a PTS-carbohydrate (or anaerobic conditions), e.g., the number of EI, HPr and EIIA^{Crr} is 5-fold higher than during growth on a non-PTS carbohydrate (or aerobic conditions) [43–45], and the corresponding concentrations for PEP are 0.08 and 1.0 mM [46]. For inducible systems, the number of molecules per cell may reach from a few in an uninduced to 8,000 in a fully induced cell. For *S. typhimurium* during growth on glucose, EI has been estimated to 5,000 molecules per cell (2 μM) compared to 20,000 (20–50 μM) for HPr and IIA^{Crr} [43,44,47]. For enteric bacteria, the K_M-value of phospho-HPr to IIA^{Crr} is 0.3 μM, to other IIAs about 1 μM. Different K_M values for phosphorylation of various IIAs or IIA domains by P~HPr could lead to a preferential utilization of one PTS substrate over another. Such a hierarchical utilization, usually with glucose (and sucrose) at the top, has in fact been observed.

During uptake of a few substrate molecules, whose threshold is set basically by the binding affinity of the IIC transporter for its substrates, the excess of P~HPr molecules is probably sufficient to 'buffer' the triggering of a signal. During fast uptake of a substrate, e.g. α-methyl-glucoside through II^{Glc}, the level of P~IIA^{Crr} drops from 80% to 10% [48]. This should cause a rapid dephosphorylation of HPr and of nearly all EI molecules together with a 10–30-fold drop in PEP concentration. In *in vitro* tests, the dimerization of the dephosphorylated EI monomers to dimers becomes under these conditions the limiting step [41,42]. If we assume that this is similar in whole cells, dephosphorylated EI monomers would accumulate. A concomitant drop in PEP concentration would further slow the rephosphorylation of EI and accentuate the dephosphorylation of HPr and IIA^{Crr} as well. Replenishment of the PEP pool through metabolism of the substrate-phosphate molecules taken up during group translocation may finally reset a new equilibrium [49] and cause adaptation of the cell. According to this scenario, transient changes in the phosphorylation level of EI, HPr and IIA^{Crr}, as well as in intracellular PEP concentrations, should constitute the clearest signals emanating from the PTS to be sensed and communicated to global regulators and other sensory systems.

3.3. The PTS signals external stimuli and the metabolic state of a cell to global regulatory and sensory systems

In this section we will discuss four examples which illustrate how cells use the PTS as an elaborate signal transduction system with a central role in the control of the quest of food. The four have been described individually and in detail in recent reviews [8,9,14–16]. Here we will restrict their description to the general principles according to which the information gathered by the PTS is used by a cell to adapt its metabolic activity. The regulatory systems chosen communicate either with IIA^{Crr}, with HPr or with EI itself.

3.3.1. IIA^{Crr} dependent carbon catabolite repression in enteric bacteria

In Fig. 7, the glucose-PTS is drawn in analogy to other complex chemosensors which contain a receptor and a transmitter module, e.g., the well-known two-component systems of bacteria. According to this analogy, the IIC IIB domains constitute the membrane-bound and specific receptor for glucose. Uptake and phosphorylation of a substrate then correspond to the stimulus. It triggers a change in the phosphorylation state of the transmitter, made up by the protein kinase ('catalytic subunit') EI with its phosphocarrier ('universal joint') HPr and its phospho-acceptor ('targeting subunit') IIA^{Crr}. The phosphorylation state of IIA^{Crr} is determined by the rate of phosphorylation through P~HPr, the rate of dephosphorylation through IICB^{Glc} during glucose transport, and (because of the reversibility of the system) through HPr during transport of any PTS carbohydrate through any of the 16 Enzymes II of a cell.

In enteric bacteria, P~IIA^{Crr} is apparently a major activator for the enzyme adenylate cyclase (gene *cyaA*) which catalyzes the synthesis of cAMP from ATP [50–54]. In contrast to the communication between PTS proteins ('input' side of Fig. 5) which is through covalent protein phosphorylation, no IIA^{Crr}-dependent phosphorylation of the adenylate cyclase has been found. In analogy to other (two-component) sensory systems, the output module IIA^{Crr} of the transmitter seems to communicate with its receiver adenylate cyclase and to modulate its activity. Although there can be no doubt that IIA^{Crr} is crucial in this regulation, no details on its mechanism can be given. P~IIA^{Crr} in the phosphorylated state activates the cyclase, thus causing high cAMP-levels under 'famine conditions'. Free IIA^{Crr} lacks the stimulating effect or even inhibits the cyclase, while high levels of PEP

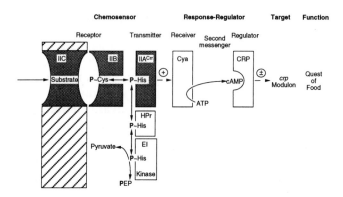

Fig. 7.   IIA^{Crr}-dependent carbon catabolite repression in enteric bacteria. A PTS (symbols as in Fig. 4) is drawn as a chemosensor which comprises a substrate-specific receptor (IIC) and a transmitter (protein kinase) which upon stimulation (uptake and phosphorylation of a substrate) activates the receiver adenylate cyclase (Cya). This enzyme generates the alarmone (or 'second-messenger') 3'-5' cyclic-adenosine-monophosphate (cAMP) which activates the cAMP receptor protein CRP. CRP corresponds to a global activator which controls the expression of all genes which are a member of the crp modulon, and hence the quest of food.

seem to favor cAMP synthesis. The alarmone or, in analogy to eukaryotic systems, 'second-messenger' cAMP binds to the cAMP-accepting-protein (CAP), also called cAMP-receptor-protein (CRP, gene crp). CRP corresponds to a global regulatory protein for the crp-modulon (references in Refs. [55–57]). The cAMP.CRP complex, but not free CRP, binds to a specific consensus present in all promoters of the crp-modulon and allows efficient transcription rates from promoters which are otherwise 'closed' to RNA polymerase binding. This includes the crp gene which is thus autoregulated by its gene product CRP [58]. The addition of glucose to cells growing on a poor carbon source stimulates through an unknown process (deactivation of CRP?) a rapid excretion of cAMP, and through a transient dephosphorylation of IIA^{Crr} the deactivation of the adenylate cyclase. This rapid fall in cAMP (up to 100 fold) and later in CRP level (up to 5 fold) seems to be the major cause for the regulatory phenomena known as transient and as permanent catabolite repression.

3.3.2. IIA^{Crr}-dependent inducer exclusion in enteric bacteria

In enteric bacteria, free but not phosphorylated IIA^{Crr} has been shown to bind to several non-PTS transporters and metabolic enzymes, primarily those involved in the uptake and synthesis of inducers (references in Refs. [7–9]). This binding inhibits the corresponding activities as has been shown with purified IIA^{Crr} and several transporters as well as with purified glycerol kinase ('inducer exclusion'). The 3D-structure of a complex between the tetrameric glycerol kinase and four IIA^{Crr} monomers has been determined [59,60]. The contact between the molecules is confined again to a limited number of hydrophobic residues and resembles the unspecific binding sites between HPr and the IIA domains mentioned before. The contact site includes His90 of IIA^{Crr}. It is conceivable that introduction of a negatively charged phosphate at His90 prevents the complexing of IIA^{Crr} with the glycerol kinase.

Binding of free IIA^{Crr} to the various non-PTS proteins requires the presence of a substrate for the target protein [48,61]. This arrangement prevents the non-productive binding of IIA^{Crr} to proteins and transporters for which no substrates are available and which need not to be inhibited. It guarantees, on the other hand, that the 15,000 odd molecules of IIA^{Crr} in a cell are sufficient to inhibit, if necessary, more than one non-PTS system. The regulation, as expected, is most efficient during partial induction of the target systems or under high catabolite repression caused by a PTS carbohydrate [47]. Under these conditions, the concentration of free IIA^{Crr} is high and the number of target proteins low. Similarly, systems encoded by genes with a low uninduced basal level and a high requirement for cAMP.CRP (e.g., those for glucitol, L-arabinose, glycerol, Krebs cycle intermediates, maltose), or systems with low affinity transport systems (e.g., those for lactose, melibiose, D-xylose) are most sensitive to the concerted action of catabolite repression and inducer exclusion. Systems, however, with a (semi-) constitutive expression of their genes and a high substrate affinity (e.g., those for glucose, N-acetyl-glucosamine, mannose, D-mannitol or gluconate) are more resistant.

3.3.3. HPr-dependent carbon catabolite repression in Bacillus subtilis

B. subtilis ptsHI mutants are pleiotropically negative for several non-PTS carbohydrates and thus resemble such mutants from enteric bacteria [62]. Furthermore, growth in the presence of glucose causes phenomena which resemble carbon catabolite repression in enteric bacteria, i.e., genes for growth on, e.g., xylose; maltose and glycerol are not transcribed despite the presence of an inducer (references in Refs. [15,16]). From mutant and enzyme studies, a regulatory system begins to emerge whose essential features are summarized in Fig. 8. According to these data, carbon catabolite repression in Gram-positive bacteria is complex and probably involves more than one mechanism. Neither a IIA^{Crr}-equivalent function of IIA^{Glc}, nor intracellular cAMP or CRP-like proteins have been found in these bacteria. Instead, HPr and a repressor CcpA, which controls the expression of a larger group of genes are central to the systems described in Fig. 8. The corresponding model is highly speculative, not well supported by data. Some of its features, however, fit remarkably well into our picture of the PTS as a signal system for global regulatory networks, and only these features will be discussed here.

The model assumes that many operons and regulons of *B. subtilis* are repressed through a repressor CcpA (mnemonic for catabolite control protein) [63–65]. Members of this group encode mostly enzymes for peripheral catabolite pathways, e.g., these which become negative in Pts$^-$ mutants. The model (see Fig. 8) assumes furthermore that during growth on poor (non-PTS) carbohydrates, the CcpA repressor is inactive and the genes consequently fully derepressed. Under such growth conditions ('famine'), all PTS proteins are in the phosphorylated state at the usual histidine and cysteine residues. During uptake of a PTS carbohydrate, in particular of glucose ('feast'), pools of dephosphorylated IIAs, HPr and EI molecules

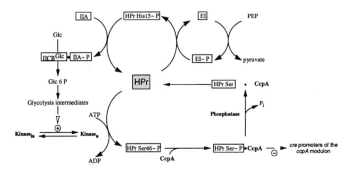

Fig. 8.   HPr-dependent carbon catabolite repression in *B. subtilis*. During growth on non-PTS
carbohydrates ('famine', upper part), the PTS proteins are in a fully phosphorylated state, in particular
HPr containing the usual phospho-His15 residue. During growth on PTS substrates ('feast', lower
part) HPr is dephosphorylated and glycolytic intermediates accumulate. This activates an ATP-dependent protein kinase which phosphorylates HPr at Ser46. In this form, HPr complexes to a global
regulator CcpA which then represses all genes which are member of the *ccpA* modulon ('catabolite
repression'). A phosphatase completes the cycle.

accumulate transiently and glycolytic intermediates (in particular Glc6P, Fru6P,
Fru1,6BP, PEP) build up. These intermediates apparently activate an ATP-dependent protein kinase, which phosphorylates HPr molecules with a free, non-phosphorylated His15 at the highly conserved Ser46 residue [62,66]. HPr Ser-P then seems
to activate the repressor CcpA through non-covalent complexing and triggers its
binding to a consensus sequence of various promoters (*cre*, mnemonic for catabolite responsive element) ([67,68; and references in Ref. [15]). As a consequence,
catabolite repression prevents the transcription of *cre* promoters. The system mirrors the regulation of the *crp* modulon in enteric bacteria in which feast conditions
cause deactivation of the CRP activator (i.e. catabolite repression), while the same
conditions here cause the activation of the CcpA repressor and hence catabolite
repression. Accordingly, all operons and regulons containing the *cre* consensus
sequence in their promoters are members of a modulon which is controlled at the
transcription level by the global regulator CcpA. We propose to call this group the
ccpA modulon of *B. subtilis* in analogy to the *crp* modulon of enteric bacteria. In
both systems, carbon catabolite repression is eventually caused by a non-covalent
protein-protein interaction between a PTS protein and a global regulator ('output'
in Fig. 5) and thus represents faithfully the metabolic state of a cell.

3.3.4. EI-dependent chemotaxis towards PTS-carbohydrates

In enteric bacteria, positive gradients of PTS-carbohydrates elicit smooth swim-
ming or prolonged runs and counter-clockwise (CCW) flagellar rotation. Attractants cause runs by suppressing tumbles or random movements which redirect a cell
in a new direction. Tumbles are caused by clockwise (CW) flagellar rotation. As a
result of prolonged runs and suppressed tumbles ('phobotaxis'), attractants elicit a
positive chemotaxis (references in Refs. [69,70]). As summarized in Fig. 9, a
phospho relay is involved in tumble generation [71,72]. This relay involves first a

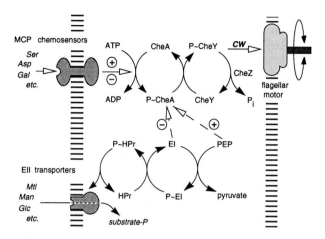

Fig. 9. EI dependent chemotaxis towards PTS-carbohydrates. The upper part of the figure shows the protein phosphorylation reactions modulated by methyl-accepting-protein (MCP) chemosensors, the protein kinase CheA, the receiver CheY and a phosphatase CheZ to elicit flagellar responses (cw, clock-wise rotation) and chemotaxis. The lower part shows in simplified form the reactions involved in the uptake and phosphorylation of carbohydrates by the PTS as described in detail in Fig. 6. The chemotactic model postulates a negative (−) cross-circuiting signal between unphosphorylated EI and CheA, i.e. the generation of phospho-CheY and of tumble movements is prevented ('positive chemotaxis'). A weak activation of CheA by PEP (+) could constitute a second cross-circuiting signal.

protein kinase ('transmitter') CheA and second a phosphate receiver CheY. CheA autophosphorylates in the dimeric form at a histidine residue, using ATP as the phosphodonor, and subsequently donates the phosphate group to an aspartate residue in CheY. Phosphorylation of CheY induces a conformational change that enables it to interact with the flagellar switch and trigger CW rotation, CCW being the default state. P-CheY is short-lived because of a self-catalyzed hydrolysis, a reaction augmented by the protein phosphatase CheZ.

The flux of phosphate groups through this signalling pathway is controlled by modulation of the CheA autophosphorylation rate in response to changes in ligand occupancy at a set of transmembrane receptors known as methyl-accepting chemotaxis proteins (MCPs). These receptors do not transport carbohydrates into the cell, but rather measure their external levels as the cell swims about. Stimulus information is signalled across the membrane to a cytoplasmic domain which communicates through an ancillary protein CheW with the transmitter CheA. An increase in attractant concentration at an MCP causes inhibition of CheA and smooth-swimming, whereas a drop in attractant level (or the presence of a repellent) stimulates CheA and thus triggers a tumble movement [73,74].

PTS carbohydrates are sensed as chemoeffectors during the uptake process by a different process (references in Refs. [8,69,70]). Neither the binding of a substrate to an Enzyme II, nor the generation of intracellular carbohydrate-phosphate, nor its subsequent degradation, are sufficient to trigger a chemotactic response; nor are MCPs required for PTS chemotaxis. In contrast, the general PTS proteins EI and

HPr are necessary for uptake of PTS carbohydrates and also required, together with CheA and CheY, for a chemotactic response to them. Recent *in vivo* and *in vitro* studies with purified PTS and Che proteins [14,75] confirm an older model [76,77] according to which the flagellar signal in PTS chemotaxis derives from an uptake driven change in phosphate flux through EI and HPr. They indicate that during uptake of a PTS-carbohydrate through any Enzyme II, EI is dephosphorylated more rapidly by HPr than it is rephosphorylated at the expense of PEP. The effect is reinforced (as described in Section 3.2) by a concomitant drop in PEP concentration. Consequently, unphosphorylated EI builds up. Tests with purified proteins revealed that free EI inhibited CheA autophosphorylation, whereas phosphorylated EI, HPr or P~HPr did not. Uncoupled HPr mutants described before, which were able to transport PTS substrates but unable to respond chemotactically to them, exhibited *in vitro* reduced phosphotransfer rates from EI. This also indicates that the building up of free EI consequent to PTS transport and phosphorylation could be the signalling link between the PTS and the MCP phospho relays. As expected according to the model, overexpression of such uncoupled HPr mutants restored *in vivo* the chemotactic response by reincreasing the dephosphorylation rate of EI during transport. In the CheA autophosphorylation assays there was no detectable transfer of [^{32}P] to either EI or HPr, thus discounting the possibility that phosphates are transferred from CheA to EI, nor was there any indication for such a transfer between P-CheY and general PTS proteins. In terms of this model, adaptation to PTS-mediated stimuli would be either the restoration of pre-stimulus levels of unphosphorylated EI through the PEP cycle or a change of CheA to offset EI inhibition.

These results suggest that free EI may inhibit the autophosphorylation of CheA directly in a manner analogous to MCP signalling. There are two functional domains of EI. The amino-terminal phosphotransfer domain of EI with its autophosphorylation site (His189), apparently interacts with HPr through non-specific ('hydrophobic–hydrophobic') binding, whereas the carboxy-terminal domain seems to be involved in dimerization [78].

The CheA molecule consists of at least four functional domains with intervening linkers [79]. The carboxy-terminal part of CheA is responsible for the coupling to the MCP chemoreceptors, whereas the amino-terminal P1 domain which contains the autophosphorylation site (His48) interacts with the centrally located transmitter domain. These two domains are separated by the P2 domain, which seems to be involved in the recognition of the various phosphorylation targets by CheA. It is tempting to speculate that EI can block the interaction (steric hindrance) between the autophosphorylation site and the transmitter domain by binding itself to P1 or P2. This mechanism would be the fourth example of a communication between PTS- and non-PTS-proteins ('output' in Fig. 5) through non-covalent interaction rather than by protein phosphorylation as seems to be usual among PTS-proteins ('input' in Fig. 5). An apparent exception to this rule seems to be the cross-regulation between the IIBA^{Bgl} domains of the β-glucoside PTS with the anti-terminator BglG. It has been claimed that its regulatory activity is modulated by phosphorylation/dephosphorylation (references in Ref. [8]).

4. Concluding remarks on pro- and eukaryotic signal transduction systems

It is generally believed that the typical eukaryotic signal transduction system contains as its central part ('catalytic subunit') a protein kinase (usually a tyrosine or a serine/threonine kinase) and a protein phosphatase (references in Refs. [80, 81]). Both modulate through 'targeting subunits' the activities of usually a large variety of 'target proteins'. The system often responds to extracellular stimuli, e.g., chemicals, hormones, neurotransmitters, growth factors or morphopoetic regulators, which are sensed by membrane-bound sensors and signalled to the catalytic units in the form of second-messengers such as cAMP and cGMP. Typical target loci are ion channels involved in sensory procedures and transcription regulators. The genes regulated by such systems often correspond to genes which control cellular differentiation at a higher hierarchical level, e.g., 'maternal', 'gap' and 'homeobox' genes of *Drosophila* or cell division genes of many plants and fungi.

Only now, more than 30 years after its discovery, do we realize how exactly the PTS with its many target loci fits into the 'eukaryotic' scheme. For a free-living unicellular organism, 'outside' normally signifies the physical universe and not a neighboring cell as in the eukaryotes. The PTS with its many transport systems thus is an optimal sensory system to monitor the environment. The information gathered through it is integrated and communicated by means of phospho-histidines (rarely -cysteines) to the general proteins IIA^{Crr}, HPr and the protein kinase EI, and from these to a group of target proteins. These include the global regulators CRP for the *crp* and CcpA for the *ccpA* modulon which control a plethora of cellular functions. The similarities with the eukaryotic systems cannot be purely accidental. Tyrosyl- and seryl/threonyl-specific kinases with a role in bacterial differentiation seem to be common [82–84], while histidyl-kinases which closely resemble bacterial enzymes have been detected in fungi and in green plants [81]. The similarities rather seem to reflect the existence of an old and universal biochemistry present throughout the pro- and the eukaryotic world which allows the coupling between complex sensory reception systems and global regulatory networks through hierarchical signal transduction systems. This ensures the optimal adaptation of a cell to its environment.

Abbreviations

cAMP, 3'-5', cyclic-adenosine-monophosphate
EI, Enzyme I
EII, Enzyme II
HPr, Histidine protein
PEP, phospho*enol*pyruvate
PTS, phosphotransferase system.

Acknowledgements

We would like to thank E. Placke for help in preparing the manuscript. We also thank the Deutsche Forschungsgemeinschaft and the Volkswagen Stiftung for financial support.

References

1. Kundig, W., Gosh, S. and Roseman, S. (1964) Proc. Natl. Acad. Sci. USA **52**, 1067–1074.
2. Tanaka, S. and Lin, E.C.C. (1967) Proc. Natl. Acad. Sci. USA **57**, 913–919.
3. Egan, J.B. and Morse, M.L. (1966) Biochim. Biophys. Acta **112**, 63–73.
4. Murphy, W.H. and Rosenblum, E.D. (1964) Proc. Soc. Exp. Biol. Med. **116**, 544–548.
5. Kundig, W., Kundig, F.D., Anderson, B. and Roseman, S. (1966) J. Biol. Chem. **241**, 3243–3246.
6. Danchin, A, (1989) FEMS Microbiol. Rev. **63**, 1–200.
7. Meadow, N.D., Fox, D.K. and Roseman, S. (1990) Annu. Rev. Biochem. **59**, 497–542.
8. Postma, P.W., Lengeler, J.W. and Jacobson, G.J. (1993) Microbiol. Rev. **57**, 543–594.
9. Saier Jr., M.H. (1993) J. Cell Biochem. **51**, 1–90.
10. Lengeler, J.W., Jahreis, K. and Wehmeier, U.F. (1994) Biochim. Biophys. Acta **1188**, 1–28.
11. Postma, P.W. and Roseman, S. (1976) Biochim. Biophys. Acta **457**, 213–257.
12. Saier Jr., M.H. and Roseman, S. (1972) J. Biol. Chem. **247**, 972–975.
13. Roseman, S. and Meadow, N.D. (1990) J. Biol. Chem. **265**, 2993–2996.
14. Lengeler, J.W., Bettenbrock, K. and Lux, R. (1994) in: Phosphate in Microorganisms. Cellular and Molecular Biology, eds A.M. Torriani-Gorini, E. Yagil and S. Silver. ASM Press, Washington D.C., pp. 192–188.
15. Hueck, C.J. and Hillen, W. (1995) Molec. Microbiol. **15**, 395–401.
16. Saier Jr., M.H., Chauvaux, S., Deutscher, J., Reizer, J. and Ye, J-J. (1995) TIBS **20**, 267–272.
17. Saier Jr., M.H. (1994) Microbiol. Rev. **58**, 71–93.
18. Kyte, J. and Doolittle, R.F. (1982) J. Mol. Biol. **157**, 105–132.
19. von Heijne, G. and Nilsson, I. (1990) Cell **62**, 1135–1141.
20. Sugiyama, J.E., Mahmoodian, S. and Jacobson, G.R. (1991) Proc. Natl. Acad. Sci. USA **88**, 9603–9607.
21. Buhr, A. and Erni, B. (1993) J. Biol. Chem. **268**, 11599–11603.
22. Grisafi, P.L., Scholle, A., Sugiyama, J., Briggs, C., Jacobson, G.R. and Lengeler, J.W. (1989) J. Bacteriol. **171**, 2719–2727.
23. Lolkema, J.S., Dijkstra, D.S., ten Hoeve-Duurkens, R.H. and Robillard, G.T. (1990) Biochemistry **29**, 10659–10663.
24. Briggs, C.E., Khandekar, S.S. and Jacobson, G.R. (1992) Res. Microbiol. **143**, 139–149.
25. Postma, P.W. (1981) J. Bacteriol. 147, 382–389.
26. Ruijter, G.J.G., Postma, P.W. and van Dam, K. (1990) J. Bacteriol. **172**, 4783–4789.
27. Ruijter, G.J.G., van Meurs, G., Verwey, M.A., Postma, P.W. and van Dam, K. (1992) J. Bacteriol. **174**, 2843–2850.
28. Pas, H.H., ten Hoeve-Duurkens, R.H. and Robillard, G.T. (1988) Biochemistry **27**, 5520–5525.
29. Lolkema, J.S., Dijksstra, D.S. and Robillard, G.T. (1992) Biochemistry, **31**, 5514–5521.
30. Lolkema, J.S., ten Hoeve-Duurkens, R.H., Dijkstra, D.S. and Robillard, G.T. (1991) Biochemistry, **30**, 6716–6721.
31. Elferink, M.G.L., Driessen, A.J.M. and Robillard, G.T. (1990) J. Bacteriol. **172**, 7119–7125.
32. Lolkema, J.S., Wartna, E.S. and Robillard, G.T. (1993) Biochemistry, **32**, 5848–5854.
33. Gottesmann, S. (1984) Ann. Rev. Genet. 18, 415–441.
34. Iuchi, S. and Lin, E.C.C. (1988) Proc. Natl. Acad. Sci. USA **85**, 1888–1892.
35. Neidhardt, F.C., Ingraham, J.L. and Schaechter, M. (1990) Physiology of the Bacterial Cell. A Molecular Spproach. Sinauer Associates Inc., Sunderland, Mass.
36. Sharma, S., Georges, F., Delbaere, L.T.J., Lee, J.S., Klevit, R.E. and Waygood, E.B. (1991) Proc. Natl. Acad. Sci. USA **88**, 4877–4881.
37. Presper, K.A., Wong, C.Y., Liu, L., Meadow, N.D. and Roseman, S. (1989) Proc. Natl. Acad. Sci. USA **86**, 4052–4055.
38. Worthylake, D., Meadow, N.D., Roseman, S., Liao, D.-I., Herzberg, O. and Remington, S.J. (1991) Proc. Natl. Acad. Sci. USA **88**, 10382–10386.
39. Van Nuland, N.A.J., Kroon, G.J.A., Dijkstra, K., Wolters, G.K., Scheek, R.M. and Robillard, G.T. (1993) FEBS Lett. **315**, 11–15.

40. Lengeler,, J.W., Titgemeyer, F., Vogler, A.P. and Wöhrl, B.M. (1990) Phil. Trans. R. Soc. Lond. B **326**, 489–504.
41. Weigl, N., Kukuruzinska, M.A., Nakazawa, A., Waygood, E.B. and Roseman, S. (1982) J. Biol. Chem. **257**, 14477–14491.
42. Chauvin, F., Brand, L. and Roseman, S. (1994) J. Biol. Chem. **269**, 20270–20274.
43. Scholte, B.J., Schuitema, A.R.J. and Postma, P.W. (1982) J. Bacteriol. **149**, 576–586.
44. Matoo, R.L. and Waygood, E.B. (1983) Can. J. Biochem. **61**, 29–37.
45. De Reuse, H. and Danchin, A. (1988) J. Bacteriol. **170**, 3827–3837.
46. Lowry, O.H., Carter, J., Ward, J.B. and Glaser, L. (1971) J. Biol. Chem. **246**, 6511–6521.
47. Van der Vlag, J., Van Dam, K. and Postma, P.W. (1994) J. Bacteriol. **176**, 3518–3526.
48. Nelson, S.O., Schuitema, A.R.J. and Postma, P.W. (1986) Eur. J. Biochem. **154**, 337–341.
49. Pertierra, A.G. and Cooper, R.A. (1977) J. Bacteriol. **129**, 1208–1214.
50. Harwood, J.P., Gazdar, C., Prasad, C., Peterkofsky, A., Curtis, S.J. and Epstein, W. (1976) J. Biol. Chem. **251**, 2464–2468.
51. Feucht, B,U. and Saier Jr., M.H. (1980) J. Bacteriol. **141**, 603–610.
52. Reddy, P., Meadow, N., Roseman, S. and Peterkofsky, A. (1985) Proc. Natl. Acad. Sci. USA **82**, 8300–8304.
53. Liberman, E., Saffen, D., Roseman, S. and Peterkofsky, A. (1986) Biochem. Biophys. Res. Commun. **141**, 1138–1144.
54. Peterkofsky, A., Svenson, I. and Amin, N. (1989) FEMS Microbiol. Rev. **63**, 103–108.
55. Pastan, I. and Adhya, S. (1976) Bacteriol. Rev. **40**, 527–551.
56. Botsford, J.L. and Harman, J.G. (1992) Microbiol. Rev. **56**, 100–122.
57. Kolb, A., Busby, S., Buc, H., Garges, S. and Adhya, S. (1993) Annu. Rev. Biochem. **62**, 749–795.
58. Ishizuka, H., Hanamura, A., Inada, T. and Aiba, H. (1994) EMBO J. **13**, 3077–3082.
59. Hurley, J.H., Worthylake, D., Faber, H.R., Meadow, N.D., Roseman, S., Pettigrew, D.W. and Remington, S.J. (1993) Science **259**, 673–677.
60. Pettigrew, D.W., Frese, M., Meadow, N.D., James Remington, S. and Roseman, S. (1994) in: Phosphate in Microorganisms, Cellular and Molecular Biology, eds A. Torriani-Goriini, E. Yagil and S. Silver, pp. 335–342, ASM Press, Washington, D.C.
61. Osumi, T. and Saier Jr., M.H. (1982) Proc. Natl. Acad. Sci. USA **79**, 1457–1461.
62. Deutscher, J., Reizer, J., Fischer, C., Galinier, A., Saier Jr., M.H. and Steinmetz, M. (1994) J. Bacteriol. **176**, 3336–3344.
63. Henkin, T.M., Grundy, F.J., Nicholson, W.L. and Chambliss, G.H. (1991) Mol. Microbiol. **5**, 575–584.
64. Chambliss, G.H. (1993) in: Bacillus subtilis and Other Gram-positive Bacteria: Biochemistry, Physiology and Molecular Genetics, eds A.L. Sonenshein, J.A. Hoch and R. Losick, pp. 213–219, ASM, Washington, D.C.
65. Miwa, Y., Saikawa, M. and Fujita, Y. (1994) Mikrobiology **140**, 2567–2575.
66. Deutscher, J., Küster, E., Bergstedt, U., Charrier, V. and Hillen, W. (1995) Molec. Microbiol. **15**, 1049–1053.
67. Weickert, M.J. and Chambliss, G.H. (1990) Proc. Natl. Acad. Sci. USA **87**, 6238–6242.
68. Hueck, C.J., Hillen, W. and Saier Jr., M.H. (1994) Res. Microbiol. **145**, 503–518.
69. Armitage, J.P. (1993) in: Signal Transduction. Prokaryotic and Simple Eukaryotic Systems, eds J. Kurjan and B.L. Taylor, pp. 43–65, Academic Press Inc., London.
70. Titgemeyer, F. (1993) J. Cell. Biochem. **51**, 69–74.
71. Bourret, R.B., Borkovich, K.A. and Simon, M.I. (1991) Annu. Rev. Biochem. **60**, 401–441.
72. Parkinson, J.S. (1993) Cell **73**, 857–871.
73. Borkovitch, K.A., Kaplan, N., Hess, J.F. and Simon, M.I. (1989) Proc. Natl. Acad. Sci. USA **86**, 1208–1212.
74. Ninfa, E.G., Stock, A., Mowbray, S. and Stock, J. (1991) J. Biol. Chem. **266**, 9764–9770.
75. Lux, R., Jahreis, K., Bettenbrock, K., Parkinson, J.S. and Lengeler, J.W. (1995) Proc. Natl. Acad. Sci. USA **92**, 11583–11587.
76. Lengeler, J.W., Auburger, A-M., Mayer, R. and Pecher, A. (1981) Mol. Gen. Genet. **183**, 163–170.
77. Grübl, G., Vogler, A.P. and Lengeler, J.W. (1990) J. Bacteriol. **172**, 5871–5876.

78. LiCalsi, C., Crocenzi, T.S., Freire, E. and Roseman, S. (1991) J. Biol. Chem. **266**, 19519–19527.
79. Parkinson, J.S. and Kofoid, E.C. (1992) Annu. Rev. Genet. **26**, 71–112.
80. Hubbard, M.J. and Cohen, P. (1993) Trends Biochem. Sci. **18**, 172–177.
81. Alex, L.A. and Simon, M.I. (1994) Trends Genet. **10**, 133–138.
82. South, S.L., Nichols, R. and Montie, T.C. (1994) Molec. Microbiol. **12**, 903–910.
83. Matsumoto, A., Hong, S-K., Ishizuka, H., Horinouchi, S. and Beppu, T. (1994) Gene **146**, 47–56.
84. Freestone, P., Grant, S., Toth, I. and Norris, V. (1995) Molec. Microbiol. **15**, 573–580.

CHAPTER 26

Ion Selectivity and Substrate Specificity: The Structural Basis of Porin Function

J.P. ROSENBUSCH

Department of Microbiology, Biozentrum,
University of Basel, Klingelbergstr. 70, Basel, Switzerland

© *1996 Elsevier Science B.V.*
All rights reserved

Handbook of Biological Physics
Volume 2, edited by W.N. Konings, H.R. Kaback and J.S. Lolkema

Contents

1. Introduction

Porins are proteins that form water-filled channels across outer membranes of Gram-negative bacteria. These porins are unusually stable, a significant factor in their early crystallization [1]. Porins solved to high resolution (1.8–3.1 Å) include two examples from photosynthetic bacteria [2,3], and three from *Escherichia coli* [4,5]. All of them are homotrimers, although this does not necessarily pertain to all other proteins of the bacterial outer membrane [6]. Some members of the porin family reveal high sequence homology [7], while others do not: the porin from *Rhodobacter capsulatus* [8], as well as maltoporin from *E. coli* [9] do not exhibit significant homologies between themselves, nor with the porins belonging to a group with high sequence conservation (60–70%). The latter include matrix porin (the product of the *ompF* gene [10]), osmoporin and phosphoporin (the products of the *ompC* and *phoE* genes, respectively [7]. Nonetheless, their architecture is similar overall [11], a finding which is surprising also in view of their rather distinct function.

For the present consideration, which is concerned with the structural basis of the functional characteristics of porins, we focus on two examples that were investigated in Basel for over twenty years: matrix porin [12], an apparently nonspecific pore that appears to allow facilitated diffusion of small (<600 Da), polar molecules (for a review, see Ref. [13]), and maltoporin [14], the function and topology of which has been studied extensively [15]. In addition to its general pore function, the former is voltage-gated [16], a property it shares with homologous porins. Its channels are moderately cation-selective, while phosphoporin has a preference for anions [17], such as phosphates. Maltoporin fulfils its porin function by allowing the passage of small nutrients through its molecular sieve [18], yet it also facilitates the diffusion of malto-oligosaccharides (maltodextrins) which exceed in size the nominal exclusion limit [19]. Matrix porin may thus be taken as a paradigm of an unspecific porin, facilitating diffusion at rates proportional to solute concentration, while the rates of the specific substrates of maltoporin follow saturation kinetics [13]. Their respective structures, to be discussed presently, reveal that their properties, and some more subtle characteristics not yet mentioned, can be understood on the basis of the structures, and their modes of action can be simulated and the resulting models subjected to rigorous test by tailor-designed mutants.

2. Overall folding pattern and protein sequences

The sequences, shown in Fig. 1 as two-dimensional representations based on the topologies of the respective proteins, reveal that matrix porin (panel A) and maltoporin (panel B) exhibit similar arrays of β-strands that cross the hydrophobic core

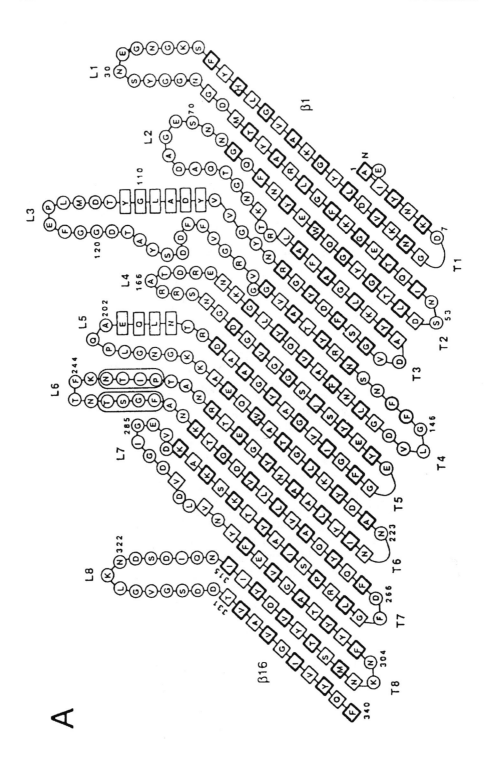

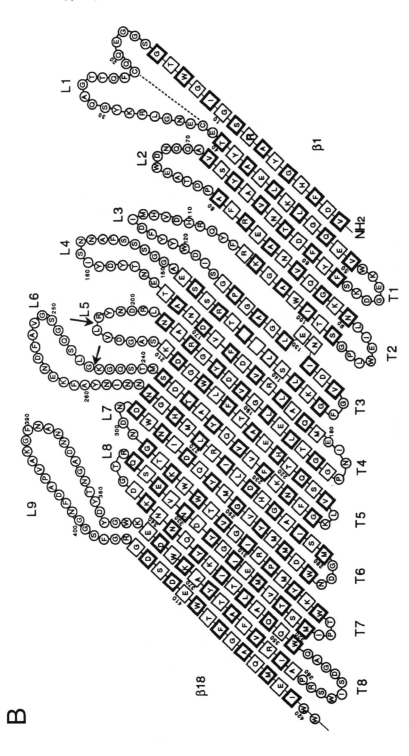

Fig. 1. The sequences of the polypeptides of matrix porin (A) and maltoporin (B) unrolled as 2-dimensional arrays.

of the membrane by 8–9 residues (squares), with aromatic rings demarcating the membrane boundaries. These arrangements should be imagined as folding into a barrel whose sheets are oriented perpendicular to the plane of the membrane. Individual strands are tilted by 40° within the sheet. The side-chains of every other residue in the transmembrane zone are exposed to a medium with a low dielectric constant (bold squares) and are themselves hydrophobic: most of the residues involved are also devoid of a hydrogen bonding potential. Hydrogen bond donors and acceptors in the backbones are saturated by bonds to the corresponding groups in both neighbouring strands, similar as it is found in silk (in a planar array). The intervening residues (light squares) are oriented towards the inside of the barrel where they either contribute to the channel wall, and hence interact with water, or where van der Waals contacts exist with residues present within the channel lumen. The intervening residues may thus be either hydrophilic or hydrophobic, further reducing the constraints for search algorithms of folding patterns. Similarly, the frame may be displaced by so-called β-bulges (residues 2; 81–82; 290–291 in panel A; 130–131 in panel B), causing further impediment for simple structure predictions. The individual β-strands are linked by short turns (T) at the periplasmic end (bottom of the panels), and longer loops (L) which are exposed at the cell surface and vary somewhat in length. Loops 3 in both porins bend into the channel and contribute critically to the constriction of the lumen. By this bending, a gap is formed in the barrel wall which is filled in both porins by loops 2 (#L2) from a neighbouring subunit which latches into the notch.

```
          --β1-       . .-----β2----........--β3-------.        ------β4---
LamB      VDFHG--32--CETYAELKLGQEVWKEGDKSFYFDTNVAYSVA--10--AFREANVQGKN
OmpF  --10--VDLYG--21--DMTYARLGFKGETQINSDLTGYGQWEYNFQGN--11--KTRLAFAGLKY
          --β1-       .------β2------.-----β3-----...         ...---β4---


          ---β5---.      ....    .----β6--    ---β7---     ----β8----
LamB  -8--TIWAGKRFY--3--DVHM--8--SGPGAGLEN--5--GKLSLAAT--24--NDVFDVRLAQ-
OmpF  -3--GSFDYGRNY-10--DMLP-16--VGGVATYRN--9--LNFAVQYL--12--GDGVGGSISY-
          ---β5--..    . ....    ...---β6--    ---β7---     ----β8----


          -----β9-----     .-----β10-----    .-----β11-----
LamB  -6--GTLELGVDYGRA--12--SKDGWLFTAEHTQS--2--KGFNKFVVQYATDS--28-
OmpF  -4--FGIVGAYGAADR--11--GKKAEQWATGLKYD--2--NIYLAANYGETRNA--14-
          -----β9-----     ...---β10-----    ------β11----.


      .----β12----      ------β13-----...    --β14--     -β15-
LamB  -GHMLRILDHGAI--6--DMMYVGMYQDINWDNDN--3--WWTVGIR--15--GYDNV--89--
OmpF  -NKTQDVLLVAQY--6--RPSIAYTKSKAKDVEGI--5--VNYFEVG--15--IINQI--21--
      .----β12----      ------β13---.....    ..-β14-     β15...
```

Fig. 2. Sequence comparison of maltoporin (LamB) and matrix porin (OmpF). (Courtesy T. Schirmer, unpublished.)

Apart from these similarities, distinct differences exist between the two molecules. There are 18 transmembrane strands in maltoporin (panel B), as compared to 16 in matrix porin (panel A). In porin, the carboxyterminal end, located on strand β16, forms a salt-bridge to the amino-terminus on the first half-strand β1*, thus forming a quasi-closed barrel. In maltoporin, the amino- and the carboxytermini are also in proximity, but on strands which are adjacent to each other. In the latter protein, a disulfide bridge exists at the exterior surface of the cell which links residues Cys_{22} and Cys_{38} in L1. A peculiarity of matrix porin is that in the loops, several short peptide segments exhibit limited periodic structures. Thus, α-helical conformations (rectangles) exist in L3 and L5, while two peptide segments occur in a short β-structure in L6. Despite these differences, the similarities in both porins suggest divergent evolution, albeit rather distant. The protein sequences in Fig. 2 show the parts of the two sequences which are superimposable on the basis of their tertiary structure. Fifteen strands (β1–15) show analogous segments, but interstrand lengths (small figures between β-strands) often deviate significantly. A number of prediction algorithms have been devised [20–22] but are not immediately relevant in the present context.

3. The architecture of porin molecules

The space-filling models of matrix porin (Fig. 3, panel a) and maltoporin (panel b) are based on their X-ray structures and exhibit once more striking resemblance, even though the differences are obvious. Both molecules are encircled by hydrophobic belts in areas where contacts to the membrane core are close. These zones are approximately 25 Å in width, and aromatic bands demarcate the lipid-water interphase (for better visualization, the carbon atoms in aromatic residues are shown in white). The domains in the periplasmic space are polar and small, reflecting short turns. The extracellular domains in both porins are larger and more polar. Some anionic groups may be linked to phospholipids or lipopolysaccharides (glycolipids) by bridges formed by divalent cations. Positive charges may be linked directly by salt-bridges to the negative charges of the lipopolysaccharides, the lipid moiety of which constitutes the outer leaflet of the bilayer. Outside of the membrane core, and located on β-strands in register with residues that are exposed to lipids within membrane boundaries, are ionizable; they occur rather frequently. Typical examples are the three lysyl residues which exist next to each other outside the membrane boundary, seen in the β13 strand of matrix porin (Fig. 1A). In the crevices between subunits, several aromatic residues are visible, particularly in matrix porin. They contribute, in part, to the multiple hydrophobic interactions between neighbouring subunits, which, in the center of the trimers, form a globular, closely packed protein core with a mass of about 15 kDa. These interactions contribute significantly to the stability of porins.

Figure 4 shows a schematic representation of a single monomer of maltoporin. Rather unexpectedly, it reveals three residues (N228/D274/Y288) that interact directly with the acyl chains of the lipid (encircled in the figure). The hydrogen bonding potential of their side-chains is saturated by mutual H-bonds. This arrangement is similar to the H-bonding in a periodically arrayed backbone and appears

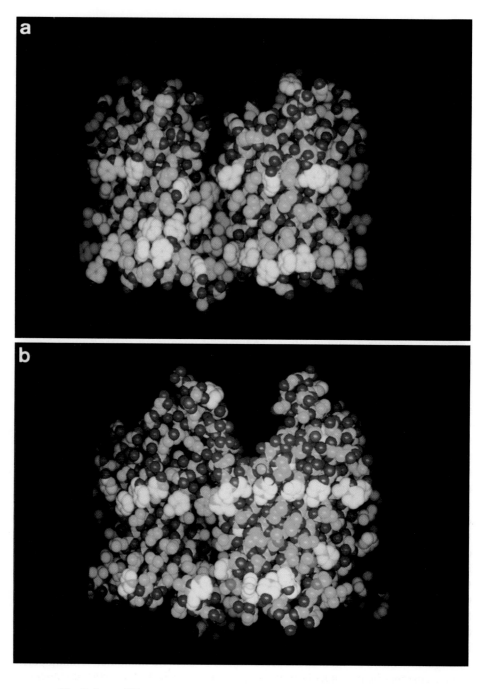

Fig. 3. Space filling models of matrix porin (a) and maltoporin (b) trimers.

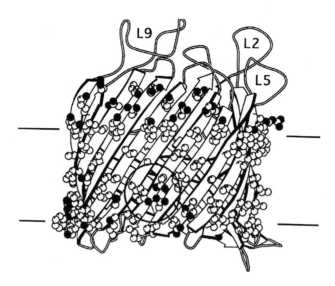

Fig. 4. Ribbon representation of a maltoporin monomer.

hydrophobic to its surroundings [23].

The representation in Fig. 5a shows the comparison of sections of matrix porin (red) and maltoporin (black). The close similarity of the outline of the β-barrels is once more remarkable. Due to two additional β-strands in each maltoporin monomer, an extension of the elliptical shape of matrix porin to the kidney-bean shape in maltoporin is notable (marked with an *asterisk* in the topmost subunit in the figure). The side-chains of the three subunits interact closely around the molecular 3-fold axis (*triangle*) of the trimer. This, as well as the contacts between subunits that are more remote from the symmetry axis also contribute to the protein's stability. In each subunit, the constricting loop (L3) near the central plane of the membrane compartmentalize the barrel into two sections: the water-filled channel proper (open diamond), and the compartment between L3 and the barrel wall (around the arrow), filled with side-chains. The tips of loop 3 are nearly superimposed in matrix porin and in maltoporin, despite the fact that the latter contains 18 residues, while the former consists of 33. In matrix porin, L3 contains an α-helical segment which closely follows the barrel wall. In either case, the space between L3 and the wall is filled by residues which form both hydrophobic contacts as well as hydrogen bonds (not drawn). The tips of L3 (to the right of the arrow) are in close contact with the barrel walls. An interesting feature exists in matrix porin near the tip of L3 (at the position of the arrow): a hydroxyl group of a serine initiates a series of hydrogen bonds by linkage to two consecutive carboxylate groups (S272–E296–D312). D312 is in turn hydrogen-bonded to two main-chain amide groups in L3 (E117, F118). The solid sphere indicates the position of the tip (#D74) of the latch in #L2 from the neighbouring subunit, which forms a salt-bridge to the cationic cluster on the barrel wall.

The stereo-representation in Fig. 5b illustrates many of the observations outlined so far. It also reveals (though it is difficult to see) that in maltoporin, three

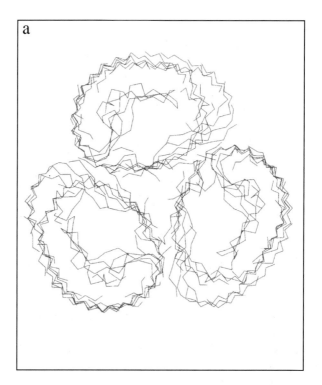

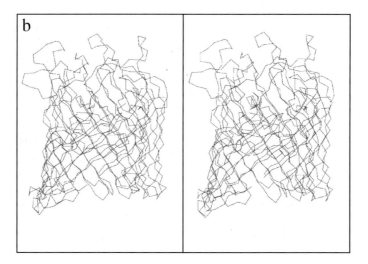

Fig. 5.   Structural comparison of matrix porin (red) and maltoporin (black). Panel (a): a section across the trimers approximately in the center of the membrane. Panel (b): Stereo-view of the two superimposed monomers. (Courtesy T. Schirmer, unpublished.)

loops, in addition to L3, also contribute to the channel constriction: L1 and L6 from the same subunit, and the latch (#L2) from a neighbouring subunit. The area in the constriction site, with its diameter of 5×5 Å, is thus much smaller compared to that in matrix porin (7×11 Å). This accounts largely for the five-fold reduced conductance of maltoporin, of 0.15 nS as compared to 0.85 nS in matrix porin.

4. Functional properties of porins

Can the functional characteristic of the two types of channels, slightly ion-selective but non-specific (matrix porin), and rather specific for malto-oligosaccharides (maltoporin) now be explained on the basis of their structure? The dimensions of the constriction in matrix porin (Fig. 6a) appear large enough to allow unhindered passage of ions (flux about 10^8 ions/sec/channel), comparable to that observed in gramicidin [24]. The rate of glucose permeation (50 molecules/channel/sec; see [16]) though obviously much smaller due to its modest dipole moment [25], appears hardly retarded relative to the diffusion coefficient of this sugar in bulk solvent. This indicates that collisions with the channel wall, and exchange of water of hydration are rare events. In this regard, it is interesting to observe that the strong electrostatic field, originating primarily from the charge segregation in the constriction (the cationic cluster on the channel wall and the carboxylate groups on the constriction loop, see Fig. 6a) actually continues in screw-shape all along the channel [26]. It seems also clear that hydrophobic solutes, with ordered water clathrates covering the hydrophobic surfaces, are considerably retarded [13]. Mutants are available either from nutrient selection [27], or from colicin N resistance [28], and a series of site-directed mutants have been constructed [29]. Their properties are in agreement with expectation with regard to single channel conductance, critical transmembrane voltage closing channel closure, as well as the changed ion selectivities [30,31]. In an evolutionary variant, phospohoporin from *E. coli*, a single lysyl side-chain that protrudes into the channel and near the center of the constriction site [4] reverses the ion selectivity to a preference for anions [32]. At least two questions remain: does the conformation, as it is present in the crystal and as it is shown in the figure, correspond to the open state? Although intuitively seemingly obvious, it is, in the absence of another crystal form that may be identified as closed state, not possible to answer this question unequivocally. Experiments measuring the water flux, normalized to that of water permeation across pure lipid bilayers, indicates that flux across closed channels (monitored by voltage clamping) is not significantly reduced [33]. The second, and related question concerns the mechanism of channel closing. Does the constricting loop L3 move? Does a change in water structure occur? Are there minor changes in the positions of ionized groups?

The channel constriction in maltoporin (Fig. 6b) appears much narrower (5×5 Å) and more crowded by side-chain residues than in the corresponding positions of matrix porin. There exists a segregation of aromatic residues (shown in gray), while the ionizable groups do not show the pattern observed in matrix porin. One tyrosyl residue is found opposite to the aromatic cluster. A longitudinal cross-section of

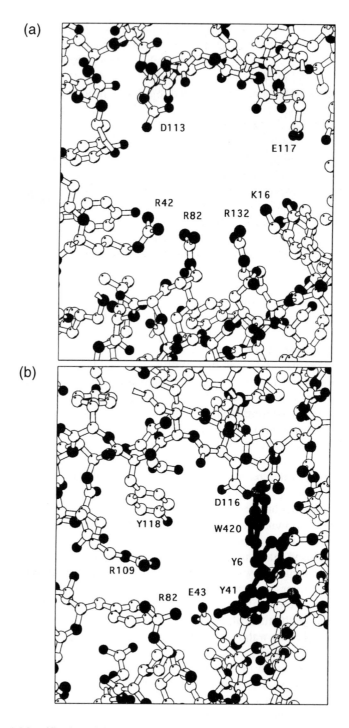

Fig. 6. Magnification of the constriction sites of matrix porin (a) and maltoporin (b).

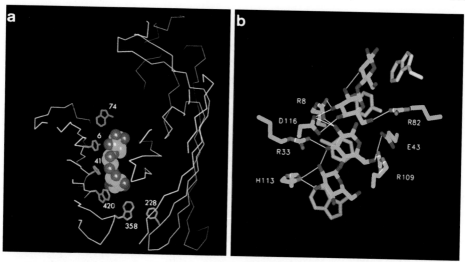

Fig. 7.   The channel constriction in maltoporin. Panel (a): Section perpendicular to the membrane plane. The residues of the greasy slide are indicated by their number in the sequence. The space-filling model of maltotriose is positioned according to the electron density. Panel (b): Two ionic tracks (left and right) and the 'greasy slide' (background) surround two maltose molecules (stick model).

maltoporin (Fig. 7a) resolves the arrangement of the aromatic residues. The barrel (white lines) is represented with the parts occluding the view of the channel (removed for clarity). The aromatic residues appear arranged in a helical pathway across the narrows of the channel, with a curvature comparable to that of maltodextrin in solution [34]. The contiguous array of aromatic rings thus forms a 'greasy slide'. The tryptophanyl residue (#W74) at the top of the slide is contributed by #L2 of the neighbouring subunit, which latches into the gap left by the bending of L3 into the channel. Two tyr (Y6, Y41) and two trp residues (W420, W358) belong to the pathway, while contacts with phenylalanine F228 have not been observed. Tyrosine 118 (not shown) is on the (white) loop 3, approaching the ligand closely. Soaking experiments with maltotriose have yielded a density in close contact with the three aromatic residues, Y6, Y41, and W420. The ligand is represented here as space filling model to show that its hydrophobic face (yellow, on the left of the model) is in van der Waals distance from the hydrophobic residues. We therefore consider the greasy slide as providing a mechanism that could contribute to the rate enhancement (k_{cat}) of the translocation. In addition to the slide, there exist other interactions with the sugars. Fig. 7b shows two tracks of ionizable residues (R8, D116, R33, H113, and R82, E43, and R109) on both sides of the oligosaccharide. In the background (in blue), the greasy slide can be seen (#W74, Y6, Y41, W420). The ionizable amino acid side-chain residues appear in hydrogen-bonding distance with all sugar hydroxyl groups (here shown are the positions of two maltose residues (yellow/red) from soaking experiments of crystals [35]). These bonds (white lines) are likely to determine both affinity and specificity of the channel for any given substrate as follows. If a solute approaches the channel constriction, it may interact with the first residue(s) of the greasy slide, which guides it into the channel, provided it exhibits a hydrophobic face. The hydrogen bonds formed determine its affinity at the entrance of the channel narrows (K_M). If it moves along the slide, the rate is determined by the interaction with the aromatic residues on the one side, and by hydrogen bondings to the ionizable residues in two tracks on the other. In Fig. 7b, two maltose molecules are shown which, according to the electron density map, overlap slightly. The occupancy of each site (#W74, Y6, and Y41, W420) is about 50%. It can also be seen that the binding of the glucose units is slightly off-register relative to the aromatic rings in the greasy slide. It thus appears that the positioning of the molecules is determined by the ionized groups rather than by the stacking of glycosyl rings on aromatic residues. The crystal structure of maltoporin-ligand complexes do provide explanations why certain sugars, such as sucrose, are bound but not translocated [36], but the contributions to the rate enhancement of sugar translocation is not clear in quantitative terms. On the whole, the analogy of the translocator to an enzyme is suggestive, and it will be interesting to see to what extent the concepts described here apply to more specific permeases. But detailed mechanistic questions remain unresolved. The formation of a hydrogen bonding networks appears to explain specificity. How is this related, in detail, to the observed rates of translocation? Are there conformational changes in the side-chain residues involved directly in translocation? Or does a concerted motion effectuate translocation? Can the diffusion be conceived of as one-dimensional? From the crystallographic data, all that is clear is that the β-barrels are not distinguishable in the liganded and the unliganded states. Changes in the conformation of the side-chain residues were not observed [35], but higher resolution will required for definitive conclusions.

5. Aqua incognita: hydration and solvation

The topology of maltoporin channels reveals unequivocally that there is no space for water molecules in the channel if it is occupied by specific ligands. The situation in matrix porin is drastically different. There clearly exists space for a large number of water molecules, of which some have been found ordered in crystallographic analyses [37,38]. Yet nothing is known about the hydration of the various ionizable residues. Clearly, the pK-values of the three arginyl residues in the cationic cluster cannot correspond to the values found in bulk solution. Calculations have revealed, for instance, that the *in situ* pK value of R82 is close to neutral [26]. The guanidinium groups are in close (π–π) contact, however, so that caution must rule when interpreting such results. Similar calculations have revealed that the buried carboxyl groups between L3 and the barrel wall are uncharged. Again this appears intuitively attractive, but we know as yet little about the local effect of an external electric field on the state of ionization of these residues. This is clearly a significant problem when the question is raised whether the constriction loop L3 moves during channels closure [39]. Another question will require further investigation: so far, it has been tacitly assumed that no shielding of fixed charges occur by mobile ions. This could not have been observed in the X-ray analysis, due to the poor scattering of counter-ions such as sodium or potassium. It also depends on the residence times of such ions in the vicinity of their counterions. The use of other ions, such as Cs^+, Ba^{+2}, or Tb^{+3} may help to resolve these questions. Terbium may be followed also by fluorescence, and may thus open different time windows.

6. Porins as models for other channels

The detailed results from the structural and the functional studies, given in outline above, illustrate that the results from high resolution structural studies and single channel recordings of channel-forming proteins, allow solute translocation to be understood at a molecular level. The contributions of molecular pathology, be that by the application of strong selective pressures (growth on carbon sources which, due to the constriction size, are not normally taken up by *E. coli* [27], or by the exposure to bacterial toxins [28]) has clearly proven valuable. Mutations that are tailor-designed [29] on the basis of structural information at atomic resolution may be more intriguing, as problems can be addressed much more specifically. Answering questions concerning the trajectories of solutes requires further simulations of the electrostatic properties of porin, the prediction of the effects of changes in charge distributions, and detailed molecular dynamics simulations [39]. The manoeuvrability of bacteria such as *E. coli* has the obvious advantage that the models proposed by computational approaches can be subjected to rigorous testing. In addition, mapping the topology of the surface by means of atomic force microscopy allows protein-lipid interactions to be studied in detail [40] and may give information on conformational changes in the loops. Interatomic distance measurements by solid-state NMR spectroscopy may, moreover, prove a very powerful tool in assessing channel dynamics, conformational changes, and solute flux.

Matrix porin may also be viewed as a model for other channels. Nanotubes have recently been studied by allowing self-assembly of small synthetic peptides [41]. In view of the fact that much attention has been focused on mimicking biological function, it may be intriguing to consider porin as natural nanotubes which, moreover, are voltage-gated (a porister?). These proteins can be prepared in large quantities, and the genetics makes them versatile, with changes engineered at specific sites, and the results accessible to analysis by crystallographic methods. It will be a challenging task for the future to determine to what extent such ideas may be realistic.

Acknowledgments

Supported by grants of the Swiss National Science Foundation. I thank Lucienne Letellier for allowing me to peruse her contribution to this volume in advance. The critical reading of this manuscript by Tilman Schirmer, and his providing me generously with illustrations is highly appreciated. The constructive comments, and the criticism of the manuscript by Dr. Robert S. Eisenberg, Rush University, Chicago IL, are most gratefully acknowledged.

References

1. Garavito, R.M. and Rosenbusch, J.P. (1980) J. Cell. Biol. **86**, 327–329.
2. Weiss, M.S., Abele, U., Weckesser, J., Welte, W., Schiltz, E. and Schulz, G.E. (1991) Science **254**,

1627–1630.
3. Kreusch, A., Neubüser, E., Schiltz, E., Weckesser, J. and Schulz, G.E. (1994) Protein Sci. **3**, 58–63.
4. Cowan, S.W., Schirmer, T., Rummel, G., Steiert, M., Ghosh, R., Pauptit, R.A., Jansonius, J.N. and Rosenbusch, J.P. (1992) Nature **358**, 727–733.
5. Schirmer, T., Keller, T.A., Wang, Y.-F. and Rosenbusch, J.P. (1995) Science **264**, 914–916.
6. Letellier, L. and Bonhivers, M. (1996) in: Transport Processes in Membranes, eds W. Konings, H.R. Kaback and J.S. Lolkema. Handbook of Biological Physics, Vol. 2, pp. 615–636. Elsevier, Amsterdam.
7. Mizuno, T., Chou, M.-Y. and Inouye, M. (1983) J. Biol. Chem. **258**, 6932-6940.
8. Schiltz, E., Kreusch, A., Nestel, U. and Schulz, G.E. (1991) Eur. J. Biochem. **199**, 587–594.
9. Clément, J.M. and Hofnung, M. (1981). Cell **27**, 507–514.
10. Chen, R., Krämer, C., Schidmayr, W., Chen-Schmeisser, U. and Henning, U. (1982) Biochem. J. **203**, 33–43.
11. Pauptit, R.A., Schirmer, T., Jansonius, J.N., Rosenbusch, J.P., Parker, M.W., Tucker, A.D., Tsernoglou, D., Weiss, M.S. and Schulz, G.E. (1991) J. Struct. Biol. **107**, 136–145.
12. Rosenbusch, J.P. (1974) J. Biol. Chem. **249**, 8019–8029.
13. Nikaido, H. (1992) Mol. Microbiol. **6**, 435–442.
14. Neuhaus, J.-M., Schindler, H. and Rosenbusch, J.P. (1983) EMBO J. **2**, 1987–191
15. Charbit, A., Gehring, K., Nikaido, H., Ferenci, T. and Hofnung, M. (1988) J. Mol. Biol. **201**, 487–496.
16. Schindler, H. and Rosenbusch, J.P. (1978) Proc. Natl. Acad. Sci. USA **75**, 3751–3755.
17. Benz, R., Schmid, A. and Hancock, R.E.W. (1985) J. Bacteriol. **162**, 722–727.
18. von Meyenburg, K. and Nikaido, H. (1977) Biochem. Biophys. Res. Commun. **78**, 1100–1107.
19. Wandersman, C., Schwartz, M. and Ferenci, T. (1979) J. Bacteriol. **140**, 1–13.
20. Paul, C. and Rosenbusch, J.P. (1985) EMBO J. **4**, 1593–1597.
21. Jeanteur, D., Lakey, J.H. and Pattus, F. (1991) Mol. Microbiol. **5**, 2153–2164
22. Schirmer, T. and Cowan, S.W. (1993) Prot. Sci. **2**, 1361–1363.
23. Rosenbusch, J.P. (1990) Experientia **46**, 167–173.
24. Anderson, O. (1983) Biophys. J. **53**, 119–133.
25. Tait, M.J., Suggett, A., Franks, F., Ablett, S. and Quickenden, P.A. (1972) J. Solution Chem. **1**, 131–151.
26. Karshikoff, A., Spassov, V., Cowan, S.A., Ladenstein, R. and Schirmer, T. (1994) J. Mol. Biol. **240**, 372–384.
27. Benson, S.A., Occi, J.L.L. and Sampson, B.A. (1988) J. Mol. Biol. **203**, 961–970.
28. Jeanteur, D., Schirmer, T., Fourel, D., Simonet, V., Rummel, G., Widmer, C., Rosenbusch, J.P., Pattus, F. and Pagès, J.-M. (1994) Proc. Natl. Acad. Sci. USA **91**, 10675–10679.
29. Prilipov, A. and Rosenbusch, J.P. (1996) In preparation.
30. Saint, N., Widmer, C., Prilipov, A., Luckey, M., Lou, K.-L., Schirmer, T. and Rosenbusch, J.P. (1996) J. Biol. Chem., in press.
31. Lou, K.-L., Saint, N., Rummel, G., Benson, S.A., Rosenbusch, J.P. and Schirmer, T. (1996) J. Biol. Chem., in press.
32. Bauer, K., Struyve, M., Bosch, D., Benz, R. and Tommassen, J. (1989) J. Biol. Chem. **264**, 16393–16398.
33. Steiert, M. (1993). Structure and function of phosphoporin: Properties of a porin from the outer membrane of *Escherichia coli*. Ph.D. Thesis (University of Basel).
34. Goldsmith, E., Sprang, S. and Flettrick, R. (1982) J. Mol. Biol. **156**, 411–427.
35. Dutzler, R., Wang, Y.-F., Rosenbusch, J.P. and Schirmer, T. (1996) Structure **4**, 127–134.
36. Andersen, C., Jordy, M. and Benz, R. (1995) J. Gen. Physiol. **105**, 385–401.
37. Weiss, M.S. and Schulz, G.E. (1992) J. Mol. Biol. **227**, 493–509.
38. Cowan, S.W. and Schirmer, T. (1996) Personal communication.
39. Watanabe, M., Rosenbusch, J.P., Schirmer, T. and Karplus, M. (1996) Computer simulations of the OmpF porin from the outer membrane from Escherichia coli. Biophys. J., in press.
40. Schabert, F.A., Henn, C. and Engel, A. (1995) Science **268**, 92–94.
41. Ghadiri, M.R., Granja, J.R. and Buehler, L.K. (1994) Nature **369**, 301–304.

CHAPTER 27

Intrinsic and Extrinsic Channels in Bacteria

L. LETELLIER and M. BONHIVERS

Laboratoire des Biomembranes, URA CNRS 1116,
Université Paris Sud, 91405 Orsay cedex, France

© 1996 Elsevier Science B.V.
All rights reserved

Handbook of Biological Physics
Volume 2, edited by W.N. Konings, H.R. Kaback and J.S. Lolkema

Contents

1. Introduction

Channels of Gram-negative bacteria were until recently generally thought to be mainly located in the outer membrane. The biochemistry, biophysics and genetics of the proteins forming these channels have been extensively studied and the determination of the 3D structure of some has led to a substantial progress in their characterization. The literature on these outer membrane proteins has increased exponentially during the last few years and numerous reviews have been published [1–9]. The structural properties of these proteins are reviewed in another chapter of this book. Therefore, we will only briefly summarize some of their properties with reference to their channel-forming ability.

Classically, channels and carriers are distinguished on the basis of the mechanism of transport. Channels are generally viewed as tunnels in which the binding sites for the transported species are accessible from both sides of the membrane at the same time. Although a channel can undergo a conformational change from the closed state to the open state, no alteration of conformation is necessary for transport of the solute once it has entered the channel. In contrast, solutes transported by carriers first bind at one side of the membrane and a conformational change is required for their release on the other side. Thus the solute binding site of a carrier is accessible only on one side of the membrane at a given time [10]. However, as it is unlikely that membrane proteins can undergo major conformational changes, it is often assumed that carriers possess some kind of inner channel which allows the translocation of the transported solute. The recent discovery that two *Escherichia coli* outer membrane proteins, FepA and FhuA, possess the dual function of transporters and channels [11–13] opens the field to new exciting mechanistic studies.

In contrast to the extensive investigations of ion channels in the outer membrane there have been little attention given to possible channel activities in the cytoplasmic membrane. The reason for this lack of interest was mainly due to Mitchell's chemiosmotic theory that implies the maintenance of a proton gradient across the inner membrane and consequently argues against the presence of ion channels. Indeed, high rates of ion transport though ion channels would be expected to dissipate the electrochemical proton gradient. This situation changed in 1987 with the finding that the *E. coli* envelope contains pressure-sensitive ion channels [14]. which were later found to be located in the inner membrane. In the same year a patch clamp study revealed the presence of a voltage-dependent ion channel in the energy-transducing inner mitochondrial membrane [15]. Subsequently pressure-dependent channels were also found in Gram-positive bacteria [16]. Recently, one of these channels belonging to *E. coli*, has been characterized at the molecular level

[17]. A paragraph will be devoted to their description and to their possible physiological functions.

It is now widely believed that translocated segments of presecretory proteins cross the cytoplasmic membranes via specific channels formed by one or several proteins of the protein translocation machineries (reviewed in Refs. [18,19]). We will present the experiments supporting this hypothesis [20]. Besides these intrinsic channel proteins, extrinsic peptides or proteins might transiently form channels in the bacterial envelope. These molecules mostly belong to the family of toxins and the channels formed are responsible for killing of the bacteria. The pore-forming colicins are among the best characterized toxins. Their properties are described in another chapter. Antimicrobial peptides have been found in a large number of diverse organisms ranging from bacteria to man. Interest in such molecules has been increasing during the last few years because many of them have potentially valuable antibiotic properties. It is generally believed that some of these peptides kill bacteria by forming channels. The demonstration of their channel activity is however not always simple. We will present some of the experiments which argue in favor of this mode of action.

Extrinsic channels are also transiently formed in Gram-negative bacteria upon transfer of phage DNA through the envelope (reviewed in Ref. [21]). We will present evidence for a mechanism of channel-forming for passage of phage DNA.

The study of channel activity in bacteria relies on various direct and indirect experimental approaches which will be briefly presented.

2. Experimental approaches to the study of channel activity

The pore-forming activity of outer and inner membrane proteins can be studied both in intact cells and *in vitro* systems. *In vitro* assays use either isolated outer and inner membrane vesicles or purified proteins. Several *E. coli* outer membrane proteins have now been purified. A general protocol for the first steps of purification of most of these proteins has been described by Nikaido [22] and reviewed by Benz and Bauer [2]. Membrane vesicles are generally prepared by passage of the bacteria through a French pressure cell. Detergents solubilizes those proteins of these membrane vesicles not associated to the peptidoglycan and precipitates those bound ·to the peptidoglycan. Peptidoglycan-linked proteins can be released by murein digestion or salt extraction. Proteins may be further purified by gel filtration and/or affinity chromatography.

2.1. Estimating pore size from vesicle permeability and liposome swelling assays

The vesicle permeability assay was introduced by Nakae to identify the pore-forming proteins of *E. coli* and *Salmonella typhimurium*: Radioactively labeled solutes of various molecular masses are entrapped in proteoliposomes. The diameter of the pore can then be estimated from the Stokes radius of the largest solute able to diffuse through the channel [23,24].

A second technique is to measure the rate of solute diffusion through the channel [25,26]. A large molecular mass solute such as dextran or stachyose that is not able to diffuse through channels is entrapped in the channel-containing liposomes. The proteoliposomes are added under rapid mixing to an isotonic solution of the solute whose permeability is to be assayed. The osmotic gradient which is formed promotes the influx of water together with that of the tested solute. This results in a swelling of the liposomes which is monitored by light scattering. The initial swelling rate is a measure of the rate of penetration of the solute through the channels. The pore diameter is estimated from the measurements of the relative permeability towards different solutes as calculated according to the model developed by Renkin [27]. However there are some limitations in the interpretation of the results of such assays [1–3].

2.2. In vivo assay of porin function

Porin activity can be estimated from the measurement of the rate of enzymatic hydrolysis in the periplasm of compounds diffusing through the outer membrane pores [28]. The rate of hydrolysis can be generally described by a combination of the rates of passive diffusion through the porins (which obeys Fick's law) and of enzymatic hydrolysis in the periplasm according to Michaelis–Menten kinetics [29,30]. The specificity of diffusion through particular porins can be established by using mutants deficient in these porins. The usefulness and limits of such assays are reviewed by Nikaido and Vaara [1].

2.3. Indirect in vivo assay of channel activity in whole cells

E. coli cells accumulate K^+ to a high concentration (a few hundred mM) in the cytoplasm due to the action of specific constitutive and inducible transport systems [31]. The primary effect of various extrinsic peptides and proteins is to induce a leakage of K^+. The amplitude of this efflux is generally sufficiently large for the changes in the K^+ concentration in the external medium to be directly measured with a K^+-selective electrode. This technique allows continuous measurements without requiring any manipulation of the cell suspension such as centrifugation or filtration. Provided that this efflux can be dissociated from that induced by the K^+ transporters and that one knows the number of peptides/proteins susceptible to form channels, it is possible to determine if the K^+ efflux is channel-mediated. This technique has proven useful in the characterization of the channel activity induced by phage and colicins [32,33], particularly because the membrane permeability changes could be induced by only one protein molecule per bacterium. The interpretation of the data with peptides is more difficult since the number of peptides inducing permeability changes are often ill-defined and/or high ($>10^6$ per bacterium) [34,35].

2.4. Electrophysiological techniques

2.4.1. Lipid bilayer techniques

Three different lipid bilayer techniques are currently used. The first uses solvent-containing lipid membranes formed by the Mueller–Rudin method. The detergent-solubilized protein is added to one of the compartments bathing the lipid bilayer membrane [36]. It spontaneously inserts into the bilayer. A voltage is applied through electrodes connected with salt bridges on both sides of the bilayer. Channel activity is detectable by a stepwise increase of the current. This technique applies essentially for membrane proteins which like porins withstand to be diluted in the absence of detergent since the presence of detergent in the compartments bathing the lipid bilayer is precluded. The second technique uses generally solvent-free membranes [37]. Vesicles reconstituted from lipid and protein are spread on the surface of the aqueous phase on both sides of a thin Teflon foil in which there is a small hole (100 μm diameter) initially above the water level. The surface of both compartments are covered with lipid layers containing the protein. The water level on both sides of the membrane is then raised and a lipid bilayer containing the protein is formed across the hole. Asymmetric insertion of the proteins is therefore possible if the protein is added only on one side of the membrane. In the third technique membrane vesicles containing the protein to be studied are added to the *cis* side of the bilayer. Fusion of the vesicles with the lipid bilayer is induced by maintaining the *cis* compartment hyperosmotic with regard to the *trans* compartment [38].

2.4.2. Preparation of bacteria, membranes and patch clamp methodologies

Direct recording of electrical activity in bacteria was not practical until recently due to the size of almost all bacteria. These difficulties were overcome with the successful application of the patch clamp technique [39,40] to giant bacteria, spheroplasts and protoplasts and to proteoliposomes containing purified membrane fractions (reviewed in Ref. [41]).

(a) *Preparation of giant bacteria, spheroplasts and protoplasts*: Giant Gram-negative bacteria are obtained by growing cells in the presence of cephalexin, an inhibitor of septation [14]. Cells thus obtained appear as long filaments. A lysozyme-EDTA treatment partially destroys the cell wall to yield giant sphero-plasts. A combination of the lysozyme-EDTA treatment with a mild osmotic shock is supposed to improve the release of the outer membrane and peptidoglycan [42]. Alternatively, cell wall biosynthesis can be inhibited with penicillin, yielding large spheroplasts with diameters up to 15 μm [43]. Recently an *E. coli lpp ompA* double mutant which has a round phenotype was used. In the presence of cephalexin it grows as giant round cells which are directly amenable to patch clamp recording [44,45].

Gram-positive bacteria can be grown in the presence of lysozyme to digest the cell wall and to prevent cell division. For *Streptococcus faecalis* and *Bacillus subtilis* this leads to protoplasts with diameters up to 4 μm [16,46].

(b) *Preparation of giant proteoliposomes*: This technique offers an interesting alternative to *in vivo* studies since it allows the manipulation of purified material and modification of the composition of the proteoliposomes. Proteoliposomes are obtained by the fusion of membrane vesicles with azolectin liposomes by a cycle of dehydration–rehydration [47]. Their diameters varies between 5 and 100 μm [48,49].

(c) *Patch clamp methodology*: A glass pipette is filled with an electrolyte solution bathing a wire electrode connected to a low-noise electronic device. The tip of the pipette is pressed onto the surface of the cell or of the proteoliposome. Suction is applied to the inside of the pipette, resulting in the formation of a seal between the membrane and the glass. Suction is then released and the current can flow between the pipette electrode and the bath electrode providing that open channels are present in the membrane patch formed. In the case of excised patch, the inner surface of the membrane is exposed to the bath solution. The electronic device allows the voltage to be clamped at the desired value and the current to be amplified and recorded [41].

3. Channels from the outer membrane

The outer membrane provides a barrier to the influx of deleterious compounds from hostile environments such as the bile salts and digestive enzymes found in the intestinal tract of animals. It also acts as a barrier against antibiotics and allows the efflux of waste products [1]. Half of the mass of the outer membrane is protein which has been subdivided into three classes: nonspecific porins, specific porins and high affinity, energy-dependent transport proteins. The nonspecific porins form large water-filled channels with a size exclusion limit for hydrophilic solutes which differs between bacterial species with a maximum of 800 Da. Specific porins have little selectivity for small solutes but selectively recognize larger molecules such as oligosaccharides for LamB or nucleosides for Tsx (reviewed by Nikaido [6]). The presence of specific binding sites inside these channels accelerates the diffusion of solutes present at low concentration. Diffusion through these porins however exhibits a saturation kinetics, unlike that through nonspecific porins, which is linearly dependent on the concentration difference across the membrane [3]. Nutrients like siderophores and vitamin B_{12} which are present in very low concentrations in the environment are taken up by the high affinity, energy-dependent transport systems (reviewed in Refs. [5–7]).

3.1. Nonspecific porins

Enteric proteins including *Escherichia coli* produce three major porins: OmpF (or matrix porin), OmpC and PhoE (or phosphoporin). OmpF is preferentially synthesized in medium of low osmolarity and OmpC in high salt [50]. PhoE is produced in conditions of phosphate starvation. In addition to their permeation properties these proteins which form a homologous family, also serve as receptors for various phages and colicins [51]. All three porins are found as extremely stable trimers which resist the action of proteases, chaotropic agents and surfactants. The 3D structure of the major porin from *Rhodobacter capsulatus* (crystal structure at 1.8 Å resolution) [52], of OmpF from *E. coli* (resolution 2.4 Å) [53] and of PhoE (resolution 3 Å) [54] were solved by X-ray crystallography. A common feature of these structures is that each subunit consists of antiparallel β-barrels. In all cases there is a loop located on the external surface and connecting two β-strands inside the barrel, narrowing the channel to a small eyelet (reviewed in Ref. [54]).

Permeability studies with reconstituted proteoliposomes containing purified porins have been used to define the molecular size exclusion limits of porins for a variety of hydrophilic and hydrophobic substances [24,55] (reviewed in Ref. [1]). However, even below the size exclusion limits the diffusion rates of solutes may differ by several orders of magnitude reflecting differences in hydrophobicity. A clear example in the case of the porin OmpF is L-arabinose (MW = 150) which diffuses 100-fold faster than dissacharides (MW = 342). Permeation rates of β-lactam antibiotics through given porins have also been determined in intact cells [6].

Single channel conductance and ionic selectivity of porins have been deduced from studies in planar lipid bilayers (reviewed in Refs. [2,56]). Pore radii were in some cases deduced from the single channel conductance. However, the calculations were often based on inappropriate values of the single channel conductance and may thus have led to incorrect conclusions. Each subunit of the porins trimers forms a channel. Most porins are cation selective, a property which may be related to the fact that their environment contains bile salts. Despite high sequence homology (63% identity) the channel properties of OmpF and PhoE are different. The conductance of the OmpF pore is larger than that of PhoE (0.8 nS compared to 0.6 nS in 1 M salt). OmpF is weakly cation selective whereas PhoE has a preference for the transport of anions [2]. This difference in selectivity is probably due to residue 125, a lysine in PhoE replaced by a glycine in OmpF and positioned in the narrowest portion of the channel [57]. The eyelet region of the different pores is probably responsible for channel selectivity. Conflicting results have been published concerning the voltage-gating of porins, some authors suggesting that it is an artefact of reconstitution [58]. To clarify this situation Lakey and Pattus used three different protocols for porins purification and reconstitution in bilayers. They conclude that porins show voltage-sensitivity following all methods of reconstitution [59]. OmpF and PhoE channels close above –90 mV and –100 mV, respectively [60,61]. Further proofs of the voltage-dependence of OmpF and OmpC were obtained using the patch clamp technique on membrane vesicles fused with giant liposomes [62] and with purified OmpF and OmpC reconstituted in azolectin liposomes [48,63]. However the experiments of Sen et al. strongly suggest that voltage does not control porin activity in intact cells [64]. Therefore, it remains unclear whether the voltage-gating observed in vitro reflects physiological behaviour in vivo.

Since their discovery in E. coli and S. typhymirium, porins have been found in almost all Gram-negative bacteria studied. There is an increasing interest in the porins of pathogens because of their possible use as vaccines and in bacterial typing and their involvement in antibiotic resistance. A phylogenetic tree of porins belonging to 14 different bacterial species has been constructed from their sequences [65,66]. A list of the channel properties of some of these porins can be found in [2].

OmpA is one of the most abundant outer membrane proteins in E. coli. OmpA plays an essential role in the structural stability of the outer membrane, in maintenance of cell morphology and in conjugation [67]. It is also a receptor for phage and colicin [51]. Despite some debate, there is evidence that this class of proteins forms channels in vesicles [68] and in planar lipid bilayers [69]. However, unlike the other porins there is no indication that the protein oligomerizes.

The outer membrane of *Pseudomonas aeruginosa* is characterized by a very low nonspecific permeability making this organism highly resistant to a wide variety of antibiotics [70,71]. *P. aeruginosa* lacks the 'classical' trimeric porins [72] but contains several specific porins involved in the uptake of sugars, amino acids and peptides (reviewed in Ref. [58]). The major nonspecific porin is OrpF which shows homology to OmpA from *Escherichia coli* [73,74]. Although OrpF has a large exclusion limit [70], it has a low channel-forming activity both *in vivo* and *in vitro*.

3.2. Specific porins

The *E. coli* maltoporin LamB is an inducible protein encoded by the maltose regulon which also encodes the other proteins required for the uptake and metabolism of maltose and linear maltooligosaccharides [75]. The structure at 3.1 Å resolution of the LamB trimer has recently been solved [76]. The scaffolding of the monomer is an 18-stranded, antiparallel β-barrel forming a channel. Although the maltoporin does not show sequence homology with the other porins, their folding show similarities. In particular, all contain loops exposed on the cell surface which fold into the barrel and contribute to a constriction inside the channel. The LamB porin forms water-filled channels responsible for the specific transport of maltose and maltodextrins [77–79]. and possesses a specific binding site for maltose within the channel. Deletion of one of the loops of LamB transforms the maltoporin into a nonspecific channel [80]. Residues presumed to be responsible for the sugar specificity of the channel have been identified in the 3D structure of the protein [76].

Tsx protein is a minor *E. coli* outer membrane protein the synthesis of which is coregulated with systems for nucleoside uptake and metabolism. At low concentration of substrate Tsx mediates the transport of nucleosides whereas at high concentration nucleosides diffuse through OmpF. Tsx is also the receptor for colicin K and phage T6 (reviewed in [58]). Tsx shows pore-forming activity in lipid bilayer membranes. The channel, which contains a binding site for nucleosides [81] has a very small single channel conductance (10 pS in 1 M KCl compared to 1500 pS and 155 pS for OmpF and LamB, respectively, under similar ionic conditions) [82].

TolC is also a minor *E. coli* protein which is believed to play a role in the expression of outer membrane proteins and in import and secretion of toxins and of Haemolysin A [83–85]. The purified protein shows channel-forming activity in lipid bilayers. High concentrations (>10 mM) of a tripeptide cause a decrease in the channel conductance which led the authors to conclude that the channel is specific for amino acids and peptides [86]. However the specificity of TolC channel for peptides needs to be confirmed.

3.3. Porins in Gram-positive bacteria

Contrary to the cell wall of most Gram-positive bacteria, the cell wall of Mycobacteria contains a substantial amount of lipids of unusual structure (mycolic acids with very long hydrocarbon chains of up to 50 carbons). These organisms are also characterized by intrinsic resistance to antimicrobial agents. Permeability experiments with *Mycobacterium chelonae* have demonstrated that the permeation rate of cephaloridine into this organism is about 10^4 fold lower than through the outer membrane of *E. coli* and suggested that hydrophilic solutes might cross the cell wall through water-filled channels [87]. A protein of 59 kDa that forms a water-filled channel in the cell envelope of *M. chelonae* has been recently characterized by reconstitution methods [88].

3.4. Proteins with the dual function of carrier and channel

The transport through the outer membrane of nutrients present at very low concentration in the environment is catalyzed by several proteins: FepA, BtuB and FhuA catalyze the transport of ferric enterobactin [89], vitamin B12 [90], ferrichrome and albomycin respectively [91]. Further transport through the envelope is ensured by various proteins in the periplasmic space and in the cytoplasmic membrane [92]. Transport of these nutrients also requires the participation of TonB, a protein anchored in the cytoplasmic membrane and extending into the periplasm and of ExbB and ExbD, two proteins that form a complex with TonB [93,94]. TonB is implicated in energy transduction between the cytoplasmic and outer membrane (for a recent review see [95] and Chapter 28).

These proteins, which like the porins are assumed to fold as β-barrels, have long time been considered to belong to a class separate from that of porins. Recently, however, it was shown that introducing internal deletions within one of the 'loop' regions on the external surface connecting the transmembrane domains involved in binding of the substrates of FepA and FhuA converted these receptors into nonspecific diffusion channels [11–13,96]. In the case of FepA from which a segment of 140 amino acids was removed this conclusion was deduced from *in vivo* transport experiments and swelling assays with proteoliposomes [11]. Both *in vivo* and planar lipid bilayer assays supported this proposal in the case of a FhuA derivative (FhuA Δ(322–355)) from which the loop extending from residues 322 to 355 had been removed [12]. The main criticism of these experiments is that they do not indicate whether the wild type proteins behave also like channels. Data from our laboratory recently showed that wild type FhuA has indeed the dual function of a transporter and of a channel [13]. FhuA, which is also the receptor for phage T1, T5, Φ80 and colicin M (reviewed in Ref. [97]), was purified. The protein is monomeric in neutral detergents. It has a 51% β-sheet content [98] and is fully active as judged by its capacity to bind phage T5 and to induce the release of the phage DNA [99]. Interaction of phage T5 with FhuA converted the carrier into a channel, the electrophysiological characteristics of which were similar to those of FhuA Δ(322–355). These data suggest that binding of T5 to loop 322–355 of FhuA triggers a conformational change of the loop that unmasks an inner channel in FhuA which is used for the transport of ferrichrome [13].

In conclusion, these data strongly support the model that these transport proteins are gated porins possessing a high affinity ligand binding domain in the loop that controls the transport through the channel.

4. Channels in the inner membrane

The observations of rapid effluxes of solutes induced by osmotic downshock [100,101] suggested the existence of channel(s) activity in the cytoplasmic membrane. Application of the patch clamp technique has indeed revealed the presence in the bacterial envelope of mechanosensitive stretch-activated channels (MS channels) that might be involved in osmoregulation. Nonmechanosensitive channels are also present in bacteria but investigation of their properties is only now beginning. Their participation in translocation and/or secretion of macromolecules (proteins and genetic material) has been postulated.

4.1. Mechanosensitive channels

Since the first report of MS channels in chick skeletal muscle [102] MS channels have been found in phylogenetically diverse organisms including animals, plants, fungi and bacteria. The ubiquity of MS channels suggests that they have important physiological functions. It has been proposed that MS channels play a role in sensing (touch and hearing), motility, cell division, smooth muscle contraction as well as in chemotaxis and regulation of osmolarity according to cell type. Recent reviews describe their properties [103,104].

4.1.1. E. coli harbors a variety of mechanosensitive, stretch-activated channels
The mechanical stimulus used to activate the MS channels in most experiments is hydrostatic pressure or suction applied through a patch clamp pipette. Pressures of a few kPa are generally sufficient to stretch the cell membrane and to induce the opening of MS channels also called stretch-activated (SA) channels. The first SA channel to be characterized was a large (950 pS in 0.3 M salt) channel in *E. coli* spheroplasts [14]. Subsequent studies by Berrier et al. [48] on giant proteoliposomes revealed the presence in *E. coli* of at least 6 conductances activated by stretch, ranging from 100 to 1500 pS (in 0.1 M KCl). Subsequently these MS channels were classified into different families which can be grouped into three different subfamilies, each subfamily being activated at a different threshold of applied pressure [45]. Bacterial MS channels are the only one to date that current methodology is able to functionally reconstitute [48] and that have been purified [105].

4.1.2. MS channels in Gram-positive bacteria
Stretching the plasma membrane of giant protoplasts for *Streptococcus faecalis* or *Bacillus subtilis* results in the activation of a whole array of conductances, up to a few nS in 0.35 M NaCl. This multiplicity of conductances may reflect cooperative gating of aggregates of channels with varying stoichiometries [16,41,106–108].

4.1.3. Localization of the MS channels in the cell envelope
Localizing the MS channels in Gram-negative bacteria is important for understanding their function but their localization has long been a matter of debate. Kung and coworkers favored the outer membrane [44] whereas evidence in favor of MS channels being located in the inner membrane was reported by Ghazi and coworkers

[45,48,109]. This localization was confirmed by recent whole-cell patch-clamp recording of *E. coli* protoplasts [42]. The fact that MS channels are found in Gram-positive bacteria which only contain a plasma membrane also strongly favors a localization in the cytoplasmic membrane of Gram-negative bacteria.

4.1.4. Pharmacology of the MS channels

The physiological study of MS channels has been hindered by the absence of specific blockers. The only currently available agent which can be considered as a blocker of most MS channels is gadolinium, a lanthanide which blocks the *E. coli* MS channels in the micromolar concentration range [109,110].

Various amphipaths are slow activators of MS channels in *E. coli* spheroplasts [111]. This effect was explained according to the bilayer couple hypothesis of Sheetz and Singer [112] which predicts that, given the asymmetry in electric charge between the two leaflets of the membrane, differential insertion of cationic and anionic amphipaths in the two leaflets would result in stress in the membrane. This interpretation was proposed when it was thought that the MS channels were located in the highly asymmetric outer membrane. As MS channels are likely to be located in the inner membrane, the lipid composition of which is symmetric, this interpretation is questionable.

4.1.5. Identification, purification and cloning of MS channels

Proteins from an octyl-glucoside extract of *E. coli* membrane vesicles were separated by gel filtration chromatography and reconstituted into asolectin liposomes. Each fraction was tested for MS channel activity using the patch clamp technique [113]. Two MS channels of 500 pS and 1500 pS conductance (in 0.1 M KCl) and termed MscS and MscL were characterized. By a series of chromatographic enrichment steps a protein of MM 17 kDa was identified, and its gene *mscL* cloned [17]. *mscL* is the first identified gene coding for a MS channel. Recombinant MscL protein was recently produced and purified protein shows electrophysiological activity [105]. This opens the field for mutagenesis and hence the identification of the regions of the protein involved in mechanotransduction.

4.1.6. Physiological role of MS channels

The osmotic pressure difference between the inside and the outside of the bacteria is enormous. Gram-negative bacteria have to deal with a pressure of about 2–5 atm (200–500 kPa) while the cell wall of Gram-positive bacteria withstands pressures of up to 25 atm (2.5 MPa) [114,115]. Bacteria have developed complex mechanisms to adapt to variations in the osmolarity of their environment. In response to an increase of the osmolarity of the medium bacteria accumulate potassium, synthesize osmoprotectants including proline or glycine betaïne or pump them from the external medium. Several active transport systems contribute to this regulation (see Refs. [31,116,117] for recent reviews). Upon an hypotonic challenge bacteria release ions (mainly potassium), small metabolites and osmoprotectants [100, 118]. The high rates and the nonspecificity of the efflux of solutes released upon an hypoosmotic shock led Berrier et al. [109] to propose that stretch-activated channels may be the exit pathway for these solutes. This was supported by the observation that gadolinium inhibited both MS channels and the efflux of metabolites (such as lactose and ATP) from osmotically shocked cells [109]. The electrophysiological responses of *E. coli* giant protoplasts to a pressure applied to the interior of the cell and to osmotic downshock were recently investigated using whole-cell patch clamp recording [42]. The mechanosensitive channels previously characterized in proteoliposomes [109], were sensitive to an osmotic gradient across the membrane. The fact that bacteria possess MS channels of different sizes and thresholds of activation that can be activated according to the intensity of the downshock may be an advantage for a species confronted with variable osmotic conditions [45].

4.2. Secretion of proteins through the bacterial envelope: a possible involvement of protein-conducting channels in E. coli

There are various machineries of export or secretion for proteins synthesized in the cytoplasm which have to be transported to the envelope or to be exported. In bacteria export is post-translational and the newly-synthesized proteins are generally translocated in an unfolded conformation (Chapter 32). One proposal, supported by *in vitro* experiments, supposes that proteins translocate through the lipid bilayer: signal sequences would partition into the lipid bilayer to initiate protein translocation [119]. To account for the efficiency of the transport (the ABC transporters for example are capable of transporting often hydrophilic proteins of over 1000 residues in an unfolded conformation [120,121]), it is more widely believed that translocation takes place through protein-conducting channels. The only reported electrophysiological data consistent with this hypothesis are those of Simon and Blobel who detected a large channel (115 pS in 45 mM potassium glutamate) in inverted vesicles of *E. coli* fused with planar lipid bilayers. Rough microsomes like *E. coli* membranes share the ability to translocate secretory proteins. They also contain a similar channel. Thus the authors concluded that this channel may represent a protein-conducting channel [38]. Addition of nM concentrations of the signal sequence of the LamB protein to *E. coli* membrane vesicles fused with a lipid bilayer results in an increase of the membrane conductance. Signal sequences may therefore open protein-conducting channels and opening of the channel may be coupled to the initiation of translocation [20]. These data corroborate the observation that large aqueous channels open in the endoplasmic reticulum upon the release of nascent translocating peptides from membrane-bound ribosomes [122]. Although very attractive there is currently no confirmation of this model in any of the translocating systems in bacteria. Further investigations will certainly be needed before definitive conclusions can be drawn.

5. Extrinsic channels

5.1. Antimicrobial peptides and polypeptides

Antimicrobial peptides and polypeptides are primeval elements of host defence produced by a large variety of animal groups, by insects, plants and by bacteria.

They have a broad diversity of structure and mode of action. In some cases, however there are structural similarities which make them interesting tools for studies of evolution. This paragraph will only deal with those molecules which present characteristics reminiscent of those of channels.

Cecropins, magainins and defensins found in animals and insects have evoked considerable interest in recent years as they have therapeutic potential interest. These peptides kill both Gram-positive and Gram-negative bacteria but cause little or no harm to eukaryotic cells [123].

The first characterized antimicrobial polypeptides produced by bacteria were of the colicin type. We will only get a general view of their mechanism of action as their properties are described in this book. Gram-positive bacteria produce bacteriocin-like molecules which are active against a broad range of bacterial species and particularly against lactic acid bacteria. This has led to an increasing interest in these molecules because of their potential application as food preservatives and prevention of bacterial infection. Their origin, biosynthesis and biological activity have been described in recent reviews [124,125]. We will therefore focus on their channel activity.

5.1.1. Cecropins

Cecropins are 31 to 39 residue, positively-charged peptides. They were first identified as part of the immune response of certain silkworm species and were also found in pig intestines. The 3D structure as deduced from NMR studies consists of two α-helices joined by a hinge region containing a glycine-proline doublet. The N-terminal α-helix is amphipathic and the C-terminus hydrophobic. They contain no cysteine. Cecropins are among the most potent antibacterial peptides and unlike most of them very active against Gram-negative bacteria. Their target is believed to be the inner membrane which they permeabilize (reviewed in Ref. [126]). They also permeabilize liposomes and form voltage-dependent anionic channels in lipid bilayers (single channel conductance 2.5 nS in 0.1 M NaCl). A comparative study of their properties and those of synthetic analogs has allowed to determine the structural requirements for binding and pore formation. A model has been proposed: the first step involves oligomerization followed by electrostatic adsorption onto the bilayer-water interface. Then the hydrophobic C-terminal region inserts into the bilayer leaving the amphipathic helix at the interface. The final step is formation of a channel by cecropin oligomers [127]. Computer models predict that dimers of cecropin would form channel-containing regular lattices in the membrane [128]. Using *E. coli* with well defined mutations affecting the outer membrane, Vaara and Vaara [129] concluded that the mode of action of cecropins on the outer membrane resembles that of cationic quaternary amine surfactants. If this is the case, it seems less likely that they form channels in the inner membrane as such detergent-like effect is likely to occur also in this membrane. Gazit et al. [130] studying the binding of fluorescent derivatives of cecropins to liposomes and bacteria concluded that membrane disintegration rather than a pore-forming mechanism is the major cause of membrane permeation. Therefore, whether or not these peptides form channels is still a matter of debate.

5.1.2. Defensins

Defensins are small cationic peptides that contain 29 to 35 amino acid residues including six invariant cysteines forming three intramolecular disulfide bridges. They belong to three different groups, the mammalian α- or 'classical' defensins, the β-defensins and the insect defensins, which differ from each other in the spacing and connectivity of the cysteines and by their 3D structure [131]. Their antimicrobial spectrum includes Gram-positive and Gram-negative bacteria [132] but insect defensins are predominantly active against Gram-positive organisms [123]. Mammalian defensins induce a sequential permeabilization of the outer and inner membrane followed by simultaneous cessation of respiration and of macromolecule synthesis. Permeabilization of the inner membrane can be inhibited by metabolic inhibitors or by a decrease of the protonmotive force [133]. Planar lipid bilayer studies of the rabbit neutrophil defensin NP1 indicated that the peptide forms voltage-dependent and weakly anion selective channels [134] in lipid bilayers. Release of entrapped dextrans from liposomes upon addition of HNP-2, suggested that this human neutrophil defensin forms multimeric pores with a diameter of 2.5 nm. A speculative model was proposed in which an annular pore is formed by a hexamer of dimers [131,135], Formation of voltage-dependent ion channels by defensins was also demonstrated in the case of the insect defensin A [34]. *In vivo* experiments showed that defensin A induced permeability changes of the inner membrane of *Micrococcus luteus* likely to be due to the formation of channels: cytoplasmic potassium was lost, the inner membrane was partially depolarized, cytoplasmic ATP was decreased (but ATP was not found in the external medium) and respiration was inhibited. These effects were prevented or arrested in the presence of divalent cations or by a decrease of the membrane potential below a threshold of 110 mV. Patch clamp experiments with defensin A incorporated into giant liposomes revealed the presence of channels. However, the channels were not well-behaved, showing heterogeneity both in conductances and in opening and closing durations. This behavior was also reported for the mammalian defensins. There are at least two not necessarily exclusive explanations for this: the pores may be formed by multimers of variable numbers of defensin molecules, or each multimer may have different conductance levels.

5.1.3. Magainins

Magainins are a class of antimicrobial peptides secreted from the skin of the frog *Xenopus laevis*. They are 23 residues in length, positively charged and amphipathic. They disrupt the permeability barrier of the *E. coli* inner membrane [136]. Ion channel formation in planar lipid bilayers was demonstrated in the case of magainin 1 [137]. Liposomes permeability assays have not led to a clear model of the mode of action of the peptide. Grant and collaborators summarize their experiments by saying: magainin 2a can permeabilize membranes via bilayer destabilization or by channels. Which mechanism predominates will depend on the membrane system, composition and degree of polarization [138]. Whereas it is reasonably accepted that magainins form a multimeric channel, the characteristics of the channel and the mechanism by which it forms remain less defined. The experiments of Juretic et al.

[139] with cytochrome oxidase-containing liposomes have led to the conclusion that at room temperature most of the magainin is in the membrane-bound monomeric form and that the membrane permeability complex may be a pentamer or hexamer or at least a trimer but that at low temperature some of the membrane-bound magainin might be a dimer. Recently the following model was proposed: (a) the peptide binds to the outer surface of the bilayer and lies parallel to the surface in a monomer-dimer equilibrium; (b) transient pores of multimers are formed; (c) upon closing the peptide translocates across the membrane and is again in the monomer–dimer equilibrium [140].

5.1.4. Colicins

These bacteriocins are toxins of high molecular mass (40–70 kDa) produced by and active against *E. coli* and closely related bacteria. Colicin A, B, E1 and Ia belong to the group of colicins which form voltage-dependent ionic channels in planar lipid bilayers [141,142]. The primary effects on bacteria of these colicins are a leakage of internal K^+ and of phosphate, a collapse of the electrochemical proton gradient $\Delta\tilde{\mu}H^+$ and a decrease in the internal ATP level [33,143,144]. On the basis of these *in vitro* and *in vivo* properties, it is generally believed that the killing activity of these colicins results from the formation of channels in the inner membrane. Colicins have been investigated by molecular biology, biochemical and biophysical methods such that their structure, organization in the bacterial envelope and function is characterized at the molecular level.

5.1.5. Bacteriocins from Gram-positive bacteria

Most of the bacteriocins-like peptides are small (MM in the range of a thousand Daltons), heat stable, cationic and amphiphilic molecules. They show little or no receptor specificity. The best characterized peptides belong to two subclasses according to whether or not they contain lanthionine. Those containing the thioether amino acids are called lantibiotics. Many bacteriocins have two or more cysteines connected by a disulfide bridge.

Most of the bacteriocins of Gram-positive bacteria are membrane active. They cause an efflux of ions and small solutes, hydrolysis and partial efflux of cytoplasmic ATP, inhibition of the respiratory activity and depolarization of the cytoplasmic membrane. This has been shown for Pep5, nisin, subtilin, epidermin, gallidermin, streptococcin and lactacin (reviewed in Ref. [125]). The divalent cations Mg^{2+} and Ca^{2+} have an inhibitory effect on the *in vivo* action of nisin Z and lactacin F against *Listeria monocytogenes* and *lactobacillus* strains. Gadolinium is even more efficient (in the μM range) in preventing or arresting the permeabilizing effect of these bacteriocins [35,145]. The inhibitory effect of gadolinium is not specific to these bacteriocins since it was also observed with defensin A [34] and colicin A and N [146]. The permeabilizing effect of these peptides was confirmed from studies on membranes vesicles and liposomes [147–149]. Planar lipid bilayer experiments indicate that these bacteriocins form pores. Most of these pores show features common to those of the antibacterial peptides described above: conductance fluctuations among different levels with a burstlike character were more frequently found than steplike current increments. This is examplified in the case of Pep5 [150]. Models for the formation of the pores have also been derived from *in vivo* and *in vitro* data and from analysis of the 3D structure of some of the peptides (reviewed in Ref. [125]).

Although of different phylogenetic origins, the antimicrobial peptides described here show common features: they are all cationic and amphiphilic. They permeabilize membranes probably by forming multi-conductance pores. The insertion/ opening of these pores requires in most cases an energized membrane. Liposome permeability assays using fluorescence probes have been extensively used to investigate their mode of action. The peptide-induced permeability of liposomes towards carboxyfluorescein and other fluorescent probes has often been considered as a criterion for channel formation. These data have to be interpreted with caution since leakage of vesicle content is often best explained by membrane destabilization. This abundant literature has unfortunately not allowed a clarification of the situation. The questions with which we are still faced are the following: Do the peptides interact with membranes as monomers or multimers? Do they lie parallel to the membrane surface or not? Is the membrane potential responsible for membrane insertion and/or change in the state of oligomerization? Is membrane destabilization or channel formation responsible for permeabilization?

5.2. Bacteriophage channels

The T-even phage T4 is a very large phage containing a double-stranded DNA molecule of approximately 160 kbp. When the phage attaches to the receptor, the tail contracts, the tip of the internal tube of the phage tail sheet is brought close to the cytoplasmic membrane and the DNA is injected through the cytoplasmic membrane. During the first minutes of infection T4 as well as T2 induce a rapid and transient leakage of ^{32}P and ^{35}S and of cytosolic Mg^{2+} and K^+. There is however no ion loss if the transfer of phage DNA is prevented (for a recent review see Ref. [151]). We have quantitatively analyzed the K^+ efflux induced by phage T4 using a K^+-valinomycin selective electrode and shown that a single phage induces the release of 10^6 ions per second from its host indicating that the efflux is channel-mediated. The initial rate of efflux increases linearly with the number of phages suggesting that each phage induces the formation of one channel [32]. Channel activity was associated with phage-DNA penetration. This observation led us to propose that the main function of these channels is to mediate the transfer of the double-stranded phage DNA.

The T-odd phage T5 possesses a noncontractile tail and a double stranded DNA of 121 kbp. The process of DNA transfer is very unusual: first 8% of the DNA (First Step Transfer or FST DNA) is transferred to the cytoplasm. Then, there is a pause of about 4 minutes during which proteins encoded by this fragment are synthesized. Two of these proteins, A1 and A2, are required for the transfer of the remaining 92% of the DNA or second step transfer (SST) of DNA (reviewed in Ref. [151]). Phage T5 also causes a K^+ efflux but in two steps and the timing of which correlated to the timing of DNA penetration The amplitude of the K^+ efflux also suggested that it is channel-mediated [152].

The very different characteristics of the channel induced by phage T4 and T5 (rate of efflux, dependence for their opening on the transmembrane electrical potential etc.) suggests that the two types of channels are different and presumably originate from the phage. To identify phage protein(s) putatively involved in the formation of these channels, *E. coli* cells were infected with radioactively labeled phage T5. The envelope of these bacteria was fractionated on successive sucrose gradients. A protein belonging to the phage tail, pb2, was recovered in the contact sites between the inner and the outer membrane. Since this protein was only present in the envelope when the DNA was transferred, this suggested that pb2 might be involved in the translocation of phage genome [153]. The only protein of T5 straight fiber, pb2, is a multimer, the length of which would be sufficient to span the whole envelope and thus to gain access to the inner membrane. The putative DNA channel activity of the purified protein is under investigation.

Is channeling a general mechanism for phage DNA penetration? It is probable that the tailed phages at least containing a double stranded DNA share a common mode of DNA penetration. The transient permeabilization of the cell envelope at the beginning of the infectious process by many of these phages probably results from the formation of channels [21].

Although much effort has been devoted to the study of the small single stranded DNA phages, the mechanism by which their DNA traverses the cell envelope is unknown. The closely related F-specific filamentous (Ff) phages fd, f1, and M13 infect *E. coli* strains which harbor the conjugative F plasmid. Infection is initiated by the binding of one end of the phage to the tip of the F pilus. This binding is mediated by the N-terminus of the gene 3-encoded viral protein (g3p). The gene 3-encoded protein has been purified; the oligomeric form of g3p forms large pores, when reconstituted into lipid bilayers, the estimated diameter of which is large enough to allow the passage of the single-stranded DNA [154].

This review has tried to point out the great diversity of channel-mediated functions in bacteria. Some of these channel proteins have been purified and electrophysiological tools can be used in *in vivo* studies such that the understanding of channel function can progress rapidly.

Acknowledgements

C. Berrier, A. Ghazi and C. Hétru are kindly acknowledged for critical reading of the manuscript.

References

1.   Nikaido, H. and Vaara, M. (1985) Microbiol. Rev. **49**, 1–32.
2.   Benz, R. and Bauer, K. (1988) Eur. J. Biochem. **176**, 1–19.
3.   Jap, B.K. and Wallan, P.J. (1990) Quart. Rev. Biophys. **23**, 367–403.
4.   Schirmer, T. and Rosenbusch, J.P. (1991) Curr. Opin. Struct. Biol. **1**, 539–545.
5.   Nikaido, H. and Saier, M.H.J. (1992) Science **258**, 936–942.
6.   Nikaido, H. (1992) Mol. Microbiol. **6**, 435–442.
7.   Nikaido, H. (1993) Trends. Microbiol. **1**, 5–8.

8. Cowan, S.W. (1993) Curr. Opin. Struct. Biol. **3**, 501–507.

9. Cowan, S.W. and Schirmer, T. (1994) in: Bacterial Cell Wall, Vol. 27, eds J.-M. Ghuysen and R. Hakenbeck. Elsevier, London, pp. 353–361.

10. Gennis, R.B. (1989) in: Biomembranes: Molecular Structure and Function, ed C.C. Cantor. Springer-Verlag, New York.

11. Rutz, J.M., Liu, J., Lyons, J.A., Goranson, J., Armstrong, S.K., McIntosh, M.A., Feix, J.B. and Klebba, P.E. (1992) Science **258**, 471–475.

12. Killmann, H., Benz, R. and Braun, V. (1993) EMBO. J. **12**, 3007–3016.

13. Bonhivers, M., Ghazi, A., Boulanger, P. and Letellier, L. (1996) EMBO J. **15**, 1850–1856.

14. Martinac, B., Buechner, M., Delcour, A.H., Adler, J. and Kung, C. (1987) Proc. Natl. Acad. Sci. USA **84**, 2297–2301.

15. Sorgato, M.C., Keller, B.H. and Stühmer, W. (1987) Nature **330**, 498–500.

16. Zoratti, M. and Petronilli, V. (1988) FEBS. Lett. **240**, 105–109.

17. Sukharev, S.I., Blount, P., Martinac, B., Blattner, F.R. and Kung, C. (1994) Nature **368**, 265–268.

18. Salmond, G. and Reeves, J. (1993) TIBS **18**, 7–12.

19. Pugsley, A. (1993) Microbiol. Rev. **57**, 50–108.

20. Simon, S.M. and Blöbel, G. (1992) Cell **69**, 677–684.

21. Letellier, L. and Boulanger, P. (1989) Biochimie **71**, 167–174.

22. Nikaido, H. (1983) Meth. Enzym. **97**, 85–100.

23. Nakae, T. (1976) Biochem. Biophys. Res. Commun. **71**, 877–884.

24. Nakae, T. (1976) J. Biol. Chem. **251**, 2176–2178.

25. Nikaido, H. and Rosenberg, E.Y. (1981) J. Gen. Physiol. **77**, 121–135.

26. Nikaido, H. and Rosenberg, E.Y. (1983) J. Bacteriol. **153**, 241–252.

27. Renkin, E.M. (1954) J. Gen. Physiol. **38**, 225–253.

28. Zimmermann, W. and Rosselet, A. (1977) Antimicrob. Agents. Chemother. **12**, 368–372.

29. Yoshimura, F. and Nikaido, H. (1985) Antimicrob. Agents Chemother **27**, 84–92.

30. Nikaido, H., Rosenberg, E.Y. and Foulds, J. (1983) J. Bacteriol. **153**, 232–240.

31. Bakker, E.P. (1993) in: Alkali Cation Transport Systems in Prokaryotes, ed E.P. Bakker. CRC Press, Boca Raton, pp. 205–224.

32. Boulanger, P. and Letellier, L. (1988) J. Biol. Chem. **263**, 9767–9775.

33. Bourdineaud, J.P., Boulanger, P., Lazdunski, C. and Letellier, L. (1990) Proc. Natl. Acad. Sci. USA. **87**, 1037–41.

34. Cocianchich, S., Ghazi, A., Hetru, C., Hoffmann, J. and Letellier, L. (1993) J. Biol. Chem. **268**, 19239–19245.

35. Abee, T., Klaenhammer, T. and Letellier, L. (1994) J. Environm. Microbiol. **60**, 1006–1013.

36. Benz, R., Janko, K., Boos, W. and Laüger, P. (1978) Biochim. Biophys. Acta **511**, 305–319.

37. Schindler, H. and Rosenbusch, J.P. (1978) Proc. Natl. Acad. Sci. USA **75**, 3751–3755.

38. Simon, S.M., Blobel, G. and Zimmerberg, J. (1989) Proc. Natl. Acad. Sci. USA **86**, 6176–6180.

39. Hamill, O.P., Marty, A., Neher, E., Sakmann, B. and Sigworth, F.J. (1981) Pflüger Arch. **391**, 85–100.

40. Sackmann, B. and Neher, E. (1983) in: Single Channel Recording. Plenum Press, New York.

41. Zoratti, M. and Ghazi, A. (1993) in: Alkali cation transport systems in procaryotes, ed E.P. Bakker. CRC Press, Boca Raton, pp. 349–358.

42. Cui, C., Smith, D.O. and Adler, J. (1995) J. Membr. Biol. **144**, 31–42.

43. Szabo', I., Petronilli, V., Guerra, L. and Zoratti, M. (1990) Biochem. Biophys. Res. Comm. **171**, 280–286.

44. Buechner, M., Delcour, A.H., Martinac, B., Adler, J. and Kung, C. (1990) Biochim. Biophys. Acta **1024**, 111–121.

45. Berrier, C., Besnard, M., Ajouz, B., Coulombe, A. and Ghazi, A. (1996) J. Membr. Biol. **151**, in press.

46. Zoratti, M., Petronilli, V. and Szabo', I. (1990) Biochem. Biophys. Res. Comm. **168**, 443–450.

47. Criado, M. and Keller, B. (1987) FEBS Lett. **224**, 172–176.

48. Berrier, C., Coulombe, A., Houssin, C. and Ghazi, A. (1989) FEBS Lett. **259**, 27–31.

49. Delcour, A.H., Martinac, B., Adler, J. and Kung, C. (1989) Biophys. J. **56**, 631–636.

50. Mizuno, T., Wurtzel, E.T. and Inouye, M. (1982) J. Biol. Chem. **257**, 13692–13698.

51. Lugtenberg, B. and van Alphen, L. (1983) Biochim. Biophys. Acta **737**, 51–115.

52. Weiss, M.S. and Schulz, G.E. (1992) J. Mol. Biol. **227**, 493–509.
53. Cowan, S.W., Schirmer, T., Rummel, G., Steiert, M., Ghosh, R., Pauptit, R.A., Jansonius, J.N. and Rosenbusch, J.P. (1992) Nature **358**, 727–733.
54. Schulz, G. (1993) Curr. Op. Cell. Biol. **5**, 701–707.
55. Nakae, T. and Nikaido, H. (1975) J. Biol. Chem. **250**, 7359–7365.
56. Benz, R. (1988) Ann. Rev. Microbiol. **42**, 359–393.
57. Bauer, K., Struyve, M., Bosch, D., Benz, R. and Tomassen, J. (1989) J. Biol. Chem. **264**, 16393–16398.
58. Benz, R. (1993) in: Bacterial Cell Wall, Vol. 27, eds J.-M. Ghuysen and R. Hakenbeck. Elsevier, London, pp. 397–420.
59. Lakey, J.H. and Pattus, F. (1989) Eur. J. Biochem. **186**, 303–308.
60. Dargent, B., Hofmann, W., Pattus, F. and Rosenbusch, J.P. (1986) EMBO J. **5**, 773–778.
61. Buehler, L.K., Kusumoto, S., Zhang, H. and Rosenbusch, J. (1991) J. Biol. Chem. **266**, 24446–24450.
62. Delcour, A.H., Martinac, B., Adler, J. and Kung, C. (1989) J. Memb. Biol. **119**, 267–275.
63. Berrier, C., Coulombe, A., Houssin, C. and Ghazi, A. (1993) J. Memb. Biol. **133**, 119–127.
64. Sen, K., Hellman, J. and Nikaido, H. (1988) J. Biol. Chem. **263**, 1182–1187.
65. Jeanteur, D., Lakey, J.H. and Pattus, F. (1991) Mol. Microbiol. **5**, 2153–2164.
66. Jeanteur, D., Lakey, J.H. and Pattus, F. (1994) in: Bacterial Cell Wall, Vol. 27, eds J.-M. Ghuysen and R. Hakenbeck. (eds.), Elsevier, London, pp. 363–378.
67. Morona, R., Tomassen, J. and Henning, U. (1985) Eur. J. Biochem. **150**, 161–169.
68. Sugawara, E. and Nikaido, H. (1992) J. Biol. Chem. **267**, 2507–2511.
69. Saint, N., De, E., Julien, S., Orange, N. and Molle, G. (1993) Biochim. Biophys. Acta **1145**, 119–123.
70. Yoshimura, F. and Nikaido, H. (1982) J. Bacteriol. **152**, 636–642.
71. Hancock, R.E.W., Siehnel, R. and Martin, N. (1990) Mol. Microbiol. **4**, 1069–1075.
72. Nikaido, H., Nikaido, K. and Harayama, S. (1991) J. Biol. Chem. **266**, 770–779.
73. Duchene, M., Schweizer, A., Loospeich, F., Krauss, G., Marget, M., von Specht, B.U. and Domdey, H. (1988) J. Bacteriol. **170**, 155–162.
74. deMot, R. (1992) Mol. Gen. Genet. **231**, 489–493.
75. Schwartz, M. (1987) in: E. coli and S. typhimurium: Cellular and Molecular Biology, ed AMS. F.C. Neidhart, Washington, DC, pp. 1482–1502.
76. Schirmer, T., Keller, T.A., Wang, Y.-F. and Rosenbusch, J.P. (1995) Science **267**, 512–514.
77. Werts, C., Charbit, A., Bachellier, S. and Hofnung, M. (1992) Mol. Gen. Genet. **233**, 372–378.
78. Luckey, M. and Nikaido, H. (1980) Proc. Natl. Acad. Sci. USA **77**, 167–171.
79. Benz, R., Schmidt, A. and Vos-Scheperkeuter, H. (1987) J. Membr. Biol. **100**, 21–29.
80. Klebba, P.E., Hofnung, M. and Charbit, A. (1994). EMBO J. **13**, 4670–4675.
81. Benz, R., Schmid, A., Maier, C. and Bremer, E. (1988) Eur. J. Biochem. **176**, 699–705.
82. Maier, C., Bremer, E., Schmid, A. and Benz, R. (1988) J. Biol. Chem. **263**, 2493–2499.
83. Webster, R.E. (1991) Mol. Microbiol. **5**, 1005–1011.
84. Fath, M.J., Skvirsky, R.C. and Kolter, R. (1991)J. Bacteriol. **172**, 7549–7556.
85. Wandersman, C. and Delepaire, P. (1991) Mol. Microbiol. **5**, 1005–1011.
86. Benz, R., Maier, E. and Gentschev, I. (1993) Zbl. Bakt. **278**, 187–196.
87. Jarlier, V. and Nikaido, H. (1990) J. Bacteriol. **172**, 1418–1423.
88. Trias, J., Jarlier, V. and Benz, R. (1992) Science **258**, 1479–1481.
89. McIntosh, M.A. and Earhart, C.F. (1977) J. Bacteriol. **131**, 331–339.
90. DeVaux, L.C. and Kadner, R.J. (1985) J. Bacteriol. **162**, 888–896.
91. Hantke, K. and Braun, V. (1978) J. Bacteriol. **135**, 190–197.
92. Guerinot, M.L. (1994) Annu. Rev. Microbiol. **48**, 743–772.
93. Kampfenkel, K. and Braun, V. (1992) J. Bacteriol. **174**, 5485–5487.
94. Kampfenkel, K. and Braun, V. (1993) J. Biol. Chem. **268**, 6050–6057.
95. Postle, K. (1993) J. Bioenerget. Biomemb.t **25**, 591–601.
96. Liu, J., Rutz, J.M., Feix, J.B. and Klebba, P.E. (1993) Proc. Natl. Acad. Sci. USA **90**, 10653–10657.
97. Braun, V. (1995) FEMS Microbiol. Rev. **16**, 295–307.
98. Boulanger, P., Le Maire, M., Dubois, S., Desmadril, M. and Letellier, L. (submitted).
99. Letellier, L., Bonhivers, M. and Boulanger, P. (submitted).

100. Tsapis, A. and Képès, A. (1977) Biochim. Biophys. Acta **469**, 1–12.
101. Meury, J., Robin, A. and Monnier-Champeix, P. (1985) Eur. J. Biochem. **151**, 613–619.
102. Guharay, F. and Sachs, F. (1984) J. Physiol. (Lond.) **352**, 685–701.
103. Morris, C.E. (1990) J. Membr. Biol. **113**, 93–107.
104. Martinac, B. (1993) in: CRC Thermodynamics of Membrane Receptors and Channels, ed M.B. Jackson. CRC Press, Boca Raton, FL, pp. 327–352.
105. Häse, C., Le Dain, A.C. and Martinac, B. (1995) J. Biol. Chem. **270**, 1–6.
106. Zoratti, M. and Szabo', I. (1991) Biochem. Biophys. Res. Comm. **168**, 443–450.
107. Szabo', I., Petronilli, V. and Zoratti, M. (1992) Biochim. Biophys. Acta **1112**, 29–38.
108. Szabo', I., Petronilli, V. and Zoratti, M. (1993) J. Membr. Biol. **131**,. 203–218.
109. Berrier, C., Coulombe, A., Szabo', I., Zoratti, M. and Ghazi, A. (1992) Eur. J. Biochem. **206**, 559–565.
110. Martinac, B., Adler, J. and Kung, C. (1991) in: Congr. Comp. Physiol. Tokyo, eds Abstr., B. pp. S13.
111. Martinac, B., Adler, J. and Kung, C. (1990) Nature **348**, 261–263.
112. Sheetz, M. and Singer, S.J. (1974) Proc. Natl. Acad. Sci. USA **71**, 4457–4461.
113. Sukharev, S.I., Martinac, B., Arshavsky, V. and Kung, C. (1993) Biophys. J. **65**, 177–183.
114. Koch, A.L. and Pinette, M.F.S. (1987) J. Bacteriol. **169**, 3654–3663.
115. Pinette, M.F.S. and Koch, A.L. (1987) J. Bacteriol. **169**, 4737–4742.
116. Czonka, L.N. and Hanson, A.D. (1991) Ann. Rev. Microbiol. **45**, 569–606.
117. Booth, I.R. (1993) in: Bakker, E.P. (eds.), Alkali Cation Transport Systems in Prokaryotes, ed E.P. Bakker. CRC Press, Boca Raton, FL, pp. 309–331.
118. Schleyer, M., Schmid, R. and Bakker, E.P. (1993) Arch. Microbiol. **160**, 424–431.
119. Killian, J.A., De Jong, A.M.P., Bijvelt, J., Verkleij, A.J. and de Kruijff, B. (1990) EMBO J. **9**, 815–819.
120. Blight, M.A. and Holland, I.B. (1990) Mol. Microbiol. **4**, 873–880.
121. Higgins, C.F. (1992) Ann. Rev. Cell. Biol. **8**, 67–113.
122. Simon, S.M. and Blobel, G. (1991) Cell **65**, 371–380.
123. Hoffmann, J.A. and Hetru, C. (1992) Immunol. Today **13**, 411–415.
124. Jack, R.W., Tagg, J.R. and Ray, B. (1995) Microbiol. Rev. **59**, 171–200.
125. Sahl, H.-J., Jack, R.W. and Bierbaum, G. (1995) Eur. J. Biochem. **230**, 827–853.
126. Boman, H.G. (1991) Cell **65**, 205–207.
127. Christensen, B., Fink, J., Merrifield, R.B. and Mauzerall, D. (1988) Proc. Natl. Acad. Sci. USA **85**, 5072–5076.
128. Durell, S.R., Raghunathan, G. and Guy, H.R. (1992) Biophys. J. **63**, 1623–1631.
129. Vaara, M. and Vaara, T. (1994) Antimicrobiol. Agents Chemother. **38**, 2498–2501.
130. Gazit, E., Lee, W.-J., Brey, P.T. and Shai, Y. (1994) Biochemistry, **33**, 10681–10692.
131. White, S.H., Wimley, W.C. and Selsted, M.E. (1995) Curr. Opin. Struct. Biol. **5**, 521–527.
132. Ganz, T. and Lehrer, R.I. (1994) Curr. Opin. Immunol. **6**, 584–589.
133. Lehrer, R.I., Lichtenstein, A.K. and Ganz, T. (1993) Ann. Rev. Immunol. **11**, 105–128.
134. Kagan, B.L., Selsted, M.E., Ganz, T. and Leherer, R.I. (1990) Proc. Natl. Acad. Sci. USA **87**, 210–214.
135. Wimley, W.C., Selsted, M.E. and White, S.H. (1994) Protein Sci. **3**, 1362–1373.
136. Westerhoff, H., Juretic, D., Hendler, R.W. and Zasloff, M. (1989) Proc. Natl. Acad. Sci. USA **86**, 6597–6601.
137. Duclohier, H., Molle, G. and Spach, G. (1989) Biophys. J. **56**, 1017–1021.
138. Grant, J., E., Beeler, T.J., Taylor, K.M.T., Gable, K. and Roseman, M.A. (1992) Biochemistry **31**, 9912–9918.
139. Juretic, D., Hendler, R.W., Kamp, F., Caughey, W.S., Zasloff, M. and Westerhoff, H. (1994) Biochemistry **33**, 4562–4570.
140. Matsuzaki, K., Murase, O., Fujii, N. and Miyajima, K. (1995) Biochemistry **34**, 6521–6526.
141. Schein, S.J., Kagan, B.L. and Finkelstein, A. (1978) Nature **276**, 159–163.
142. Pattus, F., Martinez, M.C., Dargent, B., Cavard, D., Verger, R. and Lazdunski, C. (1983) Biochemistry **22**, 5698–5707.
143. Kordel, M., Benz, R. and Sahl, H.G. (1988) J. Bacteriol. **170**, 84–88.

144. Guihard, G., Bénédetti, H., Besnard, M. and Letellier, L. (1993) J. Biol. Chem. **268**, 17775–17780.
145. Abee, T., Rombouts, F.M., Hugenholtz, J., Guihard, G. and Letellier, L. (1994) Appl. Environ. Microbiol. **60**, 1962–1968.
146. Bonhivers, M., Guihard, G., Pattus, F. and Letellier, L. (1995) Eur. J. Biochem. **229**, 155–163.
147. Gao, F., Abee, T. and Konings, W. (1991) Appl. Environ. Microbiol. **57**, 2164–2170.
148. Garcera, M., Elferink, M., Driessen, A. and Konings, W. (1993) Eur. J. Biochem. **212**, 417–422.
149. Driessen, J.M., van den Hooven, H., Kuiper, W., van de Kamp, M., Sahl, H.-G., Konings, R. and Konings, W. (1995) Biochemistry **34**, 1606–1614.
151. Letellier, L. (1993) in: Alkali Cation Transport Systems in Procaryotes, ed E.P. Bakker. CRC Press, Boca Raton, FL, pp. 359–379.
152. Boulanger, P. and Letellier, L. (1992) J. Biol. Chem. **267**, 3168–3172.
153. Guihard, G., Boulanger, P. and Letellier, L. (1992) J. Biol. Chem. **267**, 3173–3178.
154. Wuttke, G., Knepper, J. and Rasched, I. (1989) Biochim. Biophys. Acta **985**, 239–249.

CHAPTER 28

Communication Between Membranes in TonB-Dependent Transport Across the Bacterial Outer Membrane

R.J. KADNER, C.V. FRANKLUND and J.T. LATHROP

Department of Microbiology, School of Medicine,
University of Virginia, Charlottesville, VA, USA

© *1996 Elsevier Science B.V.*
All rights reserved

Handbook of Biological Physics
Volume 2, edited by W.N. Konings, H.R. Kaback and J.S. Lolkema

Contents

1. Overview of outer membrane structure and transport properties

Gram-negative bacteria comprise the majority of the bacterial world. The defining structural feature of Gram-negative bacteria, and a major factor in their evolutionary success, is the presence of an outer membrane surrounding the cytoplasmic membrane and peptidoglycan layer. Compared to other biological membranes, the outer membrane characteristically has a very unusual composition, lipid asymmetry, protein content, and function. Outer membranes usually exhibit surprisingly low permeability by hydrophobic compounds and high permeability by polar compounds, owing to their unusual lipid content and the presence of porin channels, respectively. They protect the cell from toxic enzymes and nonpolar compounds, including detergents and many antibiotics, and thus have profound influence on the cell's physiological, medical and ecological capabilities. This chapter reviews our current understanding of the structure, mechanism and interactions of the components of a family of active nutrient transport systems that are located in the outer membrane, but whose action is dependent on communication between both cellular membranes.

1.1. Outer membrane structure

The composition and structure of the bacterial outer membrane has been extensively reviewed [1,2]. The lipid components of this membrane are atypical. The lipids of the inner leaflet are the common glycerol phosphatides, phosphatidyl glycerol, phosphatidyl ethanolamine, and cardiolipin. Most or all of the lipid in the outer leaflet is composed of the complex glycolipid, lipopolysaccharide (LPS), whose membrane-embedded portion, called lipid A, is a diglucosamine moiety with 6 to 8 fatty acyl chains (usually 14:0 or 3-OH 14:0) with additional substitutions on three positions. The presence of the very bulky and rigid lipid A moiety and the linkage of each LPS molecule to other molecules of LPS and protein by divalent cation-mediated and direct salt bridges greatly reduces the fluidity of the outer membrane leaflet. Biophysical studies of the permeability and other properties of isolated outer membrane lipids are complicated by their rapid reorientation and loss of transmembrane asymmetry that occurs when the outer membrane is disrupted [3]. Physiological measurements of the entry into intact cells of several series of antibiotics of similar mode of action and basic structure, but differing oil:water partitioning properties, reveal that the outer membrane exhibits unusually low penetrability by nonpolar molecules that pass readily across most biological membranes [4]. A simplistic explanation for this behavior is that nonpolar molecules dissolve poorly in the very rigid outer lipid layer of the outer membrane.

The unusual behavior of the outer membrane is also influenced by the nature of most outer membrane proteins. They form stable structures that resist denaturation by detergents and digestion by proteases; many also form stable trimeric complexes that resist denaturation. The deduced amino acid sequences of most outer membrane proteins are fairly polar and lack the extensive stretches of highly hydrophobic character that are prominent features of the hydropathy plots of most other membrane proteins [5]. Spectroscopic studies of isolated outer membrane proteins indicate a high content of β-sheet conformation. These observations led to the conclusion that the membrane-spanning segments of outer membrane proteins are amphipathic β-sheets of 8 to 12 residues, rather than the nonpolar or amphipathic α-helical membrane spanners in other proteins. This proposal was clearly validated by the crystallographic structures of several outer membrane proteins (see below).

The outer membrane contains a relatively small number of proteins: 5–10 major proteins, and 10–40 minor proteins in lesser amounts. A few outer membrane proteins, such as murein lipoprotein and OmpA, may play structural roles but they are not essential for cell viability or for outer membrane assembly. The majority of outer membrane proteins participate in transport activities to allow the inward passage of nutrients or the outward movement of surface or secreted proteins and other macromolecules.

1.2. Types of outer membrane transport systems

The high permeability of the outer membrane for hydrophilic solutes is explained by the presence of three basic types of transport systems [6]. The first type is responsible for the movement of most of the molecules that cross the outer membrane and is represented in *Escherichia coli* by the nonspecific or general porins, OmpF and OmpC, and the phosphoporin, PhoE. These general porins distinguish among solutes only on the basis of their molecular dimensions, charge, and polar character [7]. OmpF porin favors the passage of polar molecules with cationic character and molecular weight below 600 Da. There is no indication of stereoselectivity of solute passage or of specific binding of solute to the porin. The porin channels allow only equilibration of solute concentrations across the membrane and show no apparent regulatory control other than a voltage gating that is probably not relevant in the intact cell. The OmpF and OmpC porins are weakly selective for cations, whereas PhoE is weakly selective for anions. The OmpF, OmpC and PhoE porins, when reconstituted in phospholipid bilayers, exhibit channel conductances of 0.8, 0.5 and 0.6 nS (measured at 1 M KCl), and are closed by transmembrane voltages of 90, 160 and 100 mV, respectively [7,8].

The second type of outer membrane transport process is carried out by substrate-specific porins. These channels possess specific binding sites which confer a strong preference for passage of their transport substrates. As channels, they only allow equilibration of their substrate across the outer membrane. Examples include the maltoporin LamB [9] and the nucleoside porin Tsx in *E. coli* [10]. The *E. coli* FadL protein, involved in fatty acid transport across the outer membrane, may belong to this class, although its structure differs from that of the porin family [11]. Many of the porins in the outer membrane of *Pseudomonas aeruginosa* exhibit preferences for specific groups of amino acids or for monosaccharides. As described below, the three-dimensional structure of LamB provides insight into the basis for this substrate preference.

The third type of outer membrane transport system uses the TonB-dependent transporters. These outer membrane transport proteins bind their substrates with high affinity and specificity, and seem to catalyze energy-dependent substrate accumulation, or uphill transport, into the periplasmic space. These transport systems are dependent on the function of the TonB protein complex. This complex appears to couple metabolic energy to the transporters in the outer membrane, which is separated from the cell's supply of energy-rich compounds and is unable to sustain an appreciable transmembrane ion gradient. This chapter addresses the structure and mechanism of the TonB-dependent transporters and the fascinating properties of the TonB energy coupling complex.

1.3. Structure of outer membrane proteins

Although there is no high-resolution structural information about TonB-dependent transporters, valuable guides can be found in the structures of several porin proteins. The structures of porins from several purple photosynthetic bacteria were determined by Schulz and colleagues [12], and more recently the structures of several *E. coli* porins (OmpF, PhoE, LamB) were solved by Rosenbusch's group [8,13] and are discussed in the chapter by Rosenbusch (Chapter 26).

Each porin forms a water-filled transmembrane channel of a defined size and electrostatic charge. The aqueous interior of the channels restricts passage of nonpolar compounds which form micellar structures. The OmpF and PhoE porin monomers take the shape of a barrel composed of 16 anti-parallel, amphipathic β-strands tilted about 45° relative to the plane of the membrane [8]. At the ends of the transmembrane strands are short β-turns on the periplasmic face and longer, irregular but tightly packed loops on the external face. One of these long loops, called the eyelet, folds into the barrel interior to constrict the size of the channel. The amino acid residues in this loop are major determinants of the size and charge discrimination properties of the porin channel, which are altered in response to sequence changes in some *ompF* mutants. A striking feature of the structure is the high prevalence of aromatic residues at the level of the membrane head groups, where they interdigitate with other subunits in the porin trimer and help stabilize the protein at the membrane:water interface. The porin barrel forms a very stable structure, in part because a hydrophobic core forms at the trimer interface.

The structure of maltoporin LamB, which binds sugars with an affinity around 0.1 mM [9], is superficially similar to that of the general porins [13]. LamB exists as a trimer whose subunits form a barrel with 18 anti-parallel, amphipathic β-strands, which are terminated by the aromatic collars and by short turns on the periplasmic face and long, tightly packed loops on the external face. Four loops extend into the barrel and constrict the channel about halfway down its length. The channel opening is thus smaller than in the general porins, and the placement of aromatic residues along the constriction inside the channel is proposed to comprise a 'greasy slide' for the binding, orientation, and preferential passage of oligo-saccharides. On the outer surface, the external loops form tightly packed protuber-ances around half of the top of the barrel, but they do not appear to restrict or affect solute entry into the channel [14]. Thus, all of the porins share the common characteristic structure of a very stable, interdigitated, multimeric complex formed by amphipathic, anti-parallel β-sheets, and it is likely that the TonB-dependent transporters are built on the same basic structure.

2. TonB-dependent transport systems

As far as is known, the only physiological substrates for the TonB-dependent transport systems are iron-siderophore and heme complexes, iron-containing pro-teins, and vitamin B_{12} (CN-Cbl) or other corrin derivatives. The recent description of the sequence relatedness of the chondroitin sulfate transporter, CsuF, from *Bacteroides thetaiotaomicron* suggests that the substrate range may be broader [15]. Furthermore, this protein is the first example of a possible TonB-dependent system outside the beta and gamma subdivision of purple proteobacteria. Siderophores are the low-molecular-weight iron(III) chelators that are produced by microorganisms and extract iron from the insoluble ferric hydroxide complexes that normally form under nonacidic, aerobic conditions. Most microorganisms require iron for growth and many can acquire iron by using the siderophores produced and excreted by themselves or by other microorganisms. The possession of multiple siderophore uptake systems is indicative of the importance of iron transport for bacterial growth. In addition, many pathogenic bacteria can acquire iron directly from heme or the host's iron-binding and transport proteins, such as transferrin, lactoferrin, or hemo-globin [16]. Recent studies suggest that transport of iron from these macromolecu-lar sources is also TonB dependent.

There are several features common to the TonB-dependent transport substrates, i.e., chelated iron and cobalamin. Most are too large to pass through the porin channels at rates capable of sustaining growth. They are metabolically valuable or essential nutrients which are normally present in the environment in low and limiting concentrations, such that competition for them with other microorganisms is likely to be important for cellular growth in mixed culture situations. To compen-sate for the poor diffusion of these substrates through the porin channels, bacteria possess complex transport systems that can acquire these nutrients at the low concentrations normally present in the environment.

The outer membrane transporters and the TonB protein complex appear to be involved solely in solute transport across the outer membrane into the periplasmic space. Once in the periplasm, substrates are transported into the cytoplasm by separate transport systems that employ periplasmic substrate-binding proteins and the ATP-binding cassette (ABC) components typical of periplasmic permeases. The iron-siderophore complex is taken from the periplasm into the cell by the action of ABC periplasmic permeases specific for each group of substrates. Once in the cell, the iron is removed from the siderophore, reduced to Fe(II) state and incorporated into cellular iron proteins. The siderophore can be chemically modified and is then released back into the medium. In the case of transport of iron taken from host proteins, the carrier protein remains outside, and the iron is transferred to a bacterial surface or periplasmic iron-binding protein and internalized. Details on this process are not yet available.

2.1. TonB-dependent transporters

Previous work on TonB-dependent transport has been reviewed [6,17–19]. Table 1 lists properties of some of the TonB-dependent transporters. In some cases, assignment as a TonB-dependent system is based only on sequence homologies, rather than analysis of *tonB* mutants in that strain. *E. coli* possesses at least 8 TonB-dependent transporters, which range in size from 594 to 917 residues, at least twice that of the porins. The most extensively studied in terms of structure–function relationships are BtuB, FepA and FhuA, which mediate the transport of cobalamin, ferric enterobactin, and ferrichrome, respectively.

Typical of outer membrane proteins, the sequences of TonB-dependent transporters exhibit a generally hydrophilic character, with few or no long stretches of nonpolar residues. All are synthesized with a signal sequence which is removed by leader peptidase I during translocation to the outer membrane. Secondary structure predictions indicate a high content of β-sheet conformations. Some proteins (e.g., FepA) contain cysteine residues and may possess disulfide bonds, whereas other proteins (e.g., BtuB) do not, indicating that disulfide bonds are not a conserved structural feature or necessary for assembly.

Most of these transporters serve for the binding and uptake of toxic bacteriocin proteins and bacteriophages, as well as their physiological substrates. Colicins are bactericidal proteins produced by *E. coli* strains and are divided into two groups: group A colicins are dependent on TolA function for entry, and group B colicins are dependent on TonB function. In most cases, the receptors used by the group B colicins require TonB for transport of their natural ligands and share homology regions, called TonB boxes, with the other TonB-dependent transporters. The group B colicins themselves contain sequences related to TonB box-1, which are needed for their uptake [20,21]. Not all of the lethal ligands of the TonB-dependent transporters require TonB function for entry. For example, uptake via BtuB of cobalamin requires TonB, but entry of the E colicins and phage BF23 does not. Similarly, uptake via FhuA of ferrichrome, colicin M, microcin 25 [22] and phages T1 and φ80 require TonB, but entry of phage T5 does not. Conversely, entry of colicin-10 is TonB dependent, although its receptor, the nucleoside porin Tsx, does not carry out any other TonB-dependent reactions and lacks the TonB boxes [23]. The mechanism of colicin entry is beyond the scope of this chapter, but it is clear that these processes can provide insights into TonB function.

2.2. Conserved sequence elements

Comparison of the sequences of TonB-dependent transport proteins reveals the presence of several regions of substantial and characteristic sequence conservation,

Table 1
Some TonB-dependent outer membrane transporters

Prot	Organism	Genbank	Size[a]	Substrates [b]	CM Transp.	TonB Boxes 1.0	TonB Boxes 4.0
BsrA	B. pertussis	U13950	711	Fe-Enterobactin		VQQMATVQ	YGSGAMGGVVNIITKR
BtuB	E. coli	P06129	594	Cobalamins, E colicins*, BF23*	BtuCD	DTLVVTAN	YGSDAIGGVVNIITTR
BtuB	S. typhimurium	P37409	594	Cobalamins, E colicins*, BF23*	BtuCD	DTLVVTAN	YGSDAIGGVVNIITTR
CirA	E. coli	P17315	638	Fe-DHB, ColIa, ColIb, ColV		ETMVVTAS	YGSDALGGVVNIITKK
CsuF	B. thetaiotaomicron	L42372	1036	Chondroitin sulfate		KPMVITLK	YGARGANGVVIITTKK
FatA	V. anguillarum	P11461	691	Fe-Anguibactin	FatBCD	ESITVYGQ	PPNGSVGGSINLVTKR
FcuA	Y. enterocolitica	S30948	722	Ferrichrome		DGQVANGG	PSGSGVGGMINLEPKR
FecA	E. coli	P13036	741	Fe-Dicitrate	FecBCD	DALTVVGD	YGPQSVGGVVMFVTRA
FepA	E. coli	P05825	723	Fe-Enterobactin, ColB, ColD	FepBCDG	DTIVVTAA	YGNGAAGGVVNIITKK
FepA	P. aeruginosa	Q05098	721	Fe-Enterobactin	FepBCDG	QTVVATAQ	YGNGAASSVVNIITKQ
FhuA	E. coli	P06971	714	Ferrichrome, ColM, T1, φ80, T5*	FhuCDB	DTITVTAA	YGKSSPGGLLNMVSKR
FhuE	E. coli	P16869	693	Coprogen, Rhodotorulate	FhuCDB	ETVIVEGS	TGTGNPSAAINMVRKH
FoxA	Y. enterocolitica	Q01674	684	Ferrioxamine		DTIEVTAK	YGQSIPGGVVMMTSKR
FptA	P. aeruginosa	U03161	683	Fe-Pyrochelin		ETELPDMV	HGTGNPAATVNLVRKR
FpvA	P. aeruginosa	U07359	770	Fe-Pyoverdine		NAITISVA	TGAGSLGATINLIRKK
FyuA	E. coli	Z35105	652	Fe-Enterobactin		STLVVTAS	YGKSAQGGIINIVTQQ
FyuA	Y. enterocolitica	Z35487	651	Fe-Yersiniabactin, Pesticin		STLEVTAS	TGKSAQGGIINIVTQQ
HemR	Y. enterocolitica	P31499	658	Heme		DTMVVTAT	YGSGALGGVISYETVD
HutA	V. cholerae	L27149	672	Heme		DEVVVSTT	QGSDAIGGIVAFETKD
HxuC	H. influenzae	U09840	704	Heme/Hemopexin		DSINVIAT	WGSGALGGVVAMRTPN
IrgA	V. cholerae	P27772	627	Unknown		ETMVVTAA	YGSDAIGGVINIITRK
IutA	E. coli	P14542	700	Fe-Aerobactin, Cloacin DF13*	FhuCDB	ETFVVSAN	YGGGSTGGLINIVTKK
LbpA	N. meningitis	X79838	917	Lactoferrin		DTIQVKAK	HGSGALGGAVAFRTKE
NosA	P. stutzerii	M60717	639	Copper complexes		APMVITGV	WGPGNSAATILLERDP
PsnA	Y. pestis	U09530	654	Fe-Yersiniabactin		STLVVTAS	YGKSAQGGIINIVTQQ
PupA	P. putida	P25184	772	Fe-Pseudobactin		NTVTVTAS	TGAGDPSAVVNVIRKR
PupB	P. putida	P38047	764	Fe-Pseudobactin		GTGLYTTY	SGMGNPSATINLIRKR
Tbp1	N. meningitis	JN0819	884	Transferrin		DTIQVKAK	YGNALAGSVAFQTKT
ViuA	V. cholerae	U11759	651	Fe-Vibriobactin		PVLVVIGE	QGRNTSAGAIVMKSND

[a] Polypeptide length, less signal peptide, given in amino acid residues. [b] Substrates marked with asterisk * do not require the function of TonB for entry.

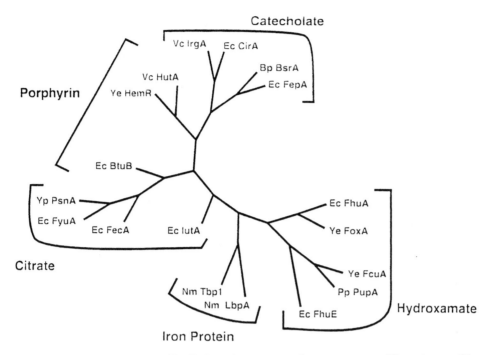

Fig. 1.   Phylogenetic analysis of TonB-dependent outer membrane transporters. The mature peptide sequences of the indicated proteins were aligned using the Pileup program in the University of Wisconsin GCG package version 8.0.1 [108] with a gap creation penalty of 5.0 and a gap extension penalty of 0.5. A bootstrapped protein parsimony analysis was performed on the resulting alignment using 100 data sets with the Seqboot and Protpars programs from the Phylip package of Felsenstein [109]. A consensus tree was then produced using Consense and Drawtree programs of Phylip.

interspersed with regions of little or no obvious relatedness. When compared pairwise, some transporters are more closely related to one another in these variable regions than are others. A phylogenetic relationship of some TonB-dependent transporters is presented in Fig. 1. Transporters that carry chemically related substrates, such as catechols versus hydroxamates, tetrapyrroles, citrate-based sidero-phores, or iron-binding proteins show greater sequence relatedness in the variable regions, as would be expected for segments of proteins that bind structurally similar substrates. However, this correlation of protein sequence and substrate structure is not invariant, as should be expected for sets of proteins that diverged at different times [24].

The regions that are strongly conserved among TonB-dependent proteins are called TonB boxes. Their locations in BtuB, FepA and FhuA are indicated in Fig. 2. They may be involved in interaction with and response to TonB, and are not likely to comprise substrate recognition regions. Some regions could act inde-pendently of TonB, such as in subunit interactions, binding to LPS or other outer membrane proteins, or in targeting to the outer membrane. The TonB box-4 region [25], owing in part to its greater length, is the most strikingly conserved region and

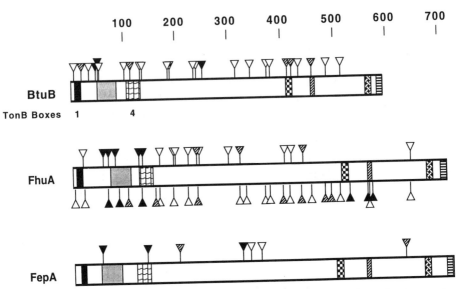

Fig. 2.   Location and effect of oligonucleotide linker insertions in TonB-dependent transporters. The mature lengths of BtuB, FhuA, and FepA are illustrated. Conserved domains are indicated by the various filled boxes. TonB boxes 1 and 4 are specifically labelled. Positions of linkers in BtuB [32], FhuA [31] upper set, [33] lower set, and FecA [30] are indicated by the symbols. The phenotype of the transporters with the linker insertions is indicated by the shading of the symbol. Open symbols designate positions where linker insertions had no discernable effect upon transporter function. Linker insertions which resulted in partial loss of function, or loss of specific functions are shown as striped symbols. Positions of linker insertions resulting in complete loss of transporter functions are shown as black symbols.

represents the signature sequence that defines the family of the TonB-dependent transporters. Only one region, TonB box-1, has been subjected to extensive genetic analysis. For sake of comparison, the sequences of TonB boxes -1 and -4 are listed in Table 1.

2.3. Genetic analysis of transporter function

The BtuB, FepA and FhuA proteins have been examined by a variety of genetic analyses to probe the involvement of specific regions in transporter function and membrane localization. Numerous amino acid substitutions in the TonB box-1 region (which is also present in the group B colicins) indicate the involvement of this region in interaction with TonB. The most extensive study of amino acid substitutions in TonB box-1 was performed with BtuB [26]. Changes in this region which introduce residues likely to distort secondary structure (Pro or Gly) conferred a TonB-uncoupled phenotype, in which ligand binding activity was normal but TonB-dependent transport by that transporter, but not by other transporters, was abolished. Other substitutions in this region, even at positions at which substitution of proline or glycine eliminated transport activity and those which resulted in a radical change of residue character at each position in TonB box-1, had little to no detectable effect on transport activity. It was concluded that the participation of TonB box-1 in energy coupling does not require recognition of specific amino acid side chains, a surprising conclusion in light of the strong conservation of sequence in this region. Instead, higher order protein structure seems to be crucial. Comparable results were seen with other TonB-dependent proteins [21,27,28]. It may be that TonB box-1 acts in a complex with other parts of the transporter protein. As will be described later, the defective phenotype resulting from mutations in TonB box-1 can be corrected by amino acid substitutions in TonB in an allele-specific manner, indicating a direct interaction between these proteins. This interaction may involve other regions of the transporters besides just TonB box-1.

A related study of the role of individual amino acid residues in the function of BtuB involved suppression of 13 nonsense mutations throughout the *btuB* gene using a panel of amber suppressor mutations which introduce different amino acids [29]. Most amino acid substitutions at most sites had no obvious effect on the level of BtuB in the outer membrane or its activity towards any of its substrates. Suppression with a lysine-inserting tRNA at most positions resulted in defective function, primarily with respect to adsorption of phage BF23. It is possible that each of the amber mutations occurred in a region that contributes to BF23 binding. However, the lysine-inserting suppressor was the least efficient suppressor of the panel, and it is conceivable that the resistance to BF23 was related to the low amount of intact BtuB that was produced rather than the presence of a lysine at each of these positions.

A mutagenesis approach that has been useful and informative involves the insertion of short 2 to 16 amino acid peptide sequences at specific sites in the BtuB, FepA, and FhuA transporter proteins [30–33]. The positions of these insertions and their effect on transporter activity are represented in Fig. 2. It was surprising that the majority of proteins carrying short linker insertions exhibited no apparent defect in their level in the outer membrane or in any of their transport activities. This finding indicates a striking degree of plasticity in the sequence requirements for membrane insertion and transport function. Although cytoplasmic membrane proteins can tolerate a considerable degree of amino acid replacement without effect, they are much less tolerant of peptide insertion, especially those that place charged residues in the transmembrane segments.

Some linker insertions resulted in partial or total loss of outer membrane transport functions. As summarized in Fig. 2, most of these disruptive insertions fell in or near the TonB boxes. Some insertions of 12 or 16 amino acids in FhuA possessed transport activity for all substrates but rendered the protein susceptible to site-specific cleavage when the protease subtilisin was added to intact cells or spheroplasts [33]. Since the native protein is resistant to protease digestion unless denatured, these insertions in the C-terminal half of FhuA must have distorted the structure of the protein in regions nonessential for transport activity to allow subtilisin action.

The availability of the linker insertions facilitated *in vitro* construction of in-frame deletions that remove defined portions of the coding sequences of BtuB, FepA and FhuA [30,34,35]. The length and position of the deletions relative to putative transmembrane segments was determined by the location of the restriction sites created by the linker insertions. With a sufficient panel of insertions, it is expected that short deletions can be obtained which could reveal the contribution of a region to membrane insertion and transport function. The results vary among different proteins. In FhuA, most deletions from residue 21 to 440 (out of 714 residues in the mature polypeptide) resulted in substantial or complete loss of transport function, although the proteins were incorporated in the outer membrane, as detected with sensitive immunological reagents [36]. Deletions beyond residue 100 were present in much lower than normal amounts when Coomassie blue staining was used for detection, suggesting that proteins with deletions in residues 100 to 300 can be incorporated in the outer membrane, but are not stably maintained. An exception involves proteins with deletions removing segments around residue 180 and 430, which were not detected in the outer membrane. Similar results were seen with deletions in BtuB, except that most proteins with deletions removing sequences past residue 50 were not stably incorporated in the outer membrane and were thus completely defective for transport function [35].

As discussed further below, some deletions in FepA and FhuA that remove a large extracellular loop resulted in a substantial increase in nonspecific outer membrane permeability [37–39]. Cells expressing these proteins exhibit susceptibility to antibiotics and detergents to which the wild-type cell is resistant. It was proposed that the TonB-dependent transporters form a transmembrane barrel structure analogous to that of the porins, but this channel is closed on the outside by a dome consisting of interacting external protein loops. The deleted loop contains residues that contribute to the substrate-binding sites (see below) and is thought to serve as a regulated gate allowing selective access to the underlying porin channel. The deletion of this loop presumably prevents the formation of the gate and allows access of unrelated solutes to the channel. The deletion variants exhibit increased channel permeability when purified and reconstituted into membranes [37,38].

To test the generality of this response, in-frame deletions were constructed covering portions along most of the length of BtuB [110]. Many of these deletions, even ones that are not stably incorporated in the outer membrane, resulted in increased outer membrane permeability, as indicated by increased susceptibility to antibiotics that are normally excluded by the outer membrane. Some of these deletion variants might possess ungated channels, but it is possible that some variant proteins affect the local structure of the outer membrane. This may allow the nonpolar solutes to diffuse around the variant protein, rather than through its opened channel, owing to changes in the local lipid asymmetry.

Another surprising example of plasticity was provided by the properties of in-frame duplications, in which segments of BtuB were repeated in tandem manner [35,110]. Unlike the deletions, which removed the same segments as were duplicated, the duplication proteins were incorporated stably in the outer membrane, and almost all of them functioned normally for all transport activities. If the deletion derivatives were unstably incorporated in the outer membrane because of the improper progression of transmembrane segments, then the duplications should exhibit the same phenotype, since they possess the same improper progression of transmembrane segments. The fact that this was not the case suggests that stable membrane incorporation requires the presence of the normal number of transmembrane segments, but that additional ones are tolerated, either by expanding the size of the barrel structure or by forming an extra appendage that does not, however, interfere with transport function.

2.4. Transmembrane topology of transport proteins

Many models for transmembrane orientation have been proposed, in most of which the transporters have between 16 and 30 membrane spanning regions. Unfortunately, at present there are no three-dimensional structures of TonB-dependent proteins, and there are no convincing guides for prediction of the orientation of transmembrane segments or of extramembranous loops, analogous to the positive-inside rule for other membrane proteins. The topological reporter PhoA, which has been widely used with cytoplasmic membrane proteins, does not seem to be able to cross the outer membrane and has not been helpful [40]. Thus, the transmembrane orientation of these proteins has been addressed by immunological, biochemical, and genetic approaches.

As mentioned above, some linker insertions render FhuA susceptible to subtilisin added to one or both sides of the outer membrane, which is interpreted as indicating the orientation of the region carrying the inserted peptides [33]. Peptide insertions of a foreign epitope recognized by a monoclonal antibody have been obtained in FhuA and FepA [41,42]. Since the transporters carrying these epitope insertions retained transport function, it is assumed that they possess the native conformation and orientation. Furthermore, several panels of monoclonal antibodies directed against native epitopes in FepA and FhuA have been prepared and the locations of their epitopes on the primary sequence roughly mapped [43]. Some epitopes that are predicted to lie in external loops are reactive in intact cells, whereas others are not. It is likely that many linear epitopes are buried in tightly packed loops and cannot be recognized by an antibody unless the protein is denatured. These approaches have clearly identified several sites as exposed on the cell surface, but there are many other sites for which the evidence about their transmembrane orientation or exposure is equivocal.

2.5. Substrate-binding sites

The locations of putative substrate-binding sites in the TonB-dependent transporters have been proposed based on the loss of specific transport activities in some insertion and deletion mutants, and the inhibition of transport activities by the binding of certain monoclonal antibodies. A large extracellular loop covering residues 258 to 340 in FepA and 316 to 356 in FhuA appears to play an important role in substrate binding. This role was examined by testing the ability of overlapping hexapeptides with the sequence of parts of this loop in FhuA to compete for substrate binding with the intact loop. Several of these hexapeptides from two or three distinct portions of the external loop were able to bind and inactivate phages T1, T5, and φ80, as well as colicin M, all of which bind to the intact loop [44]. The presence of these domains in such close proximity to each other suggests that they may interact with each other to form a single functional domain in the intact protein. In addition, three other hexapeptides separated in location from the above-mentioned regions also inhibited ligand binding. The presence of tyrosine residues in most of these peptides, as well as the abrogation of ligand binding by imidazole suggests the involvement of one or more tyrosine residues in FhuA ligand binding. Definitive assignment of these residues awaits the construction of the appropriate mutants.

3. TonB complex

Mutations in the *tonB* gene were first described in 1943 by Luria and Delbruck [45] as one class of mutations that confer resistance to phage T1. The other T1-resistance locus, *tonA*, encodes the receptor for the phage and has been renamed *fhuA*, in light of its physiological function in ferric hydroxamate uptake [46]. Mutants defective at *tonB* grow poorly on many media and are inhibited by the presence of chromium salts, which complex ferric ions [47]. Phage ϕ80 can bind reversibly to FhuA, but TonB is required for irreversible adsorption of the phage and DNA injection [48]. The amount of reversibly versus irreversibly adsorbed phage can be titered, and a quantitative measure of TonB function thus obtained. When Exb mutants were selected for resistance to colicin B due to the hyper-excretion of the siderophore enterobactin, mutations were found in *tonB* as well as two adjacent genes, *exbB* and *exbD*. These three gene products are used only for energy coupling to the TonB-dependent outer membrane transporters and are required only for transport across the outer membrane. This was shown by the observations that TonB is not required when the outer membrane is permeabilized, either in spheroplasts [49] or by osmotic shock [48]. It is the mechanism of the communication and energy transduction by TonB between the cytoplasmic and outer membranes that is of considerable interest.

3.1. TonB sequence characteristics

The *tonB* gene has been cloned and sequenced from *Escherichia coli* [50], *Salmonella typhimurium* [51], *Serratia marcescens* [52], *Yersinia entercolitica* [53], *Klebsiella pneumonia* [54], *Enterobacter aerogenes* [55], *Haemophilus influenzae* [56], and *Pseudomonas putida* [57]. The heterologous TonB proteins can substitute for the *E. coli* homolog to varying degrees, ranging from complete restoration of all TonB transport functions by *S. typhimurium* [51], to marginal complementation by *Y. entercolitica* [53] to no detectable complementation by *P. putida* [57]. The degree of sequence identity to the *E. coli* protein is related to the ability to complement transport activities by the *E. coli* transporters [54,55].

Inspection and comparison of the TonB sequences identified three potentially interesting regions: a hydrophobic segment near the amino terminus, the unusual X-Pro repeat region near the middle, and a highly conserved region near the carboxyl terminus. The *E. coli* TonB protein is 239 residues long, and is translocated to its intermembrane site without proteolytic processing. Despite its actual size of 26 kDa, TonB migrates on SDS-PAGE as a protein of 36–40 kDa. This anomalous migration results from its unusually high proline content (16%), which is particularly concentrated in the central X-Pro region, which is composed of repeats of Glu-Pro followed shortly by an equal number of Lys-Pro repeats. There are 6 such repeats in the *E. coli* homolog and 7 in *S. typhimurium*, but they are absent from the *P. putida* homolog. As expected, the isolated X-Pro region has an unusual and highly rigid conformation, as indicated by NMR analysis [58]. Similar NMR experiments indicated that a synthetic, 33-residue X-Pro peptide interacts with the FhuA protein [59]. However, Larsen et al. [60] found that deletion of part of the X-Pro repeat did not impair TonB function. They suggested that the rigid structure formed by the X-Pro region contributes to the extension of TonB across the periplasmic space to allow contact with the outer membrane transporters, but that it itself is not involved in direct contact.

The highly conserved carboxyl terminal region of TonB is strongly hydrophobic in *E. coli*; but not in some other bacteria. A hybrid protein in which the C-terminal region of the *E. coli* homolog was replaced with the weakly hydrophobic region from the *K. pneumoniae* homolog possessed the same activity as wild-type *E. coli* version, indicating that the strong hydrophobicity of this region is not required for TonB function [54]. Deletion of the 15 C-terminal residues abolished TonB activity, but deletion of the C-terminal eight residues did not [61].

3.2. Accessory proteins

Although null mutations in *tonB* result in the loss of transport function by the TonB-dependent transport systems, mutations in *exbB* and *exbD* conferred a partial defect, such as reduced substrate transport and reduced susceptibility to colicins and phages. No other mutations that specifically affect the TonB-dependent transport process have yet been described. The *exb* genes in *E. coli* are organized in an operon at 65 min that is separated from the *tonB* gene at 28 min. In other organisms, such as *Pseudomonas putida* and *Haemophilus influenzae* [57,62], all three genes are organized in the same operon in the order *exbB-exbD-tonB*. When the sequence of the 26.1 kDa ExbB and the 15.5 kDa ExbD proteins were deduced from the nucleotide sequence, they were seen to share about 25% sequence identity to the *tolQR* gene products, respectively, which function together with the closely linked *tolA* gene in the uptake of the group A colicins [63–65]. The low amount of TonB-dependent transport activity seen in *exbB* or *exbD* mutants was abolished by the presence of additional mutations in *tolQ* or *tolR*, respectively [66]. These results suggest that, since the related homologues of the accessory proteins can partly substitute for each other, TolA and TonB might carry out similar functions.

3.3. Transmembrane topology

The transmembrane topologies of TonB and the Exb proteins have been investigated with the use of the topological reporters PhoA or BlaM, and by susceptibility to proteolytic digestion in spheroplasts. PhoA fusion analysis of TonB indicated that the bulk of the protein is translocated across the cytoplasmic membrane and resides in the periplasmic space [67]. The strongly hydrophobic and highly conserved amino terminal region from residues 9 to 31 is the transmembrane segment, while the amino-terminal residues 1 to 8 lie in the cytoplasm. The presence of the transmembrane segment is required for translocation of the remainder of TonB from the cytoplasm, since its deletion blocks export [68,69]. Translocation is also blocked by introduction of a charged residue (Gly 26 to Asp) in the middle of this region [70]. The highly conserved sequence of the transmembrane segment suggests that it serves as more than just a membrane anchor and translocation signal. Replacement of the TonB amino terminus with a transmembrane segment from the TetA protein allowed membrane translocation and anchorage, but eliminated TonB transport function, indicating that the transmembrane segment makes sequence-specific functional interactions with some cellular component [68,70].

Further indication of the role of the transmembrane segment and the necessity for membrane anchorage of TonB was provided by a mutant form of TonB in which a leader peptidase cleavage site was placed at the end of the transmembrane segment [70]. In this mutant, the bulk of the TonB protein was translocated across the cytoplasmic membrane and then released into the periplasm. This cleaved protein was nonfunctional for transport activity.

The ExbB protein is predicted to contain three transmembrane domains, oriented so that the amino terminus is outside the cytoplasm and there are two large cytoplasmic domains: an internal loop and the C-terminal end [71]. ExbD is predicted to possess a single transmembrane segment (residues 23–43) oriented so that the bulk of the protein (residues 44–141) resides in the periplasm [72]. Essentially identical topologies are predicted for TolQ and TolR, the homologs of ExbB and ExbD.

3.4. Interaction of TonB with transporters and Exb proteins

The transmembrane topology of TonB is consistent with models in which it interacts with cytoplasmic membrane components, including ExbB or ExbD, and also with its outer membrane transporters. Several lines of biochemical and genetic evidence support this hypothesis. When TonB protein is overexpressed from multicopy plasmids or using a strong T7 RNA polymerase-driven promoter, the amplified protein is unstable, with a half-life of <10 min [73,74]. This instability appears to result from deranged stoichiometry of interacting proteins. The excess overexpressed protein is stabilized substantially when either ExbB or FhuA, but not ExbD, is simultaneously overexpressed. Simultaneous overexpression of ExbB and TonB stabilized TonB by decreasing its sensitivity to proteases (primarily OmpT) [74,75]. TonB expressed at normal levels in a wild-type cell is quite stable, with a half-life >90 min [76], but it is unstable in the absence of either ExbB or ExbD [77]. ExbB could protect TonB from protease degradation by physically covering protease sensitive sites or by causing TonB to adopt a protease-resistant conformation [75].

The physical stability of TonB, indicated by its long half-life in wild-type cells, is in contrast to its functional lability. The level of TonB function, assayed by the uptake of several independent substrates, decreases with a half-time of 15–20 min upon inhibition of general protein synthesis or of TonB synthesis alone [78,79]. The rate of loss is accelerated by the presence of substrates for TonB-dependent transport systems. This functional lability in conjunction with physical stability could reflect that only newly synthesized TonB or TonB that is at a specific location in the cell can be functional.

The existence of allele-specific suppressors is consistent with the formation of complexes between TonB and ExbB or the transporter proteins. As described above, certain mutations in TonB box-1 in BtuB, Cir, or FhuA [26–28] result in a TonB-uncoupled phenotype, in which ligand binding is normal, but all TonB-dependent transport activities of the transporter are lost. Mutations in multicopy *tonB* have been isolated as weak suppressors of this transport defect in *btuB* and in *fhuA* mutants [27,55,80]. Almost all of these suppressor mutations resulted in the change of a single residue on TonB, Gln-160, to Leu, Lys or Pro. The Gln-160 residue is located near the C-terminal conserved region, but is itself not strongly conserved. The Gln-160 substitutions exhibited allele-specific activity, in that each suppressor had a different ability to suppress different TonB box substitutions in the transporters. These results suggest that TonB and the transporters interact, although they do not require that the Gln-160 residue itself is making direct contact, rather that the residue replacing Gln-160 affects the conformation of the interacting surface in distinct ways.

A similar approach made use of the finding that deletion of Val-17 in TonB resulted in complete loss of TonB function, without affecting its level or translocation to the periplasm. Partial restoration of transport activity was obtained with a mutation (A39E) near the first transmembrane segment of ExbB [81]. Although the allele specificity of this suppression is not resolved, this phenomenon is consistent with the existence of a TonB-ExbB complex.

An intriguing result that is consistent with the interaction of TonB box-1 with TonB is the finding that treatment of cells with the pentapeptide Glu-Thr-Val-Ile-Val, which is part of the TonB box-1 sequence of FhuE, inhibited several TonB-dependent processes, including growth on low-iron media and killing by TonB-dependent phage and colicins [82].

Biochemical evidence for the interaction of TonB with other proteins was obtained by K. Postle and colleagues through the use of formaldehyde crosslinking, combined with immunoprecipitation of TonB and any covalently associated proteins [69,75]. Four specific TonB-containing crosslinked complexes were seen with wild-type cells. One complex migrating at the position of 59 kDa was assigned as TonB-ExbB complex. It was not found in cells that were TonB⁻, ExbB⁻, or contained the deletion of Val-17 of TonB.

Formaldehyde treatment also resulted in formation of a 195 kDa complex that was recognized by antibodies specific for TonB or FepA [75]. The complex was absent in both FepA⁻ and TonB⁻ strains but it occurred in the forms of TonB lacking the transmembrane anchor. There were two crosslinked complexes of 77 and 43.5 kDa containing TonB associated with unknown proteins. No complexes were seen between TonB and FhuA or any other outer membrane transporter.

4. Energetics and regulation of TonB function

Considerable progress is being made in defining the function, location and interactions of the components that participate in TonB-dependent transport. To understand the significance of their interactions, it is necessary to define the physiological parameters of these transport processes and the nature and consequences of energy coupling; specifically, whether TonB mediates active transport.

4.1. Regulation of synthesis of TonB and the transporters

The levels of TonB and its client and accessory proteins are regulated by growth conditions. The levels of TonB protein and mRNA are repressed about three-fold by iron supplementation, relative to iron-limited cells [83]. The levels of the iron-siderophore transport proteins are also derepressed to various degrees by iron deprivation. A major regulatory factor in the response to iron supply is the Fur (ferric uptake regulatory) protein, a typical transcriptional repressor that is activated for binding to a specific DNA sequence when it is complexed with divalent cations, such as Mn(II) or Fe(II) [84]. The DNA sequence to which Fur protein binds is present, often in multiple copies, in the promoter region of iron-regulated genes [85]. The degree of repression by iron of the TonB-dependent ferric-siderophore transporters ranges from about 3-fold for FhuA to >10-fold for FepA [86].

Under anaerobic conditions, when iron is readily available as Fe(II), TonB is not necessary for growth [87], but is still required for transport of vitamin B_{12}. Nevertheless, *tonB* mRNA expression under anaerobic conditions is repressed ten-fold by iron in Fur-dependent manner [85]. TonB transcription has been reported to be growth-stage dependent, and regulated by oxygen through effects on DNA supercoiling [88]; however, chemicals, pH, and temperature, all of which affect DNA supercoiling, had no effect on TonB transcription [85].

The level of BtuB in the outer membrane is not affected by the iron supply, but is repressed >20-fold by cobalamin supplementation [89, 90]. The *btuC* locus for transport across the cytoplasmic membrane does not appear to be subject to transcriptional regulation [91].

TonB-dependent transport activity could be limited by the level of the outer membrane transporter, of TonB or its accessory proteins, of the ABC system for transport across the cytoplasmic membrane, or, in the case of iron, of the components for iron release from siderophores and reduction and incorporation into iron-containing proteins. Iron supplementation affects the levels of each of those components for iron uptake, but it affects only the TonB component for CN-Cbl uptake. CN-Cbl uptake in wild-type cells is not affected by iron supplementation, indicating that even the 3-fold repressed level of TonB is sufficient for maximal CN-Cbl uptake. When the level of BtuB and the rate of CN-Cbl uptake is amplified by multicopy overexpression, the repressed level of TonB is no longer sufficient, as shown by the repression of CN-Cbl uptake by iron [79,92]. This finding is the first demonstration that the repression of TonB can affect transport activity.

4.2. Energetics of transport

The uptake of ferric-enterobactin and CN-Cbl in wild-type cells is inhibited by conditions that diminish either the proton motive force (pmf) or the ATP pool, even when the interchange of the pmf and ATP pools is blocked by *unc(atp)* mutations. These results indicate that transport employs both pmf- and ATP-dependent processes. Transport across the cytoplasmic membrane is likely to be ATP dependent, like other systems that use ABC periplasmic permeases. Arsenate is an inhibitor of the maintenance of the cellular ATP pool. When CN-Cbl transport is assayed in *btuC* mutants (which are blocked in transport across the cytoplasmic membrane and thus isolate outer membrane transport) no effect by arsenate is seen [93]. This demonstrates that ATP is not required for transport across the outer membrane under aerobic conditions. CN-Cbl uptake across the outer membrane in *btuC* mutants is as susceptible to inhibition by protonophores or inhibitors of pmf generation as it is in the wild-type cell, and roughly as susceptible as known pmf-driven symport systems [93,94]. These results indicate that the pmf is necessary for CN-Cbl transport across the outer membrane, but there has not been a demonstration of a direct role of the pmf as energy source. One consequence of depletion of the pmf is loss of ion gradients that create the turgor pressure that keeps the cytoplasmic membrane and the outer membrane in apposition. If proper positioning between the two membranes is needed to allow TonB to contact its client transporters, if only so that TonB can span the periplasm, the loss of the pmf could indirectly inhibit transport by disrupting proper membrane contacts.

There have been contradictory reports on whether TonB-dependent transporters carry out active transport, defined as the accumulation of unbound substrate inside the periplasm to concentrations substantially higher than the external level. These experiments must be performed in mutant strains defective in transport across the cytoplasmic membrane so that only entry into the periplasm is measured. Elimination of the periplasmic substrate-binding protein will eliminate the appearance of accumulation through protein-mediated retention of substrate in the periplasm and thus give a more accurate reflection of the free concentration of substrate. Reynolds et al. [95] estimated that transported CN-Cbl accumulated to a concentration of 25 μM in the periplasm of a *btuC* strain blocked in transport across the cytoplasmic membrane, corresponding to a concentration ratio of at least 1000-fold. Accumulation was prevented by treatment of the cells with protonophores or genetic loss of TonB function. The demonstration of periplasmic accumulation during iron transport is more difficult, owing to the uncertainty of the time of the release of the labeled iron from the siderophore. In a *fepB* mutant defective in the putative periplasmic binding protein for ferric enterobactin, little accumulation was seen [96]. Similarly, in *fhuB* mutants blocked in cytoplasmic membrane transport, little accumulation of ferrichrome was obtained unless the FhuD periplasmic binding protein was overexpressed [97]. However, substantial accumulation of aerobactin was described in strains defective for cytoplasmic membrane transport [98]. In these systems, the external concentration of siderophore (in the μM range) was much higher than in the cobalamin transport assays (in the nM range) and accumulation to 25 μM would not have appeared to be a substantial gradient.

4.3. Interactions among TonB-dependent transporters and substrates

All of the evidence presented above indicates that TonB interacts with the outer membrane transporter proteins, and not with their substrates, with the possible exception of the colicins that possess TonB box-1 sequences. It is most likely that each TonB molecule acts in conjunction with numerous transporters. The stoichiometry of TonB relative to its client proteins is not definite, but it is estimated that under derepressed conditions, there are 3000 molecules of TonB per cell, but more than 10,000 molecules of each of 4 to 8 iron transporters [99]. How can so few TonB molecules service so many transporters, if the mobility of each TonB molecule is restricted by its anchorage to a protein complex in the cytoplasmic membrane and its need to poke through the peptidoglycan mesh? Rough calculations of the outer membrane surface area that could be contacted by a cytoplasmic membrane-embedded protein with an extended arm of 100 amino acids show that 3000 randomly distributed molecules could reach the entire cell surface of 6 μm^2, but only if the 100-residues were in a β-sheet conformation. Thus, each transporter molecule might be reached by a TonB molecule. Given the slow rate of transport (<1 molecule per sec per receptor), the amount of TonB could be adequate to activate each transporter.

Even if each transporter can be contacted by TonB, there might still be competition between transporters for the limiting amounts of TonB, which would be indicated by the inhibition of uptake of one TonB-dependent substrate by the presence of another. Mutual inhibition of uptake has been seen under certain conditions. The uptake of CN-Cbl in wild-type cells is inhibited by the presence of siderophores, so long as the transporter for that siderophore is functional [79]. The converse situation showed no effect of CN-Cbl on ferrichrome uptake, presumably owing to the very low number of BtuB molecules relative to FhuA molecules. When the number of BtuB molecules was increased by multicopy expression to a level comparable to that of the derepressed iron transporters, CN-Cbl now inhibited ferrichrome uptake to a substantial degree [92]. The degree of inhibition increased further in cells that were grown with excess iron to decrease the levels of both FhuA and TonB, and the inhibition by CN-Cbl was eliminated when the cells were grown with CN-Cbl to repress the level of BtuB.

If there is competition between transporters for limiting amounts of TonB, then overexpression of TonB should suppress the competition. Testing of this prediction is complicated by the finding that multicopy overexpression of TonB results in about a 3-fold decrease in activity of all TonB-dependent activities. The basis for this decrease is not clear, and could reflect competition between active and inactive forms of TonB for binding to the transporters [100], or inhibition by the degradation fragments that are produced from the TonB made in excess of ExbB levels (suggested by K. Postle). Overexpression of both TonB and BtuB resulted in lower levels of CN-Cbl and ferrichrome uptake than when TonB was made in single gene copy number, but there was no mutual inhibition of uptake of one substrate by the other. These results are consistent with the competition for TonB by its client transporters.

The interaction of the transporters with TonB presumably results in a conformational change in them, but evidence for such changes is indirect. The clearest example came from the study of the ability of ferrichrome, whose uptake via FhuA is TonB-dependent, to inhibit the binding of phage T5, whose uptake is TonB-independent. Hantke and Braun [101] showed that in a metabolically active *tonB*[+] strain, ferrichrome is a poor inhibitor of T5 adsorption, whereas in the absence of TonB function or cellular energy, ferrichrome is an effective inhibitor, able at low concentrations to block completely T5 adsorption. They suggested that the presence of active TonB changes the conformation of FhuA so that the binding sites for T5 and ferrichrome do not overlap. In analogous experiments with the Btu system, CN-Cbl could completely block phage BF23 adsorption, but about 100 times higher concentration of CN-Cbl was needed in cells with active TonB [100]. Also, certain *tonB* mutations suppress certain *fhuA* mutations to restore sensitivity to phage T5 [102]. There should be considerable interest in delineating the nature and extent of this putative conformational change with the aid of biophysical probes [103].

If TonB transmits metabolic energy to the outer membrane, there must exist a mechanism to prevent the futile dissipation of energy upon coupling of TonB with empty transporters. No matter what the form of the energy-dependent modification of TonB, whether chemical modification or change in its conformation, nonproductive discharge of this activation by interaction with empty transporters would represent an energetic drain that must be prevented. A likely possibility is that TonB only binds to transporters with bound ligand. This possibility would require that ligand binding results in a transmembrane conformational change that affects the structure of the TonB-binding sites, which are presumably exposed to the periplasm. The formaldehyde crosslinking approach might allow a test of whether contact with TonB requires the presence of substrates.

4.4. Channel properties of transporters

Reconstitution of TonB-dependent active transport in membrane vesicles or lipid bilayers is unlikely to be accomplished in the near future, owing to the need for proper apposition of distinct membranes and the many uncertainties about the nature of energy coupling. However, it is possible to examine some properties of the outer membrane transporters in reconstituted systems. The transport activities of the wild-type proteins can be compared to those of certain deletion derivatives. As described above, expression of FepA or FhuA variants carrying in-frame deletions of a predicted, large extracellular loop resulted in deranged outer membrane permeability properties. These cells were susceptible to a number of structurally unrelated detergents and antibiotics which did not affect cells expressing the intact proteins. These variants could allow utilization of siderophores as the iron source, but in a TonB-independent manner. It was suggested that removal of the extracellular loop removed a component of the gating function that normally prevents solute passage through the channel.

The deletion variant of both FepA [38] and FhuA [37] formed diffusion channels in reconstituted membranes that were substantially larger than the channels formed by the general porins. For example, the single-channel conductance of the FhuA Δ322-355 was 3.0 nS (in 1 M KCL), which is 3–4 times greater than the OmpF channel [37]. Wild-type FhuA formed no large ion channels under these conditions.

5. Models for TonB action

It may be helpful to discuss several models for the mechanism of TonB-dependent transport in light of the findings described above. No precedent exists for energy-coupling by interaction of proteins located in two separate membranes, and there remains uncertainty about the source of metabolic energy and the extent of substrate accumulation that results.

Holroyd and Bradbeer in 1984 [104] proposed that TonB and the pmf generate a diffusible periplasmic molecule that would mediate ligand release from the transporters into the periplasm. This model overcomes the stoichiometric quandary resulting from the difference between the numbers of TonB and transporter molecules. However, it does not account for the evidence of a direct interaction of TonB and the transporters, and no mutants have been isolated that could represent such component.

There is evidence that Ca^{2+} is specifically required for high affinity binding of CN-Cbl to BtuB [105,106]. Treatment of wild-type cells with EDTA buffers resulted in decreased CN-Cbl binding with half-maximal binding at 30 nM Ca^{2+}. The dipeptide insertion at BtuB residue 50 resulted in a decreased affinity for Ca^{2+} in its ability to stimulate CN-Cbl binding (half-maximal binding at 10 μM Ca^{2+}). It was suggested that perhaps TonB binding decreases the affinity of BtuB for Ca^{2+}, resulting in release of calcium and thus of CN-Cbl bound to BtuB. The involvement of calcium in transporter action is an interesting question, but still requires a change in protein affinity as a consequence of a functional interaction of activated TonB with BtuB.

Postle has proposed [69] that TonB should be viewed as a member of a complex containing TonB, ExbB/D, and an as yet unidentified proton translocator. TonB could be synthesized in an 'active', protease sensitive conformation, capable of interacting with the receptors. TonB would undergo a conformational change in response to the pmf, which is transmitted to the receptors via the X-Pro region. This change would cause the release of the ligand. ExbB would mediate a return by TonB to an 'inactive', protease-resistant conformation. TonB's protease sensitivity could act as a mechanism for preventing unproductive TonB-receptor interactions. This theory does not resolve the stoichiometric quandary, as the degradation of TonB bound to receptors, while preventing unproductive interactions, results in nonfunctional TonB. Thus, the amount of TonB available for interaction with the transporters would still be limited.

Klebba [18] resolves the stoichiometric discrepancy by proposing that TonB actually organizes the transporters into a higher-order conformation, as the hub of a wheel organizes the spokes. Rather than interacting with each receptor in turn, a TonB conformational change would stimulate the release of ligand from several receptors simultaneously.

5.1. Gated pore or active transporter

Klebba [39] and Braun [37,107] have demonstrated that deletion of a large, extracellular loop in FepA and FhuA opens a channel which allows passage of a variety of unrelated substrates. They propose the transporters possess a transmembrane channel analogous to the porins, but that the channel is closed on the extracellular face by the interweaving of the external loops, forming a gate that blocks passage of solutes and also forming part of the binding site for the specific substrates. Deletion of one loop thus opens the gate and allows nonspecific passage through the underlying porin channel.

It was proposed that TonB acts to open the gate to allow bound substrate to be released into the channel. This model of simple gate opening does not easily account for accumulation of substrate in the periplasm against a concentration gradient. Since CN-Cbl uptake across the outer membrane is reversible, it is most likely that BtuB can mediate bi-directional transport. It is not known whether exit is affected by TonB function, but there is clear evidence that TonB function changes the affinity of BtuB for CN-Cbl in the reverse direction [95]. Thus, a gated porin model cannot easily account for the action of BtuB and probably of the other transporters in substrate accumulation, bi-directional flux, and change in substrate affinity. These considerations suggest that BtuB undergoes a conformational change, likely dependent on energized TonB action, that results in changed affinity for CN-Cbl on opposite sites of the outer membrane. This behavior is characteristic of an active transporter, which could nonetheless operate on the structural background of a porin channel. We do not favor the term gated porin, which implies a process of downhill equilibration alone.

5.2. A model

We propose the following model, which addresses most of the described properties relating to TonB function (Fig. 3). There is considerable support for the proposal that the transporters form a transmembrane porin-like barrel, with a larger diameter than that in the porins owing to the presence of more transmembrane segments. The topological analyses indicate the presence of large extramembranous loops on the exterior face, which form a tightly interwoven dome-like structure that comprises the high-affinity substrate-binding sites. We suggest that substrate binding to the dome complex on the outer face causes a conformational change in the transporter. This results in substantial reorientation of the protein so that the high affinity substrate-binding site is exposed to the inside of the porin-like channel. This protein reorientation also exposes or alters access of the N-terminal TonB boxes. The existence of a substrate-induced reorientation could explain the observed competitive interaction among the transporter's substrates without requiring that their binding regions physically overlap, so long as the binding of one substrate alters the protein so that the binding sites for the other substrates are no longer accessible. Release of the bound substrate occurs when TonB, in an interaction-competent conformation, contacts the transporter at the newly exposed TonB-interaction region. Attachment of TonB to the exposed TonB-interaction region drives the trans-

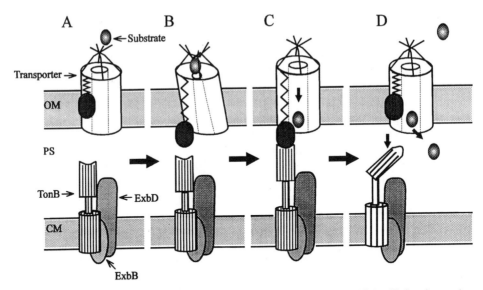

Fig. 3. A model for TonB-dependent transport. The substrate binds to a high affinity site on the transporter (A). This binding causes a conformational change in the transporter, affecting the access to ligand-binding sites on the exterior and presenting a domain for interaction with TonB (B). The binding of TonB to the substrate-loaded transporter reduces the affinity of the substrate-binding site, resulting in the release of the substrate (C). The proton-motive force across the cytoplasmic membrane generates a conformational change in TonB which breaks the TonB-transporter interaction, returning the transporter to a binding-competent conformation (D). OM, outer membrane; PS, periplasmic space; CM, cytoplasmic membrane.

porter to a new conformation with lowered substrate affinity that leads to release of substrate from the high affinity binding site. This allosteric effect could cause the release of Ca^{2+}, which would result in the release of CN-Cbl, as Ca^{2+} is required for high-affinity binding of CN-Cbl to BtuB [105,106]. Release of substrate allows the transporter to relax into a conformation with an outward-facing substrate-binding site, but with low substrate affinity. In a metabolic energy-dependent process (probably pmf) mediated by ExbB, ExbD, or an as yet unknown protein, a conformational change in TonB triggers its release from the transporter and TonB then returns to its original conformation. This process returns the transporter to its original high-affinity conformation, capable of binding external substrate. TonB would also be returned to an interaction-competent state, able to participate in additional transport events.

This model requires a ligand–receptor complex to already be formed in order for TonB to bind. Postle has shown that a TonB-BtuB complex is only formed in cross-linking experiments when CN-Cbl is present in the reaction (personal communication). This model also supposes that both substrate binding and TonB binding cause the conformational changes in the transporters which alter their binding affinities.

References

1. Lugtenberg, B. and Van Alphen, L. (1983) Biochim. Biophys. Acta **737**, 51–115.
2. Nikaido, H. and Vaara, M. (1987) in: Outer Membrane, eds F.C. Neidhardt, J.L. Ingraham, K.B. Low, B. Magasanik, M. Schaechter and H.E. Umbarger. American Society for Microbiology, Washington D.C. pp. 7–22.
3. Muhlradt, P.F. (1975) Eur. J. Biochem. **51**, 343–352.
4. Nikaido, H. and Vaara, M. (1985) Microbiol. Rev. **49**, 1–32.
5. Nikaido, H. and Saier, M.H. (1992) Science **258**, 936–942.
6. Nikaido, H. (1994) J. Biol. Chem. **269**, 3905–3908.
7. Benz, R. (1988) Ann. Rev. Microbiol. **42**, 359–393.
8. Cowan, S.W., Schirmer, T., Rummel, G., Steiert, M., Ghosh, R., Pauptit, R.A., Jansonius, J.N. and Rosenbusch, J.P. (1992) Nature **358**, 727–733.
9. Gehring, K., Cheng, C.-H., Nikaido, H. and Jap, B.K. (1991) J. Bacteriol. **173**, 1873–1878.
10. Fsihi, H., Kottwitz, B. and Bremer, E. (1993) J. Biol. Chem. **268**, 17495–17503.
11. Black, P.N. (1991) J. Bacteriol. **173**, 435–442.
12. Weiss, M.S., Abele, U., Weckesser, J., Welte, W., Schiltz, E. and Schulz, G.E. (1991) Science **254**, 1627–1630.
13. Schirmer, T., Keller, T. A., Wang, Y.-F. and Rosenbusch, J.P. (1995) Science **267**, 512–514.
14. Lepault, J., Dargent, B., Tihelaar, W., Rosenbusch, J.P., Leonard, K. and Pattus, F. (1988) EMBO J. **7**, 261–268.
15. Cheng, Q., Yu, M.C., Reeves, A.R. and Salyers, A.A. (1995) J. Bacteriol. **177**, 3721–3727.
16. Cornelissen, C.N. and Sparling, P.F. (1994) Molec. Microbiol. **14**, 843–850.
17. Kadner, R.J. (1990) Molec. Microbiol. **4**, 2027–2033.
18. Klebba, P.E., Rutz, J.M., Liu, J. and Murphy, C.K. (1993) J. Bioenerget. Biomembr. **25**, 603–611.
19. Postle, K. (1990) Molec. Microbiol. **4**, 2019–2025.
20. Braun, V., Pilsl, H. and Gross, P. (1994) Arch. Microbiol. **161**, 199–206.
21. Mende, J. and Braun, V. (1990) Molec. Microbiol. **4**, 1523–1533.
22. Salomon, R.A. and Farias, R.N. (1995) J. Bacteriol. **177**, 3323–3325.
23. Pilsl, H. and Braun, V. (1995) Molec. Microbiol. **16**, 57–67.
24. Koebnik, R., Hantke, K. and Braun, V. (1993) Molec. Microbiol. **7**, 383–393.
25. Lundrigan, M.D. and Kadner, R.J. (1986) J. Biol. Chem. **261**, 10797–10801.
26. Gudmundsdottir, A., Bell, P.E., Lundrigan, M.D., Bradbeer, C. and Kadner, R.J. (1989) J. Bacteriol. **171**, 6526–6533.
27. Bell, P.E., Nau, C.D., Brown, J.T., Konisky, J. and Kadner, R.J. (1990) J. Bacteriol. **172**, 3826–3829.
28. Schoffler, H. and Braun, V. (1989) Mol. Gen. Genet. **217**, 378–383.
29. Hufton, S.E., Ward, R.J., Bunce, N.A.C., Armstrong, J.T., Fletcher, A.J.P. and Glass, R.E. (1995) Molec. Microbiol. **15**, 381–393.
30. Armstrong, S.K., Francis, C.L. and McIntosh, M.A. (1990) J. Biol. Chem. **265**, 14536–14543.
31. Carmel, G., Hellstern, D., Henning, D. and Coulton, J.W. (1990) J. Bacteriol. **172**, 1861–1869.
32. Gudmundsdottir, A., Bradbeer, C. and Kadner, R.J. (1988) J. Biol. Chem. **263**, 14224–14230.
33. Koebnik, R. and Braun, V. (1993) J. Bacteriol. **175**, 826–839.
34. Killmann, H. and Braun, V. (1992) J. Bacteriol. **174**, 3479–3486.
35. Koster, W., Gudmundsdottir, A., Lundrigan, M.D., Seiffert, A. and Kadner, R.J. (1991) J. Bacteriol. **173**, 5639–5647.
36. Carmel, G. and Coulton, J.W. (1991) J. Bacteriol. **173**, 4394–4403.
37. Killmann, H., Benz, R. and Braun, V. (1993) EMBO J. **12**, 3007–3016.
38. Liu, J., Rutz, J.M., Feix, J.B. and Klebba, P.E. (1993) Proc. Natl. Acad. Sci. USA **90**, 10653–10657.
39. Rutz, J.M., Liu, J., Lyons, J.A., Goranson, J., Armstrong, S.K., McIntosh, M.A., Feix, J.B. and Klebba, P.E. (1992) Science **258**, 471–475.
40. Murphy, C.K. and Klebba, P.E. (1989) J. Bacteriol. **171**, 5894–5900.
41. Armstrong, S.K. and McIntosh, M.A. (1995) J. Biol. Chem. **270**, 2483–2488.
42. Moeck, G.S., Bazzaz, B.S.F., Gras, M.F., Ravi, T.S., Ratcliffe, M.J.H. and Coulton, J.W. (1994) J. Bacteriol. **176**, 4250–4259.

43. Murphy, C.K., Kalve, V.I. and Klebba, P.E. (1990) J. Bacteriol. **172**, 2736–3746.
44. Killmann, H., Videnov, G., Jung, G., Schwarz, H. and Braun, V. (1995) J. Bacteriol. **177**, 694–698.
45. Luria, S.E. and Delbruck, M. (1943) Genetics **28**, 491–511.
46. Kadner, R.J., Heller, K., Coulton, J.W. and Braun, V. (1980) J. Bacteriol. **143**, 256–264.
47. Wang, C.C. and Newton, A. (1971) J. Biol. Chem. **246**, 2147–2151.
48. Braun, V., Frenz, J., Hantke, K. and Schaller, K. (1980) J. Bacteriol. **142**, 162–168.
49. Weaver, C.A. and Konisky, J. (1980) J. Bacteriol. **143**, 1513–1518.
50. Postle, K. and Good, R.F. (1983) Proc. Natl. Acad. Sci. USA **80**, 5235–5239.
51. Hannavy, K., Barr, G.C., Dorman, C.J., Adamson, J., Mazengera, L.R., Gallagher, M.P., Evans, J.S., Levine, B.A., Trayer, I.P. and Higgins, C.F. (1990) J. Mol. Biol. **216**, 897–910.
52. Gaisser, S. and Braun, V. (1991) Mol. Microbiol. **5**, 2777–2787.
53. Koebnik, R., Baumler, A.J., Heesemann, J. and Braun, V. (1993) Molec. Gen. Genet. **237**, 152–160.
54. Bruske, A.K., Anton, M. and Heller, K. J. (1993) Gene **131**, 9–16.
55. Bruske, A.K. and Heller, K. J. (1993) J. Bacteriol. **175**, 6158–6168.
56. Jarosik, G.P., Sanders, J.D., Cope, L.D., Muller-Eberhard, U. and Hansen, E.J. (1994) Infec. Immun. **62**, 2470–2477.
57. Bitter, W., Tommassen, J. and Weisbeek, P.J. (1993) Molec. Microbiol. **7**, 117–130.
58. Evans, J.S., Levine, B.A., Trayer, I.P., Dorman, C.J. and Higgins, C.F. (1986) FEBS Lett. **208**, 211–216.
59. Brewer, S., Tolley, M., Trayer, I.P., Barr, G.C., Dorman, C.J., Hannavy, K., Higgins, C.F., Evans, J.S., Levine, B.A. and Wormald, M.R. (1990) J. Mol. Biol. **216**, 883–895.
60. Larsen, R.A., Wood, G.E. and Postle, K. (1993) Molec. Microbiol. **10**, 943–953.
61. Anton, M. and Heller, K.J. (1991) Gene **105**, 23–29.
62. Jarosik, G.P. and Hansen, E.J. (1995) Gene **152**, 89–92.
63. Eick-Helmerich, K. and Braun, V. (1989) J. Bacteriol. **171**, 5117–5126.
64. Sun, T.P. and Webster, R.E. (1987) J. Bacteriol. **169**, 2667–2674.
65. Webster, R.E. (1991) Molec. Microbiol. **5**, 1005–1011.
66. Braun, V. and Herrmann, C. (1993) Molec. Microbiol. **8**, 261–268.
67. Roof, S.K., Allard, J.D., Bertrand, K.P. and Postle, K. (1991) J. Bacteriol. **173**, 5554–5557.
68. Karlsson, M., Hannavy, K. and Higgins, C.F. (1993) Molec. Microbiol. **8**, 379–388.
69. Postle, K. (1993) J. Bioenerget. Biomembr. **25**, 591–601.
70. Jaskula, J.C., Letain, T.E., Roof, S.K., Skare, J.T. and Postle, K. (1994) J. Bacteriol. **176**, 2326–2338.
71. Kampfenkel, K. and Braun, V. (1993) J. Biol. Chem. **268**, 6050–6057.
72. Kampfenkel, K. and Braun, V. (1992) J. Bacteriol. **174**, 5485–5487.
73. Braun, V., Gunter, K. and Hantke, K. (1991) Biol. Metals **4**, 14–22.
74. Fischer, E., Gunter, K. and Braun, V. (1989) J. Bacteriol. **171**, 5127–5134.
75. Skare, J.T., Ahmer, B.M.M., Seachord, C.L., Darveau, R.P. and Postle, K. (1993) J. Biol. Chem. **268**, 16302–16308.
76. Skare, J.T. and Postle, K. (1991) Molec. Microbiol **5**, 2883–2890.
77. Ahmer, B.D., Thomas, M.G., Larsen, R.A. and Postle, K. (1995) J. Bacteriol. **177**, 4742–4747.
78. Kadner, R.J. and Bassford, P.J., Jr. (1977) J. Bacteriol. **129**, 254–264.
79. Kadner, R.J. and McElhaney, G. (1978) J. Bacteriol. **134**, 1020–1029.
80. Heller, K.J., Kadner, R.J. and Gunther, K. (1988) Gene **64**, 147–153.
81. Larsen, R.A., Thomas, M.G., Wood, G.E. and Postle, K. (1994) Molec. Microbiol. **13**, 627–640.
82. Tuckman, M. and Osburne, M.S. (1992) J. Bacteriol. **174**, 320–323.
83. Postle, K. (1990) J. Bacteriol. **172**, 2287–2293.
84. Bagg, A. and Neilands, J.B. (1987) Microbiol. Revs. **51**, 509–518.
85. Young, G.M. and Postle, K. (1994) Molec. Microbiol. **11**, 943–954.
86. McIntosh, M.A. and Earhart, C.F. (1977) J. Bacteriology **131**, 331–339.
87. Hantke, K. (1987) FEMS Microbiol. Lett. **44**, 53–57.
88. Dorman, C.J., Barr, G.C., Bhrian, N.N. and Higgins, C.F. (1988) J. Bacteriol. **170**, 2816–2826.
89. Kadner, R.J. (1978) J. Bacteriol. **136**, 1050–1057.
90. Lundrigan, M.D., Koster, W. and Kadner, R.J. (1991) Proc. Natl. Acad. Sci. USA **88**, 1479–1483.
91. DeVeaux, L.C. and Kadner, R.J. (1985) J. Bacteriol. **162**, 888–896.

92. Kadner, R.J. and Heller, K.J. (1995) J. Bacteriol. **177,** 4829–4835.
93. Bradbeer, C. (1993) J. Bacteriol. **175,** 3146–3150.
94. Bradbeer, C. and Woodrow, M.L. (1976) J. Bacteriol. **128,** 99–104.
95. Reynolds, P.R., Mottur, G.P. and Bradbeer, C. (1980) J. Biol. Chem. **255,** 4313–4319.
96. Matzanke, B.F., Ecker, D.J., Yang, T.-S., Huynh, B.H., Muller, G. and Raymond, K.N. (1986) J. Bacteriol. **167,** 674–680.
97. Koster, W. and Braun, V. (1990) J. Biol. Chem. **260,** 21407–21410.
98. Woolridge, K.G., Morrisey, J.A. and Williams, P.H. (1992) J. Gen. Microbiol. **138,** 597–603.
99. Earhart, C.F. (1996) in: *Escherichia coli* and *Salmonella*, eds F. Neidhardt et al. ASM Press.
100. Mann, B. J., Holroyd, C.D., Bradbeer, C. and Kadner, R.J. (1986) FEMS Lett **33,** 255–260.
101. Hantke, K. and Braun, V. (1978) J. Bacteriol. **135,** 190–197.
102. Killmann, H. and Braun, V. (1994) FEMS Microbiol. Lett. **119,** 71–76.
103. Liu, J., Rutz, J.M., Klebba, P.E. and Feix, J B. (1994) Biochemistry **33,** 13274–13283.
104. Holroyd, C. and Bradbeer, C. (1984) in: Cobalamin transport in *Escherichia coli*, ed D. Schlessinger. ASM Press, Washington, D.C.
105. Bradbeer, C., Reynolds, P.R., Bauler, G.M. and Fernandez, M.T. (1986) J. Biol. Chem. **261,** 2520–2523.
106. Bradbeer, C. and Gudmundsdottir, A. (1990) J. Bacteriol. **172,** 4919–4926.
107. Braun, V., Killmann, H. and Benz, R. (1994) FEBS Lett. **346,** 59–64.
108. (1994) Program Manual for the Wisconsin Package, Version 8. Genetics Computer Group, Madison, WI.
109. Felsenstein, J. (1989) Cladistics **5,** 164–166.
110. Lathrop, J.T., Wei, B.-Y., Touchie, G.A. and Kadner, R.J. (1995) J. Bacteriol. **177,** 6810–6819.

CHAPTER 29

Colicin Transport, Channel Formation and Inhibition

H. BÉNÉDETTI and V. GÉLI

*Laboratoire d'Ingénérie et Dynamique des Systèmes Membranaires,
CNRS, 31 Chemein J. Aiguier, 13402 Marseille Cedex 20, France*

© 1996 Elsevier Science B.V.
All rights reserved

*Handbook of Biological Physics
Volume 2, edited by W.N. Konings, H.R. Kaback and J.S. Lolkema*

Contents

1. Introduction

Colicins are toxic proteins produced by and active against *E. coli* and closely related bacteria. These proteins are plasmid encoded, are produced in large amounts, and often released into the extracellular medium. To kill target cells, colicins first bind to the cell outer membrane by parasitizing receptors whose physiological function is the transport of metabolites. Subsequently, the toxins have to reach their cellular targets on which they will exert their lethal activity.

Pore-forming colicins constitute the main group of colicins and form ion channels in the cytoplasmic membrane of sensitive cells, thereby destroying their energy potential. Nuclease-type colicins degrade the DNA of target cells or specifically cleave their 16S ribosomal RNA. Finally, one colicin, colicin M, has been shown to inhibit the biosynthesis of murein.

Therefore, to reach their cellular targets, nuclease-type colicins have to translocate through the outer membrane and the inner membrane of sensitive cells. In contrast, pore-forming colicins have to translocate through the outer membrane and to insert into the inner membrane. This insertion implies a molecular rearrangement of the colicin pore-forming domain and its transition from a water-soluble state to a membrane-inserted state.

In view of these properties, colicins provide a useful means of investigating the underlying mechanisms and the energetics of protein translocation across and into membranes. They are also useful tools to study voltage-gated channels and ion channel structure.

It is worth noting that the mechanism of colicin release into the extracellular medium is completely different from that involved in the import process. The release involves no topogenic export signal in the polypeptide chain of colicins [1] and is mediated by the so-called lysis proteins. These lysis proteins are encoded by a gene which forms an operon with the colicin structural gene and the expression of both genes is co-regulated under SOS control [2–5]. The lysis proteins are lipoproteins [6–10] and their mature forms have been found to be localized in both outer and inner membranes [2,6,11,12] except for that of colicin N which has only been found in the outer membrane [9]. These proteins are thought to promote a non-specific increase in the permeability of the cell envelope by exerting a direct effect on the inner membrane and an indirect effect (mediated by phospholipase A activation) on the outer membrane. This phenomenon induces the quasi-lysis of the cell culture and the death of the colicin producing cells.

Colicin producing cells are protected from the lethal activity of the toxins by a specific protein, called immunity protein. The gene of this protein is adjacent to the colicin gene on the same plasmid. The immunity protein has been shown to be an integral membrane protein in the case of pore-forming colicins. It has been demonstrated, for colicin A, that the immunity protein interacts with the pore-forming domain of the colicin as it inserts into the inner membrane and that it prevents it from opening its channel as normally in the presence of membrane potential [13]. The transmembrane helices of the immunity protein might somehow interact with membrane inserted portions of the colicin A channel in order to block any further conformational changes necessary for the channel opening.

With this very specific way of inhibiting channel opening, pore-forming colicins and their immunity proteins constitute a useful model system to understand how amino acid residues can interact in a lipidic environment.

This review explores the recent advances in our understanding of the molecular mechanisms involved in colicin translocation across the outer membrane, the insertion of pore-forming colicins into the inner membrane and the inhibition of their lethal activities by the corresponding specific immunity proteins. We will mainly focus on colicin A, which is the colicin we have been working on in our laboratory.

2. The mode of action of colicins

The mode of action of colicin is divided into three steps. They first bind to a specific receptor at the cell surface. For that purpose, some colicins have parasitized proteins of the outer membrane whose function is dedicated to the transport of iron siderophores (FepA, FhuA, FhuE, Cir), of vitamin B12 (BtuB), or nucleotides (Tsx). Others have parasitized the major porin OmpF through which small hydrophilic solutes with MW of up to 650 Daltons can pass [14–16].

During the second step of their mode of action, called 'translocation step', colicins cross the outer membrane, the periplasmic space, and the inner membrane in the case of nuclease-type colicins, in order to reach their cellular targets. Mutant strains possessing functional receptors but insensitive to a particular colicin have been isolated and described. Such mutants are usually denoted as 'tolerant' and are affected in a specific set of envelope proteins [17–20]. Colicins and a variety of bacteriophages have taken advantage of these resident proteins to enter bacteria. Group A colicins (A, E1 to E9, K, L, N, and Cloacin DF13) and filamentous phages f1, fd, and M13, need the Tol proteins (TolA, TolB, TolQ, and TolR) to penetrate into the cells [18, 21]. Group B colicins (B, D, Ia, Ib, M, 5 and 10) and phages T1 and Φ80 need TonB and its associated proteins ExbB and ExbD [17].

The last step of the mode of action consists in the membrane potential-dependent opening of the ionic-channel in the cytoplasmic membrane for pore-forming colicins (A, B, E1, Ia, Ib, N, and K), the cleavage of DNA or 16S ribosomal RNA in the cytoplasm for nuclease-type colicins (E2 to E9), and the inhibition of murein biosynthesis in the cytoplasmic membrane for colicin M.

3. The molecular organization of colicins in three domains

Consistent with the three steps in the mode of action of colicins, their polypeptide chains comprise three domains, linearly organized, corresponding to each of these steps. The N-terminal domain is involved in the translocation step, the central domain is responsible for receptor binding, and the C-terminal domain carries the lethal activity.

3.1. Determination of the domain boundaries

Different methods have been used to define more precisely the boundaries of these domains in different colicins. Genetic engineering techniques have allowed us to construct chimeric colicins from which specific regions had been deleted [22]. Hybrid colicins have also been constructed by exchanging the functional domains of colicin A and colicin E1 (two group A colicins) [23,24] and those of colicin A and colicin B (a group B colicin) [25]. Polypeptide chain fragments have been obtained by limited proteolysis and their properties have been tested [26–31]. Furthermore, it has been possible to obtain information on domain boundaries on the basis of sequence homologies: between colicins with the same type of lethal activity (A, B, E1, Ia, Ib, and N for pore-forming colicins), (E2, E7, E8, and E9 for DNase-type colicins) (E3, E6, and CloDF13 for RNase-type colicins); between colicins using the same receptors; and between colicins using the same translocation pathways (B, D),(Ia, Ib),(E2, E3, E6 and E9).

Colicins having the same type of killing activities display sometimes high levels of sequence similarities in their C-terminal domains. However, these domains are divergent enough to specifically interact with the corresponding immunity proteins thereby leading to a specific inhibition of their lethal activities. Therefore, colicins are sometimes homologous but these homologies are restricted to particular domains which correspond to the functional domains. This observation suggests that colicins have evolved through recombination of their functional domains (for a review, see Riley [32]).

3.2. Sequence homologies between different colicins

As mentioned above, some colicins which share the same receptor and the same translocation pathway display high levels of sequence homologies in their N-terminal and central domains. In contrast, colicins which only share the same receptor (BtuB for colicin A, E1, and E2) present only short regions or no region at all of sequence similarities. Therefore, these colicins might interact with different regions of the same receptor.

In the same way, colicins which share the same translocation pathway only display short stretches of similarity. Group B colicins have been shown to contain a consensus sequence of 5 to 6 aa called the TonB-box. Mende and Braun [33] and Pilsl et al.[34] have shown that this sequence plays a role in colicins B and M translocation because a colicin mutated in this region is unable to be translocated. This consensus sequence is also present in high affinity outer membrane receptors of iron siderophores and vitamin B12, and, as discussed further, this box has been shown to play an important role in the coupling of these receptors to the TonB protein.

Group A colicins do not present extensive similarities in their N-terminal domains except that they are rich in glycine residues. Recently, Pilsl and Braun [35] identified a glycin-rich pentapeptide region conserved in all Tol-dependent colicins and called it TolA-box. This sequence is important in the case of colicin E3 because a point mutation in this consensus sequence prevents the colicin to be translocated [36,37]. However, also other regions of the N-terminal domains of colicins seem to be important for translocation because a colicin A mutant deleted from aa 15 to 30 but with an intact TolA-box is also unable to translocate through the cell envelope [22].

3.3. Overall structure of colicins

Pore-forming colicins have been reported to be highly asymmetrical molecules with axial ratios of 8 to 10 [38]. Recently, the X-ray crystal structure of an entire colicin, colicin Ia, has been determined at a resolution of 4 Å [39]: the protein is highly α-helical and has an unusually elongated 'Y'-shape. The stem and the two arms of the 'Y' form three distinct structural domains which are likely to correspond to the three functional domains of the colicin. Typical structural characteristics indicate that the two arms of the 'Y' shape correspond to the N-terminal and C-terminal domains while the stem would correspond to the central domain [39].

4. The proteins of the bacterial cell envelope involved in colicin translocation

4.1. The TonB system

Since the TonB system is extensively described in a chapter of this book, we will only summarize the most recent data on the system in the light of the role it plays in the translocation step of group B colicins.

4.1.1. Components of the TonB system
The TonB system is composed of three proteins: TonB, ExbB, and ExbD (Fig. 1). ExbB is an integral membrane protein with three transmembrane domains and its

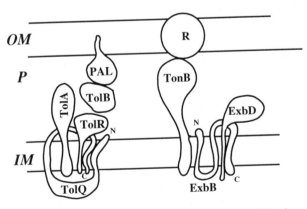

Fig. 1.   Tol and TonB systems. OM = outer membrane; P = periplasm; IM = inner membrane; R = TonB-dependent receptor; C = C-terminus; N = N-terminus.

N-terminus faces the periplasmic space [40]. TonB and ExbD are anchored in the inner membrane by a single transmembrane domain localized in their N-terminal parts and the rest of their polypeptide chains protrudes into the periplasmic space [41–43].

4.1.2. The TonB system forms a complex

TonB has been shown to interact with ExbB. The first insights indicating such an interaction have been indirect and inferred from the TonB instability in *exbB/exbD* mutants [44,45], and the fact that the requirement of ExbB and ExbD proteins for phage Φ80 sensitivity was specified by TonB transmembrane domain [46].

The interaction was then directly demonstrated by *in vivo* cross-linking experiments [47] and was shown to occur inside the membrane, and to involve the first transmembrane domain of ExbB [48,49] (Fig. 1). The evidence in favour of an interaction between TonB and ExbD is still indirect and remains based on the instability of TonB in the absence of ExbD [50].

4.1.3. The physiological role of the TonB system

TonB has a particular characteristic. A particularly proline-rich region of 33 aa is located in its amino-terminal third. This segment, composed of the (Glu-Pro)n-(Lys-Pro)m repeat motif, appears to adopt an extended constrained conformation that might allow TonB to span the periplasmic space entirely.

The transport of vitamin B12 and iron siderophores through the outer membrane is dependent on their specific receptors but also on TonB and its auxiliary proteins, ExbB and ExbD. These receptors all share the TonB-box consensus sequence. When mutated in this box, BtuB, FhuA, FhuE, and Cir receptors still bind their ligand but they cannot transport them across the outer membrane [51–54]. Mutations in the TonB-box of different receptors (BtuB, FhuA, and Cir) can be suppressed by mutations in two adjacent amino acids in TonB [52,54,55]. Furthermore, a pentapeptide corresponding to the TonB-box has been shown to inhibit the TonB activity when added to the external medium [56]. These pieces of evidence were in favour of a direct interaction between this consensus sequence and TonB. They suggested a direct interaction between TonB and these receptors and supported a model whereby TonB served as an energy transducer coupling the cytoplasmic membrane proton-motive force to the transport of ligands across the outer membrane [57] (Fig. 1). Recently, *in vivo* cross-linking experiments have demonstrated that TonB and FepA directly interact [47]. Furthermore, it has recently been proposed that the high affinity outer membrane receptors function as gated pores [58, 59]. The current model emerging from these results is that the proton motive force would drive a conformational change in TonB that would open the gated receptors and allow ligands to move through the pores [58]. Cycles of conformational changes in TonB might result in active pumping of ligands. However, the mechanism of energy transduction is still unclear. Besides their roles in the correct assembly of TonB in the cytoplasmic membrane [60], ExbB and ExbD might play a role in TonB reenergization in between rounds of energy transduction [49,61]. It may also be that other not yet identified proteins which have been shown to interact with TonB play a role in energy transduction [47].

4.2. The Tol system

The Tol system is composed of five proteins: TolA, TolB, TolQ, TolR, and PAL (Fig. 1). The corresponding genes form a cluster at 16.8 min on the chromosomal map of *E. coli* [62] and the complete sequence of the cluster has been determined [63–65].

4.2.1. Components of the Tol system

TolQ and TolR are 75% homologous with ExbB and ExbD, respectively [66]. The topologies of these proteins are also similar because TolQ has three transmembrane domains with its N-terminus facing the cytoplasm [67] and TolR is anchored to the inner membrane by a single transmembrane domain at its N-terminus [68]. Consistent with these similarities, TolQ/TolR and ExbB/ExbD have been shown to cross-complement each other [69]. It has been proposed that these two-protein systems may originate from a common ancestor [70].

TolA is anchored in the inner membrane by a N-terminal 21-aa hydrophobic region. The rest of the protein is periplasmic and can be degraded when trypsin gains access to this compartment. The TolA central region, comprising 223 residues is very rich in alanine, lysine, glutamate and aspartate residues because it contains repeated ten times the unit $ED(K)_{1-2}(A)_{2-4}$. This repeated motif might strongly stabilize a αhelical structure [64] and might provide TolA with an extended conformation in its central domain. It has been suggested that this motif might allow TolA, like TonB, to span the periplasm entirely [71].

TolB is a periplasmic protein. It is synthesized in a precursor form and its 21 aa-signal sequence is cleaved when it is exported across the cytoplasmic membrane [72].

PAL (Peptidoglycan Associated Lipoprotein) is an outer membrane protein. Its N-terminal part includes the signal sequence and the consensus cleavage site for the lipoprotein-specific signal peptidase. Furthermore, Mizuno [73–75] demonstrated that the first cysteine residue of the PAL mature form is modified by a glyceride and a lipid moiety. Therefore, as already demonstrated for other lipoproteins of the *E. coli* outer membrane, PAL would undergo several modification steps before assembling into the outer membrane. The cysteine residue at the cleavage site of the PAL signal peptide would be modified with a glyceride group and this modification would be essential for the subsequent cleavage of the signal peptide by the lipoprotein-specific signal peptidase II. The free amino group of the modified cysteine would then be further acylated with a fatty acid and the resultant mature lipoprotein would be assembled into the outer membrane (for a review, see Pugsley [76]).

PAL is non covalently associated with the peptidoglycan [77]. A positively charged α-helical motif in the C-terminal part of the protein has been shown to be important for such an interaction [78]. This motif has recently been found in a number of bacterial cell surface proteins [79].

4.2.2. The Tol system forms a complex

By analogy with the TonB system, the Tol system was also suspected to form a complex. This hypothesis was strengthened by the existence of homologies between the transmembrane anchors of TolA and TonB [46] and the fact that TonB interacts with ExbB via its transmembrane domain (and probably with TolQ because TolQ partially cross-complements ExbB). Furthermore, two indirect pieces of evidence further supported this hypothesis. First, the TolA transmembrane domain was found to specify the dependence on TolQ and TolR [46]. This result argued in favour of an interaction between TolA and TolQ/TolR. Second, the Tol/PAL proteins were shown to co-fractionate with contact sites between the inner and outer membranes of *E. coli* [80–82]. When colicin A was added to the cells, the relative amount of Tol proteins in contact sites was increased. This suggested that TolA, TolB, TolQ and TolR proteins formed a complex of definite stoichiometry [82]. The isolation of suppressor TolR mutants of a missense mutation in the third transmembrane domain of TolQ recently demonstrated a direct interaction between the third transmembrane domain of TolQ and the transmembrane domain of TolR [83]. Furthermore, *in vivo* cross-linking experiments showed that TolA transmembrane domain interacts with TolQ and TolR [84]. Therefore, the components of the Tol system localized in the inner membrane form a complex and all of them interact with each other inside the lipid bilayer via their transmembrane domains (Fig. 1). Interactions between the first and the third transmembrane domains of a single protein, TolQ, have even been reported [83]. Components of the Tol system not only form a complex in the inner membrane. Using *in vivo* cross-linking experiments and co-immunoprecipitation techniques, Bouveret et al. [85] have shown that TolB interacts with PAL (Fig. 1). Furthermore, TolB cofractionates with PAL and the other Tol proteins at the contact sites [82]. This latter result further suggests that the inner membrane complex of Tol proteins is associated with the outer membrane complex formed by TolB and PAL. In other words, the Tol system would establish contact sites between the inner and outer membranes of *E. coli*.

4.2.3. The hypothetical physiological role of the Tol system

Like the TonB system, the Tol system would therefore connect the two membranes of the bacteria. In the case of the TonB system, the role of this connection is to allow an energy transduction of the proton motive force of the inner membrane to high-affinity outer membrane receptors. In contrast, the physiological role of the contact sites formed by the Tol system is not yet known. The *tol* and *pal* mutants are hypersensitive to drugs and detergents and they release periplasmic proteins in the extracellular medium [21,86]. This pleiotropic phenotype suggests that the Tol/PAL proteins are involved in maintaining the integrity of the outer membrane of *E. coli*. What is the exact role played by the Tol/PAL proteins that may account for the outer membrane stabilization? An attractive hypothesis would be that the Tol/PAL proteins play a structural role by linking the two membranes to the peptidoglycan, via PAL. A mutation in a component of the Tol system might break this link and this would result in the destabilization of the cell envelope. Another explanation might be that the Tol system is involved in the transport and the correct assembly of specific components of the outer membrane. Mutations in Tol/PAL proteins would then block the transport and the correct organization of the cell envelope.

5. Colicin translocation step

Colicins and some phages use the two contact site systems described above to cross the target cells envelope. What are the molecular mechanisms involved? Do the colicins unfold as in the case of proteins imported into organelles and exported from bacteria? What is the mechanism which triggers the colicins inside the cells? What is the environment of their polypeptide chain during their movement across the membrane?

5.1. Group B colicins

Little is known about the mechanism of translocation of group B colicins. Until recently, all the group B colicins studied have been reported to use TonB-dependent receptors, that is to say receptors implicated in the TonB-dependent transport of iron siderophores and vitamin B12. Recently, a new colicin, colicin 5 has been shown to use the TonB system but to bind to a TonB-independent receptor, Tsx [87]. Therefore, the interaction between the colicin receptor and TonB does not seem to be a prerequisite for an efficient translocation. All group B colicins have a TonB-box in their N-terminal domain and, by analogy to the TonB-dependent receptors, they are supposed to interact with TonB. In the case of colicins B and M, Mende and Braun [33] and Pilsl et al. [34] have shown that their TonB-box and that of their receptor are important for these toxins to be translocated. To reconcile this result with the idea developed above that colicin receptors do not need to interact with TonB, we might argue that the TonB-box of the receptors might not only be necessary to interact with TonB but that the box might also be important for the receptor to be competent for the colicin to be translocated. It might be that TonB-dependent receptors have to open their gates to be competent for colicin translocation. This event might be dependent on the TonB-box. In contrast, in the case of a TonB-independent receptor like Tsx, the competent state would be obtained independently of TonB.

The importance of the TonB-box of colicins B and M also suggests that the colicins need to interact with TonB to be translocated.

Membrane protection experiments performed on colicin Ia have shown that both the N and C-terminal domains of colicin Ia are able to interact with the inner membrane and therefore are protected from proteolytic degradation [88]. In view of the 'Y' shape crystal structure of colicin Ia, in which the two arms would correspond to the N and C-terminal domains and the stem to the receptor-binding domain, this result suggests that the polypeptide chain of this colicin traverses the outer membrane twice and that the receptor-binding domain remains outside the cells.

It is noteworthy that no group B colicin has been reported to exert a nuclease-type activity. However, the so far unknown cellular target of colicin D, a colicin reported to inhibit protein synthesis [89], might be localized in the cytoplasm. The localization of this target and the discovery of other group B colicins might help us to know if the TonB system is able to promote the translocation of colicins across the inner membrane, as might be the case for the Tol system (because all nuclease-type colicins so far reported are Tol-dependent).

5.2. Group A colicins

Although the translocation pathway of group A colicins remains unclear in many aspects, much progress has been made in our understanding of the key mechanisms involved, especially in the case of pore-forming colicins and colicin A.

Group A colicins need the Tol proteins to be translocated through the cell envelope. It has to be noted that PAL, although belonging to the Tol system, is completely dispensable for colicins to enter into cells. Similarly, the leakiness and colicin tolerance phenotypes have been dissociated in many *tol* mutations. Therefore, a destabilized cell envelope does not prevent group A colicins from being translocated.

Tol proteins are not always indispensable for the entry of each colicin. TolA and TolQ are necessary for the entry of all group A colicins. In contrast, TolB and TolR are necessary for the entry of colicins A, E2, E3, and K only [18,21,62,63]. TolC, a minor outer membrane protein involved in hemolysin secretion [90] is only required for the entry of colicin E1 [21]. The requirement for different Tol proteins might reflect some differences in the mechanisms of translocation.

5.2.1. The unfolding of colicins

Pore-forming colicins kill the cells by depolarizing their cytoplasmic membrane and by inducing a phosphate efflux which leads to a depletion of the cytoplasmic ATP [91]. The opening of the pore also induces an efflux of cytoplasmic K^+. The K^+ efflux can be continuously recorded. In this way one can monitor, 'on-line', the progression of the toxin from its receptor (in the outer membrane) to its target, the inner membrane. The kinetics of K^+ efflux have been extensively studied in the case of colicin A [92]. The K^+ efflux corresponds to the opening of the colicin A channel. It is preceded by a lag time which corresponds to the binding of colicin A to its receptor and its translocation through the cell envelope. Under definite conditions, the time needed for the translocation step is the rate limiting factor of the lag time measured. Therefore, the lag time measured is a good approximation of the translocation time. This technique allowed to demonstrate that the translocation step of colicin A is accelerated if the colicin is urea-denatured before being added to the cells [93]. This result has suggested an unfolding of colicin A during its translocation step. Two different results confirmed this idea. First, externally added trypsin was shown to cleave the colicin polypeptide chain and thereby to induce the closure of the colicin A pore while having access neither to the periplasmic space nor to the inner membrane of sensitive cells [93]. Second, a disulfide bond-engineered colicin A mutant able to be translocated but unable to open its pore has been shown to prevent wild-type colicin A from binding to its receptor and from being translocated [94]. These results both indicate that colicin A maintains an extended conformation across the cell envelope and is still in contact with its receptor and its translocation machinery when its pore has been formed in the inner membrane. Therefore, when its pore opens in the inner membrane, colicin A would be in a conformation very similar to that hypothesized for the group B colicin, colicin Ia, on the basis of its 'Y' shape crystal structure [39] and the protease protection experiments [88]. Duché et al. [94] have shown that the unfolding of colicin A is initiated very early in the translocation process by the binding of the colicin to its receptor.

5.2.2. The role of porins in group A colicin translocation

Except for colicin E1 which is the only colicin dependent on TolC for its translocation step, all group A colicins need outer membrane porins (OmpF, OmpC, or PhoE) to reach their cellular targets [95–97]. An attractive idea is that colicins might traverse the outer membrane by going through the porins pore. Colicin E1 would traverse the membrane through the pore formed by TolC (because this protein has been shown to form channels in planar lipid bilayers [98]). The X-ray crystal structure of OmpF has been determined at 2.4 Å resolution and an internal loop has been shown to protrude inside the lumen of the pore and to constrict its diameter to a dimension of 7×11 Å [99]. This diameter seems too narrow to allow a polypeptide chain to go through. However, a OmpF point mutant resistant to colicins has recently strengthened this latter hypothesis because it has been shown that the mutation subdivides the pore into two intercommunicating compartments [100]. Furthermore, Jeanteur et al.[100] have suggested that colicin N has an N-terminal specific binding site within the OmpF pore lumen. Therefore, group A colicins (except colicin E1) or at least a part of their polypeptide chains may go through porins. This would greatly contribute to the further unfolding of the colicin molecules. However, these ideas are only speculative because they are not yet supported by direct evidence.

5.2.3. Colicin A and E1 interact with TolA

As mentioned above, the N-terminal domains of colicins contain all the information needed for the translocation step. It is therefore likely that this domain of group A colicins directly interacts with one or several components of the Tol system. Indeed, the N-terminal domain of colicins A and E1 and the C-terminal domain of TolA have been found to interact *in vitro* [101, and unpublished data]. A TolA protein mutated in its C-terminal domain and unable to promote colicin translocation is also unable to interact *in vitro* with the N-terminal domain of colicins A and E1 [101]. Therefore this interaction seems to reflect an interaction occurring *in vivo* when colicins are translocated. A TolA-box has recently been identified in the N-terminal domain of all Tol-dependent colicins [35]. As already mentioned, this box seems to be important for colicin E3 (a RNase-type colicin) translocation [36,37]. By analogy with the TonB-box which seems to promote binding with TonB, an attractive idea is that the TolA-box might promote binding with TolA. However, there is no evidence yet, that the TolA-box allows colicins to interact with TolA. Although this conserved motif is important for colicin E3 translocation, it is not yet known whether colicin E3 and TolA interact and in the case they would interact, it is not known if the colicin E3 mutated in its TolA-box is affected in its binding to TolA.

Colicin A seems to have to interact with TolA to be translocated. However, this interaction is not sufficient because a colicin A mutant deleted from aa 15 to 30 (and with an intact TolA-box) still binds to TolA *in vitro* but is unable to translocate through the cell envelope [22]. This result suggests that before or after binding to the C-terminal domain of TolA, colicin A undergoes a key event for its translocation step.

5.2.4. A hypothetical model for colicin translocation

At this point of the discussion, we can summarize what we know on the translocation step of our model colicin, colicin A, and we can try to draw a hypothetical model of the major molecular events which take place (Fig. 2).

(i) Colicin A starts to unfold as soon as it binds to its receptor (BtuB). (2i) Its N-terminal domain probably interacts with the OmpF porin (or OmpC with a lower

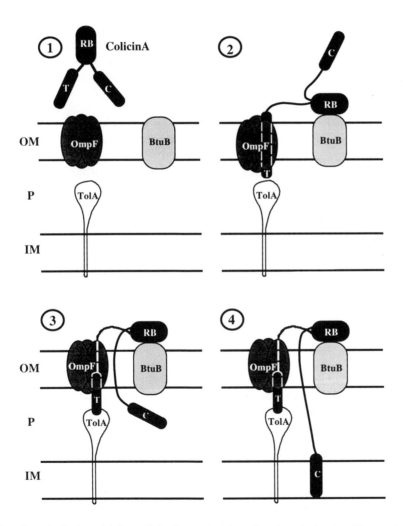

Fig. 2.   Hypothetical model for colicin A translocation. T = domain involved in translocation (N-terminal); RB = receptor-binding domain (central domain); C = pore-forming domain (C-terminal domain); OM = outer membrane; P = periplasm; IM = inner membrane.

efficiency) and this domain might traverse the outer membrane by going through the lumen of its pore. (3i) The colicin A N-terminal domain interacts with the C-terminal domain of TolA in the periplasm. (4i) While the central domain of the colicin is still in contact with the receptor, the C-terminal domain of the colicin is driven across the cell envelope, and reaches the inner membrane. This event results in a complete unfolding of the polypeptide chain in the cell envelope. (5i) Even when the colicin has opened its pore in the inner membrane, the molecule stays in contact with its receptor and its translocation machinery.

Many aspects of the process stay obscure and speculative. We do not know, for instance, if the colicin polypeptide chain traverses twice the outer membrane (Fig. 2), and if so how this could be performed. We do not know what drives the C-terminal domain through the cell envelope and we do not know either what triggers the unfolding of the colicin in the envelope. The energetics of the transport process are also unknown. A better knowledge of the Tol system, its organization and its physiological role, and an investigation of all the interactions, even transient, that might take place between the colicins and the Tol proteins, are now necessary to answer these questions.

Little is known about the translocation step of nuclease-type colicins. The translocation across the outer membrane and the periplasmic space might be similar to that of pore-forming colicins because a hybrid colicin having the N-terminal and central domains of colicin A and the C-terminal domain of colicin E3 has been shown to be active on sensitive cells [102]. However, we do not know whether the nuclease-type colicins are still accessible from the outside once their C-terminal domain has gained access to the cytoplasm. It might be that the C-terminal domain be cleaved after crossing the inner membrane. However, experiments which report the rescuing of bacteria treated with the colicin E2 DNase by trypsin added in the extracellular medium [103, 104] argue in favour of colicins acting in the cytoplasm while still anchored to the outer membrane via their receptors.

Nothing is known on how these colicins traverse the inner membrane. Do the Tol proteins play a role? Are other proteins needed?

6. *In vivo* activity of pore-forming colicins

Jacob et al. [105] have plotted the number of survivors as a function of time when bacteria were treated with various concentrations of colicin. From the exponential form of the curve in its initial portion, they have suggested that a single event was responsible for the death of the target-bacteria. The relationship between the colicin concentration and the initial rate of its bacteriocidal action has indicated that this event corresponded to the action of a single particle. A lethal unit has been defined as the amount of colicin required to kill one bacteria. However, as pointed out by Pattus et al. [106], single hit kinetics do not necessarily correspond to one molecule but rather to a single event even if it is produced by a group of molecules.

Pore-forming colicins dissipate the membrane potential [107,108]. This depolarization causes a series of metabolic effects such as inhibition of the active transport and of protein and nucleic acid synthesis, decrease of the internal ATP

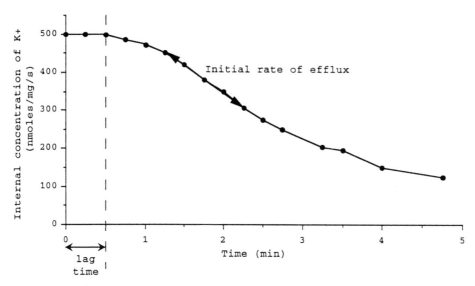

Fig. 3.   Potassium efflux induced by colicin A. At time t = 0, colicin A is added to the cells. The variation of the internal concentration of potassium is measured as a function of time. The lag time is the time comprised between the addition of colicin A and the beginning of the potassium efflux. This efflux corresponds to the opening of colicin A pore in the inner membrane. The initial rate of efflux is measured in the linear part of the curve (adapted from Bourdineaud et al. [92]).

concentration, and leak of potassium [109–113]. Colicins are not protonophores, i.e., they do not increase the permeability of the membrane to protons. It has been proposed by Plate et al. [112] that ATP was used in order to reenergize the inner membrane affected by the colicin. However, Guihard et al. [91] showed that ATP was hydrolyzed as a consequence of a shift in the ATP equilibrium due to the efflux of phosphate through the channels. This hydrolysis occurred even in a mutant defective in the F1-Fo-ATPase. Cell death results from all the events described above.

The fact that colicin A indeed formed a proteinaceous pore *in vivo* has strongly been suggested by its ability to form voltage-dependent channels that were able to be closed and reopened [92]. The channel activity of colicin A was measured by analyzing the kinetics of K^+ efflux induced by the toxin (Fig. 3). The channels formed *in vivo* are dependent on the voltage and on the pH. After colicin addition the membrane potential drops to −85 mV; above this value, the channel closes [92].

7. Properties of the channel

Schein et al. [115] have shown that colicin A formed well defined channels in planar lipid bilayer. Since this observation, pore-forming colicins have been extensively studied with lipid planar bilayers [116–122]. Pore-forming colicins create channels through the lipid bilayer of conductance of 10–30 ps which correspond to 10^7 ions/sec/channel. These channels are characterized by their sensitivities to the

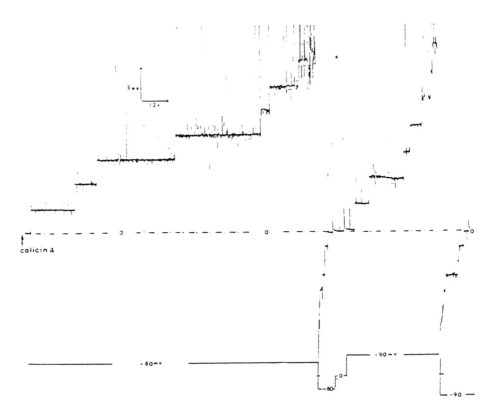

Fig. 4.   Stepwise increase of conductance of soybean phospholipid planar bilayer after injection of colicin A. A voltage of 90 mV is applied to the membrane bilayer. The steps of current increase correspond to the opening of individual pores. After the formation of 8 channels, the applied voltage is inverted (–80 mV) leading to the closing of the channels. Subsequent inversion of the applied voltage induces the opening of the channels. (Adapted from Martinez et al. [117].)

difference of the electrical potential across the membrane: when a transmembrane potential of +90 mV is applied, the opening of the channel is induced, whereas when the voltage is inverted to –90 mV, the same channels close (Fig. 4). A gating voltage corresponding to the activation or to the inactivation of 50% of the channels has been determined. The gating voltage varies for a given colicin, its value is +21, +50 and +70 mV for colicins N, A, and E1, respectively.

The pores are permeant to cations and anions but the rates at which ions are transported are slow [123]. There is a relative preference for cations versus anions which is modulated by the pH and the lipid composition [124–126]. From the permeability of the colicin E1 channel to various probes, Bullock et al. [125] have estimated the size of the pore to 9 Å. However, Cramer et al. [127] have mentioned that if the asymmetric and elongated shape of the ionic probes is taken into account, the calculated channel diameter would be smaller, around 4–5 Å. The size of the lumen is therefore small implying strong interactions between the permeant ions and the side chains of residues exposed in the lumen of the channel [128].

The pH affects the voltage-dependence [116, 123]; for instance, the gating voltages of colicin A and N channels are, at pH > 5, +50 and +21 mV and, at pH <4, −47 mV and −70 mV, respectively [120, 122]. The gating voltage of the isolated pore-forming domain does not depend on the pH. This result suggests that regions outside the pore-forming domain influence the channel behavior. For colicin A, the pore-forming domain has been suggested to interact with the receptor binding domain in a pH-dependent way [129]. Interestingly, the properties of the channel formed by the proteolytic fragment of colicin Ia (containing the channel domain) differ markedly from those formed by the intact molecule [39]. The structure of the intact colicin Ia indicates that its pore-forming domain contains an additional sub-domain which would be responsible for the functional differences between the proteolytic fragment and the whole molecule [39].

8. Structure of the soluble channel domain

The polypeptide sequence of colicins (A, B, N, E1, Ia and Ib) has been determined [16 and 130–133]. The pore-forming domains of these colicins are homologous. On the basis of the homologies, pore-forming domains have been classified into two groups: the group of colicins A, N, and B and that of colicins E1, Ia, and Ib.

The crystal structure of the channel domain of colicin A has been refined at 2.4 Å resolution [134, 135]. The domain is constituted of 10 alpha-helices of 9 to 24 residues in length. Parker and his colleagues [135] have described the structure as a bundle of 10 alpha helices arranged in three layers. The N-terminal layer (helices 1 and 2) is connected to another layer which consists of two pairs of amphipatic helices (helices 3–4 and 6–7) with a single helix (helix 5) connecting the two helix pairs. The middle layer is composed of helices 8 and 9 which form an hydrophobic hairpin buried in the core of the molecule. NMR studies of the corresponding fragment from colicin E1 indicate that it presents the same fold as the colicin A fragment [136]. Crystals of the colicin E1 fragment diffracting at 2.2 Å have been obtained but the structure of the fragment is not yet resolved [137]. The X-ray crystal structure of the entire colicin Ia has been obtained at a resolution of 4 Å [39]. Surprisingly the pore-forming domain of colicin Ia contains, in addition to the colicin A-like fragment (domain I2), an additional domain of eight helices (domain I1) which are oriented parallel to those of the domain I2. As indicated above, the domain I1 might modulate the properties of the channel formed by I2.

These channel domains have a very hydrophobic core which must therefore rearrange in order to insert into the membrane.

9. An unfolded state as a prerequisite for membrane insertion

Since the work of Eilers and Schatz [138] and that of Randall and Hardy [139], it is established as a paradigm that proteins must unfold before their insertion into or passage across biological membranes. Kinetic studies on the membrane insertion of the pore-forming domain of colicin A at low pH have shown that the intermediate

form for membrane insertion into anionic membranes was a molten globular
conformation [140]. The molten globule is characterized by the loss of tertiary
structure but conservation of the native secondary structure [141]. The role of the
negatively charged lipids would be to orientate the molecule at the surface of the
bilayer [134] or to lower the surface pH with respect to the bulk pH and thus to
promote the molten globular state and hence the membrane insertion [142]. For the
thermolytic fragment of colicin E1, the attainment of a competent state for membrane insertion at acidic pH was accompanied by an increased accessibility to
probes and sensitivity to proteases [143]. *In vitro*, other factors can promote the
conformation of the molten globule: for instance, elevated temperature or addition
of detergents result in the same effect as acidic pH [143, 144]. Spectroscopic studies
have revealed an unfolded state, somehow more compact than a molten globule,
associated with activity at acidic pH, for the colicin E1 channel [145]. However,
Schendel and Cramer [145] doubt whether, in physiological membranes, a local low
pH at the membrane surface is responsible for the formation of the unfolded
intermediate. Whatever it may be, before being membrane inserted, colicin channels lose their stable tertiary structure. As mentioned above, colicin A unfolds *in
vivo* at an early stage of the translocation step, i.e. upon interaction with its receptor
[93,94]. The interaction of the N-terminal and central domains with the translocation machinery might trigger the appearance of a molten globular conformation of
the channel domain *in vivo* [94].

10. Membrane insertion and channel formation

The pore-forming domain undergoes a conformational change from a soluble state
to a transmembrane state. The membrane associated colicin channel has several
configurations finely regulated by the transmembrane potential: a membrane-
bound closed channel and an integral conductive channel. In the absence of potential, the pore-forming domain spontaneously inserts into the lipid bilayer. Various
models of the potential-independent insertion have been proposed. Elegant studies
of energy fluorescence transfer and engineering of disulfide bonds [146–148],
proteolysis [149] and neutron scattering studies [98] have indicated that in the
absence of potential, all the helices of the colicin A pore-forming domain lie parallel
to the membrane plane at the membrane surface. Helices 1 and 2 move away from
the rest of the other helix cluster when the channel insert into the membrane
(reviewed by Lakey et al. [128]). In contrast, the membrane anchor domain of the
colicin E1 channel has been shown to be a transmembrane helical hairpin structure
(the region corresponding to the colicin A hairpin H8 + H9). Various experimental
approaches support this latter statement: saturation site directed mutagenesis and
topology experiments [150], proteolysis data [151], time-resolved spin labelling
studies [152] and fluorescence quenching experiments [153]. Do the two alternative
models reflect differences in the investigator's systems or a different behavior of
each channel fragment? This remains an open question.

As mentioned above, the colicin channel is opened when a trans-negative
potential is applied over a certain threshold voltage, implying the existence of a

voltage-responsive segment. Experiments with planar bilayers have indicated that the voltage gating of colicin E1 involved movements of the molecule across the bilayer thereby showing that colicin E1 exposes different domains on the *cis* and *trans* side of the membrane solutions in the open and closed states [118,154]. A voltage-responsive segment lying within residues Lys420/Lys461 of the colicin E1 channel has then been identified by hydrophobic photolabelling of the channel peptide in liposomes [155]. This gating peptide has further been tested using electrophysiological methods. Asp473 and His440 have been mapped on the *cis* side and on the *trans* side of the bilayer, respectively, in the open conformation [12,156]. In a clever study, Slatin et al. [157] mutated unique amino-acids of colicin Ia to cysteine and conjugated them to biotin. The biotinylated colicin was inserted into a lipid bilayer and the channels were opened or closed by varying the applied membrane potential. Streptavidin was then added on the *cis* or *trans* side of the membrane. When the channel was closed but not when it was open, streptavidin added on the *cis* side bound to a stretch of 37 residues. When streptavidin was added on the *trans* side, no binding was observed when the channel was closed but a region of 31 residues was labelled when the channel was opened. Binding of streptavidin to the open channel prevented its correct closing. A region of at least 31 amino acids can reversibly flip across the membrane implying a large conformational change when the potential is applied. Mutations in the hinge region between helices 5 and 6 of the colicin A channel [1,158] as well as restricting the movement of the same helices by a disulfide bond [148], prevent pore-formation but not membrane insertion. It has been proposed by Slatin et al. [157] that the voltage-independent insertion of the helical hairpin and the voltage-dependent insertion of the large voltage-responsive segment which reconfigures itself, when it closes and opens, leads to channel formation. Other regions may influence the properties of the channel as mentioned above. However, the structure of the channel in its membrane inserted state remains unknown. Do the helices forming the open channel interact with each other in the lipid bilayer? Genetic experiments designed to select mutations able to suppress mutations abolishing the pore-forming activity may help modelling the structure of the open channel conformation.

11. An alternative pathway for channel formation

Pore-forming colicins do not kill sensitive bacteria from the inside [159]. Only external colicin, which reaches the inner membrane via the receptors and the Tol proteins, is active. Similarly, the channel domain of colicin A produced in the cytoplasm of *E. coli* is devoid of any cytotoxicity.

The pore-forming domain of colicin A has been fused to a signal sequence. We have found that the signal sequence promotes the functional insertion of the colicin A channel into the membrane [13,160]. The hybrid protein had a cytotoxic activity that was specifically inhibited by the immunity protein. The cytotoxicity of the exported colicin A domain was abolished by a point mutation known from studies on planar bilayer and lipid vesicles to prevent channel formation but not membrane insertion [1,158]. In addition, the cytotoxicity of the chimeric protein was independent of the colicin A uptake machinery. This result suggested that the Tol proteins which are required for the transport of the wild-type colicin A from the surface of the bacterium to the inner membrane but not for channel formation. Espesset et al. [13] have then introduced a cysteine residue pair in the exported pore-forming domain, allowing a disulfide-bond to form between helices 5 and 6 belonging to the voltage-responsive segment. The channel was inactivated by the formation of a disulfide bond at the periplasmic side of the *E. coli* inner membrane. The cytotoxicity of such a double-cystein mutant was abolished upon export, but its cytotoxicity was specifically activated when dithiotreitol was added in the culture medium. These results indicate that the cytoxicity of the fusion protein is indeed related to a colicin channel and not to a hydrophobic loading of the membrane. The question arises as to whether the channel domain fused to a signal sequence is fully exported to the periplasm and then immediately inserts into the membrane or whether its translocation is stopped by the hydrophobic hairpin

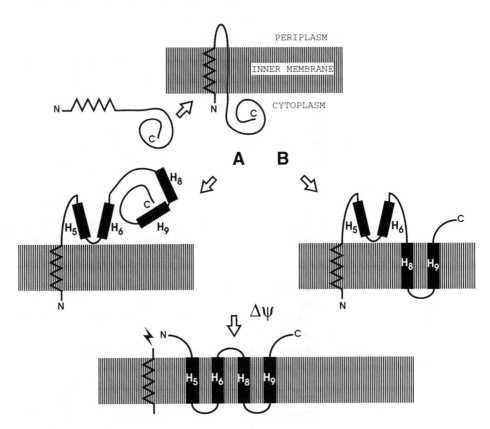

Fig. 5.   Two alternative pathways for the membrane insertion of the colicin A channel domain fused to a bacterial signal sequence. (A) The channel domain is fully exported and then inserts into the membrane. (B) The translocation of the channel domain is stopped by the hydrophobic hairpin (H8 + H9). The membrane potential $\Delta\psi$ causes insertion of the amphiphilic hairpin (H5 + H6).

(helices 8 and 9) which remains in the interior of the bilayer (Fig. 5). These *in vivo* studies underline the requirements for channel formation: the channel must be unfolded (in this case by the export machinery) and the molecule has to sense the membrane potential in the right orientation. If both requirements are fulfilled, the helices that form the channel are able to assemble whatever the pathway used for their insertion.

12. Killer toxin versus immunity: the fine art of pore inhibition

To prevent the colicin they release in the extracellular medium from killing them, colicin-producing strains synthesize immunity proteins. Immunity is a highly specific insensitivity of bacteria to the action of a particular colicin [161]. Genes encoding the colicin and the related immunity protein are carried by the same plasmid. As mentioned above, immunity proteins to pore-forming colicins protect the cell against external colicin only. It is thought that pore-forming colicins cannot function correctly from the inside because the polarity of the transmembrane potential is opposite to that required for channel formation [159]. This result explains why the immunity to pore-forming colicins is constitutively produced at a low level, despite the high level of expression of the corresponding colicins. The amount of colicins that reach the inner membrane depends on the number of import sites (approximately 400 per cell, [162]). Thus, a low amount of immunity protein is sufficient to protect the cell. A breakdown of the immunity is observed when the inner membrane is loaded with an amount of colicin in excess as compared to that of immunity protein [163,164]. The high specificity of the immunity function suggests that a molecular complex forms between the colicin and its cognate immunity protein (see below). The stoichiometry of such a complex is not yet known.

Very early, immunity proteins have been thought to act in the cytoplasmic membrane because membrane vesicles prepared from cells producing the colicin Ia immunity protein were depolarized by colicins E1 and Ib but not Ia [165]. Immunity proteins have then been shown to be integral membrane proteins [9,133,166–168]. The topology of colicin A and E1 immunity proteins (Cai and Cei) has been determined by using fusion to alkaline phosphatase and epitope mapping [168–170]. Cai contains four transmembrane segments with its N- and C-terminal regions directed towards the cytoplasm whereas Cei has only 3 membrane spanning segments with its N- and C-termini directed towards the cytoplasm and the periplasm, respectively.

Experiments involving hybrid colicins and truncated colicin E1 have indicated that immunity proteins are specific to colicin pore-forming domains [24,171]. These results have suggested that immunity proteins act in the cytoplasmic membrane by specifically preventing pore formation. The region of the channel domain responsible for the recognition of the immunity protein has then been identified by testing the killing activity of a series of colicin A/colicin B chimeric pore-forming domains on immune indicator strains [25]. The main specific determinant for immunity recognition has been found to be the hydrophobic hairpin of the channel domain. The same conclusion has been reached by analyzing colicin E1 mutants

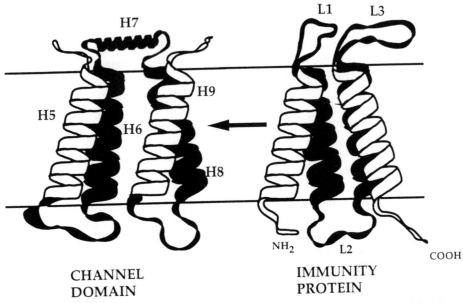

Fig. 6.   The colicin A immunity protein inhibits channel formation by interaction within the lipid bilayer with the helical hairpin (H8 and H9) of the channel.

that bypassed the protective effect of the colicin E1 immunity protein [172]. Tight and weak bypass mutants have been found in the hydrophobic and amphipathic hairpins of the colicin E1 channel, respectively. We think that the immunity protein diffuses freely in the membrane and then recognizes the channel domain by interacting with the hydrophobic hairpin of the channel.

As mentioned in the previous paragraph, we have recently shown that the colicin A pore-forming domain fused to signal sequence formed a channel in the inner membrane of the cells. The channel formed independently of the Tol proteins and it was inhibited by the colicin A immunity protein [13,161]. It has been suggested that immunity proteins interact with the channel in the open state within the core of the lipid bilayer. We have then co-produced the exported colicin A channel with an epitope-tagged immunity protein and characterized the interaction of both proteins by co-immunoprecipitation using a monoclonal antibody specific for the immunity protein tag [13]. The channel domain was specifically immunoprecipitated with the immunity protein. This interaction required a functional immunity protein. We have also shown that the immunity protein was able to interact with a mutant form of the pore-forming domain carrying a mutation located in its voltage-gated region. This mutant was devoid of pore-forming activity but still inserted into the lipid bilayer via its hydrophobic hairpin. These results indicate that the immunity protein interact with the membrane anchored pore-forming domain; this interaction requires the immunity protein helix-bundle to functionally assemble but does not require the channel to be in the open-state (Fig. 6). These results also explain why the immunity function cannot not be reconstituted into liposomes [173].

Mutagenesis studies on the colicin A immunity protein have indicated that its function is very sensitive to mutations. Various single point mutations in the hydrophobic segments as well as in the polar loops of the colicin A immunity protein decrease the immunity function [169,173] whereas the colicin E1 immunity protein tolerates a higher degree of substitution [170]. Although the protective mechanism certainly involves intramembrane helix/helix interactions, we cannot exclude a role for the polar regions of the colicin A immunity protein. Two roles can be proposed for the hydrophilic loops of the colicin A immunity protein: (i) they may stabilize the interaction of the immunity protein with the channel on both sides of the cytoplasmic membrane; (ii) they may be required for the functional assembly of the transmembrane helices of the immunity protein. These two roles may not be mutually exclusive. Whatever it may be, the immunity function requires high structural constraints. Slight conformational alterations may prevent its function. This may account for the striking conservation of prolines between related colicin immunity proteins [169]. The colicin/immunity relationships constitute a unique model system to study protein/protein interaction in a lipid bilayer.

Acknowledgements

We are grateful to Claude Lazdunski for his constant support and to Alain Rigal and David Espesset for careful reading of the manuscript.

References

1. Baty, D., Knibiehler, M., Verheij, H., Pattus, F., Shire, D., Bernadac, A. and Lazdunski, C. (1987) Proc. Natl. Acad. Sci. USA **84**, 1152–1156.
2. Cole, S.T., Saint-Joanes, B. and Pugsley, A.P. (1985) Mol. Gen. Genet. **198**, 465–472.
3. Hakkart, M.J., Veltkamp, E. and Nijkamp (1981) Mol. Gen. Genet. **183**, 318–325.
4. Lloubès, R., Baty, D. and Lazdunski, C. (1986) Nucleic Acids Res. **14**, 2621–2636.
5. Toba, M., Masaki, H. and Ohta, T. (1986) J. Biochem. **99**, 591–596.
6. Oudega, B., Ykema, A., Stegehmis, F. and De Graaf, F.K. (1984) FEMS Microbiol. Lett. **22**, 101–109.
7. Cavard, D., Baty, D., Howard, S., Verheij, H.M. and Lazdunski, C. (1987) J. Bacteriol. **169**, 2187–2194.
8. Pugsley, A.P. and Cole, S.T. (1987) J. Gen. Microbiol. **133**, 2411–2420.
9. Pugsley, A.P. (1988) Mol. Gen. Genet. **211**, 325–341.
10. Cavard, D. (1991) J. Bacteriol. **173**, 191–196.
11. Howard, S.P., Cavard, D. and Lazdunski, C. (1991) J. Gen. Microbiol. **137**, 81–89.
12. Jakes, R. and Zinder, N.D. (1984) J. Bacteriol. **157**, 582–590.
13. Espesset, D., Duché, D., Baty, D. and Géli, V. (1996) EMBO J. **15**, in press
14. Cavard, D. and Lazdunski, C. (1981) FEMS Microbiol. Lett. **12**, 311–316.
15. Chai, T., Wu, V. and Foulds, J. (1982) J. Bacteriol. **151**, 983–988.
16. Pugsley, A.P. (1987) Mol. Microbiol. **1**, 317–325.
17. Davies, J.K. and Reeves, P. (1975a) J. Bacteriol. **123**, 96–101.
18. Davies, J.K. and Reeves, P. (1975b) J. Bacteriol. **123**, 102–117.
19. Nomura, M. (1964) Biochemistry **52**, 1514–1521.
20. Nomura, M. and Witten, C. (1967) J. Bacteriol. **94**, 1093–1111.
21. Nagel de Zwaig, R. and Luria, S.E. (1967) J. Bacteriol. **94**, 1112–1123.
22. Baty, D., Frenette, M., Lloubès, R., Géli, V., Howard, S.P., Pattus, F. and Lazdunski, C. (1988) Mol. Microbiol. **2**, 807–811.

23. Frenette, M., Bénédetti, H., Bernadac, A., Baty D. and Lazdunski, C. (1991) J. Mol. Biol. **217**, 421–428.

24. Bénédetti, H., Frenette M., Baty D., Knibiehler M., Pattus F. and Lazdunski, C. (1991a) J. Mol. Biol., **217**, 429–439.

25. Géli, V. and Lazdunski, C. (1992) J. Bacteriol. **17**, 6432–6437.

26. De Graaf, F.K. and Oudega, B. (1986) Curr. Top. Microbiol. Immunol. **125**, 183–205.

27. Ohno-Iwashita, Y. and Imahori, K. (1982) J. Biol. Chem. **257**, 6446–6451.

28. Liu-Johnson, N.H., Gartenberg, M.R. and Crothers, D.M. (1986) Cell **47**, 995–1005.

29. Cleveland, M., Slatin, J., Finkelstein, A. and Levinthal, C. (1983) Proc. Natl. Acad. Sci. USA **80**, 3706–3710.

30. Brunden, K.R., Cramer, W.A. and Cohen, F.S. (1984) J. Biol. Chem. **259**, 190–196.

31. Dreher, R., Braun, V. and Wittmann-Liebold, B. (1985) Arch. Microbiol. **140**, 343–346.

32. Riley, M.A. (1993) Mol. Biol. Evol. **10**, 1380–1395.

33. Mende, J. and Braun, V. (1990) Mol. Microbiol. **4**, 1523–1533.

34. Pilsl, H., Glaser, C., Gross, P., Killman, H., Olschläger, T. and Braun, V. (1993) Mol. Gen. Genet. **240**, 103–113.

35. Pilsl, H. and Braun, V. (1995) Mol. Microbiol. **16**, 57–67.

36. Mock, M. and Schwarz, M. (1980) J. Bacteriol. **142**, 384–390.

37. Escuyer, V. and Mock, M. (1987) Mol. Microbiol. **1**, 82–85.

38. Cavard, D., Sauve, P., Heitz, H., Pattus, F., Martinez, C., Djikman, R. and Lazdunski, C. (1988) Eur. J. Biochem. **172**, 507–512.

39. Ghosh, P, Mel,S.F.and Stroud, R.M. (1994) Nature Struct. Biol. **1**, 597–604.

40. Kampfenkel, K. and Braun, V. (1993) J. Bacteriol. **175**, 4485–4491.

41. Kampfenkel, K. and Braun, V. (1992a) J. Bacteriol. **174**, 5485–5487.

42. Kampfenkel, K. and Braun, V. (1992b) J. Biol. Chem. **268**, 6050–6057.

43. Roof, S.K., Allard, J.D., Bertrand, K.P. and Postle, K. (1991) J. Bacteriol. **173**, 5554–5557.

44. Fisher, E., Günter, K. and Braun, V. (1989) J. Bacteriol. **171**, 5127–5134.

45. Skare, J.T. and Postle K. (1991) Mol. Microbiol. **5**, 2883–2890.

46. Karlsson, M., Hannavy, K. and Higgins, C.F. (1993) Mol. Microbiol. **8**, 379–388.

47. Skare, J.T., Ahmer, B.M.M., Seachord, C.L., Darveau, R.P. and Postle, K. (1993) J. Biol. Chem. **268**, 16302–16308.

48. Jaskula, J.C., Letain, T.E., Roof, S.K., Skare, J.T. and Postle. K. (1994) J. Bacteriol. **175**, 2326–2338.

49. Larsen, R.A., Thomas, M.G., Wood, G.E. and Postle, K.(1994) Mol. Microbiol. **13**, 627–640.

50. Ahmer, B.M.M., Thomas, M.G., Larsen, R.A. and Postle, K. (1995) J. Bacteriol. **177**, 4742–4747.

51. Bassford, P.J. Jr. and Kadner, R.J. (1977) J. Bacteriol. **132**, 796–805.

52. Schšffler, H. and Braun, V. (1989) Mol. Gen. Genet. **217**, 378–383.

53. Sauer, M., Hantke, K., and Braun, V. (1990) Mol. Microbiol. **4**, 427–437.

54. Bell, P.E., Nau, C.D., Brown, J.T., Konisky, J. and Kadner, R.J. (1990) J. Bacteriol. **172**, 3826–3829.

55. Heller, K., Kadner, R.J. and Günther, K. (1988) Gene **64**, 147–153.

56. Tuckman, M. and Osburne, M.S. (1992) J. Bacteriol. **174**, 320–323.

57. Hannavy, K., Barr, G.C., Dorman, C.J., Adamson, J., Mazengera, L.R., Gallagher, M.P., Evans, J.S., Levine, B.A., Trayer, I.P. and Higgins, C.F. (1990) J. Mol. Biol. **216**, 897–910.

58. Rutz, J.M., Liu, J., Lyons, J.A., Gotanson, J., Amstrong, S.K., Mc Intosh, M.A., Feix, J.B. and Klebba, P.E. (1992) Science **258**, 471–475.

59. Killmann, H., Benz, R. and Braun, V. (1993) EMBO J. **12**, 3007–3016.

60. Karlsson, M., Hannavy, K. and Higgins, C.F. (1993) Mol. Microbiol. **8**, 389–396.

61. Postle, K. (1993) J. Bioenerg. Biomembr. **25**, 591–601.

62. Sun, T.P. and Webster, R.E. (1986) J. Bacteriol. **165**, 107–115.

63. Sun, T.P. and Webster, R.E. (1987) J. Bacteriol. **169**, 2667–2674.

64. Levengood, S.K. and Webster, R.E. (1989) J. Bacteriol. **171**, 6600–6609.

65. Lazzaroni, J.C., Fognini-Lefebvre, N. and Portalier, R. (1989) Mol. Gen. Genet. **218**, 460–464.

66. Eick-Helmerich, K. and Braun, V. (1989) J. Bacteriol. **171**, 5117–5127.

67. Vianney, A., Lewin, T.M., Beyer, W.F.Jr, Lazzaroni J.C., Portalier, R. and Webster, R.E. (1994) J. Bacteriol. **176**, 822–829.

68. Müller, M.M., Vianney, A., Lazzaroni, J.C., Webster, R.E. and Portalier, R. (1993) J. Bacteriol. **175**, 6059–6061.
69. Braun, V. (1989) J. Bacteriol. **171**, 6387–6390.
70. Braun, V. and Hermann, C. (1993) Mol. Microbiol. **8**, 261–268.
71. Webster, R.E. (1991) Mol. Microbiol. **5**, 1005–1011.
72. Isnard, M., Rigal, A., Lazzaroni, J.C., Lazdunski, C. and Lloubès, R. (1995) J. Bacteriol. **176**, 6392–6396.
73. Mizuno, T. (1981) J. Biochem. **89**, 1039–1049.
74. Mizuno, T. (1981) J. Biochem. **89**, 1051–1058.
75. Mizuno, T. (1981) J. Biochem. **89**, 1059–1066.
76. Pugsley, A.P. (1993) Microbiol. Rev. **57**, 50–108.
77. Mizuno, T. (1979) J. Biochem. **86**, 991–1000.
78. Lazzaroni, J.C. and Portalier, R. (1981) J. Bacteriol. **145**, 1351–1358.
79. Koebnik, R. (1995) Mol. Microbiol. **16**, 1269–1270.
80. Bourdineaud, J.P., Howard, S.P. and Lazdunski, C. (1989) J. Bacteriol. **171**, 2458–2465.
81. Leduc, M., Ishidate, K., Shakibai, N. and Rothfield, L. (1992) J. Bacteriol. **174**, 7982–7988.
82. Guihard, G., Boulanger, P., Bénédetti, H., Lloubès, R., Besnard, M. and Letellier, L. (1993a) J. Biol. Chem. **269**, 5874–5880.
83. Lazzaroni J.C., Vianney, A., Popot, J.L., Bénédetti, H., Samatey, F., Lazdunski, C., Portalier, R. and Géli, V. (1995) J. Mol. Biol. **246**, 1–7.
84. Derouiche, R., Bénédetti, H., Lazzaroni, J.C., Lazdunski, C. and Lloubès, R. (1995) J. Biol. Chem. **270**, 11078–11084.
85. Bouveret, E., Derouiche, R., Rigal, A., Lloubès, R., Lazdunski, C. and Bénédetti, H. (1995) J. Biol. Chem. **270**, 11071–11077.
86. Fognini-Lefebvre, N., Lazzaroni, J.C. and Portalier, R. (1987) Mol. Gen. Genet. **209**, 391–395.
87. Bradley, D.E. and Howard, S.P. (1992) J. Gen. Microbiol. **135**, 1857–1863.
88. Mel, S.F., Falick, A.M., Burlingame, A.L. and Stroud, R.M. (1993) Biochemistry **312**, 9473–9479.
89. Timmis, K.N. and Hedges, A.J. (1972) Biochim. Biophys. Acta **262**, 200–207.
90. Wandersman, C. and Delepaire, P. (1990) Proc. Natl. Acad. Sci. USA **87**, 4776–4780.
91. Guihard, G., Bénédetti, H., Besnard, M. and Letellier, L. (1993b) J. Biol. Chem. **268**, 17775–17780.
92. Bourdineaud, J.P., Boulanger, P., Lazdunski, C. and Letellier, L. (1990) Proc. Natl. Acad. Sci. USA **87**, 1037–1041.
93. Bénédetti, H., Lloubès, R., Lazdunski, C. and Letellier, L. (1992) EMBO J. **11**, 441–447.
94. Duché, D, Baty, D., Chartier, M. and Letellier, L. (1994) J. Biol. Chem. **269**, 24820–24825.
95. Bénédetti, H., Frénette, M., Baty, D., Lloubès, R., Géli, V. and Lazdunski, C. (1989) J. Gen. Microbiol. **135**, 3413–3420.
96. Mock, M. and Pugsley, A.P. (1982) J. Bacteriol. **150**, 1069–1076.
97. El Kouhen, R., Hoenger, A., Engel, A. and Pagès, J.M. (1994) Eur. J. Biochem. **224**, 723–728.
98. Benz, R., Maier, E. and Gentschev, I. (1993) Zbl. Bakt. **278**, 187–196.
99. Cowan, S.W., Schirmer, T., Rummel, G., Steiert, M., Ghosh, R, Pauptit, R.A., Jansonius, J.N. and Rosenbuch, J.P. (1993) Nature **358**, 727–733.
100. Jeanteur, D., Schirmer, D., Fourel, D., Simonet, V.,Rummel, G., Widmer, C., Rosenbuch, J.P., Pattus, F. and Pagès, J.M. (1994) Proc. Natl. Acad. Sci. USA **91**, 10675–10679.
101. Bénédetti, H., Lazdunski, C. and Lloubès, R. (1991b). EMBO J., **10**, 1989–1995.
102. Bénédetti, H., Letellier, L., Lloubès, R., Géli, V., Baty, D., Pagès, J.M. and Lazdunski, C. (1992) in: Bacteriocins, Microcins and Lantibiotics, eds R. James, C. Lazdunski and F. Pattus. NATO ASI series, Vol. H 65 pp. Springer-Verlag, Berlin, Heidelberg.
103. Nose, K. and Mizuno, D. (1968) J. Biochem. **64**, 1.
104. Beppu, T., Kawabata, K. and Arima, K. (1972) J. Bacteriol. **110**, 485–493.
105. Jacob, F., Siminovitch, L. and Wollman, E. (1952) Ann. Inst. Pasteur **83**, 295–315.
106. Pattus, F., Massotte, D., Wilmsen, H.U., Lakey, J., Tsernoglou, D., Tucker, A. and Parker, M. (1990) Experientia, **46**, 180–192.
107. Weiss, M.J. and Luria, S.E. (1978) Proc. Natl. Acad. Sci. USA **75**, 2483–2487.
108. Tokuda, H. and Konisky, J. (1978) J. Biol. Chem. **253**, 7731–7737.

109. Fields, K.L. and Luria, S.E. (1969) J. Bacteriol. **97**, 57–63.
110. Fields, K.L. and Luria, S.E. (1969) J. Bacteriol. **97**, 64–77.
111. Kopecky, A.L., Copeland, D.P. and Lusk, J.E. (1975) Proc. Natl. Acad. Sci. USA **72**, 4631–4634.
112. Plate, C.A., Suit, J.L., Jetten, A.M. and Luria, S.E. (1974) J. Biol. Chem. **249**, 6138–6143.
113. Gould, J.M. and Cramer, W.A. (1977) J. Biol. Chem. **252**, 5491–5497.
14. Letellier, L. (1992) in: Alkali Cation Transport Systems in Procaryotes, ed E.P. Bakker. pp. 359–379, CRC Press, Inc., Boca Raton, FL.
115. Schein, S.J., Kagan, B.L. and Finkelstein, A. (1978) Nature **276**, 159–163.
116. Pattus, F., Cavard, D., Verger, R., Lazdunski, C., Rosenbuch, J. and Schindler, H., (1983) in: Physical Chemistry of Transmembrane Ion Motions, ed G. Spach. Elsevier, Amsterdam, pp. 407–413.
117. Martinez, M.C., Lazdunski, C. and Pattus, F. (1983) EMBO J. **2**, 1501–1507.
118. Slatin, S.L., Raymond, L. and Finkelstein, A. (1986) J. Membr. Biol. **92**, 247–254.
119. Pressler, V., Braun, V., Wittmann-Liebold, B. and Benz, R. (1986) J. Biol. Chem., **261**, 2654–2659.
120. Collarini, M., Amblard, G., Lazdunski, C. and Pattus, F. (1987) Eur. Biophys. J. **14**, 147–153.
121. Nogueira, R.A. and Varanda, W.A. (1988) J. Membr. Biol. **105**, 143–153.
122. Wilmsen, H.U., Pugsley, A.P., and Pattus, F. (1990) Eur. Biophys. J. **18**, 149–158.
123. Raymond, L., Slatin, S.L., and Finkelstein, A. (1985) J. Membr. Biol. **84**, 173–181.
124. Bullock, J.O. (1992) J. Membr. Biol. **125**, 255–271.
125. Bullock, J.O., Cohen, F.S., Dankert, J.R. and Cramer, W.A. (1983) J. Biol. Chem. **258**, 9908–9912.
126. Bullock, J.O., Kolen, E.R. and Shear, J.L. (1992) J. Membr. Biol. **128**, 1–16.
127. Cramer, W.A., Heymann, J.B., Schendel, S.L., Deriy, B.N., Cohen, F.S., Elkins, P.A. and Stauffacher, C.V. (1995) Annu. Rev. Biophys. Biomol. Struct. **24**, 611–641.
128. Lakey, J.H., Van Der Goot, F.G. and Pattus, F. (1994) Toxicology **87**, 85–108.
129. Frenette, M., Knibiehler, M., Baty, D., Géli, V., Pattus, F., Verger, R. and Lazdunski, C. (1989) Biochemistry **28**, 2509–2514.
130. Morlon, J., Lloubès, R., Varenne, S., Chartier, M. and Lazdunski, C. (1983) J. Mol. Biol. **170**, 271–285.
131. Schramm, E., Mende, J., Braun, V. and Kamp, R. (1987) J. Bacteriol. **169**, 3350–3357.
132. Chan, P.T., Ohmori, H., Tomizawa, J. and Leibovitz, J. (1986) J. Biol. Chem. **260**, 8925–8935.
133. Mankovich, J.A., Hsu, C.M. and Konisky, K. (1986) J. Bacteriol. **168**, 228–236.
134. Parker, M.W., Pattus, F., Tucker, A.D. and Tsernoglou, D. (1989) Nature **337**, 93–96.
135. Parker, M.W., Pastman, J.P.M., Pattus, F., Tucker, A.D. and Tsenoglou, D. (1992) J. Mol. Biol. **224**, 639–657.
136. Wormald, M.R., Merrill, A.R., Cramer, W.A. and Williams, R.J.P. (1990) Eur. J. Biochem. **191**, 155–161.
137. Elkins, P., Song, H.Y., Cramae, W.A. and Stauffacher, C.V. (1994) Proteins Struct. Funct. Genet. **19**, 150–157.
138. Eilers, M. and Schatz, G. (1986) Nature **322**, 228–232.
139. Randall, L.L. and Hardy, S.J. (1986) Cell **46**, 921–928.
140. Van der Goot, F.G., Gonzàlez-Manas, J.M., Lakey, J.H. and Pattus, F. (1991) Nature **354**, 408–411.
141. Ptitsyn, O.B. (1992) in: Protein Folding, ed T.E. Creighton. pp. 243–300. Freeman, New York.
142. Gonzalez-Manas, J.M., Lakey, J.H. and Pattus, F. (1992) Biochemistry **31**, 7294–7300.
143. Merrill, A.R., Cohen, F.S. and Cramer, W.A. (1990) Biochemistry **129**, 8529–8534.
144. Muga, A., Gonzalez-Manas, J.M., Lakey, J.H., Pattus, F. and Surewicz, W.S. (1993) J. Biol. Chem. **268**, 1553–1557.
145. Schendel, S.L. and Cramer, W.A. (1994) Protein Sci. **3**, 2272–2279.
146. Lakey, J.H., Baty, D. and Pattus, F. (1991) J. Mol. Biol. **218**, 639–653.
147. Lakey, J.H., Duché, D., Gonzalez-Manas, J.M., Baty, D. and Pattus, F. (1993) J. Mol. Biol. **230**, 1055–1067.
148. Duché, D., Parker, M.W., Gonzalez-Manas, J.M., Pattus, F. and Baty, D. (1994) J. Biol. Chem. **269**, 6332–6339.).
149. Massotte, D., Yamamoto, M., Scianimanico, S., Sorokine, O., van Dorsselaer, A., Nakatani, Y., Ourisson, G. and Pattus, F. Biochemistry **32**, 13787–13794.
150. Song, H.Y., Cohen, F.S. and Cramer, W.A. (1991) J. Bacteriol. **173**, 2927–2934.

151. Zhang, Y.L. and Cramer, W.A. (1992) Protein Sci. **1**, 1666–1676.
152. Shin, Y.K., Levinthal, C., Levinthal, F. and Hubbell, W.L. (1993) Science **259**, 960–963.
153. Palmer, L.R. and Merrill, A.R. (1994) J. Biol. Chem. **169**, 4187–4193.
154. Raymond, L., Slatin, S.L., Finkelstein, A., Liu, Q.R. and Levinthal, C. (1986) J. Membr. Biol. **92**, 255–268.
155. Merrill, A.R. and Cramer, W.A. (1990) Biochemistry **29**, 8529–8534.
156. Abrams, C.K., Jakes, K.S., Finkelstein, A. and Slatin, S.L. (1991) J. Gen. Phys. **98**, 77–93.
157. Slatin, S.L., Qiu, X.Q., Jakes, K.S. and Finkelstein, A. (1994) Nature **371**, 158–161.
158. Pattus, F. unpublished results.
159. Géli, V., Baty, D., Crozel, V., Morlon, J., Lloubès, R., Pattus, F. and Lazdunski, C. (1986) Mol. Gen. Genet. **202**, 455–460.
160. Espesset, D, Corda, Y., Cunningham, K., Bénédetti, H., Lloubès, R., Lazdunski, C. and Géli, V. (1994) Mol. Microbiol. **13**, 1121–1131.
161. Konisky, J. (1978) in: The Bacteria, Vol. 6, eds L.N. Omston and Sokatch, pp. 71–136. Academic Press, London.
162. Duché, D., Letellier, L., Géli, V., Bénédetti, H. and Baty, D. (1995) J. Bacteriol. **177**, 4935–4939.
163. Levisohn, R., Konisky, J. and Nomura, (1968) J. Bacteriol. **96**, 811–821.
164. Luria, S.E. (1982) J. Bacteriol. **149**, 386.
165. Weaver, C., Kagan, B., Finkelstein, A. and Konisky, J. (1981) Biochim. Biophys. Acta **645**, 137–142.
166. Goldman, K., Suit, J. and Kayalar, C. (1985) FEBS Lett. **190**, 319–323.
167. Schramm, E., Ölschläger, T., Träger, W. and Braun, V. (1988) Mol. Gen. Genet. **211**, 176–182.
168. Géli, V., Baty, D. and Lazdunski, C. (1988). Proc. Natl. Acad. Sci. USA **85**, 689–693.
169. Géli, V., Baty, D., Pattus, F. and Lazdunski, C. (1989) Mol. Microbiol. **3**, 679–687.
170. Song, H.Y. and Cramer, W.A. (1991) J. Bacteriol. **173**, 2935–2943.
171. Bishop, L.J., Bjes, E.S., Davidson, V.L. and Cramer, W.A. (1985) J. Bacteriol. **164**, 237–244.
172. Zhang, Y.L. and Cramer, W.A. (1993) J. Biol. Chem. **268**, 1–9.
173. Géli, V., Knibiehler, M., Bernadac, A. and Lazdunski, C. (1989) FEMS Microbiol. Lett. **60**, 239–244.
174. Espesset, D., Piet, P., Lazdunski, C. and Géli, V. (1994) Mol. Microbiol. **13**, 1111–1120.

CHAPTER 30

Bacterial Ion Channels

I.R. BOOTH, M.A. JONES, D. MCLAGGAN, Y. NIKOLAEV, L.S. NESS,
C.M. WOOD, S. MILLER, S. TÖTEMEYER and G.P. FERGUSON

Department of Molecular and Cell Biology, University of Aberdeen,
Marischal College, Aberdeen, UK

© 1996 Elsevier Science B.V.
All rights reserved

Handbook of Biological Physics
Volume 2, edited by W.N. Konings, H.R. Kaback and J.S. Lolkema

Contents

1. General introduction

The word 'channel' has great antiquity and originally meant a small water-filled tube or pipe [1]. Applied to biological membranes this original meaning is invaluable since it emphasizes the continuity between compartments provided by the channel protein. When a channel opens the lumen is accessible to ions from either side of the membrane and the direction and the amount of net ion flow is determined by the magnitude of the chemical gradient for the ion, the magnitude of the membrane potential and by the intrinsic properties of the channel protein. Channels are usually multi-subunit, integral membrane proteins with the ion-conducting pathway enclosed by the subunits. Similarly, many transporters, particularly those energised by phosphate bond energy, are multimeric and may translocate the solute through a pathway formed in a similar way to that found in ion channels. Both channels and transporters mediate solute movement across the membrane as a result of conformational cycles within the protein complex. A major distinction that can be drawn between a channel and a transporter, however, is that while there is a relatively fixed stoichiometry of solute movements per conformational cycle of the transporter, there is no strict linkage between the cycle of opening and closing of a channel and the number of ions moving. Relatively speaking the binding of the ion to a channel is usually very weak and consequently a large number of ions can move through the channel when it is open. In contrast, the binding of solutes to transporters is generally strong and the release on the opposite side of the membrane will require a further conformational change in the protein. As a consequence transporters can create solute gradients using energy, provided by either the proton or Na^+ motive force or the hydrolysis of ATP, to generate or modify the conformational cycle. Given the diversity of channels and transporters some of the above distinctions will disappear as molecular mechanisms are determined. However, the major distinction will remain that, while both transporter and channel undergo conformational cycles, only in the case of the transporter is the stoichiometry of solutes moving per cycle relatively fixed.

Opening an ion channel, even for a brief period, will perturb the cell: the membrane potential ($\Delta\psi$), ion gradients and energy balances may all be affected. The specificity, the number of channels, frequency and duration of the open state are all factors that will influence the degree of perturbation of cellular parameters. Bacterial cells have very small volumes (approx. 10^{-15} l) and an open, potassium-specific channel could have major consequences for the cell. Ion channels are capable of conducting greater than 10^7 ions/s. Potassium pools are of the order of 300–400 mM, corresponding to approx. 10^8 ions per cell. Opening a single potassium channel, capable of allowing the movement of 10^7 ions/s, for 1 ms would have a negligible effect on the potassium pool, but after 10 s the pool would be completely dissipated. Coordinated opening of several channels, in response to a particular stimulus, would greatly magnify these effects leading to more rapid potassium pool depletion. This does not occur because firstly, channels tend to open for relatively short durations of the order of ms rather than seconds, and secondly, because the movement of cations across the membrane affects the membrane potential. Movement of potassium out of the cell down its gradient would lead to an increase in the membrane potential; given the capacitance of the lipid bilayer, the movement of as few as 10^4 ions could cause changes of up to 200 mV in the membrane potential [2]. Assuming that the membrane potential was already at a high (negative inside) value the hyperpolarisation would prevent further ion flux through the channel (Fig. 1). Thus, the actual effect of sustained opening of a channel on the potassium pool, for example, would be determined by the permeability of the membrane to other ions, which could move across the membrane and depolarise the membrane. For channels with low or no ion specificity, e.g. the stretch-activated channels of E. coli [3], the problem is greatly exacerbated since the counterflow of ions and solutes will certainly lead to transient collapse of the membrane potential and perturbation of the cytoplasmic pH and of the pools of small molecular weight solutes. The precise consequences of channel operation should fit with the specific physiological role of the ion channel. Thus, for example, the large changes in solute pools caused by the operation of stretch-activated channels (see below) is central to their proposed role in reducing cell tugor. In contrast the small changes in the potassium pool consequent upon millisecond operation of an outwardly-rectifying potassium channel are unlikely to be sensed directly, but the consequent changes in membrane potential may be used by the cell as a sensory mechanism.

It is possible to calculate the approximate number of ion channels per bacterial cell and thus, the potential for perturbation. An average patch attached to an electrode has a diameter of 2 μm (S. Sukharov, personal communication), giving rise to an area of 3–6 μm^2 (depending upon the degree of curvature of the patch within the pipette), which is equivalent to the original surface area of 1–2 bacterial cells (assuming a spherical cell 1 μm diameter, surface area approx. 3.2 μm^2). In considering the probable number of channels of a particular type in a cell, the relationship between the size of the patch and the area of the cell becomes very important. Data only exist for the stretch-activated MscL channel of E. coli. Here 4–10 channels were detected reproducibly per patch from live spheroplasts [3], each of which is capable of completely depolarizing the cell when open. The E. coli literature supports the existence of at least four channels, MscS, MscL, KefB and KefC [3,4], and the possibility of at least one other, Kch [5,6] (see below). Consequently, the data for MscL can be multiplied four- to five-fold to give approximately 20–50 channel molecules per cell assuming similar levels of expression for each. It follows that to retain cytoplasmic solute pools and the capacity for Δψ-driven work, tight control of the opening of the channels is essential. That animal channels are tightly controlled is well established, but the smaller volume of bacterial cells makes such control even more important.

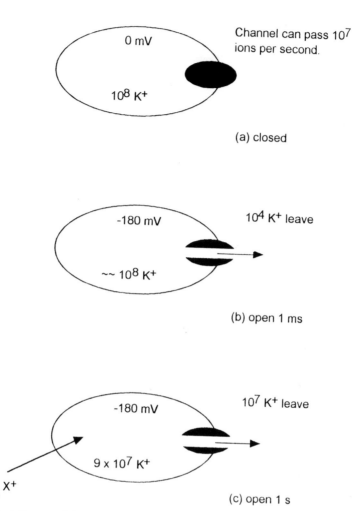

Fig. 1.   The influence of channel activity on potassium pools and the membrane potential; For the purposes of illustration a cell volume of 10^{-15} l and a cytoplasmic potassium concentration of 200 mM, a low external potassium concentration, and a channel with peak activity 10^7 potassium ions per second were assumed. At rest (a) the channel is closed and the membrane is assumed to be depolarized. Opening the channel for 1 ms (b) will elicit the exit of 10^4 K$^+$ ions, which makes little impact on the potassium pool, but raises the membrane potential by approx. -180 mV [2]. If the channel remains open for longer periods (c) the membrane potential hinders further potassium egress. Potassium exit will then be coupled with the inward movement of other cations either through the same channel or by other routes. In the normal case of a cell with a membrane potential established by proton extrusion, the activation of the channel will further hyperpolarize the membrane and potassium loss will be conditional on the uptake of other ions.

A number of channel activities have been proposed in bacteria and the degree of confidence in the supporting evidence varies (Table 1). This chapter reviews the evidence for channels in bacteria and evaluates their roles in cell physiology.

Table 1
Ion channel activities in bacterial cells

Channel class	Gene	Organism	Function	Electrophysiological characterisation?
Mechanosensitive	*mscL*	various[1]	osmoprotection	yes
Mechanosensitive	n.k.	*E. coli*	osmoprotection	yes
Glutathione-gated	*kefB, kefC*	*E. coli* and various others[2]	survival exposure to electrophiles	Yes[3]
Inwardly-rectified	*kcsA*	*S. lividans*	Potassium uptake[4]	yes
Shaker family	*kch*	*E. coli*	not known	no
K/Na	n.k.	*Rh. rubrum*	not known	yes
Porin	*ompX*[5]	various	pore	yes

The above channels have reached the stage where there is reasonable grounds for stating that they are ion channels.

[1] MscL channels have been detected in many Gram-negative and Gram-positive species, some by electro-physiological characterisation *in vivo* and others by gene sequencing and expression in *E. coli* (P. Blount personal communication).

[2] The glutathione-gated channels have been detected in a number of Gram-negative bacteria by analysis of NEM-activated efflux, and from others by gene sequencing.

[3] Electrophysiological analysis of KefC is at an early stage, but patch clamp analysis of several preparations of *E. coli* protoplasts have revealed potassium channel activity that can be eliminated by mutations affecting KefB and KefC (A. Kubalski, personal communication).

[4] The precise function of KscA in potassium uptake is unclear, but it is clearly a potassium channel that facilitates potassium uptake and thus is capable of equilibrating the potassium gradient with the membrane potential.

[5] The letter 'X' is used to denote that there at least four porins in *E. coli* and that there may be multiple porins in other bacteria.

n.k. = not known.

2. Essential background

2.1. The structure of eukaryotic ion channels

Ion channels are extremely diverse in their detailed structure and properties, reflecting the roles that they have evolved to fill in organisms as different as *Arabidopsis, Drosophila,* man and bacteria. Although the structural genes for many different types of channels have now been isolated, electrophysiological analysis of different cell types continues to reveal channel activities for which the corresponding gene product has not been identified. Therefore, we should not assume that the full scale of the diversity of channel structures has yet been revealed. Two basic types of ion channels are the ligand-gated[1], which open on binding of a molecule to the exposed

---

[1] For a broad view of ion channels and their associated issues the reader is directed to an excellent book designed for the beginner in this field: D.G. Nicholls "Proteins, Transmitters and Synapses", Blackwell Scientific Publications 1994. A more thorough treatment can be found in B. Hille "Ionic Channels of Excitable Membranes", Sinauer Associates Inc. 1992.

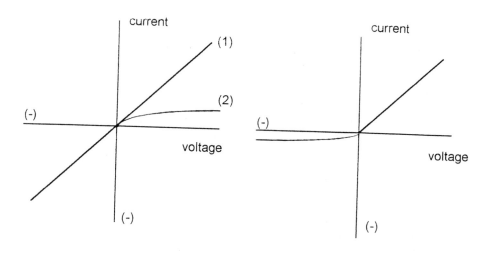

(a) inwardly-rectifying current

(b) outwardly-rectifying current

Fig. 2.   Inward and outward rectification of single ion channels; The graphs show the current voltage relationships found with strongly rectifying ion channels. (a) Inwardly-rectifying channels; in the absence of Mg^{2+} the IRK1-type of channel shows freely reversible ion flow (1), but in the presence of Mg^{2+} (2) at the cytoplasmic face of the channel, reversal of the current is diminished when the potential is positive. (b) Outwardly-rectifying channels; the channel displays outward current when the membrane potential is positive inside, but little current when the membrane potential is negative inside. In each example the voltage is in the usual convention for bacterial cells (i.e. negative inside the cell).

segments of the channel, and voltage-gated, which open in response to changes in membrane potential. Some channels are hybrids, requiring both depolarisation and a ligand before they are activated.

Complex properties are associated with different channels. This is reflected in their relative kinetics upon activation (to open), their rates of closure, changes in their sensitivity to the regulatory stimulus after activation/inactivation (desensitisation), and in their degree of rectification. In addition many mammalian channels are subject to complex regulation through the binding of allosteric effectors and by covalent modification, either of which can alter the properties of the channel. The property of rectification is a particularly important one. For many channels ion flow is determined by the relative magnitudes of the ion gradients and the membrane potential; if the membrane is polarised (cytoplasm negative) cations will flow inward through the open channel. (Direction of current is always defined as the direction of positive charge movement even if the current is actually carried by anions.) If subsequently the membrane is depolarised above the equilibrium potential of that cation its flow will reverse. Rectified channels show departures from this simple reversible operation (Fig. 2). Rectification at the macroscopic level has to do with changes in the open probability with voltage, and is not dealt with here. At the level of single channel behaviour two classes of rectifying channels have been described. *Inwardly-rectifying* potassium channels carry ions inward when the membrane potential (inside negative) is greater than the chemical potassium gradient [11]. However, when the membrane is depolarised, such channels show limited reversal. *Outwardly-rectifying* channels are functionally the opposite of inwardly-rectifying channels, carrying ions outwards under depolarising conditions, but not inwards under hyperpolarisation. All of these differences in properties are the result of small differences in sequence within the channels since there is a commonality in the structures that underlies the extreme diversity of known ligand-gated and voltage-gated channels [7–12].

2.1.1. Ligand-gated channels

Ligand-gated channels are generally activated by the binding of a small molecular mass metabolite, e.g. glutamate, acetylcholine, and glycine, to a domain located on the outside of the membrane. The binding of the metabolite activates the opening of the channel [9]. The properties of the channel are dictated both by the transmembrane domain, generally comprising four transmembrane spans, M1–M4, and by a domain located towards the carboxy-terminus and which is located on the cytoplasmic face of the membrane. The transmembrane segments may cross the membrane as α helices or as β sheets; for example the acetylcholine receptor channel is believed to be formed by the M2 segment of the transmembrane domain folded as an α helix and by M1, M3 and M4 folded into β sheets [10]. Five subunits with the stoichiometry $\alpha_2\beta\gamma\delta$, which share extensive sequence homology, are assembled to make a cation-conducting pore of approximately 7 Å diameter. Ion preference is provided by rings of negatively charged amino acids (each helix contributing amino acid side chains to the ring of negative charge), which attract cations and repel anions. The M2 segment is an amphipathic helix and the pore characteristics are believed to be contributed by serines, threonines and leucines, which form rings of hydrophilicity and hydrophobicity respectively, due to the pentameric nature of the pore [9,10]. Mutagenesis of residues in M1, M2 and M4 have confirmed the central importance of M2 in determining the channel permeation properties. Mutations affecting residues of M2 that would face into the pore alter ion conduction and alterations of the acidic residues that are located at the inner and outer mouth of the pore reduce single channel conductance. Consistent with this proposition of the importance of M2 for channel formation, the affinity of the channel for compounds that block channel conductivity is reduced by specific mutations to the M2 segment [9,10]. Finally, single channel conductance is proportional to the molecular volume of the side chain at position 4 in the helix: larger residues and smaller residues decrease and increase conductance, respectively [10,10a]. Although these details have been worked out for the acetylcholine receptor they are believed to reflect the structural organization of other ligand-gated channels and enshrine some of the fundamental principles of channel formation.

2.1.2. Voltage-gated channels

The analysis of voltage-gated channels has been dominated by the discovery and investigation of the *Shaker* potassium channels, but these channels reflect many of the structural features of the major calcium and sodium channels also [7,8,12]. The basic channel motif is considered to be the six transmembrane spans, S1–S6, in which the loop between S5 and S6 forms a hairpin sequence that enters and leaves the membrane from the same side. This sequence, called H5 or P loop, defines many of the characteristics of the channel. The S4 span contains the voltage sensor, a motif $(R/Kxx)_n$ followed by leucine residues spaced approximately seven residues apart and extending into the S5 transmembrane span [7,8]. Both the positively charged amino acids and the leucine repeat make important contributions to the voltage sensitivity of the channel. These channels may also possess extensive carboxy- and amino-terminal domains, but unlike the ligand-gated channels both of these domains are believed to be located at the cytoplasmic side of the membrane. Calcium and sodium channels have four of the S1–S6/H5 motifs covalently linked, whereas it is believed that the potassium channels are formed from the non-covalent association of four such subunits. Different subunits are believed to form heteromeric channels, but homooligomers are also formed. The sequences of S5 and S6 also affect channel conductance, as does the linker region between S4 and S5 and consequently it is thought that the voltage-gated channels are formed by four sets of S5 and S6 helices forming the pore, which is then lined with the P loop sequence. Genetic, electrophysiological and pharmacological analyses support this proposed importance of the P loop in pore formation; for example, Na^+ channels can be converted to Ca^{2+} channels and K^+ channels converted to be like cyclic nucleotide-gated channels by mutagenesis of the P loop regions [7,8,12]. Thus, the P loop is the basic determinant of selectivity and the adjacent transmembrane spans provide some degree of fine tuning.

Channels can be constituted from much simpler proteins, e.g. minK consists of a subunit with a single transmembrane span [13], while the K_{IR} family of inwardly rectifying potassium channels possess only two transmembrane spans separated by a P loop [11]. Mutagenesis has been used to delete the S1–S4 sequences of a *Shaker*-type channel leaving only the structural elements of the inwardly-rectifying channel. This mutated channel then functioned as an inwardly-rectified channel [13a].

Not all the subunits that form the channel complex are associated with pore formation. An excellent example of this is the β subunit of the voltage-gated potassium channels [14]. There are two major classes of channels in the *Shaker* family: the 'A type', which are activated rapidly by membrane depolarisation and are also rapidly inactivated ($t^{1/2}$ for closing = 1–2 mS), and the delayed rectifier type, which activate more slowly and may remain active for 1 s or more. Fast inactivation of the A type *Shaker* potassium channels is mediated by an 'inactivation ball', which consists of the amino-terminal 20 amino acids containing three critical basic amino acids attached to the pore-forming domain by a 60 residue flexible linker [15,16]. Free (untethered) peptides with the same amino acid sequence as the amino-terminus are effective blockers of channels from which the sequence has been deleted [16]. This 'ball' is believed to interact with the short loop between helices 4 and 5 to effect the closing of the channel [17]. The β1 subunits carry at their amino-terminus sequences equivalent to the inactivation ball found in the 'A type' *Shaker* channels [22], which bring about fast inactivation of the channels with which they associate. Thus, binding of a β1 subunit to a delayed rectifier channel confers the fast inactivation kinetics typical of 'A type' channels due to the presence of an 'inactivation ball' similar to that found at the amino terminus of the 'A type' channels. The β subunits bind to the channel protein with sufficiently high affinity to enable them to co-purify with the detergent-solubilised pore-forming α subunits. Sequence analysis places the β subunits in the family of oxidoreductases of the medium chain oxidoreductase/dehydrogenase [19,19a]. It is not known whether the β subunits possess any enzyme activity or whether the oxidoreductase fold may simply provide a convenient scaffold that stabilizes the inactivation peptide and enables the β subunit to move between channels. The presence of separate regulatory β subunits enables the same channel protein to be either a delayed rectifier (no β1 or inactivation ball at the amino terminus) or an 'A type', fast inactivator (ball or β1 subunit attached). Sequences similar to the β subunit have been found in bacteria, which given the existence of *Shaker*-like sequences in *E. coli*, suggests that there may be similar mechanisms regulating the activity of ion channels in bacteria (see below: discussion of Kch and KefC) [14, B. Ganetzsky, personal communication]. Thus, the diversity of channels can be greatly enhanced by the presence of different β subunits.

Eukaryotic channels are diverse on many levels. Initially this diversity was revealed by electrophysiological analysis of cell membranes, but has subsequently been borne out by molecular genetic and biochemical analysis. In bacteria analysis of channels has been pursued by analysis of physiological phenomena, which has been then been supported by genetic and biochemical analyses. Revealing the existence of ion channels in bacterial cells required the development of electrophysiological techniques, principally patch clamp technology, but also the study of the electrical properties of bacterial membranes fused with bilayer lipid membranes. Of these patch clamp has proved to be the more powerful method of analysis, revealing the presence of several channel types in *E. coli* [20,21].

2.2. Patch clamp analysis of ion channels

The recognition of the importance of ion channels in bacteria has come about through the development of methods for the application of patch clamp technology to bacterial cells [3,6,20–24]. Patch clamp technology was developed for the analysis of isolated channels in small pieces of membranes derived from animal cells [25]. The essence of the technique is that the section of membrane circumscribed by the tip of the patch pipette is electrically isolated and consequently ion flux through *individual* protein channels can be detected as current flow. For bacteria, removal of the cell wall with the consequent formation of naked protoplasts is essential, since the patch electrode must form a tight seal (giga-ohm seal) with the membrane so that all of the current flows through the channels and not through the junction between the membrane and the pipette. Bacterial protoplasts are generally too small to be patched directly, and methods for the generation of giant bacterial cells enabling the formation of large protoplasts have been essential for the development of this field [20–23]. Inhibition of cross wall formation with cephalexin leads *E. coli* cells to form 'snakes' 50–150 μm in length that form spheroplasts 5–10 μm in diameter when treated with lysozyme in an osmotically-stabilizing buffer [6]. Such preparations can be stored at –20°C for several months and retain the capacity to form good seals with the electrode. Protoplasts from *Bacillus subtilis* [22] and *Streptococcus faecalis* [23], suitable for patch clamping, have been achieved by simply removing the cell wall with lysozyme. For these organisms the protoplast continues to grow after removal of the cell wall and diameters of 2–5 μm can be achieved. Patch clamp analysis can also be carried out on liposomes to which bacterial membranes have been fused or into which solubilised membrane proteins have been reconstituted [24]; although whether a channel can be functionally reconstituted has to be individually established.

The formation of patches requires that suction (20–100 mm Hg) be applied via the electrode to draw a portion of the protoplast or liposome onto the open tip of the pipette to form a tight giga-ohm seal, so that ions passing through the channels at the tip of the electrode can be detected as changes in current [25]. Open channels are recorded as a stepwise increase in electrical conductance through the membrane patch and the specificity of the channels can be analyzed by varying the ionic composition either of the bathing medium or of the pipette contents. Once the seal has been formed the pressure can be released. Brief exposure of the attached protoplast to the air causes it to burst and leaving a single patch of membrane on the electrode. It should be noted that although text-books present the patch as a neat seal across the end of the pipette, in reality the patch may form some distance inside the pipette and may have extended contact with the walls of the pipette. Four different configurations are generally used (Fig. 3) [26]: cell-attached patches, inside-out patches, outside-out patches and whole-cell. Each of the methods has advantages. In the cell-attached, inside-out patch and outside-out patch modes only the channels in the region circumscribed by the electrode are analyzed. In the cell-attached mode metabolic pathways and signal generating systems are intact and can modify the channel in response to appropriate environmental cues. In the isolated inside-out patch the cytoplasmic face of the membrane is exposed to the bathing solution and it is much easier to modify the incubation conditions and analyze the properties of the channel under very tightly controlled conditions. However, peripheral membrane proteins and small metabolites can readily dissociate from the channel and change the properties, sometimes leading to a decay of channel activity over time if an essential activator is lost. The whole cell method involves forming the patch and then destroying the patch while retaining the seal and the attachment to the cell. In this configuration all of the channels of the cell can be analyzed simultaneously, but the drawback can be that the experimental system is very fragile and may not allow extreme conditions to be approached. An additional method, giant membrane patches, has been applied mainly to eukaryotic membrane systems. Here the electrode is wider and reads the channels over a patch of membrane up to 40 μm in diameter, which is well above the size of currently available bacterial protoplasts. However, if the bacterial channel can be expressed in oocytes then the possibility of using giant patches can be exploited. The advantage of the method is that it allows low abundance channels to be analysed, but it also allows the currents through the most active electrogenic transporters to be analysed [27].

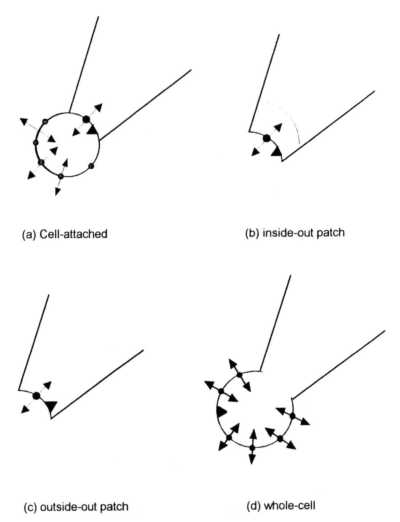

(a) Cell-attached   (b) inside-out patch

(c) outside-out patch   (d) whole-cell

Fig. 3.   Patch clamp configurations; the inside out patch, whole-cell and outside-out patch modes are all derived from the cell-attached mode by manipulation of the pressure on the cell or patch. Membrane orientation is shown by a black triangle depicting the cytoplasmic face of the membrane. The electrical circuit consists of an electrode inside the glass pipette connected with the membrane surface by an electrolyte solution and a bath reference electrode (usually connected by saturated KCl set in agar in a fine glass tube that sits within the bath). The potential across the membrane is determined by the potential applied to the pipette electrode using the patch clamp amplifier and can be arranged to have the same or the opposite orientation to that found in the cell. The channel(s) being investigated are shown in black. Other channels present in the membrane are shown hatched; note that in (a) the patch is isolated from the bath electrode by the presence of the cytoplasmic membrane (compare with (b) and (c), and thus completion of the electrical circuit is dependent upon ion flow through other transport systems and the passive permeability of the membrane. Note that in real life the patch is not usually so neat and may have a pronounced curvature into the pipette. In the case of patches derived from Gram-negative bacteria there may be perforated outer membrane located within the pipette above the seal formed with the cytoplasmic membrane.

3. Bacterial channels

Several types of channels have been characterized in bacterial cells (Table 1). The bacterial porin resides in the outer membrane of most Gram-negative bacteria and its structure has been resolved through the formation of crystals and analysis by X-ray crystallography [28–30]. It is more specifically a pore, since the open state is believed to be the most frequently encountered (see below), but it is the only channel-forming protein for which a structure and a defined physiological role are available. Mechanosensitive channels are the best documented class of bacterial channels located in the cytoplasmic membrane [3,6,20–24], but strong evidence also exists for inwardly rectifying K^+ channels in at least one Gram-positive organism [31] and for glutathione-gated K^+ channels in *E. coli* [4]. One protein, Kch [5], with a structural homology to the major *Shaker* family of eukaryotic potassium channels has been discovered in *E. coli*. The sections that follow review the evidence on the structure and function of these channels in bacterial cells.

3.1. Bacterial porins

Porins are integral membrane proteins that ensure the rapid transfer of solutes from the environment into the periplasmic space of Gram-negative bacteria [32]. The outer membrane is impermeant to most hydrophilic molecules and thus both nutrients for growth and waste products produced by the organism must pass through the pathway provided by the porins. Porins exclude molecules above approx. 600 Da, but generally exhibit poor discrimination between charged species. In *E. coli* the three major porins are OmpC, OmpF and PhoE. They show 63% sequence identity and while OmpC and OmpF are weakly cation selective, PhoE is anion selective [33] and is generally induced under conditions of phosphate limitation. Reconstitution of the porins into lipid bilayers has enabled the determination of their channel conductances in 1 M salt, which are 800 pS, 500 pS and 600 pS, respectively for OmpF, OmpC and PhoE [34,35]. The channels show voltage gating with threshold potentials of –90, –160 and –100 mV, for OmpF, OmpC and PhoE, respectively [36]. A more specific porin, maltoporin (LamB) has a much lower conductance, 150 pS, and this correlates with the smaller pore size exhibited in the crystal structure [29] (see below). Patch clamp analysis has been performed with reconstituted outer membranes of *E. coli* fused to liposomes [37]. Porin channels show alterations in the conductance state with time, which is generally assumed to reflect different conformations of the protein that exhibit altered conductivity. The channels are open for the majority of the time, but transient closures of one or more channels take place at random intervals. It was observed that an anionic molecule, membrane-derived oligosaccharide (MDO), could induce a change from the open to the closed state [38]. MDO is accumulated in the periplasm when cells are growing at low osmolarity [39]. There may be an inherent paradox here, since low osmolarity is often equated with an environment that is low in nutrients, precisely the condition when the porins are required to open to facilitate the capture of solutes required for growth. Theoretical considerations have suggested that when the external concentration of nutrients is low, the transfer of solutes through the porins is rate limiting for their acquisition by the cell [40]. Thus, if closure were more frequent under these conditions, as is indicated by the patch clamp analysis, this might be contrary to the health of the organism. However, a counter-argument is that cells leak metabolites across the cytoplasmic membrane and these are re-cycled by active transport [41]. Closure of the porins, by MDO, in low osmolarity environments might limit the loss of nutrients to the environment.

The structures of the bacterial porins have been resolved by X-ray crystallography and essentially confirms the basic features of earlier models based on structure predictions and biophysical measurements [28–30,42,43]. X-ray crystallography of three porins and molecular replacement for another has confirmed the following structural features (see Chapter 26 in this volume):

(a) The core of the molecule is composed of a multi-stranded anti-parallel β barrel (16 strands for the general porins [28,29] and 18 for LamB [29]) with the strands orientated at an angle between 35° and 50° to the axis of the barrel. The OmpF, OmpC and PhoE molecules show a marked segregation of aromatic residues (tyrosine and phenylalanine) to the boundary area of the protein where it meets the phosphate head groups. Trimers of the porin subunits are formed by hydrophobic interactions between the side chains of the β sheets and are stabilized by the insertion of the L2 loop of one subunit into the barrel of an adjacent subunit [28].

(b) Short β turns define the periplasmic face of the barrel, but the loops at the outer surface are more extensive. One or more of the loops at the outer surface fold into the barrel. In OmpF the L3 loop forms a short α helix within the lumen of the barrel and is a major contributor to size selection. The presence of the loops narrows the pore at its entrance, but the amino acid side chains that project into the lumen of the barrel provide a further narrowing [28–30]. At its narrowest point the pore in OmpF is considered to be approx. 7×11 Å compared with 15×22 Å at the periplasmic end of the barrel. At the narrowest point a molecule of glucose in its hydrated form can pass through without steric hindrance. The presence of further loops in the lumen of the LamB porin narrow this pore further to approx. 5×6 Å [29], which is consistent with the observed lower conductance of this pore.

The character of the pore is further affected by the residues that line the lumen of the barrel. For maltoporin these residues are a mixture of aromatic and ionizable residues. The latter are too far apart to make salt bridges, but are believed to be linked by water molecules. The aromatic residues are arranged along a left-hand helical path extending from above the constriction of the pore to the periplasmic surface of the protein [29]. These residues are believed to be spaced to form a series of weak sugar-binding sites by analogy with the role of aromatic residues in sugar-binding proteins. The binding of the sugar substrate to the aromatic residues above the pore may align the substrate down the axis of the pore and facilitate 'guided diffusion' through the channel [29]. This mechanism may provide a mechanistic analogue for the mechanism of ion transfer through channels. The OmpF porin channel constriction zone is delimited on one side by the main chain carbonyl groups and the side chains of an aspartate and a glutamate residue on loop L3 and on the opposite side by the side chains of three arginines and a lysine residue.

Replacement of a glycine residue found in OmpF by arginine in PhoE is believed to account for the anion selectivity of the latter, since the residue is exposed in the pore [28]. Other positively charged amino acids at the mouth of the pore are also believed to effect anion selection.

Porins conform to the generality that the channel is made up from a scaffold composed of the more structured elements of the protein, in this case β sheets, and the pore is defined by the loops that fold into the lumen of the barrel. The characteristics of the channel (cation/anion selectivity, size exclusion, substrate specificity) are defined by the properties of the amino acid side chains that extend into the lumen of the barrel.

3.2. Mechanosensitive ('stretch-activated') channels

3.2.1. Introduction

The activities of mechanosensitive (MS) channels in bacterial membranes were discovered several years ago in *E. coli*, but have since also been documented in the Gram-positive organisms *S. faecalis* and *B. subtilis* [22–24]. Mechanosensitive channels are defined as channels which open when suction (equivalent to 30–300 mm Hg; [44]) is applied to the membrane patch. Suction generates tension within the plane of the membrane and hence these channels are also referred to as 'stretch-activated' channels. When the pressure is released the channel closes. Even when the membrane is under continuous negative pressure the channels open and close indicating that the channel reverts to the closed state with a defined time course that is relatively independent of the stretch applied to the membrane (Fig. 4) [3]. Different channels can be identified by their conductance, by the pressure required to activate them, by characterisation of the protein complexes responsible for activity and by the identification of their structural genes. Such channels may also have a number of intermediate states that are revealed when the membrane stretch is sustained by suction on the patch. Given that the proposed role of the mechanosensitive channels is to relieve excessive pressure on the cell membrane, generated by high turgor, they may not be exposed to sustained tension under physiological conditions. Consequently the physiological significance of these intermediate states must be in doubt. Mechanistically, the intermediate forms of the channel are very interesting, but they are not understood at the present time.

Activation of MS channels by membrane tension has been investigated by perturbing membrane structure with amphipaths [44a]. Amphipaths insert asymmetrically into biological membranes, whose inner and outer leaflets are intrinsically different in terms of phospholipid head groups. The positive amphipaths tend to insert more readily into the more negative inner leaflet of the lipid bilayer and negatively charged amphipaths enter the less negatively charged outer leaflet of the membrane. This asymmetric insertion in one or other side of the bilayer is assumed to generate tension within the membrane. The treatment of *E. coli* spheroplasts with amphipaths showed that they could slowly activate MS channels with the effectiveness of the amphipath being linked to its lipid solubility [44a]. Cationic amphipaths could be used to counter the effects of anionic amphipaths, by inserting in the

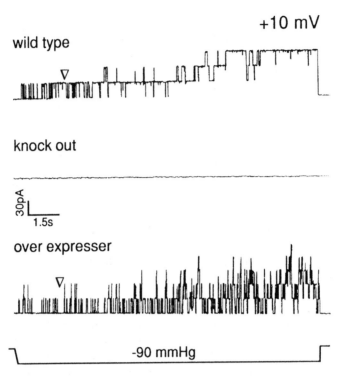

Fig. 4.   Stretch-activated channel activity associated with MscL of *E coli.*; the traces show channel activity elicited by gentle suction to the isolated membrane patch. The activity of MscL is indicated by increased current (inverted open triangle). The activity of smaller MscS channels is not shown. The top trace shows MscL activity in patches derived from the wild-type *E. coli*; the middle trace shows the trace obtained with a MscL null mutant; and the bottom trace shows the trace obtained with the knockout transformed with a plasmid carrying the wild-type *mscL* gene. The pressure profile required to elicit the change in channel activity is shown at the foot of the diagram. (Reproduced from Ref. [3] with permission).

opposite side of the polarized membrane. Sudden cooling of membranes to 1°C has also been reported to increase the activity of stretch activated channels indicating that they are sensitive to the physical state of the lipids [24]. Thus, the channels do appear to respond to membrane deformation and excessive turgor pressure in the cell may be transmitted to Msc channels through increased tension in the plane of the lipid bilayer.

3.2.2. Characteristics of different channels and their activation

The number of kinds of MS channels in *E. coli* is still a matter of debate [3,6,24], but a minimum of three can be postulated on the basis of electrophysiological and genetic data. A relatively small weakly anion-selective channel, MscS, with a conductance of about 1000 pS and a large conductance channel, MscL, 3000 pS, which is non-selective, were characterized by Kung and colleagues in detergent-extracted membrane protein fractions reconstituted into asolectin liposomes [3,44]. In the whole-cell patch clamp mode Adler's group also concluded that *E. coli* possesses stretch-activated channels with conductances of 1100 pS and 350 pS, but in their study [6] mutant strains lacking MscL were found still to exhibit both conductances. The whole-cell patch clamp experimental system is less robust than isolated patches and consequently, insufficient pressure may have been exerted such that MscL was not activated. Hence this study would be consistent with three MS channels. Several different conductances (100–150 pS, 300 pS, 500 pS and 1000 pS) were seen by Ghazi and colleagues in giant asolectin liposomes into which *E. coli* inner membranes had been fused [24]. All but the smallest of these conductances could be inhibited by 100 μM gadolinium ions (Gd^{3+}), a well-known inhibitor of eukaryotic MS channels. Recently, the purified and reconstituted MscL has been shown to be sensitive to Gd^{3+} (see below) [44,45] and the sensitivity of the larger stretch-activated channel was also confirmed in the whole-cell patch clamp studies [6]. The channels characterized in *B. subtilis and S. faecalis* are of similar conductance to the MscL channel of *E. coli* and are also sensitive to Gd^{3+} [22,23]. Given the recently discovered sequence conservation among the large mechanosensitive channels (see below), this similarity in properties is not surprising.

To perform rapid turgor adjustment the MS channels should be located in the cytoplasmic membrane. For the Gram-positive bacteria there can be no equivocation since these organisms possess only the cytoplasmic membrane. Despite some confusion in the literature the MS channels of *E. coli* are now considered to be located in the inner membrane. Thus, MS channel activity was found to be associated with cytoplasmic membranes [24]. Western blots performed with antibodies specific to MscL have now specifically located the channel protein to the inner membrane of *E.coli* (S. Sukharev, personal communication).

Physiological data and measurements of the current through MscL suggest a very large pore in the open state. It is difficult to make an accurate correlation between conductivity and the size of the pore, but the magnitude of the current passed by MS channels, and the apparent ability to transfer disaccharides (see below) across the membrane, suggest a large aperture. The bacterial porin OmpC also transfers disaccharides and exhibits conductance values that are slightly smaller than for MscL. The pore size for OmpC, established from the crystal structure is 7×11 Å [28], which is considered close to the minimum pore required for the transfer of a hydrated sugar across the membrane. The pore size of MscL again emphasizes the need for control over the opening of the stretch-activated channels. Solute release from this size channel would, of course, be indiscriminate, but would allow the cell to reduce its turgor rapidly. Indeed it is known that *E. coli* cells that are synthesizing the osmoprotectant trehalose, a disaccharide, release it into the periplasm continuously throughout growth [47]. Brief openings of the MS channels may be sufficient to cause this loss of the osmoprotectant.

The isolation of detergent-solubilised protein fractions from *E. coli* membranes and reconstitution into liposomes has allowed the identification of a protein, and its corresponding gene, for the largest of the *E. coli* stretch-activated channels, MscL [3]. MscL is a protein of 136 amino acids with two hydrophobic stretches, each large enough to span the membrane as alpha helices, separated by a long loop

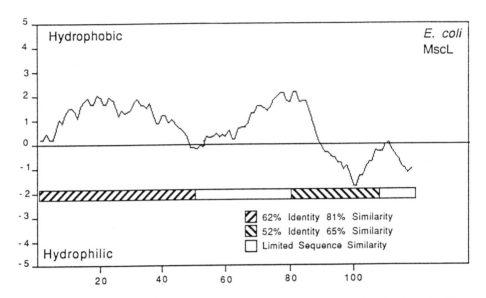

Fig. 5. Hydrophobicity plot of MscL. The plot shows the two regions of hydrophobicity (Kyte-Doolittle, w = 19; [86]), each sufficient to cross the membrane once, in the MscL protein. The central loop and the tail region are essentially hydrophilic. Also shown is the degree of sequence conservation among the published MscL sequences (*E. coli, H. influenzae,* and *Cl. perfringens*), which indicates that the two transmembrane spans are highly conserved, but that the loop and tail are variable.

sequence enriched in neutral or hydrophilic amino acids (Fig. 5). It is thought that the protein may be an oligomer and gel filtration data support a molecular mass close to 70 K, which could indicate a pentamer or greater. Similar sequences to the *E. coli* MscL have now been identified in *Haemophilus influenzae, Erwinia cara-tovora, Clostridium perfringens* and in *Pseudomonas fluorescens* (P. Blount, S. Sukharev, P. Moe and C. Kung, personal communication). A high degree of sequence conservation is observed, especially in the transmembrane stretches (Fig. 5). The amino terminal region of MscL carries three positive charges, two of which are highly conserved. According to von Heijne's rules [46] the N-terminus should be located at the cytoplasmic face of the membrane and this would also place the carboxy-terminus in the cytoplasm. Two regions of MscL show relatively poor conservation of sequence: the loop region and the carboxy-terminal region. The latter can be truncated without apparent effect on the kinetics of channel activation. However, mutations in the loop region alter channel characteristics as might be expected if this forms the pore (P. Blount, personal communication).

The possible need for the cell wall or other cellular structures for Msc activity has been eliminated by the reconstitution experiments with purified proteins. The original isolation and characterization of MscL was based upon reconstitution of relatively complex protein mixtures into liposomes [3]. Recently, the protein was purified as a protein fusion with glutathione-S-transferase, cleaved with thrombin to release the native protein and reconstituted into artificial liposomes [45]. The

purified protein was found to be fully functional, with electrophysiological characteristics similar to those of the native channel. The channel activity could be
blocked with Gd^{3+} ions and polyclonal antibodies raised against the protein were
able to abolish channel activity when pre-incubated with the MscL protein. Thus,
artificial liposomes containing purified MscL displayed normal activation when the
membrane was placed under tension. Thus, without doubt MscL activity requires
only the product of the *mscL* gene and there is a strict correlation between the
activity of this protein and the formation of large stretch-activated channels. However, the resistive force of the cell wall on the expansion of the cell must also be
considered when analysing the physiological role of the channels in controlling
membrane tension.

3.2.3. The possible physiological role of stretch-activated channels
Rigorous analysis of the role of MS channels in cell physiology is impeded by the
lack of mutants that are entirely without MS activities. However, studies with
gadolinium ions [24] have implicated these channels in the response of bacterial
cells to hypoosmotic shock (Fig. 6). It is well established that a large hypoosmotic
shock will cause *E. coli* to lose numerous solutes including amino acids, compatible
solutes, potassium, ATP, lactose and D-galactose [48–52], but not the macromole-

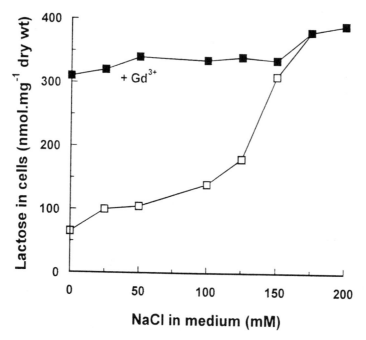

Fig. 6. Inhibition of solute efflux by gadolinium ions. *E. coli* ML308-225 cells were pre-loaded
with ^{14}C-lactose in the presence of 200 mM NaCl and the cells were then diluted into media with
progressively lower NaCl concentrations in the presence (closed symbols) or absence (open symbols) of
1 mM gadolinium ions. (Reproduced from reference [24] with permission).

cules. During hypoosmotic shock, greater than 90% viability is retained by *E. coli* cells and loss of the cytoplasmic enzyme β galactosidase is less than 5% [24,50]. If an energy source is available bacteria rapidly re-accumulate potassium, showing that they still have cellular integrity [24,52]. Thus, hypoosmotically-induced efflux is a controlled phenomenon and not cell lysis. The extent of solute loss is related to the severity of hypoosmotic shock [50]. Solute loss is independent of the final osmolarity of the medium and of the solute that makes up the osmotic pressure (provided cations are present to facilitate exchange with bound potassium ions). Cells accumulate sodium ions and exhibit transient cytoplasmic acidification after hypoosmotic shock [52]. Thus, hypoosmotic shock results in major changes in the cytoplasm consistent with a transient loss of selective permeability.

It has been proposed by several groups that the stretch-activated channels facilitate the loss of cytoplasmic solutes during hypoosmotic shock. Ghazi and colleagues showed that Gd^{3+} inhibited the release of solutes from *E. coli* and *S. faecalis* subjected to hypoosmotic stress [24]. The presence of 1 mM Gd^{3+} in the medium used for osmotic down shock was almost sufficient to inhibit the loss of lactose (Fig. 6) and ATP from *E. coli* cells. The inhibition of lactose efflux was related both to the concentration of Gd^{3+} present and to the intensity of osmotic shock applied to the cells [24]. Gadolinium was also shown to inhibit the hypoosmotically-induced efflux of ATP from *S. faecalis* [24]. However, potassium and glutamate efflux were only partially sensitive to gadolinium ions, indicating that other pathways also participate in hypoosmotic shock [24]. In patch clamp experiments, 100 μM Gd^{3+} ions inhibited the large mechanosensitive channels of *E. coli*, while channels of smaller conductances were unaffected [24]. Thus, there is a good correlation between the inhibition of the larger stretch-activated channels and the prevention of solute release during hypoosmotic shock.

3.2.4. Genetic studies of mechanosensitive channels

Genetic studies support the presence of multiple mechanosensitive efflux pathways. An *E. coli mscL* knock-out mutant has been created [3]. Patch clamp analysis confirmed that this strain lacked MscL activity, but retained other stretch-activated channels of lower conductance. However, the strain was quite able to survive hypoosmotic shock (C. Kung et al., personal communication). Work in the authors' laboratory has shown that *E. coli* AW405 lacking MscL can grow as well as the parental strain in both NaCl and KCl concentrations from 20 mM to 0.5 M salt concentration. Similarly, the mutant recovered from hypoosmotic shock with similar kinetics to the parental strain. These data indicate that the MscS and MscM channels may compensate for MscL in the mutant strain and the isolation of a mutants lacking MscS must be a priority for the further analysis of the proposed physiological roles of these channels.

The activity of the smaller conductance MS channels has been reported to be affected by a genetic locus, *kefA* [6]. Patch clamp analysis in the whole-cell mode showed that a channel, activated by membrane tension imposed by increased positive pressure from the electrode into the cell (see Fig. 3), was altered by the *kefA2* mutation carried by strain RQ2. The mutant strain RQ2, carrying the lesion *kefA2*, was isolated by Epstein as a mutant specifically-sensitive to 0.6 M K$^+$ at high osmolarity in the presence of betaine. In contrast the mutant and parent strains were indistinguishable at high osmolarity if the potassium concentration was kept low (10–20 mM). MS channels in the mutant strain, analysed in the whole-cell configuration, were activated normally, but appeared not to close spontaneously as the wild-type channels did [6]. Recent analyses of this strain, and of a *kefA* null mutant, in the isolated patch configuration has revealed no change in the stretch activated channels. However, the analysis of this genetic locus is at an early stage.

The selection protocol for *kefA* mutants was designed to identify mutations that affected potassium efflux systems other than the already identified and characterized KefB and KefC. The strain is sensitive to potassium, even at low osmolarity, but exhibits a complete cessation of growth upon the addition of betaine in the presence of 0.6 M KCl. The cells remain viable and accumulate betaine, but appear unable to regulate potassium release correctly. The *kefA* gene has been cloned and a null mutant created. This mutant does not display the phenotype associated with the RQ2 strain, grows normally at high osmolarity and displays no significant change in the activity of stretch-activated channels when analysed in isolated inside-out patches. Clearly the *kefA* gene is not the structural gene for a major non-discriminatory stretch-activated channel.

The lack of phenotype associated with the null mutant may reflect the presence of a homologue (KefD) detected by the *E. coli* genome sequencing project at approximately 94 min on the genetic map. Preliminary characterization of the putative KefA protein product suggests a large protein with domains typical of inner membrane proteins and of porins (M.A. Jones, D. McLaggan, W. Epstein and I.R. Booth, unpublished data), but otherwise with no known homologues other than KefD. KefA may be a minor, potassium-specific, stretch-sensitive channel, but it is not one of the major mechanosensitive channels detected by previous studies [3,6,20,21].

Bacteria are, therefore, likely to cope with excessive turgor by a combination of mechanisms. During small shifts in the external osmolarity and during the accumulation of compatible solutes small decreases in cell turgor might be effected by the potassium specific channels (or transport systems) detected by Ghazi and his colleagues [24]. However, large reductions in turgor can only be effected by the range of stretch-activated channels: MscS, MscM and MscL.

3.3. Inwardly rectifying potassium channels

Inwardly rectifying potassium channels, K_{IR}, channels are defined as those that show a current voltage relationship that is asymmetrical (Fig. 2), i.e. current (here carried by potassium ions) flows into the cell when the membrane potential is negative inside, but when the cell is depolarized or has a positive membrane potential the current does not readily reverse. This property makes the channel essentially non-reversible. In mammalian cells inwardly rectifying channels can be classified as either strongly or weakly inwardly rectifying, depending on the current carried through the channel when the membrane potential is positive (inside). In the case of IRK1, a mammalian channel [11], inward rectification has at least two components: tight binding of cytoplasmic Mg^{2+} ions in a voltage-dependent manner and intrinsic gating, which is more poorly understood [11]. In this case, the strength of rectification is predominantly determined by the substitution of an aspartate residue (strong rectification) by asparagine (weak rectification) in the second transmembrane span. The presence of aspartate increases the affinity for magnesium ions, which close the channel. Such channels usually show inwardly rectifying current when analyzed as cell-attached patches, but linear current-voltage relationships when the patch is detached from the cell (Fig. 2) due to the diffusion of Mg^{2+} away from the channel [11]. Addition of Mg^{2+} (or the polyamines spermine and spermidine) to the bathing solution re-establishes in the inward rectification. In eukaryotic cells two major types of inwardly rectifying potassium channels have been identified: those with six transmembrane spans and a P loop, e.g. KST1, and the simpler K_{IR} two transmembrane span class separated by a P loop, e.g. IRK1 [11,54,55].

Among bacteria an inwardly-rectifying channel, KcsA, in *Streptomyces lividans* [31] has recently been characterized electrophysiologically, purified and the corresponding gene sequenced. The purified protein is relatively small at 17.6 kDa, but nothing is known of the possible multimeric state of the protein in the membrane. The deduced KcsA protein is smaller than members of the mammalian K_{IR} family, but like these proteins it is thought to consist of two membrane-spanning regions (S1 and S2) linked by a P loop-like region. Multiple alignments of the P loop sequence reveals, however, that the deduced KcsA has closer kinship to deduced *Shaker* proteins and is more distantly related to mammalian K_{IR} channels (Table 2) [31]. However, the second putative transmembrane span lacks both aspartate and asparagine residues that might lead to Mg^{2+}-dependent rectification. Reconstitution into bilayer membranes of KcsA has revealed that the channel has several sub-conductance states, with the largest corresponding to 90 pS. The channel can be blocked by the potassium homologue, cesium ions. Channel activity requires Mg^{2+} and inward rectification is only seen in protoplasts fused with liposomes and is not seen when the purified protein is reconstituted into unilamellar liposomes and fused with planar bilayers. Inward rectifying current could not be restored by Mg^{2+} [31] and this suggests that there may be other protein components required for full channel function. The recent discovery of the association of mammalian K_{IR} channels with other membrane proteins [55a] could mean that further subunits are required for the full activity and regulation of the KcsA channel to be seen. It has been speculated that the channel has a role in potassium acquisition, but this would limit the accumulation of potassium to the magnitude of the membrane potential. The relationship between $\Delta\psi$ and the potassium gradient has not been investigated in detail. Mutants lacking this channel have been constructed and found to grow slightly poorer than the wild type [31].

During sequencing of the gene for lactate dehydrogenase in *Bacillus caldotenax* and *Bacillus stearothermophilus* a protein, LctB, that resembles the K_{IR} channel family was discovered [5,56]. The *lctB* gene encodes a protein of 134 amino acids,

Table 2
Alignment of probable P loop sequences

Kch	M	T	A	F	Y	F	S	I	E	T	M	S	T	V	G	Y	G	D	I	V	P	V
LctB	E	D	S	L	Y	L	S	G	M	T	V	L	S	V	G	Y	G	D	V	T	P	V
KcsA	P	R	A	L	W	W	S	V	E	T	A	T	T	V	G	Y	G	D	L	Y	P	V
TOK1	G	N	A	L	Y	F	C	T	V	S	L	L	T	V	G	L	G	D	I	L	P	K
slo	W	T	C	V	Y	F	L	I	V	T	M	S	T	V	G	Y	G	D	V	[Y]	C	V
Shaker	P	D	A	F	W	W	A	V	[V]	T	[M	T]	T	[V]	G	Y	G	D	M	T	P	V
IRK1	T	A	A	F	L	F	S	I	E	T	Q	T	T	I	G	Y	G	F	R	C	V	T
AKT1	V	T	S	M	Y	W	S	I	T	T	L	T	T	V	G	Y	G	D	L	H	P	V
KST1	I	T	S	L	Y	W	S	I	V	T	L	T	T	T	G	Y	G	D	L	H	A	E

The above table shows the P loop sequences for six well-characterised eukaryotic channel proteins (*Shaker*, *slo*, IRK1, AKT1, TOK1, and KST1 [7,11,18,54,55,59] and the sequences from the putative bacterial channels Kch [5], LctB [56] and KcsA [31]. In highlighting the conservation the KcsA protein is used as the reference point. Boxed residues (single line) indicate amino acids of the *Shaker* sequence that are implicated in TEA sensitivity at the inner face of the membrane. Boxed residues (double line) indicates the tyrosine residue strongly implicated in the sensitivity to TEA at the outer surface of the *Drosophila slo* channel.

significantly smaller than KcsA, with two predicted transmembrane spans separated by a P loop sequence with homology to that observed in the *Shaker* family of potassium channels (45% identity and 64% similarity; Table 2). The deduced protein shares 45% sequence identity (73% identity plus similarity) in the P loop region with KcsA (Table 2), but otherwise sequence similarity is limited. The LctB protein has a glutamate (which can substitute for aspartate in Mg^{2+}-gating of IRK1 [57]) within the predicted transmembrane span S2, but it is not clear whether this adopts the correct position to serve in gating. So far the protein has not been purified nor have any electrophysiological studies been carried out. Further experiments are required before LctB can definitely be added to the growing family of bacterial ion channels.

3.4. Kch — a putative homologue of the Shaker potassium channels

The *kch* gene product was identified as an *E. coli* sequence with potential organization similar to the *Shaker* potassium channel super-family [5]. At the present time relatively little is known about the gene product and it has not been analyzed by electrophysiology. Preliminary analysis of Kch null mutants has not revealed any obvious phenotype associated with loss of function (W. Epstein, personal communication). Therefore, the possibility that this is an active channel relies wholly on comparisons of the organisation of the protein. The organisation of Kch is very similar to that for the *Shaker* family, with six potential transmembrane spans (S1-S6), a 'P loop' (H5) between the S5 and S6, followed by an extensive carboxy-terminal hydrophilic domain separated from S6 by a linker peptide [7,8]. Kch shows strongest sequence similarity to eukaryotic potassium channels in the P loop (Table 2). However, the sequence similarity extends across the S6 span and the carboxy-terminal domain when compared with the outwardly-rectifying *slo* channel (*Drosophila* and mouse) [18,53] and inwardly rectifying plant channels (AKT1 and KST1, of *Arabidopsis* and tobacco, respectively) [54,55]. A major difference between Kch and *Shaker* lies in the putative S4 segment. Kch lacks the repeating pattern of basic amino acids in S4 that is associated with voltage-sensitivity. Recently a yeast potassium channel, TOK1 [58], also called YKC1 [58a], has been described that also lacks the S4 voltage sensor, but is otherwise related to the *Shaker* family. This channel may have a related function to that of Kch in *E. coli.* A possible function of Kch is in potassium efflux during compatible solute accumulation [5] or during other phases of potassium cycling. In *E. coli* potassium efflux elicited by hypoosmotic shock has been shown to be sensitive to 1 mM tetraethylammonium ion (TEA) [59], which would correspond to the concentrations required to inhibit the least sensitive of the eukaryotic *Shaker*-type channels [7,8]. The P loop region of the *Shaker* family is known to affect the sensitivity to TEA, which inhibits *Shaker* potassium channels by binding at the inner and outer surface of the channel pore [7] (see Table 2). Comparison of Kch with other channels suggests that Kch would be only moderately sensitive to TEA. The possibility remains that Kch could be the TEA-sensitive route for potassium efflux, but at the present time Kch remains something of an enigma.

A potassium channel activity has recently been described in *Rhodospirillum rubrum* that displays the trypsin sensitivity of inactivation typical for some *Shaker* channels [60]. *R. rubrum* chromatophores were fused with bilayer lipid membranes and the channel activities contributed by chromatophore proteins investigated. At an applied voltage of −100 mV a single channel activity was detected that appeared to be able to conduct both potassium and sodium ions and which displayed a conductance of approximately 230 pS in the presence of 100 mM KCl. The channel showed voltage-dependent inactivation and this could be overcome by trypsin treatment. Voltage-dependent inactivation of 'A' type *Shaker* channels is trypsin sensitive due to proteolytic cleavage that releases the amino-terminal inactivation ball [7,8]. Whether this activity in *R. rubrum* is analogous to Kch and/or *Shaker* will be resolved once the structural gene has been identified and the protein sequence is available.

Some further support for the existence of *Shaker*-type channels in bacteria arises from the discovery of bacterial homologues of the β subunits [14,19]. Genes encoding proteins of the appropriate oxidoreductase family [19a] have been identified by the genome sequencing project in *E. coli* and *Pseudomonas.* In addition two oxidoreductase genes have been found to be associated on the genome with the KefB and KefC potassium channel genes (see below). Whether any of these proteins has the function of β subunits or whether they are true dehydrogenases is not known at this time. However, the presence of β subunits would generate support for the presence of *Shaker* type channels, such as Kch, in bacteria.

3.5. The KefB and KefC glutathione-gated ion channels

3.5.1. Background

The *kefB* and *kefC* loci (formerly *trkB* and *trkC;* [61]) were identified during the analysis of potassium transport in *E. coli* as genes that when mutated provoked rapid potassium loss from the cells [61]. Subsequent analysis has shown that the original mutations were gain-of-function mutations, such that potassium efflux via the transport system was no longer tightly controlled [62,63]. Investigation of null mutants lacking both KefB and KefC has shown that these systems are responsible for glutathione-gated potassium efflux from *E. coli* and are important for cell survival during exposure to toxic metabolites.

From the analysis of glutathione-deficient mutants, Kepes first proposed a role for glutathione in the control of potassium retention [64]. Glutathione-deficient mutants exhibited less control over potassium retention than the wild-type parent strain. The consequence was that the mutants were unable to grow in low potassium (0.1 mM) media. This phenotype could be reversed with either glutathione or its analogue, ophthalmic acid (γ-glutamylaminobutyrylglycine). Since this peptide does not contain cysteine it was clear that redox events were not central to the regulation. Kepes proposed that glutathione, the major soluble thiol of the cell, is a tight binding ligand for one or more potassium efflux systems. Treatment of *E. coli* cells with N-ethylmaleimide (NEM) elicits rapid potassium efflux [64] and Kepes proposed that this phenomenon was explained by the depletion of glutathione by reaction with NEM. The isolation of null mutants affecting the *kefB* and *kefC* genes of *E. coli* established that both the leak observed in glutathione-deficient mutants (unpublished data) and NEM-elicited efflux takes place via these two efflux systems [63]. The efflux phenomenon is widespread in Gram-negative bacteria, which suggested its importance for cell survival [65]. In the discussion of these systems, presented below, the evidence that KefB and KefC are ligand-gated and voltage-gated channels will be reviewed.

3.5.2. The mode of action of NEM

NEM is an extremely active electrophile and will rapidly modify sulphydryl groups on proteins and glutathione, as well as attacking oxygen and nitrogen atoms in DNA. NEM is extremely toxic to bacterial cells and the compound is rapidly detoxified by cells. Detoxification is predominantly achieved through the formation of the glutathione adduct and the subsequent breakdown of this adduct regenerates reduced glutathione and a non-toxic derivative of NEM. Recovery of *E. coli* and *Pseudomonas putida* cells from exposure to low concentrations of NEM is associated with the almost complete destruction of the NEM and the elimination from the cell of its non-toxic product (S. Tötemeyer, D. McLaggan and I.R. Booth, unpublished data). However, the NEM-modification of proteins remains.

NEM elicits potassium efflux primarily through the formation of glutathione adducts that activate KefB and KefC. NEM has two other modes of action that may contribute to the observed potassium efflux: firstly, potassium uptake via the Trk system is inhibited [67] and secondly, there is some direct activation of the KefB and KefC proteins. The former is a relatively minor effect since mutants lacking both KefB and KefC exhibit only a slow leak of potassium after treatment with NEM. Direct activation of KefB and KefC is interesting and appears to take place from the periplasmic face of the membrane and leads to moderately fast potassium efflux (L. Ness and I.R. Booth, unpublished data). Direct channel activation after NEM addition to isolated membrane patches has also been observed (A. Kubalski, personal communication). However, the direct activation is inhibited by cytoplasmic glutathione and by the glutathione analogue, ophthalmic acid.

Other electrophiles act in a similar way. NEM reacts with glutathione to form N-ethylsuccinimido-S-glutathione (ESG) and methylglyoxal forms hemithiolacetal and lactoyl-S-glutathione. Less toxic electrophiles (i.e. those that do not cause a significant loss of cell viability), for example iodoacetate, form adducts that do not activate the channels strongly (L. Ness and I.R. Booth, unpublished data). The channels can also be activated by phenylmaleimide, p-chloromercuribenzoate, chlorodinitrobenzene and by diamide. The latter is an oxidant, which can convert reduced glutathione to the oxidised disulphide form. However, diamide also reacts with glutathione to form adducts and this may be the mechanism by which the channels are activated.

3.5.3. KefC as an ion channel

The analysis of the channel properties of KefC is at an early stage. Patch clamp analysis has shown that there are glutathione-gated potassium channels in *E. coli* membranes and that these are encoded by KefC (A. Kubalski, personal communication). A similar analysis of KefB has not yet been conducted, but the similarity of the primary sequence and the predicted secondary structure suggests that similar data will be obtained. Major channel activation only occurs in the presence of both glutathione and NEM. Gain-of-function mutations that cause a potassium leak through KefC (see below) have been found to alter the frequency of opening of the ion channels (A. Kubalski, B. Martinac, S. Miller and I.R. Booth, unpublished data).

Several gene sequences are now available for *kefC* (*E. coli*, *Klebsiella aerogenes* and *Haemophilus influenzae*) [68,69] and the *E. coli kefB* gene has been sequenced [70] as part of the *E. coli* genome sequencing project. All the gene products show a similar organization and a high degree of sequence similarity, (approximately 42% identity and greater than 70% similarity between the *E. coli* KefB and KefC proteins; the *Klebsiella aerogenes* KefC protein is 84% identical with the *E. coli* KefC protein). *Haemophilus influenzae* has a single KefC homologue [69], the sequence of which is equidistant from the *E. coli* KefB and KefC proteins. Cumulatively the four sequences show high similarity over the amino-terminal channel domain and substantially less conservation in the carboxy-terminal domain. KefB and KefC share similarities with eukaryotic potassium channels, but the proteins do not fall into any of the already identified eukaryotic channel families and do not possess the P loop characteristic of voltage-gated channels. They are large proteins, over 600 amino acids composed of a channel-forming amino-terminal hydrophobic domain and a substantial carboxy-terminal domain. The two domains are linked by a potential 'Q' linker, an extremely hydrophilic motif with little specific sequence

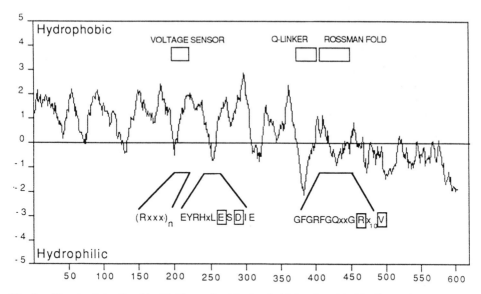

Fig. 7.   Structural details of the KefC protein. A hydrophobicity plot of the KefC protein is presented (Kyte-Doolittle, w = 19; [86]). The conserved region including the putative 'voltage sensor' and the location of mutations in the channel domain of the protein, the Q linker and the Rossman fold are indicated by horizontal bars. Elements of the conserved sequences are shown below the hydrophobicity plot. Boxed residues are those that when mutated cause enhanced potassium loss from the cells via the KefC protein.

conservation between the four KefB/C proteins. Similar sequences are found between domains in regulatory proteins and in the membrane components of the bacterial phosphotransferase system and may form flexible connections that enable the two domains to interact [71,72]. For KefB and KefC the hydrophobic region could encode several (8–10) transmembrane alpha helices rather than the six found in most eukaryotic potassium channels (Fig. 7). A possible voltage sensor, $(Rxxx)_5$ followed by regularly spaced leucine residues, similar in organisation to the voltage gate of eukaryotic voltage-gated potassium channels [7], is highly conserved in the four protein sequences available. KefB from *E. coli* and KefC from *Haemophilus influenzae* have an incomplete conservation of this motif with one of the arginine residues replaced with neutral residues. Preliminary characterization of the channel in patch clamp experiments has indicated a strong voltage-dependence (A. Kubalski, personal communication). Thus, KefB and KefC show some structural properties common to eukaryotic channels.

Three regions have been found that affect control over the activity of KefB and KefC. These have been identified as residues that when mutated alter the regulation of the protein such that the cells leak potassium [73]. Patch clamp analysis of one such mutant has revealed that the mutated KefC channel shows an increased frequency of opening (A. Kubalski, B. Martinac, S. Miller and I.R. Booth, unpublished data). In KefB, mutation of a conserved leucine (L75) enhances the rate of spontaneous leakage. For KefC two sites of mutations have been identified: firstly, two highly conserved acidic residues (E262 and D264) predicted to form àn acidic patch (ExDxE) between two transmembrane spans of the amino terminal region at the cytoplasmic face of the membrane. This sequence lies within a region that is more than 90% conserved between KefC proteins. One of the mutations observed to cause deregulation in KefC, D264A, is found in the native KefB protein as the preferred amino acid. When KefB is over-expressed it causes a spontaneous leak of potassium, whereas cells over-expressing KefC remain potassium-tight. Unfortunately, this KefB-associated potassium leak cannot be wholly ascribed to the D264A change since mutagenesis to restore the aspartate seen in KefC does not eliminate the leak (L. Ness and I.R. Booth, unpublished data). The acidic patch may form the site of interaction with other domains or proteins that are involved in controlling KefC activity (Fig. 7).

The second region in which mutations affecting control are found lies within the carboxy-terminal domain of KefB and KefC and affects a sequence homologous to the NAD-binding, Rossman fold sequences of dehydrogenases [68]. This is the most highly conserved region of the carboxy-terminal domain, which otherwise shows less conservation between the four genes sequenced to date than is observed for the amino-terminus. The importance of the carboxy-terminal domain is underlined by the finding that mutations that cause a rapid potassium leak via KefC map to conserved residues in the Rossman fold of the carboxy-terminus (Fig. 7). This indicates that this carboxy-terminal region may play a significant role in regulating channel activity and would thus be functionally analogous to the carboxy-terminal region of eukaryotic channels.

Bakker has pointed out that the TrkA, and therefore, probably KefB and KefC, sequence can be mapped onto the structural motif of the β-α-β fold of the Rossman fold of dehydrogenases [74] (see Chapter 21). It may simply be that the Rossman fold motif has proved to be a particularly stable protein conformation that can be utilized to present regulatory sequences in such a way to inactivate KefB and KefC channels. TrkA and the dehydrogenases are known to bind NAD. A possible role for NAD in regulating KefB and KefC activity has been investigated using nicotinic acid auxotrophs (*nadA*). A *nadA* auxotroph degrades its pool of NAD when starved of nicotinic acid. After 24 h starvation the NAD pool is in the μM range rather than mM in normal growing cells. Starvation of a *nadA* mutant did not cause any change in the rate of potassium leakage of *E. coli* cells and activation of KefB and KefC by the addition of NEM was unaffected (D. Mclaggan and I.R.Booth, unpublished data). These data suggest that either KefB and KefC have a high affinity for NAD, or the binding of this nucleotide is not important for regulation of activity. Finally, it has been noted that the carboxy-terminal region of the Kch channel shares sequence similarity with TrkA and to a lesser extent with KefC and KefB [75]. It may be, therefore, that these regions share a common protein-folding pattern and have regulatory features in common.

A further dimension has recently been added to the regulation of KefB and KefC. Analysis of the *kefC* gene revealed that another gene located 5′ to *kefC* overlaps physically with it. This gene, *yabF*, overlaps *kefC* by eight bases and has a homologue (*yhaH*) that precedes the *kefB* gene (C.M. Wood, S. Miller, L. Ness

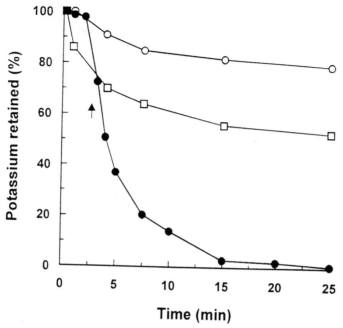

Fig. 8. Potassium efflux elicited with NEM in *E. coli* cells. *E. coli* strain Frag56 (*gshA::kan*) was grown in the presence (circles) or absence (squares) of 1 mM glutathione. Cells were harvested and resuspended in 60 mM sodium phosphate buffer pH 7 (in the absence of glucose) [61] and incubated in the presence (filled symbols) or absence (open symbols) of NEM (0.5 mM) (data unpublished; L. Ness and I.R. Booth). Potassium loss was analyzed as described previously [67].

and I.R. Booth, in preparation). Deletion of the *yabF* sequences from the cloned *yabF-kefC* region reduces the expression of KefC protein by 20- to 40-fold, suggesting that the transcription, and or translation, of the two genes is linked. A construct that carries the 150 bp 5′ to the translational start site for the *kefC* gene is transcriptionally silent suggesting that the promoter for *kefC* transcription either lies well within the *yabF* gene or *yabF* and *kefC* form an operon that is transcribed from a promoter located 5′ to *yabF*. The mRNA for the intergenic region is characterized by a strong stem-loop structure, which could give rise to translational coupling. Thus, the expression of YabF and KefC in the enteric bacteria may be strongly coordinated.

The deduced protein sequences of YabF and YhaH show high similarity to human quinone oxidoreductases. Preliminary data suggest that YabF is required for KefC activity. We have recently constructed deletions that remove *yabF-kefC* from the chromosome and analyzed the expression of plasmid constructs that either express the whole *yabF-kefC* region or express only *kefC*. In cells that lack YabF the level of KefC activity is reduced both in terms of the initial rate and the final extent of potassium efflux (Fig. 8). The residual activity may reflect the activity of YhaH, since this protein shows 58% similarity to YabF and could substitute

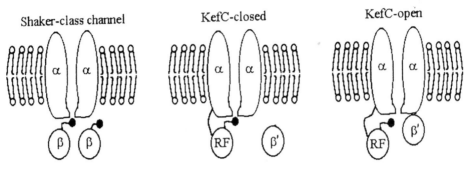

Fig. 9.   Model for KefC activation; The structure of the *Shaker* class of channels is compared with the known elements of KefC. *Shaker* channels have the α subunit that forms the channel and β subunits that regulate inactivation [7,8,14]. In the case of KefC we know that the channel is formed by the product of the *kefC* gene and is probably a homo-oligomer [76], and this subunit is considered equivalent to the α subunit. However, the carboxy-terminal domain (RF; one shown only for simplicity) plays a role in closing the channel. The YabF protein is essential for activation of the channel and given its sequence homology to oxidoreductases we suggest that this protein (β' subunit) displaces the RF module to open the channel. Whether the glutathione and glutathione adducts interact directly with channel (α subunit) or with YabF (β' subunit) is not known at the present time.

functionally. This implies either that YabF protein is required for translation of *kefC* mRNA and/or for KefC channel activity. A similar protein to YabF has been located on the *H. influenzae* genome (HU32829_13), but is not genetically linked to the *kefC* gene. It is not yet known whether this is a component of the KefC system. Although these oxidoreductases and the β subunits of the *Shaker* family belong to different dehydrogenase/oxidoreductase families, it is tempting to speculate that the glutathione-gated channels may have evolved a similar two component structure to facilitate regulation of the channel (Fig. 9).

A final feature that links KefC with eukaryotic channels is the formation of multimers [76]. The evidence in support of this is purely genetic. The mutations described above that cause potassium leakage via KefC can be suppressed by co-expression in the same cell of the wild-type gene from a multi-copy plasmid. Control experiments are consistent with suppression arising from the interaction between wild-type and mutant monomers in an oligomeric complex. Low level expression of KefC from this plasmid led to intermediate suppression of the mutant, suggesting that equal numbers of wild-type and mutant subunits still formed some leaky channels. These data are consistent with KefC forming oligomers, but unfortunately supporting biochemical evidence is not yet available.

From the above description it is evident that the general architecture of KefC resembles channels of higher organisms, but it also shows unique features that relate to the regulation of this novel class of channels. Our current model for the channel is:

(a) the channel protein forms homo-oligomers;
(b) the channel is maintained in a closed state by both the membrane potential (negative inside) and by reduced glutathione;

(c) the channel is opened on binding of glutathione adducts;

(d) expression and/or opening of the channel requires a 'β' subunit;

(e) the closure of the channel requires the interaction of the carboxy-terminal domain with a sequence in the amino-terminal domain;

(f) the channel is outwardly rectified.

Evidence to support this model has been outlined above with the exception of the expected outward rectifying nature of the channel. This property is predicted on the basis of the properties of the *kefB* and *kefC* gain-of-function mutations, which allow potassium to leak from the cell against the prevailing membrane potential. A completely symmetrical channel (i.e. showing no rectification) would tend to equilibrate the membrane potential and the potassium gradient in such mutants. The model is now being tested through genetic and electrophysiological analysis.

3.5.4. Physiological role for KefB and KefC channels

Glutathione is utilized by cells as part of the mechanism for detoxifying toxic metabolites, such as electrophiles through the formation of glutathione adducts that are subsequently metabolized by the cells. Methylglyoxal (MG), a natural metabolite produced whenever there is an imbalance between flux in the upper (from hexoses to triose phosphates) and lower (from triose phosphates to pyruvate) parts of the glycolytic sequence, serves as a potent illustration [77]. Methylglyoxal is a potent inhibitor of bacterial growth [78]. Externally-supplied methylglyoxal activates KefB and KefC (Fig. 10), with the former being the major site of activation (KefC activation is most readily seen when *kefC* is over-expressed from a multi-copy plasmid) [79]. The possession and activity of KefB and KefC protects cells during exposure to methylglyoxal, whether the electrophile is added externally or generated in the cytoplasm. Mutants lacking KefB and KefC exhibit enhanced sensitivity when exposed to methylglyoxal (Fig. 10). This does not appear to be due to altered metabolism of MG, since the rate of detoxification is unaffected by the mutations [80]. Similarly, if the channel activity is reduced by raising the external potassium concentration cell viability in the presence of MG is reduced. Activation of KefB and KefC with methylglyoxal causes the cytoplasmic pH to fall (Fig. 11) and the final value of pHi is a major determinant of cell survival [80]. Thus, the action of the channels can be mimicked by the addition of weak acids to cells. Similarly, the protective effects of the channel can be overcome by manipulations that poise the initial value of pHi at such a value that the fall engendered by channel activation does not bring the cytoplasmic pH into the range where protection is effective. Similar data show that the channels protect the cells against the toxic electrophile, NEM, by essentially the same mechanism (G.P. Ferguson, Y. Nikolaev and I.R. Booth, unpublished data). Thus the principal function of the channels appears to be potassium efflux to effect controlled acidification of the cytoplasm as ions (H^+, Na^+) move in to balance the potassium efflux.

A model to explain the physiological role of KefB and KefC is shown in Fig. 11. Methylglyoxal is detoxified by the actions of two pathways: the glutathione-dependent glyoxalase I and II and the glutathione-independent glyoxalase III [78,81].

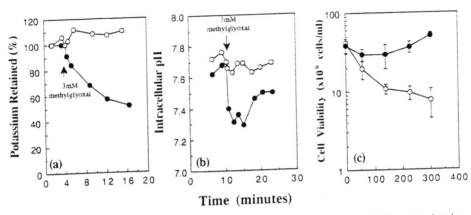

Fig. 10. Activation of KefB and KefC channels in *E. coli* protects against methylglyoxal poisoning. Log phase cells of strains MJF274 (KefB⁺KefC⁺) (closed symbols) and MJF276 (KefB⁻KefC⁻) (open symbols) were prepared as described [79,80] and resuspended in K₀.₂ medium supplemented with 0.2% glucose. The K⁺ content (a), intracellular pH (b) and viability (c) of the cells were monitored in the presence of methylglyoxal. (Reproduced from Refs. [79,80] with permission).

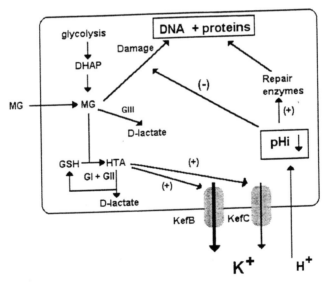

Fig. 11. Overview of the protection afforded by KefB and KefC activation. Methylglyoxal, either generated by metabolism or added exogenously, can react with proteins and DNA causing damage. Detoxification of methylglyoxal proceeds via two main pathways: glutathione-dependent glyoxalase I and II (GI and GII) or via glutathione-independent glyoxalase III (GIII), in each case generating D-lactate as the end-product. Intermediates in the glutathione-dependent pathway activate KefB and KefC leading to potassium efflux that is partially compensated by proton influx. The reduction of cytoplasmic pH causes either a reduction in the rate of damage and/or activates repair processes. When detoxification is complete, free glutathione is regenerated leading to closure of the potassium channels, potassium uptake and restoration of the cytoplasmic pH to the normal range.

The glutathione-dependent pathway also generates a signal for activation of KefB and KefC, namely the formation of glutathione adducts. Methylglyoxal also attacks proteins, DNA and lipids and consequently causes damage leading to the death of cells. The activation of the KefB and KefC channels elicits potassium efflux and the lost potassium is replaced by protons and Na$^+$ ions. The lowering of cytoplasmic pH protects the cell either by enhancing the rate of repair of damage, for example to DNA, or by slowing the rate of chemical reactions involving methylglyoxal, such that less damage is done. When methylglyoxal has been detoxified the adduct disappears, free glutathione is generated leading to closure of the channels, recovery of the potassium pool and restoration of an alkaline cytoplasmic pH. Thus, the KefB and KefC channels offer the cell a way to lower the cytoplasmic pH to effect protection against cytotoxic chemicals.

4. Concluding remarks

The analysis of bacterial channels is in its infancy. It can be seen from the above sections that ion channels may play important roles in protection against excessive turgor pressure across the membrane (MscL, MscS, and possibly KefA) and during detoxification of electrophilic compounds (KefB, KefC). Other channels, e.g. KcsA, may play important roles in potassium acquisition or in balancing the potassium gradient with the membrane potential. The reality of Kch as a channel can only be a subject for speculation, but if it were to function as an outwardly-rectified channel it might participate in membrane re-polarization during periods of starvation or, given its resemblance to the Ca^{2+}-regulated, *Drosophila slo* channel, there may be a role for this channel in signaling the Ca^{2+} status of the cell.

Other well-studied phenomena in bacteria may have their source in transport systems with channel-like properties. Most prominent among these is the Trk potassium transport system in *E. coli* (see Chapter 21). This system consists of the products of the *trkA, trkG* or *trkH* loci and the *trkE* locus. The system shows complex energetics and regulation of activity [74,75,82–84], including feedback inhibition by intracellular potassium and activation by reduced cell turgor. TrkG and TrkH are membrane proteins with the potential to form twelve membrane spans and are essentially duplicate gene functions [82]. Both proteins are predicted to have sequences that exhibit weak similarity to the P loop of the voltage-gated potassium channels of mammalian cells [12]. The TrkA protein is a peripheral membrane protein [84], which associates quite strongly with the TrkG and TrkH proteins, that may act as a major regulator of the transporter. Its structural homology to the lactate dehydrogenase family of proteins makes it analogous to the β subunits of mammalian ion channels, even though it is derived from a different dehydrogenase/oxidoreductase family. The Trk transporter has long been thought of as a possible K$^+$-H$^+$ symport [83], however, re-evaluation of the possibility that this is in fact an inwardly rectified potassium channel is warranted. The recent discovery that TrkE is actually a component of a separate ATP-dependent transporter [75], suggests that potassium flux through such a channel is regulated by the

ATP-dependent complex of TrkE. The complexity of the energy coupling to the Trk complex has been evident for many years and may find its resolution through analysis of the channel activity of the Trk complex. The situation is reminiscent of the sulphonylurea receptor of mammalian cells, which is a member of the ATP-binding cassette super-family with a two nucleotide binding folds, and which forms a channel by association with an inwardly-rectifying potassium channel, Kir6.2 [55a]. If the TrkE complex and the TrkA protein provide regulatory elements for a channel protein some of the complexity of the Trk system may be clarified.

For each of the proteins discussed above their open state has different implications for the cell and consequently tight regulation of opening is very important. The porins are believed to be open most of the time and are more correctly designated pores. This does not detract from their study as models for the structural prerequisites for the design of channels [85]. The mechanosensitive channels must open for very short periods of time, sufficient to release solutes and thereby reduce cell turgor. Extended periods in the open state will lead to such substantial energy and nutrient loss that the viability of the cell might be more greatly imperiled than from the excess turgor the channels are designed to relieve. In contrast KefB and KefC must remain open for extended periods of time as their role is the release of bulk quantities of potassium ions facilitating acidification of the cytoplasm by the counterflow of protons. As was indicated at the outset a channel with conductivity 10^7 ions/s will only cause 10% loss of the potassium pool if constantly open for 1s (provided that there is a cation influx of sufficient velocity to prevent hyperpolarisation of the membrane). The actual rate of depletion may be set by the rate of entry of other ions, consequently active KefB and KefC take many minutes to complete depletion of the potassium pool. Equally for any channel it must be borne in mind that the channel will spend extensive periods closed and may have desensitized states in which the presence of the activator no longer effects channel opening. The extended, cumulative open period for KefB and KefC is achieved by coupling the presence of the activator, and the absence of the inhibitor, to the threat to cell viability. Thus, this is a process in which the detoxification process and the channel activation are close-coupled.

As in animal cells ion channels play diverse roles in bacterial cells. In the past there has been a marked reluctance to accept their role in cell physiology because of their potential impact on energy transduction. The recognition that cells might sacrifice energetic efficiency to effect survival of stress through the operation of ion channels poses no threat to microbial bioenergetics. The latter is more appropriate to considerations of the efficiency of individual processes rather than global summations of myriad pathways, not all of which can be monitored to provide adequate basis for data comparison. In contrast, the analysis of the role of ion channels in cellular physiology adds to our understanding of the complex compromises that cells have evolved as the best overall strategy for survival. Their study not only advances our understanding of cell physiology, but it also opens up new avenues that can be explored for the design of antibiotic screens that will be needed with some urgency in the next century.

Acknowledgements

The authors wish to thank Mike Ashford, Paul Blount, Frank Harold, Ching Kung, Wolf Epstein and Sergei Sukharev for their critical reading of this manuscript and for sharing their insights in membrane bioenergetics and channel function. The authors would like to thank the Wellcome Trust and the BBSRC for financial support for their research programme. In addition the group is indebted to the following who have provide helpful comments, encouragement, vital strains and technological expertise: Evert Bakker, Wolf Epstein, Andrei Kubalski, Barry Holland, Hilde Schrempf, Julius Adler and Changhai Cui. In addition the group is indebted to the efforts of previous members of the group who have helped the story of KefA, KefB and KefC to emerge.

References

1. Sakman, B. and Neher, E. (1995) in: Single Channel Recording, 2nd Edn, ed B. Sakman and E. Neher. Plenum, New York and London, pp. vii–viii.
2. Booth, I.R. (1985) Microbiol. Rev. **49**, 359–378
3. Sukharev, S.I., Blount, P., Martinac, B., Blattner, F.R. and Kung, C. (1994) Nature **368**, 265–268.
4. Booth, I.R., Douglas, R.M., Ferguson, G.P., Lamb, A.J., Munro, A.W. and Ritchie, G.Y. (1993) in: Alkali Cation Transport Systems in Prokaryotes, ed. E.P. Bakker, CRC Press, Boca Raton, FL, pp. 291–308.
5. Milkman, R. (1994) Proc. Natl. Acad. Sci. USA **91**, 3510–3514.
6. Cui, C., Smith, D.O. and Adler, J. (1995) J. Membr. Biol. **144**, 31–42.
7. Pongs, O. (1992) Physiol. Rev. **4**, S69–S88.
8. Pongs, O. (1993) J. Memb. Biol. **136**, 1–8
9. Montal, M.(1995) Ann. Rev. Biophys. Biomol. Struct. **24**, 31–57.
10. Unwin, N. (1995) Nature (Lond.) **373**, 37–43.
10a. Villaroel, A., Herlitze, S., Koenen, M. and Sakmann, B. (1991) Nature (Lond.) **243**, 69–74.
11. Doupnik, C.A., Davidson, N. and Lester, H.A. (1995) Curr. Opin. Neurobiol. **5**, 268–277.
12. Jan, L.Y. and Jan, Y.N. (1994) Nature (Lond.) **371**, 119–122.
13. Takumi, T., Ohkobo, H. and Nakanishi, S. (1988) Science **242**, 1042–1045.
13a. Tytgat, J., Vereecke, J. and Carmeliet, E. (1994) J. Physiol. **481**, 7–13.
14. Pongs, O. (1995) Sem. Neurosci. **7**, 137–146.
15. Hoshi, T., Zagotta, W.N. and Aldrich, R.W. (1990) Science **250**, 533–538.
16. Zagotta, W.N., Hoshi, T. and Aldrich, R.W. (1990) Science **250**, 568–571.
17. Isacoff, E.Y., Jan, Y.N. and Jan, L.Y. (1991) Nature (Lond.) **353**, 86–90.
18. Atkinson, N.S., Robertson, G.A. and Ganetzky, B. (1991) Science **253**, 551–555.
19. Chounard, S.W., Wilson, G.F., Schlimgen, K. and Ganetzky, B. (1995) Proc. Natl. Acad. Sci. USA **92**, 6763–6767.
19a. Persson, B., Zigler, J.S. and Jornvall, H. (1994) Eur. J. Biochem. **226**, 15–22.
20. Martinac, B., Buechner, M., Delcour, A.H., Adler, J. and Kung, C. (1987) Proc. Natl. Acad. Sci. USA **84**, 2297–2301.
21. Delcour, A.H., Martinac, B., Adler, J. and Kung, C. (1989) Biophys. J. **56**, 631–636.
22. Zoratti, M. and Petronelli, V. (1988) FEBS Lett. **240**, 105–109.
23. Zoratti, M., Petronelli, V. and Szabo, I. (1990) Biochem. Biophys. Res. Comm. **168**, 443–450.
24. Berrier, C., Coulombe, A., Szabo, I., Zoratti, M. and Ghazi, A. (1992) Eur. J. Biochem. **168**, 443–450.
25. Hamill, O.P., Marty, A., Neher, E., Sakmann, B. and Sigworth, F.J. (1981) Plügers Arch. Eur. J. Physiol. **391**, 85–100.
26. Penner, R. (1995) in: Single Channel Recording, 2nd Edn, ed B. Sakman and E. Neher. Plenum, New York and London, pp. 3–30.
27. Hilgemann, D.W. (1995) Single Channel Recording, 2nd Edn, ed B. Sakman and E. Neher. Plenum,

New York and London, pp. 307–327.

28. Cowan, S.W., Schirmer, T., Rummel, G., Steiert, M., Ghosh, R., Pauptit, R.A., Jansonius, J.N. and Rosenbusch, J.P. (1992) Nature (Lond.) **358**, 727–733.

29. Schirmer, T., Keller, T.A., Wang, Y-F. and Rosenbusch, J.P. (1995) Science **267**, 512–514.

30. Weiss, M.S. and Schulz, G.E. (1992) J. Mol. Biol. **227**, 493–509.

31. Schrempf, H., Schmidt, O., Kümmerlen, R., Hinnah, S., Müller, Betzler, M., Steinkamp, T. and Wagner, R. (1995) EMBO J. **14**, 5170–5178.

32. Nikaido, H. (1993) Bioenerg. Biomembr. **25**, 581–589.

33. Benz, R., Schmid, A. and Hancock, R.E.W. (1985) J. Bacteriol. **162**, 722–727.

34. Dargent, B., Hofmann, W., Pattus, F. and Rosenbusch, J.P. (1986) EMBO J. **5**, 773–778.

35. Bühler, L.K., Kusuomoto, S., Zhang, H. and Rosenbusch, J.P. (1991) J. Biol. Chem. **266**, 24446–24450.

36. Schindler, H. and Rosenbusch, J.P. (1978) Proc. Natl. Acad. Sci. USA **75**, 3751–3755.

37. Buechner, M., Delcour, A.H., Martinac, B., Adler, J. and Kung, C. (1990) Biochem. Biophys. Acta **1024**, 111–121.

38. Delcour, A., Adler, J., Kung, C., Martinac, B. (1992) FEBS Lett. **304**, 216–220.

39. Kennedy, E.P. (1982) Proc. Natl. Acad. Sci. USA **79**, 1092–1095.

40. West, I.C. and Page, M.G.P. (1984) J. Theor. Biol. **110**, 11–19.

41. Goodell, E.W. and Higgins, C.F. (1987) J. Bacteriol. **169**, 3861–3865.

42. Paul, C. and Rosenbusch, J.P. (1985) EMBO J. **4**, 1593–1597.

43. Vogel, H. and Jähnig, F. (1986) J. Mol. Biol. **190**, 191–199.

44. Sukharev, S.I., Martinac, B., Arshavsky, V.Y. and Kung, C. (1993) Biophys. J. **65**, 177–183.

44a. Martinac, B., Adler, J. and Kung, C. (1990) Nature (Lond.) **348**, 261–263.

45 Hase, C.C., Ledian, A.C. and Martinac, B. (1995) J. Biol. Chem. **270**, 18329–18334.

46. von Heijne, G. (1992) J. Mol. Biol. **225**, 487–494.

47. Syrvold, O.B. and Stro, A.T. (1991) J. Bacteriol. **173**, 1187–1192.

48. Epstein, W. and Scultz, S.G. (1965) J. Gen. Physiol. **49**, 221–234.

49. Britten, R.J. and McClure, F.T. (1962) Bacteriol. Rev. **26**, 292–335.

50. Tsapis, A. and Kepes, A. (1977) Biochem. Biophys. Acta **469**, 1–12.

51. Koo, S.-P., Higgins, C.F. and Booth, I.R. (1991) J. Gen. Microbiol. **137**, 2617–2625.

52. Schleyer, M., Schmid, R. and Bakker, E.P. (1993) Arch. Microbiol. **160**, 424–431.

53. Butler, A., Tsunoda, S., McCobb, D.P., Wei, A. and Salkoff, L. (1993) Science **261**, 221–224.

54. Sentenoc, H., Bonneaud, N., Minet, M., Lacroute, F., Salmon, J-M., Gaymard, F. and Grignon, C. (1992) Science **256**, 663–665.

55. Muller-Rober, B., Ellenberg, J., Provart, N., Wilmitzer, L., Busch, H., Becker, D., Dietrich, P., Hoth, S. and Hedrich, R. (1995) EMBO J. **14**, 2409–2416.

55a. Inagaki, N., Gonoi, T., Clement, J.P., Namba, N., Inazawa, J., Gonzalez, G., Aguilar-Bryan, L., Seino, S. and Bryan, J. (1995) Science **270**, 1166–1170.

56. Barstow, D., Murphy, J., Sharman, A., Atkinson, T. and Minton, N. (1987) Nucleic Acids Res. **15**, 1331.

57. Stanfield, P.R., Davies, N.W.,Shelton, P.A., Sutcliffe, M.J., Khan, I.A., Brammar, W.J. and Conley, E.C. (1994) J. Physiol. **478**, 1–6.

58. Ketchum, K.A., Jolner, W.J., Sellers, A.J., Kaczmarek, L.K. and Goldsetin, S.A.N. (1995) Nature **376**, 690–695.

58a. Zhou, X-L., Vaillant, B., Loukin, S.H., Kung, C., Saimi, Y. (1995) FEBS Lett. **373**, 170–176.

59. Meury, J., Robin, A. and Monneir-Champeix, P. (1985). Eur. J. Biochem. **151**, 613–619.

60. Antonencko, Y.N., Rokitskaya, T.I., Kotova, E.A. and Tasiova, A.S. (1994) FEBS Lett. **337**, 77–80.

61. Epstein, W. and Kim, B.S. (1971) J. Bacteriol. **108**, 639–644.

62. Booth, I.R., Epstein, W., Giffard, P.M. and Rowland, G.C. (1985) Biochemie, **67**, 83–90.

63. Bakker, E.P., Booth, I.R., Dinnbier, U., Epstein, W. and Gajewska, A. (1987) J. Bacteriol. **169**, 3743–3749.

64. Meury, J. and Kepes, A. (1982) EMBO J. **1**, 339–343.

65. Douglas, R.M., Roberts, J.A., Munro, A.W., Ritchie, G.Y., Lamb, A.J. and Booth, I.R. (1991) J. Gen Microbiol. **137**, 1999–2005

66. Meury, J. Lebail, S. and Kepes, A. (1980) Eur. J. Biochem. **113**, 33–38.
67. Elmore, M.J., Lamb, A.J., Ritchie, G.Y., Douglas, R.M., Munro, A., Gajewska, A. and Booth, I.R. (1990) Molec. Microbiol. **4**, 405–412
68. Munro, A., Ritchie, G.Y., Lamb, A.J., Douglas, R.M. and Booth, I.R. (1991) Molec. Microbiol. **5**, 607–616
69. Fleischman, R.B., Adams, M.D., White, O., Clayton, R.A., Kirkness, E.F., Kerlavage, A.R., Bult, C.J., Tomb, J-F., Dougherty, B.A., Merrick, J.M., Mckenney, K., Sutton, G., Fitzhugh, W., Fields, C., Gocayne, J.D., Scott, J., Shirley, R., Liu, L-I., Glodek, A., Kelley,J.M., Weidman, J.F., Phillips, C.A., Spriggs, T., Hedblom, E., Cotton, M.D., Utterback, T.R., Hanna, M.C., Nguyen, D.T., Saudek, D.M., Brandon, R.C., Fine, J., Fritchman, J.L., Fuhrman, J.L., Geoghagen, N.S.M., Gnehm, C.L., MacDonald, L.A., Small, K.V., Fraser, C.M., Smith, H.O. and Venter, J.C. Science **269**, 496–512.
70. Genome entry: GB ECOW67_274
71. Wooton, J. and Drummond, M.H. (1989) Protein Eng. **2**, 535–543.
72. Schunk, T. Rhiel, E., de Meyer, R., Buhr, A., Hummel, U., Wehrli, C. Flukinger, K. and Erni, B. (1992) in: Molecular Mechanisms of Membranes and Transport, eds E. Quaglierello and F. Palmieri. Elsevier, Amsterdam, New York, London, pp. 87–96.
73. Rhoads, D.B., Waters, F.G. and Epstein, W. (1976) J. Gen. Physiol. **67**, 325–341.
74. Schlosser, A., Hamann, A. Bossemeyer, D., Schneider, E. and Bakker, E.P. (1993) Molec. Microbiol. **9**, 533–544.
75. Parra-Lopez, C., Lin, R., Apedon, A. and Groisman, E.A. (1994) EMBO J. **13**, 4053–4062.
76. Douglas, R.M., Ritchie, G.Y., Munro, A.W., McLaggan, D. and Booth, I.R. (1994) Molec. Membr. Biol. **11**, 55–63.
77. Cooper, R.A. (1984) Ann. Rev. Microbiol. **38**, 49–68.
78. Fraval, H.N.A. and McBrien, D.C.H. (1980) J. Gen. Microbiol. **117**, 127–134.
79. Ferguson, G.P., Munro, A.W., Douglas, R.M., McLaggan, D. and Booth, I.R. (1993) Molec. Microbiol. **9**, 1297–1304.
80. Ferguson, G.P., McLaggan, D. and Booth, I.R. (1995) Molec. Microbiol. **17**, 1025–1033.
81. Misra, K., Banerjee, A.B., Ray, S. and Ray, M. (1995) Biochem. J., **305**, 999–1003.
82. Dosch, D.C., Helmer, G.L., Sutton, S.H., Salvacion, F.F. and Epstein, W. (1991) J. Bacteriol. **173**, 687–696.
83. Bakker, E.P. (1993) in: Alkali Cation Transport Systems in Prokaryotes, ed. E.P. Bakker, CRC Press, Boca Raton, FL, pp. 253–276.
84. Bossemeyer, D., Borchard, A., Dosch, D.C., Helmer, G.C., Epstein, W. Booth, I.R. and Bakker, E.P. (1989) J. Biol. Chem. **264**, 16403–16410.
85. Mackinnon, R. (1995) Neuron **14**, 889–892.

CHAPTER 31

Transport of DNA Through Bacterial Membranes

K.J. HELLINGWERF

Department of Microbiology,
E.C. Slater Institute, University of Amsterdam,
Amsterdam, The Netherlands

R. PALMEN

CHU Rangueil, Bacteriologie,
University Paul Sabatier,
Toulouse, France

© *1996 Elsevier Science B.V.*
All rights reserved

Handbook of Biological Physics
Volume 2, edited by W.N. Konings, H.R. Kaback and J.S. Lolkema

Contents

1. General introduction

1.1. Uptake into Gram-positive and Gram-negative bacteria

Living cells face the problem that they have to exchange many solutes with their environment. The reactions in the cytoplasm, that transform substrates into new cellular biomass, necessitate the transport of a large array of compounds through the cytoplasmic membrane. In addition, several waste products of cellular metabolism, and solutes that passively leak into the cells, have to be actively extruded. For the transport of low molecular weight solutes, like for instance the building blocks of the major types of biopolymers (i.e. amino acids, sugars, nucleosides, etc.) a large array of transport systems has evolved. These transport systems often transduce free energy into a solute gradient, allowing cells to accumulate solutes in their cytoplasm to much higher concentrations than what is available extracellularly.

For both Gram positive- and Gram-negative bacteria the cytoplasmic membrane is the most relevant selective barrier for solute entry into the cytoplasm. Gram-negative bacteria, in addition, contain an outer membrane that surrounds the cytoplasmic membrane (and a thin peptodoglycan layer). The outer membrane generally does not function as a selective barrier for the passage of low molecular weight solutes. It contains several pore proteins that allow the passage of solutes with a molecular weight smaller than approximately 800. That selectivity in solute uptake is mediated by the transport systems from the cytoplasmic membrane in both types of organism, is also evidenced by the many similarities between the latter across the Gram border.

Macromolecules, however, cannot freely pass through the pores of the outer membrane, nor through the cytoplasmic membrane. As a consequence, transport systems have evolved also for their translocation through the different barriers of the cell envelope. Of the three major types of biopolymers (i.e. proteins, nucleic acids and polysaccharides), only for the export of proteins has this translocation process been resolved into molecular detail (see e.g. [1]): Either proteins are excreted in a *single-step* excretion process or a *two-step* translocation mechanism is operational. For the former possibility two mechanisms have been resolved, the so-called 'haemolysin route' [2] and the 'flagellar pathway' [3]. The two-step secretion process is used by several Gram-negative bacteria for degradative enzymes, like proteases and lipases. It involves the general system for protein secretion ('Sec') to translocate proteins across the cytoplasmic membrane, a periplasmic intermediate of the exported protein (often subject to processing) and a dedicated (second) excretion system that mediates the translocation of the protein across the outer membrane [4].

For DNA uptake (and for that matter also for polysaccharide excretion) much less is known about the molecular details of the translocation process. It has been proposed that a similar distinction can be made between organisms that translocate (i.e. take up) DNA in a *one-* and a *two-*step process, respectively. DNA uptake in natural transformation of *Bacillus subtilis* and *Haemophilus influenzae* would then be the two archetypes of these reactions. However, for DNA uptake this dichotomy does not follow the Gram border: Several Gram negative bacteria (like *Acinetobacter calcoaceticus*) show DNA uptake characteristics that are much more similar to those of *B. subtilis* than to *H. influenzae* [5]. The two-step DNA-uptake pathway has clearly been demonstrated only in *H. influenzae* (see further below).

Whether or not the peptidoglycan layer forms a significant barrier for the translocation of DNA (or other biopolymers), in either Gram positive- or Gram-negative organisms, is not known.

1.2. Energy sources available for DNA transport

With reference to the mechanism of energy coupling in the transport reaction, solute transport systems can be divided into primary- and secondary-transport systems [6]. *Secondary* transport systems couple the movement of one translocated species (usually the primary ion used in energy conversion in that particular organism, i.e. a proton or a sodium ion), to the translocation of a second solute. Depending on the direction of translocation this can take the form of symport or antiport. The maximal level of solute accumulation via a secondary transport system hardly ever exceeds three orders of magnitude. These systems usually consist of one integral hydrophobic membrane protein. Their archetype is the β-galactoside permease from *Escherichia coli* [7], which has been characterized in great detail. To this class also belongs the very large family of secondary proton-symport systems that contains members from both the prokaryotic and the eukaryotic kingdom [8], as well as several antiporters.

In the *primary* transport systems, the free-energy that is released in a (photo)chemical reaction is coupled to the accumulation of a substrate. Primary examples of this class are the ion-translocating ATPase-complexes, the proton-translocating complexes of a respiratory chain, light-driven ion pumps (like bacteriorhodopsin), etc., all involved in one of the most fundamental reactions of free energy transduction, the generation of ATP. To this class, however, also belong the phosphotransferase systems (for review see Ref. [9]) and the binding-protein dependent transport systems [10]. In a phosphotransferase system the free-energy of hydrolysis of phosphoenolpyruvate is invested in the accumulation of a sugar in the cytoplasm of a prokaryotic cell, in a reaction in which the phosphate group is covalently attached to the sugar during the translocation process, so that actually a sugar-phosphate accumulates in the cytoplasm. In a binding-protein dependent transport system a binding-protein from the periplasm of a Gram-negative bacterial cell binds a solute and transfers it to a complex of three polypeptides in the cytoplasmic membrane. Free-energy, most likely in the form of ATP (but see below) is subsequently invested to accumulate the solute to up to five orders of magnitude in the cytoplasm of the cell. Characteristically, the proteins involved in this uptake do not become stably phosphorylated during the transport process. The free-energy released upon hydrolysis of ATP is sufficient to energize the translocation reaction in such systems (although the stoichiometry has not yet been resolved, see Refs. [11,12]). For long, the energy coupling for this type of transport reaction has been controversial, because of claims that the proton motive force (Δp) be involved. Even though this cannot yet entirely be excluded (see discussion in Ref. [1]) it is likely that ATP hydrolysis suffices. Nevertheless, it is obvious, of course that the Δp will be involved when charged solutes are translocated through such a system [13]. Furthermore, the Δp may alter the structure of the cell envelope, and thereby affect translocation reactions (see discussion in Ref. [1]).

It is feasible that the energization of translocation of macromolecules, like nucleic acids, through biomembranes makes use of similar primary- and/or secondary translocation systems. Both forms have been claimed to be involved (for review see Ref. [14]). However, in theory DNA transport can be energized in several additional ways. An old speculation is the possibility that hydrolysis of one strand of DNA would provide the energy for the uptake of the complementary strand [15]. An additional model has been proposed by Reusch and Sadoff [16]. They have reported the presence of a cylinder, composed of polyphosphate, calcium ions and polyhydroxybutyrate in the cytoplasmic membrane of competent cells. These authors propose that this cylinder allows DNA to enter a cell, driven by the degradation of the polyphosphate. In this model no specific prediction is made about the role of ATP and/or the Δp in DNA uptake. Dubnau [17] has put forward a model based on a parallel between DNA uptake and the exchange of sugar-phosphates against inorganic phosphate, catalyzed by UhpT from *E. coli* [18]. This model combines the model of Lacks and Greenberg [15] with the reported dependence of DNA uptake into *B. subtilis* on the pH-gradient across the cytoplasmic membrane [19]. However, this model does not explicitly go into a possible role of ATP in DNA uptake. For entry of phage DNA into bacteria it has recently been proposed that transcription can be the driving force for DNA uptake [20]. This mechanism requires dsDNA and therefore can not be operational in many of the natural transformation systems because in those DNA enters in single-stranded form.

1.3. Traffic ATPases

As sequence data on the binding-protein dependent transport systems came available, it turned out that a large superfamily of proteins can be distinguished, of which most of its members are involved in transport across biomembranes. All its members share a characteristic A-type ATP-binding consensus sequence in a hydrophilic domain with a conserved secondary structure [21]. This protein family too, has members from both the prokaryotic and the eukaryotic kingdom. In addition to the binding-protein dependent transport systems, well-known examples are the multiple-drug resistance protein from tumour cells, a protein involved in the excretion of a yeast pheromone (the α-factor) and the product of the cystic fibrosis gene. On this basis, this class of proteins has recently been renamed as: M-type ATPases [22], ABC-proteins [21] and 'traffic ATPases' [23]. We will adhere to the latter term.

Several dozens of proteins have now been identified as belonging to this superfamily of traffic ATPases. Some of its members are not involved in transport reactions, like GroEL, UvrA and RecA. However, by far the majority of its members are. Strikingly, among these are components involved in transport of all three major classes of biopolymers. Common in the translocation of these biopolymers is that a more or less extended, possibly linear (but see Ref. [24]) form of the polymer is translocated. This suggests that similarities in the mechanism of energy coupling of all these types of biopolymers exist and that ATP hydrolysis energizes each of them.

2. Bioenergetics of nucleic acid translocation

Nucleic acids can be transferred across biological membranes in various processes. In bacteria the best-known examples are: injection of phage DNA, conjugation (of which transfer of Ti-DNA from *Agrobacterium* into plant cells can be seen as a differentiated form), and natural transformation. In these processes the DNA, with the exception of many phage DNAs, is transported in single-stranded form. Among these systems, natural transformation offers the most straightforward opportunity to directly assay the kinetics and stoichiometry of DNA uptake into bacteria. Mutual similarities in these processes can and have been exploited to resolve the molecular mechanism of DNA transport. This is an important motive to study these processes in parallel.

Despite the fact that the above mentioned DNA transport reactions are well known, detailed knowledge about the bioenergetic reactions driving DNA uptake is still restricted. Surprisingly, a common theme in several reports on DNA transport is the combined involvement of ATP *and* a Δp. The details of the separate processes will be discussed below.

2.1. Injection of phage DNA

We take the tailed *E. coli* phages as the primary example, because predominantly these have been studied with respect to the mechanism and bioenergetics of DNA transport. Uptake of phage DNA into recipient cells is fist of all dependent on binding of one of the (dsDNA containing) phages to a dedicated receptor in the cell envelope of the recipient bacterium. Most often this binding initially is reversible, but subsequently becomes essentially irreversible. Various components in the cell envelope can function as a phage receptor: outer-membrane pore-proteins, lipo-polysaccharides, flagella, pili, etc. After binding of the phage to the host, the injection of dsDNA is initiated. Phages have developed mechanisms enabling them to translocate their DNA across the cell envelop of the host. This can be protrusion of a contractible phage tail through outer membrane pores (like LamB or OmpF) allowing them to reach the cytoplasmic membrane (most T-even phages). Or by the generation of a channel through *both* the inner and the outer membrane, so that the phage DNA can be injected directly into the cytoplasm [25,26]. General ion leakage from the cytoplasm through this channel may be prevented by the hydrophobic tail proteins (phage T4 and T5, and possibly other phages without a contractible tail).

Translocation of phage T4-DNA, initially, was supposed to be driven by the Δp without involvement of ATP [27,28]. However, reports claiming a threshold value of the Δp of about −100 mV gave rise to the assumption that the Δp plays a regulatory role in the opening and closing of a putative pore or channel in the cytoplasmic membrane [27,29]. Furukawa et al. [30] showed that artificially contracted phage T4 can inject its DNA into spheroplasts of *E. coli* without the presence of a Δp. In their view, the only energy-demanding step is the formation of an adhesion zone between the phage tail and the cytoplasmic membrane after which DNA injection occurs, independent of energization. In their model the Δp is required for the induction of the membrane fusion between the outer and cytoplasmic membrane [25].

For other phages, like phage λ, phage T5 and phage T1 [31–33], DNA injection was independent of energization of the membrane. Phage T5 possesses a distinct mechanism via which DNA is injected. First, independent of energization, a small part (FST-DNA) of the phage DNA is injected into the cytoplasm. During the following pause in DNA injection, proteins are synthesized from the FST-DNA. Two proteins, A1 and A2, turned out to be essential for uptake of the remaining part of the phage DNA (SST-DNA). After these proteins have been synthesized and integrated into the cell envelope [34], the SST-DNA can be taken up without a further requirement for exogenous free energy input [31]. Maltouf [31] proposed that DNA translocation in phage T5 may be driven by the coiling activity of the DNA and/or histone-like proteins after binding of the incoming DNA to the inner side of the cytoplasmic membrane. No function has been proposed yet for the essential proteins A1 and A2. Alternatively, translocation of the phage T5-DNA may also be driven by transcription, as is suggested for phage T7 [20]. T7 DNA, as T5 DNA, enters via a two-stage mechanism. Initially, uptake is driven by transcription using the *E. coli* RNA polymerase, to be followed by transcription using its own RNA polymerase encoded within the first 19% of the incoming DNA.

The tail-tip protein pb2 of phage T5 forms a large water filled transmembrane channel when reconstituted into a lipid bilayer [35]. From the channel conductivity a pore diameter of 2 nm was calculated, which is sufficient to allow passage of dsDNA [35]. Recent results indicate that this protein is likely to be involved in penetration of phage T5 DNA into *E. coli* cells [36]. Infection by phage T5 induces a contact site between the inner and outer membrane, or stabilizes existing sites [36]. The contact sites of infected cells are enriched in pb2 protein and it is suggested pb2 forms the channel through which the phage DNA is translocated [35,36]. On the other hand the supposed channel could also function to depolarize the cytoplasmic membrane, thus facilitating DNA uptake [26].

Phage λ absorbs to a maltose transport protein of the outer membrane of the *E. coli* cell envelope (LamB). For the injection of DNA into the cytoplasm, phage λ requires the presence of a protein from the cytoplasmic membrane [32,37]. So-called *pel* mutants (*p*enetration of *l*ambda) of *E. coli* are resistant to infection by phage λ. This mutation mapped at an identical position as the *ptsM* gene, encoding the Enzyme II^{man} of the PTS. Further experiments demonstrated that for the injection of phage-λ DNA, no functional activity of PtsM was necessary and energy deprivation of the cells did not prevent entry of λ DNA [32].

These observations suggest that the DNA translocation reaction of phage λ is independent of Δp and/or the ATP pool of the host. More recently, Letellier and Boulanger [38] reviewed the possible existence of a channel or pore through which phage DNA is able to enter the cytoplasm. These authors proposed a model in which phage DNA is translocated across the cytoplasmic membrane via facilitated diffusion. Upon attachment of the phage to the outer membrane of the cell and protrusion of the phage tail into the periplasm, a pore or channel is opened in the cytoplasmic membrane. This opening induces an efflux of ions down their electrochemical gradient. As a consequence, the membrane becomes partially or completely depolarized and phage DNA is able to enter the cytoplasm via facilitated diffusion. Depolarization results in an inhibition of the transport of a number of metabolites that are coupled to the Δp. Closure of the pore or channel after entry causes membrane repolarization.

From these experiments the general picture emerges that translocation of phage DNA over the cytoplasmic membrane does not require energization by the Δp or ATP. Only in the case of phage T4, the Δp is essential for injection, possibly by playing a regulatory role as an energy source to keep the pore or channel open or enabling the formation of an attachment site.

2.2. Conjugation

Conjugal transfer of (plasmid) DNA between cells is defined as the DNAse insensitive transfer of DNA between donor and recipient cell. It has best been studied in enterobacteria, most notably for the F- plasmid. This plasmid codes for approximately 25 *tra* genes [39], required for conjugation. These include the genes required for pili synthesis and assembly, initiation of conjugal DNA replication and at least part of the DNA transport system.

For conjugal DNA transfer a stable mating aggregate has to be formed between a donor and an acceptor cell. The sex pili promote recognition and the initial formation of the mating aggregate. In *E. coli*, stabilization of the mating aggregate requires the functioning of the conjugal proteins TraN, TraG and outer membrane pore OmpA [39]. The transfer of DNA is subsequently initiated by the formation of a single-strand nick at the origin of transfer of the plasmid and takes place in single-stranded form, possibly guided by a pilot protein.

The actual DNA transport is generally thought to take place through a transmembrane pore, near or at the base of the sex pilus [40]. However, conjugal transfer of the loci *thr* and *his* was observed between strains separated by a 6 μm thick membrane with straight through pores of 0.015, 0.10 or 0.30 μm in diameter, preventing cell–cell contact [41]. No transfer occurred in controls in which membranes were used containing tortuous pores, or omitting the donor or recipient, suggesting that conjugal DNA can be translocated through the sex pilus.

Once cell to cell contact has been established, replication of conjugal DNA is initiated at the *oriT*. A single-stranded copy of the conjugal DNA is displaced into the acceptor cell. The entering ssDNA is then converted into dsDNA via synthesis of the complementary strand. Replication of chromosomal DNA, neither in the donor or in the recipient cell, is necessary for the translocation process [42].

Extensive studies, summarized by Grinius [14], indicated that both Δp and ATP are involved in conjugation: Experiments with EDTA-treated donor cells in which the effect of the uncoupler CCCP on the donor cells during conjugation was studied, showed that the formation of the mating aggregate was dependent on energization. Further experiments, using arsenate, suggested that formation of the mating aggregate required the presence of ATP. After the formation of the mating aggregate, the transfer of DNA is initiated and is dependent on ATP from the donor cell. Probably because ATP-consuming proteins, like Helicase I (encoded by *traI*), are involved in unwinding DNA prior to replication and transport. When arsenate was added after translocation was initiated, conjugal DNA transport could not be inhibited by arsenate, indicating that ATP hydrolysis in the recipient cells nor in the donor cells provides the driving force for DNA translocation. Experiments with CCCP- and EDTA-treated recipient cells showed that ongoing DNA transfer depended on the Δp of the recipient cell [14]. These experiments, however, were not performed with so called *unc* mutants. Such mutants lack the membrane-bound ATPase and provide a unambiguous *in vivo* system to study the effects of a separate or combined involvement of ATP and/or Δp.

The experiments of Grinius and co-workers give a rather detailed picture of the energization of conjugal DNA transport. Nevertheless, a clear bioenergetic picture of DNA translocation through the cell envelop (via some kind of intercellular contact site or sex pilus) of the donor and recipient is still missing. If conjugal DNA is transported through the sex-pilus, as indicated by Harrington and Rogerson [41], a Δp-driven uptake system on the recipient's side of the pilus seems conceivable. However, even then the question remains how the DNA is transported into the pilus on the donor side. The experiments of Grinius suggest that such a supposed export system does not require energization by Δp or ATP. Since the donor cells in these experiments did not belong to the class of the *unc* mutants [14] the involvement of ATP in energization of DNA export cannot be ruled out yet. Thus it is unclear whether conjugal DNA is exported via an ATP-driven pump or via facilitated diffusion.

If conjugal DNA is transported through a DNA-translocation complex, which spans the cell envelopes of both the donor- and the recipient cell [40], the question arises how the Δp of both bacteria influence each other. How could the Δp drive DNA translocation, since the direction of DNA transfer in donor and recipient is opposite towards the Δp of the cytoplasmic membrane of the donor and recipient. Although other explanations could be suggested, it is tempting to speculate that DNA is taken up by the recipient via an ATP driven DNA *translocase*. This is not necessarily in conflict with the results obtained by Grinius and co-workers, as it is known that inhibitors of oxidative phosphorylation can alter the intracellular pH and many ATP-dependent transport systems strictly depend on intracellular pH homeostasis [43]. If DNA is taken up via a Transport-ATPase, alterations of the intracellular pH could also explain the inhibition of DNA transport. The inner membrane protein TraD, complexed with the TraI,Y helicase/nickase proteins, might constitute an ATP driven translocation system [41], although the proposed mechanism for transport then differs from the Transport-ATPases in that the unwinding of the DNA via the action of the ATP-consuming helicase would drive the translocation of DNA [44].

The process of transfer of Ti-DNA from *Agrobacterium* to a plant cell has attracted a lot of interest in recent years because of the possibility to genetically modify plant cells. These studies have revealed a striking parallel in the mechanism of DNA transfer between this process and bacterial conjugation, partially on basis of similarity of the genetic components involved in the two processes (for review see Ref. [45]). A large number of operons (*virA* to -*G*) is involved in virulence of *Agrobacterium*. Of these, particularly the *virB* operon may be involved in DNA transport. Surprisingly, however, so far only little biochemical evidence for the presence of pili-related genes has been obtained for this operon (note that *virB*2 and -*B*3 show sequence similarity to such products [46]).

2.3. Natural transformation

Some 43 species of 24 separate bacterial genera have at least one representative that has been reported to be naturally transformable [5]. This implies that these strains are able to take up 'naked' DNA from the environment into physiologically uncompromised cells and express genes that are coded in the incoming DNA (for review see Stuart and Carlsson [47]). Many species can also be transformed *artificially*, for instance with calcium ions, in combination with a temperature shock. However, in the following we concentrate on *natural* transformation, taking *Bacillus subtilis* as the archetype. This is the best characterized model system to study DNA uptake in bacteria.

The process of natural transformation takes place in four distinct phases (see Fig. 1). DNA is taken up only in cells that are in the so-called competent state. For the development of this competence a complex regulatory network is responsible, in which — amongst many others — several early competence genes are involved (like *comA*, *comP*, *comQ*, *comX*, *degU* and *degS* in *B. subtilis*, see for instance Ref. [48]). Under appropriate conditions (i.e. absence of certain amino acids and excess glucose) these early competence genes lead to the expression of the late competence genes, coding for the DNA-binding and -uptake machinery. Subsequently, DNA can be taken up via a first step in which dsDNA is bound to cell-surface receptors and double-stranded fragments are formed of a few kilobases length. Most often these will be linear fragments of (chromosomal) DNA, but circular, or even multimeric, plasmid DNA can also be internalized. In some organisms, however (like *H. influenzae*), a distinct route for the uptake of plasmid DNA appears to exist [49].

In the next step of the uptake process, one of the two strands of such fragments is taken up into the competent cell, with simultaneous hydrolysis of the complementary strand. DNA uptake in *S. pneumoniae* proceeds in 3' to 5' direction [50]. In *H. influenzae* the same directionality of uptake was observed [51]. Uptake of DNA in *B. subtilis* did not show any directionality. Radioactive label at the 3' and the 5' end of a single strand of a labeled plasmid DNA was taken up [52] at equal rates. In *S. pneumoniae* the incoming ssDNA is bound by the eclipse protein. This is a binding protein, expressed during competence, that binds the entering ssDNA and protects it against the action of intracellular nucleases [53]. The transformation process is

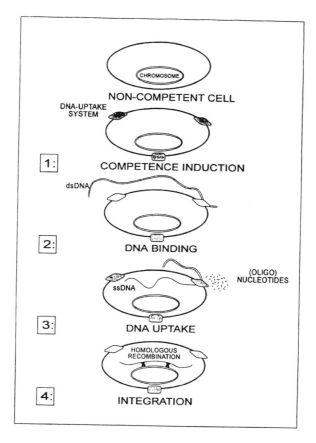

Fig. 1.   Schematic presentation of the four phases in the process of DNA uptake in natural transformation. For further explanation see text.

completed after integration of the incoming (single stranded) DNA into the chromosome and expression of genes encoded on this fragment. Among the proteins that are induced during competence induction in *B. subtilis* are the major recombinational protein RecE (*E. coli* RecA-analogue [54–56]) and the AddAB complex (*E. coli* RecBCD analogue, B. Hayema and J. Kooistra, personal communication). Also in *S. sanguis* and *S. pneumoniae* RecA is induced upon competence induction [57–59]. In other organisms, however, like *A. calcoaceticus* and *Ps. stutzeri*, RecA expression is not affected by competence development (B. Vosman, P. Rauch and K.J. Hellingwerf, unpublished experiments). Particularly when multimeric plasmid DNA is offered as transforming DNA, an intact replicon may be formed even in the absence of RecA, which subsequently may give rise to gene expression from this replicon.

DNA uptake in natural transformation in *H. influenzae* differs from this description in important respects: a vesicular membrane structure, called a transformasome, is involved in the passage of DNA through the outer membrane of this organism. Nevertheless, there are good reasons to suppose that the mechanism of translocation of DNA through the cytoplasmic membrane in the two organisms is quite analogous [14]. Furthermore, the uptake pathway of *Haemophilus* is selective for endogenous chromosomal DNA. Only DNA in which a 9 bp consensus sequence is present is taken up. The chromosome of *H. influenzae* contains 1465 of these sequences [60]. Specificity in DNA uptake has also been described for transformation of *N. gonorrhoeae* [61,62]. This latter organism recognizes a 10 bp sequence which, however, does not show any similarity with the *Haemophilus* sequence.

Studies on the energetics of DNA transport are mainly confined to experiments with *S. pneumoniae*, *B. subtilis* and *H. influenzae*. Uptake of DNA only occurs in competent cells, possessing a high energy state. Addition of an inhibitor of oxidative phosphorylation (like 2,4-dinitrophenol) blocks DNA uptake in competent cells [63–65]. Different models have been proposed to explain the energy dependence of DNA uptake.

After establishment that dsDNA in *S. pneumoniae* is converted into a ssDNA during uptake, Lacks proposed a model suggesting that the hydrolysis of the one strand could provide the energy required for the uptake of the complementary strand [66,67]. However, experiments using a mutant with a reduced activity of the membrane located nuclease (exo-5 and endo-1; required for DNA transport) demonstrated that ssDNA can readily be translocated by *S. pneumoniae* [68]. Nuclease activity is not strictly coupled to DNA uptake as suggested by Lacks, but required for conversion of dsDNA into ssDNA with is the substrate for the actual translocase. Also, effects of dissipation of components of Δp and inhibitors of the energy metabolism on the uptake of DNA, as observed in other bacteria (see below), are difficult to reconcile with this model.

In 1980, Grinius reviewed experiments on the effects of different inhibitors of Δp and oxidative phosphorylation [69]. He concluded that uptake of DNA in bacteria is dependent on Δp. A general model for uptake of DNA was proposed in which extracellular DNA binds both proteins and protons, thus acquiring a net positive charge. The positively charged complex of DNA, protein and protons then electrophoreses through the cytoplasmic membrane, thus explaining that DNA uptake is dependent on both components of Δp, e.g. $\Delta \psi$ as well as ΔpH. Pore formation is induced by interaction of the DNA with membrane constituents and does not pre-exist. Upon arrival in the relatively alkaline cytoplasm, protons are released, thus giving rise to an electrogenic mode of transport, dependent on both the $\Delta \psi$ and the ΔpH.

After routine *in situ* measurement of the magnitude of Δp and its components became possible, quantitative relationships between DNA transport and Δp were established. These studies showed that DNA uptake in *B. subtilis* was driven mainly by ΔpH, suggesting that DNA is taken up electroneutrally in symport with protons [19]. This is not in accordance with the model of Grinius who predicted an electrogenic mode of uptake. Similar experiments were performed for *H. influenzae* [70]. The results indicated that in *H. influenzae*, both components of Δp are able to drive DNA uptake and that DNA uptake is dependent on the total Δp. DNA uptake in these experiments was assayed via the determination of the transformation frequency. This means that the effects of Δp and its components on the uptake of DNA, in fact represents the effects of Δp on the whole process of natural transformation, i.e. DNA uptake but also competence development, DNA binding and DNA integration. By looking more specifically into the effects of the components of Δp on the uptake of DNA only, using radiolabeled DNA, it was shown that valinomycin and nigericin directly interacted with the DNA binding sites at the cell envelope. Other inhibitors of Δp did not inhibit binding of DNA [70]. A restriction has to be made since both experiments, on *Bacillus* as well as on *Haemophilus*, were performed with cells still possessing an active membrane bound ATPase (i.e. not in *unc* mutants). Involvement of ATP in DNA uptake can thus not be ruled out yet, just like possible effects of the uncouplers used on the intracellular pH. Such changes in intracellular pH could also inhibit uptake of DNA, as observed in *S. pneumoniae* [71].

After discovery of a polyhydroxybutyrate (pHB) cylinder in the cytoplasmic membrane of competent (for artificial transformation) *E. coli* cells, Reusch and Sadoff [72] proposed that this may function as a DNA translocase. The pHB cylinder is stabilized via Ca^{2+} ions and contains a polyphosphate core complexed to the pHB cylinder, again via Ca^{2+} ions. The dimensions of the cylinder are just large enough to allow passage of a ssDNA or RNA molecule, but not of dsDNA. However, recently it has been demonstrated that this pHB complex functions as a calcium uptake system [73]. The fact remains that several naturally transformable organisms accumulate pHB when acquiring competence [74].

Dubnau proposed a model based on a possible parallel between DNA uptake and the exchange of sugar-phosphates against inorganic phosphate, catalyzed by UhpT from *E. coli* [75]. The latter transport system exchanges two monovalent glucose-6-phosphate anions against one divalent glucose-6-phosphate^{2-} anion, resulting in the electroneutral uptake of one glucose-6-phosphate anion and one proton. To remain active, the incoming protons have to be released again and thus this system is indirectly dependent on proton recycling, i.e. on the Δp. Analogously, monovalently negatively charged (per nucleotide) dsDNA could enter via a DNA transport system, encountering a nuclease which hydrolyses one strand of the dsDNA. The nucleotides of the hydrolysed strand donate a proton to the complementary strand rendering it electroneutral. Subsequent expulsion of divalent negatively charged nucleotides results in the net uptake of ssDNA and protons.

The model of Dubnau combines the model of Lacks and the ΔpH dependence of DNA uptake found in *B. subtilis* by van Nieuwenhoven et al. [19]. On the other hand, this model does not seem to be valid for *H. influenzae*, since in this organism DNA uptake can also be driven by the $\Delta \psi$ only [70].

The models discussed above address a mode of energization of DNA uptake, exclusively via Δp or the intramolecular energy of the DNA. None of the models proposed so far, predicts the energization of DNA uptake by ATP. Competence induction in *S. pneumoniae* results in an increase of intracellular pH, a stimulation of glycolysis and an increase in ATP concentration [68]. From inhibitor studies at different pH values and studies with a mutant having a reduced transmembrane electrical potential gradient, it was concluded that DNA uptake in *S. pneumoniae* is not driven by the Δp [71]. Additional experiments showed that DNA uptake in *S. pneumoniae* depends on the intracellular pH [71], which might indicate that DNA uptake could be driven by ATP. Poolman et al. [43] have shown that the intracellular pH in prokaryotes is capable of regulating ATP-dependent solute-transport systems, belonging to the Transport-ATPases. To demonstrate ATP-dependent uptake of DNA in *S. pneumoniae*, the effects of arsenate treatment on competence induction and DNA uptake were studied [68]. It was demonstrated that competence induction is dependent on a high intracellular ATP concentration. Reduction of the ATP pool with arsenate by 20 to 30% during competence acquisition resulted in a 90% reduction of transformants formed [68]. Similarly, reducing the ATP pool, again by 20 to 30% after competence was fully induced, inhibited formation of transformants by 60%, indicating that DNA uptake may be ATP-driven. However, the ATP pool of a competent cell and of a non-competent cell is respectively 9.5 mM and 6.3 mM. The small reduction of the ATP pool could argue against a direct role of a Transport ATPase in DNA uptake because it is likely that such enzymes possess a high affinity for ATP (K_m in the order of 100 μM).

3. The DNA-translocase

A key question in studies of the mechanism of DNA transport is the identification and characterization of a (proteinaceous) complex that catalyzes the transfer of highly charged (anionic) DNA molecules through the hydrophobic interior of the cytoplasmic membrane, without causing a dramatic loss of metabolites from the cytoplasm simultaneously. This question is even more pressing for DNA translocation than for instance for protein transport, because (i) generally proteins tend to be more hydrophobic than nucleic acids and (ii) the hypothesis has been advanced that the proteins — through their signal sequences — first can spontaneously insert their N-terminus into lipidic domains of the cytoplasmic membrane, followed by entry of the export protein into the translocation channel from the side [4]. In this model the aqueous channel is immediately filled during opening, thus preventing metabolite loss.

3.1. Does it exist?

Through the years, quite some sepsis has been expressed on the assumption that a proteinaceous complex would *catalyze* the translocation of DNA into bacterial cells. As an alternative, phase boundaries between laterally separate(d) domains of the cytoplasmic membrane could function as sites of passage for DNA. The notion of the involvement of polyhydroxybutyrate and/or polyphosphate in the DNA-uptake process might support this notion. Furthermore, the specific model of Reusch and Sadoff does not have a role for catalytic activity of an enzyme during the actual translocation process either. Nevertheless, quite some evidence has accumulated that indeed a *translocase* is involved in DNA uptake.

Via the analyses in particular in *B. subtilis* [76] (but also in *H. influenzae*, see e.g. Ref. [77]) more and more genes have been identified and characterized that are required for the process of natural transformation. A subgroup of these is likely involved in transport processes. These are those that either encode intrinsic membrane proteins and/or genes whose products show homology to the traffic-ATPases. However, a crucial problem has to be solved before one can conclude that such gene products are really involved in DNA transport. As an alternative, these proteins may be required for the *assembly* (i.e. the transmembrane transport) *of the structure that forms the DNA-translocase*. That such transport of DNA-translocase subunits is necessary, is evidenced by the requirement for periplasmic processing enzymes in several transformable bacteria, like in *A. calcoaceticus* [78]. Such a translocase-assembly function may be equated to for instance the ComG protein from *B. subtilis*. Additional criteria have therefore been used to identify gene products involved in DNA transport directly, like the unimpaired (or even increased [79]) binding of DNA to cells of mutants in which the relevant genes had been inactivated.

3.2. A translocase sequenced

With the availability of the complete sequence of the genome of *H. influenzae* we are sure that, if a DNA translocase exists, we now know its sequence. Nevertheless, identification of the corresponding proteins from this data has not yet been possible. For concrete data about the proteins that may form the DNA translocase we have to rely on the strategy of analysing *com* genes and combine arguments as outlined above.

For *B. subtilis* this already allows one to draw a tentative scheme of the proteins involved in DNA processing and entry in the transformation process (see Fig. 2). ComGC proteins are processed by ComC and involved in binding of DNA and/or the passage of DNA through the outer part of the cell-envelope. Subsequently, the transforming DNA is bound by ComEA, translocated into a pore formed by ComEC and translocated and processes by ComFA, a protein that is homologous to helicases. In several organisms homologues for these gene products have been identified. The ComEC and ComFA proteins are homologous to Rec2 and Com101A of *H. influenzae* and ComEC and Rec2 are both homologous to ComA of *Neisseria*. During translocation the membrane located entry nuclease converts dsDNA into ssDNA. Biochemical characterization and further comparisons among these proteins may provide further insight into the properties of the DNA translocase.

In conjugation VirB11 and/or VirB4 have been assigned the role of translocase in *Agrobacterium* and the homologue Tra2 of the RP4 plasmid. No homology has been detected yet between these two classes of translocases.

3.3. Further strategies for identification of the translocase

In view of the uncertainties inherent in the characterization of the DNA translocase via the strategy outlined above, it is of crucial importance that biochemical evidence is added to its characterization. One of the primary goals in this respect is the reconstitution of DNA translocation activity in subcellular systems. In view of the complex nature of the DNA-uptake machinery in many bacteria, it may be necessary to use procedures similar to the ones used for *in vitro* reconstitution of flagellar

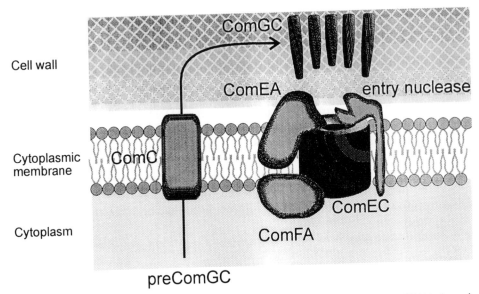

Fig. 2.   Diagrammatic representation of the components involved in the transport of DNA through
the cell envelope of *B. subtilis*.

control [80], i.e. to isolate a subcellular structure with an intact envelope. Secondly,
it is important to isolate and purify the components involved in DNA uptake. In
view of the low abundance of these proteins, this is a challenging task. Various
strategies can be thought of to label the proteins involved in DNA uptake. One
example is the specific labelling of the proteins that are expressed upon competence
induction. Alternatively, the DNA in the translocase may be crosslinked through
bifunctional reagents. For this purpose it is important that DNA can be stably
positioned in the translocase, by covalently linking a protein molecule to the
transforming DNA [5].

4. Methods for assaying DNA translocation

Several different methods for assaying DNA uptake can be deployed, each with its
own specific strength and/or disadvantages. Depending on the objective of the
experiment, any of these can be selected.

4.1. Determination of the transformation frequency after transformation with
DNA containing a selectable genetic marker

This method is commonly used to characterize the effects of a large array of
biotic and abiotic parameters (e.g. temperature, nature of DNA, incubation time,
DNA concentration, effect of inhibitors, etc.) on the transformation system of a
specific bacterium, allowing a comparison with transformation characteristics of
other naturally transformable organisms. However, it has also been used to assay

DNA uptake. A major disadvantage of this method is that the transformation frequency is the resultant of all steps involved in the formation of a transformant (notably recombination and expression) and not only of DNA uptake. This may lead to erroneous interpretation of data.

4.2. Determination of DNA uptake using radioactively labeled DNA

Most commonly [^{3}H] labeled chromosomal or plasmid DNA is used. [^{3}H] labeled DNA is easily obtained from thymidine deficient strains grown in the presence of [^{3}H]thymidine. After isolation and purification, the labeled DNA can be used for uptake studies. The amount of internalized DNA can be quantified after separation of cells from the non-internalized label, for instance by a DNAse treatment and filtration step, and correction for non-specific binding. This method is one of the most commonly used to assay DNA uptake and to study mechanistic and bioenergetic aspects of DNA uptake. In *S. pneumoniae* degradation of [^{3}H]DNA is used to assay competence as a quick and easy alternative instead of determination of the transformation frequency. The degradation products are separated from the cells and non-degraded DNA via TCA precipitation.

4.3. Quantitative assays of DNA uptake using radiolabeled PCR fragments

To study the quantitative relationship between DNA uptake and formation of transformants, measurement of uptake of [^{3}H] labeled chromosomal DNA is not sufficient. It does not allow quantification of the amount of fragments taken up due to the large size distribution of fragments of chromosomal DNA. Selection for only a single marker present on the chromosome, considerably reduces the sensitivity of the determination of the transformation frequency. These problems can be overcome by [^{35}S] labelling of a specific PCR fragment targeted to a mutated gene on the chromosome of the recipient strain. This allows quantitative determination of the number of fragments taken up per cell, combined with a highly sensitive determination of the transformation frequency. This method was used to determine the ratio between fragments taken up and the transformants formed during transformation of *A. calcoaceticus* (see below). Furthermore, this approach allows one to characterize the kinetics of DNA uptake, like determination of its K_m, V_m, etc.

5. DNA uptake in *A. calcoaceticus*

Acinetobacter calcoaceticus BD413, a typical soil organism, is highly transformable. Transforming DNA may be incorporated into its genome, thus changing its genotype and, after expression of the acquired sequences, its phenotype.

Only dsDNA is a substrate for the DNA-uptake system of *A. calcoaceticus*. It is converted into single-stranded form upon entrance into the cytoplasm, which requires energization of the cell. Dinitrophenol-treated (i.e. de-energized) cells do not form transformants [81]. Transformation is sensitive towards removal of divalent cations, suggesting the involvement of a nuclease in DNA uptake. This putative (exo)nuclease may be responsible for breakdown of one of the two DNA strands during uptake [81]. The estimated rate of DNA uptake is 60 bps s^{-1} [81]. These characteristics resemble those determined for DNA uptake in *B. subtilis* and *S. pneumoniae*. This is quite surprising since competence induction in *A. calcoaceticus* is regulated very differently from *B. subtilis* and *S. pneumoniae*. *A. calcoaceticus* induces competence maximally after a nutrient upshift of a stationary culture [82].

To be able to study the bioenergetic requirements of this DNA uptake process and to determine the efficiency of incorporating the incoming DNA fragments into the host's chromosome, a strain has been constructed (AAC360) containing an inactivated heterologous kanamycin resistance gene (*aphA3*) inserted into the *estA* gene, encoding an endogenous cytoplasmic esterase [83]. Both the *aphA3* and the *estA* have been sequenced. Hence detailed information is available about the chromosomal region under study [83,84]. After uptake of defined fragments of transforming DNA (e.g. specific PCR fragments) into this strain, one can calculate the number of DNA fragments taken up per cell. Via parallel determination of the transformation frequency, it is then possible to calculate the efficiency with which transformants are formed, in other words, how many fragments have to be taken up by competent cells, to form a single transformant.

Below, we first describe the most essential steps in the construction of AAC360, after which the transport experiments, aimed at determining the efficiency of DNA-uptake in natural transformation, are discussed in detail.

5.1. Construction of A. calcoaceticus AAC360

The *estA* gene of *A. calcoaceticus* BD413 was isolated from pAKA20 [83] via *HpaI/PvuII* digestion, ligated into the *SmaI* site of pUC18 and cloned in *E. coli* JM83 via α-complementation. Plasmids with the insert in either orientation were detected and named pAPA300-1 and pAPA300-2, respectively. All clones proved positive in esterase assays.

The *aphA3* gene, encoding kanamycin resistance, as a 1687 bp *PvuII* fragment from pPJ1 was ligated into the *EcoRV* site of the *estA* gene of pAPA300-2. This plasmid contains the esterase gene in opposite direction with respect to the ampicillin-resistance gene of pUC18. The ligation mixture was transformed into JM83. Selection of Kan^R, Amp^R clones resulted in the isolation of pAPA300-2-1 (Fig. 3A), containing the *aphA3* gene in the same direction as the esterase gene.

To inactivate the kanamycin marker of pAPA300-2-1, this plasmid was partially restricted with *Cfr10*I (plasmid pAPA300-2-1 contains two *Cfr10*I sites, one in the Kan^R marker and one in the Amp^R marker). Fragments of 6185 bp (i.e. with a single *Cfr10*I-restriction), were isolated and their 5′ end was filled-in with Klenow, creating 4 additional base pairs and an additional *Cfr10*I site next to the original site (Fig. 3B). After ligation and transformation into JM83, transformants resistant to ampicillin were selected. Subsequently, 100 ampicillin resistant colonies were screened for kanamycin sensitivity. Plasmid DNA from twelve sensitive clones was

A

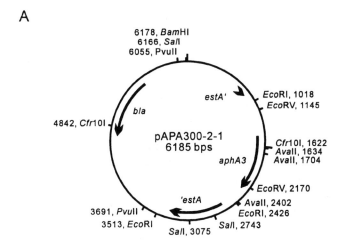

B

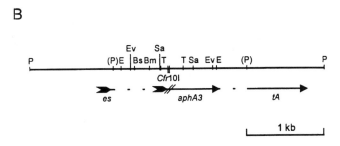

Fig. 3. (A) Restriction map of plasmid pAPA300-2-1. (B) Restriction map of the mutated *aphA3* gene region of plasmid pAPA300-2-1b. Bs, *Bss*HII; Bm, *Bsm*I; E, *Eco*RI; Ev, *Eco*RV; P, *Pvu*II; Sa, *Sau*3A; T, *Taq*I.

isolated and restricted with *Cfr10*I. Eleven of the twelve tested clones had the correct size and number of restriction sites (3, only 2 detectable) for *Cfr10*I. The kanamycin sensitive clone was named pAPA300-2-1b.

A. calcoaceticus AAC360 was obtained via transformation of BD413 with pAPA300-2-1 (selection for Kan^R and Amp^S) and subsequent transformation with pAPA300-2-1b (selection for loss of Kan^R, Fig. 3B). As a control, a representative of these strains, named AAC360, was tested for the ability to repair the mutated *aphA3* gene via transformation with an intact copy of the gene. For that purpose AAC360 was transformed with the 1687 bps *Pvu*II fragment encoding *aphA3* of pPJ1. The transformation frequency in a standard assay was in the order of 10^{-2}, indicating that AAC360 contains a copy of the inactivated *aphA3* gene and that repair proceeds with high efficiency. Spontaneous mutants have not been observed, indicating a reversion frequency of less than 1×10^{-8}.

5.2. Transformation efficiency as a function of the length of the transforming fragment

To test how transformation is dependent, quantitatively, upon the size of the transforming fragment, *A. calcoaceticus* AAC360 was used as recipient in transformations with fragments of different size containing the wild-type sequence of the mutated part of the *aphA3* gene of AAC360 (Table 1). Transformants were selected for kanamycin resistance. The highest transformation frequency was found with the largest fragment used, which holds 1750 bps upstream and 2071 bps downstream from the mutated *Cfr10*I site. The transformation frequency decreases upon a decrease in fragment size in a biphasic manner (Fig. 4). Decreasing the size from 3821 to 1025 bp results in a linear decrease of the logarithm of the transformation frequency, with a slope of 3.6. Reducing the fragment size even further, results in an increased slope of 9.3. Transformation was still detectable with a fragment of 294 bp (*Taq*I), bordering the mutation by 104 bp up- and 190 bp downstream. Apparently, these arms are long enough to allow homologous recombination with a measurable frequency.

A similar biphasic pattern has also been observed in *H. influenzae* [85] except that in this organism the size dependence of transformation frequency shows a discontinuity with fragments of 3.5 kb. Transformation in *H. influenzae* becomes increasingly inefficient with fragments smaller than 3.5 kb, whereas this point is reached in *Acinetobacter* with fragments smaller than 1 kb. The increased slope of the curve relating the formation of transformants with the fragment length, at fragment-lengths smaller than 1 kb, can be explained by: (1) A decreased efficiency of uptake of these smaller fragments; however, it was shown for *H. influenzae* that the efficiency of uptake for small fragments was the same as for large fragments [85]. (2) Partial degradation of the incoming transforming DNA, which may result in removal of the sequence that repairs the mutation. For *H. influenzae* 1.5 kb was estimated to be degraded, before the fragment is integrated into the chromosome

Table 1

The transformation frequency of AAC360 depends on the size of the transforming fragment containing the wild-type sequence of the mutated site on the chromosomal *aphA3* gene

Fragment	Size (bp)	Flanking region (upstream)*	Flanking region (downstream)*	Transformation frequency
*Pvu*II (pAPA300-2-1)	3821	1750	2071	1.8×10^{-1}
*Pvu*II (pPJ1)	1676	696	980	1.2×10^{-2}
*Eco*RI (pPJ1)	1408	604	804	7.5×10^{-3}
*Eco*RV (pPJ1)	1025	477	548	1.7×10^{-3}
*Sau*3A (pPJ1)	409	106	303	1.9×10^{-7}
*Taq*I (pPJ1)	294	104	190	1.9×10^{-8}
*Bss*HII (pPJ1)	1414	434	980	3.5×10^{-3}
*Bsm*I (pPJ1)	1215	235	980	1.8×10^{-3}
*Ava*II (pPJ1)	616	604	12	8.0×10^{-7}

* Up- and downstream have been defined with respect to the mutation in the chromosomal *aphA3* gene.

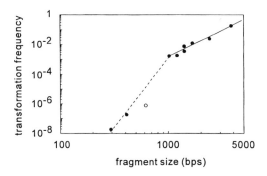

Fig. 4. Transformation of AAC360 with DNA fragments of different size (O, fragment with 12 bps downstream border).

[85]. This equals 0.5 times the minimal fragment size (minus 200 to 400 bps required for integration) that still transforms with high efficiency. Analogously, we assume that 500 bps of the incoming DNA in *A. calcoaceticus* is degraded via exonuclease activity, before the fragment is integrated. Further degradation is probably prevented by binding of RecA or Eclipse protein (single-stranded DNA binding protein) to the internalized DNA, although the presence of the latter protein has not been investigated in this organism.

The inactivated kanamycin-resistance gene can be restored via (i) two homologous recombination events, one on each bordering fragment, with an intact kanamycin cassette, or (ii) via a single recombination event, followed by integration of the single stranded fragment via displacement of one of the mutated strands. The 4 bps mismatch thus created is resolved after chromosome duplication. If the first would be the only possible mechanism, it is expected that transformation with a fragment practically devoid of one of its bordering sequences, would no longer result in the formation of a transformant. However, a significant number of transformants is formed (with a frequency of 8×10^{-7}), using a fragment (*Eco*RI-*Ava*II) with borders of 604 bps (upstream) and 12 bps (downstream) in size. The frequency obtained is lower than expected from the total size of the fragment (Fig. 4), but this may be explained by (exo)nuclease activity on the 12 bps downstream border, rendering this fragment unable to repair the mutation of the *aphA3* gene. This indicates that the transforming DNA can also be incorporated via a single homologous recombination event. This mechanism has also been proposed as a possible mechanism for integration of DNA in *Haemophilus influenzae* transformation [85].

5.3. Uptake of radiolabeled DNA

Transformation of *Acinetobacter* with fragments larger than 1 kbp results in high transformation frequencies. To be able to study radioactively labeled DNA uptake directly, PCR was used to amplify a 2402 bps fragment from pAPA300-2-1, containing the intact kanamycin resistance gene. The fragment was labeled during amplification by addition of 50 µCi $\alpha[^{35}S]$dATP. The specific activity of the fragment obtained was 1.2×10^{6} cpm/µg DNA. Competent AAC360 cells (600 µl;

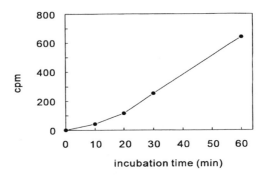

Fig. 5. Time-dependent uptake of radioactively labeled DNA by *A. calcoaceticus* AAC360.

$\pm 4 \times 10^8$ cells ml^{-1}) were incubated for selected time intervals with 2.5 ml PCR product. Uptake of DNA was determined after a DNAse treatment (20 μl of a 5 mg ml^{-1} stock solution) and separation of the cells from the medium by silicon-oil centrifugation into 7% perchloric acid. As a time-zero control, DNAse was added prior to addition of DNA.

Under these conditions uptake of DNA proceeds linearly in time after an initial lag phase of about 10 minutes (Fig. 5). This confirms earlier results, obtained by determining the time dependence of the formation of transformants [81]. The amount of label taken up after 60 minutes, corresponds to a mean uptake of 2 single-stranded 2402-base fragments per viable cell. The transformation frequency after 60 minutes of incubation was 2.7×10^{-3}. This implies that 2.7 cells out of every 1000 have been transformed into kanamycin resistance. With this frequency, and the number of fragments taken up per cell, we can calculate that 1 transformant is formed for each 750 fragments taken up. Extrapolation to transformation with saturating amounts of 2402 bps DNA and an incubation time of 3 hours (after which the transformation frequency reaches its maximum), results in a predicted maximal uptake of approximately 65 single-stranded PCR fragments per cell. This amount is equivalent to about 150 kb DNA.

Integration of the incoming DNA strongly competes with degradation of this DNA. Only 1 of 750 fragments taken up results in the formation of a transformant, due to high endonuclease activity in the cytoplasm. Evidence for significant restriction/modification activity has not been obtained (as would be apparent from the use of DNA derived from different sources). Studies on integration of chromosomal DNA in *H. influenzae*, *S. pneumoniae* and *B. subtilis* suggest that 25–50% of the DNA taken up is subsequently incorporated into the chromosome [86–88]. However, the values calculated for *B. subtilis* were obtained by comparing the DNA fraction internalized with the fraction integrated, via DNA reisolation and sucrose gradient centrifugation experiments. It is conceivable that some of the incoming DNA is already degraded during entry into the cytoplasm and will not be detected by reisolation studies. *H. influenzae*, maximally takes up 4 to 8 DNA fragments regardless of the size of the fragment, which is much lower than observed for *A. calcoaceticus*. Uptake of this limited amount of chromosomal DNA results in a maximal transformation frequency of 5×10^{-2}, whereas transformation of *A. calcoaceticus* with chromosomal DNA results in a maximal transformation frequency of 4×10^{-3} [81]. This indicates that transformation of *A. calcoaceticus* is less efficient than of *H. influenzae*, although not as inefficient as is suggested by the DNA uptake experiments. As an explanation for the limited number of DNA fragments taken up by *H. influenzae* and the independence of this number of the size of the fragments, it has been suggested that competent *H. influenzae* cells contain only 4 to 8 uptake sites which are thought to be used only once [89]. If this would be the case for *A. calcoaceticus* then a competent cell contains approximately 750 uptake sites. This is probably an overestimate because *B. subtilis* and *S. pneumoniae* are estimated to contain about 50 uptake sites per cell [90–92]. The large number of fragments taken up therefore suggests that the uptake sites of *A. calcoaceticus* can be used more than once.

DNA uptake by *A. calcoaceticus* proceeds linearly for 2 to 3 hours [81], which is relatively long compared with *H. influenzae* and *S. pneumoniae*, which are saturated for DNA uptake after a few minutes of incubation with a saturating amount of DNA [86,89]. During the 2 to 3 hours of incubation of *A. calcoaceticus* with DNA, new DNA-uptake machinery may be synthesized, which then would become available for DNA uptake. However, the addition of chloramphenicol did not inhibit the formation of transformants. In conclusion, *A. calcoaceticus* has a low efficiency of formation of transformants. Nevertheless, a high transformation frequency still is obtained in this organism, due to the fact that every competent cell can take up many DNA fragments.

6. Discussion and concluding remarks

6.1. The rate of DNA uptake

DNA translocation during injection of phage DNA into prokaryotic cells proceeds with an extraordinary high rate of 10.000 base-pairs per second (estimated for phage T4 [14]). DNA translocation rates in conjugation and natural transformation are estimated to be much slower: in the order of a 1000 and 100 bases per second respectively. Except for *H. influenzae* for which much higher rates were estimated (500–30,000 nucleotides s^{-1} [14]). The high translocation rate is in fact the rate with which the DNA (as a double stranded molecule) is taken up into the transformasome (the fist translocation step during uptake).

It is of interest to find out whether the latter rates are dependent on the size of Δp and whether or not this large difference in rates can be related to the free energy dependence of the uptake processes. Also, a difference in mechanism between translocation of double- and single-stranded DNA may be responsible for the large differences in translocation rates.

6.2. Interesting aspects of coupling and slip

Besides the rate of translocation through the DNA-uptake system, many interesting questions also remain regarding the stoichiometry of the coupling of free-energy to this translocation process (i.e. number of protons and/or ATP molecules consumed per translocated base(pair), and whether or not these stoichiometries are fixed. Furthermore, it is important to get insight in the specificity of DNA uptake systems for their substrates (i.e. aspects like: sequence, minimal length, single- vs. double stranded forms, DNA/RNA hybrids, etc.) and the orientation in which these molecules enter the cell. This may be of great importance in subsequent experiments on the purification of the DNA translocation system.

It would also be of interest to know how large the diameter is of the DNA translocation system. We can envision at least two approaches through which this question can be tackled: (i) attachment of bulky (protein) molecules to the DNA substrate with an increasing diameter and (ii) single-channel recordings of ion fluxes through the DNA translocation channel. The latter approach has recently successfully been used in the identification of a reconstituted mitochondrial peptide channel [93], whereas the former approach has successfully been applied to the *Acinetobacter* DNA-translocase [5]. Results from such studies will also reveal whether or not it is feasible to suppose that DNA is translocated in a form in which it is covered with DNA-binding proteins.

Once most of these questions have been solved, interesting aspects on an even higher level of sophistication can be addressed. This is due to the fact that nucleic acids can, depending on the specific sequence that is present, take very specific conformations. Many of these structures can accurately be predicted and their energy content calculated. This will allow one to study the relation between the energy content of secondary structure elements in DNA substrate and the rate or extent at which they are taken up. Such experiments will deepen our insight into the free energy relations of coupling between osmotic (i.e. directional) processes and the free energy content of macromolecular structure.

6.3. Impact of insight into DNA uptake

In addition to the basic scientific aspects of the nature and stoichiometry of the free energy coupling in DNA transport, a number of recent developments underline the urgency of a better understanding of DNA translocation. The first is the need to understand horizontal gene transfer in the environment. This subject has sparked a strong increase in the interest of the scientific community in understanding biochemical aspects of DNA translocation (see e.g. Refs. [94,95]).

However, it is known that DNA is taken up not only by prokaryotic cells but also by eukaryotes [96], be it that molecular details about the mechanism of this uptake process are virtually absent. In eukaryotes this process may be exploited for new vaccination strategies, particularly when it turns out to be possible to optimize the DNA uptake into eukaryotic cells [97]. Model systems to characterize DNA transport in the Eukarya are not yet available. However, DNA transport in pro- and eukaryotic organisms may show striking parallels [98]. Therefore, studies of the DNA translocase in organisms like *Bacillus* or *Acinetobacter* may become an essential stepping stone in the ultimate goal of understanding DNA uptake into plant and/or animal cells.

For rapid progress in our insight into the mechanism of and energy coupling in DNA uptake, the comparative approach (i.e. learning about DNA transport through parallel studies on DNA translocation in widely different systems) will remain an essential element, just like it has been in the past 20 years. It is important in this respect to compare as many different systems as is possible. Relevant information may come from fully unexpected fields [99].

Abbreviations

Δp, proton motive force
dsDNA, double-stranded DNA
ssDNA, single-stranded DNA

Acknowledgements

R.P. is supported by a grant from the E.C. The authors thank Dr. M.-C. Trombe for her interest in this work.

References

1. Palmen, R., Driessen, A.J.M. and Hellingwerf, K.J. (1994) Biochim. Biophys. Acta **1183**, 417–451.
2. Holland, I.B., Kenny, B. and Blight, M. (1990) Biochimie **72**, 131–141.
3. Van Gijsegem, F., Genin, S. and Boucher, C. (1993) Trends Microbiol. **1**, 175–180.
4. Pugsley, A.P. (1993) Microbiol. Rev. **57**, 50–108.
5. Palmen, R. (1994) Natural transformation in *Acinetobacter calcoaceticus*, PhD-thesis, University of Amsterdam.
6. Konings, W.N. and Michels, P.A.M. (1980) in: Diversity of Bacterial Respiratory Systems, ed C.J. Knowles. pp. 33–87, CRC Press, Boca Raton, FL.
7. Kaback, H.R. (1990) Biochim. Biophys. Acta **1018**, 160–162.
8. Maiden, M.C.J., Davis, E.O., Baldwin, S.A., Moore, D.C.M. and Henderson, P.J.F. (1987) Nature **325**, 641–643.
9. Postma, P.W. and Lengeler, J.W. (1985) Microbiol. Rev. **49**, 232–269.
10. Berger, E.A. and Heppel, L.A. (1974) J. Biol. Chem. **249**, 7747–7755.
11. Bishop, L., Agbayani, Jr.,R., Ambudkar, S.V., Maloney, P.C. and Ames, G.F.-L. (1989) Proc. Natl. Acad. Sci. USA **86**, 6953–6957.
12. Mimmack, M.L., Gallagher, M.P., Pearce, S.R., Hyde, S.C., Booth, I.R. and Higgins, C.F. (1989) Proc. Natl. Acad. Sci. USA **86**, 8257–8261.
13. Dey, S. and Rosen, B.L. (1995) J. Bacteriol. **177**, 385–389.
14. Grinius, L. (1987) in: Energy Transduction and Gene Transfer in Chemotrophic Bacteria. Macromolecules on the Move. Soviet Scientific Reviews Vol. 6. Harwood Academic Publishers GmbH, Switzerland.
15. Lacks, S. and Greenberg, B. (1976) J. Mol. Biol. **101**, 255–275.
16. Reusch, R.N. and Sadoff, H.L. (1988) Proc. Natl. Acad. Sci. USA **85**, 4176–4180.
17. Dubnau, D. (1989) in: Regulation of Procaryotic Development, eds I. Smith, R.A. Slepecky and P. Setlow, Ch. 7, pp. 147–166, American Society for Microbiology, Washington, D.C.
18. Ambudkar, S.V., Sonna, L.A. and Maloney, P.C. (1986) Proc. Natl. Acad. Sci. USA **83**, 280–284.
19. Nieuwenhoven, van, M.H., Hellingwerf, K.J., Venema, G. and Konings, W.N. (1982) J. Bacteriol. **151**, 771–776.
20. Garcia, L.R. and Molineux, I.J. (1995) J. Bacteriol. **177**, 4066–4076.
21. Hyde, S.C., Emsley, P., Hartshorn, M.J., Mimmack, M.M., Gileadi, U., Pearce, S.R., Gallagher, M.P., Gill, D.R., Hubbard, R.E. and Higgins, C.F. (1990) Nature **346**, 362–365.

22. Goffeau, A., Ghislain, M., Navarre, C., Purnelle, B. and Supply, P. (1990) Biochim. Biophys. Acta **1018**, 200–202.

23. Ames, G.F.-L., Mimura, C.S. and Shyamala, V. (1990) FEMS Microbiol. Rev. **75**, 429–446.

24. Lecker, S.H., Driessen, A.J.M. and Wickner, W. (1990) EMBO J. **9**, 2309–2314.

25. Tarahovsky, Y.S., Khusainov, A.A., Deev, A.A. and Kim, Y.V. (1991) FEBS Lett. **289**, 18–22.

26. Boulanger, P. and Lettelier, L. (1992) J. Biol. Chem. **267**, 3168–3172.

27. Kalasauskaite, E.V., Kadašait, D.L., Daugelaviius, R.J., Grinius, L. and Jasaitis, A.A. (1983) Eur. J. Biochem. **130**, 123–130.

28. Labedan, B. and Goldberg, E.B. (1979) Proc. Natl. Acad. Sci. USA **76**, 4669–4674.

29. Labedan, B., Heller, K.B., Jasaitis, A.A., Wilson, T.H. and Goldberg, E.B. (1980) Biochim. Biophys. Res. Comm. **93**, 625–630.

30. Furukawa, H., Kuroiwa, T. and Mizushima, S. (1983) J. Bacteriol. **154**, 938–945.

31. Maltouf, A.F. and Labedan, B. (1983) J. Bacteriol. **153**, 124–133.

32. Maltouf, A.F. and Labedan, B. (1985) Biochem. Biophys. Res. Comm. **130**, 1093–1101.

33. Labedan, B. and Letellier, L. (1984) J. Bioenerg. Biomembr. **16**, 1–9.

34. Duckworth, D.H. and Dunn, G.B. (1976) Arch. Biochem. Biophys. **172**, 319–328.

35. Feucht, A., Schmid, A., Benz, R., Schwarz, H. and Heller, K.J. (1990) J. Biol. Chem. **265**, 18561–18567.

36. Guihard, G., Boulanger, P. and Lettelier, L. (1992) J. Biol. Chem. **267**, 3173–3178.

37. Elliott, J. and Arber, W. (1978) Mol. Gen. Genet. **161**, 1–8.

38. Letellier, L. and Boulanger, P. (1989) Biochimie **71**, 167–174.

39. Willetts, N.S. and Skurray, R. (1987) in *Escherichia coli* and *Salmonella typhimurium*. Cellular and Molecular Biology, Vol. 2, eds F.C. Neidhardt, J.L. Ingraham, K.B. Low, B. Magasanik, M. Schaechter and H.E. Umbarger, pp. 1110–1133. American Society of Microbiology, Washington, D.C.

40. Willetts, N.S. and Wilkins, B. (1984) Microbiol. Rev. **48**, 24–41.

41. Harrington, L.C. and Rogerson, A.C. (1990) J. Bacteriol. **172**, 7263–7264.

42. Sarathy, P.V. and Siddiqi, O. (1973) J. Mol. Biol. **78**, 443–451.

43. Poolman, B., Driessen, A.J.M. and Konings W.N. (1987) Microbiol. Rev. **51**, 498–508.

44. Panicker, M.M. and Minkley, E.G., jr. (1985) J. Bacteriol. **162**, 584–590.

45. Zambryski, P. (1988) Annu. Rev. Genet. **22**, 1–30.

46. Thompson, D.V., Melchers, L.S., Idler, K.B., Schilperoort, R.A. and Hooykaas, P.J.J. (1988) Nucl. Acids Res. **16**, 4621–4636.

47. Stewart, G.J. and Carlson, C.A. (1986) Ann. Rev. Microbiol. **40**, 211–235.

48. Dubnau, D. (1991) Microbiol. Rev. **55**, 395–424.

49. Karudapuram, S., Zhao, X. and Barcak, G.J. (1995) J. Bacteriol. **177**, 3235–3240.

50. Méjean, V. and Claverys, J.-P. (1988) Mol. Gen. Genet. **213**, 444–448.

51. Barany, F., Kahn, M.E. and Smith, H.O. (1983) Proc. Natl. Acad. Sci. USA **80**, 7274–7278.

52. Vagner, V., Claverys, J.-P., Ehrlich, S.D. and Méjean, V. (1990) Mol. Microbiol. **4**, 1785–1788.

53. Morrison, D.A. and Mannarelli, B. (1979) J. Bacteriol. **140**, 655–665.

54. Vos de, W.M. and Venema, G. (1982) Mol. Gen. Genet. **187**, 439–445.

55. Vos de, W.M. and Venema, G. (1983) Mol. Gen. Genet. **190**, 56–64.

56. Vos de, W.M., de Vries, S.C. and Venema, G. (1983) Gene **25**, 301–308.

57. Martin, B., Garcia, P., Castanié, M-P. and Claverys, J-P. (1995) Mol. Microbiol. **15**, 367–379.

58. Pearce, B.J., Naughton, A.M., Campbell, E.A. and Masure, H.R. (1995) J. Bacteriol. **177**, 86–93.

59. Raina, J.L. and Macrina, F.L. (1982) Mol. Gen. Genet. **185**, 21–29.

60. Smith, H.O., Tomb, J-F., Dougherty, B.A., Fleischmann, R.D. and Venter, J.C. (1995) Science **269**, 538–540.

61. Dougherty, T.J., Asmus, A. and Tomasz, A. (1979) Biochem. Biophys. Res. Commun. **86**, 97–104.

62. Goodman, S.D. and Scocca, J.J. (1988) Proc. Natl. Acad. Sci. USA **85**, 6982–6986.

63. Barnhart, B.J. and Herriot, R.M. (1963) Biochim. Biophys. Acta **76**, 25–39.

64. Young, F.E. and Spizizen, J. (1963) J. Bacteriol. **86**, 392–400.

65. Stuy, J.H. and Stern, D. (1964) J. Gen. Microbiol. **35**, 391–400.

66. Lacks, S. (1962) J. Mol. Biol. **5**, 119–131.

67. Rosenthal, A.L. and Lacks, S.A. (1980) J. Mol. Biol. **141**, 133–146.
68. Clavé, C., Martin, F. and Trombe, M.-C. (1989) in: Genetic Transformation and Expression, eds L.O. Butler, C. Harwood and B.E.B. Moseley, pp. 27–40. Intercept, Andover.
69. Grinius, L. (1980) FEBS Lett. **113**, 1–10.
70. Bremer, W., Kooistra, J., Hellingwerf, K.J. and Konings, W.N. (1984). J. Bacteriol. **157**, 868–873.
71. Clavé, C., Morrison, D.A. and Trombe, M.-C. (1987) Bioelectrochem. Bioenerg. **17**, 269–276.
72. Reusch, R.N. and Sadoff, H.L. (1988) Proc. Natl. Acad. Sci. USA **85**, 4176–4180.
73. Reusch, R.N., Huang, R. and Bramble, L.L. (1995) Biophys. J. **69**, 754–766.
74. Reusch, R.N. and Sadoff, H.L. (1983) J. Bacteriol. **156**, 778–788.
75. Dubnau, D. (1989) in: Regulation of Procaryotic Development, eds I. Smith, R.A. Slepecky and P. Setlow, Ch. 7, pp. 147–166, American Society for Microbiology, Washington, D.C.
76. Dubnau, D. (1991) Microb. Rev. **55**, 395–424.
77. Tomb, J.F., Barcak, G.J., Chandler, M.S., Redfield, R.J. and Smith, H.O. (1989) J. Bacteriol. **171**, 3796–3802.
78. Kowalchuck, G.A., Rauch, P. and Hellingwerf, K.J. (1994) Abstract to the 3rd International Symposium on the Biology of *Acinetobacter*, Edinburgh, Scotland, September 14–16.
79. Inamine, G.S. and Dubnau, D. (1995) J. Bacteriol. **177**, 3045–3051.
80. Ravid, S. and Eisenbach, M. (1984) J. Bacteriol. **158**, 222–230.
81. Palmen, R., Vosman, B., Buijsman, P., Breek, C.K.D. and Hellingwerf, K.J. (1993) J. Gen Microbiol. **139**, 295–305.
82. Palmen, R., Buijsman, P. and Hellingwerf, K.J. (1994) Arch. Microbiol. **162**, 344–351.
83. Kok, R.G., Christoffels, V.M., Vosman, B. and Hellingwerf, K.J. (1993) J. Gen. Microbiol. **139**, 2329–2342.
84. Trieu-Cuot, P. and Courvalin, P. (1983) Gene **23**, 331–341.
85. Pifer, M.L. and Smith, H.O. (1985) Proc. Natl. Acad. Sci. USA **82**, 3731–3735.
86. Lacks, S. (1962) J. Mol. Biol. **5**, 119–131.
87. Piechowska, M. and Fox, M.S. (1971) J. Bacteriol. **108**, 680–689.
88. Kahn, M.E. and Smith, H.O. (1984) J. Membr. Biol. **81**, 89–103.
89. Deich, R.A. and Smith, H.O. (1980) Mol. Gen. Genet. **177**, 369–374.
90. Dubnau, D. and Cirigliano, C. (1972) J. Mol. Biol. **64**, 31–46.
91. Singh, R.N (1972) J. Bacteriol. **110**, 266–272.
92. Fox, M.S. and Hotchkiss, R.D. (1957) Nature **179**, 1322–1324.
93. Fèvre, F., Henry, J.-P. and Thieffry, M. (1993) J. Bioenerg. Biomembr. **25**, 55–62.
94. Dreiseikelmann, B. (1994) Microbiol. Rev. **58**, 293–316.
95. Lorenz, M.G. and Wackernagel, W. (1994) Microbiol. Rev. **58**, 563–602.
96. Braciale, T.J. (1993) Trends Microbiol. **1**, 323–324.
97. Fynan, E.F., Webster, R.G., Fuller, D.H., Haynes, J.R., Santoro, J.C. and Robinson, H.L. (1993) Proc. Natl. Acad. Sci. USA **90**, 11478–11482.
98. Muller, H.-M. and Seebach, D. (1993) Angew. Chem. Int. Ed. Engl. **32**, 477–502.
99. Wu, L.J., Lewis, P.J., Allmansberger, R., Hauser, P.M. and Errington, J. (1995) Genes Dev. **9**, 1316–1326.

CHAPTER 32

Translocation of Proteins Across the Bacterial Cytoplasmic Membrane

A.J.M. DRIESSEN

*Department of Microbiology, Groningen Biomolecular Sciences and Biotechnology Institute,
University of Groningen, 9751 NN Haren, The Netherlands*

© *1996 Elsevier Science B.V.*
All rights reserved

Handbook of Biological Physics
Volume 2, edited by W.N. Konings, H.R. Kaback and J.S. Lolkema

Contents

1. General introduction

Bacteria export proteins to the periplasmic space and outer membrane via a mechanism that involves a multisubunit integral membrane protein complex termed *translocase*. Proteins to be secreted are synthesized in the cytosol as cleavable signal sequence bearing precursor proteins. The signal sequence acts as a targeting and recognition signal for the *translocase*. The *translocase* mediates the energy-dependent translocation of the precursor protein across the cytoplasmic membrane. Our understanding of the mechanism of protein translocation has rapidly advanced due to the ability to use powerful genetic and biochemical techniques in *Escherichia coli*. Extensive genetic screening methods have identified the genes, termed *sec*(retion)-genes, that code for the components involved in this process. These are *secA*, *secB*, *secD*, *secE*, *secF*, *secG* and *secY*, each encoding one of the core-components of a complex translocation apparatus that consists of soluble, peripheral and membrane integrated proteins (Fig. 1) [1,2]. Many of these genes are essential for viability, and conditional lethal mutants have been selected that cause pleiotropic protein export defects. The products of these genes have been overproduced and purified to homogeneity. Authentic energy-dependent protein translocation as it occurs *in vivo*, has been reconstituted into proteoliposomes with purified components [3]. Homologues of the Sec-proteins have been found in many other bacteria and even in Archaea, indicating that this pathway is widely distributed. In addition, the translocation of proteins across the thylakoid of higher plants occurs via a mechanism largely identical to the bacterial protein translocation reaction [4]. The translocation of proteins into the endoplasmic reticulum (ER) of mammals and yeast involves components that are homologous to the bacterial Sec-proteins [5,6], suggesting that they also occur by similar mechanisms.

In this chapter, I will discuss the biochemical role of the individual components that catalyze the actual translocation reaction in bacteria. The enzymology of signal peptidases, and alternative and dedicated protein translocation pathways have been discussed elsewhere [7–9].

2. Enzymology of protein translocation

Protein translocation occurs via a cascade of protein–protein interactions, that involves cytosolic, peripheral and integral membrane proteins (Fig. 1). In the early stages, precursor proteins interact as nascent polypeptide chains with SecB. SecB stabilizes the precursor protein in a soluble and translocation-competent state, and targets it to the SecA protein. SecA is an ATPase, and interacts at the membrane-surface with a multisubunit integral membrane protein complex comprising the

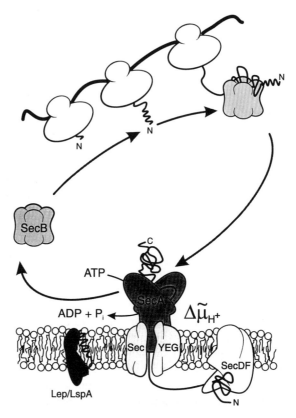

Fig. 1.   Schematic overview of the Sec-components involved in the general protein export pathway in
Escherichia coli.

SecY, SecE and SecG polypeptides. Together these proteins form a complex that is termed the *translocase*. *Translocase* mediates the ATP- and Δp-dependent translocation of the precursor protein across the membrane. The SecD and SecF polypeptides are needed for a late stage of translocation.

During the translocation reaction, the signal peptide is removed from the precursor protein by the membrane-bound signal peptidase SP-I or SP-II. SP-I is a general signal peptidase, whereas SP-II is specific for lipoproteins [7]. Other proteins are involved in the folding of the protein in the periplasmic space, the assembly of outer membrane proteins, and covalent modification of lipoproteins with lipids. In this section, the enzymology of protein translocation will be discussed with an emphasis on the *translocase* subunits.

2.1. SecB

Unlike the co-translational translocation of polypeptides into the mammalian ER [10], bacterial protein translocation is not driven by polypeptide chain elongation at the ribosome *per se* [11]. Precursor proteins can be translocated as completed polypeptide chains [11]. Some, but not all, precursor proteins interact with cytosolic molecular chaperones. SecB is the only chaperone protein with a specific function in protein translocation [12–14]. It forms a stoichiometric complex with precursor proteins [15], and stabilizes the unfolded state of the precursor protein. SecB is a folding catalyst, and retards folding sufficiently to promote a productive interaction of the precursor protein with other Sec-proteins in order to become translocated. *In vitro* studies demonstrate that precursor proteins can spontaneously form a translocation-competent state when diluted into solution from denaturant, while they loose their translocation competence during prolonged incubation in the absence of SecB [12,15–17]. The physicochemical characteristics of this translocation-competent state is obscure. Folding of the precursor of maltose binding protein (pre-MBP) involves a slow *cis-trans* isomerization of a proline residue(s), and SecB has been shown to retard this step in the folding process [18,19]. With the fast-folding precursors of the outer membrane proteins OmpA [20] and PhoE [16], SecB has been shown to prevent the formation of insoluble aggregates. The SecB-bound form of these precursor has native like secondary structure and tertiary fold, but presumably does not have the tight packing of the side-chains that is typical of the native proteins [20]. SecB associates with the mature domain of precursor proteins [18], but the specificity of this binding reaction has remained elusive. Some precursor proteins, such as pre-galactose binding protein (pre-GBP) do not require SecB for export, but bind to SecB with low affinity [21]. In the case of pre-ribose binding protein (pre-RBP), a single point mutations has been shown to be sufficient to convert it into a SecB-dependent precursor protein [19]. It has been suggested that SecB recognizes the polypeptide backbone, and that it binds preferentially at sites exhibiting β-conformation. OmpA and PhoE bear a high amount of β-structure both in their final conformations and in the SecB bound precursor forms [16,20]. The signal sequence domain of precursor proteins is, however, not essential for recognition by SecB [18,22], but facilitates binding to SecB by retarding the folding of the mature domain [18]. Biochemical mapping of the SecB binding site on pre-MBP, suggests that binding occurs in the middle region of the protein [19], in line with observations that SecB binds only larger nascent chains [23]. Mutational analysis of the pre-MBP binding region of SecB indicates the presence of two classes of *secB* mutants [24,25]. The first class of mutants has a reduced ability to form stable complexes with pre-MBP, but causes only mild defects in the rate of pre-MBP export. The pattern of mutations suggests that the primary binding site is hydrophobic and presumably contains β-sheet secondary structure. The hydrophobic character of the binding site has been substantiated with fluorescently tagged model polypeptides [26]. The second class of mutants causes a severe translocation defect, but does not disrupt the SecB:pre-MBP complex formation. These mutants are largely in acidic residues and may be involved in the SecA targeting function of SecB (see Section 2.3.2).

SecB binds a variety of unfolded proteins *in vitro*, and it has been suggested that the specificity of binding is determined by kinetic partitioning that would exclude fast-folding proteins from interacting with SecB [27]. In this model, it is hypothesized that cytosolic proteins differ from precursor proteins in that their rate of folding is higher. A kinetic analysis of the polypeptide-SecB interaction demonstrates that the rate of association is limited only by collision [26], which in solution can be faster that the rate of protein folding. This suggests that SecB is optimized to interact with unfolded polypeptide segments with high rate but low specificity. The specificity of this interaction is dictated by the rate of dissociation, since *in vivo*, SecB binding is highly selective. In the cell, SecB primarily associates with ribosome-bound nascent polypeptide chains of pre-MBP and a number of outer membrane proteins [23]. This important distinction with the *in vitro* data emphasizes the need for specificity in the binding reaction, while the observation that SecB is also needed for co-translation translocation [28], further suggests that binding of SecB to the precursor protein takes place prior to completion and stable folding of the polypeptide chain (or domains thereof). Another aspect that has received little attention is the timing of SecB-nascent chain interaction relative to that of other chaperone proteins such as DnaK and signal recognition particle (SRP) (see Section 2.2). Such studies may more precisely specify the folding state of the SecB-bound nascent chain in the cell.

SecB is only indispensable for protein translocation at high growth rates [19], or when translocation is impaired by mutations causing a decrease in the translocation efficiency [29,30]. This suggests that SecB is essential only when precursor proteins tend to queue for translocation. The specific function of SecB in protein export may relate to its binding affinity for SecA that allows a rapid and specific targeting of bound precursor proteins to the membrane surface [31,32] (see Section 2.3.2). SecB binds SecA in solution [32,33], and during this interaction it presents the signal sequence of the precursor protein to SecA in a proper conformation. This defines the role of SecB as a coupling factor that binds to nascent polypeptide chains when they emerge from the ribosome, stabilizes the completely synthesized precursors in a translocation-competent conformation, and directs them to the SecA subunit of the *translocase*.

Various soluble chaperone proteins are involved in stabilizing newly synthesized precursor proteins. Induction of the heat-shock response substitutes for a defect in SecB function, but this effect appears to be due to an, as yet unidentified, heat-shock protein other than GroEL/ES and DnaK [34]. These latter chaperones cannot replace SecB, and are involved in specific functions only [35–37].

2.2. Signal recognition particle and trigger factor

Signal recognition particle (SRP) performs a central role in protein translocation across the ER membrane of eukaryotes [38,39]. It consists of one RNA (7 SL RNA) and six protein subunits. The N-terminal domain of the 54K subunit of SRP (SRP54) contains a GTP binding site, whereas the C-terminal domain binds signal sequences and SRP RNA. Binding of SRP to the signal sequence as it emerges from the ribosome creates a cytosolic targeting complex containing the nascent polypeptide chain, the translating ribosome, and SRP. This complex is targeted to the SRP receptor at the ER membrane. The SRP receptor is a two subunit protein, with SRβ as a transmembrane GTPase that anchors SRα, a peripheral membrane GTPase, to the membrane [40]. SRP54 is stabilized by signal sequences, and binding of SRP to the SRP receptor activates SRP54 for GTP hydrolysis [41]. SRP then dissociates from both the signal sequence and the ribosome, that bind to SEC61p [42]. SRP is finally released from the SRP receptor and recycled in the cytosol. The identification of a SRP homologue in *E. coli*, several other bacteria and Archaea suggests that this pathway is not only confined to higher organisms [43]. The 4.5 S RNA of *E. coli* (the *ffs* gene product) has sequence homology to the 7 SL RNA of SRP, Fth (the *ffh* gene product), is homologous to SRP54, and FtsY is homologous to SRα [44–46]. While evidence indicates that the 4.5 S RNA of *E. coli* is involved in general protein synthesis [47], it may also function in protein translocation as it associates with Fth into a ribonucleoprotein complex that interacts with the signal sequence of nascent chains [48] and binds tightly to FtsY in a GTP-dependent manner [49]. The Fth protein can substitute for SRP54 in a mixed reconstitution with eukaryotic SRP components with respect to the particle formation and the translation arrest activity [44]. This chimeric particle cannot interact with the eukaryotic SRP receptor and cannot facilitate translocation activity. *In vivo* depletion or overexpression of Ffs causes retardation of pre-β-lactamase (pre-Bla) export but not of other proteins [50,51], Interestingly, pre-Bla shows the slowest posttranslational export kinetics in *E. coli*, and does not require SecB but uses GroEL instead [35,36]. Depletion of Fth [52] or depletion [43,53] and overexpression [53] of FtsY causes growth arrest and export retardation of several precursor proteins. *In vitro*, FtsY depletion has only a minor effect on pre-Bla export [53]. FtsY localizes at or near to the cytoplasmic membrane in *E. coli*, suggesting that it may interact with other membrane components, perhaps an as yet unidentified homologue of the mammalian SRβ. A homology search of the *Haemophilus influenza* chromosome using the yeast SRβ, however, fails to identify such a homologue (A. Driessen, unpublished data).

The precise role of SRP in the bacterial system is less well understood [43,54]. The SRP pathway may functions as an extension to the Sec pathway, in particular, during the early stages of translocation. Cross-linking approaches demonstrate that SRP binds the signal sequence of any precursor protein and the hydrophobic domains of integral membrane proteins [55]. However, once a major portion of the nascent chain has emerged from the ribosome, i.e., 80–100 amino acids, little cross-linking occurs, which suggests that SRP fulfils an early chaperone role, and possibly transfers larger nascent chains to SecB. Alternatively, SRP may guide nascent chains to the *translocase* to permit their co-translational translocation. This would require a translational arrest, docking of the entire SRP-nascent chain-ribosome complex at the membrane translocation sites, and possibly binding of the ribosome to SecY must akin Sec61α [42] (see Section 2.4). Such activities have not yet been demonstrated, but one may argue that co-translation translocation is especially relevant for the insertion of the hydrophobic domains of membrane proteins that are difficult to stabilize in the cytosol in a soluble form [56]. It is far from clear as to whether the Sec-pathway is involved in the membrane insertion of polytopic membrane proteins [57,58]. Some membrane proteins, among which SecY, seem to require SecY and not SecA [59], opening the exciting possibility that

SecA can be bypassed and that the SecYEG protein complex can be directly accessed. The SRP and chaperone pathways may co-exist for general protein export, converge at the SecYEG protein complex and/or differ in timing of nascent chain association.

Recent cross-linking approaches have identified trigger factor as a protein that interacts with short nascent chains [55]. Trigger factor was originally implicated in protein translocation due to its interaction with purified pro-OmpA upon the removal of urea by dilution [60,61]. This interaction was found to stimulate the translocation of pro-OmpA into inner membrane vesicles [62]. However, cells depleted of trigger factor did not exhibit pleiotropic secretion defects [63]. The binding characteristics of trigger factor are reminiscent of the mammalian NAC which also binds to short ribosome bound nascent chains [64]. NAC is considered to confer specificity on the subsequent interaction of SRP with signal sequences, since in the absence of NAC, SRP interacts with non-signal sequence bearing proteins and promotes their targeting to the ER, albeit inefficiently. In the absence of trigger factor, the specificity of the bacterial SRP remains unaltered suggesting that its role differs from that of NAC [55]. Trigger factor has also been implicated as a *cis–trans* proline isomerase [65], and is known to bind in a one-to-one stoichiometry to the GroEL double donut [66].

2.3. SecA

SecA is a central component of the general translocation pathway in *E. coli* and has been found in a wide variety of Gram-positive and negative bacteria, primitive algae and cyanobacteria, and in the chloroplasts of higher plants [4,67]. SecA is a 102 kDa protein which is functional as a dimer [68,69]. It is the only ATPase that is needed and sufficient to drive *in vitro* protein translocation [70]. SecA is involved in a multitude of interactions, and these are discussed in this section. SecA is also involved in the autogenous regulation of its own expression which has been reviewed elsewhere [71].

2.3.1. ATP-binding sites

SecA exhibits three distinct ATPase activity, i.e. (i) a low endogenous ATPase activity, (ii) lipid-stimulated ATPase activity, and (iii), translocation ATPase activity. The latter is the activation of the endogenous ATPase activity when SecA interacts with precursor proteins, the SecYEG protein complex, and anionic phospholipids [3,70,72]. The protein can be photoinactivated by the nucleotide-analogue 8-azido-ATP that binds to two distinct sites on the SecA protein [70]. 8-azido-ATP treated inner membranes are inactive for translocation, and activity can be restored by adding fresh SecA. 8-azido-ATP inactivated SecA exhibits an elevated endogenous ATPase activity, suggesting the presence of a third ATP binding site. Two nucleotide binding sites on SecA have been localized by cross-linking studies and analyzed by site-directed mutagenesis [73–77]. The high affinity binding site (K_D, 0.13 µM) is confined to the amino-terminal domain of the protein (NBS-I), and a low-affinity binding site (K_D, 340 µM) is located in the

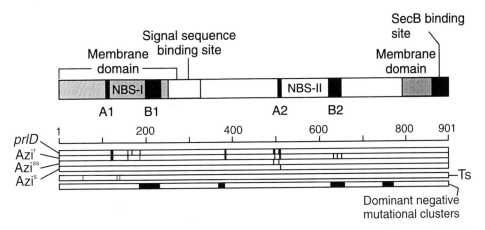

Fig. 2. Domain-structure of the SecA protein. Indicated are the interacting regions that bind the membrane, signal sequence, and SecB. NSB-I and NBS-II are the high and low affinity nucleotide binding sites, respectively. A1 and B1, and A2 and B2 correspond to the regions with similarity to the Walker A- and B-motif as found in NTP-binding proteins. Marked are clusters of dominant-negative mutations and deletions, and suppress signal sequence defects (*PrlD*), azide-resistance (Azi[r]), sensitivity (Azi[s]), super sensitivity (Azi[ss]), and temperature-sensitivity (Ts) mutations.

second half of the protein (NBS-II) (Fig. 2). Residues 102–109 of SecA exhibit the amino acid sequence motif, $G(X_4)GKT$ (the Walker A motif), characteristic of a major class of nucleotide binding sites [78]. Mutations in SecA leading to substitutions at the invariable Lys of NBS-I block the translocation activity of SecA, consistent with this sequence being part of the phosphate binding site [74,75] (Fig. 3). These mutations interfere with the release of SecA from the membrane [75], suggesting a coupling between translocation and the temporal insertion/de-insertion of SecA into the membrane [79] (see Section 2.3.2). ATP hydrolysis at NBS-I is thus essential for the de-insertion or membrane release of SecA. The second Walker motif (B-motif), hXXhhD, that in combination with the A motif completes the Mg^{2+}-phosphate protein interaction, is found in SecA at residues 205–209. This B motif is atypical as it is repeated at amino acid residues 211–215. The first region is homologous to the DEAD box found in RNA binding proteins, such as helicases [80], while the second region aligns with the B-site of the ATP-binding site of ABC proteins. Both aspartate residues of the duplicated motif are needed for the coordination shell of the Mg^{2+}-ion and are indispensable for SecA function [10,74]. Glutamate at position 181 presumably acts as catalytic base in the hydrolytic attack of the γ-phosphate bond of bound ATP. This residue clusters in a well-conserved region in between the A- and B-site, and is thought to couple the binding and hydrolysis of ATP to conformational changes in SecA (Fig. 3). Substitution of this residue for a Gln impairs the function of SecA, and locks it in an altered conformation (J. van der Wolk, unpublished results).

Residues 503–511 contain the Walker A motif of NBS-II. Mutations leading to substitutions at Arg509 block the translocation activity of SecA [74]. This mutant

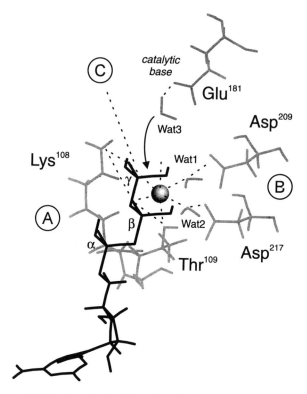

Fig. 3.   Model of the high affinity ATP-binding site (NBS-I) of SecA.

is also impaired in translocation ATPase activity, and binds ADP only in NBS-I. The amino acid region 631–653 has been tentatively assigned as second Walker B motif, but its identity as such has not yet been confirmed. Both NBS-I and NBS-II are indispensable for the translocation activity of SecA. In contrast to their ability to bind ATP, they appear to function in a cooperative manner. Many of the dominant-negative mutants in SecA cluster around the two ATP binding motifs that may cooperatively interact [74] (Fig. 2).

The effects of ATP and ADP binding on the conformational stability of SecA has been studied by differential scanning calorimetry. SecA unfolds thermally as a two domain protein of approximately equal size (T. den Blaauwen, unpublished results). Saturation of NBS-I by ADP increases the stability of the protein, and promotes interaction between both domains. Saturation of NBS-II as well, further pronounces these effects, and dramatically increases the compactness of the protein. This conformation of SecA could correspond with the membrane de-inserted state as these are the conditions at which SecA preferentially associates with the membrane-surface [81]. Saturation of NBS-I with the non-hydrolysable ATP analog AMP-PNP, does not stabilize SecA at all (T. den Blaauwen unpublished results). Under these conditions, SecA tends to penetrate into the membrane [81]. The ATP-bound and membrane inserted state of SecA could, therefore, correspond with a more extended conformation. Rotational diffusion experiments suggest that occupance of NBS-II with ADP stabilizes the SecA subunit interaction, and suggests that NBS-II is at the subunit interface.

Deletion of 66 carboxy-terminal amino acids or replacement of some of the cysteine residues of SecA by serine residues increases the level of ATP hydrolysis and impairs protein translocation [82,83]. This suggests that the carboxy-terminal region of SecA facilitates the coupling of the ATPase activity to cycles of protein translocation [82]. Site-direct tryptophane fluorescence spectroscopy studies demonstrate that the binding of ADP at NBS-I (Fig. 2), changes the conformation of the carboxy-terminus of SecA [T. Den Blaauwen, unpublished results]. These data indicate that communication between NBS-I and a carboxy-terminal domain of SecA is essential for its function.

2.3.2. Interaction with precursor proteins and SecB

SecA is present in the cell in an approximately 10-fold molar excess compared to the other components of the *translocase*. A large portion of the SecA is found in the cytosol and some associates with ribosomes [84]. Variable amounts of SecA are found to be membrane-associated [85] (see Section 2.3.3). The weak interaction between precursor proteins with SecA in solution is promoted by SecB [32,83]. These proteins form isolable binary and ternary complexes in solution [32,63]. The cellular localization of SecB and SecA suggest that they may both act early in the protein transl(oc)ation cascade.

The *in vitro* analysis of the interaction of SecA with precursor proteins has provided detailed insight into the series of events that finally leads to the initiation of translocation. Several *PrlD* (*prl* stands for *protein localization*) mutations, that suppress export defects caused by signal sequence defects, have been found in the *secA* gene [72,86]. The genetic evidence for a SecA interaction with the signal sequence domain of precursor proteins is discussed in Section 2.3.4. Biochemical evidences demonstrates that SecA interacts with precursor proteins though recognition of the positive charge at the amino terminus of the signal peptide [87], and through binding of the mature domain of the precursor protein [71,86–88]. Crosslinking experiments have demonstrated that the signal sequence associates with the 267–340 region of SecA (Fig. 2) [88]. A typical signal sequence consists of a positively charged amino-terminus, a central hydrophobic region and a hydrophilic region containing the signal peptidase cleavage site [89]. Precursor proteins with a larger number of positively charged residues at the amino-terminus bind with higher affinity to SecA, and require a lower concentration of SecA to translocate efficiently *in vitro* [88]. Synthetic signal peptides known to inhibit translocation [90] compete with precursor proteins for binding to SecA [86]. High levels of ATP antagonize this inhibitory effect [86]. Synthetic signal peptides also activate the ATPase activity of SecA when added in conjunction, and thus physically separated, with the mature domain of a precursor protein [72]. Soluble SecA binds precursors with higher affinity when liganded with ADP, and this binding reaction provokes a conformational change in the SecA protein when it occurs at NBS-II [76,91]. On the other hand, soluble SecA releases the bound ADP at NBS-I when it interacts with precursor proteins [91], and at the membrane translocation sites, this event is followed by ATP binding to SecA, i.e., ADP-ATP exchange (see Section 2.3.3).

SecA functions as a receptor for the SecB–precursor protein complex [32], and at the membrane binds SecB with high affinity (K_D 150–250 nM). The affinity is increased by precursor proteins. The site on SecA that interacts with SecB is confined in the carboxy-terminus (Fig. 2). Removal of the carboxy-terminal 66 or 70 amino acids of SecA interferes with the ability of SecA to bind SecB [83], while SecB-dependent precursor protein translocation is severely impaired [82,83]. Moreover, this region is essential for viability of E. coli cells. High affinity SecB binding activity is confined in the proximal 20 amino acids of SecA [P. Fekkes, unpublished results], a region that is highly conserved among the bacterial SecA proteins. The sequence PC[PH]CGSGKK[YF]KxC[CH]G has been proposed as consensus SecB binding motif. This sequence is likely highly flexible as it contains several glycine and proline residues. On the other hand, it may harbour specific structural attributes as any amino acid substitution that prevents the formation of the putative disulphide-bond between C896 and C885 or C887 inactivate SecA [82]. The spacing of these cysteines makes intramolecular disulphide-bond formation energetically favourable, stabilizing a flexible loop bearing several basic amino acids and exposing a strongly electropositive surface with a predicted pI of 9.5–10.8. Interestingly, SecB will also bind with low affinity to short peptides carrying a net positive charge (i.e., arginine and lysine residues) [92]. The positively charged carboxy-terminal region of SecA may interact with an β-structured acidic bristle on SecB that has been identified by mutational analysis [25].

2.3.3. Membrane interaction and topology

About 10–40% of the cellular SecA associates with the membrane [85] at sites of low and high affinity [32,93]. The lipid–SecA interaction plays an important role in translocation, as cells that are depleted of acidic phospholipids are severely blocked in protein translocation [94], but also pleiotropically altered in other membrane-associated processes [95]. In vitro experiments established that acidic phospholipids are required for membrane binding [93] and translocation ATPase activity of SecA [72]. The translocation defect of inner membranes or reconstituted proteoliposomes depleted from acidic phospholipids can be restored by the re-introduction of these lipids [96]. The amount of SecA associated with the cytoplasmic membrane or liposomes also increases with increasing acidic phospholipid content of the membrane [97]. The interaction of SecA with the lipid surface is presumably electrostatic and of poor affinity. Unabated interaction of SecA with anionic phospholipids renders the protein thermolabile [72]. ATP, translocation-competent precursor proteins, and the SecYEG protein complex stabilize the lipid-bound SecA. This is the lipid-ATPase activity of SecA [72] that is suppressed by divalent cations such as Mg^{2+}, and under normal translocating conditions concealed. SecA penetrates efficiently into a lipid monolayer containing acidic phospholipids [81], entering far into the hydrophobic acyl chain region of the membrane [98]. ATP-hydrolysis prevents this insertion and favours a soluble state of the SecA protein, whereas it does not prevent membrane-insertion of a SecA ATPase mutant [75]. In the ADP-bound state, SecA associates with the membrane surface, while in the presence of nonhydrolysable ATP analogues the inserted state dominates. At the lipid surface, SecA unfolds [99] and readily aggregates into an inactive state (A. Driessen, unpublished data).

With liposomes, SecA will only bind with low affinity to the membrane [93]. The presence of SecYEG allows high affinity binding of SecA (K_D 40 nM) [27,98]. Overexpression of the SecYEG protein results in a dramatic elevation of the amount of membrane-associated SecA (C. Van der Does and E. Manting, unpublished data). SecA bound to these high affinity sites exposes domains to lipid [99]. Part of the SecA molecules is intimately associated with the cytoplasmic membrane [85], and can only be partly removed by treatment with urea or carbonate [100,101]. Proteolysis studies with right-side-out (RSO) membrane vesicles of *E. coli* suggest that SecA penetrates the cytoplasmic membrane [102]. Under those conditions, it exposes a carboxy-terminal domain consisting of at least part of the amino acid sequence 850 up to 901, to the periplasmic face [103]. Studies with a SecYEG overproducing strain demonstrate that it is the *translocase*-bound form of SecA that exposes its carboxy-terminus to the periplasm (C. Van der Does and E. Manting, unpublished data). This carboxy-terminus has also been implicated in lipid binding [83], and in SecB binding (see Section 2.3.2). In *E. coli* SecA, the SecB binding domain is preceded by a hydrophobic sequence. It is possible that the positively charged sequence not only associates with SecB, but also with the negatively charged phospholipids. Subsequently insertion of the hydrophobic region in the membrane at the initiation of the translocation would then dislocate the bound SecB, and recycle it to the cytoplasm (see Section 4.1) During *in vitro* precursor protein translocation in the presence of ATP, a 30-kDa SecA fragment (actual size based on its mobility on SDS-PAGE is 34 kDa) was found to be resistant to protease K digestion of inside-out (ISO) membrane vesicles [79]. This fragment interacts is bound at SecYEG, and albeit in lower yields, is also found under conditions that no translocation takes place. Monoclonal antibody mapping defined its origin as amino-terminal [103]. In analogy, an amino-terminal SecA peptide (1–239), that is able to complement the *secA51(ts)* mutant, which has an altered amino acid residue 43 [104], has been found to be entirely associated with the membrane [85]. Amino-terminal truncates (1–275, [105] and 1–234, [106]) of the *Bacillus subtilis* SecA also complement the *secA51(ts)* mutant, and these fragments become membrane-bound due to the interaction with SecA51. In the absence of translocation, but in the presence of nucleotides, a trypsin-protected amino-terminal 23 kDa fragment can be formed [103]. This implies that the amino-terminus of SecA, that harbours NBS-I, is an independent domain that changes its conformation upon nucleotide binding and during translocation.

Cross-linking studies have shown that precursor protein segments that are initially associated with SecA move into contact with SecY as they traverse the membrane [107]. The intermediates, however, remain bound to SecA until they have emerged on the periplasmic face of the membrane [107]. About 20–30 amino acid residues of the precursor protein are driven across the membrane when a non-hydrolysable ATP analogues binds to the SecA subunit of the *translocase* bearing a defined translocation intermediate [108,109]. The current hypothesis is

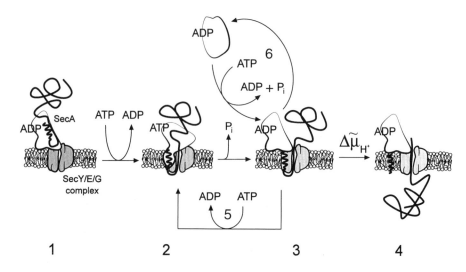

Fig. 4. Model for the initial and intermediate stages of protein translocation. SecA with bound precursor protein at the SecYEG complex (Step 1), is activated for ADP-ATP exchange. Binding of ATP facilitates the insertion of SecA into the membrane or SecYEG complex (step 2), and allows membrane insertion of an amino-terminal loop of the signal sequence and mature domain. Subsequent hydrolysis of ATP by SecA drive the release of the bound precursor protein and promotes the exclusion of SecA from the membrane bound state (step 3). In the absence of SecA association, translocation is further driven by the Δp (step 4), or SecA rebinds the partial translocated polypeptide chain, and a repeated cycle of ATP binding and hydrolysis allows the translocation of a small polypeptide domain of the intermediate through co-insertion (step 5). SecA with bound ADP is surface-localized and can be replaced by cytosolic SecA in a reaction that requires the hydrolysis of another ATP molecule (step 6).

that this is the amount of precursor protein that is carried into and partially across the membrane by the ATP-dependent insertion of SecA into the membrane (Fig. 4). ATP hydrolysis causes the dissociates the SecA–precursor protein complex bound at the SecYEG protein [108], and allows a rapid displacement of the SecA from these high affinity binding sites by cytosolic SecA [75]. This hydrolysis seems to occur at NBS-I, as a corresponding mutation in this region that prevents the hydrolysis of ATP, not only allows a tight binding of the precursor protein, but also prevents the release of SecA from the membrane [75]. Addition of apyrase to deplete the available ATP in an ongoing *in vitro* translocation reaction immediately arrests this reaction, presumably by inhibiting the de-insertion of SecA [79]. Translocation can be resumed after isolation of the vesicles and addition of ATP. A mutation of NBS-I that is still able to bind ATP, also insert into the membrane, and allows limited translocation of a polypeptide domain [75]. This mutant is deficient in membrane de-insertion, indicating that NBS-I is needed for this reaction. In summary, these data suggest that SecA undergoes nucleotide-modulated cycles of membrane-insertion and exclusion [69,75,81], and by co-insertion of the bound precursor protein drive the translocation of small polypeptide domains across the membrane as outlined in Fig. 4.

2.3.4. Signal sequence proofreading

Sodium azide is an inhibitor of the Sec-dependent protein translocation pathway in *E. coli* [110] and *B. subtilis* [111]. Inhibition of protein translocation by sodium azide is due to a block of the translocation ATPase activity of the SecA protein. Azide resistant mutants show an elevated translocation ATPase activity. The majority of the azide resistant mutants are found to be a replacement of amino acid 134 in *E. coli* [112,113], and amino acid 128 in *B. subtilis* [111]. These residues are found in the region between the Walker A and B site of NBS-I (Fig. 2). An accompanying effect of these *azi* mutations is that they, like the *prlD* mutations, enable the translocation of proteins with a defective signal sequence [112]. The *prl* class of mutants are all isolated as suppressors of signal sequence mutations, and have been found in the *secA* (*prlD*) [112–114], the *secY* (*prlA*) [19,30,115], and in the *secE* (*prlG*) [30] genes (see Section 2.4). It has been proposed that *prl* suppressors function not by restoring the recognition of altered signal sequences but rather by preventing the rejection of defective precursor proteins from the export pathway [116]. According to this hypothesis, SecA, SecY, and SecG would have a proofreading activity (see also Section 2.4). Another class of *prlD* mutants in SecA are the azide super sensitive mutants (Az^{ss}), which are clustered in the vicinity of the Walker A site of NBS-II (Fig. 2) [112]. The signal sequence defect-suppressing phenotype and the Az^{ss} phenotype caused by these *prlD* mutants are dominant in diploid analysis [112]. A possible explanation could be that in the presence of azide these mutants are unable to hydrolyse ATP once they have inserted in the membrane. Thus, they would block access to translocation channels for the *in trans* wild type SecA. Az^r is also conferred by suppressor mutations within, or in the vicinity of either subsite of NBS-II, rather than in the intervening sequence. The suppressor phenotype of these mutants is either weak or unknown, indicating that NBS-II might not be essential for the signal sequence proofreading.

2.4. SecYEG complex

Most of the genes that have been identified as components required for protein export code for integral membrane proteins [1]. The membrane-integrated domain of the *translocase* consists of three polypeptides, i.e., SecY, SecE, and SecG [3], and there function is discussed in this section.

2.4.1. Structure

SecY and SecE are essential for the viability of *E. coli*, and are needed for *in vitro* and *in vivo* protein translocation [1,117]. SecG is not essential as disruption of its gene has been shown to only confer cold-sensitivity to protein export [118]. SecY is a polytopic membrane protein and spans the membrane ten times [117]. SecE contains three transmembrane segments (TMS) [119], but only the carboxy-terminal TMS plus attached sequences suffice for activity [115]. In most other bacteria, SecE comprises only this carboxy-terminal TMS [120,121]. SecG is a glycine-rich membrane protein and comprises two TMS. SecY and SecE were genetically identified as *translocase* subunits, while SecG, previously termed 'band 1' or 'p12'

[122], was identified biochemical. It co-purifies and co-immunoprecipitates with SecY and SecE as a heterotrimeric membrane protein complex [3]. This complex is labile at ambient temperature [123]. In addition, SecG has been purified from a trichloroacetic acid soluble fraction of detergent-solubilized membranes [124], and shown to dramatically enhance the translocation activity of SecY and SecE containing proteoliposomes [125–127]. The failure to identify the *secG* gene through the extensive genetic screening of mutants may be due to the fact that SecG is not essential for viability. Suppressor mutations have been mapped in the *secG* gene that allow the translocation of a mammalian precursor protein in *E. coli* [128]. SecG function can be compensated for by an increase in the level of acidic phospholipids, which are essential for the association of SecA with the membrane [129]. Therefore, is it thought that SecG might facilitate the insertion of SecA at the *translocase*, especially at low temperatures when the membrane fluidity is low [129]. The SecYEG complex has been purified to homogeneity from a wild-type *E. coli* strain by virtue of its ability to support SecA translocation ATPase [3,130]. In reconstituted form, it mediates multiple turnovers of SecA-dependent precursor–protein translocation and is as active for protein translocation as inner membrane vesicles [131]. The heterotrimeric SecYEG complex has been functionally reconstituted from the individually isolated subunits [132], and from a SecYEG overproducing strain [133, C. van der Does, unpublished results]

Biochemical evidence indicates that SecE stabilizes SecY [134,135], and by complex formation protects SecY against proteolysis by FtsH, a membrane-bound ATP-dependent protease [136]. In wild-type cells the newly synthesized SecY immediately associates with SecE to form a complex that does not dissociate during translocation [135,137]. This contrasts genetic data that argues that SecE and SecY are dissociatable subunits [138]. SecG appears to be a less stable subunit of the *translocase*. Several pairs of *prlA* (SecY) and *prlG* (SecE) alleles that exhibit synthetic defects based on complementation screening were found [139]. These pairs suggest an interaction between the first amino-terminal periplasmic loop of SecY and the carboxy-terminal loop of SecE, and between TMS3 of SecE and TMS7 and TMS10 of SecY [139]. None of these *prl* mutations belong to the class of strong signal peptide defects suppressors, indication that the SecY–SecE interacting amino acid residues might not necessarily correspond with the residues that are involved in the proofreading activity. Other mutations in SecY that affect the association between SecY and SecE are clustered in the second cytosolic loop (C4 domain) of SecY [140,141]. The vast majority of known strong signal sequence defect suppressing *prl* mutations map to *prlA* (SecY), whereas the weaker suppressing mutation map to *prlG* (SecE) and to *prlD* (SecA) [30,112]. Most *prl* mutations cluster in the domains of SecY and SecE described above [30], and in particular in TMS7 of SecY, suggesting that this helix is intimately involved in the proofreading process.

The presence of *prl* mutations in the *secY* (*prlA*), and *secE* (*prlG*) genes suggests that SecY and SecE interact with the signal sequence domain. Most conditional lethal mutations in *secY* and *secE* are cold-sensitive. Temperature-sensitive mutations are known only for *secA* and *secY*. Protein translocation may include some intrinsic cold-sensitive steps, and it has been hypothesized that a lowering of the activity is a step that follows the cold-sensitive step invariably leads to the cold-sensitive export phenotypes [142]. A potential cold-sensitive step is the insertion of SecA into the membrane to expose the signal sequence domain to the SecYEG complex.

Several factors have been identified that stabilize overexpressed SecY and suppress the dominant negative phenotype of *secY*[d]*1*, a gene that codes for an inactive SecY protein with an internal deletion [143,144]. When expressed from a multicopy plasmid, SecY[d]1 sequesters SecE and competes with wild-type SecY for the formation of functional translocation complexes [144]. One of these suppressors is Ydr, a small hydrophilic protein that is able to partially stabilize the wild-type SecY. The function of Ydr is unclear as the disruption of its gene has no effect on protein translocation. Ydr does not functionally replace SecE, but may only mimic SecE by stabilizing SecY. The amino-terminus of Ydr bears homology to the conserved cytosolic region of SecE [144]

It has been suggested that SecY is not required for translocation [145]. SecY is not soluble in the detergent cholate, and this property has been used to obtain reconstituted proteoliposomes depleted from SecY. These proteoliposomes contain an assembly of unspecified membrane proteins and translocate a carboxy-truncated form of pre-MBP in a SecB- and SecA-dependent manner [103,145]. This contrasts the overwhelming genetic and biochemical data that SecY is required for translocation (see reviews [1,2]) and evolutionary conserved (see Section 2.4.2). Studies with crude and entirely purified reconstituted systems [3,121,125–127,133,146, 147] firmly demonstrate that SecY is essential.

The integral membrane domain of the *translocase* functions as a binding site for SecA. SecA binds at or near to SecY [32,148,149]. The number of high affinity SecA binding site increases with the amount of SecYEG complex [133, C. van der Does, unpublished data]. Overproduction of SecY and SecE alone does not result in increased SecA binding [150], whereas co-reconstitution of purified SecG with SecY and SecE also does not result in an increase in the binding of SecA [132]. The latter is unexpected, but since a kinetic analysis of SecA binding was not carried out, high affinity binding could have been obscured by the high level of low affinity binding of SecA to the lipid surface.

2.4.2. Evolutionary conservation

The heterotrimeric organization of the integral domain of the *translocase* is conserved in Bacteria, Archaea and Eukaryotes. Both the mammalian [5] and yeast [151] Sec61p complex of the ER have been purified and functionally reconstituted into proteoliposomes. The mammalian Sec61p complex mediates co-translational translocation of precursor proteins into the ER, and consists of Sec61α and γ which are homologous to SecY and SecE [6,42], and Sec61β which may be functionally homologous to SecG (see Section 2.4.1). In yeast, two *translocase* complexes have been identified. The Ssh1p complex mediates the co-translational translocation and consists of Ssh1 (a Sec61α homologue), Sss1 (a Sec61γ homologue), Sbh2 (a Sec61β homologue), and Sec62. The Sec61p complex comes in two flavours, i.e., one that catalyzes co-translational translocation of precursor proteins, and consists of the Sec61α, Sbh1, and Sss1, and one that catalyzes post-translational translocation that in addition to the previously mentioned proteins consists of Sec62 and Sec63. The Ssh1p complex is not essential in contrast to the Sec61p complex. Strikingly, neither Sbh1 nor Sbh2 were found to be essential. Cells that carry a double deletion of the *sbh1* and *sbh2* genes grow at 30°C but not at 38°C. At the latter temperature the mutant shows a severe, but not complete, translocation defect. In this respect, Sbh1 and Sbh2 show some resemblance to SecG, which is also not essential for protein translocation. Sss1 is a multi-copy suppressor of a Sec61α mutant, and has been found to stabilize Sec61α [152], analogous to the stabilization of the bacterial SecY by SecE (see Section 2.4.1). Interestingly, SecY of Archaea is more similar to Sec61α, than the bacterial SecY, as predicted from their position on the phylogenetic tree. The conserved nature of the molecular structure of the *translocase* complex strongly suggests that the basic mechanism of protein translocation in bacteria and the ER of eukaryotes is similar.

2.5. SecD and SecF

SecD and SecF are both membrane proteins that were identified by cold-sensitive mutations that cause the accumulation of precursors of exported proteins [153]. None of the *prl* mutations mapped in the *secD* and *SecF* genes, and it has been suggested that they are not involved in signal sequence recognition [1,154]. Both proteins have large periplasmic domains positioned between the first two of six TMS [154], and are not required for the biochemically reconstituted translocation reaction [3]. *secD* and *secF* null mutants are viable only at high temperatures and show a severe, cold-sensitive export defect [156]. Overexpression of SecD and SecF improves the export of proteins with defective signal sequences [156]. Genetic evidence indicates that SecD and SecF are dissociable subunits of the *translocase* at a late stage of translocation [154]. They are, however, not associated with the isolable SecYEG complex. The cold-sensitivity of their mutant phenotypes, may suggest that they promote insertion of SecA into the membrane which is thought to be a critical cold-sensitive step (see Section 2.4.1). On the other hand, protein export in spheroplasts is inhibited when incubated with an antibody against SecD [157], and the data suggests that SecD is needed or improves the efficiency of the release of the proteins in the periplasm. SecD and SecF are not required for clearing of the precursor protein from the translocation site *per se*, since the reconstituted *translocase* mediates multiple rounds of translocation in the absence [131]. Depletion of SecD and SecF severely affects the ability of cells to maintain a Δp [158]. This may be the reason for the translocation defect in *secD* and *secF* mutants. *In vitro* studies demonstrate that SecD and SecF are not needed for ATP-dependent translocation, but that they kinetically affect the Δp-dependent portion of the translocation reaction [158]. SecD and SecF are, however, not essential for coupling of the Δp to translocation as shown with the reconstituted system [3,159]. Why are *secD* and *secF* null mutants leaky for protons? Since translocation is slow in these mutant strains, leakiness could possibly arise from *translocase* units stuffed with translocation intermediates (see Section 2.4). These units would present a major energy-sink in cells causing the dissipation of Δp. By rapidly clearing the *translocase*, SecD and SecF may prevent the occurrence of such undesired ion-fluxes. In many aspects, the *in vivo* dissipation of Δp resembles the effect of the depletion of SecD and SecF. In the presence of a uncoupler, a membrane-bound processed intermediate of pre-MBP accumulates at the periplasmic side of the inner membrane [160]. Δp may thus be involved in the release of MBP, and possibly also other precursor proteins. Precursor proteins are affected in different ways in their translocation in SecD and SecF mutant strains. This effect may be related to their Δp-dependency of translocation.

Several other factors have been found that affect folding and/or release of exported proteins on the external face of the membrane. Signal peptidase I and II are needed to remove the signal peptide domain of precursor proteins and pre-lipoproteins [161]. The periplasm of *E. coli* contains at least two functional homologues of the eukaryotic protein disulphide isomerases (PDI). DsbA [162,163] and DsbC [164,165] are periplasmic proteins that are believed to be involved in the oxidation of cysteine residues and disulphide-bond rearrangements in exported proteins. DsbA, whose structure has been solved [166], is a periplasmic protein that acts in concert with the membrane-bound DsbB [167]. DsbB may specifically re-oxidizes DsbA via the electron transfer chain, thus enabling it to recycle [164,167]. In the absence of DsbA, precursor proteins remain bound at the periplasmic membrane surface in a reduced and unfolded state.

Unique as yet to Gram-positive bacteria are PrsA and PrtM, both small, membrane-anchored, lipoproteins that are involved in the maturation of exported proteins in *B. subtilis* and *Lactococcus lactis*, respectively [168–170]. These proteins presumably act as folding factors to catalyze the proline *cis–trans* isomerization when the protein emerges from the *translocase* at the outer surface of the membrane.

3. Nature of the translocation path

A central question in the study of protein translocation is the nature of the translocation pathway. Does the protein translocate through the lipid bilayer, through a proteinaceous channel, or at the interface of the lipid and protein (Fig. 5)? The answer to this question may depend on the stage of translocation.

3.1. Along a Greasy Path

There are several lines of evidence that indicate that signal peptide–lipid interactions are important for the initial stages of translocation. The positively charged amino-terminus of the signal sequence of precursor proteins may interact electrostatically with anionic phospholipids in the membrane, possibly favouring the insertion of the hydrophobic core into the bilayer via the formation of a loop with the mature amino-terminus [89,171]. A tentative correlation exists between the ability of synthetic signal peptides to insert into model membranes and to act as

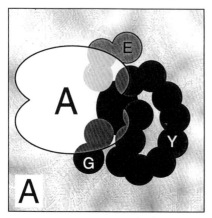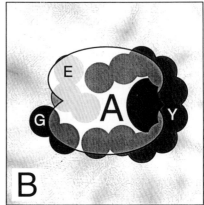

Fig. 5.   Top-view of a tentative arrangement of the SecA, SecY, SecE, and SecG subunits of the *translocase*. A. Interface model, with a U-shaped organization of the integral subunits, forming a pore that is laterally open to the lipid bilayer and that is gated by SecA, B. Pore model, with a ring-form organization of the SecYEG protein gated by SecA. The black ellipse represents the translocation pathway.

efficient targeting signals in protein translocation [172]. The total hydrophobicity of the core region is an important determinant for signal sequence function [173]. For instance, the requirement for a positive charge at the amino-terminus can be compensated for by a longer central hydrophobic stretch [174]. Under those conditions, translocation becomes less dependent on anionic phospholipids [175], suggesting that precursor proteins with a 'classic' signal sequence indeed encounters anionic phospholipids at an early stage of the translocation reaction. Anionic lipids are also required for the initial Sec-independent insertion of M13 procoat [176], possibly reflecting a primordial route of translocation. Recent evidence with the mammalian Sec61p complex indicate that transmembrane segments, and to a less extent the hydrophobic core of the signal peptide, of a nascent chains can be cross-linked to phospholipids while in transit [177]. It has been suggested that the Sec61p complex is laterally open to the lipid bilayer, and that the membrane-inserted signal sequence enters the translocation channel laterally from the membrane. An analogous situation may exist with the SecYEG complex (see Section 3.2). It is not known if lipids are functionally involved in later stages of translocation.

3.2. Or a proteinaceous orifice?

The SecYEG complex may act as a pore allowing precursor proteins to traverse the membrane, either through its centre or along its surface (Fig. 5). Translocation intermediates can be specifically photocross-linked to the SecA and SecY proteins [107]. Neither SecE nor SecG are identified with this technique, while phospholipids are nearly completely protected from cross-linking. Cross-linking experiments with early intermediates, i.e., the insertion of the signal sequence domain into the membrane, have not been carried out with the bacterial system (see Section 3.1). A systematic probing of the environment of a translocating secretory protein during translocation through the ER membrane, demonstrates that Sec61α surrounds the translocating polypeptide chain [178]. Moreover, by the use of fluorescently tagged nascent chains, the polarity of the environment of the translocating polypeptide segment has been analyzed, and these data demonstrate that the protein pass the membrane through an aqueous environment [120,179].

Electrophysiological measurements with microsomes derived from the mammalian ER show the appearance of large ion-channels when the polysomes are dissociated from the nascent chains and the membrane surface [180]. The permeability barrier between the cytoplasm and the ER lumen is thought to be maintained by a tight junction between the ribosome and the SEC61p complex [120,179]. Similar experiments have been performed with everted inner membrane vesicles or spheroplasts of *E. coli* [181]. Addition of a synthetic signal peptide provokes a large increase in the conductivity of the membrane. These observations have been taken to suggest that precursor proteins are translocated through a proteinaceous aqueous pore that opens upon the binding of the signal sequence domain. These experiments, however, suffer from the fact that it is not clear if the signal peptide-induced pore truly reflects a protein conducting pore comprised of Sec-proteins. Ambiguously, the phenomenon depends on a non-physiological polarity of the transmembrane potential ($\Delta\psi$), i.e., negative on the *trans*-side instead of positive, which makes little sense [181]. *E. coli* inner membrane vesicles show an enhanced halide-permeability when conditions are employed that result in the accumulation of translocation intermediates [182]. Similar observations have been made for the proton-permeability of membrane vesicles containing elevated levels of SecY and SecE protein [183]. An increased halide- and water-permeability of the membrane is also observed with liposomes reconstituted with the SecYEG complex (A. Driessen, unpublished results). This phenomenon is suppressed by SecA, suggesting that SecYEG is a SecA-gated channel. The ion-permeability increases dramatically when translocation-intermediate are trapped in the *translocase*. Both in the bacterial and eukaryotic systems, it appears that hydrophilic polypeptide stretches of precursor proteins are translocated across the membrane along a proteinaceous surface [120].

The SecYEG protein is only poorly discriminative, as it allows the translocation of a precursor protein that bears an internal non-polypeptide stretch, suggesting that peptide-backbone recognition is not required throughout the translocation reaction [184]. The *translocase* must, however, be able to recognize stop-transfer sequences (or TMS) in translocating membrane proteins in order to release these proteins into the cytoplasmic membrane. Recognition of such sequences could be confined to SecA and be part of the proofreading. An attractive model is that SecA inserts into the membrane and translocates the polypeptide segments along the SecYEG surface. When a stop-transfer sequence or transmembrane segment is encountered, SecA may be forced to de-insert and releases the polypeptide chain that will spontaneously inserts into the membrane according to the 'positive-inside rule' [185]. When the TMS is flanked by a positively charged region, mimicking a signal peptide domain, SecA may re-engage the protein translocation reaction by binding of the polypeptide segment and the cycle will repeat. Translocation of long polar domains requires the Sec-system, whereas short domains do not [186–189]. This model does not require the lateral opening of a putative SecYEG pore (Fig. 5), but only requires that SecA slides away from the translocation channel mouth, that would be open from the lateral side as proposed the Sec61p complex [177].

5. Energetics of protein translocation

The translocation of precursor proteins depends on two different sources of free energy: ATP hydrolysis and the Δp [159,160,190–196]. This dual energy requirement is unique for the bacterial system. ATP (or GTP) is required for most protein translocation processes that occur across biological membranes, and generally drives the release of precursor proteins from their association with molecular chaperones. Protein import into mitochondria requires $\Delta\psi$ only for the initial electrophoretic movement of the signal sequence across the inner membrane [196, 197], whereas ΔpH is required for matrix-directed membrane integration of subunit 9 of the F_o-ATPase [198]. Some proteins imported into the thylakoid of chloroplasts specifically utilize ΔpH for translocation and membrane insertion [199–201], while import of proteins into the chloroplast or the ER appears to be totally independent of a Δp [38]. Detailed biochemical studies with the bacterial system have led to the definition of a catalytic cycle of the *translocase*. This model (Fig. 4) proposes that translocation proceeds in a stepwise manner with discreet translocation intermediates [108,202,203], and is discussed in the following paragraphs. The targeting cascade has been discussed in sections 2.1, 2.2, 2.3.2 and 2.3.3.

4.1. ATP

Translocation commences with the binding of the SecA loaded with precursor protein at SecYEG complex [32]. Several scenarios can be envisioned based on the possible existence of a lipid surface-bound, freely diffusible intermediate composed of a ternary complex of SecA, SecB, and precursor protein [27]. So far, there is no evidence that a lipid-bound ternary complex is a true intermediate in the targeting cascade, although SecA is known to bind to low affinity phospholipid sites at the membrane surface [32]. The ternary complex may laterally diffuse to the SecYEG complex, and displace the SecA that already is present at this site. This process requires the hydrolysis of ATP at NBS-I [75,79]. At the SecYEG complex, SecA is activated for ADP:ATP exchange and the binding of ATP to SecA provokes a conformational change that allows the protein to insert into the membrane [79,81].This elicits the release of the signal sequence domain of the bound precursor protein [108, J. van der Wolk, unpublished results]. The signal sequence may interact with anionic phospholipids at that stage [175], and in the case of pre-PhoE, it has been shown that conformational flexibility of the signal sequence region is required to allow insertion [204]. It may adopt a loop-like structure with the mature amino-terminal region. Formation of this loop may be facilitated by the transmembrane electrical potential, $\Delta\psi$, inside negative (see Section 3.2). A critical feature of this stage of the translocation reaction is that the ATP-dependent insertion of SecA into the membrane may serve to properly expose the signal sequence domain to binding sites on SecY and SecE to allow proofreading (see also Section 2.3.4). Signal sequences (10–15 amino acid residues) are too short to span the entire membrane in α-helical conformation [89], and penetration of SecA into the lipid bilayer may shorten the distance for the amino-terminus to get across, and facilitate the stable interaction of the signal sequence with SecY and SecE. In case the signal sequence is rejected, no functional *translocase* is formed, and instead ATP hydrolysis forces SecA to release the defective precursor protein and translocation is aborted [112]. When the signal sequence is recognized, a functional *translocase* is formed, and processing of the signal domain by signal peptidase can take place [108]. Since the active site of signal peptidase is exposed to the periplasmic face of the inner membrane, about 40–45 residues must have been translocated. In an *in vitro* system, this state can be detected as the most early intermediate in the translocation of pro-OmpA as a 5/7-kDa protected proteolytic fragment termed I_{5-7} (J. van de Wolk, unpublished results). Interaction with SecYEG activates SecA for ATP hydrolysis, and this results in the release of the precursor protein, and concomitantly the de-insertion of SecA from the membrane [75,79]. Formation of a functional *translocase* also implies that the mature region of the precursor protein anchors at the SecYEG complex, in order to prevent it from diffusing away. This succession of events ensure that the precursor protein is released by SecA at the translocation site, and precludes an unproductive reactions with the lipid bilayer [108]. *In vitro*, SecB is already redundant when the ternary complex has docked at the membrane translocation site [27]. As discussed in section *II-A*, SecB may be released from the membrane when SecA binds ATP and inserts into the membrane.

Once translocation has been initiated at the expense of ATP, further ATP-dependent translocation of pro-OmpA proceeds stepwise through a series of defined transmembrane intermediates when translocation is limited by the amount of available ATP. The next stable intermediate in the cycle is translocation progress by another 5 kDa to yield a protected fragment of 10 kDa, followed by intermediates that have progressed in steps of 4–5 kDa, until pro-OmpA has translocated completely (J. van de Wolk, unpublished results). When a translocation intermediate is depleted from SecA, freshly added SecA can rebind the intermediate, insert into the membrane and through co-insertion drive the forward membrane translocation of a 2 to 2.5-kDa polypeptide stretch, i.e., one *quantum* [108]. Limited translocation can also be effected by ATP binding to SecA, and it appears that this is the amount, or *quantum*, of polypeptide chain that is translocated by a cycle of SecA membrane-insertion and de-insertion. Since stable intermediates occur every two cycles, it appears that the *translocase* freezes at discrete intervals. It is unclear why the system stalls at these steps, but the existence of SecA as a dimer [87] suggests that the two *quanta* of translocation progress requires the alternate, and synchronized, movement of the individual SecA subunits before the system comes to a stop.

4.2. Protonmotive-force

Multiple cycles of SecA-dependent translocation steps ultimately lead to the complete translocation of the polypeptide chain [108]. This process is very inefficient and slow and requires the hydrolysis of numerous ATP molecules [159]. Truncated pro-OmpA derivatives require less ATP for translocation [205]. *In vivo*, it seems more likely that the Δp drives further translocation once SecA has released the precursor protein to the SecYEG complex. Δp permits rapid and efficient translocation of intermediates provided that they are not associated with SecA [108,159]. Δp-driven translocation is completely blocked when the precursor protein associates tightly with SecA, i.e., with non-hydrolysable ATP analogues [108] or when SecA is unable to hydrolyse ATP through a mutation in NBS-I [75]. Excess SecA suppresses the Δp-dependency of precursor protein translocation [101,108,195] as it favours rapid rebinding of released precursor proteins. The observation that translocation is more dependent on Δp when SecA is limiting is consistent with this explanation [206]. *In vitro*, a collapse of the Δp results in the accumulation of early intermediates of pro-OmpA, suggesting that the latter stages in translocation progress are driven by Δp (J. van de Wolk, unpublished results). It is important to emphasize that a detailed analysis of the intermediate translocation steps has only been carried out for pro-OmpA. For pre-PhoE, it appears that the initial membrane insertion is critically dependent on Δp [122]. Mutations introduced into the mature amino-terminal region of pro-OmpA [207,208] can also render the processing of pro-OmpA Δp-dependent. In both cases, it is not known if translocation is effected by Δp, or only by Δψ (see Section 4.4).

Intermediate stages of translocation are readily reversible and SecA-mediated ATP hydrolysis is not strictly coupled to net precursor protein movement along the translocation path [108]. Futile cycles of ATP hydrolysis occur when translocation is prevented by a stable tertiary structure in the precursor protein [108, 209], or when a Δp of reversed polarity is imposed [159]. In the absence of SecA association, hysteresis movement of translocation intermediates takes place [108] that is prevented by Δp [159]. These futile cycles presumably are the cause the poor coupling between translocation and ATP hydrolysis *in vitro*. Phenomenologically, Δp increases the coupling ratio between ATP hydrolysis and precursor protein translocation [159]. Mechanistically, these processes are completely distinct [108]. *In vivo*, the main catalytic role of SecA may be to initiate translocation in order to allow further Δp-driven translocation [108,210]. Release of translocated proteins into the periplasmic space strictly requires the Δp [160]. Alternatively, SecA may be needed throughout the translocation reaction, and Δp would function merely to prevent backward translocation or other futile cycles that result from stable folded tertiary structures in the polypeptide chain [209,211]. The observation that Δp drives an extremely fast, and efficient translocation of the precursor protein once SecA has released the polypeptide chain [108] may not reflect the process as it occurs *in vivo* as it requires the physical removal of SecA from these sites, which is unlikely to occur *in vivo*.

4.3. Folding and unfolding

Precursor proteins need to be in a so-called 'loosely' folded state in order to be translocated [212]. Translocation-competent precursor proteins show significant elements of native secondary and tertiary structure, but are usually devoid of stable tertiary structures [16,20] (see Section 2.1). Nevertheless, the *translocase* is able to mediate the translocation of short segments of pro-OmpA with a stable tertiary fold, such as segments that are stabilized by a disulphide-bridge [209,211]. However, this translocation requires both ATP and Δp. This disulphide-bridge can span a 2-kDa region, i.e., a single *quantum*, without interfering with translocation [212a]. In the absence of a Δp, translocation stops at this stable fold. Δp may drive the translocation of large, folded domains, by widening the size of the translocation channel. A reversed Δp would restrict its size thereby preventing 'forward' translocation [159]. In *prlA* suppressor strains, translocation appears to be less dependent on the Δp, and a stable tertiary fold in pro-OmpA translocates in a Δp-independent manner [212b]. As discussed in Section 2.4.1, *prlA* suppressors may function by a relaxed proof-reading, and one may argue that the translocation channel in *prlA* suppressors is already locked in an open conformation.

Translocation of stably folded domains may only be of academical value as the reducing environment of the cytosol makes it unlikely that stable disulphide bonds exist. The translocation reaction has the ability to unfold polypeptide domains as demonstrated by studies with a fusion protein consisting of the cytosolic dihydro-folate reductase (Dhfr) linked to the carboxyl-terminus of pro-OmpA [109]. This fusion protein translocates up to the folded Dhfr moiety when stabilized by NADPH and methotrexate, but removal of these ligands allows a spontaneous translocation of 20–30 amino acid residues concomitant with unfolding of the Dhfr. Apparently, these residues are pulled into the *translocase*, and this seems to be sufficient to drives the unfolding of Dhfr [109]. Alternatively, one can explain these results by assuming that the folded moiety of Dhfr prevents SecA from completely inserting into the membrane. Once the ligands are removed, SecA may plunge into the membrane and force unfolding on Dhfr. The same principle may apply to the 'backward' and 'forward' translocation of unstable intermediates as for instance late pro-OmpA intermediates that translocate to full length when freed from a synthetically imposed translocation arrest [109]. This would imply that SecA contributes to unfolding when it inserts into the membrane. In addition, other protein binding or folding factors and covalent modifications of the polypeptide chain at the periplasmic face of the membrane may shift the folding equilibrium so as to allow further translocation. It has been suggest that protein translocation occurs via 'ratcheting', i.e., translocation through a proteinaceous pore is driven solely by diffusion and Brownian motion in polypeptides and that the proteinaceous components of the translocation apparatus and energetic parameters serve only to direct translocation [213]. Since translocation is driven by SecA membrane-inser-tion, it must be concluded that the 'ratchet' model is inadequate to describe protein translocation in the bacterial system. This also applies for posttranslational translo-cation of proteins into the ER and mitochondria, that are driven by a pulling force created by chaperones in the lumen or matrix [214].

4.4. Protons

In vivo and *in vitro* studies have demonstrated that $\Delta\psi$ and ΔpH are equivalent forces in translocation [159,210,215], suggesting that protons are directly involved in protein translocation. In the marine bacterium *Vibrio alginolyticus,* protein translocation appears to be coupled to the sodiummotive-force [216]. Experiments to demonstrate that protein translocation is coupled to vectorial proton movements have failed sofar. Detection of these fluxes is complicated by the appearance of proton-leaks that accompany the *in vitro* translocation reaction when intermediates jam the channel [182,183]. On the other hand, the $\Delta\psi$-driven chase of a transloca-tion intermediate of pro-OmpA is retarded more than 3-fold in deuterium oxide relative to the rate in water [193]. Such a kinetic solvent isotope effect is indicative for critical proton-transfer reactions in a rate-limiting step. It is uncertain if this step reflects vectorial proton-translocation. Scalar protons are involved in translocation, as the activity of the *translocase* is adversely affected by a lowering of the pH at the cytosolic face of the membrane [193].

Precursor proteins vary in their requirement for Δp for translocation [197]. This may be related to the presence or absence of stable-folded tertiary structure ele-ments, the number and affinity of SecA-binding sites in the polypeptide chain assuming that there is some specificity in these interactions, and the charge distri-bution along the polypeptide chain. Precursor proteins bearing a mature domain devoid of ionizable residues still require Δp for translocation [217], implying that Δp performs a mechanistic function rather than acting upon the precursor protein itself as for instance through protonation/deprotonation or electrophoresis. This latter process may be relevant to the initial insertion of the signal peptide domain into the membrane [204,207,208,218]. Other observations indicate that Δp modu-lates the activity of SecA by reducing the apparent K_m of the translocation reaction for ATP thereby allow Δp-driven translocation at low ATP concentration [219]. This phenomenon has been attributed to an accelerating effect of Δp on the rate of ADP release from SecA. An indirect modulating role of Δp seems more evident. For instance, relaxation of the translocation channel by the Δp (see Section 4.3) may ease the membrane insertion of SecA, and thereby allow faster recycling of the translocation site.

5. Concluding remarks

Protein translocation is a vital process in cell differentiation and division, biogene-sis of organelles, and plays an important role in survival and nutrient acquisition. During the last decade, extensive genetic and biochemical studies have provided a detailed view on the mechanism of protein transport across the bacterial membrane. The complicated process of protein translocation as it occurs in the cell can now be reconstituted with purified Sec-proteins. As Sec-proteins can be purified in large quantities, the bacterial system will be amendable to biophysical studies and crystallization techniques, that will reveal the innermost details of the molecular mechanism of protein translocation. In addition, it will be important to understand the molecular details of the processes that connect protein translocation and more comprehensive cellular processes such as cell division and sporulation.

Abbreviations

Δp, protonmotive force
$\Delta \psi$, transmembrane electrical potential
ΔpH, transmembrane pH gradient
Az^r, azide-resistant
Az^s, azide sensitive
Az^{ss}, azide super sensitive
Dhft, dihydrofolate reductase.
ER, endoplasmic reticulum
TMS, transmembrane segment

Acknowledgements

The author thanks Koreaki Ito, Bill Wickner, Hajime Tokuda, Kunio Yamane, Tom Rapoport, Jan Tomassen, Tom Silhavy, Janet Huie, Shoji Mizushima, Carol Kumamoto, Matti Sarvas, Colin Robinson, Chris Murphy, Ben de Kruijff, Joen Luirink, Jan-Maarten van Dijl, and Roland Freudl, and all members of his laboratory for discussion. This work was supported by grants from the Netherlands Organization for Scientific Research (N.W.O.), the Netherlands Foundation for Chemical Research (S.O.N.), Foundation of Life Sciences (S.L.W.), and the BIOTECH programm BIO2-CT-930254 of the European Community (E.C.).

References

1. Schatz, P.J. and Beckwith, J. (1990) Annu. Rev. Genet. **24**, 215–248.
2. Wickner, W., Driessen, A.J.M. and Hartl, F.-U. (1991) Annu. Rev. Biochem. **60**, 101–124.
3. Brundage, L., Hendrick, J.P., Schiebel, E., Driessen, A.J.M. and Wickner, W. (1990) Cell **62**, 649–657.
4. Yuan, J., Henry, R., McGaffery, M. and Cline, K. (1994) Science **266**, 796–798.
5. Görlich, D. and Rapoport, T.A. (1993) Cell **75**, 615–630.
6. Hartmann, E., Sommer, T., Prehn, S., Görlich, D., Jentsch, S. and Rapoport, T.A. (1994) Nature (Lond.) **367**, 654–657.
7. Dalbey, R.E. (1994) in: Signal Peptidase, ed G. von Heijne. pp. 5–15, R.G. Landes Company.
8. Pugsley, A. (1993) Microbiol. Rev. **57**, 50–108.
9. Tomassen, J., Filloux, A., Bally, M., Murgier, M. and Lazdunski, A. (1992) FEMS Microbiol. Rev. **103**, 73–90.
10. Walter, P. and Johnson, A.E. (1994) Annu. Rev. Cell Biol. **10**, 87–119.
11. Randall, L.L. (1983) Cell **33**, 231–240.
12. Collier, D.N., Bankaitis, V.A., Weiss, J.B. and Bassford, P.J., Jr. (1988) Cell **53**, 273–283.
13. Kumamoto, C.A. (1991) Mol. Microbiol. **5**, 19–22.
14. Weiss, J.P., Ray, P.H. and Bassford, P.J. Jr. (1988) Proc. Natl. Acad. Sci. USA **85**, 8978–8982.

15. Lecker, S., Lill, R., Ziegelhoffer, T., Bassford, P.J. Jr., Kumamoto, C.A. and Wickner, W. (1989) EMBO J. **8**, 2703–2709.
16. Breukink, E.J., Kusters, R. and de Kruijff, B. (1992) Eur. J. Biochem. **208**, 419–425.
17. Liu, G., Topping, T.B. and Randall, L.L. (1989) Proc. Natl. Acad. Sci. USA **86**, 9213–9217.
18. Randall, L.L., Topping, T.B. and Hardy, S.J.S. (1990) Science **248**, 860–864.
19. Randall, L.L. and Hardy, S.J.S. (1995) Trends Biochem. Sci. **20**, 65–69.
20. Lecker, S., Driessen, A.J.M. and Wickner, W. (1990) EMBO J. **9**, 2309–2314.
21. Powers, J. and Randall L.L. (1995) J. Bacteriol. **177**, 1906–1907.
22. Watanabe, M. and Blobel, G. (1989) Cell **58**, 695–705.
23. Kumamoto, C.A. and Francetic, O. (1992) J. Bacteriol. **175**, 2184–2188.
24. Gannon, P.M. and Kumamoto, C.A. (1993) J. Biol. Chem. **268**, 1590–1595.
25. Kimsey. H.H., Dagarag, M.D. and Kumamoto, C.A. (1995) J. Biol. Chem. In press.
26. Fekkes, P., den Blaauwen, T. and Driessen, A.J.M. (1995) Biochemistry **34**, 10078–10085.
27. Hardy, S.J.S. and Randall, L.L. (1991) Science **251**, 439–443.
28. Kumamoto, C.A. and Gannon, P.M. (1988) J. Biol. Chem. **263**, 11554–11558.
29. Derman, A.I., Puziss, J.W., Bassford, P.J. and Beckwith, J. (1993) EMBO J. **3**, 879–888.
30. Flower, A.M., Doebele, R.C. and Silhavy, T.J. (1994) J. Bacteriol. **176**, 5607–5614.
31. de Cock, H. and Tommassen, J. (1992) Mol. Microbiol. **6**, 599–604.
32. Hartl, F.-U., Lecker, S., Schiebel, E., Hendrick, J.P. and Wickner, W. (1990) Cell **63**, 269–279.
33. Hoffschulte, H.K., Drees, B. and Müller, M. (1994) J. Biol. Chem. **269**, 12833–12839.
34. Altman, E., Kumamoto, C.A. and Emr, S. (1991) EMBO J. **10**, 239–245.
35. Kusukawa, N., Yura, T., Ueguchi, C., Akiyama, Y. and Ito, K. (1989) EMBO J. **8**, 3517–3521.
36. Laminet, A.A., Kumamoto, C.A. and Plückthun, A. (1991) Mol. Microbiol. **5**, 117–122.
37. Phillips, G.J. and Silhavy, T.J. (1990) Nature (Lond.) **344**, 882–884.
38. Rapoport, T.A. (1992) Science **258**, 931–936.
39. Wunderlich, M., Otto, A., Seckler, R. and Glockshuber, R. (1993) Biochemistry **32**, 12251–12256.
40. Miller, J.D., Tajima, S., Lauffer, L. and Walter, P. (1995) J. Cell Biol. **128**, 273–282.
41. Miller, J.D., Wilhelm, H., Gierasch, L., Gilmore, R. and Walter, P. (1993) Nature (Lond.) **366**, 351–354.
42. Görlich, D., Prehn, S., Hartmann, E., Kalies, K.-U. and Rapoport, T.A. (1992) Cell **71**, 489–503.
43. Luirink, J. and Dobberstein, B. (1994) Mol. Microbiol. **11**, 9–13.
44. Bernstein, H.D., Poritz, M.A., Strub, K., Hoben, P.J., Brenner, S., and Walter, P. (1989) Nature (Lond.) **340**, 482–486.
45. Poritz, M.A., Strub, K. and Walter, P. (1988) Cell **55**, 4–6.
46. Römisch, K., Webb, J., Herz, J., Prehn, S., Frank, R., Vingron, M and Dobberstein, B. (1989) Nature (Lond.) **340**, 478–482.
47. Brown, S. (1989) J. Mol. Biol. **209**, 79–90.
48. Luirink, J., High, S., Wood, H., Giner, A., Tollervey, D. and Dobberstein, B. (1992) Nature (Lond.) **359**, 741–743.
49. Miller, J.D., Bernstein, H.D. and Walter, P. (1994) Nature (London) **367**, 657–659.
50. Poritz, M.A., Bernstein, H.D., Strub, K., Zopf, D., Wilhelm, H. and Walter, P. (1990) Science **250**, 1111–1117.
51. Ribes, V., Römisch, K., Giner, A., Dobberstein, B. and Tollervey, D. (1990) Cell **63**, 591–600.
52. Phillips, G.J. and Silhavy, T.J. (1992) Nature (Lond.) **359**, 744–746.
53. Luirink, J., ten Hagen-Jongman, C.M., van der Weijden, C.C., Oudega, B., High, S., Dobberstein, B. and Kusters, R. (1994) EMBO J. **13**, 2289–2296.
54. Bassford, P., Beckwith, J., Ito, K., Kumamoto, C., Mizushima, S., Oliver, D., Randall, L., Silhavy, T., Tai, P.C. and Wickner, W. (1991) Cell **65**, 367–368.
55. Valent, Q.A., Kendall, D.A., High, S., Kusters, R., Oudega, B. and Luirink, J. (1995) EMBO J. **14**, 5494–5505.
56. MacFarlane, J. and Müller, M. (1995) Eur. J. Biochem. **233**, 766–771.
57. Ahrem, B., Hoffschulte, H.K. and Müller, M. (1989) J. Cell Biol. **108**, 1637–1646.
58. Ito, K. and Akiyama, Y. (1991) Mol. Microbiol. **5**, 2243–2253.
59. Swidersky, U.E., Rienhöfer-Schweer, Werner, P.K., Ernst, F., Benson, S.A., Hoffschulte, H.K. and

Müller, M. (1992) Eur. J. Biochem. **207**, 803–811.
60. Crooke, E. and Wickner, W. (1987) Proc. Natl. Acad. Sci. USA **84**, 5216–5220.
61. Lill, R., Crooke, E., Guthrie, B. and Wickner, W. (1988) Cell **54**, 1013–1018.
62. Crooke, E., Guthrie, B., Lecker, S., Lill, R. and Wickner, W. (1988) Cell **54**, 1003–1011.
63. Guthrie, B. and Wickner, W. (1990) J. Bacteriol. **172**, 5555–5562.
64. Wiedmann, B., Sakai, H., Davis, T.A. and Wiedmann, M. (1994) Nature **370**, 434–440.
65. Stoller, G., Rücknagel, K.P., Nierhaus, K.H., Schmid, F.X., Fischer, G. and Rahfeld, J.-U. (1995) EMBO J. **14**, 4939–4948.
66. Kandror, O., Sherman, M., Rhode, M. and Goldberg, A.L. (1995) EMBO J. **14**, 6021–6027.
67. Oliver, D.B., Cabelli, R.J. and Jarosik, G.P. (1990) J. Bioenerg. Biomembr. **22**, 311–338.
68. Akita, M., Sasaki, S., Matsuyama, S. and Mizushima, S. (1989) J. Biol. Chem. **265**, 8164–8169.
69. Driessen, A.J.M. (1993) Biochemistry **32**, 13190–13197.
70. Lill, R., Cunningham, K., Brundage, L. A., Ito, K., Oliver, D. and Wickner, W. (1989) EMBO J. **8**, 961–966.
71. Oliver, D.B. (1993) Mol. Microbiol. **7**, 159–165.
72. Lill, R., Dowhan, W. and Wickner, W. (1990) Cell **60**, 271–280.
73. Matsuyama, S., Kimura, E. and Mizushima, S. (1990) J. Biol. Chem. **265**, 8760–8765.
74. Mitchell, C. and Oliver, D. (1993) Mol. Microbiol. **10**, 483–497.
75. van der Wolk, J., Klose, M., Breukink, E., Demel, R.A., de Kruijff, B., Freudl, R. and Driessen, A.J.M. (1993) Mol. Microbiol. **8**, 31–42.
76. Van der Wolk, J.W., Klose, M., de Wit, J.G., den Blaauwen, T., Freudl, R. and Driessen, A.J.M. (1995) J. Biol. Chem. **270**, 18975–18982.
77. Klose, M., Schimz, K.-L., van der Wolk, J., Driessen, A.J.M. and Freudl, R. (1993) J. Biol. Chem. **268**, 4504–4510.
78. Walker J.E., Saraste, M., Runswick, M.J. and Gay, N.J. (1982) EMBO J. **1**, 945–951.
79. Economou, A. and Wickner, W. (1994) Cell **78**, 835–843.
80. Koonin, E.V. and Gorbalenya, A.E. (1992) FEBS Lett. **298**, 6–8.
81. Breukink, E.J., van Demel, R., de Korte-Kool G. and de Kruijff, B. (1992) Biochemistry **31**, 1119–1124.
82. Radapandi, T. and Oliver, D. (1994) Biochem. Biophys. Res. Commun. **200**, 1477–1483.
83. Breukink, E., Nouwen, N., van Raalte, A., Mizushima, S., Tommassen, J. and De Kruijff, B. (1995) J. Biol. Chem. **270**, 7902–7907.
84. Liebke, H.H. (1987) J. Bacteriol. **169**, 1174–1181.
85. Cabelli, R.J., Dolan, K.M., Qian, L. and Oliver, D.B. (1991) J. Biol. Chem. **266**, 24420–24427.
86. Cunningham, K. and Wickner, W. (1989) Proc. Natl. Acad. Sci. **86**, 8630–8634.
87. Akita, M., Shinkai, A., Matsuyama, S.-I. and Mizushima, S. (1991) Biochem. Biophys. Res. Commun. **174**, 211–216.
88. Kimura, E., Akita, M., Matsuyama, S. and Mizushima, S. (1991) J. Biol. Chem. **266**, 6600–6606.
89. von Heijne, G. (1990) J. Membr. Biol. **115**, 195–201.
90. Chen, L., Tai, P.C., Briggs, M. S. and Gierasch, L. M. (1987) J. Biol. Chem. **262**, 1427–1429.
91. Shinkai, A., Mei, L.H., Tokuda, H. and Mizushima, S. (1991) J. Biol. Chem. **266**, 5827–5833.
92. Randall, L.L. (1992) Science **257**, 241–245.
93. Hendrick, J.P. and Wickner, W. (1991) J. Biol. Chem. **266**, 24596–24600.
94. de Vrije, T., de Swart, R.L., Dowham, W., Tommassen, J. and de Kruijff, B. (1988) Nature (Lond.) **334**, 173–175.
95. Van der Goot, F., Didat, N., Pattus, F., Dowham, W. and Letellier, L. (1993) Eur. J. Biochem. **213**, 217–221.
96. Kusters, R., Dowham, W. and de Kruijff, B. (1991) J. Biol. Chem. **266**, 8659–8662.
97. Kusters, R., Huijbregts, R. and de Kruijff, B. (1992) FEBS Lett. **308**, 97–100.
98. Keller, R.C.A., Snel, M.E., De Kruijff, B. and Marsh, D. (1995) FEBS Lett. **331**, 19–24.
99. Ulbrandt, N.D., London, E. and Oliver, D.B. (1992) J. Biol. Chem. **267**, 15148–15192.
100. Cunningham, K., Lill, R., Crooke, E., Rice, M., Moore, K., Wickner, W. and Oliver, D.B. (1989) EMBO J. **8**, 955–959.
101. Watanabe, M. and Blobel, G. (1993) Proc. Natl. Acad. Sci. USA **90**, 9011–9015.

102. Kim, Y.L., Rajapandi, T. and Oliver, D.B. (1994) Cell **78**, 845–853.

103. den Blaauwen, T., de Wit, J.G., de Ley, L., Breukink, E.J. and Driessen, A.J.M. (1996) submitted.

104. Schmid, M.G., Dolan, K.M. and Oliver, D. (1991) J. Bacteriol. **173**, 6605–6611.

105. Overhoff, B., Klein, M., Spies, M. and Freudl, R. (1991) Mol. Gen. Genet. **228**, 417–423.

106. Takamatsu, H., Nakane, A., Oguro, A., Saidaie, Y., Nakamura, K. and Yamane, K. (1994) J. Biochem. **116**, 1287–1294.

107. Joly, J.C. and Wickner, W. (1993) EMBO J. **12**, 255–263.

108. Schiebel, E., Driessen, A.J.M., Hartl, F.-U. and Wickner, W. (1991) Cell **64**, 927–939.

109. Arkowitz, R.A., Joly, J.C. and Wickner, W. (1993) EMBO J. **12**, 243–253.

110. Oliver, D.B., Cabelli, R.J., Dolan, K.M. and Jarosik, G.P. (1990) Proc. Natl. Acad. Sci. USA **87**, 8227–8231.

111. Nakane, A., Takamatsu, H., Oguro, A., Sadaie, Y., Nakamura, K. and Yamane, K. (1995) Microbiol. **141**, 113–121.

112. Huie, J.L. and Silhavy, T.J. (1995) J. Bacteriol. **177**, 3518–3526.

113. Fortin, Y., Phoenix, P. and Drapeau, G.R. (1990) J. Bacteriol. **172**, 607–6610.

114. Fikes, J.D. and Bassford, P.J., Jr. (1989) J. Bacteriol. **171**, 402–409.

115. Schatz, P.J., Bieker, K.L., Ottemann, K.M., Silhavy, T.J. and Beckwith, J. (1991) EMBO J. **10**, 1749–1757.

116. Osborne, R.S. and Silhavy, T.J. (1993) EMBO J. **12**, 3391–3398.

117. Ito, K. (1992) Mol. Microbiol. 6, 2423–2428.

118. Nishiyama, K., Hanada. M. and Tokuda, H (1994) EMBO J. **13**, 3272–3277.

119. Schatz, P.J., Riggs, P.D., Jacq, A., Fath, M.J. and Beckwith, J. (1989) Genes Dev. **3**, 1035–1044.

120. Johnson, A.E. (1993) Trends Biochem. Sci. **18**, 456–458.

121. Murphy, C.K. and Beckwith, J. (1994) Proc. Natl. Acad. Sci. USA **91**, 2557–2561.

122. Douville, K., Leonard, M., Brundage, L.A., Nishiyama, K., Tokuda, H., Mizushima, S. and Wickner, W. (1994) J. Biol. Chem. **269**, 18705–18707.

123. Brundage, L., Fimmel, C.J., Mizushima, S. and Wickner, W. (1992) J. Biol. Chem. **267**, 4166–4170.

124. Nishiyama, K.-I., Misuzhima, S. and Tokuda, H. (1993) EMBO J. **12**, 3409–3415.

125. Akimaru, K., Matsuyama, S.-I., Tokuda, H. and Mizushima, S. (1991) Proc. Natl. Acad. Sci. USA **88**, 6545–6549.

126. Nishiyama, K.-I., Misuzhima, S. and Tokuda, H. (1992) J. Biol. Chem. **267**, 7170–7176.

127. Tokuda, H., Akimaru, J., Matsuyama, S., Nishuyama, K. and Mizushima, S. (1991) FEBS Lett. **279**, 233–236.

128. Bost, S. and Belin, D. (1995) EMBO J. **14**, 4412–4421.

129. Kontinen, V.P. and Tokuda, H. (1995) FEBS Lett. **364**. 157–160.

130. Driessen, A.J.M. and Wickner, W. (1990) Proc. Natl. Acad. Sci. USA **87**, 3107–3111.

131. Bassilana, M. and Wickner, W. (1993) Biochemistry **32**, 2626–2630.

132. Hanada, M., Nishiyama, K.-I., Mizushima, S. and Tokuda, H. (1994) J. Biol. Chem. **269**, 23625–23631.

133. Douville., K., Price, A., Eichler, J., Economou, A. and Wickner, W. (1995) J. Biol. Chem. **270**, 20106–20111.

134. Matsuyama, S., Akimaru, J. and Mizushima, S. (1990) FEBS Lett. **269**, 96–100.

135. Taura, T., Baba, T., Akiyama, Y. and Ito, K. (1993) J. Bacteriol. **175**, 7771–7775.

136. Kihara, A., Akiyama, Y. and Ito, K. (1995) Proc. Natl. Acad. Sci. USA **92**, 4532–4536.

137. Joly, J.C., Leonard, M. and Wickner, W. (1994) Proc. Natl. Acad. Sci. USA **91**, 4703–4707.

138. Bieker, K.L. and Silhavy, T.J. (1990) Cell **61**, 833–842.

139. Flower, A.M., Osborne, R.S. and Silhavy, T.J. (1995) EMBO J. **14**, 884–893.

140. Baba, T., Taura, T., Shimoike, T., Akiyama, Y., Yoshihisa, T. and Ito, K. (1994) Proc. Natl. Acad. Sci. USA **91**, 4539–4543.

141. Taura, T., Ito, K. and Akiyama, Y. (1994) Mol. Gen. Genet. **231**, 261–269.

142. Pogliano, J.A. and Beckwith, J. (1993) Genetics **133**, 763–773.

143. Shimoike, T., Akiyama, Y., Baba, T., Taura, T. and Ito, K. (1992) Mol. Microbiol. **6**, 1205–1210.

144. Shimoike, T., Taura, T., Kihara, A., Yoshihisa, A., Akiyama, Y., Cannon, K. and Ito, K. (1995) J. Biol. Chem. **270**, 5549–5526.

145. Watanabe, M., Nicchitta, C.V. and Blobel, G. (1990) Proc. Natl. Acad. Sci. USA **87**, 1990–1964.
146. Tokuda, H., Shiozuka, K. and Mizushima, S. (1990) Eur. J. Biochem. **192**, 583–589.
147. Matsuyama, S.-I., Fujita, Y., Sagara, K. and Mizushima, S. (1992) Biochim. Biophys. Acta **1122**, 77–84.
148. Emr, S.D., Hanley-Way, S. and Silhavy, T.J. (1981) Cell **23**, 79–88.
149. Watanabe, M. and Blobel, G. (1989) Proc. Natl. Acad. Sci. USA **86**, 1895–1899.
150. Kim, Y.L. and Oliver, D.B. (1994) FEBS Lett. **339**, 175–180.
151. Panzner, S., Dreier, L., Hartmann, E., Kostka, S. and Rapoport, T.A. (1995) Cell **81**, 561–570.
152. Esnault, Y., Blondel, M.-O., Deshaies, R.J., Scheckman, R. and Kepes, F. (1993) EMBO J. **12**, 4083–4093.
153. Gardel, C., Johnson, K., Jacq, A. and Beckwith, J. (1990) EMBO J. **9**, 3209–3216.
154. Bieker-Brady, K.L. and Silhavy, T.J. (1992) EMBO J. **11**, 3165–3174.
155. Pogliano, J.A. and Beckwith, J. (1994) EMBO J. **13**, 554–561.
156. Pogliano, J.A. and Beckwith, J. (1994) J. Bacteriol. **176**, 804–814.
157. Matsuyama, S.-I., Fujita, Y. and Mizushima, S. (1993) EMBO J. **12**, 265–270.
158. Arkowitz, R.A. and Wickner, W. (1994) EMBO J. **13**, 954–963.
159. Driessen, A.J.M. (1992) EMBO J. **11**, 847–853.
160. Geller, B.L. (1990) J. Bacteriol. **172**, 4870–4876.
161. Dalbey, R.E. and von Heijne, G. (1992) Trends Biochem. Sci. **17**, 474–478.
162. Bardwell, J.C.A., McGovern, K. and Beckwith, J. (1991) Cell **67**, 581–589.
163. Kamitani, S., Akiyama, Y. and Ito, K. (1992) EMBO J. **11**, 57–62.
164. Missiakes, D., Georgopoulos, C. and Raina, S. (1994) EMBO J. **13**, 2013–2020.
165. Shevchik, V.E., Condemine, G. and Robert-Baudouy, J. (1994) EMBO J. **13**, 2007–2012.
166. Martin, J.L., Bardwell, J.C.A. and Kuriyan, J. (1993) Nature (Lond.) **365**, 464–468.
167. Bardwell, J.C.A., Lee, J.-O., Jander, G., Martin, N., Belin, D. and Beckwith, J. (1993) Proc. Natl. Acad. Sci. USA **90**, 1038–1042.
168. Jacobs, M., Andersen, J.B., Kontinen, V.P. and Sarvas, M. (1993) Mol. Microbiol. **8**, 957–966.
169. Kontinen, V.P., Saris, P. and Sarvas, M. (1991) Mol. Microbiol. **5**, 1273–1283.
170. Haandrikman, A.J., Kok, J., Soemitro, S., Ledeboer, A.M., Konings, W.N. and Venema, G. (1989) J. Bacteriol. **171**, 2789–2794.
171. Inouye, M. and Halegoua, S. (1980) CRC Crit. Rev. Biochem. **7**, 339–371.
172. Briggs, M.S. and Gierash, L.M. (1986) Adv. Prot. Chem. **38**, 109–180.
173. Hikita, C. and Mizushima, S. (1992) J. Biol. Chem. **267**, 4882–4888.
174. Hikita, C. and Mizushima, S. (1992) J. Biol. Chem. **267**, 12375–12379.
175. Phoenix, D, Kusters, R., Hikita, C., Mizushima, S. and de Kruijff, B. (1993) J. Biol. Chem. **268**, 17069–17073.
176. Kusters, R., Breukink, E., Gallusser, A., Kuhn, A. and de Kruijff, B. (1994) J. Biol. Chem. **269**, 1560–1563.
177. Martoglio, B., Hofmann, M.W., Brunner, J. and Dobberstein, B. (1995) Cell **81**, 207–214.
178. Mothes, W., Prehn, S. and Rapoport, T.A. (1994) EMBO J. **13**, 3973–3982.
179. Crowley, K.S., Reinhart, G.D. and Johnson, A.E. (1993) Cell **73**, 1101–1115.
180. Simon, S.M. and Blobel, G. (1991) Cell **65**, 371–380.
181. Simon, S.M. and Blobel, G. (1992) Cell **69**, 677–684.
182. Schiebel, E. and Wickner, W. (1992) J. Biol. Chem. **267**, 7505–7510.
183. Kawasaki, S., Mizushima, S. and Tokuda, H. (1993) J. Biol. Chem. **268**, 8193–8198.
184. Kato, M. and Mizushima, S. (1993) J. Biol. Chem. **268**, 3586–3593.
185. von Heijne, G. (1989) Nature (Lond.) **341**, 456–458.
186. Andersson, H. and von Heijne, G. (1993) EMBO J. **12**, 683–691.
187. von Heijne, G. (1994) FEBS Lett. **346**, 69–72.
188. Kuhn, A. (1988) Eur. J. Biochem. **177**, 267–271.
189. Kuhn, A., Zhu, H.-Y. and Dalbey, R.E. (1990) EMBO J. **9**, 2385–2389.
190. Chen, L. and Tai, P.C. (1985) Proc. Natl. Acad. Sci. USA **82**, 4384–4388.
191. Chen, L. and Tai, P.C. (1986) J. Bacteriol. **168**, 828–832.
192. Date, T., Goodman, J.M. and Wickner, W.T. (1980) Proc. Natl. Acad. Sci. USA **77**, 4669–4673.

193. Driessen, A.J.M. and Wickner, W. (1991) Proc. Natl. Acad. Sci. USA **88**, 2471–2475.

194. Geller, B.L. and Wickner, W. (1985) J. Biol. Chem. **260**, 13281–13285.

195. Yamada, H., Tokuda, H. and Mizushima, S. (1989). J. Biol. Chem. **264**, 1723–1728.

196. Zimmermann, R., Watts, C. and Wickner, W. (1982) J. Biol. Chem. **257**, 6529–6536.

197. Pfanner, N. and Neupert, W. (1985) EMBO J. **4**, 2819–2825.

198. Rojo, E.E., Stuart, R.A., and Neupert, W. (1995) EMBO J. **14**, 3445–3451.

199. Chaddock, A.M., Mant, A., Karnauchov, I., Brink, S., Herrmann, R.G., Klösgen, R.B. and Robinson, C. (1995) EMBO J. **14**, 2715–1722.

200. Creighton, A.M., Hulford, A., Mant, A., Robinson, D. and Robinson, C. (1995) J. Biol. Chem. **270**, 1663–1669.

201. Cline, K., Ettinger, W.F. and Theg, S.M. (1992) J. Biol. Chem. **267**, 2688–2696.

202. Geller, B.L. and Green, H.M. (1989) J. Biol. Chem. **264**, 16465–16469.

203. Tani, K., Shiozuka, K., Tokuda, H. and Mizushima, S. (1989) J. Biol. Chem. **264**, 18582–18588.

204. Nouwen, N., Tommassen, J. and De Kruijff, B. (1994) J. Biol. Chem. **269**, 16029–16033.

205. Bassilana, M., Arkowitz, R.A. and Wickner, W. (1992) J. Biol. Chem. **267**, 25246–25250.

206. Yamada, H., Matsuyama, S.-I., Tokuda, H. and Mizushima, S. (1989) J. Biol. Chem. **264**, 18577–18581.

207. Geller, B.L., Zhu, H.-Y., Cheng, S., Kuhn, A. and Dalbey, R. (1993) J. Biol. Chem. **268**, 9442–9447.

208. Lu, K., Yamada, H. and Mizushima, S. (1991) J. Biol. Chem. **266**, 9977–9982.

209. Tani, K., Tokuda, H. and Mizushima, S. (1990) J. Biol. Chem. **265**, 17341–17347.

210. Driessen, A.J.M. (1992) Trends Biochem. Sci. **17**, 219–223.

211. Tani, K. and Mizushima, S. (1991) FEBS Lett. **285**, 127–131.

212. Randall, L.L. and Hardy, S.J.S. (1986) Cell **46**, 921–928.
    a. Uchida, K., Mori, H. and Mizushima, S. (1995) J. Biol. Chem. **270**, 30862–30868.
    b. Nouwen, N., de Kruijff, B. and Tommassen, J. (1996) Proc. Natl. Acad. Sci. USA, in press.

213. Simon, S.M., Peskin, C.S. and Oster, G.F. (1992) Proc. Natl. Acad. Sci. USA **89**, 3770–3774.

214. Wickner, W.T. (1994) Science **266**, 1197–1198.

215. Bakker, E.P. , Randall, L.L. (1984) EMBO J. **3**, 895–900.

216. Tokuda, H., Kim, Y.L. and Mizushima, S. (1990) FEBS Lett. **264**, 10–12.

217. Kato, M., Tokuda, H. and Misushima, S. (1992) J. Biol. Chem. **267**, 413–418.

218. Daniels, C.J., Bole, D.G., Quay, S.C. and Oxender, D.L. (1981) Proc. Natl. Acad. Sci. USA **78**, 5396–5400.

219. Shiozuka, K., Tani, K., Mizushima, S. and Tokuda, H. (1990) J. Biol. Chem. **31**, 18843–18847.

CHAPTER 33

Protein Transport Across the Outer and Inner Membranes of Mitochondria

M.F. BAUER and W. NEUPERT

Adolf Butenandt Institut für Physiologische Chemie,
Physikalische Biochemie und Zellbiologie der Universität München,
Goethestrasse 33, 80336 München, Germany

© 1996 Elsevier Science B.V.
All rights reserved

Handbook of Biological Physics
Volume 2, edited by W.N. Konings, H.R. Kaback and J.S. Lolkema

Contents

1. Introduction

Eukaryotic cells are characterized by extensive subcellular compartmentation whose structural basis is the existence of a number of highly specialized membrane-bounded organelles. Each of these organelles is equipped with a specific subset of proteins allowing them to fulfil specific tasks in cellular metabolism. Mitochondria are present in virtually all eukaryotic cells. They are made up by two highly specialized membrane systems, the outer and inner membrane, and two aqueous compartments, the matrix and the intermembrane space. They provide energy for the cell in the form of ATP by coupling oxidation to the synthesis of ATP (for review see Ref. [170]).

Mitochondria possess an own genome (mtDNA). In mammals the mitochondrial genome encodes 13 polypeptides, all of which are components of the oxidative phosphorylation cascade. In addition, the sequences for RNA species (two rRNAs and 22 tRNAs) necessary for mitochondrial protein biosynthesis are found in mammalian mtDNA. The mitochondrial genomes of lower eukaryotes encode, in addition to subunits of oxidative phosphorylation complexes, proteins acting in mtDNA maintenance and protein synthesis. Also in these cases, only a small number of proteins is encoded by the mitochondrial genome. The total number of different proteins making up a mitochondrion can presently only be guessed; this number may account 1000–5000. Most mitochondrial proteins are nuclear encoded and synthesized in the form of precursor proteins on cytosolic ribosomes (Fig. 1) [5,50]. These precursor proteins are imported into mitochondria in a multi-step process which is facilitated by the coordinated action of two translocation systems in the mitochondrial outer and inner membranes (Fig. 1) [37,124].

Recent research in the area of protein translocation has focused mainly on the specificity of targeting of precursor proteins to the organelles and on the molecular nature of the translocation machineries which facilitate the transfer of macromolecules across the mitochondrial membranes. The use of *in vitro* systems and of molecular genetics has helped considerably to unravel the structure and function of components which mediate recognition, insertion, translocation and intramitochondrial sorting of precursor proteins. In addition, recent studies have provided insights into the energetic requirements of the import processes and into the subsequent folding of the imported proteins to their native state. In this contribution we will provide an overview on the current knowledge on the molecular mechanisms underlying protein import into and across the mitochondrial membranes.

2.1. The general pathway of protein import

Most, but not all, of the nuclear encoded mitochondrial precursor proteins are synthesized with a positively charged presequence at the NH_2-terminus that is

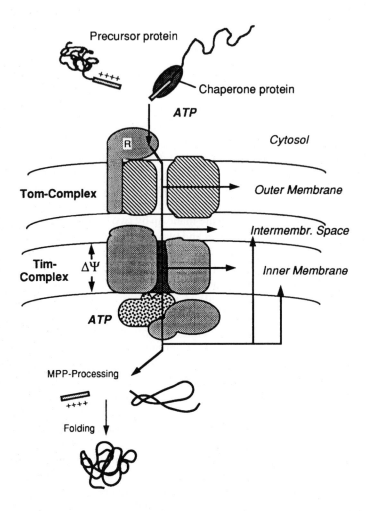

Fig. 1.   The general pathway of protein import. General pathway of protein translocation across the mitochondrial inner and outer membrane to the various mitochondrial compartments (see Section 1.1). R = receptor.

responsible for their targeting to the organelle. The mitochondrial preproteins are believed to be maintained in a translocation competent state by cytosolic factors, in particular members of the heat shock protein family of 70 kDa [25] and by binding factors specific for presequences [105,40]. Receptors on the outer surface of the mitochondrial outer membrane specifically recognize and bind the precursors prior to their translocation (R; Fig. 1). The translocation across the mitochondrial membranes is then mediated by the import machineries of the outer (Tom complex) and the inner membrane (Tim complex) at sites of close contact between both membranes [124,37]. Recognition of the polypeptide chain at the surface of the outer membrane and insertion into the translocation site is mediated by the Tom complex. The further movement within the contact site, namely the transfer of the presequence across the inner membrane strictly requires energy in form of an electrochemical potential ($\Delta\psi$) and ATP in the matrix (Fig. 1) [144,33,28]. Completion of translocation into the inner membrane or matrix is independent of $\Delta\psi$, but requires the action of the mitochondrial heat shock protein mt-Hsp70 which becomes recruited by the Tim complex through the peripheral membrane component Tim44. Tim44 and mt-Hsp70 interact with the incoming precursor protein during ATP-hydrolysis driven cycles of binding and release, thereby driving import. Upon entering the mitochondrial matrix, cleavable signal sequences, if present, are proteolytically removed by the mitochondrial processing peptidase (MPP) (Fig. 1). Once inside the matrix, precursors interact with molecular chaperones, which support their folding and assembly into functional complexes or facilitate sorting to their final destination. Additional maturation steps can occur during import, which include covalent and non-covalent modifications. A number of variations to this general import pathway exist. In particular, proteins of the outer membrane, inner membrane and the intermembrane space use only parts of the general pathway or take additional steps in order to reach their final destinations.

2. Nuclear encoded mitochondrial preproteins

2.1. Mitochondrial targeting sequences

Many different proteins destined for various cellular locations are synthesized on cytosolic polysomes. This implies that the information necessary for targeting to the respective organelle must be contained in the protein itself. It turned out that small portions of the polypeptide chains of these proteins, designated as signal or targeting sequences, carry this information. The majority of the nuclear encoded mitochondrial proteins, but not all, carry this informative stretch in form of a cleavable presequence located at the NH_2-terminus. Our present understanding of the functional role of presequences was initiated by experiments of Hurt et al. [64] and Horwich et al. [63], who demonstrated that a presequence of a mitochondrial protein when fused to a cytosolic protein could direct this 'passenger' into mitochondria.

The mitochondrial presequences are usually called 'targeting sequences' or 'matrix targeting signals'. Importantly, the targeting sequences of different precursors do not share sequence similarities, however, they have distinct common structural characteristics: (i) presequences are rich in positively charged residues (arginines and lysines), (ii) in almost all cases they lack acidic amino acid residues, (iii) they have an unusually high content of hydroxylated residues, (iv) their chain length usually ranges between 10 to 60 amino acid residues, and (v) many of them are predicted to form amphipathic α-helices with the hydrophobic face exposed to one side and the positively charged face to the opposite side of the helix [175] (for review see Refs. [13,134]). The potential to form a polarized α-helical structure is thought to be an essential prerequisite for the function of the presequence [35].

The amphipathic organization of α-helices is thought to prevail in a hydrophobic environment. It has been proposed that this structure is responsible for the initial interaction with the lipid bilayer of the outer membrane on the basis mainly of experiments with artificial lipid vesicles [24,23]. However, the significance of such a reaction for the signal deciphering *in vivo* is a matter of debate. It has become clear that the specific recognition of the targeting signals of precursor proteins occurs via proteinaceous receptor components on the mitochondrial surface. These receptors have been identified during the past few years and their role in translocation could be demonstrated (see below) (for review see Refs. [89,125]). The delivery to these receptors might be mediated by specific cytosolic signal recognition factors at least in some cases [42]. The targeting signals are recognized not only on the mitochondrial surface but also on the inner face of the outer membrane and, in a further step, by components located at the surface of the inner membrane (for details see below).

A number of precursor proteins, including all outer membrane proteins and many of those destined for the inner membrane, the intermembrane space or the matrix do not contain cleavable targeting signals. They obviously carry signal sequences within the mature part of the protein. In principle these preproteins can be subdivided into two classes according to the localization of their targeting signals: (i) precursors with non-cleavable targeting signals located at or close to the NH_2-terminus, similar to those of cleavable precursors; and (ii) precursors which do not contain NH_2-terminal targeting signals and therefore must be targeted to mitochondria by means of internal signals.

Examples with NH_2-terminal non-cleavable presequences include outer membrane proteins, e.g. Tom70 of yeast as well as matrix proteins, such as 3-oxoacyl-CoA thiolase (reviewed in Ref. [159]). In the case of Tom70 information for targeting and sorting was found in the first 41 aminoterminal amino acid residues. While residues 1 through 12 contains the targeting signal, the sorting to the correct topology is determined by the hydrophobic segment (residue 10–37) which anchors the protein in the outer membrane [51,52,53,65,133,159].

A prominent group of inner membrane proteins targeted by means of internal signal(s) is the family of carriers of the type of the ADP/ATP carrier (AAC). These proteins are characterized by three domains of about 100 amino acids which are suggested to have evolved by triplication of one ancestral gene [141]. Within these repeats, stretches of about 20 amino acids are present in the COOH-terminal half of each domain which are predicted to form α-helices [2] and resemble the classical mitochondrial presequences [71,175,160].

Still only little is known about the structural characteristics and mode of action of internal targeting signals. Recent studies have just begun to address the questions of whether such precursors use the same import machinery as those targeted by NH_2-terminal sequences and how they traverse the outer and inner membranes. One recent example of this group is the BCS1 protein which is anchored to the inner membrane by a single transmembrane domain in an N_{out}–C_{in} orientation. A positively charged sequence located immediately COOH-terminally to this domain has been shown to function as an internal targeting signal. During import this internal signal seems to cooperate with the transmembrane domain thereby forming a hairpin loop structure during translocation [30]. It is still unknown as to how these internal signals act in conjunction with further signals to accomplish both the specific targeting to the mitochondria and specific sorting to mitochondrial subcompartments.

2.2. Cytosolic factors involved in mitochondrial protein import

The mechanistic independence of synthesis of mitochondrial proteins in the cytosol and their translocation into mitochondria has been demonstrated early on [46]. Since then, posttranslational transport of virtually all proteins investigated has been observed *in vitro*. Recent work strongly supports a posttranslational mode of protein translocation also *in vivo* [180,83], although it must be stated that a co-translational mechanism is not excluded. Therefore, the discussion is still ongoing as to whether *in vivo* transport may occur co-translationally [31,173]. There is, however, no important mechanistic differences between a post- and co-translational pathway, unless a strictly coupled mechanism is postulated which appears extremely unlikely.

A major problem with any posttranslational pathway is that most precursor proteins pass through the cytosol as not natively folded species which have the tendency to aggregate and which may be subject to rapid proteolytic degradation. Several components have been found to exist in the cytosol which are believed to maintain precursor proteins in an import-competent state (Table 1). One of these components is Hsp70 in the cytosol. Deletion of three of the four Hsp70 species encoded by the ssa genes, ssa1$^-$, ssa2$^-$ and ssa4$^-$ [26] resulted in accumulation of mitochondrial precursor proteins. Furthermore, protein import into isolated mitochondria was stimulated by cytosolic Hsp70 (ct-Hsp70) and another unidentified factor [19].

Maintenance of import competence is believed to reflect a dynamic situation in which formation of inappropriate intra- and intermolecular contacts is prevented by cycles of binding and release of Hsp70 [26]. This process requires hydrolysis of ATP and further cytosolic chaperones. One of these is Ydj1p, a 45 kDa yeast homolog of the prokaryotic chaperone DnaJ [15,4]. Ydj1p is required to assist Hsp70 in stabilizing preproteins in a non-aggregated, only partially folded state [16]. Ydj1p seems to interact with Hsp70 (Ssa-1) in a manner similar to that observed for *E. coli* DnaK and DnaJ [21,47].

Two additional binding factors have been described, both of which are thought to participate in transferring import-competent precursors to mitochondria. First, MSF (*m*itochondrial import *s*timulating *f*actor) is a heterodimer found in the liver cytosol [40,112,113]. MSF belongs to the 14–3–3 protein family which is known to play a role in signal transduction [104]. It consists of subunits of 32 kDa (MSFL) and 30 kDa (MSFS). MSFL and MSFS are highly homologous to each other and to a cytosolic factor, Bmh1p, recently found in the yeast cytosol [1]. Five basic amino acid residues in the presequence seem to be essential for the recognition by and binding to MSF [84]. MSF is believed to facilitate the dissociation and unfolding of aggregated polypeptide chains in an ATP-dependent manner by binding to the presequence [1]. Moreover, MSF may target precursor proteins to mitochondria [40,41]. A specific interaction of the precursor-MSF complex with specific receptor components (Tom37/Tom70) of the outer membrane was suggested [42].

Another cytosolic factor, PBF (*p*resequence *b*inding *f*actor) [105] has been described to stimulate the import of certain precursor proteins *in vitro*, but so far has not been characterized in detail [106]. The different factors found may interact with distinct parts of a precursor protein, and may serve different purposes in stabilizing preproteins and in targeting them to the mitochondria [31].

Table 1
Components of the mitochondrial protein translocation machinery

Compartment	Name[4]	Former Name & Synonyms	Organism	Gene (cloned)	Essential for viability	Molecular mass	Function & Topology	Cleavable Pre-sequence	No. of TMS	Orientation	References
Cytosol											
	ct-Hsp70		*S. cerevisiae*	*SSA1-4*	+	70 kDa	Molecular chaperone, keeps precursors in a translocation competent conformation	No	--	--	26,19
	Ydj1p Mas5p		*S. cerevisiae*	*YDJ1 MAS5*	not essential[1]	45 kDa	Molecular chaperone; unfolding; partly membrane associated	No	--	--	15,4
	MSF				n.d.	30/32 kDa Dimer	Molecular chaperone specific for mitochondrial preproteins	No	--	--	40
Outer Membr.											
Tim-Complex	Tom20	MOM19 Mas20p	*N.crassa S. cerevisiae*	*TOM20 MOM19 MAS20*	-	19 kDa 20 kDa	Receptor component of the outer membrane translocation complex (Tom complex); interacts with Tom22	No	1	C out, N in	161,128
	Tom70	MOM72 Mas70p	*N. crassa S. cerevisiae*	*TOM70 MOM72 MAS70*	-	72 kDa 70 kDa	Receptor component in the Tom complex; interacts with Tom22	No	1	C out, N in	162,164,57
	Tom22	MOM22 Mas22p Mas17p	*N. crassa S. cerevisiae*	*TOM22 MOM22 MAS22, MAS17*	+ +	22 kDa 17 kDa	Proposed function as receptor and part of GIP; interacts with Tom20 and Tom70; transfer from membrane receptors to GIP.	No	1	N out, C in	80,108,90,107
	Tom40	MOM38 Isp42p	*N. crassa S. cerevisiae*	*TOM40 - ISP42*	+	38 kDa 42 kDa	Major component of GIP	No	> 2		81,6
	Tom5	MOM8	*N. crassa S. cerevisiae*	-	n.d.	8 kDa	Component of the GIP interacts with Tom40	No	n.d.	n.d.	102,163
	Tom7	MOM7	*N. crassa S. cerevisiae*	-	n.d.	7 kDa	Component of the GIP; interacts with Tom40	No	n.d.	n.d.	102,163
	Tom6	Isp6p	*S. cerevisiae*	*TOM6 ISP6*	-	6 kDa	Component of GIP; intacts with Tom40	No	n.d.	n.d.	78
	Tom37	Mas37p	*S. cerevisiae*	*TOM37 MAS37*	-	37 kDa	Receptor component; interacts with Tom70	No	n.d.	n.d.	39

Compartment	Name[4]	Former Name & Synonyms	Organism	Gene (cloned)	Essential for viability	Molecular mass	Function & Topology	Cleavable Pre-sequence	No. of TMS	Orientation	References
Inner Membr.											
Tim-Complex (Membrane Components)	Tim17	Mim17 Sms1p, Mpi2	*S. cerevisiae*	*TIM17 MIM17 SMS1*	essential	17 kDa	Integral component of the Tim complex; is found in complex with Tim23: Proposed component of the channel	No	2-3	N out, C in	94,139,86,7
	Tim23	Mim23 Mas6p,Mpi3	*S. cerevisiae*	*TIM23 MIM23 MAS6*	essential	23 kDa	Integral component of the inner membrane translocation channel; is found in complex with Tim17; Proposed component of the channel	No	3-4	N out, C in	22,29
		Mim33 Mim14 Mim20 Mim55	*S. cerevisiae*	-	n.d.	33 kDa 14 kDa 20 kDa 55 kDa	Co-immunoprecipitate with the Tim complex	n.d.	n.d.	n.d.	7,9
(Matrix Components)	Tim44	Mim44 Isp45, Mpilp	*S. cerevisiae*	*TIM44 MIM44, ISP45, MPI1*	essential	44 kDa	Peripheral membrane protein; Binds incoming preproteins at the IM inner surface; acts in a complex with mt-Hsp70	Yes	--	--	93,61,143,8,153
Intermembr. Space		Imp1p	*S. cerevisiae*	*IMP1*	non-essential[3]	21 kDa	Located at the outer surface of the IM; second processing of bipartite pre-sequences of IMS and IM targeted precursors. Homology to *E. coli* leader peptidase	Yes	-	-	149
		Imp2p	*S. cerevisiae*	*IMP2*	non-essential[3]	20 kDa	Second cleavage of bipartite presequence. Heterodimer with Imp1p.	Yes	-	-	111,150

Table 1 (*continuation*)

Compartment	Protein & Synonyms	Organism	Gene (cloned)	Essential for viability	Molecular mass (kDa)	Function & Topology	Cleavable Pre-sequence	No. TMS	Orientation	References
Matrix										
	mt-Hsp70	S. cerevisiae	SSC-1	essential	70 kDa	Acts in complex with Tim44 in driving the translocation reaction	Yes	--	--	20,77,153
	MGE GrpE	S. cerevisiae	MGE1 GRPE	essential	21-24 kDa	Cooperates with mt-Hsp70 in translocation of preproteins	Yes	--	--	12,87,177,179
	Mdj1p	S. cerevisiae	MDJ1	not-essential[2]	49 kDa	Co-chaperone involved in protein folding; interacts with mt-Hsp70	Yes	--	--	138
	Hsp60 Mif4p	N. crassa S. cerevisiae	MIF4	essential	58 kDa 60 kDa	Chaperonin; mediates folding and assembly of proteins together with cpn10	Yes	--	--	18,116,131
	Hsp10 Cpn10	S. cerevisiae	CPN10	essential	11 kDa	Co-chaperonin; mediates folding and assembly of proteins together with Hsp60	Yes	--	--	92,135,136
Processing Peptidase **MPP**	α-MPP Mas2p, Mif2p	N. crassa S. cerevisiae	MAS2/ MIF2	essential	57 kDa 51 kDa	Subunit of Matrix Processing Peptidase; heterodimer with β-MPP	Yes	--	--	54,181,72
	β-MPP Mas1p, Mif1p	N. crassa S. cerevisiae	MAS1/ MIF1	essential	52 kDa 48 kDa	Subunit of Matrix Processing Peptidase; heterodimer with α-MPP	Yes	--	--	54,181,127
	MIP OPP	S. cerevisiae Rat liver	--	n.d.	75 kDa	Cleavage of octapeptide of bipartite presequence	Yes	--	--	76,69,70

[1] viable at 24 °C, but not at 37 °C.
[2] viable at 24 °C, but not at 37 °C. Nonviable on nonfermentable carbon sources.
[3] Nonviable on nonfermentable carbon sources
[4] According to the uniform nomenclature for the protein transport machinery of the mitochondrial membranes published in TIBS (1996) 21, 51.

3. Translocation across the mitochondrial outer membrane

3.1. The Tom complex of the mitochondrial outer membrane: receptors and general insertion pore

During the past years several different approaches have led to the identification of components of the protein import apparatus of the mitochondrial outer membrane. These components form a complex (the 'Tom complex') composed of at least 8 membrane proteins (the Tom proteins, Fig. 2) which are involved in mediating the insertion and translocation of the preproteins [163]. The Tom components can be

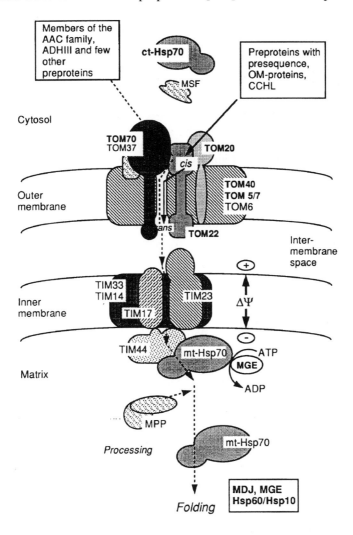

Fig. 2. The import machineries of the mitochondrial outer and inner membranes. Schematic diagram of the components involved in import. Arrows indicate the route of transit.

subdivided into two classes (Table 1): (i) 'receptors', which expose large protease-sensitive domains at the surface; and (ii) constituents of the 'general insertion pore' (GIP) that show resistance to attack by external proteases (Fig. 2) [89].

Early studies with *Neurospora crassa* and *Saccharomyces cerevisiae* have identified two distinct mitochondrial outer membrane receptors: one termed Tom20 (former MOM19 in *N. crassa* and Mas20p in *S. cerevisiae* (molecular mass ca. 19–20 kDa)); and a second component of about 70 kDa, termed Tom70 (former MOM72 (*N. crassa*) or Mas70p (*S. cerevisiae*)). To avoid confusion the various components of the import apparatus in different organisms were recently uniformed; their former names and synonyms are listed in Table 1.

Further insight into the structure of the Tom translocation machinery was gained mainly by the analysis of the protein complex that was found by co-immunoprecipitation with antibodies against Tom20. In *N. crassa* at least six proteins were observed to be associated with Tom20. According to their apparent molecular masses they were termed Tom7, Tom5, Tom22, Tom40, and Tom70 (Fig. 2) [163].

The yeast homologue of *N. crassa* Tom40 was detected by a different biochemical approach, namely by crosslinking of precursor proteins arrested as translocation intermediates spanning both mitochondrial membranes [174,6]. Co-immunoprecipitation experiments with *S. cerevisiae* mitochondria revealed a protein complex of comparable composition as found with *N. crassa* [102]. Tom6, a constituent of the yeast complex was identified by means of a genetic approach, namely as a high copy suppressor of a conditional mutation in the *TOM40* gene (Fig. 2) [78]. An additional component, termed Tom37, was initially identified by screening a collection of *S. cerevisiae* mutants temperature sensitive for growth on a non-fermentable carbon source (Fig. 2) [39].

3.1.1. Mitochondrial import receptors

Tom20 and Tom70

Both Tom70 and Tom20 are anchored via an NH_2-terminal hydrophobic segment to the membrane, while the hydrophilic parts of the proteins are exposed to the cytosol [161,151]. The cytosolic domain of Tom70 is about 65 kDa and contains a series of a consensus motif called tetratricopeptide repeat previously found in a number of proteins required for mitosis and RNA synthesis [156]. These sequences are known to be involved in protein-protein interaction. Therefore, it was speculated that they play a role in the interaction with the other Tom components or/and with presequences [162,146]. Remarkably, also Tom20 contains such a motif.

The nature of the specificity of these receptors was mainly concluded from antibody inhibition studies [161,128]. In support of this, mitochondria isolated from *N. crassa* mutants defective in Tom20 show a decrease in import for the same subset of proteins identified in Tom20 antibody inhibition studies [45]. The function of Tom20 is crucial for normal growth of *N. crassa* cells, as Tom20 disruption led to almost complete growth stop [44]. In *S. cerevisiae* the encoding genes (*TOM20, TOM70*) are not essential for the cell viability, however, cells with deleted Tom20 grow only on fermentable carbon sources. Deletion of both receptor genes together is lethal.

The interaction of preproteins with Tom70 was demonstrated by crosslinking and co-immunoprecipitation experiments. Tom70 is a receptor with a restricted substrate specificity mainly for preproteins with non-cleavable signal sequences, including the ADP/ATP carrier, AAC. The limited specificity of binding of Tom70 is supported by studies on the recognition of a Tom70 fragment to polyA$^+$ RNA [147] which only recognized AAC and cytochrome c_1. On the other hand, also a few proteins containing presequences such as the phosphate carrier, cytochrome c_1, ADHIII (Fig. 2) [27,146] seem to be influenced by Tom70 in their import efficiency. This preference, however, is not absolute, since a variety of other precursors without NH$_2$-terminal presequences (like porin and cytochrome c heme lyase) do not require Tom70 for their import. It was even proposed that Tom70 has the ability to recognize some features of the mature domains of precursor proteins [56].

When Tom70 in *N. crassa* or in yeast were selectively depleted, AAC was still targeted to mitochondria, albeit with reduced efficiency. In this situation Tom20 could function as a receptor for the AAC [164]. On the other hand, Tom70 seems to play only a little if any role in the import of preproteins containing presequences (such as cytochrome b$_2$), or artificial fusion proteins containing dihydrofolate reductase (DHFR) and of cytochrome c heme lysases [57,165].

Unlike Tom70, Tom20 functions as a receptor for most preproteins (including proteins of the outer membrane and presequence-bearing preproteins). These precursors are directly recognized by Tom20 which functions in close cooperation with Tom22 (see below). Also, artificial fusion proteins composed of a mitochondrial targeting sequence fused to a non-mitochondrial polypeptide (e.g. DHFR) interact with Tom20 [49]. Therefore, Tom20 can be regarded as the main preprotein entry point into mitochondria. The function of Tom70 may then be viewed as a means to accumulate those preproteins which are recognized by Tom20 (Fig. 2).

Tom22

Tom22, a protease-sensitive constituent of the protein import complex, was initially identified as a component co-precipitating with antibodies against Tom20 [80]. Tom22 deficient *N. crassa* mutants are unable to import any precursor known to utilize the general import route except Tom20, Tom70 and Tom22 itself [108]. In yeast, the equivalent *TOM22* gene was isolated as a suppressor of *TOM20*-deficiency and, independently, by selection of yeast mutants whose growth deficiency could be complemented by *N. crassa* Tom22 [91,107]. In both organisms *N. crassa* and *S. cerevisiae* Tom22 is essential for the growth of cells [60,91,108].

Tom22 is anchored in the outer membrane by a hydrophobic segment located roughly in the middle of the protein with the NH$_2$-terminal domain exposed to the cytosol and the COOH-terminal domain to the intermembrane space [80]. Thus, the protein has an opposite orientation as compared to Tom20 and Tom70. The cytoplasmic domain of *Neurospora* Tom22 carries a cluster of 18 negatively charged residues; yeast Tom22 has a similar distribution of negative charges. This region was proposed to interact with the positive charges of the presequences [80]. In *Neurospora* Tom22 has been proposed to play a role in the transfer of the ADP/ATP carrier (AAC) from the Tom70-bound state into the translocation pore [80]. From these data, the functional role of Tom22 appeared to be downstream of the Tom70-receptor-bound state. However, it remained unclear whether Tom22 plays a similar role for preproteins which enter the mitochondria in a Tom20-dependent fashion [79]. Recently, a novel *in vitro* system using purified outer membranes was established for direct measurement of binding of presequence containing preproteins. With this approach it could be demonstrated that both Tom22 and Tom20 are needed simultaneously for efficient interaction of the mitochondrial targeting sequence with the outer membrane [100]. This specific interaction can be abolished by inactivation of either Tom22 or Tom20. Tom22 appears to be in direct contact with Tom20, thereby forming a complex which functions as the presequence receptor on the mitochondrial surface [100,103]. The salt sensitivity of presequence binding to Tom22/Tom20 suggests that it is mediated largely by electrostatic forces. Lability of this binding may be important for allowing efficient transfer into the translocation channel or for rapid removal of non-cognate proteins from the receptor. Furthermore, the levels of Tom22 and Tom20 appear to be coordinated in the cell [45] which may be a reflection of complex formation between these two proteins. In yeast disruption of the *TOM20* gene leads to greatly reduced levels of Tom22.

Since Tom22 apparently plays a role also in the transfer of AAC the question arises as to how the presequence receptor Tom22/Tom20 may functionally interact with the more specialized receptor Tom70. Neither Tom22 nor Tom20 are involved in high affinity binding of AAC, yet the transfer of Tom70-bound AAC across the outer membrane requires the function of Tom22 [80]. Tom20, however, appears to play a minor role in this reaction or is even dispensable [45,146,162].

Based on these data a model was proposed in which Tom70 facilitates the recognition of the AAC internal targeting signal by Tom22 (Fig. 2, arrows) [100]. It is assumed that Tom70 allows AAC to adopt a conformation which permits its faster and more efficient introduction into the general import pathway via Tom22. This model would be in line with the observation that the Tom70-bound stage is not an obligatory step in AAC transport [164]. Inactivation of Tom70 by biochemical or genetic means results in a decrease in the import efficiency of AAC, but import remains dependent on Tom22/Tom20. This is in agreement with the observations described above that AAC is able to enter the general import pathway directly via the Tom22/Tom20.

Tom37

Tom37, in yeast was cloned by the complementation of a respiratory deficient mutant with defects in the lipid composition of mitochondrial membranes [39]. A direct involvement of Tom37 in lipid metabolism or lipid import could not be defined, however, deficiency in Tom37 led to a weak protein import defect. Deletion of both *TOM70* and *TOM37* genes is lethal, whereas disruption of *TOM37* alone is tolerated by the yeast cells. Tom37 and Tom70 have been proposed to form a heterodimer and functionally interact in recognizing preproteins [39,42].

3.1.2. General insertion pore (GIP)

Tom40

Tom40 is the major constituent of the protease resistant portion of the protein import complex which was co-immunoprecipitated by anti-Tom20 antibodies (Fig. 2) [80]. *TOM40* is an essential gene in *S. cerevisiae* [6]. The topology of Tom40 cannot unequivocally be predicted on the basis of the amino acid sequence; the protein is an integral membrane protein as it is resistant against alkaline extraction. The exact membrane orientation of Tom40 remains elusive so far, but recent data seem to indicate that both termini face the intermembrane space (M. Kiebler, unpublished).

The precise function of Tom40 is not yet clear. In both yeast and *Neurospora*, it can be crosslinked to a preprotein spanning both mitochondrial membranes at translocation contact sites [163,174]. A crosslink of the precursor of AAC to Tom40 could be observed only when the precursor was transferred into a protease-resistant location within the Tom-complex, a site called GIP ('general insertion pore') [120]. These observations indicate a direct interaction between preproteins in transit and Tom40, when the preprotein is inserted into the outer membrane translocation pore [124] and support the view of Tom40 being part of the translocation pore or channel (Fig. 2).

Tom7/Tom5/ITom6

N. crassa Tom7 and Tom5 were identified by co-immunoprecipitation and chemical crosslinking as constituents of the protein import complex [163]. Both proteins are believed to be part of GIP, together with Tom (Fig. 2). The membrane topology of these small Tom proteins is not clear, yet. Tom7 or Tom5 may be homologous to Tom6 which was identified as a high copy suppressor of a Tom40 temperature-sensitive (Tom40^{ts}) mutant [78]. Tom6 is anchored to the outer membrane by its COOH-terminus with the NH$_2$-terminal domain facing the cytosol. It appears to be targeted to the outer membrane by a signal in the COOH-terminal part of the protein. Thus, the import process of Tom6 is similar to that of Tom20/Tom70 [147,151], and therefore most likely is distinct from that of most other mitochondrial preproteins, which are recognized by protein receptors on the mitochondrial surface [14]. A direct interaction of yeast Tom40 with Tom6 is suggested by co-immunoprecipitation and it seems very likely that the association of Tom6 to Tom40 is functionally related to the interaction of *Neurospora* Tom7/Tom5 with Tom40.

As has been observed early on, protein import can still occur at low efficiency, even if the mitochondrial import receptors Tom20 and Tom70 are removed from the outer membrane surface of mitochondria by means of gene disruption or by treatment of isolated mitochondria with proteases [45,90,119]. This low efficient 'bypass import' shows similar characteristics as normal mitochondrial protein import with respect to the requirement for targeting sequences and for a membrane potential; however, it appears to lack the high degree of receptor specificity.

3.2. Mechanisms and energy requirements of protein translocation across the mitochondrial outer membrane

At this point the question arises as to how targeting sequences initiate translocation and what the driving forces are for the movement of preproteins across the outer membrane. Recently, an *in vitro* system was developed using purified outer membrane vesicles (OMV) of *Neurospora crassa* [98]. The availability of this system together with the possibility to genetically manipulate the components of the Tomcomplex has allowed a detailed study of these questions.

3.2.1. Initiation of protein translocation by presequence recognition and binding

Translocation of preproteins is initiated by the specific binding to the protease-sensitive receptor proteins at the outer surface of mitochondria (see above). The receptor components necessary for the translocation of presequence containing precursor proteins are Tom20 and Tom22. This interaction is specific for the NH_2-terminal presequence, since purified presequence peptides as well as complete preproteins compete for the binding site equally well [100]. Binding to this so-called *cis*-site (Fig. 2) was abolished by prebinding of specific antibodies to either Tom20 or Tom22 [99]. Likewise, a strong reduction of binding as compared to wild-type OMV was seen using OMV isolated from Tom20 or Tom22 null mutants [45,108]. Thus, both Tom20 and Tom22 contribute to this initial step and they cooperate in the recognition and binding of the targeting sequences. Furthermore, both components can be crosslinked by various chemical crosslinking reagents suggesting their presence as a protein complex in the outer membrane. This complex therefore may be viewed as a two-subunit receptor providing the recognition site for presequence bearing-precursors at the mitochondrial surface.

Since different presequences have different lengths and positions of positively charged segments vary among various presequences, the *cis*-site receptor may exhibit a binding 'area' rather than a well defined binding site [100]. Binding to the *cis*-site is readily reversible and weakened in its stability at increasing salt concentrations. This observation may reflect the involvement of weak electrostatic forces acting between the presequences and the receptor components. Salt bridges may be formed between the positively charged residues within the targeting signal and the cytosolic domain of Tom22.

3.2.2. Transfer of the presequence to the trans-site of the outer membrane

Binding of preproteins to the *cis*-site is followed by transfer into the general insertion pore (GIP) (Fig. 2) (for review see Refs. [79,99]). Transfer through the putative translocation channel allows interaction of the NH_2-terminal targeting sequence with a second specific binding site located at the inner face of the outer membrane, called *trans*-site (Fig. 2). The affinity of binding to the *trans*-site is higher than to the *cis*-site. Thereby, a translocation intermediate is formed with the presequence already translocated across the outer membrane, the NH_2-terminal portion of the mature protein spanning the membrane, and the remaining COOH-terminal part of the preprotein still residing in the cytosol (Fig. 2). The molecular nature of the *trans*-site is not known. Tom40 and Tom22 would be possible candidates for forming this site (Fig. 2).

Binding of the presequence to the *trans*-site and further insertion of the NH_2-terminal part of the mature protein into GIP are accompanied by the unfolding of the following segments of preproteins. Neither a membrane potential nor ATP was found to be required for directing the presequence to the *trans*-site. Possibly, the energy derived from presequence binding constitutes the driving force for transfer across the outer membrane. A model was proposed in which the *cis*- and *trans*-binding sites are connected by a 'passive' channel which facilitates reversible movement of NH_2-terminal segments of the preprotein. Such a 'passive channel' is characterized by the property not to undergo any specific interaction with the preprotein; it may just see the polypeptide chain backbone during sliding. Directionality of movement would be due to the relative affinity of the binding sites [99]. Reports describing reversible sliding of a polypeptide chain in transit within putative channels of the outer and the inner mitochondrial membrane [171] would be in line with such a model. Reversible sliding has also been reported for presecretory proteins in the endoplasmic reticulum and *E. coli* [157], therefore, the concept of a 'passive' channel could be a general theme in intracellular protein translocation [114].

4. Translocation across the inner mitochondrial membrane

4.1. Components of the inner membrane translocation machinery

Proteins of the inner membrane translocation machinery have been identified in *S. cerevisiae* in a genetic approach screening for mutants defective in mitochondrial import (for review see Ref. [126]). A number of *trans*-acting recessive nuclear mutations were found and classified into four complementation groups [22]. One complementation group contained mutations in the *SSC1* gene encoding for the 70kDa heat shock protein (mt-Hsp70) located in the mitochondrial matrix [20] while three novel genes could be shown to encode proteins of the mitochondrial inner membrane. Some of these proteins have been discovered by different groups at more or less the same time using different experimental approaches. Therefore, various synonyms came up for the identical genes and their products. These are summarized in Table 1. We are using here the recently uniformed terminology which is in analogy to that of the import components of the outer membrane (Toms). According to their apparent molecular masses these gene products were named Tim44, Tim23 and Tim17 (Table 1; Fig. 2). All three components are essential for yeast viability [20,22,29,93]. Their involvement in mitochondrial import was demonstrated in a number of biochemical approaches. In particular, crosslinking to preproteins spanning both mitochondrial membranes *in vitro* was used to probe the spatial relationship of preproteins in transit to the Tim components [8,86,140, 143,153]. Import defects were analyzed *in vivo* by following the accumulation of precursor proteins after depletion of Tim components [29,93,139].

Tim44 (Mpi1p, Isp45p) is a hydrophilic protein which is peripherally associated with the inner face of the inner membrane [61,93,153]. No membrane spanning segments were predicted from the primary sequence; on the other hand the COOH-terminus was proposed to be inserted into the inner membrane [61,93].

Tim23 (Mas6p, Mpi3p) was identified by characterizing a mutant defective in mitochondrial protein import which had a temperature sensitive growth phenotype (*mas6*) [29]. Tim17 (Sms1p, Mpi2p) was found by its ability to function as a high copy-suppressor of the temperature-sensitive growth-defect of the yeast *mas6* mutant and by gene complementation [94,139]. Both, Tim17 and Tim23 are integral inner membrane proteins synthesized without a cleavable presequence and three to four predicted membrane spanning segments [22,29,94,139]. These proteins share significant sequence similarity in regions that contain the putative transmembrane segments. However, they cannot substitute for each other functionally [94,139]. Tim23 carries a hydrophilic domain of ca. 100 amino acid residues at the NH_2-terminus which is absent in Tim17.

Most recently, further inner membrane proteins of 14 kDa, 20 kDa, 33 kDa and 55 kDa were described as components of the Tim complex by means of co-immunoprecipitations (Table 1) [7,9]. A functional role of these components in protein import across the inner membrane has not yet been established.

4.2. Molecular mechanisms and energy requirements of translocation across the mitochondrial inner membrane

Current biochemical data suggest that the Tim17, Tim23, Tim44, and mt-Hsp70 interact with each other thereby forming a complex. At least Tim44 and mt-Hsp70 at the same time appear to directly interact with preproteins in transit across the inner membrane. It seems likely that the inner membrane complex forms a dynamic translocation machinery. This machinery was proposed to consist of two substructures with different functions: (i) An integral inner membrane complex (Tim complex) consisting of Tim17 and Tim23, and probably several other components which forms a translocation channel; and (ii) a complex at the matrix-side consisting of Tim44 and mt-Hsp70 which binds the incoming preprotein chain (Figs. 2 and 3). The ATP dependence of protein translocation across the inner membrane appears to be directly related to the action of these latter two components.

4.2.1. Role of the Membrane Potential ($\Delta\psi$)
Energization of the inner membrane is an absolute requirement for the translocation of preproteins targeted to the inner membrane or the matrix [33,144]. It is the electrical component $\Delta\psi$ of the total proton-motive force that drives translocation [97,123]. $\Delta\psi$ is required for the transfer of the targeting sequence of a preprotein across the inner membrane; it is, however, not necessary for the movement of the mature part of the preprotein through the import channel of the inner membrane [145]. The mechanism of $\Delta\psi$ action is not clear. The membrane potential may exert an electrophoretic effect on the positively charged presequences [97]. On the other hand, $\Delta\psi$ may influence the conformation of domains of the Tim complex or other components of the inner membrane in such a manner that translocation of the targeting signal is triggered or that a gating effect is exerted. It is interesting in this

regard that translocation of preproteins with differently charged presequences is affected to different degrees upon decreasing $\Delta\psi$ by addition of CCCP (carbonyl cyanide *m*-chlorophenylhydrazone) in increasing concentrations [97].

4.2.2. The Tim44/mt-Hsp70 complex drives translocation by ATP-hydrolysis

Currently two ATP-dependent steps in the import reaction have been identified. (i) There is requirement for ATP in the cytosol; the reversible interaction of cytosolic Hsp70 and MSF with preproteins requiring ATP-hydrolysis has been discussed above. (ii) A second ATP requiring step is in the matrix and is related, at least in part, to the role of mt-Hsp70 in threading preproteins through the inner membrane [77,32]. Studies on protein translocation at experimentally defined ATP levels suggested that import of matrix and intermembrane-space targeted preproteins require ATP in the matrix (in addition to a membrane potential $\Delta\psi$). This requirement is for securing the presequence on the matrix side of the inner membrane and for the completion of translocation [167].

First evidence for the key role of mt-Hsp70 in providing energy for the translocation process was obtained by the characterization of a temperature-sensitive yeast strain (ssc1-2) harbouring a mutation in the mt-Hsp70 gene *SSC1* [77,117]. It turned out that translocation of matrix-targeted precursor proteins across the inner membrane strictly depends on mt-Hsp70. The induction of the phenotype by shifting cells to the non-permissive temperature led to the accumulation of preproteins in the cytosol. Upon import into isolated ssc1-2 mitochondria preproteins became arrested as intermediates with the NH_2-terminal part exposed in the matrix and processed by MPP, whilst the mature portions still exposed on the surface of the mitochondria. This analysis of the mutant ssc1 phenotype was complemented by studies on the interaction of mt-Hsp70 with partially translocated polypeptide chains. Complex formation between mt-Hsp70 and preproteins in transit was demonstrated by co-immunoprecipitation [77,176] and by chemical crosslinking of these two components [96,142].

Interestingly, part of mt-Hsp70 was found in close proximity to the inner membrane by immunoelectron microscopy [17]. Tim44, located at the matrix side of the inner membrane recruits mt-Hsp70 by forming a dynamic complex at the sites where the unfolded chain enters the matrix space (Fig. 2) [85,129,153]. This interaction is regulated by ATP and unfolded polypeptide substrate [153] and has been proposed to energetically drive the translocation reaction [109,166]. Current evidence suggests a mechanism by which Tim44 first binds the unfolded polypeptide chain emerging from the putative translocation channel, then transfers it to mt-Hsp70, thereby stimulating the ATP-dependent dissociation of the complex (Fig. 3) [153]. The Tim44/mt-Hsp70 complex was found to be stabilized when ATP-hydrolysis is inhibited by added EDTA or non-hydrolysable nucleotide analogs (e.g. AMP-PNP) [153]. On the other hand, unfolded proteins and synthetic peptides stimulate the ATP-dependent dissociation of the Tim44/Hsp70 complex. Thus, cycles of binding and release of Tim44, mt-Hsp70 and preprotein in transit could be regulated by ATP-hydrolysis (Fig. 3).

Mge1p, a mitochondrial GrpE homologue, has been identified as a matrix located component in *S. cerevisiae* involved in the regulation of mt-Hsp70 dependent precursor import and folding [12,68,87,177]. In contrast to the gene encoding for the mitochondrial DnaJ homologue MDJ1 which triggers the ATP-hydrolysis of DnaK [138] the gene encoding Mge1p, MGE1 (also termed YGE1) is essential. Mge1p functions as a nucleotide exchange factor for mt-Hsp70 which facilitates release of bound ADP, thereby allowing the formation of mt-Hsp70 in the ATP form. In this way it regenerates the competence of mt-Hsp70 for substrate binding in the matrix by exchange of ADP versus ATP and also the formation of the complex of mt-Hsp70 with Tim44 [179].

How could such cycles of binding and release be used to exert a unidirectional driving force on the incoming polypeptide chain? It has been proposed that mt-Hsp70 targeted by Tim44 binds to the incoming preprotein with high affinity and thereby locks the preprotein in a membrane-spanning fashion (Fig. 3). After dissociation of the Tim44/mt-Hsp70 complex the incoming chain can further slide through the import channel exclusively in the inward direction. Tim44 then can interact with a segment of the preprotein more COOH-terminal and this initiates a new cycle of mt-Hsp70 binding. Thus, repeated cycles of dissociation and rebinding would complete the import by favouring movement into the matrix (Fig. 3).

The Tim44/mt-Hsp70 complex therefore may function as a ratchet such that the movement of the translocating chain would be driven by Brownian motion and

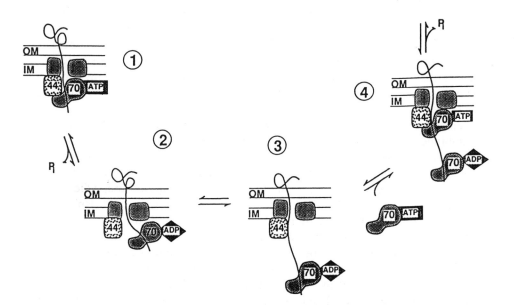

Fig. 3.   Working model for the ATP-dependent interaction of mt-Hsp70 and Tim44 in preprotein translocation across the mitochondrial inner membrane. For further explanation see text. Abbreviations: 44, Tim44; 70, mt-Hsp70; OM, outer membrane; IM, inner membrane.

segments of the incoming chains are trapped in a stepwise manner when emerging from the import channel at the matrix side. This mechanism has been termed 'molecular ratchet' or 'Brownian ratchet' [158]. Alternatively or additionally, a nucleotide dependent conformational change of mt-Hsp70 could facilitate the transfer reaction by exerting a 'pulling force' on the incoming chain [36,121]. Binding of incoming peptides to the COOH-terminal domain of mt-Hsp70 seems to be allosterically regulated by the binding of nucleotide to the NH_2-terminal binding site [101]. It is possible that the two domains undergo significant conformational rearrangements during the reaction cycles or can move relatively to one another thereby generating a pulling force.

4.2.3. The Tim complex is coupled to the Tim44/mt-Hsp70 driving system

The identification of the Tim proteins provided a basis for the structural and functional analysis of the inner membrane import machinery. Tim17 and Tim23 form a stable complex in the membrane that can be solubilized by mild detergents such as digitonin. They are present in the complex in a 1:1 stoichiometry, as are probably also Tim14 and Tim33. Tim44 is associated with the Tim complex, but its interaction is rather labile and sensitive to salt and detergent. Tim44, Tim23 and Tim17 could be crosslinked to membrane-spanning translocation intermediates suggesting that these proteins are in vicinity of incoming polypeptide chains [129,153]. Synthetic lethality between different mutant alleles of these three Tim proteins provides further evidence for their functional interaction [9]. Tim44 may interact with the membrane embedded components in a dynamic fashion, but could possibly also constitute an integral component of the Tim complex.

In order to investigate early steps in import, chimaeric preproteins consisting of DHFR fused to short segments of matrix targeted precursors were arrested by stabilizing the DHFR domain so that only the matrix targeting sequence was spanning the two membranes. Specific crosslinks to Tim17, Tim23, Tim44 and mt-Hsp70 were found with these intermediates in the same way as with longer constructs which were already processed in the matrix [7,86,140]. The components of the inner membrane import machinery are apparently in close proximity to the presequence at an early stage of translocation and remain so throughout the import process [7]. The crosslinking efficiency was strongly reduced when the mitochondrial matrix was depleted of ATP prior to the crosslinking procedure, indicating that the chain in transit is held in the translocation channel mainly by mt-Hsp70. Under ATP depletion conditions the translocation intermediates were found to slide back in the import channel and, if not too long to even fall out of the mitochondria [171].

In the presence of a translocating chain, the Tim complex forms 'translocation contact sites' with the import machinery of the outer membrane, the Tom complex (Fig. 4). Matrix targeted membrane spanning translocation intermediates, when bound to mt-Hsp70 are obviously not able to laterally leave the translocation contact sites [7]. Under these conditions Tom40 together with the Tim proteins and mt-Hsp70 can be co-immunoprecipitated with the intermediates. The individual components of the Tom and Tim complexes, however, do not bind to the translo-

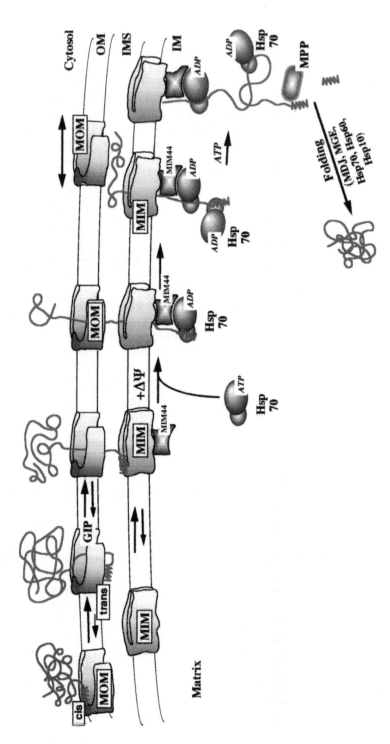

Fig. 4. Dynamic cooperation of protein import machineries in mitochondrial outer and inner membranes.

cating chain with high affinity. Association between the Tom and Tim complexes could only be observed when the intermediates were arrested on the outer face by stabilizing the DHFR domain and on the matrix face by binding to mt-Hsp70 [7,62]. Thus, the two translocation machineries may constitute a translocation channel that provides a proteinaceous environment for an unfolded polypeptide in transit.

Recent observations suggest a critical role of Tim proteins in the recognition of preproteins after transfer across the outer membrane and in the initiation of translocation across the inner membrane. Preprotein binding and translocation can be inhibited by pretreating mitoplasts with antibodies against Tim17 and Tim23, but not against Tim44. When added after the preprotein was arrested in an inner membrane spanning fashion, the antibodies did not interfere with the translocation process. Thus, portions of the Tim complex exposed to the intermembrane space might be involved in the initiation of protein translocation across the inner membrane [7].

To describe the complete process of translocation across both mitochondrial membranes the following scenario can be envisaged: The presequence of a preprotein is translocated first across the outer membrane and held at the *trans*-site (Figs. 2, 3 and 4) [100]. The Tom and Tim complexes diffusing laterally in the two membranes make contact when a Tom contains a preprotein at the *trans*-site (Fig. 4). Then, the presequence switches over to a recognition component of the Tim complex facing the intermembrane space. The electrical membrane potential then drives insertion into and translocation of the targeting sequence across the inner membrane. At the inner side, mt-Hsp70 in complex with Tim44 then binds to the targeting signal trapping the preprotein in a spanning fashion and thereby initiates translocation of the rest of the preprotein [85,129,153]. The association of Tim44 with the Tim complex would link a passive channel to the molecular machinery that energetically drives the import process (Figs. 3 and 4) [7].

5. Independent function of the outer and inner translocation machineries

In support of this hypothetical mechanism of protein translocation matrix-targeted proteins were found to be able, under certain conditions, to traverse the two membranes in consecutive steps. This suggests independent function of the two machineries [67,98,155].

An example of the function of the Tom complex independent of the Tim machinery is the import of the intermembrane space protein cytochrome c heme lyase (CCHL). CCHL precursor only requires receptor and GIP functions to reach its final destination. Fusion proteins consisting of a matrix targeting signal and CCHL, as well as large portions of AAC precursor, can be translocated into the intermembrane space in the absence of a membrane potential [155,122].

Furthermore, yeast mitochondria with a disrupted outer membrane (mitoplasts) directly import preproteins across the inner membrane, thereby bypassing the Tom complex [172]. On the other hand, coupling of the two translocation machineries seems to be necessary for efficient and coordinated translocation. Nevertheless, a sealed channel spanning the two membranes probably does not exist. Instead, the import systems of the outer and inner membrane appear to form translocation contact sites by dynamic and transient interactions.

6. Processing and assembly of imported preproteins

6.1. Processing in the matrix

Most of the precursor proteins with cleavable presequences are initially targeted to the matrix. During or after their translocation across the inner membrane the targeting signals are removed by the mitochondrial processing peptidase in the matrix (MPP) [11]. In a number of cases the initial processing, performed by MPP, is followed by an additional proteolytic maturation step in the matrix. A second portion is removed either by MPP, as with the precursor of F_0-ATPase subunit 9 [148] or by a monomeric metalloprotease named MIP (*m*itochondrial *i*ntermediate *p*eptidase) which removes an octapeptide from the NH_2-termini generated by MPP [69,75]. MIP was purified by its ability to specifically bind to a polypeptide containing an artificially generated octapeptide [76,70].

The MPP consists of two structurally related subunits, α-MPP and β-MPP, which both are required for processing. β-MPP is at least related to the core I protein of complex III (ubiquinol:cytochrome c oxidoreducatse; bc_1-complex) of the respiratory chain. In some species β-MPP and core I are identical proteins [154].

The isolated processing peptidase has a neutral pH-optimum [3]. The processing activity is stimulated by addition of divalent cations such as Mn^{2+} and Mg^{2+} with *N. crassa* MPP and by Zn^{2+}, Co^{2+} and Mn^{2+} with *S. cerevisiae* MPP [10]. MPP activity is inhibited by divalent cation chelators such as EDTA, EGTA and 1,10-phenanthroline, and is insensitive towards other inhibitors (e.g. serine proteinase inhibitors) indicating that MPP is a metal dependent peptidase. However, it is not clear whether it contains metal in the catalytic center. A potential metal binding site with the signature HXXEH known to play a crucial role in metal binding and catalytic activity of insulinases is conserved in all β-MPP subunits but is degenerated in the α-MPP proteins [13]. This motif is an inversion of the active site motif of the thermolysin superfamily, HEXXH [74].

Both, α-MPP and β-MPP are synthesized with cleavable presequences. The active proteins located in the matrix perform the proteolytic maturation of their own precursors proteins [54,152]. This is in agreement with the hypothesis that mitochondria arise by growth and division of preexisting mitochondria.

Despite great efforts it is still unclear how the MPP recognizes the highly degenerated cleavage sites. There is no obvious consensus sequence, besides the presence of a positively charged amino acid (usually arginine) in the −2 or −3 position at the cleavage site [34,82,110]. Recent mutational analysis of matrix targeting signals and the MPP processing sites of cleavable precursor proteins suggest that the targeting signal is, or can be, different from the signal recognized by MPP. The information for matrix targeting is concentrated to a stretch of 10 to 20 amino acid residues within the presequence. This portion of the polypeptide chain has the potential to form an amphiphilic α-helix [175]. The location of the MPP cleavage sites is variable with respect to the targeting sequence. While MIP cleavage sites are always found COOH-terminal to the amphiphilic stretch, MPP processing sites can also be located within the targeting signal, but both are never found NH$_2$-terminal to the targeting sequence [3,115]. Recent experiments suggest that amphiphilicity is not a prerequisite for MPP cleavage, on the other hand formation of an α-helix does not interfere with MPP processing (Klaus, Brunner and Neupert, in preparation).

One class of mitochondrial precursors destined for the inner membrane or the intermembrane space carry a bipartite presequences which become processed in a two-step reaction [132,168]. The NH$_2$-terminal portion of these presequences direct the preprotein into the matrix where they are cleaved off by MPP. Within the second part such presequences contain a sorting signal for the intermembrane space characterized by a hydrophobic stretch flanked by positive charges at the NH$_2$-terminal side. This portion shows similarities with a prokaryotic signal sequence for the export of proteins to the periplasmic space [50].

Sorting signals are removed by the membrane-bound intermembrane space peptidase (IMP). IMP is a heterodimer which resembles the bacterial leader peptidase [149,150]. The subunits, named IMP1 and IMP2 are structurally related and both are catalytically active with differences in their substrate specificity [111].

Imp1 cleaves the sorting signal of the nuclear-encoded, imported cytochrome b$_2$ and, at the same time the NH$_2$-terminal leader sequence of the precursor of cytochrome oxidase subunit II which is encoded in the mitochondrial genome and synthesized on mitochondrial ribosomes [55]. The way how these sorting signals work is currently a matter of debate.

Sorting of yeast NADH-cytochrome b$_5$-reductase occurs by an interesting mechanism differing from the general import pathway. The precursor form of this protein carries an NH$_2$-terminal matrix-targeting signal followed by a hydrophobic sequence. One third of the precursor becomes anchored in the outer membrane by the hydrophobic stretch with the COOH-terminal major part of the protein facing the cytosol. The other precursor molecules become initially translocated into the inner membrane with the help of the matrix targeting sequence where they become arrested and then cleaved by the inner membrane protease IMP1 [43]. This process of incomplete translocation arrest in the outer membrane represents a novel mechanism by which the product of a single gene is sorted into different compartments of the same organelle.

6.2. The folding machinery of the mitochondrial matrix

Following the vectorial movement across the mitochondrial inner membrane the imported proteins must undergo a folding process in order to achieve their functional, active conformation. The Tim44 dependent targeting of mt-Hsp70 to the polypeptide chain emerging at the inner face of the inner membrane appears to prevent aggregation or unproductive folding. Some mt-Hsp70 initially bound may remain with the imported preprotein until folding to the native state has occurred.

Two pathways of folding have been proposed. Some preproteins may be able to acquire the native state directly upon release of bound mt-Hsp70. This appears to require cycles of release and rebinding of mt-Hsp70. The matrix co-chaperone Mdj1p and the nucleotide exchange factor Mge1p cooperate with mt-Hsp70 in this reaction (see above) [88]. Similar to the situation found in *E. coli*, disruption of the MDJ1 gene is not lethal under normal growth conditions but a deficiency in the folding of some newly imported proteins is observed [138].

Many, if not most, preproteins have to undergo a more complex folding pathway. Here, after release of mt-Hsp70 the preprotein is transferred to the heat shock protein Hsp60, the mitochondrial homologue of *E. coli* GroEL [48]. Hsp60 belongs to the family of 'chaperonins' which are highly conserved throughout evolution in prokaryotes, and in chloroplasts and mitochondria of eukaryotic cells. It is essential for the viability of yeast cells and its expression can be increased by heat shock treatment [73,131]. As a classical chaperone Hsp60 exhibits ATPase activity modulated by a cofactor Hsp10 (or cpn10). Hsp60 is a homo-oligomeric protein composed of 14 subunits of 60 kDa arranged in two heptameric stacked rings forming a barrel-like structure with a large central cavity [66,178]. Each of the subunits has a substrate binding site.

The mechanism of chaperonin-mediated folding of imported chains is not completely understood at present. Upon ATP hydrolysis conformational changes in the substrate binding domain are supposed to regulate the interaction with the bound substrate. Repeated cycles of release and rebinding favour conformational changes of the folding substrate, thereby enabling the formation of the correct tertiary structure. As shown for the bacterial GroEL/GroES system mt-cpn10 forms a complex with mt-Hsp60 able to bind unfolded substrate. A current focus of Hsp60/GroEL function is whether folding occurs within the central cavity and only the folded 'product' is set free or whether folding occurs in an equilibrium of bound and free substrate in solution. Hsp60/GroEL does not appear to play a direct role in assembly of protein complexes [18].

The regulation of the ATPase activities of Hsp60 subunits by the mt-Hsp10 is also an issue of considerable interest. In *S. cerevisiae* Hsp10/mt-cpn10 is an essential component [135,58,59]. Purification of the small heat shock protein was achieved by a functional assay enriching for an re-folding activity conferred to an unfolded protein bound to Hsp60 [92,135]. It has been identified within mitochondria of several organisms as a protein highly homologous to bacterial GroES [38] indicating that the basic mechanisms of chaperonin mediated folding appears to be conserved from prokaryotes to mitochondria. Recent investigations confirmed the involvement of mitochondrial cyclophilin (CyP20) localized in the matrix within the mt-Hsp70/mt-Hsp60 mediated folding process [169].

7. Summary and perspectives

The identification of quite a number of protein components involved in mitochondrial protein import has contributed significantly to our insights into the mechanism of this complex process. Mitochondrial preprotein import is a multi-step process which is facilitated by the sequential and coordinated action of two separate import machineries located in the outer and inner mitochondrial membrane (Tom and Tim complex). The Tom machinery translocates the charged presequences of matrix targeted preproteins across the outer membrane but not the entire protein.

Tom20/Tom22 and perhaps Tom70, integral components of the outer membrane, act as receptors for the targeting signals on the surface of the mitochondrial surface (*cis*-site) (Fig. 2). Binding of the signal to the *cis*-site is followed by its transfer across the outer membrane where it interacts with the *trans*-site (Fig. 2). The transfer presumably occurs via a general translocation pore (GIP) formed by components of the Tom complex, in particular Tom40. As the presequence binds to the *trans*-site, the unfolding of adjacent structures is promoted [100]. The Tom machinery cannot support complete translocation of a matrix targeted precursor across the outer membrane.

Preproteins destined for the outer membrane and the intermembrane space can be imported and inserted by the Tom machinery independently of the inner membrane [99]. The driving forces for these reactions are unknown; the free energies of membrane insertion, membrane association and folding could represent such forces. From the *trans*-site the matrix targeting sequence is passed on to the inner membrane complex (Tim) (Figs. 2 and 4) [7]. The components on the inner membrane which receive the presequence and probably initiate the transfer into the matrix remain to be identified. Translocation of the targeting sequence but not of the mature part of the preprotein across the inner membrane requires an electrical potential $\Delta\psi$. Upon emerging from the putative inner membrane channel, presumably formed by Tim23 and Tim17 (and a number of so far uncharacterized other components), the polypeptide chain becomes locked by binding to the Tim44/mt-Hsp70 complex, thereby the reversible movement of the preprotein in the Tom and Tim channels is prevented (Fig. 3) [153]. Unidirectional movement into the matrix then is facilitated by multiple cycles of binding and dissociation of the Tim44/mt-Hsp70 complex to the unfolded preprotein chain emerging from the import channel. This process has been proposed to occur by a 'molecular ratchet mechanism' since Tim44/mt-Hsp70 appear to work like a ratchet. The source of the energy for driving unfolding is not entirely clear. One possibility is that spontaneous limited unfolding and 'Brownian motion' allows reversible sliding in the channel and both unfolding and preprotein movement is made irreversible by ATP-driven Tim44/mt-Hsp70 binding (Fig. 3) [121,153,158]. Spontaneous unfolding of preprotein domains on the outside of the mitochondria appears to be a rate limiting step of import *in vitro* (C. Klaus, W. Neupert and M. Brunner, unpublished]. Alternatively, a conformational change of mt-Hsp70 upon ATP-hydrolysis or binding could exert a driving force to 'pull' a preprotein into the matrix. Upon reaching the matrix presequence containing preproteins are processed by the matrix processing peptidase (MPP). They are transferred to a folding device composed of mt-Hsp70, MDJ, Mge1p, Hsp60 and Hsp10 which facilitates the acquisition of the functional tertiary structure. Complex pathways and components involved in sorting to the mitochondrial subcompartments and/or assembly to protein complexes then complete the import of nuclear encoded mitochondrial proteins.

Despite the considerable progress made during the last years in understanding these complex processes, quite a number of fundamental questions remain to be answered. For instance, it is not clear what the exact structures within the targeting sequences are that are recognized by the receptor components and how the receptors interact with the signals. It will also be interesting to learn more about the chaperone function in the cytosol and whether these have a specific role, beyond preprotein stabilization, in triggering binding to the receptors.

One of the most interesting questions concern the structure and detailed function of the translocation machineries in both outer and inner membranes. Certainly not all of the components involved have been discovered so far. Mechanistic aspects of how these machineries work are most intriguing. This includes the energetics of the translocation process. In particular, the precise function of the Tim44/mt-Hsp70 complex, an important part of the mechanism, needs to be studied in detail. Further, the cooperation of the two transport machineries upon translocating proteins across both mitochondrial membranes by a single step mechanism deserves much attention.

Finally, much has to be done to understand the process of sorting, folding and assembly within the mitochondrion. Since there are multiple pathways in all these processes their elucidation will be a challenging task.

References

1. Alam, R., Hachiya, N., Sakaguchi, M., Kawabata, S., Kitajima, M., Mihara, K. and Omura, T. (1994) J. Biochem. **116**, 416.
2. Aquila, H., Link, T.A. and Klingenberg, M. (1985) EMBO J. **4**, 2369.
3. Arretz, M., Schneider, H., Guiard, B., Brunner, M. and Neupert, W. (1994) J. Biol. Chem. **269**, 4959.
4. Atencio, D.P. and Yaffe, M.P. (1992) Mol. Cell Biol., **12**, 283.
5. Attardi, G. and Schatz, G. (1988) Ann. Rev. Cell Biol. **4**, 289.
6. Baker, K.P., Schaniel, A., Vestweber, D. and Schatz, G. (1990) Nature **348**, 605.
7. Berthold, J., Bauer, M.F., Schneider, H.C., Klaus, C., Brunner, M. and Neupert, W. (1995) Cell **81**, 1085.
8. Blom, J., Kuebrich, M., Rassow, J., Voos, W., Dekker, P.J.T., Maarse, A.C., Meijer, M. and Pfanner, N. (1993) Mol. Cell Biol. **13**, 7364.
9. Blom, J., Dekker, P.J.T. and Meijer, M. (1995) Eur. J. Biochem. **232**, 309.
10. Boehni, P.C., Daum, G. and Schatz, G. (1983) J. Biol. Chem. **258**, 4937.
11. Boehni, P., Gasser, S., Leaver, C. et al. (1980) in: The Organization of the Mitochondrial Genome, eds A.M. Kroon and C. Saccone. Elsevier/North-Holland, Amsterdam, p. 423.
12. Bolliger. L., Deloche, O., Glick, B.S., Georgopoulos, C., Jeno, P., Kronidou, N., Horst, M., Morishima, N. and Schatz, G. (1994) EMBO J. **13**, 1998.
13. Brunner, M., Klaus, C. and Neupert, W. (1994) in: Signal Peptidases, ed R.G. von Heijne, Landes Company.
14. Cao, W. and Douglas, M.G. (1995) J. Biol. Chem. **270**, 5674.
15. Caplan, A.J. and Douglas, M.G. (1991) J. Cell Biol. **114**, 609.
16. Caplan, A.J., Cyr, D.M. and Douglas, M.G. (1992) Cell **71**, 1143.
17. Carbajal, E.M., Beaulieu, J.F., Nicole, L.M. and Tanguay, R.M. (1993) Exp. Cell Res. **207**, 300.
18. Cheng, M.Y., Hartl, F.-U., Martin, J., Pollock, R.A., Kalousek, F., Neupert, W., Hallberg, E.M., Hallberg, R.L. and Horwich, A.L. (1989) Nature **337**, 620.
19. Chirico, W.J., Waters, M.G. and Blobel, G. (1988) Nature **332**, 805.
20. Craig, E.A., Kramer, J., Shilling, J., Werner-Washburne, M., Holmes, S., Kosic-Smithers, J. and Nicolet, C. (1989) Mol. Cell Biol. **9**, 3000.

21. Cyr, D.M., Lu, X. and Douglas, M.G. (1992) J. Biol. Chem. **267**, 2092.
22. Dekker, P.J.T., Keil, P., Rassow, J., Maarse, A.C., Pfanner, N. and Meijer, M. (1993) FEBS Lett. **330**, 66.
23. DeKruijff, B. (1994) FEBS Lett. **346**, 78.
24. DeKroon, A.I.P.N., DeGier, J. and DeKruijff, B. (1991) Biochim. Biophys. Acta **1068**, 111.
25. Deshaies, R.J., Koch, B.D. and Schekman, R. (1988) Trends Biochem. Sci. **13**, 384.
26. Deshaies, R.J., Koch, B.D., Werner-Washburne, M., Craig, E.A. and Schekman, R. (1988) Nature **332**, 800.
27. Dietmeier, K., Zara, V., Palmisano, A., Palmieri, F., Voos, W., Schlossmann, J., Moczko, M., Kispal, G. and Pfanner, N. (1993) J. Biol. Chem. **268**, 25958.
28. Emr, S.D., Vassarotti, A., Garrett, J., Geller, B.L., Takeda, M. and Douglas, M.C. (1986) J. Cell Biol. **102**, 523.
29. Emtage, J.L. and Jensen, R.E. (1993) J. Cell Biol. **122**, 1003.
30. Foelsch, H., Guiard, B., Neupert, W. and Stuart, R.A. (1995) EMBO J. **15**, 479–487.
31. Fujiki, M. and Verner, K. (1993) J. Biol. Chem. **268**, 1914.
32. Gambill, B.D., Voos, W., Kang, P.J., Miao, B., Langer, T., Craig, E.A. and Pfanner, N. (1993) J. Cell Biol. **123**, 109.
33. Gasser, S.M., Daum, G. and Schatz, G. (1982) J. Biol. Chem. **257**, 13034.
34. Gavel, Y. and von Heijne, G. (1990) Protein Eng. **4**, 33.
35. Gavel, Y., Nilsson, L. and von Heijne, G. (1988) FEBS Lett. **235**, 173.
36. Glick, B. (1995) Cell **80**, 11.
37. Glick, B., Wachter, C. and Schatz, G. (1991) Trends Cell Biol. **1**, 99.
38. Goloubinoff, P. Gatenby, A.A. and Lorimer, G.H. (1989) Nature **337**, 44.
39. Gratzer, S., Lithgow, T., Bauer, R.E., Lamping, E., Paltauf, F., Kohlwein, S.D., Haucke, V., Junne, T., Schatz, G. and Horst, M. (1995) J. Cell Biol. **129**, 25.
40. Hachiya, N., Alam, R., Sakasegawa, Y., Sakaguchi, M., Mihara, K. and Omura, T. (1993) EMBO J. **12**, 1579.
41. Hachiya, N., Komiya, T., Alam, R., Iwahashi, J., Sakaguchi, M., Mihara, K. and Omura, T. (1994) EMBO J. **13**, 5146.
42. Hachiya, N., Mihara, K., Suda, K., Host, M., Schatz, G., Lithgow, T. (1995) Nature **376**, 705.
43. Hahne, K., Haucke, V., Ramage, L. and Schatz, G. (1994) Cell **79**, 829.
44. Harkness, T.A., Metzenberg, R.L., Schneider, H., Lill, R., Neupert, W. and Nargang, F.E. (1994) Genetics **136**, 107.
45. Harkness, T.A., Nargang, F.E., van der Klei, I., Neupert, W. and Lill, R. (1994) J. Cell Biol. **124**, 637.
46. Harmey, M.A., Hallermayer, G., Korb, H. and Neupert, W. (1977) Eur. J. Biochem. **81**, 533.
47. Hartl, F.U., Hlodan, R. and Langer, T. (1994) Trends Biochem. Sci. **19**, 20.
48. Hartl, F.U., Martin, J. and Neupert, W. (1992) Annu. Rev. Biophys. Biomol. Struct. **21**, 293.
49. Hartl, F.U., Pfanner, N., Nicholson, D.W. and Neupert, W. (1989) Biochim. Biophys. Acta **988**, 1.
50. Hartl, F.U. and Neupert, W. (1990) Science **247**, 930.
51. Hase, T., Mueller, U., Riezman, H. and Schatz, G. (1984) EMBO J. **3**, 3157.
52. Hase, T., Riezman, H., Suda, K. and Schatz, G. (1983) EMBO J. **2**, 2169.
53. Hase, T., Nakai, M. and Matsubara, H. (1986) FEBS Lett. **197**, 199.
54. Hawlitschek, G., Schneider, H., Schmidt, B., Tropschug, M., Hartl, F.U. and Neupert, W. (1988) Cell **53**, 795.
55. Herrmann, J.M., Koll, H., Cook, R.A., Neupert, W., Stuart, R.A. (1995), J. Biol. Chem. **270**, 27079–27086.
56. Hines, V. and Schatz, G. (1993) J. Biol. Chem. **268**, 449.
57. Hines, V., Brändt, A., Griffith, G., Horstmann, H., Broetsch, H. and Schatz, G. (1990) EMBO J. **9**, 3191.
58. Hohfeld, J. and Hartl, F.U. (1994) J. Cell Biol. **126**, 305.
59. Hohfeld, J. and Hartl, F.U. (1994) Curr. Opin. Cell Biol. **6**, 499.
60. Honlinger, A., Kubrich, M., Moczko, M., Gartner, F., Mallet, C., Bussereau, F., Eckerskorn, C., Lottspeich, T., Dietmeier, K., Jacquet, M. et al. (1995) Mol. Cell Biol. **15**, 3382.
61. Horst, M., Jeno, P., Kronidou, N.G., Bolliger, L., Oppliger, W., Scherer, P., Manning, Krieg, U.,

Jascur, T. and Schatz, G. (1993) EMBO J. **12**, 3035.

62. Horst, M., Hilfiker-Rothenfluh, S., Oppliger, W. and Schatz, G. (1995) EMBO J. **14**, 2293.
63. Horwich, A.L., Kalousek, F., Mellmann, I. and Rosenberg, L.E. (1985) EMBO J. **4**, 1129.
64. Hurt, E.C., Pesold-Hurt, B. and Schatz, G. (1984) FEBS Lett. **178**, 306.
65. Hurt, E.C., Pesold-Hurt, B., Suda, K., Oppliger, W. and Schatz, G. (1985) EMBO J. **4**, 2061.
66. Hutchinson, E.G., Tichelaar, W., Hofhaus, G., Weiss, H. and Leonard, K.R. (1989) EMBO J. **8**, 1485.
67. Hwang, S.T. and Schatz, G. (1989) J. Cell Biol. **86**, 8432.
68. Ikeda, E., Yoshida, S., Mitsizawa, H., Uno, I. and Toh-e, A. (1994) FEBS Lett. **339**, 265.
69. Isaya, G., Kalousek, F., Fenton, W.A. and Rosenberg, L.E. (1991) J. Cell Biol. **113**, 65.
70. Isaya, G., Miklos, D. and Rollins, R.A. (1994) Mol. Cell Biol. **14**, 5603.
71. Ito, A., Pgishima, T., Ou, W., Omura, T., Aoyagi, H., Lee, S., Mihara, H. and Izumiya, N. (1985) J. Biochem. **98**, 1571.
72. Jensen, R.E. and Yaffe, M.P. (1988) EMBO J. **7**, 3863.
73. Johnson, C., Chandrasekhar, G.N., Georgopoulos, C. (1989) J. Bacteriol. **171**, 1590.
74. Jongeneel, C., Bouvier, J. and Bairoch, A. (1989) FEBS Lett. **242**, 211.
75. Kalousek, F., Hendrick, J.P. and Rosenberg, L.E. (1988) Proc. Natl. Acad. Sci. USA **85**, 7536.
76. Kalousek, F., Isaya, G. and Rosenberg, L.E. (1992) EMBO J. **11**, 2803.
77. Kang, P.J., Ostermann, J., Shilling, J., Neupert, W., Craig, E.A. and Pfanner, N. (1990) Nature **348**, 137.
78. Kassenbrock, C.K., Cao, W. and Douglas, M.G. (1993) EMBO J. **12**, 3023.
79. Kiebler, M., Becker, K., Pfanner, N. and Neupert, W. (1993) J. Membr. Biol. **135**, 191.
80. Kiebler, M., Keil, P., Schneider, H., van der Klei, I., Pfanner, N. and Neupert, W. (1993) Cell **74**, 483.
81. Kiebler, M., Pfaller, R., Soellner, T., Griffiths, G., Horstmann, H., Pfanner, N. and Neupert, W. (1990) Nature **348**, 610.
82. Kitada, S., Shimokata, K., Niidome, T., Ogishima, T. and Ito, A. (1995) J. Biochem. Tokyo **117**, 1148.
83. Kleene, R., Pfanner, N., Neupert, W. and Tropschug, M. (1987) Biol. Chem. Hoppe-Seyler **368**, 1260.
84. Komiya, T., Hachiya, N., Sakaguchi, M., Omura, T. and Mihara, K. (1994) J. Biol. Chem. **269**, 30893.
85. Kronidou, N.G., Oppliger, W., Bollinger, L., Hannavy, K., Glick, B.S., Schatz, G. and Horst, M. (1994) Proc. Natl. Acad. Sci. USA **91**, 12818.
86. Kuebrich, M., Keil, P., Rassow, J., Dekker, P.J.T., Blom, J., Meijer, M. and Pfanner, N. (1994) FEBS Lett. **349**, 222.
87. Laloraya, S., Gambill, B.D. and Craig, E.A. (1994) Proc. Natl. Acad. Sci. USA **91**, 6481.
88. Langer, T., Lu, C., Echols, H., Flanagan, J., Hayer-Hartl, M.K. and Hartl, F.U. (1992) Nature **356**, 683.
89. Lill et al. (1994) in: Advances in Molecular and Cell Biology, ed F. Hartl. Academic Press, New York.
90. Lithgow, T., Junne, T., Wachter, C. and Schatz, G. (1994) J. Biol. Chem. **269**, 15325.
91. Lithgow, T., Junne, T., Suda, K., Gratzer, S. and Schatz, G. (1994) Proc. Natl. Acad. Sci. USA **91**, 11973.
92. Lubben, T.H., Gatenby, A.A., Donaldson, G.K., Lorimer, G.H. and Viitanen, P.V. (1990) Proc. Natl. Acad. Sci. USA **87**, 7683.
93. Maarse, A.C., Blom, J., Grivell, L.A. and Meijer, M. (1992) EMBO J. **11**, 3619.
94. Maarse, A.C., Blom, J., Keil, P., Pfanner, N. and Meijer, M. (1994) FEBS Lett. **349**, 215.
95. Mahlke, K., Pfanner, N., Martin, J., Horwich, A.L., Hartl, F.U. and Neupert, W. (1990) Eur. J. Biochem. **192**, 551.
96. Manning-Krieg, U., Scherer, P.E. and Schatz, G. (1991) EMBO J. **10**, 3273.
97. Martin, J., Mahlke, K. and Pfanner, N. (1991) J. Biol. Chem. **266**, 18051.
98. Mayer, A., Lill, R. and Neupert, W. (1993) J. Cell Biol. **121**, 1233.
99. Mayer, A., Neupert, W. and Lill, R. (1995) Cell **80**, 127.
100. Mayer, A., Nargang, F.E., Neupert, W. and Lill, R. (1995) EMBO J. **14**, 4204.
101. McKay, D.B., Wilbanks, S.M., Flaherty, K.M., Ha, J.H., O'Brien, M.C. and Shirvanee, L.L. (1994) in: The Biology of Heats Shock Proteins and Molecular Chaperones, eds R.I. Morimoto, A. Tissieres and C. Georgopoulos. Cold Sping Harbor Laboratory Press, pp. 153–177.
102. Moczko, M., Dietmeier, K., Sollner, T., Segui, B., Steger, H.F., Neupert, W. and Pfanner, N. (1992) FEBS Lett. **310**, 265.

103. Moczko, M., Ehmann, B., Gartner, F., Honlinger, A., Schafer, E. and Pfanner, N. (1994) J. Biol. Chem. **269**, 9045.
104. Morrison, D. (1994) Science **266**, 56.
105. Murakami, K. and Mori, M. (1990) EMBO J. **9**, 3201.
106. Murakami, K, Tanase, S., Morino, Y. and Mori, M. (1992) J. Biol. Chem. **267**, 13119.
107. Nakai, M. and Endo, T. (1995) FEBS Lett. **357**, 202.
108. Nargang, F.E., Kuenkele, K.P., Mayer, A., Ritzel, R.G., Neupert, W. and Lill, R. (1995) EMBO J. **14**, 1099.
109. Neupert, W., Hartl, F.U., Craig, E.A. and Pfanner, N. (1990) Cell **63**, 447.
110. Niidome, T., Kitada, S., Shimokata, K., Ogishima, T. and Ito, A. (1994) J. Biol. Chem. **269**, 24719.
111. Nunnari, J., Fox, D.F. and Walter, P. (1993) Science **262**, 1997.
112. Ono, H., Tuboi, S. (1990) Arch. Biochem. Biophys. **280**, 299.
113. Ono, H. and Tuboi, S. (1988) J. Biol. Chem. **263**, 3188.
114. Ooi, C.E. and Weiss J. (1992) Cell **71**, 87.
115. Ou, W.J., Kumamoto, T., Mihara, K., Kitada, S., Niidome, T., Ito, A. and Omura, T. (1994) J. Biol. Chem. **269**, 24673.
116. Ostermann, J., Horwich, A.L., Neupert, W. and Hartl, F.U. (1989) Nature **341**, 125.
117. Ostermann, J., Voss, W., Kang, P.J., Craig, E.A., Neupert, W. and Pfanner, N. (1990) FEBS Lett. **277**, 281.
118. Palleros, D.R., Reid, K.L., Shi, L., Welch, W.J. and Fink, A.L. (1993) Nature **365**, 664.
119. Pfaller, R., Pfanner, N. and Neupert, W. (1989) J. Biol. Chem. **264**, 34.
120. Pfaller, R., Steger, H.F., Rassow, J., Pfanner, N. and Neupert, W. (1988) J. Cell Biol. **107**, 2483.
121. Pfanner, N. and Meijer M. (1995) Curr. Biol. **5**, 132.
122. Pfanner, N. and Neupert W. (1987) J. Biol. Chem. **262**, 7528.
123. Pfanner, N. and Neupert W. (1985) EMBO J. **4**, 2819.
124. Pfanner, N., Rassow, J., van der Klei, I. and Neupert, W. (1992) Cell **68**, 999.
125. Pfanner, N., Soellner, T. and Neupert, W. (1991) Trends Biochem Sci. **16**, 63.
126. Pfanner, N., Craig, E.A. and Meijer, M. (1994) Trends Biochem. Sci. **19**, 368.
127. Pollock, R.A., Hartl, F.U., Cheng, M.Y., Ostermann, J., Horwich, A.L. and Neupert, W. (1988) EMBO J. **7**, 3493.
128. Ramage, L., Junne, T., Hahne, K., Lithgow, T. and Schatz, G. (1993) EMBO J. **12**, 4115.
129. Rassow, J., Maarse, A.C., Krainer, E., Kuebrich, M., Mueller, H., Meijer, M., Craig, E.A. and Pfanner, N. (1994) J. Cell Biol. **127**, 1547.
130. Rassow, J., Mohrs, K., Koidl, S., Barthelmess, I.B., Pfanner, N. and Tropschug, M. (1995) Mol. Cell Biol. **15**, 2654.
131. Reading, D.S., Hallberg, R.L. and Myers, A.M. (1989) Nature **337**, 655.
132. Reid, G.A., Yonetani, T. and Schatz, G. (1982) J. Biol. Chem. **257**, 13068.
133. Riezman, H., Hay, R., Witte, C., Nelson, N. and Schatz, G. (1983) EMBO J. **2,** 1113.
134. Roise, D. and Schatz, G. (1988) J. Biol. Chem. **263**, 4509.
135. Rospert, S., Glick, B.S., Jeno, P., Schatz, G., Todd, M.J., Lorimer, G.H. and Viitanen, P.V. (1993) Proc. Natl. Acad. Sci. USA **90**, 10967.
136. Rospert, S., Junne, T., Glick, B.S. and Schatz, G. (1993) FEBS Lett. **335**, 358.
137. Rospert, S., Mueller, S., Schatz, G. and Glick B.S. (1994) J. Biol. Chem. **269**, 17279.
138. Rowley, N., Prip Buus, C., Westermann, B., Brown, C., Schwarz, E., Barrell, B. and Neupert, W. (1994) Cell **77**, 249.
139. Ryan, K.R., Menold, M.M., Garrett, S. and Jensen, R.E. (1994) Mol. Biol. Cell **5**, 529.
140. Ryan, K.R. and Jensen, R.E. (1993) J. Biol. Chem. **268**, 23743.
141. Saraste, M. and Walker J.E. (1982) FEBS Lett. **144**, 250.
142. Scherer, P.E., Krieg, U.C., Hwang, S.T., Vestweber, D. and Schatz, G. (1990) EMBO J. **9**, 4315.
143. Scherer, P.E., Manning Krieg, U., Jeno, P., Schatz, G. and Horst, M. (1992) Proc. Natl. Acad. Sci. USA **89**, 11930.
144. Schleyer, M., Schmidt, B. and Neupert, W. (1982) Eur. J. Biochem. **125**, 109.
145. Schleyer, M. and Neupert, W. (1985) Cell **43**, 330.
146. Schlossmann, J., Dietmeier, K., Pfanner, N. and Neupert, W. (1994) J. Biol. Chem. **269**, 11893.

147. Schlossmann, J. and Neupert, W. (1995) J. Biol. Chem. **270**, 27116–27121.
148. Schmidt, B., Wachter, E., Sebald, W. and Neupert, W. (1984) Eur. J. Biochem. **144**, 581.
149. Schneider, A., Behrens, M., Scherer, P., Pratje, E., Michaelis, G. and Schatz, G. (1991) EMBO J. **10**, 247.
150. Schneider, A., Opplinger, W. and Jenoe, P. (1994) J. Biol. Chem. **269**, 8635.
151. Schneider, H., Soellner, T., Dietmeier, K., Eckerskorn, C., Lottspeich, F., Troelzsch, B., Neupert, W. and Pfanner, N. (1991) Science **254**, 1659.
152. Schneider, H., Arretz, M., Wachter, E. and Neupert, W. (1990) J. Biol. Chem. **265**, 9881.
153. Schneider, H.C., Berthold, J., Bauer, M.F., Dietmeier, K., Guiard, B., Brunner, M. and Neupert, W. (1994) Nature **371**, 768.
154. Schulte, U., Arretz, M., Schneider, H., Tropschug, M., Wachter, E., Neupert, W. and Weiss, H. (1989) Nature **339**, 147.
155. Segui-Real, B., Kispal, G., Lill, R. and Neupert, W. (1993) EMBO J. **12**, 2211.
156. Sikorski, R.S., Boguski, M.S., Goebl, M. and Hieter, P. (1990) Cell **60**, 307.
157. Simon, S.M. and Blobel, G. (1991) Cell **65**, 371.
158. Simon, S.M., Peskin, C.S. and Oster, G.F. (1992) Proc. Natl. Acad. Sci. USA **89**, 3770.
159. Shore, G.C., McBride, H.M., Millar, D.G., Steenaart, Nae, Nguyen, M. (1995) Eur. J. Biochem. **227**, 1.
160. Smagula, C. and Douglas, M.G. (1988) J. Biol. Chem. **263**, 6783.
161. Soellner, T., Griffiths, G., Pfaller, R., Pfanner, N. and Neupert, W. (1989) Cell **59**, 1061.
162. Soellner, T., Pfaller, R., Griffiths, G., Pfanner, N. and Neupert, W. (1990) Cell **62**, 107.
163. Soellner, T., Rassow, J., Wiedmann, M., Schlossmann, J., Keil, P., Neupert, W. and Pfanner, N. (1992) Nature **355**, 84.
164. Steger, H.F., Soellner, T., Kiebler, M., Dietmeier, K.A., Pfaller, R., Troelzsch, K.S., Tropschug, M., Neupert, W. and Pfanner, N. (1990) J. Cell Biol. **111**, 2353.
165. Steiner, H., Zollner, A., Haid, A., Neupert, W. and Lill, R. (1995) J. Biol. Chem. **270**, 22842.
166. Stuart, R.A., Cyr, D.M., Craig, E.A. and Neupert, W. (1994) Trends Biochem Sci. **19**, 87.
167. Stuart, R.A., Gruhler, A., van der Klei, I., Guiard, B., Koll, H. and Neupert, W. (1994) Eur. J. Biochem. **220**, 9.
168. Teintze, M., Slaugther, M., Weiss, H. and Neupert, W. (1982) J. Biol. Chem. **257**, 10364.
169. Tropschug, M., Barthelmess, I.B. and Neupert, W. (1989) Nature **342**, 953.
170. Tzagoloff, A. (1982) in: Mitochondria. Plenum Press, New York.
171. Ungermann, C., Neupert, W. and Cyr, D.M. (1994) Science **266**, 1250.
172. Vale, R.D. (1994) Cell **78**, 733.
173. Verner, K. (1993) Trends Biochem. Sci. **18**, 366.
174. Vestweber, D., Brunner, J., Baker, A. and Schatz, G. (1989) Nature **341**, 205.
175. von Heijne, G. (1986) EMBO J. **5**, 1335.
176. Voos, W., Gambill, B.D., Guiard, B., Pfanner, N. and Craig, E.A. (1993) J. Cell Biol. **123**, 119.
177. Voos, W., Gambill, B.D., Laloraya, S., Ang, D., Craig, E.A. and Pfanner, N. (1994) Mol. Cell Biol. **14**, 6627.
178. Weissmann, J.S., Sigler, P.B. and Horwich, A.L. (1995) Science **268**, 523.
179. Westermann, B., Prip-Buus, C., Neupert, W. and Schwarz, E. (1995) EMBO J. **14**, 3452.
180. Wienhues, U., Becker, K., Schleyer, M., Guiard, B., Tropschug, M., Horwich, A.L., Pfanner, N. and Neupert, W. (1991) J. Cell Biol. **115**, 1601.
181. Yang, M., Jensen, E.R., Yaffe, M.P., Opplinger, W. and Schatz, G. (1988) EMBO J. **7**, 3857.
182. Zheng, X., Rosenberg, L.E., Kalousek, F. and Fenton, W.A. (1993) J. Biol. Chem. **268**, 7489.

Author Index

Akita, M., *see* Kimura, E., 787

Akiyama, S.I., 158

Akiyama, S.I., *see* Kamiwatari, M., 158

Akiyama, S.I., *see* Richert, N., 158

Akiyama, S.I., *see* Yoshimura, A., 159

Akiyama, Y., 135

Akiyama, Y., *see* Baba, T., 788

Akiyama, Y., *see* Ito, K., 786

Akiyama, Y., *see* Kamitani, S., 789

Akiyama, Y., *see* Kihara, A., 788

Akiyama, Y., *see* Kusukawa, N., 786

Akiyama, Y., *see* Shimoike, T., 788

Akiyama, Y., *see* Taura, T., 788

Al-Awqati, Q., 471

Al-Awqati, Q., *see* Adelsberg, J.S. v., 309

Al-Awqati, Q., *see* Barasch, J., 471

Al-Awqati, Q., *see* Young, G.P., 472

Al-Shawi, M.K., 161

Al-Shawi, M.K., *see* Senior, A.E., 72, 161, 183

Al-Shawi, M.K., *see* Nakamoto, R.K., 73

Al-Shawi, M.K., *see* Urbatsch, I.L., 161

Alam, R., *see* Hachiya, N., 819

Alam, R., 818

Alarcón, T., *see* López-Brea, M., 183

Alberti, M., *see* Ma, D., 184

Albrecht, S.P.J., *see* Pfenninger-Li, X.D., 46

Albright, L.M., *see* Jiang, J., 445

Alcantara, R., *see* Zafra, F., 445

Alden, J.C., 20

Aldrich, R.W., *see* Zagotta, W.N., 727

Aldrich, R.W., *see* Hoshi, T., 727

Alex, L.A., 598

Alexander-Bowman, S.J., *see* Bañueles, M.A., 499

Alfonso, A., 430

Alger, J.R., *see* Slonczewski, J.L., 530

Alger, S.R., *see* Slonczewski, J.L., 531

Allard, J.D., 183, 201

Allard, J.D., *see* Roof, S.K., 662, 688

Allard, W.J., *see* Mueckler, M., 90

Allard, W.J., *see* Mueckler, M., 342

Allen, D.P., *see* Drenckhahn, D., 308

Allen, D.P., *see* Low, P.S., 308

Allen, D.P., *see* Verlander, J.W., 309

Allen, R., *see* Bowman, B.J., 108

Allikmets, R., 185

Allison, M.J., *see* Anantharam, V., 259

Allison, M.J., *see* Baetz, A., 260

Allison, W.S., 71

Allison, W.S., *see* Andrews, W.W., 72

Allison, W.S., *see* Bullough, D.A., 72, 73

Allison, W.S., *see* Yoshida, M., 72

Allison, W.S., *see* Zhuo, S., 73

Allisson, M.J., *see* Ruan, Z.-S., 259

Allmansberger, R., *see* Wu, L.J., 757

Almquist, K.C., *see* Cole, S.P.C., 18, 161

Almquist, K.C., *see* Grant, C.E., 161

Alon, N., *see* Riordan, J.R., 159

Alon, T., *see* Karpel, R., 530

Alouf, J.E., *see* Launay, J., 404, 444

Alow, N., *see* Riordan, J.R., 160

Alper, S.L., 305, 308, 309, 471

Alper, S.L., *see* Brosius III, F.C., 306

Alper, S.L., *see* Erocolani, L., 308

Alper, S.L., *see* Humphreys, B.D., 309

Alper, S.L., *see* Jarolim, P., 307

Alper, S.L., *see* Jiang, L., 309

Alpern, R.J., *see* Cano, A., 380

Alpern, R.J., *see* Horie, S., 380

Alpert, C.A., *see* Deutscher, J., 547

Altenbach, C., 201, 226

Altenbach, C., *see* Farahbakhsh, Z.T., 227

Altenbach, C., *see* Ujwal, M.L., 201, 225

Altenberg, G.A., 162, 186

Altendorf, K., 498

Altendorf, K., *see* Abee, T., 498

Altendorf, K., *see* Birkenhäger, R., 73

Altendorf, K., *see* Bowman, E.J., 108, 472

Altendorf, K., *see* Burkovski, A., 71

Altendorf, K., *see* Fischer, S., 71

Altendorf, K., *see* Hesse, J., 498

Altendorf, K., *see* Laimins, L.A., 498

Altendorf, K., *see* Laubinger, W., 45, 74

Altendorf, K., *see* Möllenkamp, T., 498

Altendorf, K., *see* Puppe, W., 498

Altendorf, K., *see* Schneider, E., 45, 70, 73

Altendorf, K., *see* Siebers, A., 496, 498

Altendorf, K., *see* Voelkner, P., 498

Altendorf, K., *see* Walderhaug, M.O., 498

Altendorf, K., *see* Zanotti, F., 70

Altendorf, K., *see* Zimman, P., 498

Altman, E., 786

Altschul, S.F., *see* Schuler, G., 431, 531

Amaber, T., *see* Whitley, P., 201

Subject Index

ABC, 145
ABC family, 146
ABC protein, 121
ABC transporter, 174, 491
ABC type ATPases, 4, 115
access channel, 41
accessory protein, 170
acetate kinase, 254
acetoacetate, 462
Acetobacterium woodii, 42
acetylcholine transporter, 414
AChR, 198
acridine orange, 463
activation by Na^+, 30
active transport, 655
adaptation, 587
adenylate cyclase, 538–541
ADHIII, 803
ADP/ATP carrier, 796
adrenal chromaffin granules, 105
β_2-adrenergic receptor, 196
AE1, 283, 287, 461
AE2, 283, 287
AE3, 283, 288
Ag^+, 3
alarmones, 585
Alicyclobacillus acidocaldarius, 483
alkaline phosphatase (AP), 192, 194, 195
alkaliphiles, 519
alkaliphilic bacilli, 43
alkaliphilic bacteria, 43
alternative promoters, 287
amiloride, 507
9-aminoacridine, 96, 463
ammonium chloride, 468
ammonium prepulse technique, 455

AMP-PNP, 809
amphetamines, 383, 396, 435
amphipathic α-helices, 795
amphipaths, 626, 707
animal lysosomes, 105
anion exchange, 261
anion exchanger (AE) gene family, 283
ankyrin, 297, 298, 303
antibiotics, 167
antibody, 295
antibody binding, 197, 198
antidepressants, 393
antimicrobial peptides, 627
antimonite, 5
antiparallel β-barrels, 621
antiporter mutants, 510
antiporter Nhe-1, 508
antiporters, 233, 263, 313
antiporters, molecular physiology of, 522
apyrase, 772
Arabidopsis, 698
aromatic residues, 297
ArsA, 7
ArsB, 12
arsenite, 5
arsenite-translocating ATPase, 6
As(+3), 3
Ascidia sydneiensis samea, 106
asialoorosomucoid, 463
asymmetry, 319
atebrin, 463
ATP, 629, 655, 711
ATP binding cassette, 174
ATP binding proteins, 491
ATP dependence, 808
ATP hydrolysis, 795